3rd EDITION

VOYAGES
THROUGH THE UNIVERSE

Andrew Fraknoi
Foothill College

David Morrison
NASA Ames Research Center

Sidney Wolff
National Optical Astronomy Observatories

THOMSON

BROOKS/COLE

Australia • Canada • Mexico • Singapore • Spain • United Kingdom • United States

THOMSON
BROOKS/COLE

Astronomy Editor: *Chris Hall*
Development Editor: *Marie Carigma-Sambilay*
Assistant Editor: *Carol Benedict*
Editorial Assistant: *Melissa Newt*
Technology Project Manager: *Sam Subity*
Marketing Manager: *Kelley McAllister*
Marketing Assistant: *Leyla Jowza*
Advertising Project Manager: *Nathaniel Bergson-Michelson*
Project Manager, Editorial Production: *Hal Humphrey*
Print/Media Buyer: *Barbara Britton*
Permissions Editor: *Joohee Lee*

Production Service: *Nancy Shammas, New Leaf Publishing Service*
Cover and Text Designer: *Roy Neuhaus Design*
Art Editor: *Kathy Joneson*
Photo Researcher: *Jane Sanders*
Copy Editor: *Carol Reitz*
Illustrator: *Rolin Graphics*
Cover Photo: *T. A. Rector (NRAO/AUI/NSF and NOAO/AURA/NSF) and B. A. Wolpa (NOAO/AURA/NSF)*
Compositor: *G & S Typesetters, Inc.*
Cover Printer: *Transcontinental Printing/Interglobe*
Printer: *Transcontinental Printing/Interglobe*

For more information about our products, contact us at:
Thomson Learning Academic Resource Center
1-800-423-0563
For permission to use material from this text, contact us by:
Phone: 1-800-730-2214 **Fax:** 1-800-730-2215
Web: http://www.thomsonrights.com

Student Edition: ISBN 0-495-01789-2
Instructor's Edition: ISBN 0-534-40906-7

Brooks/Cole — Thomson Learning
10 Davis Drive
Belmont, CA 94002-3098
USA

Asia
Thomson Learning
5 Shenton Way #01-01
UIC Building
Singapore 068808

Australia/New Zealand
Thomson Learning
102 Dodds Street
Southbank, Victoria 3006
Australia

Canada
Nelson
1120 Birchmount Road
Toronto, Ontario M1K 5G4
Canada

Europe/Middle East/Africa
Thomson Learning
High Holborn House
50/51 Bedford Row
London WC1R 4LR
United Kingdom

Latin America
Thomson Learning
Seneca, 53
Colonia Polanco
11560 Mexico D.F.
Mexico

Spain/Portugal
Paraninfo Thomson Learning
Calle/Magallanes, 25
28015 Madrid, Spain

About the Authors

Andrew Fraknoi is the Chair of the Astronomy Department at Foothill College near San Francisco, where his courses are taken by about 900 students per year. He is also Director of Project ASTRO at the Astronomical Society of the Pacific—a national program that forms partnerships between volunteer astronomers and school teachers in their communities and develops astronomy activity kits for families to use outside of school. From 1978 to 1992 he was Executive Director of the Society, as well as Editor of *Mercury* Magazine and the *Universe in the Classroom* Newsletter. He has taught astronomy and physics at San Francisco State University, Cañada College, and the University of California Extension Division. He is co-author and editor of *The Universe at Your Fingertips* and *More Universe at Your Fingertips*, two widely used collections of astronomy teaching activities and resources. For five years he was the lead author of a nationally syndicated newspaper column on astronomy, and he appears regularly on radio and television explaining astronomical developments. With Sidney Wolff, he is co-editor of *Astronomy Education Review,* a new on-line journal/magazine for those working in space science education (see: aer.noao.edu). In addition, he has organized three national symposia on teaching introductory astronomy at the college level, and over 20 national workshops on improving the way astronomy is taught in earlier grades. He has received the Annenberg Foundation Prize of the American Astronomical Society and the Klumpke-Roberts Prize of the Astronomical Society of the Pacific for his contributions to the public understanding of astronomy. Asteroid 4859 was named Asteroid Fraknoi in 1992 in recognition of his work in astronomy education.

David Morrison is the Senior Scientist at the NASA Astrobiology Institute, where he participates in a variety of research programs in astrobiology—the study of the living universe. From 1996–2001 he was the Director of Astrobiology and Space Research at NASA Ames Research Center, managing basic and applied research programs in the space, life, and Earth sciences. Dr. Morrison received his Ph.D. in astronomy from Harvard University, and until he joined NASA he was Professor of Astronomy at the University of Hawaii. Internationally known for his research on small bodies in the solar system, Dr. Morrison is the author of more than 130 technical papers and has published a dozen books. He chaired the official NASA study of impact hazards that recommended that a Spaceguard Survey be carried out to search for potentially threatening asteroids and comets and in 1995 received the NASA Outstanding Leadership Medal for this work. He is also the recipient of the Dryden Medal for research from the American Institute of Aeronautics and Astronautics, and of the Klumpke-Roberts award of the Astronomical Society of the Pacific for contributions to science education. He has served as President of the Astronomical Society of the Pacific, Chair of the Astronomy Section of the American Association for the Advancement of Science, and President of the Planetary Commission of the International Astronomical Union. Asteroid 2410 Morrison is named in his honor.

Sidney C. Wolff received her Ph.D. from the University of California at Berkeley and then joined the Institute for Astronomy at the University of Hawaii. During the 17 years she spent in Hawaii, the Institute for Astronomy developed Mauna Kea into the world's premier international observatory. She became Associate Director of the Institute for Astronomy in 1976 and Acting Director in 1983. During that period, she earned international recognition for her research, particularly on stellar atmospheres and how they can help us understand the evolution, formation, and composition of stars. She is currently working on problems relating to how stars form. In 1984, Wolff was named Director of the Kitt Peak National Observatory and in 1987 became Director of the National Optical Astronomy Observatories. She was the first woman to head a major observatory in the United States. As Director of NOAO, she and her staff oversaw facilities used annually by nearly 1000 visiting scientists. During its early phases, she was Director of the Gemini Project, which is an international program to build two state-of-the-art 8m telescopes. Wolff is currently involved with the design of a telescope that would scan the whole visible sky every week or so; conventional telescopes would take decades to complete a similar survey. Such a facility would open up the systematic study of time variable and moving objects. She has served as President of the American Astronomical Society, and is also a member of the Board of Trustees of Carleton College, a liberal arts school that excels in science education. The author of more than 70 professional articles, she has written a monograph, *The A-Type Stars: Problems and Perspectives,* as well as several astronomy textbooks. She is the editor of the new on-line journal, *Astronomy Education Review.*

Preface for the Student

IN COLLEGE TEXTBOOKS, there is a long tradition that the preface of the book is read by the instructor and the rest of the book by the student. Still, many students start reading the preface (it does come first) and then wonder why it doesn't say much to them.

So we begin our book with a preface specifically for student readers. It's not about the subject matter of astronomy, which is introduced in the Prologue. Rather, we want to tell you a little about the book and give you some hints for the effective study of astronomy. (Your professor will probably have other, more specific suggestions for doing well in your class.)

THE WORLD OF ASTRONOMY

Astronomy, the study of the universe beyond the borders of our planet, is one of the most exciting and rapidly changing branches of science. Even scientists from other fields often confess to having had a lifelong interest in astronomy, though they may now be doing something earth-bound like biology, chemistry, engineering, or writing software. There are fewer than 10,000 professional astronomers in the entire world. However, astronomy has a large group of amateur devotees who spend many an evening with a telescope under the stars observing the sky, and who occasionally make a discovery, such as a new comet or a developing storm on Saturn.

Many other people are "armchair astronomers"—fascinated just to read about bizarre worlds and processes that astronomers are uncovering. Some are intrigued by the scientific search for planets or life in other star systems. Others like to follow the challenges of space exploration, such as upgrading the Hubble Space Telescope by the shuttle astronauts or the results from the Cassini mission to probe the giant planet Saturn and its intriguing moon Titan. Hearing about an astronomical event in the news media may be what first sparked your interest in taking an astronomy course.

But some of the things that make astronomy so interesting also make it a challenge for the beginning student.

The universe is a big place, full of objects and processes that do not have familiar counterparts here on Earth. Like a visitor to a new country, you will need time to feel familiar with the territory or the local customs. Astronomy, like other sciences, has its own special vocabulary, some of which you will have to learn if you want to have a good voyage with us. (Since the title of this text includes the word *Voyages*, we will use voyaging metaphors from time to time. Humor your authors; we mean well.)

FEATURES OF THIS BOOK

To assist students taking their first college-level course in astronomy, we have built a number of special features into this book, and we invite you to make use of them:

- All technical terms are printed in **boldface** type the first time they are used and clearly defined in the text; their definitions are listed alphabetically in Appendix 3 (the Glossary), so you can refer to them at any time. The summaries at the end of each chapter also include these boldface terms as a review.

- The book begins with a historical summary of astronomy and then surveys the universe, starting at home and finishing with the properties of the entire cosmos. But don't worry if your instructor doesn't assign all the chapters or doesn't assign the chapters in order. Throughout the book "directional signs" lead you to any earlier material you need to know before tackling the current section.

- We use tables to bring together numerical data for your convenience. For example, some tables summarize the important properties of each planet in the solar system (including the Earth). Students who want to see more of the data that astronomers use regularly can investigate the appendices at the back of the book.

- Figure captions clearly and fully describe what phenomena or objects students are looking at. We have tried to look at each figure with student eyes and ask what would help clarify the diagram or image for a reader just being introduced to astronomy.

- Each chapter ends with a summary of the essential points in the chapter, plus small-group activities (more about these in a minute), review questions, thought questions, and numerical problems to help you "process" what you have learned. We have put one or more worked out examples into a "Figuring for Yourself" box in most chapters to help you with these problems. Some simple multiple-choice quiz questions to help you review are also available on our Web site: http://info.brookscole.com/voyages.

- Suggestions for further reading in print and on the Web are included at the end of each chapter, in case you want to learn more about a particular topic. These books, articles, and Web sites are written at the same introductory level as this text. Appendices 1 and 2 contain a listing of general sources of astronomical information. Appendix 1 also includes Web sites that can guide you to the best new astronomical images (many of which we could not fit into the text).

- Each chapter features a series of small-group activities—mostly questions that take off in new directions from the material in the chapter and are meant to foster debate and discussion among students. Your instructor or discussion section leader may use these in class, or you may want to discuss questions that interest you with the other members of your study group. (What study group? you may be asking. See below.)

STUDYING ASTRONOMY

How can you increase your chances of doing well in your astronomy class? Here are a few suggestions for studying astronomy that come from good teachers and good students from around the country:

- First, the best advice we can give you is to be sure to leave enough time in your schedule to study the material in this class regularly. It sounds obvious, but it is not very easy to catch up with a subject like astronomy by trying to do everything just before an exam. Try to put aside some part of each day, or every other day, when you can have uninterrupted time for reading and studying astronomy.

- Try to read each assignment in the book twice, once before it is discussed in class and once afterward. Take notes or use a highlighter to outline ideas that you may want to review later. Also, take some time to coordinate the notes from your reading with the notes you take in class. Many students start college without good note-taking habits. If you are not a good note-taker, try to get some help. Many colleges and universities have student learning centers that offer short courses, workbooks, or videos on developing good study habits. Good note-taking skills will also be useful for many jobs or activities you may get involved with after college.

- Form a small astronomy study group with people in your class; get together as often as you can and discuss the topics that may be giving group members trouble. Make up sample exam questions and make sure everyone in the group can answer them confidently. If you have always studied alone, you may at first resist this idea, but don't be too hasty to say no. Study groups are a very effective way of digesting new information, learning a foreign language (which astronomy words sometimes resemble), or studying a broad field like law or astronomy.

- Before each exam, do a concise outline of the main ideas discussed in class and presented in your text. Compare your outline with those of other students as a check on your own study habits.

- If you find a topic in the text or in class especially difficult or interesting, don't hesitate to make use of the resources in your library or on the Web for additional study (each chapter has suggestions for you).

- Don't be too hard on yourself! If astronomy is new to you, many of the ideas and terms in this book may be unfamiliar. Astronomy is like any new language; it may take a while to become a good conversationalist. Practice as much as you can, but also realize that it is natural to be overwhelmed by the vastness of the universe and the variety of things that are going on in it.

We hope you enjoy reading this text as much as we enjoyed writing it. We are always glad to hear from students who have used the text and invite you to send us your reactions to the book and suggestions for how we can improve future editions. We promise you we will read and consider every serious letter we receive. You can send your comments to Andrew Fraknoi, Astronomy Department, Foothill College, 12345 El Monte Road, Los Altos Hills, CA 94022, USA. (Please note that we will not send you the answers to the chapter problems or do your homework for you, but all other thoughts are most welcome.)

Andrew Fraknoi
David Morrison
Sidney Wolff

Preface for the Instructor

WE'RE DELIGHTED THAT YOU ARE using or considering this book. *Voyages* is an astronomy text written with today's students in mind—it is designed for non-science majors who may even be a little intimidated by science, and who approach astronomy with more interest than experience. With features that we hope will appeal to everyone from university economics majors to first-time community college students, *Voyages* is written to draw in and engage readers while preserving the accuracy and timeliness that our colleagues expect of us.

In thinking through how best to present astronomy to students whose science and math background may be limited, we have tried to make the language friendly and inviting and to use examples and analogies drawn from everyday experiences. Yet we have not hesitated to include the very latest research results and to explain the latest thinking in the many disciplines that make up our field. We very much hope our ways of drawing in readers—including vignettes from the lives of astronomers, references to other fields that are influenced by astronomy, and occasional touches of humor—can help make this a book that your students will actually enjoy reading. There is, we believe, nothing less useful than a textbook the instructor opens more often than the student!

ORGANIZATION AND SPECIAL FEATURES

Voyages is designed to serve both a one-semester overview course and a two-semester introduction to the field. (It is available in one full volume and two shorter, separate books, one on the solar system and one on stars and galaxies. As befits its special role in astronomy, the Sun's chapter is included in both books, together with the introductory material on telescopes, radiation, and history that many astronomy courses cover.)

We realize that some instructors will want to teach topics in a different order than we have presented or leave some out altogether. It's fair to say that in astronomy—much more than in any other introductory science class—there are as many approaches to the basic course as there are instructors teaching it. Accordingly, we have tried to write the text so that the chapters and even main sections within chapters often stand on their own. Colleagues with very different course outlines from ours have told us that *Voyages* is, in fact, a book they use without difficulty.

Since we do not expect students to remember every concept introduced in previous chapters, we have inserted unobtrusive verbal "sign posts" throughout to help students find where a concept was defined or explained in detail. We sometimes review a key idea that may have been introduced many chapters ago. Our aim is that your student should be able to use the book as an easy navigational tool, no matter how you teach your course. A complete glossary (even including terms not used in the book but often used by instructors) is supplied in Appendix 3.

Our book is not an encyclopedia of astronomy (which is why students do not need to work out for six weeks in a gym before they can carry it around). We tried not to overwhelm the reader with detail, but instead to focus on the major threads and overarching ideas of astronomy. We probably have a bit less jargon than most textbooks, but we have worked hard not to sacrifice any of the key concepts that we believe students should learn in a basic course.

The third edition includes many of the latest ideas and discoveries in astronomy, not merely for their novelty but for their value in advancing the quest for a coherent understanding of the universe. In each case, we have tried to fit the latest research results into a wider context and to explain clearly what they mean. Among the recent topics included in this edition are:

- a complete update on cosmology, including early results from the Wilkinson MAP satellite;
- a new chapter on astrobiology, with careful attention to recent ideas about the origin and early evolution of life on Earth;
- a full report on the Shoemaker NEAR spacecraft mission and its exploration of the asteroid Eros;
- updated discussions of the discoveries of extrasolar planets and their significance;
- a new section on gravitational waves in astronomy;

- the latest on the relationship between solar activity and Earth climate.

We also portray astronomy as a human endeavor and have tried to include descriptions and images of some of the key men and women who have created our science over the years. Illustrations also include many of the latest images from the Hubble Space Telescope and other space instruments, as well as an up-to-date collection of color images from ground-based observatories around the world. Full-color diagrams are used as teaching tools, not as cosmetic devices. Figure captions contain full explanations of what the student should be seeing and understanding, not just the cryptic one-liners featured in so many other books.

Among the other features of the book that we hope will particularly help students and instructors are:

- suggestions for accurate, accessible Web sites and readings at the end of each chapter and in the appendices for those who want to explore further;
- three levels of end-of-chapter questions: review questions, thought-provoking exercises that ask students to apply what they have learned, and numerical problems;
- *Inter-Activity* questions in each chapter, consisting of a series of interactive group exercises (see box) that can be a change of pace during a lecture, form the basis of a discussion section, or serve as homework to teams of students.

At the request of a number of our adopters, our third edition features a new series of "Figuring for Yourself" boxes in each chapter. These demonstrate how to solve problems relating to the chapter and prepare students for doing the numerical questions. Our aim is to give instructors who use mathematics more extensively in their lectures both examples and questions to assign, and yet not to break up the main flow of our narrative for those who don't. We have significantly increased the number of numerical problems in this edition as well.

All our teaching materials are kept updated at the text Web site: info.brookscole.com/voyages.

The chapters in *Voyages* also feature highlighted boxes designed to help non-science students appreciate the breadth of astronomy without distracting them from the main narrative.

- *Making Connections:* These special boxes show how astronomy connects to students' experiences with other fields of human activity, from poetry to engineering, from popular culture to natural disasters.
- *Voyagers in Astronomy:* These profiles of noted astronomers focus not only on their work but also on their lives and times.
- *Astronomy Basics:* In these short sections we try to explain fundamental science ideas and terms that other texts often assume (incorrectly) that students know.
- *Seeing for Yourself:* Here, students become familiar with the sky and everyday astronomical phenomena through observations using simple equipment.

ANCILLARY MATERIALS

We are happy to offer instructors who adopt *Voyages* a growing series of learning and teaching aids to make their lives a little bit easier. These include:

1. *The* Voyages *Instructor's Manual/Test Bank*
   Our unique instructor's manual contains answers to thought questions and numerical problems in the textbook, hints for selected *Inter-Activity* exercises, various possible course outlines, and annotated guides to printed and Web resources for astronomy instructors. It's especially aimed at instructors new to teaching but will, we think, still contain much of interest to veteran professors. The manual also includes a large number of multiple-choice test questions for use in a variety of classroom settings (also provided on disk). We're proud to say that, unlike most texts, our multiple-choice questions have five answers, and are far more than just regurgitation of factoids. Many of the questions require applying the principles students learned in new situations, or finding similarities or differences among disparate phenomena. There is also a liberal sprinkling of humor among the answers to cheer up the suffering students during the exam.

2. *ExamView® Computerized Test Bank for Windows and Macintosh*
   ExamView features all the questions from the printed test bank in a format that allows instructors to edit them, add questions, and compose many versions of the same test, in print or online.

3. *The* Voyages *Transparency Collection*
   This set contains 100 of the best figures and images in the book as full-page transparencies for your use in the classroom.

4. *The* Voyages *Multimedia Manager*
   This CD-ROM includes virtually all the diagrams, photos, and paintings from the text, together with a number of other images that we wanted to use in the book but couldn't find room for (especially recent images from planetary exploration and from large telescopes in orbit and on the ground). The software allows you to include them in your own presentations and add your own files or graphics in any order.

5. *The* Sky™ *Student Edition on CD-ROM (packaged free with every text)*
   See the sky from anywhere over a period of 12,000 years with this well-reviewed planetarium program. Comes with a database of 118,000 stars and 13,000 deep-sky objects and prints star charts for labs or observing sessions.

6. *InfoTrac® College Edition (free 4-month access code packaged with the book)*
   This online library of magazines and journals lets your students read or print complete articles from nearly 4,000 publications. Great for instructors who assign pa-

Collaborative Group Activities as a Teaching Tool

Recent research has shown that listening to an hour or hour-and-a-half lecture on some topic in astronomy is not always conducive to effective student learning. Students seem to learn better when they are active participants rather than passive listeners. Unfortunately, many of us teach huge classes, where it might seem that student participation could not possibly be practical. Yet, science instructors in many fields are beginning to come up with ways that even large-lecture classes can do small-group activities. (Readings and Web sites to assist you with trying such activities can be found in the *Voyages* Instructors' Manual.)

In the text, we focus on one approach, which involves small groups and collaborative group activities either during lecture or in discussion section. The instructor divides the class (or asks the students to divide themselves) into groups of three or four. The smaller you make the groups, the less likely it is that some students will be able to remain uninvolved, but groups of two are typically not sufficient for good discussions to evolve.

Each group is given a task and a time limit. Each chapter of *Voyages* presents a series of *Inter-Activity* exercises that are good for group discussion and typically do not require any outside materials or data. In addition, some instructors like to present each group with some simple data to analyze (on a handout or an overhead). After group members discuss or do the activity, they elect a spokesperson, who shares their results with the whole class. If your class is really large,

perhaps only a representative number of groups will report on any given day.

Based on what they hear in the reports, groups can then return to their discussion and change or refine their answers. Some instructors require each person or the whole group to write a short summary of their conclusions; others feel that the experience of interacting with their peers is sufficient.

This approach works well even with showing slides. Put up a high-resolution Galileo image of the surface of Europa, for example, and ask the groups to discuss what they are seeing, how large an area they think it is, and how such features might originate. The important thing is not for students to get a "right answer" but to get them thinking, talking, interacting, and participating in their own education. Students are so used to being passive in lecture classes that this approach may take a few tries before they get fully involved, but an instructor (and section leaders) walking around to facilitate discussion and get those who are congenitally uninvolved to participate can work wonders after a while.

We should warn you that many of the suggested group activities in *Voyages* are open-ended and may engender a great deal of discussion. You should probably have a strategy for coming to closure in a way that still leaves students feeling that their opinions are valued. A short paper (perhaps for extra credit), an interactive Web site, or a chance for further debate in a discussion section are some ways of capturing student enthusiasm while still being able to move on in the main lecture.

---

pers or require outside research but have students who whine about going to a library.

7. *Virtual Laboratories (free access code packaged with the book)*
Michael Guidry, of the University of Tennessee, has created a series of online interactive lab exercises on 20 important concepts in astronomy. Students can print out or e-mail completed lab assignments to the instructor.

8. *WebTutor (with WebCT and Blackboard)*
Software to help you set up course-specific and text-specific course management tools, with syllabi, office hours, student discussions, quizzes, and much more.

In addition, your Brooks/Cole–Thomson Learning representative can discuss with you many other materials available for adopters, including *RedShift*™ software, lab workbooks for both *The Sky* and *RedShift*, an *InfoTrac College Edition* student guide, a collection of short astronomy videos from CNN, and *My Course 2.0* Web software.

THE REGISTERED ADOPTERS PROGRAM

With many textbooks, the time you place your order is the last time you hear from the publisher or the authors. In fact, many authors and publishers have no idea who adopted their book and never communicate with the instructors except through sales reps. We hope to be a little bit different; we'd like to know who our adopters are, stay in touch with them, and provide new materials, updated information, and so on. Over the years, the community of *Voyages* adopters has not only grown but has been very important to us in shaping the development of the texts.

Accordingly, we have created a "Registered Adopters" program for *Voyages*. To join, you merely send us an e-mail with your name, e-mail, college mailing address, the course you are teaching, and which text you are using. We will verify the adoption with your bookstore and then be in touch with you from time to time, mostly to let you know when we are publishing something (or giving away something free) that may help your teaching. There is no cost involved, and we promise not to release the names for any other purpose.

If you are not yet a registered adopter, we urge you to get on our list. You can do so by e-mailing the lead author of the book at:

fraknoiandrew@fhda.edu

(Note that the last name does not have an "n" before the "k"—that misspelling is the number one reason e-mails get returned.) If you prefer, you can send the information by particle mail to: Andrew Fraknoi, Astronomy Dept., Foothill College, 12345 El Monte Road, Los Altos Hills, CA 94022.

LET US HEAR FROM YOU

Unlike stars and planets, textbooks in astronomy do not exist in a vacuum. All three authors have benefited tremendously over the years from the advice of colleagues and students who teach astronomy. We want to be sure that we continue to make changes and updates in *Voyages* that will be the most useful to you. We welcome your comments and suggestions for improving the *Voyages* texts, our Web site, and our ancillary materials.

Please address your comments to Andrew Fraknoi (see addresses above), who will share them with the other two authors.

ACKNOWLEDGMENTS

We would like to thank the many colleagues and friends who have provided information, images, and encouragement, including: Charles Avis, Charles Bailyn, Bruce Balick, Roger Bell, Roy Bishop, Clark Chapman, Chip Clark, Grace Deming, David DesMarais, Richard Dreiser, Doug Duncan, George Djorgovski, Alex Filippenko, Debra Fischer, Dale Frail, Alex Fraknoi, Lola Fraknoi, Kathy Fransham, Michael Friedlander, Alan Friedman, Ian Gatley, Paul Geissler, Margaret Geller, Cheryl Gundy, Heidi Hammel, Alan Harris, Todd Henry, George Jacoby, Bruce Jakosky, William Keel, Jonathan Lunine, Geoff Marcy, Jeff McClintock, John McNeel, Michael Merrill, Jeff Moore, Janet Morrison, Jerome Orosz, Donald Osterbrock, Tobias Owen, Michael Perryman, Carle Pieters, Axel Quetz, Dennis Schatz, Rudy Schild, Maarten Schmidt, William Schopf, Joe Term, Ray Villard, Adrienne Wasserman, Richard Wolff, and Don Yeomans.

We are grateful to Lynette Cook, Bill Hartmann, John Spencer, Don Davis, and Don Dixon for permission to reproduce their wonderfully illustrative astronomical paintings, and to David Malin for his assistance and his superb astronomical photographs.

We appreciate the guidance of colleagues who reviewed all or part of the manuscript.

For the First Edition

Grady Blount
Texas A&M University

Michael Briley
University of Wisconsin, Oshkosh

David Buckley
East Strohdsburg University

John Burns
Mt. San Antonio College

Paul Campbell
Western Kentucky University

Eugene R. Capriotti
Michigan State University

George L. Cassiday
University of Utah

John Cunningham
Miami-Dale Community College

Grace Deming
University of Maryland

Miriam Dittman
DeKalb College

Gary J. Ferland
University of Kentucky

George Hamilton
Community College of Philadelphia

Ronald Kaitchuck
Ball State University

William C. Keel
University of Alabama

Steven L. Kipp
Mankato State University

Jim Lattimer
SUNY, Stonybrook

Robert Leacock
University of Florida

Terry Lemley
Heidelberg College

Bennett Link
Montana State University

Charles H. McGruder III
Western Kentucky University

Stephen A. Naftilan
Claremont Colleges

Anthony Pabon
DeAnza College

Cynthia W. Peterson
University of Connecticut

Andrew Pica
Salisbury State University

Terry Richardson
College of Charleston

Margaret Riedinger
University of Tennessee, Knoxville

Jim Rostirolla
Bellevue Community College

Michael L. Sitko
University of Cincinnati

John Stolar
West Chester University

Charles R. Tolbert
University of Virginia

Steve Velasquez
Heidelberg College

David Weinrich
Moorhead State University

David Weintraub
Vanderbilt University

Mary Lou West
Montclair State University

Dan Wilkins
University of Nebraska, Omaha

J. Wayne Wooten
Pensacola Junior College

For the Second Edition

Mitchell C. Begelman
University of Colorado

Stephen Danford
University of North Carolina, Greensboro

Richard French
Wellesley College

Catharine Garmany
University of Colorado, Boulder

Owen Gingerich
Harvard University

Edward Harrison
University of Massachusetts

Scott Johnson
Idaho State University

Mark Lane
Palomar College

John Patrick Lestrade
Mississippi State University

Michael C. LoPresto
Henry Ford Community College

Anthony Marston
Drake University

Ronald A. Schorn
Texas A&M University

Vernon Smith
University of Texas, El Paso

Ronald Stoner
Bowling Green State University

Jack W. Sulentic
University of Alabama

Stephen Walton
California State University, Northridge

Warren Young
Youngstown State University

For the Third Edition

Paul DeVries
Miami University

Steven Doty
Denison University

John F. Hawley
University of Virginia

Michael Kaufman
San Jose State University

Robbie F. Kouri
Our Lady of the Lake University

Michael C. Lopresto
Henry Ford Community College

Ed Oberhofer
Lake Sumter Community College

No book project of this complexity could succeed without the diligent efforts of many people at Brooks/Cole–Thomson Learning. We very much appreciate the assistance of:

Keith Dodson—Acquisitions Editor
Marie Carigma-Sambilay—Development Editor
Carol Benedict—Assistant Editor
Melissa Newt—Editorial Assistant
Samuel Subity—Technology Project Manager
Kelley McAllister—Marketing Manager
Sandra Perin—Marketing Assistant
Hal Humphrey—Production Project Editor

Nancy Shammas, New Leaf Publishing—Production Service
Nathaniel Bergson-Michelson—Executive Advertising Project Manager

But most of all, we appreciate the hundreds of instructors and many thousands of students around the United States, Canada, and the world, who have honored us by using this textbook on their own voyages through the universe.

Andrew Fraknoi
David Morrison
Sidney Wolff

Brief Contents

Contents

APPENDICES

Distant Galaxies A mosaic of images of remote galaxies of stars, taken with four cameras aboard the repaired Hubble Space Telescope in March 1994. The area of the sky shown in this image is extremely small—about the size of President Roosevelt's eye on a U.S. dime held at arm's length. The brilliant galaxy that fills the small image (upper left) lies on the outskirts of the Coma Cluster, a rich group of galaxies so far away that light from it takes 300 million years to reach us. Everything on this image that is not a round dot is a galaxy of billions of stars (the round dots are stars in our own Milky Way Galaxy). Except for the galaxy we mentioned above and the spiral-shaped galaxy at right, all the other galaxies in this view are significantly farther away than the Coma Cluster, and are shown with great clarity from the Hubble's position above the Earth's atmosphere. They present a marvelous array of shapes and colors, challenging astronomers to understand their birth and evolution.

Prologue
Science and the
Universe: A Brief Tour

Everybody, at last, is getting nosy. I predict that in our time astronomy will become the gossip of the marketplace.

Galileo speaking in Bertolt Brecht's play _Galileo._

We invite you to come with us on a series of voyages to explore the universe as astronomers understand it today. Out there we will find vast, awesome, and magnificent realms—full of objects that have no counterparts on Earth. Nevertheless, we hope you will learn to call the universe our home, and that you will see that its evolution has been directly responsible for your presence on planet Earth today.

Along our journey, we will encounter:

- a grand canyon system so large that on Earth it would stretch from Los Angeles to Washington, D.C. (Figure P.1).
- a scar and other evidence on Earth that tells us that the dinosaurs (and many other creatures) died because of a cosmic collision
- a tiny moon whose gravity is so weak that one good throw from its surface could put a baseball into orbit.
- a collapsed star so dense that to duplicate its interior we would have to squeeze every human being on Earth into a single raindrop.
- exploding stars whose violent end could wipe a planet orbiting a neighboring star clean of all its life-forms (Figure P.2).
- a "cannibal galaxy" with billions of stars that has already consumed a number of its smaller galaxy neighbors and is not yet finished finding new victims.
- the radio echo that is the faint but unmistakable signal of the creation event itself.

Ace Astronomy™ The AceAstronomy icon throughout the text indicates an opportunity for you to test yourself on key concepts and to explore animations and interactions on the AceAstronomy website at **http://ace.brookscole.com/voyages**

Virtual Laboratories

 Units, Measure and Unit Conversion

Mars Mosaic   This image of Mars was constructed at the U.S. Geological Survey in Flagstaff, Arizona, by combining many individual images taken by the Viking spacecraft starting in 1976. The view is centered on the Valles Marineris (Mariner Valley) complex of canyons, which is as long as the United States is wide.

It is these kinds of discoveries that make astronomy such an exciting field for both scientists and people in other walks of life. But we will explore more than just the objects in our universe and the latest discoveries about them. We will pay equal attention to the *process* by which we have come to understand the realms beyond our Earth and the tools we need to increase that understanding.

Astronomy today must still be carried out mostly from Earth. We gather information about the cosmos from the messages that the universe is kind enough to send our way. And because the stars are the fundamental building blocks of the universe, decoding the message of starlight has been a central challenge and triumph of modern astronomy. By the time you finish reading this text, you will know a bit about how to read that message and how to understand what it is telling us.

1 THE NATURE OF ASTRONOMY

Astronomy is defined as the study of the objects that lie beyond our planet Earth, and of the processes by which these objects interact with one another. But, as we will see, it is much more than that. It is also humanity's attempt to organize what we learn into a clear history of the universe, from the instant of its birth in the Big Bang to the present moment in which you are finishing this sentence.

Putting a cosmic history together takes some nerve—especially for a little creature living on a rocky ball circling a nondescript star in the suburbs of the Milky Way—and we are surely far from finished with it. Throughout the book, we emphasize how much science is a *progress report*—one that constantly changes as new techniques and instruments allow us to probe the universe more deeply.

In considering the history of the universe, we will see again and again that the cosmos *evolves;* it changes in profound ways over long periods of time. Although you may have read in the popular press that the concept of evolution is controversial in science, this is simply not true. Few ideas are better established than the notion that the universe is not the same today as it was long ago, and that it has existed and been evolving for billions of years.

You will see, for example, that the universe could not have produced the readers of this book during the first generation of stars after the Big Bang. The ingredients in the recipe for an astronomy student simply did not exist then in sufficient numbers. The universe still had to make the carbon, the calcium, and the oxygen needed to construct something as interesting and complicated as you. Today, many billions of years later, the universe has evolved into a more hospitable place for life. Tracing the evolutionary processes that continue to shape the universe is one of the most important (and satisfying) parts of modern astronomy.

European Southern Observatory

■ **FIGURE P.2**

A Stellar Corpse We see the remains of a star that was seen to explode in our skies in 1054 (and was briefly bright enough to be visible during the daytime). Today, the remnant is called the Crab Nebula, and its central region is seen here in a new image taken with the Very Large Telescope in Chile. Such exploding stars were crucial to the development of life in the universe.

2 THE NATURE OF SCIENCE

Unlike religion or parts of philosophy, science accepts nothing on faith. The ultimate judge in science is always the experiment or observation: what nature itself reveals. Science, then, is not merely a body of knowledge, but a *method* by which we attempt to understand nature and how it behaves. This method begins with many observations over a period of time. From the trends in the observations, scientists may come up with a *model* of the particular phenomenon we want to understand. Such models are always approximations of nature itself, subject to further testing.

To take a concrete astronomical example, ancient astronomers constructed a model (partly from observations, partly from philosophical beliefs) that the Earth was the center of the universe and that everything moved around it. At first the available observations of the Sun, Moon, and planets could be fit to this model, but eventually better observations required the model to add circle after circle to the movements of the planets to keep the Earth at the center. As the centuries went by and improved instruments were developed for keeping track of objects in the sky, the old model (even with a huge number of circles) could no longer explain all the observed facts. As we will see in Chapter 1, a new model, with the Sun at the center, fit the experimental evidence better. After a period of philosophical struggle, it became accepted as our view of the universe.

When they are first proposed, new models or ideas are sometimes called *hypotheses.* Many students today think that there can be no new hypotheses in a science such as astronomy, that everything important has already been learned. Nothing could be further from the truth. Throughout this book you will find discussions of recent, and occasionally still controversial, hypotheses in astronomy—concerning, for instance, the significance for life on our planet of huge chunks of rock and ice that hit the Earth; the existence of vast quantities of invisible "dark energy," which could make up the bulk of the universe; and the presence of a strange "black hole" at the center of the Milky Way. All such hypotheses are based on difficult observations done at the forefront of our technology, and all require further testing before we fully incorporate them into our standard astronomical models.

This last point is crucial: If there is no possible way of *testing* a hypothesis, it does not belong in the realm of science. The most straightforward approach to such testing in science is to perform an experiment. If the experiment is properly conducted, either its results will agree with the predictions of the hypothesis or they will contradict it. If the experimental result is truly inconsistent with the hypothesis, a scientist must discard the hypothesis (no matter how fond of it she may have grown) and try to develop an alternative. If the experimental result agrees with predictions, that is satisfying but does not prove that the hypothesis is correct. Perhaps later experiments will contradict

crucial parts of the hypothesis. But the more experiments a hypothesis survives, the more likely we are to accept it as a useful description of nature.

When you read about experiments, you probably have a mental picture of a scientist in a laboratory, making some careful test or measurement. This is certainly the case for a biologist or a chemist, but what can astronomers do when our laboratory is the universe? It's impossible to put a group of stars into a test tube or order another comet from a scientific supply company.

As a result, astronomy is sometimes called an *observational* science; we often make our tests by observing many samples of the kind of object we want to study and carefully noting how different samples vary. The invention of new instruments sometimes lets us look at astronomical objects from new perspectives or in greater detail. Our hypotheses are then judged in the light of this new information, and they pass or fail in the same way that we would evaluate the result of a laboratory experiment. Because the objects we want to study are far away and hard to see, many years may pass before the appropriate observational tests can be made in astronomy. We must simply wait patiently until either nature or better instruments make some deciding test possible.

Much of astronomy is also a *historical* science—meaning that what we observe has already happened in the universe, and we can do nothing to change it. In the same way, a geologist cannot alter what has happened to our planet, and a paleontologist cannot bring an ancient animal back to life. While this sometimes makes astronomy challenging, it also gives us fascinating opportunities to discover the secrets of our cosmic past.

You might compare an astronomer to a detective trying to solve a crime that occurred before the detective ever got to the scene. There is all sorts of evidence, but both the detective and the scientist must sift through it carefully and organize it to test various hypotheses about what really happened. And there is another way in which the scientist is like a detective: They both must prove their case. The detective must convince the district attorney, the judge, and perhaps ultimately the jury that his hypothesis is correct. Similarly the scientist must convince colleagues, editors of journals, and ultimately a broad cross section of other scientists that his hypothesis is provisionally correct. In both cases, one can only ask for evidence "beyond a reasonable doubt." And sometimes new evidence will force both the detective and the scientist to revise their last hypotheses.

It is this self-correcting aspect of science that sets it off from most human activities. Scientists spend a great deal of time questioning and criticizing one another. No project is funded and no report is published without extensive *peer review*—that is, without careful examination by other scientists in the same field. In nonscience areas, young people are often taught to accept the authority of their elders without question, but in science (after proper training), everyone is encouraged to try new and better experiments and to challenge any and all hypotheses. Young scientists know that one of the best ways to advance their careers is to find a weakness in our current understanding and to correct it with a new or modified hypothesis.

This is one of the reasons science has made such dramatic progress. An undergraduate science major today knows more science and math than did Isaac Newton, one of the most brilliant scientists who ever lived. Even in this introductory astronomy course you will learn about objects and processes that a few generations ago no one even dreamed existed. While the domain of science is limited, within that domain its achievements have been glorious.

3 THE LAWS OF NATURE

Over the centuries, scientists have extracted from countless observations certain fundamental principles, called *scientific laws*. These are, in a sense, the rules of the game nature plays. One remarkable discovery about nature—one that underlies everything you will read in this book—is that the same laws apply everywhere in the universe. The rules that govern the behavior of gravity on Earth, for example, are the same rules that determine the motion of two stars in a system so far away that your eye cannot find them in the sky.

Note that without the existence of such universal laws, we could not make much headway in astronomy. If each pocket of the universe had not only different objects but also different rules, we would have little chance of interpreting what happened in other "neighborhoods." But the consistency of the laws of nature gives us enormous power to understand distant objects without traveling to them and learning the local laws. In the same way, if every U.S. state or Canadian province had completely different laws, it would be very difficult to carry out commerce or even to understand the behavior of people in different regions. But a consistent set of laws allows us to apply what we learn or practice in one state to any other state.

You might ask whether the laws of the natural world could (under the right circumstances) be suspended. This is an enticing fantasy, but, despite many attempts, not a shred of scientific evidence has been found to support such an idea. While it would be nice if we could suspend the law of gravity and float away from our porch through effort of will alone, in real life such experiments generally result in broken bones.

This is not to say that our models or rules cannot change. New experiments and observations can lead to new or more sophisticated models—models that can include even new phenomena and new laws about their behavior. The general theory of relativity proposed by Albert Einstein is a perfect example of such a transformation, one that took place less than a century ago. It led us to predict, and more recently to catch, indirect glimpses of a strange new class of objects astronomers call black holes. But wishing isn't going to bring such new models into existence; only the patient process of observing nature ever more finely can reap such rewards.

One important problem about describing scientific models has to do with the limitations of language. When we try to describe complex phenomena in everyday terms, the words themselves may not be adequate to do the job. For example, you may have heard the structure of the atom likened to a miniature solar system. While some aspects of our modern model of the atom do remind us of planetary orbits, many other of its aspects are fundamentally different.

This problem is the reason scientists often prefer to describe their theories using equations rather than words. In this book, designed to introduce the field of astronomy, we use mainly words to discuss what scientists have learned. We avoid math beyond basic algebra. But if this course piques your interest and you go on in science, more and more of your studies will involve the precise language of mathematics.

4 NUMBERS IN ASTRONOMY

You may have heard a television reporter refer to a large number (such as the national debt) as "astronomical." In astronomy, we do deal with distances on a scale you may never have thought about before, and with numbers larger than any you may have encountered. Most students take a while to learn to navigate among the millions and billions that astronomers tend to throw about in their everyday discussions. But with some practice, you can become just as good at it!

By the way, if you sometimes have trouble sorting out millions and billions, take heart. Even so distinguished a group as the Presidential Advisory Committee on the Future of the U.S. Space Program, in a 1990 report, listed the distance to the planet Uranus as 1.7 million miles, when in fact Uranus is 1.7 billion miles away. A million (1,000,000) is a thousand times less than a billion (1,000,000,000). If Uranus were suddenly that close, we would all have noticed it because it would be bigger and brighter than the Moon in our skies.

In this book we adopt two approaches that make dealing with astronomical numbers a little bit easier. First, we use a system for writing large and small numbers called *powers-of-ten notation* (or sometimes *scientific notation*). This system is very appealing because it does away with the huge number of zeros that can seem overwhelming to the reader.

In scientific notation, if you want to write a figure like $490,000 (which is definitely not the starting salary for an astronomer but might be for a television star), you write:

$$4.9 \times 10^5$$

The small, raised number after the 10, called an *exponent*, keeps track of the number of tens you have to multiply together to get the number you want. In our example, five tens are multiplied together, so that $10 \times 10 \times 10 \times 10 \times 10$ gives 100,000. Multiply 100,000 by 4.9 and you get our astronomical starting salary. Another way to remember the basics of this notation is to note that 5 is the number of places you have to move the decimal point to the right to convert 4.9 to 490,000. If you are encountering this system for the first time or would like a refresher, we suggest you look at Appendix 4 for more information and examples.

Small numbers are written with negative exponents. Three millionths (0.000003) is expressed as:

$$3.0 \times 10^{-6}$$

One reason this notation is so popular among scientists—trust us, it is, even if you at first don't like it—is that it makes arithmetic a lot easier. To multiply two numbers in scientific notation, you need only add their exponents. Thus, you can multiply a thousand times a billion:

$$10^3 \times 10^9 = 10^{12}$$

To divide numbers, just subtract exponents.

The second way we try to keep numbers simple is to use a consistent set of units—the metric International System of Units. Unlike the system used in the United States, in which a completely arbitrary number such as 5280 feet equals 1 mile, metric units are related by powers of ten; a kilometer, for example, equals a thousand (10^3) meters. If you are used to U.S. units, the kilometer is about ⁶⁄₁₀ of a mile. The metric system, which has been adopted by every major country in the world except the United States, is summarized in Appendix 5. You may want to make it part of your vocabulary if you want to be ready for the future (when everyone will be using metric units).

5 LIGHT YEARS

To give you a chance to practice scientific notation, and to set the scene for the tour of the universe in the next section, let's define a common unit astronomers use to describe distances in the universe. A *light year* (LY) is the distance that light travels during 1 year. Since light always travels at the same speed, and since its speed turns out to be the fastest possible speed in the universe, it makes a good standard for keeping track of distances. Some students complain about this name for a unit of distance; light years seem to imply that we are measuring time. But this mix-up of time and distance is common in everyday life as well—for example, when we tell a friend to meet us at a movie theater that's 20 minutes away.

So, how many kilometers are there in a light year? If you are new to scientific notation, we suggest you work through all the steps of the following example for yourself on a separate sheet of paper. Light travels at the amazing pace of 3×10^5 kilometers per second (km/s). Think about that: Light covers 300,000 km every single second. In 1 second, it can travel seven times around the circumference of the Earth; a commercial airplane, in contrast, would take about two days to go around once, not counting time to refuel.

Now that we know how far light goes in a second, we can calculate how far it goes in a year:

1. There are 60 (6×10^1) seconds in each minute and 6×10^1 minutes in every hour.
2. Multiply those numbers and you get that there are 3.6×10^3 seconds per hour.
3. Thus light covers 3×10^5 km/s $\times$ 3.6×10^3 s/h = 1.08×10^9 km/h.
4. There are 24 or 2.4×10^1 hours in a day and 365.24 (3.65×10^2) days in a year.
5. The product of those two numbers is 8.77×10^3 h/year.
6. Multiplying that product by 1.08×10^9 km/h gives 9.46×10^{12} km in a light year.

That's almost 10 trillion km that light covers in a year. A string 1 LY long could fit around the circumference of the Earth 236 million times!

You might think that such a long unit would more than reach to the nearest star. But the stars are far more remote than our imaginations (or episodes of *Star Trek*) might lead us to believe. Even the nearest star is 4.3 LY away—more than 40 trillion km. Other stars visible to the unaided eye are hundreds or even thousands of light years away (Figure P.3). This is why astronomers are skeptical that UFOs are extraterrestrial spacecraft coming here across vast distances, briefly picking up two rural fishermen or loggers, and then going straight home. It seems like such a small reward for such a large investment.

6 CONSEQUENCES OF LIGHT TRAVEL TIME

There is another reason the speed of light is such a natural unit of distance for astronomers. Information about the universe comes to us almost exclusively through various forms of light, and all such light travels at the speed of light—that is, 1 LY every year. This sets a limit on how quickly we can learn about events in the universe. If a star is 100 LY away, the light we see from it tonight left that star 100 years ago and is just now arriving in our neighborhood. The soonest we can learn about any changes in that star—its blowing up, for example—is 100 years after the fact. For a star 500 LY away, the light we detect tonight left 500 years ago and is carrying 500-year-old news.

Some students, accustomed to CNN and other news media known for instant world coverage, at first find this frustrating. "You mean when I see that star up there," they ask, "I won't know what's actually happening there now for another 500 years?" But that's not really the right way to think about the situation. For astronomers, *now* is when the light reaches us here on Earth. There is no way for us to know anything about that star (or other object) until its light reaches us; despite the fondest dreams of science-fiction writers, instant communication through the universe is not possible.

■ **FIGURE P.3**

The Orion Nebula This beautiful cloud of cosmic raw material (gas and dust from which new stars and planets are being made) called the Orion Nebula is about 1500 LY away. That's a distance of roughly 1.4×10^{16} km—a pretty big number! The picture we see is actually a seamless mosaic of 15 smaller images taken with the Hubble Space Telescope. The field of view is about 2.5 LY wide and shows only a small part of a vast reservoir of mostly dark material. The gas and dust in this region are illuminated by the intense light from a few extremely energetic adolescent stars in the neighborhood.

But what at first may seem a great frustration is actually a tremendous benefit in disguise. If astronomers really want to piece together what has happened in the universe since its beginnings, they need to find evidence about each epoch of the past. Where can we find evidence today about cosmic events that occurred billions of years ago? Unfortunately, the universe doesn't leave written records (or even decent videotapes) of its main activities!

The delay in the arrival of light provides such evidence automatically. The farther out in space we look, the longer the light has taken to get here, and the longer ago it left its place of origin. By looking billions of light years out into space, astronomers are actually seeing billions of years into the past. In this way, we can reconstruct the history of the cosmos and get a sense of how it has evolved over time.

This is one reason astronomers strive to build telescopes that can collect more and more of the faint light the universe sends us. The more light we collect, the fainter the objects we can make out. On average, fainter objects are farther away and can thus tell us about periods of time even deeper in the past. New instruments, such as the Hubble Space Telescope (Figure P.4) and Hawaii's Keck Telescope (see Chapter 5), are giving astronomers views of deep space and deep time better than any we have had before.

7   A TOUR OF THE UNIVERSE

Let us now take a brief introductory tour of the universe as astronomers understand it today, just to get acquainted with the sorts of objects and distances we will encounter throughout the text. We begin at home with the Earth, a

■ **FIGURE P.4**

Telescope in Orbit The Hubble Space Telescope, shown here being repaired aboard the Space Shuttle *Endeavour* in December 1993, is an example of the new generation of astronomical instruments in space.

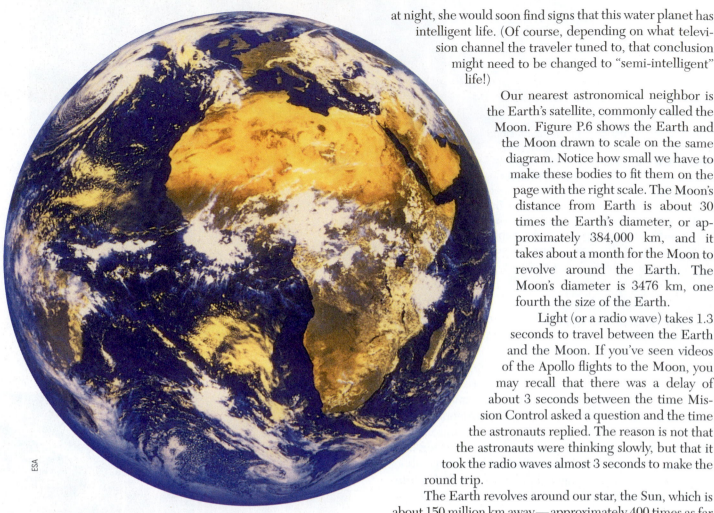

FIGURE P.5

Humanity's Home Base Planet Earth as viewed from a perspective in space. This image was taken on May 3, 1990, by the Meteosat weather satellite high above the Earth's equator.

at night, she would soon find signs that this water planet has intelligent life. (Of course, depending on what television channel the traveler tuned to, that conclusion might need to be changed to "semi-intelligent" life!)

Our nearest astronomical neighbor is the Earth's satellite, commonly called the Moon. Figure P.6 shows the Earth and the Moon drawn to scale on the same diagram. Notice how small we have to make these bodies to fit them on the page with the right scale. The Moon's distance from Earth is about 30 times the Earth's diameter, or approximately 384,000 km, and it takes about a month for the Moon to revolve around the Earth. The Moon's diameter is 3476 km, one fourth the size of the Earth.

Light (or a radio wave) takes 1.3 seconds to travel between the Earth and the Moon. If you've seen videos of the Apollo flights to the Moon, you may recall that there was a delay of about 3 seconds between the time Mission Control asked a question and the time the astronauts replied. The reason is not that the astronauts were thinking slowly, but that it took the radio waves almost 3 seconds to make the round trip.

The Earth revolves around our star, the Sun, which is about 150 million km away—approximately 400 times as far as the Moon. We still call the average Earth–Sun distance an astronomical unit (AU) because in the early days of astronomy it was the most important measuring standard. Light takes a little more than 8 minutes to travel 1 AU, which means our latest news from the Sun is always 8 minutes old.

It takes the Earth 1 year (3×10^7 seconds) to go around the Sun at our distance; to make it around, we must travel at approximately 110,000 km/h. (If you, like many students in the United States, still prefer miles to kilometers, you might find the following trick helpful. To convert kilometers to miles, you can multiply kilometers by 0.6. Thus 110,000 km/h becomes 66,000 mi/h—a fast clip no matter what units you use.) Since gravity holds us firmly to the Earth and there is no resistance to the Earth's motion in

nearly spherical planet about 13,000 km in diameter (Figure P.5). A space traveler entering our planetary system would easily distinguish the Earth by the large amount of liquid water that covers some two thirds of its crust. If the traveler had equipment to receive radio or television signals, or came close enough to see the lights of our cities

Earth

Moon

FIGURE P.6

The Earth and Moon, Drawn to Scale

the vacuum of space, we participate in this breakneck journey without being aware of it day by day.

The diameter of the Sun is about 1.5 million km; our Earth could fit comfortably inside one of the minor eruptions that occur on the surface of our star. If the Sun were reduced to the size of a basketball, the Earth would be a small apple seed some 30 m from the ball.

The Earth is only one of nine planets that revolve around the Sun. These planets, along with their satellites and swarms of smaller bodies, make up the solar system, what we might call the family of the Sun (Figure P.7). A planet is defined as a body of significant size that orbits a star and does not produce its own light. (If a large body consistently produces its own light, it gets to be called a star.) We'll see later in the book that this simple definition, perfectly fine as we begin our voyages, will need to be modified a bit. Astronomers have discovered a class of objects intermediate between stars and planets, which we call *brown dwarfs*. These are, in a sense, failed stars—balls of hot gas that do not have what it takes to keep producing significant amounts of light. Our technology has just recently gotten good enough to detect these.

We are able to see the nearby planets in our skies only because they reflect the light of our local star, the Sun. If the planets were much farther away, the tiny amount of light that they manage to reflect would not be visible to us. The planets we have so far been able to discover orbiting other stars were found from the pull their gravity exerts on their parent stars, not from their light.

Jupiter, the largest planet in the solar system, is about 143,000 km in diameter, 11 times the size of the Earth (Figure P.8). Its distance from the Sun is five times that of the Earth, or 5 AU. On the scale where the Sun is a basketball, Jupiter is about the size of a grape located about 150 m from the basketball. The orbits of the planets are shown schematically in Figure P.9. Pluto has a significantly non-circular orbit, but its average distance from the Sun is about 40 AU, or 5.9 billion km. On our basketball scale, Pluto is a grain of sand about 1 km from the ball.

The Sun is our local star, and all the other stars are also suns—enormous balls of glowing gas that generate vast amounts of energy by nuclear reactions deep within. We will discuss the fascinating processes that make the stars shine in more detail later in the book. The other stars look faint only because they are so very far away. If we continue our basketball analogy, Proxima Centauri, the nearest star beyond the Sun, which is 4.3 LY away, is almost 7000 km from the basketball.

When we look up at the star-studded country sky on a clear night, all the stars visible to the unaided eye turn out to be part of a single collection of stars we call the Milky

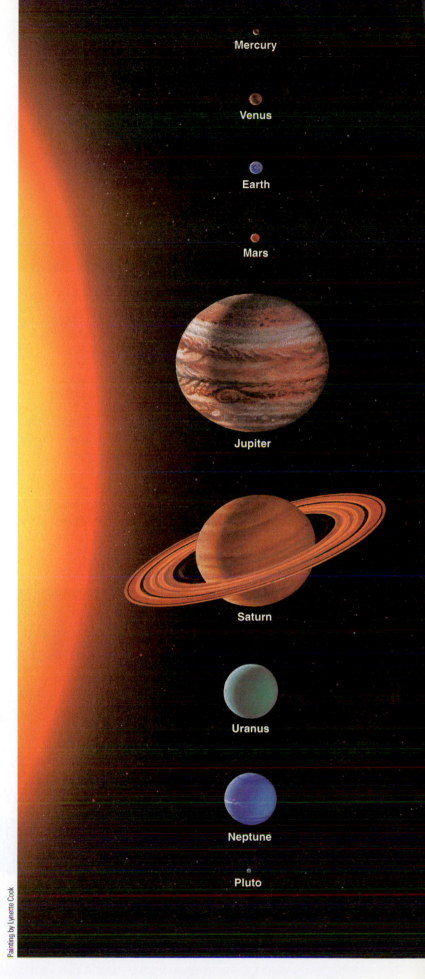

■ **FIGURE P.7**

Our Solar Family The Sun and the planets shown to scale. Notice the size of the Earth compared to the giant planets.

■ **FIGURE P.8**

Jupiter The largest planet in our solar system is Jupiter. We could fit almost 11 Earths side by side into its equator, and it contains as much mass as all the other planets combined. This image, which also shows one of its large satellites, was taken in February 1979 when the Voyager 1 spacecraft flew by.

Way Galaxy, or simply the Galaxy. (When referring to the Milky Way, we capitalize *Galaxy;* when talking about other galaxies of stars, we use lowercase *galaxy*.) The Sun is one of hundreds of billions of stars that make up the Galaxy; its extent, as we will see, staggers the human imagination.

Let's make a rough scale drawing showing the stars within 10 LY of the Sun (Figure P.10). The small circle la-

beled (a) represents a sphere 10 LY in radius centered on the Sun. We find roughly ten stars in this sphere. Now we change the scale: The circle labeled (b) represents a sphere 100 LY in radius. Note that all of (a) is now just a small circle in the center. Sphere (b) contains about 10,000 (10^4) stars, far too many to make the job of counting them pleasant or the job of naming them reasonable. And yet in going out to a distance of even 200 LY, we have traversed only a tiny part of the Milky Way Galaxy.

Now let's draw a circle of radius 1000 LY as (c), in which once again, the circle in (b) is just a small center. Within the 1000-LY sphere, we find some 10 million (10^7) stars. In Figure P.10d, we change the scale and examine the entire Galaxy, a wheel-shaped system whose visible diameter is roughly 100,000 LY (seen edge-on in our figure). Our Galaxy looks like a giant frisbee with a small ball in the middle. If we could move outside our Galaxy and look down on the disk of the Milky Way from above, it would probably resemble the galaxy in Figure P.11, its spiral structure outlined by the blue light of hot adolescent stars.

The Sun is somewhat less than 30,000 LY from the center of the Galaxy, in a location with nothing much to distinguish it. From our position inside the Milky Way Galaxy, we cannot see through to its far rim (at least not with ordinary light) because the space between the stars is not completely empty. It contains a sparse distribution of gas (mostly the simplest element, hydrogen) intermixed with tiny solid particles that we call interstellar dust. This gas and dust collect into enormous clouds in many places in the Galaxy, becoming the raw material for future generations of stars. Figure P.12 shows a beautiful image of the disk of the Galaxy as seen from our vantage point.

■ **FIGURE P.9**

The Orbits of the Planets in Our Solar System (Left) The inner planets. (Right) The outer planets (note the change of scale).

■ FIGURE P.10

Counting Stars   The distribution of stars around the Sun within (a) 10 LY, (b) 100 LY, (c) 1000 LY, and (d) the Galaxy (with its disk seen edge-on).

European Southern Observatory

■ FIGURE P.11

A Spiral Galaxy   This galaxy of billions of stars, called by its catalog number NGC 1232, is about 100 million LY away and is thought to be similar to our own Milky Way Galaxy. Here we see the giant wheel-shaped system face-on, looking down on the colorful disk of stars with the spiral arms outlined by hot, young, bluish stars.

Roger Angel, Steward Observatory/University of Arizona

■ **FIGURE P.12**

The Milky Way   Because we are inside the Galaxy, we
see its disk in cross section flung across the sky like a great
white avenue of stars. This full-sky view, taken with a special
lens from the summit of Mount Graham in southern Arizona,
shows the Milky Way, with its myriad stars and dark "rifts"
of dust. The outer circle is the horizon, and you can see, in
addition to fir trees and test towers, the lights of several
Arizona cities around it.

Anglo-Australian Observatory/David Malin Images

■ **FIGURE P.13**

A Stellar Nursery   In a cloud of cosmic raw material
called the Rosette Nebula, a cluster of bright, hot stars can
be seen forming in the upper left.

Typically, the interstellar material is so extremely sparse
that the space between stars is a far, far better vacuum than
anything we can produce in terrestrial laboratories. Yet, the
dust in space, building up over thousands of light years, can
block the light of more distant stars. Like the distant buildings that disappear from our view on a smoggy day in Los
Angeles, the more distant regions of the Milky Way cannot
be seen behind the layers of interstellar smog. Luckily, astronomers have found that stars and raw material shine
with various forms of *invisible* light that do penetrate the
smog, and so we have been able to build up a pretty good
map of the Galaxy.

Recent observations, however, have also revealed a
rather surprising and disturbing fact. There appears to be
more—much more—to the Galaxy than meets the eye (or
the telescope). From various investigations, we have evidence that much of our Galaxy is made of material that we
cannot at present observe directly with our instruments
(not even with those that detect the invisible forms of light
we just mentioned). We therefore call this component of
the Galaxy *dark matter*. We know the dark matter is there
by the pull its gravity exerts on the stars and raw material
we can observe, but what this dark matter is made of and

how much of it exists remain a mystery. (As you will see,
this dark matter is not confined to our Galaxy; it appears to
be an important part of other star groupings as well.)

By the way, not all stars live by themselves, as the Sun
does. Many are born in double or triple systems, with two
or three stars revolving about each other (and even larger
numbers of partners possible). Because the stars influence
each other in such close systems, multiple stars allow us to
measure characteristics that we cannot discern from observing single stars. In a number of places, enough stars
have formed together that we recognize them as star clusters (Figure P.13). Some of the largest of the more than a
thousand star clusters that astronomers have cataloged
contain hundreds of thousands of stars and take up volumes
of space hundreds of light years across.

Because stars live a long time (compared to the people
who like to watch them), you often hear them referred to
as "eternal." But in fact no star can last forever. Since the
"business" of stars is making energy, and energy production
requires some sort of fuel to be used up, eventually all stars
will run out of fuel and go out of business. (This news should

not make you rush out to stock up on thermal underwear; our Sun still has at least 7 billion years to go.) Ultimately, the Sun and all stars will die, and it is in their death throes that some of the most intriguing and important processes of the universe stand revealed. For example, we now know that many of the atoms in our bodies were once inside stars. These stars exploded at the ends of their lives, recycling their material back into the reservoir of the Galaxy. In this sense, all of us are literally made of "star dust."

Photo taken with the U.K. Schmidt Telescope; Anglo-Australian Telescope Board

■ **FIGURE P.14**

A Neighbor Galaxy   The Large Magellanic Cloud, at a distance of 160,000 LY, is one of the Milky Way's closest neighbor galaxies and contains billions of stars. The reddish regions are nebulae, places where gas and dust are being illuminated by the energetic light of young stars. The brightest nebula, at left center, is nicknamed the Tarantula.

<table><tr><td>8</td></tr></table>

THE UNIVERSE ON THE LARGE SCALE

In a very rough sense, you could think of the solar system as your house or apartment and the Galaxy as your town, made up of many houses and buildings. In the 20th century, astronomers were able to show that just as our world is made up of many, many towns, so the universe is made up of enormous numbers of galaxies. (We define the universe to be everything that exists that is accessible to our observations.) Galaxies stretch as far in space as our telescopes can see, many billions of them within the reach of modern instruments. When they were first discovered, some astronomers called galaxies "island universes," and the term is aptly descriptive: Galaxies do look like islands of stars in the vast, dark seas of intergalactic space.

The nearest galaxy, just discovered in 1993, is a small one that lies 75,000 LY from the Sun in the direction of the constellation Sagittarius, where the smog in our own Galaxy makes it especially difficult to discern. (A constellation, we should note, is one of the 88 sections into which astronomers divide the sky, each named after a prominent star pattern within it.) The existence of this Sagittarius dwarf galaxy is, in fact, still controversial and will require other observations before all astronomers are convinced. Beyond it lie two other small galaxies, about 160,000 LY away. First recorded by Magellan's crew as he sailed around the world, these are called the Magellanic Clouds (Figure P.14). All three of these small galaxies are satellites of the Milky Way, interacting with it through the force of gravity. Ultimately, all three may even be swallowed by our much larger Galaxy.

The nearest large galaxy is a spiral quite similar to our own, located in the constellation of Andromeda and thus often called the Andromeda Galaxy; it is also known by one of its catalog numbers, M31 (Figure P.15). Given the number of galaxies, no reasonable person would suggest giving all of them proper names. M31 is about 2 million LY away and, along with the Milky Way, is part of a small cluster of over 40 galaxies that we call the Local Group.

At distances of 10 to 15 million LY we find other small galaxy groups, and then at about 50 million LY there is a more impressive system, with thousands of member galaxies, called the Virgo Cluster. We have discovered that galaxies occur mostly in clusters, both large and small

Tony Hallas

■ **FIGURE P.15**

Closest Spiral Galaxy   The Andromeda Galaxy (M31) is a spiral-shaped collection of stars similar to our own Milky Way. Its flat disk of stars looks blue because its light is dominated by the vigorous light output of hot blue stars. The bulge in its center, on the other hand, contains older stars, which are predominantly yellow and red. Two smaller galaxies can be seen on either side of M31.

Image taken with the U.K. Schmidt Telescope, Anglo-Australian Observatory

FIGURE P.16

Fornax Cluster of Galaxies On this image, you can see a cluster of galaxies located about 60 million LY away, in the constellation of Fornax. All the objects that are not pinpoints of light on the picture are galaxies of billions of stars.

(Figure P.16). Some of the clusters themselves form into larger groups called superclusters.

Our Local Group, as well as the Virgo Cluster, is part of such a supercluster, which stretches over a diameter of at least 60 million LY. We are just beginning to explore the structure of the universe at these enormous scales and are already finding some unexpected results (which will be discussed in Chapter 27.)

At even greater distances, where many ordinary galaxies are too dim to see, we find the quasars. These are brilliant centers of galaxies, glowing with the light of some extraordinarily energetic process. One theory suggests that perhaps a giant black hole is swallowing whole neighborhoods of raw material. (We will describe the bizarre objects called black holes once we have discussed the lives of the stars. If you are not a black hole fan now, we predict you will be by the time you've finished this book.) Whatever the quasars are, their brilliance makes them the most distant beacons we can see in the dark oceans of space. They allow us to probe the universe 10 billion or more LY away, and thus 10 billion or more years in the past.

With the quasars we see a substantial way back to the Big Bang explosion that marks the beginning of time. Beyond the quasars, we can detect only the feeble glow of the explosion itself, filling the universe and thus coming to us from all directions in space. The discovery of this "afterglow of creation" was one of the most exciting events in 20th-century science, and we are still exploring the many things it has to tell us about the earliest times of the universe.

For now, we might just note that such ideas are far easier to state than to discover. Measurements of the properties of galaxies and quasars in remote locations require large telescopes, sophisticated light-amplifying devices, and painstaking labor. Every clear night, at observatories around the world, astronomers and students are at work on such mysteries as the birth of new stars and the large-scale structure of the universe, mostly observing one star or one galaxy at a time and fitting their results into the tapestry of our understanding.

9 THE UNIVERSE OF THE VERY SMALL

The foregoing discussion should impress on you that the universe is extraordinarily large and extraordinarily empty. The universe on average is 10,000 times emptier than our Galaxy. Yet, as we have seen, even the Galaxy is mostly empty space. The air we breathe has about 10^{19} atoms in each cubic centimeter—and we think of air as pretty empty stuff. In the interstellar gas of the Galaxy, there is about one atom in every cubic centimeter. Intergalactic space is so sparsely filled that to find one atom, on average, we must search through a cubic meter of space! Most of the universe is fantastically empty; places that are dense, such as the bodies of our readers, are tremendously rare.

Yet even the familiar solids, such as this textbook, are mostly space. If we could take such a solid apart, piece by piece, we would eventually reach the molecules of which it is formed. Molecules are the smallest particles into which matter can be divided while still retaining its chemical properties. A molecule of water (H_2O), for example, consists of two hydrogen atoms and one oxygen atom, bonded together.

Molecules, in turn, are built up of atoms, which are the smallest particles of an element that can still be identified

TABLE P.1 *The Cosmically Abundant Elements*

Element°	Symbol	Number of Atoms per Million Hydrogen Atoms
Hydrogen	H	1,000,000
Helium	He	80,000
Carbon	C	450
Nitrogen	N	92
Oxygen	O	740
Neon	Ne	130
Magnesium	Mg	40
Silicon	Si	37
Sulfur	S	19
Iron	Fe	32

° Our list of elements is arranged in order of the atomic number, which is the number of protons in each nucleus.

as that element. For example, an atom of gold is the smallest possible piece of gold (although one atom of gold won't impress your sweetheart very much). Nearly 100 different kinds of atoms (elements) exist in nature, but most of them are rare, and only a handful account for more than 99 percent of everything with which we come in contact. The most abundant elements in the cosmos today are listed in Table P.1; think of this table as the "greatest hits" of the universe when it comes to elements.

All atoms consist of a central, positively charged nucleus surrounded by negatively charged electrons. The bulk of the matter in each atom is found in the nucleus, which consists of positive protons and electrically neutral neutrons all tightly bound together in a very small space. Each element is defined by the number of protons in its atoms: Thus, any atom with 6 protons in its nucleus is called carbon, any with 50 protons is called tin, and any with 70 protons is called ytterbium. (Ytterbium, as you can probably guess, is not big on the cosmic hit parade of elements, but we like its name. For a list of the elements, see Appendix 13.)

The distance from an atomic nucleus to its electrons is typically 100,000 times the size of the nucleus itself. This is why we say that even solid matter is mostly space. The typical atom is far emptier than the solar system out to Pluto. (The distance from the Earth to the Sun, for example, is only 100 times the size of the Sun.) This is another reason atoms are not like miniature solar systems.

Remarkably, physicists have discovered that everything that happens in the universe, from the smallest atoms to the largest superclusters of galaxies, can be explained through the action of only four forces: gravity, electromagnetism (which combines the actions of electricity and magnetism), and two forces that act at the nuclear level. The fact that there are four forces (and not a million, or just

one) has puzzled physicists and astronomers for many years and has led to a quest for a unified picture of nature.

10 A CONCLUSION AND A BEGINNING

If you are typical of students new to astronomy, you have probably reached the end of our tour in this Prologue with mixed emotions. On the one hand, you may be fascinated by some of the new ideas you've read about, and eager to learn more. On the other, you may be feeling overwhelmed by the number of topics we have covered, and the number of new words and ideas we have introduced. Learning astronomy is a little like learning a new language: At first it seems there are so many new expressions that you'll never learn them all, but with practice, you soon develop facility with them.

At this point you may also feel a bit small and insignificant, dwarfed by the cosmic scales of distance and time. Such a feeling is not a bad thing from time to time (sometimes we wish more of our politicians and film stars felt this way). And just before a difficult exam, or when you've ended a treasured relationship, it can certainly help to see your problems in a cosmic perspective. But there is another way to look at what we've learned from our first glimpses of the cosmos.

Let us consider the history of the universe from the Big Bang to today and compress it, for easy reference, into a single year. (We have borrowed this idea from Carl Sagan's Pulitzer Prize-winning book, *The Dragons of Eden*, published in 1977 by Random House.) On this scale, the Big Bang happened at the first moment of January 1, the solar system formed around September 10, and the oldest rocks we can date on Earth go back to the third week in September (Figure P.17).

Where in this "cosmic year" does the origin of human beings fall? The answer turns out to be the evening of December 31! The invention of the alphabet doesn't occur until the 50th second of 11:59 P.M. on December 31. And the beginnings of modern astronomy are a mere fraction of a second before the New Year. Seen in a cosmic context, the amount of time we have had to study the stars is minute, and our success at piecing together as much of the story as we have is remarkable.

Certainly, our attempts to understand the universe are not complete. As new instruments and new ideas allow us to gather even better data about the cosmos, our present picture of astronomy will very likely undergo many changes. In fact, we would not be surprised if your grandchildren's grandchildren found some of the contents of this book a bit primitive. Still, as you read our current progress report on the exploration of the universe, take a few minutes every once in a while just to savor how much we have already learned.

December 19
The first vertebrates
(animals with a
backbone)

December 30
The extinction
of dinosaurs

December 26
The first
mammals

December 31
Late evening:
The first humans

December 31
20 seconds
before
midnight:
The invention
of agriculture

December 31
9 seconds
before
midnight:
The invention
of the
alphabet

September 10
The formation
of the Earth
(and other
planets in our
solar system)

September 23
The end of
heavy asteroid
and comet
bombardment

September 25
The origin of life
on Earth (estimated)

January 1
The Big Bang

■ **FIGURE P.17**

Charting Cosmic Time On a cosmic calendar, where the time since the Big Bang is compressed into one year, creatures we would call human do not emerge on the scene until the evening of December 31.

Ace◐Astronomy™ Log into AceAstronomy and select this chapter to see the Active Figure called "Earth Calendar."

SURFING THE WEB

If you have enjoyed some of the spectacular photographs in this Prologue and have access to the World Wide Web, we want to recommend the following Web sites to you. They will allow you to check out some of the best images astronomers have of cosmic objects and to keep track of new images as they come in:

Anglo-Australian Observatory Image Collection:
www.aao.gov.au/images.html
A marvelous library of images (focusing on nebulae and galaxies) taken using large Australian telescopes. Many are by David Malin, one of the finest astronomical photographers of our time.

Hubble Space Telescope Images:
hubblesite.org/newscenter/archive
All the magnificent Hubble Space Telescope images, with captions, detailed background information, and new research results. Check the latest images, browse the Hubble's "greatest hits," or search for objects of interest to you.

National Optical Astronomy Observatories Image Gallery: www.noao.edu/noaosci.html
NOAO includes a number of major telescopes in the United States and the Southern Hemisphere. Some of the best images from NOAO instruments are collected at this site.

Planetary Photojournal: photojournal.jpl.nasa.gov
This site features thousands of the best images from planetary exploration, with detailed captions and excellent indexing.

Astronomy Picture of the Day:
antwrp.gsfc.nasa.gov/apod/astropix.html
Astronomers Robert Nemiroff and Jerry Bonell feature one relatively new celestial image each day with a brief nontechnical caption. Over the years, some of the best astronomical images have been featured here.

European Southern Observatory:
www.eso.org/outreach/gallery/
This growing album contains images from the large telescopes in the Southern Hemisphere run by a consortium of European countries.

The constellation of Orion (a star grouping with a distinctive belt of three stars, part of the figure of the great hunter of ancient Greek mythology) can be seen above and to the right of the University of Toronto's telescope dome on Las Campanas mountain in Chile. To find our way among the stars, we rely on signposts like groups of stars that make a recognizable pattern we can point out to one another.

1 Observing the Sky: The Birth of Astronomy

If the Lord Almighty had consulted me before embarking upon Creation, I should have recommended something simpler.

Statement attributed to Alfonso the tenth, King of Castile, after having the Ptolemaic system of planet motions explained to him

THINKING AHEAD

Much to your surprise, a member of the Flat Earth Society moves in next door. He believes that the Earth is flat and that all the NASA images of a round Earth are either faked or simply show the round (but flat) disk of the Earth from above. How could you prove to this neighbor that the Earth is really a sphere? (You can find some of our answers in the chapter and a summary in a box later in the chapter, but try to figure out some ways by yourself before peeking at our suggestions! You might want to discuss ideas with other students in the class.)

Today, few people really spend much time looking at the night sky. When we see patterns in the dark, we're much more likely to be looking at a television or movie screen than at the heavens over our heads. But in ancient days, before electric lights and television robbed so many people of the beauty of the sky, the stars and planets were an important aspect of everyone's daily life. All the records that we have—on paper and in stone—show that ancient civilizations around the world noticed, worshipped, and tried to understand the lights in the sky and fit them into their own view of the world. These ancient observers found both majestic regularity and never-ending surprise in the motions of the heavens.

The Babylonians and Greeks, for example, believing the planets to be gods, studied their motions in hopes of understanding the presumed influences of those planet-gods on human affairs; thus developed the religion of astrology. Through their careful study of the planets, however, the ancient Greeks and later the Romans were also laying the foundations of the science of astronomy. In this chapter we take a look at the night sky as seen with the unaided eye and also examine some of the interesting history of how we came to understand the realm above our heads.

1.1 THE SKY ABOVE

Our senses suggest to us that the Earth is the center of the universe—the hub around which the heavens turn. This **geocentric** (Earth-centered) view was what almost everyone believed until the Renaissance. After all, it is simple, logical, and seemingly self-evident. Furthermore, the geocentric perspective reinforced those philosophical and religious systems that taught the unique role of human beings as the central focus of the cosmos. However, the geocentric view happens to be wrong. One of the great themes of our intellectual history is the overthrow of the geocentric perspective. Let us therefore take a look at the steps by which we reevaluated the place of our world in the cosmic order.

1.1.1 The Celestial Sphere

If you go on a camping trip or live far from city lights, your view of the sky on a clear night is pretty much identical to that seen by people all over the world before the invention of the telescope. Gazing up, you get the impression that the sky is a great hollow dome, with you at the center (Figure 1.1). The top of that dome, the point directly above your head, is called the **zenith,** and where the dome meets the Earth is called the **horizon.** It's easy to see the horizon as a circle around you from the sea or a flat prairie, but from most places where people live today, the horizon is hidden by mountains, trees, buildings, or smog.

If you lie back in an open field and observe the night sky for hours, as ancient shepherds and travelers regularly did, you will see stars rising on the eastern horizon (just as the Sun and Moon do), moving across the dome of heaven in the course of the night, and setting on the western horizon. Watching the sky turn like this night after night, you might eventually get the idea that the dome of the sky is really part of a great sphere that is turning around you, bringing different stars into view as it turns. The early Greeks regarded the sky as just such a **celestial sphere** (Figure 1.2). Some thought of it as an actual sphere of transparent crystalline material, with the stars embedded in it like tiny jewels.

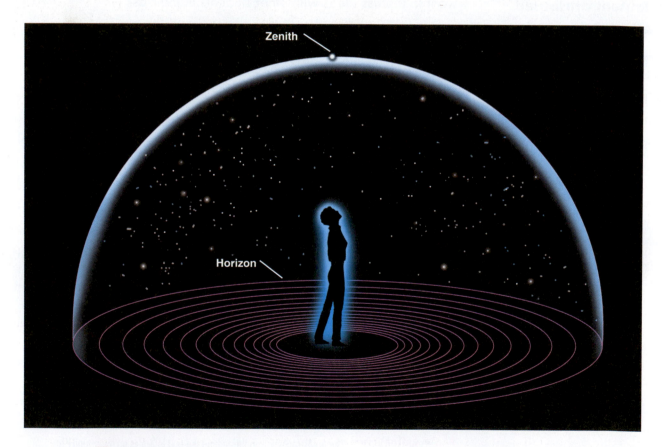

■ **FIGURE 1.1**

The Sky Around Us *The dome of the sky, as it appears to a naive observer. The horizon is where the sky meets the ground, and the observer's zenith is the point directly overhead.*

Ace⊘Astronomy™ Log into AceAstronomy and select this chapter to see the Active Figure called "Celestial Sphere."

■ **FIGURE 1.2**

Circles on the Celestial Sphere
Here we show the (imaginary) celestial sphere around the Earth, on which objects in the sky can turn. Note the axis through the Earth. In reality, it is the Earth that turns around this axis, creating the illusion that the sky revolves around us. Note that the Earth in this picture has been tilted so that your location is at the top and the North Pole is where the N is.

Today, we know that it's not the celestial sphere that turns as the night and day proceed, but rather the planet on which we live. We can put an imaginary stick through the Earth's North and South Poles; we call this stick our planet's **axis.** It is because the Earth turns on this axis every 24 hours that we see the Sun, Moon, and stars rise and set with clockwork regularity. These celestial objects are not on a dome, but at greatly varying distances from us in space. Nevertheless, it is sometimes still convenient to talk about the celestial dome or sphere to help us keep track of objects in the sky. There is even a special theater, called a *planetarium*, in which we project a simulation of the stars and planets onto a white dome.

As the celestial sphere rotates, the objects on it maintain their positions with respect to one another. A grouping of stars like the Big Dipper has the same shape during the course of the night, although it turns with the sky. During a single night, even objects that we know to have significant motions of their own, such as the nearby planets, seem fixed relative to the stars. (Only the *meteors*—brief "shooting stars" that flash into view for just a few seconds—move appreciably with respect to the celestial sphere. This is because they are not stars at all. Rather, they are small pieces of cosmic dust, burning up as they hit the Earth's atmosphere.) We can use the fact that the entire celestial sphere seems to turn together to help us set up systems for keeping track of what things are visible in the sky and where they happen to be at a given time.

1.1.2 Celestial Poles and Celestial Equator

To help orient us in the turning sky, astronomers use a system that extends our Earth's special points into the sky. We extend the line of the Earth's axis outward: The points where the axis meets the celestial sphere are defined as the **north celestial pole** and the **south celestial pole.** As the Earth rotates about its axis, the sky appears to turn in the opposite direction about those celestial poles (Figure 1.3). We also (in our imaginations) throw the Earth's equator onto the sky and call this the **celestial equator.** It lies halfway between the celestial poles, just as the Earth's equator lies halfway between our planet's poles.

Now let's imagine how riding on different parts of the spinning Earth affects our view of the sky. The apparent motion of the celestial sphere depends on your latitude (position north or south of the equator). Bear in mind that the Earth's axis is pointing at the celestial poles, so those two points on the sky do not appear to turn.

If you stood at the North Pole of the Earth, for example, you would see the north celestial pole overhead, at your zenith. The celestial equator, 90° from the celestial

■ **FIGURE 1.3**

Circling the North Celestial Pole Time exposure showing trails left by stars as a result of the apparent rotation of the celestial sphere. (In reality, it is the Earth that rotates.) The bright trail at top center was made by Polaris (the North Star), which is very close to the north celestial pole.

National Optical Astronomy Observatories

poles, lies along your horizon. As you watched them during the course of the night, the stars would all circle around the celestial pole, with none rising or setting (Figure 1.4). Only that half of the sky that is north of the celestial equator is ever visible to an observer at the North Pole. Similarly, an observer at the South Pole would see only the southern half of the sky.

If you were at the Earth's equator, on the other hand, you would see the celestial equator (which after all is just an "extension" of the Earth's equator) pass overhead through your zenith. The celestial poles, being 90° from the celestial equator, must then be at the north and south points on your horizon. As the sky turns, all stars rise and set; they move straight up from the east side of the horizon and set straight down on the west side. During a 24-hour period, all stars are above the horizon exactly half the time. (Of course, during some of those hours, the Sun is too bright for us to see them.)

What would an observer in the latitudes of the United States or Europe see? Remember, we are neither at the Earth's pole nor at the equator, but in between them. In our case, the north celestial pole is neither overhead nor on the horizon, but in between. It appears above the northern horizon at an angular height, or altitude, equal to the observer's latitude (see Figure 1.4). In San Francisco, for example, where the latitude is 38° N, the north celestial pole is 38° above the northern horizon.

For an observer at 38° N latitude, the south celestial pole is 38° below the southern horizon and thus never visible. As the Earth turns, the whole sky seems to pivot about the north celestial pole. For this observer, stars within 38° of the North Pole can never set. They are always above the horizon, day and night. This part of the sky is called the **north circumpolar zone.** To observers in the continental United States, the Big and Little Dippers and Cassiopeia are examples of star groups in the north circumpolar zone. On the other hand, stars within 38° of the south celestial pole never rise. That part of the sky is the **south circumpolar zone.** To most U.S. observers, the Southern Cross is in that zone. (Don't worry if you are not familiar with the star groups we just mentioned. We will introduce them more formally later on.)

At this particular time in Earth history, there happens to be a star very close to the north celestial pole. It is called Polaris, the pole star, and has the distinction of being the star that moves the least amount as the northern sky turns each day. (Because it moved so little while the other stars moved much more, it played a special role in the mythology of several Native American tribes, for example. Some called it the "fastener of the sky.")

1.1.3 Rising and Setting of the Sun

We have described the movement of stars in the night sky. The stars continue to circle during the day, but the brilliance of the Sun makes them difficult to see. (The Moon is

North celestial pole
Zenith
Observer
N
S
At North Pole

Zenith
S N
North celestial pole
At equator

Altitude of pole = latitude of the observer
North celestial pole
N
S
At intermediate latitude

■ FIGURE 1.4

Star Circles at Different Latitudes   The turning of the sky looks different depending on your latitude on Earth. (a) At the North Pole, the stars circle the zenith and do not rise and set. (b) At the equator, the celestial poles are on the horizon, and the stars rise straight up and set straight down. (c) At intermediate latitudes, the north celestial pole is at some position between overhead and the horizon. Its angle turns out to be equal to the observer's latitude. Stars rise and set at an angle to the horizon.

Ace◑Astronomy™   Log into AceAstronomy and select this chapter to see the Active Figure called "Daily Motion in the Sky."

still easily seen in the daylight, however.) On any given day, we can think of the Sun as being located at some position on the hypothetical celestial sphere. When the Sun rises—that is, when the rotation of the Earth carries the Sun above the horizon—sunlight is scattered by the molecules of our atmosphere, filling our sky with light and hiding the stars that are above the horizon.

For thousands of years, astronomers have been aware that the Sun does more than just rise and set. It gradually changes position on the celestial sphere, moving each day about 1° to the east relative to the stars. Very reasonably, the ancients thought this meant that the Sun was slowly moving around the Earth, taking a period of time we call one **year** to make a full circle. Today, of course, we know that it is the Earth that is going around the Sun, but the effect is the same: The Sun's position in our sky changes day to day. We have a similar experience when we walk around a campfire at night; we see the flames appear in front of each person seated about the fire in turn.

The path the Sun appears to take around the celestial sphere each year is called the **ecliptic** (Figure 1.5). Because of its motion on the ecliptic, the Sun rises about 4 minutes later each day with respect to the stars. The Earth must make just a bit more than one complete rotation (with respect to the stars) to bring the Sun up again. As the months go by and we look at the Sun from different places in our orbit, we see it projected against different stars in the background, or we would, at least, if we could see the stars in the daytime. In practice, we must deduce what stars lie behind and beyond the Sun by observing the

stars visible in the opposite direction at night. After a year, when the Earth has completed one trip around the Sun, the Sun will appear to have completed one circuit of the sky along the ecliptic.

The ecliptic does not lie along the celestial equator but is inclined to it at an angle of about 23°. In other words, the Sun's annual path in the sky is not lined up with the Earth's equator. This is because our planet's axis of rotation is tilted by about 23° from a vertical line sticking out of the plane of the ecliptic (Figure 1.6). Being tilted from "straight up" is not at all unusual among planets; Uranus and Pluto are actually tilted so much that they orbit the Sun "on their side." The inclination of the ecliptic is the reason the Sun moves north and south in the sky as the seasons change. In Chapter 3 we will discuss the progression of the seasons in more detail.

1.1.4 Fixed and Wandering Stars

The Sun is not the only object that moves among the fixed stars. The Moon and each of the five planets that are visible to the unaided eye—Mercury, Venus, Mars, Jupiter, and Saturn—also slowly change their positions from day to day. During a single day, the Moon and planets all rise and set as the Earth turns, just as the Sun and stars do. But like the Sun, they have independent motions among the stars, superimposed on the daily rotation of the celestial sphere. Noticing these motions, the Greeks of two thousand years ago distinguished between what they called the fixed stars, those that maintain fixed patterns among themselves

Constellation on the Ecliptic	Dates When the Sun Crosses It	Constellation on the Ecliptic	Dates When the Sun Crosses It
Capricornus	Jan 21–Feb 16	Leo	Aug 10–Sept 16
Aquarius	Feb 16–Mar 11	Virgo	Sept 16–Oct 31
Pisces	Mar 11–Apr 18	Libra	Oct 31–Nov 23
Aries	Apr 18–May 13	Scorpius	Nov 23–Nov 29
Taurus	May 13–June 22	Ophiuchus	Nov 29–Dec 18
Gemini	June 22–July 21	Sagittarius	Dec 18–Jan 21
Cancer	July 21–Aug 10		

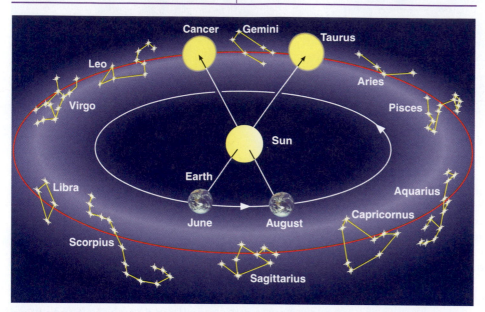

■ **FIGURE 1.5**

Constellations on the Ecliptic
As the Earth revolves around the Sun, we sit on "platform Earth" and see the Sun moving around the sky. The circle in the sky that the Sun appears to make around us in the course of a year is called the ecliptic. This circle (like all circles in the sky) goes through a set of constellations; the ancients thought that these constellations, which the Sun (and the Moon and planets) visited, must be special and incorporated them into their system of astrology (see Section 1.3). Note that at any given time of the year, some of the constellations crossed by the ecliptic are visible in the night sky and others are in the day sky and thus hidden by the brilliance of the Sun. As we discuss later in the chapter, today the zodiac constellations in which we "find" the Sun each month are no longer lined up with the signs the astrologers use.

Ace Astronomy™ Log into AceAstronomy and select this chapter to see the Active Figures called "Constellations from Different Latitudes" and "Constellations in Different Seasons."

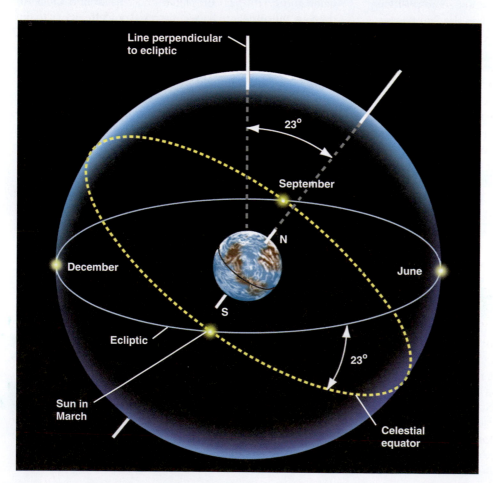

■ **FIGURE 1.6**

The Celestial Tilt The celestial equator is tilted by 23° to the ecliptic. As a result, North Americans and Europeans see the Sun north of the celestial equator and high in our sky in June, and south of the celestial equator and low in the sky in December.

through many generations, and the wandering stars, or **planets.** The word *planet,* in fact, means "wanderer" in Greek.

Today we do not regard the Sun and Moon as planets, but the ancients applied the term to all seven of the moving objects in the sky. Much of ancient astronomy was devoted to observing and predicting the motions of these celestial wanderers. The Moon, being the Earth's nearest celestial neighbor, has the fastest apparent motion; it completes a trip around the sky in about one month (*moonth*). To do this, the Moon moves about 12°, or 24 times its own apparent width on the sky, each day.

The individual paths of the Moon and planets in the sky all lie close to the ecliptic, although not exactly on it. This is because the paths of the planets about the Sun, and of the Moon about the Earth, are all in nearly the same plane, as if they were circles on a huge sheet of paper. The planets, the Sun, and the Moon are thus always found in the sky within a narrow 18°-wide belt called the **zodiac** that is centered on the ecliptic (see Figure 1.5). (The root of the term *zodiac* is the same as that of the word *zoo* and means a collection of animals; many of the patterns of stars within the zodiac belt reminded the ancients of animals, such as a fish or a goat.)

How the planets appear to move in the sky as the months pass is a combination of their actual motions plus the motion of the Earth about the Sun; consequently, their paths are somewhat complex. As we will see, this complexity has fascinated and challenged astronomers for centuries.

1.1.5 Constellations

The backdrop for the motions of the "wanderers" in the sky is the canopy of stars. If there were no clouds in the sky and we were on a flat plain with nothing to obstruct our view, we could see about 3000 stars with the unaided eye. To find their way around such a multitude, the ancients found groupings of stars that made some familiar geometric pattern or (more rarely) resembled something they knew. Each civilization found its own patterns in the stars, much like a modern Rorschach test in which you are asked to discover patterns or pictures in a set of inkblots. The ancient Chinese, Egyptians, and Greeks, among others, found their own groupings or **constellations** of stars helpful in navigating among them and in passing their star-lore on to their children.

You may be familiar with some of the star patterns we still use today, such as the Big and Little Dippers or Orion, the hunter, with his distinctive belt of three stars (Figure 1.7). However, many of the stars we see are not part of a distinctive star pattern at all, and a telescope can reveal millions of stars too faint for the eye to see. Therefore, in the early decades of the 20th century, astronomers from many countries decided to establish a more formal system for organizing the sky.

Today, we use the term *constellation* to mean one of 88 sectors into which we divide the sky, much as the United States has been divided into 50 states. The modern bound-

■ **FIGURE 1.7**

Orion The winter constellation of Orion, the hunter, as illustrated in the 17th-century atlas by Hevelius.

J. M. Pasachoff and the Chapin Library

aries between the constellations are imaginary lines in the sky running north–south and east–west, so that each point in the sky falls in a specific constellation. All the constellations are listed in Appendix 14. Whenever possible, we have named each modern constellation after the Latin translation of one of the ancient Greek star patterns that lies within it. Thus, the constellation of Orion is a kind of box on the sky, which includes, among many other objects, the stars that made up the ancient picture of the hunter. Some people use the term *asterism* to denote an especially noticeable star pattern within a constellation (or sometimes spanning parts of several constellations). For example, the Big Dipper is an asterism within the constellation of Ursa Major, the Big Bear.

Students are sometimes puzzled because the constellations seldom resemble the people or animals for which they were named. In all likelihood, the Greeks themselves did not name groupings of stars because they looked like actual people or objects (any more than the outline of Washington State resembles George Washington). Rather, they named sections of the sky in honor of the characters in their mythology and then fit the star configurations to the animals and people as best they could.

1.2 ANCIENT ASTRONOMY

Let us now look briefly back into history. Much of modern Western civilization is derived in one way or another from the ideas of the ancient Greeks and Romans, and this is true in astronomy as well. However, many other ancient cultures also developed sophisticated systems for observing and interpreting the sky.

1.2.1 Astronomy Around the World

Ancient Babylonian, Assyrian, and Egyptian astronomers knew the approximate length of the year. The Egyptians of 3000 years ago, for example, adopted a calendar based on a 365-day year. They kept careful track of the rising time of the bright star Sirius in the predawn sky, whose yearly cycle corresponded with the flooding of the river Nile. The Chinese also had a working calendar and determined the length of the year at about the same time. They recorded comets, bright meteors, and dark spots on the Sun. (Many types of astronomical objects were introduced in the Prologue of this book. If you are not familiar with terms like *comets* and *meteors,* you may want to look at the Prologue.) Later, Chinese astronomers kept careful records of "guest stars"—those that are normally too faint to see but suddenly flare up to become visible to the naked eye for a few weeks or months. We still use some of these records in studying stars that exploded a long time ago.

The Mayan culture in Central America developed a sophisticated calendar based on the planet Venus, and they made astronomical observations from sites dedicated to this purpose a thousand years ago. The Polynesians learned to navigate by the stars over hundreds of kilometers of open ocean, a skill that enabled them to colonize new islands far away from where they began.

In the British Isles, before the widespread use of writing, ancient people used stones to keep track of the motions of the Sun and Moon. We still find some of the great stone circles they built for this purpose, dating from as far back as 2800 B.C.E. The best known of these is Stonehenge, which is discussed in Chapter 3.

1.2.2 Early Greek and Roman Cosmology

Our concept of the cosmos—its basic structure and origin—is called **cosmology,** a Greek word. Before the invention of telescopes, humans had to depend on the simple evidence of their senses for a picture of the universe. The ancients developed cosmologies that combined their direct view of the heavens with a rich variety of philosophical and religious symbolism.

At least 2000 years before Columbus, educated people in the eastern Mediterranean region knew the Earth was round. Belief in a spherical Earth may have stemmed from the time of Pythagoras, a philosopher and mathematician who lived 2500 years ago. He believed circles and spheres to be "perfect forms" and suggested that the Earth should therefore be a sphere. As evidence that the gods liked spheres, the Greeks cited the fact that the Moon is a sphere, using evidence we will describe later.

The writings of Aristotle (384–322 B.C.E.), the tutor of Alexander the Great, summarize many of the ideas of his day. They describe how the progression of the Moon's phases—its changing shape—results from our seeing different portions of the Moon's sunlit hemisphere during the month (see Chapter 3). Aristotle also knew that the Sun has to be farther away from the Earth than is the Moon, be-

■ **FIGURE 1.8**

The Earth's Round Shadow A lunar eclipse, with the Moon moving into and out of the Earth's shadow. Note the curved shape of the shadow—evidence for a spherical Earth that has been recognized since antiquity.

cause occasionally the Moon passes exactly between the Earth and the Sun and temporarily hides the Sun from view. We call this a solar eclipse (see Chapter 3).

Aristotle cited two convincing arguments that the Earth must be round. First is the fact that as the Moon enters or emerges from the Earth's shadow during an eclipse of the Moon, the shape of the shadow seen on the Moon is always round (Figure 1.8). Only a spherical object always produces a round shadow. If the Earth were a disk, for example, there would be some occasions when the sunlight would be striking it edge-on and its shadow on the Moon would be a line.

As a second argument, Aristotle explained that travelers who go south a significant distance are able to observe stars that are not visible farther north. And the height of the North Star—the star nearest the north celestial pole—decreases as a traveler moves south. On a flat Earth, everyone would see the same stars overhead. The only possible explanation is that the traveler must have moved over a curved surface on the Earth, showing stars from a different angle. (See the box *How Do We Know the Earth Is Round?* for more ideas on proving the Earth is round.)

Ace⟳Astronomy™ Log into AceAstronomy and select this chapter to see Astronomy Exercise "Aristarchus's Measurement."

One brave Greek thinker, Aristarchus of Samos (310–230 B.C.E.), even suggested that the Earth was moving around the Sun, but Aristotle and most of the ancient Greek scholars rejected this idea. One of the reasons for

■ **FIGURE 1.9**

Light Rays from Space   The more distant an object, the more nearly parallel are the rays of light coming from it.

their conclusion was the thought that if the Earth moved about the Sun, they would be observing the stars from different places along Earth's orbit. As the Earth moved along, nearby stars should shift their positions in the sky relative to more distant stars. (In a similar way, we see foreground objects appear to move against a more distant background whenever we are in motion. When we ride on a train, the trees in the foreground appear to shift their position relative to distant hills as the train rolls by. Unconsciously, we use this phenomenon all of the time to estimate distances around us.)

The apparent shift in the direction of an object as a result of the motion of the observer is called *parallax*. We call the shift in the apparent direction of a star due to the Earth's orbital motion *stellar parallax*. The Greeks made dedicated efforts to observe stellar parallax, even enlisting the aid of Greek soldiers with the clearest vision, but to no avail. The brighter (and presumably nearer) stars just did not seem to shift as the Greeks observed them in the spring and then again in the fall.

Ace ◐ **Astronomy**™   Log into AceAstronomy and select this chapter to see Astronomy Exercise "Parallax I."

This meant either that the Earth was not moving or that the stars had to be so tremendously far away that the parallax shift was immeasurably small. A cosmos of such enormous extent required a leap of imagination that most ancient philosophers were not prepared to make, so they retreated to the safety of the Earth-centered view, which would dominate Western thinking for nearly two millennia.

1.2.3   Measurement of the Earth by Eratosthenes

The Greeks not only knew the Earth was round, they were also able to measure its size. The first fairly accurate determination of the Earth's diameter was made in about 200 B.C.E. by Eratosthenes, a Greek living in Alexandria, Egypt. His method was a geometrical one, based on observations of the Sun.

The Sun is so distant from us that all the light rays that strike our planet approach us along essentially parallel lines. To see why, look at Figure 1.9. Take a source of light near

How Do We Know the Earth Is Round?

In addition to the two ways (from Aristotle's writings) discussed in this chapter, you might also reason as follows:

1. Let's watch a ship leave its port and sail into the distance on a clear day. On a flat Earth, we would just see the ship get smaller and smaller as it sailed away. But that's not what we actually observe. Instead, ships sink below the horizon, with the hull disappearing first and the mast remaining visible for a while longer. Eventually only the top of the mast can be seen, as the ship sails around the curvature of the Earth; then finally the ship disappears.

2. The Space Shuttle circles the Earth once every 90 minutes or so. Photographs taken from the shuttle and other satellites show that the Earth is round from every perspective.

3. Suppose you made a friend in each time zone of the Earth. You could call all of them in the same hour and ask, "Where is the Sun?" On a flat Earth, each caller would give you roughly the same answer. But on a round Earth you would find that for some, the Sun would be high in the sky, while for others it would be rising, or setting, or completely out of sight (and this last group of friends would be upset with you for waking them up).

the Earth—say, at position A. Its rays strike different parts of the Earth along diverging paths. From a light source at B, or at C still farther away, the angle between rays that strike opposite parts of the Earth is smaller. The more distant the source, the smaller is the angle between the rays. For a source infinitely distant, the rays travel along parallel lines.

Of course, the Sun is not infinitely far away, but light rays striking the Earth from a point on the Sun diverge from one another by an angle far too small to be observed with the unaided eye. As a consequence, if people all over the Earth who could see the Sun were to point at it, their fingers would all be essentially parallel to one another. (The same is also true for the planets and stars, an idea we will use in our discussion of how telescopes work.)

Eratosthenes noticed that on the first day of summer at Syene, Egypt (near modern Aswan), sunlight struck the bottom of a vertical well at noon. This indicated that the Sun was right over the well (that Syene was on a direct line from the center of the Earth to the Sun). At the corresponding time and date in Alexandria, he observed that the Sun was not directly overhead but was slightly south of the zenith, so that its rays made an angle with the vertical equal to about 1/50 of a circle (7°). Since the Sun's rays striking the two cities are parallel to one another, why would the two rays not make the same angle with the Earth's surface? Eratosthenes reasoned that the curvature of the round Earth meant that "straight up" was not the same in the two cities. And the measurement of the angle in Alexandria, he realized, allowed him to figure out the size of the Earth. Alexandria, he saw, must be 1/50 of the Earth's circumference north of Syene (Figure 1.10). Alexandria had been measured to be 5000 stadia north of Syene (the *stadium* was a Greek unit of length, derived from the length of the racetrack in a stadium). Eratosthenes thus found that the Earth's circumference must be 50 × 5000, or 250,000 stadia.

It is not possible to evaluate precisely the accuracy of Eratosthenes' solution because there is doubt about which of the various kinds of Greek stadia he used as his unit of distance. If it was the common Olympic stadium, his result was about 20 percent too large. According to another interpretation, he used a stadium equal to about 1/6 km, in which case his figure was within 1 percent of the correct value of 40,000 km. Even if his measurement was not exact, his success at measuring the size of our planet by using only shadows, sunlight, and the power of human thought was one of the greatest intellectual achievements in history.

1.2.4 Hipparchus and Precession

Perhaps the greatest astronomer of pre-Christian antiquity was Hipparchus, born in Nicaea in what is present-day Turkey. He erected an observatory on the island of Rhodes in the period around 150 B.C.E., when the Roman Republic was increasing its influence throughout the Mediterranean region. There he measured as accurately as possible the directions of objects in the sky, compiling a pioneering star catalog with about 850 entries. He designated celestial coordinates for

■ FIGURE 1.10

How Eratosthenes Measured the Size of the Earth
The Sun's rays come in parallel, but because the Earth's surface curves, a ray at Syene comes straight down, while a ray at Alexandria makes an angle of 7° with the vertical. That means, in effect, that at Alexandria the Earth's surface has curved away from Syene by 7° out of 360°, or 1/50 of a full circle. Thus the distance between the two cities must be 1/50 the circumference of the Earth.

Ace◉Astronomy™ Log into AceAstronomy and select this chapter to see the Active Figure called "Erastothenes's Experiment" and Astronomy Exercise "Erastothenes's Calculations."

each star, specifying its position in the sky, just as we can specify the position of a point on the Earth by giving its latitude and longitude. He also divided the stars into **magnitudes** according to their apparent brightnesses. He called the brightest ones "stars of the first magnitude," the next brightest group "stars of the second magnitude," and so forth. This system, in modified form, still remains in use today.

By observing the stars and comparing his data with older observations, Hipparchus made one of his most remarkable discoveries: The position in the sky of the north celestial pole had altered over the previous century and a half. Hipparchus correctly deduced that this had happened not only during the period covered by his observations, but was in fact happening all the time: The direction around which the sky appears to rotate changes slowly but continuously.

You will recall from Section 1.1.2 that the north celestial pole is just the projection of the Earth's North Pole into the sky. If the north celestial pole is wobbling around, then the Earth itself must be doing the wobbling. Today we understand that the direction in which the Earth's axis points does indeed change slowly but regularly, a motion we call **precession.** If you've ever watched a child's spinning top wobble, you have observed a similar kind of motion. The

■ FIGURE 1.11

Precession Just as the axis of a rapidly spinning top slowly wobbles in a circle, so the axis of the Earth wobbles in a 26,000-year cycle. Today the north celestial pole is near the star Polaris, but about 5000 years ago it was close to a star called Thuban, and in 14,000 it will be closest to the star Vega.

top's axis describes a path in the shape of a cone, as the Earth's gravity tries to topple it over (Figure 1.11).

Because our planet is not an exact sphere, but bulges a bit at the equator, the pulls of the Sun and Moon cause it to wobble like a top. It takes about 26,000 years for the Earth's axis to complete one circle of precession. As a result of this motion, the point where our axis points in the sky changes as time goes on. While Polaris is the star closest to the north celestial pole today (it will reach its closest point around the year 2100), the star Vega in the constellation of Lyra will be the north star in the year 14,000.

1.2.5 Ptolemy's Model of the Solar System

The last great astronomer of the Roman era was Claudius Ptolemy (or Ptolemaeus), who flourished in Alexandria in about the year 140. He wrote a mammoth compilation of astronomical knowledge, which today is called by its Arabic name, *Almagest* (meaning "The Greatest"). *Almagest* does not deal exclusively with Ptolemy's own work; it includes a discussion of the astronomical achievements of the past, principally those of Hipparchus. Today it is our main source of information about the work of Hipparchus and other Greek astronomers.

Ptolemy's most important contribution was a geometrical representation of the solar system that predicted the

positions of the planets for any desired date and time. Hipparchus, not having enough data on hand to solve the problem himself, had instead amassed observational material for posterity to use. Ptolemy supplemented this material with new observations of his own and produced a cosmological model that endured more than a thousand years, until the time of Copernicus.

The complicating factor in explaining the motions of the planets is that their apparent wanderings in the sky result from the combination of their own motions with the Earth's orbital revolution. As we watch the planets from our vantage point on the moving Earth, it's a little like watching a car race from inside one of the cars. Sometimes other cars pull ahead, while at other times we might pass opponents' cars, making them seem to move backward for a while with respect to our car.

Figure 1.12a shows the motion of the Earth and a planet farther from the Sun, such as Mars or Jupiter. The Earth travels around the Sun in the same direction as the other planet and in nearly the same plane, but its speed in orbit is faster. As a result, it periodically overtakes the planet, like a faster race car on the inside track. The figure shows where we see the planet in the sky at different times. In Figure 1.12b, we show what the path of the planet looks like among the stars.

Normally, planets move eastward in the sky over the weeks and months as they orbit the Sun, but from positions

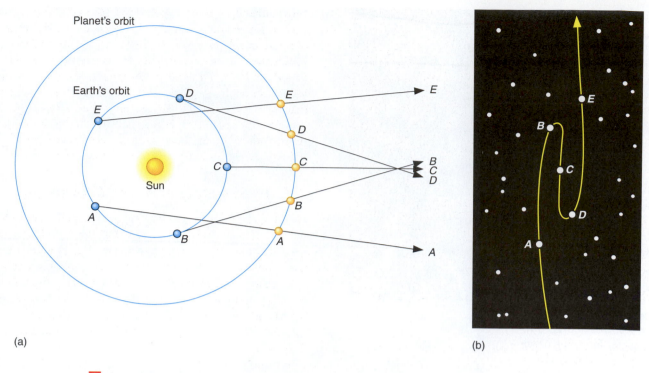

(a)

(b)

■ FIGURE 1.12

Retrograde Motion of a Planet Beyond the Earth's Orbit   The letters on the diagram show where the Earth (blue) and the other planet (yellow) are at different times. (a) Actual positions of the planet and the Earth. (b) What the path of the planet looks like in the Earth's skies against the background of stars.

B to *D* in Figure 1.12, as the Earth passes the planet, it appears to drift backward, moving west in the sky. Even though it is actually moving to the east, the faster Earth has overtaken it and seems, from our perspective, to be leaving it behind. As the Earth rounds its orbit toward position *E,* the planet again takes up its apparent eastward motion in the sky. The temporary apparent westward motion of a planet as the Earth swings between it and the Sun is called **retrograde motion.** Such backward motion is much easier for us to understand today, now that we know the Earth is one of the moving planets and not the unmoving center of all creation. But Ptolemy was faced with the far more complex problem of explaining such motion while assuming a stationary Earth.

Furthermore, because the Greeks believed that celestial motions had to be circles, Ptolemy had to construct a model using circles alone. To do it, he needed dozens of circles, some moving around other circles, in a complex structure that makes a modern viewer dizzy. But we must not let our modern judgment cloud our admiration for Ptolemy's achievement. In his day, a complex universe centered on the Earth was perfectly reasonable and, in its own way, quite beautiful. (On the other hand, see the quote that starts this chapter.)

Ptolemy solved the problem of explaining the observed motions of planets by having each planet revolve in a small orbit called an **epicycle.** The center of the epicycle then

revolved about the Earth on a circle called a *deferent* (Figure 1.13). When the planet is at position *x* in Figure 1.13 on the epicycle orbit, it is moving in the same direction as the center of the epicycle; from Earth the planet appears to be moving eastward. When the planet is at *y,* however, its motion is in the direction opposite to the motion of the epicycle's center around the Earth. By choosing the right combination of speeds and distances, Ptolemy succeeded in having the planet moving westward at *y* at the correct speed and for the correct interval of time, thus replicating retrograde motion with his model.

However, we shall see in the next chapter that the planets, like the Earth, travel about the Sun in orbits that are ellipses, not circles. Their actual behavior cannot be represented accurately by a scheme of uniform circular motions. In order to match the observed motions of the planets, Ptolemy had to center the deferent circles not on the Earth but at points some distance from the Earth. In addition, he introduced uniform circular motion around yet another axis, called the *equant point.* All of this considerably complicated his scheme.

It is a tribute to the genius of Ptolemy as a mathematician that he was able to develop such a complex system to account successfully for the observations of planets. It may be that Ptolemy did not intend his cosmological model to describe reality, but merely to serve as a mathematical representation that allowed him to predict the positions of the

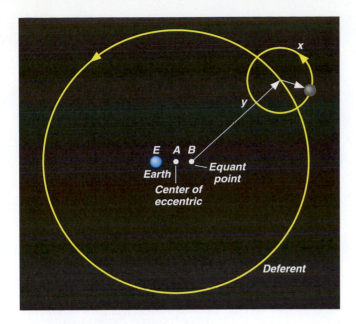

■ **FIGURE 1.13**

Ptolemy's Cosmological System Each planet orbits around a small circle called an epicycle. The system is not exactly centered on the Earth but on a point called the equant. The Greeks needed all this complexity to explain the motions in the sky because they believed that the Earth was stationary.

Ace⊛Astronomy™ Log into AceAstronomy and select this chapter to see the Active Figure called "Epicycles."

planets at any time. Whatever his thinking, his model, with some modifications, was accepted as authoritative in the Muslim world and (later) in Christian Europe.

Many ancient cultures regarded the planets and stars as representatives or symbols of the gods or other supernatural forces that controlled their lives. For them, the study of the heavens was not an abstract subject; it was directly connected to the life-and-death necessity of understanding the actions of the gods and currying favor with them. Before the time of our scientific perspective, everything that happened in nature—from the weather, to diseases and accidents, to celestial surprises like eclipses or new comets—was thought to be an expression of the whims or displeasures of the gods. Any signs that helped people understand what these gods had in mind were considered extremely important.

The movements of the seven objects that had the power to "wander" through the realm of the sky—the Sun, the Moon, and five planets visible to the unaided eye—clearly must have special significance in such a system of thinking.

Most ancient cultures associated these seven objects with various supernatural rulers in their pantheon and kept track of them for religious reasons. Even in the comparatively sophisticated Greece of antiquity, the planets had the names of gods and were credited with having the same powers and influences as the gods whose names they bore. From such ideas was born the ancient system called **astrology,** still practiced by some people today, in which the positions of these bodies among the stars of the zodiac are thought to hold the key to understanding what we can expect from life.

1.3.1 The Beginnings of Astrology

Astrology began in Babylonia about two and a half millennia ago. The Babylonians, believing that the planets and their motions influenced the fortunes of kings and nations, used their knowledge of astronomy to guide their rulers. When the Babylonian culture was absorbed by the Greeks, astrology gradually came to influence the entire Western world and eventually spread to the Orient as well.

By the 2nd century B.C.E. the Greeks democratized astrology by developing the idea that the planets influence every individual. In particular, they believed that the configuration of the Sun, Moon, and planets at the moment of birth affected a person's personality and fortune, the doctrine called *natal astrology.* Natal astrology reached its zenith with Ptolemy 400 years later. As famous for his astrology as for his astronomy, Ptolemy compiled the *Tetrabiblos,* a treatise on astrology that remains the "bible" of the subject. It is essentially this ancient religion, older than Christianity or Islam, that is still practiced by today's astrologers.

1.3.2 The Horoscope

The key to natal astrology is the **horoscope,** a chart showing the positions of the planets in the sky at the moment of an individual's birth. When a horoscope is charted, the planets (including the Sun and Moon, classed as wanderers or planets by the ancients) must first be located in the zodiac. At the time the system of astrology was set up, the circle of the zodiac was divided into 12 sectors called *signs*, each 30° long. Each sign was named after a constellation in the sky through which the Sun, Moon, and planets were seen to pass—the sign of Virgo after the constellation of Virgo, for example.

When someone today casually asks you your "sign," what they are asking for is your "sun-sign"—which zodiac sign the Sun was in at the moment you were born. But more than 2000 years have passed since the signs received their names from the constellations. Because of precession, the constellations of the zodiac slide westward along the ecliptic, going once around the sky in about 26,000 years. Thus today the real stars have slipped around by about 1/12 of the zodiac—about the width of one sign (see Figure 1.5).

In most forms of astrology, however, the signs have remained assigned to the dates of the year they had when astrology was first set up. This means that the astrological signs and the real constellations are out of step; the sign of Aries, for example, now occupies the constellation of Pisces. When you look up your sun-sign in a newspaper astrology column, the name of the sign associated with your birthday is no longer the name of the constellation in which the Sun was actually located when you were born. To know that constellation, you must look for the sign *before* the one that includes your birthday.

A complete horoscope shows the location of not only the Sun but also the Moon and each planet in the sky by indicating its position in the appropriate sign of the zodiac. However, as the celestial sphere turns (owing to the rotation of the Earth), the entire zodiac moves across the sky to the west, completing a circuit of the heavens each day. Thus the position in the sky (or "house" in astrology) must also be calculated. There are more or less standardized rules for the interpretation of the horoscope, most of which (at least in Western schools of astrology) are derived from the *Tetrabiblos* of Ptolemy. Each sign, each house, and each planet, the last acting as a center of force, is supposed to be associated with particular matters in a person's life.

The detailed interpretation of a horoscope is a very complicated business, and there are many schools of astrological thought on how it should be done. Although some of the rules may be standardized, how each rule is to be weighed and applied is a matter of judgment—and "art." It also means that it is very difficult to tie astrology down to specific predictions or to get the same predictions from different astrologers.

1.3.3 Astrology Today

Astrologers today use the same basic principles laid down by Ptolemy nearly 2000 years ago. They cast horoscopes (a process much simplified by the development of appropriate computer programs) and suggest interpretations. Sun-sign astrology (which you read in the newspapers and many magazines) is a recent simplified variant of natal astrology. Although even professional astrologers do not place much trust in such a limited scheme, which tries to fit everyone into just 12 groups, sun-sign astrology is taken seriously by many people (perhaps because it is so commonly discussed in the media). In a recent poll of teenagers in the United States, more than half said they "believed in astrology."

Today, we know much more about the nature of the planets as physical bodies, as well as about human genetics, than the ancients could. It is hard to imagine how the positions of the Sun, Moon, or planets in the sky at the moment of our birth could have anything to do with our personality or future. There are no known forces, not gravity or anything else, that could cause such effects. (For example, a simple calculation shows that the gravitational pull of the obstetrician delivering a newborn baby is greater than that of Mars.) Astrologers thus have to argue that there are unknown forces exerted by the planets that depend on their configurations with respect to one another and that do not vary according to the distance of the planet—forces for which there is no shred of evidence.

Another curious aspect of astrology is its emphasis on planet configurations at birth. What about the forces that might influence us at conception? Isn't our genetic makeup more important for determining our personality than the circumstances of our birth? Would we really be a different person if we had been born a few hours earlier or later, as astrology claims? (Back when astrology was first conceived, birth was thought of as a moment of magic significance, but today we understand a lot more about the long process that precedes it.)

Actually, very few thinking people today buy the claim that our entire lives are predetermined by astrological influences at birth, but many people apparently believe that astrology has validity as an indicator of affinities and personality. A surprising number of Americans make judgments about people—whom they will hire, associate with, and even marry—on the basis of astrological information. To be sure, these are difficult decisions, and you might argue that we should use any relevant information that might help us to make the right choices. But does astrology actually provide any useful information on human personality? This is the kind of question that can be tested using the scientific method (see Making Connections: *Testing Astrology*).

The results of hundreds of tests are all the same: There is no evidence that natal astrology has any predictive power, even in a statistical sense. Why then do people often seem to have anecdotes about how well their own astrologer advised them? Effective astrologers today use the language of the zodiac and the horoscope only as the outward trappings of their craft. Mostly they work as amateur therapists, offering simple truths that clients like or need to hear. (Recent studies have shown that just about any sort of short-term therapy makes people feel a little better. This is because the very act of talking about our problems with someone is in itself beneficial.)

The scheme of astrology has no basis in scientific fact, however. It is an interesting historical system, left over from prescientific days and best remembered for the impetus it gave people to learn the cycles and patterns of the sky. From it grew the science of astronomy, which is our main subject for discussion.

1.4 THE BIRTH OF MODERN ASTRONOMY

Astronomy made no major advances in strife-torn medieval Europe. The birth and expansion of Islam after the 7th century led to a flowering of Arabic and Jewish cultures that preserved, translated, and added to many of the astronomical ideas of the Greeks. (Many of the names of the brightest stars, for example, are today taken from the Arabic, as are such astronomical terms as *zenith*.)

MAKING CONNECTIONS

Testing Astrology

In response to modern public interest in astrology, scientists have carried out a wide range of statistical tests to assess its predictive power. The simplest of these examine sun-sign astrology to determine whether—as astrologers assert—some signs are more likely than others to be associated with some objective measure of success, such as winning Olympic medals, earning high corporate salaries, or achieving elective office or high military rank. (You can make such a test yourself by looking up the birth dates of all members of Congress, for example, or all members of the U.S. Olympic Team.) Are our political leaders somehow selected at birth by their horoscopes and thus more likely to be Leos, say, than Scorpios?

You don't even need to be specific about your prediction in such tests. After all, many schools of astrology disagree about which signs go with which personality characteristics. To demonstrate the validity of the astrological hypothesis, it would be sufficient if the birthdays of all our leaders clustered in any one or two signs in some statistically significant way. Dozens of such tests have been performed, and all have come up completely negative: The birth dates of leaders in all fields tested have been found to be randomly distributed among *all* the signs. Sun-sign astrology does not predict anything about a person's future occupation or strong personality traits.

In a fine example of such a test, two statisticians examined the reenlistment records of the Marine Corps. We suspect our readers will agree that it takes a certain kind of personality to not only enlist but also reenlist in the Marines. If sun-signs can predict strong personality traits—as astrologers claim—then the reenlistees (with similar personalities) should have been distributed preferentially in those one or few signs that matched the personality of someone who loves being a Marine. However, the reenlistees were distributed randomly among all the signs.

More sophisticated studies have also been done, involving full horoscopes calculated for thousands of individuals. The results of all of these studies are also negative: None of the systems of astrology has been shown to be at all effective in connecting astrological aspects to personality, success, or finding the right person to love.

Other tests show that it hardly seems to matter what a horoscope interpretation says, as long as it is vague enough, and as long as each subject feels it was personally prepared just for him or her. The French statistician Michel Gaugelin, for example, sent the horoscope interpretation for one of the worst mass-murderers in history to 150 people, but told each recipient that it was a "reading" prepared exclusively for him or her. Ninety-four percent of the readers said they recognized themselves in the interpretation of the mass-murderer's horoscope.

Geoffrey Dean, an Australian researcher, reversed the astrological readings of 22 subjects, substituting phrases that were the opposite of what the horoscope actually said. Yet his subjects said that the resulting readings applied to them just as often (95 percent) as the people to whom the original phrases were given. Apparently, those who seek out astrologers just want guidance, and almost any guidance will do.

As European culture began to emerge from its long dark age, trading with Arab countries led to a rediscovery of ancient texts such as *Almagest* and to a reawakening of interest in astronomical questions. This time of rebirth (in French, *Renaissance*) in astronomy was embodied in the work of Copernicus (Figure 1.14).

1.4.1 Copernicus

One of the most important events of the Renaissance was the displacement of Earth from the center of the universe, an intellectual revolution initiated by a Polish cleric in the 16th century. Nicolaus Copernicus (1473–1543) was born in Torun, a mercantile town along the Vistula River. His training was in law and medicine, but his main interests were astronomy and mathematics. His great contribution to science was a critical reappraisal of the existing theories of planetary motion and the development of a new Sun-centered, or **heliocentric,** model of the solar system. Copernicus concluded that the Earth is a planet and that all the planets circle the Sun. Only the Moon orbits the Earth (Figure 1.15).

Copernicus described his ideas in detail in his book *De Revolutionibus Orbium Coelestium (On the Revolution of Celestial Orbs),* published in 1543, the year of his death. By this time, the old Ptolemaic system (like a cranky old machine) needed significant adjustments to predict the positions of the planets correctly. Copernicus wanted to develop an improved theory from which to calculate planetary positions, but in doing so, he was himself not free of all traditional prejudices.

He began with several assumptions that were common in his time, such as the idea that the motions of the heavenly bodies must be made up of combinations of uniform

■ **FIGURE 1.14**

Copernicus Nicolaus Copernicus (1473–1543), cleric and scientist, played a leading role in the emergence of modern science. While he could not prove that the Earth revolves about the Sun, he presented such compelling arguments for this idea that he turned the tide of cosmological thought and laid the foundations upon which Galileo and Kepler so effectively built in the following century.

■ **FIGURE 1.15**

Copernicus' System Heliocentric plan of the solar system in the first edition of Copernicus' *De Revolutionibus.* Notice the word "Sol" for Sun in the middle.

circular motions. But he did not assume (as most people did) that the Earth had to be in the center of the universe, and he presented a defense of the heliocentric system that was elegant and persuasive. His ideas, although not widely accepted until more than a century after his death, were much discussed among scholars and ultimately had a profound influence on the course of world history.

One of the objections raised to the heliocentric theory was that if the Earth were moving, we would all sense or feel this motion. Solid objects would be ripped from the surface, a ball dropped from a great height would not strike the ground directly below it, and so forth. But a moving person is not necessarily aware of that motion. We have all experienced seeing an adjacent train, car, or ship appear to move, only to discover that it is we who are moving.

Copernicus argued that the apparent motion of the Sun about the Earth during the course of a year could be represented equally well by a motion of the Earth about the Sun. He also reasoned that the apparent rotation of the celestial sphere could be explained by assuming that the Earth rotates while the celestial sphere is stationary. To the ob-

jection that if the Earth rotated about an axis, it would fly into pieces, Copernicus answered that if such motion would tear the Earth apart, the still faster motion of the much larger celestial sphere required by the geocentric hypothesis would be even more devastating.

1.4.2 The Heliocentric Model

The most important idea in Copernicus' *De Revolutionibus* is that the Earth is one of six (then known) planets that revolve about the Sun. Using this concept, he was able to work out the correct general picture of the solar system. He placed the planets, starting nearest the Sun, in the correct order: Mercury, Venus, Earth, Mars, Jupiter, and Saturn. Further, he deduced that the nearer a planet is to the Sun, the greater is its orbital speed. With his theory, he was able to explain the complex retrograde motions of the planets without epicycles and to work out a roughly correct scale for the solar system.

Copernicus could not *prove* that the Earth revolves about the Sun. In fact, with some adjustments the old Ptolemaic system could have accounted as well for the motions of the planets in the sky. But Copernicus pointed out that the Ptolemaic cosmology was clumsy and lacking the beauty and symmetry of its successor.

In Copernicus' time, in fact, few people thought there were ways to prove whether the heliocentric or the older geocentric system was correct. A long philosophical tradition, going back to the Greeks and defended by the Catholic Church, held that pure human thought combined with divine revelation represented the path to truth. Nature, as revealed by our senses, was suspect. For example, Aristotle had reasoned that heavier objects (having more of the quality that made them heavy) must fall to Earth faster than lighter ones. This is absolutely incorrect, as any simple experiment dropping two balls of different weights will show. But in Copernicus' day, experiments did not carry much weight (if you will pardon the expression); Aristotle's brilliant reasoning was more convincing.

In this environment, there was little motivation to carry out observations or experiments to distinguish between competing cosmological theories (or anything else). It should not surprise us, therefore, that the heliocentric idea was debated for more than half a century without any tests being applied to determine its validity. (In fact, in the American colonies, the older geocentric system was still taught at Harvard University in the first years after it was founded in 1636.)

Contrast this with the situation today, when scientists rush to test each new hypothesis and do not accept any ideas until the results are in. For example, when two researchers at the University of Utah announced in 1989 that they had discovered a way to achieve nuclear fusion (the process that powers the stars) at room temperature, other scientists at more than 25 laboratories around the United States attempted to duplicate "cold fusion" within a few weeks—without success, as it turned out. The cold fusion theory soon went down in flames!

How would we look at Copernicus' model today? When a new hypothesis or theory is proposed in science, it must first be checked for consistency with what is already known. Copernicus' heliocentric idea passes this test, for it allows planetary positions to be calculated at least as well as does the geocentric theory. The next step is to see what predictions the new hypothesis makes that differ from those of competing ideas. In the case of Copernicus, one example is the prediction that, if Venus circles the Sun, the planet should go through the full range of phases just as the Moon does, whereas if it circles the Earth, it should not (Figure 1.16). But in those days, before the telescope, no one imagined testing this prediction.

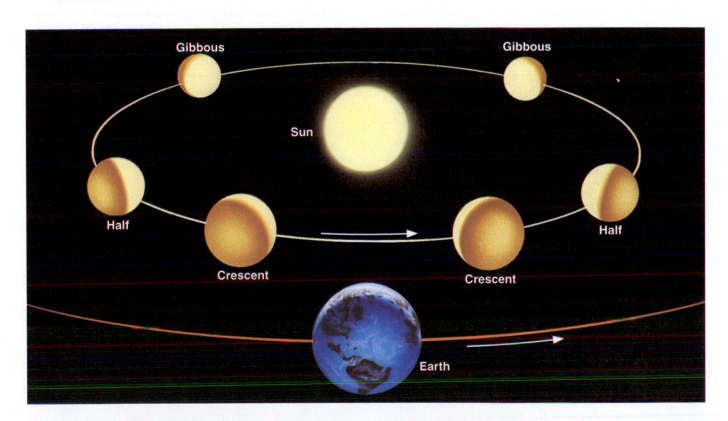

■ **FIGURE 1.16**

Venus Shows Phases   The phases of Venus according to the heliocentric theory. As Venus moves around the Sun, we see changing illumination of its surface, just as we see the face of the Moon illuminated differently in the course of a month.

1.4.3 Galileo and the Beginning of Modern Science

Many of the modern scientific concepts of observation, experimentation, and the testing of hypotheses through careful quantitative measurements were pioneered by a man who lived nearly a century after Copernicus. Galileo Galilei (1564–1642; Figure 1.17), a contemporary of Shakespeare, was born in Pisa. Like Copernicus, he began training for a medical career, but he had little interest in the subject and later switched to mathematics. He held faculty positions at the Universities of Pisa and Padua, and eventually he became mathematician to the Grand Duke of Tuscany in Florence.

Galileo's greatest contributions were in the field of mechanics, the study of motion and the actions of forces on bodies. It was familiar to all persons then, as it is to us now, that if something is at rest, it tends to remain at rest and requires some outside influence to start it in motion. Rest was thus generally regarded as the natural state of matter. Galileo showed, however, that rest is no more natural than motion.

If an object is slid along a rough horizontal floor, it soon comes to rest, because friction between it and the floor acts as a retarding force. However, if the floor and the object are both highly polished, the object, given the same initial speed, will slide farther before stopping. On a smooth layer of ice, it will slide farther still. Galileo reasoned that if all resisting effects could be removed, the object would continue in a steady state of motion indefinitely. He argued that a force is required not only to start an object moving from rest but also to slow down, stop, speed up, or change the direction of a moving object. You will appreciate this if you have ever tried to stop a rolling car by leaning against it, or a moving boat by tugging on a line.

Galileo also studied the way objects **accelerate**—change their speed or direction of motion. Galileo watched objects as they fell freely or rolled down a ramp. He found that such objects accelerate uniformly; that is, in equal intervals of time they gain equal increments in speed. Galileo formulated these newly found laws in precise mathematical terms that enabled future experimenters to predict how far and how fast objects would move in various lengths of time.

Sometime in the 1590s Galileo adopted the Copernican hypothesis of a heliocentric solar system. In Roman Catholic Italy, this was not a popular philosophy, for the Church authorities still upheld the ideas of Aristotle and Ptolemy, and they had powerful political and economic reasons for insisting that the Earth was the center of creation. Galileo not only challenged this thinking but also had the audacity to write in Italian rather than scholarly Latin, and to lecture publicly on these topics. For him, there was no contradiction between the authority of the Church in matters of religion and morality, and the authority of nature

Yerkes Observatory

■ FIGURE 1.17

Galileo Galileo Galilei (1564–1642) advocated that we perform experiments or make observations to ask nature its ways. When Galileo turned the telescope to the sky, he found that things were not the way philosophers had supposed.

Ace✷Astronomy™ Log into AceAstronomy and select this chapter to see Astronomy Exercise "Galileo's Experiment."

(revealed by experiments) in matters of science. It was primarily because of Galileo and his "dangerous" opinions that in 1616 the Church issued a prohibition decree stating that the Copernican doctrine was "false and absurd" and not to be held or defended.

1.4.4 Galileo's Astronomical Observations

It is not certain who first conceived of the idea of combining two or more pieces of glass to produce an instrument that enlarged images of distant objects, making them appear nearer. The first such telescopes that attracted much notice were made by the Dutch spectacle maker Hans Lippershey in 1608. Galileo heard of the discovery and, without ever having seen an assembled telescope, constructed one of his own with a three-power magnification, which made distant objects appear three times nearer and larger (Figure 1.18).

On August 25, 1609, Galileo demonstrated a telescope with a magnification of 9× to government officials of the city-state of Venice. By a magnification of 9×, we mean

■ **FIGURE 1.18**

Telescopes Used by Galileo The longer telescope has a wooden tube covered with paper and a lens 26 mm across.

that the linear dimensions of the object being viewed appeared nine times larger or, alternatively, that the objects appeared nine times closer than they really were. There were obvious military advantages associated with a device for seeing distant objects. For his invention, Galileo's salary was nearly doubled, and he was granted lifetime tenure as a professor. (His university colleagues were outraged, particularly because the invention was not even original.)

Others had used the telescope before Galileo to observe things on Earth. But, in a flash of insight that changed the history of astronomy, Galileo realized that he could turn the power of the telescope toward the heavens. Before using his telescope for astronomical observations, Galileo had to devise a stable mount, and he improved the optics to provide a magnification of 30×. Galileo also needed to acquire confidence in the telescope.

At that time, human eyes were believed to be the final arbiter of truth about sizes, shapes, and colors. Lenses, mirrors, and prisms were known to distort distant images by enlarging, reducing, or even inverting them. Galileo undertook repeated experiments to convince himself that what he saw through the telescope was identical to what he saw up close. Only then could he begin to believe that the miraculous phenomena the telescope revealed in the heavens were real.

Beginning his astronomical work late in 1609, Galileo found that many stars too faint to be seen with the unaided eye became visible with his telescope. In particular, he found that some nebulous blurs resolved into many stars, and that the Milky Way—the strip of whiteness across the night sky—was also made up of a multitude of individual stars.

Examining the planets, Galileo found four satellites, or moons, revolving about Jupiter in times ranging from just under 2 days to about 17 days. This discovery was particularly important because it showed that not everything has to revolve around the Earth! Furthermore, it demonstrated that there could be centers of motion that are themselves in motion. Defenders of the geocentric view had argued that if the Earth were in motion, then the Moon would be left behind because it could hardly keep up with a rapidly moving planet. Yet here were Jupiter's satellites doing exactly that! (To recognize this discovery and honor his work, NASA named a spacecraft that explored the Jupiter system Galileo.)

With his telescope, Galileo was able to carry out the test of the Copernican theory mentioned earlier, based on the phases of Venus. Within a few months he had found that Venus goes through phases like the Moon, showing that it must revolve about the Sun, so that we see different parts of its daylight side at different times (see Figure 1.16). These observations could not be reconciled with any model in which Venus circled about the Earth.

Galileo also observed the Moon and saw craters, mountain ranges, valleys, and flat dark areas that he thought might be water. These discoveries showed that the Moon might be not so dissimilar to the Earth—suggesting that the Earth, too, could belong to the realm of celestial bodies.

After Galileo's work, it became increasingly difficult to deny the Copernican view, and slowly the Earth was dethroned from its central position in the universe and given its rightful place as one of the planets attending the Sun. Galileo himself had to appear before the Inquisition to answer charges that his work was heretical, and he was condemned to house arrest. His books were on the Church's forbidden list until 1836, although in countries where the Roman Catholic Church held less sway, they were widely read and discussed. Not until 1992 did the Catholic Church publicly admit that it had erred in the matter of censoring Galileo's ideas.

The new ideas of Copernicus and Galileo began a revolution in our conception of the cosmos. It eventually became evident that the universe is a vast place and that the Earth's role in it is relatively unimportant. The idea that the Earth moves around the Sun like the other planets raised the possibility that they might be worlds themselves, perhaps even supporting life. As the Earth was demoted from its position at the center of the universe, so too was humanity. The universe, despite what we may wish, does not revolve around us, and we must find a more modest place for ourselves in the scheme of the cosmos.

Most of us take these things for granted today, but four centuries ago such concepts were frightening and heretical for some, immensely stimulating for others. The pioneers of the Renaissance started the European world along the path toward science and technology that we still tread today. For them, nature was rational and ultimately knowable, and experiments and observations provided the means to reveal its secrets.

Observing the Planets

At almost any time of the night, and at any season, you can spot one or more bright planets in the sky. All five of the planets known to the ancients—Mercury, Venus, Mars, Jupiter, and Saturn—are more prominent than any but the brightest stars, and they can be seen even from urban locations if you know where and when to look. One way to tell planets from bright stars is that planets twinkle less.

Venus, which stays close to the Sun from our perspective, appears either as an "evening star" in the west after sunset or as a "morning star" in the east before sunrise. It is the brightest object in the sky after the Sun and Moon. It far outshines any real star, and under the most favorable circumstances it can even cast a visible shadow. Some young military recruits have tried to shoot Venus down as an approaching enemy craft or UFO!

■ ■ ■ ■ ■ ■ ■ ■ ■ ■ ■ ■ ■ ■ ■ ■ ■ ■

One way to tell planets from bright stars is that planets twinkle less.

■ ■ ■ ■ ■ ■ ■ ■ ■ ■ ■ ■ ■ ■ ■ ■ ■ ■

Mars, with its distinctive red color, can be nearly as bright as Venus is when close to the Earth, but normally it remains much less conspicuous. Jupiter is most often the second-brightest planet, approximately equaling in brilliance the brightest stars. Saturn is dimmer, and it varies considerably in brightness, depending on whether its large rings are seen nearly edge-on (faint) or more widely opened (bright).

Mercury is quite bright, but few people ever notice it because it never moves very far from the Sun (it's never more than 28° away in the sky) and is always seen against bright twilight skies.

True to their name, the planets "wander" against the background of the "fixed" stars. Although their apparent motions are complex, they reflect an underlying order, upon which the heliocentric model of the solar system, as described in this chapter, was based. The positions of the planets are often listed in newspapers (sometimes on the weather page), and clear maps and guides to their locations can be found each month in such magazines as *Sky & Telescope* and *Astronomy* (available at most libraries). There are also a number of computer programs (including shareware and commercial programs) that allow you to calculate and display where the planets are on any night.

Activities

1. If either Venus or Mercury is visible, plot its position with respect to the background stars and note the date. Use a map of the relevant portion of the sky, or sketch the location of the brighter stars. Wait a few days and repeat the observation. Estimate how many degrees the planet moves per day. (The width of your thumb, held straight out at arm's length, covers about 1° of sky; your fist at arm's length covers about 10°.) If you make enough observations, you can estimate each planet's maximum distance in degrees from the Sun.

2. Use binoculars to try to determine the shape of Venus. This experiment will be easier when Venus is on the same side of the Sun as the Earth.

3. Determine which of the outer planets—Mars, Jupiter, and Saturn—are visible. For each visible planet, draw its position with respect to the background stars and note the date. Repeat this experiment every few days for two months or as long as each planet is visible. Is the planet moving east or west? Did it appear to change direction? Did you observe retrograde motion for any of the planets?

SURFING THE WEB

The Constellations and Their Stars:
www.astro.wisc.edu/~dolan/constellations/
Chris Dolan has assembled handy information about all the constellations and the most important stars, nebulae, clusters, and galaxies that can be found in them.

The Galileo Project: es.rice.edu/ES/humsoc/Galileo/
Historian of astronomy Albert VanHelden of Rice University maintains this rich lode of information about Galileo's life, work, and times, with text, images, maps, timelines, and so on.

SUMMARY

1.1 The direct evidence of our senses supports a **geocentric** perspective, with the **celestial sphere** pivoting on the **celestial poles** and rotating about a stationary Earth. We see only half of this sphere at one time, limited by the **horizon;** the point directly overhead is our **zenith.** The Sun's annual path on the celestial sphere is the **ecliptic,** a line that runs through the center of the **zodiac,** the 18°-wide strip of sky within which we always find the Moon and planets. The celestial sphere is organized into 88 **constellations,** or sectors.

1.2 Ancient Greeks such as Aristotle recognized that the Earth and Moon are spheres and understood the phases of the Moon, but because of their inability to detect stellar parallax, they rejected the idea that the Earth moves. Eratosthenes measured the size of the Earth with surprising precision. Hipparchus carried out many astronomical observations, making a star catalog, defining the system of stellar **magnitudes,** and discovering **precession** from the apparent shift in the position of the north celestial pole. Ptolemy of Alexandria summarized classical astronomy in his *Almagest;* he explained planetary motions, including **retrograde motion,** with remarkably good accuracy using a model centered on the Earth. This geocentric model, based on combinations of uniform circular motion using **epicycles,** was accepted as authority for more than a thousand years.

1.3 The ancient religion of **astrology,** whose main contribution to civilization was a heightened interest in the heavens, began in Mesopotamia. It reached its peak in the Greco-Roman world, especially as recorded in the *Tetrabiblos* of Ptolemy. Natal astrology is based on the assumption that the positions of the planets at the time of our birth, as described by a **horoscope,** determine our future. However, modern tests clearly show that there is no evidence for this, even in a broad statistical sense, and there is no verifiable theory to explain what might cause such an astrological influence.

1.4 Nicolaus Copernicus introduced the **heliocentric cosmology** to Renaissance Europe in his book *De Revolutionibus*. Although he retained the Aristotelian idea of uniform circular motion, Copernicus suggested that the Earth is a planet and that the planets all circle about the Sun, dethroning the Earth from its position at the center of the universe. Galileo Galilei was the father of both modern experimental physics and telescopic astronomy. He studied the **acceleration** of moving objects, and in 1610 began telescopic observations, discovering the nature of the Milky Way, the large-scale features of the Moon, the phases of Venus, and four satellites of Jupiter. Although he was accused of heresy for his support of the heliocentric cosmology, Galileo is credited with observations and brilliant writings that convinced most of his scientific contemporaries of the reality of the Copernican theory.

INTER-ACTIVITY

(This section in each chapter will be devoted to activities that small groups of students can collaborate on, either during class, in a discussion section, or as independent projects. Your instructor may assign some of these or you can use them in a study group to extend your understanding of astronomy.)

A With your group, consider the question with which we began this chapter: How many ways can you think of to prove to a member of the Flat Earth Society that our planet is indeed round?

B Make a list of ways in which a belief in astrology (the notion that your life path or personality is controlled by the position of the Sun, Moon, and planets at the time of your birth) might be harmful to an individual or to society at large.

C Members of the group should compare their experiences with the night sky. Have you seen the Milky Way? Can you identify any constellations? Make a list of reasons you think so many fewer people know the night sky today than at the time of the ancient Greeks. What reasons can you think of that a person today may want to be acquainted with the night sky?

D Constellations commemorate great heroes, dangers, or events in the legends of the people who name them. Suppose we had to start from scratch today, naming the patterns of stars in the sky. What would you choose to commemorate by naming a constellation after it/him/her and why (begin with people from history; then, if you have time, include living people as well)? Can the members of your group agree on any choices?

E Although astronomical mythology no longer holds a powerful sway over the modern imagination, we still find proof of the power of astronomical images in the number of products in the marketplace that have astronomical names. How many can your group come up with? (Think of things like "Milky Way" candy bars, "Saturn" cars, and "Comet" cleanser.)

REVIEW QUESTIONS

Ace Astronomy™ Assess your understanding of this chapter's topics with additional quizzing and animations at **http://ace.brookscole.com/voyages**

1. From where on Earth could you observe all of the stars during the course of a year? What fraction of the sky can be seen from the North Pole?
2. Describe a practical way to determine in which constellation the Sun is found at any time of the year.
3. What is a constellation as astronomers define it today? What does it mean when an astronomer says, "I saw a comet in Orion last night"?

4. Give four ways to demonstrate that the Earth is round.
5. Explain why we see retrograde motion of the planets, according to both geocentric and heliocentric cosmologies.
6. Draw a picture that explains why Venus goes through phases the way the Moon does, according to the heliocentric cosmology. Does Jupiter also go through phases as seen from the Earth? Why?
7. In what ways did the work of Copernicus and Galileo differ from the traditional views of the ancient Greeks and of the Catholic Church?
8. What were five of Galileo's discoveries that were important to science?

THOUGHT QUESTIONS

9. Show with a simple diagram how the lower parts of a ship disappear first as it sails away from you on a spherical Earth. Use the same diagram to show why lookouts on old sailing ships could see farther from the masthead than from the deck. Would there be any advantage to posting lookouts on the mast if the Earth were flat? (Note that these nautical arguments for a spherical Earth were quite familiar to Columbus and other mariners of his time.)

10. Parallaxes of stars were not observed by ancient astronomers. How can this fact be reconciled with the heliocentric hypothesis?

11. Design an experiment to test whether or not the planets and their motions influence human behavior.

12. Why do you think so many people believe in astrology? What psychological needs does such a belief system satisfy?

13. Consider three cosmological perspectives: the geocentric perspective, the heliocentric perspective, and the modern perspective in which the Sun is a minor star on the outskirts of one galaxy among billions. Discuss some of the cultural and philosophical implications of each point of view.

FIGURING FOR YOURSELF

In this section of the book, we will be offering (or reviewing) some basic formulas for doing astronomical calculations and then giving you some examples to figure out on your own.

A circle consists of 360 degrees (°). When we measure the angle in the sky that something moves, we can use the formula: speed = distance covered/time taken. This is true whether the motion is measured in kilometers per hour or degrees per hour; we just need to use consistent units.

14. The Moon moves relative to the background stars. Go outside at night and note the position of the Moon relative to nearby stars. Repeat the observation a few hours later. How far has the Moon moved? (For reference, the diameter of the Moon is about 1/2°.) Based on your estimate of its motion, how long will it take for the Moon to return to the position relative to the stars in which you first observed it?

15. The north celestial pole appears at an altitude above the horizon that is equal to the observer's latitude. Identify Polaris, the North Star, which lies very close to the north celestial pole. Measure its altitude. (This can be done with a protractor. Alternatively, your fist, extended at arm's length, spans a distance approximately equal to 10°.) Compare this estimate with your latitude. (Note that this experiment cannot be easily performed in the Southern Hemisphere because Polaris itself is not visible in the south and no bright star is located near the south celestial pole.)

16. Suppose Eratosthenes had found that in Alexandria at noon on the first day of summer, the line to the Sun makes an angle of 30° with the vertical. What then would he have found for the Earth's circumference?

17. Suppose Eratosthenes' results for the Earth's circumference were quite accurate. If the diameter of the Earth is 12,740 km, what is the length of his stadium in kilometers?

18. A small asteroid makes a "near miss" of the Earth. At its closest, it is seen to move across the sky at a speed (rate) of 1° per minute. Approximately how long will it take to cross the sky, from horizon to horizon?

19. Suppose you are on a strange planet and observe, at night, that the stars do not rise and set but circle parallel to the horizon. Now you walk in a constant direction for 8000 miles, and at your new location on the planet you find that all stars rise straight up in the east and set straight down in the west, perpendicular to the horizon.
a. How could you determine the circumference of the planet without any further observations?
b. What evidence is there that the Greeks could have done what you suggest?
c. What is the circumference, in miles, of that planet?

SUGGESTIONS FOR FURTHER READING

Culver, B. and Ianna, P. *Astrology: True or False.* 1988, Prometheus Books. The best skeptical book about astrology.

Ferris, T. *Coming of Age in the Milky Way.* 1988, Morrow. A general history of our understanding of the organization of the universe.

Fraknoi, A. "Your Astrology Defense Kit" in *Sky & Telescope,* Aug. 1989, p. 146. A review of the tenets and tests of astrology, along with some skeptical questions.

Gingerich, O. "From Aristarchus to Copernicus" in *Sky & Telescope,* Nov. 1983, p. 410.

Gingerich, O. "How Galileo Changed the Rules of Science" in *Sky & Telescope,* Mar. 1993, p. 32.

Gurshtein, A. "In Search of the First Constellations" in *Sky & Telescope,* June 1997, p. 47. This and an earlier article in the Oct. 1995 issue of *Sky & Telescope* present intriguing ideas about the origins of the ancient star groupings.

Krupp, E. *Beyond the Blue Horizon: Myths and Legends of the Sun, Moon, Stars, and Planets.* 1991, HarperCollins. Superb introductory book on the sky stories of ancient cultures.

Krupp, E. *Skywatchers, Shamans, and Kings.* 1997, Wiley. An excellent primer on ancient monuments and astronomical systems, and the terrestrial purposes they served.

Kuhn, T. *The Copernican Revolution.* 1957, Harvard U. Press. Classic study of the changes wrought by Copernicus' work.

Reston, J. *Galileo: A Life.* 1994, HarperCollins. A well-written, popular-level biography by a journalist.

Sagan, C. *Cosmos.* 1980, Random House. The chapter titled "Backbone of Night" focuses on the ancient Greek astronomers.

In February 1984, astronaut Bruce McCandless, having let go of the tether to the Space Shuttle, became the first human satellite of planet Earth. Here he is shown riding the Manned Maneuvering Unit, which had nitrogen gas jets that enabled him to move around in orbit and ultimately return to the shuttle. As he circled our planet, he was subject to the same basic laws of motion that govern all other satellites.

NASA

2 Orbits and Gravity

One had to be a Newton to notice that the Moon is falling, when everyone sees that it doesn't fall.

Paul Valery in "Analects" (*Collected Works*, Vol. 14, 1966)

THINKING AHEAD

How would you find a new planet that is too dim to be seen with the unaided eye and is so far away that it moves very slowly among the stars? This was the problem confronting astronomers during the 19th century as they tried to pin down a full inventory of our solar system.

If we could look down on the solar system from somewhere out in space, far above the plane of the planets' orbits, interpreting planetary motions would be much simpler. But the fact is, we must observe the positions of all the other planets from our own moving planet. Scientists of the Renaissance did not know the details of the Earth's orbit any better than the orbits of the other planets. Their problem, as we saw in Chapter 1, was that they had to deduce the nature of all planetary motion using only their observations of the other planets' positions in the sky as seen from the Earth. To solve this complex problem more fully, better observations and models of the planetary system were needed.

2.1 THE LAWS OF PLANETARY MOTION

At about the time that Galileo was beginning his experiments with falling bodies, the efforts of two other scientists dramatically advanced our understanding of the motions of the planets. These two astronomers were the observer Tycho Brahe and the mathematician Johannes Kepler. Together, they placed the speculations of Copernicus on a sound mathematical basis and paved the way for the work of Isaac Newton in the next century.

2.1.1 Tycho Brahe's Observatory

Three years after the publication of Copernicus' *De Revolutionibus*, Tycho Brahe (1546–1601) was born to a family of Danish nobility. He developed an early interest in astronomy and as a young man made significant astronomical observations. Among these was a careful study of what we now know was an exploding star that flared up to great brilliance in the night sky. His growing reputation gained him the patronage of the Danish King Frederick II, and at the age of 30 Tycho was able to establish a fine astronomical observatory on the North Sea island of Hven (Figure 2.1). Tycho was the last and greatest of the pre-telescopic observers in Europe.

At Hven, Tycho made a continuous record of the positions of the Sun, Moon, and planets for almost 20 years. His extensive and precise observations enabled him to note that the positions of the planets varied from those given in published tables, which were based on the work of Ptolemy. But Tycho was an extravagant and cantankerous fellow, and he accumulated enemies among government officials. When his patron, Frederick II, died in 1597, Tycho lost his political base and decided to leave Denmark. He took up residence in Prague, where he became court astronomer to the Emperor Rudolf of Bohemia. There, in the year before his death, Tycho found a most able young mathematician, Johannes Kepler, to assist him in analyzing his extensive planetary data.

2.1.2 Kepler

Kepler (1571–1630) (Figure 2.2) was born into a poor family in the German province of Württemberg and lived much of his life amid the turmoil of the Thirty Years' War. He attended university at Tübingen and studied for a theological career. There he learned the principles of the Copernican system and became converted to the heliocentric hy-

Granger Collection, New York

■ **FIGURE 2.1**

Tycho at Hven   A stylized engraving showing Tycho Brahe using his instruments to measure the altitude of a celestial object above the horizon. Note that the scene includes hints of the grandeur of Tycho's observatory.

AIP/Niels Bohr Library

■ **FIGURE 2.2**

Kepler   Johannes Kepler (1571–1630) was a German mathematician and astronomer. His discovery of the basic laws that describe planetary motion placed the heliocentric cosmology of Copernicus on a firm mathematical basis.

pothesis. Eventually, Kepler went to Prague to serve as an assistant to Tycho, who set him to work trying to find a satisfactory theory of planetary motion—one that was compatible with the long series of observations made at Hven.

Tycho was reluctant to provide Kepler with much material at any one time, for fear that Kepler would discover the secrets of the universal motions by himself, thereby robbing Tycho of some of the glory. Only after Tycho's death in 1601 did Kepler get full possession of the priceless records. Their study occupied most of Kepler's time for more than 20 years.

Kepler's most detailed study was of Mars, for which the observational data were the most extensive. In 1609 he published the first results in *The New Astronomy,* where we find his first two laws of planetary motion. Their discovery was a profound step in the development of modern science.

2.1.3 The Orbit of Mars

Kepler began his research under the assumption that the orbits of planets were circles, but the observations contradicted this idea. Working with data for Mars, he eventually discovered that the orbit of that planet had the shape of a somewhat flattened circle, or **ellipse.** Next to the circle, the ellipse is the simplest kind of closed curve, belonging to a family of curves known as *conic sections* (Figure 2.3).

You might recall from math classes that in a circle, the distance from a special point called the center to anywhere on the circle is exactly the same. In an ellipse, the *sum* of the

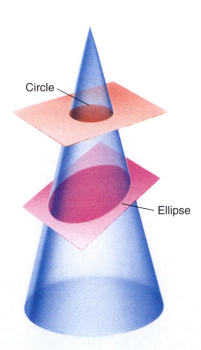

■ FIGURE 2.3

Conic Sections   The circle and the ellipse are both formed by the intersection of a plane with a cone. This is why both curves are called conic sections.

■ FIGURE 2.4

Drawing an Ellipse   We can construct an ellipse with two tacks and a string, where the tacks are the foci of the ellipse. Note that the length of the string remains the same, so that the sum of the distances from any point on the ellipse to the foci is always constant.

distances from two special points inside the ellipse to any point on the ellipse is always the same. These two points inside the ellipse are called its **foci** (singular: **focus**), a word invented for this purpose by Kepler.

This property suggests a simple way to draw an ellipse (Figure 2.4). We wrap the ends of a loop of string around two tacks pushed through a sheet of paper into a drawing board, so that the string is slack. If we push a pencil against the string, making the string taut, and then slide the pencil against the string all around the tacks, the curve that results is an ellipse. At any point where the pencil may be, the sum of the distances from the pencil to the two tacks is a constant length—the length of the string. The tacks are at the two foci of the ellipse.

The widest diameter of the ellipse is called its **major axis.** Half this distance—that is, the distance from the center of the ellipse to one end—is the **semimajor axis,** which is usually used to specify the size of the ellipse. For example, the semimajor axis of the orbit of Mars, which also turns out to be the planet's average distance from the Sun, is 228 million kilometers (km).

The shape (roundness) of an ellipse depends on how close together the two foci are, compared with the major axis. The ratio of the distance between the foci to the length of the major axis is called the **eccentricity** of the ellipse.

If the foci (or tacks) are in the same place, the eccentricity is zero and the ellipse is just a circle; thus, a circle is an ellipse of zero eccentricity. In a circle, the semimajor axis would be the radius. We can make ellipses of various shapes by varying the spacing of the tacks (as long as they are not farther apart than the length of the string). The greater the eccentricity, the more elongated is the ellipse, up to a maximum eccentricity of 1.0.

The size and shape of an ellipse are completely specified by its semimajor axis and its eccentricity. Kepler found that Mars has an elliptical orbit, with the Sun at one focus (the other focus is empty). The eccentricity of the orbit of Mars is only about 0.1; its orbit, drawn to scale, would be practically indistinguishable from a circle. Yet the difference is critical for understanding planetary motions.

Kepler generalized this result in his first law and said that *the orbits of all the planets are ellipses.* Here was a decisive moment in the history of human thought: It was not necessary to have only circles in order to have an acceptable cosmos. The universe could be a bit more complex than the Greek philosophers had wanted it to be.

Kepler's second law deals with the speed with which each planet moves along the ellipse. Working with Tycho's observations of Mars, Kepler discovered that the planet speeds up as it comes closer to the Sun and slows down as it pulls away from the Sun. He expressed the precise form of this relationship by imagining that the Sun and Mars are connected by a straight, elastic line (Figure 2.5). When Mars is closer to the Sun (positions 1 and 2 in Figure 2.5), the elastic line is not stretched as much, and the planet moves rapidly. Farther from the Sun, as in positions 3 and 4, the line is stretched a lot, and the planet does not move so fast. As Mars travels in its elliptical orbit around the Sun, the elastic line sweeps out areas of the ellipse (the colored regions in our figure). Kepler found that in equal intervals of time, the *areas* swept out in space by this imaginary line are always equal; that is, the area of the region from 1 to 2 is the same as that from 3 to 4.

This is also a general property of the orbits of all planets. A planet in a circular orbit always moves at a constant speed, but in an elliptical orbit the planet's speed varies considerably, following the rule Kepler found.

2.1.4 Laws of Planetary Motion

Kepler's first two laws of planetary motion describe the shape of a planet's orbit and allow us to calculate the speed of its motion at any point in the orbit. Kepler was pleased to have discovered such fundamental rules, but they did not satisfy his quest to understand planetary motions. He wanted to know why the orbits of the planets were spaced as they are, and to find a mathematical pattern in their orbital periods—a "harmony of the spheres," as he called it. For many years he worked to discover mathematical relationships governing planetary spacings and the time each planet took to go around the Sun.

In 1619, Kepler succeeded in finding a basic relationship that links the semimajor axes of the planets' orbits and their periods of revolution. The relationship is now known as Kepler's third law. It applies to all of the planets, including the Earth, and it provides a means for calculating their relative distances from the Sun.

For the solar system, Kepler's third law takes its simplest form when the period is expressed in years (the revo-

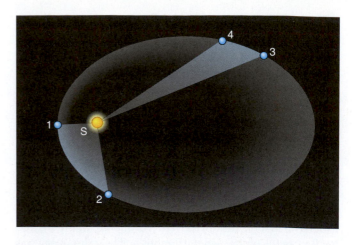

■ **FIGURE 2.5**

Kepler's Second Law: The Law of Equal Areas
A planet moves most rapidly on its elliptical orbit when at position 1, nearest the Sun, which is at one focus of the ellipse. The orbital speed of the planet varies in such a way that in equal intervals of time a line between the Sun and the planet sweeps out equal areas. Thus, the area swept out from 1 to 2 is the same as that from 3 to 4. Note that the eccentricities of the planets' orbits are substantially less than shown here.

lution period of the Earth) and the semimajor axis of the orbit is expressed in terms of the Earth's average distance from the Sun, called the **astronomical unit (AU).** One astronomical unit is equal to 1.5×10^8 km. With these units, Kepler's third law is

$$(\text{distance})^3 = (\text{period})^2$$

For example, this relationship tells us how to calculate Mars' average distance from the Sun (the semimajor axis of its orbit) from its period of 1.88 years. The period squared is $1.88 \times 1.88 = 3.53$, and this equals the cube of the semimajor axis. So what number must be cubed to give 3.53? The answer is 1.52 ($1.52 \times 1.52 \times 1.52 = 3.53$). Thus, the planet's semimajor axis in astronomical units must be 1.52. In other words, to go around the Sun in a little less than 2 years, Mars must be about 50 percent farther from the Sun than Earth is.

Kepler's three laws of planetary motion can be summarized as follows:

Kepler's First Law Each planet moves about the Sun in an orbit that is an ellipse, with the Sun at one focus of the ellipse.

Kepler's Second Law The straight line joining a planet and the Sun sweeps out equal areas in space in equal intervals of time.

Kepler's Third Law The squares of the planets' periods of revolution are in direct proportion to the cubes of the semimajor axes of their orbits.

These three laws provided a precise geometric description of planetary motion within the framework of the

Copernican system. With these tools, it was possible to calculate planetary positions with undreamed-of precision. But Kepler's laws are purely descriptive; they do not help us understand what forces of nature constrain the planets to follow this particular set of rules. That step was left to Newton.

2.2 NEWTON'S GREAT SYNTHESIS

It was the genius of Isaac Newton (1643–1727) that found a conceptual framework that completely explained the observations and rules assembled by Galileo, Tycho, Kepler, and others. Newton was born in Lincolnshire, England, in the year after Galileo's death (Figure 2.6). Against the advice of his mother, who wanted him to stay home and help with the family farm, he entered Trinity College at Cambridge in 1661 and eight years later was appointed Professor of Mathematics there. Among Newton's contemporaries in England were architect Christopher Wren, authors Samuel Pepys and Daniel Defoe, and composer G. F. Handel.

2.2.1 Newton's Laws of Motion

As a young man in college, Newton became interested in natural philosophy, as science was then called. He worked out some of his first ideas on mechanics and optics during the plague years of 1665 and 1666, when students were sent home from college. Newton, a moody and often difficult man, continued to work on his ideas in private, even inventing new mathematical tools to help him deal with the complexities involved. Eventually, his friend Edmund Halley (profiled in Chapter 12) prevailed on him to collect and publish the results of his remarkable investigations on motion and gravity. The result was a volume that set out the underlying system of the physical world, *Philosophiae Naturalis Principia Mathematica.* The *Principia,* as the book is generally known, was published at Halley's expense in 1687.

At the very beginning of the *Principia,* Newton proposes three laws that he presumes to govern the motions of all objects:

Newton's First Law Every body continues doing what it is already doing—being in a state of rest, or moving uniformly in a straight line—unless it is compelled to change by an outside force.

Newton's Second Law The change of motion of a body is proportional to the force acting on it, and is made in the direction in which that force is acting.

Newton's Third Law To every action there is an equal and opposite reaction (*or* the mutual actions of two bodies upon each other are always equal and act in opposite directions).

In the original Latin, the three laws contain only 59 words, but those few words set the stage for modern science. Let us examine them more carefully.

AIP/Niels Bohr Library

■ FIGURE 2.6

Newton Isaac Newton (1643–1727), whose work on the laws of motion, gravity, optics, and mathematics laid the foundations of much of physical science.

2.2.2 Interpretation of Newton's Laws

Newton's first law is a restatement of one of Galileo's discoveries, called the **conservation of momentum**—where momentum is a measure of a body's motion. The law states that in the absence of any outside influence, a body's momentum remains unchanged. We use the word *momentum* in everyday expressions as well, as in "This bill in Congress has a lot of momentum: It seems unstoppable." In other words, a stationary object stays put, and a moving object keeps moving unless some force intervenes. Momentum depends on three factors. The first is *speed*—how fast a body moves (zero if it is stationary). The second is the direction in which the body is moving. Scientists use the term *velocity* to describe both speed and direction. For example, 20 kilometers per hour (km/h) due south is velocity; 20 km/h is speed. The third factor in momentum is what Newton called *mass.* Mass is a measure of the amount of matter in a body, as we will discuss later. Momentum then is mass × velocity.

As mentioned in Chapter 1, it is not so easy to see this rule in action in the everyday world. On Earth, objects in motion do not remain in motion because outside forces are always acting. One important force is friction, which slows things down. If you roll a ball down the sidewalk, it eventually comes to a stop because the rubbing of the ball against the sidewalk exerts a force. But out in space between the stars, where friction is negligible, objects could in fact continue to move (coast) indefinitely. Newton's first law is sometimes called the *law of inertia,* inertia being the tendency of objects (and legislatures) to keep doing what they are already doing.

The momentum of a body can change only under the action of an outside influence. Newton's second law defines *force* in terms of its ability to *change momentum.* A force (a push or a pull) has both size and direction. When force is applied to a body, the momentum changes in the direction of the applied force. This means that a force is required to change either the speed or the direction of a body, or both—that is, to start it moving, to speed it up, to slow it down, to stop it, or to change its direction.

The rate of change in an object's velocity is called **acceleration.** Newton showed that the acceleration of a body was proportional to the force being applied to it. Suppose, after a hard exam, you push your astronomy textbook away from you in disgust on a long, smooth table. The harder you keep pushing the book, the more it will speed up. How much a force will accelerate an object is also determined by the object's mass. If you kept pushing a pen with the same force with which you pushed the textbook, the pen—having less mass—would be accelerated to a greater speed. (Note that we said above that the table was smooth because we wanted to minimize the effect of friction. You'd have even better luck accelerating the textbook on ice or out in space.)

Newton's third law is the most profound. Basically, it is a generalization of the first law, but it also gives us a way to define mass. If we consider a system of two or more objects isolated from outside influences, Newton's first law says that the total momentum of the system of objects should remain constant. Therefore, any change of momentum within the system must be balanced by another change that is equal and opposite to it, so that the momentum of the entire system is not changed.

This means that forces in nature do not occur alone: We find that in each situation there is always a pair of forces that are equal to and opposite each other. If a force is exerted on an object, it must be exerted by something else, and the object will exert an equal and opposite force back on that something.

Suppose that during that same difficult exam, another student runs screaming from the room and jumps out a (not very high) window. The force pulling him down after he jumps (as we will see in the next section) is the gravitational force between him and the Earth. Both he and the Earth must experience the same total change of momentum because of the influence of this mutual force. So, both the student and the Earth are accelerated by each other's pull. However, the student does much more of the moving! Because the Earth has enormously greater mass, it can experience the same change of momentum by accelerating only a very small amount. Things fall toward the Earth all the time, but the acceleration of our planet as a result is far too small to be noticed or measured.

A more obvious example of the mutual nature of forces between objects is familiar to all who have batted a baseball. The recoil you feel as you swing your bat shows that the ball exerts a force on it during the impact, just as the bat does on the ball. Similarly, when a rifle is discharged, the force pushing the bullet out of the muzzle is equal to the

NASA

■ FIGURE 2.7

Demonstrating Newton's Third Law   The U.S. Space Shuttle at launch, powered by three liquid fuel engines burning liquid oxygen and liquid hydrogen, with two solid fuel boosters.

force pushing backward upon the gun and the shoulder of the person shooting it.

This is the principle behind jet engines and rockets: The force that discharges the exhaust gases from the rear of the rocket is accompanied by force that shoves the rocket forward. The exhaust gases need not push against air or the Earth; a rocket actually operates best in a vacuum (Figure 2.7).

2.2.3 Mass, Volume, and Density

Before we go on to discuss Newton's other work, we want to take a brief look at some terms that will be important to sort out clearly. We begin with *mass,* which is a measure of the amount of material in an object.

Volume is a measure of the physical space occupied by a body—say, in cubic centimeters or liters. In short, the volume is the "size" of an object; it has nothing to do with its mass. A penny and an inflated balloon may both have the same mass, but they have very different volumes.

The penny and the balloon are also very different in **density,** which is a measure of how much mass we have per unit volume. Specifically, density is the ratio of mass to volume. Note that often in everyday language we use "heavy"

TABLE 2.1	*Densities of Materials*
Material	**Density (g/cm³)**
Gold	19.3
Lead	11.4
Iron	7.9
Earth (bulk)	5.6
Rock (typical)	2.5
Water	1.0
Wood (typical)	0.8
Insulating foam	0.1
Silica gel	0.02

■ **FIGURE 2.8**

The Conservation of Angular Momentum   When a spinning figure skater brings her arms in, their distance from the spin center is small, so her speed increases. When her arms are out, their distance from the spin center is greater, so she slows down.

and "light" as indications of density (rather than weight), as, for instance, when we say that iron is heavy or that puff pastry is light.

The units of density that will be used in this book are grams per cubic centimeter (g/cm³).[1] If a block of some material has a mass of 300 g and a volume of 100 cm³, its density is 3 g/cm³. Familiar materials span a considerable range in density, from artificial materials such as plastic insulating foam (less than 0.1 g/cm³) to gold (19.3 g/cm³) (Table 2.1). In the astronomical universe, much more remarkable densities can be found, all the way from a comet's tail (10^{-16} g/cm³) to a neutron star (10^{15} g/cm³).

To sum up, then, mass is "how much," volume is "how big," and density is "how tightly packed."

2.2.4 Angular Momentum

The concept of **angular momentum** is a bit more complex, but it is important for understanding many astronomical objects. Angular momentum is a measure of the momentum of an object as it rotates or revolves about some fixed point. Whenever we deal with the revolution of spinning objects, from planets to galaxies, we have to consider angular momentum. The angular momentum of an object is defined as the product of three quantities: its mass, its velocity, and its distance from the fixed point around which it turns.

If these three quantities remain constant—that is, if the motion of a particular object takes place at a constant speed and at a fixed distance from the spin center—then the angular momentum is also a constant. So, ignoring air resistance, if you tie a string to your astronomy textbook and twirl it around your head at constant speed, you will have a system with constant angular momentum.

More generally, angular momentum is constant, or conserved, in any rotating system in which no external forces act, or in which the only forces are directed toward or away from the spin center. An example of such a system is a planet orbiting the Sun. Kepler's second law is an example of the conservation of angular momentum. When a planet approaches the Sun on its elliptical orbit, the distance to the spin center decreases; the planet speeds up to keep the angular momentum the same. Similarly, when the planet is farther from the Sun, it revolves more slowly.

The concept is illustrated by figure skaters, who bring their arms and legs in to spin more rapidly, and extend their arms and legs to slow down (Figure 2.8). (You can duplicate this yourself on a well-oiled piano stool, by starting yourself spinning slowly with your arms extended and then pulling your arms in.) And just as a planet speeds up when it approaches the Sun, a shrinking cloud of dust or a star collapsing on itself (both situations you will encounter as you read on) increases its spin rate as it contracts. There is less distance to the spin center, so the speed increases to keep the angular momentum the same.

2.3 UNIVERSAL GRAVITY

2.3.1 The Law of Gravity

Newton's laws of motion show that, left to themselves, objects at rest stay at rest, and those in motion continue moving uniformly in a straight line. Thus, it is the *straight line,* not the *circle,* that defines the most natural state of motion. But the planets move in ellipses, not straight lines. Some

[1]Generally we use the standard metric (or SI) units in this book. The proper metric unit of density is kg/m³. But to most people g/cm³ provides a more meaningful unit because the density of water is exactly 1 g/cm³. Density expressed in g/cm³ is sometimes called specific density or specific weight.

force must be bending their paths. That force, Newton proposed, is **gravity.**

In Newton's time gravity was something associated with the Earth alone. Everyday experience shows us that the Earth exerts a gravitational force upon objects at its surface. If you drop something off a leaning tower in Pisa, it falls toward the Earth, accelerating as it falls. Newton's insight was that the Earth's gravity might extend as far as the Moon and produce the acceleration required to curve the Moon's path from a straight line and keep it in its orbit. He further hypothesized that gravity is not limited to the Earth, but that there is a general force of attraction between all material bodies. If so, the attractive force between the Sun and each of the planets could keep each in its orbit.

Once Newton boldly hypothesized that there is a *universal attraction* among all bodies everywhere in space, he had to determine the exact nature of the attraction. The precise mathematical description of that gravitational force had to dictate that the planets move exactly as Kepler had observed them to (as expressed in Kepler's three laws). Also, the law of gravity had to predict the correct behavior of falling bodies on the Earth, as observed by Galileo. How must gravitational force depend on distance in order for these conditions to be met?

The answer to this question required mathematical tools that had not yet been developed. But this did not deter Isaac Newton, who invented what we today call *calculus* to deal with this problem. Eventually he was able to conclude that the force of gravity must drop off with increasing distance between the Sun and a planet (or between any two objects) in proportion to the *inverse square* of their separation. In other words, if a planet were twice as far from the Sun, the force would be $1/2^2$ or 1/4 as large. Put the planet three times farther away, and the force is $1/3^2$ or 1/9 as strong.

Newton also concluded that the gravitational attraction between two bodies must be proportional to their masses. The more mass an object has, the stronger the pull of its gravity. The gravitational attraction between any two objects is given by one of the most famous formulas in all of science:

$$\text{force} = GM_1M_2/R^2$$

where M_1 and M_2 are the masses of the two objects and R is their separation. The number represented by G is called the constant of gravitation. With such a force and the laws of motion, Newton was able to show mathematically that the only orbits permitted were exactly those described by Kepler's laws.

Newton's law of gravity works for the planets, but is it really universal? The gravitational theory should also predict the observed acceleration of the Moon toward the Earth, falling around the Earth at a distance of about 60 times the radius of the Earth, as well as of an object (say, an apple) dropped near the Earth's surface. The falling of an apple is something we can measure quite easily; can we use it to predict the motions of the Moon?

Newton's theory says that the force on (and therefore the acceleration of) an object toward the Earth should be inversely proportional to the square of its distance from the center of the Earth. Objects like apples at the surface of the Earth (R = 1 Earth-radius, the distance from its center) are observed to accelerate downward at 9.8 meters per second per second (9.8 m/s^2).

Therefore, if the law holds, the Moon, 60 Earth-radii from its center, should experience an acceleration toward the Earth that is $1/60^2$, or 3600 times less—that is, about 0.00272 m/s^2. This is precisely the observed acceleration of the Moon in its orbit. What a triumph! Imagine the thrill Newton must have felt to realize he had discovered and verified a law that holds for the Earth, apples, the Moon, and, as far as we know, everything in the universe!

Gravity is a "built-in" property of mass. Wherever masses occur, they will interact via the force of gravity. The more mass there is, the greater the force it can exert. Here on Earth, the largest concentration of mass is, of course, the planet we stand on, and its pull dominates the gravitational interactions we experience. But everything with mass attracts everything else with mass anywhere in the universe.

For example, you and this textbook (each having mass) attract each other via gravity. (We certainly hope it's not the *only* way you find our book attractive!) If you let the book fall, however, it is pulled much more strongly by the Earth, so it falls to the Earth, not toward you. But if you and the textbook somehow found yourselves out in space, far from any stars or planets, you would discover that gravitational attraction would start to pull you together.

Newton's law also suggests that gravity never becomes zero. It quickly gets weaker with distance, but it continues to act to some degree no matter how far away you get. The pull of the Sun is stronger at Mercury than at Pluto, but it can be felt far beyond Pluto, where (as we shall see) we have good evidence that it continues to make enormous numbers of smaller bodies move around huge orbits. And its pull joins with the pull of billions of other stars to create the gravitational pull of the Milky Way Galaxy in which we live. That force, in turn, can make other smaller galaxies orbit around the Milky Way, and so forth.

Why is it then, you may ask, that the astronauts aboard the Space Shuttle appear "weightless" and we see images on television of people and objects floating in the spacecraft? After all, the astronauts in the shuttle are only a few hundred kilometers above the surface of the Earth, not a significant distance compared to the size of the Earth. Gravity is certainly not a great deal weaker that much farther away. The astronauts feel weightless for the same reason that passengers in an elevator whose cable has broken or in an airplane whose engines no longer work feel weightless[2]

---

[2] In the film *Apollo 13*, the scenes where the astronauts were "weightless" were actually filmed in a falling airplane. As you might imagine, the plane fell for only short periods before the engines cut in again.

■ **FIGURE 2.9**

Astronaut in Free Fall   Aboard the Space Shuttle *Columbia* in July 1997, astronaut Susan Still floats before the Spacelab control center. Both she and the shuttle are falling freely around the Earth, and so she experiences no force relative to the shuttle (just as people in a freely falling elevator would feel weightless for the brief time they had left to live).

(Figure 2.9). They are *falling,* and in free fall, they accelerate at the same rate as everything around them, including their spacecraft or a camera with which they are taking photographs of Earth. Thus, *relative* to the shuttle, the astronauts experience no additional force and so feel "weightless." Unlike the falling elevator passengers, however, the astronauts are falling *around* the Earth, not *to* the Earth (and will continue falling).

2.3.2   Orbital Motion and Mass

Kepler's laws are descriptions of the orbits of objects moving according to Newton's laws of motion and the law of gravity. Knowing that gravity is the force that attracts planets toward the Sun, however, allowed Newton to rethink Kepler's third law. Recall that Kepler had found a relationship between the period of a planet's revolution and its distance from the Sun. Newton's law of gravity can be used to show mathematically that this relationship is actually

$$D^3 = (M_1 + M_2) \times P^2$$

As explained in Section 2.1.4, we express distances in the solar system in astronomical units and periods in years. But Newton's formulation introduces the additional factor of the masses of the Sun (M_1) and planet (M_2), both expressed in units of the Sun's mass.

How did Kepler miss this factor? In units of the Sun's mass, the mass of the Sun is 1; in units of the Sun's mass, the mass of a typical planet is a negligibly small fraction, and so ($M_1 + M_2$) is very, very close to 1. This makes Newton's formula look almost the same as Kepler's. The tiny mass of the planets is the reason that Kepler did not realize that both masses had to be included in the calculation. There are many cases in astronomy, however, in which we do need to include the two mass terms—for example, when two stars or two galaxies orbit each other.

Including the mass term allows us to use this formula in a new way. If we can measure the motions (distances and periods) of objects acting under their mutual gravity, then the formula will permit us to deduce their masses. For example, we can calculate the mass of the Sun by using the distances and periods of the planets, or the mass of Jupiter by noting the motions of its moons. Indeed, Newton's reformulation of Kepler's third law is one of the most powerful concepts in astronomy. Our ability to deduce the masses of objects from their motions is key to understanding the nature and evolution of many astronomical bodies. We will use this law repeatedly throughout this text in calculations that range from the orbits of comets to the interactions of galaxies.

2.4 ORBITS IN THE SOLAR SYSTEM

2.4.1   Description of an Orbit

The path of an object through space is called its **orbit,** whether that object is a spacecraft, planet, star, or galaxy. An orbit, once determined, allows the future positions of the object to be calculated.

Two points in any orbit have been given special names. The place where the planet is closest to the Sun (*helios* in Greek) is called the **perihelion** of its orbit, and the place where it is farthest away and moves the most slowly is the **aphelion.** For a satellite orbiting the Earth (*geos* in Greek), the corresponding terms are **perigee** and **apogee.**

2.4.2   Orbits of the Planets

Today, Newton's work enables us to calculate and predict the orbits of the planets with marvelous precision. We

Astronomy and the Poets

When Copernicus, Kepler, Galileo, and Newton formulated the fundamental rules that underlie everything in the physical world, they changed much more than the face of science. For some, they gave humanity the courage to let go of old superstitions and see the world as rational and manageable; for others, they upset comforting, ordered ways that had served humanity for centuries, leaving only a dry, mechanical clockwork universe in their wake.

Poets of the time reacted to such changes in their own work and debated whether the new world picture was an appealing or frightening one. John Donne (1573–1631), in a poem called *Anatomy of the World,* laments the passing of the old certainties:

> The new philosophy [science] calls all in doubt,
> The element of fire is quite put out;
> The Sun is lost, and th' earth, and no man's wit
> Can well direct him where to look for it.

(Here the "element of fire" refers also to the sphere of fire, which medieval thought placed between the Earth and the Moon.)

By the next century, poets like Alexander Pope were celebrating Newton and the Newtonian world view. Pope's famous couplet, written upon Newton's death, goes:

> Nature, and nature's laws lay hid in night.
> God said, Let Newton be! and all was light.

In his 1733 poem, *An Essay on Man,* Pope revels in the complexity of the new views of the world, incomplete though they are:

> Of man, what see we, but his station here,
> From which to reason, to which refer? . . .
> He, who thro' vast immensity can pierce,
> See worlds on worlds compose one universe,
> Observe how system into system runs,
> What other planets circle other suns,
> What vary'd being peoples every star,
> May tell why Heav'n has made us as we are . . .
> All nature is but art, unknown to thee;
> All chance, direction, which thou canst not see;
> All discord, harmony not understood;
> All partial evil, universal good:
> And, in spite of pride, in erring reason's spite,
> One truth is clear, whatever is, is right.

Poets and philosophers continued to debate whether humanity was exalted or debased by the new views of science. The 19th-century poet Arthur Hugh Clough (1819–1861) cries out in his poem *The New Sinai:*

> And as of old from Sinai's top God said that God is one,
> By science strict so speaks He now to tell us, there is None!
> Earth goes by chemic forces; Heaven's a Mécanique Celeste!
> And heart and mind of humankind a watchwork as the rest!

(A "mécanique celeste" is a clockwork model to demonstrate celestial motions.)

The 20th-century poet Robinson Jeffers (whose brother was an astronomer) saw it differently in a poem called *Star Swirls:*

> There is nothing like astronomy to pull the stuff out of man.
> His stupid dreams and red-rooster importance:
> Let him count the star-swirls.

---

know nine planets, beginning with Mercury closest to the Sun and extending outward to Pluto. The average orbital data for the planets are summarized in Table 2.2. (Ceres is the largest of the *asteroids;* see the next section.)

According to Kepler's laws, Mercury must have the shortest period of revolution (88 Earth-days); thus it has the highest orbital speed (averaging 48 km/s). At the opposite extreme, Pluto has a period of 249 years and an average orbital speed of just 5 km/s.

All of the planets have orbits of rather low eccentricity. The most eccentric orbits are those of Mercury (0.21) and Pluto (0.25); all of the rest have eccentricities less than 0.1. It is fortunate for the development of science that Mars has an eccentricity greater than many of the other planets. Otherwise the pre-telescopic observations of Tycho would not have been sufficient for Kepler to deduce that its orbit had the shape of an ellipse rather than a circle.

The planetary orbits are also confined close to a common plane, which is near the plane of the Earth's orbit (the

Ace Astronomy™ Log into AceAstronomy and select this chapter to see Astronomy Exercise "Orbital Motion."

TABLE 2.2 *Orbital Data for the Planets*

Planet	Semimajor Axis (AU)	Period (yr)	Eccentricity
Mercury	0.39	0.24	0.21
Venus	0.72	0.62	0.01
Earth	1.00	1.00	0.02
Mars	1.52	1.88	0.09
(Ceres)	2.77	4.60	0.08
Jupiter	5.20	11.86	0.05
Saturn	9.54	29.46	0.06
Uranus	19.19	84.07	0.05
Neptune	30.06	164.80	0.01
Pluto	39.60	248.60	0.25

ecliptic). The strange orbit of Pluto is inclined about 17° to the average, but all of the other planets lie within 10° of the common plane of the solar system.

2.4.3 Orbits of Asteroids and Comets

In addition to the nine planets, there are many smaller objects in the solar system. Some of these are natural satellites that orbit all of the planets except Mercury and Venus. In addition, there are two classes of smaller objects in heliocentric orbits: the *asteroids* and the *comets*. Both asteroids and comets are believed to be small chunks of material left over from the formation process of the solar system.

In general, the asteroids have orbits with smaller semimajor axes than do the comets (Figure 2.10). The great majority of them lie between 2.2 and 3.3 AU, in the region known as the **asteroid belt.** As you can see in Table 2.2, the asteroid belt (represented by its largest member, Ceres) is in the middle of a gap between the orbits of Mars and Jupiter. It is because these two planets are so far apart that stable orbits of small bodies can exist in the region between them.

Comets generally have orbits of larger size and greater eccentricity than those of the asteroids. Typically, the eccentricities of their orbits are 0.8 or higher. According to Kepler's second law, therefore, they spend most of their time far from the Sun, moving very slowly. As they approach perihelion, the comets speed up and whip through the inner parts of their orbits more rapidly.

Ace⌬Astronomy™   Log into AceAstronomy and select this chapter to see the Active Figure called "Orbital Motion."

2.5 MOTIONS OF SATELLITES AND SPACECRAFT

2.5.1 Space Flight and Satellite Orbits

The law of gravity and Kepler's laws describe the motions of Earth satellites and interplanetary spacecraft as well as the planets. Sputnik, the first artificial Earth satellite, was launched by what was then called the Soviet Union on October 4, 1957. Since that time, thousands of satellites have been placed into orbit around the Earth, and spacecraft have also orbited the Moon, Venus, Mars, Jupiter, and the asteroid Eros.

Once an artificial satellite is in orbit, its behavior is no different from that of a natural satellite, such as our Moon. If the satellite is high enough to be free of atmospheric friction, it will remain in orbit forever, following Kepler's laws in a perfectly respectable way. However, although there is no difficulty in maintaining a satellite once it is in orbit, a great deal of energy is required to lift the spacecraft off the Earth and accelerate it to orbital speed.

To illustrate how a satellite is launched, imagine a gun firing a bullet horizontally from the top of a high mountain (Figure 2.11a—adapted from a similar diagram by Newton, shown in Figure 2.11b). Imagine, further, that the friction of the air could be removed and that nothing can get in the bullet's way. Then the only force that acts on the bullet after it leaves the muzzle is the gravitational force between the bullet and Earth.

If the bullet is fired with some velocity we can call v_a, it continues to have that forward speed, but meanwhile the gravitational force acting upon it pulls it downward toward the Earth, where it strikes the ground at point a. However, if it is given a higher muzzle velocity, v_b, its higher forward speed carries it farther before it hits the ground. This is because, regardless of its forward speed, the downward gravitational force is the same. Thus, this faster-moving bullet strikes the ground at point b.

If our bullet is given a high enough muzzle velocity, v_c, the curved surface of the Earth causes the ground to tip out from under it so that it remains the same distance above the ground and falls around the Earth in a complete circle. The speed needed to do this—called the **circular satellite velocity**—is about 8 km/s, or about 17,500 miles per hour (mph) in more familiar units.

Each year more than 50 new satellites are launched into orbit by such nations as Russia, the United States, China, Japan, India, and Israel, as well as by the European Space Agency (ESA), a consortium of European nations (Figure 2.12). Most satellites are launched into low Earth orbit,

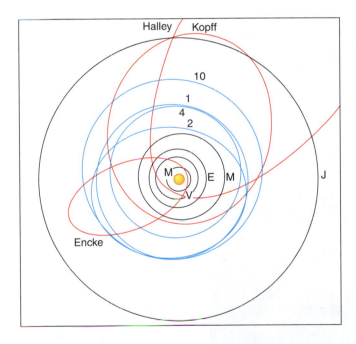

■ FIGURE 2.10

Solar System Orbits   We see the orbits of typical comets and asteroids compared with those of the planets Mercury, Venus, Earth, Mars, and Jupiter (black circles). Shown in red are three comets: Halley, Kopff, and Encke. In blue are the four largest asteroids: 1 Ceres, 2 Pallas, 4 Vesta, and 10 Hygeia.

(a)

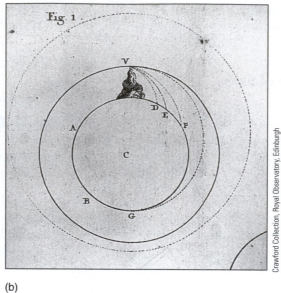

Crawford Collection, Royal Observatory, Edinburgh

(b)

■ **FIGURE 2.11**

Firing a Bullet into Orbit
(a) For paths *a* and *b*, the velocity is not enough to prevent gravity from pulling the bullet back to Earth; in case *c*, the velocity allows the bullet to fall completely around the Earth. (b) A diagram by Newton in his *De Mundi Systematic*, 1731 edition, illustrating the same concept shown in panel (a).

since this requires the minimum launch energy. At the orbital speed of 8 km/s, they circle the planet in about 90 minutes. Low Earth orbits are not stable indefinitely because the ldrag generated by friction with the thin upper atmosphere eventually leads to a loss of energy and "decay" of the orbit.

2.5.2 Interplanetary Spacecraft

The exploration of the solar system has been carried out largely by robot spacecraft sent to the other planets. To escape Earth, these craft must achieve **escape velocity,** the

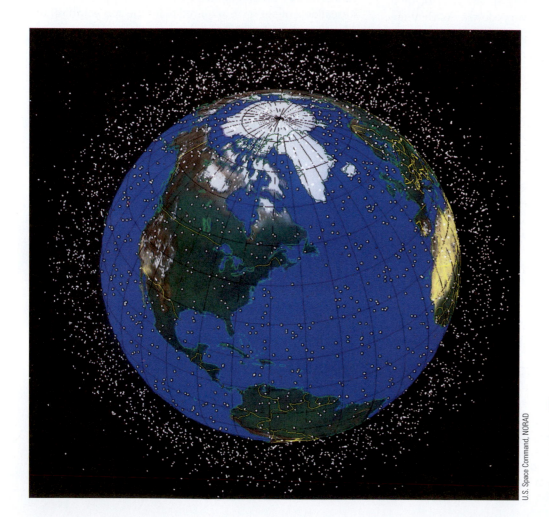

U.S. Space Command, NORAD

Ace⊗Astronomy™ Log into AceAstronomy and select this chapter to see the Active Figure called "Gravity and Orbits" and Astronomy Exercise "Escape Velocity."

■ **FIGURE 2.12**

Satellites in Earth Orbit
A plot of all known unclassified satellites and satellite debris in Earth orbit larger than about the size of a softball.

speed needed to move away from the Earth forever, which is about 11 km/s (about 25,000 mph). After this they coast to their targets, subject only to minor trajectory adjustments provided by small thruster rockets on board. In interplanetary flight, these spacecraft follow Keplerian orbits around the Sun, modified only when they pass near one of the planets.

As it comes close to its target, a spacecraft is deflected by the planet's gravitational force into a modified orbit, either gaining or losing energy in the process. By carefully choosing the aim point in a planetary encounter, controllers have actually been able to use a planet's gravity to redirect a flyby spacecraft to a second target. Voyager 2 used a series of gravity-assisted encounters to yield successive flybys of Jupiter (1979), Saturn (1980), Uranus (1986), and Neptune (1989). The Galileo spacecraft, launched in 1989, flew past Venus once and Earth twice to gain the energy required to reach its ultimate goal of orbiting Jupiter.

If we wish to orbit a planet, we must slow the spacecraft with a rocket when the spacecraft is near its destination, allowing it to be captured into an elliptical orbit. Additional rocket thrust is required to bring a vehicle down from orbit for a landing on the surface. Finally, if a return trip to Earth is planned, the landed payload must include enough propulsive power to repeat the entire process in reverse.

NASA/ARC

■ **FIGURE 2.13**

Modern Computing Power Supercomputers at NASA Ames Research Center are capable of tracking the motions of more than a million objects under their mutual gravitation.

2.6 GRAVITY WITH MORE THAN TWO BODIES

Until now we have considered the Sun and a planet (or a planet and one of its satellites) as nothing more than a pair of bodies revolving around each other. In fact, all the planets exert gravitational forces upon one another as well. These interplanetary attractions cause slight variations from the orbits that would be expected if the gravitational forces between planets were neglected. Unfortunately, the problem of treating the motion of a body that is under the gravitational influence of two or more other bodies is very complicated and can be handled properly only with large computers. Fortunately, astronomers have such computers at their disposal in universities and government research institutes.

2.6.1 The Interactions of Many Bodies

As an example, suppose you have a cluster of a thousand stars all orbiting a common center (such clusters are quite common, as we shall see). If we know the exact position of each star at any given instant, we can calculate the combined gravitational force of the entire group on any one member of the cluster. Knowing the force on the star in question, we can therefore find how it will accelerate. If we know how it was moving to begin with, we can then calculate how it will move in the next instant of time, thus tracking its motion.

However, the problem is complicated by the fact that the other stars are also moving and thus changing the effect they will have on our star. Therefore, we must simultaneously calculate the acceleration of each star produced by the combination of the gravitational attractions of all the others in order to track the motions of all of them, and hence of any one. Such complex calculations have been carried out with modern computers to track the evolution of hypothetical clusters of stars with up to a million members (Figure 2.13).

Within the solar system, the problem of computing the orbits of planets and spacecraft is somewhat simpler. We have seen that Kepler's laws, which do not take into account the gravitational effects of the other planets on an orbit, really work quite well. This is because these additional influences are very small in comparison with the dominant gravitational attraction of the Sun. Under such circumstances, it is possible to treat the effects of other bodies as small *perturbations* (or disturbances). During the 18th and 19th centuries, mathematicians developed many elegant techniques for calculating perturbations, permitting them to predict very precisely the positions of the planets. Such calculations eventually led to the prediction and discovery of a new planet in 1846.

2.6.2 The Discovery of Neptune

The discovery of the eighth planet, Neptune, was one of the high points in the development of gravitational theory. In 1781, William Herschel, a musician and unpaid astronomer,

(a)

(b)

a, Yerkes Observatory; b, Corbis/Bettmann

■ **FIGURE 2.14**
Mathematicians Who Discovered a Planet (a) John Couch Adams (1819–1892) and (b) Urbain J. J. Leverrier (1811–1877) share the credit for discovering the planet Neptune.

accidentally discovered the seventh planet, Uranus. It happens that Uranus had been observed a century before, but in none of those earlier sightings was it recognized as a planet; rather, it was simply recorded as a star. Herschel's discovery showed that there could be planets in the solar system too dim to be visible to the unaided eye, but ready to be discovered with a telescope if we just knew where to look.

By 1790, an orbit had been calculated for Uranus using observations of its motion in the decade following its discovery. Even after allowance was made for the perturbing effects of Jupiter and Saturn, however, it was found that Uranus did not move on an orbit that exactly fit the earlier observations of it made since 1690. By 1840, the discrepancy between the positions observed for Uranus and those predicted from its computed orbit amounted to about 0.03°—an angle barely discernible to the unaided eye but still larger than the probable errors in the orbital calculations. In other words, Uranus just did not seem to move on the orbit predicted from Newtonian theory.

In 1843, John Couch Adams, a young Englishman who had just completed his studies at Cambridge, began a detailed mathematical analysis of the irregularities in the motion of Uranus to see whether they might be produced by the pull of an unknown planet. His calculations indicated the existence of a planet more distant than Uranus from the Sun. In October 1845, Adams delivered his results to George Airy, the British Astronomer Royal, informing him where in the sky to find the new planet. We now know that Adams' predicted position for the new body was correct to within 2°, but for a variety of reasons, Airy did not follow up right away.

Meanwhile, French mathematician Urbain Jean Joseph Leverrier, unaware of Adams or his work, attacked the same problem and published its solution in June 1846. Airy, noting that Leverrier's predicted position for the unknown planet agreed to within 1° with that of Adams, suggested to James Challis, Director of the Cambridge Observatory,

that he begin a search for the new object. The Cambridge astronomer, having no up-to-date star charts of the Aquarius region of the sky where the planet was predicted to be, proceeded by recording the positions of all the faint stars he could observe with his telescope in that location. It was Challis' plan to repeat such plots at intervals of several days, in the hope that the planet would distinguish itself from a star by its motion. Unfortunately, he was negligent in examining his observations; although he had actually seen the planet, he did not recognize it.

About a month later, Leverrier suggested to Johann Galle, an astronomer at the Berlin Observatory, that he look for the planet. Galle received Leverrier's letter on September 23, 1846, and, possessing new charts of the Aquarius region, found and identified the planet that very night. It was less than a degree from the position Leverrier predicted. The discovery of the eighth planet, now known as Neptune (the Latin name for the god of the sea), was a major triumph for gravitational theory, for it dramatically confirmed the generality of Newton's laws. The honor for the discovery is properly shared by the two mathematicians, Adams and Leverrier (Figure 2.14).

We should note that the discovery of Neptune was not a complete surprise to astronomers, who had long suspected the existence of the planet based on the "disobedient" motion of Uranus. On September 10, 1846, two weeks before Neptune was actually found, John Herschel, son of the discoverer of Uranus, remarked in a speech before the British Association, "We see [the new planet] as Columbus saw America from the shores of Spain. Its movements have been felt trembling along the far-reaching line of our analysis with a certainty hardly inferior to ocular demonstration."

This discovery was a major step forward in combining Newtonian theory with painstaking observations. Such work continues in our own times with the discovery of planets around other stars—work that we will discuss in chapters to come.

SURFING THE WEB

Images of Tycho Brahe:
www.mhs.ox.ac.uk/tycho/index.htm
Virtual exhibit from the Museum of the History of Science in Oxford, England, with images of and information about Tycho and his contemporaries.

Mathematical Discovery of Planets Site:
www-groups.dcs.st-and.ac.uk/~history/HistTopics/Neptune_and_Pluto.html
Has the fuller story of how Neptune was predicted and discovered, and who did what to whom. A spirited defense of George B. Airy (who comes across as the "villain" in many versions of the story), written by historian Alan Chapman, can be found at: www.u-net.com/ph/lassell/adams-airy.htm

Isaac Newton Sites:

- *Biography on the Math Tutor Site:*
  www-groups.dcs.st-and.ac.uk/~history//Mathematicians/Newton.html

- *Luminarium Newton Pages:*
  www.luminarium.org/sevenlit/newton/index.html

Sites for Observing Earth Satellites:

- *Visual Satellite Observer's Home Page:*
  www.satellite.eu.org/satintro.html
  Has lots of instructions and links.

- *The Satellite Observing Resources Page:*
  www.znark.com/sat/sattrack.html
  Also has links and tutorials.

- *Orbitessera:* www.mindspring.com/~n2wwd/
  A more advanced site.

- *Satellite Passes over North American Cities:*
  www.bester.com/satpasses.html

SUMMARY

2.1 Tycho Brahe was the most skillful of the pre-telescopic astronomical observers. His accurate observations of planetary positions provided the data used by Johannes Kepler to derive the three fundamental laws of planetary motion that bear his name: (1) Planetary orbits are **ellipses** (a figure described by its **semimajor axis** and **eccentricity**) with the Sun at one **focus;** (2) in equal intervals, a planet's orbit sweeps out equal areas; and (3) if times are expressed in years and distances in **astronomical units,** the relationship between the period (P) and semimajor axis (D) of an orbit is given by $D^3 = P^2$.

2.2 In his *Principia,* Isaac Newton established the three laws that govern the motion of objects: (1) Bodies continue at rest or in uniform motion unless acted upon by an outside force; (2) an outside force causes an acceleration (and changes the **momentum**) of an object; and (3) for each action there is an equal and opposite reaction. Momentum is a measure of the motion of an object and depends on both its mass and its velocity. **Angular momentum** is a measure of the motion of a spinning or revolving object. The **density** of an object is its mass divided by its volume.

2.3 Gravity, the attraction of all mass for all other mass, is the force that keeps the planets in orbit. Newton's law of gravity relates gravitational force to mass and distance ($F = GM_1M_2/R^2$). Newton was able to show the equivalence of gravitational force (weight) on Earth to the gravitational force between objects in space. When Kepler's laws are reexamined in the light of gravitational theory, it becomes clear that the masses of both Sun and planet are important for the third law, which becomes $D^3 = (M_1 + M_2) \times P^2$. Mutual gravitational effects permit us to calculate the masses of astronomical objects, from comets to galaxies.

2.4 The lowest point in a satellite **orbit** around the Earth is its **perigee,** and the highest point is its **apogee** (corresponding to **perihelion** and **aphelion** for an orbit about the Sun). The planets follow orbits about the Sun that are nearly circular and in the same plane. Most asteroids are found between Mars and Jupiter in the **asteroid belt,** whereas comets generally follow orbits of high eccentricity.

2.5 The orbit of an artificial satellite depends on the circumstances of its launch. The **circular satellite velocity**

at the Earth's surface is 8 km/s, and the **escape velocity** is 11 km/s. There are many possible interplanetary trajectories, including those that use gravity-assisted flybys of one object to redirect the spacecraft toward its next target.

2.6 Gravitational problems that involve more than two interacting bodies are much more difficult to deal with than two-body problems. They require large computers for accurate solutions. If one object dominates gravitationally, it is possible to calculate the effects of a second object in terms of small perturbations. This approach was used by Adams and Leverrier to predict the position of Neptune from its perturbations of the orbit of Uranus and thus discover a new planet mathematically.

INTER-ACTIVITY

A An eccentric, but very rich, alumnus of your college makes a bet with the dean that if you drop a marble and a bowling ball from the top of a tall water tower, the bowling ball would hit the ground first. Have your group discuss whether you would make a side bet that the alum is right. How would you decide who is right?

B Suppose a member of your group felt unhappy about his or her weight. Where could a person go to weigh one-fourth as much as he or she does now? Where would the person weigh even less? Would changing the unhappy person's weight have any effect on his or her mass?

C When the Apollo astronauts landed on the Moon, some commentators complained that it ruined the mystery and "poetry" of the Moon forever (and that lovers could never gaze at the full moon in the same way again). Others felt that knowing more about the Moon could only enhance its interest to us as we see it from Earth. How does your group feel? Why?

D Figure 2.12 shows an impressive swarm of satellites in orbit around the Earth. What do you think all these satellites do? How many categories of functions for Earth satellites can your group come up with?

REVIEW QUESTIONS

Ace ◯ Astronomy™ Assess your understanding of this chapter's topics with additional quizzing and animations at **http://ace.brookscole.com/voyages**

1. State Kepler's three laws in your own words.
2. Why did Kepler need Tycho Brahe's data to formulate his laws?
3. Which has more mass: an armful of feathers or an armful of lead? Which has more volume: a kilogram of feathers or a kilogram of lead? Which has higher density: a kilogram of feathers or a kilogram of lead?
4. Explain how Kepler was able to find a relationship (his third law) between the periods and distances of the planets that did not depend on the masses of the planets or the Sun.
5. Write out Newton's three laws of motion in terms of what happens with the momentum of objects.
6. What planet has the largest:
   a. semimajor axis
   b. speed of revolution around the Sun
   c. period of revolution around the Sun
   d. eccentricity
7. Why do we say that Neptune was the first planet to be discovered through the use of mathematics?

THOUGHT QUESTIONS

8. Is it possible to escape the force of gravity by going into orbit about the Earth? How does the force of gravity in the International Space Station (orbiting 500 km above the Earth's surface) compare with that on the ground? (*Hint:* The Earth's gravity acts as if all the mass were concentrated at the center of the Earth. Is the station significantly farther from the Earth's center than the Earth's surface is?)

9. What is the momentum of a body whose velocity is zero? How does Newton's first law of motion include the case of a body at rest?

10. Evil space aliens drop you and your astronomy instructor 1 km apart out in space, very far from any star or planet. Discuss the effects of gravity on each of you.

11. A body moves in a perfectly circular path at constant speed. Are there forces acting in such a system? How do you know?

12. As air friction causes a satellite to spiral inward closer to the Earth, its orbital speed increases. Why?

13. Use a history book or an encyclopedia to find out what else was happening in England during Newton's lifetime, and discuss what trends of the time might have contributed to his accomplishments and the rapid acceptance of his work.

FIGURING FOR YOURSELF

We have given several formulas in this chapter, and in this section we will ask you to apply them. Go back to the appropriate sections of the chapter if you need a reminder of such formulas as Kepler's third law.

14. What is the semimajor axis of a circle of diameter 24 cm? What is its eccentricity?

15. If 24 g of material fills a cube 2 cm on a side, what is the density of the material?

16. Draw an ellipse by the procedure described in the text, using a string and two tacks. Arrange the tacks so that they are separated by 1/10 the length of the string. Comment on the appearance of your ellipse. This (if you have been careful in your construction) is approximately the shape of the orbit of Mars.

17. The Earth's distance from the Sun varies from 147 million to 152 million km. What is the eccentricity of its orbit? (*Hint:* The distance between the foci of the ellipse is twice the distance between the Sun and the center of the ellipse. To find this, you need to first find the center of the ellipse by dividing the major axis in two. It helps to draw a diagram for yourself.)

18. Look up the revolution periods and distances from the Sun for Venus, Earth, Mars, and Jupiter in Table 2.2. Calculate D^3 and P^2 (with D in units of AU and P in units of years), and verify that they obey Kepler's third law.

19. What would be the period of a planet whose orbit has a semimajor axis of 4 AU? Of an asteroid with a semimajor axis of 10 AU?

20. What is the distance from the Sun (in astronomical units) of an asteroid with a period of revolution of 8 years? What is the distance of a planet with a period of 45.66 days?

21. In 1996 astronomers discovered an ice dwarf beyond Pluto that now bears the unromantic name of 1996 TL 66. It has a semimajor axis of 84 AU. What is its period according to Kepler's third law?

22. Newton showed that the periods and distances in Kepler's third law depend on the masses of the objects. What would be the period of revolution of the Earth (at 1 AU from the Sun) if the Sun had twice its present mass?

23. By what factor would a person's weight at the surface of the Earth be reduced if the Earth had its present mass but eight times its present volume? What if it had its present size but only one-third its present mass?

SUGGESTIONS FOR FURTHER READING

Christianson, G. "Newton's *Principia:* A Retrospective" in *Sky & Telescope,* July 1987, p. 18.

Christianson, G. "The Celestial Palace of Tycho Brahe" in *Scientific American,* Feb. 1961, p. 118.

Cohen, I. "Newton's Discovery of Gravity" in *Scientific American,* Mar. 1981, p. 166.

Gingerich, O. *The Eye of Heaven: Ptolemy, Copernicus and Kepler.* 1993, American Institute of Physics Press.

King-Hele, D. and Eberst, R. "Observing Artificial Satellites" in *Sky & Telescope,* May 1986, p. 457.

Koestler, A. *The Sleepwalkers: A History of Man's Changing Vision of the Universe.* 1959, Macmillan. A journalist's recreation of Renaissance astronomical developments.

Standage, T. *The Neptune File: Planet Detectives and the Discovery of Worlds Unseen.* 2000, Walker.

Thoren, V. *The Lord of Uraniborg.* 1990, Cambridge U. Press. Definitive modern study of Brahe's life and work.

Wilson, C. "How Did Kepler Discover His First Two Laws?" in *Scientific American,* Mar. 1972.

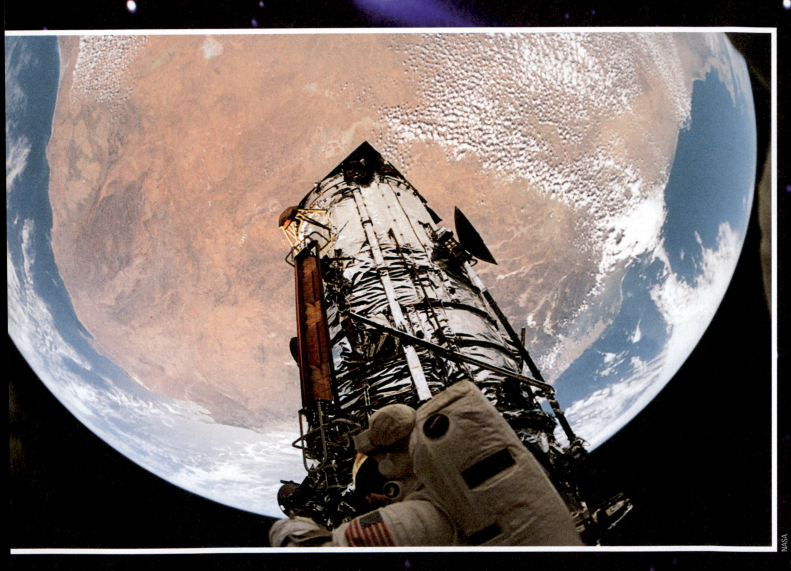

Southern Summer As captured with a fish-eye lens aboard the Space Shuttle on December 9, 1993, the Earth hangs above the Hubble Space Telescope as it is repaired. The reddish continent is Australia, its size and shape distorted by the special lens. Because the seasons in the Southern Hemisphere are opposite ours, it is summer in Australia on this December day.

NASA

3 Earth, Moon, and Sky

To witness a total eclipse of the Sun is privilege that comes to but a few people. Once seen, however, it is a phenomenon never to be forgotten. . . . There is something in it all that affects even the strongest nerves. . . .

Isabel Lewis in A Handbook of Solar Eclipses (1924)

THINKING AHEAD

If the Earth's orbit is nearly a perfect circle (as we saw in earlier chapters), why is it hotter in summer and colder in winter in many places around the globe? And why are the seasons in Australia or Peru the opposite of those in the United States or Europe?

The story is told that Galileo, as he left the Hall of the Inquisition following his retraction of the doctrine that the Earth rotates and revolves about the Sun, said under his breath, "But nevertheless it moves." Historians are not sure whether the story is true, but certainly Galileo knew that the Earth was in motion, whatever church authorities said.

It is the motions of the Earth that produce the seasons and give us our measures of time and date. The Moon's motions around us provide the concept of the month and the cycle of lunar phases. In this chapter we examine some of the basic phenomena of our everyday world in their astronomical context.

Ace○**Astronomy**™ The AceAstronomy icon throughout the text indicates an opportunity for you to test yourself on key concepts and to explore animations and interactions on the AceAstronomy website at **http://ace.brookscole.com /voyages**

Virtual Laboratories

 Tides and Tidal Forces in Astronomy

3.1 EARTH AND SKY

3.1.1 Locating Places on the Earth

Let's begin by fixing our position on the surface of planet Earth. As we discussed in Chapter 1, the Earth's axis of rotation defines the locations of its North and South Poles and of its equator, halfway between. Two other directions are also defined by the Earth's motions: East is the direction toward which the Earth rotates, and west is its opposite. At almost any point on the Earth, the four directions—north, south, east, and west—are well defined, despite the fact that our planet is round rather than flat. The only exceptions are exactly at the North and South Poles, where the directions east and west are ambiguous (because the poles do not turn).

We can use these ideas to define a system of coordinates attached to our planet. Such a system, like the layout of streets and avenues in Manhattan or Salt Lake City, helps us find where we are or want to go. Coordinates on a sphere, however, are a little more complicated than those on a flat surface. We must define circles on the sphere that play the same role as the rectangular grid that you see on city maps.

A **great circle** is any circle on the surface of a sphere whose center is at the center of the sphere. For example, the Earth's equator is a great circle on the Earth's surface, halfway between the North and South Poles. We can also imagine a series of great circles that pass through the North and South Poles. These circles are called **meridians;** they each cross the equator at right angles.

Any point on the surface of the Earth will have a meridian passing through it (Figure 3.1). This meridian

■ FIGURE 3.2

The Royal Greenwich Observatory in England
At the internationally agreed-upon zero point of longitude, tourists can stand and straddle the exact line where longitude "begins."

specifies the east–west location, or *longitude,* of that place. By international agreement (and it took many meetings for the world's countries to agree), your longitude is defined as the number of degrees of arc along the equator between your meridian and the one passing through Greenwich, England.

Why Greenwich, you might ask? Every country wanted 0° longitude to pass through its own capital. Greenwich, the site of the old Royal Observatory (Figure 3.2), was selected because it was between continental Europe and the United States, and because it was the site for much of the development of a way to measure longitude at sea. Longitudes are measured either to the east or to the west of the Greenwich meridian from 0° to 180°. As an example, the longitude of the clock-house benchmark of the U.S. Naval Observatory in Washington, D.C., is 77.066° W.

Your *latitude* (or north–south location) is the number of degrees of arc you are away from the equator along your meridian. Latitudes are measured either north or south of the equator from 0° to 90°. As an example, the latitude of the previously mentioned Naval Observatory benchmark is 38.921° N. The latitude of the South Pole is 90° S.

3.1.2 Locating Places in the Sky

Positions in the sky are measured in a way that is very similar to the way we measure positions on the surface of the Earth. Instead of latitude and longitude, however, astronomers use coordinates called **declination** and **right ascension.** To denote positions of objects in the sky, it is of-

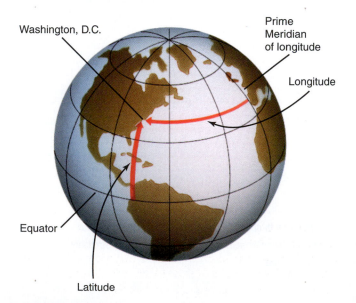

■ FIGURE 3.1

The Latitude and Longitude of Washington, D.C.

ten convenient to make use of the fictitious celestial sphere. We saw in Chapter 1 that the sky appears to rotate about points above the North and South Poles of the Earth—points in the sky called the north celestial pole and the south celestial pole. Halfway between the celestial poles, and thus 90° from each, is the celestial equator, a great circle on the celestial sphere that is in the same plane as the Earth's equator. We can use these markers in the sky to set up a system of celestial coordinates.

Declination on the celestial sphere is measured the same way that latitude is measured on the sphere of the Earth: from the celestial equator toward the north (positive) or south (negative). So Polaris, the star near the north celestial pole, has a declination of almost +90°.

Right ascension (RA) is like longitude, except that instead of Greenwich, its arbitrarily chosen point where we start counting is the *vernal equinox*, a point in the sky where the ecliptic (the Sun's path) crosses the celestial equator. RA can be expressed either in units of angle (degrees) or in units of time. This is because the celestial sphere appears to turn around the Earth once a day as our planet turns on its axis. Thus the 360° of RA that it takes to go once around the celestial sphere can just as well be set equal to 24 hours. Then each 15° of arc is equal to 1 hour of time. The hours can be further subdivided into minutes. For example, the celestial coordinates of the bright star Vega are RA 18 h 36.2 m (= 279.05°) and declination +38.77°.

One way to visualize these circles in the sky is to imagine the Earth as a transparent sphere with the terrestrial coordinates (latitude and longitude) painted on it with dark paint. Imagine the celestial sphere around us as a giant ball, painted white on the inside. Then imagine yourself at the center of the Earth, with a bright light bulb in the middle, looking out through its transparent surface to the sky. The terrestrial poles, equator, and meridians will be projected as dark shadows on the celestial sphere, giving us the system of coordinates in the sky.

3.1.3 The Turning Earth

We have seen that the apparent rotation of the celestial sphere could be accounted for either by a daily rotation of the sky around a stationary Earth, or by the rotation of the Earth itself. Since the 17th century, it has been generally accepted that it is the Earth that turns, but not until the 19th century did the French physicist Jean Foucault provide an unambiguous demonstration of this rotation. In 1851, he suspended a 60-meter (m) pendulum weighing about 25 kilograms (kg) from the dome of the Pantheon in Paris and started the pendulum swinging evenly. If the Earth had not been turning, there would have been no force to alter the pendulum's plane of oscillation, and so it would have continued tracing the same path. Yet after a few minutes Foucault could see that the pendulum's plane of motion was turning. Foucault explained that it was not the pendulum that was shifting, but rather the Earth that was turning underneath it (Figure 3.3). You can now find such

© Bob Ernett, Photographer

■ **FIGURE 3.3**
The Foucault Pendulum

pendulums in many science centers and planetaria around the world.

Can you think of other ways to prove that it is the Earth and not the sky that is turning? (See Inter-Activity problem A at the end of this chapter.)

3.2 THE SEASONS

One of the fundamental facts of life at Earth's mid-latitudes, where most of this book's readers live, is that there are significant variations in the heat we receive from the Sun during the course of the year. We thus divide the year into *seasons*, each with its different amount of sunlight. The difference between seasons gets more pronounced the farther north or south from the equator we travel, and the seasons in the Southern Hemisphere are the opposite of what we find in the northern half of the Earth. With these observed facts in mind, let us ask what causes the seasons.

According to recent surveys, many people believe that the seasons are the result of the changing distance between the Earth and the Sun. This sounds reasonable at first: It

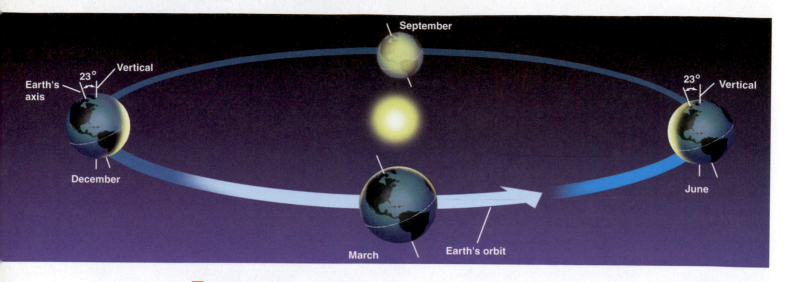

■ FIGURE 3.4

Seasons We see the Earth at different seasons as it circles the Sun. During our winter in the north, the Southern Hemisphere "leans into" the Sun and is illuminated more directly. In summer, it is the Northern Hemisphere that is leaning into the Sun and has longer days. In spring and autumn, the two hemispheres receive more equal shares of sunlight.

Ace◯Astronomy™ Log into AceAstronomy and select this chapter to see the Active Figure called "Seasons" and Astronomy Exercise "The Seasons."

should be colder when the Earth is farther from the Sun. But the facts don't bear out this hypothesis. Although the Earth's orbit around the Sun is an ellipse, its distance from the Sun varies by only about 3 percent. That's not enough to cause significant variations in the Sun's heating. To make matters worse for people in North America who hold this hypothesis, the Earth is actually closest to the Sun in January, when the Northern Hemisphere is in the middle of winter. And if distance were the governing factor, why would the two hemispheres have opposite seasons? As we shall show, the seasons are actually caused by the 23° tilt of the Earth's axis.

3.2.1 The Seasons and Sunshine

Figure 3.4 shows the Earth's annual path around the Sun, with the Earth's axis tilted by 23°. Note that our axis con-

tinues to point in the same direction in the sky throughout the year. As the Earth travels around the Sun, in June the Northern Hemisphere "leans into" the Sun and is more directly illuminated. In December, the situation is reversed: The Southern Hemisphere leans into the Sun, and the Northern Hemisphere leans away. In September and March, the Earth leans "sideways"—neither into the Sun nor away from it—so the two hemispheres are equally favored with sunshine.

How does the Sun's favoring one hemisphere translate into making it warmer for us down on the surface of the Earth? There are two effects we need to consider. When we lean into the Sun, sunlight hits us at a more direct angle and is more effective at heating the Earth's surface (Figure 3.5). You can get a similar effect by shining a flashlight onto a wall. If you shine the flashlight straight on, you get an intense spot of light on the wall. But if you hold the flashlight

(a) (b)

■ FIGURE 3.5

The Sun's Rays in Summer and Winter (a) In summer the Sun appears high in the sky and its rays hit the Earth more directly, spreading out less. (b) In winter the Sun is low in the sky and its rays spread out over a much wider area, becoming less effective at heating the ground.

North celestial pole — Sun's path June 22 — N — E — S — W — Celestial equator

Sun's path March 21 Sept. 23 — N — E — S — W — Celestial equator

Sun's path Dec. 22 — N — E — S — W — Celestial equator

■ **FIGURE 3.6**

The Sun's Path in the Sky for Different Seasons   On June 22, the Sun rises north of east and sets north of west; for observers in the Northern Hemisphere of the Earth, the Sun spends about 15 hours above the horizon. On December 22, the Sun rises south of east and sets south of west; it spends only 9 hours above the horizon, which means short days and long nights in northern lands (and a strong need for people to hold celebrations to cheer themselves up). On March 21 and September 23, the Sun spends equal amounts of time above and below the horizon.

Ace Astronomy™   Log into AceAstronomy and select this chapter to see the Active Figure called "Shadow and Seasons" and Astronomy Exercise "Sunrise Through the Seasons."

at an angle (if the wall "leans out" of the beam), then the spot of light is more spread out. In the same way, the sunlight in June is more direct and intense on the Northern Hemisphere, and hence more effective at heating.

The second effect has to do with the length of time the Sun spends above the horizon (Figure 3.6). Even if you've never thought about astronomy before, we're sure you have observed that the days get longer in summer and shorter in winter. Let's see why this happens.

As we saw in Chapter 1, an equivalent way to look at our path around the Sun each year is to pretend that the Sun moves around the Earth (on a circle called the ecliptic). Because the Earth's axis is tilted, the ecliptic is tilted by about 23° relative to the celestial equator (review Figure 1.6), and where we see the Sun in the sky changes as the year wears on. In June, the Sun is north of the celestial equator and spends more time with those who live in the Northern Hemisphere. It rises high in the sky and is above the horizon in the United States for as long as 15 hours. Thus, the Sun not only heats us with more direct rays, but it also has more time to do it each day. (Notice in Figure 3.6 that our gain is the Southern Hemisphere's loss. There, the June Sun is low in the sky, meaning fewer daylight hours. In Chile, for example, June is a colder, darker time of year.) In December, when the Sun is south of the celestial equator, the situation is reversed.

Let's look at what the Sun's illumination on Earth looks like at some specific dates of the year, when these effects

are at their maximum. On about June 22 (the date we who live in the Northern Hemisphere call the *summer solstice* or sometimes the first day of summer), the Sun shines down most directly upon the Northern Hemisphere of the Earth. It appears 23° north of the equator and thus on that date passes through the zenith of places on the Earth that are at 23° N latitude. The situation is shown in detail in Figure 3.7. To a person at latitude 23° N (near Hawaii, for example), the Sun is directly overhead at noon. This latitude, where the Sun can appear at the zenith at noon on the first day of summer, is called the *Tropic of Cancer*.

We also see in Figure 3.7 that the Sun's rays shine down all around the North Pole at the solstice. As the Earth turns on its axis, the North Pole is continuously illuminated by the Sun; all places within 23° of the pole have sunshine for 24 hours. The Sun is as far north on this date as it can get; thus, $90° - 23°$ (or 67° N) is the southernmost latitude where the Sun can be seen for a full 24-hour period (the "land of the midnight Sun"). That circle of latitude is called the *Arctic Circle*.

Many early cultures scheduled special events around the summer solstice to celebrate the longest days and thank their gods for making the weather warm. This required people to keep track of the lengths of the days and the northward trek of the Sun in order to know the right day for the "party." (You can do the same thing by watching for several weeks, from the same observation point, where the Sun rises or sets relative to a fixed landmark. In spring, the

Sun's rays

N
Arctic Circle
Tropic of Cancer
23°
Equator
Tropic of Capricorn
Antarctic Circle
S

■ **FIGURE 3.7**

The Earth on June 22 This is the date of the summer solstice in the Northern Hemisphere. Note that as the Earth turns on its axis (the line connecting the North and South Poles), the North Pole is in constant sunlight while the South Pole is veiled in 24-hour darkness. The Sun is at the zenith for observers on the Tropic of Cancer.

Sun will rise farther and farther north of east, and set farther and farther north of west, reaching the maximum around the summer solstice.)

Now look at the South Pole in Figure 3.7. On June 22, all places within 23° of the South Pole—that is, south of what we call the *Antarctic Circle*—do not see the Sun at all for 24 hours.

The situation is reversed 6 months later, about December 22 (the date of the *winter solstice* or the first day of winter in the Northern Hemisphere), as shown in Figure 3.8. Now it is the Arctic Circle that has the 24-hour night and the Antarctic Circle that has the midnight Sun. At lati-

tude 23° S, called the *Tropic of Capricorn*, the Sun passes through the zenith at noon. Days are longer in the Southern Hemisphere and shorter in the north. In the United States and southern Europe, we might get only 9 or 10 hours of sunshine during the day. It is winter in the Northern Hemisphere and summer in the Southern Hemisphere.

Many cultures that developed some distance north of the equator have a celebration around December 22 to help people deal with the depressing lack of sunlight and the often dangerously cold temperatures. Originally, this was often a time for huddling with family and friends, for sharing the reserves of food and drink, for rituals asking the gods to return the light and heat and turn the cycle of the seasons around. Many cultures constructed elaborate devices for anticipating when the shortest day of the year was coming. Stonehenge in England, built long before the invention of writing, is probably one such device (see Section 3.4). In our own time, we continue the winter solstice tradition with various holiday celebrations (although few of us now feel the need to pause while eating the sacrificial turkey to ask the sun god to make the days longer).

Halfway between the solstices, on about March 21 and September 23, the Sun is on the celestial equator. From Earth, it appears above our planet's equator and favors neither hemisphere. Every place on the Earth then receives exactly 12 hours of sunshine and 12 hours of night. The points where the Sun crosses the celestial equator are called the *vernal* (spring) and *autumnal* (fall) *equinoxes*.

3.2.2 The Seasons at Different Latitudes

The seasonal effects are different at different latitudes on Earth. Near the equator, for instance, all seasons are much the same. Every day of the year, the Sun is up half the time, so there are always 12 hours of sunshine and 12 hours of

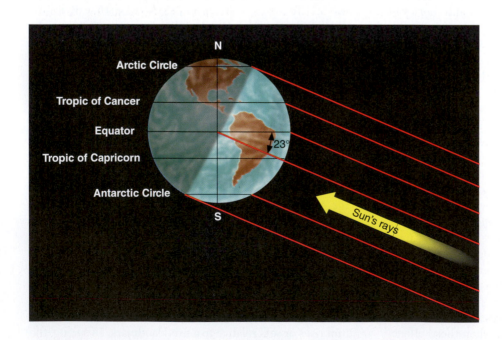

N
Arctic Circle
Tropic of Cancer
Equator
Tropic of Capricorn
Antarctic Circle
S
23°
Sun's rays

■ **FIGURE 3.8**

The Earth on December 22 This is the date of the winter solstice in the Northern Hemisphere. Now the North Pole is in darkness for 24 hours and the South Pole is illuminated. The Sun is at the zenith for observers on the Tropic of Capricorn and thus is low in the sky for the residents of the United States and Canada (just the right time to sacrifice a ritual turkey and huddle with your loved ones).

night. Local residents define the seasons by the amount of rain rather than by the amount of sunlight. As we travel north or south, the seasons become more pronounced, until we reach extreme cases in the Arctic and Antarctic.

At the North Pole, all celestial objects that are north of the celestial equator are always above the horizon and, as the Earth turns, circle around parallel to it. The Sun is north of the celestial equator from about March 21 to September 23, so at the North Pole, the Sun rises when it reaches the vernal equinox and sets when it reaches the autumnal equinox. Each year there are 6 months of sunshine at each pole, followed by 6 months of darkness.

3.2.3 Clarifications About the Real World

In our discussions so far, we have been describing the rising and setting of the Sun and stars as they would appear if the Earth had little or no atmosphere. In reality, however, the atmosphere has the curious effect of allowing us to see a little way "over the horizon." This effect is a result of *refraction,* the bending of light passing through air or water, something we will discuss in Chapter 5. Because of this atmospheric refraction, the Sun appears to rise earlier and to set later than it would if no atmosphere were present.

In addition, the atmosphere scatters light and provides some twilight illumination even when the Sun is below the horizon. Astronomers define morning twilight as beginning when the Sun is 18° below the horizon, and evening twilight extends until the Sun sinks more than 18° below the horizon.

These atmospheric effects require small corrections in many of our statements about the seasons. At the equinoxes, for example, the Sun appears to be above the horizon for a few minutes longer than 12 hours, and below the horizon for less than 12 hours. These effects are most dramatic at the Earth's poles, where the Sun actually rises more than a week before it reaches the celestial equator.

You probably know that the summer solstice (June 22) is not the warmest day of the year, even if it is the longest. The hottest months in the Northern Hemisphere are July and August. This is because our weather involves the air and water covering the Earth's surface, and these large reservoirs do not heat up instantaneously. After all, a swimming pool does not get warm the moment the Sun rises but is warmest late in the afternoon, after it has had time to absorb the Sun's heat. In the same way, the Earth gets warmer after it has had a chance to absorb the extra sunlight that is the Sun's summer gift to us. And the coldest times of winter are a month or more after the winter solstice.

3.3 KEEPING TIME

The measurement of time is based on the rotation of the Earth. Throughout most of human history, time has been reckoned by the positions of the Sun and stars in the sky.

Only recently have mechanical and electronic clocks taken over this function in regulating our lives.

3.3.1 The Length of the Day

The most fundamental astronomical unit of time is the day, measured in terms of the rotation of the Earth. There is, however, more than one way to define the day. Usually, it is the rotation period of the Earth with respect to the Sun, called the **solar day.** After all, for most people sunrise is more important than the rising time of Arcturus or some other star, so we set our clocks to some version of sun-time. However, astronomers also use a **sidereal day,** which is defined in terms of the rotation period of the Earth with respect to the stars.

A solar day is slightly longer than a sidereal day because (as you can see from Figure 3.9) the Earth moves along its path around the Sun in a day. Suppose we start the day when the Earth's orbital position is at A, with both the Sun and some distant star (located in direction C) being above an observer at point O on the Earth. When the Earth has completed one rotation with respect to the distant star, C is again above O. However, notice that because of the movement of the Earth along its orbit from A to B, the Sun has not yet reached a position above O. To complete a solar day, the Earth must rotate an additional amount, equal to 1/365 of a full turn. The time required for this extra rotation is 1/365 of a day, or about 4 minutes, so the solar day is about 4 minutes longer than the sidereal day.

Because our ordinary clocks are set to solar time, stars rise 4 minutes earlier each day. Astronomers prefer sidereal time for planning their observations because in that system, a star rises at the same time every day.

3.3.2 Apparent Solar Time

Apparent solar time is reckoned by the actual position of the Sun in the sky (or, during the night, its position below the horizon). This is the kind of time indicated by sundials, and it probably represents the earliest measure of time used by ancient civilizations. Today we adopt the middle of the night as the starting point of the day and measure time in hours elapsed since midnight.

During the first half of the day, the Sun has not yet reached the meridian (the great circle in the sky that goes from the north point on the horizon through our zenith to the south point.). We designate those hours as before midday (ante meridiem, or A.M.). We then start numbering the hours after noon over again and designate them by P.M. (post meridiem).

Although apparent solar time seems simple, it is not really very convenient to use. The exact length of an apparent solar day varies slightly during the year. The eastward progress of the Sun in its annual journey around the sky is not uniform because the speed of the Earth varies slightly in its elliptical orbit. Another reason is that the Earth's axis of rotation is not perpendicular to the plane of its revolution. Thus, apparent solar time does not advance

C   C
                 To remote point on
                 celestial sphere

1°

$\frac{1}{365}$ day
(4 min.)

1°   O

        O   O

A   B

■ FIGURE 3.9

The Difference Between a Sidereal Day and a Solar Day This is a top view, looking down as the Earth orbits the Sun. Because the Earth moves around the Sun (roughly 1° per day), after one complete rotation of the Earth relative to the stars, we do not see the Sun in the same position.

at a uniform rate. After the invention of mechanical clocks that ran at a uniform rate, it became necessary to abandon the apparent solar day as the fundamental unit of time.

3.3.3 Mean Solar Time and Standard Time

Mean solar time is based on the average value of the solar day over the course of the year. A mean solar day contains exactly 24 hours, and it is what we use in our everyday time-keeping. Although mean solar time has the advantage of progressing at a uniform rate, it is still inconvenient for practical use because it is determined by the position of the Sun. For example, noon occurs when the Sun is overhead. But because we live on a round Earth, the exact time of noon is different as you change your longitude by moving east or west.

If mean solar time were strictly observed, people traveling east or west would have to reset their watches continually as the longitude changed, just to read the local mean time correctly. For instance, a commuter traveling from Oyster Bay on Long Island to New York City would have to adjust the time on the trip through the East River tunnel because Oyster Bay time is actually about 1.6 min more advanced than that of Manhattan. (Imagine an airplane trip in which the flight attendant gets on the intercom every minute, saying, "Please reset your watch for local mean time.")

Until near the end of the last century, every city and town in the United States did keep its own local mean time. With the development of railroads and the telegraph, however, the need for some kind of standardization became evident. In 1883, the nation was divided into four standard time zones (now five, including Hawaii and Alaska). Within each zone, all places keep the same *standard time,* with the local mean solar time of a standard line of longitude running more or less through the middle of each zone. Now travelers reset their watches only when the time change has amounted to a full hour. Pacific standard time is 3 hours earlier than eastern standard time, a fact that becomes painfully obvious in California when someone on the East Coast forgets and calls you at 5:00 A.M.

For local convenience, the boundaries between the U.S. time zones are chosen to correspond to divisions between states. Since 1884, standard time has been in use around the world by international agreement (with 24 time zones circling the globe). Almost all countries have adopted one or more standard time zones, although one of the largest nations, India, has settled on a half-zone, being 5½ hours from Greenwich standard.

Daylight saving time is simply the local standard time of the place plus 1 hour. It has been adopted for spring and summer use in most states in the United States, as well as in many other countries, to prolong the sunlight into evening hours, on the apparent theory that it is easier to change the time by government action than it would be for individuals or businesses to adjust their own schedules to

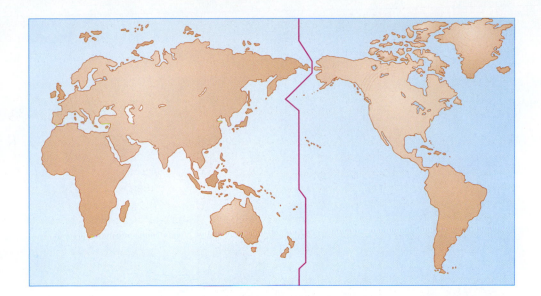

■ **FIGURE 3.10**

Where the Date Changes
The international date line is an arbitrarily drawn line on the Earth where the date changes. So that neighbors do not have different days, the line is located where the Earth's surface is mostly water.

produce the same effect. It does not, of course, "save" any daylight at all—because the amount of sunlight is not determined by what we do with our clocks.

3.3.4 The International Date Line

The fact that time is always more advanced to the east presents a problem. Suppose you travel eastward around the world. You pass into a new time zone, on the average, about every 15° of longitude you travel, and each time you dutifully set your watch ahead an hour. By the time you have completed your trip, you have set your watch ahead through a full 24 hours and thus gained a day over those who stayed at home.

The solution to this dilemma is the **international date line,** set by international agreement to run approximately along the 180° meridian of longitude. The date line runs about down the middle of the Pacific Ocean, although it jogs a bit in a few places to avoid cutting through groups of islands and through Alaska (Figure 3.10). By convention, at the date line the date of the calendar is changed by one day. Crossing the date line from west to east, thus advancing your time, you compensate by decreasing the date; crossing from east to west, you increase the date by one day. To maintain our planet on a rational system of timekeeping, we simply must accept that the date will differ in different cities at the same time. A good example is the date when the Imperial Japanese Navy bombed Pearl Harbor in Hawaii, known in the United States as Sunday, December 7, 1941, but taught to Japanese students as Monday, December 8.

3.4 THE CALENDAR

3.4.1 The Challenge of the Calendar

"What's today's date?" is one of the most common questions you can ask (usually when writing a check, or worrying about the next exam). Long before the era of digital watches that tell the date, people used calendars to help measure the passage of time.

There are two traditional functions of any calendar. First, it must keep track of time over the course of longer spans, allowing people to anticipate the cycle of the seasons and to honor special religious or personal anniversaries. Second, to be useful to a large number of people, a calendar must use natural time intervals that everyone can agree on—those defined by the motions of the Earth, Moon, and sometimes even the planets. The natural units of our calendar are the *day,* based on the period of rotation of the Earth; the *month,* based on the period of revolution of the Moon about the Earth; and the *year,* based on the period of revolution of the Earth about the Sun. Difficulties have resulted from the fact that these three periods are not commensurable; that is, one does not divide evenly into any of the others.

The rotation period of the Earth is, by definition, 1.0000 day. The period required by the Moon to complete its cycle of phases, called the *lunar month,* is 29.5306 days. The basic period of revolution of the Earth, called the *tropical year*, is 365.2422 days. The ratios of these numbers are not convenient for calculations. This is the historic challenge of the calendar, dealt with in various ways by different cultures.

3.4.2 Early Calendars

Even the earliest cultures were concerned with the keeping of time and the calendar. Particularly interesting are monuments left by Bronze Age people in northwestern Europe, especially the British Isles. The best preserved of the monuments is Stonehenge, about 13 kilometers (km) from Salisbury in southwest England (Figure 3.11). It is a complex array of stones, ditches, and holes arranged in concentric circles. Carbon dating and other studies show that Stonehenge was built during three periods ranging

David Morrison

■ **FIGURE 3.11**

Part of Stonehenge This ancient monument was built between 2800 and 1500 B.C.E. and used to keep track of the motions of the Sun and Moon. Today, heedless tourists and vandals have disturbed and chipped away at the stones to such a degree that the site is now fenced in and entry is restricted.

David Morrison

■ **FIGURE 3.12**

Ruins of the Caracol This Mayan observatory at Chichén Itza in the Yucatan, Mexico, dates from around the year 1000.

from about 2800 to 1500 B.C.E. Some of the stones are aligned with the directions of the Sun and Moon during their risings and settings at critical times of the year (such as the summer and winter solstices), and it is generally believed that at least one function of the monument was connected with the keeping of a calendar.

The Maya in Central America, who thrived more than a thousand years ago, were also concerned with the keeping of time. Their calendar was as sophisticated as, and perhaps more complex than, contemporary calendars in Europe. The Maya did not attempt to correlate their calendar accurately with the length of the year or lunar month. Rather, their calendar was a system for keeping track of the passage of days and for counting time far into the past or future. Among other purposes, it was useful for predicting astronomical events—for example, the positions of Venus in the sky (Figure 3.12).

The ancient Chinese developed an especially complex calendar, largely limited to a few privileged hereditary court astronomer–astrologers. In addition to the motions of the Earth and Moon, they were able to fit in the approximately 12-year cycle of Jupiter, which was central to their system of astrology. The Chinese still preserve some aspects of this system in their cycle of 12 "years"—the Year of the Dragon, the Year of the Pig, and so on—that are defined by the position of Jupiter in the zodiac.

Our Western calendar derives from Greek calendars dating back to at least the 8th century B.C.E. They led, eventually, to the *Julian calendar,* introduced by Julius Caesar, which approximated the year at 365.25 days, fairly close to the actual value of 365.2422. The Romans achieved this approximation by declaring years to have 365 days each, with the exception of every fourth year. The *leap year* was to have one extra day, bringing its length to 366 days and thus making the average length of the year in the Julian calendar 365.25 days.

The Romans had dropped the almost impossible task of trying to base their calendar on the Moon as well as the Sun, although a vestige of older lunar systems can be seen in the fact that our months have an average length of about 30 days. However, lunar calendars remained in use in other cultures, and Islamic calendars are still primarily lunar rather than solar.

3.4.3 The Gregorian Calendar

Although the Julian calendar (which was adopted by the early Christian Church) represented a great advance, its average year still differed from the true year by about 11 minutes, an amount that accumulates over the centuries to an appreciable error. By 1582, that 11 minutes per year had added up to the point where the first day of spring was occurring on March 11, instead of March 21. If the trend were allowed to continue, eventually the Christian celebration of Easter would be occurring in early winter. Pope Gregory XIII, a contemporary of Galileo, felt it necessary to institute further calendar reform.

The Gregorian calendar reform consisted of two steps. First, 10 days had to be dropped out of the calendar to bring the vernal equinox back to March 21; by proclamation, the day following October 4, 1582, became October 15. The second feature of the new Gregorian calendar was a change in the rule for leap year, making the average length of the year more closely approximate the tropical year. Gregory decreed that three of every four century years, all leap years under the Julian calendar, would be common years henceforth. The rule was that only century years divisible by 400 would be leap years. Thus, 1700, 1800, and 1900—all divisible by 4 but not by 400—were not leap years in the Gregorian calendar. On the other hand, the years 1600 and 2000, both divisible by 400, were leap years. The average length of this Gregorian year, 365.2425 mean solar days, is correct to about 1 day in 3300 years.

The Catholic countries immediately put the Gregorian reform into effect, but countries under control of the Eastern Church and most Protestant countries did not adopt it until much later. It was 1752 when England and the American colonies finally made the change. By parliamentary decree, September 2, 1752, was followed by September 14. Although special laws were passed to prevent such abuses as landlords collecting a full month's rent for September, there were still riots, and people demanded their 12 days back. Russia did not abandon the Julian calendar until the time of the Bolshevik revolution. The Russians then had to omit 13 days to come into step with the rest of the world.

3.5 PHASES AND MOTIONS OF THE MOON

After the Sun, the Moon is the brightest and most obvious object in the sky. Unlike the Sun, it does not shine under its own power but merely glows with reflected sunlight. If you were to follow its progress in the sky for a month, you would observe a cycle of **phases,** with the Moon starting dark and getting more and more illuminated by sunlight over the course of about two weeks. After the Moon's disk becomes fully bright, it begins to fade, returning to dark about two weeks later. These changes fascinated and mystified many early cultures, which came up with marvelous stories and legends to explain the cycle of the Moon. Even in the modern world, many people don't understand what causes the phases, thinking that they are somehow related to the shadow of the Earth. Let us see how the phases can be explained by the motion of the Moon relative to the bright light source in the solar system, the Sun.

3.5.1 Lunar Phases

Although we know the Sun moves 1/12 of its path around the sky each month, for purposes of explaining the phases, we can assume that the Sun's light comes from roughly the same direction during the course of a four-week lunar cycle. The Moon, on the other hand, moves completely around the Earth in that time. As we watch the Moon from our vantage point on Earth, how much of its face we see illuminated by sunlight depends on the angle the Sun makes with the Moon.

Here is a simple experiment to show you what we mean: Stand about 6 feet in front of a bright electric light in a completely dark room (or outdoors at night) and hold in your hand a small round object such as a tennis ball or an orange. Your head can then represent the Earth, the light the Sun, and the ball the Moon. Move the ball around your head (making sure you don't cause an eclipse by blocking the light with your head). You will see phases just like those of the Moon on the ball. (Another good way to get acquainted with the phases and motions of the Moon is to follow our satellite in the sky for a month or two, recording its shape, its direction from the Sun, and when it rises and sets.)

Let's examine the Moon's cycle of phases using Figure 3.13, which depicts the Moon's behavior for the entire month. The trick to this figure is that you must imagine yourself standing on the Earth, *facing* the Moon in each of its

Sun's rays

■ **FIGURE 3.13**

The Phases of the Moon   The appearance of the Moon during the course of a complete monthly cycle. The upper part shows a perspective from space, with the Sun off to the right in a fixed position. Imagine yourself standing on the Earth, facing the Moon in each part of its orbit around the Earth. In position A, for example, you are facing the Moon from the right side of the Earth in the middle of the day. The strip below shows the appearance of the Moon from Earth as you would see it from each lettered position. (Please note that the distance of the Moon from the Earth is not to scale in this diagram: The Moon is roughly 30 Earth-diameters away from us, but we would have needed a big expensive foldout to show you the diagram to scale!)

Ace🌑Astronomy™   Log into AceAstronomy and select this chapter to see the Active Figure called "Lunar Phases" and Astronomy Exercises "Phases of the Moon" and "Moon Phase Calendar."

phases. So for the position labeled *A*, you are on the right side of the Earth and it's the middle of the day; for position *E*, you are on the left side of the Earth in the middle of the night. Note that in every position on Figure 3.13, the Moon is half illuminated and half dark (as a ball in sunlight should be). The difference at each position has to do with what part of the Moon faces Earth.

The Moon is said to be *new* when it is in the same general direction in the sky as the Sun (position A). Here its illuminated (bright) side is turned away from us and its dark side is turned toward us. You might say that the Sun is shining on the "wrong" side of the Moon from our perspective. In this phase the Moon is invisible; its dark, rocky surface does not give off any light of its own. Because the new moon is in the same part of the sky as the Sun, it rises at sunrise and sets at sunset.

But the Moon does not remain in this phase long because it moves eastward each day in its monthly path around us. Since it takes about 30 days to orbit the Earth and there are 360° in a circle, the Moon will move about 12° in the sky

each day (or about 24 times its own diameter). A day or two after the new phase, the thin *crescent* first appears, as we begin to see a small part of the Moon's illuminated hemisphere. It has moved into a position where it now reflects a little sunlight toward us along one side. The bright crescent increases in size on successive days as the Moon moves farther and farther around the sky away from the direction of the Sun (position *B*). Because the Moon is moving eastward away from the Sun, it rises later and later each day, like a student during summer vacation.

After about one week, the Moon is one quarter of the way around its orbit (position *C*) and so we say it is at the *first quarter* phase. Half of the Moon's illuminated side is visible to Earth observers. Because of its eastward motion, the Moon now lags about one quarter of the day behind the Sun, rising around noon and setting around midnight.

During the week after the first quarter phase, we see more and more of the Moon's illuminated hemisphere (position *D*), a phase that is called *waxing* (or growing) *gibbous*. Eventually, the Moon arrives at position *E* in our

MAKING CONNECTIONS

Astronomy and the Days of the Week

The week seems independent of celestial motions, although its length may have been based on the time between quarter phases of the Moon. In Western culture, the seven days of the week are named after the seven "wanderers" that the ancients saw in the sky: the Sun, the Moon, and the five planets visible to the unaided eye (Mercury, Venus, Mars, Jupiter, and Saturn).

In English, we can easily recognize the names Sun-day, Moon-day, and Saturn-day, but the other days are named after the Norse equivalents of the Roman gods that gave their names to the planets. In languages more directly re-

lated to Latin, the correspondences are clearer. Wednesday, Mercury's day, for example, is *Mercoledi* in Italian, *Mercredi* in French, and *Miercoles* in Spanish. Mars gives its name to Tuesday (*Martes* in Spanish), Jupiter or Jove to Thursday (*Giovedi* in Italian), and Venus to Friday (*Vendredi* in French).

There is no reason that the week has to have seven days rather than five or eight. It is interesting to speculate that if we had lived in a planetary system where more planets were visible, the Beatles could have been right and we might well have had "Eight Days a Week."

figure, where it and the Sun are opposite each other in the sky. The side of the Moon turned toward the Sun is turned toward the Earth, and we have *full* phase.

When the Moon is full, it is opposite the Sun in the sky. The Moon does the opposite of what the Sun does, rising at sunset and setting at sunrise. Note what that means in practice. The completely illuminated (and thus very noticeable) Moon rises just as it gets dark, remains in the sky all night long, and sets as the Sun's first rays are seen at dawn. And when is it highest in the sky and most noticeable? At midnight, a time made famous in generations of horror novels and films. (Note how the behavior of a vampire like Dracula parallels the behavior of the full Moon: Dracula rises at sunset, does his worst mischief at midnight, and must be in his coffin by sunrise. The old legends were a way of personifying the behavior of the Moon, which was a much more dramatic part of people's lives in the days before city lights and television.)

Folk wisdom has it that more crazy behavior is seen during the time of the full moon (the Moon even gives a name to crazy behavior—"lunacy"). But, in fact, statistical tests of this "hypothesis," involving thousands of records from hospital emergency rooms and police files, do not reveal any correlation with the phases of the Moon. For example, as many homicides occur during the new moon or the crescent moon as during the full moon. Most investigators believe that the real story is not that more crazy behavior happens on nights with a full moon, but rather that we are more likely to notice or remember such behavior with the aid of a bright celestial light that is up all night long.

During the two weeks following the full moon, the Moon goes through the same phases again in reverse order (points *F, G,* and *H* in Figure 3.13), returning to new phase after about 29.5 days. About a week after full moon, for example, the Moon is now at *third quarter*—meaning that it is three quarters of the way around, not that it is three-fourths illuminated. In fact, half of the visible side of the

Moon is again dark. At this phase, the Moon is now rising around midnight and setting around noon.

Note that there is one thing quite misleading about Figure 3.13. If you look at the Moon in position *E,* although it is full in theory, it appears as if its illumination would in fact be blocked by a big fat Earth, and hence we would not see anything on the Moon except the Earth's shadow. In reality, the Moon is nowhere near as close to the Earth (nor is its path so identical with the Sun's in the sky) as this diagram (and the diagrams in every school textbook) might lead you to believe. The Moon is actually *30 Earth-diameters* away from us; the Prologue contains a diagram that shows the two objects to scale. And, since the Moon's orbit is tilted relative to the path of the Sun in the sky, the Earth's shadow misses the Moon most months. That's why we regularly get treated to a full moon. The times when the Earth's shadow does fall on the Moon are called lunar eclipses and are discussed in Section 3.7.

You can see from this tour of the Moon's phases and times of rising and setting that the writer of the old song that went "I've got the Sun in the morning, and the Moon at night" did not quite remember his college astronomy course. It is just during the full phase that the Moon is up only at night. At other times of the month, it may be visible all morning (third quarter) or all afternoon (first quarter) in the daytime sky.

3.5.2 The Moon's Revolution and Rotation

The Moon's sidereal period—that is, the period of its revolution about the Earth measured with respect to the stars—is a little over 27 days: 27.3217 days to be exact. The time interval in which the phases repeat—say, from full to full—is 29.5306 days. The difference is the fault of the Earth's motion around the Sun. The Moon must make more than a complete turn around the moving Earth to get back to the same phase with respect to the Sun. As we saw, the Moon

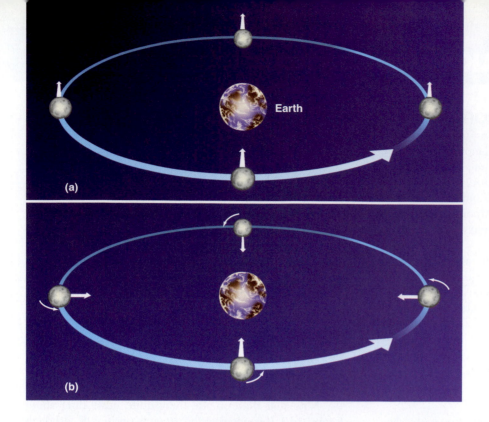

The Moon Without and With Rotation In this figure we stuck a white arrow into a fixed point on the Moon to keep track of its sides. (a) If the Moon did not rotate as it orbited the Earth, it would present all of its sides to our view; hence the white arrow would point directly toward the Earth only in the bottom position on the diagram. (b) Actually, the Moon rotates in the same period that it revolves, so we always see the same side (the white arrow keeps pointing to the Earth).

changes its position on the celestial sphere rather rapidly: Even during a single evening, the Moon creeps visibly eastward among the stars, traveling its own width in a little less than 1 hour. The delay in moonrise from one day to the next, caused by this eastward motion, averages about 50 minutes.

The Moon *rotates* on its axis in exactly the same time that it takes to *revolve* about the Earth. As a consequence, the Moon always keeps the same face turned toward the Earth (Figure 3.14). You can try this yourself by "orbiting" your roommate or another volunteer. Start by facing your roommate. If you make one rotation (spin) with your shoulders in the exact same time that you revolve around him or her, you will continue to face your roommate during the whole "orbit."

The differences in the Moon's appearance from one night to the next are due to changing illumination by the Sun, not to its own rotation. You sometimes hear the back side of the Moon (the side we never see) called the "dark side." This is a misunderstanding of the real situation: Which side is light and which is dark changes as the Moon moves around the Earth. The back side is dark no more frequently than the front side. Since the Moon rotates, the Sun rises and sets on all sides of the Moon. With apologies to Pink Floyd, there is simply no regular "Dark Side of the Moon."

3.6 OCEAN TIDES AND THE MOON

Anyone living near the sea is familiar with the twice-daily rising and falling of the *tides*. Early in history it was clear that tides must be related to the Moon because the daily

delay in high tide is the same as the daily delay in the Moon's rising. A satisfactory explanation of the tides, however, awaited the theory of gravity, supplied by Newton.

3.6.1 The Pull of the Moon on the Earth

The gravitational forces exerted by the Moon at several points on the Earth are illustrated in Figure 3.15. These forces differ slightly from one another because the Earth is not a point, but has a certain size: All parts are not equally distant from the Moon, nor are they all in exactly the same direction from the Moon.

Moreover, the Earth is not perfectly rigid. As a result, the differences among the forces of the Moon's attraction on different parts of the Earth (called *differential forces*) cause the Earth to distort slightly. The side of the Earth nearest the Moon is attracted toward the Moon more strongly than is the center of the Earth, which in turn is attracted more strongly than is the side opposite the Moon. Thus, the differential forces tend to stretch the Earth slightly into a *prolate spheroid* (a football shape), with its long diameter pointed toward the Moon.

If the Earth were made of water, it would distort until the Moon's differential forces over different parts of its surface came into balance with the Earth's own gravitational forces pulling it together. Calculations show that in this case the Earth would distort from a sphere by amounts ranging up to nearly 1 m. Measurements of the actual deformation of the Earth show that the solid Earth does distort, but only about one-third as much as water would, because of the great rigidity of the Earth's interior.

■ FIGURE 3.15

The Pull of the Moon The Moon's differential attraction of different parts of the Earth. (Note that the differences have been exaggerated for educational purposes.)

Because the tidal distortion of the solid Earth amounts at its greatest to only about 20 cm, the Earth does not distort enough to balance the Moon's differential forces with its own gravity. Hence, objects at the Earth's surface experience tiny horizontal tugs, tending to make them slide about. These *tide-raising forces* are too insignificant to affect solid objects like astronomy students or rocks in the Earth's crust, but they do affect the waters in the oceans.

3.6.2 The Formation of Tides

The tide-raising forces, acting over a number of hours, produce motions of the water that result in measurable tidal bulges in the oceans. Water on the side of the Earth facing the Moon flows toward it, with the greatest depths at the point just below the Moon. On the side of the Earth opposite the Moon, water flows to produce a tidal bulge there as well (Figure 3.16).

Note that the tidal bulges in the oceans do not result from the Moon's compressing or expanding the water, nor from the Moon's lifting the water "away from the Earth." Rather, they result from an actual flow of water over the Earth's surface toward the two regions below and opposite the Moon, causing the water to pile up to greater depths at those places (Figure 3.17).

In the idealized (and, as we shall see, oversimplified) model just described, the height of the tides would be only a few feet. The rotation of the Earth would carry an observer at any given place alternately into regions of deeper and shallower water. An observer being carried toward the regions under or opposite the Moon, where the water was deepest, would say, "The tide is coming in"; when carried away from those regions, the observer would say, "The

■ FIGURE 3.17

Minas Basin Seen at Low and High Tides

tide is going out." During a day, the observer would be carried through two tidal bulges (one on each side of the Earth) and so would experience two high tides and two low tides.

The Sun also produces tides on the Earth, although it is less than half as effective as the Moon at tide raising. The actual tides we experience are a combination of the larger effect of the Moon and the smaller effect of the Sun. When the Sun and Moon are lined up (at new moon or full moon), the tides produced reinforce each other and so are greater than normal (Figure 3.18). These are called spring tides (the name is connected not to the season but to the idea that higher tides "spring up"). Spring tides are approximately the same, whether the Sun and Moon are on the same or opposite sides of the Earth, because tidal bulges

■ FIGURE 3.16

Tidal Bulges in an "Ideal" Ocean

VOYAGERS IN ASTRONOMY

George Darwin and the Slowing of the Earth

The rubbing of water over the face of the Earth involves an enormous amount of energy. Over long periods of time, the friction of the tides is slowing down the rotation of the Earth. Our day gets longer by about 0.002 second each century. That seems very small, but such tiny changes can add up over millions and billions of years.

Although the Earth's spin is slowing down, the angular momentum (see Chapter 2) in a system such as the Earth–Moon system cannot change. Thus some other spin motion must speed up to take the extra angular momentum. The details of what happens were worked out over a century ago by George Darwin (1845–1912), the son of naturalist Charles Darwin. George Darwin had a strong interest in science but studied law for six years and was admitted to the bar. However, he never practiced law, returning to science instead and eventually becoming a professor at Cambridge University. He was a protégé of Lord Kelvin, one of the great physicists of the 19th century, and be-

G. Darwin

© Royal Society

came interested in the long-term evolution of the solar system. He specialized in making detailed (and difficult)

■ ■ ■ ■ ■ ■ ■ ■ ■ ■ ■ ■ ■

Darwin calculated that the Moon will slowly spiral outward, away from the Earth.

■ ■ ■ ■ ■ ■ ■ ■ ■ ■ ■ ■ ■

mathematical calculations of how orbits and motions will change over geologic time.

What Darwin calculated for the Earth–Moon system was that the Moon will slowly spiral outward, away from the Earth. As it moves farther away, it will orbit less quickly (just as planets farther from the Sun move more slowly in their orbits). Thus, the month will get longer. Also, because the Moon will be more distant, total eclipses of the Sun (see Section 3.7) will no longer be visible from Earth.

The day and the month will both continue to get longer, although bear in mind that the effects are very gradual. The calculations show that ultimately—billions of years in the future—the day and the month will be the same length (about 47 of our present days), and the Moon will be stationary in the sky over the same spot on the Earth. This kind of alignment is already true for Pluto's moon Charon (among others). Its rotation and orbital period are the same length as a day on Pluto.

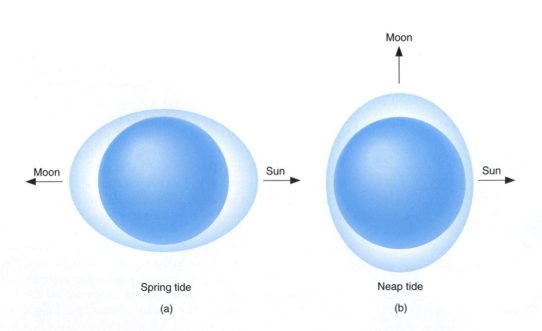

■ **FIGURE 3.18**

Tides Caused by Different Alignments of the Sun and Moon
(a) In spring tides, the Sun's and Moon's pulls reinforce each other. (b) In neap tides, the Sun and the Moon pull at right angles to each other and the resulting tides are lower than usual.

Spring tide
(a)

Neap tide
(b)

Moon

Moon

Sun

Sun

occur on both sides. When the Moon is at first quarter or last quarter (at right angles to the Sun's direction), the tides produced by the Sun partially cancel the tides of the Moon, making them lower than usual. These are called *neap tides*.

The "simple" theory of tides, described in the preceding paragraphs, would be sufficient if the Earth rotated very slowly and were completely surrounded by very deep oceans. However, the presence of land masses stopping the flow of water, the friction in the oceans and between oceans and the ocean floors, the rotation of the Earth, the wind, the variable depth of the ocean, and other factors, all complicate the picture. This is why in the real world, some places have very small tides while in other places huge tides become tourist attractions.

3.7 ECLIPSES OF THE SUN AND MOON

One of the fortunate coincidences of living on Earth at the present time is that the two most prominent astronomical objects, the Sun and the Moon, have nearly the same apparent size in the sky. Although the Sun is about 400 times larger in diameter than is the Moon, it is also about 400 times farther away, so both the Sun and the Moon have the same angular size—about 1/2°. As a result, the Moon, as seen from the Earth, can appear to cover the Sun, producing one of the most impressive events in nature.

Any solid object in the solar system casts a shadow by blocking the light of the Sun from a region behind it. This shadow in space becomes apparent whenever another object moves into it. In general, an **eclipse** occurs whenever any part of either the Earth or the Moon enters the shadow of the other. When the Moon's shadow strikes the Earth, people within that shadow see the Sun at least partially covered by the Moon; that is, they witness a **solar eclipse.** When the Moon passes into the shadow of the Earth, people on the night side of the Earth see the Moon darken in what is called a **lunar eclipse.**

The shadows of the Earth and the Moon consist of two parts: a cone where the shadow is darkest, called the *umbra,* and a lighter, more diffuse region of darkness called the *penumbra.* As you can imagine, the most spectacular eclipses occur when an object enters the umbra. Figure 3.19 illustrates the appearance of the Moon's shadow and what the Sun and Moon would look like from different points within the shadow.

If the path of the Moon in the sky were identical to the path of the Sun (the ecliptic), we might expect to see an eclipse of the Sun and the Moon each month—whenever the Moon got in front of the Sun or into the shadow of the Earth. However, as we mentioned, the Moon's orbit is tilted relative to the plane of the Sun's orbit by about 5° (imagine two hula hoops with a common center, but tilted a bit). As a result, most months the Moon is sufficiently above or below the Sun to avoid an eclipse. But when the

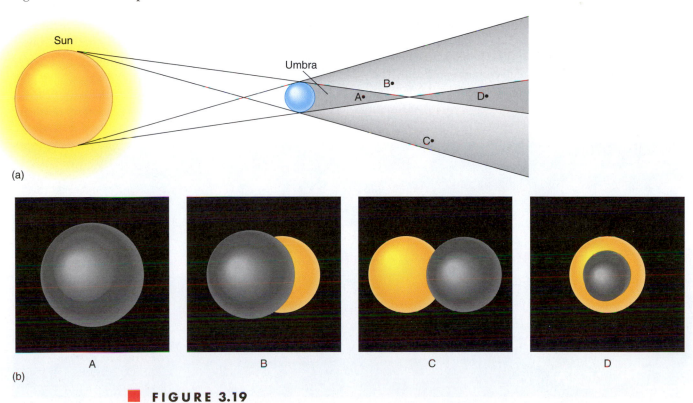

■ FIGURE 3.19

Explaining Solar Eclipses   (a) The shadow cast by a spherical body (the Moon, for example). Notice the dark umbra and the lighter penumbra. Four points in the shadow are labeled with letters. (b) What the Sun and Moon would look like in the sky at the four labeled points.

two paths cross (twice a year), it is then "eclipse season" and eclipses are possible.

3.7.1 Eclipses of the Sun

The apparent or angular sizes of both the Sun and Moon vary slightly from time to time as their distances from the Earth vary. (Figure 3.19 shows the distance of the observer varying at points A–D, but the idea is the same.) Much of the time, the Moon looks slightly smaller than the Sun and cannot cover it completely, even if the two are perfectly aligned. However, if an eclipse of the Sun occurs when the Moon is somewhat nearer than its average distance, the Moon can completely hide the Sun, producing a *total* solar eclipse. A total eclipse of the Sun occurs at those times when the umbra of the Moon's shadow reaches the surface of the Earth.

The geometry of a total solar eclipse is illustrated in Figure 3.20. If the Sun and Moon are properly aligned, then the Moon's darkest shadow intersects the ground at a small point on the Earth's surface. Anyone on the Earth within the small area covered by the tip of the Moon's shadow will, for a few minutes, be unable to see the Sun and will witness a total eclipse. At the same time, observers on a larger area of the Earth's surface who are in the penumbra will see only a part of the Sun eclipsed by the Moon: We call this a *partial* solar eclipse.

As the Moon moves eastward in its orbit, the tip of its shadow sweeps eastward at about 1500 km/h along a thin band across the surface of the Earth. The thin zone across the Earth within which a total solar eclipse is visible (weather permitting) is called the *eclipse path*. Within a region about 3000 km on either side of the eclipse path, a partial solar eclipse is visible. It does not take long for the Moon's shadow to sweep past a given point on Earth. The duration of totality may be only a brief instant; it can never exceed about 7 minutes.

Because a total eclipse of the Sun is so spectacular, it is well worth trying to see one if you can. There are people whose hobby is "eclipse chasing" and who brag about how many they have seen in their lifetimes. Because much of the Earth's surface is water, eclipse chasing can involve lengthy boat trips (and often requires air travel as well). As a result, eclipse chasing is rarely within the budget of a typical college student. Nevertheless, a list of future eclipses is given for your reference in Appendix 9, just in case you strike it rich early!

3.7.2 Appearance of a Total Eclipse

What can you see if you are lucky enough to catch a total eclipse? A solar eclipse starts when the Moon just begins to silhouette itself against the edge of the Sun's disk. A partial phase follows, during which more and more of the Sun is covered by the Moon. About an hour after the eclipse begins, the Sun becomes completely hidden behind the Moon. In the few minutes immediately before this period of totality begins, the sky noticeably darkens, some flowers close up, and chickens may go to roost. As an eerie twilight suddenly descends during the day, other animals (and people) may get disoriented. During totality the sky is dark enough that planets become visible in the sky, and usually the brighter stars do as well.

As the bright disk of the Sun becomes entirely hidden behind the Moon, the Sun's remarkable **corona** flashes into view (Figure 3.21). The corona is the Sun's outer atmosphere, consisting of sparse gases that extend for millions of miles in all directions from the apparent surface of the Sun. It is ordinarily not visible because the light of the corona is feeble compared with the light from the underlying layers of the Sun. Only when the brilliant glare from the Sun's visible disk is blotted out by the Moon during a total eclipse is the pearly white corona visible.

■ **FIGURE 3.20**

Geometry of a Total Solar Eclipse Note that our diagram is not to scale. The Moon blocks the Sun as seen from some parts of the Earth and casts a shadow on our planet.

■ **FIGURE 3.21**

Sun's Corona The corona (thin outer atmosphere) of the Sun, visible during the July 11, 1991, total solar eclipse, photographed from near La Paz, Mexico, by a dedicated eclipse chaser.

The total phase of the eclipse ends, as abruptly as it began, when the Moon begins to uncover the Sun. Gradually the partial phases of the eclipse repeat themselves, in reverse order, until the Moon has completely uncovered the Sun.

3.7.3 Eclipses of the Moon

A lunar eclipse occurs when the Moon enters the shadow of the Earth. The geometry of a lunar eclipse is shown in Figure 3.22. The Earth's dark shadow is about 1.4 million km long, so at the Moon's distance (an average of 384,000 km) it could cover about four full moons. Unlike a solar eclipse, which is visible only in certain local areas on the Earth, a lunar eclipse is visible to everyone who can see the Moon. Because a lunar eclipse can be seen (weather permitting) from the entire night side of the Earth, lunar eclipses are observed far more frequently from a given place on Earth than are solar eclipses.

An eclipse of the Moon is total only if the Moon's path carries it through the Earth's umbra. If the Moon does not enter the umbra completely, we have a partial eclipse of the Moon.

A lunar eclipse can take place only when the Sun, Earth, and Moon are in a line. The Moon is opposite the Sun, which means the Moon will be in full phase before the eclipse, making the darkening even more dramatic. About 20 minutes or so before the Moon reaches the dark shadow, it dims somewhat as the Earth partly blocks the sunlight. As the Moon begins to dip into the shadow, the curved shape of the Earth's shadow upon it soon becomes apparent.

Even when totally eclipsed, the Moon is still faintly visible, usually appearing a dull coppery red. The illumination on the eclipsed Moon is sunlight that has been bent into the Earth's shadow by passing through the Earth's atmosphere.

After totality the Moon moves out of the shadow and the sequence of events is reversed. The total duration of the eclipse depends on how closely the Moon's path approaches the axis of the shadow. For an eclipse where the Moon goes through the center of the Earth's shadow, each partial phase consumes at least 1 hour, and totality can last as long as 1 hour and 40 minutes. Total eclipses of the Moon occur, on the average, about once every two or three years.

Thanks to our understanding of gravity and motion (see Chapter 2), eclipses can now be predicted centuries in advance. We've come a long way since humanity stood frightened by the darkening of the Sun or the Moon, fearing the displeasure of the gods. Today we enjoy the sky show with an appreciation of the majestic forces that keep our solar system running.

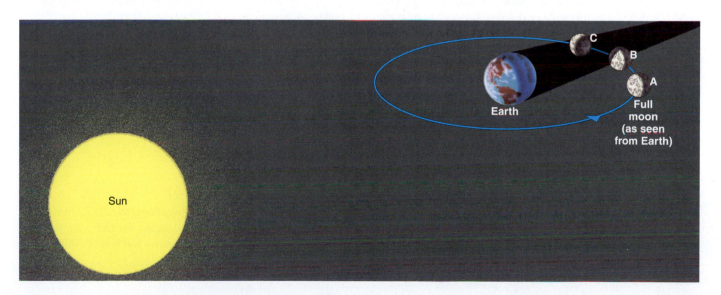

■ **FIGURE 3.22**

Geometry of a Lunar Eclipse The Moon is full as seen from Earth at A; at position B, the Moon has begun to move into the Earth's shadow; at position C, the eclipse is over and the Moon is moving out of the Earth's shadow. (Note that this diagram is also not to scale and that the distance the Moon moves in its orbit during the eclipse has been exaggerated here for educational purposes.)

How to Observe Solar Eclipses

It is extremely dangerous to look at the Sun: Even a brief exposure can damage your eyes. Normally, few sane people are tempted to do this because it is painful (and something your mother told you never to do!). But during a solar eclipse, the temptation to take a look is strong. Think before you give in. The fact that the Moon is covering part of the Sun doesn't make the uncovered part any less dangerous to look at. Still, there are perfectly safe ways to follow the course of a solar eclipse, if you are lucky enough to be in the path of the shadow.

The easiest technique is to make a pinhole projector. Take a piece of cardboard with a small (1-mm) hole punched in it and hold it several feet above a light surface, such as a concrete sidewalk or a white sheet of paper, so that the hole is "aimed" at the Sun. The hole produces a fuzzy but adequate image of the eclipsed Sun (see diagram). Alternatively, you can let the tiny spaces between leaves form multiple pinhole images against a wall or sidewalk. Watching hundreds of little crescent Suns dancing in the breeze can be captivating.

It is safe to look at the Sun directly when it is totally eclipsed, even through binoculars or telescopes.

Although there are safe filters for looking at the Sun directly, people have suffered eye damage by looking through improper filters (or no filter at all). For example, neutral-density photographic filters are not safe because they transmit infrared radiation that can cause severe damage to the retina. Also unsafe are smoked glass, completely exposed color film, and many other homemade filters.

It is safe to look at the Sun directly when it is totally eclipsed, even through binoculars or telescopes. Unfortunately the total phase, as we discussed, is all too brief. But if you know when it is coming (and going), be sure you look, for it's an unforgettably beautiful sight. And, despite the ancient folklore that presents them as dangerous times to be outdoors, the partial phases of eclipses—as long as you are not looking directly at the Sun—are certainly not any more dangerous than being out in sunlight.

During past eclipses, unnecessary panic has been created by uninformed public officials acting with the best intentions. There were two marvelous total eclipses in Australia in the 20th century during which townspeople held newspapers over their heads for protection and schoolchildren cowered indoors with their heads under their desks. What a pity that all those people missed what would have been one of the most memorable experiences of their lifetimes!

How to watch a solar eclipse safely during its partial phases using a pinhole projector.

SUMMARY

3.1 The terrestrial system of latitude and longitude makes use of the **great circles** called **meridians.** An analogous celestial coordinate system is called **right ascension** (RA) and **declination,** with the vernal equinox serving as reference point (like the prime meridian at Greenwich on the Earth). These coordinate systems help us locate any object on the celestial sphere. The Foucault pendulum is a way to demonstrate that the Earth rather than the sky is turning.

3.2 The familiar cycle of the seasons results from the 23° tilt of the Earth's axis of rotation. At the *summer solstice,* the Sun is higher in the sky and its rays strike the Earth more directly. The Sun is in the sky for more than half the day and can heat the Earth longer. At the *winter solstice,* the Sun is low in the sky and up for fewer than 12 hours. At the *vernal* and *autumnal equinoxes,* the Sun is on the celestial equator and we get 12 hours of day and night. The seasons are different at different latitudes.

3.3 The basic unit of astronomical time is the day (either the **solar day** or the **sidereal day**). **Apparent solar time** is based on the position of the Sun in the sky, and **mean solar time** is based on the average value of a solar day during the year. By international agreement, we define 24 time zones around the world, each with its own standard time. The convention of the **international date line** is necessary to reconcile times in different parts of the Earth.

3.4 The fundamental problem of the calendar is to reconcile the incommensurable lengths of the day, the month, and the year. Most modern calendars, beginning with the Roman (Julian) calendar of the 1st century B.C.E., neglect the problem of the month and concentrate on achieving the correct number of days in a year by using such conventions as the leap year. Today, most of the world has adopted the Gregorian calendar established in 1582.

3.5 The Moon's monthly cycle of **phases** results from the changing angle of its illumination by the Sun. The full moon is visible in the sky only during the night; other phases are visible during the day as well. Because its period of revolution is the same as its period of rotation, the Moon always keeps the same face toward the Earth.

3.6 The twice-daily ocean *tides* are primarily the result of the Moon's differential gravitational force on the material of the Earth's crust and ocean. These tidal forces cause ocean water to flow into two tidal bulges on opposite sides of the Earth; each day the Earth rotates through these bulges. Actual ocean tides are complicated by the additional effects of the Sun, and by the shape of the coasts and ocean basins.

3.7 The Sun and Moon have nearly the same angular size (about 1/2°). A **solar eclipse** occurs when the Moon moves between the Sun and the Earth, casting its shadow on a part of the Earth's surface. If the eclipse is total, the observer is in the Moon's *umbra,* the light from the bright disk of the Sun is completely blocked, and the solar atmosphere (the **corona**) comes into view. Solar eclipses take place rarely in any one location, but they are among the most spectacular sights in nature. A **lunar eclipse** takes place when the Moon moves into the Earth's shadow; it is visible (weather permitting) from the entire night hemisphere of the Earth.

INTER-ACTIVITY

A Have your group brainstorm about other ways (besides the Foucault pendulum) you could prove that it is our Earth that is turning once a day, and not the sky turning around us. (*Hint:* How does the spinning of the Earth affect the oceans and the atmosphere?)

B What would the seasons on Earth be like if the Earth's axis were not tilted? How many things about life on Earth can you think of that would be different in this case?

C After college and graduate training, members of your group are asked to set up a school in New Zealand. Describe some ways your school schedule in the Southern Hemisphere would have to differ from what we are used to in the North.

D During the traditional U.S. Christmas vacation weeks, you are sent to the vicinity of the South Pole on a research expedition (depending on how well you did on your astronomy midterm, either as a research assistant or as a short-order cook!). Have your group discuss how the days and nights will be different there and how these differences might affect you.

E Discuss with your group all the stories you have heard about the Moon and crazy behavior. Why do you think people associate crazy behavior with the full Moon? What other legends besides vampire stories are connected with the phases of the Moon?

F Your college town becomes the founding site for a strange new cult that worships the Moon. These true believers gather regularly around sunset and do a dance in which they must extend their arms in the direction of the Moon. Have your group discuss which way their arms will be pointing *at sunset* when the Moon is new, first quarter, full, and third quarter.

REVIEW QUESTIONS

Ace⊙Astronomy™ Assess your understanding of this chapter's topics with additional quizzing and animations at **http://ace.brookscole.com/ voyages**

1. Discuss how latitude and longitude on Earth are similar to declination and right ascension in the sky.
2. What is the latitude of the North Pole? The South Pole? Why does longitude have no meaning at the North and South Poles?
3. Make a table showing each main phase of the Moon and roughly when the Moon rises and sets for each phase. During which phase can you see the Moon in the middle of the morning? In the middle of the afternoon?
4. What are the advantages and disadvantages of apparent solar time? How is the situation improved by introducing mean solar time and standard time?
5. What are the two ways that the tilt of the Earth's axis causes the summers in the United States to be warmer than the winters?
6. Why is it difficult to construct a practical calendar based on the Moon's cycle of phases?
7. Explain why there are two high tides and two low tides every day. Strictly speaking, should the period during which there are two high tides be 24 hours? If not, what should the interval be?
8. What is the phase of the Moon during a total solar eclipse? During a total lunar eclipse?

THOUGHT QUESTIONS

9. Where are you on the Earth if you experience the following? (Refer back to Chapter 1 as well as this chapter.)
a. The stars rise and set perpendicular to the horizon.
b. The stars circle the sky parallel to the horizon.
c. The celestial equator passes through the zenith.
d. In the course of a year, all stars are visible.
e. The Sun rises on September 23 and does not set until March 21 (ideally).

10. In countries at far northern latitudes, the winter months tend to be so cloudy that astronomical observations are nearly impossible. Why can't good observations of the stars be made at those places during the summer months?

11. What is the phase of the Moon if it
a. rises at 3:00 P.M.?
b. is highest in the sky at 7:00 A.M.?
c. sets at 10:00 A.M.?

12. A car accident occurs around midnight on the night of a full moon. The driver at fault claims he was blinded momentarily by the Moon rising on the eastern horizon. Should the police believe him?

13. The secret recipe to the ever-popular veggie burgers in the college cafeteria is hidden in a drawer in the director's office. Two students decide to break in and get their hands on it, but they want to do it a few hours before dawn

on a night when there is no Moon, so they are less likely to be caught. What phases of the Moon would suit their plans?

14. Your great uncle, who often exaggerates events in his own life, tells you about a terrific adventure he had on February 29, 1900. Why would this story make you suspicious?

15. One year, when money is no object, you enjoy your birthday so much that you want to have another one right away. You get into your supersonic jet. Where should you and the people celebrating with you travel? From what direction should you approach? Explain.

16. Suppose you lived in the crater Copernicus on the side of the Moon facing the Earth.
a. How often would the Sun rise?
b. How often would the Earth set?

c. During what fraction of the time would you be able to see the stars?

17. In a lunar eclipse, does the Moon enter the shadow of the Earth from the east or west side? Explain why.

18. Describe what an observer at the crater Copernicus would see while the Moon is eclipsed. What would the same observer see during what would be a total solar eclipse as viewed from the Earth?

19. The day on Mars is 1.026 Earth-days long. The martian year lasts 686.98 Earth-days. The two moons of Mars take 0.32 Earth-day (for Phobos) and 1.26 Earth-days (for Deimos) to circle the planet. You are given the task of coming up with a martian calendar for a new Mars colony. What might you do?

FIGURING FOR YOURSELF

Now that you understand a bit more about why our systems of time and calendar work the way they do, you can ask what would happen if things on Earth had turned out differently.

20. Suppose the tilt of the Earth's axis were only 16°. What, then, would be the difference in latitude between the Arctic Circle and the Tropic of Cancer? What would be the effect on the seasons compared with that produced by the actual tilt of 23°?

21. Suppose the Earth took exactly 300.0 days to go around the Sun, and everything else (the day, the month) was the same? What kind of calendar would we have? How would this affect the seasons?

22. Consider a calendar based entirely on the day and the month (the Moon's period from full phase to full phase). How many days are there in a month? Can you figure out a scheme analogous to leap year to make this calendar work? Can you also incorporate the idea of a week into your lunar calendar?

23. a. If a star rises at 8:30 P.M. tonight, approximately what time will it rise two months from now?
b. What is the altitude of the Sun at noon on December 22, as seen from a place on the Tropic of Cancer?

24. Show that the Gregorian calendar will be in error by 1 day in about 3300 years.

SUGGESTIONS FOR FURTHER READING

Aveni, A. *Empires of Time: Calendars, Clocks, and Cultures.* 1989, Basic Books.

Bartky, I. and Harrison, E. "Standard and Daylight Saving Time" in *Scientific American,* May 1979.

Brunier, S. and Luminet, J. *Glorious Eclipses: Their Present, Past, and Future.* 2000, Cambridge U. Press.

Coco, M. "Not Just Another Pretty Phase" in *Astronomy,* July 1994, p. 76. Moon phases explained.

Gingerich, O. "Notes on the Gregorian Calendar Reform" in *Sky & Telescope,* Dec. 1982, p. 530.

Harris, J. and Talcott, R. *Chasing the Shadow: An Observer's Guide to Eclipses.* 1994, Kalmbach.

Kluepfel, C. "How Accurate Is the Gregorian Calendar?" in *Sky & Telescope,* Nov. 1982, p. 417.

Krupp, E. "Calendar Worlds" in *Sky & Telescope,* Jan. 2001, p. 103. On how the days of the week got their names.

Krupp, E. "Behind the Curve" in *Sky & Telescope,* Sept. 2002, p. 68. On the reform of the calendar by Pope Gregory XIII.

Pasachoff, J. "Solar Eclipse Science: Still Going Strong" in *Sky & Telescope,* Feb. 2001, p. 40. On what we have learned and are still learning from eclipses.

Pasachoff, J. and Ressmeyer, R. "The Great Eclipse" in *National Geographic,* May 1992, p. 30. About the July 1991 solar eclipse, with spectacular photographs.

Rey, H. *The Stars: A New Way to See Them.* 1976, Houghton Mifflin. Good introduction to time, the seasons, and celestial coordinates by the author of the "Curious George" children's stories.

Sobel, D. *Longitude.* 1995, Walker. Story of John Harrison, who solved the problem of how to measure longitude at sea.

Steel, D. *Marking Time: The Epic Quest to Invent the Perfect Calendar.* 2001, Wiley.

The Very Large Array of Radio Telescopes in New Mexico Seen with a Rainbow Just as water droplets in our atmosphere can disperse sunlight into a rainbow of color, so instruments called spectrometers allow astronomers to break the light from planets, stars, or galaxies into their component colors. The details of these colors can help us decipher what conditions are like in the objects from which they originate.

4 Radiation and Spectra

In the last hundred years the power of interpreting the messages brought by light has increased greatly. We are in touch with the stars.

Sir William Bragg in *The Universe of Light* (1933, G. Bell and Sons)

THINKING AHEAD

The nearest star is so far away that the fastest spacecraft humans have built would take almost 100,000 years to get there. Yet we very much want to know what material this neighbor star is composed of and how it differs from our own Sun. How can we learn about the chemical makeup of stars that we cannot hope to visit or sample?

In astronomy, most of the objects that we study are completely beyond our reach. The temperature of the Sun is so high that a spacecraft would be fried long before it reached the solar surface, and the stars are much too far away to visit in our lifetimes with the technology now available. Even light, which travels at a speed of 300,000 kilometers/second (km/s), takes more than four years to reach us from the *nearest* star. If we want to learn about the Sun and stars, or even about most members of our solar system, we must rely on techniques that allow us to analyze them from a distance.

Here we are in luck. Coded into the light and other kinds of radiation that reach us from objects in the universe is a wide range of information about what those objects are like and how they work. If we can decipher this "cosmic code" and read the messages it contains, we can learn an enormous amount about the cosmos without ever having to leave Earth or its immediate environment.

Ace◉Astronomy™ The AceAstronomy icon throughout the text indicates an opportunity for you to test yourself on key concepts and to explore animations and interactions on the AceAstronomy website at **http://ace.brookscole.com/voyages**

Virtual Laboratories

 Properties of Light and Its Interaction with Matter

 The Doppler Effect

The light and other radiation we receive from the stars and planets is generated by processes at the atomic level—by changes in the way the parts of the atom interact and move. Thus, to appreciate how light is generated, we must first explore how atoms work. There is a bit of irony in the fact that in order to understand some of the largest structures in the universe, we must become acquainted with some of the smallest.

Notice that we have twice used the phrase *light and other radiation.* One of the key ideas explored in this chapter is that visible light is not unique; it is merely the most familiar example of a much larger family of radiation that can carry information to us.

The word *radiation* will be used frequently in this book, so it is important to understand what it means. In everyday language, *radiation* is often used to describe certain kinds of energetic subatomic particles released by radioactive materials in our environment. (An example is the kind of radiation used to treat cancer.) But this is not what we mean when we use the word *radiation* in an astronomy text. *Radiation,* as used in this book, is a general term for the waves (including light) that provide our primary link with the universe beyond our own solar system.

American Institute of Physics, Niels Bohr Library

■ **FIGURE 4.1**

James Clerk Maxwell   Maxwell (1831–1879) unified the rules governing electricity and magnetism into a coherent theory.

 THE NATURE OF LIGHT

As we saw in earlier chapters, Newton's theory of gravity accounts for the motions of the planets as well as the objects on Earth. Application of this theory to a variety of problems dominated the work of scientists for nearly two centuries. In the 19th century, many physicists turned to the study of electricity and magnetism, which—as we shall see in a moment—are intimately connected with the production of light.

The scientist who played a role in this field comparable to Newton's role in the study of gravity was physicist James Clerk Maxwell (1831–1879), born and educated in Scotland (Figure 4.1). Inspired by a number of ingenious experiments that showed an intimate relationship between electricity and magnetism, Maxwell developed a theory that describes both with only a small number of elegant equations. It is this theory that allows us to understand the nature and behavior of light.

4.1.1 Maxwell's Theory of Electromagnetism

We will look at the structure of the atom in more detail in Section 4.4, but we begin by noting that the typical atom consists of several types of particles, a number of which have not only mass but an additional property called electric charge. In the nucleus (central part) of every atom are *protons,* which are positively charged; outside the nucleus are *electrons,* which have a negative charge.

Maxwell's theory deals with these electric charges and their effects, especially when they are moving. In the vicin-ity of an electric charge, another charge feels a force of attraction or repulsion: Opposite charges attract; like charges repel. When charges are not in motion, we observe only this electric attraction or repulsion. If charges are in motion, however (as they are inside every atom and in a wire carrying a current), then we measure another force called *magnetism.*

Magnetism was well known for much of recorded human history, but its cause was not understood until the 19th century. Experiments with electric charges demonstrated that magnetism was the result of moving charged particles. Sometimes the motion is clear, as in the coils of heavy wire that make an industrial electromagnet. Other times it is more subtle, as in the kind of magnet you buy in a hardware store, in which many of the electrons inside the atoms are spinning in roughly the same direction; it is the alignment of their motion that causes the material to become magnetic.

Physicists use the word *field* to describe the action of forces that one object exerts on other distant objects. For example, we say the Sun produces a *gravitational field* that controls Earth's orbit, even though the Sun and Earth do not come directly into contact. Using this terminology, we can say that stationary electric charges produce *electric fields,* and moving electric charges also produce *magnetic fields.*

Actually, the relationship between electric and magnetic phenomena is even more profound. Experiments showed that changing magnetic fields could produce electric currents (and thus changing electric fields), and changing electric currents could (in turn) produce changing magnetic fields. So once begun, electric and magnetic field changes could continue to trigger each other.

Maxwell analyzed what would happen if electric charges were *oscillating* (moving constantly back and forth) and found that the resulting pattern of electric and

■ FIGURE 4.2

Making Waves An oscillation in a pool of water creates an expanding disturbance called a wave.

© 1993, Comstock, Inc.

magnetic fields would spread out and travel rapidly through space. Something similar happens when your finger or a nervous frog moves up and down in a pool of water. The disturbance moves outward and creates a pattern we call a *wave* in the water (Figure 4.2). You might at first think that there must be very few situations in nature where electric charges oscillate, but this is not at all the case. As we shall see, atoms and molecules (which consist of charged particles) oscillate back and forth all the time. The resulting electromagnetic disturbances are among the most common phenomena in the universe.

Maxwell was able to calculate the speed at which an electromagnetic disturbance moves through space; he found that it is equal to the speed of light, which had been measured experimentally. On that basis, he speculated that light was one form of a family of possible electric and magnetic disturbances called **electromagnetic radiation**, a conclusion that was again confirmed in laboratory experiments. When light (reflected from the pages of an astronomy textbook, for example) enters a human eye, its changing electric and magnetic fields stimulate nerve endings, which then transmit the information contained in these changing fields to the brain. The science of astronomy is primarily about analyzing radiation from distant objects to understand what they are and how they work.

4.1.2 The Wave-Like Characteristics of Light

The changing electric and magnetic fields in radiation are similar (as hinted earlier) to the waves that can be set up in a quiet pool of water. In both cases the disturbance travels rapidly outward from the point of origin and can use its energy to disturb other things farther away. (For example, in water, the expanding ripples moving away from our twitching frog could disturb the peace of a grasshopper sleeping on a leaf in the same pool.) In the case of electromagnetic waves, the radiation generated by a transmitting antenna full of charged particles at your local radio station can, sometime later, disturb a group of electrons in your home radio antenna and bring you the news and weather while you are getting ready for class or work in the morning.

The waves generated by charged particles differ from water waves in some profound ways, however. Water waves require water to travel in. Sound waves, to give another example, are pressure disturbances that require air to travel through. But electromagnetic waves do not require water or air: The fields generate each other and so can move through a vacuum (such as outer space). This was such a disturbing idea to 19th-century scientists that they actually made up a substance to fill all of space—one for which there was not a single shred of evidence—just so light waves could have something to travel through: They called it the *aether*. Today we know that there is no aether and that electromagnetic waves have no trouble at all moving through empty space (as all the starlight visible on a clear night must surely be doing).

The other difference is that *all* electromagnetic waves move at the same speed in space (the speed of light), which turns out to be the fastest possible speed in the universe. No matter where electromagnetic waves are generated and no matter what other properties they have, when they are moving (and not interacting with matter), they move at the speed of light. Yet you know from everyday experience that waves like light are not all the same. For example, we perceive that light waves differ from one another in a property we call color. Let's see how we can denote the differences among the whole broad family of electromagnetic waves.

■ FIGURE 4.3

Characterizing Waves Electromagnetic radiation has wave-like characteristics. The wavelength (λ) is the distance between crests, the frequency (f) is the number of cycles per second, and the speed (c) is the distance the wave covers over time.

The nice thing about a wave is that it is a repeating phenomenon. Whether it is the up-and-down motion of a water wave or the changing electric and magnetic fields in a wave of light, the pattern of disturbance repeats in a cyclical way. Thus, any wave motion can be characterized by a series of crests and troughs (Figure 4.3). Moving up a full crest and down a full trough completes one cycle. The horizontal length covered by one cycle—the distance between one crest and the next, for example—is called the **wavelength.** Ocean waves provide an analogy: The wavelength is the distance that separates successive wave crests.

For visible light, our eyes (and minds) perceive different wavelengths as different colors: Red, for example, is the longest visible wavelength and violet is the shortest. The main colors of visible light from longest to shortest wavelength can be remembered using the mnemonic ROY G. BIV—for red, orange, yellow, green, blue, indigo, violet. Other invisible forms of electromagnetic radiation have different wavelengths, as we will see in the next section.

We can also characterize different waves by their **frequency,** the number of wave cycles that pass per second. If you count ten crests moving by each second, for example, then the frequency is 10 cycles per second (cps). In honor of Heinrich Hertz, the physicist who—inspired by Maxwell's work—discovered radio waves, a cps is also called a *hertz (hz).* Take a look at the dial of your radio, for example, and you will see the channel assigned to each radio station on the dial characterized by its frequency, usually in units of Khz (kilohertz, or thousands of hertz) or Mhz (megahertz, or millions of hertz).

Wavelength and frequency are related because all electromagnetic waves travel at the same speed. To see how this works, imagine a parade in which everyone is forced by prevailing traffic conditions to move at exactly the same speed. You stand on a corner and watch the waves of marchers come by. First you see row after row of very skinny fashion models. Because they are not very wide, a good number of the models can fit by you each minute; we can say they have a high frequency. Next, however, come several rows of circus elephants. The elephants are large and marching at the

same speed as the models, so far fewer of them can march past you per minute: Because they have a wider spacing, they must be content with a lower frequency.

Mathematically, we can express this relationship as

$$c = \lambda f$$

where the Greek letter for "l"—lambda, λ—is used to denote wavelength and c is the scientific symbol for the speed of light. The formula can be expressed as follows: For any wave motion, the speed at which a wave moves equals the frequency times the wavelength. Waves with longer wavelengths have lower frequencies.

4.1.3 Light as a Photon

The electromagnetic wave model of light (as formulated by Maxwell) was one of the great triumphs of 19th-century science. When Heinrich Hertz actually made invisible electromagnetic waves (what today are called radio waves) on one side of a room and detected them on the other in 1887, it ushered in a new era that led to the modern age of telecommunications. However, by the beginning of the 20th century, more sophisticated experiments had revealed that light behaves in certain ways that cannot be explained by the wave model. Reluctantly, physicists had to accept that sometimes light (and all other electromagnetic radiation) behaves more like a "particle"—or at least a self-contained packet of energy—than a wave. We call such packets of electromagnetic energy **photons.**

The fact that light behaves like a wave in certain experiments and like a particle in others was a very surprising and unlikely idea; at first, it was as confusing to physicists as it seems to students. After all, our common sense says that waves and particles are opposite concepts. On one hand, a wave is a repeating disturbance that by its very nature is not in only one place, but spreads out. A particle, on the other hand, is something that can be in only one place at any given time. Strange as it sounds, though, countless experiments now confirm that electromagnetic radiation can sometimes behave like a wave and at other times like a particle.

Then, again, perhaps we shouldn't be surprised that something that always travels at the "speed limit" of the universe and doesn't need a substance to travel through might not obey our everyday commonsense ideas. The confusion that this wave–particle duality of light caused in physics was eventually resolved by the introduction of a more complicated theory of waves and particles, now called *quantum mechanics.* (This is one of the most interesting fields of modern science, but it is mostly beyond the scope of our book. If you get interested in it, see some of the references at the end of this chapter.)

In any case, you should now be prepared when scientists (or the authors of this book) sometimes discuss electromagnetic radiation as if it consisted of waves and at other times refer to it as a stream of photons. A photon (being a packet of energy) carries a specific amount of energy. We can use the idea of energy to connect the photon and wave models. How much energy a photon has depends on its fre-

■ **FIGURE 4.4**

The Inverse-Square Law for Light
As light energy radiates away from its source, it spreads out in such a way that the energy decreases as the square of the distance from its source.

quency when you think about it as a wave. A low-energy radio wave has a low frequency as a wave, while a high-energy x ray at your dentist's office is a high-frequency wave. Among the colors of visible light, violet-light photons have the highest energy and red has the lowest.

Test whether the connection between photons and waves is clear to you. In the above example, which photon would have the longer wavelength as a wave: the radio wave or the x ray? If you answered the radio wave, you are correct. Radio waves have lower frequency, so the wave cycles are longer (they are elephants, not fashion models).

4.1.4 Propagation of Light

Let's think for a moment about how light from a light bulb moves through space. The expansion of light waves is extremely democratic: They travel away from the bulb, not just toward *your* eyes but in all directions. They must therefore cover an ever-widening space. Yet the total amount of light available can't change once the light has left the bulb. This means that, as the same expanding shell of light covers a larger and larger area, there must be less and less of it in any given place. Light (and all other electromagnetic radiation) gets weaker and weaker as it gets farther from its source.

The increase in the area that the radiation must cover is proportional to the square of the distance that the radiation has traveled (Figure 4.4). If we stand twice as far from the source, our eyes will intercept 2-squared (2×2), or 4 times less light. If we stand 10 times farther from the source, we get 10-squared, or 100 times less light. You can see how this weakening means trouble for sources of light at astronomical distances! One of the nearest stars, called Alpha Centauri, emits about the same total energy as does the Sun. But it is about 270,000 times farther away, and so it appears about 73 billion times fainter. No wonder that the stars, which close-up would look more or less like the Sun, look like faint pinpoints of light from far away.

This idea—that the apparent brightness of a source (how bright it looks to us) gets weaker with distance in the way we have described—is known as the **inverse-square law** of light propagation. In this respect, the propagation of radiation is similar to the effects of gravity. Remember that the force of gravity between two attracting masses is also inversely proportional to the square of their separation.

4.2 THE ELECTROMAGNETIC SPECTRUM

Objects in the universe send us an enormous range of electromagnetic radiation. Scientists call this range the **electromagnetic spectrum,** which they have divided into a number of categories. The spectrum is shown in Figure 4.5, with some information about the waves in each part or band.

4.2.1 Types of Electromagnetic Radiation

Electromagnetic radiation with the shortest wavelengths, no longer than 0.01 nanometer (nm), is called **gamma rays** ($1 \text{ nm} = 10^{-9}$ m; see Appendix 5). The name *gamma* comes from the third letter of the Greek alphabet; gamma rays were the third kind of radiation discovered coming from radioactive atoms when physicists first investigated their behavior. Because gamma rays carry a lot of energy, they can be dangerous for living tissues. We shall see that gamma radiation is generated deep in the interior of stars. They are also generated by some of the most violent phenomena in the universe, such as the deaths of stars and the merging of stellar corpses. Gamma rays coming to Earth are absorbed by our atmosphere before they reach the ground (which is a good thing for our health); thus, they can only be studied using instruments in space.

Electromagnetic radiation with wavelengths between 0.01 nm and 20 nm is referred to as **x rays.** These are now familiar to all of us from visits to the doctor and dentist; being more energetic than light, x rays are able to penetrate soft tissues and so allow us to make images of the shadows of the bones within them. While x rays can penetrate a short length of human flesh, they are stopped by the large numbers of atoms in the Earth's atmosphere with which they interact. Thus, x-ray astronomy (like gamma-ray astronomy) could not develop until we invented ways of sending instruments above our atmosphere (Figure 4.6).

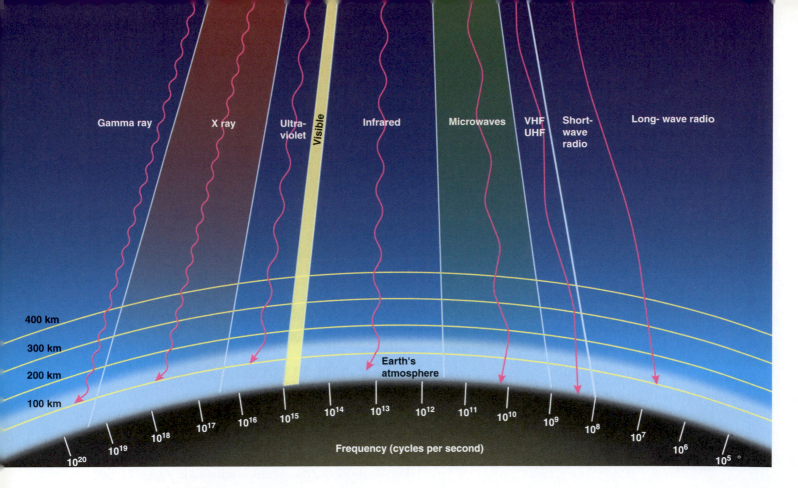

Gamma ray X ray Ultra-violet Visible Infrared Microwaves VHF UHF Short-wave radio Long-wave radio

400 km
300 km
200 km
100 km

Earth's atmosphere

10^{20} 10^{19} 10^{18} 10^{17} 10^{16} 10^{15} 10^{14} 10^{13} 10^{12} 10^{11} 10^{10} 10^{9} 10^{8} 10^{7} 10^{6} 10^{5}

Frequency (cycles per second)

■ **FIGURE 4.5**

Radiation and the Earth's Atmosphere The bands of the electromagnetic spectrum and how well Earth's atmosphere transmits them. Note that high-frequency waves from space do not make it to the surface and must therefore be observed from space. Some infrared and microwaves are absorbed by water and thus are best observed from high altitudes. Low-frequency radio waves are blocked by Earth's ionosphere.

■ **FIGURE 4.6**

The X-Ray Sky What the sky would look like if we had eyes tuned to certain types of x rays (and were standing above Earth's atmosphere). This map of the sky (with the disk of our Milky Way Galaxy running across its center) was constructed from data gathered by the European ROSAT satellite. Each color (red, yellow, and blue) shows x rays of different frequencies or energies. The red outlines the glow from a hot local bubble of gas all around us, blown by one or more exploding stars in our cosmic vicinity. The yellow and blue show more distant sources of x rays, such as the remnants of other exploded stars or the center of our Galaxy (in the middle of the picture), as well as a diffuse background of x rays that result from the blending of x rays from many more distant sources in the universe.

Max Planck Institute for Extraterrestrial Physics

■ **FIGURE 4.7**
The Action of a Prism
When we pass a beam of white sunlight through a prism, we see a rainbow-colored band of light that we call a continuous spectrum.

700 nm
600 nm
500 nm
400 nm

Radiation intermediate between x rays and visible light is **ultraviolet** (meaning higher energy than violet). Outside the world of science, ultraviolet light is sometimes called "black light" because our eyes cannot see it. Ultraviolet radiation is mostly blocked by a section of Earth's atmosphere called the *ozone* layer, but a small fraction of the ultraviolet rays from our Sun do penetrate to cause sunburn or, in extreme cases, skin cancer in human beings. Ultraviolet astronomy is also best done from space.

Electromagnetic radiation with wavelengths between roughly 400 and 700 nm is called **visible light** because these are the waves that human vision can perceive. As we shall see, this is also the band of the electromagnetic spectrum where the Sun gives off the greatest amount of radiation. These two observations are not coincidental: Human eyes evolved to see the kinds of waves that the Sun produced most effectively. The more light there was to see by, the more efficiently we could evade predators and make babies before we got stepped on or eaten (which, in a crude way, is what life is all about from the point of view of your genetic material). Visible light penetrates the Earth's atmosphere effectively, except when it is temporarily blocked by passing clouds.

In 1672, in the first paper that he submitted to the Royal Society, Newton described an experiment in which he permitted sunlight to pass through a small hole and then through a prism. Newton found that sunlight, which gives the impression of being white, is actually made up of a mixture of all the colors of the rainbow (Figure 4.7). Although all the colors of light may impinge on objects in our world, not all of them are reflected. When white light from the Sun or from a light fixture hits a pair of blue jeans, the dye in the jeans absorbs all the colors except blue. Some of the blue colors are reflected back into our eyes, and our brain interprets that particular set of wavelengths as blue color.

Between visible light and radio waves are the wavelengths of **infrared** or heat radiation. Astronomer William Herschel first discovered infrared in 1800 while trying to measure the temperatures of different colors of sunlight spread out into a spectrum. He noticed that when he accidentally pushed his thermometer beyond the reddest color, it still registered heating due to some invisible energy coming from the Sun. This was the first hint about the existence of the other (invisible) bands of the electromagnetic spectrum, although it would take many decades for our full understanding to develop.

A "heat lamp" radiates mostly infrared radiation, and the nerve endings in our skin are sensitive to this band of the electromagnetic spectrum. Infrared waves are absorbed by water and carbon dioxide molecules in Earth's atmosphere, and for this reason, infrared astronomy is best done from high mountaintops, high-flying airplanes, and spacecraft.

All electromagnetic waves longer than infrared are called **radio waves,** but this is so broad a category that we generally divide it into several subsections. Among the most familiar of these are *microwaves,* used in short-wave communication and microwave ovens; radar waves, used at airports and by the military; FM and television waves, which carry the news and entertainment without which it is now hard to imagine modern culture; and AM radio waves, which were the first to be developed for broadcasting. The wavelengths of these different categories range from a few millimeters to hundreds of meters. Other radio radiation can have wavelengths as long as several kilometers.

With such a wide range of wavelengths, you will not be surprised to learn that not all radio waves interact with Earth's atmosphere the same way. Microwaves are absorbed by water vapor (which makes them effective in heating foods with high water contents in microwave ovens). FM and TV waves are not absorbed and can travel easily through our atmosphere. AM radio waves are absorbed or reflected by a layer in Earth's atmosphere called the *ionosphere.* Given some of the truly dumb things people are saying these days on AM radio, maybe it's just as well that this type of radiation can not get out of our planet's shielding atmosphere!

We hope this brief survey has left you with one strong impression. Although visible light is what most people associate with astronomy, the light that our eyes can see is only a tiny fraction of the broad range of waves that the universe sends us. Today we understand that judging some astronomical phenomenon by using only the light we can see is like hiding under the table at a big dinner party and judging all the guests by nothing but their shoes. There's a lot more to each person than meets our eye under the table. It is very important for those who study astronomy today to avoid being *visible-light chauvinists*—to respect only the information seen by their eyes while ignoring the

TABLE 4.1 *Types of Electromagnetic Radiation*

Type of Radiation	Wavelength Range (nm)	Radiated by Objects at This Temperature	Typical Sources
Gamma rays	Less than 0.01	More than 10^8 K	Produced in nuclear reactions; requires very-high-energy processes
X rays	0.01–20	10^6–10^8 K	Gas in clusters of galaxies; supernova remnants; solar corona
Ultraviolet	20–400	10^4–10^6 K	Supernova remnants; very hot stars
Visible	400–700	10^3–10^4 K	Stars
Infrared	10^3–10^6	10–10^3 K	Cool clouds of dust and gas; planets; satellites
Radio	More than 10^6	Less than 10 K	No astronomical objects this cold; radio emission produced by electrons moving in magnetic fields (synchrotron radiation)

Ace⊛Astronomy™ Log into AceAstronomy and select this chapter to see Astronomy Exercise "The Electromagnetic Spectrum."

information gathered by instruments sensitive to other bands of the electromagnetic spectrum.

Table 4.1 summarizes the bands of the electromagnetic spectrum and indicates the temperatures and typical astronomical objects that emit each kind of electromagnetic radiation. Many of the objects in the table will not be familiar to you this early in the course, but you can return to this table as you learn more about the types of objects astronomers study.

4.2.2 Radiation and Temperature

Some astronomical objects emit mostly infrared radiation, others mostly visible light, and still others mostly ultraviolet radiation. What determines the type of electromagnetic radiation emitted by the Sun, stars, and other astronomical objects? The answer often turns out to be their *temperature*.

At the microscopic level, everything in nature is in motion. A *solid* is composed of molecules and atoms that are in continuous vibration: They move back and forth in place, but their motion is much too small for our eyes to make out. A *gas* consists of molecules that are flying about freely at high speed, continually bumping into one another and bombarding the surrounding matter. The energy of these random molecular and atomic motions is called *heat.* The hotter the solid or gas, the more rapid the motion of its molecules or atoms. The temperature of something is thus a measure of the average motion energy of the particles that make it up.

This motion at the microscopic level is responsible for much of the electromagnetic radiation on Earth and in the universe. As atoms and molecules move about and collide, or vibrate in place, their charged particles generate electromagnetic radiation (as we saw in Section 4.1). The characteristics of this radiation are determined by the temperature of those atoms and molecules. In a hot solid or gas, for example, the individual particles vibrate in place or move rapidly from collision to collision, so the emitted waves are, on average, more energetic. In very cool material, the particles have low-energy atomic and molecular motions and thus generate lower-energy waves.

4.2.3 Radiation Laws

To understand in more quantitative detail the relationship between temperature and electromagnetic radiation, we may consider an idealized object that (unlike your sweater or your astronomy instructor's head) does not reflect or scatter any radiation but absorbs all the electromagnetic energy that falls on it. The energy that is absorbed causes the atoms and molecules in it to vibrate or move around at increasing speeds. As it gets hotter, this object will radiate electromagnetic waves until absorption and radiation are in balance. We want to discuss such an idealized object because, as you will see, stars behave in very nearly the same way.[1]

The radiation from our ideal object has several characteristics, as illustrated in Figure 4.8. The graphs show the power emitted at each wavelength by objects of different temperatures. In science, the word *power* means the energy coming off per second (and it is typically measured in *watts,* which you are probably familiar with from buying light bulbs.)

First of all, notice that the white curves show that at each temperature our object emits a *continuous spectrum* of waves; that is, it emits radiation at all wavelengths. This is because in any solid or denser gas, some molecules or atoms vibrate or move between collisions slower than average, and some move faster than average. So when we look at the electromagnetic waves emitted, we find a broad range or spectrum of energies and wavelengths. More waves are emitted at the *average* vibration or motion rate (the tall part of each curve), but if we have a large number of atoms or molecules, some waves will be detected at each wavelength.

Second, note that an object at a higher temperature emits *more* power at all wavelengths than does a cooler one. In a hot gas (the taller curves in Figure 4.8), for example, the atoms have more collisions and give off more waves at each possible energy or wavelength. In the real

---

[1] Such an object is called a *blackbody* in more advanced courses, but since blackbodies in nature—like stars—do not look black, we prefer not to use the term.

FIGURE 4.8

Radiation Laws Illustrated
These graphs show a count of
the number of photons per
second (or the total amount
of power) given off at different
wavelengths for objects at three
different temperatures (three
white curves). Note that at
hotter temperatures, more
energy is emitted at all
wavelengths. The higher the
temperature, the shorter the
wavelength at which the peak
amount of energy is radiated
(this is called Wien's law).

Ace★Astronomy™ Log into AceAstronomy and select this chapter to see Astronomy Exercise "Blackbody & Wien's Law."

world of stars, this means that hotter stars give off more energy at every wavelength than do cooler stars.

Third, the graph shows us that the higher the temperature, the shorter the wavelength at which the maximum power is emitted. Remember that shorter wavelength means higher frequency and energy. It makes sense, then, that hot objects give off their average waves at shorter wavelengths (higher energies) than do cool objects. You may have observed examples of this rule in everyday life. When a burner on an electric stove is turned on low, it emits only heat, which is infrared radiation, but does not glow with visible light. If the burner is set to a higher temperature, it starts to glow a dull red. At a still higher setting, it glows a brighter orange-red (shorter wavelength). At even higher temperatures, which cannot be reached with ordinary stoves, metal can appear brilliant yellow or even blue-white.

We can use these ideas to come up with a rough sort of "thermometer" for measuring the temperatures of stars. Because many stars give off most of their waves in visible light, the color of light that dominates a star's appearance is a rough indicator of its temperature. If one star looks red and another looks blue, which one has the higher temperature? Because blue is the shorter-wavelength color, it is the sign of a hotter star. (Note that the temperatures we associate with different colors in science are not the same as the ones artists use. In art, red is often called a "hot" color and blue a "cool" color, but in nature, it's the other way around.)

We can develop a more precise star thermometer by measuring how much energy a star gives off at each wavelength and by constructing diagrams like Figure 4.8. The location of the peak (or maximum) in the power curve of each star can tell us its temperature. The temperature at the surface of the Sun, which is where the radiation that we see is emitted, turns out to be 5800 K. (Throughout this text we use the Kelvin or absolute temperature scale. On

this scale, water freezes at 273 K and boils at 373 K. All molecular motion ceases at 0 K. The various temperature scales are described in Appendix 5.) There are stars cooler than the Sun and stars hotter than the Sun; there are even some stars so hot that most of their energy is given off at ultraviolet wavelengths.

The wavelength at which maximum power is emitted can be calculated according to the equation

$$\lambda_{max} = \frac{3 \times 10^6}{T}$$

where the wavelength is in nanometers (one billionth of a meter) and the temperature is in Kelvins. This relationship is called **Wien's law.** For the Sun, the wavelength at which the maximum energy is emitted is 520 nm, which is near the middle of that portion of the electromagnetic spectrum called visible light. Characteristic temperatures of other astronomical objects, and the wavelengths at which they emit most of their power, are listed in Table 4.1.

We can also describe our observation that hotter objects radiate more power at all wavelengths in a mathematical form. If we sum up the contributions from all parts of the electromagnetic spectrum, we obtain the total energy emitted by a blackbody. What we usually measure from a large object like a star is the **energy flux,** the power emitted per square meter. The word *flux* means "flow" here: We are interested in the flow of power into an area (like the area of a telescope mirror). It turns out that the energy flux from a blackbody at temperature T is proportional to the fourth power of its absolute temperature. This relationship is known as the **Stefan-Boltzmann law** and can be written in the form of an equation as

$$F = \sigma T^4$$

where F stands for the energy flux and σ (Greek letter sigma) is a constant number.

Notice how impressive this result is. Increasing the temperature of a star would have a tremendous effect on the power it radiates. If the Sun, for example, were twice as hot—that is, if it had a temperature of 11,600 K—it would radiate 2^4, or 16, times more power than it does now. Tripling the temperature would raise the power output 81 times! Hot stars really shine away a tremendous amount of energy.

Ace⟲Astronomy™ Log into AceAstronomy and select this chapter to see Astronomy Exercises "Stefan-Boltzmann I," "Stefan-Boltzmann II," and "Stefan-Boltzmann, Wien's Law, and Black Body Radiation."

4.3 SPECTROSCOPY IN ASTRONOMY

As we hinted at the beginning of the chapter, electromagnetic radiation carries a lot of information about the nature of stars and other astronomical objects. To extract this information, however, astronomers must be able to study the amounts of energy we receive at different wavelengths of visible light (and other radiation) in fine detail. Let's examine how we can do this and what we can learn.

4.3.1 Optical Properties of Light

Visible light and other forms of electromagnetic energy exhibit certain behaviors that are important to the design of telescopes and other instruments. For example, light can be *reflected* from a surface. If the surface is smooth and shiny, as in a mirror, the direction of the reflected light beam can be calculated accurately from a knowledge of the shape of the reflecting surface. Light is also bent, or *refracted,* when it passes from one kind of transparent material into another—say, from the air into a glass lens.

Reflection and refraction of light are the basic properties that make possible all optical instruments—from eyeglasses to giant astronomical telescopes. Such instruments are generally combinations of glass lenses, which bend light according to the principles of refraction, and curved mirrors, which depend on the properties of reflection. Small optical devices, such as eyeglasses or binoculars, generally use lenses, while large telescopes depend almost entirely on mirrors for their main optical elements. In Chapter 5 we will discuss astronomical instruments and their uses. For now, we turn to another behavior of light, one that is essential for the decoding of light.

When light passes from one transparent material to another, an interesting effect occurs in addition to simple refraction. Because the bending of the beam depends on the wavelength of the light as well as the properties of the material, different wavelengths or colors of light are bent by different amounts and are therefore separated. This phenomenon is called **dispersion.**

Figure 4.7 shows how light is separated into different colors with a *prism*—a piece of glass in the shape of a triangle. Upon entering one face of the prism, light is refracted once, the violet light more than the red, and upon leaving the opposite face, the light is bent again and further dispersed. If the light leaving the prism is focused on a screen, the different wavelengths or colors that make up white light are lined up side by side just like a rainbow (Figure 4.9). (In fact, a rainbow is formed by the dispersion of light through raindrops; see Making Connections: *The Rainbow.*) Because this array of colors is a spectrum of light, the instrument used to disperse the light and form the spectrum is called a **spectrometer.**

4.3.2 The Value of Stellar Spectra

If the spectrum of the white light from the Sun and stars were simply a continuous rainbow of colors, astronomers would have little interest in the detailed study of a star's spectrum once they had learned its approximate temperature. When Newton, who first described the laws of refraction and dispersion in optics, observed the solar spectrum, all he could see was a continuous band of colors.

In 1802, however, William Wollaston built an improved spectrometer that included a lens to focus the Sun's spectrum on a screen. With this device Wollaston saw that the colors were not spread out uniformly, but instead some ranges of color were missing, appearing as dark bands in the solar spectrum. He mistakenly attributed these lines to natural boundaries between the colors. In 1815 the German physicist Joseph Fraunhofer, upon a more careful examination of the solar spectrum, found about 600 such dark lines (missing colors), which pretty quickly ruled out the boundary hypothesis (Figure 4.10).

These physicists called the missing colors *dark lines* because the rainbow-colored (continuous) spectrum as they viewed it was interrupted at various wavelengths, and in their spectrometers such interruptions looked like dark lines. Later, researchers found that similar dark lines could be produced in the spectra (*spectra* is the plural of *spectrum*) of artificial light sources. They did this by passing their light through various apparently transparent substances—usually containers with just a bit of thin gas in them.

These gases turned out not to be transparent at *all* colors: They were quite opaque at a few sharply defined wavelengths. Something in each gas had to be absorbing just a

■ FIGURE 4.9

A Continuous Spectrum When white light passes through a prism, its dispersion forms a continuous spectrum of all the colors. Although it is hard to see in this printed version, in a well-dispersed spectrum, many subtle gradations in color are visible as your eye scans from one end (the violet) to the other (the red).

The Rainbow

Rainbows are an excellent illustration of the refraction of sunlight. You have a good chance of seeing a rainbow anytime you are between the Sun and a rain shower; this situation is illustrated in Figure 4A. The raindrops act like little prisms and break white light into the spectrum of colors. Suppose a ray of sunlight encounters a raindrop and passes into it. The light changes direction—is refracted (Figure 4B)—when it passes from air to water; the blue and violet light are refracted more than the red. Some of the light is then reflected at the backside of the drop and reemerges from the front, where it is again refracted. As a result, the white light is spread out into a rainbow of colors.

Note that in Figure 4B violet light lies above the red light after it emerges from the raindrop. When you look at a rainbow, however, the red light is higher in the sky. Why? Look again at Figure 4A. If the observer looks at a raindrop that is high in the sky, the violet light passes over her head and the red light enters her eye. Similarly, if the observer looks at a raindrop that is low in the sky, the violet light reaches her eye and the drop appears violet, whereas the red light from that same drop strikes the ground and is not seen. Colors of intermediate wavelengths are refracted to the eye by drops that are intermediate in altitude between the drops that appear violet and the ones that appear red. Thus, a single rainbow always has red on the outside and violet on the inside.

For an even simpler example of refraction, put a pencil at a slanted angle in a glass of water. What do you see? Can you offer an explanation?

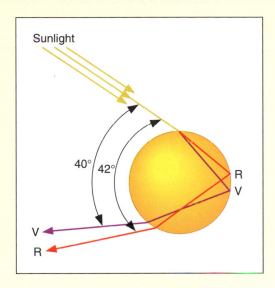

■ **FIGURE 4B**

A diagram showing the path of light passing through a raindrop. Refraction separates white light into its component colors.

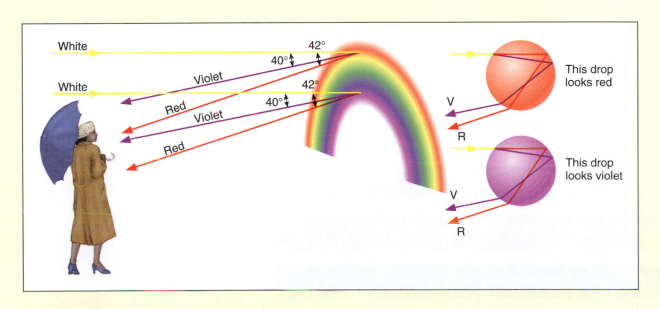

■ **FIGURE 4A**

This diagram shows how light from the Sun, which is located behind the observer, can be refracted by raindrops to produce a rainbow.

■ **FIGURE 4.10**

The Visible Spectrum of the Sun   Our star's spectrum is crossed by dark lines
produced by atoms in the solar atmosphere that absorb light at certain wavelengths.

few colors of light and no others. All gases did this, but each
different element absorbed a different set of colors and
thus showed different dark lines. If the gas in a container
consisted of two elements, then light passing through it was
missing the colors (showing dark lines) that go with both of
the elements. So it became clear that certain lines in the
spectrum "go with" certain elements. This discovery, as we
shall see, was one of the most important steps forward in
the history of astronomy.

What would happen if there were no continuous spectrum for our gases to remove light from? What if, instead,
we heated the same thin gases until they were hot enough
to glow with their *own light?* When the gases were heated,
a spectrometer revealed no continuous spectrum, but several separate *bright lines.* That is, these hot gases *emitted*
light only at certain specific wavelengths or colors. The colors at which they emitted when they were heated were the
very same colors at which they had absorbed when a continuous source of light was behind them.

When the gas was pure hydrogen, it would absorb or
emit one pattern of colors; when it was sodium, it would absorb or emit a different pattern. A mixture of hydrogen and
sodium emitted both sets of spectral lines. Thus, scientists
began to see that different substances showed distinctive
spectral signatures by which their presence could be detected (Figure 4.11). Just as your signature allows the bank

to identify you, the unique pattern of colors for each type
of atom can help us identify which element or elements are
in the gas.

Ace⊗Astronomy™   Log into AceAstronomy and select this chapter
to see Astronomy Exercise "Emission and Absorption Spectra" and the Active
Figure called "Kirchoff's Laws."

4.3.3   Types of Spectra

In such experiments, then, we can distinguish among the
three different types of spectra. A **continuous spectrum**
(formed when a solid or very dense gas gives off radiation)
is an array of all wavelengths or colors of the rainbow. A continuous spectrum can serve as a backdrop from which the
atoms of a much less dense gas can absorb light. A dark line,
or **absorption line spectrum,** consists of a series or pattern of dark lines—missing colors—superimposed upon
the continuous spectrum of a source. A bright line, or **emission line spectrum,** appears as a pattern or series of bright
lines; it consists of light in which only certain discrete wavelengths are present. (Figure 4.18—a little further ahead in
this chapter—is a picture of what each of these looks like.)

When it's a gas, each particular chemical element or
compound produces its own characteristic pattern of spectral lines—its spectral signature. No two patterns are alike.

Ace⊗Astronomy™   Log into AceAstronomy and
select this chapter to see Astronomy Exercise "Stellar Atomic
Absorption Lines."

■ **FIGURE 4.11**

Line Spectra from Different Elements
Each type of hot glowing gas (each element)
produces its own unique pattern of lines, so the
composition of a gas can be identified by its
spectrum.

96   CHAPTER 4   RADIATION AND SPECTRA

In other words, each particular gas can absorb or emit only certain wavelengths of light peculiar to that gas. The temperature and other conditions determine whether the lines are bright or dark, but the wavelengths for any element are the same in either case. It is the precise pattern of wavelengths that makes the signature of each element unique. (Liquids and solids can also generate spectral lines or bands, but they are broader and less well defined and hence more difficult to interpret.)

The dark lines in the solar spectrum thus give evidence of certain chemical elements between us and the Sun, absorbing those wavelengths of sunlight. Because the space between us and the Sun is pretty empty, astronomers quickly realized that the atoms doing the absorbing must be in a thin atmosphere of cooler gas around the Sun. Since the Sun is made entirely of gas, this outer atmosphere is not all that different from the rest of the Sun, just thinner and cooler. Thus, we can use what we learn about its composition as an indicator of what the whole Sun is made of. Similarly, we can use the presence of absorption and emission lines to analyze the composition of other stars and clouds of gas in space.

Such analysis of spectra is the key to modern astronomy. Only in this way can we "sample" the stars, which are too far away for us to visit. Even if we could visit them, they are so hot that it would be very difficult to find graduate students willing to go take a sample. But there is no need to ask for such volunteers: Encoded in the electromagnetic radiation from celestial objects is clear information about the chemical makeup of these objects. Only by understanding what the stars were made of could astronomers begin to form theories about what made them shine and how they evolved.

In 1860 the German physicist Gustav Kirchhoff became the first person to use spectroscopy to identify an element in the Sun when he found the spectral signature of sodium gas. In the years that followed, astronomers found many other chemical elements in the Sun and stars. In fact, the element helium was found first in the Sun from its spectrum and only later identified on Earth. (The word *helium* comes from *helios*, the Greek name for the Sun.)

And why are there specific lines for each element? The answer to that question was not found until the 20th century; it required the development of a model for the atom. We therefore turn next to a closer examination of the atoms that make up all matter.

4.4 THE STRUCTURE OF THE ATOM

The idea that matter is composed of tiny particles called atoms is at least 25 centuries old. It took until the 20th century, however, for scientists to invent instruments that permitted them to probe inside an atom and find that it is not, as had been thought, hard and indivisible. Instead, the atom is a complex structure composed of still smaller particles.

4.4.1 Probing the Atom

The first of these smaller particles was discovered by British physicist James Thomson in 1897. Named the *electron,* this particle is negatively charged. (It is the flow of these particles that produces currents of electricity, whether in lightning bolts or in the wires leading to your hair dryer.) Because an atom in its normal state is electrically neutral, each electron in an atom must be balanced by the same amount of positive charge.

The next problem was to determine where in the atom the positive and negative charges are located. In 1911 British physicist Ernest Rutherford devised an experiment that provided part of the answer to this question. He bombarded an extremely thin piece of gold foil, only about 400 atoms thick, with a beam of *alpha particles* emitted from a radioactive material (Figure 4.12). We now know that alpha particles are helium atoms that have lost their electrons and thus are positively charged. Most of the alpha particles passed through the gold foil just as if it and the atoms in it were nearly empty space. About 1 in 8000 of the alpha particles, however, completely reversed direction and bounced backward from the foil. Rutherford wrote, "It was quite the most incredible event that has ever happened to me in my life. It was almost as incredible as if you fired a 15-inch shell at a piece of tissue paper and it came back and hit you."

The only way to account for the alpha particles that reversed direction when they hit the gold foil was to assume that nearly all of the mass as well as all of the positive charge in each individual gold atom was concentrated in a tiny center or **nucleus.** When an alpha particle strikes a nucleus, it reverses direction, much as a cue ball reverses direction when it hits another billiard ball. Rutherford's model placed the other type of charge—the negative electrons—in orbit around this nucleus.

Rutherford's model required that the electrons be in motion. Positive and negative charges attract each other, so stationary electrons would fall into the nucleus. Also, because both the electrons and the nucleus are extremely small, most of the atom is empty, which is why nearly all of Rutherford's alpha particles were able to pass right through the gold foil without colliding with anything. Rutherford's model was a very successful explanation of the experiments he conducted, although eventually scientists would discover that the nucleus has structure as well.

4.4.2 The Atomic Nucleus

The simplest possible atom (and the most common one in the Sun and stars) is hydrogen: The nucleus of ordinary hydrogen contains a single positively charged particle called a *proton.* Moving around this proton is a single *electron.* The mass of an electron is nearly 2000 times smaller than the mass of a proton; the electron carries an amount of charge exactly equal to that of the proton but opposite in sign (Figure 4.13). Opposite charges attract each other, so it is electromagnetic force that holds the proton and electron

■ FIGURE 4.12

Rutherford's Experiment (a) When Rutherford allowed alpha particles from a radioactive source to strike a target of gold foil, he found that, although most of them went straight through, some rebounded back in the direction from which they came. (b) From this experiment he concluded that the atom must be constructed like a miniature solar system, with the positive charge concentrated in the nucleus and the negative charge orbiting in the large volume around the nucleus.

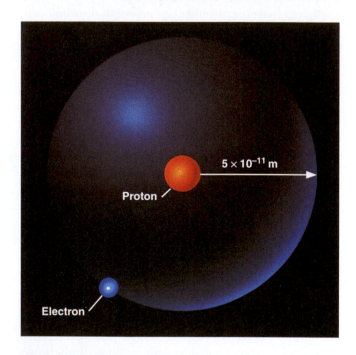

■ FIGURE 4.13

The Hydrogen Atom This is a schematic diagram of a hydrogen atom in its lowest energy state, also called the *ground state*. The proton and electron have equal but opposite charges, which exert an electromagnetic force that binds the hydrogen atom together. In this sketch the size of the particles is exaggerated for educational purposes; they are not to scale.

■ FIGURE 4.14

The Helium Atom Here we see a schematic diagram of a helium atom in its lowest energy state. Two protons are present in the nucleus of all helium atoms. In the most common variety of helium, the nucleus also contains two neutrons, which have nearly the same mass as the proton but carry no charge. Two electrons can be seen orbiting the nucleus.

together, just as gravity is the force that keeps the planets in orbit around the Sun.

There are, of course, other types of atoms. Helium, for example, is the second most abundant element in the Sun. Helium has two protons in its nucleus, instead of the single proton that characterizes hydrogen. In addition, the he-

lium nucleus contains two *neutrons*, particles with a mass comparable to that of the proton but with no electric charge. Moving around this nucleus are two electrons, so the total net charge of the helium atom is also zero (Figure 4.14).

From this description of hydrogen and helium, perhaps you have guessed the pattern for building up all the

■ **FIGURE 4.15**

Isotopes of Hydrogen A single proton in the nucleus defines the atom to be hydrogen, but there may be zero, one, or two neutrons. The most common isotope is the one with only a single proton. A hydrogen nucleus with one neutron is called deuterium; one with two neutrons is called tritium. The symbol for hydrogen is H; the superscript to the left indicates the total number of particles (protons plus neutrons) in the nucleus.

elements (different types of atoms) that we find in the universe. The type of element is determined by the *number of protons* in the nucleus of the atom. For example, any atom with 6 protons is the element carbon, with 8 protons is oxygen, with 26 is iron, and with 92 is uranium. On Earth, a typical atom has the same number of electrons as protons, and these electrons follow complex orbital patterns around the nucleus. Deep inside stars, however, it is so hot that the electrons get loose from the nucleus and (as we shall see) live separate, productive lives.

The number of neutrons in the nucleus is usually approximately equal to the number of protons, but the number of neutrons is not necessarily the same for all atoms of a given element. For example, most hydrogen atoms contain no neutrons at all. There are, however, hydrogen atoms that contain one proton and one neutron, and others that contain one proton and two neutrons. The various types of hydrogen nuclei with different numbers of neutrons are called **isotopes** of hydrogen (Figure 4.15), and other elements have isotopes as well. You can think of isotopes as siblings in the same element "family"—closely related but with different characteristics and behaviors.

4.4.3 The Bohr Atom

Rutherford's model for atoms has one serious problem. Maxwell's theory of electromagnetic radiation says that when electrons change either their speed or the direction of motion, they must emit energy. Orbiting electrons constantly change their direction of motion, so they should emit a constant stream of energy. According to Maxwell's theory, all electrons should spiral into the nucleus of the atom as they lose energy, and this collapse should happen very quickly—in about 10^{-16} seconds.

It was the Danish physicist Niels Bohr (1885–1962) who solved the mystery of how electrons remain in orbit. He was trying to develop a model of the atom that would also explain certain regularities observed in the spectrum

of hydrogen (see Section 4.5). He suggested that the spectrum of hydrogen can be understood if we assume that orbits of only certain sizes are possible for the electron. Bohr further assumed that as long as the electron moves in only one of these allowed orbits, it radiates no energy. Its energy would change only if it moved from one orbit to another.

This suggestion, in the words of science historian Abraham Pais, was "one of the most audacious hypotheses ever introduced in physics." If something equivalent were at work in the everyday world, you might find that, as you went for a walk after astronomy class, nature permitted you to walk 2 steps per minute, 5 steps per minute, and 12 steps per minute, but no speeds in between. No matter how you tried to move your legs, only *certain* walking speeds would be permitted. To make things more bizarre, it would take no effort to walk at one of the allowed speeds, but it would be difficult to change from one speed to another. Luckily, no such rules apply at the level of human behavior, but at the microscopic level of the atom, experiment after experiment has confirmed the validity of Bohr's strange idea. Bohr's suggestions became one of the foundations of the new (and much more sophisticated) model of the subatomic world called *quantum mechanics.*

In Bohr's model, if the electron moves from one orbit to another closer to the atomic nucleus, it must give up some energy in the form of electromagnetic radiation. If the electron goes from an inner orbit to one farther from the nucleus, however, it requires some additional energy. One way to obtain the necessary energy is to absorb electromagnetic radiation that may be streaming past the atom from an outside source.

A key feature of Bohr's model is that each of the permitted electron orbits around a given atom has a certain energy value. To move from one orbit to another (which will have *its* own specific energy value) requires a change in the electron's energy—a change determined by the difference between the two energy values. If the electron goes to a lower level, the energy difference will be given off; if the electron goes to a higher level, the energy difference must be obtained from somewhere else. Each jump (or transition) to a different level has a fixed and definite energy change associated with it.

A crude analogy for this situation might be life in a tower of luxury apartments where the rent is determined by the quality of the view. Such a building has certain definite numbered levels or floors on which apartments are located. No one can live on floor 5.37 or 22.5. In addition, the rent gets higher as you go up to higher floors. If you want to exchange an apartment on the 20th floor for one on the 2nd floor, you will not owe as much rent and you will get a refund from the landlord. But if you want to move from the 3rd floor to the 25th floor, you'd better be prepared to find some extra resources because your rent will increase. In the atom, too, the "cheapest" place for an electron to live is the lowest possible level, and energy is required to move to a higher level.

Here we have one of the situations where it is easier to think of electromagnetic radiation as photons than as waves (see p. 88). As electrons move from one level to another, they give off or absorb little packets of energy. When an electron moves to a higher level, it absorbs a photon of just the right energy (provided one is available). When it moves to a lower level, it emits a photon with the exact amount of energy it no longer needs in its "lower-cost living situation."

The photon and wave perspectives must be equivalent: Light is light, no matter how we look at it. Thus each photon carries a certain amount of energy that is proportional to the frequency of the wave it represents. The value of its energy is given by the formula

$$E = h \times f$$

where the constant of proportionality, h, is called Planck's constant. It is named for Max Planck, the German physicist who was one of the originators of the quantum theory. If metric units are used (that is, if energy is measured in joules and frequency in hertz), then Planck's constant has the value $h = 6.626 \times 10^{-34}$ joules. Higher-energy photons correspond to higher-frequency waves (which have a shorter wavelength); lower-energy photons are waves of lower frequency.

To take a specific example, consider a calcium atom inside the Sun's atmosphere in which an electron jumps from a lower level to a higher level. To do this, it needs about 5×10^{-19} joules of energy, which it can conveniently obtain by absorbing a passing photon of that energy coming from deeper inside the Sun. This photon is equivalent to a wave of light whose frequency is about 7.5×10^{14} hz and whose wavelength is about 3.9×10^{-7} m (393 nm), in the deep violet part of the visible light spectrum. Although it may seem strange at first to switch from picturing light as a photon (or energy packet) to picturing it as a wave, such switching has become second nature to astronomers and can be a handy tool for doing calculations about spectra.

4.5 FORMATION OF SPECTRAL LINES

We can use Bohr's model of the atom to understand how spectral lines are formed. The concept of energy levels for the electron orbits in an atom leads naturally to an explanation of why atoms absorb or emit only specific energies or wavelengths of light.

4.5.1 The Hydrogen Spectrum

Let's look at the hydrogen atom from the perspective of the Bohr model. Suppose a beam of white light (which consists of photons of all wavelengths) shines through a gas of atomic hydrogen. A photon of wavelength 656 nm has just the right energy to raise an electron in a hydrogen atom from the second to the third orbit. Thus, as all the photons of different energies (or wavelengths or colors) stream by the hydrogen atoms, photons with *this* particular wave-

length can be absorbed by those atoms whose electrons are orbiting on the second level. When they are absorbed, the electrons on the second level will move to the third level, and a number of the photons of this wavelength and energy will be missing from the general stream of white light.

Other photons will have the right energies to raise electrons from the second to the fourth orbit, or from the first to the fifth orbit, and so on. Only photons with exactly these correct energies can be absorbed. All of the other photons will stream past the atoms untouched. Thus, the hydrogen atoms absorb light at only certain wavelengths and produce dark lines at those wavelengths in the spectrum we see.

Suppose we have a container of hydrogen gas through which a whole series of photons is passing, allowing many electrons to move up to higher levels. When we turn off the light source, these electrons "fall" back down from larger to smaller orbits and emit photons of light—but, again, only light of those energies or wavelengths that correspond to the energy difference between permissible orbits. The orbital changes of hydrogen electrons that give rise to spectral lines are shown in Figure 4.16. (Note that in a real hydrogen atom the electron orbits are not as evenly spaced as those shown in this diagram.)

Similar pictures can be drawn for atoms other than hydrogen. However, because these other atoms ordinarily have more than one electron each, the orbits of their electrons are much more complicated, and the spectra are more complex as well. For our purposes, the key conclusion is this: *Each type of atom has its own unique pattern of electron orbits, and no two sets of orbits are exactly alike.* This means that each type of atom shows its own unique set of spectral lines, produced by electrons moving between its unique set of orbits.

Astronomers and physicists have worked hard to learn the lines that go with each element by studying the way atoms absorb and emit light in laboratories here on Earth. Then they can use this knowledge to identify the elements in celestial bodies. In this way, we now know the chemical makeup of not just any star, but even galaxies of stars so distant that their light started on its way to us long before Earth had even formed.

4.5.2 Energy Levels and Excitation

Bohr's model of the hydrogen atom was a great step forward in our understanding of the atom. However, we know today that atoms cannot be represented by quite so simple a picture; even the concept of sharply defined electron orbits is not really correct. We can still retain the concept that only certain discrete energies are allowable for an atom. These energies, called **energy levels,** can be thought of as representing certain average distances of the electron's possible orbits from the atomic nucleus.

Ordinarily, an atom is in the state of lowest possible energy, its **ground state.** In the Bohr model, the ground state corresponds to the electron being in the innermost orbit.

FIGURE 4.16

The Bohr Model for Hydrogen Here we follow the emission or absorption of photons by a hydrogen atom according to the Bohr model. Several different series of spectral lines are shown, corresponding to transitions of electrons from or to certain allowed orbits. Each series of lines that terminates on a specific inner orbit is named for the physicist who studied it. At the top, for example, you see the Balmer series; arrows show electrons jumping from the second orbit ($n = 2$) to the third, fourth, fifth, and sixth orbits. Each time a "poor" electron from a lower level wants to rise to a higher position in life, it must absorb energy to do so. It can absorb the energy it needs from passing waves (or photons) of light. The next set of arrows (Lyman series) shows electrons falling down to the first orbit from different (higher) levels. Each time an electron goes downward toward the nucleus, it can afford to give off (emit) some energy it no longer needs.

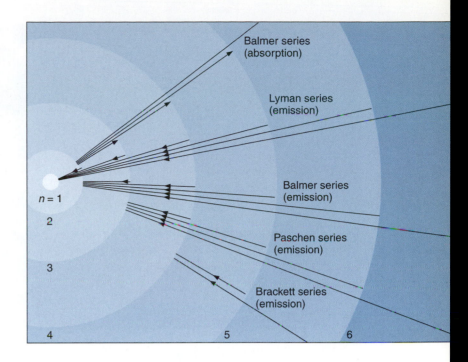

An atom can absorb energy, which raises it to a higher energy level (corresponding, in the Bohr picture, to an electron's movement to a larger orbit). The atom is then said to be in an **excited state.** Generally, an atom remains excited for only a very brief time. After a short interval, typically a hundred-millionth of a second or so, it drops back spontaneously to its ground state, with the simultaneous emission of light. The atom may return to its lowest state in one jump, or it may make the transition in steps of two or more jumps, stopping at intermediate levels on the way down. With each jump it emits a photon of the wavelength that corresponds to the energy difference between the levels at the beginning and end of that jump.

An energy-level diagram for a hydrogen atom and several possible atomic transitions are shown in Figure 4.17; it shows the same series of emission lines as the diagram of the Bohr model in Figure 4.16. When we measure the energies involved as the atom jumps between levels, we find that the transitions to or from the ground state, called the *Lyman series* of lines, result in the emission or absorption of ultraviolet photons. But the transitions to or from the first excited state (labeled $n = 2$ in Figure 4.16), called the *Balmer series,* produce emission or absorption in visible light. In fact, it was to explain this Balmer series that Bohr first suggested his model of the atom.

We mentioned that atoms that have absorbed specific photons from a passing beam of white light and have thus become excited generally de-excite themselves and emit that light again in a very short time. You might wonder, then, why *dark* spectral lines are ever produced. In other words, why doesn't this reemitted light quickly "fill in" the darker absorption lines?

Imagine a beam of white light coming toward you through some cooler gas. Some of the reemitted light *is* ac-

tually returned to the beam of white light you see, but this fills in the absorption lines only to a slight extent. The reason is that the atoms in the gas reemit light *in all directions* and only a small fraction of the reemitted light is in the direction of the original beam (toward you). In a star, much of the reemitted light actually goes in directions leading

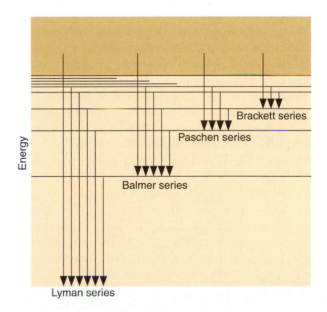

FIGURE 4.17

Energy-Level Diagram for Hydrogen At higher and higher energy levels, the levels become more and more crowded together, approaching a limit. The shaded region represents energies at which the atom is ionized (the electron is no longer attached to the atom). Each series of arrows represents electrons falling from higher levels to lower ones, releasing photons or waves of energy in the process.

Cloud of gas

Source of
continuous
spectrum

(b)
Continuous spectrum with dark lines

(c)

(a)
Continuous spectrum

Bright line spectrum

■ **FIGURE 4.18**

Three Kinds of Spectra When we see a light bulb or other source of continuous radiation (a), all the colors are present. When the continuous spectrum is seen through a thinner gas cloud, the cloud's atoms produce absorption lines in the continuous spectrum (b). When the excited cloud is seen without the continuous source behind it, its atoms produce emission lines (c). We can learn which types of atoms are in the cloud from the pattern of the absorption or emission lines.

back into the star, which does observers outside the star no good whatsoever.

Figure 4.18 summarizes the different kinds of spectra we have discussed. A light bulb produces a continuous spectrum (a in the figure). When that continuous spectrum is viewed through a thinner cloud of gas, an absorption line spectrum can be seen superimposed on the continuous spectrum (b). If we look only at a cloud of excited gas atoms (with no continuous source seen behind it), we see that the excited atoms give off an emission line spectrum (c).

Atoms in a hot gas are moving at high speeds and continually colliding with one another and with any loose electrons. They can be excited and de-excited by these collisions as well as by absorbing and emitting light. The speed of atoms in a gas depends on the temperature. When the temperature is higher, so are the speed and energy of the collisions. The hotter the gas, therefore, the more likely that electrons will occupy the outermost orbits, which correspond to the highest energy levels. This means that the level where electrons *start* their upward jumps in a gas can serve as an indicator of how hot that gas is. In this way, the absorption lines in a spectrum give astronomers information about the temperature of the regions where the lines originate.

4.5.3 Ionization

We have described how certain discrete amounts of energy can be absorbed by an atom, raising it to an excited state

and moving one of its electrons farther from its nucleus. If enough energy is absorbed, the electron can be completely removed from the atom. The atom is then said to be **ionized.** The *minimum amount* of energy required to remove one electron from an atom in its ground state is called its *ionization energy.*

Still greater amounts of energy must be absorbed by the now-ionized atom (called an **ion**) to remove an additional electron deeper in the structure of the atom. Successively greater energies are needed to remove the third, fourth, fifth, and so on, electrons from the atom. If enough energy is available, an atom can become completely ionized, losing all of its electrons. A hydrogen atom, having only one electron to lose, can be ionized only once; a helium atom can be ionized twice; and an oxygen atom up to eight times. When we examine regions of the cosmos where there is a great deal of energetic radiation, such as the neighborhoods where hot young stars have recently formed, we see a lot of ionization going on.

An atom that has become ionized has lost a negative charge—the missing electron—and thus is left with a net positive charge. It therefore exerts a strong attraction on any free electron. Eventually, one or more electrons will be captured and the atom will become neutral (or ionized to one less degree) again. During the electron capture process, the atom emits one or more photons. Which photons are emitted depends on whether the electron is captured at once to the lowest energy level of the atom or stops at one or more intermediate levels on its way to the lowest available level.

Just as the excitation of an atom can result from a collision with another atom, ion, or electron (collisions with electrons are usually most important), so can ionization. The rate at which such collisional ionizations occur depends on the speeds of the atoms and hence on the temperature of the gas.

The rate at which ions and electrons recombine also depends on their relative speeds—that is, on the temperature. In addition, it depends on the density of the gas: The higher the density, the greater the chance for recapture because the different kinds of particles are crowded more closely together. From a knowledge of the temperature and density of a gas, it is possible to calculate the fraction of atoms that have been ionized once, twice, and so on. In the Sun, for example, we find that most of the hydrogen and helium atoms in its atmosphere are neutral, whereas most of the calcium atoms, as well as many other heavier atoms, are ionized once.

The energy levels of an ionized atom are entirely different from those of the same atom when it is neutral. Each time an electron is removed from the atom, the energy levels of the ion, and thus the wavelengths of the spectral lines it can produce, change. This helps astronomers differentiate the ions of a given element. Ionized hydrogen, having no electron, can produce no absorption lines.

4.6 THE DOPPLER EFFECT

Although the last two sections contain many new concepts, we hope you have seen one major idea emerge. Astronomers can learn about the elements in stars and galaxies by de-coding the information in their spectral lines. There is a complicating factor in learning how to decode the message of starlight, however. If a star is moving toward or away from us, its lines will be in a slightly different place in the spectrum from where they would be in a star at rest. And most objects in the universe do have some motion relative to the Sun.

4.6.1 Motion Affects Waves

In 1842 Christian Doppler pointed out that if a light source is approaching or receding from the observer, the light waves will be, respectively, crowded more closely together or spread out. (This is actually true for all kinds of waves, including sound: Doppler first measured the effect of motion on waves by hiring a group of musicians to play on an open railroad car as it was moving along the track.) The general principle, now known as the **Doppler effect,** is illustrated in Figure 4.19.

In part (a) of the figure, the light source (S) is at rest with respect to the observer. The source gives off a series of waves with crests labeled 1, 2, 3, and 4. The light waves spread out evenly in all directions, like the ripples from a splash in a pond. The crests are separated by a distance, λ, where λ is the wavelength. The observer, who happens to be located in the direction of the bottom of the page, sees the light waves coming nice and evenly, one wavelength apart. Observers located anywhere else would see the same thing.

On the other hand, if the source is moving with respect to the observer, as in part (b), the situation is more complicated. Between the time one crest is emitted and the next is ready to come out, the source has moved a bit—in our case, toward the bottom of the page and toward our original observer. By moving, the source has decreased the

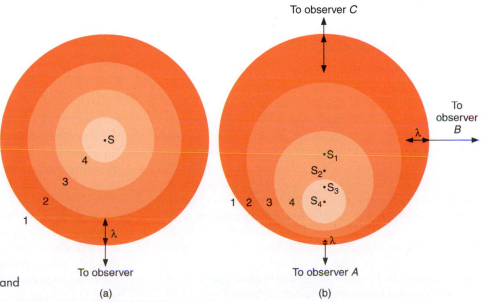

Ace Astronomy™   Log into AceAstronomy and select this chapter to see Astronomy Exercise "Doppler Shift."

■ FIGURE 4.19

The Doppler Effect (a) A source S makes waves whose numbered crests (1, 2, 3, 4) wash over a stationery observer. (b) The source S now moves toward observer A and away from observer C. Wave crest 1 was omitted when the source was at position S1, crest 2 at position S2, and so forth. Observer A sees the waves compressed by this motion and sees a blueshift (if the waves are light). Observer C sees the waves stretched out by the motion and sees a redshift. Observer B, whose line of sight is perpendicular to the source's motion, sees no change in the waves (and feels left out).

distance between crests—squeezed the crests together a bit, you might say.

In part (b), the sources are in four positions, S_1, S_2, S_3, and S_4, each corresponding to the emission of one wave crest. To our original observer, labeled observer A, the waves seem to follow one another more closely, at a decreased wavelength and thus increased frequency. (Remember, all light waves travel at the speed of light, no matter what. This means that motion cannot affect the speed, but only the wavelength and the frequency. As the wavelength decreases, the frequency must increase. If the waves are shorter, more will be able to fit by during each second.)

The situation is not the same for other observers. Let's look at the situation from the point of view of observer C, located opposite observer A in the figure. For her, the source is moving away from her location. As a result, the waves are not squeezed together but instead are spread out by the motion of the source. The crests arrive with an increased wavelength and decreased frequency. To observer B, in a direction at right angles to the motion of the source, no effect is observed. The wavelength and frequency remain the same as they were in part (a) of the figure.

We can see from this illustration that the Doppler effect is produced only by a motion toward or away from the observer, a motion called **radial velocity.** Observers between A and B and between B and C would observe, respectively, some shortening or lengthening of the light waves for that part of the motion of the source that is along their line of sight.

You have probably heard the Doppler effect with sound waves. When a train whistle or police siren approaches you and then moves away, you will notice a decrease in the pitch (or frequency) of the sound waves. They have changed from slightly more frequent than at rest when coming toward you, to slightly less frequent than at rest when moving away from you. (A nice example of this change in the sound of a train whistle can be heard at the end of the Beach Boys song "Caroline, No" on their album *Pet Sounds.*)

4.6.2 Color Shifts

When the source of waves moves toward you, the wavelengths decrease a bit. If the waves involved are visible light, then the colors of the light change slightly. As wavelengths decrease, they shift toward the blue end of the spectrum: Astronomers call this a *blueshift.* (Since the end of the spectrum is really violet, the term should probably be *violetshift,* but blue is a more common color.) When the source moves away from you and the wavelengths get longer, we call the change in colors a *redshift.* Because the Doppler effect was first used with visible light in astronomy, the terms *blueshift* and *redshift* became well established. Today astronomers use these words to describe changes in radio wavelengths or x-ray wavelengths as comfortably as they use them to describe changes in visible light.

The greater the motion toward or away from us, the greater the Doppler shift. If the relative motion is entirely along the line of sight, the formula for the Doppler shift of light is

$$\frac{\Delta\lambda}{\lambda} = \frac{v}{c}$$

where λ is the wavelength emitted by the source, $\Delta\lambda$ is the difference between λ and the wavelength measured by the observer, c is the speed of light, and v is the relative speed of the observer and the source in the line of sight. The variable v is counted as positive if the velocity is one of recession, and negative if it is one of approach. Solving this equation for the velocity, we find

$$v = c \times \frac{\Delta\lambda}{\lambda}$$

If a star approaches or recedes from us, the wavelengths of light in its continuous spectrum appear shortened or lengthened, respectively, as do those of the dark lines. However, unless its speed is tens of thousands of kilometers per second, the star does not appear noticeably bluer or redder than normal. The Doppler shift is thus not easily detected in a continuous spectrum and cannot be measured accurately in such a spectrum. The wavelengths of the absorption lines can be measured accurately, however, and their Doppler shift is relatively simple to detect.

This may sound like a horrible note on which to end the chapter. If all the stars are moving and motion changes the wavelength of each spectral line, won't this be a disaster for astronomers trying to figure out what elements are present in the stars? After all, it is the precise wavelength (or color) that tells astronomers which lines belong to which element. And we first measure these wavelengths in containers of gas in our laboratories, which are not moving. If every line in a star's spectrum is now shifted by its motion to a different wavelength (color), how can we be sure which lines and which elements we are looking at in a star whose speed we do not know?

Take heart: This situation sounds worse than it really is. Astronomers rarely judge the presence of an element in an astronomical object by a single line. It is the *pattern* of lines unique to hydrogen or calcium that enables us to determine that those elements are part of the star or galaxy we are observing. The Doppler effect does not change the pattern of lines from a given element—it only shifts the whole pattern slightly toward redder or bluer wavelengths. The shifted pattern is still quite easy to recognize. Best of all, when we do recognize a familiar element's pattern, we get a bonus: The amount the pattern is shifted now tells us the speed of the object in our line of sight.

We hope you can see why the training of new astronomers includes so much work on learning to decode light (and other electromagnetic radiation). We have seen that a skillful "decoder" can learn the temperature of a star, what elements are in it, and even its speed in a direction toward us or away from us. That's really an impressive bit of decoding for stars that are light years away.

SUMMARY

4.1 James Clerk Maxwell showed that whenever charged particles change their motion, as they do in every atom and molecule, they give off waves of energy. Light is one form of this **electromagnetic radiation.** The **wavelength** of light determines the color of visible radiation. Wavelength (λ) is related to **frequency** (f) and the speed of light (c) by the equation $c = \lambda f$. Electromagnetic radiation sometimes behaves like waves, but at other times it behaves as if it were a little packet of energy, called a **photon.** The apparent brightness of a source of electromagnetic energy decreases with increasing distance from that source in proportion to the square of the distance, a relationship known as the **inverse-square law.**

4.2 The **electromagnetic spectrum** consists of **gamma rays, x rays, ultraviolet radiation** (all forms of electromagnetic radiation with wavelengths shorter than that of visible light), **visible light, infrared,** and **radio radiation** (the last two with wavelengths longer than that of visible light). Many of these wavelengths cannot penetrate the layers of the Earth's atmosphere and must be observed from space. The emission of electromagnetic radiation is intimately connected to the temperature of the source. The higher the temperature of an idealized emitter of electromagnetic radiation, the shorter is the wavelength at which the maximum amount of radiation is emitted. The mathematical equation describing this relationship ($\lambda_{max} = 3 \times 10^6/T$) is known as **Wien's law.** The total power emitted per square meter increases with increasing temperature. The relationship between emitted **energy flux** and temperature ($F = \sigma T^4$) is known as the **Stefan–Boltzmann law.**

4.3 A **spectrometer** is a device that forms a spectrum, often utilizing the optical phenomenon of **dispersion.** The light from an astronomical source can consist of a **continuous spectrum,** a bright line or **emission line spectrum,** or a dark line or **absorption line spectrum.** Because each element leaves its spectral "signature" in the pattern of lines we observe, spectral analyses reveal the composition of the Sun and stars.

4.4 Atoms consist of a **nucleus** containing one or more positively charged protons. All atoms except hydrogen also contain one or more neutrons in the nucleus. Negatively charged electrons orbit the nucleus. The number of protons defines an element (hydrogen, helium, and so on) of the atom. Nuclei with the same number of protons but different numbers of neutrons are different **isotopes** of the same element.

4.5 When an electron moves from one **energy level** to another, a photon is emitted and a spectral emission line is formed. Absorption lines are formed when an electron moves to a higher energy level. Since each atom has its own characteristic set of energy levels, each is associated with a unique pattern of spectral lines. An atom in its lowest energy level is in the **ground state.** If an electron is in an orbit other than the least energetic one possible, the atom is said to be **excited.** If an atom has lost one or more electrons, it is called an **ion** and is said to be **ionized.** The spectra of different ions look different and can tell astronomers about the temperatures of the source they are observing.

4.6 If an atom is moving toward us when an electron changes orbits and produces a spectral line, we see that line shifted slightly toward the blue of its normal wavelength in a spectrum. If the atom is moving away, we see the line shifted toward the red. This shift is known as the **Doppler effect** and can be used to measure the **radial velocities** of distant objects by the formula $v = c(\Delta\lambda/\lambda)$.

INTER-ACTIVITY

A Have your group make a list of all the electromagnetic wave technology you use during a typical day.

B How many applications of the Doppler effect can your group think of in everyday life? For example, why would the highway patrol find it useful?

C Have members of your group go home and "read" the face of your radio set and then compare notes. What do all the words and symbols mean? What frequencies can your radio tune in on? What is the frequency of your favorite radio station? What is its wavelength?

D If your instructor gave you a spectrometer, what kind of spectra does your group think you would see from each of the following: (1) a household lightbulb, (2) the Sun, (3) the "neon lights of Broadway," (4) an ordinary household flashlight, and (5) a streetlight on a busy shopping street?

E Suppose astronomers wanted to send a message to an alien civilization that is living on a planet with an atmosphere very similar to that of Earth's. This message must travel through space, make it through the other planet's atmosphere, and be noticeable to the residents of that planet. Have your group discuss what band of the electromagnetic spectrum might be best for this message and why. (Some people, including an earlier Congress, have warned scientists not to send such messages and reveal the presence of our civilization to a possible hostile cosmos. Do you agree with this concern?)

REVIEW QUESTIONS

Ace Astronomy™ Assess your understanding of this chapter's topics with additional quizzing and animations at **http://ace.brookscole.com/voyages**

1. What distinguishes one type of electromagnetic radiation from another? What are the main categories (or bands) of the electromagnetic spectrum?

2. What is a wave? Use the terms *wavelength* and *frequency* in your definition.

3. Is your textbook the kind of idealized object (described in Section 4.2.3) that absorbs all the radiation falling on it? Explain. How about the black sweater worn by one of your classmates?

4. Where in an atom would you expect to find electrons? Protons? Neutrons?

5. Explain how emission lines and absorption lines are formed. In what sorts of cosmic objects would you expect to see each?

6. Explain how the Doppler effect works for sound waves, and give some familiar examples.

7. What kind of motion for a star does not produce a Doppler effect? Explain why.

8. Describe how Bohr's model used the work of Rutherford and Maxwell. Why was Bohr's model considered a radical notion?

THOUGHT QUESTIONS

9. Make a list of some of the many practical consequences of Maxwell's theory of electromagnetic waves (television is one example).

10. Explain how you would calculate the total amount of power radiated into space by the Sun. What information about the Sun would you need to make this calculation?

11. What type of electromagnetic radiation is best suited to observing a star with:
a. a temperature of 5800 K
b. a gas heated to a temperature of 1 million K
c. a person on a dark night

12. Why is it dangerous to be exposed to x rays but not (or at least much less) dangerous to be exposed to radio waves?

13. Go outside on a clear night and look carefully at the brightest stars. Some should look slightly red and others slightly blue. The primary factor that determines the color of a star is its temperature. Which is hotter: a blue star or a red one? Explain.

14. Water faucets are often labeled with a red dot for hot water and a blue dot for cold. Given Wien's law, does this labeling make sense?

15. The planet Jupiter appears yellow and Mars is red. Does this mean that Mars is cooler than Jupiter? Explain your answer.

16. Suppose you are standing at the exact center of a park surrounded by a circular road. An ambulance drives completely around this road, with siren blaring. How does the pitch of the siren change as it circles around you?

17. How could you measure Earth's orbital speed by photographing the spectrum of a star at various times throughout the year? (*Hint:* Suppose the star lies in the plane of Earth's orbit.)

FIGURING FOR YOURSELF

The formula on page 88 for the relationship between the speed and other characteristics of a wave can be derived from our basic understanding of motion. The speed of anything that is moving is

$$\text{speed} = \frac{\text{distance traveled}}{\text{time taken}}$$

(So, for example, a car on the highway traveling at a speed of 100 km/h covers 100 km during the time of 1 h.) For an electromagnetic wave to travel the distance of one of its wavelengths, λ, at the speed of light, c, we have $c = \lambda/t$. The frequency of a wave is the number of cycles per second. If a wave has a frequency of a million cycles per second, then the time for each cycle to go by is a millionth of a second. So $t = 1/f$. Substituting into our wave equation, we get $c = \lambda \times f$.

18. What is the wavelength of the carrier wave of a campus radio station, broadcasting at a frequency of 97.2 Mhz (million cps)?

19. What is the frequency of a red laser beam, with a wavelength of 670 nm, that your astronomy instructor might use to point to slides during a lecture on galaxies?

20. You go to a dance club to forget how hard your astronomy midterm was. What is the frequency of a wave of ultraviolet light coming from a blacklight in the club, if its wavelength is 150 nm?

21. What is the energy of the photon with the frequency you calculated in problem 20?

22. "Tidal waves," or tsunamis, are waves caused by earthquakes that travel rapidly through the ocean. If tsunamis travel at the speed of 600 km/h and approach a shore at the rate of one wave crest every 15 min, what would be the distance between those wave crests at sea?

23. How many times brighter or fainter would a star appear if it were moved to:
a. twice its present distance
b. ten times its present distance
c. half its present distance

24. Two stars with identical diameters are the same distance away. One has a temperature of 5800 K; the other has a temperature of 2900 K. Which is brighter? How much brighter is it?

25. If the emitted infrared radiation from Pluto has a wavelength of maximum intensity at 50,000 nm (50 μm), what is the temperature of Pluto (assuming it follows Wien's law)?

26. What is the temperature of a star whose maximum light is emitted at a wavelength of 290 nm?

27. Suppose a spectral line of some element, normally at 500 nm, is observed in the spectrum of a star to be at 500.1 nm. How fast is the star moving toward or away from the Earth?

SUGGESTIONS FOR FURTHER READING

Augensen, H. and Woodbury, J. "The Electromagnetic Spectrum" in *Astronomy*, June 1982, p. 6.

Bova, B. *The Beauty of Light*. 1988, Wiley. A readable introduction to all aspects of the production and decoding of light by a science writer.

Connes, P. "How Light Is Analyzed" in *Scientific American*, Sept. 1968.

Darling, D. "Spectral Visions: The Long Wavelengths" in *Astronomy*, Aug. 1984, p. 16; "The Short Wavelengths" in *Astronomy*, Sept. 1984, p. 14.

Gingerich, O. "Unlocking the Chemical Secrets of the Cosmos" in *Sky & Telescope*, July 1981, p. 13.

Gribbin, J. *In Search of Schroedinger's Cat*. 1984, Bantam; *Schroedinger's Kittens and the Search for Reality*. 1995, Little, Brown. Clear, basic introductions to the fundamental ideas of quantum mechanics by a British physicist/science writer.

Hearnshaw, J. *The Analysis of Starlight*. 1986, Cambridge U. Press. A history of spectroscopy.

Sobel, M. *Light*. 1987, U. of Chicago Press. An excellent nontechnical introduction to all aspects of light.

Stencil, R. et al. "Astronomical Spectroscopy" in *Astronomy*, June 1978, p. 6.

NASA

The Hubble Space Telescope (HST)   Astronauts Story Musgrave and Jeffrey Hoffman are seen refurbishing the orbiting telescope in space. Australia's west coast can be seen in the background.

5 Astronomical Instruments

Beyond the [faintest stars the eye can see] you will behold through the telescope a host of other stars, which escape the unassisted sight, so numerous as to be almost beyond belief . . .

Galileo Galilei in *Siderius Nuncius*, 1610, reporting on his first observations of the night sky with a telescope

THINKING AHEAD

When you look at the night sky far from city lights on a camping trip, there seem to be an overwhelming number of stars up there. In reality, only about 6000 stars are visible to the unaided eye. The light from most stars is so weak by the time it reaches Earth that it cannot register on human vision. How can we learn about the vast majority of objects in the universe that our unaided eyes simply cannot see?

In this chapter we describe the tools astronomers use to extend their vision. We have learned almost everything we know about the universe from studying electromagnetic radiation (see Chapter 4). In the 20th century, access to space made it possible to detect electromagnetic radiation at all wavelengths from gamma rays to radio waves. The different wavelengths carry different kinds of information, and the appearance of any given object often depends on the wavelength at which the observations are made (Figure 5.1).

There are three basic components in a modern system for measuring radiation from astronomical sources. First, there is a **telescope,** which serves as a "bucket" for collecting visible light (or radiation at other wavelengths). Just as you can catch more rain with a garbage can than with a coffee cup, large telescopes gather much more light than your eye can. Second, there is an instrument attached to the telescope that sorts the incoming radiation according to wavelength. Sometimes the sorting is fairly crude. For example, we might simply want to measure blue light separately from red light so that we

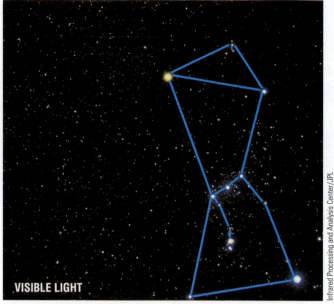

VISIBLE LIGHT

(a)

Infrared Processing and Analysis Center/JPL

INFRARED

(c)

Infrared Processing and Analysis Center/JPL

(b)

Max Planck Institut für Extraterrestriche Physik

■ **FIGURE 5.1**

**The Orion Region of the Sky at Different
Wavelengths**   The same part of the sky looks different
when observed with instruments sensitive to different bands
of the spectrum. (a) Visible light. This shows part of the Orion
region as the human eye sees it. Note the stars that form the
constellation of Orion, the hunter. (b) X rays. Here the view
emphasizes the pointlike x-ray sources in the vicinity. The
colors change from yellow to white to blue with increasing
energy of the x rays. The bright, hot stars in Orion are still
seen in this image, but so are many other objects located at
very different distances, including other stars, white dwarfs,
galaxies, and quasars at the edge of the observable
universe. (The belt and sword of Orion are shown for
reference.) *(ROSAT)* (c) Infrared radiation. Here we see the
glowing dust in this region.

can determine the color of a star and estimate its temperature. But at other times we want to look at very narrow
wavelength regions so that we can see individual spectral
lines (see Chapter 4) and determine what an object is made
of. Third, there is some kind of **detector,** a device that can
sense the radiation in the wavelength regions we have chosen and make a permanent record of the observations.

The history of the development of astronomical telescopes is really just the story of how new technologies have
been applied to improve the efficiency of each of these three
basic components: the telescopes, the wavelength-sorting
devices, and the detectors. Let's look at each in turn.

5.1   TELESCOPES

Many ancient cultures constructed special sites for observing the sky (Figure 5.2). At these ancient *observatories* they
could measure the positions of celestial objects, mostly to
keep track of time and date. Many of these ancient observatories had religious and ritual functions as well. The eye

(a)

(b)

■ **FIGURE 5.2**

Two Pretelescopic Observatories   (a) The Jantar Mantar, built in 1724 by Maharaja Jai Singh in Delhi, India. (b) Seventeenth-century bronze instruments from the old Chinese imperial observatory, Beijing.

was the only device available to gather light, all of the colors in the light were observed at once, and the only permanent record of the observations was made by human beings writing down or sketching what they saw.

In 1610 Galileo first used a simple tube with lenses that he could hold in his hand to observe the sky and gather more light than his eyes alone could. Even this small telescope revolutionized ideas about the nature of the planets and the position of the Earth, and it got Galileo into a lot of trouble with church authorities (as we explained in Chapter 1).

5.1.1   How Telescopes Work

Telescopes have come a long way since Galileo's time. Now they tend to be huge devices; the most expensive cost hundreds of millions of dollars. The reason astronomers keep building bigger and bigger telescopes is that celestial objects, such as planets, stars, and galaxies, send much more light to Earth than any human eye (with its tiny opening) can catch. There is plenty of starlight to go around: If you have ever watched the stars with a group of friends, you know that each of you can see each of the stars. If a thousand more people were watching, each of them would also catch a bit of each star's light. Yet, as far as you are concerned, the light not shining into your eye is wasted. It would be great if some of this "wasted" light could also be captured and brought to your eye. This is precisely what a telescope does.

The most important functions of a telescope are (1) to *collect* the faint light from an astronomical source and (2) to *focus* all the light gathered up into an image. Most objects of interest to astronomers are extremely faint; the

more light we can collect, the better we can study such objects. (Although we use the word *light,* some types of telescopes collect not visible light, but each of the various forms of electromagnetic radiation. We want no light-chauvinists among our readers!)

Telescopes that collect visible radiation use a lens or mirror (which we will discuss in a moment) to gather up the light. Other types of telescopes may use collecting devices that look very different from the lenses and mirrors that we are familiar with, but they serve the same function. In all types of telescopes, the light-gathering ability is determined by the area of the device acting as the light-gathering bucket. Since most telescopes have round mirrors or lenses, we can compare their light-gathering power by comparing their **apertures** (or diameters).

The amount of light a telescope can collect increases with the square of the aperture. A telescope with a mirror that is 4 meters (m) in diameter can collect 16 times as much light as a telescope that is 1 m in diameter. (The diameter is squared because the area of a circle equals $\pi d^2/4$, where d is the diameter of the circle.)

After the telescope forms an image, we need to have some way to detect and record it so that we can measure, reproduce, and analyze the image in a host of ways. Before the 19th century, astronomers simply viewed images with their eyes and wrote descriptions of what they saw. This was very inefficient and did not lead to a very reliable long-term record; eyewitness testimony is notoriously inaccurate!

In the 19th century, the use of photography became widespread. In those days, photographs were a chemical record of an image on a specially treated glass plate. Today the image is generally detected with sensors similar to those in television cameras, recorded electronically, and stored in computers. This permanent record can then be used for detailed studies. Professional astronomers rarely look through the large telescopes that they use for their research.

5.1.2 Formation of an Image by a Lens or a Mirror

Whether or not you wear glasses, you see the world through lenses; they are key elements of your eyes. A lens is a transparent piece of material that bends the rays of light passing through it. If the light rays are parallel, the lens brings them together in one place to form an image (Figure 5.3). If the curvatures of the lens surfaces are just right, all parallel rays of light (say, from a star) are bent, or *refracted*, in such a way that they converge toward a point, called the **focus** of the lens. At the focus, an image of the light source appears. In the case of parallel light rays, the distance of the focus, or image, behind the lens is called the *focal length* of the lens.

As you look at Figure 5.3, you may ask why two rays of light from the same star would be parallel to each other. After all, if you draw a picture of a star shining in all directions, the rays of light coming from the star don't look parallel at all. But remember that the stars (and other astronomical objects) are all outrageously far away. By the time the rays of light headed our way actually arrive at Earth, they are—for all practical purposes—parallel to each other. Put another way, any rays that were *not* more or less parallel to the ones pointed at Earth are now heading in some very different direction in the universe.

To view the image formed by the primary lens in a telescope, we can install another lens called an *eyepiece* (Fig-

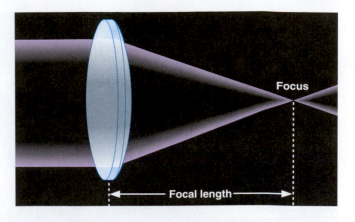

■ FIGURE 5.3

Formation of an Image by a Simple Lens Parallel rays from a distant source are bent by the convex lens so that they all come together in a single place (the focus) to form an image.

Ace Astronomy™ Log into AceAstronomy and select this chapter to see Astronomy Exercises "Snell's Law," "Lenses: Focal Length," and "Telescopes: Objective Lens & Eyepiece."

ure 5.4). This lens can magnify the image. Stars are points of light and magnifying them makes little difference, but the image of a planet or galaxy, which has structure, can often benefit from magnification, just as a tiny image in a magazine can sometimes look clearer through a magnifying glass. Using different eyepieces, we can change the magnification of the image.

A telescope can also be built by using a concave *mirror*—one curved like the inner surface of a sphere—to form an image (see Figure 5.4). Telescope mirrors are coated with a shiny metal, usually silver or aluminum or oc-

Ace Astronomy™ Log into AceAstronomy and select this chapter to see the Active Figure called "Different Telescopes."

(a) Refractor

(b) Reflector

■ FIGURE 5.4

Refracting and Reflecting Telescopes (a) Light enters a *refracting* telescope through a lens at the upper end, which focuses the light near the bottom of the telescope. An eyepiece then magnifies the image so that it can be viewed by the eye, or a detector like a photographic plate can be placed at the focus. (b) The upper end of a *reflecting* telescope is open, and the light passes through to the primary mirror, located at the bottom of the telescope. The mirror then focuses the light at the top end, where it can be detected. Alternatively, as in this sketch, a second mirror may reflect the light to a position outside the telescope structure, where an observer can have easier access to it.

casionally gold, to make them highly reflecting. If the mirror has the correct shape, all parallel rays are reflected back to the same point, the focus of the mirror. Thus, images are produced by a mirror exactly as they are by a lens.

Many people, when thinking of a telescope, picture a long tube with a large glass lens at one end. This design is called a **refracting telescope.** Galileo's telescopes were refractors, as are binoculars and opera glasses today (Figure 5.5). There is, however, a limit to the size of refracting telescopes. The largest one ever built is the 40-in. refractor at Yerkes Observatory in Wisconsin. The main problem is that the light must pass *through* the lens of a refractor. That means the glass must be perfect all the way through, but it has proven to be impossible to make large pieces of glass without flaws and bubbles in them.

Also, since the light must pass through the lens, it can be supported only around the edges. The force of gravity will cause a large lens to sag and distort the path of the light rays as they pass through it. Finally, again because the light passes through it, both sides of the lens must be manufac-

tured to precisely the right shape in order to produce sharp images.

Telescopes designed with mirrors avoid all of these problems. Because the light is reflected from the front surface only, flaws and bubbles within the glass do not affect the path of the light. Only the front surface has to be manufactured to a precise shape, and the mirror can be supported from the back. For these reasons, most astronomical telescopes (both amateur and professional) use a mirror rather than a lens as their primary optical part; these are called **reflecting telescopes** (see Figure 5.4).

The first successful reflecting telescope was built by Newton in 1668. The concave mirror is placed at the bottom of a tube or open framework. The mirror reflects the light back up the tube to form an image near the front end at a location called the *prime focus.* The image can be observed at the prime focus, or additional mirrors can intercept the light and redirect it to a position where the observer can get at it easily (Figure 5.6). Since an astronomer at the prime focus can block quite a bit of the light coming to the main mirror, the use of a small *secondary mirror* allows more light to get through the system.

Eyepiece

Prism

Objective

Light path in binoculars

■ **FIGURE 5.5**

Binoculars Binoculars are common examples of refracting telescopes. The main light-collecting element in each side is the initial (objective) lens. After following a path through several prisms (used to bend the light and shorten the length of the instrument), the light is viewed through a magnifying eyepiece. Binoculars make the image both brighter and larger, helping you see what is happening on stage or on the field—even if you can't afford really good tickets.

(a) Prime focus   (b) Newtonian focus   (c) Cassegrain focus

■ **FIGURE 5.6**

Focus Arrangements for Reflecting Telescopes
Standard reflecting telescopes have three options for where the light is brought to a focus. (a) Prime focus: Light is detected where it comes to a focus after reflecting from the primary mirror. (b) Newtonian focus: Light is reflected by a small secondary mirror off to one side, where it can be detected (see also Figure 5.4b). (c) Cassegrain focus: Light is reflected by a small secondary mirror back down through a hole in the primary mirror and detected at a station below.

5.2 TELESCOPES TODAY

Since the time of Newton, when the sizes of the mirrors in telescopes were measured in inches, reflecting telescopes have grown ever larger. In 1948, U.S. astronomers built a telescope with a 5-m (200-in.) diameter mirror on Palomar Mountain in southern California. It remained the largest visible-light telescope in the world for several decades. The giants of today, however, have primary mirrors (i.e., the largest mirrors in the telescope) that are 8 to 10 m in diameter (Figure 5.7).

5.2.1 Modern Visible-Light and Infrared Telescopes

The decade starting in 1990 saw telescope building around the globe at an unprecedented rate (see Table 5.1, which also includes Web sites for each telescope in case you want to visit or learn more about them). Breakthroughs in technology made it possible at long last to build telescopes significantly larger than the 5-m telescope at Palomar for a reasonable cost. All of these new telescopes have been designed to work well in the infrared as well as visible wavelengths.

The differences between the Palomar telescope and the modern Gemini North telescope are easily seen in Figure 5.8. The Palomar telescope is a massive steel structure designed to hold the 14.5-ton primary mirror. Glass tends to sag under its own weight, and a sagging telescope mirror will distort the image. The Palomar telescope was designed to minimize the sagging with a mirror so thick that it is stiff and rigid. Thick also means heavy, however, and hence a huge steel structure is needed to hold the mirror. A mirror 8 m in diameter built using the same technology would have to weigh at least eight times as much and would require an enormous steel structure to support it.

The 8-m Gemini North telescope looks like a featherweight by contrast, and indeed it is. The mirror is only about 8 in. thick and weighs 24.5 tons, less than twice as much as the Palomar mirror. This mirror does sag, but with modern computers it is possible to measure that sag many times each second and apply forces at 120 different locations on the back of the mirror to correct the sag. This *active control* of the mirror shape makes it possible to build huge lightweight telescopes; 17 telescopes with mirrors 6.5 m in diameter and larger have been constructed since 1990.

The twin 10-m Keck telescopes on Mauna Kea, which were the first of these new-technology instruments, use precision control in an entirely novel way. Instead of a single primary mirror 10 m in diameter, each Keck telescope achieves its large aperture by combining the light from 36 separate hexagonal mirrors, each 1.8 m wide (Figure 5.9). Computer-controlled actuators constantly adjust these 36 mirrors so that the overall reflecting surface acts like a single mirror with just the right shape to collect and focus the light into a sharp image.

In addition to holding the mirror, the steel structure of a telescope is designed so that the entire telescope can be pointed quickly toward any object in the sky. Since the Earth is rotating, the telescope must have a motorized drive system that moves it backward very smoothly at

Gemini Observatory

■ **FIGURE 5.7**

A Large Telescope Mirror This image shows the primary mirror of the Gemini North telescope just after it was coated with aluminum. The mirror is 8 m in diameter; note the person in the center hole in the mirror.

TABLE 5.1 *Large Optical Telescopes Being Built or in Operation*

Aperture (m)	Telescope Name	Location	Status	Web Address
16.4	Very Large Telescope (four 8.2-m telescopes)	Cerro Paranal, Chile°	Four telescopes complete	www.eso.org/vlt/
11.8	Large Binocular Telescope (two 8.4-m telescopes)	Mount Graham, Arizona	First light 2004	medusa.as.arizona.edu/lbtwww/lbt.html
10.0	Keck I and II (two 10-m telescopes)	Mauna Kea, Hawaii	Completed 1993–96	www2.keck.hawaii.edu
10.4	Gran Telescopio Canarias (GTC)	La Palma, Canary Islands	First light 2005 (planned)	www.gtc.iac.es/home.html
9.1	Hobby–Eberly (HET)	Mount Locke, Texas	Completed 1997	www.as.utexas.edu/mcdonald/het/het.html
9.1	Southern African Large Telescope	Sutherland, South Africa	First light 2004	www.salt.ac.za
8.3	Subaru (Pleiades)	Mauna Kea, Hawaii	First light 1998	www.naoj.org/
8.0	Gemini (North)	Mauna Kea, Hawaii [†]	First light 1999	www.gemini.edu
8.0	Gemini (South)	Cerro Pachon, Chile [†]	First light 2000	www.gemini.edu
6.5	Multi-Mirror (MMT)	Mount Hopkins, Arizona	First light 1998	sculptor.as.arizona.edu/foltz/www/mmt.html
6.5	Magellan (twin telescopes)	Las Campanas, Chile	First light 1997 and 2002	www.ociw.edu/magellan_lco/
6.0	Bolshoi Teleskop	Mount Pastukhov, Russia	Completed 1976	—
5.0	Hale Telescope	Mount Palomar, California	Completed 1948	www.astro.caltech.edu/palomar/
4.2	Herschel Telescope	Canary Islands, Spain	Completed 1987	www.ing.iac.es/Astronomy/telescopes/wht/
4.2	SOAR	Cerro Pachon, Chile	First light 2004	www.soartelescope.org
4.0	Blanco Telescope	Cerro Tololo, Chile [†]	Completed 1974	www.ctio.noao.edu/telescopes/4m/base4m.html
4.0	Visible and Infrared Survey Telescope (Vista)	Cerro Paranal, Chile	First light 2006 (planned)	www.vista.ac.uk/
3.9	Anglo-Australian (AAT)	Siding Spring, Australia	Completed 1975	www.aao.gov.au/index.html
3.8	Mayall Telescope	Kitt Peak, Arizona [†]	Completed 1973	www.noao.edu/kpno
3.8	United Kingdom Infrared	Mauna Kea, Hawaii	Completed 1979	www.jach.hawaii.edu/JACpublic/UKIRT/home.html
3.7	Advanced Electro-Optical System Telescope	Haleakala, Hawaii	Completed 2000	—
3.6	Telescopio Nazionale Galileo	La Palma, Canaries	Completed 1998	www.tng.iac.es
3.6	Canada–France–Hawaii	Mauna Kea, Hawaii	Completed 1979	www.cfht.hawaii.edu/
3.6	ESO Telescope	Cerro La Silla, Chile°	Completed 1976	www.ls.eso.org/
3.6	ESO New Technology	Cerro La Silla, Chile°	Completed 1989	www.ls.eso.org/
3.5	Max Planck Institut	Calar Alto, Spain	Completed 1983	www.mpia-hd.mpg.de/Public/CAHA/index.html
3.5	WIYN	Kitt Peak, Arizona [†]	Completed 1993	www.noao.edu/wiyn/wiyn.html
3.5	Astrophysical Research Corporation	Apache Point, New Mexico	Completed 1993	www.apo.nmsu.edu/
3.5	Starfire Optical Range	Kirtland AFB, New Mexico	Completed 1994	www.de.afrl.af.mil/SOR/

° Part of the European Southern Observatory (ESO).
† Part of the U.S. National Optical Astronomy Observatories (NOAO).

(a)

California Institute of Technology

■ **FIGURE 5.8**

Modern Reflecting Telescopes (a) The Palomar 5-m reflector. The Hale telescope on Palomar Mountain has a complex mounting structure that enables the telescope (in the open "tube" pointing upward in this photo) to swing easily into any position. The people in the bottom foreground give you a sense of scale. (b) The Gemini North 8-m telescope. The Gemini North telescope has a mirror twice the diameter of the Palomar mirror, but note how much less massive the whole instrument seems. The Gemini North telescope was completed about 50 years after the Palomar telescope. Engineers took advantage of new technologies to build a telescope that is much lighter in weight relative to the size of the primary mirror.

(b)

Gemini Observatory

California Association for Research in Astronomy

■ **FIGURE 5.9**

Thirty-Six Eyes Are Better Than One A close-up of the 10-m Keck telescope's mirror as it was being assembled. The finished mirror is composed of 36 hexagonal sections. In this view, 18 sections have been installed. To get a sense of the mirror's size, note the person working at the center of the mirror assembly.

exactly the same rate the Earth is moving forward, so it can continue to point at the object being observed. All this machinery must be housed in a dome to protect it from the elements. The dome has an opening in it that can be positioned in front of the telescope and moved along with it so that the light from the objects being observed is not blocked.

5.2.2 Picking the Best Observing Sites

A telescope like the Gemini or Keck telescope costs about $100 million to build. That kind of investment simply demands that the telescope be placed in the best possible site. Since the end of the 19th century, astronomers have realized that the best observatory sites are on mountains far from the lights and pollution of cities. Although a number of urban observatories remain, especially in the large cities of Europe, they have become administrative centers or museums. The real action takes place far away, often on desert mountains or isolated peaks in the Atlantic and Pacific Oceans, where we find the living quarters, computers, electronic and machine shops, and of course the telescopes themselves. A large observatory today requires a supporting staff of 20 to 100 people in addition to the astronomers.

The performance of a telescope is determined not only by the size of its mirror but also by its location. The Earth's atmosphere, so vital to life, is a big headache for the observational astronomer. In at least four ways our air imposes limitations on the usefulness of telescopes:

1. The most obvious limitation is weather—clouds, wind, rain, and the like. At the best sites the weather is clear as much as 75 percent of the time.

2. Even on a clear night, the atmosphere filters out a certain amount of starlight, especially in the infrared, where the absorption is due primarily to water vapor. Astronomers therefore prefer dry sites, generally found at the highest altitudes.

3. The sky should also be dark. Near cities the air scatters the glare from lights, producing an illumination that hides the faintest stars and limits the distances that can be probed by telescopes. (Astronomers call this effect *light pollution.*) Observatories are best located at least a hundred miles from the nearest large city.

4. Finally, the air is often unsteady; light passing through this turbulent air is disturbed, resulting in blurred star images. Astronomers call these effects "bad **seeing.**" When seeing is bad, images of celestial objects are distorted by the constant twisting and bending of light rays by turbulent air.

Ace🌀Astronomy™ Log into AceAstronomy and select this chapter to see Astronomy Exercises "Particulate," "Heat," and "Light Pollution."

The best observatory sites are therefore high, dark, and dry. The world's largest telescopes are found in such remote mountain locations as the Andes Mountains of Chile (Figure 5.10), the desert peaks of Arizona, the Canary Islands in the Atlantic Ocean, and Mauna Kea in Hawaii, a dormant volcano 13,700 ft (4200 m) high.

5.2.3 The Resolution of a Telescope

In addition to gathering as much light as they can, astronomers also want to have the sharpest images possible. **Resolution** refers to the fineness of detail present in the image. As you can imagine, astronomers are always eager to make out more detail in the images they study, whether they are following the weather on Jupiter or trying to peer into the violent heart of a cannibal galaxy that recently ate its neighbor for lunch.

One factor that determines how good the resolution will be is the size of the telescope. Larger apertures produce sharper images. Until very recently, however, optical and infrared telescopes on the Earth's surface never produced images as sharp as optics theory said they should. The problem—as we saw above—is the atmosphere. The light from an astronomical source has to move through many blobs or cells of gas, ranging from inches to several feet in size, each having a slightly different temperature from its neighbor. Each cell acts like a lens, bending (refracting) the path of the light by a small amount. This slightly changes the position where each light ray finally reaches the detector in a telescope. The cells of air are in motion, constantly be-

■ **FIGURE 5.10**

A High and Dry Site Cerro Paranal, a mountain summit 2.7 km above sea level in Chile's Atacama Desert, is the site of the European Southern Observatory's Very Large Telescope. This photograph shows the four 8-m telescope buildings on the site and vividly illustrates that astronomers prefer high and dry sites for their instruments.

Ace🌀Astronomy™ Log into AceAstronomy and select this chapter to see the Active Figure called "Resolution and Telescopes," and Astronomy Exercises "Telescopes and Resolution I" and "Telescopes and Resolution II."

ing blown through the light path of the telescope by winds, often in different directions at different altitudes. As a result, the path followed by the light is constantly changing.

For an analogy, think about watching a ticker tape parade from a window high up in a skyscraper. You decide to throw some confetti down toward the marchers. Even if you drop a handful all at the same time and in the same direction, air currents will toss the pieces around and they will reach the ground in different places. As we described earlier, we can think of the light from the star as a series of parallel beams, each making its way through the atmosphere. Each path will be slightly different, and each will reach the detector of the telescope in a slightly different place. The result is a blurred image, and because the cells are being blown by the wind, the nature of the blur will change many times each second. (You have probably noticed this effect as the "twinkling" of stars seen from Earth. The light beams are bent enough that part of the time they reach your eye, and part of the time some of them miss, thereby making the star seem to vary in brightness. In space, the light of the stars is steady.)

Astronomers scour the world for locations where this atmospheric blurring is as small as possible. It turns out the best sites are in coastal mountain ranges and on isolated volcanic peaks in the middle of an ocean. Air that has flowed long distances over water before it encounters land is especially stable.

The resolution of an image is measured in units of angle on the sky, typically in units of *arcseconds*. One arcsecond is 1/3600 degree, and there are 360 degrees in a full circle. So we are talking about tiny angles on the sky. To give you a sense of just how tiny, we might note that 1 arcsecond is how big a quarter would look when seen from a distance of *5 kilometers*. The best images obtained from the ground with traditional techniques reveal details as small as several tenths of an arcsecond across. This image size is remarkably good. Astronomers are never satisfied, however, and one of the major reasons for launching the Hubble Space Telescope was to escape the Earth's atmosphere and obtain even sharper images.

Recently, thanks to fast computers, astronomers have devised a technique called **adaptive optics** that can beat the atmosphere at its own game of blurring. This technique (which is most effective in the infrared region of the spectrum with our current technology) makes use of a small flexible mirror placed in the beam of a telescope. A sensor measures how much the atmosphere has distorted the image, and as often as 500 times each second sends instructions to the flexible mirror on how it should change shape in order to compensate exactly for the distortions produced by the atmosphere. The light is thus brought back to a nearly perfectly sharp focus at the detector. Figure 5.11

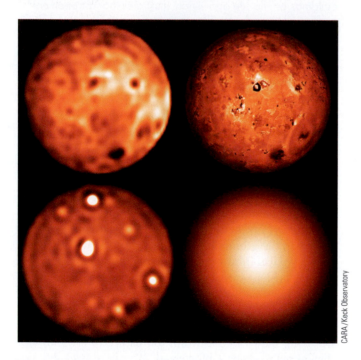

CARA/Keck Observatory

■ **FIGURE 5.11**

The Power of Adaptive Optics The lower right image shows Jupiter's satellite Io as it appears through the 10-m Keck telescope without adaptive optics; the image is blurred by the Earth's atmosphere. For comparison, the upper right picture was taken by the Galileo spacecraft when it was passing close to the surface of Io. The two left-hand images were taken in the infrared with the Keck telescope plus adaptive optics, which makes corrections for the blurring by the Earth's atmosphere. The top one looks very much like the image taken by Galileo. The lower was taken at a wavelength that is very sensitive to heat, and we see the hot lava erupted by Io's volcanoes.

shows just how effective this new technique is. With adaptive optics, ground-based telescopes can achieve resolutions of 0.1 arcsecond or a little better in the infrared. This impressive figure is the equivalent of the resolution that the Hubble Space Telescope achieves in the visible-light region of the spectrum.

ASTRONOMY BASICS
How Astronomers Really Use Telescopes

In the popular view, an astronomer spends most nights in a cold observatory peering through a telescope, but this picture, which may have had some validity in previous centuries, is not very accurate today. Most astronomers do not live at observatories, but near the universities or laboratories where they work. A typical astronomer might spend only a week or so each year observing at the telescope and the rest of the time measuring or analyzing the data acquired during that week. Many astronomers use radio telescopes or space experiments, which work just as well during the day. Still others work at purely theoretical problems (often using high-speed computers) and never observe at a telescope of any kind.

Even when astronomers are observing with a large telescope, they seldom peer through it. Electronic detectors record the data permanently for detailed analysis after the observations are completed. At some observatories, it is now possible to conduct observations remotely, with the astronomer sitting at an office computer terminal, which can be thousands of miles away from the telescope.

Time on the major telescopes is at a premium, and an observatory director will typically receive many more requests for telescope time than can be accommodated during the year. Astronomers must therefore write a convincing proposal explaining how they would like to use the telescope and why their observations will be important to the progress of astronomy. A committee of astronomers is then asked to judge and rank the proposals, and time is assigned to those with the greatest merit. Even if your proposal is among the high-rated ones, you may have to wait many months for your turn. If the skies are cloudy on the nights you have been assigned, it may be more than a year before you get another chance.

Some older astronomers still remember long, cold nights spent alone in an observatory dome, with only music from a tape recorder or all-night radio station for company. The sight of the stars shining brilliantly hour after hour through the open slit in the observatory dome was unforgettable. So was the relief as the first pale light of dawn announced the end of a 12-hour observing session. Astronomy is much easier today, with teams of observers working together, perhaps without ever leaving a computer workstation in a warm room. But some of the romance has gone from the field, too.

5.3 VISIBLE-LIGHT DETECTORS AND INSTRUMENTS

After a telescope captures its share of the radiation from an astronomical source, the radiation must be *detected* and measured. The first detector used for astronomical observations was the human eye, but it suffers from being connected to an imperfect recording and retrieving device, the human brain. Photography and modern electronic detectors have eliminated the foibles and quirks of human memory by making a permanent record of the information from the cosmos.

The eye also suffers from having a very short *integration time*; it takes only a fraction of a second to add light energy together before sending the image to the brain. One important advantage of modern detectors is that the light from astronomical objects can be collected by the detector over longer periods of time; this technique is called "taking a long exposure." Exposures of several hours are required to detect very faint objects in the cosmos.

Before the light reaches the detector, astronomers normally use some instrument to sort the light according to wavelength. The instrument may be as simple as colored filters, which transmit light within a specified range of wavelengths. Red cellophane is an everyday example of a filter that transmits only red light. After the light passes through the filter, it forms an image, which astronomers can then use to measure the *brightness* and *color* of objects. We will show you many examples of such images in the later chapters of this book, and we will describe what can be learned from them.

Alternatively, the instrument may be a complicated set of optics that spread the light out into its rainbow of colors so that the astronomer can measure individual lines in the spectrum. Such an instrument is called a *spectrometer*, since it allows astronomers to measure (meter) the spectrum of a source of radiation. Both types of wavelength-sorting instruments still have to use detectors to record and measure the properties of light.

5.3.1 Photographic and Electronic Detectors

Throughout most of the 20th century, photographic film or *plates* served as the prime astronomical detectors, whether for direct imaging or for photographing spectra. In a plate, a light-sensitive chemical coating is applied to a piece of glass that, when developed, provides a lasting record of the image. At observatories around the world, vast collections of photographs preserve what the sky has looked like during the past 100 years. Photography represents a huge improvement over the human eye, but it still has serious limitations. Photographic films are inefficient: Only about 1 percent of the light that actually falls on the film contributes to the image; the rest is wasted.

Astronomers today are delighted to have much more efficient electronic detectors to record astronomical images. Most often, these are **charge-coupled devices (CCDs),** which are similar to the detectors used in video camcorders or in digital cameras. In a CCD, photons of radiation generate a stream of charged particles (electrons) that are stored on the detector and counted at the end of the exposure. Because CCDs record as much as 60 to 70 percent of all the photons that strike them, we can detect much fainter objects. Among these are a host of small moons around the outer planets, icy miniplanets beyond Pluto, and dwarf galaxies of stars. CCDs also provide more accurate measurements of the brightness of astronomical objects than photography, and their output can go directly into a computer for analysis.

5.3.2 Infrared Observations

Observing the universe in the infrared band of the spectrum presents some additional challenges. The infrared extends from wavelengths near 1 micrometer (μm), which is about the longwave sensitivity limit of both CCDs and photography, out to 100 μm or longer. Recall from Chapter 4 that infrared radiation is heat radiation. The main challenge to infrared astronomers is to distinguish the tiny amount of heat that reaches the Earth from stars and galaxies from the much greater heat radiated by the telescope itself and our planet's atmosphere.

Typical temperatures on the Earth's surface are near 300 K, and the atmosphere through which observations are made is only a little cooler. According to Wien's law (see Section 4.2), the telescope, the observatory, and even the sky are radiating infrared energy with a peak wavelength of about 10 μm. To infrared eyes, everything on Earth is brightly aglow. The challenge is to detect faint cosmic sources against this sea of light. The infrared astronomer must always contend with the situation that a visible-light observer would face if working in broad daylight with a telescope and optics lined with bright fluorescent lights.

To solve this problem, astronomers must protect the infrared detector from nearby radiation, just as you would shield photographic film from bright daylight. Since anything warm radiates infrared energy, the detector must be isolated in very cold surroundings; often it is held near absolute zero (1 to 3 K) by immersing it in liquid helium. The second step is to reduce the radiation emitted by the telescope structure and optics, and to block this heat from reaching the infrared detector.

Like the infrared, each band of the electromagnetic spectrum presents its own challenges to the astronomer. It is a challenge, for example, to design "mirrors" to reflect such penetrating radiation as x rays and gamma rays, which normally pass straight through matter. However, although the technical details of design are complicated, the three basic components of an observing system are the same at all wavelengths: a telescope to gather up the radiation, filters or an instrument to sort the radiation according to wavelength, and some method for detecting and making a permanent record of the observations.

5.3.3 Spectroscopy

Spectroscopy is one of the astronomer's most powerful tools, providing information about the composition, temperature, motion, and other characteristics of celestial objects. More than half the time spent on most large telescopes is used for spectroscopy.

The many different wavelengths present in light can be separated by passing it through a spectrometer to form a spectrum. The design of a simple spectrometer is illustrated in Figure 5.12. Light from the source (actually, the image of a source produced by the telescope) enters the instrument through a small hole or narrow slit and is collimated (made into a beam of parallel rays) by a lens. The light then passes through a prism, producing a spectrum: Different wavelengths leave the prism in different directions because each wavelength is bent by a different amount when it enters and leaves the prism. A second lens placed behind the prism focuses the many different images of the slit or entrance hole onto a CCD or other detecting device. This collection of images (spread out by color) is the spectrum that astronomers can then analyze at leisure.

In practice, astronomers today are more likely to use a different device, called a *grating*, to disperse the spectrum. A grating is a piece of material with thousands of grooves in its surface. Like a prism, a grating also spreads light out into a spectrum.

In addition to visible and infrared radiation, radio waves from astronomical objects can also be detected from the surface of the Earth (see Chapter 4). In the early 1930s Karl G. Jansky, an engineer at Bell Telephone Laboratories, was experimenting with antennas for long-range radio communication when he encountered some mysterious static—radio radiation coming from an unknown source (Figure 5.13). He discovered that this radiation came in strongest about 4 min earlier on each successive day and correctly concluded that since the Earth's sidereal rotation period is 4 min shorter than a solar day (see Section 3.3), the radiation must be originating from some region fixed on the celestial sphere. Subsequent investigation showed that the source of this radiation was part of the Milky Way; Jansky had discovered the first source of cosmic radio waves.

In 1936 Grote Reber, who was an amateur astronomer interested in radio communications, built from galvanized iron and wood the first antenna specifically designed to receive cosmic radio waves. Over the years, Reber built several such antennas and used them to carry out pioneering surveys of the sky for celestial radio sources; he remained active in radio astronomy for more than 30 years. During the first decade, he worked practically alone because

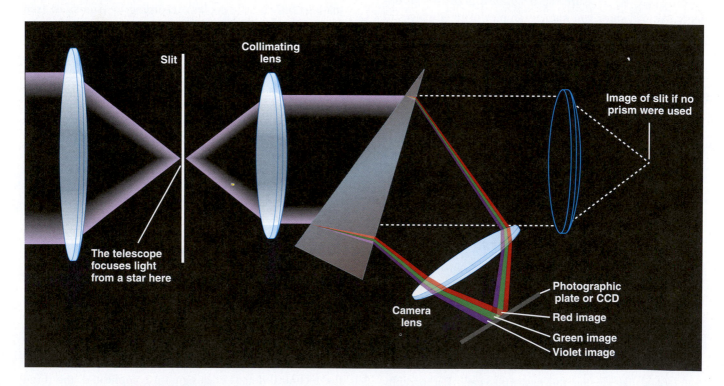

■ FIGURE 5.12

A Prism Spectrometer The light from the telescope is focused on a slit, so the light of one object at a time enters the spectrometer. A prism (or grating) disperses the light into a spectrum, which is then photographed or recorded electronically.

George Ellery Hale: Master Telescope Builder

All of the major research telescopes in the world today were begun in the 20th century. The giant among early telescope builders was George Ellery Hale (1868–1938). Not once but four times he initiated projects that led to the construction of what was the world's largest telescope at the time. And he was a master at finding and winning over wealthy benefactors to underwrite the construction of these new instruments.

Hale's training and early research were in solar physics. In 1892, at age 24, he was named Associate Professor of Astral Physics and director of the astronomical observatory at the University of Chicago. At that time, the largest telescope in the world was the 36-in. refractor at the Lick Observatory near San Jose, California. Taking advantage of an existing glass blank for a 40-in. telescope, Hale set out to raise the money for a larger telescope than the one at Lick. One prospective donor was Charles T. Yerkes, who, among other things, ran the trolley system in Chicago.

Hale wrote to Yerkes, encouraging him to support construction of the giant telescope by saying that "the donor could have no more enduring monument. It is certain that Mr. Lick's name would not have been nearly so widely known today were it not for the famous observatory established as a result of his munificence." Yerkes agreed, and the new telescope was completed in May 1897; it remains the largest refractor in the world.

Even before the completion of the Yerkes refractor, Hale was not only dreaming of building a still larger telescope but was also taking concrete steps to achieve that goal. In the 1890s there was a major controversy about the relative quality of refracting and

George Ellery Hale

reflecting telescopes. Hale realized that 40 in. was close to the maximum feasible aperture for refracting telescopes. If telescopes with significantly larger apertures were to be built, they would have to be reflecting telescopes.

Using funds borrowed from his own family, Hale set out to construct a 60-in. reflector. For a site, he left the Midwest for the much better conditions on Mount Wilson, at that time a wilderness peak above the small city of Los Angeles. In 1904, at the age of 36, Hale received funds from the Carnegie Foundation to establish the Mount Wilson Observatory. The 60-in. mirror was placed in its mount in December 1908.

Two years earlier, in 1906, Hale had already approached John D. Hooker, who had made his fortune in hardware and steel pipe, with a proposal to build a 100-in. telescope. The technological risks were substantial. The 60-in. telescope was not yet complete, and the usefulness of large reflectors for astronomy had yet to be demonstrated. George Ellery Hale's brother called him "the greatest gam-

bler in the world." Once again, Hale was successful in obtaining funds, and the 100-in. telescope was completed in November 1917. (It was with this telescope that Edwin Hubble was able to establish that the spiral nebulae were separate islands of stars—or galaxies—quite removed from our own Milky Way; see Chapter 25.)

Hale was not through dreaming. In 1926, he wrote an article in *Harper's Magazine* about the scientific value of a still larger telescope. This article came to the attention of the Rockefeller Foundation, which granted $6 million (a very large sum in those days) for the construction of a 200-in. telescope. Hale died in 1938, but the 200-in. (5-m) telescope on Palomar Mountain was dedicated ten years later and is now named in Hale's honor.

The Yerkes 40-in. (1-m) refracting telescope.

Bell Laboratories

■ FIGURE 5.13

The First Radio Telescope
The rotating radio antenna used by Jansky in his serendipitous discovery of radio radiation from the Milky Way. On June 8, 1998, a memorial sculpture was placed at the site of this telescope in Holmdel, New Jersey.

professional astronomers had not yet recognized the vast potential of radio astronomy.

5.4.1 Detection of Radio Energy from Space

It is important to understand that radio waves cannot be "heard": They are not the sound waves you hear coming out of the radio receiver in your home or car. Like light, radio waves are a form of electromagnetic radiation, but unlike light, we cannot detect them with our senses—we must rely on electronic equipment to pick them up. In commercial radio broadcasting, we encode sound information (music or the voice of a newscaster) into the radio waves. These must be decoded at the other end and then turned back into sound by speakers or headphones. There are two familiar ways to *modulate* (encode information into) radio waves: by changing the frequency of the waves (i.e., by changing how often they arrive each second), which is called *frequency modulation* or FM, or by broadcasting at a constant frequency but changing the details of the shape of the wave, which is called *amplitude modulation* or AM.

The radio waves we receive from space do not, of course, have rock music or other program information encoded in them (although astronomers would love to tune in on a radio broadcast from some distant civilization; see Chapter 29). If cosmic radio signals were translated into sound, they would sound like the static you hear when dialing between stations. Nevertheless, there is information in the radio waves we receive—information that can tell us about the chemistry and physical conditions in the sources of the waves.

Just as vibrating charged particles can produce electromagnetic waves (see Chapter 4), electromagnetic waves can make charged particles move back and forth. Radio waves can produce a current in conductors of electricity such as metals. An antenna is such a conductor: It intercepts radio waves, which induce a feeble current in it. The current is then amplified in a radio receiver until it is strong enough to measure or record. Like your television or radio set, receivers can be tuned to select a single frequency (channel). In astronomy today, however, it is more common to use sophisticated data-processing techniques that allow thousands of separate frequency bands to be detected simultaneously. Thus, the astronomical radio receiver operates much like a spectrometer on an optical telescope, giving information about how much radiation we receive at each wavelength or frequency. After computer processing, the radio signals are recorded on magnetic disks for further analysis.

Radio waves are reflected by conducting surfaces just as light is reflected from a shiny metallic surface, and according to the same laws of optics. A radio-reflecting telescope consists of a concave metal reflector (called a *dish*), analogous to a telescope mirror. The radio waves collected by the dish are reflected to a focus, where they can then be directed to a receiver and analyzed. Because humans are such visual creatures, radio astronomers often construct a pictorial representation of the radio sources they observe. Figure 5.14 shows such a radio image of a distant galaxy, where radio telescopes reveal vast jets and regions of radio emission that are completely invisible in photographs taken with light.

Radio astronomy is a young field compared with optical astronomy, but it has experienced tremendous growth in recent decades. The world's largest radio reflectors that can be pointed to any direction in the sky have apertures of 100 m. One of these has recently been constructed at the U.S. National Radio Astronomy Observatory in West Virginia (Figure 5.15). Table 5.2 lists some of the major radio telescopes of the world.

FIGURE 5.14

A Radio Image This image has been constructed of radio observations of a galaxy called Cygnus A. Colors have been added to help the eye sort out regions of different radio intensity. Red regions are the most intense, blue the least. The visible galaxy would be a small dot in the center of the image. The radio image reveals jets of expelled material (more than 160,000 LY long) on either side of the galaxy.

FIGURE 5.15

The Robert C. Byrd Green Bank Telescope This new, fully steerable radio telescope in West Virginia went into operation in August 2000. Its dish is about 100 m across.

5.4.2 Radio Interferometry

As we discussed earlier, a telescope's ability to show us fine detail (its resolution) depends on its aperture, but it also depends on the wavelength of the radiation the telescope is gathering. The longer the waves, the harder it is to resolve fine detail in the images or maps we make. Because radio waves have such long wavelengths, they present tremendous challenges for astronomers who need good resolution. In fact, even the largest radio dishes on Earth—operating alone—cannot make out as much detail as the typical small visible-light telescope used in a college astronomy lab. To overcome this difficulty, radio astronomers have learned to sharpen their images by linking two or more radio telescopes together electronically.

Telescopes linked together in this way are called an **interferometer.** This may seem like a strange term because the telescopes in an interferometer work cooperatively; they don't interfere with each other. *Interference,* however, is a technical term for the way waves that arrive in our instruments at slightly different times interact with each other, and this interaction allows us to coax more detail out of our observations. The resolution of an interferometer depends on the separation of the telescopes, not on their individual apertures. Two telescopes separated by 1 km provide the same resolution as would a single dish 1 km across (although they are not, of course, able to collect as much radiation as a radio-wave bucket 1 km across).

To get even better resolution, astronomers combine a large number of radio dishes into an **interferometer array.** In effect, such an array works like a large number of two-dish interferometers, all observing the same part of the sky together. Computer processing of the results permits the reconstruction of a high-resolution radio image. The most extensive such instrument is the National Radio Astronomy Observatory's Very Large Array (VLA) near Socorro, New Mexico (see the opening figure in Chapter 4). It consists of 27 movable radio telescopes (on railroad tracks), each having an aperture of 25 m, spread over a total span of about 36 km. By electronically combining the signals from all of its individual telescopes, this array permits the radio astronomer to make pictures of the sky at radio wavelengths comparable to those obtained with a visible-light telescope, with a resolution of about 1 arcsec.

Initially, the size of interferometer arrays was limited by the requirement that all of the dishes be accurately wired together. The maximum dimensions of the array were thus only a few tens of kilometers. However, larger interferometer separations can be achieved if the telescopes do not require a physical connection. Astronomers have learned to time the arrival of electromagnetic waves coming from space very precisely at each telescope and combine the data later. If the telescopes are as far apart as

TABLE 5.2 *Major Radio Observatories of the World*

Observatory	Location	Description	Web Site
Individual Radio Dishes			
Arecibo Telescope (National Astronomy & Ionospheric Center)	Arecibo, Puerto Rico	305-m fixed dish	www.naic.edu
Greenbank Telescope (National Radio Astronomy Observatory)	Green Bank, West Virginia	110 × 100-m steerable dish	www.gb.nrao.edu/GBT/GBT.html
Effelsberg Telescope (Max Planck Institute für Radioastronomie)	Bonn, Germany	100-m steerable dish	www.mpifr-bonn.mpg.de/index_e.html
Lovell Telescope (Jodrell Bank Radio Observatory)	Manchester, England	76-m steerable dish	www.jb.man.ac.uk/
Goldstone Tracking Station (NASA/JPL)	Barstow, California	70-m steerable dish	gts.gdscc.nasa.gov/
Australia Tracking Station (NASA/JPL)	Tidbinbilla, Australia	70-m steerable dish	www.cdscc.nasa.gov/
Parkes Radio Observatory	Parkes, Australia	64-m steerable dish	www.parkes.atnf.csiro.au/
Arrays of Radio Dishes			
Australia Telescope	Several sites in Australia	8-element array (seven 22-m dishes plus Parkes 64-m)	www.atnf.csiro.au
MERLIN	Cambridge, England, and other British sites	Network of 7 dishes (the largest is 32 m)	www.jb.man.ac.uk/merlin/
Westerbork Radio Observatory	Westerbork, the Netherlands	12-element array of 25-m dishes (1.6-km baseline)	www.astron.nl/wsrt
Very Large Array (NRAO)	Socorro, New Mexico	27-element array of 25-m dishes (36-km baseline)	www.aoc.nrao.edu/vla/html/
Very Long Baseline Array (NRAO)	Ten U.S. sites, Hawaii to the Virgin Islands	10-element array of 25-m dishes (9000-km baseline)	www.aoc.nrao.edu/vlba/html
Millimeter-Wave Telescopes			
IRAM	Granada, Spain	30-m steerable mm-wave dish	iram.fr/
James Clerk Maxwell Telescope	Mauna Kea, Hawaii	15-m steerable mm-wave dish	www.jach.hawaii.edu/JCMT/pages/intro.html
Nobeyama Cosmic Radio Observatory	Minamimaki-Mura, Japan	6-element array of 10-m mm-wave dishes	www.nro.nao.ac.jp/~nma/index-e.html
Hat Creek Radio Observatory (University of California)	Cassel, California	6-element array of 5-m mm-wave dishes	bima.astro.umd.edu/bima

California and Australia, or as West Virginia and Crimea in Ukraine, the resulting resolution far surpasses that of visible-light telescopes.

The United States operates the Very Long Baseline Array (VLBA), made up of ten individual telescopes stretching from the Virgin Islands to Hawaii (Figure 5.16). The VLBA, completed in 1993, can form astronomical images with a resolution of 0.0001 arcsec, permitting features as small as 10 astronomical units (AU) to be distinguished at the center of our Galaxy.

5.4.3 Radar Astronomy

Radar is the technique of transmitting radio waves to an object in our solar system and then detecting the radio radiation that the object reflects back. The time required for

National Radio Astronomy Observatory

■ **FIGURE 5.16**

The Very Long Baseline Array A map showing the distribution of the ten antennas that constitute an array of radio telescopes stretching across the United States.

the round trip can be measured electronically with great precision. Because we know the speed at which radio waves travel (the speed of light), we can determine the distance to the object or a particular feature on its surface (such as a mountain).

Radar observations have been used to determine the distances to the planets and have played an important role in navigating spacecraft throughout the solar system. In addition, as will be discussed in later chapters, radar observations have determined the rotation periods of Venus and Mercury, probed tiny Earth-approaching asteroids, and allowed us to investigate the surfaces of Mercury, Venus, Mars, and the large satellites of Jupiter.

Any radio dish can be used as a radar telescope if it is equipped with a powerful transmitter as well as a receiver. The most spectacular facility in the world for radar astronomy is the 1000-ft (305-m) telescope at Arecibo in Puerto Rico (Figure 5.17). The Arecibo telescope is too large to be pointed directly toward different parts of the sky. Instead, it is constructed in a huge natural "bowl" (more than a mere dish!) formed by several hills, and it is lined with reflecting metal panels. A limited ability to track astronomical sources is achieved by moving the receiver system, which is suspended on cables 100 m above the surface of the bowl.

5.5 OBSERVATIONS OUTSIDE THE EARTH'S ATMOSPHERE

The Earth's atmosphere blocks most radiation at wavelengths shorter than visible light, so we can make ultraviolet, x-ray, and gamma-ray observations only from space. Getting above the distorting effects of the atmosphere is also an advantage at visible and infrared wavelengths. The stars don't "twinkle" in space, so the amount of detail you can observe is limited only by the size of your instrument. On the other hand, it is expensive to place telescopes into space and repairs can present a major challenge. This is why astronomers continue to build telescopes for use on the ground as well as for launching into space.

5.5.1 Airborne and Space Infrared Telescopes

Water vapor, the main source of atmospheric interference for making infrared observations, is concentrated in the lower part of the Earth's atmosphere. For this reason a gain of even a few hundred meters in elevation can make an important difference in the quality of an observatory site. Given the limitations of high mountains, most of which attract clouds and violent storms, it was natural for astronomers to investigate the possibility of observing infrared waves from airplanes and ultimately from space.

Infrared observations from airplanes have been made since the 1960s, starting with a 15-cm telescope on board a Learjet. From 1974 through 1995, NASA operated a 0.9-m airborne telescope flying regularly out of the Ames Research Center south of San Francisco. Observing from an altitude of 12 km, the telescope was above 99 percent of the atmospheric water vapor. NASA is now constructing (in partnership with the German government) a much larger 2.5-m telescope, called the Stratospheric Observatory for Infrared Astronomy (SOFIA), to fly in a modified Boeing 747.

Getting even higher and making observations from space itself have important advantages for infrared astronomy. First is the elimination of all interference from the atmosphere. Equally important is the opportunity to cool the entire optical system of the instrument in order to nearly eliminate infrared radiation from the telescope itself. If we tried to cool a telescope within the atmosphere, it would quickly become coated with condensing water vapor and other gases, making it useless. Only in the vacuum of space can optical elements be cooled to hundreds of degrees below freezing and still remain operational.

The first orbiting infrared observatory, launched in 1983, was the Infrared Astronomical Satellite (IRAS), built as a joint project among the United States, the Netherlands,

FIGURE 5.17

The Largest Radio and Radar Dish The Arecibo Observatory, with its 1000-ft radio dish filling a valley in Puerto Rico, is part of the National Astronomy and Ionosphere Center, operated by Cornell University under a cooperative agreement with the National Science Foundation.

National Astronomy and Ionosphere Center

Choosing Your Own Telescope

If the astronomy course you are taking whets your appetite for exploring the sky further, you may be thinking about buying your own telescope. The good news is that many excellent amateur telescopes are now on the market, and prices are much more reasonable than they were 20 years ago. The bad news is that a good telescope, like a good camera or video recorder, is still not inexpensive, and some research is required to find the model that is best for you. The best sources of information about personal telescopes are the two popular-level magazines published for amateur astronomers: *Sky & Telescope* and *Astronomy*. Both carry regular articles of advice, reviews, and ads from reputable telescope dealers around the country. (See also some Web sites listed at the end of the chapter.)

In some ways, choosing a telescope is like choosing a car: Personal preference can play a major role in the decision. A certain shape or brand name may be important to some car buyers but no big deal to others. In the same way, some of the factors that determine which telescope is right for you depend on your preferences:

- Will you be setting up the telescope in one place and leaving it there, or do you want an instrument that is portable and can come with you on camping trips? Does it have to be carried some distance by a person, or can it get to its destination in a car?
- Do you want to observe the sky with your eyes only, or do you want to take photographs? (Long-exposure photography, for example, requires a good clock-drive to turn your telescope to compensate for the Earth's rotation.)
- What sorts of objects will you be observing? Are you interested primarily in comets, planets, star clusters, or galaxies? Or do you want to observe all kinds of celestial sights?

You may not know the answers to some of these questions yet. For this reason, our number-one recommendation is that you "test-drive" some telescopes first. Most communities have amateur astronomy clubs that sponsor star parties open to the public. The members of these clubs often know a lot about telescopes and can share their ideas with you. Your instructor may know where the nearest amateur astronomy club meets; many of the clubs in the country are also listed on the Web sites for *Sky & Telescope* (skyandtelescope.com) and *Astronomy* magazine (www.astronomy.com).

Furthermore, you may already have an instrument like a telescope at home (or have access to one through a relative or friend). Many amateur astronomers recommend starting your survey of the sky with a good pair of binoculars. These are easily carried around and can show you many objects not visible (or clear) to the unaided eye.

When you are ready to purchase a telescope, you might find the following thoughts useful:

- The key characteristic of a telescope is the aperture of the main mirror or lens; when someone says they have a 6-in. or 8-in. telescope, they mean the diameter of the collecting surface. The larger the aperture, the more light you can gather and the fainter the objects you can see or photograph.
- Telescopes of a given aperture that use lenses (refractors) are typically more expensive than those using mirrors (reflectors) because both sides of a lens must be polished to great accuracy. And, because the light passes through it, the lens must be made of high-quality glass throughout. In contrast, only the front surface of a mirror must be accurately polished.
- Magnification is not one of the criteria on which to base your choice of a telescope. As we discussed in the main text, the magnification of the image is done by a smaller eyepiece, so the magnification can be changed by changing eyepieces. However, a telescope will magnify not only the astronomical object you are viewing but also the turbulence of the Earth's atmosphere. If the magnification is too high, your image will shimmer and shake and be difficult to view. A good telescope will come with a variety of eyepieces that stay within the range of useful magnifications.
- The mount of a telescope is one of its most critical elements. Because a telescope shows a tiny field of view, which is magnified significantly, even the smallest vibration or jarring of the telescope can move the object you are viewing around or out of your field of view. A sturdy and stable mount is essential for serious viewing or photography (although it clearly affects how portable your telescope can be).
- A telescope requires some practice to set up and use effectively. Don't expect everything to go perfectly on your first try. Take some time to read the instructions. If a local amateur astronomy club is nearby, use it as a resource.

and Britain. IRAS was equipped with a 0.6-m telescope cooled to a temperature of less than 10 K. For the first time the infrared sky could be seen as if it were night, rather than through a bright foreground of atmospheric and telescope emissions. IRAS carried out a rapid but comprehensive survey of the entire infrared sky over a ten-month period, cataloging more than 250,000 sources of infrared radiation. Since then, several other infrared telescopes have operated in Earth orbit with much better sensitivity and resolution due to improvements in infrared detectors. The most powerful of these infrared instruments is the Spitzer Space Telescope, which was launched in 2003.

5.5.2 Hubble Space Telescope

In April 1990, a great leap forward (or should we say upward) in astronomy was made with the launch of the Hubble Space Telescope (HST). This telescope has an aperture of 2.4 m, the largest put into space so far. (Its aperture was limited by the size of the payload bay in the Space Shuttle, which was its launch vehicle.) It was named for Edwin Hubble, the astronomer who discovered the expansion of the universe in the 1920s (see Chapter 25).

HST is operated jointly by the NASA Goddard Space Flight Center and the Space Telescope Science Institute in Baltimore. It is the first orbiting observatory designed to be serviced by shuttle astronauts, and they have made several visits to improve or replace its initial instruments and to repair some of the systems that operate the spacecraft (see the opening figure for this chapter).

HST's mirror was ground and polished to a remarkable degree of accuracy. If we were to scale up its mirror to the size of the entire continental United States, there would be no hill or valley larger than about 6 cm in its smooth surface. Unfortunately, after it was launched, scientists discovered that the primary mirror had a slight error in its shape equal to roughly 1/50 the width of a human hair. Small as that sounds, it was enough to ensure that much of the light entering the telescope did not come to a clear focus and that all the images were blurry. In a misplaced effort to save money, a complete test of the optical system had not been carried out before launch, so the error was not discovered until HST was in orbit.

The solution was to do something very similar to what we do for astronomy students with blurry vision: put corrective optics in front of their eyes. In December 1993, in one of the most exciting and difficult space missions ever flown, astronauts captured the orbiting telescope and brought it back into the shuttle payload bay. There they installed a package containing compensating optics as well as a new improved camera before releasing HST back

into orbit. The telescope now works as it was designed to, and we have sprinkled the new images from HST throughout this book. HST continues to need repairs, however— gyros fail, batteries run down, etc. The explosion of the Space Shuttle Columbia in 2003 and the subsequent new safety requirements for manned flights have left future repair missions, and hence the life expectancy of HST, in serious doubt.

5.5.3 High-Energy Observatories

Ultraviolet, x-ray, and gamma-ray observations can be made only from space. Such observations first became possible in 1946, with V2 rockets captured from the Germans. The U.S. Naval Research Laboratory put instruments on these rockets for a series of pioneering flights used initially to detect ultraviolet radiation from the Sun. Since then, many other rockets have been launched to make x-ray and ultraviolet observations of the Sun, and later of other celestial objects as well.

Beginning in the 1960s, a steady stream of high-energy observatories has been launched into orbit to reveal and explore the universe at short wavelengths. The most recent of these is the Chandra x-ray telescope, which was launched in 1999 (Figure 5.18). It is producing x-ray images with

■ FIGURE 5.18

The Chandra X-Ray Satellite This painting shows NASA's Chandra X-ray Observatory in orbit. Chandra, the world's most powerful x-ray telescope, was deployed in July 1999.

NASA

TABLE 5.3 *Some Recent Observatories in Space*

Observatory	Dates of Operation	Bands of the Spectrum	Notes	Web Site
Einstein (HEAO-2)	1978–1981	x rays	first to take x-ray images	heasarc.gsfc.nasa.gov/docs/einstein/heao2.html
International UV Explorer (IUE)	1978–1996	UV	UV spectroscopy	www.vilspa.esa.es/iue/iue.html
Infrared Astron. Satellite (IRAS)	1983–1984	IR	mapped 250,000 sources	irsa.ipac.caltech.edu/IRASdocs/
Hipparcos	1989–1993	visible light	measured over 100,000 precise positions	astro.esa.int/Hipparcos/hipparcos.html
Cosmic Background Explorer (COBE)	1989–1993	IR, mm	observed the 3-degree background radiation	space.gsfc.nasa.gov/astro/cobe/
Compton Gamma-Ray Observatory	1990–2000	gamma rays	gamma-ray sources and spectra	cossc.gsfc.nasa.gov/
Hubble Space Telescope (HST)	1990–	visible, UV, IR	2.4-m mirror; images and spectra	hubblesite.org
Roentgen Satellite (ROSAT)	1990–1998	x rays	x-ray images and spectra	wave.xray.mpe.mpg.de/rosat
Infrared Space Observatory (ISO)	1995–1998	IR	infrared images and spectra	isowww.estec.esa.nl/
Rossi X-Ray Timing Explorer	1995–	x rays	variability of x-ray sources	heasarc.gsfc.nasa.gov/docs/xte/XTE.html
BeppoSAX	1996–2002	x rays	observes over wide spectral range	bepposax/gsfc.nasa.gov/bepposax
HALCA	1997–2000	radio	8-m radio dish, made array with dishes on the ground	www.vsop.isas.ac.jp/
Far UV Spectroscopic Explorer (FUSE)	1998–	UV	far-UV spectroscopy	fuse.pha.jhu.edu/
Chandra X-Ray Astronomy Facility	1999–	x rays	x-ray images and spectra	xrtpub.harvard.edu/
XMM Newton	1999–	x rays	x-ray spectroscopy	xmm.vilspa.esa.es
Microwave Anisotropy Probe (MAP)	2001–	microwaves	examine the background radiation from the Big Bang	map.gsfc.nasa.gov
Internat'l Gamma Ray Astrophys. Laboratory (INTEGRAL)	2002–	x and gamma rays	higher resolution gamma-ray images	sci.esa.int/home/integral
Spitzer Space Telescope	2003–	IR	0.85-m telescope in space	www.spitzer.caltech.edu
Stratospheric Observatory for IR Astron. (SOFIA)	2004–	IR	airborne 2.5-m IR telescope	sofia.arc.nasa.gov/

unprecedented resolution and sensitivity. The 2002 Nobel Prize in physics was awarded to Riccardo Giacconi, who was a pioneer in the field of building and launching sophisticated x-ray instruments. Table 5.3 lists some of the most important space observatories.

5.6 THE FUTURE OF LARGE TELESCOPES

Each new generation of large telescopes, either in space or on the ground, simply whets the appetite of both scientists and the public. When we go on a hike, we are eager to see what lies just around the next bend in the path. Researchers are no different, and astronomers and engineers are now working on the technologies that would allow us to explore even more distant parts of the universe and to see them more clearly.

The premier space facility planned for the next decade is the James Webb Space Telescope, which (in a departure from tradition) is named after one of the early administrators of NASA instead of a scientist. This telescope will have a mirror 6 m in diameter and made up, like the Keck telescopes, of 36 small hexagons. The telescope is scheduled for launch in 2010 and should have the sensitivity needed to detect the very first generation of stars, formed when the universe was only a few hundred million years old.

■ **FIGURE 5.19**

Artist's Conception of the Overwhelmingly Large Telescope
The mirror in this telescope, which is currently being designed in Europe, is 100 m across—slightly wider than a football field is long. The telescope is being designed to operate in the open air without a protective enclosure. Protective mirror covers deploy from four surrounding buildings when the telescope is not in use.

On the ground, radio astronomers are planning the Atacama Large Millimeter Array (ALMA), which will consist of 64 radio telescopes, each 12 m in diameter. ALMA will be an interferometer and will be able to achieve spatial resolution ten times better than the Hubble Space Telescope but at millimeter wavelengths. The array will be located at an altitude of 16,400 ft on a plateau in the Atacama Desert of northern Chile, one of the driest spots on Earth. ALMA will be a particularly good tool for studying molecules, which are found in cool regions of gas where star formation is taking place.

And finally, several groups of infrared and optical astronomers around the globe are exploring the feasibility of building ground-based telescopes with mirrors 30–100 m across. Stop and think what this means: 100 m is longer than a football field! It is technically impossible to build and trans-port a single mirror that is 30 m or larger in diameter. These giant telescopes will follow the Keck design and be made up of a thousand or more smaller hexagonal mirrors, all held precisely in position so that they form a continuous surface.

The most ambitious of these projects is the OWL project (Figure 5.19), which is being developed in Europe. OWL stands for Overwhelmingly Large Telescope, and the design calls for a 100-m primary mirror. The cost is likely to exceed a billion dollars. These giant telescopes will combine light-gathering power with high-resolution imaging and will be able to attack many important astronomical problems. For example, they should be able to tell us when, where, and how often planets form around other stars. They should even be able to provide us images and spectra of planets, and thus perhaps give us the first real evidence that life exists elsewhere (see Chapter 29).

SURFING THE WEB

Web sites for the major telescopes and observatories on the ground and in space are given in Tables 5.1, 5.2, and 5.3.

Multiwavelength Astronomy:
www.ipac.caltech.edu/Outreach/Multiwave/multiwave.html
Good images of the same celestial object taken with telescopes collecting different wavelength bands. You can see radio, visible-light, and x-ray images side by side.

Early Radio Astronomy:
www.nrao.edu/intro/ham.connection.html
A brief summary of the history of radio astronomy, with emphasis on the connection with amateur (or "ham") radio.

💻 Whispers from the Cosmos:
www.ncsa.uiuc.edu/Cyberia/Bima/BimaHome.html
Online "exhibit" about radio astronomy in general and the Berkeley–Illinois–Maryland Association array of radio telescopes in particular.

💻 Orbiting Astronomical Observatories:
www.seds.org/~spider/oaos/oaos.html
A good annotated list, with fine links, of the telescopes in Earth orbit.

💻 History of High-Energy Astrophysics:
heasarc.gsfc.nasa.gov/docs/heasarc/headates/heahistory.html
A good chronology with links detailing all major experiments and space missions relevant to the study of gamma rays, x rays, or ultraviolet waves from space.

💻 Sites with Information on Buying Your Own Telescope:

- Astronomy magazine:
www.astronomy.com/content/static/beginners/scopebuyingguide/sbg_1.asp

- Sky & Telescope magazine:
http://skyandtelescope.com/howto/scopes/article_241_1.asp

- The Telescope Review Site:
www.weatherman.com/BEGINNER.HTM

- Weber State Planetarium Buying Your First Telescope:
physics.weber.edu/planet/telescope.html
What we like about this site is that it recommends NOT buying a telescope at first.

- The Heretic's Guide to Buying Your First Telescope:
www.findascope.com

SUMMARY

5.1 A **telescope** collects the faint light from astronomical sources and brings it to a **focus,** where an instrument (filters or a *spectrometer*) sorts the light according to wavelength. Light is then directed to a **detector,** where a permanent record is made. The light-gathering power of a telescope is determined by its **aperture**—that is, by the area of its largest or primary lens or mirror. The aperture is a convex lens in a **refracting telescope** or a concave mirror in a **reflector** that brings the light to a focus. Most large telescopes are reflectors; it is easier to manufacture and support large mirrors because the light does not have to pass through the glass.

5.2 New technologies for supporting lightweight mirrors have led to the construction of a number of new large telescopes since 1990. The site for an astronomical observatory must be carefully chosen for clear weather, dark skies, low water vapor, and excellent atmospheric **seeing** (low atmospheric turbulence). The **resolution** (the fineness of detail that can be seen) of a visible-light or infrared telescope is degraded by turbulence in the Earth's atmosphere. The new technique of **adaptive optics,** however, can make corrections for this turbulence in real time and produce exquisite images.

5.3 Visible-light detectors include the eye, photographic film, and the **charge-coupled device (CCD).** Detectors sensitive to infrared radiation must be cooled to very low temperatures. A spectrometer disperses the light into a spectrum to be recorded for detailed analysis.

5.4 In the 1930s, radio astronomy was pioneered by Jansky and Reber. A radio telescope is basically a radio antenna (often a large curved dish) connected to a receiver. Significantly enhanced resolution can be obtained with **interferometers,** including **interferometer arrays** like the 27-element VLA. Expanding to very long baseline interferometers, radio astronomers can achieve resolutions as good as 0.0001 arcsec. **Radar** astronomy involves transmitting as well as receiving. The largest radar telescope is the 305-m bowl at Arecibo.

5.5 Infrared observations are made with telescopes aboard aircraft and in space, as well as from ground-based facilities on dry mountain peaks. Ultraviolet, x-ray, and gamma-ray observations must be made from above the atmosphere. Many orbiting observatories have been flown to observe in these bands of the spectrum in the last few decades. The largest-aperture instrument in space is the Hubble Space Telescope (HST). The premier x-ray satellite is called Chandra. The Space Infrared Telescope Facility, launched in 2003, will be delivering its results while you are reading this text.

5.6 New and even larger telescopes are on the drawing boards. The James Webb Space Telescope, a 6-m successor to Hubble, is currently scheduled for launch in 2010. Radio astronomers are planning to build an interferometric array for millimeter astronomy called ALMA in the Atacama Desert of northern Chile. And several groups around the world are developing designs for ground-based optical and infrared telescopes with primary mirrors 30–100 m in diameter.

INTER-ACTIVITY

A Most large telescopes get many more proposals for observing projects than there is night observing time available in a year. Suppose your group is the telescope time allocation committee reporting to an observatory director. What criteria would you use in deciding how to give out time on the telescope? What steps could you take to make sure all your colleagues thought the process was fair and people would still talk to you at future astronomy meetings?

B Your group is a committee of nervous astronomers about to make a proposal to the government of a small European country to chip in to build the world's largest telescope in the high dry desert of the Chilean Andes Mountains. You expect the government ministers to be very skeptical about supporting this project. What arguments would you make to convince them to participate?

C The same government ministers we met in activity B ask you to draw up a list of the pros and cons of having the world's largest telescope in the mountains of Chile (instead of a mountain in Europe). What would your group list in each column?

D Make a list of all the ways in which an observing session at a large optical telescope and a large radio telescope might differ. (*One hint:* Bear in mind that because the Sun is not especially bright at many radio wavelengths, observations with radio telescopes can often be done during the day.)

E Another "environmental threat" to astronomy (besides light pollution) comes from the spilling of terrestrial communications into the "channels"—wavelengths and frequencies—previously reserved for radio astronomy. For example, the demand for cellular phones means that more and more radio channels will be used for this purpose. The faint signals from cosmic radio sources could be drowned in a sea of earthly conversation (translated and sent as radio waves). Assume your group is a congressional committee being lobbied by both radio astronomers who want to save some clear channels for doing astronomy and the companies that stand to make a lot of money from expanding cellular phone use. What arguments would sway you to each side? [For a real-world example of where this issue is being debated, see *Science* magazine, Nov. 28, 1997, p. 1569.]

REVIEW QUESTIONS

Ace ☯ Astronomy™ Assess your understanding of this chapter's topics with additional quizzing and animations at **http://ace.brookscole.com/voyages**

1. Name the two spectral windows through which electromagnetic radiation easily reaches the surface of the Earth, and describe the largest-aperture telescope currently in use for each window.
2. List the six bands into which we commonly divide the electromagnetic spectrum, and list the largest-aperture telescope currently in use in each band.
3. When astronomers discuss the apertures of their telescopes, they say bigger is better. Explain why.
4. Why are the largest visible-light telescopes in the world made with mirrors rather than lenses?
5. Compare the eye, photographic film, and CCDs as detectors for light. What are the advantages and disadvantages of each?
6. Radio and radar observations are often made with the same antenna, but otherwise they are very different techniques. Compare and contrast radio and radar astronomy in terms of the equipment needed, the methods used, and the kind of results obtained.

7. Why do astronomers place telescopes in Earth orbit? What are the advantages for the different regions of the spectrum?
8. What was the problem with the Hubble Space Telescope and how was it solved?
9. Describe the techniques radio astronomers use to obtain a resolution comparable to what astronomers working with visible light can achieve.
10. What kind of visible-light and infrared telescopes on the ground are astronomers planning for the future? Why are they building them on the ground and not in space?
11. What kind of visible-light and infrared telescopes are astronomers planning to launch into space in the future?

12. What happens to the image produced by a lens if the lens is "stopped down" with an iris diaphragm—a device that covers its periphery?

13. What would be the properties of an *ideal* astronomical detector? How closely do the actual properties of a CCD approach this ideal?

14. Fifty years ago, the astronomers on the staff of Mount Wilson and Palomar Observatories each received about 60 nights per year for their observing programs. Today an astronomer feels fortunate to get 10 nights per year on a large telescope. Can you suggest some reasons for this change?

15. The largest observatory complex in the world is on Mauna Kea in Hawaii, at an altitude of 4.2 km. This is by no means the tallest mountain on Earth. What are some factors astronomers consider when selecting an observatory site? Don't forget practical ones. Should astronomers, for example, consider building an observatory on Mount McKinley (Denali) or Mount Everest?

16. Another site recently developed for astronomy is the Antarctic plateau. Discuss its advantages and disadvantages.

17. Suppose you are looking for sites for an optical observatory, an infrared observatory, and an x-ray observatory. What are the main criteria of excellence for each? What sites are actually considered the best today?

18. Radio astronomy involves wavelengths much longer than those of visible light, and many orbiting observatories have probed the universe for radiation of very short wavelengths. What sorts of objects and physical conditions would you expect to be associated with radiation emissions of very long and very short wavelengths?

19. The dean of a university located near the ocean (who was not a science major in college) proposes building an infrared telescope right on campus and operating it in a nice heated dome so that astronomers will be comfortable on cold winter nights. Criticize this proposal, giving your reasoning.

The primary reason for building large telescopes is to gather light from a larger area and hence to see fainter objects. The human eye actually works on the same principle. Your pupil expands under dark conditions to capture more light. It takes a few seconds for the pupil to expand in response to the change in illumination as you go from a fully lighted room to the dark outdoors at night. Far more important for dark adaptation is the gradual release of a chemical called rhodopsin, which makes the cones in your eye more sensitive to low light levels. Reaching maximum dark adaption takes about 30 minutes. Have you noticed how your eyes gradually get better at seeing in the dark as the minutes pass—in a movie theater, for example?

The larger the area over which light is gathered, the more light can be captured and brought to a focus. Remember that the area A of a circle is given by

$$A = \frac{\pi d^2}{4}$$

where d is the diameter of the circle.

20. In broad daylight, the size of your pupil is typically 3 mm. In dark situations, it expands to about 7 mm. How much more light can it gather?

21. How much more light can be gathered by a telescope that is 8 m in diameter than by your fully dark-adapted eye (neglect the increased sensitivity because of rhodopsin)?

22. How much more light can the Keck telescope (with its 10-m-diameter mirror) gather than an amateur telescope whose mirror is 25 cm (0.25 m) across?

23. People are often bothered when they discover that reflecting telescopes have a second mirror in the middle to bring the light out to an accessible focus where big instruments can be mounted. "Don't you lose light?" people ask. Well, yes, you do, but there is no better alternative. You can estimate how much light is lost by such an arrangement. The primary mirror (the one at the bottom in Figure 5.6) of the Gemini North telescope is 8 m in diameter. The secondary mirror at the top is about 1 m in diameter. Use the formula for the area of a circle to estimate what fraction of the light is blocked by the secondary mirror.

24. Telescopes can now be operated remotely from a warm room, but until about 25 years ago, astronomers worked at the telescope to guide it so that it remained pointed in exactly the right place. In a large telescope, like the Palomar 200-in. telescope, astronomers sat in a cage at the top of the telescope, where the secondary mirror is located in Figure 5.6. Assume for the purpose of your calculation that the diameter of this cage was 40 in. What fraction of the light is blocked? Now you can see why these cages were made very cramped and uncomfortable—but what a ride!

25. The HST cost about $1.7 billion for construction and $300 million for its shuttle launch, and it costs $250 million per year to operate. If the telescope lasts for 20 years, what is the total cost per year? Per day? If the telescope can be used just 30 percent of the time for actual observations, what is the cost per hour and per minute for the as-tronomer's observing time on this instrument? What is the cost per person in the United States? Was your investment in Hubble worth it?

SUGGESTIONS FOR FURTHER READING

Bartusiak, M. "The Next Generation Space Telescope: Predicting the Past" in *Astronomy,* Feb. 2001, p. 40. On what the instrument might detect.

Chaisson, E. *The Hubble Wars.* 1994, HarperCollins. Controversial story of the building of the HST.

Dyer, A. "What's the Best Telescope for You?" in *Sky & Telescope,* Dec. 1997, p. 28.

Florence, R. *The Perfect Machine: Building the Palomar Telescope.* 1994, Harper Perennial. Highly readable account of how the 200-in. telescope was built.

Gehrels, N. et al. "The Compton Gamma-Ray Observatory" in *Scientific American,* Dec. 1993, p. 68.

Graham, R. "Astronomy's Archangel" in *Astronomy,* Nov. 1998, p. 56. On master mirror-maker Roger Angel of the University of Arizona.

Graham, R. "High Tech Twin Towers" in *Astronomy,* Oct. 2000, p. 36. On the Gemini telescopes.

Grunsfeld, J. "Remaking the Hubble" in *Sky & Telescope,* Mar. 2002, p. 30. On the spring 2002 upgrade mission, by one of the astronauts.

Harrington, P. "Back to Basics with Binoculars" in *Astronomy,* Aug. 2002, p. 64. Brief introduction to how binoculars work and what to look for.

Irion, R. "Prime Time" in *Astronomy,* Feb. 2001, p. 46. On how time is allotted on the major research telescopes.

Janesick, J. and Blouke, M. "Sky on a Chip: The Fabulous CCD" in *Sky & Telescope,* Sept. 1987, p. 238.

Jayawardhana, R. "The Age of the Behemoths" in *Sky & Telescope,* Feb. 2002, p. 30. On current and planned large telescopes.

Junor, B. et al. "Seeing the Details of the Stars with Next Generation Telescopes" in *Mercury,* Sept./Oct. 1998, p. 26. On adaptive optics.

Krisciunas, K. *Astronomical Centers of the World.* 1988, Cambridge U. Press. History of and guide to major observatories.

Krisciunas, K. "Science with the Keck Telescope" in *Sky & Telescope,* Sept. 1994, p. 20. What it's like to use the telescope, and what work is being done with it.

Pilachowski, C. and Trueblood, M. "Telescopes of the 21st Century" in *Mercury,* Sept./Oct. 1998, p. 10. On the telescopes now being built or planned on the ground.

Schilling, G. "Adaptive Optics Comes of Age" in *Sky & Telescope,* Oct. 2001, p. 30. Excellent 11-page review.

Shore, L. "VLA: The Telescope That Never Sleeps" in *Astronomy,* Aug. 1987, p. 15.

Tarenghi, M. "Eyewitness View: First Sight for a Glass Giant" in *Sky & Telescope,* Nov. 1998, p. 47. On the first observations with the Very Large Telescope.

Whitt, K. "Destination: La Palma" in *Astronomy,* Jan. 2001, p. 62. Photo album and brief introduction to the Newton, Kapteyn, Herschel, and other telescopes there.

Exploring the Surface of Mars The view from the Mars Pathfinder lander, part of which is visible at the bottom of the picture. Some distance away is the little six-wheeled Sojourner rover, about the size of a microwave oven. During the summer of 1996, the public avidly followed the adventures of this little robot on the martian surface.

NASA/JPL

6

Other Worlds: An Introduction to the Solar System

In all of the history of mankind, there will be only one generation that will be the first to explore the Solar System, one generation for which, in childhood, the planets are distant and indistinct discs moving through the night sky, and for which, in old age, the planets are places, diverse new worlds in the course of exploration.

Carl Sagan in _The Cosmic Connection_ (1973, Doubleday)

THINKING AHEAD

Surrounding the Sun is a family of worlds, large and small: nine planets, dozens of moons, and smaller pieces too numerous to count. It is a complex system with a wide range of conditions on the various worlds. Yet the system has some remarkable regularities to it. Most objects circle the Sun in the same direction and in the same plane, for example. What caused the orbits of all these worlds to share a common motion?

During the past 40 years we have learned more about the solar system than anyone imagined before the space age. In addition to gathering information with powerful new telescopes, we have sent spacecraft directly to many members of the planetary system. (Planetary astronomy is the only branch of our science in which we can—at least vicariously—travel to the objects we want to study.) With evocative names like Voyager, Pioneer, and Pathfinder, our robot explorers have flown past or around every planet except Pluto, returning images and data that have dazzled both astronomers and the public. In the process, we have also investigated dozens of fascinating moons, four ring systems, four asteroids, and two comets.

Probes have penetrated the atmosphere of Jupiter and landed on the surfaces of Venus, Mars, the Moon, and the asteroid Eros. Humans have set foot on the Moon and returned samples of its surface soil for analysis in the laboratory (Figure 6.1). We have even discovered other places in our solar system that might be able to support life.

Virtual Laboratories

Planetary Geology

Planetary Atmospheres and Their Retention

■ **FIGURE 6.1**

Astronauts on the Moon The lunar lander and surface rover from the *Apollo 15* mission are seen in this view of the one place beyond Earth that has been explored directly by humans.

The chapters about planets in today's textbooks are nothing like the ones earlier generations of students were reading. We can now envision the members of the solar system as other worlds like our own, each with its own chemical and geological history, and each with unique sights that interplanetary tourists may someday want to visit. Some have called these last few decades the golden age of planetary exploration, comparable to another golden age of exploration—the period starting in the 15th century, when great sailing ships plied the Earth's oceans and humanity began to get familiar with our own planet's surface.

In this chapter we discuss our planetary system and introduce the idea of comparative planetology—studying how the planets work by comparing them with one another. We want to get to know the planets not only for their own sakes but also to see what they can tell us about the origin and evolution of the entire solar system. In the next seven chapters, we will describe the better-known members of the solar system in some detail, returning to the theme of our origins in Chapter 13.

6.1 OVERVIEW OF OUR PLANETARY SYSTEM

The solar system[1] consists of the Sun and many smaller objects: the planets, their satellites and rings, and such "debris" as asteroids, comets, and dust. Decades of observation and spacecraft exploration have revealed that most of these objects formed together with the Sun about 4.5 billion years ago. They represent clumps of material that condensed from an enormous cloud of gas and dust. The central part of this cloud became the Sun, and a small fraction of the material in the outer parts eventually formed the other objects.

6.1.1 An Inventory

The Sun, a rather ordinary star, is by far the most massive member of the solar system, as shown in Table 6.1. It is an enormous ball of incandescent gas, 1.4 million kilometers (km) in diameter, with an interior temperature of millions of degrees. The Sun will be discussed later in this book, as our first (and best-studied) example of a star.

Table 6.1 also shows that most of the material of the planets is actually concentrated in the largest one, Jupiter, which is more massive than all the rest of the planets com-

---

[1]The generic term for a group of planets and other bodies circling a star is *planetary system*. Ours is called the *solar system* because our Sun is sometimes called *Sol*. Strictly speaking, then, there is only one solar system; planets circling other stars are planetary systems.

TABLE 6.1	*Mass of Members of the Solar System*
Object	**Percentage of Total Mass**
Sun	99.80
Jupiter	0.10
Comets	0.05
All other planets	0.04
Satellites and rings	0.00005
Asteroids	0.000002
Cosmic dust	0.0000001

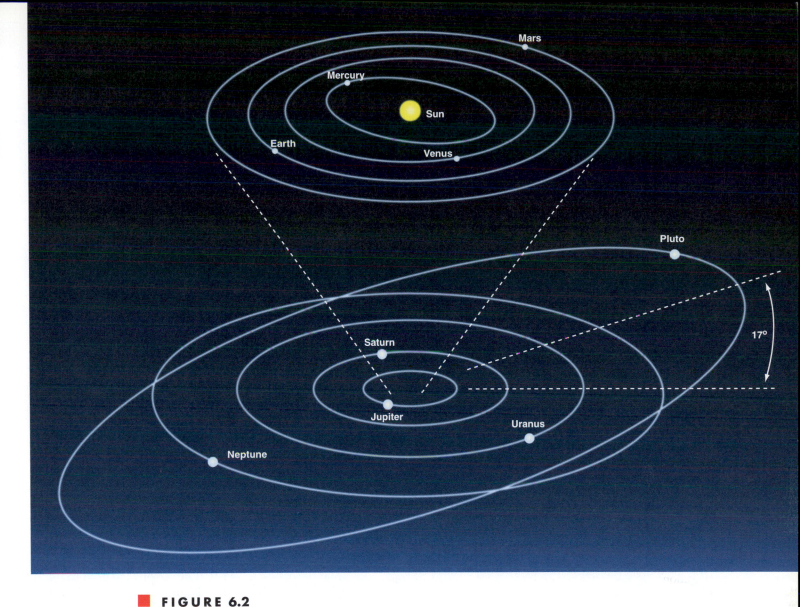

■ **FIGURE 6.2**

The Orbits of the Planets   All the planets except Pluto orbit the Sun in roughly the same plane.

bined. Astronomers were able to measure the masses of the planets centuries ago, using Newton's laws of gravity, by measuring the planets' gravitational effects on one another or on moons that orbit about them. Today we make even more precise measurements of their mass by tracking their gravitational effects on the motion of spacecraft that pass near them.

Besides the Earth, five other planets (Mercury, Venus, Mars, Jupiter, and Saturn) were known to the ancients, and three (Uranus, Neptune, and Pluto) were discovered since the invention of the telescope. The nine planets all revolve in the same direction around the Sun (Figure 6.2). They orbit in approximately the same plane, like cars traveling on concentric tracks on a giant, flat racecourse. Each planet stays in its own traffic lane, following a nearly circular orbit about the Sun and obeying the "traffic laws" discovered by Galileo, Kepler, and Newton.

Each of the planets also rotates (spins) about an axis running through it, and in most cases the direction of rota-

tion is the same as the direction of revolution about the Sun. The exception is Venus, which rotates very slowly backward (that is, in a retrograde direction). Uranus and Pluto also have strange rotations, each spinning about an axis tipped nearly on its side.

The four planets closest to the Sun (Mercury through Mars) are called the inner or **terrestrial planets;** often the Moon is also discussed as a part of this group, bringing the total of terrestrial objects to five. (We generally call the Earth's satellite the *Moon,* with a capital *M,* and the other satellites *moons,* with lowercase *m*'s.) The terrestrial planets are relatively small worlds, composed primarily of rock and metal. All of them have solid surfaces that bear the records of their geological history in the forms of craters, mountains, and volcanoes (Figure 6.3).

The next four planets (Jupiter through Neptune) are much larger and are composed primarily of lighter ices, liquids, and gases. We call these four the jovian (after "Jove," another name for Jupiter in mythology) or **giant planets,**

FIGURE 6.3

The Surface of Mercury The pockmarked face of a terrestrial world, more typical of the inner planets than the watery surface of Earth. This black-and-white image, taken with the Mariner 10 spacecraft, shows a region more than 400 km wide.

a name they richly deserve (Figure 6.4). More than 1400 Earths could fit inside Jupiter, for example. These planets do not have solid surfaces on which future explorers might land. They are more like vast, spherical oceans with much smaller, dense cores.

At the outer edge of the system lies little Pluto, which is neither a terrestrial nor a jovian planet; it is similar to the satellites of the outer planets and a number of icy objects that orbit the Sun beyond Pluto (see Chapter 12). It may have a common origin with them. Pluto was the last planet to have its mass measured—something astronomers could not do until Pluto's satellite was discovered in 1978. Table 6.2 summarizes some of the main facts about these nine planets.

6.1.2 Smaller Members of the Solar System

Most of the planets are accompanied by one or more satellites or moons; only Mercury and Venus move through space alone. There are more than 100 known satellites (see Appendix 8 for a listing of the larger ones), and undoubtedly many other small ones remain undiscovered. The largest of the satellites are as big as small planets and just as interesting; these include, in addition to our Moon, the four largest moons of Jupiter (called the Galilean satellites after their discoverer) and the largest moons of Saturn and Neptune (confusingly named Titan and Triton).

Each of the jovian planets also has rings made up of countless small bodies ranging in size from mountains down to mere grains of dust, all in orbit about the equator of the planet. The bright rings of Saturn are the widest and by far the easiest to see; they are among the most beautiful sights in the solar system (Figure 6.5). But all of the four

EARTH

FIGURE 6.4

The Four Giant Planets A montage showing the four giant planets (as photographed by the Voyager cameras) and Earth to scale.

TABLE 6.2 *The Planets*

Name	Distance from Sun (AU)°	Revolution Period (years)	Diameter (km)	Mass (10^{23} kg)	Density (g/cm^3)†
Mercury	0.39	0.24	4,878	3.3	5.4
Venus	0.72	0.62	12,102	48.7	5.3
Earth	1.00	1.00	12,756	59.8	5.5
Mars	1.52	1.88	6,787	6.4	3.9
Jupiter	5.20	11.86	142,984	18,991	1.3
Saturn	9.54	29.46	120,536	5,686	0.7
Uranus	19.18	84.07	51,118	866	1.3
Neptune	30.06	164.82	49,660	1,030	1.6
Pluto	39.44	248.60	2,200	0.01	2.1

° An AU (or astronomical unit) is the distance from Earth to the Sun.
† We give densities in units where the density of water is 1. To get densities in units of kg/m^3, multiply the value we give by 1000.

ring systems are interesting to scientists because of their complicated forms, influenced by the pull of the satellites that also orbit these giant planets.

The solar system has many other less conspicuous members. The **asteroids** are rocky bodies that orbit the Sun like miniature planets, mostly in the space between Mars and Jupiter (Figure 6.6). Most asteroids are remnants of the initial population of the solar system that existed before the planets themselves formed. Some of the smallest satellites of the planets, such as the moons of Mars, are very likely captured asteroids.

Another class of small bodies is composed in part of ice, made of frozen gases such as water, carbon dioxide, carbon monoxide, and methyl alcohol; these objects are called **comets.** Comets also are remnants from the formation of the solar system, but they were formed and continue (with rare exceptions) to orbit the Sun in distant, cooler regions—stored in a sort of cosmic deep freeze.

Finally, there are countless grains of broken rock, which we call cosmic dust, scattered throughout the solar system. When these particles collide with Earth's atmosphere (as millions do each day), they burn up, producing brief flashes of light that we see as shooting stars or **meteors.** Occasionally, some larger chunk of material from space survives the passage through the atmosphere to become a **meteorite,** a piece of the solar system that has landed on Earth. (You can see meteorites on display in many natural history museums and can sometimes even purchase pieces of them from gem and mineral dealers.)

6.1.3 A Scale Model of the Solar System

Astronomy often deals with dimensions and distances that far exceed our ordinary experience. What does 1.4 billion kilometers—the distance from the Sun to Saturn—really mean to anyone? Sometimes it helps to visualize such large systems in terms of a scale model.

■ **FIGURE 6.5**

Saturn and Its Rings   Voyager 1 image of a crescent Saturn and its complex system of rings, taken from a distance of about 1.5 million km. Notice that you can see the edge of the planet through the rings at left.

■ FIGURE 6.6

Asteroid Eros Close-up view of this small Earth-crossing asteroid, taken by the NEAR-Shoemaker spacecraft from an altitude of about 100 km. This view of the heavily cratered surface is about 10 km wide. The spacecraft orbited Eros for a year before gently landing on its surface.

In our imagination, let us build a scale model of the solar system, adopting a scale factor of 1 billion (10^9)—that is, reducing the actual solar system by dividing every dimension by a factor of 10^9. Earth then has a diameter of 1.3 cm, about the size of a grape. The Moon is a pea orbiting this grape at a distance of 40 cm, or a little more than a foot away. The Earth-Moon system fits into a bookbag.

In this model the Sun is nearly 1.5 m in diameter, about the average height of an adult, and our Earth is at a distance of 150 m—about one city block—from the Sun. Jupiter is five blocks away from the Sun, and its diameter is 15 cm, about the size of a very large grapefruit. Saturn is 10 blocks from the Sun, Uranus 20 blocks, and Neptune 30 blocks. Pluto, whose distance varies quite a bit during its 249-year orbit, is currently just beyond 30 blocks and getting farther away with time. Most of the satellites of the outer solar system are the sizes of various kinds of seeds orbiting the grapefruit, oranges, and lemons that represent the outer planets.

In our scale model, a human is reduced to the dimensions of a single atom, and cars and spacecraft to the size of molecules. Sending the Voyager spacecraft to Neptune involves navigating a single molecule from the Earth-grape

toward a lemon 5 km away with an accuracy equivalent to the width of a thread in a spider's web.

If that model represents the solar system, where would the nearest stars be? If we keep the same scale, the closest stars would be tens of thousands of kilometers away. If you built our scale model in the city where you live, you would have to place the representations of these stars on the other side of the Earth or beyond.

By the way, model solar systems like the one we just presented are being built in cities throughout the world. In Sweden, for example, Stockholm's huge Globe Arena has become a model for the Sun, and Pluto is represented by a 12-cm sculpture in the small town of Delsbo 300 km away. Another model solar system has been built on the Mall between the White House and Congress in Washington, D.C. (perhaps proving they are worlds apart?).

6.2 COMPOSITION AND STRUCTURE OF PLANETS

The fact that there are two distinct kinds of planets—the rocky terrestrial planets and the gas-rich jovian planets—leads us to believe that they formed under different conditions. Certainly their compositions are dominated by different elements. Let us look at each type in more detail.

6.2.1 The Giant Planets

The two largest planets, Jupiter and Saturn, have nearly the same chemical makeup as the Sun itself; they are composed primarily of the two elements hydrogen and helium, with 75 percent of the mass being hydrogen and 25 percent helium. On Earth, both hydrogen and helium are gases, so Jupiter and Saturn are sometimes called gas planets. But this name is misleading. Jupiter and Saturn are so large that the gas is compressed in their interior until the hydrogen becomes a liquid. Because the bulk of both planets consists of compressed, liquefied hydrogen, we should really call them liquid planets (Figure 6.7).

Under the force of gravity, the heavier elements sink toward the inner parts of a liquid or gaseous planet. Both Jupiter and Saturn thus have cores composed of heavier rock, metal, and ice, but we cannot see these regions directly. In fact, when we look down from above, all we see is the atmosphere with its swirling clouds. We must infer the existence of the denser core inside these planets from studies of each planet's gravity.

Uranus and Neptune are much smaller than Jupiter and Saturn, but each also has a core of metal, rock, and ice. In one way, however, Uranus and Neptune are very similar to Jupiter and Saturn: All four of these giant planets have cores of about the same mass—roughly ten times the entire mass of the Earth. Uranus and Neptune were less efficient at attracting hydrogen and helium gas, so they have much smaller atmospheres in proportion to their cores.

MAKING CONNECTIONS

Names in the Solar System

We humans just don't feel comfortable until something has a name. Types of butterflies, new elements, and mountains of Venus all need names for us to feel we are acquainted with them. How do we give names to objects and features in the solar system?

Planets and satellites are named after gods and heroes in Greek and Roman mythology (with a few exceptions among the moons of Uranus that are drawn from English literature). When William Herschel, a German immigrant to England, first discovered the planet we now call Uranus, he wanted to name it Georgium Sidus (George's star) after King George III of his adopted country. This caused such an outcry among astronomers in other countries, however, that the classical tradition was resumed and has been maintained ever since. Luckily, there were a lot of minor gods in the ancient pantheon, so plenty of names are left for the many small moons we are discovering around the giant planets. (Appendix 8 lists the larger satellites; how many names do you recognize?)

Comets are named after their discoverers (giving an extra incentive to comet hunters); asteroids are named by their discoverers after just about anyone or anything they want. Recently, asteroid names have been used to recognize people who have made significant contributions to astronomy, including two of the authors of this book.

That was pretty much all the naming that was needed while our study of the solar system was confined to Earth. But now our robot probes have surveyed and photographed many worlds in great detail, and each world has a host of features that suddenly need names. To make sure that naming things in space remains multinational, rational, and somewhat dignified, astronomers have given the responsibility of approving names to a special committee of the International Astronomical Union, the body that includes scientists from every country that does astronomy.

This committee has developed a set of rules for naming features on other worlds. For example, craters on Venus are named for women who have made significant contributions to human knowledge and welfare. Volcanic features on Jupiter's moon Io, which is in a constant state of volcanic activity, are named after gods of fire and thunder from the mythologies of many cultures. Craters on Mercury commemorate famous novelists, playwrights, artists, and composers. On Saturn's satellite Tethys, all the features are named after characters in Homer's great epic poem, *The Odyssey*. As we explore further, it may well turn out that more places in the solar system need names than Earth history can provide. Perhaps by then, explorers and settlers on these worlds will be ready to develop their own names for the places that (if only for a while) they will call home.

■ FIGURE 6.7

Saturn This image of Saturn and its ring system was taken by the Cassini spacecraft on May 7, 2004 as it approached the planet. It shows the planet in its natural color. Like the other giant planets, Saturn is made mostly of gas and liquid. Its rings are made up of many small icy particles, which — like the planet — reflect sunlight back to our eyes or spacecraft cameras.

Chemically, each of the giant planets is dominated by hydrogen and its many compounds. Nearly all of the oxygen present is chemically combined with hydrogen to form H_2O (water). Chemists call such a hydrogen-dominated composition *reduced*. Throughout the outer solar system, we find abundant water (mostly in the form of ice) and reducing chemistry.

6.2.2 The Terrestrial Planets

The terrestrial planets are quite different from the giants. In addition to being much smaller, they are composed primarily of rocks and metals. These, in turn, are made of elements that are less common in the universe as a whole. The most abundant rocks, called silicates, are made of silicon and oxygen, while the most common metal is iron. We can tell from their densities (see Table 6.2) that Mercury has the highest proportion of metals (which are denser) and the Moon has the lowest. (To refresh your memory about the definition of density, see Section 2.2.)

Earth, Venus, and Mars all have roughly similar bulk compositions: About one third of their mass consists of iron–nickel or iron–sulfur combinations; two thirds is made of silicates. Because these planets are largely composed of oxygen compounds (such as the silicate minerals of their crusts), their chemistry is said to be *oxidized*.

When we look at the internal structure of each of the terrestrial planets, we find that the densest metals are in a central core, with the lighter silicates near the surface. If these planets were liquid, like the giant planets, we could understand this effect as the result of the sinking of heavier materials due to the pull of gravity. Thus we believe that, although the terrestrial planets are solid today, they must once have been hot enough to melt.

Differentiation is the name given to the process by which gravity helps separate a planet's interior into layers of different compositions and densities. The heavier metals sink to form a core, while the lightest minerals float to the surface to form a crust. Later, when the planet cools, this layered structure is preserved. In order for a rocky planet to differentiate, it must be heated to the melting point of rocks, typically above 1300 K. (Temperature was defined in Chapter 4; see Appendix 5 for a discussion of the different temperature scales.)

6.2.3 Moons, Asteroids, and Comets

Chemically and structurally, Earth's Moon is like the terrestrial planets, but it is an exception among satellites. Most of the moons are in the outer solar system, and they have compositions similar to those of the cores of the giant planets around which they orbit. The three largest satellites, Ganymede and Callisto in the jovian system and Titan in the saturnian system, are composed half of frozen water, half of rocks and metals. Most of these moons differentiated during formation, and today they have cores of rock and metal, with upper layers and crusts of very cold and thus very hard ice (Figure 6.8).

The asteroids and comets, as well as the smallest moons, were probably never heated to the melting point. They therefore retain their original composition and structure. This is one of their main attractions for astronomers: They represent relatively unmodified material dating back to the time of the formation of the solar system. In a sense, they are ancient chemical "fossils," helping us to learn about a time whose traces have long been erased on larger worlds.

6.2.4 Temperatures: Going to Extremes

Generally speaking, the farther a planet or satellite is from the Sun, the cooler it is. The planets are heated by the radiant energy of the Sun, which gets weaker as the square of the distance. You know how rapidly the heating effect of a fireplace or an outdoor "radiant heater" diminishes as you walk away from it—the same effect applies to the Sun. Mercury, the closest planet to the Sun, has a blistering surface

JPL/NASA

■ **FIGURE 6.8**

Ganymede A view of Jupiter's satellite Ganymede from the Galileo spacecraft, taken in June 1996. The brownish-gray color of the surface indicates a dusty mixture of rocky material and ice. The bright spots are places where recent impacts have uncovered fresh ice from underneath.

temperature of more than 500 K (230°C), as hot as the cleaning cycle on an electric oven, whereas the surface temperature on Pluto is only about 50 K, colder than liquid air.

Mathematically, the temperatures (expressed on the Kelvin scale) decrease approximately in proportion to the square root of the distance from the Sun. Pluto is 30 AU from the Sun, or 100 times the distance of Mercury. Thus, Pluto's temperature is less than that of Mercury by the square root of 100, or a factor of 10—from 500 K to 50 K.

Because of this strong dependence of temperature on distance from the Sun, Earth is the only planet where surface temperatures generally lie between the freezing and boiling points of water. Our planet is the only one with liquid water on its surface and therefore (as far as we know) the only one to support life.

ASTRONOMY BASICS
There's No Place Like Home

In the classic film *The Wizard of Oz*, Dorothy, our heroine, concludes after her many adventures in "alien" environments that "there's no place like home." The same can be said of the other worlds in our solar system. There are many fascinating places, large and small, that we might like to visit, but on none of them could human beings survive without a great deal of artificial assistance.

A thick atmosphere keeps the surface temperature on our neighbor Venus at a sizzling 700 K (near 900°F), hot enough to melt several metals. And the pressure of its mostly carbon dioxide atmosphere equals that found almost a kilometer underwater on Earth. Mars, on the other hand, has temperatures generally below freezing, with air (also mostly CO_2) so thin, it's like that found at an altitude of 30 km (100,000 ft) in Earth's atmosphere. And the red planet is so dry that it has not had any rain for billions of years.

In the outer layers of the jovian planets, it is neither warm enough nor solid enough for human habitation. Any bases we build in the systems of the giant planets may well have to be in space or on one of their satellites—none of which is particularly hospitable to a luxury hotel with swimming pool and palm trees. Perhaps we will find warmer havens deep inside the clouds of Jupiter or in the ocean under the frozen ice of its moon Europa.

All of this suggests that we had better take good care of the Earth because it is the only site where large numbers of our species can survive unaided. Recent human activity may be reducing the habitability of our planet by adding pollutants to the atmosphere or reducing the fraction of the Earth's surface where vegetation can thrive and continue to supply us with oxygen. In a solar system that seems unready to receive us, making the Earth less hospitable to life may be a grave mistake.

■ ■ ■ ■ ■ ■ ■ ■ ■ ■ ■ ■ ■ ■

6.2.5 Geological Activity

The crusts of all of the terrestrial planets, as well as of the larger satellites, have been modified over their histories by both internal and external forces. Externally, each has been battered by a slow rain of projectiles from space, leaving surfaces pockmarked by impact craters of all sizes (see Figure 6.3). We have good evidence that this bombardment was far greater in the early history of the solar system, but it certainly continues to this day. (The collision of more than 20 large pieces of Comet Shoemaker-Levy 9 with Jupiter in the summer of 1994 is one recent example of this process. Figure 6.9 shows the aftermath of these collisions, when debris clouds larger than the Earth could be seen in Jupiter's atmosphere.)

At the same time, internal forces on the terrestrial planets have buckled and twisted their crusts, built up mountain ranges, erupted as volcanoes, and generally reshaped the surfaces in what we call geological activity. (The prefix *geo* means "Earth," so this is a bit of an "Earth-chauvinist" term, but it is so widely used that your authors bow to tradition.) Among the terrestrial planets, Earth and Venus have experienced the most geological activity over their histories,

STScI/NASA

■ **FIGURE 6.9**

Jupiter with Huge Dust Clouds The Hubble Space Telescope took this sequence of images of Jupiter in summer 1994, when the fragments of Comet Shoemaker-Levy 9 collided with the giant planet. Here we see the site hit by fragment G, from 5 minutes to five days after impact. Several of the dust clouds generated by the collisions became larger than the Earth.

although some of the satellites in the outer solar system are also surprisingly active. In contrast, our own Moon is a dead world where geological activity ceased billions of years ago.

Geological activity on a planet is the result of a hot interior. The forces of volcanism and mountain building are driven by heat escaping from the cores of the planets. As we will see, each of the planets was heated at the time of its birth, and this primordial heat initially powered extensive volcanic activity, even on our Moon. But small objects like the Moon soon cooled off. The larger the planet or satellite, the more likely it is to retain its internal heat, and therefore the more we expect to see surface evidence of continuing geological activity. The effect is similar to our own experience with a baked potato: The larger the potato, the more slowly it cools. If we want a potato to cool quickly, we cut it into small pieces.

For the most part, the history of volcanic activity on the terrestrial planets conforms to the predictions of this simple theory. The Moon, smallest of these objects, is a geologically dead world. Although we know much less about Mercury, it seems likely that this planet, too, ceased most volcanic activity about the same time the Moon did. Mars represents an intermediate case: It has been much more active than the Moon, but less so than the Earth. Earth and Venus, the largest terrestrial planets, still have molten interiors and high levels of geological activity even today, some 4.5 billion years after their birth.

NASA

■ **FIGURE 6.10**
Our Cratered Moon   This image of the Moon's surface was taken during the *Apollo 16* mission and shows craters of many different sizes.

6.3   DATING PLANETARY SURFACES

How do we know the age of the surfaces we see on planets and satellites? If a world has a surface (as opposed to being mostly gas and liquid), we have developed some techniques for estimating how long ago that surface solidified. Note that the age of these surfaces is not necessarily the age of the planet as a whole. On geologically active objects (including the Earth), vast outpourings of molten rock or the erosive effects water and ice have erased evidence of earlier epochs and present us with only a relatively young surface for investigation.

6.3.1   Counting the Craters

One way to estimate the age of a surface is by counting the number of impact craters. This technique works because the rate at which impacts have occurred in the solar system has been roughly constant for several billion years. Thus, in the absence of forces to eliminate craters, the number of craters is simply proportional to the length of time the surface has been exposed. This technique has been successfully applied to planets, satellites, and even asteroids (Figure 6.10).

Bear in mind that crater counts can tell us only the time since the surface experienced a major change. Esti-

mating ages from crater counts is a little like walking along a sidewalk in a snowstorm after the snow has been falling steadily for a day or more. You may notice that in front of one house the snow is deep, while next door the sidewalk may be almost clear. Do you conclude that less snow has fallen in front of Ms. Jones's house than Mr. Smith's? More likely, you conclude that Jones has recently swept the walk clean and Smith has not. Similarly, the numbers of craters indicate how long it has been since a planetary surface was last "swept clean" by lava flows or by molten materials ejected when a large impact happened nearby.

Astronomers can thus use the numbers of craters on different parts of the same world to provide important clues about how regions on that world evolved. On a given planet or moon, the more heavily cratered terrain will generally be older (i.e., more time will have elapsed there since something swept the region clean).

6.3.2   Radioactive Rocks

Another way to trace the history of a solid world is to measure the ages of individual rocks. After samples were brought back from the Moon by the Apollo astronauts, the techniques that had been developed to date rocks on Earth could also be applied to establish a chronology for the Moon. In addition, a few samples of material from the Moon, Mars, and the large asteroid Vesta have fallen to Earth as meteorites and can also be examined directly (see Chapter 13).

Scientists measure the ages of rocks using the properties of natural **radioactivity.** Around the beginning of the

20th century, physicists began to understand that some atomic nuclei are not stable but can spontaneously split apart (decay) into smaller nuclei. The process of radioactive decay involves the emission of particles such as electrons, or of radiation in the form of gamma rays (see Chapter 4).

For any one radioactive nucleus, it is not possible to predict when the decay process will happen. Such decay is random in nature, just as the throw of dice is: As gamblers have all too often found out, it is impossible to say just when dice will come up 7 or 11. But for a very large number of dice tosses, we can calculate the odds that 7 or 11 will come up. Similarly, if you have a very large number of radioactive atoms of one type (say, uranium), there is a specific time period, called its **half-life,** during which the chances are fifty–fifty that decay will occur for any of the nuclei.

A particular nucleus may last a shorter or longer time than its half-life, but in a large sample almost exactly half will have decayed after a time equal to one half-life. Half of those remaining will have decayed in two half-lives, leaving only one half of a half—or one quarter—of the original sample (Figure 6.11).

If you had 1 gram of pure radioactive nuclei whose half-life was 100 years, then after 100 years you would have 1/2 gram, after 200 years 1/4 gram, after 300 years only 1/8 gram, and so forth. However, the material does not disappear. Instead, the radioactive atoms are replaced with their decay products. Sometimes the radioactive atoms are called *parents*, and the decay products are called *daughter* elements.

In this way, radioactive elements whose half-lives we have determined can provide accurate nuclear clocks. By comparing how much of a radioactive parent element is left in a rock with how much of its daughter products have accumulated, we can learn how long the decay process has been going on and hence how long ago the rock formed. Table 6.3 summarizes the decay reactions used most often to date lunar and terrestrial rocks.

When astronauts first flew to the Moon, one of their most important tasks was to bring back lunar rocks for ra-

dioactive age dating. Until then, astronomers and geologists had no real idea of the age of the lunar surface. Counting craters had given relative ages (for example, the heavily cratered lunar highlands were older than the dark lava plains), but scientists could not measure the actual age in years. Some thought that the ages were as young as those of Earth's surface, implying an active geology on the Moon. Only in 1969, when the first Apollo samples were dated, did we learn that the Moon is an ancient, geologically dead world. Using such dating techniques, we can determine the ages of both the Earth and the Moon: Each was formed about 4.5 billion years ago. (We show this date in the "Cosmic Calendar" at the end of the Prologue.)

By the way, note that the decay of radioactive nuclei generally releases energy in the form of heat. Although the energy from a single nucleus is not very large (in human terms), the enormous numbers of radioactive nuclei in a planet or satellite (especially early on) can be a significant source of internal energy for that world.

TABLE 6.3 *Radioactive Decay Reactions Used to Date Rocks*

Parent	Daughter	Half-Life (billion yr)
Samarium-147*	Neodymium-143	106
Rubidium-87	Strontium-87	48.8
Thorium-232	Lead-208	14.0
Uranium-238	Lead-206	4.47
Potassium-40	Argon-40	1.31

*The number after each element is its atomic weight, equal to the number of protons plus neutrons in its nucleus. This specifies the *isotope* of the element; different isotopes of the same element differ in the number of neutrons.

Ace⊙Astronomy™ Log into AceAstronomy and select this chapter to see the Active Figure called "Radioactive Decay."

■ **FIGURE 6.11**

Radioactive Decay This graph shows (in red) the amount of a radioactive sample that remains after several half-lives have passed. After one half-life, half the sample is left; after two half-lives, one half of the remainder (or 1/4) is left; and after three half-lives, one half of that (or 1/8) is left. Note that in reality, the decay of radioactive materials in a rock sample would not cause any visible change in the appearance of the rock; the splashes of color are shown here for educational purposes only.

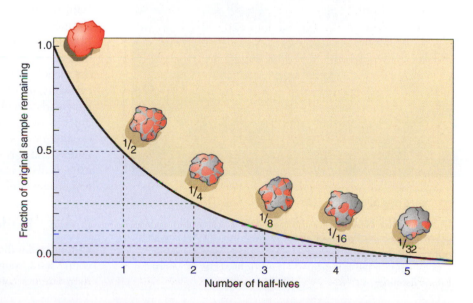

6.4 ORIGIN OF THE SOLAR SYSTEM

Much of astronomy is motivated by a desire to understand the origin of things—to find at least partial answers to the age-old questions of where the universe, the Sun, the Earth, and we ourselves came from. Each planet and satellite is a fascinating place that may stimulate our imaginations as we try to picture what it would be like to visit there. But taken together, the members of the solar system also preserve patterns that can tell us about the formation of the entire system. As we begin our exploration of the planets, we want to introduce our modern picture of how the solar system formed. We will return to this question in more detail in Chapter 13.

6.4.1 Looking for Patterns

One way to approach the question of origins is to look for regularities among the planets. We found, for example, that all the planets revolve in the same direction around the Sun and that they lie in nearly the same flat plane. The Sun also spins in the same direction about its own axis. Astronomers interpret this pattern as evidence that the Sun and planets formed together from a spinning system of gas and dust that we call the **solar nebula** (Figure 6.12).

The composition of the planets gives another clue. Spectroscopic analysis (see Chapter 4) allows us to determine what elements are present in the Sun and the planets.

Art by planetary astronomer William Hartmann

■ FIGURE 6.12

In the Solar Nebula An expert artist's conception of the solar nebula, the flattened cloud of gas and dust from which our planetary system formed. Icy and rocky planetesimals can be seen in the foreground. The bright center is where the Sun is forming.

The Sun has the same hydrogen-dominated composition as Jupiter and Saturn and therefore appears to have been formed from the same reservoir of material. In comparison, each of the terrestrial planets and satellites is relatively deficient in the light gases and the various ices that form out of the common elements oxygen, carbon, and nitrogen. Instead, on Earth and its neighbors we see mostly the rarer heavy elements like iron and silicon. This pattern suggests that the processes that led to planet formation in the inner solar system must somehow have rejected much of the lighter materials. These must have escaped, leaving just a residue of heavy stuff.

The reason for this is not hard to guess, bearing in mind the heat of the Sun. The inner planets (which are closer to the Sun) are now made of rock and metal, which can survive heat, but they contain very little ice or gas, which evaporate when the temperatures are high. (To see what we mean, just compare how long a rock and an ice cube survive when they are placed in the sunlight.) In the outer solar system, where it has always been cooler, the planets and their satellites are mostly composed of ice and gas.

6.4.2 The Evidence from Far Away

A second approach to understanding the origin of the solar system is to look outward for evidence that other, similar systems of planets are forming elsewhere. We cannot look back in time to the formation of our own system, but many stars in space are much younger than the Sun. In these systems, the process of planet formation might still be accessible to direct observation. One of the important astronomical discoveries of recent years is the presence of just such planetary nurseries. What we actually observe are other "solar nebulas" or *circumstellar disks*—flattened, spinning clouds of gas and dust surrounding young stars. These disks resemble our own solar system's initial stages of formation billions of years ago (Figure 6.13).

In addition to discovering these stellar nurseries, astronomers have recently developed tools to let them detect planets in orbit around other stars. So far we can find only giant planets, so we know nothing about the presence or absence of terrestrial planets in other planetary systems. But already we know that other planetary systems are not necessarily configured like our own. For example, many of the planets found around distant stars are in highly elliptical orbits, unlike the planets in the solar system, which follow nearly circular orbits. In other cases, giant planets are much closer to their stars than Mercury is to the Sun. An exciting new opportunity is opening as astronomers try to understand this bewildering variety of planetary systems, instead of focusing exclusively on our own solar system, which may not be typical at all.

6.4.3 Building Planets

Circumstellar disks are a common occurrence around very young stars, suggesting that disks and stars form together. Astronomers can use theoretical calculations to see how

C. R. O'Dell / Rice U. / NASA

■ **FIGURE 6.13**

Planetary Nurseries Hubble Space Telescope photos of the Orion nebula, a region of current star formation, and of an imbedded circumstellar disk orbiting a very young star. This disk is a contemporary analogue of our own solar nebula—a location where planets are probably being formed today.

solid bodies might form from the gas and dust in these disks as they cool. These models show that material begins to coalesce by first forming smaller objects, precursors of the planets, which we call **planetesimals.**

Fast computers simulate the way millions of planetesimals, each no larger than 100 km in diameter, might have gathered together under their mutual gravity to form the planets we see today. We are beginning to understand that this process was a violent one, with planetesimals crashing into each other and sometimes even disrupting the growing planets themselves. As a consequence of these violent impacts (as well as the heat from radioactive materials in them), all of the planets were heated and differentiated—which helps explain their present internal structures.

The process of impacts and collisions in the early solar system was complex and often apparently random. The so-lar nebula model can explain many of the regularities we find in the solar system, but the random collisions of massive collections of planetesimals could be the reason for some exceptions to the "rules" of solar system behavior. For example, why do the planets Uranus and Pluto spin on their sides? Why does Venus spin slowly and in the opposite direction from all the other planets? Why does the composition of the Moon resemble the Earth in many ways and yet exhibit substantial differences? Although we cannot know for sure, the answers to such questions probably lie in enormous collisions that took place in the solar system long before life on Earth began.

Today, some 4.5 billion years after its origin, the solar system is—thank goodness—a much less violent place. As we will see, however, some planetesimals have continued to interact and collide, and their fragments move about the

Carl Sagan: Solar System Advocate

The best-known astronomer in the world during the 1970s and 1980s, Carl Sagan devoted most of his professional career to studying the planets, and much of his considerable energy to raising public awareness of what we can learn from exploring the solar system. Born in Brooklyn, New York, in 1934, Sagan became interested in astronomy as a youngster; he also credits science fiction stories for sustaining his fascination with what's out there. After earning both his bachelor's and PhD degrees from the University of Chicago, he worked at the University of California and Harvard before becoming professor of astronomy at Cornell University in 1968.

In the early 1960s, when many scientists still thought that Venus might turn out to be a hospitable place, Sagan demonstrated that the atmosphere of our neighborhood planet could act like a giant greenhouse, keeping the heat in and raising the temperature enormously. He showed that the seasonal changes astronomers had seen on Mars were caused not by vegetation but by wind-blown dust. He was a member of the scientific teams for many of the robotic missions that explored the solar system and was instrumental in getting NASA to put a message-bearing plaque aboard the Pioneer spacecraft, as well as audio-video records on the Voyager probes (as explained in Chapter 29).

To encourage public interest and public support of planetary exploration, Sagan helped found The Planetary Society, now the largest space-interest organization in the world. He was a tireless and eloquent exponent of the need to study the solar system close-up, and of the value of learning about other worlds in order to take better care of our own.

Carl Sagan

Sagan did experimental work simulating conditions on the early Earth and demonstrated how some of life's fundamental building blocks might have formed from the primordial "soup" of natural compounds on our

> . . . He was a tireless and eloquent exponent of the need to study the solar system close-up . . .

planet. He and his students built "Jupiter boxes" to demonstrate how conditions in the giant planet's atmosphere may in some ways resemble what it was like on Earth before life began. In addition, he and his colleagues developed computer models showing that the consequences of nuclear war for the Earth would be even more devastating than anyone had thought (this is now called the *nuclear winter* hypothesis) and demonstrating some of

the serious consequences of continued pollution of our atmosphere.

Sagan was perhaps best known, however, as a brilliant popularizer of astronomy and the author of many books on science, including the best-selling *Cosmos,* the Pulitzer-Prize-winning *Dragons of Eden,* and several evocative tributes to solar system exploration such as *The Cosmic Connection* and *Pale Blue Dot.* His book *The Demon Haunted World,* finished just before his death in 1996, is perhaps the best antidote to fuzzy thinking about pseudo-science and irrationality in print today. His intriguing science fiction novel, *Contact,* also a successful film, is recommended by many science instructors as a scenario for making contact with life elsewhere that is much more reasonable than most science fiction.

An appearance on *The Tonight Show* in 1973 (where his lucid explanations caught the attention of then-host Johnny Carson, an amateur astronomer) began a long series of television appearances that culminated in Sagan's 13-part public television series, *Cosmos.* It has been seen by an estimated 500 million people in 60 countries and has become one of the most-watched series in the history of public broadcasting. Sagan, by the way, was quick to point out that not once in the episodes of *Cosmos* did he say "billions and billions," a phrase now indelibly associated with him (although he did use it in other media appearances over the years). A few astronomers scoffed at a scientist who spent so much time in the public eye, but it is probably fair to say that Sagan's enthusiasm and skill as an explainer won more friends for the science of astronomy than anyone or anything else in the second half of the 20th century.

solar system as roving "transients" that can make trouble for the established members of the Sun's family, such as our own Earth. In Chapter 7 we will examine some of the devastation caused by impacts from space in the past along with some of the dangers that may face us in the future.

SURFING THE WEB

The Nine Planets: A Multimedia Tour of the Solar System:
nineplanets.org
Impressive amounts of information about each planet and moon in the solar system, with sections on space exploration, images, and links.

Views of the Solar System:
www.solarviews.com/eng/homepage.htm
Another good site of information and images about the worlds in our solar system. The site is similar to *The Nine Planets,* but more focused on graphics. There is useful background information, a data browser, a glossary, and more.

Planetary Sciences Site:
nssdc.gsfc.nasa.gov/planetary/planetary_home.html
NASA's National Space Science Data Center has a somewhat more technical, but still quite accessible, site with information and images on each world our spacecraft have explored, with concise descriptions of the various missions to the planets.

Gazetteer of Planetary Nomenclature:
planetarynames.wr.usgs.gov
A helpful resource if you want to find out how objects and features in the solar system get their names.

Planetary Photojournal: photojournal.jpl.nasa.gov/
This NASA site features thousands of the best images from planetary exploration, with detailed captions and excellent indexing. You can dial up images by world, feature name, or date, and download them in a number of formats.

The Thousand-Yard Model: Earth as a Peppercorn:
www.noao.edu/education/peppercorn/pcmain.html
Another scale model of the solar system, with careful instructions for doing a "walking tour" and other activities.

SUMMARY

6.1 Our solar system consists of the Sun, nine planets, nearly 100 known satellites, and a host of smaller objects. The planets can be divided into two groups: the inner **terrestrial planets** and the outer **giant planets.** Pluto does not fit into either category. The giant planets are composed mostly of liquids and gases. Smaller members of the solar system include **asteroids,** rocky and metallic objects found mostly between Mars and Jupiter; **comets,** made mostly of frozen gases, which generally orbit far from the Sun; and countless smaller grains of cosmic dust. When a **meteor** survives its passage through our atmosphere and falls to Earth, we call it a **meteorite.**

6.2 The giant planets have dense cores roughly ten times the mass of Earth, surrounded by layers of hydrogen and helium. The terrestrial planets consist mostly of rocks and metals. They were once molten, which allowed their structures to **differentiate** (i.e., their denser materials sank to the center). The Moon resembles the terrestrial planets in composition, but most of the other satellites—which orbit the giant planets—have a large quantity of frozen ices within them. Generally, worlds closer to the Sun have higher temperatures. The surfaces of terrestrial planets have been modified by impacts from space and by varying degrees of geological activity.

6.3 The ages of the surfaces of objects in the solar system can be estimated by counting craters: On a given world, a more heavily cratered region will always be older. We can also use samples of rocks with **radioactive** elements in them to obtain the time since the layer in which the rock formed last solidified. The **half-life** of a radioactive element is the time it takes for half the sample to decay; we determine how many half-lives have passed by how much of a sample remains the radioactive element and how much has become the decay product.

6.4 Regularities among the planets lead astronomers to hypothesize that the Sun and the planets formed together in a giant, spinning cloud of gas and dust called the **solar nebula.** Recent observations show tantalizingly similar circumstellar disks around other stars. We have also found giant planets orbiting a number of nearby stars. Within the solar nebula, material first coalesced into **planetesimals;** many of these gathered together to make the planets and satellites. The remainder can still be seen as comets and asteroids.

INTER-ACTIVITY

A Your group should discuss and make a list of the reasons humanity might have to explore other worlds. Does the group think such missions of exploration are worth the investment? Why?

B Your instructor will assign each group a world. Your task is to think about what it would be like to be there. Feel free to look ahead in the book to the relevant chapters. Discuss where on or around your world we would establish a foothold and what we would need to survive there.

C In Astronomy Basics: *There's No Place Like Home* in this chapter, your authors discuss briefly how human activity is transforming our planet's overall environment. Can you think of other ways that this is happening?

D Some scientists criticized Carl Sagan for "wasting his research time" popularizing astronomy. Do you think scientists should spend their time interpreting their field for the public? Why or why not? Are there ways that scientists who are not as eloquent or charismatic as Carl Sagan can still contribute to the public understanding of science?

E Your group has been named a special committee by the International Astronomical Union to suggest names of features (like craters, trenches, etc.) on a newly explored asteroid. Given the restrictions that any people after whom features are named must no longer be alive, what names or types of names would you suggest? (You are not restricted to names of people, by the way.)

F A member of your group has been kidnapped by a little-known religious cult that worships the planets. They will release him only if your group can tell which of the planets are currently visible in the sky during the evening and morning. You are forbidden from getting your instructor involved. How else could you find out the information you need? (Be as specific as you can; if your instructor says it's OK, feel free to continue answering this question in a library.)

REVIEW QUESTIONS

Ace⊙Astronomy™ Assess your understanding of this chapter's topics with additional quizzing and animations at **http://ace.brookscole.com/voyages**

1. List some reasons that the study of the planets has progressed more in the last few decades than has any other branch of astronomy.

2. Imagine you are a travel agent in the next century. An eccentric billionaire asks you to arrange a "Guinness Book of Solar System Records" kind of tour. Where would you direct him to find the following (use this chapter and Appendices 7 and 8):
   a. the least-dense planet
   b. the densest planet
   c. the largest moon in the solar system
   d. the planet at whose surface you would weigh the most (*Hint:* Weight is directly proportional to surface gravity.)
   e. the smallest planet
   f. the planet that takes the longest time to rotate
   g. the planet that takes the shortest time to rotate
   h. the planet whose diameter is the closest to Earth's
   i. the moon that takes the shortest time to orbit its planet
   j. the densest moon
   k. the moon that orbits the farthest from its planet
   l. the most massive moon
   m. the most spectacular ring system

3. How do terrestrial and giant planets differ? List as many ways as you can think of.

4. Why are there so many craters on the Moon and so few on Earth?

5. How do asteroids and comets differ?

6. How and why is the Earth's Moon different from the larger satellites of the giant planets?

7. Where would you look for some "original" planetesimals left over from the formation of our solar system?

8. Describe how we use radioactive elements and their decay products to find the age of a rock sample. Is this necessarily the age of the entire world from which the sample comes? Explain.

9. What was it like in the solar nebula? Why did the Sun form at its center?

10. What can we learn about the formation of our solar system by studying other stars? Explain.

THOUGHT QUESTIONS

11. Using some of the astronomical resources in your college library or the World Wide Web, find five names of features on each of three other worlds that are named after people. In a sentence or two, describe who each of these people was.

12. Explain why the planet Venus is differentiated, but asteroid Fraknoi, a very boring and small member of the asteroid belt, is not.

13. Would you expect as many impact craters per unit area on the surface of Venus as on the surface of Mars? Why or why not?

14. Interview a sample of 20 people who are not taking an astronomy class and ask them if they can name a living astronomer. What percentage of those interviewed were able to name one? Typically, the one living astronomer the public knows these days is Stephen Hawking. Why is he better known than most astronomers? How would your results have differed if you had asked the same people to name a movie star or a professional football player?

FIGURING FOR YOURSELF

Let's compare the densities of several members of the solar system. The density of an object equals its mass divided by its volume. The volume of a sphere (like a planet) is calculated with the formula:

$$V = \frac{4}{3}\pi R^3$$

where π (Greek pi) has a value of approximately 3.14. Although planets are not perfect spheres, this formula works well enough. The masses and diameters of the planets are given in Table 6.2. For the data on the satellites, look in Appendix 8. We will select Saturn's satellite Mimas as our example, with a mass of 4×10^{19} kg and a diameter of approximately 400 km (radius 200 km = 2×10^5 m). The volume of Mimas is $4/3 \times 3.14 \times (2 \times 10^5)^3 = 3.3 \times 10^{16}$ m^3. Density is mass divided by volume: $4 \times 10^{19} / 3.3 \times 10^{16} = 1.2 \times 10^3$ kg/m^3. Note that the density of water in these units is 1000 kg/m^3, so Mimas must be mainly made of ice, not rock. (Note that the density of Mimas given in Appendix 8 is 1.2, but the units used there are different. In that table, we give density in units of g/cm^3, for which the density of water equals 1. Can you show, by converting units, that 1 g/cm^3 is the same as 1000 kg/m^3?)

15. Calculate the average density of our own planet, Earth. Show your work. (See Table 6.2 for data.)

16. Calculate the density of Jupiter. Show your work. Is it more or less dense than Earth? Why?

17. Calculate the density of Saturn. Show your work. How does it compare with the density of water? Explain how this can be?

18. What is the density of Jupiter's moon Europa (see Appendix 8 for data on moons)? Show your work.

19. Look at Appendices 7 and 8 and find the satellite whose diameter is the largest fraction of the diameter of the planet it orbits.

20. Barnard's Star, the second closest star to us, is about 56 trillion (5.6×10^{12}) km away. Calculate how far it would be using the scale model in Section 6.1.3.

21. A radioactive nucleus has a half-life of 5×10^8 years. Assuming that a sample of rock (say, in an asteroid) solidified right after the solar system formed, approximately what fraction of the radioactive element should be left in the rock today?

SUGGESTIONS FOR FURTHER READING

Barnes-Svarney, P. "The Chronology of Planetary Bombardments" in *Astronomy,* July 1988, p. 21.

Beatty, J. et al. *The New Solar System,* 4th ed. 1999, Sky Publishing. A compendium of articles on planetary astronomy by leading scientists in the field. Occasionally technical.

Burrows, W. E. *Exploring Space: Voyages in the Solar System and Beyond.* 1990, Random House. A science journalist looks at the broad political and social context of planetary exploration.

Greeley, R. and Batson, R. *The NASA Atlas of the Solar System.* 1997, Cambridge U. Press. Over-sized, coffee-table book with the best maps and images of the planets beautifully displayed.

Hartmann, W. "Piecing Together Earth's Early History" in *Astronomy,* June 1989, p. 24.

Jayawardhana, R. "Spying on Planetary Nurseries" in *Astronomy,* Nov. 1998, p. 62.

Kross, J. "What's In a Name?" in *Sky & Telescope,* May 1995, p. 28. Discusses the naming of features in the solar system.

Morrison, D. and Owen, T. *The Planetary System,* 3rd ed. 2003, Addison-Wesley. A thorough introduction to our modern knowledge.

O'Meara, S. "The World's Largest Solar System Scale Model" in *Sky & Telescope,* July 1993, p. 99. At the Lakeview Museum in Peoria, IL. (See an update in the March 1998 issue, p. 80, on a wide range of such models.)

Sagan, C. *Pale Blue Dot.* 1994, Random House. Eloquent introduction to planetary exploration and a sequel to Sagan's *Cosmos.*

Sheehan, W. *Worlds in the Sky.* 1992, U. of Arizona Press. History of how we came to understand the planets.

Stern, S. and Levinson, H. "Toward a Planet Paradigm" in *Sky & Telescope,* Aug. 2002, p. 42. On how to define a planet.

Molten lava fountaining from a volcanic vent in Hawaii—testament to the dynamic nature of the Earth's crust.

Earth as a Planet

By coming face to face with alternative fates of worlds more or less like our own, we have begun to better understand the Earth. Every one of these worlds is lovely and in-structive. But, so far as we know, they are also, every one of them, desolate and barren. Out there, there are no 'better places.' So far, at least.

Carl Sagan in *Pale Blue Dot* (1994, Random House)

THINKING AHEAD

Airless worlds in our solar system seem peppered with craters large and small. The Earth, on the other hand, has few craters, but a thick atmo-sphere and much surface activity. We believe that in the past the Earth, too, had many craters that have since been erased by forces in its crust and atmosphere. What can the persistent cratering on so many other worlds tell us about the history of our planet?

As our first step in exploring the solar system in more detail, we turn to the most familiar planet, our own Earth. The first humans to see the Earth as a blue sphere floating in the blackness of space were the *Apollo 8* as-tronauts who made the first voyage around the Moon in 1968. For many people, the historic images showing our world as a small, distant globe repre-sent a pivotal moment in human history, when it became difficult for educated human beings to avoid viewing our world with a global perspective.

In earlier chapters we discussed our planet's size and orbit, and we de-scribed the seasons and tides, which are related to the motions of the Earth. In this chapter we examine the composition and structure of our planet with its envelope of ocean and atmosphere. We ask how our terrestrial environ-ment came to be the way it is today. Because we live on it, we know the Earth much better than any other planet, and we can perform experiments on its surface that we cannot yet do elsewhere with any regularity. This direct knowl-edge helps us to use the Earth as a reference point for understanding other worlds, where we are mostly limited to remote observations.

Virtual Laboratories

Planetary Geology

Planetary Atmospheres and Their Retention

THE GLOBAL PERSPECTIVE

The Earth is a medium-sized planet with a diameter of approximately 13,000 km (Figure 7.1). As one of the inner or *terrestrial* planets, it is composed primarily of heavy elements such as iron, silicon, and oxygen—very different from the composition of the Sun and stars, which are dominated by the light elements hydrogen and helium. The Earth's orbit is nearly circular, and Earth is warm enough to support liquid water on its surface. It is the only planet in our solar system that is neither too hot nor too cold, but "just right" for the development of life as we know it. Some of the basic properties of the Earth are summarized in Table 7.1.

7.1.1 Earth's Interior

The interior of a planet—even our own Earth—is difficult to study, and its composition and structure must be determined indirectly. Our only direct experience is with the outermost skin of the Earth's crust, a layer no more than a few kilometers deep. It is important to remember that in many ways we know less about our own planet 5 km beneath our feet than we do about the surfaces of Venus and Mars.

The Earth is composed largely of metal and silicate rock (see Section 6.2). Most of this material is in a solid

Ace◐**Astronomy**™   Log into AceAstronomy and select this chapter to see the Active Figure called "Seismic Waves."

TABLE 7.1	*Some Properties of the Earth*
Semimajor axis	1.00 AU
Period	1.00 year
Mass	5.98×10^{24} kg
Diameter	12,756 km
Escape velocity	11.2 km/s
Rotation period	23 h 56 m 4 s
Surface area	5.1×10^8 km^2
Atmospheric pressure	1.00 bar

state, but some of it is hot enough to be molten. The structure of the material in the Earth's interior has been probed in considerable detail by measuring the transmission of **seismic waves** through the Earth. These are waves that spread through the interior of the Earth from earthquakes or explosion sites.

Seismic waves travel through a planet rather like sound waves through a struck bell. Just as the sound frequencies vary depending on what material the bell is made of and how it is constructed, so a planet's response depends on its composition and structure. By monitoring the seismic waves in different locations, scientists can learn about the layers the waves have traveled through. Some of these vibrations travel along the surface; others pass directly through the interior. Seismic studies have shown that our planet's interior consists of several distinct layers with different compositions, illustrated in Figure 7.2.

The top layer is the **crust,** the part of the Earth we know best (Figure 7.3). The crust under the oceans, which cover 55 percent of the surface, is typically about 6 km thick and is composed of volcanic rocks called **basalt.** Produced by the cooling of volcanic lava, basalts are made primarily of the elements silicon, oxygen, iron, aluminum, and magnesium. The continental crust, which covers 45 percent of the surface, is 20 to 70 km thick and is predominantly composed of a different volcanic class of silicates (rocks made of silicon and oxygen) called **granite.** These crustal rocks typically have densities of about 3 g/cm^3. (For comparison, the density of water is 1 g/cm^3.) The crust is the easiest layer for geologists to study, but it makes up only about 0.3 percent of the total mass of the Earth.

■ **FIGURE 7.1**

Planet Earth   The Earth from space, taken by the Apollo 17 astronauts in December 1972. This is one of the rare images of a full Earth taken during the Apollo program; most images show only part of the Earth's disk in sunlight.

NASA

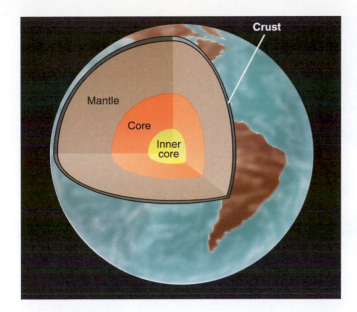

FIGURE 7.2

The Interior Structure of the Earth The crust, mantle, and inner and outer cores (liquid and solid, respectively) as revealed by seismic studies.

The largest part of the solid Earth, called the **mantle,** stretches from the base of the crust down to a depth of 2900 km. The mantle is more or less solid, but at the temperatures and pressures found there, the mantle rock can deform and flow slowly. The density in the mantle increases downward from about 3.5 g/cm^3 to more than 5 g/cm^3 as a result of the compression produced by the weight of overlying material. Samples of upper-mantle material are occasionally ejected from volcanoes, permitting a detailed analysis of its chemistry.

Beginning at a depth of 2900 km, we encounter the dense metallic **core** of the Earth. With a diameter of 7000 km, our core is substantially larger than the entire planet Mercury. The outer core is liquid, but the innermost part of the core (about 2400 km in diameter) is probably solid. In addition to iron, the core probably also contains substantial quantities of nickel and sulfur, all compressed to a very high density.

The separation of the Earth into layers of different densities is an example of *differentiation* (a concept explained in Chapter 6). The fact that the Earth is differentiated is evidence that it was once warm enough for the mantle rocks to melt, permitting the heavier metals to sink to the center and form the dense core.

7.1.2 Magnetic Field and Magnetosphere

We can find additional clues about the Earth's interior from its magnetic field. Our planet behaves in some ways as if a giant bar magnet were inside it, approximately aligned with the rotational poles of the Earth. This magnetic field is actually generated by moving material in the Earth's liquid

J. Frawley and NASA

FIGURE 7.3

Earth's Crust This computer-generated globe shows the surface of the Earth's crust with the oceans removed, as determined from satellite images and ocean floor radar mapping. The amounts of vegetation on the continents, as recorded by the instruments aboard the NOAA satellites, are indicated by color, ranging from light brown for the least vegetation, through yellow and orange, to green for the most.

Ace Astronomy™ Log into AceAstronomy and select this chapter to see Astronomy Exercise "Convection and Magnetic Fields I: Planet."

metallic core. As the liquid metal inside the Earth circulates, it sets up a circulating electric current. And, as discussed in Chapter 4, whenever charged particles are moving, they produce a magnetic field.

The Earth's magnetic field extends into surrounding space. When a charged particle encounters a magnetic field, it becomes trapped in the magnetic zone. Above the Earth's atmosphere, our field is able to trap small quantities of electrons and other atomic particles. This region, called the **magnetosphere,** is defined as the zone within which the Earth's magnetic field dominates over the weak interplanetary magnetic field that extends outward from the Sun (Figure 7.4).

Where do the charged particles trapped in our magnetosphere come from? They flow outward from the hot surface of the Sun; astronomers call this flow the *solar wind* (see Chapter 14). It not only provides particles for the Earth's magnetic field to trap; it also elongates our field in the direction pointing away from the Sun. Typically, the Earth's magnetosphere extends about 60,000 km, or ten Earth radii, in the direction of the Sun. But pointing away from the Sun, the magnetic field is shaped like a wind sock and can reach as far as the orbit of the Moon.

The magnetosphere was discovered in 1958 by instruments on the first U.S. Earth satellite, Explorer 1, which

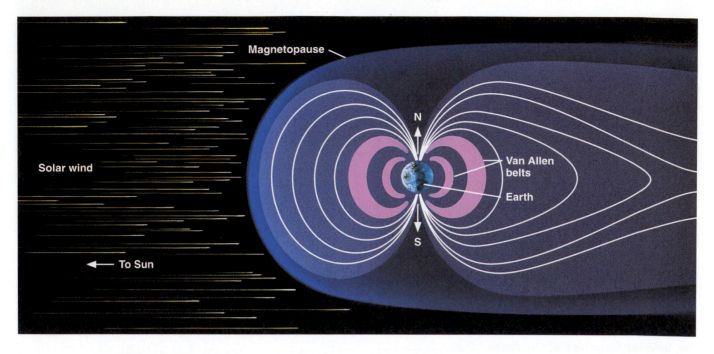

FIGURE 7.4

Earth's Magnetosphere A cross-sectional view of our magnetosphere (or zone of magnetic influence), as revealed by numerous spacecraft missions. Note how the wind of charged particles from the Sun "blows" the magnetic field outward like a wind sock.

recorded the ions (charged particles) trapped in its inner part. The regions of high-energy ions in the magnetosphere are often called the *Van Allen belts* in recognition of the University of Iowa professor who built the scientific instrumentation for Explorer 1 and correctly interpreted the satellite measurements. Since 1958, hundreds of spacecraft have explored various regions of the magnetosphere. In Chapter 14 we will discuss how the magnetic field and Van Allen belts of the Earth give rise to the brilliant curtains of light called *auroras*, visible in our atmosphere near the poles.

7.2 THE CRUST OF THE EARTH

Let us now examine our planet's outer layers in more detail. The Earth's crust is a dynamic place. Volcanic eruptions, erosion, and large-scale movements of the continents constantly rework the surface of our planet. Geologically, ours is the most active planet. Many of the geological processes described in this section have taken place on other planets as well, but usually in their distant pasts. Some of the satellites (moons) of the giant planets do have impressive activity (as we will see in Chapter 11). For example, Jupiter's satellite Io has a remarkable number of active volcanoes.

7.2.1 Composition of the Crust

The Earth's crust is largely made up of oceanic basalt and continental granite. These are both examples of **igneous** rock, the term used for any rock that has cooled from a

molten state. All volcanically produced rock is igneous (Figure 7.5).

Two other kinds of rock are familiar to us on the Earth, although it turns out that neither is common on other planets. **Sedimentary** rocks are made of fragments of igneous rocks or the shells of living organisms deposited by wind or water and cemented together without melting. On Earth, these rocks include the common sandstones, shales, and limestones. **Metamorphic** rocks are produced when high temperature or pressure alters igneous or sedimentary rocks physically or chemically (the word *metamorphic* means "changed in form"). Metamorphic rocks are produced on Earth because geological activity carries surface rocks down to considerable depths and then brings them back up to the surface. Without such activity, these changed rocks would not exist.

There is a fourth very important category of rock that can tell us much about the early history of the planetary system. This is **primitive** rock, which has largely escaped chemical modification by heating. Primitive rock represents the original material out of which the planetary system was made. No primitive material is left on the Earth because the entire planet was heated early in its history. To find primitive rocks, we must look to smaller objects such as comets, asteroids, and small planetary satellites.

A block of marble on Earth is composed of materials that have gone through all four of these states. Beginning as primitive material before the Earth was born, it was heated in the early Earth to form igneous rock, chemically transformed and redeposited (perhaps many times) to form sedimentary rock, and finally changed several kilometers

■ **FIGURE 7.5**

Igneous Rocks The formation of igneous rock as liquid lava cools and freezes. This is a lava flow from a basaltic eruption in Hawaii.

below the Earth's surface into the hard, white metamorphic stone we see today.

7.2.2 Plate Tectonics

Geology is the study of the Earth's crust and the processes that have shaped its surface throughout history. (Although *geo-* means related to the Earth, you will see in future chapters that astronomers also talk about the geology of other planets.) Heat escaping from the Earth's interior provides the energy for the formation of mountains, valleys, volcanoes, and even the continents and ocean basins themselves. But not until the middle of the 20th century did geologists succeed in understanding just how these landforms are created.

Plate tectonics is a theory that explains how slow motions within the mantle of the Earth move large segments of the crust, resulting in a gradual "drifting" of the continents as well as the formation of mountains and other large-scale geological features. Plate tectonics is now a concept as basic to geology as evolution by natural selection is to biology or gravity is to understanding the orbits of planets.

Ace ⊘**Astronomy**™ Log into AceAstronomy and select this chapter to see Astronomy Exercise "Convection and Plate Tectonics."

The Earth's crust and upper mantle (to a depth of about 60 km) are divided into about a dozen major plates that fit together like the pieces of a jigsaw puzzle (Figure 7.6). These plates are also capable of moving slowly with respect to each other. In some places, such as the Atlantic Ocean, the plates are moving apart; in others, they are being forced together. The power to move the plates is provided by slow **convection** of the mantle, a process by which heat escapes from the interior through the upward flow of warmer material and the slow sinking of cooler material. (Convection, in which energy is transported from a warm region such as the interior of the Earth to a cooler region such as the upper mantle, is a process we encounter often in astronomy—in stars as well as planets. It is also important in boiling water for coffee while studying for astronomy exams.)

As the plates slowly move, they bump into each other and cause dramatic changes in the Earth's crust over time. Four basic kinds of interactions between crustal plates are possible at their boundaries: (1) They can pull apart; (2) one plate can burrow under another; (3) they can slide alongside each other; or (4) they can jam together. Each of these activities is important in determining the geology of the Earth, as we now explain.

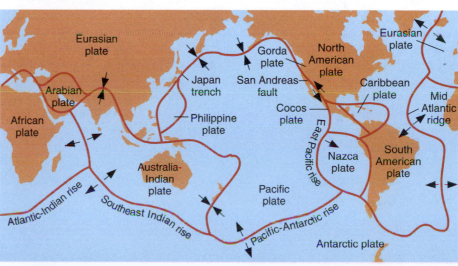

■ **FIGURE 7.6**

Earth's Continental Plates The major plates into which the crust of the Earth is divided. Arrows indicate the motion of the plates at average speeds of 4 to 5 cm per year.

Alfred Wegener: Catching the Drift of Plate Tectonics

When studying maps or globes of the Earth, many students notice that the coast of North and South America, with only minor adjustments, could fit pretty well against the coast of Europe and Africa. It seems as if these great land masses could once have been together and then were somehow torn apart. The same idea had occurred to others (including Francis Bacon as early as 1620), but not until the 20th century could such a proposal be more than speculation. The man who made the case for continental drift in 1912 was a German meteorologist and astronomer named Alfred Wegener.

Born in Berlin in 1880, Wegener was from an early age fascinated by Greenland, the world's largest island, which he dreamed of exploring. He studied at the universities in Heidelberg, Innsbruck, and Berlin, receiving a doctorate in astronomy by reexamining 13th-century astronomical tables. But his interests turned more and more toward the Earth, particularly its weather. He carried out experiments using kites and balloons, becoming so accomplished that he and his brother set a world record in 1906 by flying for 52 hours in a balloon.

Wegener first conceived of continental drift in 1910 while examining a world map in an atlas. But it took two years for him to assemble sufficient data to propose the idea in public: He published the results of his work in book form in 1915. Wegener's evidence went far beyond the congruence in the shapes of the continents. He proposed that the similarities between unusual fossils found only in certain parts of South America and

Alfred Wegener
(1880–1930)

Historical Pictures Service, Chicago

Africa indicated that these two continents were joined at one time. He also showed that resemblances among living animal species on different continents could best be explained by assuming that the continents were once connected in a supercontinent he called *Pangaea* (from Greek elements meaning "all land").

■ ■ ■ ■ ■ ■ ■ ■ ■ ■ ■ ■ ■

Wegener's suggestion first met with hostile reaction from scientists.

Wegener's suggestion met with a hostile reaction from most scientists. Although he had marshaled an impressive list of arguments for his theory, he was missing a *mechanism:* No one could explain *how* solid continents could drift over thousands of miles. A few scientists were sufficiently impressed by Wegener's work to con-

tinue searching for additional evidence, but many found the notion of moving continents too different from the accepted view of the Earth's surface to take seriously. Developing an understanding of the mechanism (plate tectonics) would take decades of further progress in geology, oceanography, and geophysics.

Wegener was disappointed in the reception to his suggestion, but he continued his research, and in 1924 he was appointed to a special meteorology and geophysics professorship created especially for him at the University of Graz (where he was, however, ostracized by most of the geology faculty). Four years later, on his fourth expedition to his beloved Greenland, he celebrated his 50th birthday with colleagues and then set off toward a different camp on the island. He never made it; he was found a few days later, dead of an apparent heart attack.

Critics of science often point to the resistance to the continental drift hypothesis as an example of the flawed way that scientists regard new ideas. (Many people who have advanced crackpot theories have claimed that they are being unjustly ridiculed, just as Wegener was.) But we think there is a more positive light in which to view the story of Wegener's suggestion. Scientists in his day maintained a skeptical attitude because they needed more evidence and a clear mechanism that would fit what they understood about nature. Once the evidence and the mechanism were clear, Wegener's hypothesis quickly became the centerpiece of our new view of a dynamic Earth.

7.2.3 Rift and Subduction Zones

Plates pull apart from each other along **rift zones,** such as the Mid-Atlantic ridge, driven by upwelling currents in the mantle (Figure 7.7a). A few rift zones are found on land. The best known is the central African rift, an area where the African continent is slowly breaking apart. Most rift zones, however, are in the oceans. Molten rock (called **magma**) rises from below to fill the space between the receding plates; this is basaltic lava, the kind of igneous rock that forms most of the ocean basins.

From a knowledge of how the sea floor is spreading, we can calculate the average age of the oceanic crust. About 60,000 km of active rifts have been identified, with average separation rates of about 4 cm per year. The new area added to the Earth each year is about $2\ km^2$, enough to renew the entire oceanic crust in a little more than 100 million years. This is a very short interval in geological time, less than 3 percent of the age of the Earth. The present ocean basins thus turn out to be among the youngest features on our planet.

As new crust is added to the Earth, the old crust must go somewhere. When two plates come together, one plate is often forced down beneath another in what is called a **subduction zone** (Figure 7.7b). Generally, the thick continental masses cannot be subducted, but the thin oceanic plates can be rather readily thrust down into the upper mantle. Often a subduction zone is marked by an ocean trench, a fine example being the deep Japan trench along the coast of Asia. The subducted plate is forced down into regions of high pressure and temperature, eventually melting several hundred kilometers below the surface. Its material is recycled back into a downward-flowing convection current, ultimately balancing the flow of material that rises along rift zones. The amount of crust destroyed at subduction zones is approximately equal to the amount formed at rift zones.

All along the subduction zone, earthquakes and volcanoes mark the death throes of the plate. Some of the most destructive earthquakes in history have taken place along subduction zones. These include the 1923 Yokohama earthquake and fire, which killed 100,000 people, and the 1976 earthquake that leveled the Chinese city of Tangshan and resulted in nearly half a million deaths.

7.2.4 Fault Zones and Mountain Building

Along much of their lengths, the crustal plates slide parallel to each other. These plate boundaries are marked by cracks or **faults.** Along active fault zones, the motion of one plate with respect to the other is several centimeters per year, about the same as the spreading rates along rifts.

(a) RIFT ZONE

(b) SUBDUCTION ZONE

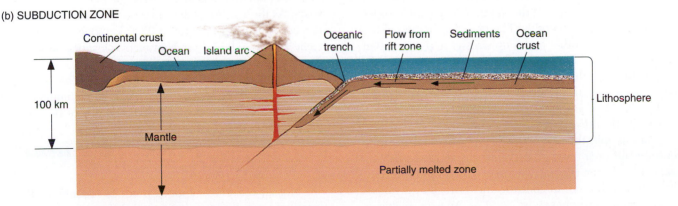

■ **FIGURE 7.7**

Rift Zone (a) and Subduction Zone (b) The regions (mostly beneath the oceans) where new crust is formed and old crust is destroyed as part of the cycle of plate tectonics.

David Porter/Science Photo Library/Photo Researchers, Inc.

■ **FIGURE 7.8**

The San Andreas Fault We see part of a very active region in California where one crustal plate is sliding sideways with respect to the other. The fault is marked by the valley running up the center of the photo. Major slippages along this fault can produce extremely destructive earthquakes.

One of the most famous faults is the San Andreas fault, lying on the boundary between the Pacific plate and the North American plate (Figure 7.8). This fault runs from the Gulf of California to the Pacific Ocean northwest of San Francisco. The Pacific plate, to the west, is moving northward, carrying Los Angeles, San Diego, and parts of the southern California coast with it. In several million years, Los Angeles will be an island off the coast of San Francisco.

Unfortunately for us, the motion along fault zones does not take place smoothly. The creeping motion of the plates against each other builds up stresses in the crust that are released in sudden, violent slippages to generate earthquakes. Because the average motion of the plates is constant, the longer the interval between earthquakes, the greater the stress and the more energy released when the surface finally moves.

For example, the part of the San Andreas fault near the central California town of Parkfield has slipped every 25 years or so during the past century, moving an average of about 1 m each time. In contrast, the average interval between major earthquakes in the Los Angeles region is about 150 years, and the average motion is about 7 m. The last time the San Andreas fault slipped in this area was in

1857; tension has been building ever since, and sometime soon it is bound to be released. Sensitive instruments placed within the Los Angeles basin show that the basin is distorting and contracting in size as these tremendous pressures build up beneath the surface.

When two continental masses are moving on a collision course, they push against each other under great pressure. The Earth buckles and folds, dragging some rock deep below the surface and raising other folds to heights of many kilometers. This is the way most of the mountain ranges on Earth were formed; the Alps, for example, are a result of the African plate's bumping into Europe. As we will see, however, quite different processes produced the mountains on other planets.

Once a mountain range is formed by upthrusting of the crust, its rocks are subject to erosion by water and ice. The sharp peaks and serrated edges have little to do with the forces that initially make mountains. Instead, they result from the processes that tear mountains down. Ice is an especially effective sculptor of rock (Figure 7.9). In a planet without moving ice or running water (such as the Moon or Mars), mountains remain smooth and dull.

7.2.5 Volcanoes

Volcanoes mark locations where magma rises to the surface. One example is the mid-ocean ridges, which are long undersea mountain ranges formed by magma rising from the Earth's mantle at plate boundaries. A second major kind of volcanic activity is associated with subduction zones, and volcanoes sometimes also appear in regions where continental plates are colliding. In each case, the volcanic activity gives us a way to sample some of the material from deeper within our planet.

Other volcanic activity occurs above mantle "hot spots"—areas far from plate boundaries where heat is nevertheless rising from the interior of the Earth. One of the best-known such hot spots is under the island of Hawaii, where it currently supplies the magma to maintain three active volcanoes, two on land and one under the ocean. The Hawaiian hot spot has been active for at least 100 million years. As the Earth's plates have moved during that time, the hot spot has generated a 3500-km-long chain of volcanic islands. The tallest Hawaiian volcanoes are among the largest individual mountains on Earth, more than 100 km in diameter and rising 9 km above the ocean floor. As we saw in Chapter 5, one of these volcanic mountains, the now-dormant Mauna Kea, has become one of the world's great sites for doing astronomy.

Ace⬙Astronomy™ Log into AceAstronomy and select this chapter to see the Active Figure called "Hotspot Volcanoes."

Not all volcanic eruptions produce mountains. If the lava flows rapidly from long cracks, it can spread out to form lava plains. The largest known terrestrial eruptions, such as those that produced the Snake River basalts in the

■ FIGURE 7.9

Mountains on Earth The Alps, a young region of the Earth's crust where sharp mountain peaks are being sculpted by glaciers. We owe the beauty of our young, steep mountains to the erosion of ice and water.

northwestern United States or the Deccan plains in India, are of this type. As we will see later, similar lava plains are found on the Moon and the other terrestrial planets.

7.3 THE EARTH'S ATMOSPHERE

We live at the bottom of the ocean of air that envelops our planet. The atmosphere, weighing down upon the Earth's surface under the force of gravity, exerts a pressure at sea level that scientists define as 1 **bar** (a term that comes from *barometer*, an instrument used to measure atmospheric pressure). A bar of pressure means that each square centimeter of the Earth's surface has a weight equivalent to 1.03 kg pressing down on it. Humans have evolved to live at this pressure; make the pressure a lot lower or higher, and we do not function well at all.

The total mass of our atmosphere is about 5×10^{18} kg; this sounds like a big number, but it is only about a millionth of the total mass of the Earth. The atmosphere represents a smaller fraction of the Earth than the fraction of your mass represented by the hair on your head.

7.3.1 Structure of the Atmosphere

The structure of the atmosphere is illustrated in Figure 7.10. Most of the atmosphere is concentrated near the surface of the Earth, within about the bottom 10 km—where clouds form and airplanes fly. Within this region, called the **troposphere,** warm air, heated by the surface, rises and is replaced by descending currents of cooler air—another example of convection. This circulation generates clouds and wind. Within the troposphere, temperature drops rapidly with increasing elevation to values near 50°C below freez-

ing at its upper boundary, where the **stratosphere** begins. Most of the stratosphere, which extends to about 50 km above the surface, is cold and free of clouds.

Ace◉Astronomy™ Log into AceAstronomy and select this chapter to see the Active Figure called "Planetary Atmospheres."

■ FIGURE 7.10

The Structure of the Earth's Atmosphere Height increases up the left side of the diagram, and the names of the different layers are shown at the right. In the upper ionosphere, ultraviolet radiation from the Sun can strip electrons from their atoms, leaving the atmosphere ionized. (This layer can scatter or reflect AM radio waves, assuring that some of the dumb things said on AM radio never make it out of our planet's atmosphere.) The curving red line shows the temperature; see the scale for temperature at the bottom of the chart.

Near the top of the stratosphere is a layer of **ozone** (O_3), a heavy form of oxygen with three atoms per molecule instead of the usual two. Since ozone is a good absorber of ultraviolet light, it protects the surface from some of the Sun's dangerous ultraviolet radiation, making it possible for life to exist on land. Because ozone is essential to our survival, we react with justifiable concern to evidence that atmospheric ozone is being destroyed. Just such evidence has been accumulating; one especially important agent of this destruction is a series of industrial chemicals called CFCs (chlorofluorocarbons).

Each year a large ozone hole forms above the Antarctic continent, and by now the ozone loss has progressed into the temperate zones. The production of CFCs has been banned by international agreement, but these chemicals are destroyed so slowly that we can expect increasing loss of ozone well into this century.

At heights above 100 km, the atmosphere is so thin that orbiting satellites can pass right through it with very little friction. At these elevations, individual atoms can occasionally escape completely from the gravitational field of the Earth. There is a continuous, slow leaking of atmosphere—especially of lightweight atoms, which move faster than heavy ones. The Earth's atmosphere cannot, for example, hold on for long to hydrogen or helium, which escape into space.

7.3.2 Atmospheric Composition and Origin

At the Earth's surface, the atmosphere consists of 78 percent nitrogen (N_2), 21 percent oxygen (O_2), and 1 percent argon (A), with traces of water vapor (H_2O), carbon dioxide (CO_2), and other gases. Variable amounts of dust particles and water droplets are also found suspended in the air.

A complete census of the Earth's volatile materials, however, should look at more than the gas that is now present. *Volatile* materials are those that evaporate at a relatively low temperature. If the Earth were just a little bit warmer, some materials that are now liquid or solid might become part of the atmosphere. Suppose, for example, that our planet were heated to above the boiling point of water (100°C, or 373 K); that's a large change for humans, but a small change compared to the range of possible temperatures in the universe. At 100°C the oceans would boil and the water vapor would become a part of the atmosphere.

To estimate how much water vapor would be released, we note that there is enough water to cover the entire Earth to a depth of about 3000 m. Since the pressure exerted by 10 m of water is equal to about 1 bar, the average pressure at the ocean floor is about 300 bars. Since water weighs the same whether in liquid or in vapor form, if the oceans boiled away, the atmospheric pressure of the water would still be 300 bars. Water would therefore greatly dominate the Earth's atmosphere, with nitrogen and oxygen reduced to the status of trace constituents.

On a warmer Earth, another source of additional atmosphere would be found in the sedimentary carbonate rocks of the crust. These are minerals that contain abundant carbon dioxide. If all these rocks were heated, they would release about 70 bars of CO_2, far more than the current CO_2 pressure of only 0.0005 bar. Thus, the atmosphere of a warm Earth would be dominated by water vapor and carbon dioxide, with a surface pressure close to 400 bars.

Several lines of evidence convince scientists that the composition of the Earth's atmosphere has changed over our planet's history. Scientists can infer the amount of atmospheric oxygen, for example, by studying the chemistry of minerals that formed at various times. We examine this question in more detail in Section 7.4.

Today we see that CO_2, H_2O, sulfur dioxide (SO_2), and other gases are released from deeper within the Earth through the action of volcanoes. Much of this apparently new gas, however, is probably recycled material that has been subducted through plate tectonics. But where did our planet's original atmosphere come from?

Three possibilities exist for the original source of the Earth's atmosphere and oceans (since the early Earth was warmer, we can think of today's oceans as liquified early atmosphere, as just discussed): (1) The atmosphere could have been formed with the rest of the Earth, as it accumulated from the debris left over from the formation of the Sun; (2) it could have been released from the interior through volcanic activity, subsequent to the formation of the Earth; or (3) it may have been derived from impacts by comets or other icy materials from the outer parts of the solar system. Current evidence favors a combination of the interior and cometary sources.

7.3.3 Weather and Climate

All planets with atmospheres have *weather*, which is the name we give to the circulation of the atmosphere. The energy that powers the weather is derived primarily from the sunlight that heats the surface. Both the rotation of the planet and slower seasonal changes cause variations in the amount of sunlight striking different parts of the Earth. The atmosphere and oceans try to redistribute the heat from warmer to cooler areas. Weather on any planet represents the response of its atmosphere to changing inputs of energy from the Sun (see Figure 7.11 for a dramatic example).

Climate is a term used to refer to the effects of the atmosphere that last through decades and centuries. Changes in climate (as opposed to the random variations in weather from one year to the next) are often difficult to detect over short time periods, but as they accumulate, their effect can be devastating. Modern farming is especially sensitive to temperature and rainfall; for example, calculations indicate that a drop of only 2°C throughout the growing season would cut the wheat production by half in Canada and the United States.

NASA

FIGURE 7.11

A Storm from Space A large tropical storm marked by clouds swirling around a low-pressure region, photographed by astronauts orbiting the Earth. The combination of Earth's tilted axis of rotation, moderately rapid rotation, and oceans of liquid water can lead to violent weather on our planet.

The best-documented changes in the Earth's climate are the great ice ages, which have periodically lowered the temperature of the Northern Hemisphere over the past million years or so. The last ice age, which ended about 14,000 years ago, lasted some 20,000 years. At its height, the ice was almost 2 km thick over Boston and stretched as far south as New York City. Today we are in a relatively warm period, interpreted by many scientists as a fairly short-lived interglacial interval between major ice ages.

Scientists think that the ice ages are primarily the result of changes in the tilt of the Earth's rotational axis, produced by the gravitational effects of the other planets. They are less certain about recent evidence that on several occasions about a billion years ago the entire ocean froze over—a situation called "snowball Earth." As we will see in Chapter 9, there is also evidence of periodic climate changes on Mars, and modern calculations suggest that these also have their origin in slow changes of that planet's orbit and rotational axis.

7.4 LIFE AND CHEMICAL EVOLUTION

As far as we know, Earth seems to be the only planet in the solar system with life. The origin and development of life are an important part of the story of our planet. Life arose early in Earth's history, and living organisms have been interacting with their environment for billions of years. We recognize that life-forms have evolved to adapt themselves to the environment on Earth, and we are now beginning to realize that the Earth itself has been changed in important ways by the presence of living matter. The study of the co-evolution of life and our planet is one of the subjects of the modern science of *astrobiology*.

7.4.1 The Origin of Life

The record of the birth of life on Earth has been lost in the restless motions of the crust. By the time the oldest surviving rocks were formed, about 3.9 billion years ago, life already may have existed, according to chemical evidence. At 3.5 billion years ago, life had achieved the sophistication to build large colonies called *stromatolites*, a form so successful that stromatolites still grow on Earth today (Figure 7.12). But little crust survives from these ancient times, and abundant fossils have been produced only during the past 600 million years—less than 15 percent of our planet's history.

There is little direct evidence about the actual origin of life. We know that the atmosphere of the early Earth, unlike today, contained abundant carbon dioxide but no oxygen gas. In the absence of oxygen, many complex chemical reactions are possible that lead to the production of amino acids, proteins, and other chemical building blocks of life. Therefore it seems likely that these chemical building

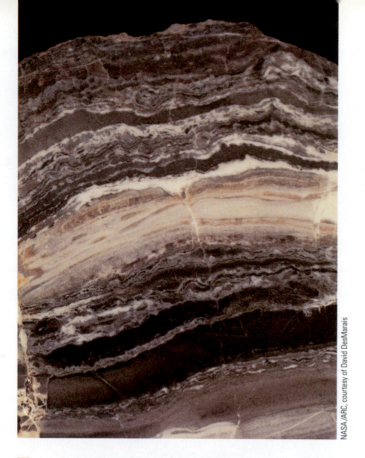

■ **FIGURE 7.12**

Cross Sections of Fossil Stromatolites The layered, domelike structures are mats of sediment trapped in shallow waters by large numbers of blue-green bacteria that can photosynthesize. Such colonies of microorganisms date back more than 3 billion years.

blocks were available very early in Earth's history and would have combined to make living organisms.

For tens of millions of years after its formation, life (perhaps little more than large molecules, like the viruses of today) probably existed in warm, nutrient-rich seas, living off accumulated organic chemicals. Eventually, however, this easily accessible food became depleted and life began the long evolutionary road that led to the vast numbers of different organisms on Earth today. As it did so, life began to influence the chemical evolution of the atmosphere.

In addition to the study of life's history as revealed by chemical and fossil evidence in ancient rocks, scientists have new tools from the rapidly advancing fields of genetics and *genomics*—the study of the genetic code that is shared by all life on Earth. While each individual has a unique set of genes (which is why genetic "fingerprinting" is so useful for the study of crime), we also have many genetic traits in common. Your *genome,* the complete map of the DNA in your body, is identical at the 99.9% level with that of Julius Caesar or Mozart. At the 99% level, the human and chimpanzee genomes are the same. By looking at the gene sequences of many organisms, we can determine that all life in Earth is descended from a common ancestor, and we can use the variations between species as a measure of how closely different species are related.

These genetic analysis tools have allowed scientists to construct what is called the "tree of life" (Figure 7.13). This diagram illustrates the way organisms are related by examining one sequence of the nucleic acid RNA that all species have in common. The figure shows that life on Earth is dominated by microscopic creatures that you have probably never heard of. Note that the plant and animal kingdoms are just two little branches at the upper right. Most of the diversity of life, and most of our evolution, has taken place at the microbial level. Indeed, it may surprise you to know that there are more microbes in a bucket of soil than there are stars in the Galaxy! You may want to keep this in mind when we turn to the search for life on other worlds later in this book. The "aliens" that are most likely to be out there are microbes.

Such genetic studies lead to other interesting conclusions as well. For example, it appears that the earliest surviving terrestrial life-forms were all adapted to live at high temperatures. Some biologists think that life might actually have begun in locations on our planet that were extremely hot. We will return to this question in Section 7.5.4. Yet another intriguing possibility is that life began on Mars rather than the Earth and was "seeded" onto our planet by meteorites traveling from Mars to Earth. Mars rocks are still making their way to Earth (as we will see in Chapter 13), but so far none has shown evidence of serving as a "spaceship" to carry microorganisms from Mars to Earth.

7.4.2 The Evolution of the Atmosphere

One of the key steps in the evolution of life on Earth was the development of tiny blue-green algae, a very successful life-form that takes in carbon dioxide from the environment and releases oxygen as a waste product. These successful microorganisms proliferated, giving rise to all the life-forms we call plants. Since the energy for making new plant material from chemical building blocks comes from sunlight, we call the process *photosynthesis.*

Studies of the chemistry of ancient rocks show that Earth's atmosphere lacked abundant free oxygen until about 2 billion years ago, in spite of the presence of plants releasing oxygen by photosynthesis. Apparently, chemical reactions with the Earth's crust removed the oxygen gas as quickly as it formed. Slowly, however, the increasing evolutionary sophistication of life led to a growth in the plant population and thus increased oxygen production. At the same time, it appears that increased geological activity led to heavy erosion on our planet's surface. This buried much of the plant carbon before it could recombine with oxygen to form CO_2.

Free oxygen began accumulating in the atmosphere about 2 billion years ago, and the increased amount of this gas led to the formation of the Earth's ozone layer, which protects the surface from deadly solar ultraviolet light. Before that, it was unthinkable for life to venture outside the protective oceans, so the land masses of Earth were barren.

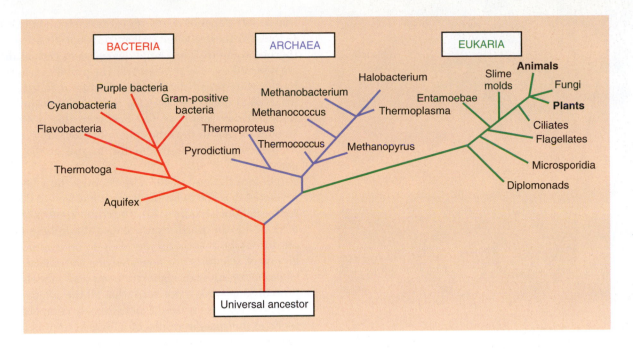

■ FIGURE 7.13

The Tree of Life This chart shows the main subdivisions of life on Earth and how they are related. The distance between any two species measured along the lines connecting them indicates the amount of evolutionary time that separates them. Note that the animal and plant kingdoms are just short branches in the upper right, along with the fungi. The most fundamental division of Earth's living things is onto three large domains called *bacteria*, *archaea*, and *eukaria*. Most of the species listed are microscopic.

The presence of oxygen and hence ozone thus allowed the colonization of the land. It also made possible a tremendous proliferation of animals, which lived by taking in and using the organic materials produced by plants as their own energy source.

As animals evolved in an environment increasingly rich in oxygen, they were able to develop techniques for breathing oxygen directly from the atmosphere. We humans take it for granted that plenty of free oxygen is available in the Earth's atmosphere to release energy from the food we take in. Although it may seem funny to think of it this way, we are life-forms that have evolved to breathe in the waste product of plants.

On a planetary scale, one of the most important consequences of life has been a decrease in atmospheric carbon dioxide. In the absence of life, Earth would probably have a much more massive atmosphere dominated by CO_2, like that of the planet Venus. But life in combination with high levels of geological activity has effectively stripped us of most of this gas.

7.4.3 The Greenhouse Effect and Global Warming

We have a special interest in the carbon dioxide content of the atmosphere because of the role this gas plays in retaining heat from the Sun through a process called the **green-house effect.** To understand how the greenhouse effect works, consider the fate of the sunlight that strikes the surface of the Earth. The light penetrates our atmosphere, is absorbed by the ground, heats the surface layers, and is then reemitted as infrared or heat radiation (Figure 7.14). However, the CO_2 in our atmosphere, which is transparent to visible light, is largely opaque to infrared energy. As a result, it acts like a blanket, trapping the heat in the atmosphere and impeding its flow back to space.

On average, as much heat reaches the surface from the atmospheric greenhouse effect as from direct sunlight. For this reason, temperatures are only slightly lower at night than in the daytime. Over time, this extra heat means that the Earth's surface regions get warmer than they would have without the effect of the CO_2. The more CO_2 there is in our atmosphere, the higher the temperature at which the Earth's surface reaches a new balance.

The greenhouse effect in a planetary atmosphere is similar to the heating of a gardener's greenhouse or the inside of a car left out in the Sun with the windows rolled up. In these examples, the window glass plays the role of carbon dioxide, letting sunlight in but reducing the outward flow of heat radiation. As a result, a greenhouse or car interior winds up much hotter than would be expected from the heating of sunlight alone. On Earth, the current greenhouse effect elevates the surface temperature by about 23°C. Without this greenhouse effect, the average surface

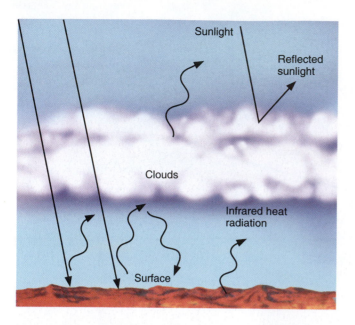

■ FIGURE 7.14

How the Greenhouse Effect Works   Sunlight that penetrates to the Earth's lower atmosphere and surface is reradiated as infrared or heat radiation, which is trapped by CO2 in the atmosphere. The result is a higher surface temperature for our planet.

temperature would be well below freezing and the Earth would be locked in a global ice age.

That's the good news; the bad news is that the heating due to the greenhouse effect is increasing. Modern indus-

trial society depends on energy extracted from burning fossil fuels. As these ancient coal and oil deposits are *oxidized* (burned using oxygen), large quantities of carbon dioxide are released into the atmosphere. The problem is increased by the widespread destruction of tropical forests, which we depend on to extract CO_2 from the atmosphere and replenish our supply of O_2. So far in this century, the amount of CO_2 in the atmosphere has increased by about 25 percent, and it is continuing to rise at 0.5 percent per year.

Before the end of the 21st century, the CO_2 level is predicted to reach twice the value it had before the industrial revolution (Figure 7.15). The consequences of such an increase for the Earth's surface and atmosphere (and the creatures who live there) are likely to be complex. Many groups of scientists are now studying the effects of such *global warming* with elaborate computer models. At the present time we are not sure what all the effects will be, since the Earth's surface is an extremely complicated system and the specific results of increased temperature will surely vary from one location to another.

Already some global warming is apparent. In North America and Europe, summer temperatures throughout the 1990s reached record highs. Glaciers are retreating, and the Arctic Sea ice is now much thinner than when it was first explored with nuclear submarines in the 1950s. The increasing greenhouse effect comes at a particularly difficult time: Our planet is already in an unusually warm interglacial period. We are rapidly entering unknown territory where human activities are contributing to the highest temperatures on Earth in more than 50 million years.

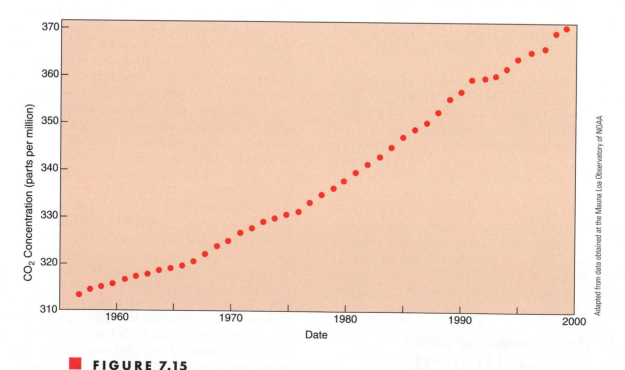

■ FIGURE 7.15

The Increase of Atmospheric Carbon Dioxide over Time   The amount of CO_2 is expected to be double its preindustrial level before the end of the 21st century.

7.5 COSMIC INFLUENCES ON THE EVOLUTION OF EARTH

7.5.1 Where Are the Craters on Earth?

In discussing the Earth's geology in Section 7.2, we dealt only with the effects of internal forces, expressed through the processes of plate tectonics and volcanism. On the Moon, in contrast, we see primarily craters, produced by the impacts of interplanetary debris (as described in Chapter 6). Why don't we see more evidence on Earth of the kinds of impact craters that are so prominent on the Moon and other planets?

It is not possible that the Earth escaped being struck by the interplanetary debris that has pockmarked the Moon. From a cosmic perspective, the Moon is almost next door. Our atmosphere does make small pieces of cosmic debris burn up (which we see as *meteors*—commonly called shooting stars). But the layers of air provide no shield against the large impacts that form craters several kilometers in diameter.

In the course of its history, the Earth must therefore have been cratered as heavily as the Moon. The difference is that on the Earth these craters are destroyed by our active geology before they can accumulate. As plate tectonics constantly renews our crust, evidence of past cratering events is destroyed. Only in the past few decades have geologists succeeded in identifying the eroded remnants of many old impact craters. Even more recent is our realization that these impacts have had an important influence on the evolution of life.

7.5.2 Recent Impacts

The collision of interplanetary debris with the Earth is not a hypothetical idea: Evidence of relatively recent impacts can be found on our planet's surface. The best-studied historic collision took place on June 30, 1908, near the Tunguska River in Siberia. In this desolate region, there was a remarkable explosion in the atmosphere about 8 km above the surface. The shock wave flattened more than a thousand square kilometers of forest (Figure 7.16). Herds of reindeer and other animals were killed, and a man at a trading post 80 km from the blast was thrown from his chair and knocked unconscious. The blast wave spread around the world, as recorded by instruments designed to measure changes in atmospheric pressure.

Despite this violence, no craters were formed by the Tunguska explosion. Shattered by atmospheric pressure, a stony projectile with a mass of approximately 100,000 tons disintegrated to create a blast equivalent to a 15-megaton nuclear bomb.[1] Had it been smaller or more fragile, the

[1] A megaton is the explosive energy released by a million tons of TNT—not a million pounds or kilograms but a million *tons!*

Novosty

■ FIGURE 7.16

Aftermath of the Tunguska Explosion   This photograph, taken 21 years after the blast, shows a part of the forest that was devastated by the 15-megaton explosion, as a stony projectile about the size of a large office building (60 m in diameter) collided with our planet.

impacting body would have dissipated its energy at high altitude and probably attracted no attention. Today such high-altitude atmospheric explosions are monitored regularly by military surveillance systems. If it had been larger or made of stronger material (such as metal), the Tunguska projectile would have penetrated all the way to the surface of the Earth and formed a crater. Instead, only the shock of the explosion reached the surface, but the devastation it left behind in Siberia bore witness to the power of such impacts. Imagine if the same rocky impactor had exploded over New York City in 1908; history books might today record it as one of the most deadly events in human history.

The best-known recent crater on Earth was formed about 50,000 years ago in Arizona. The projectile in this case was a lump of iron about 40 m in diameter. (We will discuss in Chapter 8 how the explosion of a small chunk of material can make a much bigger crater.) Now called Meteor Crater and a major tourist attraction on the way to the Grand Canyon, the crater is about a mile across and has all the features associated with similar-sized lunar impact craters (Figure 7.17). Meteor Crater is one of the few impact features on the Earth that remain relatively intact; other older craters are so eroded that only a trained eye can distinguish them. Nevertheless, more than 150 have been identified. (See our list of Web sites if you want to find out more about these other impact scars.)

Meteor Crater, Northern Arizona

■ **FIGURE 7.17**

Meteor Crater in Arizona
Here we see a 50,000-year-old impact scar made by the collision of a 40-m lump of iron with our planet. Although impact craters are common on less-active bodies such as the Moon, this is one of the very few well-preserved craters on the Earth.

7.5.3 Extinction of the Dinosaurs

The impact that produced Meteor Crater would have been dramatic indeed to any humans who witnessed it, since the energy release was equivalent to a 15-megaton nuclear bomb. But such explosions are devastating only in their local areas; they have no *global* consequences. Much larger (and rarer) impacts, however, can disturb the ecological balance of the entire planet and thus influence the course of evolution.

The best-documented large impact took place 65 million years ago, at the end of what is now called the Cretaceous era of geological history. This point in the history of life on Earth was marked by a **mass extinction,** in which more than half of the species on our planet died out. There are a dozen or more mass extinctions in the geological record, but this particular event has always intrigued paleontologists because it marks the end of the dinosaur age. For tens of millions of years these great creatures had flourished. Then they suddenly disappeared (along with many other species), and thereafter the mammals began the development and diversification that ultimately led to the readers (and authors) of this book.

The object that collided with the Earth at the end of the Cretaceous era struck a shallow sea in what is now the Yucatan peninsula of Mexico. Its mass must have been more than a trillion tons (Figure 7.18) because scientists have found a worldwide layer of sediment deposited from the dust cloud that enveloped the planet after its impact. First identified in 1979, this sediment layer is rich in the

Painting by Don Davis

■ **FIGURE 7.18**

A 20-km Asteroid Hits the Earth   This painting shows an impact similar to the one that led to a mass extinction 65 million years ago.

168   CHAPTER 7   EARTH AS A PLANET

rare metal iridium and other elements that are relatively abundant in asteroids and comets but exceedingly rare in the Earth's crust. Even though it was diluted by the material the explosion excavated from the surface of the Earth, this cosmic component is easily identified. In addition, this layer of sediment contains many minerals characteristic of the temperatures and pressures of a gigantic explosion.

The impact released energy equivalent to 5 billion Hiroshima-sized nuclear bombs and excavated a crater 200 km across and deep enough to penetrate through the Earth's crust. This large crater, named Chicxulub for a small town near its center, has subsequently been buried in sediment, but its outlines can still be identified (Figure 7.19). The explosion that created the Chicxulub crater lifted about *100 trillion tons* of dust into the atmosphere! We can determine this amount by measuring the thickness of the sediment layer that formed when this dust settled to the surface.

Such a quantity of airborne material would have blocked sunlight completely, plunging the Earth into a period of cold and darkness that lasted several months. Other worldwide effects included large-scale fires (started by the hot, flying debris from the explosion) that destroyed most of the planet's forests and grasslands, and a long period in which rainwater around the globe was acidic. It was these environmental effects, rather than the explosion itself, that were responsible for the mass extinction, including the death of the dinosaurs.

A demonstration of just such effects took place in July 1994, when pieces of a comet called Shoemaker–Levy 9 impacted the planet Jupiter. Some comet fragments were larger than a kilometer across. The comet fragments exploded as they penetrated the giant planet's atmosphere, and vast clouds of dark material were seen to expand over great distances (see Chapter 12).

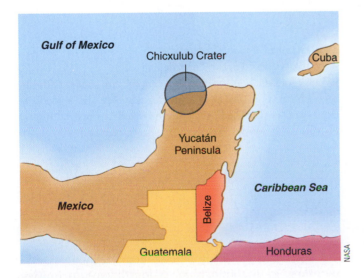

■ FIGURE 7.19

Site of the Chicxulub Crater Map showing the location of the buried crater from the impact 65 million years ago on Mexico's Yucatan peninsula. The crater is now buried under more than 500 m of sediment.

7.5.4 Impacts and the Evolution of Life

Several other mass extinctions in the geological record have been tentatively identified with large impacts, although none is so dramatic as the blast that destroyed the dinosaurs. Even without such specific documentation, it is now clear that impacts of this size do occur on all planets and that their effects can be catastrophic for life. A catastrophe for one group of living things, however, may create opportunities for another group. Following each mass extinction, there is a sudden evolutionary burst as new species develop to fill the ecological niches opened by the event.

Impacts by comets and asteroids represent the only mechanisms we know of that could cause truly global catastrophes and seriously influence the evolution of life all over the planet. According to some estimates, the majority of all species extinctions may be due to such impacts. As paleontologist Stephen Jay Gould of Harvard noted, such a perspective fundamentally changes our view of biological evolution. The central issues for the survival of a species must now include more than just its success in competing with other species and adapting to slowly changing environments, as envisioned by Darwinian natural selection. Of at least equal importance is its ability to survive random global catastrophes due to impacts.

Still earlier in its history, the Earth was subject to even larger impacts from the leftover debris of planet formation. We know the Moon was repeatedly struck by objects larger than 100 km in diameter, a thousand times more massive than the object that wiped out most terrestrial life 65 million years ago (see Chapter 8). The Earth must have experienced similar large impacts during its first 700 million years of existence. Some of them were probably violent enough to strip the planet of most of its atmosphere and to boil away its oceans. Such events would sterilize the planet, destroying any life that had begun. Life may have formed and been wiped out several times before our own primitive ancestors took hold sometime about 4 billion years ago.

The fact that the oldest surviving microbes are thermophiles (adapted to very high temperatures) can also be explained by such large impacts. An impact that was just a bit too small to sterilize the planet would still have destroyed anything that lived in what we consider "normal" environments, and only the creatures that liked high temperatures would survive. Thus the oldest surviving terrestrial life-forms are probably the remnants of a sort of evolutionary bottleneck caused by repeated large impacts early in the planet's history.

7.5.5 Impacts in Our Future?

The impacts by asteroids and comets that have had such a major influence on life are not necessarily a thing of the past. In the full scope of planetary history, 65 million years ago was just yesterday. The Earth actually orbits the Sun within a sort of cosmic shooting gallery, and though major impacts are rare, they are by no means over. Humanity can ultimately expect to suffer the same fate as the dinosaurs—unless we figure out a way to predict the next big impact and to protect our planet from such a global catastrophe.

During the 1990s a few astronomers began to analyze the cosmic impact hazard and to persuade the government to support a search for potentially hazardous asteroids. Several small but sophisticated wide-field telescopes are now used for this search, which is called the Spaceguard Survey. The objective is to find any threatening asteroid many decades before it hits.

Already we know that there are currently no asteroids on Earth-crossing orbits that are as big (10 to 15 km) as the one that killed the dinosaurs. The Spaceguard Survey now concentrates on finding asteroids 1 km or larger in diameter, which is big enough to kill hundreds of millions of people. By 2002 the search had netted more than half of the 1100 such asteroids that are estimated to exist. None of those discovered so far poses us any danger. Of course, we cannot make a similar statement about the asteroids that have not yet been discovered, but these will be found one by one and evaluated for their potential hazard. This asteroid survey is one of few really practical projects carried out by astronomers, and it could help to save our planet from another catastrophe of the sort that redirected the course of biological evolution 65 million years ago. (We will return to the subject of Earth-crossing asteroids in Chapter 12.)

SURFING THE WEB

Earth on the Solar System Web Sites:

- *The Nine Planets Site:*
  seds.lpl.arizona.edu/nineplanets/nineplanets/earth.html

- *Views of the Solar System Site:*
  www.solarviews.com/eng/earth.htm

- *Planetary Sciences Site:*
  nssdc.gsfc.nasa.gov/planetary/planets/earthpage.html

Tutorial on the Magnetosphere:
ssdoo.gsfc.nasa.gov/education/lectures/magnetosphere.html
Primer on the Earth's zone of magnetic influence, by a NASA scientist.

The Electronic Volcano: www.dartmouth.edu/~volcano
Everything you wanted to know about volcanoes, including catalogs, maps, current events, and images.

NASA Observatorium Ozone Guide:
observe.arc.nasa.gov/nasa/exhibits/ozone/Ozone1.html
Basic introduction to the ozone problem, with annoying typefaces but useful information and links.

The Earth from Space: earth.jsc.nasa.gov
A collection of photos taken by the astronauts from orbit.

National Earthquake Information Center:
www.neic.cr.usgs.gov/
A wealth of information about earthquakes from the U.S. Geological Survey.

Asteroid and Comet Impact Hazard Page:
impact.arc.nasa.gov
Voyages co-author David Morrison is behind this informative site about the dangers of asteroids and comets hitting the Earth and what we might do about it. Included are congressional and scientific position papers, a list of objects that pass near the Earth, useful background information, and annotated links.

Terrestrial Impact Craters Site:
www.solarviews.com/eng/tercrate.htm
Concise introduction to impact craters on our planet, with images, information, and an educator's guide. See also a good site from the Canadian Geological Survey called "Terrestrial Impact Structures" at: www.unb.ca/passc/ImpactDatabase.

SUMMARY

7.1 The Earth is the prototype terrestrial planet. Its interior composition and structure are probed using **seismic waves.** Such studies reveal that the Earth has a metal **core** and a silicate **mantle.** The outer layer or **crust** consists primarily of oceanic **basalt** and continental **granite.** A global magnetic field, generated in the core, produces the Earth's **magnetosphere,** which can trap charged atomic particles.

7.2 Terrestrial rocks can be classified as **igneous, sedimentary,** or **metamorphic.** A fourth type, **primitive rock,** is not found on the Earth. Our planet's geology is dominated by **plate tectonics,** in which crustal plates move slowly in response to mantle **convection.** The surface expression of plate tectonics includes continental drift, recycling of the ocean floor, mountain building, **rift zones,**

subduction zones, faults, earthquakes, and volcanic eruptions of **magma** from the interior.

7.3 The atmosphere has a surface pressure of 1 **bar** and is composed primarily of N_2 and O_2, plus such important trace gases as H_2O, CO_2, and **ozone (O_3).** Its structure consists of the **troposphere,** the **stratosphere,** and tenuous higher regions. Atmospheric oxygen is the product of photosynthetic life-forms on Earth. Atmospheric circulation (weather) is driven by seasonally changing deposition of sunlight. Longer-term climatic variations, such as the ice ages, are probably due to changes in the planet's orbit and axial tilt.

7.4 Life originated on Earth at a time when the atmosphere lacked O_2 and consisted mostly of CO_2. Later, photosynthesis gave rise to free oxygen and ozone. Modern ge-nomic analysis lets us see how the wide diversity of species on the planet are related to each other. Carbon dioxide in the atmosphere heats the surface through the **greenhouse effect;** today, increasing amounts of atmospheric CO_2 are leading to the global warming of our planet.

7.5 Earth, like the Moon and other planets, has been influenced by the impacts of cosmic debris, including such small recent examples as Meteor Crater and the Tunguska explosion. Larger past impacts are responsible for at least some **mass extinctions,** including the large extinction 65 million years ago that ended the Cretaceous era and killed many species, including the dinosaurs. Such impacts have probably played an important role in the evolution of life. Today astronomers are working to predict the next impact in advance, to give us a chance to defend our planet from such a catastrophe.

INTER-ACTIVITY

A If we can predict that lots of ground movement takes place along subduction zones and faults, then why do so many people live there? Should we try to do anything to discourage people from living in these areas? What inducement would your group offer people to move? Who would pay for the relocation? (Note that two of the authors of this book live quite close to the San Andreas and Hayward faults. If they wrote this chapter and haven't moved, what are the chances others living in these kinds of areas will move?)

B After your group reads the box on Alfred Wegener, discuss some reasons his idea did not catch on right away among scientists. From your studies in this course and in other science courses (in college and before), can you cite other scientific ideas that we now accept but that had controversial beginnings? Can you think of any scientific theories that are still controversial today? If your group comes up with some, can you also discuss ways that scientists can decide whether each theory on your list is right?

C Suppose we knew that a large chunk of rock or ice (about the same size as the one that hit 65 million years ago) will impact the Earth in about five years. What could or should we do about it? (The film *Deep Impact* dealt with this theme.) Does your group think that the world as a whole should spend more money to find and predict the orbits of cosmic debris near the Earth?

D Carl Sagan pointed out that any defensive weapon that we might come up with to deflect an asteroid *away* from the Earth could be used as an offensive weapon by an unstable government leader in the future to cause an asteroid not heading our way to come toward the Earth. The history of human behavior, he noted, has shown that most weapons that are built (even with the best of motives) seem to wind up being used. Bearing this in mind, does your group think we should be building weapons to protect the Earth from asteroid or comet impact? Can we afford not to build them? How can we safeguard them?

REVIEW QUESTIONS

Ace○Astronomy™ Assess your understanding of this chapter's topics with additional quizzing and animations at **http://ace.brookscole.com/voyages**

1. What are the Earth's core and mantle made of? Explain how we know.
2. Describe the differences among primitive, igneous, sedimentary, and metamorphic rock, and relate these differences to their origins.
3. Explain briefly how the following phenomena happen on Earth, relating your answers to the theory of plate tectonics:

    a. earthquakes
    b. continental drift
    c. mountain building
    d. volcanic eruptions
    e. creation of the Hawaiian island chain

4. Describe three ways in which the presence of life has affected the composition of the Earth's atmosphere.
5. How do impacts by comets and asteroids influence the Earth's geology, its atmosphere, and the evolution of life?
6. Why are there so many impact craters on our neighbor world, the Moon, and so few on the Earth?

7. If you wanted to live where the chances of a destructive earthquake were small, would you pick a location near a fault zone, near a mid-ocean ridge, near a subduction zone, or on a volcanic island such as Hawaii? What are the relative risks of earthquakes at each of these locations?

8. If all life were destroyed on Earth by a large impact, would new life eventually form to take its place? Explain how conditions would have to change for life to start again on our planet.

9. Why will a decrease in the Earth's ozone be harmful to life?

10. Why are we concerned about the increases in CO_2 and other gases that cause the greenhouse effect in the Earth's atmosphere? What steps can we take in the future to reduce the levels of CO_2 in our atmosphere? What factors stand in the way of taking the steps you suggest (you may include technological, economic, and political factors in your answer).

11. Is there evidence of climate change in your area over the past century? How would you distinguish a true climate change from the random variations in weather that take place from one year to the next?

12. Do you think scientists should make plans to defend the Earth from future asteroid impacts? Is it right to intervene in the same evolutionary process that made the development of mammals (including us) possible after the big impact 65 million years ago?

After scientists mapped the boundaries between tectonic plates in the Earth's crust and measured the annual rate at which the plates move (which is about 5 cm per year), we could estimate quite a lot about the rate at which the geology of the Earth is changing. As an example, let's suppose that the next slippage along the San Andreas fault in southern California takes place in the year 2007 and that it completely relieves the accumulated strain in this region. How much slippage is required for this to occur?

The speed of motion of the Pacific plate relative to the North American plate is 5 cm per year. That's 500 cm per century, or 5 m per century. The last southern California earthquake was in 1857. The time from 1857 to 2007 is 150 years, or 1.5 centuries, so the slippage to completely relieve the strain would be $5 \times 1.5 = 7.5$ m.

13. Europe and North America are also moving apart by about 5 m per century. As the continents separate, new ocean floor is created along the Mid-Atlantic Rift. If the rift is 5000 km long, what is the total area of new ocean floor created in the Atlantic each century? (Remember that 1 km equals 1000 m.)

14. Over the entire Earth there are 60,000 km of active rift zones, with average separation rates of 5 m per century. How much area of new ocean crust is created each year over the entire planet? (This area is approximately equal to the amount of ocean crust that is subducted, since the total area of the oceans remains about the same.)

15. With the information from problem 14, you can calculate the average age of the ocean floor. First find the total area of the ocean floor (equal to about 60 percent of the surface area of the Earth). Then compare this with the area created (or destroyed) each year. The average lifetime is the ratio of these numbers—the total area of ocean crust compared to the amount created (or destroyed) each year.

16. What is the volume of new oceanic basalt added to the Earth's crust each year? Assume that the thickness of the new crust is 5 km, that there are 60,000 km of active rifts, and that the average speed of plate motion is 5 cm per year. What fraction of the Earth's entire volume does this annual addition of new material represent?

17. Suppose a major impact that produces a mass extinction takes place on the Earth once every 5 million years. Further suppose that if such an event occurred today, you and most other humans would be killed (this would be true even if the human species as a whole survived). Such impact events are random, and one could take place at any time. Calculate the probability that such an impact will occur within the next 50 years (within your lifetime).

18. How do the risks of dying from the impact of an asteroid or comet compare with other risks we are concerned about, such as dying in a car accident or from heart disease or some other "natural" cause? (*Hint:* To find the annual risk, go to the library or the Web and look up the annual number of deaths from a particular cause in a particular country, and then divide by the population of that country.)

SUGGESTIONS FOR FURTHER READING

Allegre, C. and Schneider, S. "The Evolution of the Earth" in *Scientific American,* Oct. 1994, p. 66.

Becker, L. "Repeated Blows" in *Scientific American,* Mar. 2002, p. 60. On whether there have been repeated giant impacts and extinctions on Earth.

Chyba, C. "Death from the Sky: Tunguska" in *Astronomy,* Dec. 1993, p. 38. Excellent review article.

Frankel, C. *The End of the Dinosaurs: Chicxulub Crater and Mass Extinctions.* 1999, Cambridge U. Press. Approaches extinctions by focusing on the story of the crater.

Gallant, R. "Journey to Tunguska" in *Sky & Telescope,* June 1994, p. 38. A planetarium director tells of his trip to the site and fills in its history.

Hartmann, W. "Piecing Together Earth's Early History" in *Astronomy,* June 1989, p. 24.

Hartmann, W. and Miller, R. *The History of the Earth.* 1993, Workman. Lavishly illustrated chronicle of our planet's evolution.

Heppenheimer, T. "Journey to the Center of the Earth" in *Discover,* Nov. 1987, p. 86.

Jones, P. and Wigley, T. "Global Warming Trends" in *Scientific American,* Aug. 1990, p. 84. Three hundred years of data analyzed.

Lewis, J. *Rain of Iron and Ice: The Very Real Threat of Comet and Asteroid Bombardment.* 1995, Addison-Wesley. Good overview of impacts.

Lunine, J. *Earth: Evolution of a Habitable World.* 1998, Cambridge U. Press. History of Earth in a cosmic context.

Meissner, R. *The Little Book of Planet Earth.* 2002, Copernicus/Springer Verlag.

Morrison, D. and Chapman, C. "Target Earth: It Will Happen" in *Sky & Telescope,* Mar. 1990, p. 261. (See an updated version in *Astronomy,* Oct. 1995, p. 34.)

Murphy, J. and Nance, R. "Mountain Belts and the Supercontinent Cycle" in *Scientific American,* Apr. 1992, p. 84. There has been more than one supercontinent through geological time.

Orgel, L. "The Origin of Life on Earth" in *Scientific American,* Oct. 1994, p. 77.

Powell, J. *Night Comes to the Cretaceous.* 1998, Harcourt Brace. Fascinating look at the controversy surrounding the proposal that an impact killed off the dinosaurs, illustrating the reactions of scientists to a revolutionary new idea.

Steel, D. *Target Earth.* 2000, Readers Digest Books. On impacts and the people who study them.

York, D. "The Earliest History of the Earth" in *Scientific American,* Jan. 1993, p. 90.

***Apollo 11* Astronaut Buzz Aldrin on the Surface of the Moon.**
Because there is no atmosphere, ocean, or geological activity on the Moon
today, the footprints you see in this image will likely be preserved in the lunar
soil for millions of years.

NASA

8

Cratered Worlds: The Moon and Mercury

This is one small step for a man, one giant leap for mankind.

What astronaut Neil Armstrong meant to say when he became the first human being to walk on another world. (In his nervousness, he left out the word "a," making the statement a bit less effective.)

THINKING AHEAD

The Moon is the only other world human beings have ever visited. What is it like to stand on the surface of our natural satellite? And what can we learn from going there and bringing home pieces of a different world?

We begin our discussion of the planets as worlds with two relatively simple objects: the Moon and Mercury. Unlike the Earth, the Moon is geologically dead, a place that has exhausted its internal energy sources. Because its airless surface preserves the events that happened long ago, the Moon provides us with a window on the earlier epochs of solar system history. The planet Mercury is in many ways similar to the Moon, which is why the two are discussed together. Both are relatively small, lacking in atmospheres, deficient in geological activity, and dominated by the effects of impact cratering. Still, the processes that have molded their surfaces are not unique to these two worlds; they have acted on many other members of the planetary system as well.

Ace⟲Astronomy™ The AceAstronomy icon throughout the text indicates an opportunity for you to test yourself on key concepts and to explore animations and interactions on the AceAstronomy website at **http://ace.brookscole.com/voyages**

Virtual Laboratories

 Planetary Geology

 Planetary Atmospheres and Their Retention

8.1 GENERAL PROPERTIES OF THE MOON

The Moon has only 1/80 the mass of the Earth and about 1/6 the Earth's surface gravity—too low to retain an atmosphere (Figure 8.1). Moving molecules of a gas can escape from a planet just the way a rocket does, and the lower the gravity, the easier it is for the gas to leak away into space. If, early in its history, the Moon acquired an atmosphere from its hot interior or collected a temporary envelope of gases from impacting comets, such an atmosphere was lost before it could leave any recognizable evidence of its short existence. Indeed, the Moon is dramatically deficient in a wide range of *volatiles,* those elements and compounds that evaporate at relatively low temperatures. Some of the Moon's properties are summarized in Table 8.1, along with comparative values for Mercury.

TABLE 8.1 *Properties of the Moon and Mercury*

	Moon	Mercury
Mass (Earth = 1)	0.0123	0.055
Diameter (km)	3476	4878
Density (g/cm^3)	3.3	5.4
Surface gravity (Earth = 1)	0.17	0.38
Escape velocity (km/s)	2.4	4.3
Rotation period (days)	27.3	58.65
Surface area (Earth = 1)	0.27	0.38

8.1.1 Exploration of the Moon

Most of what we know about the Moon today derives from the U.S. Apollo program, which sent nine piloted spacecraft to our satellite between 1968 and 1972, landing 12 astronauts on its surface (see the photograph that opens this chapter). Before the era of spacecraft studies, astronomers had mapped the side of the Moon that faces the Earth with telescopic resolution of about 1 km, but lunar geology hardly existed as a scientific subject. All that changed beginning in the early 1960s.

Initially the Russians took the lead in lunar exploration with Luna 3, which returned the first photos of the lunar far side in 1959, and then with Luna 9, which landed on the surface in 1966 and transmitted pictures and other data to Earth. However, these efforts were overshadowed on July 20, 1969, when the first American astronaut set foot on the Moon.

Table 8.2 summarizes the nine Apollo flights—six that landed and three others that circled the Moon but did not land. The initial landings were on flat plains selected for safety reasons. But with increasing experience and confidence, NASA targeted the last three missions to more geologically interesting locales. The level of scientific exploration also increased with each mission, as the astronauts spent longer times on the Moon and carried more elaborate equipment. Finally, on the last Apollo landing, NASA included one scientist, geologist Jack Schmitt, among the astronauts (Figure 8.2).

In addition to landing on the lunar surface and studying it at close hand, the Apollo missions accomplished three objectives of major importance for lunar science. First, the astronauts collected nearly 400 kg of samples for detailed laboratory analysis on Earth (Figure 8.3). These samples, still being studied, have probably revealed more about the Moon and its history than all other lunar studies combined. Second, each Apollo landing after the first one deployed

■ FIGURE 8.1

The Moon from Earth The Moon as photographed by astronauts from a spacecraft near the Earth. The left side of the image shows part of the hemisphere that faces the Earth; several dark maria are visible. The right side shows part of the hemisphere that faces away from the Earth; it is dominated by highlands. The resolution of this image is several kilometers, similar to that of high-powered binoculars or a small telescope.

NASA

TABLE 8.2 *Apollo Flights to the Moon*

Flight	Date	Landing Site	Main Accomplishment
Apollo 8	Dec. 1968	—	First humans to fly around the Moon
Apollo 10	May 1969	—	First spacecraft rendezvous in lunar orbit
Apollo 11	July 1969	Mare Tranquillitatis	First human landing on the Moon; 22 kg of samples returned
Apollo 12	Nov. 1969	Oceanus Procellarum	First ALSEP; visit to Surveyor 3 lander
Apollo 13	Apr. 1970	—	Landing aborted due to explosion in command module
Apollo 14	Jan. 1971	Mare Nubium	First "rickshaw" on the Moon
Apollo 15	July 1971	Imbrium/Hadley	First "rover"; visit to Hadley Rille; astronauts traveled 24 km
Apollo 16	Apr. 1972	Descartes	First landing in highlands; 95 kg of samples returned
Apollo 17	Dec. 1972	Taurus Mountains	Geologist among the crew; 111 kg of samples returned

an Apollo Lunar Surface Experiment Package (ALSEP), which continued to operate for years after the astronauts departed. (The ALSEPs were turned off by NASA in 1978 as a cost-cutting measure.) Third, the orbiting Apollo Command Modules carried a wide range of instruments to photograph and analyze the lunar surface from above.

The last human left the Moon in December 1972, just a little more than three years after Neil Armstrong took his "giant leap for mankind." The program of lunar exploration was cut off in midstride due to political and economic pressures. It had cost just about $100 per American, spread over ten years—the equivalent of one large pizza per person per year. Yet for many people, the Moon landings were one of the central events in 20th-century history.

The giant Apollo rockets built to travel to the Moon were left to rust on the lawns of NASA centers in Florida, Texas, and Alabama, although recently some have at least been moved indoors to museums (Figure 8.4). Today no nation on Earth has the capability of returning people to the Moon; NASA estimates that it would require more than a decade to mount such an effort, and the financial and political will to do so is lacking. (In a bizarre piece of irony, a few people even question whether we went to the Moon at all, proposing instead that the Apollo program was a fake,

FIGURE 8.2

A Scientist on the Moon Geologist (and later U.S. Senator) Jack Schmitt in front of a large boulder in the Litrow Valley at the edge of the lunar highlands. The "rover" vehicle the astronauts used for driving around on the Moon is seen at left. Note how black the sky is on the airless Moon. No stars are visible because the surface is brightly lit by the Sun, and the exposure therefore is not long enough to reveal stars.

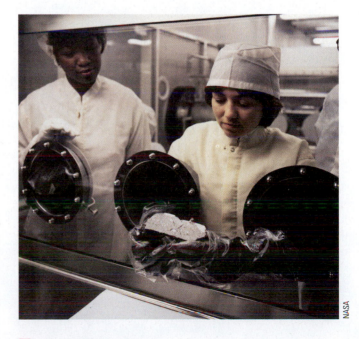

FIGURE 8.3

Handling Moon Rocks Lunar samples collected in the Apollo project are analyzed and stored in NASA facilities at the Johnson Space Center in Houston, Texas. Here a technician is examining a rock sample using gloves in a sealed environment to avoid contaminating the sample.

FIGURE 8.4

Moon Rocket on Display One of the unused Saturn 5 rockets built to go to the Moon, now a tourist attraction at NASA's Johnson Space Center in Houston.

filmed on a Hollywood sound stage! See the Moon hoax Web sites at the end of this chapter for some replies from scientists to such claims.) Having reached our nearest neighbor in space, humanity has for now retreated back to our own planet. It is hard to guess how long it will be before humans again venture out into the solar system.

The scientific legacy of Apollo remains, however, as discussed in the following sections of this chapter. The Apollo results have also been recently supplemented by three small scientific orbiters, one launched by NASA, one by Japan, and the third built by the U.S. military as a by-product of the "star wars" space defense effort (Figure 8.5).

8.1.2 Composition and Structure of the Moon

The composition of the Moon is not the same as that of the Earth. With an average density of only 3.3 g/cm^3, the Moon must be made almost entirely of silicate rock. Compared to the Earth, it is depleted in iron and other metals. It is as if the Moon were composed of the same basic silicates as the Earth's mantle and crust, with the core metals

FIGURE 8.5

Lunar South Pole This image, made of approximately 1500 individual pictures taken with the Clementine orbiter spacecraft, shows the south pole region of the Moon. Don't let the round border of the image fool you; you are looking at only a small portion of the Moon. For a sense of scale, note that the largest visible impact feature, the Schrodinger Basin at the lower right, is 320 km in diameter. Clementine gave astronomers their first close-up look at the Moon's polar regions. The south pole is of particular interest because ice was found in the permanent shadow zone (center).

and the volatiles selectively removed. These differences in composition between Earth and Moon provide important clues concerning the origin of the Moon, a topic we return to later in this chapter.

Probes of our satellite's interior carried out with seismometers taken to the Moon as part of the Apollo program confirm the absence of a large metal core. The Moon also lacks a global magnetic field like that of the Earth, a result consistent with the theory that such a field must be generated by motions in a liquid metal core. Not only is the metal lacking, but in addition the Moon is cold and has a solid interior, with very little seismic activity.

We also know from the study of lunar samples that water and other volatiles are largely absent from most of the crust. Very near the poles, however, NASA's Lunar Prospector mission measured the presence of hydrogen and interpreted this as an indication of ice mixed with the soil. The total quantity of water ice was estimated to be at least 6 billion tons. As liquid, this would only be enough water to fill a lake a few miles across, but compared with the rest of the dry lunar crust, so much water is remarkable. Presumably the water was carried to the Moon by comets and asteroids that hit our satellite. Some small fraction froze in a few extremely cold regions where the Sun never shines. One reason this discovery could be important is that it raises the possibility of future human habitations near the lunar poles, or even of a lunar base as a way station on the routes to Mars and the rest of the solar system.

<table>
<tr><td>8.2</td><td></td></tr>
</table>

8.2 THE LUNAR SURFACE

8.2.1 General Appearance

If you look at the Moon through a telescope, you can see that it is covered by impact craters of all sizes. The most conspicuous of the Moon's surface features—those that can be seen with the unaided eye and that make up the feature often called "the man in the Moon"—are vast splotches of darker lava flows.

Centuries ago, the early lunar observers thought that the Moon had continents and oceans and that it was a possible abode of life. Early names given to the lunar "seas"—Mare Nubium (Sea of Clouds), Mare Tranquillitatis (Tranquil Sea), and so on—are still in use today. In contrast, the "land" areas between the seas do not have names. Thousands of individual craters have been named, however, mostly for great scientists and philosophers. Among the most prominent are those named for Plato, Copernicus, Tycho, and Kepler. Galileo has only a very small crater, however, reflecting his low standing among the Roman Catholic scientists who made some of the first lunar maps.

We know today that the resemblance of lunar features to terrestrial ones is superficial. Even when they look somewhat similar, the origins of lunar features such as craters and mountains may be very different from their terrestrial counterparts. The Moon's relative lack of internal

activity, together with the absence of air and water, make most of its geological history unlike anything we know on Earth. And much of the lunar surface is older than the rocks of the Earth's crust, as we learned when the first lunar samples were analyzed in the laboratory.

8.2.2 Lunar History

To trace the detailed history of the Moon or of any planet, we must be able to estimate the ages of individual rocks. Once lunar samples were brought back by the Apollo astronauts, the radioactive dating techniques that had been developed for the Earth (see Section 6.3) were applied to them. The solidification ages of the samples ranged from about 3.3 to 4.4 billion years, substantially older than most of the rocks on the Earth. As we saw in Chapter 6, it now seems clear that both the Earth and the Moon were formed 4.5 billion years ago.

Most of the surface of the Moon (83 percent) consists of silicate rocks called *anorthosites;* these regions are known as the lunar **highlands.** They are made of relatively low-density rock that solidified on the cooling Moon like slag floating on the top of a smelter. Because they formed so early in lunar history (between 4.1 and 4.4 billion years ago), the highlands are also extremely heavily cratered, bearing the scars of billions of years of impacts by interplanetary debris (Figure 8.6).

The most prominent lunar features are the **maria** (Latin for "seas") that we mentioned earlier. *Maria* is the plural term, by the way; the singular is *mare* (pronounced "mah-ray"). These dark, round plains, which are much less cratered than the highlands, cover just 17 percent of the

■ **FIGURE 8.6**

Lunar Highlands The old, heavily cratered lunar highlands make up **83 percent** of the Moon's surface. The width of this image is 250 km.

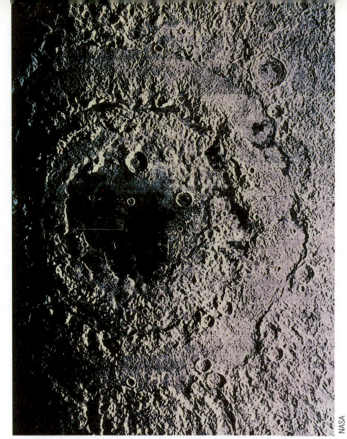

■ FIGURE 8.7

Lunar Maria About 17 percent of the Moon's surface consists of the maria—flat plains of basaltic lava. This view of Mare Imbrium also shows numerous secondary craters and evidence of material ejected from the large crater Copernicus on the upper horizon. Copernicus is an impact crater almost 100 km in diameter that was formed long after the lava in Imbrium had already been deposited.

■ FIGURE 8.8

Mare Orientale The youngest of the large lunar impact basins is Orientale, formed 3.8 billion years ago. Its outer ring is about 1000 km in diameter, roughly the distance between New York City and Detroit, Michigan. Unlike most of the other basins, Orientale has not been filled in with lava flows, so it retains its striking "bull's-eye" appearance. It is located on the edge of the Moon as seen from Earth.

lunar surface, mostly on the side facing the Earth (Figure 8.7). Today we know that the maria are volcanic plains, laid down in eruptions billions of years ago. They partly fill huge depressions called *impact basins,* which were produced by collisions of large chunks of material with the Moon relatively early in its history (Figure 8.8).

The lunar maria are all composed of basalt, very similar in composition to the crust under the oceans of the Earth or to the lavas erupted by many terrestrial volcanoes. A series of large eruptions on the Moon between 3.3 and 3.8 billion years ago (dated from laboratory measurements of returned samples) formed smooth flows typically a few meters thick but extending over distances of hundreds of kilometers. Eventually, these lava flows filled in the lowest parts of the giant impact basins to form the mare surfaces we see today (Figure 8.9).

Volcanic activity may have begun very early in the Moon's history, although most evidence of the first half-billion years is lost. What we do know is that the major mare volcanism, which involved the release of liquid magma from hundreds of kilometers below the surface, ended about 3.3 billion years ago. After that the Moon's interior cooled, and by 3 billion years ago all volcanic activity had ceased. Since then our satellite has been a geologically dead world, changing only slowly as the result of random impacts.

■ FIGURE 8.9

Rock from a Lunar Mare Sample of basalt from the mare surface. The gas bubbles are characteristic of rock formed from lava. All lunar rocks are chemically distinct from terrestrial rocks, a fact that has recently allowed scientists to identify a few lunar samples among the thousands of meteorites that reach Earth.

NASA

■ **FIGURE 8.10**

Lunar Mountain Mt. Hadley on the edge of Mare Imbrium, photographed by the *Apollo 15* astronauts. Note the smooth contours of the lunar mountains, which have not been sculpted by water or ice.

The major lunar mountains are all the result of the Moon's history of impacts. Long, arc-shaped ranges that border the maria are debris ejected from the impacts that formed these giant basins. These mountains have low, rounded profiles that resemble old, eroded mountains on Earth (Figure 8.10). But appearances can be deceiving. The mountains of the Moon have not been eroded, except for the effects of meteoritic impacts. They are rounded because that is the way they formed, and there has been no water or ice to carve them into cliffs and sharp peaks. (Compare Figure 8.10 with Figure 7.9.)

8.2.3 On the Lunar Surface

> The surface is fine and powdery. I can pick it up loosely with my toe. But I can see the footprints of my boots and the treads in the fine sandy particles.
> —Neil Armstrong, *Apollo 11* astronaut, immediately after stepping onto the Moon for the first time

The surface of the Moon is buried under a fine-grained soil of tiny, shattered rock fragments. The dark basaltic dust of the lunar maria was kicked up by every astronaut footstep, and this dust eventually worked its way into all of the astronauts' equipment. The upper layers of the surface are porous, consisting of loosely packed dust into which their boots sank several centimeters (Figure 8.11). This lunar dust, like so much else on the Moon, is the product of impacts. Each cratering event, large or small, breaks up the rock of the lunar surface and scatters the fragments. Ultimately, billions of years of impacts have reduced much of the surface layer to particles about the size of dust or sand.

In the absence of any air, the lunar surface experiences much greater temperature extremes than the surface of the Earth, even though the Earth is virtually the same distance from the Sun. Near local noon, when the Sun is highest in the sky, the temperature of the dark lunar soil rises to near the boiling point of water. During the long lunar night

NASA

■ **FIGURE 8.11**

Footprint in Moondust Apollo photo of an astronaut's bootprint in the lunar soil.

(which, like the lunar day, lasts two Earth weeks) it drops to about 100 K (−173°C). The extreme cooling is a result not only of the absence of air but also of the porous nature of the dusty soil, which cools more rapidly than solid rock would.

8.3 IMPACT CRATERS

The Moon provides an important benchmark for understanding the history of the planetary system. Most solid worlds show the effects of impacts, often extending back to the era when a great deal of debris from our system's formation process was still present. On the Earth, this long history has been erased by our active geology. On the Moon, in contrast, most of the impact history is preserved. If we can understand what has happened on the Moon, we may be able to apply this knowledge to the other planets and satellites.

8.3.1 Volcanic Versus Impact Origin of Craters

Until the middle of the 20th century, scientists did not generally recognize that lunar craters were the result of impacts. Since impact craters are extremely rare on Earth, geologists did not expect them to be the major feature of lunar geology. They reasoned (perhaps unconsciously) that since the craters we have on Earth are volcanic, the lunar craters must have a similar origin.

One of the first geologists to propose that lunar craters were the result of impacts was Grove K. Gilbert, a scientist with the U.S. Geological Survey in the 1890s. He pointed out that the large lunar craters, which are mountain-rimmed, circular features with floors generally below the level of the surrounding plains, are larger and have different shapes from known volcanic craters. Terrestrial volcanic craters are smaller and deeper and almost always occur at the tops of volcanic mountains (Figure 8.12). Gilbert therefore concluded that the lunar craters were not volcanic. The only alternative was an impact origin. His careful reasoning, although not accepted at the time, laid the foundations for the modern science of lunar geology.

Gilbert concluded that the lunar craters were produced by impacts, but he didn't understand why all of them were circular and not oval. The reason lies in the speed with which projectiles approach the Earth or Moon. Attracted by the gravity of the larger body, the incoming chunk strikes with at least escape velocity, which is 11 km/s for the Earth and 2.4 km/s (5400 mi/h) for the Moon. To this escape velocity is added whatever speed the projectile already had with respect to the Earth or Moon, typically 10 km/s or more.

At these speeds, the energy of impact produces a violent explosion that excavates a large volume of material in a symmetrical way. Photographs of bomb and shell craters on Earth confirm that explosion craters are always essen-

■ **FIGURE 8.12**

Volcanic and Impact Craters Profiles of typical terrestrial volcanic craters (a) and typical lunar impact craters (b) are quite different.

tially circular. Only following World War I did scientists recognize the similarity between impact craters and explosion craters, but Gilbert did not live to see his impact hypothesis widely accepted.

8.3.2 The Cratering Process

Let's consider how an impact at these high speeds produces a crater. When such a fast projectile strikes a planet, it penetrates two or three times its own diameter before stopping. During these few seconds its energy of motion is transferred into a shock wave, which spreads through the target body, and into heat, which vaporizes most of the projectile and some of the surrounding target. The shock wave fractures the rock of the target, while the hot silicate vapor generates an explosion similar to that of a nuclear bomb detonated at ground level (Figure 8.13). The size of the excavated crater depends primarily on the speed of impact, but generally it is 10 to 15 times the diameter of the projectile.

Ace★Astronomy™ Log into AceAstronomy and select this chapter to see the Active Figure called "The Moon's Craters," and Astronomy Exercise "Cratering."

(a)   (b)   (c)   (d)

■ **FIGURE 8.13**

Stages in the Formation of an Impact Crater (a) The impact occurs. (b) The projectile vaporizes and a shock wave spreads through the lunar rock. (c) Ejecta are thrown out of the crater. (d) Most of the ejected material falls back to fill the crater and form an ejecta blanket.

Observing the Moon

The Moon is one of the most beautiful sights in the sky, and it is the only object close enough to reveal its *topography* (mountains and valleys) to us without a visit from a spacecraft. A fairly small amateur telescope easily shows craters or mountains on the Moon as small as a few kilometers across.

Even as seen through a good pair of binoculars, the appearance of the Moon's surface changes dramatically with its phase (see Chapter 3 for an explanation of the phases). At full phase it shows almost no topographic detail, and you must look closely to see more than a few craters. This is because the sunlight illuminates the surface straight on, and in this flat lighting no shadows are cast. Much more revealing is the view near first or third quarter, when sunlight streams in from the side and topographic features cast sharp shadows. It is almost always more rewarding to study a planetary surface under such oblique lighting, when the maximum information about the surface relief can be obtained.

The flat lighting at full phase does, however, accentuate brightness contrasts on the Moon, such as those between the maria and highlands. Notice on the accompanying photo that several of the large mare craters seem to be surrounded by white material, and that the light streaks or rays that can stretch for hundreds of kilometers across the surface are clearly visible. These lighter features are ejecta, splashed out from the crater-forming impact.

By the way, there is no danger in looking at the Moon with binoculars or telescopes. The reflected sunlight is never bright enough to hurt you. In fact, the sunlit surface of the Moon has about the same brightness as a sunlit landscape on Earth. So explore away. Several guides to observing the Moon are listed in the recommendations for further reading at the end of this chapter.

One interesting thing about the Moon that you can see without binoculars or telescopes is popularly called "the new Moon in the old Moon's arms." Look at the Moon when it is a thin crescent and you can often make out the faint circle of the entire lunar disk, even though the sunlight shines on only the crescent. The rest of the disk is illuminated not by sunlight but by earthlight—sunlight reflected from the Earth. The light of the full Earth on the Moon is about 50 times brighter than that of the full Moon shining on the Earth.

Lick Observatory

The Appearance of the Moon at Different Phases   Illumination from the side (*left*) brings craters and other topographic features into sharp relief. At full phase (*right*) there are no shadows and it is more difficult to see such features. However, the flat lighting at full phase brings out some surface features, such as the bright rays of ejecta that stretch out from a few large young craters.

■ FIGURE 8.14

Typical Impact Crater King Crater on the far side of the Moon, a fairly recent lunar crater 75 km in diameter, shows most of the features associated with large impact structures.

An impact explosion of the sort described above leads to a characteristic kind of crater, as illustrated in Figure 8.14. The central cavity is initially bowl-shaped (the word *crater* comes from the Greek word for "bowl"), but the rebound of the crust partially fills it in, producing a flat floor and sometimes creating a central peak. Around the rim, landslides create a series of terraces.

The rim of the crater is turned up by the force of the explosion, so it rises above both the floor and the adjacent terrain. Surrounding the rim is an *ejecta blanket* consisting of material thrown out by the explosion. This debris falls back to create a rough, hilly region, typically about as wide as the crater diameter. Additional, higher-speed ejecta fall at greater distances from the crater, often digging small *secondary craters* where they strike the surface (see Figure 8.7).

Some of these streams of ejecta can extend from the crater for hundreds or even thousands of kilometers, creating on the Moon the bright *crater rays* that are prominent in photos taken near full phase (see Seeing for Yourself: *Observing the Moon*). The brightest lunar crater rays are associated with large young craters such as Kepler and Tycho.

8.3.3 Using Crater Counts

As discussed in Chapter 6, if a world has had little erosion or internal activity, like the Moon during the past 3 billion years, it is possible to use the number of impact craters on its surface to estimate the age of that surface. By "age" we mean the time since a major disturbance occurred on that surface (such as the volcanic eruptions that produced the lunar maria).

We can't directly measure the rate at which craters are being formed on the Earth and Moon, since (as we saw in Chapter 7) the average interval between large crater-forming impacts is longer than the span of human history. Our best-known example of such a large crater, Meteor Crater in Arizona, is 50,000 years old. However, the cratering rate can be estimated from the number of craters on the lunar maria or calculated from the number of potential "projectiles" (asteroids and comets) present in the solar system today, as will be discussed in Chapter 12. Both lines of reasoning lead to about the same answers.

For the Moon these calculations indicate that a crater 1 km in diameter should be produced about every 200,000 years, a 10-km crater every few million years, and one or two 100-km craters every billion years. If the cratering rate stayed the same over the millennia, we can figure out how long it must have taken to make all the craters we see in the lunar maria. Our calculations show that it would have taken several billion years. This result is similar to the age determined for the maria from radioactive dating—3.3 to 3.8 billion years old. The fact that these two calculations agree suggests that our assumption was right: Comets and asteroids in approximately their current numbers have been impacting planetary surfaces for at least 3.8 billion years. Calculations carried out for other planetary bodies (and their satellites) indicate that they also have been subject to about the same number of interplanetary impacts during this time.

Earlier than 3.8 billion years ago, however, we have good reason to believe that the impact rates must have been a great deal higher. This becomes immediately evident when we compare the numbers of craters on the lunar highlands with those on the maria. Typically, there are ten times more craters on the highlands than on a similar area of maria. Yet the radioactive dating of highland samples showed that they are only a little older than the maria, typically 4.2 billion years rather than 3.8 billion years. If the rate of impacts had been constant throughout the Moon's history, the highlands would have had to be at least ten times older. They would thus have had to form 38 billion years ago—long before the universe itself began.

In science, when an assumption leads to an unbelievable conclusion, we must go back and re-examine that assumption—in this case, the constant impact rate. The contradiction is resolved if the impact rate varied over time, with a much heavier bombardment earlier than 3.8 billion years ago (Figure 8.15). This "heavy bombardment" produced most of the craters we see today in the highlands.

FIGURE 8.15

Cratering Rates over Time The number of craters being made on the Moon's surface has varied with time over the past 4.3 billion years.

8.4 THE ORIGIN OF THE MOON

It is characteristic of modern science to ask how things originated. Understanding the origin of the Moon has proved to be challenging for planetary scientists, however. Part of the difficulty is simply that we know so much about our satellite (quite the opposite of our usual problem in astronomy). Put in simple terms, the Moon is both tantalizingly similar to the Earth and frustratingly different.

8.4.1 Ideas for the Origin of the Moon

Most of the hypotheses for the Moon's origin follow one of three general ideas:

1. The fission theory—the Moon was once part of the Earth but separated from it early in their history.
2. The sister theory—the Moon formed together with (but independently of) the Earth, as we believe many satellites of the outer planets formed.
3. The capture theory—the Moon formed elsewhere in the solar system and was captured by the Earth.

Unfortunately, there seem to be fundamental problems with each of these ideas. Perhaps the easiest hypothesis to reject is the capture theory. Its primary drawback is that no one knows of any way that the early Earth could have captured a large satellite from elsewhere. One body approach-

ing another cannot go into orbit around it without a substantial loss of energy; this is the reason that spacecraft destined to orbit other planets are equipped with retro-rockets. Furthermore, if such a capture did take place, the satellite would go into a very eccentric orbit rather than the nearly circular orbit the Moon occupies today. Finally, there are too many compositional similarities between the Earth and the Moon, particularly an identical fraction of the major isotopes of oxygen, to justify seeking a completely independent origin.

The fission hypothesis, which states that the Moon separated from the Earth, was suggested in the late 19th century by George Darwin, profiled in Chapter 3. More modern calculations have shown that the sort of spontaneous fission or splitting imagined by Darwin is impossible. Furthermore, it is difficult to understand how a Moon made out of terrestrial material in this way could have developed the many distinctive chemical differences now known to characterize our satellite.

Scientists have therefore been left with the sister hypothesis—that the Moon formed alongside the Earth—or with some modification of the fission hypothesis that finds a more acceptable way for the lunar material to have separated from the Earth. But the more we learn about our satellite, the less these old ideas seem to fit the bill.

8.4.2 The Giant Impact Hypothesis

In an effort to resolve these apparent contradictions, scientists developed a fourth hypothesis for the origin of the Moon, one that involves a giant impact early in the Earth's history. There is increasing evidence that large chunks of material—objects of essentially planetary mass—were orbiting in the inner solar system at the time that the terrestrial planets formed. The giant impact hypothesis envisions the Earth being struck obliquely by an object approximately one-tenth the Earth's mass—a "bullet" about the size of Mars. This is very nearly the largest impact the Earth could experience without being shattered.

Such an impact would disrupt much of the Earth by penetrating into the core, ejecting a vast amount of material into space, and releasing almost enough energy to break the planet apart (Figure 8.16). The simulations done on large computers indicate that material totaling several percent of the Earth's mass could be ejected in such an impact. Most of this material would be from the stony mantles of the Earth and the impacting body, not from their metal cores. This ejected rock vapor then cooled and formed a ring of material orbiting around the Earth. It was this ring that ultimately condensed into the Moon.

While we do not have any current way of showing that the giant impact hypothesis is the correct model of the Moon's origin, it does offer potential solutions to most of the major problems raised by the chemistry of the Moon. First, since the Moon's raw material is derived from the *mantles* of the Earth and the projectile, the absence of metals is easily understood. Second, most of the volatile elements would have been lost during the high-temperature

Courtesy of A. G. W. Cameron, Center for Astrophysics

■ **FIGURE 8.16**

How an Impact May Have Formed the Moon Four frames from a series of computer simulations showing a giant impact with the Earth. Iron is shown as spheres of blue and green. Rock is represented by red, pink, yellow, and white spheres, in order of increasing temperature. In this particular simulation, the impactor has 1/5 the mass in the system, and the Earth has 4/5. After the oblique impact, a plume of material is ejected into space. It is this plume that, under the right circumstances, can condense to form the Moon. Note that the plume eventually has much more pink (rock) than blue or green (iron), which explains why the Moon resembles the mantle of the Earth.

phase following the impact, explaining the lack of these materials on the Moon. Yet by making the Moon primarily of terrestrial mantle material, it is also possible to understand similarities such as identical abundances of various oxygen isotopes.

This idea that large impacts (especially during the early history of the solar system) have played a major role in shaping the worlds we see is not unique to our study of the Moon. As you read through the other chapters about the planets and their satellites, you will see further indications that a number of the present-day characteristics of our system may be due to our violent past.

8.5 MERCURY

The planet Mercury is similar to the Moon in many ways. Like the Moon, it has no atmosphere and its surface is heavily cratered. As described later in this chapter, it also shares with the Moon the likelihood of a violent birth history.

8.5.1 Mercury's Orbit

Mercury is the nearest to the Sun of the nine planets and, in accordance with Kepler's third law, it has the shortest period of revolution about the Sun (88 of our days) and the highest average orbital speed (48 km/s). It is appropriately named for the fleet-footed messenger god of the Romans. Because Mercury remains close to the Sun, it can be difficult to pick out in the sky. As you might imagine, it's best seen when its eccentric orbit takes it as far from the Sun as possible.

The semimajor axis of Mercury's orbit—that is, the planet's average distance from the Sun—is 58 million km, or 0.39 astronomical units (AU). However, because its orbit has the high eccentricity of 0.206, Mercury's actual distance from the Sun varies from 46 million km at perihelion to 70 million km at aphelion. (If some of these terms are new to you, you may want to look at Chapter 2 or the glossary at the end of the book.) Pluto is the only planet with a more eccentric orbit than Mercury. Furthermore, the 7° inclination of Mercury's orbit to the plane of the ecliptic is also greater than that of any other planet except Pluto.

8.5.2 Composition and Structure

Table 8.1 compares some basic properties of Mercury with those of the Moon. Mercury's mass is 1/18 that of the Earth; again Pluto is the only planet with a smaller mass. Mercury is also the second smallest of the planets, having a diameter of 4878 km, less than half that of the Earth. Mercury's density is 5.4 g/cm^3, much greater than the density of the Moon, indicating that the composition of these two objects differs substantially.

Its composition is one of the most interesting things about Mercury; it is unique among the planets. Mercury's high density tells us that it must be composed largely of heavier materials such as metals. The most likely models for Mercury's interior suggest a metallic iron–nickel core amounting to 60 percent of the total mass, with the rest of the planet made up primarily of silicates. The core has a diameter of 3500 km and extends out to within 700 km of the surface. We could think of Mercury as a metal ball the size of the Moon surrounded by a rocky crust 700 km thick (Figure 8.17). Unlike the Moon, Mercury also has a weak magnetic field. The existence of this field is consistent with a large metal core, and it suggests that at least part of the core must be liquid in order to generate the observed field. (See Section 7.1 for the explanation of how magnetic fields and liquid cores are connected.)

■ FIGURE 8.17

Mercury's Internal Structure The interior of Mercury is dominated by a metallic core about the same size as our Moon.

8.5.3 Mercury's Strange Rotation

Visual studies of Mercury's indistinct surface markings were once thought to indicate that the planet kept one face to the Sun (as the Moon does to the Earth). Thus, for many years it was widely believed that Mercury's rotation period was equal to its revolution period of 88 days, making one side perpetually hot while the other was always cold.

Radar observations of Mercury in the mid-1960s, however, showed conclusively that Mercury does not keep one side fixed to the Sun. Recall from Section 5.4 that the frequency of a transmitted radar pulse can be controlled precisely. Any motion of the target then introduces a measurable change in this frequency as a result of the Doppler effect (see Section 4.6). Such target motions can include rotation: If a planet is turning, one side seems to be approaching the Earth while the other is moving away from it. The result is to spread or broaden the precise transmitted frequency into a range of frequencies in the reflected signal (Figure 8.18). The degree of broadening provides an exact measurement of the rotation rate of the planet.

Mercury's sidereal period of rotation (how long it takes to turn with respect to the distant stars) was found to be about 59 days, which is just two thirds of the planet's period of revolution. Subsequently astronomers found that a situation where the spin and the orbit of a planet are in a 2-to-3 ratio is stable. That is, it can continue for a long time, like the situation where the time to spin equals the time to orbit (a relationship Pluto exhibits, for example).

Mercury is very hot on its daylight side, but because it has no appreciable atmosphere, it gets surprisingly cold during the long nights. The temperature on the surface climbs to 700 K at noontime. After sunset, however, the

■ FIGURE 8.18

Doppler Radar Measures Rotation When a radar beam is reflected from a rotating planet, the motion of one side of the planet's disk toward us and the other away from us causes Doppler shifts in the reflected signal. The effect is to cause both a "redshift" and a "blueshift," and thus to widen the spread of frequencies in the radio beam.

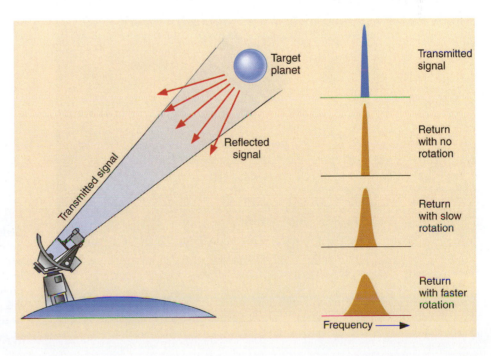

temperature drops, reaching 100 K just before dawn. The *range* in temperature on Mercury is thus 600 K (or 600°C), more than on any other planet.

8.5.4 The Surface of Mercury

The first close-up look at Mercury came in 1974, when the U.S. spacecraft Mariner 10 passed 9500 km from the surface of the planet and transmitted more than 2000 photographs to Earth, revealing details with a resolution down to 150 m. Mercury strongly resembles the Moon in appearance (Figures 8.19 and 8.20). It is covered with thousands of craters and larger basins up to 1300 km in diameter. Some of the brighter craters are rayed, like Tycho and Copernicus on the Moon, and many have central peaks. There are also *scarps* (cliffs) more than a kilometer high and hundreds of kilometers long, as well as ridges and plains.

Most of the mercurian features have been named in honor of artists, writers, composers, and other contributors to the arts and humanities, in contrast with the scientists commemorated on the Moon. Among the most prominent craters are Bach, Shakespeare, Tolstoy, Mozart, and Goethe.

There is no evidence of plate tectonics on Mercury. Distinctive long scarps, however, can sometimes be seen cutting across craters; this means the scarps must have formed later than the craters (Figure 8.21). These long cliffs appear to have their origin in the slight compression of Mercury's crust. Apparently, at some point in its history this planet shrank, wrinkling the crust, and it must have done so after most of the craters on its surface had already

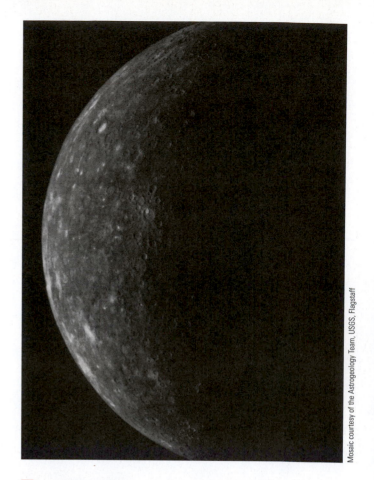

Mosaic courtesy of the Astrogeology Team, USGS, Flagstaff

■ **FIGURE 8.19**

Best Photo of Mercury The planet Mercury as photographed by Mariner 10 in 1974. The heavily cratered surface resembles that of the Moon.

NASA/JPL

■ **FIGURE 8.20**

Caloris Basin Part of the Caloris basin on Mercury, 1300 km in diameter, can be seen at the bottom. This partially flooded impact basin is the largest structural feature on Mercury seen by Mariner 10. Compare this photo with Figure 8.8, the Orientale basin on the Moon.

What a Difference a Day Makes

Mercury rotates three times for each two orbits around the Sun. It is the only planet that exhibits this relationship between its spin and its orbit, and there are some interesting consequences for any observers who might someday be stationed on the surface of Mercury.

Here on Earth, we take it for granted that days are much shorter than years. Therefore, the two ways of defining the local "day"—how long the planet takes to rotate, and how long the Sun takes to return to the same position in the sky (see Chapter 3)—are the same for most practical purposes. But this is not the case on Mercury. While Mercury rotates in 59 Earth days, the time for the Sun to return to the same place in Mercury's sky turns out to be two Mercury years, or 176 Earth days. (Note that this result is not intuitively obvious, so don't be upset with yourself if you didn't come up with it.) Thus, if one day at noon a Mercury explorer suggests to her companion that they should meet at noon the next day, this could mean a very long time apart!

To make things even more interesting, recall that Mercury has an eccentric orbit, meaning that its distance from the Sun varies significantly during each mercurian year. By Kepler's law, the planet moves fastest in its orbit when closest to the Sun. Let's examine how this affects the way we would see the Sun in the sky during one 176-Earth-day cycle. We'll look at the situation as if we were standing on the surface of Mercury in the center of a giant basin that astronomers call Caloris (see Figure 8.20).

At the location of Caloris, Mercury is most distant from the Sun at sunrise; this means the rising Sun looks smaller in the sky (although still more than twice the size it appears from the Earth). As the Sun rises higher and higher, it looks bigger and bigger; Mercury is now getting closer to the Sun in its eccentric orbit. At the same time, the apparent motion of the Sun slows down as Mercury's faster motion in orbit begins to catch up with its rotation.

At noon the Sun is now three times larger than it looks from Earth and hangs almost motionless in the sky. As the afternoon wears on, the Sun eventually looks smaller and smaller again, and moves faster and faster in the sky. At sunset, a full Mercury year (or 88 Earth days) after sunrise, the Sun is back to its smallest apparent size as it dips out of sight. Then it takes another Mercury year before the Sun rises again. (Sunrise and sunset are much more sudden on Mercury, since there is no atmosphere to bend or scatter the rays of sunlight.)

Astronomers call locations like the Caloris basin the "hot longitudes" on Mercury because the Sun is closest to the planet at noon, just when it is lingering overhead for many Earth days. This makes them the hottest places on the planet.

We bring all this up not because the exact details of this scenario are so important, but to illustrate how many of the things we take for granted on Earth are not the same on other worlds. As we've mentioned before, one of the best things about taking an astronomy class should be ridding you forever of any "Earth chauvinism" you might have. The way things are on our planet is just one of the many ways nature can arrange reality.

formed. If the standard cratering chronology applies to Mercury, this shrinkage must have taken place during the past 4 billion years. If we understood the interior structure and composition of the planet better, we could probably calculate how internal changes in temperature might have led to this global compression, which has no counterpart on the Moon or the other terrestrial planets.

Although it has become something of a forgotten planet in the two decades since the Mariner 10 flybys, two spacecraft missions are now planned to revisit the planet:

■ **FIGURE 8.21**

Discovery Scarp on Mercury This long cliff, nearly 1 km high and more than 100 km long, cuts across several craters. We conclude that the compression that made "wrinkles" like this in the planet's surface must have taken place after the craters were formed.

NASA/JPL

the Messenger probe from NASA and BepiColombo from ESA (the European Space Agency). Astronomical studies from the Earth are also still useful. For example, enhanced radar reflectivity from both poles of the planet was discovered in 1992, indicating the presence of ice just beneath the surface. Ice on a planet so close to the Sun may sound surprising, but the temperature very near the poles remains low enough for ice to survive. Still, the discovery of these icy polar caps was unexpected, since they seem to imply that the polar regions of Mercury have remained below freezing since the formation of the planet. Probably there are cold traps there similar to the deep craters that shelter ice at the poles of our Moon.

8.5.5   The Origin of Mercury

The problem with understanding how Mercury formed is the reverse of the problem posed by the composition of the Moon. We have seen that, unlike the Moon, Mercury is composed mostly of metal. However, astronomers think that Mercury should have formed with about the same ratio of metal to silicate as that found on the Earth or Venus. How did it lose so much of its rocky material?

The most probable explanation for Mercury's silicate loss may be similar to the explanation for the Moon's lack of a metal core. Mercury is likely to have experienced several giant impacts very early in its youth, and one or more of these may have torn away a fraction of its mantle and crust, leaving a body dominated by its iron core.

Today astronomers recognize that the early solar system was a chaotic place, with the final stages of planet formation characterized by impacts of great violence. Some objects of planetary mass may have been destroyed, while others could have fragmented and then reformed, perhaps more than once. Both the Moon and Mercury, with their strange compositions, bear testimony to the catastrophes that must have characterized the solar system during its youth.

SURFING THE WEB

The Moon—General Sites About the Moon, with Good Links:

- *The Nine Planets Site:*
  seds.lpl.arizona.edu/nineplanets/nineplanets/luna.html

- *Views of the Solar System Site:*
  www.solarviews.com/eng/moon.htm

- *Planetary Sciences Site:*
  nssdc.gsfc.nasa.gov/planetary/planets/moonpage.html

- *Exploring the Moon:*
  cass.jsc.nasa.gov/pub/expmoon/lunar_missions.html
  An extensive site from NASA's Johnson Space Flight Center, with information on U.S. and Russian Moon exploration and good links to many other Web resources.

- *Apollo Lunar Surface Journal:*
  www.hq.nasa.gov/office/pao/History/alsj/
  Information, interviews, maps, photos, video and audio clips, and much more on each of the Apollo landing missions.

Recent Missions to the Moon:

- *Clementine Mission Site:*
  www.nrl.navy.mil/clementine/

- *Lunar Prospector Home Page:* lunar.arc.nasa.gov/

- *Touring the Moon with Binoculars:*
  skyandtelescope.com/observing/objects/moon/
  An introductory guide, with a map, for viewing the Moon with simple equipment.

Sites That Take a Skeptical Look at Claims That the Moon Landings Were a Hoax:

- *The Bad Astronomy Site:*
  www.badastronomy.com/bad/tv/foxapollo.html

- *Science@NASA Site:*
  science.nasa.gov/headlines/y2001/ast23Feb_2.htm

- *Comments by Astronomer Jim Scotti:*
  pirlwww.lpl.arizona.edu/~jscotti/NOT_faked/

Mercury:

- *The Nine Planets Site:*
  seds.lpl.arizona.edu/nineplanets/nineplanets/mercury.html

- *Views of the Solar System Site:*
  www.solarviews.com/eng/mercury.htm

- *Planetary Sciences Site:*
  nssdc.gsfc.nasa.gov/planetary/planets/mercurypage.html

- *Mercury Unveiled:*
  www.psrd.hawaii.edu/PSRdiscoveries/Jan97/Mercury Unveiled.html
  Planetary scientist G. Jeffrey Taylor updates our current understanding of Mercury based on reanalyzed Mariner 10 images.

- *Messenger Mission:* messenger.jhuapl.edu
  Previews the NASA probe to Mercury.

SUMMARY

8.1 Most of what we know about the Moon derives from the Apollo program, including 400 kg of lunar samples still being intensively studied. The Moon has 1/80 the mass of the Earth and is severely depleted in both metals and volatile materials. It is made almost entirely of silicates like those in the Earth's mantle and crust. However, spacecraft have found evidence of a small amount of water near the lunar poles, most likely deposited by comet and asteroid impacts.

8.2 Lunar rocks can be dated by their radioactivity. Like the Earth, the Moon was formed about 4.5 billion years ago. The heavily cratered **highlands** are made of rocks more than 4 billion years old. The darker volcanic plains of the **maria** were erupted between 3.3 and 3.8 billion years ago. Generally, the surface is dominated by impacts, including continuing small impacts that produce its fine-grained soil.

8.3 A century ago Gilbert suggested that the lunar craters were of impact origin, but the cratering process was not well understood until more recently. High-speed impacts produce explosions and excavate craters with raised rims, ejecta blankets, and often central peaks. Cratering rates have been roughly constant for the past 3 billion years but earlier were much greater. Crater counts can be used to derive approximate ages for geological features on the Moon and other planets.

8.4 The three standard hypotheses for the origin of the Moon are the fission hypothesis, the sister hypothesis, and the capture hypothesis. All have problems, and recently they have been supplanted by the giant impact hypothesis, which ascribes the origin of the Moon to the impact of a Mars-sized projectile with the Earth 4.5 billion years ago.

8.5 Mercury is the nearest planet to the Sun and the fastest moving. Mercury is similar to the Moon in having a heavily cratered surface and no atmosphere, but it differs in having a very large metal core. Early in its evolution, it apparently lost part of its silicate mantle, probably due to one or more giant impacts. Long scarps on its surface testify to a global compression of Mercury's crust during the past 4 billion years.

INTER-ACTIVITY

A We mentioned that no nation on Earth now has the capability to send a human being to the Moon, even though the United States once sent 12 astronauts to land there. How does your group feel about this? Should we continue the exploration of space with human beings? Should we put habitats on the Moon? Should we go to Mars? Does humanity have a "destiny in space"? Whatever your answer to these questions is, make a list of the arguments and facts that support your position.

B When they first hear about the giant impact hypothesis for the origin of the Moon, many students are intrigued and wonder why we can't cite more evidence for it. Could your group make a list of reasons we cannot find any traces on the Earth of the great impact that formed the Moon?

C We discussed that the ice (mixed into the soil) that is found on the Moon was most likely delivered by comets. Have your group make a list of all the reasons the Moon would not have any ice of its own left from its early days.

D Can your group make a list of all the things that would be different if our Earth had no Moon? Don't restrict your answer to astronomy and geology. (You may want to review Chapter 3.)

E If, one day, humanity decides to establish a colony on the Moon, where should we put it? Make a list of the advantages and disadvantages of locating such a human habitat on the near side, the far side, or at the poles. What site would be best for visible-light and radio astronomy?

F A member of the class (but luckily, not a member of your group) suggests that he has always dreamed of building a vacation home on the planet Mercury. Can your group make a list of all the reasons such a house would be hard to build and keep in good repair?

REVIEW QUESTIONS

Ace Astronomy™ Assess your understanding of this chapter's topics with additional quizzing and animations at **http://ace.brookscole.com/voyages**

1. What is the composition of the Moon, and how does it compare to the composition of the Earth? Of Mercury?
2. Outline the main events in the Moon's geological history.
3. Explain how high-speed impacts form circular craters. How can this explanation account for the characteristic features of impact craters?
4. Explain the evidence for a period of heavy bombardment on the Moon about 4 billion years ago. What might have been the source of the impacting debris during this early era?
5. How did our exploration of the Moon differ from that of Mercury (and the other planets)?
6. Summarize the four main theories for the origin of the Moon. Give pros and cons for each.
7. What do current ideas about the origins of the Moon and Mercury have in common? How do they differ?

THOUGHT QUESTIONS

8. One of the primary scientific objectives of the Apollo program was the return of lunar material.
   a. Why was this so important?
   b. What can be learned from samples?
   c. Are they still of value now?

9. Apollo astronaut David Scott dropped a hammer and a feather together on the Moon, and both reached the ground at the same time. What are the two distinct advantages that this experiment on the Moon had over the same experiment as performed by Galileo on the Earth?

10. Galileo thought the lunar maria might be seas of water. If you had no better telescope than the one he had, could you prove that they are not composed of water?

11. Why did it take so long for geologists to recognize that the lunar craters had an impact origin rather than a volcanic one?

12. How might a crater made by the impact of a comet with the Moon differ from a crater made by the impact of an asteroid?

13. Why are the lunar mountains smoothly rounded rather than having sharp, pointed peaks (as they were almost always depicted in science-fiction illustrations and films before the first lunar landings)?

14. The lunar highlands have about ten times more craters on a given area than do the maria. Does this mean that the highlands are ten times older? Explain your reasoning.

15. At the end of Section 8.2, your authors say that lunar night and day each last about 2 Earth weeks. After looking over the information in Chapter 3 and this chapter about the motions of the Moon, can you explain why? (It helps to draw a diagram for yourself.)

16. Give several reasons Mercury would be a particularly unpleasant place to build an astronomical observatory.

17. If, in the remote future, we establish a base on Mercury, keeping track of time will be a challenge. Discuss how to define a year on Mercury, and the two ways to define a day. Can you come up with ways that humans raised on Earth might deal with time cycles on Mercury?

18. The Moon has too little iron, Mercury too much. How can both of these anomalies be the result of giant impacts? Explain how the same process can yield such apparently contradictory results.

FIGURING FOR YOURSELF

The average density of a body equals its mass divided by its volume:

$$\text{density} = \frac{\text{mass}}{\frac{4}{3}\pi \times \text{radius}^3}$$

Astronomers can measure both mass and radius accurately when a spacecraft flies by a body.

19. Using the information in this chapter, calculate the average density of the Moon. Show your work. Does your calculation agree with the figure we give in this chapter?

20. Using the information in this chapter, calculate the average density of Mercury. Show your work. Does your calculation agree with the figure we give in this chapter?

21. The Moon was once closer to the Earth than it is now. When it was at half its present distance, how long was its period of revolution? (See Chapter 2 for the formula to use.)

22. Astronomers believe that the deposit of lava in the giant mare basins did not happen in one flow but in many different eruptions spanning some time. Indeed, in any one mare we find a variety of rock ages, typically spanning about 100 million years. The individual lava flows as seen in Hadley Rill by the *Apollo 15* astronauts were about 4 m thick. Estimate the average time interval between the beginnings of successive lava flows if the total depth of the lava in the mare is 2 km.

23. The Moon requires about one month (0.08 year) to orbit the Earth. Its distance from us is about 400,000 km (0.0027 AU). Use Kepler's third law, as modified by Newton, to calculate the mass of the Earth relative to the Sun.

SUGGESTIONS FOR FURTHER READING

On the Moon

Chaikin, A. *A Man on the Moon*. 1994, Viking Press. A well-reviewed history of manned lunar exploration.

Coco, M. "Staging a Moon Shot" in *Astronomy,* Aug. 1992, p. 62. Describes photographing the Moon.

Foust, J. "NASA's New Moon" in *Sky & Telescope*, Sept. 1998, p. 48. On results from the Lunar Prospector mission.

Jayawardhana, R. "Deconstructing the Moon" in *Astronomy,* Sept. 1998, p. 40. An update on the giant impact hypothesis for the origin of the Moon.

Kitt, M. *The Moon: An Observing Guide for Backyard Telescopes*. 1992, Kalmbach. Eighty-page illustrated primer for beginners.

Ryder, G. "Apollo's Gift: The Moon" in *Astronomy,* July 1994, p. 40. Good evolutionary history of the Moon.

Schmitt, H. "Exploring Taurus–Littrow: Apollo 17" in *National Geographic,* Sept. 1973. First-person account by the only scientist to walk on the Moon.

Spudis, P. "The Giant Holes of the Moon" in *Astronomy,* May 1996, p. 50. On the results of the Clementine mission. (See also reports in *Sky & Telescope,* July 1995, p. 32; Feb. 1997, p. 24.)

Spudis, P. *The Once and Future Moon*. 1998, Smithsonian Institution Press. A geologist discusses what our exploration of the Moon has taught us, and what we might do in the future to know and use the Moon better.

Taylor, G. "The Scientific Legacy of Apollo" in *Scientific American*, July 1994, p. 40.

On Mercury

Armstrong, J. "Heading Back to the Forgotten Planet" in *Astronomy*, Oct. 2002, p. 40. Preview of Messenger Mission.

Beatty, J. "Mercury's Cool Surprise" in *Sky & Telescope*, Jan. 1992, p. 35.

Chapman, C. "Mercury's Heart of Iron" in *Astronomy*, Nov. 1988, p. 22. Good introduction to our modern view of the planet.

Nelson, R. "Mercury: The Forgotten Planet" in *Scientific American*, Nov. 1997, p. 56.

Sheehan, W. and Dobbins, T. "Mesmerized by Mercury" in *Sky & Telescope,* June 2000, p. 109. History of Mercury observations and how amateur astronomers can contribute.

Strom, R. *Mercury: The Elusive Planet*. 1987, Smithsonian Institution Press. Best book on the inner planet.

Heavily Eroded Canyonlands on Mars This Viking spacecraft view, taken from Mars orbit, looks down on a small part of the Valles Marineris canyon complex and shows an area about 60 km across.

Earth-Like Planets: Venus and Mars

One deduction from this thin air [on Mars] we must be careful not to make— that because it is thin, it is incapable of supporting intelligent life. That beings constituted physically as we are would find it a most uncomfortable habitat is pretty certain. But lungs are not wedded to logic, as public speeches show, and there is nothing in the world or beyond it to prevent, so far as we know, a being with gills, for example, from being a most superior person.

Percival Lowell, in Mars (1895)

THINKING AHEAD

For decades science-fiction writers have conjured up visions of Venus and Mars as Earth-like worlds where humans would someday visit and (perhaps) come face to face with alien life. Now that our robot probes have explored both planets, we know that neither has conditions similar to our own. How did it happen that these three neighboring terrestrial planets have turned out to be so different from each other?

The Moon and Mercury are geologically dead. In contrast, the larger terrestrial planets—the Earth, Venus, and Mars—are more active and interesting worlds. We have already discussed the Earth, and we now turn to Venus and Mars. These are the nearest planets and the most accessible to spacecraft. Not surprisingly, the greatest effort in planetary exploration has been devoted to these fascinating worlds. In this chapter we discuss some of the results of four decades of scientific exploration of Mars and Venus.

Ace Astronomy™ The AceAstronomy icon throughout the text indicates an opportunity for you to test yourself on key concepts and to explore animations and interactions on the AceAstronomy website at **http://ace.brookscole.com/voyages**

Virtual Laboratories

 Planetary Geology

 Planetary Atmospheres and Their Retention

9.1 THE NEAREST PLANETS: AN OVERVIEW

As you might expect from close neighbors, Mars and Venus are among the brightest objects in the night sky. The average distance of Mars from the Sun is 227 million km (1.52 AU), or about half again as far from the Sun as the Earth. However, Mars has a relatively large orbital eccentricity of 0.093, which means its distance varies by 42 million km as it goes around the Sun.

Venus' orbit, in contrast, is very nearly circular, at a distance of 108 million km (0.72 AU) from the Sun. Like Mercury, Venus sometimes appears as an "evening star" and sometimes as a "morning star." The closest Mars ever gets to Earth is about 56 million km. Venus approaches the Earth more closely than does any other planet: At its nearest it is only 40 million km away from us.

9.1.1 Appearance

Venus looks very bright, and even a small telescope reveals that it goes through phases like the Moon. Galileo discovered that Venus displays a full range of phases, and he used this as an argument to show that Venus must circle the Sun and not the Earth (see Figure 1.16). The planet's actual surface is not visible because it is shrouded by dense clouds that reflect about 70 percent of the sunlight that falls on them, frustrating efforts to study the underlying surface, even with cameras in orbit around the planet (Figure 9.1).

In contrast, Mars is more tantalizing as seen through a telescope (Figure 9.2). The planet is distinctly red, due (as we now know) to the presence of iron oxides in its soil. This color may account for its association with war (and bloodletting) in the legends of early cultures. The best resolution obtainable from telescopes on the ground is about 100 km, or about the same as what we can see on the Moon with the unaided eye. At this resolution, no hint of topographic structure can be detected—no mountains, no valleys, not even impact craters. On the other hand, bright polar ice caps can be seen easily, together with dusky surface markings that sometimes change in outline and intensity from season to season. Of all the planets, only Mars has a surface that can be clearly made out from the Earth.

For a few decades around the turn of the 20th century, some astronomers believed they saw evidence of an intelligent civilization on Mars. The controversy began in 1877, when the Italian astronomer Giovanni Schiaparelli announced that he could see long, faint, straight lines on Mars that he called *canale,* or channels. In English-speaking countries, the term was mistakenly translated as *canals,* implying an artificial origin.

Even before Schiaparelli's observations, astronomers had watched the bright polar caps change size with the seasons and had seen variations in the dark surface features. With a little imagination, it was not difficult to picture the

NASA/ARC

■ FIGURE 9.1

Venus as Photographed by the Pioneer Venus Orbiter This ultraviolet image shows upper-atmosphere cloud structure that would be invisible at visible wavelengths. Note that there is not a glimpse of the planet's surface.

Steve Larson, University of Arizona

■ FIGURE 9.2

Mars Seen from the Earth's Surface These are among the best Earth-based photos of Mars, taken in 1988 when the planet was exceptionally close to the Earth. The polar caps and dark surface markings are evident, but not the topographic features.

Percival Lowell: Dreaming of an Inhabited Mars

Percival Lowell (1855–1916) was born into the well-to-do Massachusetts family about whom John Bossidy made the famous toast:

> And this is good old Boston,
> The home of the bean and the cod,
> Where the Lowells talk to the Cabots
> And the Cabots talk only to God.

Percival's brother Lawrence became president of Harvard University, and his sister Amy, who liked to dress in men's clothing and smoke cigars, became a distinguished poet. Percival was already interested in astronomy as a boy; he made observations of Mars at age 13. His undergraduate thesis at Harvard dealt with the origin of the solar system, but he did not pursue this interest immediately. Instead he entered the family business and traveled extensively in Asia. In 1892, however, he decided to dedicate himself to carrying on Schiaparelli's work and solving the mysteries of the martian canals.

In 1894, with the help of astronomers at Harvard but using his own funds, Lowell built an observatory on a high plateau in Flagstaff, Arizona, where he hoped the seeing would be clear enough to show him Mars in unprecedented detail. He and his assistants quickly accumulated a tremendous number of drawings and maps, purporting to show a vast network of martian canals (see Figure 9.2). He elaborated his ideas about the inhabitants of the red planet in several books, including *Mars* (1895)

Percival Lowell in about 1910, observing with his 24-inch telescope at Flagstaff, Arizona.

and *Mars and Its Canals* (1906), and in hundreds of articles and speeches. As Lowell put it:

> A mind of no mean order would seem to have presided over the system we see—a mind certainly of considerably more comprehensiveness than that which presides over the various departments of our own public works. Party politics, at all events, have had no part in them; for the system is planet-wide. . . . Certainly what we see hints at the existence of beings who are in advance of, not behind us, in the journey of life.

Lowell's views captured the public imagination and inspired many novels and stories, the most famous of

which was H. G. Wells' *War of the Worlds* (1897). In this famous "invasion" novel, the thirsty inhabitants of a dying planet Mars (based entirely on Lowell's ideas) come to conquer the Earth with advanced technology.

Although the Lowell Observatory first became famous for its work on the martian canals, both Lowell and the observatory eventually turned to other projects as well. As we will see in Chapter 11, Lowell became interested in the search for a ninth (and then undiscovered) planet in the solar system. In 1930 Pluto was found at the Lowell Observatory, and it is not a coincidence that the name selected for the new planet starts with Lowell's initials. It was also at the Lowell Observatory that the first measurements were made of the great speed at which galaxies are moving away from us, observations that would ultimately lead to our modern view of an expanding universe.

Lowell continued to live at his observatory, marrying at age 53 and publishing extensively. He relished the debate his claims about Mars caused far more than the astronomers on the other side, who often complained that Lowell's work was making planetary astronomy a less respectable field. At the same time, the public fascination with the planets fueled by Lowell's work (and its interpreters) may, several generations later, have helped fan support for the space program and the many missions whose results grace the pages of our text.

canals as crops bordering irrigation ditches that brought water from the melting polar ice to the parched deserts of the red planet. (They assumed the polar caps were composed of water ice, which isn't exactly true, as we will see shortly).

Until his death in 1916, the most effective proponent of intelligent life on Mars was Percival Lowell, a self-made American astronomer and member of the wealthy Lowell

family of Boston (see the Voyagers in Astronomy box). A skilled author and speaker, Lowell made what seemed to the public a convincing case for intelligent Martians, who had constructed the huge canals to preserve their civilization in the face of a deteriorating climate (Figure 9.3).

The argument for a race of Martians, however, hinged on the reality of the canals, a matter that remained in serious

■ **FIGURE 9.3**

Lowell's Mars Globe One of the remarkable globes of Mars prepared by Percival Lowell, showing a network of dozens of canals, oases, and triangular water reservoirs that he claimed were visible on the red planet.

than 200 years, a period during which tens of thousands of martian days have passed. As a result, the rotation period can be calculated to within a few hundredths of a second.

The rotational axis of Mars has a tilt of about 25°, similar to the tilt of the Earth's axis. Thus Mars experiences seasons very much like those on Earth. Because of the longer martian year (almost two Earth years), however, each season there lasts about six of our months.

Since no surface detail can be seen through Venus' clouds, its rotation period can be found only by bouncing radar signals off the planet (as explained for Mercury in the last chapter). The first radar observations of the planet's rotation were made in the early 1960s. Later, topographical surface features were identified on the planet that showed up in the reflected radar signals. The rotation period of Venus, precisely determined from the motion of such "radar features" across its disk, is 243 days. Even more surprising than how *long* Venus takes to rotate is the fact that it spins in a backward or retrograde direction (east to west).

Stop for a moment, and think about how odd this slow rotation makes the calendar on Venus. The planet takes 225 Earth days to orbit the Sun and 243 Earth days to spin on its axis. As a result, the time the Sun takes to return to the same place in the sky turns out to be 117 Earth days. (If you say "See you tomorrow" on Venus, you'll have a long time to wait.) Although we do not know the reason for Venus' slow backward rotation, we can guess that it may have suffered one or more extremely powerful collisions during the formation process of the solar system (see Chapter 8).

dispute among astronomers. The canal markings were always difficult to study, glimpsed only occasionally because atmospheric conditions caused the tiny image of Mars to shimmer in the telescope. Lowell saw canals everywhere (even a few on Venus), but many other observers could not see them at all and remained unconvinced of their existence. When larger telescopes than Lowell's failed to confirm the presence of canals, the skeptics felt vindicated.

Astronomers had given up the idea of martian canals by the 1930s, although it persisted in the public consciousness until the first spacecraft photographs clearly showed that no canals existed. Now it is generally accepted that the straight lines were an optical illusion, the result of the human mind's tendency to see order in random features that are glimpsed dimly at the limits of the eye's resolution. When we see small, dim dots of surface markings, our minds tend to connect those dots into straight lines.

9.1.2 Rotation of the Planets

Astronomers have determined the rotation period of Mars with great accuracy by watching the motion of permanent surface markings; its sidereal day is 24 h 37 m 23 s, just a little longer than the rotation period of the Earth. This high precision is not obtained by watching Mars for a single rotation, but by noting how many turns it makes over a long period of time. Good observations of Mars date back more

9.1.3 Basic Properties of Venus and Mars

Before discussing each planet individually, let us compare some of their basic properties with each other and with the Earth (Table 9.1). Venus is in many ways the Earth's twin, with a mass 0.82 times the mass of the Earth and an almost identical density. The amount of geological activity is also relatively high, almost as high as on Earth. On the other hand, with a surface pressure nearly 100 times greater than ours, Venus' atmosphere is not at all like that of Earth. The surface of Venus is also remarkably hot, with a temperature of 730 K (over 850°F), hotter than the self-cleaning cycle of your oven. One of the major challenges presented by Venus is to understand why the atmosphere and surface environment of this twin have diverged so sharply from those of our own planet.

Mars, by contrast, is rather small, with a mass only 0.11 times the mass of the Earth. It is larger than either the Moon or Mercury, however, and unlike them it retains a thin atmosphere. Mars is also large enough to have supported considerable geological activity in the past. But the most fascinating thing about Mars is that long ago it probably had a thick atmosphere and seas of liquid water; the conditions we associate with the development of life. There is even a chance that some form of life persists today in protected environments below the martian surface.

TABLE 9.1 *Properties of Earth, Venus, and Mars*

	Earth	Venus	Mars
Semimajor axis (AU)	1.00	0.72	1.52
Period (year)	1.00	0.61	1.88
Mass (Earth = 1)	1.00	0.82	0.11
Diameter (km)	12,756	12,102	6,790
Density (g/cm^3)	5.5	5.3	3.9
Surface gravity (Earth = 1)	1.00	0.91	0.38
Escape velocity (km/s)	11.2	10.4	5.0
Rotation period (hours or days)	23.9 h	243 d	24.6 h
Surface area (Earth = 1)	1.00	0.90	0.28
Atmospheric pressure (bar)	1.00	90	0.007

9.2 THE GEOLOGY OF VENUS

Since Venus has about the same size and composition as the Earth, we might expect its geology to be similar. This is partly true, but Venus does not exhibit the same kind of *plate tectonics* as the Earth, and we will see that its lack of erosion results in a very different surface appearance.

9.2.1 Spacecraft Exploration of Venus

More spacecraft have been sent to Venus than to any other planet. Although the 1962 U.S. Mariner 2 flyby was the first, the Soviet Union launched most of the subsequent missions to Venus. In 1970 Venera 7 became the first probe to land and broadcast data from the surface of Venus. It operated for 23 minutes before succumbing to the high surface temperature. Additional Venera probes and landers followed, photographing the surface and analyzing the atmosphere and soil.

To understand the geology of Venus, however, we needed to make a global study of its surface, a task made very difficult by the perpetual cloud layers surrounding the planet. The problem resembles the challenge facing air traffic controllers on many an evening at San Francisco airport, when the weather is so foggy that they can't locate the incoming planes visually. The solution is similar in both cases: Use a radar instrument to probe through the obscuring layer.

The first crude global radar map was made by the U.S. Pioneer Venus orbiter in the late 1970s, followed by better maps from the twin Soviet Venera 15 and 16 radar orbiters in the early 1980s. However, most of our information on the geology of Venus is derived from the U.S. Magellan spacecraft, which mapped Venus from orbit between 1991 and 1993. With its powerful *imaging radar*, Magellan was

able to study the surface at a resolution of 100 m, much higher than that of previous missions, yielding our first detailed look at the surface of our sister planet (Figure 9.4). (The Magellan spacecraft returned more data to Earth than all previous planetary missions combined; each 100 minutes of data transmission from the spacecraft provided enough information, if translated into characters, to fill two 30-volume encyclopedias.)

Consider for a moment how good Magellan's resolution of 100 m really is. It means the radar images from Venus can show anything on the surface larger than a football field. Suddenly a whole host of topographic features on Venus became accessible to our view. As you look at the radar images throughout this chapter, bear in mind that these are constructed from radar reflections, not from visible-light photographs. For example, bright features on these images are an indication of rough terrain, while darker regions are smoother.

NASA/JPL

■ **FIGURE 9.4**

Radar Map of Venus Global radar image of Venus centered on 180° longitude. This composite image has a resolution of a few kilometers. Colors have been added to indicate elevation, with blue meaning low and red high. The large continent of Aphrodite stretches around the equator, where the bright (therefore rough) surface has been deformed by tectonic forces in the crust of Venus.

9.2.2 Probing Through the Clouds of Venus

The radar maps of Venus reveal a planet that looks much the way the Earth might look if our planet's surface were not constantly being changed by erosion and deposition of sediment. Because there is no water or ice on Venus and the surface wind speeds are low, almost nothing obscures or erases the complex geological features produced by the movements of Venus' crust and by volcanic eruptions. Having finally penetrated below the clouds of Venus, we find its surface to be naked, revealing the history of hundreds of millions of years of geological activity. Venus is in many ways a geologist's dream planet.

About 75 percent of the surface of Venus consists of lowland lava plains. Superficially, these plains resemble the basaltic ocean basins of the Earth, but they were not produced in quite the same way. There is no evidence of subduction zones (see Chapter 7) on Venus, indicating that, unlike the Earth, this planet never experienced plate tectonics. Although *convection* (the rising of hot materials) in its mantle generated great stresses in the crust of Venus, they did not start great continental plates moving. The formation of the lava plains of Venus more nearly resembles that of the lunar maria. Both were the result of widespread lava eruptions without the crustal spreading associated with plate tectonics.

Rising above the lowland lava plains are two full-scale continents of mountainous terrain. The largest continent on Venus, called Aphrodite, is about the size of Africa (you can see it stand out in Figure 9.4). Aphrodite stretches along the equator for about one third of the way around the planet. Next in size is the northern highland region Ishtar, which is about the size of Australia. Ishtar contains the highest region on the planet, the Maxwell Mountains, which rise 11 km above the surrounding lowlands. (The Maxwell Mountains are the only feature on Venus named after a man. They commemorate James Clerk Maxwell, whose theory of electromagnetism led to the invention of radar. All other features are named for women, historic or mythological.)

9.2.3 Craters and the Age of the Venusian Surface

One of the first questions astronomers addressed with the high-resolution Magellan images was the age of the surface of Venus. Remember that the age of a surface is rarely the age of the world it is on. A young age merely implies an active geology; the planet has forces at work that can resurface the world in short order. As described in the previous chapter, such ages can be derived from counting impact craters (Figure 9.5a is an example of what these craters look like on the Venus radar images). The more densely cratered the surface, the greater its age. The largest crater on Venus (called Mead) is 275 km in diameter, slightly larger than the largest known terrestrial crater (Chicxulub) but much smaller than the lunar impact basins.

(a)

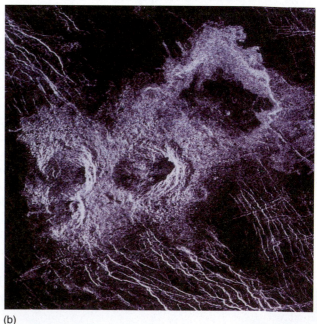
(b)

NASA/JPL

■ FIGURE 9.5

Impact Craters on Venus (a) Large impact craters in the Lavinia region of Venus. Because they are rough, the crater rims and ejecta appear brighter on these radar images than do the smoother surrounding lava plains. The largest of these craters has a diameter of 50 km. (b) The small, complex crater named after writer Gertrude Stein. The triple impact was caused by the breaking apart of the incoming asteroid during its passage through the thick atmosphere of Venus. The projectile had an initial diameter of between 1 and 2 km.

You might think that the thick atmosphere of Venus would protect the surface from impacts, burning up the projectiles long before they could reach the surface. But this is the case for only smaller projectiles. Crater statistics show very few craters less than 10 km in diameter, indicating that projectiles smaller than about 1 km (the size that typically produces a 10-km crater) were stopped by the atmosphere. Those craters with diameters from 10 to 30 km are frequently distorted or multiple, apparently because the incoming projectile broke apart in the atmosphere before it could strike the ground (Figure 9.5b). If we limit ourselves to impacts that produce craters with diameters of 30 km or larger, however, then crater counts are as useful on Venus for measuring surface age as they are on airless bodies such as the Moon.

There are typically only about 15 percent as many craters in the venusian plains as on the lunar maria, indicating a surface age only about 15 percent as great, or roughly 500 million years old. These results indicate that Venus is indeed a planet with persistent geological activity, intermediate between that of the Earth's ocean basins (which are younger and more active) and that of its continents (which are older and less active).

Almost all of the craters on Venus look fresh, with little degradation or filling in by either lava or windblown dust. This is one way we know that the rates of erosion or sediment deposition are very low. We have the impression that relatively little has happened since the venusian plains were last resurfaced by large-scale volcanic activity. Apparently Venus experienced some sort of planetwide volcanic convulsion about 500 million years ago, a mysterious event that is unlike anything we know of in terrestrial history.

9.2.4 Volcanoes on Venus

Like the Earth, Venus is a planet with widespread volcanism, and we can see many different types of volcanic features. In the lowland plains, volcanic eruptions are the principal way the surface is renewed, with large flows of highly fluid lava destroying old craters and generating a fresh surface. In addition, numerous younger volcanic mountains and other structures are associated with surface hot spots— places where convection in the planet's mantle transports the interior heat (and hot magma) to the surface.

The largest individual volcano on Venus, called Sif Mons, is about 500 km across and 3 km high—broader but lower than the Hawaiian volcano Mauna Loa. At its top is a volcanic crater or *caldera* about 40 km across, and its slopes show individual lava flows up to 500 km long. Thousands of smaller volcanoes dot the surface, down to the limit of visibility of the Magellan images, which correspond to cones or domes about the size of a shopping-mall parking lot. Most of these seem similar to terrestrial volcanoes. Other volcanoes have unusual shapes, such as the "pancake domes" illustrated in Figure 9.6.

All of this volcanism is the result of eruption of lava onto the surface of the planet. But the hot magma rising from the interior of a planet does not always make it to the surface. On

NASA/JPL

■ **FIGURE 9.6**

Pancake-Shaped Volcanoes on Venus   These remarkable circular domes, each about 25 km across and about 2 km tall, are the result of eruptions of highly viscous (sludgy) lava.

both the Earth and Venus, this upwelling magma can collect to produce bulges in the crust. Many of the granite mountain ranges on Earth, such as the Sierra Nevada in California, involve subsurface collections of magma that push up on the material above them. Such subsurface bulges are common on Venus, where they produce large circular or oval features called *coronae* (singular: *corona*) (Figure 9.7).

NASA/JPL

■ **FIGURE 9.7**

The "Miss Piggy" Corona   Fotla Corona, located in the plains to the south of Aphrodite Terra. This circular feature, about 200 km across, is a region where magma from deeper inside Venus is pushing up on the surface. Curved fracture patterns clearly show where the material beneath has put stress on the surface. A number of pancake and dome volcanoes are also visible. Fotla was a Celtic fertility goddess. Some students see a resemblance between this corona and Miss Piggy of the Muppets (her right ear is the pancake volcano in the upper center of the image).

9.2.5 Tectonic Activity

Convection currents in the mantle of Venus do more than bring magma to the surface. As on the Earth, these convection currents push and stretch the crust itself. Such forces are called **tectonic,** and the geological features that result from these forces are called *tectonic features.* On Venus' lowland plains, tectonic forces have broken the lava surface to create remarkable patterns of ridges and cracks (Figure 9.8). In a few places the crust has even torn apart to gen-

erate rift valleys. The circular features associated with coronae are tectonic ridges and cracks, and most of the mountains of Venus also owe their existence to tectonic forces.

The Ishtar continent, which has the highest elevations on Venus, is the most dramatic product of these tectonic forces. Ishtar and its tall Maxwell Mountains resemble the Tibetan Plateau and Himalayan Mountains on the Earth. Both are the product of compression of the crust, and both are maintained by the continuing forces of mantle convection.

9.2.6 On Venus' Surface

The successful Venera landers of the 1970s found themselves on an extraordinarily inhospitable planet, with a surface pressure of 90 bars and a temperature hot enough to melt lead and zinc. Despite these unpleasant conditions, the spacecraft were able to photograph their surroundings and collect surface samples for chemical analysis before their instruments gave out.

The probes found that the rock in the landing areas is igneous, primarily basalts. Examples of the Venera photographs are shown in Figure 9.9. Each picture shows a flat, desolate landscape with a variety of rocks, some of which may be ejecta from impacts. Other areas show flat, layered lava flows.

STScI/NASA

■ **FIGURE 9.8**

Ridges and Cracks   This region of the Lakshmi Plains on Venus has been fractured by tectonic forces to produce a cross-hatched grid of cracks and ridges. Be sure to notice the fainter linear features that run perpendicular to the brighter ones! This image shows a region 40 km across. Lakshmi is a Hindu goddess of prosperity.

9.3 THE MASSIVE ATMOSPHERE OF VENUS

The atmosphere of Venus produces the high surface temperature and shrouds the surface in a perpetual red twilight. The Sun does not shine directly through the heavy clouds, but the surface is fairly well lit by diffuse light—about the same as the light on Earth under a heavy overcast. The weather at the bottom of this deep atmosphere remains perpetually hot and dry, with calm winds. Because of the heavy blanket of clouds and atmosphere, one spot on the surface of Venus is similar to any other as far as weather is concerned.

USSR Academy of Science and Carle Pieters, Brown University

■ **FIGURE 9.9**

The Surface of Venus   Views of the surface of Venus from the Venera 13 spacecraft. The upper images show the scene as recorded by the spacecraft cameras; everything is orange because the thick atmosphere of Venus absorbs the bluer colors of light. Computer processing has transformed the bottom images to show what the scenes would look like under Earth lighting conditions. The horizon is visible in the upper corner of each image.

TABLE 9.2	Atmospheric Compositions of Earth, Venus, and Mars (in %)		
Gas	Earth	Venus	Mars
Carbon dioxide (CO_2)	0.03	96	95.3
Nitrogen (N_2)	78.1	3.5	2.7
Argon (Ar)	0.93	0.006	1.6
Oxygen (O_2)	21.0	0.003	0.15
Neon (Ne)	0.002	0.001	0.0003

9.3.1 Composition and Structure of the Atmosphere

The most abundant gas on Venus is carbon dioxide (CO_2), which accounts for 96 percent of the atmosphere. The second most common gas is nitrogen. The predominance of carbon dioxide over nitrogen is not surprising when you recall that the Earth's atmosphere would also be mostly carbon dioxide if this gas were not locked up in marine sediments (see Section 7.3).

Table 9.2 compares the compositions of the atmospheres of Venus, Mars, and the Earth. Expressed in this way, as percentages, the proportions of the major gases are very similar for Venus and Mars, but in total quantity their atmospheres are dramatically different. With its surface pressure of 90 bars, the venusian atmosphere is more than 10,000 times more massive than its martian counterpart. Overall, the atmosphere of Venus is very dry; absence of water is one of the important ways that Venus differs from the Earth.

The venusian atmosphere has a huge troposphere that extends up to at least 50 km above the surface (Figure 9.10). Within the troposphere, the gas is heated from below and circulates slowly, rising near the equator and descending over the poles. Being at the base of the atmosphere of Venus is something like being a kilometer or more below the ocean surface on the Earth. There the mass of water evens out temperature variations and results in a uniform environment—the same effect the thick atmosphere has on Venus.

In the upper troposphere, between 30 and 60 km above the surface, a thick cloud layer is composed primarily of sulfuric acid droplets. Sulfuric acid (H_2SO_4) is formed from the chemical combination of sulfur dioxide (SO_2) and water (H_2O). In the atmosphere of the Earth, sulfur dioxide is one of the primary gases emitted by volcanoes, but it is quickly diluted and washed out by rainfall. In the dry atmosphere of Venus, this unpleasant substance is apparently stable. Below 30 km the venusian atmosphere is clear of clouds.

9.3.2 Surface Temperature on Venus

The high surface temperature of Venus was discovered by radio astronomers in the late 1950s and confirmed by the Mariner and Venera probes. How can our neighbor planet be so hot? Although Venus is somewhat closer to the Sun than is the Earth, its surface is hundreds of degrees hotter than you would expect from the extra sunlight it receives.

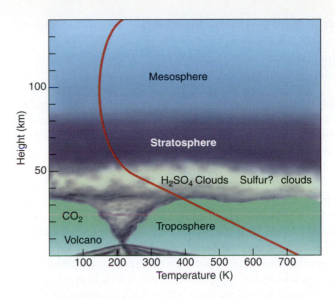

■ FIGURE 9.10

Venus' Atmosphere   The layers of the massive atmosphere of Venus, based on data from the Pioneer and Venera entry probes. Height is measured along the left axis, the bottom scale shows temperatures, and the red line allows you to read off the temperature at each height. Notice how steeply the temperature rises below the clouds thanks to the planet's huge greenhouse effect.

Ace◉Astronomy™   Log into AceAstronomy and select this chapter to see the Active Figure called "Planetary Atmospheres," and Astronomy Exercise "The Greenhouse Effect."

Scientists wondered what could be heating the surface of Venus to a temperature above 700 K. The answer turned out to be the *greenhouse effect.*

The greenhouse effect works on Venus just as it does on the Earth (see Section 7.3). But since Venus has so much more CO_2—almost a million times more—the effect is much stronger. Sunlight that diffuses through the atmosphere heats the surface just as on Earth, but the thick CO_2 acts as a blanket, making it very difficult for the infrared radiation from the ground to get back into space. As a result, the surface heats up until eventually it is emitting enough to balance the energy it receives from the Sun.

Has Venus always had such a massive atmosphere and high surface temperature, or might it have evolved to such conditions from a climate that was once more nearly Earth-like? The answer to this question is of particular interest to us as we look at the increasing levels of CO_2 in the Earth's atmosphere, as discussed in Section 7.4. Are we in any danger of transforming our own planet into a hellish place like Venus?

Let us try to reconstruct the possible evolution of Venus from an Earth-like beginning to its present state. We imagine that Venus began with moderate temperatures, water oceans, and much of its CO_2 dissolved in the ocean or chemically combined with the surface rocks, as is the case on Earth today. Then we allow for modest additional heating—by a small rise in the energy output of the Sun, for example, or by an increase in atmospheric CO_2. When we make models of how Venus' atmosphere would respond to such effects, it turns out that even a small amount of extra heat can lead to increased evaporation from the oceans and the release of gas from surface rocks.

This in turn means a further increase in the atmospheric CO_2 and H_2O, gases that would amplify the greenhouse effect. That would lead to still more heating and the release of further CO_2 and H_2O. Unless some other processes intervene, the temperature thus continues to rise. Such a situation is called the *runaway greenhouse effect*.

We want to emphasize that the runaway greenhouse effect is not just a larger greenhouse effect; it is an evolutionary process. The atmosphere evolves from having a small greenhouse effect, such as on the Earth, to a situation where greenhouse warming is a major factor, as we see today on Venus. Once the larger greenhouse conditions develop, the planet establishes a new, much hotter equilibrium near its surface. Reversing the situation is difficult because water vapor in the planet's atmosphere is not stable in the presence of ultraviolet light. Ultraviolet light from the Sun tends to break down the molecules of H_2O into their constituent parts—oxygen and hydrogen. The light element hydrogen can escape from the atmospheres of the terrestrial planets, leaving the oxygen behind to combine chemically with surface rock. The loss of water is therefore an *irreversible* process; once the water is gone, it cannot be restored. There is evidence that this is just what happened to the water once present on Venus.

Although we do not know the point at which a stable greenhouse effect breaks down and turns into a runaway greenhouse effect, Venus stands as clear testament to the fact that a planet cannot continue heating indefinitely without a major change in its oceans and atmosphere. It is a conclusion that we and our descendants will surely want to pay close attention to.

J. Bell, Cornell University; NASA/STScI

■ **FIGURE 9.11**

Mars Photographed by the Hubble Space Telescope This is one of the best photos of Mars taken from our planet, obtained by the HST in June 2001 when Mars was only 68 million km away. The resolution is about 20 km—much better than can be obtained with ground-based telescopes but still insufficient to reveal the underlying geology of Mars.

9.4 THE GEOLOGY OF MARS

Mars is more interesting to most people than Venus because it is more hospitable. Even from the distance of the Earth, we can see surface features on Mars and follow the seasonal changes in its polar caps (Figure 9.11). Although the surface today is dry and cold, evidence collected by spacecraft suggests that Mars once had blue skies and lakes of liquid water. Even today, it is the sort of place we can imagine astronauts visiting and perhaps even setting up permanent bases.

9.4.1 Spacecraft Exploration of Mars

Mars has been intensively investigated by spacecraft, and many additional missions are planned. The first visitor was the U.S. Mariner 4, which flew by Mars in 1965 and transmitted 22 photos to Earth. These pictures showed an apparently bleak planet with abundant impact craters. In those days craters were unexpected; some people who were romantically inclined still hoped to see canals or something like them. In any case, newspaper headlines sadly announced that Mars was a "dead planet."

In 1971 Mariner 9 became the first spacecraft to orbit another planet, mapping the entire surface of Mars at a resolution of about 1 km and discovering a great variety of geological features, including volcanoes, huge canyons, intricate layers on the polar caps, and channels that appeared to have been cut by running water. Geologically, Mars didn't look so dead after all. Next came the Viking spacecraft, which were among the most ambitious and successful of all planetary missions.

Two Viking *orbiters* surveyed the planet and served to relay communications for two *landers* on the surface. After an exciting and sometimes frustrating search for a safe landing spot, the Viking 1 lander touched down on the surface of Chryse Planitia (the Plains of Gold) on July 20, 1976, exactly seven years after Neil Armstrong's historic first step on the Moon. Two months later, Viking 2 landed with equal success in another plain farther north, called Utopia. The last message from the Viking landers arrived on November 5, 1982, after an enormously successful mission of studying the red planet.

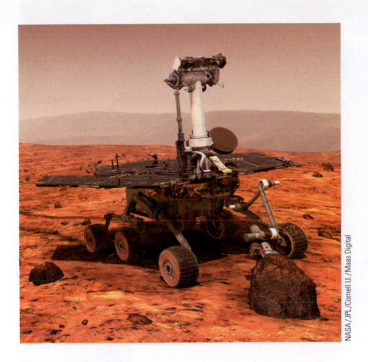

■ **FIGURE 9.12**

Mars Exploration Rover on the Surface (artist's conception). Twin rovers on opposite sides of Mars explored the planet during 2004 and 2005.

Mars then languished unvisited for two decades after Viking, until NASA initiated a new exploration program in the 1990s. Mars Pathfinder landed a small rover on the surface in 1997, followed by an orbiter (Global Surveyor) with high-resolution cameras. A second NASA orbiter arrived in 1998, followed by a European orbiter in 2004. That same year, two capable rovers, Opportunity and Spirit, landed on opposite sides of the planet in regions selected for their potential to tell us about the history of water on Mars. More missions are planned with spacecraft of increasing sophistication, focusing on the possibility of life in the past, or even surviving to the present. Within about ten years it may be possible to return samples of martian material for analysis in labs on Earth. Many NASA planners hope that humans will fly to Mars on piloted missions, arriving between 2025 and 2030.

9.4.2 Global Properties of Mars

Mars has a diameter of 6790 km, just over half the diameter of the Earth, giving it a total surface area very nearly equal to the continental area of our planet. Its density of 3.9 g/cm^3 suggests a composition consisting primarily of silicates but with a small metal core. The planet has no global magnetic field, although there are areas of strong surface magnetization that indicate a global field billions of years ago. Apparently the red planet has no liquid material in its core today that would conduct electricity.

Thanks to the Mars Global Surveyor, we have mapped the entire planet (Figure 9.13). A laser altimeter on board made millions of separate measurements of the surface topography to a precision of a few meters—good enough to show even the annual deposition and evaporation of the polar caps. Like the Earth, Moon, and Venus, the surface of Mars has continental or highland areas as well as widespread volcanic plains. The total range in elevation from the top of the highest mountain (Olympus Mons) to the bottom of the deepest basin (Hellas) is 31 km.

Approximately half the planet consists of heavily cratered highland terrain, found primarily in the southern hemisphere. The other half, which is mostly in the north, contains younger, lightly cratered volcanic plains at an average elevation about 5 km lower than the highlands. Remember that we saw a similar pattern on Earth, the Moon, and Venus. A geological division into older highlands and younger lowland plains seems to be characteristic of all the terrestrial planets except Mercury.

Lying across the north–south division of Mars is an uplifted continent the size of North America. This is the 10-km-high Tharsis bulge, a volcanically active region crowned by four great volcanoes that rise still higher into the martian sky.

9.4.3 Volcanoes on Mars

The lowland plains of Mars look very much like the lunar maria, and they have about the same density of impact craters. Like the lunar maria, they probably formed between 3 and 4 billion years ago. Apparently Mars experienced extensive volcanic activity at about the same time the Moon did, producing similar basaltic lavas.

The largest volcanic mountains of Mars are found in the Tharsis area (you can see them in Figure 9.13), although smaller volcanoes dot much of the surface. The most dramatic volcano on Mars is Olympus Mons (Mount Olympus), with a diameter larger than 500 km and a summit that towers more than 20 km above the surrounding plains (Figure 9.14). The volume of this immense volcano is nearly 100 times greater than that of Mauna Loa in Hawaii. Placed on the Earth's surface, Olympus would more than cover the entire state of Washington or Missouri.

Images taken from orbit allow scientists to search for impact craters on the slopes of these volcanoes in order to estimate their age. Many of the volcanoes show fair numbers of such craters, suggesting that they ceased activity a billion years or more ago. However, Olympus Mons has very, very few impact craters. Its present surface cannot be more than about 100 million years old; it may even be much younger. Some of the fresh-looking lava flows might have been formed a hundred years ago, or a thousand, or a million, but geologically speaking they are quite young. This leads geologists to the conclusion that Olympus Mons probably remains intermittently active today—something future Mars land developers may want to keep in mind.

Altitude [km]

■ **FIGURE 9.13**

Mars Globe from Laser Ranging These globes are highly precise topographic maps, reconstructed from millions of individual elevation measurements made with the Mars Global Surveyor. Color is used to indicate elevation. The hemisphere on the left includes Olympus Mons, the highest mountain on Mars; the hemisphere on the right includes the Hellas basin, which has the lowest elevation on Mars.

■ **FIGURE 9.14**

Olympus Mons The largest volcano on Mars, and probably the largest in the solar system, is Olympus Mons, illustrated in this computer-generated rendering based on data from the Mars Global Surveyor's Laser Altimeter. Placed on Earth, the base of Olympus Mons would completely cover the state of Missouri; the caldera, the circular opening at the top, is 65 km across, about the size of Los Angeles.

■ **FIGURE 9.15**

Martian Landslides This Viking orbiter image shows one section of the Valles Marineris canyon system. The canyon walls are about 100 km apart here. Look carefully and you can see enormous landslides whose debris is piled up underneath the cliff walls, which tower some 10 km above the canyon floor.

9.4.4 Martian Cracks and Canyons

The Tharsis bulge has many interesting geological features in addition to its huge volcanoes. In this part of the planet the surface itself has bulged upward, forced by great pressures from below, resulting in extensive tectonic cracking of the crust. Among the most spectacular tectonic features are the canyons called the Valles Marineris (or Mariner Valleys, named after Mariner 9, which first revealed them to us). They extend for about 5000 km (nearly a quarter of the way around Mars) along the slopes of the Tharsis bulge. If it were on Earth, this canyon system would stretch all the way from Los Angeles to Washington, D.C. The main canyon is about 7 km deep and up to 100 km wide, large enough for the Grand Canyon of the Colorado River to fit comfortably into one of its side canyons. The opening image for this chapter shows a spectacular overview of part of Valles Marineris.

The term *canyon* is somewhat misleading here because the Valles Marineris canyons have no outlets and were not cut by running water. They are basically tectonic cracks, produced by the same crustal tensions that caused the Tharsis uplift. However, water has played a later role in shaping the canyons, primarily by seeping from deep springs and undercutting the cliffs. This undercutting led to landslides that gradually widened the original cracks into the great valleys we see today (Figure 9.15). Today the primary form of erosion in the canyons is probably wind.

While the Tharsis bulge and Valles Marineris are impressive, in general we see fewer tectonic structures on Mars than on Venus. In part, this may reflect a lower general level of geological activity, as would be expected for a smaller planet. But it is also possible that evidence of widespread faulting has been buried by wind-deposited sediment over much of Mars. Like the Earth, Mars may have hidden part of its geological history under a cloak of soil.

9.4.5 The View on the Martian Surface

Five spacecraft have landed successfully on Mars: two Vikings (1976), Mars Pathfinder (1997), and the two recent Mars Exploration Rovers (2004) (Figure 9.16). The twin Viking landers were large spacecraft carrying a wide variety of instruments. Each lander peered at its surroundings through color stereo cameras, sniffed the atmosphere with a variety of analytical tools, and poked at nearby rocks with its mechanical arm. Each Viking also searched directly for microbes in the soil, as we will describe in more detail in Section 9.5.5. But the Viking landers were fixed in place. In contrast, the current pair of rovers is capable of driving for a total distance of several kilometers from the landing site. The next generation of spacecraft should be able to cover as much as 100 km.

NOTE TO PRODUCTION: Source info or photo credit to come. Please follow up with Chris Hall or authors.]

■ **FIGURE 9.16**

Four Martian Landing Sites The first three Mars landing sites are all similar in appearance, showing flat windswept plains littered with rocks. The most recent view from a rover is much different, including a crater that reveals layered sedimentary rocks just below the surface.

The philosophy of the current NASA Mars program is to "follow the water" in an effort to understand the ancient climate on Mars. A planet that once had water is also a planet that might have supported life. Mars Pathfinder was targeted to land in an outwash plain where ancient martian rivers might have deposited a variety of rock, but the analysis accomplished by the small Sojourner rover indicated that the nearby rocks were volcanic basalts like those at the Viking sites.

Careful analysis of potential landing sites using data from new orbiters preceded the targeting of the two landers in 2004. Each was sent to a region where there was evidence of past water—one in what appeared to be an ancient martian lake, the other in a region with evidence for the mineral hematite, which on Earth is produced in a wet environment. Unfortunately, the former lake turned out to have a thin layer of lava rocks that covered the lakebed sediments and made it impossible for the rover to reach them. On the opposite side of the planet, however, luck was with the scientists driving the other rover.

Figure 9.17 illustrates one of the craters studied by the Opportunity rover. On the walls of the crater are exposed layers of sedimentary rock—the first to be found on Mars. These layers showed clear evidence of being formed in a shallow salty sea. Ripple marks preserved the evidence of moving water. The composition also indicated formation of the rock under water—with large amounts of the mineral hematite and also sulfur and salt-rich compounds created when a shallow sea dried up, billions of years ago.

What would it be like to stand on the surface of Mars, on a wind-swept plain under a pink sky? All of the Mars landers have photographed their surroundings (Figure 9.16). They have also included "weather stations" to monitor the martian atmosphere. The temperatures on Mars vary greatly due to the absence of moderating oceans and clouds. Typically the summer maximum near the equator is 240°K (−33°C), dropping to 190°K (−83°C) just before dawn. In the winter both water frost and carbon dioxide frost can form on the surface.

In spite of the thinness of the martian air, wind plays an important role in redistributing surface material. During major dust storms, the visibility can drop almost to zero. The effects of wind can be seen well from orbit. Figure 9.18 shows streaks caused by dust devils that strip off lighter dust from a darker underlying rock surface. Figure 9.19 shows a beautiful area of dark sand dunes on top of lighter material.

9.4.6 Martian Samples

Much of what we know of the Moon, including the circumstances of its origin, comes from studies of lunar samples, but spacecraft have not yet returned martian samples to Earth for laboratory analysis. It is with great interest, therefore, that scientists have discovered that samples of

■ **FIGURE 9.17**

Evidence for Water on Mars
Two views of sedimentary rock found
by the Opportunity rover on Mars.
Endurance Crater's wall shows
multiple layers of rock deposited
in a shallow, salty sea. The inset
shows a microscopic view of the
"blueberries" composed of hematite
that provide further evidence for
liquid water at this site.

NASA/JPL/Cornell/USGS

Malin Space Science Systems/NASA

■ **FIGURE 9.18**

Tracks of Dust Devils   This high-resolution photo from the
Mars Global Surveyor shows the dark tracks of several dust
devils that have stripped away a thin coating of light-colored
dust. This view is of an area about 3 km across. Dust devils
are one of the most important ways that dust gets
redistributed by the martian winds.

NASA/JPL/Malin Space Science Systems

■ **FIGURE 9.19**

Sand Dunes on Mars   These dark dunes in the interior of
Proctor Crater overlay a lighter sandy surface. Each dune in
this high-resolution view is about 1 km across.

■ **FIGURE 9.20**

Martian Meteorite A fragment of basalt, ejected from Mars in a crater-forming impact, that eventually arrived on the Earth's surface.

martian material are nevertheless already here on Earth, available for study. About two dozen of these martian rocks exist, all members of a rare class of *meteorites* (Figure 9.20), rocks that have fallen from space (see Chapter 13 for more on meteorites).

How would rocks have escaped from Mars? Many impacts have occurred on the red planet, as shown by its heavily cratered surface. The larger impacts can eject debris at so great a speed that some fragments actually attain escape velocity from Mars. A long time later (typically a few million years), a very small fraction of these fragments collide with the Earth and survive their passage through our atmosphere, just like other meteorites. (By the way, several rocks from the Moon have also reached our planet as meteorites, although we were able to demonstrate their lunar origin only after the Apollo missions brought back samples directly from the lunar surface.)

Most of the martian meteorites are volcanic basalts; most of them are also relatively young, about 1.3 billion years old. We know from details of their composition that they are not from the Earth or the Moon. Besides, there was no volcanic activity on the Moon to form them as recently as 1.3 billion years ago. It would be very difficult for ejecta from impacts on Venus to escape through its thick atmosphere. By the process of elimination, the only reasonable origin seems to be Mars, where the Tharsis volcanoes were certainly active at that time.

The martian origin of these meteorites was confirmed by analysis of tiny gas bubbles trapped inside several of them. These bubbles match the atmospheric properties of Mars as measured directly by Viking. It appears that some atmospheric gas was trapped in the rock by the shock of the impact that ejected it from Mars and started it on its way toward Earth.

One of the most exciting results from analysis of these martian samples has been the discovery of both water and organic (carbon-based) compounds in them, which suggests that Mars may once have had oceans and perhaps even life on its surface. As we have already hinted, there is other evidence for the presence of flowing water on Mars in the remote past.

9.5 MARTIAN POLAR CAPS AND CLIMATE

The atmosphere of Mars today has an average surface pressure of only 0.007 bar, less than 1 percent that of the Earth. (This is how thin the air is about 30 km above the Earth's surface.) Martian air is composed primarily of carbon dioxide (95 percent), with about 3 percent nitrogen and 2 percent argon. The proportions of different gases are similar to those in the atmosphere of Venus (see Table 9.2), but a lot less of each gas is found on Mars.

9.5.1 Clouds on Mars

Several types of clouds can form in the martian atmosphere. First there are dust clouds, raised by winds, which can sometimes grow to produce global-scale dust storms. Second are water-ice clouds similar to those on Earth. These often form around mountains, just as happens on our planet. Finally, the CO_2 of the atmosphere can itself condense at high altitudes to form hazes of dry ice crystals. The CO_2 clouds have no counterpart on Earth, since on our planet temperatures never drop low enough (down to about 150 K or about $-125°C$) for this gas to condense.

Although the atmosphere contains water vapor and occasional clouds of water ice, liquid water is not stable under present conditions on Mars. Part of the problem is the low temperatures on the planet. But even if the temperature on a sunny summer day rises above the freezing point, the low pressure means that liquid water still cannot exist on the surface, except at the lowest elevations. At a pressure of less than 0.006 bar, only the solid and vapor forms of water are possible. In effect, the boiling point is as low or lower than the freezing point, and water changes directly from solid to vapor without an intermediate liquid state.

9.5.2 The Polar Caps

Through a telescope, the most prominent surface features on Mars are the bright polar caps, which change with the seasons, similar to the seasonal snow cover on Earth. We do not usually think of the winter snow in northern latitudes as a part of our polar caps, but seen from space, the thin snow merges with the Earth's thick, permanent ice caps to create an impression much like that seen on Mars.

The *seasonal caps* on Mars are composed not of ordinary snow but of frozen CO_2 (dry ice). These deposits condense directly from the atmosphere when the surface tem-

46103
600 m

NASA/JPL/Malin Space Science Systems

■ **FIGURE 9.21**

Layers at the Martian North Pole   The small inset in the left image shows a map of the residual north polar cap of Mars, which is about 1000 km across and composed of water ice. The small black box in the middle of the map shows the area covered in the tilted Viking orbiter image at left. The box in that image shows the area of the Mars Global Surveyor high-resolution image at right. On the right image, we see a slope on the edge of the permanent north polar cap, with dozens of layers visible—some thinner than 10 m.

perature drops below about 150 K. The caps develop during the cold martian winters and extend down to about 50° latitude by the start of spring.

Quite distinct from these thin seasonal caps of CO_2 are the *permanent* or *residual caps* that are always present near the poles (Figure 9.21). The southern permanent cap has a diameter of 350 km and is composed of frozen CO_2 deposits together with a great deal of water ice. Throughout the southern summer, it remains at the freezing point of CO_2, 150 K, and this cold reservoir is thick enough to survive the summer heat intact.

The northern permanent cap is different. It is much larger, never shrinking to a diameter less than 1000 km, and is composed of water ice. Summer temperatures in the north are too high for the frozen CO_2 to be retained. Measurements from the Mars Global Surveyor have established the exact elevations in the north polar region of Mars, showing that it is a large basin about the size of our own Arctic Ocean basin. The ice cap itself is about 3 km thick, with a total volume of about 10 million km^3 (similar to that of the Earth's Mediterranean Sea). If Mars ever had extensive liquid water, this north polar basin would have contained a shallow sea. There is some indication of ancient shorelines visible, but better images will be required to verify this suggestion.

The polar caps represent a huge reservoir of water compared with the very small amounts of water vapor in

the atmosphere. However, the caps appear to be only the tip of the iceberg. Recent measurements, made by the Mars Odyssey orbiter, indicate that about one third of the surface crust (in the top meter) at high latitudes is composed of water ice or permafrost. Unlike Venus or the Moon, Mars is not dry—it is just too cold for liquid water to be present.

Images taken from orbit also show a distinctive type of terrain surrounding the permanent polar caps, as shown in Figure 9.21. At latitudes above 80° in both hemispheres, the surface consists of recent layered deposits that cover the older cratered ground below. Individual layers are typically ten to a few tens of meters thick, marked by alternating light and dark bands of sediment. Probably the material in the polar deposits is dust carried by wind from the equatorial regions of Mars.

What do these terraced layers tell us about Mars? Some cyclic process is depositing dust and ice over long periods of time. The time scales represented by the polar layers are tens of thousands of years. Apparently the martian climate experiences periodic changes at intervals similar to those between ice ages on the Earth. Calculations indicate that the causes are probably also similar: The gravitational pull of the other planets produces variations in Mars' orbit and tilt as the great clockwork of the solar system goes through its paces.

NASA/from Mars Digital Image Map, processing by Brian Fessler, LPI

■ **FIGURE 9.22**

Runoff Channels These runoff channels in the old martian highlands are interpreted as the valleys of ancient rivers fed by either rain or underground springs. The width of this image is about 200 km.

9.5.3 Channels and Gullies on Mars

Although no liquid water exists on Mars today, evidence has accumulated that long ago rivers flowed on the red planet. Two kinds of geological features appear to be remnants of ancient watercourses, while a third class—smaller gullies—suggests intermittent outbreaks of liquid water even today. We will examine each of these features in turn.

In the highland equatorial plains there are multitudes of small, sinuous (twisting) channels—typically a few meters deep, some tens of meters wide, and perhaps 10 or 20 km long (Figure 9.22). They are called *runoff channels* because they look like what geologists would expect from troughs that carried the surface runoff of ancient rainstorms. These runoff channels seem to be telling us that the planet had a very different climate long ago. To estimate the age of these channels, we look at the cratering record. Crater counts show that this part of the planet is more cratered than the lunar maria but less cratered than the lunar highlands. Thus the runoff channels are probably older than the lunar maria, presumably about 4 billion years old.

The second set of water-related features we see are *outflow channels* (Figures 9.23 and 9.24), much larger than the runoff channels. The largest of these, which drain into the Chryse basin where Pathfinder landed, are 10 km or more wide and hundreds of kilometers long. Many features of these outflow channels have convinced geologists that they were carved by huge volumes of running water, far too great to be produced by ordinary rainfall. Where could such floodwater have come from on Mars?

As far as we can tell, the regions where the outflow channels originate contained abundant water frozen in the soil as permafrost. Some local source of heating must have

NASA/Malin Space Science Systems

■ **FIGURE 9.23**

Evidence of Liquid Water on Mars This intriguing channel, called Nanedi Valles, resembles Earth riverbeds in some (but not all) ways. The tight curves and terraces seen in the channel certainly suggest the sustained flow of a fluid like water. The channel is about 2.5 km across, and the entire Mars Global Surveyor image is 10 km wide.

released this water, leading to a period of rapid and catastrophic flooding. Perhaps this heating was associated with the formation of the volcanic plains on Mars, which date back to roughly the same time as the outflow channels.

Note that neither the runoff channels nor the outflow channels are wide enough to be visible from Earth, nor do they follow straight lines. They could not have been the "canals" Percival Lowell imagined seeing on the red planet.

The third type of water feature, the smaller *gullies*, was discovered by the camera on the Mars Global Surveyor. These images achieved a resolution of a few meters, good enough to see something as small as a truck or bus on the surface (Figures 9.25 and 9.26). On the steep walls of valleys and craters at high latitudes, there are often erosional

■ **FIGURE 9.24**

Outflow Channels Here we see a region of large outflow channels, photographed by Viking. These features appear to have been formed in the distant past from massive floods of water. The width of this image is about 150 km.

■ **FIGURE 9.26**

Stratification in the Martian Crust As many as a hundred layers can be seen in this high-resolution photo of a wind-eroded mesa within the Valles Marineris canyons. Many geologists interpret this photo as evidence for layers of sediment deposited in an ancient martian sea. The width of the image is only 1.5 km.

(a)

(b)

■ **FIGURE 9.25**

Recent Gullies on Mars (a) Gullies on a cliff near the south polar cap. (b) Gullies on the wall of Newton Crater. Each image is about 2 km across.

Astronomy and Pseudoscience: The "Face on Mars"

People like human faces. We humans have developed great skill in recognizing people and interpreting facial expressions. We also have a tendency to see faces in many natural formations, from clouds to the Man in the Moon. One of the curiosities that emerged from the Viking orbiters' global mapping of Mars was the discovery of a strangely shaped mesa in the Cydonia region that resembled a human face. Despite later rumors of a cover-up, the "Face on Mars" was, in fact, recognized by Viking scientists and included in one of the early mission press releases. At the low resolution and oblique lighting under which the Viking image was obtained, the mile-wide mesa has something of a Sphinx-like appearance.

Unfortunately, a small band of individuals decided that this formation was an artificial, carved sculpture of a human face placed on Mars by an ancient civilization that thrived there hundreds of thousands of years ago. A band of "true believers" grew around the face and tried to deduce the nature of the "sculpture" and who made it. This group linked the face to a variety of other pseudoscientific phenomena such as crop circles (patterns in fields of grain, mostly in Britain, now known to be the work of pranksters).

Members of this group accused NASA of covering up evidence of intelligent life on Mars, and they received a great deal of help in publicizing their perspective from tabloid media. Some of the believers picketed JPL at the time of the failure of the Mars Observer spacecraft, circulating stories that the "failure" of the Mars Observer was itself a fake, and that its true (now secret) mission was to photograph the face.

The high-resolution Mars Observer Camera (MOC) was reflown on the Mars Global Surveyor mission, which arrived at Mars in 1997. On April 5, 1998, in Orbit 220, the MOC obtained an oblique image of the face at a resolution of 4 m/pixel, a factor of ten improvement in resolution over the Viking image. Another image in 2001 had even higher resolution. Immediately released by NASA, the new images showed a low mesa-like hill cut crossways by several roughly linear ridges and depressions, which were misidentified in the 1976 photo as the eyes and mouth of a face. Only with a large dose of imagination can any resemblance to a face be seen in the new images, demonstrating how dramatically our interpretation of geology can change with large improvements in resolution.

After 20 years of promoting pseudoscientific interpretations and various conspiracy theories, will the Face on Mars believers now accept reality? Probably they will find a way out, perhaps by accusing NASA of faking the new picture. One suggestion they have already made is that the secret mission of the Mars Observer included a nuclear bomb used to destroy the face before it could be photographed in greater detail by the Mars Global Surveyor.

Space scientists find these suggestions incredible. NASA is spending increasing sums for research on life in the universe, and a major objective of current and upcoming Mars missions is to search for evidence of past microbial life on Mars. Conclusive evidence of extraterrestrial life would be one of the great discoveries of science and incidentally might well lead to increased funding for NASA. The idea that NASA or other government agencies would (or could) mount a conspiracy to suppress such welcome evidence is truly bizarre.

Alas, the Face on Mars story is only one example of a whole series of conspiracy theories that are kept before the

---

features that look like gullies carved by flowing water. These gullies are very young; not only are there no superimposed impact craters, but in some instances the gullies seem to cut across recent wind-deposited dunes. Perhaps there is liquid water underground that can occasionally break out to produce short-lived surface flows before the water can freeze or evaporate. These gullies are one of the most exciting and mysterious recent discoveries on Mars.

9.5.4 Climate Change on Mars

The evidence discussed so far suggests that billions of years ago martian temperatures must have been warmer and the atmosphere must have been much more substantial than it is today. That way, ancient Mars could have had river channels and even great lakes filled with water; but what could have changed the climate on Mars so dramatically?

We presume that, like the Earth and Venus, Mars formed with a much thicker atmosphere than it has now, and that this atmosphere maintained a higher surface temperature thanks to the greenhouse effect. But Mars is a smaller planet, and its lower gravity means that gases could escape more easily than from Earth and Venus. As more and more of the atmosphere escaped into space, the temperature on the surface gradually fell.

Eventually Mars became so cold that water froze out of the atmosphere, further reducing its ability to retain heat. The planet experienced a sort of *runaway refrigerator ef-*

public by dedicated believers, by people out to make a fast buck, and by irresponsible media attention. Others include the "urban legend" that the Air Force has the bodies of extraterrestrials at a secret base, the widely circulated report that UFOs crashed near Roswell, New Mexico (actually it was a balloon carrying scientific instruments to find evidence of Soviet nuclear tests), or the notion that alien astronauts helped build the Egyptian pyramids and many other ancient monuments because our ancestors were too stupid to do it alone.

In response to the increase in publicity given to these "fiction science" ideas, a group of scientists, educators, scholars, and magicians (who know a good hoax when they see one) have formed the Committee for the Scientific Investigation of Claims of the Paranormal (CSICOP). Two of the authors of your book are active on the committee. For more information about its work delving into the rational explanations for paranormal claims, see their excellent magazine, *The Skeptical Inquirer,* or check out their Web site at www.csicop.org/.

JPL/NASA

NASA/Malin Space Science Systems

The "Face on Mars" is seen at the left in low resolution from Viking and to the right with 20 times better resolution from the Mars Global Surveyor.

fect, just the opposite of the runaway greenhouse effect that occurred on Venus. Probably this loss of atmosphere took place within less than a billion years after Mars formed. The result is the cold, dry Mars we see today.

Conditions a few meters below the martian surface, however, may be much different. There liquid water might persist, kept warm by the internal heat of the planet. Even on the surface, long climate cycles may be caused by the changing orbit and tilt of the planet. At times, one or both of the polar caps might melt, releasing a great deal of water vapor into the atmosphere. Perhaps an occasional impact by a comet might produce a temporary atmosphere that is thick enough to permit liquid water on the surface for a few weeks or months. Some have even suggested that future technol-

ogy might allow us to *terraform* Mars—that is, to engineer its atmosphere and climate in ways that might make the planet more hospitable for long-term human habitation.

9.5.5 The Search for Life on Mars

If there was running water on Mars in the past, perhaps there was life as well; could life, in some form, remain in the martian soil today? Testing this possibility, however unlikely, was one of the primary objectives of the Viking landers. These landers carried, in addition to the instruments already discussed, miniature biological laboratories to test for microorganisms in the martian soil. They looked for evidence of *respiration* by living animals, *absorption of*

MAKING CONNECTIONS

Planetary Protection

When scientists begin to search for life on another planet, they must make sure that we do not contaminate the other world with life carried from the Earth. At the very beginning of spacecraft exploration on Mars, an international agreement specified that all landers were to be carefully sterilized to avoid accidentally transplanting terrestrial microbes to Mars. In the case of Viking, we know the sterilization was successful. Viking's failure to detect martian organisms also implies that these experiments did not detect hitchhiking terrestrial microbes.

As we have learned more about the harsh conditions on the martian surface, the sterilization requirements have been somewhat relaxed. It is evident that no terrestrial microbes could grow on the martian surface, with its low temperature, absence of water, and intense ultraviolet radiation. Microbes might survive in a dormant, desiccated state, but they cannot grow and proliferate on Mars.

The problem of contaminating Mars will become more serious, however, as we begin to search for life below the surface, where temperatures are higher and no ultraviolet light penetrates. The situation will be even more daunting if we consider human flights to Mars. Any humans will carry with them a multitude of terrestrial microbes of all kinds, and it is hard to imagine how we can effectively keep the two biospheres isolated from each other if Mars has indigenous life. Perhaps the best situation could be one in which the two life-forms are so different that each is effectively invisible to the other—not recognized on a chemical level as living or as potential food.

The most immediate issue of public concern is not with the contamination of Mars but with any dangers associated with returning Mars samples to the Earth. NASA is committed to the complete biological isolation of returned samples until they are demonstrated to be safe. Even though the chances of contamination are extremely low, it is better to be safe than sorry.

Most likely there is no danger, even if there is life on Mars and alien microbes hitch a ride to Earth inside some of the returned samples. In fact, Mars is sending samples to Earth all the time in the form of the Mars meteorites. Since some of these bugs (if they exist) could probably survive the trip to Earth, we may have been exposed many times over to martian microbes. Either they do not interact with our terrestrial life, or in effect our planet has already been inoculated against such alien bugs.

---

nutrients offered to organisms that might be present, and an *exchange of gases* between the soil and its surroundings for any reason whatsoever. Martian soil was scooped up by the spacecraft's long arm and placed into the experimental chambers, where it was isolated and incubated in contact with a variety of gases, radioactive isotopes, and nutrients to see what would happen. A fourth instrument pulverized the soil and analyzed it carefully to determine what organic (carbon-bearing) material it contained.

The Viking experiments were so sensitive that, had one of the spacecraft landed anywhere on Earth (with the possible exception of Antarctica), it would easily have detected life. But, to the disappointment of many scientists and members of the public, no life was detected on Mars. The soil tests for absorption of nutrients and gas exchange did show some activity, but this was most likely caused by chemical reactions that began as water was added to the soil and had nothing to do with life. In fact, these experiments showed that martian soil seems much more chemically active than terrestrial soils because of its exposure to solar ultraviolet radiation (since Mars has no ozone layer).

The organic chemistry experiment showed no trace of organic material, which is apparently destroyed by the sterilizing effect of this ultraviolet light. While the possibility of life on the surface has not been eliminated, most experts consider it negligible. Although Mars has the most Earth-like environment of any planet in the solar system, the sad fact is nobody seems to be home today, at least on the surface.

As a result, scientists increasingly focus their interest on the possibility of life underneath the surface, where liquid water might still be found today. If life once flourished on Mars, at a time when its climate was warmer and wetter, it seems reasonable to imagine that this life evolved into subsurface environments where it might even persist, just out of our sight. The key question, therefore, is whether conditions on Mars were really once favorable to life. Discoveries from the 2004 Mars rovers seem to indicate a positive answer, with their direct evidence of a salty sea near the martian equator billions of years ago.

In 1996, several scientists reported that one of the martian meteorites found in the Antarctic contained evidence of fossil microorganisms, but further examination of this evidence has left most scientists unconvinced. However, if there was once extensive life on Mars, even just microbes, there should be some sort of fossil evidence on the red planet itself. If so, samples of martian rock brought back to Earth by a future mission would be a good place to look for indications of past life. We hope we can find the right sort of rocks—sedimentary rocks from ancient seas or from deposits that formed at hot springs. Scientists are also working on better ways to identify microbial fossils in very old rock. The most powerful searches for martian life (past or present) will thus be carried out first in laboratories here on Earth.

DIVERGENT PLANETARY EVOLUTION

Venus, Mars, and our own planet Earth form a remarkably diverse triad of worlds. Although all three orbit at roughly the same distance from the Sun and all apparently started with about the same chemical mix of silicates and metals, their evolutionary paths have diverged. As a result, Venus became hot and dry, Mars became cold and dry, and only the Earth ended up with what we consider a hospitable climate.

We have discussed the runaway greenhouse effect on Venus and the runaway refrigerator effect on Mars. But we do not understand exactly what started these two planets down these evolutionary paths. Was the Earth ever in danger of a similar fate? Or might it still be diverted onto one of these paths, perhaps due to stress on the atmosphere generated by human pollutants? One of the reasons for studying Venus and Mars is to seek insight into these questions.

Some people have even suggested that if we understood the evolution of Mars and Venus better, we could possibly reverse their evolution and restore more Earth-like environments. While it seems unlikely that humans could ever make either Mars or Venus into a replica of the Earth, considering such possibilities is a useful part of our more general quest to understand the delicate environmental balance that distinguishes our planet from its two neighbors. In Chapter 13 we will return to the comparative study of the terrestrial planets and their divergent evolutionary histories.

SURFING THE WEB

 General Venus sites:

- *The Nine Planets Site:*
  seds.lpl.arizona.edu/nineplanets/nineplanets/venus.html

- *Views of the Solar System Site:*
  www.solarviews.com/eng/venus.htm

- *Planetary Sciences Site:*
  nssdc.gsfc.nasa.gov/planetary/planets/venuspage.html

- *Magellan Mission to Venus:*
  www.jpl.nasa.gov/magellan/
  Magellan radar images, the news releases, and mission background, as well as technical information and links.

General Mars sites:

- *The Nine Planets Site:*
  seds.lpl.arizona.edu/nineplanets/nineplanets/mars.html

- *Views of the Solar System Site:*
  www.solarviews.com/eng/mars.htm

- *Planetary Sciences Site:*
  nssdc.gsfc.nasa.gov/planetary/planets/marspage.html

- *Center for Mars Exploration Home Page:*
  cmex-www.arc.nasa.gov/
  A fine site with background information, maps and images, Mars mission summaries, educational resources, and lots of links.

- *Mars Missions Home Page: mars.jpl.nasa.gov*
  Images, mission overviews, updates, and educational activities.

- *Mars Explorer for Armchair Astronauts:*
  www-pdsimage.wr.usgs.gov/PDS/public/mapmaker
  Browse through a map of Mars made by the cartographers at the U.S. Geological Survey, zoom in on an area of your choice, and let the mapmaker program create a detailed map for you.

- *Face on Mars:*
  science.nasa.gov/headlines/y2001/ast24may_1.htm
  All the information on the new images of "the face," with background and even links to a climbing trail for the "face" mesa.

Planetary Protection Information:
www.spaceref.com/Directory/Astrobiology_and_Life_Science/planetary_protection/
A private site full of information and articles about all aspects of space science; this section has a good introduction to the ideas of protecting the Earth and other planets from contamination.

SUMMARY

9.1 Venus, the nearest planet, is a great disappointment through the telescope because of its impenetrable cloud cover. Mars is more tantalizing, with dark markings and polar caps. Early in the 20th century it was widely believed that the "canals" of Mars indicated intelligent life there. Mars has only 11 percent the mass of the Earth, but Venus is nearly our twin in size and mass. Mars rotates in 24 hours and has seasons like the Earth; Venus has a retrograde rotation period of 243 days. Both planets have been extensively explored by spacecraft.

9.2 Venus has been mapped by radar, especially with the Magellan spacecraft. Its crust consists of 75 percent lowland lava plains, numerous volcanic features, and many large coronae, which are the expression of subsurface volcanism. The planet has been modified by widespread **tectonics** driven by mantle convection, forming complex patterns of ridges and cracks and building high continental regions such as Ishtar. The surface is extraordinarily inhospitable, with pressure of 90 bars and temperature of 730 K, but several Russian Venera landers investigated it successfully.

9.3 The atmosphere of Venus is 96 percent CO_2. Thick clouds at altitudes of 30 to 60 km are made of sulfuric acid, and a CO_2 greenhouse effect maintains the high surface temperature. Venus presumably reached its current state from more Earth-like initial conditions as a result of a runaway greenhouse effect, which included the loss of large quantities of water.

9.4 Much of what we know about Mars is derived from six spacecraft: two Viking orbiters and the Mars Global Surveyor in orbit, and two Viking landers and the Mars Pathfinder operating on the surface. Mars has heavily cratered highlands in its southern hemisphere but younger, lower volcanic plains over much of its northern half. The Tharsis bulge, as big as North America, includes several huge volcanoes; Olympus Mons is more than 20 km high and 500 km in diameter. The Valles Marineris canyons are tectonic features widened by erosion. The Viking and Pathfinder landers revealed barren, windswept plains at Chryse and Utopia. Currently there is great interest in martian meteorites, which are samples of the martian crust and may preserve information from the time that Mars was warmer and wetter and could even have supported life.

9.5 The martian atmosphere has a surface pressure of less than 0.01 bar and is 95 percent CO_2. It has dust clouds, water clouds, and carbon dioxide (dry ice) clouds. Liquid water is not possible today, but there is subsurface permafrost at high latitudes. Seasonal polar caps are made of dry ice, but the northern residual cap is water ice. Evidence of a very different climate in the past is found in water erosion features, both runoff channels and outflow channels, the latter carved by catastrophic floods. Even more mysterious are the gullies that seem to indicate small but recent episodes of water erosion. The Viking landers searched for martian life in 1976, with negative results, but life might have flourished long ago. The search for fossil life is one of the prime motivations for future sample return missions.

9.6 Earth, Venus, and Mars have diverged in their evolution from what may have been similar beginnings. We need to understand why if we are to protect the environment of the Earth.

INTER-ACTIVITY

A Your group has been asked by high NASA officials to start planning the first human colony on Mars. Begin by making a list of what sorts of things humans would need to bring along to be able to survive for years on the surface of the red planet.

B As a publicity stunt, the mayor of Venus, Texas (there really is such a town), proposes that NASA fund a mission to Venus with humans on board. Clearly, the good mayor neglected to take an astronomy course in college. Have your group assemble a list of as many reasons as possible why it is unlikely that humans will soon land on the surface of Venus.

C Even if humans would have trouble surviving on the surface of Venus, this does not mean we could not learn a lot more about our veiled sister planet. Have your group brainstorm a series of missions (pretend cost is no object) that would provide us with more detailed information about the Venus atmosphere, surface, and interior.

D Some time late in the 21st century, when travel to Mars has become somewhat routine, a very wealthy couple asks you to plan a honeymoon tour of Mars that includes the most spectacular sights on the red planet. Constitute your group as the Percival Lowell

Memorial Tourist Agency, and come up with a list of not-to-be missed tourist stops on Mars. (If you have time to complete this assignment outside of class and have access to the Web, be sure to check out some of the Mars Web sites as part of your research.)

E Astronomers have been puzzled and annoyed about the extensive media publicity that was given the small group of "true believers" who claimed the "Face on Mars" was not a natural formation (see Making Connections: *Astronomy and Pseudoscience*). Have your group make a list of the reasons many of the media were so enchanted by this story. What do you think astronomers could or should do to get the skeptical, scientific perspective about such issues before the public?

F Assume your group is a Long-Term Planetary Environment Planning Committee at the Environmental Protection Agency during a future administration. Make a list of ways in which human activity is affecting the Earth's climate, and discuss what might be done about it.

G Your group is a special committee of scientists set up by the United Nations to specify how any Mars samples should be returned to Earth so that possible martian microbes do not harm Earth life. What precautions would you recommend, starting at Mars and going all the way to the labs that analyze the martian samples back on Earth?

REVIEW QUESTIONS

Ace ⊙ **Astronomy**™ Assess your understanding of this chapter's topics with additional quizzing and animations at **http://ace.brookscole.com/voyages**

1. List several ways that Venus, Earth, and Mars are similar, and several ways that they are different.

2. Although neither Venus nor Mars has oceans today, both are considered to have "continents" somewhat like those on the Earth. Describe the continents on each planet. What fraction of the surface area do continents take up on each world?

3. Compare the current atmospheres of Earth, Venus, and Mars in terms of composition, thickness (and pressure at the surface), and the greenhouse effect.

4. How might Venus' atmosphere have evolved to its present state through a runaway greenhouse effect?

5. The Magellan mission has taught us the most about Venus, while the Viking and Mars Global Surveyor missions have been the source of the largest amount of information about Mars. Compare the instruments on each mission, and describe how each helped us understand the planet it surveyed.

6. Describe the current atmosphere on Mars. What evidence suggests that it must have been different in the past?

7. Explain the runaway refrigerator effect and the role it may have played in the evolution of Mars.

8. What evidence do we have that there was running (liquid) water on Mars?

THOUGHT QUESTIONS

9. What are the advantages of using radar imaging rather than ordinary cameras to study the topography of Venus? What are the relative advantages of these two approaches to mapping the Earth or Mars?

10. Venus and Earth are nearly the same size and distance from the Sun. What are the main differences in the geology of the two planets? What might be some of the reasons for these differences?

11. Why is there so much more carbon dioxide in the atmosphere of Venus than in that of the Earth? Why so much more than on Mars?

12. If the Viking missions were such a rich source of information about Mars, why have we sent the Pathfinder, Global Surveyor, and other more recent missions to Mars?

Make a list of questions about Mars that still puzzle astronomers.

13. Compare Mars with Mercury and the Moon in terms of overall properties. What are the main similarities and differences?

14. Contrast the mountains on Mars and Venus with those on Earth and the Moon.

15. We believe that all of the terrestrial planets had similar histories when it comes to impacts from space. Explain how this idea can be used to date the formation of the martian highlands, the martian basins, and the Tharsis volcanoes. How certain are the ages derived for these features?

16. Is it likely that life ever existed on either Venus or Mars? Justify your answer in each case.

17. Suppose that, decades from now, astronauts are sent to Mars and Venus. In each case, describe what kind of protective gear they would have to carry, and what their chances for survival would be if their spacesuits ruptured.

18. We believe that Venus, Earth, and Mars all started with a significant supply of water. Explain where that water is now for each planet.

FIGURING FOR YOURSELF

It is interesting to estimate the amount of water (in the form of ice) on Mars and to compare this with the amount of water on Earth. In each case, we can find the total volume of a layer on a sphere by multiplying the area of the sphere ($4\pi r^2$) by the thickness of the layer. For the Earth, the ocean water is equivalent to a layer 3 km thick spread over the entire planet, and the radius of the Earth is 6.378×10^6 m (see Appendix 6). The volume of this water is therefore $4\pi(6.378 \times 10^6)^2 = 5.1 \times 10^{14}$ m^2 times the thickness of 3000 m. This gives 1.5×10^{18} cubic meters of water. Since water has a density of 1 ton per cubic meter, the mass is 5.1×10^{18} tons.

19. For Mars most of the water we are sure of is in the form of ice near the poles. We can calculate the amount of ice in one of the residual polar caps if it is (for example) 2 km thick and has a radius of 400 km (the area of a circle is $\pi \times$ radius^2). Estimate how much water might be present in the polar caps (using the dimensions given above).

Compare this with the amount of water in the Earth's oceans calculated above.

20. Estimate the amount of water there could be in a global region of subsurface permafrost on Mars (assuming two permafrost thicknesses, 1 and 10 km, and a concentration of ice in the permafrost of 10 percent by volume). Compare the two results you get with the amount of water in the Earth's oceans calculated above.

21. At its nearest approach, Venus comes within about 40 million km of the Earth. How distant is it at its farthest?

22. If you weigh 150 lb on the surface of the Earth, how much would you weigh on Venus? On Mars?

23. Calculate the relative land areas—that is, the amounts of the surface not covered by liquids—of the Earth, Moon, Venus, and Mars. (Assume that 70 percent of the Earth is covered with water.)

SUGGESTIONS FOR FURTHER READING

On Venus

Cooper, H. *The Evening Star: Venus Observed.* 1993, Farrar, Straus and Giroux. Good primer on Venus exploration, with a focus on the Magellan mission.

Grinspoon, D. "Venus Unveiled" in *Astronomy,* May 1997, p. 44.

Grinspoon, D. *Venus Revealed: A New Look Below the Clouds of Our Mysterious Twin Planet.* 1997, Addison Wesley. Excellent overview of Venus, including the fascinating geology revealed by the Magellan radar images.

Kargel, J. "Rivers of Venus" in *Sky & Telescope,* Aug. 1997, p. 32. On lava channels.

Kasting, J. et al. "How Climate Evolved on the Terrestrial Planets" in *Scientific American,* Feb. 1988. Compares Venus with the other inner planets.

Saunders, S. "Venus: A Hellish Place Next Door" in *Astronomy,* Mar. 1990, p. 18.

Robinson, C. "Magellan Reveals Venus" in *Astronomy,* Feb. 1995, p. 32.

Stofan, E. "The New Face of Venus" in *Sky & Telescope,* Aug. 1993, p. 22.

On Mars

Barile, S. "2002: A Martian Odyssey" in *Mercury* (the magazine of the Astronomical Society of the Pacific), Sept./Oct. 2002, p. 30. On the first year of observations with the spacecraft's instruments.

Bell, J. "Mars Pathfinder: Better Science?" in *Sky & Telescope,* July 1998, p. 36. Good review of mission results.

Bergreen, L. *Voyage to Mars: NASA's Search for Life Beyond Earth.* 2000, Riverhead. History of NASA's Mars exploration over the last decade or so.

Goldsmith, D. *The Hunt for Life on Mars.* 1997, Dutton. Popular discussion of life on Mars, with particular attention to the story of Mars rock ALH 84001 and its possible microbial fossils.

Golombeck, M. "The Mars Pathfinder Mission" in *Scientific*

American, July 1998, p. 40. Good review by the mission's chief scientist.

Hartmann, W. "Red Planet Renaissance" in *Astronomy,* July 2000, p. 36. Review of our view of Mars by a noted astronomer.

Hoyt, W. *Lowell and Mars.* 1976, U. of Arizona Press. Probably the best study of the canal controversy.

Kargel, J. and Strom, R. "The Ice Ages of Mars" in *Astronomy,* Dec. 1992, p. 40.

Kerr, R. "Requiem for Life on Mars? Support for Microbes [in Meteorites] Fades" in *Science,* vol. 282, p. 1398 (Nov. 20, 1998).

Morton, O. *Mapping Mars: Science, Imagination, and the Birth of a New World.* 2002, Picador. Enjoyable and authoritative account of current Mars studies from a well-known science journalist.

Newcott, W. "Return to Mars" in *National Geographic,* Aug. 1998. Report on the Pathfinder mission with great color images, some in 3D.

Pendick, D. "The Real Men in Black" in *Astronomy,* July 1999, p. 36. Profile of John Rummel, NASA's Planetary Protection Officer, whose task it is to see that microorganisms from other worlds don't contaminate the Earth.

Raeburn, P. and Golombek, M. *Mars: Uncovering the Secrets of the Red Planet.* 2000, Econo-Clad books. The story of the Pathfinder mission from the perspective of the project scientist.

The Jovian Planets This montage shows the four jovian planets to scale. Each image was recorded by one of the Voyager spacecraft that flew by these giant outer worlds.

NASA/JPL

10 The Giant Planets

What do we learn about the Earth by studying the planets? Humility.

Andrew Ingersoll in discussing the Voyager missions in 1986

THINKING AHEAD

Are you a surface chauvinist? You should be! You grew up on a planet with a solid surface and probably take it for granted that a world must have a surface. Yet, in the outer solar system, four planets do not have solid surfaces. You cannot land *on* Jupiter, for example, but only *in* it. What would it be like to visit such a world?

Beyond Mars and the asteroid belt we encounter a new region of the solar system: the realm of the giants. Here planets are much larger, the distances between them are vastly increased, and worlds are accompanied by extensive systems of satellites and rings. From many perspectives the outer solar system is where the action is, and the giant planets are the more important members of the Sun's family. When compared to these outer giants, the little cinders of rock and metal that orbit closer to the Sun can seem like insignificant afterthoughts.

10.1 EXPLORING THE OUTER PLANETS

Five planets—Jupiter, Saturn, Uranus, Neptune, and Pluto—are found in the outer solar system. The first four of these are the giant or jovian planets, the subject of this chapter. In contrast, Pluto is a small world, by far the smallest in the planetary system. Pluto is physically similar to the satellites of the giant planets and is therefore discussed with them in Chapter 11.

10.1.1 Composition and Chemistry

As we saw in Chapter 6, most of the mass in the planetary system is in the outer solar system. Jupiter alone exceeds the mass of all the other planets combined (Figure 10.1). The chemistry of the outer solar system is also different, with hydrogen rather than oxygen dominating. When the planets and the Sun were forming, it was cooler in the parts of the solar nebula farther from the Sun. This allowed water ice and other volatile compounds to condense, whereas these materials remained as gas (and eventually dissipated) in the inner solar system.

As a result, the outer planets accumulated a significantly greater mass of solid material. This meant that the core bodies for the giant planets had more gravity to capture the gases hydrogen and helium, which were abundant

in the solar nebula. Hence, the outer planets were able to grow significantly bigger than those closer to the Sun. Table 10.1 lists the main materials that were available in the outer solar nebula to form planets, based on the elements we observe making up the Sun.

With so much hydrogen available, the chemistry of the outer solar system became *reducing* (see Chapter 6). Most of the oxygen combined with hydrogen to make H_2O and was thus unavailable to form oxidized compounds with other elements. As a result, the compounds detected in the atmospheres of the giant planets are hydrogen-based gases such as methane (CH_4) and ammonia (NH_3), or more complex hydrocarbons (combinations of hydrogen and carbon) such as ethane (C_2H_6) and acetylene (C_2H_2).

TABLE 10.1 *Abundances in the Outer Solar Nebula*

Material	Percent (by mass)
Hydrogen (H_2)	75
Helium (He)	24
Water (H_2O)	0.6
Methane (CH_4)	0.4
Ammonia (NH_3)	0.1
Rock and metal	0.3

NASA/JPL/University of Arizona

■ **FIGURE 10.1**

Jupiter Jupiter as seen by the Cassini spacecraft on its way to Saturn. The storm system called the Great Red Spot is visible to the lower right. Unlike many spacecraft photos, the colors here are only slightly enhanced.

TABLE 10.2 *Missions to the Outer Solar System*

Planet	Spacecraft	Encounter Date
Jupiter	Pioneer 10	Dec. 1973 (flyby)
	Pioneer 11	Dec. 1974 (flyby)
	Voyager 1	Mar. 1979 (flyby)
	Voyager 2	July 1979 (flyby)
	Ulysses°	Dec. 1991 (flyby)
	Galileo	Dec. 1995 (orbiter and probe)
	Cassini	Dec. 2002 (flyby)
Saturn	Pioneer 11	Sept. 1979 (flyby)
	Voyager 1	Nov. 1980 (flyby)
	Voyager 2	Aug. 1981 (flyby)
	Cassini	July 2004 (orbiter)
Uranus	Voyager 2	Jan. 1986 (flyby)
Neptune	Voyager 2	Aug. 1989 (flyby)

° The Ulysses spacecraft was designed to study the polar regions of the Sun. To get away from the ecliptic, it needed a gravity boost from flying by Jupiter.

10.1.2 Exploration of the Outer Solar System

Seven spacecraft, six from the United States and one from Europe, have penetrated beyond the asteroid belt into the realm of the giants. The challenges of probing so far away from Earth are considerable. Flight times to the giant planets are measured in years to decades rather than the few months required to reach Venus or Mars. Spacecraft must be highly reliable and capable of a fair degree of independence and autonomy. This is because, even at the speed of light, messages take several hours to pass between Earth and the spacecraft. If a problem develops near Saturn, for example, the spacecraft computer must deal with it directly. To wait hours for the alarm to reach Earth and instructions to be routed back to the spacecraft could spell disaster.

These spacecraft also must carry their own electrical energy sources, since sunlight is too weak to supply energy through solar cells. Heaters are required to keep instruments at proper operating temperatures, and spacecraft must have powerful radio transmitters and large antennas if their precious data are to be transmitted to receivers on Earth a billion kilometers or more away. Table 10.2 summarizes the spacecraft missions to the outer solar system.

The first spacecraft to investigate the regions past Mars were Pioneers 10 and 11, launched in 1972 and 1973 as pathfinders to Jupiter. One of their main objectives was simply to determine whether a spacecraft could actually navigate through the belt of asteroids that lies beyond Mars without colliding with small particles in the belt. Another was to measure the radiation hazards in the enormous *mag-*

NASA/JPL

■ **FIGURE 10.2**

The Voyager Spacecraft   The camera (at bottom) is on an arm that swivels. To the left of the white main antenna for sending data back to Earth, you can see a gold disk; this is the cover for an audiovisual recording of the sights and sounds of Earth, included for the remote possibility that some other civilization might find the capsule someday. The entire spacecraft weighs about a ton on Earth.

netosphere (or zone of magnetic influence) of Jupiter. Both spacecraft passed through the asteroids without incident, but the energetic particles in Jupiter's magnetic field nearly wiped out their electronics, providing information necessary for the safe design of subsequent missions.

Pioneer 10 flew past Jupiter in 1973, after which it sped outward toward the limits of the solar system. Pioneer 11 undertook a more ambitious program, using the gravity of Jupiter during its 1974 encounter to divert it toward Saturn, which it reached in 1979. The two small craft eventually became the first human-built objects to leave the realm of the planets.

Much more advanced were Voyagers 1 and 2, launched in 1977 (Figure 10.2). The Voyagers each carried 11 scientific instruments, including cameras and spectrometers, as well as devices to measure the characteristics of planetary magnetospheres. Voyager 1 reached Jupiter in 1979 and used a gravity assist from that planet to take it on to Saturn in 1980 (see Figure 2.12). Voyager 2 arrived at Jupiter four months later and then followed a different path to complete a grand tour of the outer planets, reaching Saturn in 1981, Uranus in 1986, and Neptune in 1989. This trajectory was made possible by the approximate alignment of the four giant planets on the same side of the Sun. About once every 175 years, these planets are in a position that allows a single spacecraft to visit them all by using gravity-assisted flybys to adjust its course for each next encounter. Because of the alignment, every planet in the outer solar system except Pluto has been visited by spacecraft.

All of these missions were flybys of the giant planets: One quick look and the spacecraft moved on. For a more detailed study, we must send spacecraft that can go into orbit around a planet. This was the mission of Galileo, launched toward Jupiter in 1989, and Cassini, sent to Saturn in 1997.

10.1.3 Galileo and Cassini

The Galileo spacecraft arrived at Jupiter in 1995, beginning its investigations by deploying an entry probe into the planet for the first direct studies of Jupiter's gaseous outer layers. Like a flaming meteor, the probe plunged at a shallow angle into the jovian atmosphere, traveling at a speed of 50 kilometers per second. (That's more than 160,000 kilometers per hour—fast enough to fly from New York to San Francisco in 100 seconds!)

Atmospheric friction slowed the probe within two minutes, reaching a maximum deceleration of nearly 230 g and producing temperatures at the front of its heat shield as high as 15,000°C. This high entry speed was a result of acceleration by the strong gravitational attraction of Jupiter. As the probe speed dropped to 2500 km/h, the remains of the glowing heat shield were jettisoned, and a parachute was deployed to lower the instrumented probe spacecraft more gently into the atmosphere (Figure 10.3). The data from the probe instruments were relayed to Earth via the Galileo main spacecraft.

The probe continued to operate for nearly an hour, descending 200 km into the atmosphere. A few minutes later the polyester parachute melted, and within a few hours the main aluminum and titanium structure of the probe vaporized to become a part of Jupiter itself. About two hours after the receipt of the final probe data, the main spacecraft fired its retro-rockets so it could be captured into orbit around the planet, where its primary objectives were to study Jupiter's large and often puzzling moons (see Chapter 11).

The Cassini mission to Saturn (Figure 10.4), a cooperative venture between NASA and the European Space Agency, is similar to Galileo in its twofold approach. Arriving in 2004, Cassini will go into orbit around Saturn and begin studying its rings and satellites. In January 2005 Cassini will also deploy an entry probe. In this case the probe will not explore Saturn itself, but its intriguing moon Titan, whose surface is veiled by a thick, opaque atmosphere (see Chapter 11).

NASA/ARC

■ FIGURE 10.3

Galileo Probe Fall into Jupiter This artist's impression shows the Galileo probe descending into the clouds on its parachute just after the protective heat shield separated. The probe made its measurements of Jupiter's atmosphere on December 7, 1995.

NASA/JPL

■ FIGURE 10.4

Cassini Mission Drops Its Probe In this artist's conception, the Huygens probe drops from the Cassini orbiter into the hazy clouds of Saturn's moon Titan.

Engineering and Space Science: Teaching an Old Spacecraft New Tricks

By the time Voyager 2 arrived at Neptune in 1989, 12 years after its launch, the spacecraft was beginning to show almost human signs of old age. The arm on which the camera and other instruments were located was "arthritic"—it could no longer move easily in all directions. The communications system was "hard of hearing"—part of its radio receiver was not operating anymore. The "brains" had significant "memory loss"—some of the computer memory on board had failed. And the whole spacecraft was beginning to run out of energy—its radioactive generators had begun showing serious signs of wear.

To make things even more of a challenge, the craft's mission at Neptune was in many ways the most difficult of all four flybys. For example, since sunlight at Neptune is 900 times weaker than at Earth, the on-board camera had to take longer exposures in this light-starved environment. This was a nontrivial requirement, given that the spacecraft was shooting by Neptune at ten times the speed of a rifle bullet, or 42,000 mi/h (at this speed, it could have gone around the Earth's equator in only 36 min).

The solution was to swivel the camera backward at a rate that would exactly compensate for the forward motion of the spacecraft. But since the arm on which the camera was mounted could not move with the ease and grace it once had (and because instructions from Earth took about 4 hours to reach Neptune), engineers had to preprogram the ship's computer to execute an incredibly complex series of maneuvers for each image. Some of the beautiful illustrations you see in this chapter are a testament to their ingenuity.

Another problem that confronted the mission scientists was the sheer distance of the craft from its controllers on Earth. Voyager 2 communicated its results (including the pattern of dark and light dots that make up each image) by means of its on-board radio transmitter. By the time the radio beam arrived at the Earth, it had traveled (and spread out over) about 4.8 billion km. The power that reached us was approximately 10^{-16} W, or 20 billion times less power than it takes to operate a digital watch. Thirty-eight different antennas on four continents were used by NASA to collect the faint signals from the spacecraft and decode the precious information about Neptune that they contained.

10.2 THE JOVIAN PLANETS

Let us now examine the four jovian planets in more detail. Our approach is not just to catalog their characteristics but to compare them with each other, noting their similarities and differences, and attempting to relate their properties to their differing masses and distances from the Sun.

10.2.1 Basic Characteristics

The giant planets are much farther from the Sun than we are. The distance of Jupiter from the Sun is 778 million km—5.2 times the Earth's distance, or 5.2 AU; thus Jupiter takes just under 12 years to circle the Sun. Saturn is about twice as far away as Jupiter and takes very nearly the standard human generation of 30 years to complete one revolution.

As we saw in Chapter 2, Uranus and Neptune were not discovered until larger telescopes came into regular use. Uranus orbits at 19 AU with a period of 84 years, while Neptune at 30 AU requires 165 years for each circuit of the Sun. Not until 2010 will a full Neptune "year" have passed since its discovery in 1845.

Jupiter and Saturn have many similarities in composition and internal structure, although Jupiter is nearly four times more massive. In contrast, Uranus and Neptune are smaller worlds that differ in composition and structure from their larger siblings. Some of the main properties of these four planets are summarized in Table 10.3.

Jupiter, the giant among giants, has enough mass to make 318 Earths. Its diameter is about 11 times the Earth's (and about one tenth of the Sun's). Jupiter's average density is 1.3 g/cm^3, much lower than that of any of the terrestrial planets. (Recall that water has a density of 1 g/cm^3.) Jupiter's material is spread out over a volume so large that more than 1400 Earths could fit within it.

Saturn's mass is 95 times that of the Earth, and its average density is only 0.7 g/cm^3—the lowest of any planet. Since this is less than the density of water, Saturn would be light enough to float—if a bathtub large enough to contain it existed (but then imagine the ring it would leave)!

Uranus and Neptune each have a mass about 15 times that of the Earth and hence only 5 percent as great as

TABLE 10.3 Basic Properties of the Jovian Planets

Planet	Distance (AU)	Period (years)	Diameter (km)	Mass (Earth = 1)	Density (g/cm^3)	Rotation (hours)
Jupiter	5.2	11.9	142,800	318	1.3	9.9
Saturn	9.5	29.5	120,540	95	0.7	10.7
Uranus	19.2	84.1	51,200	14	1.3	17.2
Neptune	30.1	164.8	49,500	17	1.6	16.1

Jupiter. Their densities of 1.3 g/cm^3 and 1.6 g/cm^3, respectively, are much higher than that of Saturn, in spite of the reduced internal compression that results from a weaker gravity than that of Saturn. This tells us that their composition must be fundamentally different, consisting for the most part of heavier materials than hydrogen and helium, the primary constituents of Jupiter and Saturn. We will return to this question of their composition shortly.

10.2.2 Appearance and Rotation

When we look at the giant planets, we see only their atmospheres, composed primarily of hydrogen and helium gas (see the image that opens this chapter). The uppermost clouds of Jupiter and Saturn, the part we see when looking down at these planets from above, are composed of ammonia crystals. On Neptune the upper clouds are made of methane. On Uranus we see no obvious cloud layer at all, but only a deep and featureless haze.

Seen through a telescope, Jupiter is a colorful and dynamic planet. Distinct details in its cloud patterns allow us to determine the rotation rate of its atmosphere at the cloud level, although such atmospheric rotation may have little to do with the spin of the underlying planet. Much more fundamental is the rotation of the mantle and core; these can be determined by periodic variations in radio waves coming from Jupiter, which are controlled by its magnetic field. Since the magnetic field originates deep inside the planet, it shares the rotation of the interior. This rotation period of 9 h 56 m gives Jupiter the shortest "day" of any planet. In the same way, we can measure that the underlying rotation period of Saturn is 10 h 40 m. Uranus and Neptune have slightly longer rotation periods of about 17 hours, also determined from the rotation of their magnetic fields.

Remember from Chapters 3 and 9 that the Earth and Mars have seasons because their spin axes, instead of "standing up straight," are tilted relative to the Sun's orbital plane. This means that, as the Earth revolves around the Sun, sometimes one hemisphere and sometimes the other "leans into" the Sun.

What are the seasons like for the giant planets? The spin axis of Jupiter is tilted by only 3°, so there are no seasons to speak of. Saturn, however, does have seasons, since

its spin axis is inclined at 27° to the perpendicular to its orbit. Neptune has about the same tilt as Saturn (29°); therefore, it experiences similar seasons (only more slowly). The strangest seasons of all are on Uranus, which has a spin axis tilted by 98° with respect to the north direction. Practically speaking, we can say that Uranus orbits on its side, and its ring and satellite system follows along, orbiting about Uranus' equator (Figure 10.5).

We don't know what caused Uranus to be tipped over like this, but one possibility is a collision with a large planetary body when our system was first forming. Whatever the cause, this unusual tilt creates dramatic seasons. When Voyager 2 arrived at Uranus, its south pole was facing directly into the Sun. The southern hemisphere was having a roughly 21-year sunlit summer, while during that same period the northern hemisphere was plunged into darkness. For the next 21-year season, the Sun shines on Uranus' equator, and both hemispheres go through cycles of light and dark (Figure 10.6). Then there are 21 years of an illuminated northern hemisphere and a dark southern hemisphere. If you were to install a floating platform at the south pole of Uranus, for example, it would experience 42 years of light and 42 years of darkness. Any future astronauts crazy enough to set up camp there could spend most of their lives without ever seeing the Sun.

10.2.3 Composition and Structure

Although we cannot see into these planets, astronomers are confident that the interiors of Jupiter and Saturn are composed primarily of hydrogen and helium. Of course, these gases have been measured only in their atmospheres, but calculations first carried out 50 years ago showed that these two light gases are the only possible materials out of which a planet with the observed masses and densities of Jupiter and Saturn could be constructed.

Having said that much, we should add that the precise internal structure of the two planets is not so easy to predict. This is mainly because the planets are so big that the hydrogen and helium in their centers become tremendously compressed and behave in ways that these gases can never behave on Earth. The best theoretical models we have of Jupiter's structure predict a central pressure greater than 100 million bars and a central density of about

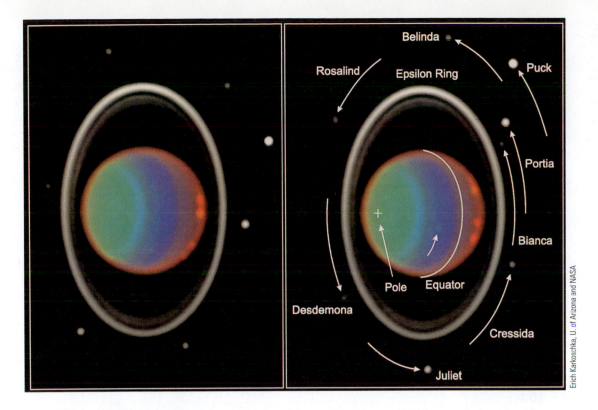

FIGURE 10.5

Infrared Image of Uranus The infrared camera on the Hubble Space Telescope took these false-color images of the planet Uranus, its ring system, and satellites in 1997. The south pole of the planet (marked with a + on the right image) faces the Sun; its green color shows a strong local haze. The two images were taken 90 min apart, and during that time the five reddish clouds can be seen to rotate around the parallel to the equator. The rings (which are very faint in visible light, but prominent in infrared) and eight satellites can be seen around the equator. This was the "bull's-eye" arrangement that Voyager saw as it approached Uranus in 1986.

31 g/cm³. (By contrast, the Earth's core has a central pressure of 4 million bars and a central density of 17 g/cm³.)

At the pressures inside the giant planets, familiar materials can take on strange forms. A few thousand kilometers below Jupiter and Saturn's visible clouds, pressures become so great that hydrogen changes from a gaseous to a liquid state. Still deeper, this liquid hydrogen is further compressed and begins to act like a metal, something it never does on Earth. (In a metal, electrons are not firmly attached to their parent nuclei but can wander around. This is why metals are such good conductors of electricity.) On Jupiter, the greater part of the interior is liquid metallic hydrogen.

Because Saturn is less massive, it has only a small volume of metallic hydrogen, but most of its interior is liquid. Uranus and Neptune are too small to reach internal pressures sufficient to liquefy hydrogen. We will return to the question of the metallic hydrogen layers when we examine the magnetic fields of the giant planets.

Each of these planets has a core composed of heavier materials, as demonstrated by detailed analyses of their gravitational fields. Presumably these cores are the original rock-and-ice bodies that formed before the capture of gas from the surrounding nebula (see Chapters 6 and 13). The cores exist at pressures of tens of millions of bars. While scientists speak of the giant planet cores being composed of rock and ice, we can be sure that neither rock nor ice assumes any familiar forms at such pressures and the accompanying high temperatures. What is really meant by *rock* is any material made up primarily of iron, silicon, and oxygen. The term *ice* denotes materials composed primarily of the elements carbon, nitrogen, and oxygen in combination with hydrogen.

Figure 10.7 illustrates the interior structures of the four jovian planets. It appears that all four have similar cores of rock and ice. On Jupiter and Saturn, the cores constitute only a few percent of the total mass, consistent with the initial composition of raw materials shown in Table 10.1. However, most of the mass of Uranus and Neptune resides in these cores, demonstrating that the two planets were unable to attract massive quantities of hydrogen and helium when they were first forming.

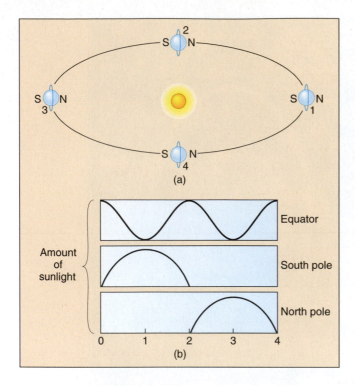

FIGURE 10.6

The Strange Seasons on Uranus (a) The orbit of
Uranus as seen from above. At the time Voyager 2 arrived
(position 1), the south pole was facing the Sun. As we move
counterclockwise in the diagram, we see the planet 21 years
later at each step. (b) The amount of sunlight seen at the
poles and the equator of Uranus over the course of its 84-
year revolution around the Sun.

10.2.4 Internal Heat Sources

Because of their large sizes, all of the giant planets were
strongly heated during their formation by the collapse of
surrounding material onto their cores. Jupiter, being the
largest, was the hottest. Some of this primordial heat can
still remain inside such large planets. In addition, it is possible for giant, largely gaseous planets to generate heat after formation by slowly contracting. (With so large a mass,
even a minuscule amount of shrinking can generate
significant heat.) The effect of these internal energy
sources is to raise the temperatures in the interiors and atmospheres of the planets higher than we would expect
from the heating effect of the Sun alone.

Jupiter has the largest internal energy source, amounting to 4×10^{17} watts (W); that is, it is glowing with the
equivalent of 4 million billion 100-W light bulbs. This is
about the same as the total solar energy absorbed by
Jupiter. The atmosphere of Jupiter is therefore something
of a cross between a normal planetary atmosphere (like the
Earth's), which obtains most of its energy from the Sun,
and the atmosphere of a star, which is entirely heated by an
internal energy source. Most of the internal energy of
Jupiter is primordial heat, left over from the formation of
the planet 4.5 billion years ago.

Saturn has an internal energy source about half as
large as that of Jupiter, which means (since its mass is only
about one quarter as great) that it is producing twice as
much energy per kilogram of material as does Jupiter.
Since Saturn is expected to have much less primordial heat,
there must be another source at work generating most of
this 2×10^{17} W of power. This source is the separation of
helium from hydrogen in Saturn's interior. In the liquid hydrogen mantle, the heavier helium forms drops that sink
toward the core, releasing gravitational energy. In effect,
Saturn is still differentiating—letting lighter material rise
and heavier material fall.

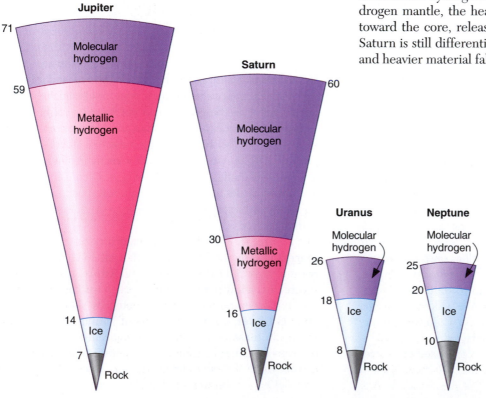

FIGURE 10.7

**Internal Structures of the
Jovian Planets** Jupiter and
Saturn are composed primarily
of hydrogen and helium, but
Uranus and Neptune consist,
in large part, of compounds of
carbon, nitrogen, and oxygen.
(The diagrams are drawn to
scale; numbers show radius in
thousands of km.)

■ FIGURE 10.8

Jupiter in Radio Waves This false-color image of Jupiter was made with the Very Large Array (of radio telescopes) in New Mexico. We see part of the magnetosphere, brightest in the middle because the largest number of charged particles are in the equatorial zone of Jupiter. The planet itself is slightly smaller than the circular green region in the center. Different colors are used to indicate different intensities of synchrotron radiation.

Imke dePater/NRAO

Uranus and Neptune are different. Neptune has a small internal energy source, while Uranus does not emit a measurable amount of internal heat. As a result, these two planets have almost the same temperature, in spite of Neptune's greater distance from the Sun. No one knows why these two planets differ in their internal heat.

10.2.5 Magnetic Fields

Each of the giant planets has a strong magnetic field, generated by electric currents in its rapidly spinning interior. Associated with the magnetic fields are the planets' *magnetospheres,* which are defined as the regions around the planet within which the planet's own magnetic field dominates over the general interplanetary magnetic field. The magnetospheres of these planets are their largest features, extending millions of kilometers into space.

In the late 1950s, astronomers discovered that Jupiter was a source of radio waves that got more intense at longer rather than at shorter wavelengths—just the reverse of what is expected from thermal radiation (radiation caused by the normal vibrations of particles within all matter; see Chapter 4). Such behavior is typical, however, of the radiation emitted when high-speed electrons are accelerated by a magnetic field. We call this **synchrotron radiation** because it was first observed on Earth in particle accelerators, called synchrotrons. This was our first hint that Jupiter must have a strong magnetic field.

Later observations showed that the radio waves are coming from a region surrounding Jupiter with a diameter several times that of the planet itself (Figure 10.8). The evidence suggested, therefore, that a vast number of charged atomic particles must be circulating around Jupiter, spiraling around the lines of force of a magnetic field associated with the planet. This is just what we observe happening, but on a smaller scale, in the Van Allen belt around the Earth (see Chapter 7). The magnetic fields of Saturn, Uranus, and Neptune, discovered by the spacecraft that first passed close to these planets, work in a similar way.

Inside each magnetosphere, charged particles are trapped into spiraling around in the magnetic field; as a result they can be accelerated to high energies. The charged particles come from the Sun or from the neighborhood of the planet itself. In Jupiter's case, Io, one of its moons,

turns out to have a constant series of volcanic eruptions that propel charged particles into space.

The axis of Jupiter's magnetic field (the line that connects the magnetic north pole with the magnetic south pole) is not aligned exactly with the axis of rotation of the planet; rather, it is tipped by about 10°. Uranus and Neptune have even greater magnetic tilts of 60° and 55°, respectively. Saturn's field, on the other hand, is perfectly aligned with its rotation axis. Why different planets have such different magnetic tilts is not well understood.

The physical processes around the jovian planets turn out to be milder versions of what astronomers find in many distant objects, from the remnants of dead stars to the puzzling distant powerhouses we call quasars. One reason to study the magnetospheres of the giant planets and the Earth is that they provide nearby accessible analogues of more energetic and challenging cosmic processes.

10.3 ATMOSPHERES OF THE GIANT PLANETS

The atmospheres of the jovian planets are the parts we can observe or measure directly. Since these planets have no solid surfaces, their atmospheres are more representative of their general compositions than is the case with the terrestrial planets. These atmospheres also present us with some of the most dramatic examples of weather patterns in the solar system. As we will see, storms on these planets can grow bigger than the entire planet Earth.

10.3.1 Atmospheric Composition

When sunlight reflects from the atmospheres of the giant planets, the atmospheric gases leave their "fingerprints" in the spectrum of light (as we saw in Chapter 4). Spectroscopic observations of the jovian planets began in the 19th century, but for a long time astronomers were not able to interpret the spectra they observed. As late as the 1930s, the most prominent features photographed in these spectra remained unidentified. Then better spectra revealed the presence of methane (CH_4) and ammonia (NH_3) in the atmospheres of Jupiter and Saturn.

James Van Allen: Several Planets Under His Belt

The career of physicist James Van Allen has spanned the birth and growth of the space age; in fact he played a major role in its development. Born in Iowa in 1914, Van Allen received his PhD from the University of Iowa. He then worked for several research institutions and served in the Navy during World War II. In 1951 he was appointed Professor of Physics at the University of Iowa, where he has remained ever since.

After the war, Van Allen and his collaborators began using the V2 rockets captured from the Germans, as well as rockets built in the United States, to explore cosmic radiation in the Earth's outer atmosphere. Van Allen designed a technique, nicknamed "the rockoon," in which a small rocket is lifted up by a balloon before being launched, enabling it to reach a higher altitude.

Over dinner one night in 1950, Van Allen and several colleagues came up with the idea of the International Geophysical Year (IGY), an opportunity for scientists around the world to coordinate their investigations of the physics of the Earth, especially research done at high altitudes. Stretched to 18 months, the IGY took place between July 1957 and December 1958. In 1955, the United States and the Soviet Union each committed themselves to launching an Earth-orbiting satellite during IGY, a competition that began what came to be known as the space race.

When the Soviet Union won the first lap of the race by launching Sputnik 1 in October 1957, the U.S. government spurred its scientists and en-

James Van Allen

Photo by D. Hokanson

gineers to even greater efforts to get something into space and thereby maintain the country's prestige. However, the main U.S. satellite program, called Vanguard, was running into all sorts of difficulties, and each of its early launches crashed or exploded. In the meantime, a second team of

After WWII, Van Allen began using captured V2 rockets to probe nearby space.

rocket engineers and scientists had quietly been working on a military launch vehicle called Jupiter-C. Van Allen spearheaded the design of the instruments aboard the small satellite, called Explorer I, that this rocket would carry. On January 31, 1958, Ex-

plorer I became the first U.S. satellite in space.

Unlike Sputnik, Explorer I was equipped to make scientific measurements. Van Allen had placed an instrument aboard Explorer I that could detect high-energy charged particles above the atmosphere, and it was this instrument that discovered what we now call the Van Allen belts around the Earth. This was the first scientific discovery of the space program and made Van Allen's name a household word around the world.

Van Allen and his colleagues continued making measurements of the magnetic and particle environment around our planet with more sophisticated spacecraft and instruments. When Pioneers 10 and 11 made the first exploratory surveys of the Jupiter and Saturn environments, they again carried instruments designed by Van Allen. Some scientists therefore refer to the charged-particle zones around those planets as Van Allen belts as well. (Once, when Van Allen was giving a lecture at the University of Arizona, the graduate students in planetary science asked him if he would leave his belt at the school. It is now proudly displayed as the "Van Allen belt.")

Van Allen has continued to be a strong supporter of space science and has been an eloquent senior spokesman for the American scientific community, warning NASA not to put all its efforts into human exploration but to remember how productive and (relatively) inexpensive its robot probes and instruments have always been in helping us learn more about the solar system.

At first astronomers thought that methane and ammonia might be the main constituents of these atmospheres, but now we know that hydrogen and helium are actually the dominant gases. The confusion arose because neither hydrogen nor helium possesses easily detected spectral features. It was not until the Voyager spacecraft measured the far-infrared spectra of Jupiter and Saturn that a reliable abundance for the elusive helium could be found. The compositions of the two atmospheres are generally similar, except that on Saturn there is less helium—the result of the precipitation of helium that contributes to Saturn's internal energy source. The most precise measurements of

NASA/JPL

■ FIGURE 10.9

Jupiter's Colorful Clouds
Also shown in this Voyager photo are two of Jupiter's satellites: Io (*left*) and Europa (*right*). The colors have been somewhat exaggerated to emphasize the differences among the zones and bands. The actual colors are a bit more muted, as shown in Figure 10.1.

composition were made on Jupiter by the Galileo entry probe in 1995, and as a result, we know the abundances of some elements in the jovian atmosphere even better than those in the Sun.

10.3.2 Clouds and Atmospheric Structure

The clouds of Jupiter are among the most spectacular sights in the solar system, much beloved by makers of science-fiction films. They range in color from white to orange to red to brown, swirling and twisting in a constantly changing kaleidoscope of patterns (Figure 10.9). Saturn shows similar but much more subdued cloud activity: Instead of vivid colors, its clouds have a nearly uniform butterscotch hue (Figure 10.10).

Different gases freeze at different temperatures. At the temperatures and pressures of the upper atmospheres of Jupiter and Saturn, methane remains a gas but ammonia can condense and freeze. (Similarly, water vapor condenses high in the Earth's atmosphere to produce clouds of ice crystals.) The primary clouds that we see around these planets,

Space Telescope Science Institute/NASA

■ FIGURE 10.10

Saturn Over Four Years These beautiful images of Saturn were recorded by the Hubble Space Telescope between 1996 and 2000. Since Saturn is tilted by 27°, we see the orientation of the rings around its equator change as the planet moves along its orbit. Note the horizontal bands in the atmosphere.

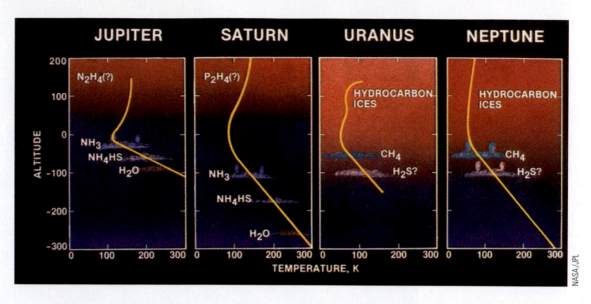

■ **FIGURE 10.11**

Atmospheric Structure for the Jovian Planets   In each diagram, the yellow line shows how the temperature (see the scale on the bottom) changes with altitude. The location of the main layers on each planet is also shown.

Ace◯Astronomy™   Log into AceAstronomy and select this chapter to see the Active Figure called "Planetary Atmospheres."

whether from a spacecraft or through a telescope, are composed of frozen ammonia crystals. The ammonia clouds mark the upper edge of the planets' *troposphere;* above that lies the *stratosphere,* the coldest part of the atmosphere. (These layers were defined for the Earth in Chapter 7.)

Figure 10.11 is a diagram showing the structure and clouds in the atmospheres of all four jovian planets. On both Jupiter and Saturn, the temperature near the cloud tops is about 140 K (only a little cooler than the polar caps of Mars). On Jupiter this cloud level is at a pressure of about 0.1 bar, but on Saturn it occurs lower in the atmosphere, at about 1 bar. Because the ammonia clouds lie so much deeper on Saturn, they are more difficult to see, and the overall appearance of the planet is much more bland than is Jupiter.

Within the tropospheres of these planets, the temperature and pressure both increase with depth. Through breaks in the ammonia clouds, we can see tantalizing glimpses of other cloud layers that can form in these deeper regions of the atmosphere—regions that were sampled directly by the Galileo probe for Jupiter.

As it descended to a pressure of 5 bars, the probe should have passed into a region of frozen water clouds, then below that into clouds of liquid water droplets, perhaps similar to the common clouds of the terrestrial troposphere. At least this is what scientists expected. But the probe saw no water clouds, and it measured a surprisingly low abundance of water vapor in the atmosphere. It soon became clear to the Galileo scientists that the probe happened to descend through an unusually dry, cloud-free region of the atmosphere—a giant downdraft of cool, dry gas.

Andrew Ingersoll of Caltech, a member of the Galileo team, called this entry site the "desert" of Jupiter. It's a pity that the probe did not enter in a more representative region, but that's the luck of the cosmic draw. The probe continued to make measurements to a pressure of 22 bars but found no other cloud layers before its instruments stopped working. It also detected lightning storms, but only at great distances, suggesting that the probe itself was in a region of clear weather.

Above the visible ammonia clouds in Jupiter's atmosphere, we find the clear stratosphere, which reaches a minimum temperature near 120 K. At still higher altitudes temperatures rise again, just as they do in the upper atmosphere of the Earth, because here the molecules absorb ultraviolet light from the Sun. This input of energy causes chemical reactions among the atmospheric gases—a process we call **photochemistry.** In Jupiter's upper atmosphere, photochemical reactions create a variety of fairly complex compounds of hydrogen and carbon that form a thin layer of smog far above the visible clouds. We show this smog as a fuzzy orange region in Figure 10.11; however, this thin layer does not block our view of the clouds beneath it.

We must mention one puzzle of the main jovian clouds: the mystery of their colors. The ammonia-condensation clouds identified on the planet should be white, like water clouds on Earth, yet we see beautiful and complex patterns of pastel red, orange, and brown. Some additional chemical or chemicals must be present to create such colors, but we do not know what they are. Various photochemically produced organic compounds have been suggested, as well as sulfur and red phosphorus. But there are no firm identifications, nor any immediate prospects of solving this mystery.

The atmospheric structure of Saturn is similar to that of Jupiter. Temperatures are somewhat colder, however, and the atmosphere is more extended, due to Saturn's lower surface gravity. Thus the layers are stretched out over a longer distance (as you can see in Figure 10.11). Overall, though, the same atmospheric regions, condensation clouds, and photochemical reactions that we see on Jupiter should be present on Saturn.

Unlike Jupiter and Saturn, Uranus is almost entirely featureless as seen at wavelengths that range from the ultraviolet to the infrared (see its rather boring image in the opening collage for this chapter). Calculations indicate that the basic atmospheric structure of Uranus should resemble that of Jupiter and Saturn, although its upper clouds (at the 1-bar pressure level) are composed of methane rather than ammonia. However, the absence of an internal heat source suppresses up-and-down movement and leads to a very stable atmosphere with little visible structure.

Neptune differs from Uranus in its appearance, although their basic atmospheric temperatures are similar. The upper clouds are composed of methane, which forms a thin cloud layer near the top of the troposphere at a temperature of 70 K and a pressure of 1.5 bars. Most of the atmosphere above this level is clear and transparent, with less haze than is found on Uranus. The scattering of sunlight by gas molecules lends Neptune a pale blue color similar to that of the Earth's atmosphere (Figure 10.12). Another cloud layer, perhaps composed of hydrogen sulfide ice particles, exists below the methane clouds at a pressure of 3 bars.

Unlike Uranus, Neptune has an atmosphere in which *convection currents*—vertical drafts of gas—emanate from the interior, powered by the planet's internal heat source. These currents carry warm gas above the 1.5-bar cloud level, forming additional clouds at elevations about 75 km higher. These high-altitude clouds form bright white patterns against the blue planet beneath. Voyager photographed distinct shadows on the methane cloud tops, permitting the altitudes of the high clouds to be calculated. (See Figure 10.13, a remarkable close-up of Neptune's outer layers that could never have been obtained from Earth.)

10.3.3 Winds and Weather

The atmospheres of the jovian planets have many regions of high pressure (where there is more air) and low pressure (where there is less). Just as it does on Earth, air flows between these regions, setting up wind patterns that are then distorted by the rotation of the planet. By observing the changing cloud patterns on the jovian planets, we can measure wind speeds and track the circulation of their atmospheres.

The atmospheric motions we see on these planets are fundamentally different from those on the terrestrial planets. The giants spin faster, and their rapid rotation tends to smear out the circulation of the air into horizontal (east–

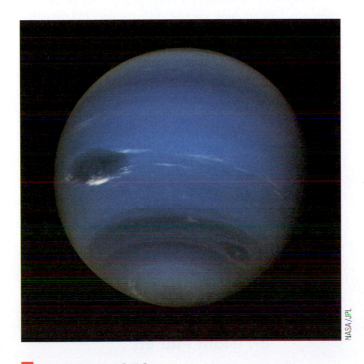

NASA/JPL

■ **FIGURE 10.12**

Neptune The planet Neptune is seen here as photographed by Voyager in 1989. The blue color, exaggerated with computer processing, is caused by the scattering of sunlight in the upper atmosphere.

NASA/JPL

■ **FIGURE 10.13**

High Clouds in the Atmosphere of Neptune
These bright, narrow cirrus clouds are made of methane ice crystals. From the shadows they cast on the thicker cloud layer below, we can measure that they are about 75 km higher than the main clouds.

Left two: NASA/JPL. Right two: Andrew Ingersoll and Don Davis, courtesy of Sky Publishing Corp.

■ **FIGURE 10.14**

Winds on the Giant Planets   The speed and direction of the east–west winds on Jupiter, Saturn, Uranus, and Neptune, as measured by Voyager, are shown on these photographic maps. Latitude is shown on the left column of each chart, with the equator being 0°. The dark line shows the wind speed at each latitude (measured relative to the planet's core). Notice that Saturn has very strong winds at the equator.

west) patterns parallel to the equator. In addition, there is no solid surface below the atmosphere against which the circulation patterns can rub and lose energy (which is how tropical storms on Earth ultimately die out when they come over land). As we have seen, on all the giants except Uranus, heat from the inside contributes about as much energy to the atmosphere as sunlight from the outside. This means that deep convection currents of rising hot air and falling cooler air circulate throughout the atmospheres of the planets in the vertical direction.

The main features of Jupiter's visible clouds (see Figures 10.1 and 10.9), for example, are alternating dark and light bands that stretch around the planet parallel to the equator. These bands are semipermanent features, although they shift in intensity and position from year to year. Consistent with the small tilt of Jupiter's axis, the pattern does not change with the seasons.

More fundamental than these bands are underlying east–west wind patterns in the atmosphere, which do not appear to change at all, even over many decades. These are displayed in Figure 10.14, which indicates how strong the winds are at each latitude for the giant planets. At Jupiter's equator, a jet stream flows eastward with a speed of about 90 m/s (300 km/h), similar to the speed of jet streams in the Earth's upper atmosphere. At higher latitudes there are alternating east- and west-moving streams, with each hemisphere an almost perfect mirror image of the other. Saturn shows a similar pattern, but with a much stronger equatorial stream flowing at a speed of 1300 km/h.

The light zones on Jupiter are regions of upwelling air capped by white ammonia cirrus clouds. They apparently represent the tops of upward-moving convection currents. The darker belts are regions where the cooler atmosphere

moves downward, completing the convection cycle; they are darker because fewer ammonia clouds mean we can see deeper into the atmosphere, perhaps down to a region of ammonium hydrosulfide (NH_4SH) clouds. The Galileo probe sampled one of the clearest of these dry downdrafts.

In spite of the strange seasons induced by the 98° tilt of its axis, Uranus' basic circulation is parallel with its equator, as is the case on Jupiter and Saturn. The mass of the atmosphere and its capacity to store heat are so great that the alternating 42-year periods of sunlight and darkness have little effect. In fact, Voyager measurements show that the atmospheric temperatures are even a few degrees higher on the dark winter side than on the hemisphere facing the Sun. Here is another indication that the behavior of such giant planet atmospheres is a complex problem: We simply do not understand the seasonal effects on Uranus in detail.

Neptune's weather is characterized by strong east–west winds generally similar to those observed on Jupiter and Saturn. The highest wind speeds near its equator reach 2100 km/h (600 m/s), nearly twice as fast as the peak winds on Saturn. The Neptune equatorial jet stream actually approaches supersonic speeds (faster than the speed of sound in Neptune's air).

10.3.4   Giant Storms on Giant Planets

Superimposed on the regular atmospheric circulation patterns we have just described are many local disturbances—weather systems or *storms*, to borrow the term we use on Earth. The most prominent of these are large, oval-shaped, high-pressure regions on both Jupiter and Neptune.

JPL/NASA

■ **FIGURE 10.15**

The Great Red Spot   This is the largest storm system on Jupiter, as seen during the Voyager spacecraft flyby. Below and to the right of the Red Spot is one of the white ovals, which are similar but smaller high-pressure features. The white oval is roughly the size of planet Earth, to give you a sense of the huge scale of the weather patterns we are seeing. The colors on the Jupiter image have been somewhat exaggerated here so astronomers (and astronomy students) can study their differences more effectively. See Figure 10.1 to get a better sense of the colors your eye would actually see near Jupiter.

The largest and most famous storm on Jupiter is the Great Red Spot, a reddish oval in the southern hemisphere. The Red Spot changes slowly; it was 25,000 km long when Voyager arrived in 1979, but it had shrunk to 20,000 km by the end of the Galileo mission in 2000 (Figure 10.15). The giant storm has persisted in Jupiter's atmosphere ever since astronomers were first able to observe it after the invention of the telescope, more than 300 years ago.

In addition to its longevity, the Red Spot differs from terrestrial storms in being a high-pressure region; on our planet such storms are regions of lower pressure. The Red Spot's counterclockwise rotation has a period of six days. Three similar but smaller disturbances (about as big as the Earth) formed on Jupiter in the 1930s. They look like white ovals and one can be seen clearly below and to the right of the Great Red Spot in Figure 10.15. In 1998 the Galileo spacecraft watched as two of these ovals collided and merged into one (Figure 10.16).

We don't know what causes the Great Red Spot or the white ovals, but we do understand how they can last so long once they form. On Earth, the lifetime of a large oceanic hurricane or typhoon is typically a few weeks, or even less when it moves over the continents and encounters friction with the land. Jupiter has no solid surface to slow down an atmospheric disturbance, and furthermore the sheer size of the disturbances lends them stability. We can calculate that on a planet with no solid surface, the lifetime of anything as large as the Red Spot should be measured in centuries, while lifetimes for the white ovals should be measured in decades, which is pretty much what we have observed.

Despite Neptune's smaller size and different cloud composition, the 1989 Voyager images showed that it had an atmospheric feature surprisingly similar to the jovian Great Red Spot. Neptune's Great Dark Spot was nearly 10,000 km long (see Figure 10.12). On both planets, the giant storms formed at latitude 20°S, had the same shape, and took up about the same fraction of the planet's diameter. The Great Dark Spot rotated with a period of 17 days, versus about 6 days for the Great Red Spot. When the repaired Hubble Space Telescope examined Neptune in the mid 1990s, however, astronomers could find no trace of the Great Dark Spot on their images.

Although many of the details of the weather on the jovian planets are not yet understood, it is clear that if you are a fan of dramatic weather, these worlds are the place to look. We study the features in these atmospheres not only for what they have to teach us about conditions in the jovian planets, but also because we hope they can help us understand the weather on Earth just a bit better.

(a)

Cornell University/Caltech /NASA

(c)

Caltech /NASA

(b)

Caltech /NASA

■ FIGURE 10.16

Storms on Jupiter   Three examples of storms on Jupiter, illustrating the use of enhanced color and contrast to bring out faint features. (a) We show two versions of the same Galileo view of the large white ovals that are the most long-lived storms after the Great Red Spot. The top image shows how the scene might appear to the human eye, while the bottom is in false color (and includes infrared information). The left white oval is about three-fourths the size of the Earth. Note the cyclone-shaped storm between the two top ovals. (b) This 1996 Galileo image shows an enhanced view of a large hole in the clouds, similar to the place where the Galileo probe entered the atmosphere. The image is about 44,000 km across, and the dark hole (which is free of clouds) is nearly as wide as the Earth. (c) Here we see a thunderhead on Jupiter—a plume of water clouds carried to high altitudes by a strong updraft. The white plume is 1000 km across and about 25 km high. Such clouds occasionally form in regions of high atmospheric turbulence, such as this site near the outer edge of the Great Red Spot.

SURFING THE WEB

General Jupiter sites:

- *The Nine Planets Site:*
  seds.lpl.arizona.edu/nineplanets/nineplanets/jupiter.html
- *Views of the Solar System Site:*
  www.solarviews.com/eng/jupiter.htm
- *Planetary Sciences Site:*
  nssdc.gsfc.nasa.gov/planetary/planets/jupiterpage.html

General Saturn sites:

- *The Nine Planets Site:*
  seds.lpl.arizona.edu/nineplanets/nineplanets/saturn.html
- *Views of the Solar System Site:*
  www.solarviews.com/eng/saturn.htm
- *Planetary Sciences Site:*
  nssdc.gsfc.nasa.gov/planetary/planets/saturnpage.html

General Uranus sites:

- *The Nine Planets Site:*
  seds.lpl.arizona.edu/nineplanets/nineplanets/uranus.html
- *Views of the Solar System Site:*
  www.solarviews.comsolar/eng/uranus.htm
- *Planetary Sciences Site:*
  nssdc.gsfc.nasa.gov/planetary/planets/uranuspage.html

General Neptune sites:

- *The Nine Planets Site:*
  seds.lpl.arizona.edu/nineplanets/nineplanets/neptune.html
- *Views of the Solar System Site:*
  www.solarviews.com/eng/neptune.htm
- *Planetary Sciences Site:*
  nssdc.gsfc.nasa.gov/planetary/planets/neptunepage.html

Galileo Mission Site: www.jpl.nasa.gov/galileo
All the new Galileo images and results can be found here, together with educational activities and background information.

Voyager Project Home Page: voyager.jpl.nasa.gov
In 1998, Voyager 1 became the most distant human-made object in space. This site chronicles the planetary flybys of the two Voyagers and gives updates on their continuing mission to explore deep space and search for the heliopause. There is a good science summary for each planet the Voyagers investigated.

Cassini Mission Site: saturn.jpl.nasa.gov/index.cfm
Information about the mission and its engineering and science objectives. There are images of mission equipment and goals and several pages for educators.

SUMMARY

10.1 The outer solar system contains the four jovian (or giant) planets: Jupiter, Saturn, Uranus, and Neptune. Their chemistry is generally reducing, and Jupiter and Saturn have overall compositions similar to that of the Sun. These planets have been explored by the Pioneer, Voyager, Galileo, and Cassini spacecraft. Voyager 2, perhaps the most successful of all space-science missions, explored Jupiter (1979), Saturn (1981), Uranus (1986), and Neptune (1989)—a grand tour of the giant planets. Galileo successfully deployed a probe into the atmosphere of Jupiter in 1995 before beginning a 5-year orbital mission.

10.2 Jupiter is 318 times more massive than the Earth. Saturn is about 25 percent, and Uranus and Neptune only 5 percent, as massive as Jupiter. All four have deep atmospheres and opaque clouds, and all rotate quickly (with periods from 10 to 17 hours). Jupiter and Saturn have exten-

sive mantles of liquid hydrogen. Uranus and Neptune are depleted in hydrogen and helium relative to Jupiter and Saturn (and the Sun). Each giant planet has a core of "ice" and "rock" of about 10 Earth masses. Jupiter, Saturn, and Neptune have major internal heat sources, obtaining as much (or more) energy from their interiors as by radiation from the Sun. Uranus has no measurable internal heat. Jupiter has the strongest magnetic field and largest magnetosphere of any planet, first discovered by radio astronomers from observations of **synchrotron radiation.**

10.3 The four giant planets have generally similar atmospheres, composed mostly of hydrogen and helium. Their atmospheres contain small quantities of methane and ammonia gas, both of which also condense to form clouds. Deeper (invisible) cloud layers consist of water and possibly ammonium hydrosulfide (Jupiter and Saturn) and hydrogen sulfide (Neptune). In the upper atmospheres, hydrocarbons and other trace compounds are produced by **photochemistry.** We do not know what colors the clouds of Jupiter. Atmospheric motions on the giant planets are dominated by east–west circulation. Jupiter displays the most active cloud patterns, with Neptune second. Saturn is generally bland, and Uranus is featureless (perhaps due to its lack of an internal heat source). Large storms (oval-shaped high-pressure systems such as the Great Red Spot on Jupiter and the Great Dark Spot on Neptune) can be found in some of the giant planet atmospheres.

INTER-ACTIVITY

A A new member of Congress has asked your group to investigate why the Galileo probe launched into the Jupiter atmosphere in 1995 survived only 57 minutes. Make a list of all the reasons the probe did not last longer, and why it was not made more durable. (Remember that the probe had to hitch a ride to Jupiter!)

B Select one of the jovian planets and get your group to write a script for an evening news weather report for the planet you chose. Be sure you specify roughly how high in the atmosphere the region lies for which you are giving the report.

C What does your group think should be the next step to learn more about the giant planets? Put cost considerations aside for a moment: What kind of mission would you recommend to NASA to learn more about these giant worlds? Which world or worlds should get the highest priority and why?

D Suppose that an extremely dedicated (and slightly crazy) astronomer volunteers to become a human probe into Jupiter (and somehow manages to survive the trip through Jupiter's magnetosphere alive). As he enters the upper atmosphere of Jupiter, would he fall faster or slower than he would fall doing the same suicidal jump into the atmosphere of the solid Earth? Groups that have some algebra background could even calculate the force he would feel compared to the force on Earth. (*Bonus question:* If he were in a capsule, falling into Jupiter feet first, and the floor of the capsule had a scale, what would the scale show as his weight?)

REVIEW QUESTIONS

Ace Astronomy™ Assess your understanding of this chapter's topics with additional quizzing and animations at **http://ace.brookscole.com/ voyages**

1. Describe the differences in the chemical makeup of the inner and outer parts of the solar system. What is the relationship between what the planets are made of and the temperature where they formed?
2. How did the giant planets grow to be so large?
3. Why is it difficult to drop a probe like Galileo's gently into the atmosphere of Jupiter? How did the Galileo engineers solve this problem?

4. What are the visible clouds on the four giant planets composed of, and why are they different from each other?
5. Describe the seasons on the planet Uranus.
6. Compare the atmospheric circulation (weather) for the four giant planets.
7. What are the main atmospheric heat sources of each of the giant planets?
8. How do the clouds on the four giant planets differ?

THOUGHT QUESTIONS

9. Jupiter is denser than water, yet composed for the most part of two light gases, hydrogen and helium. What makes Jupiter as dense as it is?

10. Would you expect to find free oxygen gas in the atmospheres of the giant planets? Why or why not?

11. Why would a tourist brochure (of the future) describing the most dramatic natural sights of the giant planets have to be revised more often than one for the terrestrial planets?

12. The water clouds believed to be present on Jupiter and Saturn exist at temperatures and pressures similar to those in the clouds of the terrestrial atmosphere. What would it be like to visit such a location on Jupiter or Saturn? In what ways would the environment differ from that in the clouds of Earth?

13. Describe the different processes that lead to substantial internal heat sources for Jupiter and Saturn. Since these two objects generate much of their energy internally, should they be called stars instead of planets? Justify your answer.

14. Use sources in your college library to learn more about the Galileo mission. What technical problems occurred between the mission launch and the arrival of the craft in the jovian system, and how did the mission engineers deal with them? (Good sources of information include *Astronomy* and *Sky & Telescope* magazines, plus the mission web site.)

FIGURING FOR YOURSELF

The wind speeds in circular storm systems can be formidable on both Earth and the giant planets. First think about our big terrestrial hurricanes. If you watch their behavior in satellite images shown on the weather channels, you will see that they require about one day to rotate. If a storm has a diameter of 400 km and rotates once in 24 hours, what is the wind speed? Speed equals distance divided by time. The distance in this case is the circumference ($\pi \times$ diameter), or approximately 1200 km, and the time is 24 hours, so the speed at the edge of the storm is 50 km/h, or about 35 mi/h. Toward the center of the storm, of course, the wind speeds can be much higher.

15. Jupiter's Great Red Spot rotates in 6 days and has a circumference equivalent to a circle with radius 10,000 km. Calculate the wind speed at the outer edge of the spot.

16. Calculate the wind speed at the edge of Neptune's Great Dark Spot, which was 2500 km wide and rotated in 17 days.

17. Calculate how many Earths would fit into the volumes of Saturn, Uranus, and Neptune.

18. As the Voyager spacecraft penetrated into the outer solar system, the illumination from the Sun declined. Relative to the situation at Earth, how bright is the sunlight at each of the jovian planets?

19. The ions in the inner parts of the jovian magnetosphere rotate with the same period as Jupiter. Calculate how fast they are moving at the orbit of Jupiter's moon Io (see Appendix 8). Will these ions strike Io from behind or in front as it moves about Jupiter?

SUGGESTIONS FOR FURTHER READING

Beatty, J. "Into the Giant" in *Sky & Telescope*, Apr. 1996, p. 20. On the Galileo probe.

Beebe, R. "Queen of the Giant Storms" in *Sky & Telescope*, Oct. 1990, p. 359. Excellent review of the Red Spot.

Cowling, T. "Big Blue: The Twin Worlds of Uranus and Neptune" in *Astronomy*, Oct. 1990, p. 42. Nice long review of the two planets.

Hanlon, M. *The Worlds of Galileo*. 2001, St. Martin's Press. Good overview of the Galileo mission and its discoveries.

Harland, D. *Jupiter Odyssey: The Story of NASA's Galileo Mission*. 2000, Praxis/Springer-Verlag. Another good mission summary.

Littmann, M. *Planets Beyond: Discovering the Outer Solar System*, 2nd ed. 1989, John Wiley. Good introduction to Uranus, Neptune, and Pluto.

Lunine, J. "Neptune at 150" in *Sky & Telescope*, Sept. 1996, p. 38.

Nice review.

McAnally, J. "A Jupiter Observing Guide" in *Sky & Telescope*, Oct. 2000, p. 124. A good introduction for serious amateurs.

Miner, E. *Uranus*. 1990, Ellis Horwood/Simon & Shuster.

Miner, E. & Wessen, R. *Neptune: The Planet, Rings, and Satellites*. 2001, Springer Verlag.

Nellis, E. "Making Metallic Hydrogen" in *Scientific American*, May 2000, p. 84. On experiments on Earth to duplicate what is found naturally in Jupiter.

Rogan, J. "Bound for the Ringed Planet" in *Astronomy*, Nov. 1997, p. 36. On the Cassini mission.

Sagan, C. *Pale Blue Dot*. 1994, Random House. Chapters 6–9 explore the outer solar system.

Smith, B. "Voyage of the Century" in *National Geographic*, Aug. 1990, p. 48. Beautiful summary of the Voyager mission to all four outer planets.

This montage, assembled from individual Galileo and Voyager images, shows a "family portrait" of Jupiter (with its giant red spot) and its four large satellites. From top to bottom, we see Io, Europa, Ganymede, and Callisto. The colors are exaggerated by image processing to emphasize contrasts. See Figure 10.1 for a better sense of the true colors of Jupiter's atmosphere, for example.

DLR/NASA

Rings, Moons, and Pluto

Our imaginations always fall short of anticipating the beauty we find in nature.

Geologist Laurence Soderblom, discussing the 1989 Voyager encounter with Neptune's moons

THINKING AHEAD

On an active world, a map is always temporary. The boundaries of towns change, new roads are built, rivers slowly alter their channels. Over geological time, even the Earth's mountains, seas, and continents change shape and form. But what kind of map would be worth making of a world where the surface landforms changed substantially in only a decade? Such a world can be found orbiting the planet Jupiter.

All of the giant planets are accompanied by moons or satellites (astronomers use the two terms interchangeably), which orbit about them like planets in a miniature solar system. More than 100 satellites are known in the outer solar system, too many to name individually or discuss in any detail. The characteristics of the more important moons are summarized in Appendix 8. Astronomers expect that many other small satellites will be found in the future.

Before the Voyager missions to the outer planets, even the largest of these satellites were mere points of light in our telescopes. Suddenly, in less than a decade, we had close-up images (and much nonvisual information as well); they then became individual worlds for us, each with unique features and peculiarities. The Galileo mission added greatly to our knowledge of the satellites of Jupiter, and the Cassini mission will, we hope, do the same for the Saturn system.

Ace⊘Astronomy™ The AceAstronomy icon throughout the text indicates an opportunity for you to test yourself on key concepts and to explore animations and interactions on the AceAstronomy website at **http://ace.brookscole.com/voyages**

In this chapter we focus our attention on a few of the more interesting satellites and on the tiny planet Pluto, which is smaller than a number of the outer planets' moons. Then we examine the ring systems that gird each of the four giant planets. The processes we see at work in these rings have applications ranging from the formation of planetary systems to the spiral structure of galaxies.

11.1 RING AND SATELLITE SYSTEMS

The rings and satellites of the outer solar system are not composed of the same materials as the mostly rocky objects in the inner solar system. We should expect this, since they formed in regions of lower temperature, cool enough so that large quantities of water ice were available as building materials. Most of these objects also contain dark, organic compounds mixed with their ice and rock. The presence of this dark, primitive material often means that the objects reflect very little light. Don't be surprised, therefore, to find many objects in the ring and satellite systems that are both icy and black.

Most of the satellites in the outer solar system are in *direct* or regular orbits; that is, they revolve about their parent planet in a west-to-east direction and in the plane of the planet's equator. In addition, a number of irregular satellites orbit in a *retrograde* (east-to-west) direction or else have orbits of high eccentricity (more elliptical than circular) or high inclination (moving in and out of the planet's equatorial plane). These are usually small satellites, located relatively far from their planet; they were probably formed far away and subsequently captured by the planet they now orbit. (Perhaps the fact that they were not born locally will excuse their ill-mannered behavior.)

11.1.1 The Jupiter System

Jupiter has 52 known satellites (that's the number as we write; probably more will have been discovered by the time you read this sentence) and a faint ring. The satellites include four large moons—Callisto, Ganymede, Europa, and Io (see the image that opens the chapter)—discovered in 1610 by Galileo and therefore often called the *Galilean satellites.* The smaller of these, Europa and Io, are about the size of our Moon, while the larger, Ganymede and Callisto, are bigger than the planet Mercury. We will discuss the Galilean satellites individually in a moment and try to understand the interesting differences among them.

Most of the jovian moons are much smaller. The majority are in retrograde orbits more than 20 million km from Jupiter; these are very likely small captured asteroids.

11.1.2 The Saturn System

Saturn has at least 30 known satellites in addition to a magnificent set of rings (Figure 11.1). The largest of the satellites, Titan, is almost as big as Ganymede in the jovian system, and

■ **FIGURE 11.1**

Saturn Family Portrait This NASA montage of Voyager images shows a perspective view of Saturn and several of its satellites. The satellites are not to scale: Giant Titan is the tiny orange ball toward the back. The moon in the foreground is called Enceladus. The color of Saturn is intensified by the image processing.

NASA/JPL

it is the only satellite with a substantial atmosphere. We will discuss it in detail in the next section. Saturn has six other large regular satellites with diameters between 400 and 1600 km, a handful of small moons orbiting in or near the rings, and a collection of captured strays similar to those of Jupiter.

The rings of Saturn, one of the most impressive sights in the solar system, are broad and flat, with only a few gaps. They are not solid, but rather a huge collection of icy fragments, all orbiting the equator of Saturn in a traffic pattern that makes rush hour in a big city look simple by comparison. Individual ring particles are composed of water ice and are typically the size of ping-pong, tennis, and basketballs.

11.1.3 The Uranus System

The ring and satellite system of Uranus is tilted at 98°, just like the planet itself. It consists of 11 rings and more than 20 known satellites. The five largest satellites are similar in size to the six regular satellites of Saturn, with diameters of 500 to 1600 km, while most of the smaller satellites and the ring particles are very dark, in contrast to those of Saturn.

Discovered in 1977, the rings of Uranus are narrow ribbons of material with broad gaps in between—very different from the broad rings of Saturn. Astronomers suppose that the ring particles are confined to these narrow paths by the gravitational effects of numerous small satellites, most of which we have not yet glimpsed. (We will discuss how satellites can do this in Section 11.4.4.)

11.1.4 The Neptune System

Neptune has eight known satellites: six regular satellites close to the planet, and two irregular satellites farther out. The most interesting of these is Triton, a relatively large moon in a retrograde orbit. Triton has an atmosphere, and active volcanic eruptions were discovered there by Voyager in its 1989 flyby. We will compare it in detail with the planet Pluto, which it resembles, later in this chapter.

The rings of Neptune are narrow and faint. Like those of Uranus, they are composed of dark materials and are thus not easy to see.

From 1996 to 1999 the Galileo spacecraft careered through the jovian system on a complex but carefully planned trajectory that provided repeated close encounters with the Galilean satellites. Although scientists had thought they understood the basics of these satellites from the two Voyager flybys of 1979, the new data brought many surprises. And, to the delight of scientists and the public, the high-resolution Galileo cameras offered stunning views that far exceeded the Voyager images in their scientific value. This wealth of new information provides us the opportunity, quite literally, to rewrite the textbooks on the Galilean satellites. Beginning in 2004, we can expect a similar bonanza of information about Titan, obtained from the Cassini spacecraft.

We discuss each of these five large satellites in turn, but to compare them with each other, we first list some basic facts about them (and our own Moon) in Table 11.1.

11.2.1 Callisto: An Ancient, Primitive World

We begin our discussion of the Galilean satellites with the outermost one, Callisto—not because it is remarkable but because it is not. This makes it a convenient object with which other, more active, worlds can be compared. Its distance from Jupiter is about 2 million km, and it orbits the planet in 17 days. Like our own Moon, Callisto rotates in the same period as it revolves, so it always keeps the same face toward Jupiter. Callisto's day thus equals its month: 17 days. Its noontime surface temperature is only 130 K (about 140°C below freezing), so that water ice is stable on its surface the year around.

Callisto has a diameter of 4820 km, almost the same as the planet Mercury (Figure 11.2). Yet its mass is only one third as great, which means its density (the mass divided by the volume) must be only one third as great as well. This tells us that Callisto has far less of the rocky and metallic materials found in the inner planets and must instead be an

TABLE 11.1 *The Largest Satellites*

Name	Diameter (km)	Mass (Moon = 1)	Density (g/cm^3)	Reflectivity (%)
Moon	3476	1.0	3.3	12
Callisto	4820	1.5	1.8	20
Ganymede	5270	2.0	1.9	40
Europa	3130	0.7	3.0	70
Io	3640	1.2	3.5	60
Titan	5150	1.9	1.9	20

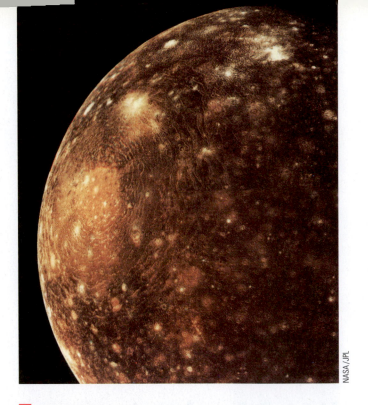

NASA/JPL

■ **FIGURE 11.2**

Callisto   Jupiter's outermost large satellite shows a heavily cratered surface of dark ice in this Voyager image. The multi-ringed feature at center left is the giant impact basin called Valhalla.

from the details of its gravitational pull on the Galileo spacecraft during several very close flybys. This fact surprised scientists, who expected that all the big icy moons would be differentiated. It should be easier for an icy body to differentiate than for a rocky one because the melting temperature of ice is so low. Only a little heating will soften the ice and get the process started, allowing the rock and metal to sink to the center while the slushy ice floats to the surface. Yet Callisto seems to have frozen solid before the process of differentiation was complete.

The surface of Callisto is covered with impact craters, like the lunar highlands. The survival of these craters tells us that an icy object can retain impact craters in its surface. An especially impressive impact feature, called Valhalla (after the home of the gods in Norse mythology), can be seen as a bull's-eye on Figure 11.2. The light material brought up by the impact in the center is as large as the state of Colorado, and the rings surrounding it have a diameter almost as large as the continental United States. Callisto is unique among the planet-sized objects of the solar system in the absence of interior forces to drive geological change. You might say that the satellite was stillborn, and it has remained geologically dead for more than 4 billion years.

In thinking about ice so far from the Sun, we must take care not to judge its behavior from the much warmer ice we know and love on Earth. At the temperatures of the outer solar system, ice on the surface is nearly as hard as rock and it behaves similarly. Ice on Callisto does not deform or flow like ice in glaciers on the Earth.

In general form, the icy craters of Callisto look very much like the rocky craters of the Moon and Mercury. Yet when we look at the close-up photographs from Galileo, the similarity with lunar craters evaporates (Figure 11.3). Some process has been at work to break down crater walls and bury parts of the surface. Although the results of this

icy body through much of its interior. Such a satellite can show us how the geology of an icy object compares with one made primarily of rock.

Unlike the worlds we have studied so far, Callisto has not fully *differentiated* (separated into layers of different density materials). We can tell that it lacks a dense core

Arizona State U./JPL/NASA

■ **FIGURE 11.3**

Worn Craters on Callisto
This pair of Galileo images shows an area about 30 km across in the Valhalla region of Callisto. Details as small as 10 m across can be seen. Sunlight is illuminating the surface from the left. Note that the impact craters are eroded even though Callisto has no atmosphere, and that white frost seems to have condensed on the highest parts of the eroded crater rims.

erosion look rather like the effect of windblown dust on Earth and Mars, this explanation cannot be correct because Callisto has no atmosphere. Instead, it must be the slow process of *sublimation* of ice (turning from solid directly into gas due to heating by the Sun) that breaks down the craters over billions of years and redistributes the dusty (non-icy) surface material.

11.2.2 Ganymede, the Largest Satellite

Ganymede, the largest satellite in the solar system, also shows a great deal of cratering (Figure 11.4). About one quarter of its surface seems to be as old and heavily cratered as that of Callisto; the rest formed more recently, as we can tell by the sparse covering of impact craters as well as the relative freshness of the craters. If we judge from crater counts, this fresher terrain on Ganymede is somewhat younger than the lunar maria or the martian volcanic plains, perhaps 2 to 3 billion years old.

The differences between Ganymede and Callisto are more than skin deep. Ganymede is a differentiated world, like the terrestrial planets. Measurements of its gravity field tell us that the rock sank to form a core about the size of our Moon, with a mantle and crust of ice "floating" above it. In

JPL/NASA

■ **FIGURE 11.5**

Close-up of Ganymede Details as small as 75 m across can be seen in this image of Jupiter's largest moon, taken by the Galileo spacecraft cameras. We see an area 55 by 35 km in the Uruk Sulscus region, with a complex assortment of ridges, grooves, craters, and evidence of crustal shifting. The linelike sets of features are parallel sunlit ridges. We can see signs that the icy surface has pulled apart and slid sideways in shaping the landscape.

addition, the Galileo spacecraft discovered that Ganymede has a magnetic field—the sure signature of a partially molten interior. There is very likely liquid water trapped within the interior. Thus, Ganymede is not a dead world but rather a place of continuing geological activity powered by an internal heat source. Some surface features could be as young as the surface of Venus (half a billion years).

The younger terrain is the result of tectonic and volcanic forces (Figure 11.5). In some places the crust apparently cracked, flooding many of the craters with water from the interior. Extensive mountain ranges were formed from compression of the crust, forming long ridges with parallel valleys spaced a few kilometers apart. In some places older impact craters were split and pulled apart. There are even indications of large-scale crustal movements that are similar to the plate tectonics of the Earth.

Why is Ganymede so different from Callisto? Possibly the small difference in size and internal heating between

Galileo Imaging Team; JPL/NASA

■ **FIGURE 11.4**

Ganymede This global view of Ganymede, the largest moon in the solar system, was assembled from images taken by the Galileo spacecraft. The large circular dark region at the upper right is called Galileo Regio. Such darker places are older, more heavily cratered regions; the lighter areas are younger. Really bright spots are sites of geologically recent impacts.

the two led to this divergence in their evolution. But more likely the gravity of Jupiter is to blame for Ganymede's continuing geological activity. Ganymede is close enough to Jupiter that *tidal forces* from the giant planet may have episodically heated its interior and triggered major convulsions on its crust.

As we discussed in Section 3.6, a tidal force results from the unequal gravitational pull on two sides of a body. In a complex kind of modern dance, the large satellites of Jupiter are caught in the varying gravity grip of both the giant planet and each other. This leads to gravitational flexing or kneading in their centers, which can heat them up. (A fuller explanation is given in the section on Io, below.) We will see as we move inward to Europa and Io that the role of jovian tides becomes more critical for satellites close to the planet.

11.2.3 Europa, a Satellite with an Ocean

Europa and Io, the inner two Galilean satellites, are not icy worlds like most of the satellites of the outer planets. With densities and sizes similar to our Moon, they appear to be predominantly rocky objects. How did they fail to acquire a majority share of the ice that must have been plentiful in the outer solar system at the time of their formation?

The most probable cause is Jupiter itself, which was hot enough to radiate a great deal of infrared energy during the first few million years after its formation. This infrared radiation would have heated the disk of material near the planet that would eventually coalesce into the closer moons. Thus any ice near Jupiter was vaporized, leaving Europa and Io with compositions more appropriate to bodies in the inner solar system.

Despite its mainly rocky composition, Europa has an ice-covered surface, as astronomers have long known from examining spectra of sunlight reflected from it (Figure 11.6). In this it resembles the Earth, which also has a layer of water on its surface, but in Europa's case most of the water may be frozen. There are very few impact craters, indicating that the surface of Europa is in a continual state of geological self-renewal. Judging from crater counts, the surface must be no more than a few million years old, and perhaps substantially less. In terms of its ability to erase impact craters, Europa is more geologically active than the Earth!

When we look at close-up photos of Europa, we see a strange, complicated surface (Figures 11.7 and 11.8). For the most part the satellite is extremely smooth, but it is crisscrossed with cracks and low ridges that often stretch for thousands of kilometers across the icy plains. Some of these long lines are single, but most are double or

JPL/NASA

■ **FIGURE 11.6**

Europa's Intriguing Surface Jupiter's moon Europa is the only place in the solar system beyond Earth that appears to have liquid water—an ocean beneath a thick crust of ice. Both the contrast and the color are highly exaggerated in this wide-angle view of a region about 1200 km across. Scientists enhance the color to bring out subtle differences in composition between the different types of terrain. Europa's crust is nearly pure water ice, with very few hills or other topographic features.

JPL/NASA

■ **FIGURE 11.7**

Evidence for an Ocean on Europa A close-up of an area called Conamara Chaos is shown here with enhanced color. This view is 70 km wide in its longer dimension. It appears that Conamara is a region where Europa's icy crust is (or recently was) relatively thin and there is easier access to the possible liquid or slushy ocean beneath. Not anchored to solid crust underneath, many of the ice blocks here seem to have slid or rotated from their original positions. In fact, the formations seen here look similar to views of floating sea-ice and icebergs in the Earth's Arctic Ocean.

■ **FIGURE 11.8**

A Faded Crater We know that the surface of Europa is geologically young because there are so few impact craters. One of the largest craters on Europa is Pwyll, about 26 km in diameter. Even though Pwyll is the result of a recent impact, its crater rim is nearly gone, fading away as the icy crust bends and readjusts over time. The darker material in and outside the crater was probably dredged up from deeper inside Europa by the force of the impact.

multiple, looking rather like the remnants of a colossal freeway system (Figure 11.9 and the first panel of Figure 11.11).

It is very difficult to make straight lines on a planetary surface. In Chapter 9, we explained that when Percival Lowell saw what appeared to him to be straight lines on Mars (the so-called martian "canals"), he attributed them to the engineering efforts of intelligent beings. We now know the lines on Mars were optical illusions, but the lines on Europa are real. If the crust consists of ice floating without much friction on an ocean of liquid water, then such long cracks would be expected. To examine this exciting possibility more closely, NASA redirected the Galileo spacecraft during the third and fourth years of its mission to pass nearer to Europa and send back very high-resolution images of its intriguing surface (Figure 11.10, for example).

The Galileo images appear to confirm the existence of a global ocean. In many places the surface of Europa looks just as we would expect for a thick layer of ice that was broken up into giant icebergs and ice flows and then refrozen in place. When the ice breaks, water or slush from below may be able to seep up through the cracks and make the ridges and multiple-line features we observe. Many episodes of ice cracking, shifting, rotating, and refreezing are required to explain the complexity we see. The icy crust might vary in thickness from a kilometer or so up to 20 km.

■ **FIGURE 11.9**

Europa's Complex Geological History In this high-resolution view, the ice is wrinkled and crisscrossed by long ridges. Where these ridges intersect, we can see which ones are older and which younger; the younger ones cross over the older ones. To some observers, this system of ridges resembles a giant freeway system on Europa, but the ridges are much wider than our freeways and are a natural result of the flexing of the moon.

■ **FIGURE 11.10**

Detail of an Ice Ridge A very high-resolution Galileo image of one young double ridge. The area in this picture is only 17 km across, and the ridge itself is about 3 km wide and 300 m high. It appears to have formed when viscous icy material was forced up through a long straight crack in the crust.

Europa   Ganymede   Callisto

■ FIGURE 11.11

Comparing Three Icy Satellites These Galileo images show the surfaces of Europa, Ganymede, and Callisto (from left to right) with the same resolution (detail). Note that the number of craters (and thus the age of the surface we see) increases as we go from Europa to Ganymede to Callisto. The Europa image is one of those where the system of cracks and ridges resembles a freeway system.

In some places the ice appears to be free-floating, in others it seems to be grounded against the rocky ocean floor. Confirmation that this liquid ocean is present today came from measurements of the small magnetic field induced by Europa's interactions with the magnetosphere of Jupiter. The "magnetic signature" of Europa is that of a liquid water ocean, not of ice or rock.

If Europa really has a large ocean of liquid water under its ice, then it is the only place in the solar system, other than Earth, with such an environment. To remain liquid, this ocean must be warmed by heat escaping from the interior of Europa. Hot (or at least warm) springs might be active there, analogous to those we have discovered in the deep oceans of the Earth. However, to maintain large quantities of liquid water, there must be substantial heating from Europa's interior—more than can be explained by the energy from radioactive rocks (the mechanism that heats the interiors of the terrestrial planets).

Such heat can be generated by the tidal heating mechanism we mentioned earlier, but the heating will vary at different times during Europa's orbit. Scientists still don't know how much tidal energy is produced inside Europa and whether this heating is steady enough to maintain a global ocean at all times. At some times the water may be in the form of an icy slush, depending on subtle changes in the orbit of Europa.

What makes the idea of an ocean with possible warm springs especially exciting is the discovery in *Earth's oceans* of large ecosystems of primitive life around subterranean hot springs. Such life derives all its energy from the mineral-laden water and thrives independent of the sunlight shining on the Earth's surface. Is it possible that similar ecosystems could exist today under the ice of Europa? Some people have suggested that Europa is the most likely place beyond the Earth to find life in the solar system. This possibility ensures that Europa will be a continuing target of planetary missions in the future.

11.2.4 Io, a Volcanic Satellite

Io, the innermost of Jupiter's Galilean satellites, is a close twin of our Moon, with nearly the same size and density. We might therefore expect it to have experienced a similar history. Its appearance, as photographed from space, tells us another story, however (Figure 11.12). Instead of being a dead, cratered world, Io turns out to have the highest level of volcanism in the solar system, greatly exceeding that of the Earth.

Io's active volcanism was discovered by the Voyager spacecraft. Eight volcanoes were actually seen erupting when Voyager 1 passed in March 1979, and six of these were still active four months later when Voyager 2 passed.

NASA/JPL/USGS

■ **FIGURE 11.12**

Two Sides of Io   A composite of Voyager images assembled by the U.S. Geologic Survey shows both sides of the volcanically active moon Io. The orange deposits are sulfur snow; the white is sulfur dioxide. (Carl Sagan once quipped that Io looks like it desperately needs a shot of penicillin.)

With the improved instruments carried by the Galileo spacecraft, more than 50 eruptions were found during the calendar year 1997 alone. Many of the eruptions produce graceful plumes that extend hundreds of kilometers out into space (Figure 11.13).

The Galileo data show that most of the volcanism on Io consists of hot silicate lava, like the volcanoes on Earth. Sometimes the hot lava encounters frozen deposits of sulfur and sulfur dioxide. When these icy deposits are suddenly heated, the result is great eruptive plumes far larger than any ejected from terrestrial volcanoes. As the rising plumes cool, the sulfur and sulfur dioxide recondense as solid particles, which fall back to the surface in colorful "snowfalls" that extend as much as a thousand kilometers from the vent. Planetary scientists, comparing Galileo images to earlier Voyager maps, have found at least a dozen areas on Io larger than the entire state of Connecticut that have been resurfaced by volcanic activity in the years between the two missions. Major new surface features were even seen to appear between Galileo orbits, as shown in Figure 11.14.

As the Galileo mission drew to a close, controllers were willing to take risks in getting close to Io. Approaching this satellite is always a dangerous maneuver because the belts of atomic particles trapped in the magnetic environment of Jupiter (see Chapter 10) are at their most intense near Io's orbit. Indeed, in its very first pass by Io, the spacecraft absorbed damaging radiation beyond its design levels. To

NASA/JPL

■ **FIGURE 11.13**

Volcanic Plume on Io   The large plume rising about 250 km above Io's surface (seen in white above the edge of the moon) erupted from the center of an enormous volcanic feature that takes up most of this image. It is named Pele, after the Hawaiian goddess of fire, and is about the size of the state of Alaska.

NASA/JPL/U. of Arizona

■ **FIGURE 11.14**

Volcanic Changes on Io   These three images were taken of the same 1700-km-square region of Io in April 1997, September 1997, and July 1999. The dark volcanic center called Pillan Patera experienced a huge eruption, producing a dark deposit some 400 km across. In the right image, however, some of it is being covered by the reddish material from the volcano Pele (see Figure 11.13), indicating that both volcanic vents probably continue to be active. Also, a small unnamed volcano to the right of Pillan has erupted since 1997, and some of its dark deposit and a yellow ring around it are visible on the right image. The color range is exaggerated in these images.

keep the system working at all, controllers had to modify or disable various fault-protection software routines in the onboard computers. In spite of these difficulties, the spacecraft achieved four successful Io flybys, obtaining photos and spectra of the surface with unprecedented resolution.

Our new maps of Io reveal more than 100 recently active volcanoes on its surface (Figure 11.15). For more that 50 of these, infrared measurements have been made of the hot lava. Some of these eruptions show temperatures as high as 1500°C, hotter than all but a few terrestrial volcanoes. Huge flows spread out from these vents, covering about 25 percent of the moon's total surface with still-warm lava. From these measurements, it seems clear that the bright surface colors that first attracted attention to Io are the result of a thin veneer of sulfur compounds. The underlying volcanism is driven by eruptions of molten silicates, just like on the Earth (Figure 11.16).

11.2.5   Tidal Heating

How can Io maintain such ongoing volcanism in spite of its small size? The answer, as we hinted earlier, lies in gravitational heating of the satellite by Jupiter. Io is about the same distance from Jupiter as our Moon is from the Earth. Yet Jupiter is more than 300 times more massive than Earth, causing forces that pull the satellite into an elongated shape, with a several-kilometer-high bulge extending toward Jupiter.

If Io always kept exactly the same face turned toward Jupiter, this bulge would not generate heat. However, Io's orbit is not exactly circular due to gravitational perturbations (tugs) from Europa and Ganymede. In its slightly ec-

centric orbit, Io twists back and forth with respect to Jupiter, at the same time moving nearer and farther from the planet on each revolution. The twisting and flexing heat Io, much as repeated flexing of a wire coat-hanger heats the wire.

After billions of years, this constant flexing and heating have taken their toll on Io, driving away water and carbon dioxide and other gases, until now sulfur and sulfur compounds are the most volatile materials remaining. The moon's inside is entirely melted, and the crust itself is constantly recycled by volcanic activity.

In moving inward toward Jupiter from Callisto to Io, we have encountered more and more evidence of geological activity and internal heating. Callisto has a surface more than 4 billion years old; Ganymede has been active as recently at the past billion years; Europa has an icy crust at most a few million years old; and parts of the visible surface of Io were deposited in the brief 16-year interval between the Voyager and Galileo missions. (Three of these surfaces are compared in Figure 11.11.) Just as the character of the planets in our solar system depends in large measure on their distance from the Sun (and on the amount of heat they receive), so it appears that distance from a giant planet like Jupiter can play a large role in the composition and evolution of its satellites (at least partly due to differences in internal heating by Jupiter's unrelenting tides).

11.2.6   Titan, a Satellite with Atmosphere

Titan, discovered in 1655 by the Dutch astronomer Christian Huygens, is the largest satellite of Saturn. It was the first satellite discovered after Galileo saw the four large moons

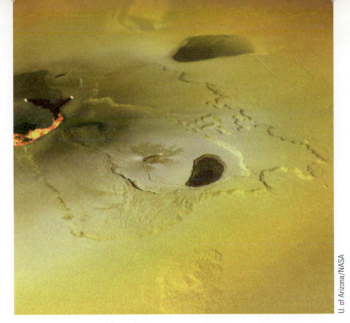

■ FIGURE 11.16

Lava Fountains on Io   Galileo captured a number of eruptions along the chain of huge volcanic calderas (or pits) on Io called Tvashtar Catena. In this false-color image, infrared and visible-light information have been combined. The orange-yellow areas at left are places where fresh, hot lava is erupting. The rounded ribbon of lava is about 60 km long.

■ FIGURE 11.15

Giant Lava Flow on Io   This mosaic of Galileo images shows a region 500 by 200 km that contains the longest active lava flow yet discovered in the solar system. We see a patchwork quilt of dark lava interacting with the colored snows of sulfur compounds. The most recent lava eruptions are the darkest because they have not yet had a chance to be covered by the sulfur dioxide plumes the lava flows cause. It is not always clear where the liquid lava is coming up from Io's interior because the fresh lava can travel under older lava deposits and emerge a long distance—hundreds of kilometers—from its point of origin.

of Jupiter. Titan has roughly the same diameter, mass, and density as Callisto or Ganymede. Presumably it also has a similar composition—about half ice and half rock, although we don't know whether this satellite has differentiated. What is most remarkable about Titan is the presence of a substantial atmosphere, discovered by Gerard Kuiper in 1944. He noticed that spectra of Titan showed absorption features indicating the presence of methane gas (Figure 11.17).

■ FIGURE 11.17

Titan's Thick Atmosphere   To Voyager, Titan looked like a fuzzy orange ball (see Figure 11.1). Cassini, getting a closer look, shows a nearly detached layer of haze in Titan's upper atmosphere. The inset shows the haze in greater detail. (The purple color has been enhanced.)

The 1980 Voyager flyby of Titan determined that the surface density of its atmosphere is four times greater than that on the Earth. The atmospheric pressure on this satellite is 1.6 bars, higher than that on any other moon and, remarkably, even higher than that of the terrestrial planets Mars and Earth. The atmospheric composition is primarily nitrogen, an important way in which Titan's atmosphere resembles Earth's.

A variety of additional compounds have been detected in Titan's upper atmosphere, including carbon monoxide (CO), hydrocarbons (compounds of hydrogen and carbon) such as methane (CH_4), ethane (C_2H_6), and propane (C_3H_8), and nitrogen compounds such as hydrogen cyanide (HCN), cyanogen (C_2N_2), and cyanoacetylene (HC_3N). Their presence indicates an active chemistry, in which sunlight interacts with atmospheric nitrogen and methane to create a rich mix of organic molecules. This is very much the way we believe organic compounds were formed on the Earth when it still had its original atmosphere. The discovery of HCN on Titan is particularly interesting, since this molecule is the starting point for the formation of some components of DNA, the fundamental genetic molecule essential to life on Earth.

Titan has multiple layers of clouds and haze. In the bottom 10 km are condensation clouds composed of methane. The much higher dark reddish haze layers (Figure 11.17) are complex organic chemicals created by the energy of sunlight falling on the molecules of Titan's atmosphere. As these particles settle to the surface, they may form tarlike organic deposits. At Titan's surface temperature of 90K, these compounds will not decompose as they do on Earth. Thus they may preserve a chemical history that goes back billions of years.

In January 2005, scientists took a major step forward with the landing on Titan of the Huygens Probe, a part of the Cassini mission to Saturn. The Huygens probe had been under development in Europe since the 1980s. It worked flawlessly, decelerating in Titan's upper atmosphere and then descending for more than two hours by parachute, making detailed analyses of the atmosphere and taking panoramic photos of the surface. Since the atmosphere is thicker than that of the Earth, the probe's descent was relatively gentle. Better yet, it landed on soft ground and continued to transmit data back from the surface for two more hours. The first photos are shown in Figure 11.18.

Several hydrocarbons such as methane and ethane could be stable at the surface of Titan, where they could form both liquid "seas" and condensation clouds in the lower atmosphere. The Huygens descent photos show complex light and dark regions with "river channels" suggesting that methane, the major hydrocarbon, behaves in the same way as water does on our planet. Although the probe was designed to float if it landed in liquid, it apparently came down on a solid surface—but perhaps one that was muddy or slushy—as shown in Figure 11.18b. At these temperatures Titan's water is frozen solid, and it is methane as both clouds and liquid rain that create the weather on this fascinating moon.

ESA/NASA/U. OF ARIZONA

■ **FIGURE 11.18**

First Views of the Surface of Titan The European Hyugens probe took these photos when it landed on Titan in January 2005. (a) View of an apparent drainage network from the altitude of 16 km. (b) Photo of the surface. The "rocks" are probably frozen water or other ices. The color is reddish because only red wavelengths are transmitted by the atmosphere of Titan.

11.3 TRITON AND PLUTO

In this section we discuss two apparently similar objects: Neptune's satellite Triton and the planet Pluto. Seeking to understand the ways in which these two worlds are similar to each other is an example of the comparative approach to studying astronomy. Pluto is the only planet not visited by a spacecraft; therefore comparing it with Triton, which was studied by Voyager in 1989, is one way we can learn more about it.

11.3.1 Triton and Its Volcanoes

Triton (don't get it confused with Titan) has a diameter of 2720 km and a density of 2.1 g/cm^3, indicating that it's probably composed of 75 percent rock mixed with 25 percent water ice. Measurements indicate that Triton's surface has the coldest temperature of any of the worlds our robot representatives have visited. Because its reflectivity is so high (about 80 percent), Triton reflects most of the solar energy that falls on it, resulting in a surface temperature between 35 and 40 K.

The surface material of Triton is made of frozen water, nitrogen, methane, and carbon monoxide. Methane and nitrogen exist as gas in most of the solar system, but they are frozen at Triton's temperatures. Only a small quantity of ni-

■ **FIGURE 11.19**

Neptune's Moon Triton This mosaic of Voyager images of Triton shows a wide range of surface features. The pinkish area at the bottom is Triton's large southern polar cap. The south pole of Triton faces the Sun here, and the slight heating effect is driving some of the material northward, where it is colder.

NASA/JPL/Image processing at USGS

trogen vapor persists to form an atmosphere. Although the surface pressure of this atmosphere is only 16 millionths of a bar, this is sufficient to support thin haze or cloud layers.

Triton's surface, like that of many other satellites in the outer solar system, reveals a long history of geological evolution (Figure 11.19). Although some impact craters are found, many regions have been flooded fairly recently by the local version of "lava" (perhaps water or water–ammonia mixtures). There are also mysterious regions of jumbled or mountainous terrain.

The Voyager flyby of Triton took place at a time when the satellite's southern pole was tipped toward the Sun, allowing this part of the surface to enjoy a period of relative warmth. (Remember that "warm" on Triton is still outrageously colder than anything we experience on Earth.) A polar cap covers much of Triton's southern hemisphere, apparently evaporating along the northern edge. This polar cap may consist of frozen nitrogen that was deposited during the previous winter. Remarkably, the Voyager images showed that the evaporation of this polar cap generates geysers or volcanic plumes of nitrogen gas. (Fountains of such gas rose about 10 km high, visible in the thin atmosphere because dust from the surface rose with them and

colored them dark.) These plumes differ from the volcanic plumes of Io in their composition and also in that they derive their energy from sunlight warming the surface rather than from internal heat.

11.3.2 Discovery of Pluto

Now we turn to Pluto, a small planet that resembles Triton in many ways and is quite different from the giant planets in the outer solar system. Pluto was discovered through a careful, systematic search, unlike Neptune, whose position was calculated from gravitational theory. Nevertheless, the history of the search for Pluto began with indications that Uranus and Neptune had slight departures from their predicted orbits (you may want to review the last section of Chapter 2 for more on what such departures mean). Early in the 20th century several astronomers—most notably Percival Lowell, then at the peak of his fame as an advocate of intelligent life on Mars (see Chapter 9)—became interested in this problem.

At the time Lowell made his calculations, Neptune had moved such a short distance since its discovery that its orbit could not be used effectively to search for perturbations by an unknown ninth planet. Therefore Lowell and his contemporaries based their calculations primarily on the tiny remaining irregularities in the motion of Uranus. Lowell's computations indicated two possible locations for a perturbing planet; the more likely of the two was in the constellation of Gemini. He predicted a mass for the planet intermediate between those of the Earth and Neptune (his calculations gave about 6.6 Earth masses). Other astronomers, however, obtained other solutions from the tiny orbital irregularities, even including one model that indicated two planets beyond Neptune.

At his Arizona observatory, Lowell searched without success for the unknown planet from 1906 until his death in 1916. Subsequently, his brother donated to the observatory a 33-cm, wide-field photographic telescope that could record a 12° x 14° area of the sky on a single photograph. The new camera went into operation in 1929.

In February 1930, a young observing assistant named Clyde Tombaugh (see the Voyagers in Astronomy box), comparing photographs he made on January 23 and 29 of that year, found an object whose motion appeared to be about right for a planet far beyond the orbit of Neptune (Figure 11.20). It was within 6° of the position Lowell had predicted for the unknown planet. The new planet was named for Pluto, the Roman god of the underworld, who dwelt in remote darkness, just like the new planet. The choice of this name, among hundreds suggested, was helped by the fact that the first two letters were Percival Lowell's initials.

Although at the time the discovery of Pluto appeared to be a vindication of gravitational theory similar to the triumph of Adams and Leverrier in predicting the position of Neptune, we now know that Lowell's calculations were wrong. When the mass of Pluto was finally measured, it was found to be much less than that of the Moon. Pluto could not possibly have exerted any measurable pull on either Uranus or Neptune. Astronomers are now convinced that the reported small anomalies in the motions of Uranus and Neptune are not, and never were, real.

11.3.3 Pluto's Motion and Satellite

Pluto's orbit has the highest inclination to the ecliptic (17°) of any planet, meaning that the plane of its orbit is significantly tilted relative to the plane in which the other planets move. Pluto also has the largest eccentricity (0.248) in our solar system, so that its orbit deviates quite a bit from a perfect circle. Its average distance from the Sun is 40 AU, or 5.9 billion km, but its perihelion distance is less than 4.5 billion km, within the orbit of Neptune. Even though the orbits of these two planets cross, there is no danger of collision because of the high inclination of Pluto's orbit. (By the way, many of the giant planets found orbiting other stars have an even greater orbital eccentricity than Pluto.)

Pluto completes its orbital revolution in a period of 248.6 years; since its discovery in 1930, it has traversed less than one quarter of its long path around the Sun. But almost 50 years after that discovery, astronomers found that it is not alone in its distant travels. In 1978, astronomers found that Pluto has a satellite, which was named Charon after the boatman who ferried the dead into the realm of Pluto in mythology. The exact nature of Charon's orbit was not confirmed until 1985, when the system had turned enough that the satellite and the planet began to go behind each other as seen from the Earth. These observations showed that the satellite was in a retrograde orbit and had a diameter of about 1200 km, more than half the size of Pluto itself. This makes Charon the satellite whose size is the largest fraction of its parent planet (Figure 11.21). Seen from Pluto, Charon would be as large as eight full moons on Earth.

Furthermore, Pluto orbits "on its side" like Uranus, and Charon circles in its equatorial plane around its rotation axis. The gravitational interaction between the two worlds has locked them in the tightest of embraces. Not only does Charon take the same time to rotate and revolve, thus keeping the same face toward Pluto, but this period—the month on Pluto—is the same length as Pluto's day. This means that if you were on the Charon-facing side of Pluto,

Lowell Observatory

■ **FIGURE 11.20**

Pluto's Motion Portions of the two photographs by which Clyde Tombaugh discovered Pluto in 1930. The left one was taken on January 23 and the right on January 29. Note that Pluto, indicated by an arrow, has moved among the stars during those six nights. If we hadn't put an arrow next to it, you probably would never have spotted the dot that moved.

Clyde Tombaugh: From the Farm to Fame

Clyde Tombaugh discovered Pluto when he was 24 years old, and his position as staff assistant at the Lowell Observatory was his first paying job. Tombaugh had been born on a farm in Illinois, but when he was 16 his family moved to Kansas. There, with his uncle's encouragement, he observed the sky through a telescope the family had ordered from the Sears catalog. Tombaugh later constructed a larger telescope on his own and devoted his nights (when he wasn't too tired from farm work) to making detailed sketches of the planets.

In 1928, after a hailstorm ruined the crop, Tombaugh decided he needed a job to help support his family. Although he had only a high school education, he thought of becoming a telescope builder. He sent his planet sketches to the Lowell Observatory, seeking advice about whether such a career choice was realistic. By a wonderful twist of fate, his query arrived just when the Lowell astronomers realized that a renewed search for a ninth planet would require a very patient and dedicated observer.

The large photographic plates (pieces of glass with photographic emulsion on them) that Tombaugh

Clyde Tombaugh at the Lowell Observatory at the time of his discovery of Pluto in 1930.

Lowell Observatory

was hired to take at night and search during the day contained an average of about 160,000 star images each. How to find Pluto among them? The technique involved taking two photographs about a week apart. During that week, a planet would move a tiny bit, while the stars remained in the same place relative to each other. A new instrument called a "blink comparator" could quickly alternate the two images in an eyepiece. The stars,

being in the same position on the two plates, would not appear to change as the two images were "blinked." But a moving object would appear to wiggle back and forth as the plates were alternated.

After examining more than 2 million stars (and many false alarms), Tombaugh found his planet on February 18, 1930. The astronomers at the observatory checked his results carefully, and the find was announced on March 13, the 149th anniversary of the discovery of Uranus. Congratulations and requests for interviews poured in from around the world. Visitors descended on the observatory by the scores, wanting to see the place where the first new planet in almost a century had been discovered, as well as the person who had discovered it.

In 1932 Tombaugh took leave from Lowell, where he had continued to search and blink, to get a college degree. Eventually he received a master's degree in astronomy and taught navigation for the Navy during World War II. In 1955, after working to develop a rocket-tracking telescope, he became a professor at New Mexico State University, where he helped found the astronomy department. He died in 1997.

■ **FIGURE 11.21**

Pluto and Charon An image of Pluto and its satellite Charon, taken with the Faint Object Camera on the Hubble Space Telescope. The clarity of the image allowed astronomers to measure the diameters of the two worlds to an accuracy of 1 percent.

A. Stern, M. Buie, NASA & ESA

you would see the moon hover in the same place all the time, while from Pluto's opposite hemisphere Charon would never be visible.

11.3.4 The Nature of Pluto

Pluto has not been visited by spacecraft, and it is so faint that the world's largest telescopes are required to study it. The diameter of Pluto is 2190 km, only 60 percent that of the Moon. From the diameter and mass, we find a density of 2.1 g/cm^3, suggesting that Pluto is a mixture of rocky materials and water ice in about the same proportions as in Triton.

Pluto's surface is highly reflective, and its spectrum demonstrates the presence of frozen methane, carbon monoxide, and nitrogen. The surface temperature ranges from about 50 K when Pluto is farthest from the Sun to 60 K when it's closest. Even this small difference is enough to cause a partial sublimation of the methane and nitrogen ice to generate an atmosphere when Pluto is close to the Sun. In fact, for the past few years Pluto has been in its warmest period, near perihelion, and its atmosphere is accordingly near maximum size and density. Observations of a distant star seen through that thin atmosphere suggest that the surface pressure is about a ten-thousandth of the Earth's. Because Pluto is a few degrees warmer than Triton, its atmospheric pressure is about ten times greater.

In February 1999, Pluto crossed the orbit of Neptune and became the ninth planet once more (it had been eighth from the Sun since 1979). Now it will slowly swing farther and farther from the Sun, and more and more of its atmosphere will slowly freeze out.

We have a long way to go to understand how Pluto arrived at its unusual place in the outer solar system. In some ways Pluto is like the odd cousin that everyone hopes will not show up at the next family reunion. Neither its path around the Sun nor its size resembles the jovian planets. Pluto is much more similar to a moon such as Triton than to a planet such as Saturn. We now know that Pluto is the largest example of a whole family of distant asteroid- or cometlike objects in what is called the Kuiper belt, which we will discuss in Chapter 12.

11.4 PLANETARY RINGS

All four of the giant planets have rings, with each ring system consisting of billions of small particles or moonlets orbiting close to their planet. Each of these rings displays a complicated structure that seems related to interactions between the ring particles and the larger satellites. However, the four ring systems are very different from each other in mass, structure, and composition (Table 11.2).

Saturn's large ring system is made up of icy particles spread out into several vast, flat rings containing a great deal of fine structure. The Uranus and Neptune ring systems, on the other hand, are nearly the reverse of Saturn's; they consist of dark particles confined to a few narrow rings with broad empty gaps in between. Jupiter's ring and at

TABLE 11.2 *Properties of Ring Systems*

Planet	Outer Radius (km)	Outer Radius (R_{planet})	Mass (kg)	Reflectivity (%)
Jupiter	128,000	1.8	10^{10}(?)	?
Saturn	140,000	2.3	10^{19}	60
Uranus	51,000	2.2	10^{14}	5
Neptune	63,000	2.5	10^{12}	5

least one of Saturn's are merely transient dust bands, constantly renewed by erosion of dust grains from small satellites. (Figure 11.22 is a diagram comparing the four ring systems.) In this section we focus on the two most massive ring systems, those of Saturn and Uranus.

11.4.1 What Causes Rings?

A ring is a collection of vast numbers of particles, each like a tiny moon obeying Kepler's laws as it follows its own orbit around the planet. Thus the inner particles revolve faster than those farther out, and the ring as a whole does not rotate as a solid body. In fact, it is better not to think of a ring rotating at all, but rather to consider the revolution of its individual moonlets.

If the ring particles were widely spaced, they would move independently, like separate small satellites. However, in the main rings of Saturn and Uranus the particles are close enough to each other to exert mutual gravitational influence, and occasionally even to rub together or bounce off each other in low-speed collisions. Because of these interactions, we see phenomena such as waves that move across the rings the way water waves move over the surface of the ocean.

There are two basic theories of how such rings come to be. First is the breakup hypothesis, which suggests that the rings are the remains of a shattered satellite. A passing comet or asteroid might have collided with the satellite, breaking it into pieces. Tidal forces then inhibited the fragments from coming together again, and they dispersed into a disk.

Ace **Astronomy**™ Log into AceAstronomy and select this chapter to see the Active Figure called "Roche Limit."

The second hypothesis, which takes the reverse perspective, suggests that the rings are made of particles that were unable to come together to form a satellite in the first place. In either theory, the gravity of the planet plays an important role. Close to the planet, tidal forces can tear bodies apart or inhibit loose particles from coming together. The rings of both Saturn and Uranus lie within the region in which large satellites are probably not stable against such forces (the edge of this region is sometimes called the *Roche limit*, after the French mathematician who first worked out the theory). We do not know which explanation holds for the rings, although many scientists have concluded that at least a few of the rings are relatively young and must therefore be the result of breakup.

■ FIGURE 11.22

Four Ring Systems A diagram showing the locations of the ring systems of the four giant planets. The left axis represents the planet's surface. The dotted vertical line is the limit inside which gravitational forces can break up satellites. Notice that each planet's system is drawn *to a different scale*, so that this stability limit lines up for all four of them. The black dots are the inner satellites of each planet on the same scale as its rings. Notice that only really small bodies survive inside the stability limit.

11.4.2 Rings of Saturn

The rings of Saturn are one of the most beautiful sights in the solar system (Figures 10.10 and 11.23). From outer to inner, the three brightest rings are labeled with the extremely romantic names of A, B, and C Rings. Table 11.3 gives the dimensions of the rings in both kilometers and units of the radius of Saturn, R_s. The B Ring is the brightest and has the most closely packed particles, whereas the A and C Rings are translucent. The total mass of the B Ring, which is probably close to the mass of the entire ring system, is about equal to that of an icy satellite 250 km in diameter (thus it could have originated in the breakup of such a satellite). Between the A and B Rings is a wide gap named the Cassini Division after Gian Domenico Cassini, who first glimpsed it in 1675.

The rings of Saturn are very broad and very thin. The width of the main rings is 70,000 km, yet their thickness is only about 20 m. If we made a scale model of the rings out of paper the thickness of the pages in this book, we would have to make them 1 km across—about eight city blocks. On this scale, Saturn itself would loom as high as an 80-story building.

The ring particles are composed primarily of water ice, and they range from grains the size of sand up to house-sized boulders. An insider's view of the rings would probably resemble a bright cloud of floating snowflakes and hailstones, with a few snowballs and larger objects, many of these loose aggregates of smaller particles (Figure 11.24).

In addition to the broad A, B, and C Rings, Saturn has a handful of very narrow rings no more than 100 km wide. The most substantial of these, which lies just outside the A Ring, is called the F Ring; its surprising appearance is discussed below. In general, Saturn's narrow rings resemble the rings of Uranus and Neptune.

TABLE 11.3 *Rings of Saturn*

| Ring Name | Outer Edge | | Width |
	(R_s)	(km)	(km)
F	2.324	140,180	90
A	2.267	136,780	14,600
Cassini Division	2.025	122,170	4,590
B	1.949	117,580	25,580
C	1.525	92,000	17,490

NASA/JPL

■ FIGURE 11.23

Saturn's Rings from Underneath Since you are looking up at sunlight filtering through the rings in this Voyager image, the brightest parts of the ring system are the gaps and the least dense regions. The denser B Ring is not transmitting much light.

NASA/JPL

■ FIGURE 11.24

Ring Pieces An artist's idealized version of the sizes of typical particles inside the A Ring of Saturn. The ball at the center is roughly the size of an adult's head. In the real rings, the particles are not spherical but are likely to have irregular shapes.

11.4.3 Rings of Uranus and Neptune

The rings of Uranus are narrow and black, making them almost invisible from the Earth. The nine main rings were discovered in 1977 from observations made of a star as Uranus passed in front of it. We call the passage of one astronomical object in front of another an *occultation*. During the 1977 occultation, astronomers expected the star's light to disappear as the planet moved across it. But in addition the star dimmed briefly several times before Uranus reached it, as each narrow ring passed between the star and the telescope. Thus the rings were mapped out in detail even though they could not be seen or photographed directly, like counting the number of cars in a train at night by watching the blinking of a light as the cars successively pass in front of it. When Voyager approached Uranus in 1986, it was able to study the rings at close range; the spacecraft also photographed two new rings (Figure 11.25).

The outermost and most massive of the rings of Uranus is called the Epsilon Ring. It is only about 100 km wide and probably no more than 100 m thick (similar to the F Ring of Saturn). The Epsilon Ring encircles Uranus at a distance of 51,000 km, about twice the radius of Uranus. This ring probably contains as much mass as all of Uranus' other ten rings combined; most of them are narrow ribbons less than 10 km wide—just the reverse of the broad rings of Saturn.

The individual particles in the uranian rings are nearly as black as lumps of coal. While astronomers do not understand the composition of this material in detail, it seems to consist in large part of black carbon and hydrocarbon compounds. Organic material of this sort is rather common in the outer solar system. Many of the asteroids and comets (discussed in Chapter 12) also are composed of dark, tarlike materials. In the case of Uranus, its ten small inner satel-

NASA/JPL

■ FIGURE 11.25

The Narrow Rings of Uranus The Voyager team had to expose this image for a long time to get a glimpse of these dark rings; you can see the grainy structure of "noise" in the electronics of the camera in the picture background.

Ace⊙Astronomy™ Log into AceAstronomy and select this chapter to see the Active Figure called "Uranus's Ring Detection."

lites have a similar composition, suggesting that one or more satellites might have broken up to make the rings.

The rings of Neptune are generally similar to those of Uranus but even more tenuous (Figure 11.26). There are only four of them, and the particles are not uniformly distributed along their lengths. Because these rings are so difficult to investigate from the Earth, it will probably be a long time before we understand them very well.

11.4.4 Interactions Between Rings and Satellites

Much of our fascination with planetary rings is a result of their intricate structures, most of which owe their existence to the gravitational effect of satellites. Without satellites the rings would be flat and featureless. Indeed, it is becoming clear that without satellites there would probably be no rings at all because, left to themselves, thin disks of matter gradually spread and dissipate.

Most of the gaps in Saturn's rings, and also the location of the outer edge of the A Ring, result from gravitational

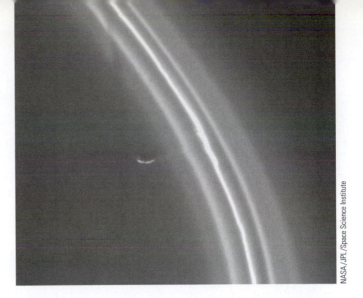

NASA/JPL/Space Science Institute

■ FIGURE 11.27

Saturn's Complex F Ring   Saturn's shepherd moon Prometheus (100 km long) glides along next to the multiple strands of the F Ring in this close-up view from Cassini.

JPL/NASA

■ FIGURE 11.26

Neptune's Rings   A long exposure of the rings of Neptune, as photographed by Voyager in 1989. Note the two denser regions of the outer ring.

resonances with small inner satellites. A **resonance** takes place when two objects have orbital periods that are exact ratios of each other, such as one-to-two or one-to-three. For example, any particle in the gap at the inner side of the Cassini Division of Saturn's rings would have a period equal to one-half that of Saturn's satellite Mimas. Such a particle would be nearest Mimas in the same part of its orbit every second revolution. The repeated gravitational tugs of Mimas, acting always in the same direction, would perturb it, forcing it into a new orbit outside the gap and therefore no longer representing a resonance. In this way, the Cassini Division became depleted of ring material over long periods of time.

One of the most interesting rings of Saturn is the narrow F Ring, which contains several apparent ringlets within its 90-km width. In places the F Ring breaks up into two or three parallel strands that sometimes show bends or kinks. Most of the rings of Uranus and Neptune are also narrow ribbons like the F Ring of Saturn. Clearly, the gravity of some objects must be keeping the particles in these rings from spreading out.

The best hypothesis is that the rings are controlled gravitationally by small satellites orbiting very close to them. This certainly seems to be the case for Saturn's F Ring: Voyager showed that it is bounded by the orbits of two satellites, now called Pandora and Prometheus (Figure 11.27). These two small objects (each about 100 km in diameter) are referred to as *shepherd satellites,* since their gravitation serves to "shepherd" the ring particles and keep them confined to a narrow ribbon. A similar situation applies to the Epsilon Ring of Uranus, which is shepherded by the satellites Cordelia and Ophelia. These two shepherds, each about 50 km in diameter, orbit about 2000 km inside and outside the ring.

Theoretical calculations suggest that the other narrow rings in the uranian and neptunian systems should also be controlled by shepherd satellites, but none has been located. The calculated diameter for such shepherds—about 10 km—was just at the limit of detectability for the Voyager cameras, so it is impossible to say whether they are present or not. (Given all the narrow rings we see, many scientists still hope to find another, more satisfactory mechanism for keeping them confined.)

Just recently, astronomers have found a relatively narrow ring of dust around another star (called HR 4796A) and have applied the same kind of analysis to suggest that there must be shepherding *planets* on either side of the ring to keep it so narrow. (At the distance of the star, such planets cannot be seen with our present telescopes—the same situation we face for the smaller shepherd moons around the outer planets.) This is a wonderful example of how techniques developed in one branch of astronomy often find application in another.

The main rings of Saturn include several narrow gaps that are the result of satellites, which clear lanes in a manner similar to that of the shepherd satellites. As each small satellite moves through its gap, it produces waves in the surrounding ring material like the wake left by a moving ship. One of these satellites, called Pan, was discovered in 1991 from its wake; calculations based on the waves it produced pinpointed its location, and when the appropriate ten-year-old Voyager picture was examined, there it was!

NOTE: Many of the sites about planets and planetary missions listed for Chapters 6 and 10 also include good information about the satellites of the planets. For example, the Nine Planets site has separate sections for many of the main moons in the solar system.

🖥 *Galileo Mission Moons Page:*
www.jpl.nasa.gov/moons/moons.html
A fine summary of what the Galileo spacecraft found out about the Galilean moons, with some of the great images from the mission.

🖥 *General Pluto sites:*

- *The Nine Planets Site:*
  seds.lpl.arizona.edu/nineplanets/nineplanets/pluto.html

- *Views of the Solar System Site:*
  www.solarviews.com/eng/pluto.htm

- *Planetary Sciences Site:*
  nssdc.gsfc.nasa.gov/planetary/planets/plutopage.html

🖥 *The Clyde Tombaugh Web Site:*
www.klx.com/clyde/index.html
The discoverer of Pluto passed away in January 1997 at age 90. Jeff Faust at MIT has brought together articles on Tombaugh and his work, plus information about Pluto.

🖥 *Pluto Home Page:* dosxx.colorado.edu/plutohome.html
A site from astronomer Fran Bagenal at the University of Colorado with good images, diagrams (with captions), brief information, and links about Pluto.

🖥 *Pluto Portal:* www.plutoportal.net
Another good site with information about Pluto.

🖥 *New Horizons Mission:* pluto.jhuapl.edu
A site describing a planned mission to Pluto.

🖥 *The Planetary Rings Node:* ringside.arc.nasa.gov/
Mark Showalter at NASA's Ames Research Center maintains this somewhat technical site devoted to the rings. It includes lots of images, movies, links, and information.

SUMMARY

11.1 The four jovian planets are accompanied by impressive systems of satellites and rings. More than 100 satellites have been discovered in the outer solar system. Most of these satellites are composed in part of water ice, with varying amounts of rock and dark primitive material mixed in. Of the four ring systems, Saturn's is the largest; Uranus and Neptune have narrow rings of dark material, and Jupiter has a tenuous ring of dust.

11.2 The largest satellites are Ganymede, Callisto, and Titan, all low-density objects that are composed of more than half water ice. Callisto has an ancient cratered surface, while Ganymede shows evidence of extensive tectonic and volcanic activity, persisting until perhaps a billion years ago. Voyager was not able to see Titan's surface because it is unique among jovian-planet moons in having a thick atmosphere (surface pressure of 1.6 bar) with an opaque haze layer. Io and Europa are denser and smaller, each about the size of our Moon. Europa is the only other object in the solar system thought to have a global ocean of liquid water (under an ice crust). Io is the most volcanically active object in the solar system; its eruptions are powered by tidal interactions with Jupiter.

11.3 Triton, Neptune's retrograde satellite, and Pluto resemble each other in size, composition, and temperature. Triton has a very thin atmosphere, and we can see eruptions of nitrogen when surface layers of its extensive polar cap evaporate while facing the Sun. The planet Pluto, discovered by Clyde Tombaugh in 1930, has never been visited by spacecraft. Observations indicate the presence of frozen methane, carbon monoxide, and nitrogen on its surface, and a thin (perhaps temporary) atmosphere. Pluto's moon Charon has a diameter more than half of Pluto's, and its and Pluto's motions are gravitationally locked.

11.4 Rings are composed of vast numbers of individual particles (ranging in size) orbiting so close to a planet that its gravitational forces could have broken larger pieces apart or kept pieces from gathering together. Saturn's rings are broad, flat, and nearly continuous, except for a handful of gaps. The particles are mostly water ice, with typical dimensions of a few centimeters. The rings of Uranus are narrow ribbons separated by wide gaps and contain much less mass. They were discovered in 1977 during a stellar occultation. Neptune's rings are similar but contain even less material. Much of the complex structure of the rings is due to waves and **resonances** induced by satellites within the rings or orbiting outside them. Voyager found that several thin rings were accompanied by shepherd moons that kept the ring material confined.

INTER-ACTIVITY

A Imagine it's the distant future and humans can now travel easily among the planets. Your group is a travel agency, with the task of designing a really challenging tour of the Galilean satellites for a group of sports enthusiasts. What kind of activities are possible on each world? How would rock climbing on Ganymede, for example, differ from rock climbing on Earth? (If you design an activity for Io, you had better bring along very strong radiation shielding. Why?)

B In the same spirit as activity A, have your agency design a tour that includes the seven most spectacular sights among any and all of the satellites or rings covered in this chapter. What are the "not to be missed" destinations that future tourists will want to visit and why? Which of the sights you pick are going to be spectacular if you are on the moon's surface or inside the ring, and which would look interesting only from far away in space?

C In this chapter we could cover only a few of the dozens of satellites in the outer solar system. Using the Web (see the Surfing the Web sections in this and the previous chapters) or your college library, organize your group into a research team and find out more about one of the moons we did not cover in any detail. Our favorites include Uranus' Miranda with its jigsaw puzzle surface, Saturn's Mimas with a "knockout" crater called Herschel, and Saturn's Iapetus whose two hemispheres differ significantly. Prepare a report to attract tourists to the world you selected.

D In a novel entitled *2010*, science fiction writer Arthur C. Clarke, inspired by the information coming back from the Voyager spacecraft, had fun proposing a life-form under the ice of Europa that was evolving toward intelligence. Suppose future missions do indeed find some sort of life (not necessarily intelligent but definitely alive) under the ice of Europa—life that evolved completely independently from life on Earth. Have your group discuss what effect such a discovery would have on humanity's view of itself. What should be our attitude toward such a life-form? Do we have an obligation to guard it against contamination by our microbes and viruses? Or, to take an extreme position, should we wipe it out before it becomes competitive with Earth life or contaminates our explorers with microorganisms we are not prepared to deal with? Who should be in charge of making such decisions?

E Work together to make a list of all the reasons it is hard to send a mission to Pluto. How would you design a mission that will get there despite your list of difficulties?

F Your group has been asked by NASA to come up with one or more missions to learn about Europa. Review what we know about this satellite so far and then design a robotic mission that would answer some of the questions we have. You can assume that budget is not a factor, but your instruments have to be realistic. (Bear in mind that Europa is cold and far from the Sun.)

G Imagine your group is the first landing party on Pluto (let's hope you remembered to bring long underwear!). You land in a place where Charon is visible in the sky and you observe Charon for one Earth week. Describe what Charon will look like during that week. Now you move your camp to the opposite hemisphere of Pluto. What will Charon look like there during the course of a week?

REVIEW QUESTIONS

Ace⬤Astronomy™ Assess your understanding of this chapter's topics with additional quizzing and animations at **http://ace.brookscole.com/voyages**

1. What are the satellites of the outer planets made of, and why is their composition different from that of the Moon?
2. Compare the geology of Callisto, Ganymede, and Titan. Why does Titan have an atmosphere while the other two large satellites do not?
3. What is the evidence for a liquid water ocean on Europa, and why is this interesting to scientists searching for extraterrestrial life?
4. Explain the energy source that powers the volcanoes of Io.
5. Compare the properties of Titan's atmosphere with those of the Earth's atmosphere.
6. How was Pluto discovered? Why did it take so long to find it?
7. How are Triton and Pluto similar?
8. Describe and compare the rings of Saturn and Uranus, including their possible origins.

9. Why do you think the outer planets have such extensive systems of rings and satellites, while the inner planets do not?

10. Which would have the longer orbital period: a satellite 1 million km from the center of Jupiter, or a satellite 1 million km from the center of Earth? Why?

11. Ganymede and Callisto were the first icy objects to be studied from a geological point of view. Summarize the main differences between their geology and that of the rocky terrestrial planets.

12. Compare the properties of the volcanoes on Io with those of terrestrial volcanoes. Give at least two similarities and two differences.

13. Would you expect to find more impact craters on Io or Callisto? Why?

14. Why is it unlikely that humans will be traveling to Io? (*Hint:* Review the information about Jupiter's magnetosphere in Chapter 10.)

15. Where did the nitrogen in Titan's atmosphere come from? Compare its origin with that of the nitrogen in our atmosphere.

16. Do you think there are many impact craters on the surface of Titan? Why or why not?

17. Why do you suppose the rings of Saturn are made of bright particles, whereas the particles in the rings of Uranus and Neptune are black?

18. Suppose you miraculously removed all of Saturn's satellites. What would happen to its rings?

FIGURING FOR YOURSELF

Using the occultations of stars by the rings of Uranus, astronomers have been able to measure details in the ring structure to a resolution of 10 km. This is a much higher resolution than can be obtained in a conventional photo of the rings. Let's figure out what angular resolution (in arcsec) a space telescope in Earth orbit would have to achieve to obtain equal resolution. To solve this problem, we use the "small-angle formula" to relate angular and linear diameter in the sky. For angles in the sky that are small, the formula is usually written as

$$\frac{\text{angular diameter}}{206{,}265} = \frac{\text{linear diameter}}{\text{distance}}$$

where angular diameter is expressed in arcsec. The distance of Saturn near opposition is about 9 AU = 1.4×10^9 km. Substituting in the above formula and solving for the angular resolution, we get $(206{,}265 \times 10)/1.4 \times 10^9$, which is about 10^{-3} arcsec, or a milliarcsec. This is not possible for our telescopes to achieve. For comparison, the best resolution from either the HST or ground-based telescopes is about 0.1 arcsec, or 100 times worse than what we would need. This is why such occultation measurements are so useful for astronomers.

19. How close to Uranus would a spacecraft have to get to obtain the same resolution as in our example above with a camera that has an angular resolution of 2 arcsec?

20. How close to Saturn would a spacecraft have to be to make out detail in its rings as small as 20 km, if its camera has an angular resolution of 5 arcsec?

21. If the HST has an angular resolution of 0.1 arcsec and Pluto is at its closest distance to the Sun, what is the smallest detail the telescope could make out on its surface? Does this tell you why astronomers really hope NASA or some other space agency will send a spacecraft to the ninth planet?

22. Saturn's A, B, and C Rings extend about 75,000 to 137,000 km from the center of the planet. Use Kepler's third law to calculate the difference between how long the inner edge and the outer edge of the three-ring system take to revolve about the planet.

23. Use the information in Appendix 8 to calculate what you would weigh on Titan, Io, and Uranus' satellite Miranda.

SUGGESTIONS FOR FURTHER READING

On Satellites

Carroll, M. "Europa: Distant Ocean, Hidden Life?" in *Sky & Telescope*, Dec. 1997, p. 50.

Elliot, J. "The Warming Wisps of Triton" in *Sky & Telescope*, Feb. 1999, p. 42.

Harland, D. M. *Jupiter Odyssey: The Story of NASA's Galileo Mission;* 2000, Springer-Praxis. Good mission overview.

Johnson, T. "The Galileo Mission to Jupiter and Its Moons" in *Scientific American*, Feb. 2000, p. 40. Ten-page review of mission results.

Johnson, T., et al. "The Moons of Uranus" in *Scientific American*, Apr. 1987.

Lorenz, R. and Mitton, J. *Lifting Titan's Veil: Exploring the Giant Moon of Saturn.* 2002, Cambridge U. Press.

Mackenzie, D. "Is There Life Under the Ice?" in *Astronomy*, Aug. 2001, p. 32. On future exploration of Europa.

Rothery, D. *Satellites of the Outer Planets: Worlds in Their Own Right.* 1999, Oxford U. Press.

Spencer, J. "Galileo's Closest Look at Io" in *Sky & Telescope*, May 2001, p. 40.

Talcott, R. "The Violent Volcanoes of Io" in *Astronomy*, May 1993, p. 41.

On Pluto

Binzel, R. "Pluto" in *Scientific American*, June 1990.

Burnham, R. "At the Edge of Night: Pluto and Charon" in *Astronomy*, Jan. 1994, p. 41.

Harrington, R. and Harrington, B. "The Discovery of Pluto's Moon" in *Mercury*, Jan./Feb. 1979, p. 1.

Hoyt, W. *Planets X and Pluto.* 1980, U. of Arizona Press. Superb history of the search for Pluto and other outer planets.

Levy, D. *Clyde Tombaugh: Discoverer of Pluto.* 1991, U. of Arizona Press.

Stern, S. "Journey to the Farthest Planet" in *Scientific American*, May 2002, p. 56. On possible missions to Pluto and the Kuiper belt.

Stern, S. and Mitton, J. *Pluto and Charon.* 1998, Wiley.

Tombaugh, C. "The Discovery of Pluto" in *Mercury*, May/June 1986, p. 66, and July/Aug. 1986, p. 98.

On Rings

Burns, J., et al. "Bejeweled Worlds" in *Scientific American*, Feb. 2002, p. 64. On rings throughout the solar system.

Elliot, J., et al. "Discovering the Rings of Uranus" in *Sky & Telescope*, June 1977, p. 412.

Esposito, L. "The Changing Shape of Planetary Rings" in *Astronomy*, Sept. 1987, p. 6.

Sobel, D. "Secrets of the Rings" in *Discover*, Apr. 1994, p. 86. Discusses the outer planet ring systems.

Comet Hale–Bopp was one of the most attractive and easily visible comets of the 20th century. It is shown here as it appeared in the sky on March 8, 1997, near the much more distant North America Nebula (the reddish glowing cloud of gas). You can see the comet's long blue ion tail and the shorter white dust tail.

Comets and Asteroids: Debris of the Solar System

The day will yet come when posterity will be amazed that we remained ignorant of things that will to them seem so plain. . . . Men will some-day be able to demonstrate in what regions comets have their paths [and] what is their size and constitution. Let us be satisfied with what we have discovered, and leave a little truth for our descendants to find out.

Lucius Annaeus Seneca, in *Natural Questions* (first century)

THINKING AHEAD

Hundreds of smaller members of the solar system, called asteroids and comets, are known to have crossed the Earth's orbit in the past, and many others will do so in the centuries ahead. What could we do if we knew a few years in advance that one of these bodies (with significant mass) would hit the Earth?

To understand the early history of life on Earth, scientists need to study ancient fossils. To reconstruct the early history of the solar system, we need cosmic fossils—material that formed when our system was very young. When it comes to answering our questions about origins, the planets themselves are largely mute. Melted, battered by giant impacts, twisted by tectonic forces, they retain little evidence of their births. Reconstructing their early history is almost as difficult as determining the circumstances of human birth by merely looking at an adult. Instead, we must turn to clues provided by the surviving remnants of the creation process—ancient but smaller objects with which we share our cosmic neighborhood.

Astronomers have rather arbitrarily divided these objects into two categories. **Asteroids** are rocky or metallic and contain little volatile (easily evaporated) material. They differ from the terrestrial planets primarily in size and are sometimes even called minor planets. **Comets** are icy objects that contain frozen water and other volatile materials (but with solid grains mixed in). Comets are most easily noticed when they get near the Sun and form a visible

Virtual Laboratories

 Kuiper Belt Objects

(but temporary) atmosphere and an extended tail of gas and dust.

12.1 ASTEROIDS

The asteroids are mostly found in the broad space between Mars and Jupiter, a region of the solar system called the *asteroid belt*. Asteroids are too small to be seen without a telescope; the first of them was not discovered until the beginning of the 19th century.

12.1.1 Discovery and Orbits of the Asteroids

In the late 1700s, many astronomers were hunting for an additional planet they thought should exist in the gap between the orbits of Mars and Jupiter. Some German astronomers even organized themselves into a group they called "the celestial police" to search for this "missing" member of the solar system. However, it was another astronomer who discovered the first asteroid as a by-product of his search for new comets.

In January 1801, the Sicilian astronomer Giovanni Piazzi initially thought he had found this missing planet when he discovered the first asteroid, which he named Ceres, orbiting at 2.8 astronomical units (AU) from the Sun. His discovery was followed the next year by that of another little planet in a similar orbit, and two more were found in 1804 and 1807. Clearly, there was not a single missing planet between Mars and Jupiter but rather a whole group of objects, each no more than 1000 km across.

By 1890 more than 300 had been discovered by sharp-eyed observers. In that year, Max Wolf of Heidelberg introduced astronomical photography to the search for asteroids, greatly accelerating the discovery of additional objects (Figure 12.1). More than 20,000 asteroids now have well-determined orbits. The pace of discovery is so fast that, although it required 197 years to find the first 10,000 asteroids, only 3 years were needed to double this number to 20,000.

Asteroids are given both a number (corresponding to the order of discovery) and a name. Originally, the names of asteroids were chosen from goddesses in Greek or Roman mythology. After exhausting these and other female names (including, later, those of wives, friends, flowers, cities, and others), astronomers turned to the names of colleagues (and other people of distinction) whom they wished to honor. For example, asteroids 2410 and 4859 are named Morrison and Fraknoi, for two authors of this text.

It would be a formidable task to discover, determine orbits for, and catalog all the asteroids bright enough to be photographed with modern telescopes. Nevertheless, the total number of such objects can be estimated by systematically sampling regions of the sky. These studies indicate that approximately a million (10^6) asteroids have diameters larger than 1 km.

■ **FIGURE 12.1**

Asteroid Trails Time exposure showing trails left by asteroids (arrows). The telescope moved backward to compensate for the rotation of the Earth; thus the stars remained fixed. But in the time the image was taken, the asteroids moved relative to the stars and left a trail.

The largest asteroid is Ceres, with a diameter just less than 1000 km. Two, Pallas and Vesta, have diameters of about 500 km, and about 15 more are larger than 250 km (Table 12.1). The number of asteroids increases rapidly with decreasing size; there are about 100 times more objects 10 km across than there are 100 km across. The total mass in all the asteroids, which probably represent just a tiny fraction of the original asteroid population, is less than that of the Moon. The orbits of the four largest asteroids were illustrated in Figure 2.10.

TABLE 12.1 *The Largest Asteroids*

Name	Year of Discovery	Semimajor Axis (AU)	Diameter (km)	Class[*]
Ceres	1801	2.77	940	C
Pallas	1802	2.77	540	C
Vesta	1807	2.36	510	[†]
Hygeia	1849	3.14	410	C
Interamnia	1910	3.06	310	C
Davida	1903	3.18	310	C
Cybele	1861	3.43	280	C
Europa	1868	3.10	280	C
Sylvia	1866	3.48	275	C
Juno	1804	2.67	265	S
Psyche	1852	2.92	265	M
Patientia	1899	3.07	260	C
Euphrosyne	1854	3.15	250	C

[*] C = carbonaceous; S = stony; M = metallic.
[†] Vesta has a very unusual (once thought unique) basaltic surface.

■ **FIGURE 12.2**

Asteroid Positions The positions of more than 6000 asteroids in February 1990, seen from above. Also shown are the planets Earth, Mars, and Jupiter. Notice the two groups of Trojan asteroids in approximately the same orbit as Jupiter.

The asteroids all revolve about the Sun in the same west-to-east direction as the planets, and most of their orbits lie near the plane in which the Earth and other planets circle. The majority of asteroids lie in a zone we call the **asteroid belt,** defined as the region that contains all asteroids with semimajor axes in the range 2.2 to 3.3 AU (Figure 12.2). (See Chapter 2 if you need a reminder of the terms that apply to orbits.) Asteroids at these distances take 3.3 to 6 years to orbit the Sun. Although more than 75 percent of the known asteroids are in the main belt, they are not closely spaced. The volume of the belt is actually very large, and the typical spacing between objects (down to 1 km in size) is several million kilometers. (This is fortunate for spacecraft like Galileo and Cassini that need to travel through the belt without a collision.)

Still, over the long history of our solar system, there must have been a good number of collisions among the asteroids themselves. In 1917, the Japanese astronomer Kiyotsuga Hirayama found that some asteroids fall into *families,* groups with similar orbital characteristics. He hypothesized that each family may have resulted from a breakup of a larger body or, more likely, from the collision of two asteroids. Slight differences in the speeds with which the various fragments left the collision scene account for the small spread in orbits now observed for the different asteroids in a given family. Several dozen such families exist, and observations have shown that individual members of the larger ones are physically similar, just as we would expect if they were fragments of a common parent.

12.1.2 Composition and Classification

Asteroids are as different as black and white. The majority are very dark, with reflectivities of only 3 to 4 percent, like a lump of coal. However, another large group has typical reflectivities of 15 to 20 percent, a little greater than that of the Moon, and still others have reflectivities as high as 60 percent. To understand more about these differences and how they are related to chemical composition, astronomers study the spectra of the light reflected from asteroids for clues about their composition.

The dark asteroids are revealed from spectral studies to be *primitive* bodies (chemically little changed since the beginning of the solar system) composed of silicates mixed with dark, organic carbon compounds. Two of the largest asteroids, Ceres and Pallas, are primitive, as are almost all of the objects in the outer third of the belt. Most of the primitive asteroids are classed as C asteroids, where C stands for carbonaceous or carbon-rich.

The second most populous asteroid group is the S asteroids, where S stands for a stony or silicate composition. Here the dark carbon compounds are missing, resulting in higher reflectivity and clearer spectral signatures of silicate minerals. At first astronomers wondered if this was evidence of subsequent heating or chemical alteration. However, it appears that most of the S-type asteroids are also primitive; their different composition indicates they were formed in a different location from the C-type asteroids.

Asteroids of a third class, much less numerous than those of the first two, are composed primarily of metal and are called M asteroids. Spectroscopically, the identification of metal is difficult, but for at least the largest M asteroid, Psyche, this identification has been confirmed by radar. Since a metal asteroid, like an airplane or ship, is a much better reflector of radar than is a stony object, Psyche appears bright when we aim a radar beam at it.

How did such metal asteroids come to be? We suspect that each came from a parent body large enough for its molten interior to settle out or differentiate. The heavier metals sank to the center. When this parent body shattered in a later collision, the fragments from the core were rich in metals. There is enough metal in even a 1-km M-type asteroid to supply the world with iron and most other industrial metals for the foreseeable future, if only we could bring one safely to Earth.

In addition to the M asteroids, a few other asteroids show signs of early heating and differentiation. These have basaltic surfaces like the volcanic plains of the Moon and Mars; the large asteroid Vesta (discussed in a moment) is in the latter category. Why only a small percentage of the total number of asteroids have these characteristics, we do not know.

The different classes of asteroids are grouped together at different distances from the Sun (Figure 12.3). By tracing how their compositions vary with distance from the Sun, we can reconstruct some of the properties of the solar nebula from which they originally formed.

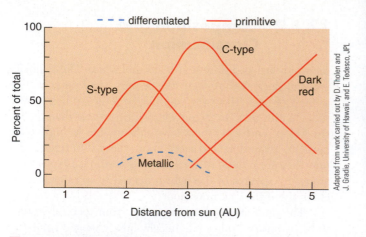

Adapted from work carried out by D. Tholen and J. Gradie, University of Hawaii, and E. Tedesco, JPL

■ **FIGURE 12.3**

Where Different Types of Asteroids Are Found
Asteroids of different composition are distributed at different distances from the Sun. The S type and C type are both primitive; the M type consists of cores of differentiated parent bodies.

NASA

■ **FIGURE 12.4**

A Piece of Vesta   Photograph of one of the meteorites believed to be a volcanic fragment from the crust of asteroid Vesta.

12.1.3   Vesta: A Volcanic Asteroid

Vesta is one of the most interesting of the asteroids. It orbits the Sun with a semimajor axis of 2.4 AU in the inner part of the main belt. Its relatively high reflectivity of almost 30 percent makes it the brightest of the main-belt objects, visible to the unaided eye if you know just where to look. But its real claim to fame is that its surface is covered with basalt, indicating that Vesta was once volcanically active in spite of its small size (about 500 km in diameter).

You may be surprised to learn that samples of Vesta's surface are available for direct study in the laboratory (and we thus know a great deal about this asteroid). Scientists have long suspected that *meteorites*—chunks of solar system material that survived their trip to the Earth's surface (see Chapter 13)—come from the asteroids, but there is generally no way to identify the particular source of a given meteorite that strikes the Earth. However, because of Vesta's unusual surface composition, the identification of its fragments seems fairly firm. (As we discussed in Chapter 9, we are also able to determine which meteorites come from Mars because we find pockets of trapped Mars gases in them.)

The meteorites believed to come from Vesta are a group of about 30 basaltic meteorites very similar in composition (Figure 12.4). Chemical analysis has shown that they cannot have come from the Earth, Moon, or Mars. On the other hand, their spectra (measured in the laboratory) perfectly match the spectrum of Vesta obtained telescopically. The age of the lava flows from which these meteorites derived has been measured at 4.4 to 4.5 billion years, very soon after the formation of the solar system. This age is consistent with what we might expect for Vesta; whatever process heated such a small object was probably intense and short-lived.

12.1.4   Asteroids Up Close

On the way to its 1995 encounter with Jupiter, the Galileo spacecraft was targeted to fly close to two main-belt asteroids called Gaspra and Ida. Gaspra, the first target, is an S-type asteroid. Astronomers knew from variations in its brightness that it had a rotation period of 7 hours. (Its name, incidentally, comes from a resort where 19th-century Russians enjoyed vacationing; the U.S. equivalent would be to call an asteroid Niagara or Mazatlan.) The Galileo camera revealed that Gaspra is 16 km long and highly irregular, as befits a fragment from a catastrophic collision (Figure 12.5).

The detailed images have allowed us to count the craters on Gaspra and to estimate the length of time its surface has been exposed to collisions. From the rather sparse number of craters, the Galileo scientists have concluded that Gaspra is only about 200 million years old (that is, the collision that formed Gaspra and other members of its family took place about 200 million years ago). Calculations suggest that an asteroid the size of Gaspra can expect another catastrophic collision sometime in the next billion years, at which time it will be disrupted to form another generation of still-smaller fragments.

The second target was Ida, a larger S-type asteroid 56 km long. The greatest surprise of the Galileo flyby of Ida, and the most important result scientifically, was the discovery of a satellite—named Dactyl—in orbit about the asteroid (Figure 12.6). Although only 1.5 km in diameter, smaller than many college campuses, Dactyl provides scientists with something otherwise beyond their reach—a measurement of the mass and density of Ida using Kepler's laws. The satellite's distance of about 100 km and its orbital period of about 24 hours indicate that Ida has a density of approximately 2.5 g/cm³, which matches the density of primitive rocks. More recently, both large visible-light telescopes and high-powered planetary radar have discovered other asteroidal satellites, so that we are now able to accumulate valuable data on asteroid masses and densities.

■ **FIGURE 12.5**

Mathilde, Gaspra, and Ida   The first three asteroids photographed from spacecraft flybys, printed to the same scale. Gaspra and Ida are S type and were investigated by the Galileo spacecraft; Mathilde is C type and was a flyby target for the NEAR-Shoemaker spacecraft.

JPL/NASA/Johns Hopkins

Mathilde   Gaspra   Ida

We might mention that Phobos and Deimos, the two satellites of Mars, are probably captured asteroids. They were first studied at close range by the Viking orbiters in 1977 and later by the Mars Global Surveyor. Both are irregular, somewhat elongated, and heavily cratered. Their largest dimensions are about 25 km and 13 km, respectively.

In 1996, NASA launched the Near Earth Asteroid Rendezvous (NEAR) spacecraft, designed to match orbit with an asteroid rather than merely flying past at high speed. On the way to Eros, its prime target (see Section 12.2.4), the NEAR spacecraft flew past a much larger C-type asteroid in the main belt, called Mathilde. It passed so close that the mass of the asteroid could be measured without a satellite, just from its influence on the path of the spacecraft. To our surprise, the density turned out to be very low—about 1.6 g/cm^3.

Since no rock has a density this low, Mathilde must be a loosely bound collection of smaller rocks with substantial spaces between them—a sort of orbiting gravel bank. Another clue came from the fact that Mathilde has huge craters (as you can see in Figure 12.5). The impacts that formed them would have shattered a denser body, but a porous collection of pieces might well have survived such hits.

NASA/JPL

■ **FIGURE 12.6**

Ida and Dactyl   Asteroid Ida and its moon Dactyl photographed by the Galileo spacecraft in 1993. Irregularly shaped Ida is 56 km in its longest dimension, while Dactyl is about 1.5 km across.

12.2 ASTEROIDS FAR AND NEAR

Not all asteroids are in the main asteroid belt. In this section we consider some special groups of asteroids that stray outside the belt's boundaries.

12.2.1 The Trojans

The Trojan asteroids are located far beyond the main belt, orbiting the Sun at 5.2 AU, about the same distance as Jupiter. The gravity of the giant planet makes most orbits of asteroids near it unstable, but calculations first carried out in the 18th century show that there are two points in the or-

bit of Jupiter near which an asteroid can remain almost indefinitely. These two points make equilateral triangles with Jupiter and the Sun (Figure 12.7). Between 1906 and 1908, four such asteroids were found; the number has now increased to several hundred. These asteroids are named for heroes from Homer's epic poem, the *Iliad,* and are collectively called the Trojans.

Measurements of the reflectivities and spectra of the Trojans show that they are dark, primitive objects like those in the outer part of the asteroid belt. They appear faint because they are so dark and far away, but actually the larger Trojans are quite sizable. Four of them—Hektor, Diomedes, Agamemnon, and Patroclus—have diameters between 150 and 200 km.

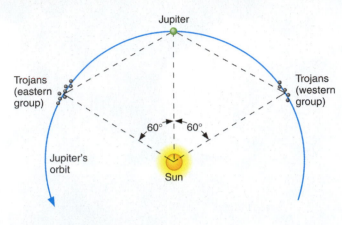

■ **FIGURE 12.7**

Trojan Asteroids   Locations of the Trojan points of the orbit of Jupiter. See also Figure 12.2.

12.2.2   Asteroids in the Outer Solar System

There may be many asteroids with orbits that carry them far beyond Jupiter, but they are difficult to detect and only a few have been discovered. The largest of these mysterious objects is Chiron, with a path that carries it from just inside the orbit of Saturn at its closest approach to the Sun out to almost the distance of Uranus. The diameter of Chiron is estimated to be about 200 km.

In 1992, a still-more-distant object named Pholus was discovered with an orbit that takes it 33 AU from the Sun, beyond the orbit of Neptune. Pholus has the reddest surface of any object in the solar system, indicating a strange (and still unknown) surface composition. As more objects are discovered in these distant reaches, astronomers have decided that they too will be given the names of centaurs from classical mythology; this is because the centaurs were half human, half horse, and these new objects display some of the properties of both asteroids and comets.

In 1988, as it was reaching its closest approach to the Sun, Chiron brightened by about a factor of two. This is not the sort of behavior we expect from self-respecting asteroids. Astronomers realized that Chiron (and perhaps the other centaurs) likely contained abundant volatile materials such as water ice or carbon monoxide ice. As the object gets closer to the Sun, these materials begin to evaporate and produce bright reflective atmospheres of gas and dust. As we will see in the next section, this is precisely the behavior we associate with *comets*. But Chiron and Pholus are much larger than any known comet. What a spectacular show they would put on if either was diverted into the inner solar system! Chiron has been shown to be in an unstable orbit, so this may yet happen sometime in the distant future.

12.2.3   Earth-Approaching Asteroids

Asteroids that stray far *outside* the main belt are of interest mostly to astronomers. But asteroids that come *inward*, especially those with orbits that come close to or cross the or-

bit of the Earth, are of interest to political leaders, military planners, and all the readers of this text. Some of these asteroids briefly become the closest celestial objects to us. In 1994 a 2-km asteroid passed within 1.1 million km of the Earth, and later that same year a 1-km object was picked up passing closer than the Moon. Some of these small asteroids have collided with the Earth in the past, and others will surely do so in the future, as we saw in Chapter 7. Together with any comets that come close to our planet, such asteroids are known collectively as **near-Earth objects (NEOs).** As the dinosaurs found out 65 million years ago, the collision of a significant-sized NEO could be a catastrophe for life on our planet.

Astronomers have been urging for some time that the first step in protecting the Earth from future impacts must be to learn what potential impactors are out there. Not until 1998, however, did NASA formally adopt what has been called the Spaceguard Survey Goal: to discover and track 90 percent of the Earth-approaching asteroids larger than 1 km in diameter within ten years—that is, by 2008. By the end of 2002, more than 640 of these 1-km NEOs had been discovered, as well as many smaller ones. Figure 12.8 shows how the pace of discoveries has been increasing over the last years.

How will astronomers know when they have discovered 90 percent of these asteroids? There are several ways to estimate the total number even before they have been individually located. One way is to look at the numbers of larger craters on the Moon. Remember that these craters were made by impacts just like the ones we are considering; it's just that the Moon preserves billions of years of impacts on its airless surface, while the Earth soon erases the imprints of past impacts. Thus the number of larger craters on the Moon allows us to estimate how often impacts have occurred on the Moon and Earth over the past several billion years. The number of impacts is directly related to the

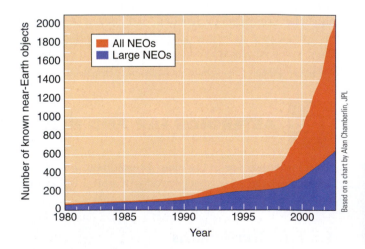

■ **FIGURE 12.8**

Discovery of Near-Earth Asteroids   The accelerating rate of discovery of NEOs is illustrated in this graph, which shows both the total number of known NEOs and the number of those 1 km in diameter or larger, the size that dominates the current impact risk on Earth.

number of asteroids and comets that are typically orbiting in the inner solar system.

Another approach is to see how often the surveys (which are mostly automated searches for points of light that move among the stars) rediscover a previously known asteroid. At the beginning of a survey, all the objects it finds will be new. But as the survey becomes more complete, more and more of the moving points the survey cameras record will be rediscoveries. Applying such reasoning, astronomers calculate that there are probably between 1000 and 1200 such Earth-approaching asteroids larger than 1 km, and we have already discovered more than half of them.

The best news is that none of the NEOs discovered so far is on a trajectory that will impact the Earth within the foreseeable future. However, we can't speak for the 500 or so larger than 1 km that have not yet been found or the much more numerous smaller ones. Researchers who work with asteroid orbits estimate that for asteroids we are not yet tracking, we will have about a 5-second warning that one is going to hit the Earth—we won't see it until it enters the atmosphere! Clearly, this estimate gives us a lot of motivation to continue the surveys and track as many of the larger asteroids as possible.

Though entirely predictable over times of a few centuries, the orbits of Earth-approaching asteroids are unstable over long time spans. These objects will meet one of two fates: Either they will impact one of the terrestrial planets, or they will be ejected gravitationally from the inner solar system due to a near-encounter with a planet. The probabilities of these two outcomes are about the same. The timescale for impact or ejection is only about a hundred million years, very short compared with the age of the solar system. Calculations show that approximately one quarter of the current Earth-approaching asteroids will eventually end up colliding with the Earth itself. The larger of these impacts will generate environmental catastrophes for our planet as described in Section 7.5.

If the current population of Earth-approaching asteroids will be removed by impact or ejection in a hundred million years, there must be a continuing source of new objects to replenish our supply. Some of them probably come from the asteroid belt, where collisions between asteroids can eject fragments into Earth-crossing orbits. Others may be dead comets that have exhausted their volatile materials (see the next section). So far we have no way of distinguishing between asteroidal fragments and the solid remnants of former comets.

If an NEO comes close enough, it can be studied with radar as well as with visible-light telescopes. Radar can even be used to produce an image of the asteroid. Steven Ostro and his colleagues at the NASA Jet Propulsion Laboratory have succeeded in imaging more than a dozen with sufficient resolution to establish their shapes and sizes. Their greatest success was with a 5-km-long NEO called Toutatis, which approached to within 3 million km of the Earth in 1992. Several radar images of Toutatis are shown in Figure 12.9, obtained with the NASA radar system in California. It appears to be a double object, consisting of two irregular lumps, with diameters of 3 km and 2 km, squashed together.

NASA/JPL, courtesy of Steven Ostro

■ **FIGURE 12.9**

Radar Images of Toutatis Reconstruction of the size and shape of asteroid Toutatis obtained from radar data collected during its close flyby of Earth in December 1992. Toutatis appears to consist of two irregular, lumpy bodies rotating in contact with each other. Its maximum dimension is about 5 km.

■ FIGURE 12.10

Eros in Color The left image shows Eros from a distance of about 1700 km, taken two days before the NEAR-Shoemaker spacecraft went into orbit around the asteroid. The asteroid's surface is heavily cratered, indicating that the surface is old. The image in the box at right was taken from a distance of 283 km and thus shows much more detail.

12.2.4 A Rendezvous with Eros

On Valentine's Day 1999, the first spacecraft to explore an Earth-approaching asteroid went into orbit around the asteroid Eros, and one year later it landed on the surface. The arrival month was dictated by the position of the asteroid in the solar system, but the specific arrival date was selected to highlight the asteroid's name, which honors the Greco-Roman god of love. Eros is highly elongated: 34 km long but only 11 km wide (Figure 12.10). It is the largest of the Earth-approaching asteroids—somewhat larger than the impacting body that killed off the dinosaurs when it collided with our planet.

Getting into an intimate relationship with Eros was the culmination of the NEAR mission, launched in 1996 as part of NASA's Discovery series. On its way to Eros, the NEAR spacecraft was renamed after planetary geologist Eugene Shoemaker, a pioneer in our understanding of craters and impacts, who died in a tragic car accident in 1997.

For a year the NEAR–Shoemaker spacecraft orbited the little asteroid at various altitudes, measuring its surface and interior composition as well as mapping Eros from all sides (Figure 12.11). An additional chemical analysis of the surface was performed after the landing. The new information shows that Eros is a fractured chip off a larger body, made of some of the most primitive materials in the solar system. Several other asteroids have been revealed as loosely bound rubble, but not Eros. Its uniform density (about the same as that of Earth's crust) and extensive global-scale grooves and ridges show that it is a cracked but solid rock.

Although it is solid inside, Eros does seem to have a good deal of loose surface material sliding down toward lower elevations. In some places the surface rubble layer is 100 m deep. The top of this rubble layer is dotted with scattered, half-buried boulders. There are so many of these boulders that at the closest distances they are more numerous than the craters (Figure 12.12). Of course, with the gravity so low on this small world, a visiting astronaut would find loose boulders rolling toward her pretty slowly and could easily leap high enough to avoid being hit.

One reason scientists are interested in the interior structure of Earth-approaching asteroids like Eros is that humans will probably need to defend themselves against an asteroid impact someday. If we ever found one of these as-

■ FIGURE 12.11

Looking Down on the North Pole of Eros Constructed from six images taken from an altitude of 200 km, this mosaic shows Eros from a different angle. The large crater at the top has been named Psyche (after the maiden who was Eros' lover in classical mythology) and is about 5.3 km wide. A saddle-shaped region can be seen directly below it. Craters of many different sizes are visible.

FIGURE 12.12

Craters and Boulders Taken from an altitude of 43 km, this scene on Eros is about 1.3 km across. You can see many boulders, some dramatically displayed on the black skyline. Because of the orientation of the spacecraft, the viewer gets the impression that the cliff on the left side of the image is steep, but in reality the slope is gentle.

NASA/JHUAPL

teroids on a collision course, we would certainly try to deflect it so it would miss the Earth. With today's technology, such deflection would be done by exploding a nuclear bomb near the asteroid and nudging it off course. But to achieve a successful deflection, we need to know about the density and interior structure of the asteroid.

A nearby explosion would have a much greater effect on a solid rocky asteroid such as Eros than on a loose rubble pile such as Mathilde. Think of climbing a sand dune compared to climbing a rocky hill with the same slope. On the dune, much of our energy is absorbed in the slipping sand, so the climb is much more difficult. In a comparable way, a nuclear explosion would have little effect on the orbit of a loosely bound asteroid.

12.3 THE "LONG-HAIRED" COMETS

Comets differ from asteroids primarily in their icy composition, a difference that also makes them become much brighter as they approach the Sun. In some early cultures, these "hairy stars" were considered omens of disaster. Today we no longer fear the comets but eagerly anticipate those that come close enough to us to put on a good sky show.

12.3.1 Appearance of Comets

Comets have been observed from the earliest times; accounts of spectacular comets are found in the histories of virtually all ancient civilizations. The typical comet is not spectacular, however, but has the appearance of a rather faint, diffuse spot of light somewhat smaller than the Moon and many times less brilliant. Comets are relatively small chunks of icy material that develop an atmosphere as they get closer to the Sun. Later there may be a very faint, nebulous **tail** extending several degrees away from the main body of the comet.

Like the Moon and planets, comets appear to wander among the stars, slowly shifting their positions in the sky from night to night. Unlike the planets, however, most comets appear at unpredictable times, which perhaps explains why they have frequently inspired fear and superstition. Comets typically remain visible for periods that vary from a couple of weeks to several months.

12.3.2 Comet Orbits

The study of comets as members of the solar system dates from the time of Isaac Newton, who first suggested that their orbits were extremely elongated ellipses. Newton's colleague Edmund Halley (see the Voyagers in Astronomy box) developed these ideas, and in 1705 he published calculations of 24 cometary orbits. In particular, he noted that the orbits of the bright comets appearing in the years 1531, 1607, and 1682 were so similar that the three could well be the same comet, returning to perihelion at average intervals of 76 years. If so, he predicted that the object should next return about 1758. Although Halley had died, when the comet appeared as he predicted, it was given the name Comet Halley (rhymes with "valley") in honor of the man who first recognized it as a permanent member of our solar system.

We now know from historical records that Comet Halley has actually been observed and recorded on every passage near the Sun at intervals from 74 to 79 years since 239 B.C.E. The period of its return varies somewhat

Edmund Halley: Astronomy's Renaissance Man

Halley, a brilliant astronomer who made contributions in many fields of science and statistics, was by all accounts a generous, warm, outgoing man. In this he was quite the opposite of his good friend Isaac Newton, whose great work, the *Principia* (see Chapter 2), Halley encouraged, edited, and helped pay to publish. Halley himself published his first scientific paper at age 20, while still in college. As a result, he was given a royal commission to go to Saint Helena (a remote island off the coast of Africa where Napoleon would later be exiled) to make the first telescopic survey of the southern sky. After returning, he received the equivalent of a master's degree and was elected to the prestigious Royal Society in England, all at the age of 22.

In addition to his work on comets, Halley was the first astronomer to recognize that the so-called "fixed" stars move relative to each other, by noting that several bright stars had changed their positions since the ancient Greek catalogs published by Ptolemy. He wrote a paper on the possibility of an infinite universe, proposed that some stars may be variable, and discussed the nature and size of *nebulae* (glow-

Edmund Halley (1656–1742)

Yerkes Observatory, University of Chicago

Halley was the first astronomer to recognize that so-called "fixed" stars move relative to each other.

ing cloudlike structures visible in telescopes). While in Saint Helena, Halley observed the planet Mercury going across the face of the Sun and devel-

oped the mathematics of how such *transits* could be used to establish the scale of the solar system.

In other fields, Halley published the first table of human life expectancies (the precursor of life-insurance statistics); wrote papers on monsoons, trade winds, and tides (charting the tides in the English Channel for the first time); laid the foundations for the systematic study of the Earth's magnetic field; studied evaporation and how inland waters become salty; and even designed an underwater diving bell. He served as a British diplomat, advising the emperor of Austria and squiring the future czar of Russia around England (avidly discussing, we are told, both the importance of science and the quality of local brandy).

For a brief time Halley was a deputy comptroller of the British Mint (where he uncovered some local graft) and served as a Navy captain for scientific voyages (once he had to deal with a mutiny). In 1703 he became a professor of geometry at Oxford and in 1720 was appointed Astronomer Royal of England. He continued observing the Earth and the sky and publishing his ideas for another 20 years, until death claimed him at age 85.

because of orbital changes produced by the pull of the giant planets. In 1910 the Earth was brushed by the comet's tail, causing much needless public concern. Comet Halley last appeared in our skies in 1986 (Figure 12.13), when it was met by several spacecraft that gave us a wealth of information about its makeup; it will return in 2061.

Only a few comets return in a time measurable in human terms, like Comet Halley. Most of the short-period comets have had their orbits changed by coming too close to one of the giant planets—most often Jupiter (they are thus sometimes called Jupiter-family comets). Most comets have long periods and will take thousands of years to return, if they return at all. As we will see later in this chapter, most Jupiter-family comets also come from a different source than Halley and the long-period comets.

Observational records exist for thousands of comets. Today, many new comets are discovered each year, not just by terrestrial observers but by spacecraft instruments as well. Most never become conspicuous and are visible only on photographs made with large telescopes. Every few years, however, a comet may appear that is bright enough to be seen easily with the unaided eye. Table 12.2 lists some well-known comets.

We are fortunate to have been visited by two bright comets in recent years; perhaps you had the chance to see at least one of them. First, in March 1996, came Comet Hyakutake, with a very long tail. A year later Comet Hale–Bopp appeared; it was as bright as the brightest stars and remained visible for several weeks, even in urban areas (see the image that opens this chapter).

Anglo-Australian Observatory

■ **FIGURE 12.13**

Comet Halley This composite of three images (one in red, one in green, one in blue) shows Comet Halley as seen with a large telescope in Australia in 1985. During the time the three images were taken in sequence, the comet moved among the stars. The telescope was moved to keep the image of the comet steady, causing the stars to become colored streaks in the background.

12.3.3 The Comet's Nucleus

When we look at a comet, all we see is its temporary atmosphere of gas and dust illuminated by sunlight. This atmosphere is called the comet's head, or *coma*. Since the gravity of such small bodies is very weak, the atmosphere is rapidly escaping all the time; it must be replenished by new material, which has to come from somewhere. The source is the small, solid **nucleus** inside, usually hidden by the glow from the much larger atmosphere surrounding it. The nucleus is the real comet, the fragment of ancient material responsible for the atmosphere and the tail (Figure 12.14).

The modern theory of the physical and chemical nature of comets was first proposed by Harvard astronomer Fred Whipple in 1950. Before Whipple's work, many astronomers thought that a comet's nucleus might be a loose aggregation of solids, the sort of orbiting "gravel bank" model we discussed for the asteroid Mathilde. Whipple proposed instead that the nucleus is a solid object a few kilometers across, composed in substantial part of water ice (but with other ices as well) mixed with silicate grains and dust. This proposal became known as the "dirty snowball" model.

The water vapor and other volatiles that escape from the nucleus when it is heated can be detected in the comet's head and tail, and we can use spectra to analyze what atoms and molecules the nucleus ice consists of. We are somewhat less certain of the non-icy component, however. No large fragments of solid matter from a comet have ever survived passage through the Earth's atmosphere.

Some very fine, microscopic grains of comet dust have been collected in the Earth's upper atmosphere, however, and have been studied in the laboratory (Figure 12.15). The spacecraft that encountered Comet Halley in March 1986 also carried dust detectors. From these various investigations it seems that much of the "dirt" in the dirty snowball

TABLE 12.2 *Some Interesting Comets*

Name	Period	Significance
Great Comet of 1577	Long	Tycho Brahe showed it was beyond the Moon
Great Comet of 1843	Long	Brightest recorded comet; visible in daytime
Daylight Comet of 1910	Long	Brightest comet of the 20th century
West	Long	Nucleus broke into pieces (1976)
Hyakutake	Long	Passed within 15 million km of Earth (1996)
Hale–Bopp	Long	Brightest recent comet (1997)
Swift–Tuttle	133 years	Parent comet of Perseid meteor shower
Halley	76 years	First comet found to be periodic; explored by spacecraft in 1986
Borrelly	6.8 years	Flyby by Deep Space 1 spacecraft (2000)
Biela	6.7 years	Broke up in 1846 and not seen again
Wild 2	6.4 years	Target of Stardust sample return mission (2004)
Tempel 1	5.7 years	Target of Deep Impact mission (2005)
Encke	3.3 years	Shortest known period

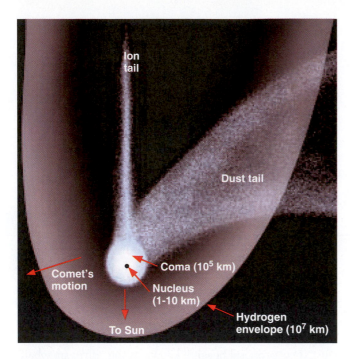

Ion tail

Dust tail

Coma (10^5 km)

Comet's motion

Nucleus (1-10 km)

To Sun

Hydrogen envelope (10^7 km)

■ **FIGURE 12.14**

Parts of a Comet This schematic illustration shows the main parts of a comet. Note that the different structures are not to scale.

Ace⊙Astronomy™ Log into AceAstronomy and select this chapter to see the Active Figure called "Build a Comet."

is tiny bits of dark, primitive hydrocarbons and silicates, rather like the material thought to be present on the primitive asteroids.

Since the nuclei of comets are small and dark, they are difficult to study from the Earth. Close-up spacecraft pho-

Donald Brownlee, University of Washington

■ **FIGURE 12.15**

Captured Comet Dust A particle that is believed to be a tiny fragment of cometary dust, collected in the upper atmosphere of the Earth.

tos have been obtained for three comets, however. Several spacecraft flew past Comet Halley at close range in 1986, with best images obtained by the European Giotto mission. In 2001 a NASA spacecraft imaged Comet Borrelly, and in 2004 a comet sample return mission called Stardust got a look at Comet Wild-2 from a distance of less than 500 km. (Figure 12.16) In spite of very different brightness and levels of activity, all thee comets are about the same size (between 5 and 10 km long) and are very dark (reflecting less than 2% of the sunlight that reaches them)—which makes these chunks of dirty ice as dark as coal!

(a)

Max Planck Institut für Aeronomie, and Ball Aerospace Corporation, courtesy of Harold Reitsema

(b)

NASA/JPL/Caltech

■ **FIGURE 12.16**

Close-up Views of Two Comets (a) Comet Halley as seen from the Giottto spacecraft as a distance of about 1000 km. Note the jets of escaping gas. (b) Comet Wild-2 from 400 km, showing deep pits perhaps formed by jets of escaping material.

12.3.4   The Comet's Atmosphere

The spectacular activity that allows us to see comets is caused by the evaporation of cometary ices heated by sunlight. Beyond the asteroid belt, where comets spend most of their time, these ices are solidly frozen. But as a comet approaches the Sun, it begins to warm up. If water (H_2O) is the dominant ice, significant quantities vaporize as temperatures rise toward 200 K, which happens somewhat beyond the orbit of Mars. The evaporating H_2O in turn releases the dust that was mixed with the ice. Since the comet's nucleus is so small, its gravity cannot hold back either the gas or the dust, both of which flow away into space at speeds of about 1 km/s.

The comet continues to absorb energy as it approaches the Sun. A great deal of this energy goes into the evaporation of its ice, as well as into heating the surface. However, observations of many comets indicate that the evaporation is not uniform and that most of the gas is released in sudden spurts, perhaps confined to a few areas of the surface. Such jets were observed directly on the surfaces of Comets Halley and Borrelly (see Figure 12.16). Most of the comet's surface is apparently inactive, with the ice buried under a layer of black silicates and carbon compounds.

Expanding into space at a speed of about 1 km/s, the comet's atmosphere can reach an enormous size. The diameter of a comet's head is often as large as Jupiter, and it can approach a diameter of a million kilometers (Figure 12.17).

Measurements show that the composition of the gas is primarily H_2O (about 80 percent in the case of Comet Halley) plus a few percent each of carbon dioxide (CO_2) and carbon monoxide (CO), along with small quantities of many additional gases, including hydrocarbons. Ultraviolet light from the Sun can break up the molecules of H_2O into oxygen and hydrogen. Modern telescopes have revealed huge hydrogen clouds around comets, up to tens of millions of kilometers across, making them, temporarily, the largest objects in the solar system.

Many comets develop tails as they approach the Sun. A comet's tail is an extension of its atmosphere, consisting of the same gas and dust that make up its head. As early as the 16th century, observers realized that comet tails always point away from the Sun (Figure 12.18), not back along the comet's orbit. Newton proposed that comet tails are formed by a repulsive force of sunlight driving particles away from the head—an idea close to our modern view. In addition to sunlight, however, we now know that cometary gas is also carried outward by streams of ions (charged particles) emitted by the Sun (Figure 12.19).

12.3.5   The Kuiper Belt

If we define a "comet" by its chemical composition (lots of water ice) rather than by its appearance (possessing an atmosphere and tail), then there is another group of "comets" that are now being found in large numbers at the edge of the solar system, beyond the orbit of Neptune. The first of these trans-Neptunian objects was discovered in 1992 by David Jewitt and Jane Luu of the University of Hawaii (Figure 12.20). They are dark (low reflectivity) and can be significantly larger than any of the comets ever seen in the inner solar system. Because of their great distance, as-

University of Arizona, courtesy of Uwe Fink

■ FIGURE 12.17
Head of Comet Halley   Here we see the cloud of gas and dust that make up the head or coma of Comet Halley on January 20, 1986. On this scale, the nucleus (hidden inside the cloud) would be a dot too small to see.

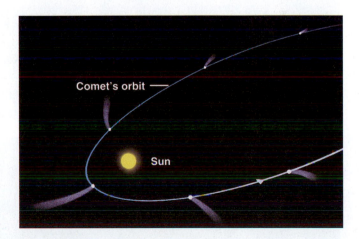

■ FIGURE 12.18
Comet Tail Points Away from the Sun   The orientation of a typical comet tail changes as the comet passes perihelion. Approaching the Sun, the tail is behind the incoming comet head, but on the way out, the tail precedes the head.

■ **FIGURE 12.19**

Comet Tails Comet Mrkos was photographed in 1957 with a wide-field telescope at Palomar Observatory. Note the smoother tail of dust curving to the right as individual dust particles spread out along the comet's orbit, and the straight ion tail pushed outward from the Sun by its wind of charged particles.

Ace⊙Astronomy™ Log into AceAstronomy and select this chapter to see Astronomy Exercise "Comets."

tronomers can detect are only the largest—generally more than 100 km in diameter.

More than 500 of these objects have now been found, and it is expected that larger survey telescopes currently under discussion could detect thousands. If any of them ever wondered into the inner solar system, they would probably develop a spectacular atmosphere and tail, but in the bitter cold beyond Neptune they are solid as a rock.

The existence of this vast reservoir of primitive icy objects—related to the comets and asteroids but not really part of either group—had been predicted theoretically 50 years earlier by the astronomer Gerard Kuiper. Today astronomers call this region beyond Neptune the **Kuiper belt** and its "inhabitants" Kuiper belt objects, or **KBOs**. We will discuss the KBOs in the next section when we look at the origin of comets.

12.4 ORIGIN AND EVOLUTION OF COMETS AND THEIR KIN

The comets and their close relatives, the KBOs, are mostly remnant ice fragments from the origin of the solar system. They are probably the most primitive objects we can study, preserved unchanged for billions of years in the deep freeze of the outer solar system.

12.4.1 The Oort Comet Cloud

Although comets are part of the solar system, observations show that most of them come initially from very great distances. By following their orbits, we can calculate that the *aphelia* (points farthest from the Sun) of new comets typi-

1992 QB1

27 Sept 92 2:35 UT

28 Sept 92 6:58 UT

ESO

■ **FIGURE 12.20**

Kuiper Belt Object Two images of object 1992QB taken in 1992 with the 3.5-m New Technology Telescope at the European Southern Observatory. The object, circled on each image as it moved among the stars, was the first member of the Kuiper belt to be found.

cally have values near 50,000 AU (more than a thousand times farther than Pluto). This clustering of aphelion distances was first noted by Dutch astronomer Jan Oort, who in 1950 proposed an idea for the origin of those comets that is still accepted today.

It is possible to calculate that a star's gravitational *sphere of influence*—the distance within which it can exert sufficient gravitation to hold onto orbiting objects—is about one third of the distance to the nearest other stars. Stars in the vicinity of the Sun are spaced in such a way that the Sun's sphere of influence extends only a little beyond 50,000 AU, or about 1 light year (LY). At such distances, however, objects in orbit about the Sun can be perturbed by the gravity of passing stars. Some of the perturbed objects can then take on orbits that bring them much closer to the Sun.

Oort suggested, therefore, that the new comets were examples of objects orbiting the Sun near the edge of its sphere of influence, whose orbits had been disturbed by nearby stars, eventually bringing them close to the Sun where we can see them. The reservoir of ancient icy objects from which such comets are derived is now called the **Oort comet cloud.**

Astronomers estimate that there are about a trillion (10^{12}) comets in the Oort cloud. In addition, we estimate that about ten times this number of potential comets could be orbiting the Sun in the volume of space between the planets and the Oort cloud at 50,000 AU. These objects remain undiscovered because they are too faint to be seen directly and their orbits are too stable to permit any of them to be deflected inward close to the Sun. The total number of cometary objects could thus be on the order of 10 trillion (10^{13}), a very large number indeed.

What is the mass represented by 10^{13} comets? We can make an estimate if we assume something about comet sizes and masses (see the Figuring for Yourself section at the end of the chapter). Let us suppose that the nucleus of Comet Halley is typical. Its observed volume is about 600 km^3. If the primary constituent is water ice with a density of about 1 g/cm^3, then the total mass of Halley's nucleus must be about 6×10^{14} kg. This is about one ten billionth (10^{-10}) of the mass of the Earth.

If our estimate is reasonable and there are 10^{13} comets with this mass out there, their total mass would be equal to about 1000 Earths—greater than the mass of all the planets put together! Therefore, cometary material could be the most important constituent of the solar system after the Sun itself.

12.4.2 KBOs and Plutinos

The Oort cloud spawns comets we can see because they are deflected into the inner solar system by passing stars. But we now know of a second source of comets. The Kuiper belt represents a relatively nearby reservoir of cometary material, in the form of a flattened disk that starts just beyond Neptune. Smaller KBOs are occasionally ejected gravitationally into the inner solar system, where they can

be captured by Jupiter and thereby contribute to the short-period or Jupiter-family comets. The Centaur asteroids are probably intermediate objects that are in transit from the Kuiper belt to the inner solar system. (You can see how awkward the traditional terminology is, when the same object is called a KBO for most of its life, then a Centaur asteroid, and finally a short-period comet.)

The biggest KBOs discovered so far have impressive diameters, approaching 1000 km. Most estimates of the sizes of KBOs are based on assumptions about how much light they reflect; the darker we assume they are, the larger they must be to reflect the amount of sunlight we measure. Recent infrared measurements with the Spitzer Space Telescope remove this ambiguity and yield real measurements of size, just as we have for main belt asteroids. To the surprise of many astronomers, the KBOs are not as dark as the comets—with surfaces that are dark grey rather than coal black. The masses of a few KBOs that are found in pairs have also been calculated from their orbits around each other, using Kepler's laws.

The three largest KBOs are called Sedna, Varuna, and Quaoar. Since many KBOs (called plutinos) have orbits that are similar to that of the planet Pluto, Pluto is sometimes also classed as the largest KBO. About 1000 KBOs have been discovered, from which astronomers infer the existence of perhaps 25,000 larger than 100 km in diameter. The orbits of the KBOs are strongly influenced by the gravity of Neptune, which confines most to a distance from the Sun between 35 and 50 AU.

The objects in the Oort cloud and the Kuiper belt probably had different origins, and they may therefore have different compositions. Astronomers are therefore very interested in making detailed measurements of the comets that are derived from these two source regions. Most of the bright comets that have been studied in the past (Halley, Hyakutake, Hale–Bopp) are Oort-cloud comets, but several of the comets targeted for spacecraft measurements in the next decade are Jupiter-family comets from the Kuiper belt (see Table 12.2).

12.4.3 The Fate of Comets

Any comet we see today will have spent nearly its entire existence in the Oort cloud or the Kuiper belt at a temperature near absolute zero. But once a comet enters the inner solar system, its previously uneventful life history begins to accelerate. It may, of course, survive its initial passage near the Sun and return to the cold reaches of space where it spent the previous 4.5 billion years. At the other extreme, it may impact the Sun or come so close that it is destroyed on its first perihelion passage (several such impacts have been observed with space telescopes that monitor the Sun). Sometimes, however, the new comet does not come that close to the Sun but instead interacts with one or more of the planets.

A comet that comes within the gravitational influence of a planet has three possible fates. It can (1) impact the planet, ending the story at once; (2) be speeded up and

MAKING CONNECTIONS

Comet Hunting as a Hobby

When amateur astronomer David Levy, the codiscoverer of Comet Shoemaker–Levy 9, found his first comet, he had already spent 928 fruitless hours searching through the dark night sky. But the discovery of that first comet only whetted his appetite. Since then he has found 7 others on his own and 13 more working with others. Despite this impressive record, he ranks only fourth in the record books for number of comet discoveries. But David is still young and hopes to break the record someday.

All around the world, dedicated amateur observers spend countless nights scanning the sky for new comets. Astronomy is one of the very few fields of science where amateurs can still make a meaningful contribution, and the discovery of a comet is one of the most exciting ways they can establish their place in astronomical history. Don

Machholz, a California amateur (and comet hunter) who has been making a study of comet discoveries, reported that between 1975 and 1995, 38 percent of all comets discovered were found by amateurs. Those 20 years yielded 67 comets for amateurs, or almost four per year. That might sound pretty encouraging to new comet hunters, until they learn that the average number of hours the typical amateur spent searching for a comet before finding one was about 420. Clearly, this is not an activity for impatient personalities.

What do comet hunters do if they think they have found a new comet? First, they must check the object's location in an atlas of the sky to make sure it really is a comet. Since the first sighting of a comet usually occurs when it is still far from the Sun and before it sports a significant tail, it will look

---

ejected, leaving the solar system forever; or (3) be perturbed into an orbit with a shorter period. In the last case, its fate is sealed. Each time it approaches the Sun, it loses part of its material and also has a significant chance of collision with a planet. Once the comet is in a short-period orbit, its lifetime starts being measured in thousands, not billions, of years. Comet Halley is the best-known example of a comet that has been reoriented in this way.

A few comets end their lives catastrophically by breaking apart (sometimes for no apparent reason; Figure 12.21).

■ **FIGURE 12.21**

Breakup of Comet LINEAR A ground-based view and a much more detailed photo with the HST showing the multiple fragments of the nucleus of Comet LINEAR. The comet disintegrated in July 2000 for no apparent reason.

Comet LINEAR
Fragments
August 5, 2000

HST

University of Hawaii

Hubble Space Telescope
WFPC2

NASA and H. Weaver (Johns Hopkins University)
STScI-PRC00-27

University of Hawaii & STScI/NASA

like only a small, fuzzy patch. And, through most amateur telescopes, so will nebulae (clouds of cosmic gas and dust) and galaxies (distant groupings of stars). Next they must check that they have not come across a comet that is already known, in which case they will only get a pat on the back instead of fame and glory. Then they must reobserve or rephotograph it some time later to see whether its motion in the sky is appropriate for comets.

Often comet hunters who think they have made a discovery get an-

Amateur astronomer David Levy

other comet hunter elsewhere in the country to confirm it. If everything checks out, the place they contact is the Central Bureau for Astronomical Telegrams at the Harvard–Smithsonian Center for Astrophysics in Cambridge, Massachusetts (Web site cfa-www. harvard.edu/cfa/ps/cbat.html). If the discovery is confirmed, the bureau will send the news out to astronomers and observatories around the world. One of the unique rewards of comet hunting is that the discoverer's name becomes associated with the new comet—a bit of cosmic fame that few hobbies can match.

Especially spectacular was the fate of a faint comet called Shoemaker–Levy 9, which broke into about 20 pieces when it passed close to Jupiter in July 1992. The fragments of Shoemaker–Levy were actually captured into a very elongated, two-year orbit around Jupiter, more than doubling the number of known jovian satellites. This was only a temporary enrichment of Jupiter's family, however, because in July 1994 the comet fragments crashed into Jupiter, releasing energy equivalent to millions of megatons of TNT.

As each cometary fragment streaked into the jovian atmosphere at a speed of 60 km/s, it disintegrated and ex-

ploded, producing a hot fireball that carried the comet dust as well as atmospheric gases to high altitudes. These fireballs were clearly visible in profile, with the actual point of impact just beyond the jovian horizon as viewed from the Earth (Figure 12.22). As each explosive plume fell back into Jupiter, a region of the upper atmosphere larger than the Earth was heated to incandescence and glowed brilliantly for about 15 min, a glow we could detect with infrared-sensitive telescopes.

Dark clouds of debris settled into the stratosphere of Jupiter, producing long-lived "bruises" (each larger than

■ FIGURE 12.22

Comet Impact on Jupiter
This sequence of images shows the last fragment of Comet Shoemaker–Levy 9 to hit Jupiter. Although the initial impact is on the edge of Jupiter as seen from Earth, the giant planet's rotation soon brings the fireball caused by the impact into view. The time of each observation and the filter through which it was taken are shown in each box.

NASA/STScI

the Earth) that could be easily seen through even small telescopes (Figure 12.23). Millions of people all over the world peered at Jupiter through telescopes or followed the event via television and the Internet. Seeing these impact explosions on Jupiter helps us to appreciate the disaster that would happen to our planet if we were hit by a comet or asteroid.

For comets that do not meet so dramatic an end, measurements of the amount of gas and dust in their atmospheres permit us to estimate the total losses during one orbit. Typical loss rates are up to a million tons per day from an active comet near the Sun, adding up to some tens of millions of tons per orbit. At that rate, a typical comet will be gone after a few thousand orbits. This will probably be the fate of Comet Halley in the long run.

■ **FIGURE 12.23**

Impact Dust Cloud on Jupiter   One group of features resulting from the impact of Comet Shoemaker–Levy 9 with Jupiter, seen with the Hubble Space Telescope 105 minutes after the impact that produced the dark rings (the compact black dot came from another fragment). The inner edge of the diffuse outer ring is about the same size as the Earth. Later, the winds on Jupiter blended these features into a broad spot that remained visible for more than a month.

SURFING THE WEB

🖥 *Asteroid and Comet Impact Hazards Page:*
impact.arc.nasa.gov/
Informative site about the dangers of asteroids and comets hitting the Earth and what we might do about it.

🖥 *Near-Earth Asteroid Rendezvous Mission:* near.jhuapl.edu/
Background information and great images from the mission that went by Mathilde and landed on Eros.

🖥 *Kuiper Belt Home Page:*
www.ifa.hawaii.edu/~jewitt/kb.html
David Jewitt of the University of Hawaii keeps track of the objects that have been discovered in the Kuiper belt and provides useful background material.

🖥 *Deep Space 1 Mission:* nmp.jpl.nasa.gov/ds1/
This was the mission that flew by Comet Borrelly.

🖥 *Stardust Mission Home Page:* stardust.jpl.nasa.gov/
This NASA mission will encounter Comet Wild 2 in 2004, collecting comet dust and returning it to Earth in 2006.

🖥 *Deep Impact Mission:* www.ss.astro.umd.edu/deepimpact
A mission to fly by and impact a probe on comet Tempel 1 in 2005.

🖥 Information About Quaoar, the Largest Known KBO:
- *At Caltech:*
  http://www.gps.caltech.edu/~chad/quaoar/
- *From the Hubble:*
  http://hubblesite.org/newscenter/archive/2002/17/

🖥 Sites About Asteroids in General:
- *Nine Planets Site:*
  seds.lpl.arizona.edu/billa/tnp/asteroids.html
- *Views of the Solar System Site:*
  www.solarviews.com/eng/asteroid.htm

SUMMARY

The solar system includes many objects that are much smaller than the planets and their larger satellites. The rocky ones are called **asteroids,** and the icy ones are called **comets.**

12.1 Ceres is the largest asteroid; about 15 are larger than 250 km and 100,000 are larger than 1 km. Most are in the **asteroid belt,** between 2.2 and 3.3 AU from the Sun. The presence of asteroid families in the belt indicates that many asteroids are the remnants of asteroid collisions and fragmentation. The asteroids include both primitive and differentiated objects. Most asteroids are classed as C type, meaning they are composed of carbonaceous materials. Dominating the inner belt are S-type (stony) asteroids, with a few M-type (metallic) ones. Vesta is rare in having a volcanic (basaltic) surface. We now have spacecraft images of several asteroids and have detected several asteroid satellites, making it possible to measure asteroid masses and densities.

12.2 The Trojan asteroids are dark, primitive objects gathered into two swarms controlled by Jupiter's gravity at 5.2 AU from the Sun. Of great interest are the Earth-approaching asteroids, called **near-Earth objects (NEOs),** estimated to number about 1100 (down to 1 km in diameter). These are on unstable orbits, and on timescales of 10^8 years they will either impact one of the terrestrial planets or be ejected. Most of them probably come from the asteroid belt, but some may be dead comets. Radar images provide us a close view of some of these without a spacecraft encounter.

12.3 Halley first showed that some comets are on closed orbits and return periodically to the Sun. The heart of a comet is its **nucleus,** a few kilometers in diameter and composed of volatiles (primarily H_2O) and solids (including both silicates and carbonaceous materials). Whipple first suggested this "dirty snowball" model in 1950; it has been confirmed by spacecraft studies of Comets Halley and Borrelly. As the nucleus approaches the Sun, its volatiles evaporate (perhaps in localized jets or explosions) to form the comet's head or atmosphere, which escapes at about 1 km/s. The atmosphere streams away from the Sun to form a long **tail.**

12.4 Oort proposed in 1950 that comets are derived from what we now call the **Oort comet cloud,** which surrounds the Sun out to about 50,000 AU (near the limit of the Sun's gravitational sphere of influence) and contains between 10^{12} and 10^{13} comets. Comets also come from the **Kuiper belt,** a disk-shaped region beyond the orbit of Neptune. Modern telescopes enable us to discover and study the largest of these **KBOs** directly. Comets are primitive bodies left over from the formation of the outer solar system. Once a comet is diverted into the inner solar system, it typically survives no more than a few thousand perihelion passages before losing all its volatiles. Some comets die spectacular deaths: Shoemaker–Levy 9, for example, broke into 20 pieces before colliding with Jupiter in 1994.

INTER-ACTIVITY

A Your group is a congressional committee charged with evaluating the funding for an effort to find all the NEOs (near-Earth objects) that are larger than 0.5 km across. Make a list of reasons it would be useful to humanity to find such objects. What should we (could we) do if we found one that will hit the Earth in a few years?

B Many cultures considered comets bad omens. Many legends associate comets with the death of kings, losses in war, ends of dynasties. Did any members of your group ever hear about such folktales? Discuss why comets may have gotten this bad reputation.

C Because asteroids have a variety of compositions and a low gravity that makes the removal of materials quite easy, some people have suggested that mining asteroids may be a way to get needed resources in the future. Make a list of materials in asteroids (and comets that come to the inner solar system) that may be valuable to a space-faring civilization. What are the pros and cons of undertaking mining operations on these small worlds?

D As discussed in the last box of the chapter, amateur comet hunters typically spend more than 400 hours scanning the skies with their telescopes to find a comet. That's a lot of time to spend (usually alone, usually far from city lights, usually in the cold, and always in the dark). Discuss with members of your group whether you can see yourself being this dedicated. Why do people undertake such quests? Do you envy their dedication?

E The largest Kuiper belt objects known are estimated to be about half the diameter of Pluto and are sometimes called miniplanets. All the planets in our solar system have so far been named after mythological gods. Have your group discuss whether they think we should continue this naming tradition with miniplanets. Why or

why not? (The group that discovered the largest KBO so far decided to go beyond Greek and Roman mythology. They named it Quaoar, after a creation god of the Tongva tribe, the Native Americans who lived in the region of Southern California where Caltech, the university where the work was done, is located.)

F The discovery of the miniplanets in the Kuiper belt again raises the issue of the nature of Pluto. Have your group review the information about Pluto and Triton (see Chapter 11) and the large Kuiper belt objects (in this chapter), and then discuss whether you think Pluto should be considered a planet, the largest KBO, or something else. Give reasons for your choice.

REVIEW QUESTIONS

Ace☯Astronomy™ Assess your understanding of this chapter's topics with additional quizzing and animations at **http://ace.brookscole.com/ voyages**

1. Why are asteroids and comets important to our understanding of solar system history?
2. Describe the main differences between C-type and S-type asteroids.
3. Compare asteroids of the asteroid belt with the Earth-

approaching asteroids. What are the main differences between the two groups?

4. Describe the nucleus of a typical comet, and compare it with an asteroid of similar size.
5. Describe the origin and eventual fate of the comets we see from the Earth.
6. What evidence do we have for the existence of the Kuiper belt? What kind of objects are found there?
7. How did Comet Shoemaker–Levy 9 end its life in 1994?

THOUGHT QUESTIONS

8. Give at least two reasons today's astronomers are so interested in the discovery of additional Earth-approaching asteroids.

9. Suppose you were designing a spacecraft that would match course with an asteroid and follow along its orbit. What sorts of instruments would you put on board, and what would you like to learn?

10. Suppose you were designing a spacecraft that would match course with a comet and move with it for a while. What sorts of instruments would you put on board, and what would you like to learn?

11. Suppose a comet were discovered approaching the Sun, one whose orbit would cause it to collide with the Earth 20 months later, after perihelion passage. (This is approximately the situation described in the science-fiction novel

Lucifer's Hammer by Larry Niven and Jerry Pournelle.) What could we do? Would there be any way to protect ourselves from a catastrophe?

12. We believe that chains of comet fragments like Comet Shoemaker–Levy 9's have collided not only with the jovian planets, but occasionally with their satellites. What sort of features would you look for on the outer planet satellites to find evidence of such collisions? (As an extra bonus, can you find any images of such features on a satellite like Callisto? You can use a Web site of planetary images, such as *Planetary Photojournal*, at: photojournal.jpl.nasa.gov.)

13. Why have we found so many KBOs in the last decade and not before?

14. Why is it hard to give exact diameters for even the larger KBOs?

FIGURING FOR YOURSELF

Suppose the Oort comet cloud contains 10^{12} comets with an average diameter of 10 km each. How can we estimate the mass of the total Oort comet cloud? We can start by assuming that typical comets are about the size of Halley and

Borrelly, with a diameter of 10 km and a density appropriate to water ice, which is 1 g/cm^3 or 1000 kg/m^3. We know that density = mass/volume, the volume of a sphere = $\frac{1}{3}\pi R^3$, and the radius $R = \frac{1}{2}D$. Therefore, for each comet,

$$\text{mass} = \text{density} \times \text{volume} = \text{density} \times \frac{4}{3}\pi\left(\frac{1}{2}D\right)^3$$

Given that $10\text{ km} = 10^4\text{ m}$, each comet's mass is

$$1000\text{kg/m}^3 \times \frac{4}{3} \times 3.14 \times \frac{1}{8} \times (10^4)^3\text{m}^3 \approx 10^{15}\text{kg}$$
$$= 10^{12}\text{ tons}$$

To calculate the total mass of the cloud, we multiply this typical mass for one comet by the number of comets:

$$10^{15} \times 10^{12} = 10^{27}\text{ kg}$$

15. How does the total mass we calculated above compare to the mass of Jupiter? To the mass of the Sun? (Give a numerical answer.)

16. How would the calculation change if a typical comet in the Oort cloud is only 1 km in diameter?

17. How would the calculation change if a typical comet in the Oort cloud is larger—say, 50 km in diameter?

18. The calculation above refers to the known Oort cloud, the source for most of the comets we see. If, as some astronomers suspect, there are ten times this many cometary objects in the solar system, how does the total mass of cometary matter compare with the mass of Jupiter?

19 If the Oort comet cloud contains 10^{12} comets and ten new comets are discovered coming close to the Sun each year, what percentage of the comets have been used up since the beginning of the solar system?

20. The mass of the asteroids is found mostly in the larger asteroids, so to estimate the total mass we need to consider only the larger objects. Suppose the three largest asteroids—Ceres (1000 km), Pallas (500 km), and Vesta (500 km)—account for half the total mass. Assume that each of these three asteroids has a density of 3 g/cm^3 and calculate their total mass. Multiply your result by 2 to obtain an estimate for the mass of the total asteroid belt. How does this compare with the mass of the Oort comet cloud?

21. Make a similar estimate for the mass of the Kuiper belt assuming that Pluto (about 2000 km), Quauar, and Varuna (each 1000 km) are the three largest (they are the biggest found so far) and that all three have Pluto's density of 2 g/cm^3. How does this mass compare with the mass of the main asteroid belt? (In fact there are probably many undiscovered large KBOs, so the true mass could be as much as ten times larger than in this example.)

22. What is the period of revolution about the Sun for an asteroid with a semimajor axis of 3 AU in the middle of the asteroid belt?

23. What is the period of revolution for a comet with aphelion at 5 AU and perihelion at the orbit of the Earth?

SUGGESTIONS FOR FURTHER READING

On Asteroids

Asphang, E. "The Small Planets" in *Scientific American*, May 2000, p. 46. On asteroids and NEAR results.

Beatty, J. "NEAR Falls for Eros" in *Sky & Telescope*, May 2001, p. 34. On the landing, with an album of photos.

Bell, J. and Mitton, J., eds. *Asteroid Rendezvous: NEAR Shoemaker's Adventures at Eros.* 2002, Cambridge U. Press.

Binzel, R. "A New Century for Asteroids" in *Sky & Telescope*, July 2001, p. 44.

Binzel, R. "Asteroids Come of Age" in *Science*, Sept. 2000, vol. 289, p. 2065.

Burnham, R. "Here's Looking at Ida" in *Astronomy*, Apr. 1994, p. 38.

Kowal, C. *Asteroids: Their Nature and Utilization*, 2nd ed. 1996, John Wiley/Praxis. Introductory book by the astronomer who discovered Chiron.

Zimmerman, R. "Ice Cream Sundaes and Mashed Potatoes" in *Astronomy*, Feb. 1999, p. 54. On the NEAR mission.

On Comets

Aguirre, E. "The Great Comet of 1997" in *Sky & Telescope*, July 1997, p. 50. On Comet Hale–Bopp.

Benningfield, D. "Where Do Comets Come From?" in *Astronomy*, Sept. 1990, p. 28.

Bortle, J. "A Halley Chronicle" in *Astronomy*, Oct. 1985, p. 98. A chronology of each pass.

Brandt, J. and Chapman, R. *Rendezvous in Space.* 1992, W. H. Freeman. Introduction by two leading comet experts.

Davies, J. *Beyond Pluto: Exploring the Outer Limits of the Solar System.* 2001, Cambridge U. Press. Discovery of minor planets in the Kuiper belt.

Gore, R. "Halley's Comet '86: Much More Than Met the Eye" in *National Geographic*, Dec. 1986, p. 758.

Levy, D. "My Life and Hard Times as a Comet Hunter" in *Sky & Telescope*, Oct. 2002, p. 72.

Levy, D. *The Quest for Comets.* 1994, Plenum Press. Personal story of comet discovery and comet science by an amateur astronomer who has found many comets.

Spencer, J. and Mitton, J., eds. *The Great Comet Crash: The Collision of Comet Shoemaker–Levy 9 and Jupiter.* 1995, Cambridge U. Press. Good nontechnical summary.

Stern, A. "Chiron: Interloper from the Kuiper Disk?" in *Astronomy*, Aug. 1994, p. 26.

Talcott, R. "Comet Borrelly's Dark Nature" in *Astronomy*, Apr. 2002, p. 42. Results of the Deep Space 1 flyby.

Weissman, P. "The Oort Cloud" in *Scientific American*, Sept. 1998, p. 84.

Collisions between planetesimals during the formative stages of the solar system. Many meteorites are fragments of these planetesimals, dating back to the origin of our planetary system.

Cosmic Samples and the Origin of the Solar System

I could more easily believe that two Yankee professors would lie than that stones would fall from the heavens.

Thomas Jefferson, responding to a report of a Connecticut meteorite fall in 1807

THINKING AHEAD

Imagine you are a scientist examining a sample of rock that had fallen from space a few days earlier and you find within it some of the chemical building blocks of life. How could you determine whether those "organic" materials were extraterrestrial or merely the result of earthly contamination?

We conclude our survey of the solar system with a discussion of its origin and evolution. Some of these ideas were introduced in Chapter 6; we now return to them, using the information we have learned about individual planets and smaller objects. In addition, astronomers have recently discovered more than a hundred planets around other stars. This is a new source of data on planetary systems, providing us a perspective that extends beyond our own particular (and perhaps atypical) solar system.

But first, we want to look at another crucial way that astronomers learn about the ancient history of the solar system: by examining samples of *primitive matter,* the debris of the processes that formed the solar system some 4.5 billion years ago. Unlike the Apollo Moon rocks, these samples of cosmic material come to us free of charge—they literally fall from the sky. We call this material cosmic dust and meteorites.

Ace Astronomy™ The AceAstronomy icon throughout the text indicates an opportunity for you to test yourself on key concepts and to explore animations and interactions on the AceAstronomy website at **http://ace.brookscole.com/voyages**

13.1 METEORS

As we saw in the last chapter, the ices in comets evaporate when they get close to the Sun, spraying millions of tons of rock and dust into the inner solar system. The Earth is surrounded by this material, sweeping through thin clouds and streams of dust in its orbit about the Sun. As each of the larger dust or rock particles enters the Earth's atmosphere, it creates a brief fiery trail; this is often called a *shooting star*, but it is properly known as a **meteor.**

13.1.1 Observing Meteors

Meteors are tiny solid particles that enter the Earth's atmosphere from interplanetary space (Figure 13.1). Since the particles move at speeds of many kilometers per second, friction with the air vaporizes them at altitudes between 80 and 130 km. The resulting flashes of light fade out within a few seconds. These "shooting stars" got their name because at night their luminous vapors look like stars moving rapidly across the sky. To be visible, a meteor must be within about 200 km of the observer. On a typical dark, moonless night, an alert observer can see half a dozen meteors per hour. Over the entire Earth, the total number of meteors bright enough to be visible must total about 25 million per day.

The typical meteor is produced by a particle with a mass of less than 1 g—no larger than a pea. But the light you see comes from the much larger region of heated, glowing gas surrounding this little grain of interplanetary material. Because of its high speed, the energy in a pea-sized meteor is as great as that of an artillery shell fired on Earth, but it is all dispersed high in the Earth's atmosphere.

If a particle the size of a golf ball strikes our atmosphere, it produces a much brighter trail called a *fireball*

P. Parviainen, Polar Image

■ FIGURE 13.1

Meteor Trails Two meteors can be seen against a starry background that includes the Milky Way.

(Figure 13.2). A piece as large as a bowling ball has a fair chance of surviving its fiery entry if its approach speed is not too high. The total mass of meteoric material entering the Earth's atmosphere is estimated to be about 100 tons per day (which seems like a lot if you imagine it all falling

Photo © 1986 John Sanford

■ FIGURE 13.2

A Fireball When a larger piece strikes the Earth's atmosphere, it can make a bright fireball. Note that during the long exposure needed to catch a fireball, the Earth's motion has made the stars into streaks.

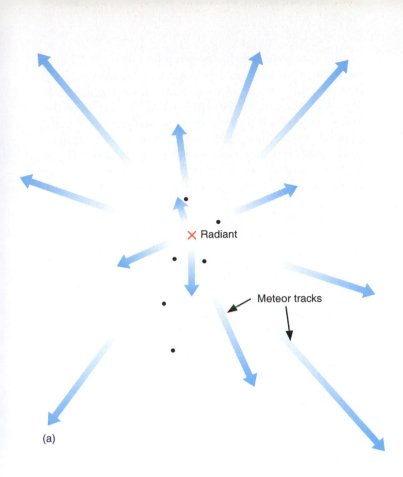

X Radiant

Meteor tracks

(a)

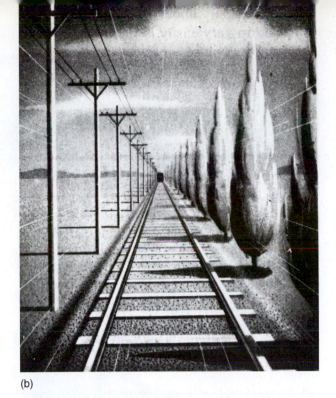

(b)

■ **FIGURE 13.3**

The Radiant of a Meteor Shower The tracks of the meteors diverge from a point in the distance, just as long, straight railroad tracks appear to do.

on your home, but it is not very significant when spread out over the whole planet).

13.1.2 Meteor Showers

Many—perhaps most—of the meteors that strike the Earth can be associated with specific comets. Some of these periodic comets still return to our view; others have long ago fallen apart, leaving only a trail of dust behind them. The many dust particles from a given comet retain approximately the orbit of their parent, continuing to move together through space (spreading out over the orbit with time). When the Earth, in its travels around the Sun, crosses such a dust stream, we see a sudden burst of meteor activity that usually lasts several hours; these events are called **meteor showers.**

The dust particles that produce meteor showers are moving together in space before they encounter the Earth. Thus, as we look up at the atmosphere, their parallel paths seem to come toward us from a place in the sky called the *radiant.* This is the direction in space from which the meteor stream is diverging, just as long railroad tracks seem to diverge from a single spot on the horizon (Figure 13.3). Meteor showers are often designated by the constellation in which this radiant is located: For example, the Perseid meteor shower has its radiant in the constellation of Perseus. The characteristics of some of the more famous meteor showers are summarized in Table 13.1.

The meteoric dust is not always evenly distributed along the orbit of the comet, so during some years more

TABLE 13.1 *Major Annual Meteor Showers*

Shower Name	Date of Maximum	Associated Comet	Comet's Period (years)
Quadrantid	January 3	Unknown	—
Lyrid	April 21	Thatcher	415
Eta Aquarid	May 4	Halley	76
Delta Aquarid	July 30	Unknown	—
Perseid	August 11	Swift–Tuttle	105
Draconid	October 9	Giacobini–Zinner	7
Orionid	October 20	Halley	76
Taurid	October 31	Encke	3
Leonid	November 16	Tempel–Tuttle	33
Geminid	December 13	Phaethon °	1.4

° Phaethon is an Earth-approaching asteroid, not a comet.

meteors are seen when the Earth intersects the dust stream, and in other years fewer. A very clumpy distribution is associated with the Leonid meteors, which in 1833 and again in 1866 (after an interval of 33 years—the period of the comet) yielded the most spectacular showers (sometimes called *meteor storms*) ever recorded (Figure 13.4). During the Leonid storm on November 17, 1966, up to a

■ **FIGURE 13.4**

A Meteor Storm   A woodcut of the great meteor shower or storm of 1833, shown here with considerable artistic license.

No shower meteor has ever survived its flight through the atmosphere and been recovered for laboratory analysis. However, there are other ways to investigate the nature of these particles and thereby to gain additional insight into the comets from which they are derived. Analysis of the flight paths of meteors shows that most of them are very light or porous, with densities typically less than 1.0 g/cm³. If you placed a fist-sized lump of meteor material on a table in the Earth's gravity, it might well fall apart under its own weight.

Such light particles break up very easily in the atmosphere, accounting for the failure of even relatively large shower meteors to reach the ground. Comet dust is apparently fluffy, rather inconsequential stuff. We can also infer this from the tiny comet particles recovered in the Earth's atmosphere (see Figure 12.15). This fluff, by its very nature, cannot reach the Earth's surface intact. However, more substantial fragments from asteroids do make it into our laboratories, as we will see in the next section.

13.2   METEORITES: STONES FROM HEAVEN

Any fragment of interplanetary debris that survives its fiery plunge through the Earth's atmosphere is called a **meteorite.** Meteorites fall only very rarely in any one locality, but over the entire Earth thousands fall each year. These rocks from the sky carry a remarkable record of the formation and early history of the solar system.

13.2.1   Extraterrestrial Origin of Meteorites

Occasional meteorites have been found throughout history, but their extraterrestrial origin was not accepted by scientists until the beginning of the 19th century. Before that, these strange stones were either ignored or considered to have a supernatural origin.

The falls of the earliest recovered meteorites are lost in the fog of mythology. A number of religious texts speak of stones from heaven, which sometimes arrived at opportune moments to smite the enemies of the authors of those texts. At least one sacred meteorite has apparently survived in the form of the Ka'aba, the holy black stone in Mecca that is revered by Islam as a relic from the time of the Patriarchs—although understandably, no chip from this sacred stone has been subject to detailed chemical analysis.

The modern scientific history of the meteorites begins in the late 18th century, when a few scientists suggested that some strange-looking stones had such peculiar composition and structure that they were probably not of terrestrial origin. The idea that indeed "stones fall from the sky" was generally accepted after a scientific team led by French physicist Jean Baptiste Biot investigated a well-observed fall in 1803. As indicated by our opening quote, however, it

hundred meteors were observed per second in some locations. The Leonid shower of 2001 was not this intense, but it peaked at nearly a thousand meteors per hour—one every few seconds—observable from any dark viewing site. Perhaps some of you saw this "once in a lifetime" event.

The most dependable meteor display is the Perseid shower, which appears each year for about three nights near August 11. In the absence of bright moonlight, you can see one meteor every few minutes during a typical Perseid shower. Astronomers estimate that the total combined mass of the particles in the Perseid swarm is nearly a billion tons; the comet that gave rise to the particles, called Swift–Tuttle, must originally have had at least that much mass. However, if its initial mass were comparable to the mass measured for Comet Halley (see Chapter 12), then Swift–Tuttle would have contained several hundred billion tons, suggesting that only a very small fraction of the original cometary material survives in the meteor stream.

Showering with the Stars

Observing a meteor shower is one of the easiest and most enjoyable astronomy activities for beginners. The best thing about it is that you don't need a telescope or binoculars—in fact, they would positively get in your way! What you do need is a site far from city lights with an unobstructed view of as much sky as possible. While the short bright lines in the sky made by individual meteors could, in theory, be traced back to a radiant point (as shown in Figure 13.3), the quick blips of light that represent the end of the meteor could happen anywhere above you.

So the key to observing meteor showers is not to restrict your field of view, but to lie back and scan the sky alertly. Try to select a good shower (see the list in Table 13.1) and a night when the Moon will not be bright at the time you are observing. The Moon, street lights, car headlamps, and bright flashlights will all get in the way of your seeing the faint meteor streaks.

■ ■ ■ ■ ■ ■ ■ ■ ■ ■ ■ ■ ■ ■ ■

The key to observing meteor showers is not to restrict your field of view, but to lie back and scan the sky alertly.

■ ■ ■ ■ ■ ■ ■ ■ ■ ■ ■ ■ ■ ■ ■

You will also see more meteors after midnight, when you are on the hemisphere of the Earth that faces forward—in the direction of the Earth's revolution around the Sun. Before midnight, you are observing from the "back side" of the Earth, and the only meteors you see will be those that traveled fast enough to catch up with the Earth's orbital motion.

When you've gotten away from all the lights, give your eyes about 15 minutes to get "dark adapted"—that is, for the pupil of your eye to open up as much as possible. (This adaptation is the same thing that happens in a dark movie theater. When you first enter, you can't see a thing, but eventually, as your pupils open wider, you can see pretty clearly by the faint light of the screen—and notice all that popcorn on the floor.)

Seasoned meteor observers find a hill or open field and make sure to bring warm clothing, a blanket, and a thermos of hot coffee or chocolate with them. (It's also nice to take along someone with whom you enjoy sitting in the dark.) Don't expect to see fireworks or a laser show: Meteor showers are subtle phenomena, best approached with a patience that reflects the fact that some of the dust you are watching burn up may first have been gathered into its parent comet more than 4.5 billion years ago, as the solar system was just forming.

Ace✪Astronomy™   Log into AceAstronomy and select this chapter to see the Active Figure called "Impacts on Earth."

took a few more years for the news to reach the young republic on the western shores of the Atlantic.

Meteorites sometimes fall in groups or showers. Such a fall occurs when a single larger object breaks up during its violent passage through the atmosphere. It is important to remember that such a *shower of meteorites* has nothing to do with a *meteor shower*. No meteorites have ever been recovered in association with meteor showers. Whatever the ultimate source of the meteorites, they do not appear to come from the comets or their associated particle streams.

13.2.2   Meteorite Falls and Finds

Meteorites are found in two ways. First, sometimes bright meteors (fireballs) are observed to penetrate the atmosphere to very low altitudes. If we search the area beneath the point where the fireball burned out, we may find one or more remnants that reached the ground. Observed *meteorite falls,* in other words, may lead to the recovery of fallen meteorites. (A few meteorites have even hit buildings or, very rarely, people; see Making Connections: *Some Striking Meteorites.*) But there are many false alarms. For example, every few months the newspapers report that a meteorite has been implicated in the start of a fire. Such stories have always proved to be wrong. The meteorite is ice-cold in space, and most of its interior remains cold even after its brief fiery plunge through the atmosphere. A freshly fallen meteorite is more likely to acquire a coating of frost than to start a fire.

Second, people sometimes discover unusual-looking rocks that turn out to be meteoritic; these rocks are termed *meteorite finds.* Now that the public has become meteorite-conscious, many unusual fragments, not all of which turn out to be from space, are sent to experts each year. Some scientists divide these objects into two categories: "meteorites" and "meteorwrongs." Outside Antarctica (see the next paragraph), genuine meteorites turn up at an average rate of 25 or so per year. Most of these end up in natural history museums or specialized meteoritical laboratories throughout the world (Figure 13.5).

Since the 1980s, sources in the Antarctic have dramatically increased our knowledge of meteorites. More than

MAKING CONNECTIONS

Some Striking Meteorites

Although meteorites fall regularly onto the Earth's surface, few of them have much of an impact(!) on human civilization. There is so much water and uninhabited land on our planet that rocks from space typically fall where no one even sees them come down. But given the number of meteorites that land each year, you may not be surprised that a few have struck buildings, cars, and even people.

In November 1982, for example, Robert and Wanda Donahue of Wethersfield, Connecticut, were watching *M*°*A*°*S*°*H* on television when a 6-lb meteorite came thundering through their roof, making a hole in the living room ceiling. After bouncing back up into the attic, it finally came to rest under their dining room table.

Eighteen-year-old Michelle Knapp of Peekskill, New York, got quite a surprise one morning in October 1992. She had just purchased her very first car, her grandmother's 1980 Chevy Malibu. But she awoke to find its rear end mangled and a crater in the family driveway—thanks to a 3-lb meteorite. Michelle was not sure whether to be devastated by the loss of her car or thrilled by all the media attention.

In June 1994, José Martín and his wife were driving from Madrid, Spain, to a golfing vacation when a fist-sized meteorite crashed through the windshield of their car, bounced off the dashboard, broke José's little finger, and then landed in the back seat. Before Martín, the most recent person known to have been struck by a meteorite was Annie Hodges of Sylacauga, Alabama. In November 1954, she was napping on a couch when a meteorite came through the roof, bounced off a large radio set, and hit her first on the arm and then on the leg.

In a recent compilation, amateur astronomer and meteorite historian Christopher Spratt found 61 recorded instances between 1790 and 1990 of meteorites striking human habitats or machinery (and we presume other cases that have gone unreported or unrecorded). Although his

John McAuliffe, deputy chief of the Wethersfield, Connecticut, Fire Department, examines the hole made by a meteorite that crashed into the home of Robert and Wanda Donahue on November 8, 1982.

compilation makes for an impressive list, two centuries are a long time and the Earth is a big place. The chance of you or your property being hit by a meteorite is so small that it is certainly not worth losing sleep over.

ten thousand meteorites have been recovered from the Antarctic as a result of the motion of the ice in some parts of that continent (Figure 13.6). Meteorites that fall in regions where ice accumulates are buried and then carried slowly to other areas where the ice is gradually worn away. After thousands of years, the rock again finds itself on the surface, along with other meteorites carried to these same locations. The ice thus concentrates the meteorites that have fallen both over a large area and over a long period of time. Once on the surface, the rocks stand out in contrast

with the ice and are thus easier to spot than in other places on our rocky world.

13.2.3 Meteorite Classification

The meteorites in our collections have a wide range of compositions and histories, but traditionally they have been placed into three broad classes. First are the **irons,** composed of nearly pure metallic nickel-iron. Second are the **stones,** the term used for any silicate or rocky meteorite.

■ **FIGURE 13.5**

Meteorite Find This 1906 photo shows a 15-ton iron meteorite found in the Willamette Valley in Oregon. Although known to Native Americans in the area, it was "discovered" by an enterprising local farmer in 1902, who proceeded to steal it and put it on display. It was eventually purchased for the American Museum of Natural History and is now on display in the museum's Rose Center in New York City as the largest iron meteorite in the United States.

J. O. Wheeler, courtesy Department of Library Services, American Museum of Natural History

Third are the much rarer **stony-irons,** made (as the name implies) of mixtures of stone and metallic iron (Figure 13.7).

Of these three types, the irons and stony-irons are the most obviously extraterrestrial because of their metallic content. Pure iron almost never occurs naturally on Earth; it is generally found here as an oxide (chemically combined with oxygen) or other mineral ore. Therefore, if you ever come across a chunk of metallic iron, it is sure to be either man-made or a meteorite.

The stones are much more common than the irons but more difficult to recognize. Often laboratory analysis is required to demonstrate that a particular sample is really of extraterrestrial origin, especially if it has lain on the ground for some time and been subject to weathering. The most

NASA

■ **FIGURE 13.6**

Antarctic Meteorite An iron meteorite lying on the Antarctic ice just before being added to our collection.

Chip Clark

■ **FIGURE 13.7**

A Variety of Meteorite Types At the top left is a piece of the Allende carbonaceous meteorite, with white inclusions that may date back to before the formation of the solar nebula. At the top right is a fragment of the iron meteorite responsible for the formation of Meteor Crater in Arizona (see Chapter 12). At bottom left is a piece of the Imilac stony-iron meteorite, a beautiful mixture of green olivine crystals and metallic iron. At bottom right is part of the Mern primitive stony meteorite.

TABLE 13.2 *Frequency of Occurrence of Meteorite Classes*

Class	Falls (%)	Finds (%)	Antarctic (%)
Primitive stones	88	51	85
Differentiated stones	8	1	12
Irons	3	42	2
Stony-irons	1	5	1

scientifically valuable stones are those collected immediately after they fall or the Antarctic samples preserved in a nearly pristine state by the ice.

Table 13.2 summarizes the frequencies of occurrence of the different classes of meteorites among the fall, find, and Antarctic categories.

13.2.4 Ages and Compositions of Meteorites

It was not until the ages of meteorites were measured and their compositions analyzed in detail that scientists appreciated their true significance. The meteorites include the oldest and most primitive materials available for direct study in the laboratory. The ages of stony meteorites can be determined from the careful measurement of radioactive isotopes and their decay products, as described in Chapter 6. Almost all meteorites have radioactive ages between 4.50 and 4.56 billion years, as old as any ages we have measured in the solar system. The few younger exceptions are igneous rocks that have been ejected from cratering events on the Moon or Mars.

The average age for the most primitive meteorites, calculated using the most accurate values now available for radioactive half-lives, is 4.55 billion years, with an uncertainty of less than 0.01 billion years. This value (which we round off to 4.5 billion years in this book) is taken to represent the *age of the solar system*—the time since the first solids condensed and began to form into larger bodies.

The traditional classification of meteorites into irons, stones, and stony-irons is easy to use because it is obvious from inspection which category a meteorite falls into (although it may be much more difficult to distinguish a meteoritic stone from a terrestrial rock). Much more significant, however, is the distinction between *primitive* and *differentiated* meteorites. The differentiated meteorites are fragments of larger parent bodies that were molten before they broke up, allowing the denser materials (such as metals) to sink to their centers. Like many rocks on Earth, they have been subject to a degree of chemical reshuffling, with the different materials sorted according to density. Differentiated meteorites include the irons, which come from the metal cores of their parent bodies, and the stony-

irons, which probably originate in regions between a metal core and a stony mantle.

13.2.5 The Most Primitive Meteorites

For information on the *earliest* history of the solar system, we turn to the most primitive meteorites—those made of materials that have *not* been subject to great heat or pressure since their formation. We can look at the spectrum of sunlight reflected from asteroids and compare their compositions with those of primitive meteorites. Such analysis indicates that their parent bodies are almost certainly asteroids. Recall from Chapter 12 that asteroids are believed to be fragments left over from the formation process of the solar system; thus it makes sense that they should be the parent bodies of the primitive meteorites.

The great majority of the meteorites that reach the Earth are primitive stones. Many of them are composed of light-colored gray silicates with some metallic grains mixed in, but there is also an important group of darker stones called *carbonaceous meteorites*. As their name suggests, these meteorites contain carbon, but we also find various complex organic molecules in them—chemicals based on carbon, which on Earth are the simple building blocks of life. In addition, some of them contain chemically bound water, and many are depleted in metallic iron. The carbonaceous meteorites probably originate among the dark, carbonaceous C-type asteroids concentrated in the outer part of the asteroid belt. Among the most useful of these meteorites have been the Allende meteorite that fell in Mexico (see Figure 13.7), the Murchison meteorite that fell in Australia (both in 1969), and the much more recent Tagish Lake meteorite that landed in a winter snowdrift on Taggish Lake, Canada, in 2000. (The fragile bits of dark material from the Tagish Lake meteorite were readily visible against the white snow, although at first they were mistaken for wolf droppings!)

The Murchison meteorite is known for the variety of organic chemicals it has yielded. Most of the carbon compounds in carbonaceous meteorites are complex, tarlike substances that defy exact analysis. Murchison also contains 16 amino acids (the building blocks of proteins), 11 of which are rare on Earth. The most remarkable thing about them is that they include equal numbers with right-handed and left-handed molecular symmetry. Amino acids can have either kind of symmetry, but all life on Earth has evolved using only the left-handed versions to make proteins. The presence of both kinds of amino acids clearly demonstrates that the ones in the meteorite had a nonterrestrial origin!

These naturally occurring amino acids and other complex organic molecules in Murchison—formed without the benefit of the sheltering environment of planet Earth—show that a great deal of interesting chemistry must have taken place when the solar system was forming. Perhaps some of the molecular building blocks of life on Earth were first delivered by primitive meteorites and comets. This is an interesting idea because our planet was probably much too hot for any organic materials to survive its earliest history. But after the Earth's surface cooled, the asteroid and comet fragments that pelted it could have refreshed its supply of organic materials.

13.3 FORMATION OF THE SOLAR SYSTEM

As we have seen, the comets, asteroids, and meteorites are surviving remnants from the processes that formed the solar system. The planets and the Sun, of course, also are the products of the formation process, although the material in them has undergone a wide range of changes. We are now ready to put together the information from all these objects to discuss what is known about the origin of the solar system.

13.3.1 Observational Constraints

There are certain basic properties of the planetary system that any theory of its formation must explain. These may be summarized under three categories: motion constraints, chemical constraints, and age constraints. We call them *constraints* because they place restrictions on our theories; unless a theory can explain the observed facts, it will not survive in the competitive marketplace of ideas that characterizes the endeavor of science.

There are many regularities to the motions in the solar system. We saw that the planets all revolve around the Sun in the same direction and approximately in the plane of the Sun's own rotation. In addition, most of the planets rotate in the same direction as they revolve and most of the satellites also move in counterclockwise orbits (when seen from the north). With the exception of the comets, the motions of the system members define a disk or frisbee shape. Nevertheless, a full theory must be prepared to deal with the exceptions to these trends, such as the retrograde rotation of Venus.

In the realm of chemistry, we saw that Jupiter and Saturn have approximately the same composition—dominated by hydrogen and helium—as the Sun and the stars. These are the two largest planets, with sufficient gravity to hold on to any gas present when and where they formed; thus we might expect them to be representative of the original material out of which the solar system formed. Each of the other members of the planetary system is, to some degree, lacking in the light elements. A careful examination of the composition of solid solar-system objects shows a striking progression from the metal-rich inner planets, through those made predominantly of rocky materials, out to objects with ice-dominated compositions in the outer solar system. The comets in the Oort cloud and the Kuiper belt are also icy objects, whereas the asteroids represent a tran-

sitional rocky composition with abundant dark, carbon-rich material.

As we saw in Chapter 6, this general chemical pattern can be interpreted as a temperature sequence: hot near the Sun and cooler as we move outward. The inner parts of the system are generally missing those materials that could not condense (form a solid) at the high temperatures found near the Sun. However, there are (again) important exceptions to the general pattern. For example, it is difficult to explain the presence of water on Earth and Mars if these planets formed in a region where the temperature was too hot for ice to condense, unless the ice or water was brought in later from cooler regions.

As far as age is concerned, we discussed that radioactive dating demonstrates that some rocks on the surface of the Earth have been present for at least 3.8 billion years, and that certain lunar samples are 4.4 billion years old. The primitive meteorites all have radioactive ages near 4.5 billion years. The age of these unaltered building blocks is considered the age of the planetary system. The similarity of the measured ages tells us that planets formed and their crusts cooled within a few hundred million years (at most) of the beginning of the solar system. Further, detailed examination of primitive meteorites indicates that they are made primarily from material that condensed or coagulated out of a hot gas; few identifiable fragments appear to have survived from before this hot-vapor stage 4.5 billion years ago.

13.3.2 The Solar Nebula

All of the foregoing constraints are consistent with the general idea, introduced in Chapter 6, that the solar system formed 4.5 billion years ago out of a rotating cloud of hot vapor and dust called the solar nebula with an initial composition similar to that of the Sun today. As the solar nebula collapsed under its own gravity, material fell toward the center, where things became more and more concentrated and hot. Increasing temperatures in the shrinking nebula vaporized most of the solid material that was originally present.

At the same time, the collapsing nebula began to rotate faster through the conservation of angular momentum (see Chapter 2). Like a figure skater pulling her arms in to spin faster, the shrinking cloud spun more quickly as time went on. Near the poles of the nebula, where orbits were slow, the nebular material fell directly into the center. Faster moving material, on the other hand, collapsed into a flat disk revolving around the central object (Figure 13.8). The existence of this disk-shaped rotating nebula explains the primary motions in the solar system that we discussed in the previous section. Formed from a rotating disk, the planets all orbit the same way.

Picture the solar nebula at the end of the collapse phase, when it was at its hottest. With no more gravitational energy (from material falling in) to heat it, most of the neb-

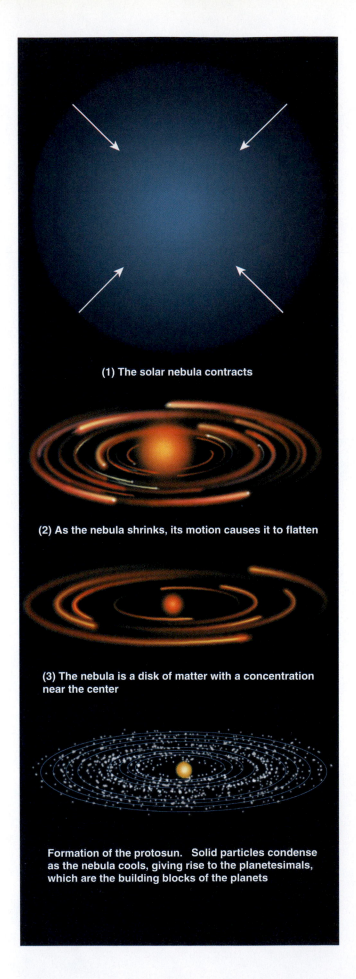

(1) The solar nebula contracts

(2) As the nebula shrinks, its motion causes it to flatten

(3) The nebula is a disk of matter with a concentration near the center

Formation of the protosun. Solid particles condense as the nebula cools, giving rise to the planetesimals, which are the building blocks of the planets

ula began to cool. The material in the center, however, where it was hottest and most crowded, formed a *star* that was able to maintain high temperatures in its immediate neighborhood by producing its own energy. Turbulent motions and magnetic fields within the disk can drain away angular momentum, robbing the disk material of some of its spin. This allowed some material to continue to fall into the growing star, while the rest of the disk gradually stabilized.

The temperature within the disk decreased with increasing distance from the Sun, much as the planets' temperatures vary with position today. As the disk cooled, the gases interacted chemically to produce compounds; eventually these compounds condensed into liquid droplets or solid grains. This is similar to the process by which raindrops on Earth condense from moist air as it rises over a mountain.

Let's look in more detail at how material condensed at different places in the maturing disk (Figure 13.9). The first materials to form solid grains were the metals and various rock-forming silicates. As the temperature dropped, these were joined throughout much of the solar nebula by sulfur compounds and by carbon- and water-rich silicates, such as those now found abundantly among the asteroids. However, in the inner parts of the disk, the temperature never dropped low enough for such materials as ice or carbonaceous organic compounds to condense, so they were lacking on the innermost planets.

Far from the Sun, cooler temperatures allowed the oxygen to combine with hydrogen and condense in the form of water (H_2O) ice. Beyond the orbit of Saturn, carbon and nitrogen combined with hydrogen to make ices such as methane (CH_4) and ammonia (NH_3). This sequence of events explains the basic chemical composition differences among various regions of the solar system.

13.3.3 Formation of the Terrestrial Planets

The grains that condensed in the solar nebula rather quickly joined into larger and larger chunks, until most of the solid material was in the form of *planetesimals* a few kilometers to a few tens of kilometers in diameter (see Section 6.4). Some planetesimals still survive today as the comets and asteroids. Others have left their imprint on the cratered surfaces of many of the worlds we studied in earlier chapters. A substantial step up in size is required, however, to go from planetesimals to planets.

■ **FIGURE 13.8**

Steps in Forming the Solar System   An artist's rough sketch of the steps in the formation of the solar system from the solar nebula. As the nebula shrinks, its rotation causes it to flatten into a disk. Much of the material is concentrated in the hot center, which will ultimately become a star. Away from the center, solid particles can condense as the nebula cools, giving rise to planetesimals, the building blocks of the planets and satellites.

FIGURE 13.9

The Chemical Condensation Sequence in the Solar Nebula The scale at left shows temperature; at right are the materials that would condense out at each temperature under the conditions expected to prevail in the nebula.

Some planetesimals were large enough to attract their neighbors gravitationally and thus to grow by the process called **accretion.** While the intermediate steps are not well understood, ultimately several dozen centers of accretion seem to have grown in the inner solar system. Each of these attracted surrounding planetesimals until it had acquired a mass similar to that of Mercury or Mars. At this stage we may think of these objects as *protoplanets*—"not quite ready for prime time" planets.

Each of these protoplanets continued to grow by the accretion of planetesimals. Every incoming planetesimal was accelerated by the gravity of the protoplanet, striking with great energy—sufficient to melt both the projectile and a part of the impact area. Soon the entire protoplanet was heated to above the melting temperature of rocks. The result was *planetary differentiation,* with heavier metals sinking toward the core and lighter silicates rising toward the surface. As they were heated, the inner protoplanets lost some of their more volatile constituents (the lighter gases), leaving more of the heavier elements and compounds behind.

13.3.4 Formation of the Giant Planets

In the outer solar system, where the raw materials included ices as well as silicates, the protoplanets grew to be much larger, with masses ten times that of the Earth. These protoplanets of the outer solar system were so large that they were able to attract and hold the surrounding gas. As the hydrogen and helium rapidly collapsed onto their cores, the giant planets were heated by the energy of contraction. Such heating also begins the process of forming stars (which we discuss in Chapter 20). But these giant planets were far too small to raise their central temperatures and pressures to the point where nuclear reactions could begin (and it is such reactions that give us our definition of a star). After glowing dull red for a few thousand years, the giant planets gradually cooled to their present state.

The collapse of gas from the nebula onto the cores of the giant planets explains how these objects acquired nearly the same hydrogen-rich composition as the Sun itself. The process was most efficient for Jupiter and Saturn; hence their compositions are most nearly "cosmic." Much less gas was captured by Uranus and Neptune, which is why these two planets have compositions dominated by the icy and rocky building blocks that made up their large cores rather than by hydrogen and helium.

13.3.5 Further Evolution of the System

All of the processes we have just described, from the collapse of the solar nebula to the formation of protoplanets, took place within a few million years. However, the story of the formation of the solar system was not complete at this stage; there were many planetesimals and other debris that did not initially accumulate to form the planets. What was their fate?

Recall from Chapter 12 that we have evidence that the comets visible to us are merely the tip of the cosmic iceberg (if you'll pardon the pun). Most comets are believed to be in the Oort cloud, far from the region of the planets. Additional comets are in the Kuiper belt, which stretches beyond the orbit of Neptune. Our models show that these icy pieces probably formed near the present orbits of Uranus and Neptune but were ejected from their initial orbits by the gravitational influence of the giant planets.

In the inner parts of the system, remnant planetesimals and perhaps several dozen protoplanets continued to whiz about. Over the vast span of time we are discussing, collisions among these objects were inevitable. Giant impacts at this stage probably stripped Mercury of part of its mantle and crust, reversed the rotation of Venus, and broke off part of the Earth to create the Moon.

Smaller-scale impacts also added mass to the inner protoplanets. Because the gravity of the giant planets could "stir up" the orbits of the planetesimals, the material impacting on the inner protoplanets could have come from almost anywhere within the solar system. In contrast to the previous stage of accretion, therefore, this new material did not represent just a narrow range of compositions (from "local" planetesimals).

As a result, much of the debris striking the inner planets was ice-rich material that had condensed in the outer part of the solar nebula. As this comet-like bombardment progressed, the Earth accumulated the water and various organic compounds that would later be critical to the formation of life. Mars and Venus probably also acquired water and organic materials from the same source.

Gradually, as the planets swept up or ejected the remaining debris, most of the leftover planetesimals disappeared. In two regions, however, stable orbits are possible where leftover planetesimals could avoid impacting the planets or being ejected from the system. These regions are the asteroid belt between Mars and Jupiter and the Kuiper belt beyond Neptune. The planetesimals (and their fragments) that survive in these special locations are what we now call the asteroids and Kuiper belt objects (KBOs).

13.4 COMPARISON WITH OTHER PLANETARY SYSTEMS

Until the middle 1990s, the practical study of the origin of planets focused on our single known example—the solar system. Although there had been a great deal of speculation about planets circling other stars, none had actually been detected. Logically enough, in the absence of data, most scientists assumed that our own system was likely to be typical. They were in for a big surprise!

13.4.1 Discovery of Other Planets

In Chapter 20 we will discuss the formation of stars and planets in some detail. About half the stars develop as binary pairs, quite different from the way our solar system formed. But as many as half the stars may have begun as our Sun did, with a protostar surrounded by a spinning circumstellar disk. The Hubble Space Telescope in orbit, as well as powerful new ground-based telescopes, enable astronomers to study directly the nearest of these circumstellar disks, in regions of space where stars are being born today.

Many of the circumstellar disks show internal structure, although the details are beyond the resolving power of current telescopes. The disks appear to be donut-shaped, with one or more gaps close to the star (Figure 13.10). Such gaps may indicate that the gas and dust in the disk have already collapsed to form a planet (Figure 13.11). The planets themselves are far too small and faint to be seen directly, but the depletion of raw materials hints at the presence of something invisible in the inner part of the circumstellar disk—and that something may well be one or more planets.

The next step, the discovery of the planets themselves, is much more difficult. Even giant planets are far too faint to be seen directly against the glare of their parent star. Observed from a great distance, the reflected light of Jupiter

M. Clampin, STScI, et al.; ACS Science Team; NASA/ESA

■ **FIGURE 13.10**

A Circumstellar Disk This Hubble Space Telescope image shows a dust disk around the young star HD 141569A, about 320 LY away. The image has been computer processed in two ways: It has been tilted to show you what it would look like if you were looking down onto the disk, and false color has been added to bring out details in the disk. The black region in the center is where a mask has been used to eclipse the bright light of the star itself, letting us see the much fainter radiation from the disk. Other observations show that the inner 4.5 billion km between the star and the disk are much clearer of dust. This clear region may have been swept out by one or more planets, which cannot be detected in such images. Two other stars in this triple star system are visible at the upper left, quite overexposed.

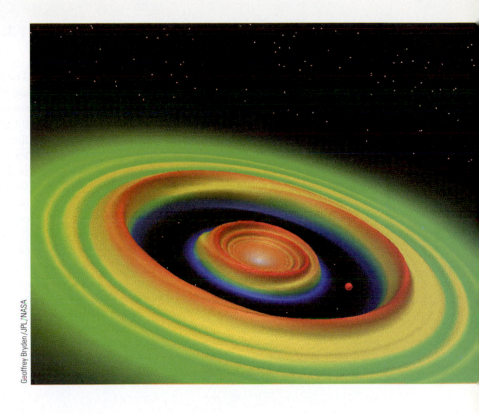

■ **FIGURE 13.11**

Model of a Circumstellar Disk with Embedded Planet This computer simulation shows the empty lane and spiral density waves that result as a giant planet is forming within the disk.

Geoffrey Bryden/JPL/NASA

is about 100 million times fainter than the light of the Sun. Planets have been detected only indirectly, mostly by their gravitational influence on the central star. As we describe in Chapter 20, the motion of the star can now be measured so precisely that we can detect the pull of any planets around them that are relatively large and close to the star.

With this technique, we can currently deduce only the presence of planets of roughly Neptune's mass or greater. Unfortunately, such techniques cannot detect terrestrial (Earth-mass) planets. Nevertheless, by the end of 2002 astronomers had discovered more than 130 giant planets, and more of them are being found all the time. We should also note that this method has been technologically feasible only for about a decade. Astronomers generally want to follow a planet's pull on its star for at least a full orbit to confirm its existence. This means that any planets that take longer than about 10 years to circle their stars would not yet have been detected. (And the farther out the planet orbits, the less it pulls, so the harder it is to detect.)

If other planetary systems were like our own, with one Jupiter-mass planet in an orbit with a 12-year period (and one Saturn mass taking 29 years), we would have found very few such systems so far. But to the great surprise of most astronomers, roughly 5 percent of observed solar-type single stars have giant planets orbiting very close—often less than the distance of Mercury from our own Sun (0.4 AU). The first such "exoplanet" to be discovered has about half the mass of Jupiter but is in an orbit just 7 million km from the solar-type star 51 Pegasi. Its orbital period is only 4.2 days (remember Mercury takes 88 days to orbit)! Such planets are being called "hot Jupiters."

13.4.2 The Configurations of Other Planetary Systems

Let's look at the first hundred external planets, all discovered between 1995 and 2002, to see what they can tell us about the way other planetary systems are configured. These hundred planets were found from a careful survey of about 700 stars similar to the Sun. Thus roughly 15 percent of these stars had planetary systems that we could detect. The other 85 percent either have no planets or (more likely) have planetary systems that do not include hot Jupiters or giant planets that take less than a decade to orbit. It is important to remember that we still cannot tell whether our solar system is typical or highly unusual. Certainly the first hundred external planets are very different from our own system, but this is at least partly a consequence of the limitations of current techniques.

One third of these hundred systems have true hot Jupiters—that is, planets of jovian mass that are closer than the orbit of Mercury. Fully half of them have giant planets with periods of less than one year—that is, lying inside the Earth's orbit. Nothing like this was expected from the theory of formation of our solar system, in which giant planets could form only in a region that was cold enough for ice to condense in the circumstellar disk.

Roughly half of the systems studied so far include one or more giant planets with eccentric orbits—eccentricities greater than that of Pluto (0.25), which has the most irregular planetary orbit in our solar system. Orbits of high eccentricity are not expected for planets that form in a spinning disk where the building blocks have nearly circular

FIGURE 13.12

Masses of External Planets This graph shows the masses of the first hundred extrasolar planets. Note that few are larger than about 13 Jupiter masses; that is, all are real planets, not failed stars. The number of planets is increasing toward small masses, in spite of the bias against finding small planets in the method we use.

orbits, but eccentricities might increase later as the result of gravitational interactions between planets. (Note that we cannot measure the inclinations of the orbits of these planets, so we do not know if they are confined to a plane as are the planets in our own system.)

Most of the planets are apparently alone; that is, only one member of a system has been found. If there are multiple planets, the others are too small or too far out to have been discovered. However, even with the limitations of our present data, about 10 percent of these systems are already known to contain two or even three planets.

One of the most interesting results from the first hundred external planets is illustrated in Figure 13.12, where we have plotted the planets according to their mass. As expected from the search techniques, most are more massive than Jupiter (since these have the biggest pull on their stars and are the easiest to find), but only three have masses in the range above 13 Jupiter masses, where the distinction between planets and very-low-mass stars becomes blurred. This result shows us that there really is a difference between planets and stars. Planets are not just small or failed stars. They form by a different process, presumably from a spinning circumstellar disk, as we have described.

Note also in Figure 13.12 that the number of detected planets increases with smaller sizes, even though small planets are much harder to detect. Although we cannot detect terrestrial planets, this result suggests that there should be many more terrestrial-mass planets than those of jovian mass.

13.4.3 Confrontation Between Theory and Data

The planetary systems discovered so far by searching for giant planets in short-period orbits do not resemble our own solar system. In particular, the presence of hot Jupiters was totally unexpected. Consequently, we have had to reassess the "standard models" for the formation of planetary systems. Science sometimes works in this way, new data contradicting our expectations. The press often talks about a scientist making experiments to "confirm" a theory. Indeed, it is comforting when new data support a hypothesis or theory and increase our confidence in an earlier result. But the most exciting and productive moments in science often come when new data contradict existing theories, forcing scientists to rethink their position and develop new and deeper insights into the way nature works.

Nothing about the new planetary systems contradicts the basic idea that planets form from the aggregation (clumping) of material within circumstellar disks. But to test this hypothesis we need to find multiplanet systems and determine whether all the planets orbit the star in the same direction and in the plane of the star's rotation. Unfortunately, the new data don't help us much in this respect, since we are missing all the smaller planets and we can't measure orbital inclinations in any case.

The biggest problem is posed by the hot Jupiters. As far as we know, a giant planet cannot be formed without the condensation of water ice, and water ice is not stable close to a star. It therefore seems likely that all the giant planets, hot or "normal," formed at a distance of several astronomical units from the star. Some of these giant planets must have subsequently migrated inward to explain the giant planets that we find orbiting in short-period orbits.

There are several suggestions for ways this migration might have occurred. Most involve interactions between the giant planets and the remnant material in the circumstellar disk from which they formed. Such interactions would have taken place when the system was very young, while material still remained in the disk. In such cases, there are processes that can transfer angular momentum from the planet to the disk, pulling the planet inward. In a way, interacting with the material in the disk "puts the brakes" on the planet, allowing it to spiral inward.

The next question is how to stop the planet from plunging into the star. Astronomers have some ideas here, but no really convincing hypothesis. Maybe a number of inward-migrating giant planets were lost (eaten by the star) and we see only the planets that were not consumed when the disk went away and the migration stopped.

One thing seems certain: Suppose a planetary system initially formed with a configuration like our own, having the giant planets on the outside and the terrestrial planets on the inside. Then such a migration by the outer planets would surely destroy any inner ones. On the other hand, we

■ **FIGURE 13.13**

Stages in the Geological History of a Terrestrial Planet In this graph, time increases downward along the left axis and the stages are labeled. Each planet is shown roughly in its present stage. The smaller the planet, the more quickly it passes through these stages.

Accretion, heating, differentiation

Formation of solid crust, heavy cratering

Widespread mare-like volcanism

Reduced volcanism, possible plate tectonics

Mantle solidification, end of tectonic activity

Cool interior, no activity

Geological history of the terrestrial planets

Venus

Earth

Mars

Mercury

Moon

are now seeing some giant planets in orbits larger than the Earth's, so such migrations may stop earlier or may not even start in some star systems. We will return to some of these ideas in the final chapter, where we consider the search for habitable planets and life beyond our own solar system.

13.5 PLANETARY EVOLUTION

While we await more discoveries of other planetary systems, let us look again at the early history of our own solar system, after the dissipation of the disk. The era of giant impacts was probably confined to the first 100 million years of solar system history, ending by about 4.4 billion years ago. Shortly thereafter, the planets cooled and began to assume their present aspects. Up until about 4 billion years ago they continued to acquire volatile materials, and their surfaces were heavily cratered from the remaining debris that hit them. However, as external influences declined, all of the terrestrial planets as well as the satellites of the outer planets began to follow their own evolutionary courses. The nature of this evolution depended on each object's composition, mass, and distance from the Sun.

13.5.1 Geological Activity

We have seen a wide range in the level of geological activity on the terrestrial planets and icy satellites. Internal sources of such activity (as opposed to pummeling from above) require energy, either in the form of primordial heat left over from the formation of a planet or from the decay of radioactive elements in the interior. The larger the planet or satellite, the more likely it is to retain its internal

heat and the more slowly it cools—this is the "baked potato effect" mentioned in Chapter 6. Therefore we are more likely to see evidence of continuing geological activity on the surface of larger worlds (Figure 13.13). Jupiter's satellite Io is an interesting exception to this rule; we saw that it has an unusual source of heat—the gravitational flexing of its interior by the tidal pull of Jupiter. Europa is probably also heated by jovian tides.

The Moon, the smallest of the terrestrial worlds, was internally active until about 3.3 billion years ago, when its major volcanism ceased. Since that time its mantle has cooled and become solid, and today even internal seismic activity has declined to almost zero. The Moon is a geologically dead world. Although we know much less about Mercury, it seems likely that this planet, too, ceased most volcanic activity about the same time the Moon did.

Mars represents an intermediate case, and it has been much more active than the Moon. The southern hemisphere crust had formed by 4 billion years ago, and the northern hemisphere volcanic plains seem to be contemporary with the lunar maria. However, the Tharsis bulge formed somewhat later, and activity in the large Tharsis volcanoes has apparently continued on and off to the present.

Earth and Venus are the largest and most active terrestrial planets. Our planet experiences global plate tectonics driven by convection in its mantle. As a result, our surface is continually reworked, and most of the Earth's surface material is less than 200 million years old. Venus has generally similar levels of volcanic activity, but unlike the Earth, it has not experienced plate tectonics. Most of its surface appears to be roughly 500 million years old. We did see that the surface of our sister planet is being modified by a kind of "blob tectonics"—where hot material from below puckers and bursts through the surface, leading to coronae, pancake volcanoes, and other such features. A better un-

Olympus Mons

Sea level

Mauna Loa   Mt. Everest   Maxwell Mountains

■ **FIGURE 13.14**

The Highest Mountains on Mars, Venus, and Earth   Mountains can rise taller
on Mars because Mars has less surface gravity and no moving plates. The vertical scale
is exaggerated by a factor of 3 to make comparison easier.

derstanding of the geological differences between Venus
and Earth is a high priority for planetary geologists.

The geological evolution of the icy satellites has been
somewhat different from that of the terrestrial planets.
Tidal energy sources have been active, and the materials nature has to work with are not the same. On these outer
worlds, we see evidence of low-temperature volcanism,
with the silicate lava of the inner planets being supplemented by sulfur and sulfur compounds on Io, and replaced
by water and other ices on other outer-planet satellites.

13.5.2 Elevation Differences

Let's look at some specific examples of how planets differ.
The mountains on the terrestrial planets owe their origins
to different processes. On the Moon and Mercury, the major mountains are ejecta thrown up by the large basin-

forming impacts that took place billions of years ago. Most
large mountains on Mars are volcanoes, produced by repeated eruptions of lava from the same vents. There are
similar (but smaller) volcanoes on Earth and Venus. However, the highest mountains on Earth and Venus are the result of compression and uplift of the surface. On the Earth,
this crustal compression results from collisions of one continental plate with another.

It is interesting to compare the maximum heights of
the volcanoes on Earth, Venus, and Mars (Figure 13.14).
On Venus and Earth, the maximum elevation differences
between these mountains and their surroundings are about
10 km. Olympus Mons, in contrast, towers more than 20
km above its surroundings and nearly 30 km above the lowest elevation areas on Mars (Figure 13.15).

One reason Olympus Mons is so much higher than its
terrestrial counterparts is that the crustal plates on Earth

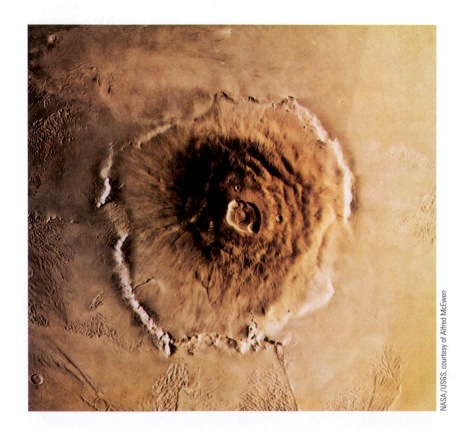

NASA/USGS, courtesy of Alfred McEwen

■ **FIGURE 13.15**

Olympus Mons   The largest martian volcano is seen from above on this spectacular
composite image created from many Viking
orbiter photographs. The volcano is nearly
500 km wide at its base and more than
20 km high.

JPL/NASA

■ **FIGURE 13.16**

Irregular Asteroid   Small objects such as asteroid Ida (shown here in multiple views taken by the Galileo spacecraft camera as it flew past) are generally irregular or elongated; they do not have strong enough gravity to pull them into a spherical shape.

never stop long enough to let a really large volcano grow. Instead, the moving plate creates a long row of volcanoes like the Hawaiian Islands. On Mars (and perhaps Venus) the crust remains stationary with respect to the underlying hot spot, and so the volcano can continue to grow for hundreds of millions of years.

A second difference relates to the strength of gravity on the three planets. The surface gravity on Venus is nearly the same as that on Earth, but on Mars it is only about one third as great. In order for a mountain to survive, its internal strength must be great enough to support its weight against the force of gravity. Volcanic rocks have known strengths, and we can calculate that on Earth 10 km is about the limit. For instance, when new lava is added to the top of Mauna Loa in Hawaii, the mountain slumps downward under its own weight. The same height limit applies on Venus, where the force of gravity is the same as Earth's. On Mars, however, with its lesser surface gravity, much greater elevation differences can be supported, which helps explain why Olympus Mons is more than twice as high as the tallest mountains of Venus or Earth.

By the way, the same kind of calculation that determines the limiting height of a mountain can be used to ascertain the largest body that can have an irregular shape. Gravity, if it can, pulls all objects into the most "efficient" shape (where all the outside points are equally distant from the center). All of the planets and larger satellites are nearly spherical, due to the force of their own gravity having pulled them into a sphere. But the smaller the object, the greater is the departure from spherical shape that the strength of its rocks can support. For silicate bodies, the limiting diameter is about 400 km; larger objects will always be approximately spherical, while smaller ones can have almost any shape (as we see in photographs of asteroids, such as Figure 13.16).

Ace◐Astronomy™   Log into AceAstronomy and select this chapter to see the Active Figure called "Planetary Atmospheres," and Astronomy Exercise "Primary Atmospheres."

13.5.3   Atmospheres

The atmospheres of the planets were formed by a combination of gas escaping from their interiors and the impacts of volatile-rich debris from the outer solar system. Each of the terrestrial planets must have originally had similar atmospheres, but Mercury was too small and too hot to retain its gas. The Moon probably never had an atmosphere, since the material composing it was depleted in volatile materials.

The predominant volatile gas on the terrestrial planets is now carbon dioxide (CO_2), but initially there were probably also hydrogen-containing gases. In this more chemically *reduced* (hydrogen-dominated) environment, there should have been large amounts of carbon monoxide (CO) and perhaps also traces of ammonia (NH_3) and methane

(CH_4). Ultraviolet light from the Sun split apart the molecules of reducing gases in the inner solar system, however. Most of the hydrogen escaped, leaving behind the oxidized atmospheres we see today on Earth, Venus, and Mars.

The fate of water was different on each of these three planets, depending on its size and distance from the Sun. Early in its history, Mars apparently had a thick atmosphere with abundant liquid water, but it could not retain those conditions. As we saw in Chapter 9, the CO_2 necessary for a substantial greenhouse effect was lost, the temperature dropped, and eventually the water froze. On Venus the reverse process took place, with a runaway greenhouse effect leading to the permanent loss of water. Only the Earth managed to maintain the delicate balance that permits liquid water to persist.

With the water gone, Venus and Mars each ended up with an atmosphere of about 96 percent carbon dioxide and a few percent nitrogen. On Earth, the presence first of water and then of life led to a very different kind of atmosphere. The CO_2 was removed and deposited in marine sediment. The proliferation of life-forms that could photosynthesize eventually led to the release of more oxygen than natural chemical reactions can remove from the atmosphere. As a result, thanks to the life on its surface, Earth finds itself with a great deficiency of CO_2, with nitrogen as the most abundant gas, and with the only planetary atmosphere that contains free oxygen.

In the outer solar system, Titan is the only satellite with a substantial atmosphere. Apparently this object contained sufficient volatiles—such as ammonia, methane, and nitrogen—to form an atmosphere. It was also large and cold enough to retain these gases. Thus, as we saw in Chapter 11, today Titan's atmosphere consists primarily of nitrogen. Compared with those on the inner planets, temperatures on Titan are too low for either carbon dioxide or water to be in vapor form. With these two common volatiles frozen solid, it is perhaps not too surprising that nitrogen has ended up as the primary atmospheric constituent.

We see that nature, starting with one set of chemical constituents, can fashion a wide range of final atmospheres appropriate to the conditions and history of each world. The atmosphere we have on Earth is the result of many eons of evolution and adaptation. And, as we saw, it can be changed by the actions of the life-forms that inhabit the planet.

13.6 CONCLUSION

We now come to the end of our study of the planetary system. Although we have learned a great deal about the other planets during the past few decades of spacecraft exploration, much remains unknown. Now that we have adopted the comparative approach to study these different worlds and the processes that form them and their environments, we see how limited our knowledge really is. More recently, we have had some surprises when we found planets around a number of nearby stars. Even such generalizations that giant planets form only in the outer parts of a planetary system may need reexamination in light of some of these new discoveries.

For now, we end by noting that we do not even understand our own planet well enough to assess the consequences of the changes humans are inadvertently inflicting on the atmosphere and oceans. The exploration of the solar system is one of the greatest human adventures, and it will continue, if only so that we can learn to appreciate and protect our own planet.

SURFING THE WEB

Gary Kronk's Meteor Showers and Comets Site:
comets.amsmeteors.org/
Amateur astronomer Gary Kronk, author of a book on meteor showers, provides lots of information on major and minor showers, meteor legends, fireballs, observing hints, and links.

The List of the Largest Meteorites "in Captivity":
www.jensenmeteorites.com/largestmeteorites.htm

Sky Online Meteor Page:
skyandtelescope.com/observing/objects/meteors/
Excellent site for serious or casual meteor observers, with lots of up-to-date data, background, information hints for viewing, and more.

Other Good Meteorite Sites:

- *Nine Planets Meteors, Meteorites, and Impacts Page:* seds.lpl.arizona.edu/nineplanets/nineplanets/meteorites.html

- *Views of the Solar System Meteorites Page:* www.solarviews.com/eng/meteor.htm

- *David Kring's Introduction to Meteorites and Their Properties:* meteorites.lpl.arizona.edu

- *Mars Meteorites Page at JPL:* www.jpl.nasa.gov/snc

For general and comparative information about the solar system, see many of the sites recommended for Chapter 6. For information on planets around other stars, see the sites we suggest in Chapter 20.

13.1 When a fragment of interplanetary dust strikes the Earth's atmosphere, it burns up to create a **meteor.** Streams of dust particles traveling through space together produce **meteor showers,** in which we see meteors diverging from a spot in the sky called the *radiant* of the shower. Many meteor showers recur each year and are associated with particular comets.

13.2 Meteorites are the debris from space (mostly asteroid fragments) that survive to reach the surface of the Earth. Meteorites are called *finds* or *falls* according to how they are found; the most productive source today is the Antarctic ice cap. Meteorites are classified as **irons, stony-irons,** or **stones** according to their composition. Most stones are primitive objects, dated to the origin of the solar system 4.5 billion years ago. The most primitive are the *carbonaceous meteorites,* such as Murchison and Allende. These can contain a number of organic molecules.

13.3 A viable theory of solar system formation must take into account motion constraints, chemical constraints, and age constraints. Meteorites, comets, and asteroids are survivors of the *solar nebula* out of which the solar system formed. This nebula was the result of the collapse of an interstellar cloud of gas and dust, which contracted (conserving its angular momentum) to form our star, the Sun, surrounded by a thin, spinning disk of dust and hot vapor. Condensation in the disk led to the formation of *planetesimals,* which became the building blocks of the planets. **Accretion** of infalling materials heated the planets, leading to their differentiation. The giant planets were also able to attract and hold gas from the solar nebula. After a few million years of violent impacts, most of the debris was swept up or ejected, leaving only the asteroids and cometary remnants surviving to the present.

All of these violent events had terminated by about 4.4 billion years ago.

13.4 The first planets circling distant solar-type stars were found in 1995, and by the end of 2002 astronomers had data on more than one hundred, all of them giant planets. Roughly 15 percent of stars like the Sun have giant planets in relatively short-period orbits, and at least 5 percent have "hot Jupiters" within 1 AU from their stars. Many of the new planets are also in eccentric orbits. These systems are fundamentally different from our own, and they suggest that giant planets can migrate inward from their place of formation where it is cold enough for ice to form. Astronomers eagerly await the development of more sensitive tools so that we can detect smaller planets and perhaps find other systems that look more like our own. Current data hint that there should be many more small (terrestrial type) planets than giants.

13.5 After their common beginning, each of the planets evolved on its own path. The different possible outcomes are illustrated by comparison of the terrestrial planets (Earth, Venus, Mars, Mercury, and the Moon). All are rocky, differentiated objects. The level of geological activity is proportional to mass: greatest for Earth and Venus, less for Mars, and absent for the Moon and Mercury. Mountains can result from impacts, volcanism, or uplift. Whatever their origin, higher mountains can be supported on smaller planets that have less surface gravity. The atmospheres of the terrestrial planets may have acquired volatile materials from comet impacts. The Moon and Mercury lost their atmospheres; most volatiles on Mars are frozen due to its greater distance from the Sun; and Venus retained CO_2 but lost H_2O when it developed a massive greenhouse effect. Only Earth still has liquid H_2O on its surface and hence can support life.

INTER-ACTIVITY

A Ever since the true (cosmic) origin of meteorites was understood, people have tried to make money selling them to museums and planetaria. More recently, a growing number of private collectors have been interested in purchasing meteorite fragments, and a network of dealers (some more reputable than others) has sprung up to meet this need. What does your group think of all this? Who should own a meteorite? The person on whose land it falls, the person who finds it, or the local, state, or federal government where it falls? Why? Should there be any limit to what people charge for meteorites? Or should all meteorites be the common property of humanity? (If you can, try to do research on what the law is now in this area.)

B Your group has been formed to advise a very rich person who wants to buy some meteorites but is afraid of being cheated and being sold some Earth rocks. How would you advise your client to make sure that the meteorites she buys are authentic?

C Your group is a committee set up to give advice to NASA about how to design satellites and telescopes in space to minimize the danger of meteor impacts. Remember that the heavier a satellite is, the harder (more expensive) it is to launch. What would you include in your recommendations?

D Discuss what you would do if you suddenly find that a small meteorite had crashed in or near your home. Whom would you call first, second, third? What would you do with the sample? (And would any damage to your home be covered by your insurance?)

E A friend of your group really wants to see a meteor shower. The group becomes a committee to assist her in fulfilling this desire. What time of year would be best? What equipment would you recommend she gets? What advice would you give her?

F Thinking that all giant planets had to be far from their stars (because the ones in our solar system are) is an example of making theories without having enough data (examples). Can your group make a list of other instances in science (and human relations) where we have made incorrect judgments without having explored enough examples?

REVIEW QUESTIONS

Ace Astronomy™ Assess your understanding of this chapter's topics with additional quizzing and animations at **http://ace.brookscole.com/voyages**

1. A friend of yours who has not taken astronomy sees a meteor shower. The next day she confides in you that she was concerned that the stars in the Big Dipper (her favorite star pattern) might be the next ones to go. How would you put her mind at ease?
2. In what ways are meteorites different from meteors? What is the probable origin of each?
3. How are comets related to meteor showers?
4. What do we mean by primitive material? How can we tell if a meteorite is primitive?
5. Describe the solar nebula, and outline the sequence of events within the nebula that gave rise to the planetesimals.
6. Why do the giant planets and their satellites have compositions different from those of the terrestrial planets?
7. How do the planets discovered so far around other stars differ from those in our own solar system? List at least two ways.
8. Explain the role of impacts in planetary evolution, including both giant impacts and more modest ones.
9. Why are some planets more geologically active than others?
10. Summarize the origin and evolution of the atmospheres of Venus, Earth, and Mars.

THOUGHT QUESTIONS

11. What methods do scientists use to distinguish a meteorite from terrestrial material?

12. Explain why we do not believe meteorites come from comets, or meteors from asteroids.

13. Why do iron meteorites represent a much higher percentage of finds than of falls?

14. Why is it more useful to classify meteorites according to whether they are primitive or differentiated rather than whether they are stones, irons, or stony-irons?

15. Which meteorites are the most useful for defining the age of the solar system? Why?

16. Suppose a new primitive meteorite is discovered (sometime after it falls in a field of soybeans) and analysis reveals that it contains a trace of amino acids, all of which show the same rotational symmetry (unlike the Murchison meteorite). What might you conclude from this finding?

17. Suppose you were to come back and take this astronomy course again in 10 years (heaven forbid!). How would our information about planets around other stars be different from the information we have today? Explain.

18. How do we know when the solar system formed? Usually we say that the solar system is 4.5 billion years old. To what does this age correspond?

19. What do you think would happen to our planetary system if Jupiter began to migrate inward to the orbit of Mercury? What would be the fate of the Earth and other terrestrial planets? Why is this not likely to happen (and thus is not necessary to worry about)?

20. We have seen how Mars can support greater elevation differences than can Earth or Venus. According to the same arguments, the Moon should have higher mountains than any of the other terrestrial planets, yet we know it does not. What is wrong with applying the same line of reasoning to the mountains on the Moon?

FIGURING FOR YOURSELF

We can use the concept of angular momentum to trace the evolution of the collapsing solar nebula. The angular momentum of an object is proportional to the square of its size (diameter) times its period of rotation ($D^2 \times P$). If angular momentum is conserved, then any change in the size of a nebula must be compensated for by a proportional change in period, in order to keep $D^2 \times P$ a constant.

Suppose that the solar nebula began with a diameter of 10,000 AU and a rotation period of 1 million years. Let's calculate its rotation period when it has shrunk to the size of Pluto's orbit, which Appendix 7 tells us has a radius of about 40 AU. This means the diameter is about 80 AU. If the subscript 1 denotes the initial state before the collapse and 2 denotes the final state at Pluto's orbit, then

$$\frac{P_2}{P_1} = \left(\frac{D_2}{D_1}\right)^2 = \left(\frac{80}{10,000}\right)^2 = (0.008)^2 = 0.000064$$

With P_1 equal to a 1,000,000 years, P_2, the new rotation period, is 64 years. This is a lot shorter than the actual time Pluto takes to go around the Sun, but it gives you a sense of the kind of speeding up the conservation of angular momentum can produce. As we noted in the chapter, other mechanisms helped the material in the disk lose angular momentum before the planets fully formed.

21. What would the rotation period of the nebula in our example be when it had shrunk to the size of Jupiter's orbit?

22. How long would material take to go around if the nebula in our example became the size of the Earth's orbit?

23. Consider the differentiated meteorites. We think the irons are from the cores, the stony-irons are from the interfaces between mantles and cores, and the stones are from the mantles of their differentiated parent bodies. If these parent bodies were like the Earth, what fraction of the meteorites would you expect to consist of irons, stony-irons, and stones? Is this consistent with the observed numbers of each?

24. Estimate the maximum height of the mountains on a hypothetical planet similar to the Earth but with twice the surface gravity of our planet.

SUGGESTIONS FOR FURTHER READING

On Meteors
Beatty, J. "Storm Chasers" in *Sky & Telescope*, June 2000, p. 42. On using multiple aircraft to observe the Leonid meteor shower. (Several other articles on meteors are in this issue.)

Bone, N. *Meteors*. 1993, Sky Publishing. A good manual and information source.

Kronk, G. "Halley's Comet Crumbs" in *Sky & Telescope*, May 2002, p. 92. On the Eta Aquarids meteor shower.

Litmann, M. *The Heavens on Fire*. 1998, Cambridge U. Press. Best introduction to meteor showers in history and science.

Reynolds, M. *Falling Stars: A Guide to Meteors & Meteorites*. 2001, Stackpole Books. A basic introduction.

On Meteorites
Alper, J. "It Came from Outer Space" in *Astronomy*, Nov. 2002, p. 36. On the analysis of organic materials in meteorites.

Bevan, A. and deLaeter, J. *Meteorites: A Journey through Space & Time*. 2002, Smithsonian Inst. Press. Introduction to meteorite science.

Gallant, R. "Saga of the Lump: The Pallas Meteorite" in *Sky & Telescope*, Jan. 1999, p. 50.

McSween, H. *Meteorites and Their Parent Planets*. 1987, Cambridge U. Press. A good general introduction by a geologist.

Norton, O. *Rocks from Space*, 2nd ed. 1998, Mountain Press Publishing. A guide for amateurs, with information on science, folklore, and collecting.

Phillips, J. "Meteorite Field Guide" in *Mercury*, Nov./Dec. 2001, p. 32. On identifying and collecting.

Schaefer, B. "Meteor[ite]s that Changed the World" in *Sky & Telescope*, Dec. 1998, p. 68. On famous cases, including Tunkuska, Chicxulub, and the Black Stone of the Ka'aba.

Spratt, C. and Stephens, S. "Against All Odds: Meteorites That Have Struck" in *Mercury*, Mar./Apr. 1992, p. 50.

For readings on the discovery of planets around other stars, see Chapter 20.

On the Evolution of the Solar System
See the readings recommended in Chapter 6, plus

Jayawardhana, R. "Spying on Planetary Nurseries" in *Astronomy*, Nov. 1998, p. 62. On dust disks around other stars that may be the birthplaces of other planetary systems.

Kasting, J. "How Climate Evolved on the Terrestrial Planets" in *Scientific American*, Feb. 1988, p. 90.

Rubin, A. "Microscopic Astronomy" in *Sky & Telescope*, July 1995, p. 24. Discusses the study of solar system history from analysis of samples.

Wood, J. "Forging the Planets: The Origin of Our Solar System" in *Sky & Telescope*, Jan. 1999, p. 36. Good overview.

Yulsman, T. "From Pebbles to Planets" in *Astronomy*, Feb. 1998, p. 56. Comparing the origin of our system to others.

SOHO; ESA & NASA

Space Weather and Its Impact on Earth   The first image shows a storm on the Sun, which blasts matter into space. This matter, which is composed of charged particles, mainly the nuclei of hydrogen atoms, speeds out into the solar system. If headed in the right direction, the particles will slam into the Earth. In the middle frame we see the northern hemisphere of the Earth, and the glow shows where charged particles are entering the Earth's atmosphere. The final image, taken by Jan Curtis, shows the visible evidence of space weather—an aurora in the sky over Alaska.

The adventure of the Sun is the great natural drama by which we live, and not to have joy in it and awe of it, not to share in it, is to close a dull door on nature's sustaining and poetic spirit.

Henry Beston, from "Midwinter" in *The Outermost House* (1928)

THINKING AHEAD

Can we trust the Sun? This may seem like a silly question—after all, the Sun has been rising in the east and setting in the west every day for billions of years. But the Sun has *not* put out exactly the same amount of heat during all that time; it has been both hotter and cooler than it is today. Since life on Earth ultimately depends on the Sun's heat and light, the way the Sun's energy output changes, and on what timescales, are questions of personal importance to those of us who share the Sun's cosmic neighborhood.

The Sun is, by many measures, a rather ordinary star—not unusually hot or cold, old or young, large or small. Indeed, we are lucky that the Sun is typical. By studying it, we learn much that helps us understand the stars in general. Just as studies of the Earth help us understand observations of the more distant planets, so too the Sun serves as a guide to astronomers in interpreting the messages contained in the light we receive from distant stars.

In this chapter we describe what the Sun looks like, how it changes with time, and how those changes affect the Earth. Some of the basic characteristics of the Sun are listed in Table 14.1, although many of the terms in that table may be unfamiliar to you until you've read further. Figure 14.1 shows what the Sun would look like if we could see all parts of it from the center to its outer atmosphere; the terms in the figure will also become familiar to you as you read on.

Ace ◯ **Astronomy** ™   The AceAstronomy icon throughout the text indicates an opportunity for you to test yourself on key concepts and to explore animations and interactions on the AceAstronomy website at **http://ace.brookscole.com /voyages**

TABLE 14.1 *Solar Data*

Datum	How Found	Value
Mean distance	Radar reflection from planets	1 AU (149,597,892 km)
Maximum distance from Earth		1.521×10^8 km
Minimum distance from Earth		1.471×10^8 km
Mass	Orbit of Earth	333,400 Earth masses (1.99×10^{30} kg)
Mean angular diameter	Direct measure	31′59″.3
Diameter of photosphere	Angular size and distance	109.3 × Earth diameter (1.39×10^6 km)
Mean density	Mass/volume	1.41 g/cm^3 (1400 kg/m^3)
Gravitational acceleration at photosphere (surface gravity)	GM/R^2	27.9 × Earth surface gravity = 273 m/s^2
Solar constant	Instrument sensitive to radiation at all wavelengths	1370 W/m^2
Luminosity	Solar constant × area of spherical surface 1 AU in radius	3.8×10^{26} W
Spectral class	Spectrum	G2 V
Effective temperature	Derived from luminosity and radius of Sun	5800 K
Rotation period at equator	Sunspots and Doppler shift in spectra taken at the edge of the Sun	24 d 16 h
Inclination of equator to ecliptic	Motions of sunspots	7°10′.5

14.1 THE VISIBLE SUN

The Sun, like all stars, is an enormous ball of extremely hot gas, shining under its own power. And we do mean enormous. The Sun has enough volume (takes up enough space) to hold about 1.3 million Earths. Most of the Sun's regions are shielded from our view by its outside layers, just as the Earth shows only its surface and atmosphere to our direct observation. The Sun has no solid surface, but it does have an extensive atmosphere—the only part of the Sun we can actually see. In this chapter we describe the solar atmosphere. The solar interior and its role in producing the energy that warms the Earth will be the subject of the next chapter.

SOHO: ESA & NASA

■ **FIGURE 14.1**

The Parts of the Sun   This drawing shows the different parts of the Sun, from the hot core where the energy is generated through regions where energy is transported outward, first by radiation and then by convection and then out through the solar atmosphere. The parts of the atmosphere are also labeled—the photosphere, chromosphere, and corona. The artist shows some typical features in the atmosphere, such as coronal holes and prominences.

14.1.1 The Composition of the Sun

Let's begin by asking what the Sun's atmosphere is made of. As explained in Chapter 4, we can use its *absorption line spectrum* to determine what elements are present. It turns out that the Sun contains the same elements as the Earth, but *not* in the same proportions. About 73 percent of the Sun's mass is hydrogen, and another 25 percent is helium. All the other chemical elements (including those we know and love in our own bodies, such as carbon, oxygen, and nitrogen) make up only 2 percent of our star. The ten most abundant gases in the Sun's visible surface layer are listed in Table 14.2. Examine that table and notice how different the composition of the Sun's outer layer is from the Earth's crust, where we live. (In our planet's crust, the three most abundant elements are oxygen, silicon, and aluminum.) But the makeup of the Sun is quite typical of the stars in general.

The fact that the Sun and the stars all have similar compositions and are made up of mostly hydrogen and helium was first shown in a brilliant thesis in 1925 by Cecilia Payne-Gaposchkin, the first woman to get a PhD in astronomy in the United States (Figure 14.2). However, the idea that the simplest light gases, hydrogen and helium, were the most abundant elements in stars was so unexpected and so shocking that she assumed her analysis of the data must be wrong. At the time, she wrote, "The enormous abundance derived for these elements in the stellar atmosphere is almost certainly not real." Even scientists sometimes find it hard to accept new ideas that do not agree with what everyone "knows" to be right!

Before Payne's work, everyone assumed that the composition of the Sun and stars would be much like that of the Earth. It was three years after her thesis that other studies proved beyond doubt that the enormous abundance of hydrogen and helium in the Sun is indeed real. (And, as we will see, the composition of the stars is much more typical

Harvard University Archives

■ FIGURE 14.2

Cecilia Payne-Gaposchkin Her 1925 PhD thesis laid the foundations for understanding the composition of the Sun and the stars. Yet she was not given a formal appointment at Harvard, where she worked, until 1938 and was not appointed a professor until 1956.

of the makeup of the universe than the odd concentration of heavier elements that characterizes our planet.)

Most of the elements found in the Sun are in the form of atoms, but several types of molecules, including water vapor and carbon monoxide, have been identified in the light from the Sun's cooler regions, such as the sunspots (see Section 14.2). The atoms and molecules contained in the Sun are all in the form of gases: The Sun is so hot that no matter can survive as a liquid or a solid.

In fact, the Sun is so hot that many of the atoms in it are *ionized*—stripped of one or more of their electrons. This separation between electrons and protons means that the Sun is an electrically charged environment—quite different from the neutral one in which you are reading this book. As we saw in Chapter 4, when charged particles flow, magnetic fields are generated. As we shall see, the strong, complicated magnetism of the Sun plays a vital role in how it appears to us.

14.1.2 The Solar Photosphere

The Earth's air is generally transparent. But on a smoggy day it can become opaque, which prevents us from seeing through it past a certain point. Something similar happens in the Sun. Its outer atmosphere is transparent, allowing us to look a short distance through it. But when we try to look through the atmosphere deeper into the Sun, our view is blocked. The **photosphere** is the layer where the Sun becomes opaque and marks the boundary past which we cannot see (Figure 14.3).

The energy that emerges from the photosphere was originally generated deep inside the Sun (more on this in Chapter 15). This energy is in the form of photons, which

| | **TABLE 14.2** | *The Abundance of Elements in the Sun* |

Element	Percentage by Number of Atoms	Percentage by Mass
Hydrogen	92.0	73.4
Helium	7.8	25.0
Carbon	0.02	0.20
Nitrogen	0.008	0.09
Oxygen	0.06	0.8
Neon	0.01	0.16
Magnesium	0.003	0.06
Silicon	0.004	0.09
Sulfur	0.002	0.05
Iron	0.003	0.14

Approx. size of Earth ➡ •

SOHO: NASA & ESA

■ **FIGURE 14.3**

Solar Photosphere plus Sunspots This photograph shows the photosphere—
the visible surface of the Sun. Also shown is an enlarged image of one of the groups of
sunspots; the size of the Earth is shown for comparison. Sunspots appear darker because
they are cooler than their surroundings. The typical temperature at the center of a large
sunspot is about 4000 K, while the photosphere has a temperature of about 5800 K.

make their way slowly toward the solar surface, taking about a million years to reach the photosphere. Inside the Sun, a photon of energy emitted by one atom travels only a short distance before it is "grabbed" (i.e., absorbed) by another. Outside the Sun, we can observe only those photons that are emitted in the solar photosphere, where the density of atoms is low and the photons can finally escape from the Sun without being captured.

As an analogy, imagine you are attending a big campus rally, and you have found a prime spot near the center of the action. Your friend arrives late and calls you on your cell phone to ask you to join her at the edge of the crowd. You decide that friendship is worth more than a prime spot, and so you work your way out through the dense crowd to meet her. You can move only a short distance before bumping into someone, changing direction, and trying again, making your way slowly to the outside edge of the crowd. Your friend can't see you until you get very close to the edge because of all the bodies in the way. So too photons making their way through the Sun are constantly bumping into atoms, changing direction, working their way slowly outward, and becoming visible only when they reach the atmosphere of the Sun, where the density of atoms is too low to block their outward progress.

Astronomers have found that the solar atmosphere changes from almost perfectly transparent to almost completely opaque in a distance of just over 400 kilometers (km); it is this thin region that we call the *photosphere,* a word that comes from the Greek for "light sphere." When astronomers speak of the "diameter" of the Sun, they mean the size of the region surrounded by the photosphere.

The photosphere looks sharp only from a distance. If you were falling into the Sun, you would not feel any surface but would just sense a gradual increase in the density of the gas surrounding you. It is much the same as falling through a cloud while skydiving. From far away, the cloud looks as if it has a sharp surface, but you do not feel a surface as you fall into it. (One big difference between these two scenarios, however, is temperature. The Sun is so hot that you would be vaporized long before you reached the photosphere. Skydiving in the Earth's atmosphere is much safer!)

We might note that the atmosphere of the Sun is not a very dense layer compared to the air in the room where you are reading this book. At a typical point in the photosphere, the pressure is less than 10 percent of the Earth's pressure at sea level, and the density is about one ten-thousandth of the Earth's atmospheric density at sea level.

Observations with telescopes show that the photosphere has a mottled appearance resembling grains of rice spilled on a dark tablecloth. This structure of the photosphere is now generally called **granulation** (see Figure 14.8, p. 319). Granules, which are typically 700 to 1000 km in diameter, appear as bright areas surrounded by narrow, darker regions. The lifetime of an individual granule is only 5 to 10 minutes.

The motions of the granules can be studied by examining the Doppler shifts in the spectra of gases just above them (see Section 4.6). The bright granules are columns of hotter gases rising at speeds of 2 to 3 km/s from below the photosphere. As this rising gas reaches the photosphere, it spreads out and sinks down again into the darker regions between the granules. The centers of the granules are hotter than the intergranular regions by 50 to 100 degrees Kelvin (K).

The granules are the tops of *convection* currents of gas rising through the photosphere. Convection is the rising of a hot liquid or gas, transferring energy from the hotter layers underneath to the cooler layers above. As the gas cools, it falls back down, and the area looks darker from above. (You can see the same sort of hot currents rising in a pot of boiling water as you make more coffee to stay awake while reading your astronomy book!)

14.1.3 The Chromosphere

The Sun's outer gases extend far beyond the photosphere (Figure 14.4). Because they are transparent to most visible radiation and emit only a small amount of light, these outer layers are difficult to observe. The region of the Sun's atmosphere that lies immediately above the photosphere is called the **chromosphere.** Until this century, the chromosphere was visible only when the photosphere was concealed by the Moon during a total solar eclipse (see Section 3.7). In the 17th century, several observers described what appeared to them as a narrow red "streak" or "fringe" around the edge of the Moon during a brief instant after the Sun's photosphere had been covered. The name *chromosphere,* from the Greek for "colored sphere," was given to this red streak.

Observations made during eclipses show that the chromosphere is about 2000 to 3000 km thick, and its spectrum consists of bright emission lines, indicating that this layer is composed of hot, transparent gases emitting light at discrete wavelengths. The reddish color of the chromosphere arises from one of the strongest emission lines in the visible part of its spectrum—the bright red line caused by hydrogen, the element that, as we have already seen, dominates the composition of the Sun.

In 1868, observations of the chromospheric spectrum revealed a yellow emission line that did not correspond to any previously known element on Earth. Scientists quickly realized they had found a new element and named it *he-*lium (after *helios,* the Greek word for "Sun"). It took until 1895 for helium to be discovered on our planet. Today, students are probably most familiar with it as the light gas used to inflate balloons, although it turns out to be the second most abundant element in the universe.

The temperature of the chromosphere is about 10,000 K. This means that the chromosphere is hotter than the photosphere, which should seem surprising. In all the situations we are familiar with, temperatures fall as one moves away from the source of heat, and the chromosphere is farther from the center of the Sun than the photosphere is. We will describe one possible explanation for its high temperature after we describe even hotter regions of the solar atmosphere.

14.1.4 The Transition Region

The increase in temperature does not stop with the chromosphere. Above it is a region in the solar atmosphere where the temperature changes from 10,000 K, typical of the chromosphere, to nearly a million degrees. The hottest part of the solar atmosphere, which has a temperature of a million degrees or more, is called the **corona.** Appropriately, the part of the Sun where the rapid temperature rise occurs is called the **transition region.** It is probably only a few tens of kilometers thick. Figure 14.5 summarizes how the temperature of the solar atmosphere changes from the photosphere outward.

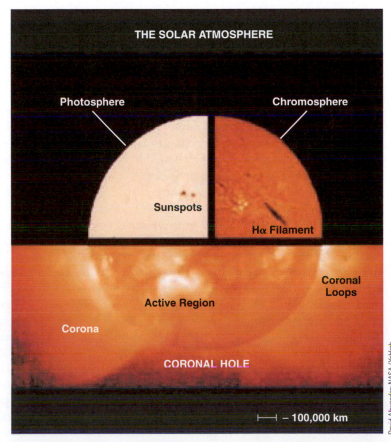

THE SOLAR ATMOSPHERE

Photosphere

Chromosphere

Sunspots

Hα Filament

Coronal Loops

Active Region

Corona

CORONAL HOLE

⊢——⊣ – 100,000 km

David Alexander, NASA/Yohkoh

■ **FIGURE 14.4**

The Sun's Atmosphere Composite image showing the three components of the solar atmosphere: the photosphere or surface of the Sun taken in ordinary light; the chromosphere, imaged in the light of the strong red spectral line of hydrogen (H-alpha); and the corona as seen with x rays.

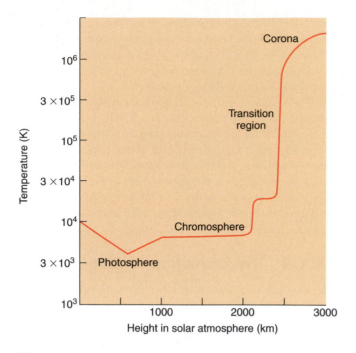

■ **FIGURE 14.5**

Temperatures in the Solar Atmosphere On this graph, temperature is shown increasing upward, and height above the photosphere is shown increasing to the right. Note the very rapid increase in temperature over a very short distance in the transition region between the chromosphere and the corona.

■ **FIGURE 14.6**

The Corona During an Eclipse This image of the Sun was taken during the solar eclipse on March 18, 1988. Since the light from the brilliant surface (photosphere) of the Sun is blocked by the Moon, it is possible to see the tenuous outer atmosphere of the Sun, which is called the corona.

Figure 14.5 makes the Sun seem rather like an onion, with smooth spherical shells, each one with a different temperature. For a long time, astronomers did indeed think of the Sun this way. However, we now know that while this idea of layers—photosphere, chromosphere, transition region, corona—describes the big picture fairly well, the Sun's atmosphere is really more complicated, with hot and cool regions intermixed. For example, clouds of carbon monoxide gas with temperatures colder than 4000 K have now been found at the same height above the photosphere as the much hotter gas of the chromosphere.

This complexity should not seem too surprising. You may have learned that temperatures in the Earth's atmosphere change as high- and low-pressure areas circulate and as the jet stream moves about. In the same way, the atmosphere of the Sun is constantly changing, with matter flowing up and down and changing temperature as it does so. Fortunately, the weather extremes on the Earth are small relative to the violent events on the Sun. Even hurricanes pale in comparison to "solar weather," as we shall see later in this chapter.

14.1.5 The Corona

The outermost part of the Sun's atmosphere is called the corona. Like the chromosphere, the corona was first observed during total eclipses (Figure 14.6). Unlike the chro-

mosphere, the corona has been known for many centuries: It was referred to by the Roman historian Plutarch and was discussed in some detail by Kepler. The corona extends millions of kilometers above the photosphere and emits half as much light as the full moon. The reason we don't see this light until an eclipse occurs is the overpowering brilliance of the photosphere. Just as bright city lights make it difficult to see faint starlight, so too the intense light from the photosphere hides the faint light from the corona. While the best time to see the corona is during a total solar eclipse, its brighter parts can now be photographed with special instruments even at other times, and it can be observed easily from orbiting spacecraft.

Studies of its spectrum show the corona to be very low in density. At the bottom of the corona there are only about 10^9 atoms per cubic centimeter, compared with about 10^{16} atoms per cubic centimeter in the upper photosphere and 10^{19} molecules per cubic centimeter at sea level in the Earth's atmosphere. The corona thins out very rapidly at greater heights, where it corresponds to a high vacuum by laboratory standards.

The corona is very hot. We know this because the spectral lines produced in it come from highly ionized atoms of iron, nickel, argon, calcium, and other elements. For example, astronomers observe lines of iron whose atoms have lost 16 electrons in the corona. Such a high degree of ionization requires a temperature of millions of degrees

■ FIGURE 14.7

Coronal Loops This image shows gas in the corona at a temperature of about 1 million K. High arches of gas reach to a height of approximately 120,000 km. The distinctive arch shape results from strong magnetic fields in the region that force the hot ionized gas to follow their loops.

Kelvin. Because of its high temperature, the corona is also very bright at x-ray wavelengths (see Figure 14.10b, p. 320).

Why is the corona so hot? Observation and theory together have identified magnetic energy as the culprit. The Sun, as we mentioned earlier, is a giant magnet, with a complicated magnetic field that can control the motion of the thin, electrically charged gas in the outer regions of the Sun. The surface of the Sun is covered with magnetic loops, much like the loops of a carpet (Figure 14.7), and the charged atoms flow along these loops.

If a loop in which the magnetic field has one polarity is jostled by turbulence into close proximity to a loop of the opposite polarity, the two loops can "reconnect," much as two bar magnets held side by side with their north poles in opposite directions will snap together. The reconnection of the solar loops releases energy, which can accelerate and heat material that rises into the corona. Magnetic fields also play the key role in heating the chromosphere.

14.1.6 The Solar Wind

One of the most remarkable discoveries about the Sun's atmosphere is that it produces a stream of charged particles (mainly protons and electrons) that we call the **solar wind.** These particles flow outward from the Sun into the solar system at a speed of about 400 km/s (about 900,000 mi/h)! The solar wind exists because the gases in the corona are so hot and moving so rapidly that they cannot be held back by solar gravity. (This wind was actually discovered by its effects on the charged tails of comets; in a sense, we can see the comet tails blow in the solar breeze the way wind socks or curtains in open windows flutter on Earth.)

Although the solar wind material is very, very rarified, the Sun has an enormous surface area. Astronomers estimate that the Sun is losing about 10 million tons of material each year. While this amount of lost mass seems large by Earth standards, it is completely insignificant for the Sun.

In visible light photographs, the solar corona appears fairly uniform and smooth. X-ray pictures, however, show that the corona has loops, plumes, and both bright and dark regions (see Figure 14.10b). Large dark regions of the corona that are relatively cool and quiet are called **coronal holes.** In these regions, magnetic field lines stretch far out into space away from the Sun, rather than looping back to the surface. The solar wind comes predominantly from coronal holes, where gas can stream away from the Sun into space unhindered by magnetic fields. Hot coronal gas, on the other hand, is present mainly where magnetic fields have trapped and concentrated it.

At the surface of the Earth, we are protected to some degree from the solar wind by our atmosphere and the Earth's magnetic field. However, the magnetic field lines come into the Earth at the north and south magnetic poles. Here charged particles accelerated by the solar wind can follow the field down into our atmosphere. As the particles strike molecules of air, they cause them to glow, producing beautiful curtains of light called the **auroras** (or the northern and southern lights). See the opening figure of this chapter for what an auroral display looks like.

Ace◐Astronomy™ Log into AceAstronomy and select this chapter to see Astronomy Exercise "Auroras."

14.2 THE CHANGING SUN

Before the invention of the telescope, the Sun was thought to be an unchanging and perfect sphere. We now know, however, that the Sun is in a perpetual state of change; its surface is a seething, bubbling cauldron of hot gas. Areas that are darker and cooler than the rest of the surface come and go. Vast plumes of gas erupt into the chromosphere and corona. Occasionally, there are even giant explosions on the Sun that send enormous streamers of charged particles and energy hurtling toward the Earth. When they

Art Walker: Doing Astronomy in Space

Since the Sun's corona is a rich source of high-energy photons, observations at x-ray wavelengths are a good way to learn more about our star. Unfortunately, x rays do not penetrate the Earth's atmosphere, so such observations must be done from rockets and satellites. Art Walker (1936–2001) was one of the pioneers in designing and flying instruments in space to learn more about the Sun.

Walker received a PhD in physics from the University of Illinois, supported in his graduate training by the Air Force Reserve officer program. After receiving his degree, he began active service at the Air Force Weapons Laboratory, which was beginning a space research program in the physics of the Sun. (See Making Connections: *Space Weather* to find out why the military is interested in the Sun.) Using a specially designed spectrometer aboard an Air Force satellite, Walker was able to take one of the first x-ray spectra of the Sun and identify several new emission lines in those spectra. Such measurements allowed him and his co-workers to probe the temperature, composition, and structure of the

Art Walker

Courtesy of Stanford University

corona, and get a much better sense of what the Sun's atmosphere is like.

After that, Walker worked with a novel design for x-ray telescopes that makes them similar to some of the telescopes astronomers use on the ground. These instruments allow scientists to take very high-resolution x-ray images of the active Sun; they are also relatively easy to build and can be flown aboard sounding rockets whose paths are spectacular arcs traveling 100 or more miles above the Earth's surface.

In an interview, Walker emphasized one of the key challenges of doing astronomy from space: "In the laboratory, if you build an experiment and it doesn't quite work right, you can always tinker with it until you get the instrumentation to operate. In the case of space observations, you have to build the instrument anticipating all the things that might go wrong and eliminating each possibility of failure. Once the instrument is launched, it is out of your control, and you just have to hope that you have anticipated everything."

After working for the Aerospace Corporation, Walker became a professor at Stanford University, where he also served as Dean of the Graduate College. He worked hard to achieve greater inclusion of women and ethnic minorities in the sciences. Perhaps his best-known student was Sally Ride, who received her PhD in astrophysics and went on to become America's first woman in space. Walker also served on the Presidential Commission that investigated the accident that destroyed the shuttle *Challenger*.

---

arrive, these can cause power outages and other serious effects on our planet.

14.2.1 Sunspots

The first evidence that the Sun varies came from studies of **sunspots,** which are large, dark features seen on the surface of the Sun (Figure 14.8). Occasionally, these spots are large enough to be visible to the unaided eye, and we have records going back over a thousand years of observers who noticed them when haze or mist reduced the Sun's intensity. (We emphasize what your mother surely told you: Looking at the Sun for even a brief time can cause permanent eye damage. This is the one area of astronomy where we don't encourage you to do your own observing without getting careful instructions or filters from your instructor!)

Today we know that sunspots are darker than the photosphere in which they are embedded because the gases in sunspots are as much as 1500 K cooler than the surround-

ing gases. Sunspots are nevertheless hotter than the surfaces of many stars. If they could be removed from the Sun, they would shine brightly. They appear dark only in contrast with the hotter, brighter photosphere around them.

Individual sunspots have lifetimes that range from a few hours to a few months. If a spot lasts and develops, it usually consists of two parts: an inner darker core, the *umbra,* and a surrounding less dark region, the *penumbra.* Many spots become much larger than the Earth, and a few have reached diameters of 50,000 km. Frequently, spots occur in groups of 2 to 20 or more. The largest groups are very complex and may have over 100 spots. Like storms on the Earth, sunspots are not fixed in position, but they drift slowly compared with the Sun's rotation.

By recording the apparent motions of the sunspots as the turning Sun carried them across its disk (Figure 14.9), Galileo in 1612 demonstrated that the Sun rotates on its axis with a rotation period of approximately one month. Modern measurements show that the Sun's rotation period

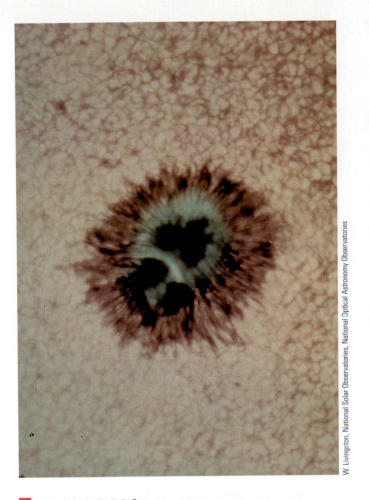

FIGURE 14.8

A Sunspot This image of a sunspot, a cooler and thus darker region on the Sun, was taken with the McMath-Pierce Solar Telescope. The spot has a dark central region (called the umbra) surrounded by a less dark region (the penumbra). Although sunspots appear dark when seen next to the hotter gases of the photosphere, an average sunspot, cut out of the solar surface and left standing in the night sky, would be about as bright as the full moon. Note that you can also see the granulation on the Sun's surface.

is about 25 days at the equator, 28 days at latitude 40°, and 36 days at latitude 80°. Our star turns in a west-to-east direction, like the orbital motions of the planets. Note that the Sun, being composed of gas, does not have to rotate rigidly, the way a solid body does.

14.2.2 The Sunspot Cycle

Between 1826 and 1850, Heinrich Schwabe, a German pharmacist and amateur astronomer, kept daily records of the number of sunspots. What he was really looking for was a planet inside the orbit of Mercury, which he hoped to find by observing its dark silhouette as it passed between the Sun and the Earth. He failed to find the hoped-for planet, but his diligence paid off with an even more important discovery—the **sunspot cycle.** He found that the number of sunspots varied systematically in cycles of about ten years.

What Schwabe observed was that, although individual spots are short-lived, the total number visible on the Sun at any one time is likely to be very much greater at certain times—the periods of sunspot maximum—than at other times—the periods of sunspot minimum. We now know that sunspot maxima occur at an *average* interval of 11 years, but the intervals between successive maxima have ranged from as short as 8 years to as long as 16 years. During sunspot maxima, more than 100 spots can often be seen at once, but even so usually less than half a percent of the surface is covered by spots (Figure 14.10). During sunspot minima, sometimes no spots are visible. The Sun's activity reached its most recent maximum in 2001.

14.2.3 Magnetism and the Solar Cycle

It is the Sun's changing magnetic field that drives the sunspot cycle. The solar magnetic field is measured using a property of atoms called the *Zeeman effect.* Recall from Chapter 4 that an atom has many energy levels and that spectral lines are formed when electrons shift from one

FIGURE 14.9

Sunspots Rotate Across Sun's Surface This sequence of photographs of the Sun's surface tracks a large group of sunspots as the Sun turns. The series of exposures follows the rotation of the sunspots across the visible hemisphere of the Sun. The top sequence shows the Sun in ordinary light; the bottom sequence shows emission from the chromosphere.

MARCH 7 MARCH 8 MARCH 9 MARCH 10 MARCH 13 MARCH 14 MARCH 15 MARCH 16 MARCH 17, 1989

Ace🌐Astronomy™ Log into AceAstronomy and select this chapter to see Astronomy Exercise "Sunspot Cycle II"

(a)

(b)

█ FIGURE 14.10 The Solar Cycle

AceAstronomy™ Log into AceAstronomy and select this chapter to see Astronomy Exercise "Zeeman Effect."

■ **FIGURE 14.11**

The Zeeman Effect These photographs show how magnetic fields in sunspots are measured by means of the Zeeman effect. The vertical black line in the picture at right indicates the position of the spectrograph slit through which light passed in order to obtain the spectrum in the picture at left. Note that the second strongest spectral line in the left-hand picture is split into three components, which indicates that a strong magnetic field is present.

National Optical Astronomy Observatories

level to another. If each energy level is precisely defined, then the difference between them is also quite precise. As an electron changes levels, the result is a sharp, narrow spectral line (either an absorption or emission line, depending on whether the electron's energy increases or decreases in the transition).

In the presence of a strong magnetic field, however, each energy level is separated into several levels very close to one another. The separation of the levels is proportional to the strength of the field. As a result, spectral lines formed in the presence of a field are not single lines, but a series of very closely spaced lines corresponding to the subdivisions of the atomic energy levels. This splitting of lines in the presence of a magnetic field is termed the Zeeman effect.

Measurements of the Zeeman effect in the spectra of the light from sunspot regions show them to have strong

■ **FIGURE 14.10**

The Solar Cycle These dramatic sequences show the activity cycle of the Sun. (a) The image shows ten maps of the magnetic field on the surface of the Sun, spanning a period of seven and a half years. The two polarities (N and S) of the magnetic field are shown against a blue disk as dark blue to black and as light blue to white, respectively. The earliest image, taken on January 8, 1992, is at the lower left and was taken just after a maximum in solar activity. The arc of images shows how the magnetic fields change during the decline of one cycle and the increase in activity associated with the next cycle. (Each image, from left to right around the arc, was taken one half to one year later than the preceding one.) The last image was taken on July 25, 1999, just as the Sun was approaching the most recent solar maximum. Note a few striking patterns in the magnetic maps: The direction from white to black polarity on the southern solar hemisphere is opposite from that on the northern hemisphere. Moreover, the direction from white to black polarity in each hemisphere alternates from one cycle to the next. The complete magnetic cycle of the Sun therefore averages around 22 years. The magnetic field heats the outer solar atmosphere to several million degrees wherever it is strong. The hot gas, forming the solar corona, emits x rays. (b) This series of images was taken by the YOHKOH satellite and shows striking changes in the intensity of coronal x-ray emission during the same period.

magnetic fields (Figure 14.11). Whenever sunspots are observed in pairs, or in groups containing two principal spots, one of the spots usually has the magnetic polarity of a north-seeking magnetic pole and the other has the opposite polarity. Moreover, during a given cycle, the leading spots of pairs (or leading principal spots of groups) in the northern hemisphere all tend to have the same polarity, while those in the southern hemisphere all tend to have the opposite polarity.

During the next sunspot cycle, however, the polarity of the leading spots is reversed in each hemisphere. For example, if during one cycle the leading spots in the northern hemisphere all had the polarity of a north-seeking pole, then the leading spots in the southern hemisphere would have the polarity of a south-seeking pole. During the next cycle, the leading spots in the northern hemisphere would have south-seeking polarity, while those in the southern hemisphere would have north-seeking polarity. Therefore, the sunspot cycle does not repeat itself in regard to magnetic polarity until two maxima have passed. The solar activity cycle, fundamentally a magnetic cycle, thus averages 22 years in length, not 11.

Why does the Sun's magnetic field change in strength and polarity in a nearly regular way? Calculations from detailed models of how the Sun works show that rotation and convection just below the solar surface can distort the magnetic fields. This causes them to grow and then decay, regenerating with opposite polarity approximately every 11 years. The calculations also show that as the fields grow stronger near solar maximum, they flow from the interior of the Sun toward its surface in the form of loops. When a large loop emerges from the solar surface, much like the loop formed by a snake as it slithers along, it creates regions of sunspot activity.

The "ends" of the loop, where they penetrate the solar surface, have different polarities. This idea of magnetic loops offers a natural explanation of why the leading and trailing sunspots in an active region have opposite polarity. The leading sunspot coincides with one end of the loop and the trailing spot with the other end. Smaller loops form the "magnetic carpet" we described earlier and appear to be responsible for heating the corona.

Magnetic fields also hold the key to explaining why sunspots are cooler and darker than the regions without strong magnetic fields. The forces produced by the magnetic field resist the motions of the bubbling columns of rising hot gases. Since these columns carry most of the heat from inside the Sun to the surface by means of convection, less heating occurs where there are strong magnetic fields. As a result, these regions are seen as darker, cooler sunspots.

Data courtesy of J. Harvey, NSO (Tucson/Kitt Peak, AZ)

14.3 ACTIVITY ABOVE THE PHOTOSPHERE

Sunspots are not the only features that vary during a solar cycle. There are extraordinary changes in the chromosphere and corona as well. To see what happens in the chromosphere, we must observe the emission lines from elements such as hydrogen and calcium. The hot corona, on the other hand, can be studied by observations of x rays.

14.3.1 Plages and Prominences

Emission lines of hydrogen and calcium are produced in the hot gases of the chromosphere. Astronomers routinely photograph the Sun through filters that transmit light only at the wavelengths that correspond to these emission lines.

■ **FIGURE 14.12**

Plages on the Sun This image of the Sun was taken with a filter that transmits only the light of the spectral line produced by singly ionized calcium. The bright cloud-like regions are the plages.

Pictures taken through these special filters show bright "clouds" in the chromosphere around sunspots; these bright regions are known as **plages** (Figure 14.12). The plages are regions within the chromosphere that have higher temperature and density than their surroundings. The plages actually contain all of the elements in the Sun, not just hydrogen and calcium. It just happens that the spectral lines of hydrogen and calcium produced by these clouds are bright and easy to observe.

Moving higher into the Sun's atmosphere, we come to the spectacular phenomena called **prominences** (Figure 14.13), which usually originate near sunspots. Eclipse observers often see prominences as red, flame-like protuberances rising above the eclipsed Sun and reaching high into the corona. Some, the quiescent prominences, are graceful loops that can remain nearly stable for many hours or even days. The relatively rare eruptive

NASA/SOHO

■ **FIGURE 14.13**

A Collage of Four Different Prominences A prominence is a huge cloud of relatively cool (about 60,000 K in this case), fairly dense gas suspended in the much hotter corona. These pictures are color coded so that white corresponds to the hottest temperatures and dark red to cooler ones. The four images were taken, moving clockwise from the upper left, on May 15, 2001; March 28, 2000; January 18, 2000; and February 2, 2001.

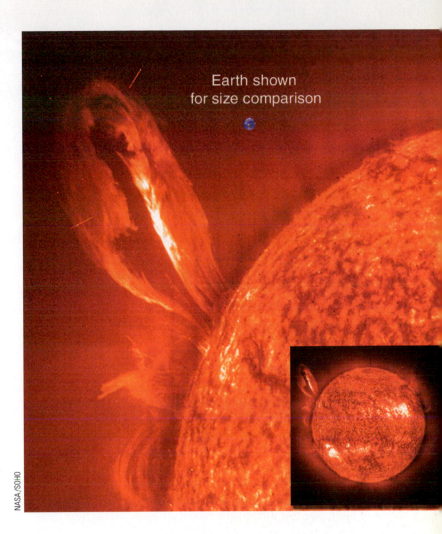

Earth shown
for size comparison

NASA/SOHO

■ **FIGURE 14.14**

An Eruptive Prominence   This image of an eruptive prominence was taken in the light of singly ionized helium in the extreme ultraviolet part of the spectrum. The prominence, which occurred on July 24, 1999, is a particularly large one. An image of the Earth is shown at the same scale for comparison.

prominences appear to send matter upward into the corona at high speeds, and the most active surge prominences may move as fast as 1300 km/s (almost 3 million mi/h). Some eruptive prominences have reached heights of more than 1 million km above the photosphere; the Earth would be completely lost inside one of these awesome displays (Figure 14.14).

14.3.2   Flares

The most violent event on the surface of the Sun is a rapid eruption called a **solar flare.** A typical flare lasts for 5 to 10 minutes and releases a total amount of energy equivalent to that of perhaps a million hydrogen bombs. The largest flares last for several hours and emit enough energy to power the entire United States at its current rate of electrical consumption for 100,000 years. Near sunspot maximum, small flares occur several times per day, and major ones may occur every few weeks.

Flares are often observed in the red light of hydrogen, but the visible emission is only a tiny fraction of the energy released when a solar flare explodes. At the moment of the explosion, the matter associated with the flare is heated to temperatures as high as 10 million K. At such high temperatures, a flood of x-ray and ultraviolet radiation is emitted.

Flares seem to occur when magnetic fields pointing in opposite directions release energy by interacting with and destroying each other—much as a stretched rubber band releases energy when it breaks. We have discussed this idea before, when we talked about heating the corona. What is different about flares is that their magnetic interactions cover a large volume in the solar corona and release a tremendous amount of electromagnetic radiation. In some cases, immense quantities of coronal material—mainly protons and electrons—may also be ejected at high speeds (500–1000 km/s) into interplanetary space. Such **coronal mass ejections** (CMEs) can affect the Earth in several ways. (See Making Connections: *Space Weather*.)

14.3.3   Active Regions

Sunspots, flares, and bright regions in the chromosphere and corona tend to occur together on the Sun. That is, they all tend to have similar longitudes and latitudes, but they are located at different heights in the atmosphere. Because they all occur together, they all vary with the sunspot cycle.

Space Weather

As if scientists did not have enough trouble trying to predict weather on Earth, they are now facing the challenge of predicting the effects of solar storms on Earth. This new field of research is called *space weather*. With thousands of satellites in orbit, astronauts taking up long term residence in the International Space Station, millions of people using cell phones and wireless communication, and nearly everyone relying on the availability of dependable electrical power, governments are now making major investments in trying to learn how to predict when solar storms will occur and how strongly they will affect the Earth.

Three solar phenomena—coronal holes, solar flares, and coronal mass ejections (CMEs)—account for most of the space weather. Coronal holes allow the solar wind to flow freely away from the Sun, unhindered by solar magnetic fields. When the solar wind reaches the Earth, it causes the Earth's magnetosphere (the region around the Earth that is filled with charged particles) to contract and then expand after the solar wind passes by. These changes can cause (usually mild) electromagnetic disturbances on Earth.

More serious are solar flares, which shower the upper atmosphere of the Earth with x rays, energetic particles, and intense ultraviolet radiation. The x rays and ultraviolet radiation can ionize atoms in the upper atmosphere, and the freed electrons can build up a charge on the surface of a spacecraft. When this static charge discharges, it can damage the electronics in the spacecraft—just as you can receive a shock when you walk across a carpet in your stocking feet in a dry climate and then touch a light switch or some other metal object.

Most disruptive are the coronal mass ejections. A CME is an erupting bubble of tens of millions of tons of gas blown away from the Sun into space. When this bubble reaches the Earth a few days after leaving the Sun, it heats the ionosphere, which expands and reaches farther into space. As a consequence, friction between the atmosphere and spacecraft increases, dragging satellites to lower altitudes. At the time of a particularly strong flare in March 1989, the system responsible for tracking some 19,000 objects orbiting the Earth temporarily lost track of 11,000 of them because their orbits were changed by the expansion of the Earth's atmosphere. During solar maximum, a number of satellites are brought to such a low altitude that they are destroyed by friction with the atmosphere. Both the Hubble Space Telescope and the International Space Station require reboosts to higher altitude so that they can remain in orbit.

When a CME reaches the Earth, it distorts the Earth's magnetic field. Since a changing magnetic field induces electrical currents, the CME accelerates electrons, sometimes to very high speeds. These "killer electrons" can penetrate deep into satellites, sometimes destroying their electronics and permanently disabling operation. This has happened with some communications satellites.

Disturbances in the Earth's magnetic field can cause disruptions in communications, especially cell phone and wireless systems. In fact, disruptions can be expected to occur several times a year during solar maximum. Changes in the Earth's magnetic field due to CMEs can also cause surges in power lines large enough to burn out transformers and cause major power outages. For example, in 1989 parts of Montreal and Quebec Province in Canada were without power for up to 9 hours as a result of a major solar storm. Electrical outages due to CMEs are more likely to occur in North America than in Europe because North America is closer to the Earth's magnetic pole where the currents induced by CMEs are strongest.

Solar storms also expose astronauts, passengers in high-flying airplanes, and even people on the surface of the Earth to increased amounts of radiation. Astronauts, for example, are limited in the total amount of radiation that they can be exposed to during their careers. A single ill-timed solar outburst could end an astronaut's career. This problem becomes increasingly serious as astronauts spend more time in space. For example, the typical daily dose of radiation aboard the Russian Mir was equivalent to about eight chest x rays. One of the major challenges in planning the human exploration of Mars is devising a way to protect astronauts from high-energy solar radiation.

Warnings of solar storms would help to minimize their disruptive effects. Power networks could be run at less than their full capacity so that they could absorb the effects of power surges. Communications networks could be prepared for malfunctions and have backup plans in place. Spacewalks could be timed to avoid major solar outbursts. Scientists are now trying to find ways to predict where and when flares and CMEs will occur and whether they will be big, fast events or small, slow ones with little consequence for Earth. The strategy is to relate changes in the appearance of small regions on the Sun and changes in local solar magnetic fields to subsequent eruptions. However, right now this predictive capability is poor, and so the only real warning we have is from actually seeing CMEs and flares occur. Since a CME travels outward at about 300 km/s, an observation of an eruption provides several days warning at the distance of the Earth. However, the severity of the impact on the Earth depends on how the magnetic field associated with the CME is oriented relative to that of the Earth. The orientation can be measured only when the

SOHO: ESA & NASA

2000/11/25 16:00 UT

2000/11/25 18:48

2000/11/26 08:06

2000/11/26 11:42.

A Flare and a Coronal Mass Ejection   This sequence of four images shows the evolution over time of a giant eruption on the Sun. The event began at the location of a sunspot group, and a flare is seen in far ultraviolet light in the upper right image. Fourteen hours later, a CME is seen blasting out into space. Three hours later, this CME has expanded to form a giant cloud of particles escaping from the Sun and beginning the journey out into the solar system. The white circle in the lower two pictures shows the diameter of the solar photosphere. The larger dark area shows where the light from the Sun has been blocked out by a specially designed instrument in order to make it possible to see the faint emission from the corona.

(*continued*)

Space Weather (continued)

CME flows past a satellite, which is located only about an hour upstream from the Earth.

Space weather predictions are available on the Web (www.noaa.gov/solar.html). Outlooks are given a week ahead, bulletins are issued when there is an event that is likely to be of interest to the public, and warnings and alerts are posted when an event is imminent or already under way.

Astronomers expected calmer space weather following the solar maximum in 2001. However, the Sun had some surprises in store—the Halloween storms of 2003. From late October through early November, three sunspots, including one the diameter of Jupiter, launched a series of solar eruptions. There were sixty solar flares, including the largest x-ray flare ever observed. A series of CMEs ejected charged particles at speeds as high as 6 million mph (2700 km/sec). Some of these particles slammed into Earth-orbiting satellites and damaged several, disabled two, and caused a radiation-monitoring instrument on a satellite orbiting Mars to fail. On Earth, there was a power outage in Sweden, and communications with airline pilots were disrupted. This series of events illustrates the urgency of developing better space weather predictions before the next solar maximum.

For example, flares are more likely to occur near sunspot maximum, and the corona is much more conspicuous at that time (see Figure 14.10). A place on the Sun where these phenomena are seen is called an **active region** (Figure 14.15). Active regions are always associated with strong magnetic fields.

14.4 IS THE SUN A VARIABLE STAR?

The Sun rises faithfully every day at a time that can be calculated precisely. Each day it deposits energy on the Earth, warming it and sustaining life. But over the last decade, scientists have accumulated evidence that the Sun is not truly constant but varies over the centuries by a small amount—probably less than 1 percent. Still, that variation is enough to have profound effects on the Earth and its climate.

14.4.1 Variations in the Number of Sunspots

You already read that the number of sunspots varies, with the time between sunspot maxima being about 11 years. However, the number of sunspots at maximum is not always the same. Considerable evidence shows that between the years 1645 and 1715, the number of sunspots, even at sunspot maximum, was much lower than it is now. This interval of extremely low activity was first noted by Gustav Spörer in 1887 and then by E. W. Maunder in 1890; it is now called the **Maunder Minimum.** The variation in the number of sunspots over the past three centuries is shown in Figure 14.16. Besides the Maunder Minimum in the 17th century, sunspot numbers were somewhat lower during the first part of the 19th century than they are now; this period is called the Little Maunder Minimum.

When the number of sunspots is high, the Sun is active in various other ways as well, and this activity affects the Earth directly. For example, there are more auroral displays when the sunspot number is high. Auroras are caused when energetic charged particles from the Sun interact with the Earth's magnetosphere, and the Sun is more likely to spew out particles when it is active and the sunspot number is high. Historical accounts indicate that auroral activity was abnormally low throughout the several decades of the Maunder Minimum.

The Maunder Minimum was a time of exceptionally low temperatures in Europe—so low that this period is described as a Little Ice Age. The river Thames in London froze at least 11 times during the 17th century, ice appeared in the oceans off the coasts of southeast England, and low summer temperatures led to short growing seasons and poor harvests (Figure 14.17).

Changes in climate can have profound impacts on human history. For example, explorers from Norway first colonized Iceland and then reached Greenland by 986. From there, they were able to make repeated visits to the northeastern coasts of North America, including Newfoundland, between about 1000 and 1350. (The ships of the time did not allow the Norse explorers to travel all the way to North America directly, but only from Greenland, which served as a station for further exploration.)

Most of Greenland is covered by ice, and the Greenland station was never self-sufficient; rather it depended on imports of food and other goods from Norway for its survival. When a little ice age began in the 13th century,

■ FIGURE 14.15

Active Regions as Seen at Different Levels in the Solar Atmosphere   If we observe the Sun at different wavelengths, we can see different temperature regions and different levels in the solar atmosphere. The central image is a three-color composite of the solar corona, observed in extreme ultraviolet light with the Transition Region and Coronal Explorer (TRACE) satellite. The colors in this image represent different temperatures: 1 million K gas is represented by green, 1.5 million K gas by blue, and 2 million K gas by red. The surrounding images are, clockwise starting from the top: a magnetic map taken by the SOHO satellite, with opposite magnetic polarities shown in black and white; an image of the solar photosphere taken in ordinary white light; two images of the chromosphere taken in the ultraviolet by the TRACE satellite; and four images of the corona taken at the three wavelengths used to construct the central image. You can see that coronal x-ray emission, chromospheric emission, strong magnetic fields, and sunspots tend to occur in the same locations—the active regions. The composite was prepared by Joe Covington.

■ **FIGURE 14.16**

The Numbers of Sunspots As They Change with Time   This diagram shows how the number of sunspots has changed with time since counts of the numbers of spots began to be recorded on a consistent scale. Note the low number of spots during the early years of the 19th century—the Little Maunder Minimum.

■ **FIGURE 14.17**

Europe's Little Ice Age   During the "Little Ice Age" in Europe, bodies of water froze every winter. This painting by Robert van den Hoecke from 1649 is entitled "Skating in the Town Moat of Brussels."

voyaging became very difficult, and support of the Greenland colony was no longer possible. The last known contact with it was made by a ship from Iceland blown off course in 1410. When European ships again began to visit Greenland in 1577, the entire colony there had disappeared.

The estimated dates for these patterns of migration follow what we know about solar activity. Solar activity was unusually high between 1100 and 1250, which includes the time the first European visits were made to North America. Activity was low from 1280 to 1340 and there was a little ice age, which was about the time regular contact with North America and between Greenland and Europe stopped.

14.4.2 The Effects of the Sun on the Earth's Climate

The historical record therefore suggests that climate and solar activity are correlated. But is this an accidental correlation? Or do changes in the Sun *cause* changes in the Earth's climate? There are two reasons that it is urgent to answer this question. First, active debate is under way about whether greenhouse gases that we humans are adding to the Earth's atmosphere are affecting the climate. If solar variability also causes changes in climate, then our estimate of the impact of human activities would be affected. Second, we can test models that predict how the atmosphere will respond to changing temperatures. We have data concerning both solar variability and climate changes that span several centuries, and we can see whether atmospheric models predict accurately the relationship between the energy output of the Sun and the climate here on Earth. If the models can explain the Earth's response to solar variability, then we will have greater confidence in the predictions of the consequences of increases in greenhouse gases.

Two steps are required to show that changes in solar luminosity affect the Earth's climate: (1) A long record of correlation between weather and the energy emitted by the Sun must be established. (2) The mechanism must be found whereby small changes in the output of solar energy are translated into significant changes in climate. Let's deal with each of these steps in turn.

14.4.3 Solar Activity and the Earth's Climate: The Correlation

It is only with the advent of the space age that we have been able to measure the luminosity of the Sun accurately enough to see whether its total energy output changes with activity level. To our surprise, it turns out that the Sun is a tenth of 1 percent or so brighter during solar maximum. We might have expected the Sun to be fainter when it is covered with a large number of dark sunspots. The extra radiation at sunspot maximum probably comes from bright regions, like plages, which are more abundant at that time.

Incidentally, we may be fortunate to live at a time when the Sun is exceptionally stable. If we look at activity cycles in stars that are similar to the Sun, we find that the Sun's current variations (just 0.1 percent of the total energy emitted) are unusually small. The luminosity of most stars varies by 0.3 percent and some by as much as 1 percent. Therefore, it seems likely that the Sun too might sometimes vary by larger amounts than it does currently, with consequently even greater effects on climate.

Since we have no historical measurements of solar luminosity (the changes were much too small to be measured until recently), we will follow the recent results from satellites and assume that the level of solar activity is an indicator of the energy emitted by the Sun. It turns out that nature provides us with an estimate of the level of solar activity going back for thousands of years.

The best quantitative evidence of long-term variations in the level of solar activity comes from studies of the radioactive isotope carbon-14. The Earth is constantly bombarded by cosmic rays—high-energy charged particles that include protons and nuclei of heavier elements (see Section 19.4). The rate at which cosmic rays from sources outside the solar system reach the upper atmosphere depends on the level of solar activity. When the Sun is active, its charged particles, streaming out into the solar system, carry the Sun's strong magnetic field with them. This magnetic field shields the Earth from incoming cosmic rays. At times of low activity, when the Sun's magnetic field is weak, cosmic rays reach the Earth in larger numbers.

When the energetic cosmic-ray particles strike the upper atmosphere, they produce several different radioactive isotopes (see Section 4.4). One such isotope is carbon-14, which is produced when nitrogen is struck by high-energy cosmic rays. The rate of carbon-14 production is higher when the activity of the Sun is lower and the solar magnetic field does not shield the Earth from bombardment by cosmic rays.

Some of the radioactive carbon atoms then form carbon dioxide molecules, which are ultimately incorporated into trees through photosynthesis. By measuring the amount of radioactive carbon in tree rings, we can estimate the historical levels of solar activity. Correlations with visual estimates of sunspot numbers over the past 300 years indicate that the carbon-14 estimates of solar activity are indeed valid. Because it takes an average of about 10 years for a carbon dioxide molecule to be absorbed from the atmosphere or ocean into plants, this technique cannot provide data on the 11-year solar cycle. It can be used, however, to look for long-term changes in the level of solar activity.

Estimates of the amount of carbon-14 in tree rings show that variations in solar activity levels have occurred throughout the last several thousand years, and at different times the Sun has been both more and less active than it is now. The measurements confirm that the amount of carbon-14 was unusually high, with solar activity correspondingly low, during both the Maunder Minimum and the Little

Observing the Sun

Looking directly at the Sun is very dangerous! Even a brief exposure can burn your retina and cause severe eye damage. Looking at the Sun directly through an unfiltered telescope is even worse because the telescope concentrates the Sun's radiation and hence damages your eye even more quickly. But there are indirect ways of observing the Sun that are both safe and instructive.

Not long after Galileo began training his first telescope on the sky, his assistant and disciple Benedetto Castelli came up with a way of observing sunspots, whose nature was a big controversy in those days. Castelli projected the image of the Sun made by the telescope onto a sheet of paper, where the sunspots could be sketched safely. Using this technique, Galileo was able to show that sunspots were not small, dark planets blocking sunlight, as some had suggested, but dark areas on the Sun itself moving around as the Sun rotated. You can try this experiment for yourself.

Set up a small telescope on a steady mount so that it points at the Sun. Remember, you have to do this by trial and error; never look through the telescope at the Sun to see how well you're doing! Instead, watch the telescope's shadow; it will be smallest when the telescope is pointing directly at the Sun. Then you'll be able to see an image of the Sun on a sheet of white cardboard or other light surface. Sketch the sunspots if there are any. Repeat the observation over the course of a week or two, and, orienting your sketches the same way, trace how the spots have moved over the face of the Sun.

Looking through a telescope at the Sun is dangerous, but you can always view the Sun safely with a small telescope by projecting its image onto a sheet of white cardboard.

Putting a cardboard collar around the telescope makes the image easier to see by casting a shadow on it and thus making the sphere of the Sun stand out. If you observe the Sun for a while, don't be surprised if its image moves across and off the paper over time. This is caused by the turning of the Earth, which also makes the Sun appear to rise and set. And don't be upset if you don't see any sunspots, especially if you are observing during the solar activity minimum.

Maunder Minimum. Activity was also low from 1410 to 1530 and from 1280 to 1340. Between about 1100 and 1250, the level of solar activity may have been even higher than it is now.

Nature also provides us with a way to estimate the average temperature of the Earth over a time period of thousands of years. Ice in the far north, near the coasts of Canada, Greenland, and Iceland, picks up microscopic rocky debris as it moves over land. This ice then floats southward into the North Atlantic, eventually melts, and drops the debris onto the ocean floor. In cold years, the ice floats farther south before melting, occasionally reaching as far south as Ireland. Studies of the material picked up from the ocean floor show that the amount of debris jumped every 1500 years or so as the ice moved farther out into the (temporarily) colder Atlantic.

In each case, peaks in the amount of debris coincided with peaks in the amount of carbon-14 measured. This result implies that times of low temperature correspond to times of low solar activity. The measurements show that the climate has warmed and cooled nine times in the last 12,000 years, each time in step with changes in the level of solar activity as determined from carbon-14. The Sun really does affect the Earth's climate.

14.4.4 Solar Variability and the Earth's Climate: The Mechanism

The evidence then is that the Earth is on average cooler when solar activity is low. But why?

The changes in the Sun's luminosity over the past several hundred years are too small to change the overall temperature of the Earth dramatically. It is also true that the temperature changes are not the same everywhere during an event like the Little Ice Age, which affected Europe more strongly than some other parts of the world. People who compute models of the Earth's atmosphere have been trying to identify mechanisms that amplify the small changes in solar luminosity. One recent model suggests that wind may be the culprit. Measurements show that the ultraviolet energy emitted by the Sun varies by 10 to 100 times more than the energy put out in the visible-light region of the spectrum. At times of high solar activity, more ultraviolet radiation is emitted and as a consequence more ozone is formed in the Earth's atmosphere. Ozone can absorb more sunlight and increase the temperature of the stratosphere. This in turn can affect the patterns of the winds aloft, modify the paths of storms, and change the temperature in lower layers of the Earth's atmosphere by differing amounts at different latitudes.

This atmospheric model shows that at times of low solar activity, the westerly winds in the northern hemisphere will be reduced. In winter, the oceans are relatively warmer than the land because water retains heat more efficiently. If winds are reduced, then less warm air will be blown from the Pacific over North America and from the Atlantic over Europe, thereby resulting in colder winters. These models predict that while global temperatures averaged over the whole Earth during the Maunder Minimum were only 0.3–0.4°C cooler, Europe and North America were 1.8–2.7°C cooler, which is what is needed to account for the Little Ice Age in the northern hemisphere.

Other factors also alter the Earth's climate. For example, the eruption of the Pinatubo volcano in June 1991 spewed out 20 million tons of sulfur dioxide (SO_2), which then spread around the globe. The presence of these particles in the atmosphere cooled the northern hemisphere by about 0.5°C on average for the next two years. This event retarded the effects of global warming, but also allowed scientists to calibrate the effects of volcanoes on climate. The 0.6°C warming of the Earth that has occurred over the past century cannot be accounted for by changes in the Sun and volcanic eruptions alone. In order to account for this warming, it is necessary to include greenhouse gases from human activities in the model.

Many of these ideas are still new and must be strengthened by further observations and modeling. Still, such studies clearly give us reason to continue to study the behavior of our star and its influences on Earth.

SURFING THE WEB

 General Information Sites About the Sun:

- *The Nine Planets Site:*
  seds.lpl.arizona.edu/nineplanets/nineplanets/sol.html

- *Views of the Solar System Site:*
  www.solarviews.com/eng/sun.htm

- *Stanford SOLAR Center:*
  solar-center.stanford.edu/

- *NASA's Sun-Earth Connection:*
  sunearth.ssl.berkeley.edu

 Sites for Space Missions Examining the Sun:

- *Ulysses:*
  ulysses.jpl.nasa.gov

- *SOHO:*
  sohowww.nascom.nasa.gov/

- *TRACE:*
  vestige.lmsal.com/TRACE/

- *Yohkoh:*
  www.lmsal.com/SXT/homepage.html

- *HESSI:*
  hesperia.gsfc.nasa.gov/hessi/

SUMMARY

14.1 The Sun, our star, is surrounded by a number of layers that make up the solar atmosphere. In order of increasing distance from the center of the Sun, they are the **photosphere,** with a temperature that ranges from 4500 K to about 6800 K; the **chromosphere,** with a typical temperature of 10^4 K; the **transition region,** a zone that may be only a few kilometers thick, where the temperature increases rapidly from 10^4 K to 10^6 K; and the **corona,** with temperatures of a few million degrees Kelvin. The Sun's surface is mottled with upwelling currents seen as hot, bright **granules. Solar wind** particles stream out into the solar system through **coronal holes.** When such particles reach the vicinity of the Earth, they produce **auroras,** which are strongest near the Earth's magnetic poles. Hydrogen and helium together make up 98 percent of the mass of the Sun, whose composition is much more characteristic of the universe at large than is the composition of the Earth.

14.2 Sunspots are dark regions where the temperature is up to 1500 K cooler than the surrounding photosphere. Their motion across the Sun's disk allows us to calculate how fast the Sun turns on its axis. The Sun rotates more rapidly at its equator, where the rotation period is about 25 days, than near the poles, where the period is slightly longer than 36 days. The number of visible sunspots varies according to a **sunspot cycle** that averages 11 years in length. Spots frequently occur in pairs. During a given 11-year cycle, all leading spots in the northern hemisphere have the same magnetic polarity, while all leading spots in the southern hemisphere have the opposite polarity. In the subsequent 11-year cycle, the polarity reverses. For this reason, the magnetic activity cycle of the Sun is often said to last for 22 years.

14.3 Sunspots, **prominences,** and bright regions, including **plages, solar flares,** and **coronal mass ejections,** all tend to occur in **active regions**—that is, in places on the Sun with the same latitude and longitude but at different heights in the atmosphere. Solar flares and coronal mass ejections can cause auroras, disrupt communications, damage satellites, and cause power outages.

14.4 Over long periods of time (100 years or more), there are changes in the level of solar activity and the number of sunspots seen at solar maximum. For example, the number of sunspots was unusually low from 1645 to 1715, a period now called the **Maunder Minimum.** Data now show that there is a correlation between the temperature on Earth and the energy output of the Sun. Models suggest that heating of the upper atmosphere by ultraviolet radiation affects wind patterns, which in turn affect weather near the Earth's surface. However, the warming of the Earth seen over the last century cannot be accounted for by changes in the Sun alone, even when combined with the effects of volcanic eruptions. The recent temperature rise is accounted for only if greenhouse gases produced by human activities are included in the models.

INTER-ACTIVITY

A Have your group make a list of all the ways the Sun affects your life on Earth. How long a list can you come up with? (Be sure you consider the everyday effects as well as the unusual effects due to high solar activity.)

B Long before the nature of the Sun was fully understood, astronomer (and planet discoverer) William Herschel (1738–1822) proposed that the hot Sun may have a cool interior and may be inhabited. Have your group discuss this proposal and come up with modern arguments against it.

C We discussed how the migration of Europeans to North America was apparently affected by climate change. If the Earth were to become significantly hotter, either because of changes in the Sun or because of greenhouse warming, one effect would be an increase in the rate of melting of the polar ice caps. How would this affect modern civilization?

D Suppose we experience another Maunder Minimum on Earth, with a drop in the average temperatures. Have your group discuss how this would affect civilization and international politics. Make a list of the most serious effects you can think of.

E Watching sunspots move across the disk of the Sun is one way to show that our star rotates on its axis. Can your group come up with other ways to show the Sun's rotation?

REVIEW QUESTIONS

Ace Astronomy™ Assess your understanding of this chapter's topics with additional quizzing and animations at **http://ace.brookscole.com/voyages**

1. Describe the main differences between the composition of the Earth and that of the Sun.
2. Make a sketch of the Sun's atmosphere showing the locations of the photosphere, chromosphere, and corona. What is the approximate temperature of each of these regions?

3. Why do sunspots look dark?
4. What is the Zeeman effect, and what does it tell us about the Sun?
5. Describe three different types of solar activity.
6. How does activity on the Sun affect the Earth?
7. Which aspects of the Sun's activity cycle have a period of about 11 years? Which vary during intervals of about 22 years?
8. Summarize the evidence indicating that over several decades or more there have been variations in the level of solar activity.

THOUGHT QUESTIONS

9. Table 14.1 indicates that the density of the Sun is 1.41 g/cm^3. What kinds of materials have similar densities? One such material is ice. How do you know that the Sun is not made of ice?

10. If the rotation period of the Sun is determined by observing the apparent motions of sunspots, must any correction be made for the orbital motion of the Earth? If so, explain what the correction is and how it arises. If not, explain why the Earth's orbital revolution does not affect the observations. Making some sketches may help answer this question.

11. Suppose an (extremely hypothetical) elongated sunspot forms that extends from a latitude of 30° to a latitude of 40° along a fixed line of longitude on the Sun. How will the appearance of that sunspot change as the Sun rotates?

12. Suppose you live in northern Canada and an extremely strong flare is reported on the Sun. What precautions might you take? What could compensate you for your troubles?

13. Give some reasons why it is difficult to determine whether or not small changes in the amount of energy radiated by the Sun have an effect on the Earth's climate.

FIGURING FOR YOURSELF

A basic equation you use almost automatically to figure out trips in your car is also useful in figuring out when events on the Sun will impact the Earth. The equation is:

distance = velocity × time   or $D = v \times t$

Dividing both sides by v, we get

$$t = \frac{D}{v}$$

If you are driving at 90 km/h and you drive 180 km, the trip will take 180/90 = 2 hours.

When you use this equation, make sure you use the same units for velocity (say, km/s) as you do for distance (km) and time (seconds).

14. Suppose you observe a major solar flare while astronauts are orbiting the Earth in the shuttle. Use the data in Section 14.3 to calculate how long it will be before the charged particles ejected from the Sun during the flare reach the shuttle.

15. Suppose an eruptive prominence rises at a speed of 150 km/s. If it does not change speed, how far from the

photosphere will it extend after 3 hours? How does this distance compare with the diameter of the Earth?

16. From the information in the figure that accompanies the Making Connections box in this chapter, estimate the speed with which the particles in the CME in the final two frames are moving away from the Sun.

17. From the Doppler shifts of the spectral lines in the light coming from the east and west edges of the Sun, astronomers find that the radial velocities of the two edges differ by about 4 km/s. Find the approximate period of rotation of the Sun. (It may help to sketch the Sun and show the velocities at the east and west edges in order to work out this problem.)

The equation for density is:

$$\text{density} = \frac{\text{mass}}{\text{volume}}$$

The volume of a sphere is given by

$$V = \left(\frac{4}{3}\right)\pi R^3$$

where R is the radius of the sphere.
Again, be sure that you use consistent units.

18. Use the data in Table 14.1 to calculate the density of the Sun. How does its density compare with the densities of other objects given in this text (e.g., planets, minerals,

ice)? Does the density alone allow you to say anything about the composition of the Sun? If so, what conclusions would you draw?

19. The edge of the Sun doesn't have to be absolutely sharp in order to look that way to us. It just has to go from being transparent to being completely opaque in a distance that is smaller than your eye can resolve. Remember from Chapter 5 on telescopes that the ability to resolve detail depends on the size of the telescope. The pupil of your eye is very small relative to the size of a telescope and therefore is very limited in the amount of detail you can see. In fact, your eye cannot see details that are smaller than 1/30 of the diameter of the Sun (about 1 arcminute). Nearly all the light from the Sun emerges from a layer that is only about 400 km thick. What fraction is this of the diameter of the Sun? How does this compare with the ability of the human eye to resolve detail? Suppose we could see light emerging directly from a layer that was 300,000 km thick. Would the Sun appear to have a sharp edge?

20. Table 14.2 shows what percentage of the atoms in the Sun consists of hydrogen and what percentage of the total mass of the Sun is composed of hydrogen. Can you explain why the two numbers are different? Show that the statement that 92 percent of the atoms are hydrogen is consistent with the statement that 73 percent of the mass is made up of hydrogen. (*Hint:* Make the simplifying assumption, which is nearly correct, that the Sun is made up entirely of hydrogen and helium.)

SUGGESTIONS FOR FURTHER READING

Akasofu, S. "The Shape of the Solar Corona" in *Sky & Telescope*, Nov. 1994, p. 24.

Baliunas, S. and Soon, W. "The Sun-Climate Connection" in *Sky & Telescope*, Dec. 1996, p. 38.

Eddy, J. "The Case of the Missing Sunspots" in *Scientific American*, May 1977, p. 80.

Emslie, A. "Explosions in the Solar Atmosphere" in *Astronomy*, Nov. 1987, p. 18. Discusses solar flares.

Frank, A. "Blowin' in the Solar Wind" in *Astronomy*, Oct. 1998, p. 60. On results from the SOHO spacecraft.

Freeman, J. *Storms in Space.* 2001, Cambridge U. Press. On solar-terrestrial relationships.

Friedman, H. *Sun and Earth.* 1986, W. H. Freeman. Contains good sections about the Sun's effects on the Earth.

Golub, L. "Heating the Sun's Million-Degree Corona" in *Astronomy*, May 1993, p. 27.

Golub, L. and Pasachoff, J. M. *Nearest Star: The Surprising Science of Our Sun.* 2001, Harvard U. Press. An introduction to the Sun, with especially good chapters on solar eclipses and space missions.

Hufbauer, K. *Exploring the Sun: Solar Science Since Galileo.* 1991, Johns Hopkins U. Press. A good history.

Jaroff, L. "Fury on the Sun" in *Time*, July 3, 1989, p. 46. Nice introduction to solar activity and the Sun's interior.

Kippenhahn, R. *Discovering the Secrets of the Sun.* 1994, John Wiley. Excellent modern introduction for the beginner.

Lang, K. "SOHO Reveals the Secrets of the Sun" in *Scientific American*, Mar. 1997.

Nichols, R. "Solar Max: 1980–1989" in *Sky & Telescope*, Dec. 1989, p. 601.

Schaefer, B. "Sunspots That Changed the World" in *Sky & Telescope*, Apr. 1997, p. 34. Historical events connected with sunspots and solar activity.

Schrijver, C. and Title, A. "Today's Science of the Sun" in *Sky & Telescope*, Feb. 2001, p. 34 and Mar. 2001, p. 34. Excellent review of recent results about the solar atmosphere.

Verschuur, G. "The Day the Sun Cut Loose" in *Astronomy*, Aug. 1989, p. 48. Examines events surrounding a huge flare.

Zirker, J. *Journey from the Center of the Sun.* 2001, Princeton U. Press. Covers both the solar interior and atmosphere.

The Sudbury Neutrino Detector   This image shows a device being built deep underground in Canada to detect neutrinos from the Sun. The detector involves a 12-meter-wide acrylic sphere that would eventually be filled with tons of heavy water. The image shows the sphere surrounded by 9600 light detectors. The passage of a neutrino through the sphere of heavy water produces a flash of light. We discuss results from this detector toward the end of this chapter.

15

The Sun: A Nuclear Powerhouse

We do not argue with the critic who urges that the stars are not hot enough for this process [nuclear fusion]; we tell him to go and find a hotter place.

Arthur Eddington in *Internal Constitution of the Stars* **(1926)**

THINKING AHEAD

Why does the Sun shine? Our star produces prodigious amounts of energy and has done so ever since human eyes first gazed upon it. And we now know that the Sun was shining for billions of years before that. How long has it been producing light? And what is the source of all that energy?

The Sun is really, really bright—the brightest thing that you have ever seen, even though it is 93 million miles away. The first question we want to ask is: How bright is it really?

With modern instruments, we have measured the Sun's output with great precision and found that our star puts out about 4×10^{26} watts (W). (See the Astronomy Basics box for more about watts.) How can we get a sense of how large that number is? Suppose for a moment that all 6 billion (6×10^9) people on Earth simultaneously turned on a thousand 100-W light bulbs. Each person would then be lit up like a Hollywood movie theater on opening night! But all those bulbs surrounding all those people would still total only 6×10^{14} W. To use as much energy as the Sun produces, we would have to find 670 billion worlds like the Earth, all doing the same stunt. In other words, on the scale of light bulbs, the Sun is unimaginably bright.

And it is not enough just to determine how much energy the Sun produces in a second. You might think of all sorts of ways to generate a huge amount of energy if you had to do it only for a second, or even for a minute. But if you had

Ace✺Astronomy™ The AceAstronomy icon throughout the text indicates an opportunity for you to test yourself on key concepts and to explore animations and interactions on the AceAstronomy website at **http://ace.brookscole.com/voyages**

Virtual Laboratories

 Helioseismology

to sustain a tremendous output of energy for a year or a billion years, then you would need a steady, reliable generator.

Just how long has the Sun been shining? As we saw in Chapter 13, we have several lines of evidence indicating that the Sun formed at roughly the same time as the planetary system—about 4.5 billion years ago. Our task, then, is to explain not only how the Sun generates so many watts now, but also what mechanism has allowed it to do so for billions of years. When geologists and biologists began to uncover clues about the great age of the Earth in the late 19th and early 20th centuries, this appeared to be an insoluble problem.

Scientists trying to explain the Sun looked first at sources of energy already familiar to them here on Earth. This seemed a reasonable strategy, since the Sun and the Earth contain the same types of atoms, albeit in different proportions. But, as we shall see, none of the energy sources known at the time could explain the Sun's longevity. It was only after scientists discovered how to tap the energy stored in the nuclei of atoms that they finally identified the source of the Sun's energy.

Visuals Unlimited/Doug Sokell

■ **FIGURE 15.1**
A Wood Fire

15.1 THERMAL AND GRAVITATIONAL ENERGY

Nineteenth-century scientists knew of two possible sources for the Sun's energy: chemical and gravitational energy. The source of chemical energy most familiar to us here on Earth is the burning (the chemical term is *oxidation*) of wood, coal, gasoline, or other fuel (Figure 15.1). We know exactly how much energy the burning of these materials can produce. We can thus calculate that even if the immense mass of the Sun consisted of a burnable material like coal or wood, our star could not produce energy at its present rate for more than a few thousand years. That's not enough time for Earth to evolve even a virus, to say nothing of a life-form as complicated as an astronomy student! Besides, geologists have found fossils in rocks that are more than 3.5 billion years old, so the temperature of the Earth (and the heat output of the Sun) must have been suitable to sustain life as long ago as that. And we also know that at the temperatures found in the Sun, nothing like solid wood or coal could survive.

15.1.1 Conservation of Energy

Other 19th-century attempts to determine what makes the Sun shine used the law of *conservation of energy*. Simply stated, this law says that energy cannot be created or destroyed, but can be transformed from one type to another—say, from heat to mechanical energy. The steam engine, which was the key to the industrial revolution, is a good example. In it, the hot steam from a boiler drives the movement of a piston, converting heat energy to motion energy.

Motion can also be transformed into heat. If you clap your hands vigorously at the end of an especially good astronomy lecture, your palms become hotter. If you rub ice on the surface of a table, the heat produced by friction melts the ice.

In the 19th century, scientists thought that the source of the Sun's heat might be the mechanical motion of mete-

Smithsonian Institution, courtesy AIP Emilio Segrè Visual Archives

■ **FIGURE 15.2**

Kelvin and Helmholtz British physicist William Thomson (Lord Kelvin) and German scientist Hermann von Helmholtz proposed that the contraction of the Sun under its own gravity might account for its energy.

orites falling into it. Their calculations showed, however, that in order to produce the total amount of energy emitted by the Sun, the mass in meteorites that would have to fall into it every 100 years would equal the mass of the Earth. The resulting increase in the Sun's mass would, according to Kepler's third law, change the period of the Earth's orbit by 2 seconds per year. Such a change would be easily measurable and was not, in fact, occurring. This source of the Sun's energy was then ruled out.

15.1.2 Gravitational Contraction as a Source of Energy

As an alternative, the German scientist Hermann von Helmholtz and the British physicist Lord Kelvin (Figure 15.2), in about the middle of the 19th century, proposed that the Sun might produce energy by the conversion of gravitational energy to heat. They suggested that the outer layers of the Sun might be "falling" inward because of the force of gravity. In other words, they proposed that the Sun could be shrinking in size and staying hot and bright as a result.

To imagine what would happen, picture the outer layer of the Sun starting to fall inward. This outer layer is a gas made up of individual atoms, all moving about in random directions. Temperature, which measures the amount of stored heat energy, depends on the speed of the atoms: Higher speeds equal higher temperatures. If this layer starts to fall, the atoms acquire an additional speed because of the falling motion. As the outer layer falls inward, it also contracts, moving the atoms closer together. Collisions become more likely, and some of them transfer the extra speed associated with the falling motion to other atoms. This, in turn, increases the speeds of those atoms and hence increases the temperature of this layer of the Sun. Other collisions excite electrons within the atoms to higher

energy orbits. When these electrons return to their normal orbits, they emit photons, which can then escape from the Sun (see Chapter 4).

Kelvin and Helmholtz calculated that a contraction of the Sun at a rate of only about 40 meters per year would be enough to produce the amount of energy that it is now radiating. Over the span of human history, the decrease in the Sun's size from such a slow contraction would be undetectable.

If we assume that the Sun began its life as a large, diffuse cloud of gas, then we can calculate how much energy has been radiated by the Sun during its entire lifetime as it has contracted from a very large diameter down to its present size. That amount of energy is on the order of 10^{42} J. Since the solar luminosity is 4×10^{26} W (J/s) or about 10^{34} J per year, contraction could keep the Sun shining at its present rate for roughly 100 million years.

In the 19th century, 100 million years at first seemed plenty long enough, since the Earth was then widely thought to be much younger than this. But toward the end of that century and into the 20th, geologists and physicists showed that the Earth (and hence the Sun) is actually much older. Contraction therefore cannot be the primary source of solar energy (although, as we shall see in Chapter 20, contraction is an important source of energy in stars that are in the process of being born).

Scientists were thus confronted with a puzzle of enormous proportions. Either an unknown type of energy was responsible for the most important energy source known to humanity or estimates of the span of time that the solar system (and life on Earth) had been around had to be seriously modified. Charles Darwin, whose theory of evolution required a longer time span than the theories of the Sun seemed to permit, was discouraged by these results and continued to worry about them until his death in 1882.

Albert Einstein

For a large part of his life, Albert Einstein was one of the most recognized celebrities of his day. Strangers stopped him on the street, and people all over the world wrote asking him for endorsements, advice, and assistance. In fact, when Einstein and the great film star Charlie Chaplin met in California, they found they shared similar feelings about the loss of privacy that comes with fame. Einstein's name was a household word despite the fact that most people did not understand the ideas that had made him famous.

Einstein was born in 1879 in Ulm, Germany. Legend has it that he did not do well in school (even in arithmetic), and thousands of students have since justified a bad grade by referring to this story. Alas, like many legends, this one is not true. Records indicate that although he tended to rebel against the authoritarian teaching style in vogue in Germany at that time, Einstein was a good student.

After graduating from the Federal Polytechnic Institute in Zurich, Switzerland, Einstein at first had trouble getting a job (even as a high-school teacher), but he eventually became an examiner in the Swiss Patent Office. Working in his spare time,

Albert Einstein in 1905

Permission granted by the Albert Einstein Archives, The Hebrew University of Jerusalem, Israel

without the benefit of a university environment but using his superb physical intuition, he wrote four papers in 1905 that would ultimately transform the way physicists looked at the world.

One of these, which earned him the Nobel Prize in 1921, set part of the foundation of *quantum mechanics*, the rich, puzzling, and remarkable theory of the subatomic realm. But his most important paper presented the *special theory of relativity*, a reexamination of space, time, and motion that added a whole new level

of sophistication to our understanding of these concepts. $E = mc^2$ was a relatively minor part of this theory, added in a later paper.

In 1916, Einstein published his *general theory of relativity*, which was, among other things, a fundamentally new description of gravity (see our chapter on black holes). When this theory was confirmed by measurements of the "bending of starlight" during a 1919 eclipse (*The New York Times* headline read, "Lights All Askew in the Heavens"), Einstein became world-famous.

In 1933, to escape Nazi persecution, Einstein left his professorship in Berlin and settled in the United States at the newly created Institute for Advanced Studies at Princeton. He remained there until his death in 1955, writing, lecturing, and espousing a variety of intellectual and political causes. For example, he agreed to sign a letter written by Leo Szilard and other scientists in 1939, alerting President Roosevelt to the dangers of allowing Nazi Germany to develop the atomic bomb first. In 1952, Einstein was offered the second presidency of Israel. In declining the position, he said, "I know a little about nature and hardly anything about men."

It was only in the 20th century that the true source of the Sun's energy was identified. The two key pieces of information required to solve the puzzle were the structure of the nucleus of the atom and the fact that mass can be converted into energy.

15.2 MASS, ENERGY, AND THE THEORY OF RELATIVITY

As we have seen, energy cannot be created or destroyed, but only converted from one form to another. One of the remarkable conclusions derived by Albert Einstein (see the Voyagers in Astronomy box), when he developed the theory

of relativity, is that the mass of an object can be considered another form of energy and could therefore be converted to energy.

15.2.1 Converting Mass to Energy

This remarkable equivalence is expressed in one of the most famous equations in all of science:

$$E = mc^2$$

where E is the symbol for energy, m is the symbol for mass, and c, the constant that relates the two, is the speed of light. Note that this equation is very similar in form to

$$\text{inches} = \text{feet} \times 12$$

or

$$\text{cents} = \text{dollars} \times 100.$$

That is, it is a conversion formula that allows you to calculate the conversion of one thing, mass, to another, energy. The conversion factor in this case turns out to be not 12 or 100 but another constant quantity, the speed of light squared. Mass does not have to travel at the speed of light (or the speed of light squared, which is impossible in nature) for this conversion to occur. The factor of c^2 is just the number that Einstein showed must be used to relate mass and energy.

Notice that this formula does not tell you *how* to convert mass into energy, just as the formula for cents does not tell you where to exchange coins for a dollar bill. The formulas merely tell you what the equivalent values are if you succeed in making the conversion. When Einstein first derived his famous formula in 1905, no one had the faintest idea how to convert mass into energy in any practical way. Einstein himself tried to discourage speculation that the conversion of atomic mass into energy would be feasible in the near future. Today, as a result of developments in nuclear physics, we regularly convert mass into energy in power plants, nuclear weapons, and high-energy physics experiments in particle accelerators.

Because c^2, the speed of light squared, is a very large quantity, the conversion of even a small amount of mass results in a very large amount of energy. For example, the complete conversion of 1 gram (g) of matter (about 1/28 ounce) would produce as much energy as the burning of 15,000 barrels of oil.

Scientists soon realized that the conversion of mass to energy is the source of the Sun's heat and light. With Einstein's equation $E = mc^2$, we can calculate that the amount of energy radiated by the Sun could be produced by the complete conversion of about 4 million tons of matter to energy inside the Sun *each second*. Destroying 4 million tons per second sounds like a lot when compared to earthly things, but bear in mind that the Sun is a *very big* reservoir of matter. In fact, we will see that the Sun contains more than enough mass to destroy such huge amounts of matter and still continue shining at its present rate for billions of years.

But this still does not tell us *how* mass can be converted to energy. To understand the process that actually occurs in the Sun, we need to explore the structure of the atom a bit further.

15.2.2 Elementary Particles

The fundamental components of matter are called **elementary particles.** The most familiar of these are the proton, neutron, and electron—the particles that make up ordinary atoms (see Section 4.4).

Protons, neutrons, and electrons are by no means all the particles that exist. First, for each kind of particle, there is a corresponding but opposite *antiparticle*. If the particle carries a charge, its antiparticle has the opposite charge. The antielectron is the *positron*, which has the same mass as the electron but is positively charged. Likewise, the antiproton has a negative charge. The remarkable thing about such *antimatter* is that when a particle comes in contact with its antiparticle, the two annihilate each other, turning into pure energy.

Since our world is made exclusively of ordinary particles of matter, any antimatter cannot survive for very long. But individual antiparticles are found in cosmic rays (particles that arrive at the top of the Earth's atmosphere from space) and can be created in particle accelerators. In fact, when we create matter from energy in our high-energy physics labs, we always get half matter and half antimatter.

Science fiction fans may be familiar with antimatter from the *Star Trek* television series and films. The Starship Enterprise is propelled by the *careful* combining of matter and antimatter in the ship's engine room. According to $E = mc^2$, the complete annihilation of matter and antimatter can produce a huge amount of energy, but keeping the antimatter fuel from touching the ship before it is needed must be a big problem. No wonder that Scotty, the chief engineer in the original TV show, always looked worried!

In 1933, physicist Wolfgang Pauli suggested that there might be another type of elementary particle. Energy seemed to disappear when certain types of nuclear reactions took place, a violation of the law of conservation of energy. Pauli was reluctant to accept the idea that one of the basic laws of physics was wrong, and he suggested a "desperate remedy." Perhaps a so-far-undetected particle, which was given the name **neutrino** ("little neutral one"), was carrying away the "missing" energy. He suggested that neutrinos were particles with zero mass that, like photons, moved with the speed of light.

The elusive neutrino was not detected until 1956. The reason it was so hard to find is that neutrinos interact very weakly with other matter and so are very difficult to capture with any kind of detector. The Earth is more transparent to a neutrino than the thinnest and cleanest pane of glass is to a photon of light. In fact, most neutrinos can pass completely through a star or planet without being absorbed. As we shall see, this "antisocial" behavior of neutrinos makes them a very important tool for studying the Sun. It also turns out that, while Pauli was right about the existence of the neutrino, he was wrong about its mass. We will return to the subject of neutrinos at the end of this chapter.

Some of the properties of the proton, electron, neutron, and neutrino are summarized in Table 15.1. (Other subatomic particles have been produced by experiments with particle accelerators, but they do not play a role in the generation of solar energy.)

TABLE 15.1	*Properties of Some Elementary Particles*	
Particle	**Mass (kg)**	**Charge**
Proton	1.67265×10^{-27}	$+1$
Neutron	1.67495×10^{-27}	0
Electron	9.11×10^{-31}	-1
Neutrino	?	0

15.2.3 The Atomic Nucleus

The nucleus of an atom is not just a loose collection of elementary particles. Inside the nucleus, particles are held together by a very powerful force called the strong nuclear force. This is a short-range force, only able to act over distances about the size of the atomic nucleus. A quick thought experiment shows how important this force is. Take a look at your little finger and think of the atoms composing it. Among them is carbon, one of the basic elements of life. Focus your imagination on the nucleus of one of your carbon atoms. It contains six protons, which have a positive charge, and six neutrons, which are neutral. Thus the nucleus has a net charge of six positives. If only the electrical force were acting, the protons in this and every carbon atom would find each other very repulsive and fly apart.

The strong nuclear force is an attractive force, stronger than the electric force, and it keeps the particles of the nucleus tightly bound together. We saw earlier that if under the force of gravity a star "shrinks"—bringing its atoms closer together—gravitational energy is released. In the same way, if particles come together under the strong nuclear force and unite to form an atomic nucleus, some of the nuclear energy is released. The energy given up in such a process is called the *binding energy* of the nucleus.

When such binding energy is released, the resulting nucleus has slightly less mass than the sum of the masses of the particles that came together to form it. In other words, the energy comes from the loss of mass. This slight deficit in mass is only a small fraction of the mass of one proton.

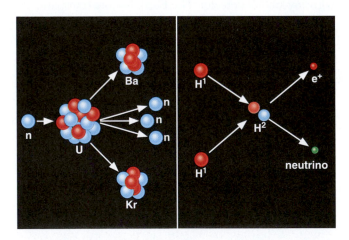

■ FIGURE 15.3

Fission and Fusion In fission, a larger nucleus breaks into two smaller components. Here a nucleus of uranium, with 92 protons and 143 neutrons, is shown undergoing fission into two smaller nuclei of barium (56 protons) and krypton (36 protons). In fusion, smaller nuclei bind together to make a larger one. We also see two nuclei of hydrogen (one proton each) fuse into a heavier hydrogen nucleus (one proton and one neutron), accompanied by the emission of a positron and a neutrino.

But because each bit of lost mass can provide a lot of energy (remember, $E = mc^2$), this nuclear energy release can be quite potent.

Measurements show that the binding energy is greatest for atoms with a mass near that of the iron nucleus (with a combined number of protons and neutrons equal to 56) and less for both the lighter and the heavier atoms. Iron, therefore, is the most stable element (since it gives up the most energy when it forms, it is the hardest nucleus to "undo").

What this means is that, in general, when light atomic nuclei come together to form a heavier one (up to iron), mass is lost and energy released. This joining together of atomic nuclei is called nuclear **fusion.**

Energy can also be produced by breaking up heavy atomic nuclei into lighter ones (down to iron); this process is called nuclear *fission* (Figure 15.3). Nuclear fission was the process we learned to use first—in atomic bombs and in nuclear reactors used to generate electric power—and it may thus be more familiar to you. It also sometimes occurs spontaneously in some unstable nuclei through the process of natural radioactivity. But fission requires big, complex nuclei, whereas we know that the stars are made up predominantly of small, simple nuclei. So we must look to fusion to explain the energy of the Sun and the stars.

15.2.4 Nuclear Attraction Versus Electrical Repulsion

So far, we seem to have a very attractive prescription for making energy: "Roll" some nuclei together and join them via nuclear fusion. This will cause them to lose some of their mass, which then turns into energy. However, every nucleus, even simple hydrogen, has protons in it—and protons all have positive charges. Since like charges repel via the electrical force, the closer we get two nuclei to each other, the more they repel. It's true that if we can get them within "striking distance" of the nuclear force, they will then come together with a much stronger attraction. But that striking distance is very tiny, about the size of a nucleus. How can we get nuclei close enough to participate in fusion?

The answer turns out to be heat—tremendous heat—which speeds the protons up enough to overcome the electrical forces that try to keep protons apart. Inside the Sun, as we saw, the most common element is hydrogen, whose nucleus is a single proton. Two protons can fuse only in regions where the temperature is greater than about 10 million degrees Kelvin (K), and the speed of the protons averages around 1000 km/s or more. (In old-fashioned units, that's over 2 million mi/h!)

In our Sun, such extreme temperatures are reached only in the regions near its center, which has a temperature of 15 million K. Calculations show that nearly all of the Sun's energy is generated within about 150,000 km of its core, or within less than 10 percent of its total volume.

Even at these high temperatures, it is exceedingly difficult to force two protons to combine. On average, a proton will rebound from other protons in the Sun's crowded core for about 14 billion years, at the rate of *100 million collisions per second,* before it fuses with a second proton. This is, however, only the *average* waiting time. Some of the enormous number of protons in the Sun's inner region are "lucky" and take only a few collisions to achieve a fusion reaction: They are the protons responsible for producing the energy radiated by the Sun. Since the Sun is about 4.5 billion years old, most of its protons have not yet been involved in fusion reactions.

Ace◉**Astronomy**™   Log into AceAstronomy and select this chapter to see Astronomy Exercise "Nuclear Fusion."

15.2.5 Nuclear Reactions in the Sun's Interior

The Sun, then, taps the energy contained in the nuclei of atoms through nuclear fusion. Let's look at what happens in more detail. Deep inside the Sun, four hydrogen atoms fuse to form a helium atom. The helium atom is slightly less massive than the four hydrogen atoms that combine to form it, and that lost mass is converted to energy.

The initial steps required to form one helium nucleus from four hydrogen nuclei are shown in Figure 15.4. First, two protons combine to make a *deuterium* nucleus, which is an *isotope* (or version) of hydrogen that contains one proton and one neutron. In effect, one of the original protons has been converted to a neutron in the fusion reaction. Electric charge has to be conserved in nuclear reactions, and it is conserved in this one. A **positron** (antimatter electron) emerges from the reaction and carries away the positive charge originally associated with one of the protons.

Since it is antimatter, this positron will instantly collide with a nearby electron, and both will be annihilated, producing pure electromagnetic energy in the form of gamma-ray photons. This gamma ray, which has been created in the center of the Sun, finds itself in a world crammed full of fast-moving nuclei and electrons. It collides with particles of matter and transfers some of its energy to them. Generally, the result of this process is to take energy away from the gamma-ray photon.

Such interactions happen to the gamma ray again and again and again as it makes its way slowly toward the outer layers of the Sun, until its energy becomes so reduced that it is no longer a gamma ray but an x ray (recall Section 4.2). Later, as it loses more energy, it becomes an ultraviolet photon. The Sun is so full of particles (targets for the photon to hit) that it takes on the order of a million years for the average photon to emerge from the Sun's photosphere. By that time, most of the photons have given up enough energy to be ordinary light—and they are the sunlight we see coming from our star. (To be precise, each gamma-ray photon is ultimately converted into many separate lower-energy photons of sunlight.)

Let's stop for a moment and really think about what we have found out. The energy you enjoy from the Sun on any

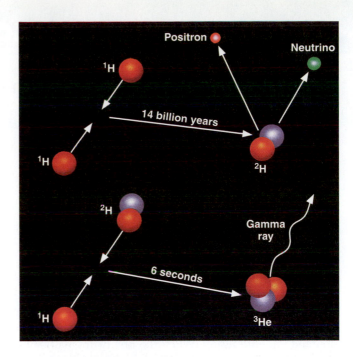

■ **FIGURE 15.4**

p-p Chain, Steps 1 and 2   These are the first two steps in the process of fusing hydrogen into helium in the Sun (the p-p chain).

given day had its origin as a gamma ray produced by nuclear reactions deep in the Sun's core roughly a million years ago. Having emerged from the Sun about 8 minutes ago, it made its way to Earth and spent its ancient energy warming you.

In addition to the positron, the fusion of two hydrogen atoms to form deuterium results in the emission of a neutrino. Because neutrinos interact so little with ordinary matter, those produced by fusion reactions near the center of the Sun travel directly to the Sun's surface and then on toward the Earth. Neutrinos move at nearly the speed of light and they get out of the Sun only about 2 seconds after they are created (see Figure 15.10, p. 348).

The next step in forming helium from hydrogen is to add another proton to the deuterium nucleus to create a helium nucleus that contains two protons and one neutron. In the process, some mass is again lost and more gamma radiation is emitted. Such a nucleus is helium because an element is defined by the number of its protons; any nucleus with two protons is called helium. But this form of helium, which we call helium-3, is not the isotope we see in the Sun's atmosphere or on Earth. That helium has two neutrons and two protons and hence is called helium-4.

To produce helium-4 in the Sun, helium-3 combines with another just like it in the third step of fusion (illustrated in Figure 15.5). Note that two energetic protons are left over from this step; they come out of the reaction ready to collide with other protons and thus to start step 1 all over again.

Fusion on Earth

Wouldn't it be wonderful if we could duplicate the Sun's energy mechanism in a controlled way on Earth? (We have already duplicated it in an uncontrolled way in hydrogen bombs, but we hope our storehouse of these will never be used.) Fusion energy would have many advantages: It would use hydrogen (or deuterium, which is heavy hydrogen) as fuel, and there is plenty of hydrogen in the Earth's lakes and oceans. Water is much more evenly distributed around the world than is oil or uranium, meaning that a few countries would no longer hold an energy advantage over the others. And unlike fission, which leaves dangerous by-products, the nuclei that result from fusion are perfectly safe.

The problem is that, as we saw, it takes very, very high temperatures for nuclei to overcome their electrical repulsion and undergo fusion. When the first hydrogen bombs were exploded in tests in the 1950s, the "fuses" to get them hot enough were fission bombs. Interactions at such temperatures are difficult to sustain and control.

Canada, Europe, Japan, and Russia are now collaborating on ITER, a project to demonstrate the feasibility of controlled fusion. ITER is based on the Tokamak design, in which a large donut-shaped container is surrounded by superconducting magnets to confine and control the hydrogen nuclei in a strong magnetic field. The goal of ITER is to build the first fusion device capable of producing thermal energy at the level equivalent to what is put out by an electric power station. The challenge is to keep the deuterium and tritium, which will participate in fusion reactions, hot enough, dense enough, long enough to produce energy. Previous fusion experiments have produced about 15 million watts of energy, but only for a second or two, and required 100 million watts to produce the conditions necessary to achieve fusion. Obviously, a process that requires more energy than it produces is not going to attract a lot of investors just yet!

In 1989, two scientists (from the University of Utah) made an announcement that simply astonished astronomers and physicists familiar with fusion. The two claimed to have achieved fusion at room temperatures, with equipment so simple it could easily be duplicated in a high-school science lab. They used an electrochemical cell in which electric current passes from one metal surface to another through a chemical solution. Even though the equipment was nowhere near the temperature physicists and astronomers had said was needed for fusion, the Utah scientists thought they had found evidence of deuterium fusion in their equipment—and their results, quickly dubbed "cold fusion," became a media sensation.

Alas, cold fusion did not hold up under the intense scrutiny to which all new scientific ideas and procedures are subjected. Other groups around the world could not duplicate the results. Even the original experiments failed to show many other characteristics associated with fusion. It turned out that the groups in Utah were probably seeing chemical reactions, not nuclear ones, and the initial results were discredited.

It appears that if we want to duplicate fusion on Earth, we have to do what the Sun does: Produce temperatures and pressures high enough to get hydrogen nuclei on intimate terms with one another. Perhaps by the time your children or grandchildren take an astronomy class in college, controlled fusion will be a reality instead of a dream.

Published with kind permission of ITER

Design of the ITER The gold areas in this diagram show where the superconducting magnets will circle the chamber within which fusion will take place. A huge magnet will keep the charged nuclei of heavy hydrogen confined. The goal is to produce 500 megawatts of energy.

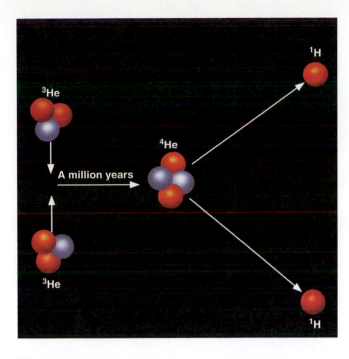

■ **FIGURE 15.5**

p-p Chain, Step 3 We see the third step in the fusion of hydrogen into helium in the Sun. Note that two of the products of the second step (see Figure 15.4) must combine before the third step becomes possible.

15.2.6 The p-p Chain

The nuclear reactions in the Sun that we have been discussing can be described succinctly through the following nuclear formulas:

$$^1H + ^1H \rightarrow ^2H + e^+ + v$$
$$^2H + ^1H \rightarrow ^3He + \gamma$$
$$^3He + ^3He \rightarrow ^4He + 2\,^1H$$

Here the superscripts indicate the total number of neutrons plus protons in the nucleus, e^+ is the positron, v is the neutrino, and γ indicates that gamma rays are emitted. Note that the first two steps must happen twice before the third step can occur, because the third step requires two helium-3 nuclei at its start.

Although, as we discussed, the first step in this chain of reactions is very difficult and generally takes a long time, the other steps happen more quickly. After the deuterium nucleus is formed, it survives an average of only about 6 seconds before being converted to ^3He. About a million years after that, the ^3He nucleus will combine with another to form ^4He.

We can compute the amount of energy these reactions generate by calculating the difference in initial and final mass. The masses of hydrogen and helium atoms in the units normally used by scientists are 1.007825u and 4.00268u, respectively. (The unit of mass, u, is defined to be 1/12 the

mass of an atom of carbon, or approximately the mass of a proton.) Here we include the mass of the entire atom, not just the nucleus, because electrons are involved as well. When hydrogen is converted to helium, two positrons are created (remember, the first step happens twice), and these are annihilated with two free electrons, adding to the energy produced.

$$4 \times 1.007825 = 4.03130u \quad \text{(mass of initial hydrogen atoms)}$$
$$\underline{-4.00268u} \quad \text{(mass of final helium atom)}$$
$$0.02862u \quad \text{(mass lost in the transformation)}$$

The mass lost, 0.02862u, is 0.71 percent of the mass of the initial hydrogen. Thus, if 1 kg of hydrogen is converted into helium, then the mass of the helium is only 0.9929 kg, and 0.0071 kg of material is converted into energy. The speed of light (c) is 3×10^8 m/s, so the energy released by the conversion of just 1 kg of hydrogen to helium is:

$$E = mc^2$$
$$E = 0.0071 \times (3 \times 10^8)^2 = 6.4 \times 10^{14}\,\text{J}$$

This amount, the energy released when a *single kilogram* (which is equivalent to 2.2 pounds) of hydrogen undergoes fusion, would supply all of the electricity used in the United States for about 2 weeks.

To produce the Sun's luminosity of 4×10^{26} W, some 600 million tons of hydrogen must be converted to helium *each second,* of which about 4 million tons turn from matter into energy. As large as these numbers are, the store of hydrogen (and thus of nuclear energy) in the Sun is still more enormous and can last a *long* time—billions of years, in fact.

At the temperatures inside the stars with masses less than about 1.2 times the mass of our Sun (a category that includes the Sun itself), most of the energy is produced by the reactions we have just described, and this set of reactions is called the **proton-proton cycle** (or sometimes, the *p-p chain*). It is called a cycle because the two protons produced in the third step can fuse with other protons to initiate the first step again. In the proton-proton cycle, protons collide directly with other protons to build into helium nuclei.

In hotter stars, another set of reactions, called the *carbon-nitrogen-oxygen* (CNO) cycle, accomplishes the same net result. In the CNO cycle, carbon and hydrogen nuclei collide to initiate a series of reactions that form nitrogen, oxygen, and ultimately helium. The nitrogen and oxygen nuclei do not survive but interact to form carbon again. Therefore, the outcome is the same as in the proton-proton cycle: Four hydrogen atoms disappear, and in their place a single helium atom is created. The CNO cycle plays only a minor role in the Sun but is the main source of energy for stars with masses greater than about twice the mass of the Sun.

Thus we have solved the puzzle that so worried scientists at the end of the 19th century. The Sun can maintain its high temperature and energy output for billions of years through the fusion of the simplest element in the universe, hydrogen. Because most of the Sun (and the other stars) is made of hydrogen, it is an ideal "fuel" for powering a star. As will be discussed in the coming chapters, we can define a star as a ball of gas capable of getting its core hot enough to initiate the fusion of hydrogen. There are balls of gas that lack the mass required to do this (Jupiter is a local example); like so many hopefuls in Hollywood, they will never, ever be stars.

15.3 THE INTERIOR OF THE SUN: THEORY

Fusion of protons can occur in the center of the Sun only if the temperature exceeds 10 million K. How do we know that the Sun is actually this hot? To determine what the interior of the Sun is like, it is necessary to resort to complex calculations. Essentially, astronomers teach a computer everything they know about the physical processes going on in the Sun's interior. The computer then calculates the temperature and pressure at every point inside the Sun and determines what nuclear reactions, if any, are taking place. The computer can also calculate how the Sun will change with time.

After all, the Sun must change. In its center, the Sun is slowly depleting its supply of hydrogen and creating helium instead. Will this composition change have measurable effects? Will the Sun get hotter? Cooler? Larger? Smaller? Brighter? Fainter? Ultimately, the changes in the center could be catastrophic, since eventually all the hydrogen fuel hot enough for fusion will be exhausted. Either a new source of energy must be found, or the Sun will cease to shine. We will describe the ultimate fate of the Sun in later chapters. For now, let's look at some of the things we must teach the computer about the Sun in order to carry out such calculations.

15.3.1 The Sun Is a Gas

The Sun is so hot that the material in it is gaseous throughout. Astronomers are grateful for this because a hot gas is easier to describe mathematically than either liquids or solids. The particles that constitute a gas are in rapid motion, frequently colliding with one another. This constant bombardment is the *pressure* of the gas (Figure 15.6). More particles within a given volume of gas produce more pressure because the combined impact of the moving particles increases with their number. The pressure is also greater when the molecules or atoms are moving faster. Since the molecules move faster when the temperature is hotter, higher temperatures produce higher pressure.

15.3.2 The Sun Is Stable

The Sun, like the majority of other stars, is stable; it is neither expanding nor contracting. Such a star is said to be in a condition of *equilibrium*. All the forces within it are

■ FIGURE 15.6

Gas Pressure The particles in a gas are in rapid motion and produce pressure through collisions with the surrounding material. Here particles are shown bombarding the sides of an imaginary container.

balanced, so that at each point within the star the temperature, pressure, density, and so on, are maintained at constant values. We will see in later chapters that even these stable stars, including the Sun, are changing as they evolve, but such evolutionary changes are so gradual that to all intents and purposes the stars are still in a state of equilibrium at any given time.

The mutual gravitational attraction between the masses of various regions within the Sun produces tremendous forces that tend to collapse the Sun toward its center. Yet we know from the history of the Earth that the Sun has been emitting roughly the same amount of energy for billions of years, and so clearly it has managed to resist collapse for a very long time. The gravitational forces must therefore be counterbalanced by some other force. That force is the pressure of the gases within the Sun (Figure 15.7). Calculations show that, in order to exert enough pressure to prevent the Sun from collapsing due to the force of gravity, the gases at its center must be maintained at a temperature of 15 million K. Think about what this tells us. Just from the fact that the Sun is not contracting, we can conclude that its temperature must be high enough at the center for protons to undergo fusion.

If the internal pressure in a star were not great enough to balance the weight of its outer parts, the star would collapse somewhat, contracting and building up the pressure inside. If the pressure were greater than the weight of the overlying layers, the star would expand, thus decreasing the internal pressure. Expansion would stop, and equilibrium

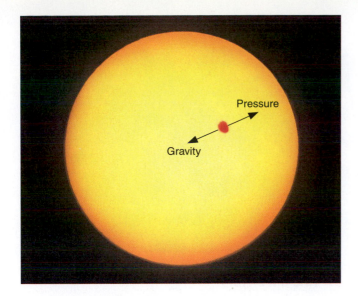

FIGURE 15.7

Hydrostatic Equilibrium In the interior of a star, the inward force of gravity is exactly balanced at each point by the outward force of gas pressure.

Ace⊙Astronomy™ Log into AceAstronomy and select this chapter to see the Active Figure called "Gravity vs. Pressure," and Astronomy Exercise "Hydrostatic Equilibrium."

would again be reached, when the pressure at every internal point again equaled the weight of the stellar layers above that point. An analogy is an inflated balloon, which will expand or contract until an equilibrium is reached between the pressure of the air inside and of that outside. The technical term for this condition is **hydrostatic equilibrium.** Stable stars are all in hydrostatic equilibrium; so are the oceans of the Earth as well as the Earth's atmosphere. The air's own pressure keeps it from falling to the ground.

15.3.3 The Sun Is Not Cooling Down

As everyone who has ever left a window open on a cold winter night knows, heat always flows from hotter to cooler regions. As energy filters outward toward the surface of a star, it must be flowing from inner, hotter regions. The temperature cannot ordinarily get cooler as we go inward in a star, or energy would flow in and heat up those regions until they were at least as hot as the outer ones. We conclude that the temperature is highest at the center of a star, dropping to lower and lower values toward the stellar surface. (The high temperature of the Sun's chromosphere and corona may therefore appear to be a paradox. But remember from Chapter 14 that these high temperatures are maintained by magnetic heating, which occurs in the Sun's atmosphere.)

The outward flow of energy through a star robs it of its internal heat, and the star would cool down if that energy were not replaced. Similarly, a hot iron begins to cool as soon as it is unplugged from its source of electric energy. There-

fore, a source of fresh energy must exist within each star. In the Sun's case, we have seen that this energy source is the fusion of hydrogen to form helium.

Ace⊙Astronomy™ Log into AceAstronomy and select this chapter to see the Active Figure called "Conduction, Convection, and Radiation."

15.3.4 Heat Transfer in a Star

Since the nuclear reactions that generate the Sun's energy occur deep within it, the energy must be transported from the center of the Sun to its surface—where we see it in the form of both heat and light. There are three ways in which energy can be transferred from one place to another. In **conduction,** atoms or molecules pass on their energy by colliding with others nearby. This happens, for example, when the handle of a metal spoon heats up as you stir a hot cup of coffee. In **convection,** currents of warm material rise, carrying their energy with them to cooler layers. A good example is hot air rising from a fireplace. In **radiation,** energetic photons move away from hot material and are absorbed by some other material to which they convey some or all of their energy. You can feel this when you put your hand close to the coils of an electric heater, allowing infrared photons to heat up your hand. Conduction and convection are both important in the interiors of planets. In stars, which are much more transparent, radiation and convection are important, while conduction can usually be ignored.

Stellar *convection* occurs as currents of hot gas flow up and down through the star (Figure 15.8). Such currents travel at moderate speeds and do not upset the overall stability of the star. Nor do they result in a net transfer of mass either inward or outward. In much the same way, heat from a fireplace can stir up air currents in a room without driving any air into or out of the room. Convection currents

FIGURE 15.8

Convection Rising convection currents carry heat from the Sun's interior to its surface, while cooler material sinks downward. Of course, nothing in a real star is as simple as diagrams in textbooks suggest.

Ace⊙Astronomy™ Log into AceAstronomy and select this chapter to see Astronomy Exercise "Convection and Magnetic Fields II: Sun."

carry heat very efficiently outward through a star. In the Sun, convection turns out to be important in the central regions and near the surface.

Unless convection occurs, the only significant mode of energy transport through a star is by electromagnetic *radiation*. Radiation is not an efficient means of energy transport in stars because gases in stellar interiors are very opaque; that is, a photon does not go far (in the Sun, typically about 0.01 m) before it is absorbed. The processes by which atoms and ions can interrupt the outward flow of photons—such as becoming ionized—were discussed in Section 4.5. The absorbed energy is always reemitted, but it can be reemitted in any direction. A photon absorbed when traveling outward in a star has almost as good a chance of being reradiated back toward the center of the star as toward its surface.

A particular quantity of energy, therefore, zigzags around in an almost random manner and takes a long time to work its way from the center of the star to the surface (Figure 15.9). In the Sun the time required is on the order of a million years. If the photons were not absorbed and reemitted along the way, they would travel at the speed of light and could reach the surface in a little over 2 seconds, just as the neutrinos do (Figure 15.10).

15.3.5 Model Stars

Scientists use the principles we have just described to calculate what the Sun's interior is like. These physical ideas are expressed as mathematical equations that are solved to determine the values of temperature, pressure, density, the efficiency with which photons are absorbed, and other physical quantities throughout the Sun. The solutions so obtained, based on a specific set of physical assumptions, are called a theoretical model for the interior of the Sun.

Figure 15.11 schematically illustrates a theoretical model for the Sun's interior. Energy is generated through fusion in the core of the Sun, which extends only about one

■ **FIGURE 15.9**

Photons Deep in the Sun A photon moving through the dense gases in the solar interior travels only a short distance before it interacts with one of the surrounding atoms. The photon usually has a lower energy after each interaction and may then travel in any random direction.

quarter of the way to the surface. This core contains about one third of the total mass of the Sun, however. At the center, the temperature reaches a maximum of approximately 15 million K and the density is nearly 150 times the density of water. The energy generated in the core is transported toward the surface by radiation until it reaches a point about 70 percent of the distance from the center to the surface. At this point convection begins, and energy is transported the rest of the way primarily by rising columns of hot gas.

Figure 15.12 shows how the temperature, density, rate of energy generation, and composition vary from the center of the Sun to its surface.

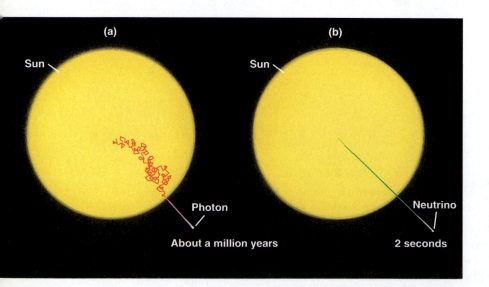

■ **FIGURE 15.10**

Photon and Neutrino Paths in the Sun
(a) Because photons generated by fusion reactions in the solar interior travel only a short distance before being absorbed or scattered by atoms and sent off in random directions, it takes about a million years for energy to make its way from the center of the Sun to its surface. (b) In contrast, neutrinos do not interact with matter but traverse straight through the Sun at the speed of light, reaching the surface in only a little more than 2 seconds.

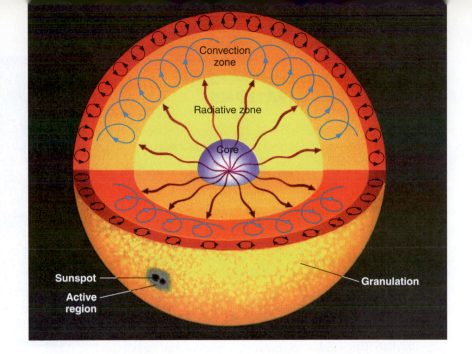

■ FIGURE 15.11

The Interior Structure of the Sun
Energy is generated in the core by the fusion of hydrogen to form helium. This energy is transmitted outward by radiation—that is, by the absorption and reemission of photons. In the outermost layers, energy is transported mainly by convection.

Ace**Astronomy**™   Log into AceAstronomy and select this chapter to see the Active Figure called "The Sun."

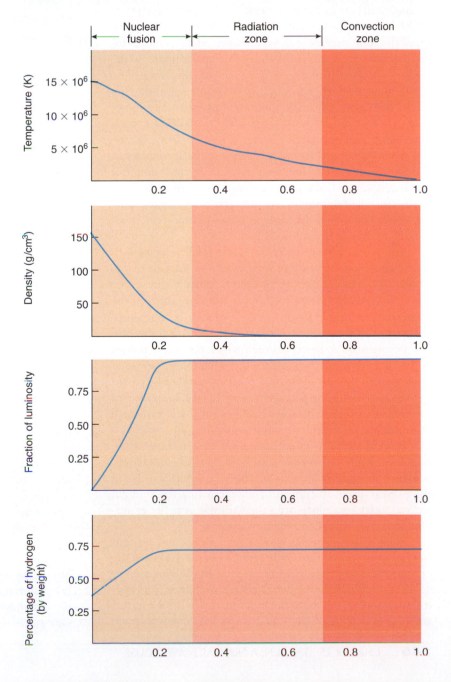

■ FIGURE 15.12

The Interior of the Sun   Diagrams showing how temperature, density, rate of energy generation, and the percentage (by mass) abundance of hydrogen vary inside the Sun. The horizontal scale shows the fraction of the Sun's radius: The left edge is the very center; the right edge, the photosphere.

15.4 THE SOLAR INTERIOR: THE OBSERVATIONS

Recall that when we observe the Sun's photosphere, we are not seeing very deeply into our star, certainly not into the regions where energy is generated. That's why the title of this section—observations of the solar interior—should seem very surprising. However, astronomers have indeed devised two types of measurements that can be used to obtain information about the inner parts of the Sun. One technique involves the analysis of tiny changes in the motion of small regions at the Sun's surface. The other relies on the measurement of the neutrinos emitted by the Sun.

15.4.1 Solar Pulsations

Astronomers have discovered that the Sun pulsates—that is, it alternately expands and contracts—just as your chest expands and contracts as you breathe. This pulsation is very slight, but it can be detected by measuring the *radial velocity* of the solar surface—the speed with which it moves toward or away from us. The velocities of small regions on the Sun are observed to change in a regular way, first toward the Earth, then away, then toward, and so on. It is as if the Sun were "breathing" through thousands of individual lungs, each having a size in the range of 4000 to 15,000 km, each fluctuating back and forth (Figure 15.13).

The typical velocity of one of the oscillating regions on the Sun is only a few hundred meters per second, and it takes only about 5 minutes to complete a full cycle from maximum to minimum velocity and back again. The change in the size of the Sun measured at any given point is no more than a few kilometers.

The remarkable thing is that these small velocity variations can be used to determine what the interior of the Sun is like. The motion of the Sun's surface is caused by waves that reach it from deep in the interior. Study of the amplitude and cycle length of the velocity changes provides information about the temperature, density, and composition of the layers through which the waves passed before they reached the surface. The situation is somewhat analogous to the use of seismic waves generated by earthquakes to infer the properties of the Earth's interior. For this reason, studies of solar oscillations (back-and-forth motions) are referred to as **solar seismology.**

It takes a little over an hour for waves to traverse the Sun from center to surface, so the waves, like neutrinos, provide information about what the solar interior is like at the present time. In contrast, remember that the sunlight we see today emerging from the Sun was actually generated in the core about a million years ago.

Solar seismology has shown that convection extends inward from the surface 30 percent of the way toward the center; we have used this information in drawing Figure 15.11. Pulsation measurements also show that the *differential rotation* that we see at the Sun's surface, with the fastest rotation occurring at the equator, persists down through

National Optical Astronomy Observatories

■ FIGURE 15.13

Oscillations in the Sun New observational techniques permit astronomers to measure small differences in velocity at the Sun's surface to infer what the deep solar interior is like. In this computer simulation, red shows surface regions that are moving away from the observer; blue marks regions moving toward the observer. Note that the velocity changes penetrate deep into the Sun's interior.

the convection zone. Below the convection zone, however, the Sun, even though it is gaseous throughout, rotates as if it were a solid body like a bowling ball. Another result from solar seismology is that the abundance of helium inside the Sun, except in the center where nuclear reactions have converted hydrogen to helium, is about the same as at the surface. That result is important to astronomers because it means we are correct when we use the abundances of the elements measured in the solar atmosphere to construct models of the solar interior.

Solar seismology also allows scientists to look below a sunspot and see how it works. In Chapter 14, we said that sunspots are cool because strong magnetic fields block the outward flow of energy. Figure 15.14 shows how gas moves around underneath a sunspot. The sunspot itself acts like a drain in a bathtub. Cool material from the sunspot flows downward at speeds of about 4000 km/h. Material surrounding the sunspot is pulled inward, carrying magnetic field with it, thus maintaining the strong field that is necessary to form a sunspot. As the new material enters the sunspot region, it too cools, becomes denser, and sinks, thus setting up a self-perpetuating cycle that can last for weeks.

The downward-flowing cool material acts as a kind of plug that blocks the upward flow of hot material, which is then diverted sideways and eventually reaches the solar surface in the region around the sunspot. This outward flow of hot material accounts for the paradox that we described in the last chapter—namely, that the Sun emits slightly more energy when more of its surface is covered by cool sunspots.

■ FIGURE 15.14

Structure of Sunspot This drawing shows our new understanding, from solar seismology, of what lies beneath a sunspot. The black arrows show the direction of the flow of material. The intense magnetic field associated with the sunspot stops the upward flow of hot material and creates a kind of plug that blocks the hot gas. As the material above the plug cools (shown in blue), it becomes denser and plunges inward, drawing more gas and more magnetic field behind it into the spot. The concentrated magnetic field causes more cooling, thereby setting up a self-perpetuating cycle that allows a spot to survive for several weeks. Since the plug keeps hot material from flowing up into the sunspot, the region below the plug, represented by red in this picture, becomes hotter. This material flows sideways and then upward, eventually reaching the solar surface in the area surrounding the sunspot.

Ace◯Astronomy™ Log into AceAstronomy and select this chapter to see Astronomy Exercise "Sunspot Cycle I."

Solar seismology is also becoming an important tool for predicting solar storms that might impact the Earth. Active regions can appear and grow to large size in only a few days. The solar rotation period is about 28 days. Therefore, regions capable of producing solar flares and coronal mass ejections can develop on the far side of the Sun where we can't see them. However, sound waves travel slightly faster in regions of high magnetic field, and waves generated in active regions traverse the Sun about 6 seconds faster than waves generated in quiet regions. By detecting this subtle difference, scientists can provide warnings of a week or more to operators of electric utilities and satellites about when a potentially dangerous active region might rotate into view. With this warning, it is possible to plan for disruptions, put key instruments into safe mode, or reschedule spacewalks in order to protect astronauts.

15.4.2 Solar Neutrinos

The second technique for obtaining information about the Sun's interior involves the detection of a few of those elusive neutrinos created during nuclear fusion. Recall from our earlier discussion that neutrinos very rarely interact with matter, and that most of the neutrinos created in the center of the Sun make their way directly out of the Sun and travel to the Earth at nearly the speed of light. As far as neutrinos are concerned, the Sun is transparent.

About 3 percent of the total energy generated by nuclear fusion in the Sun is carried away by neutrinos. So many protons react inside the Sun's core that, scientists calculate, 35 million billion (3.5×10^{16}) solar neutrinos pass through each square meter of the Earth's surface every second. If we can devise a way to detect even a few of these solar neutrinos, then we can obtain information directly about what is going on in the center of the Sun. Unfortunately for those trying to "catch" some neutrinos, the Earth and everything on it are nearly transparent to passing neutrinos, just as the Sun is.

On very, very rare occasions, however, one of the billions and billions of solar neutrinos will interact with another atom. The first successful detection of solar neutrinos made use of cleaning fluid (C_2Cl_4). The nucleus of a chlorine (Cl) atom in the cleaning fluid can be turned into a radioactive argon nucleus by an interaction with a neutrino. Because the argon is radioactive, its presence can be detected. However, since the interaction of a neutrino with chlorine happens so rarely, a huge amount of chlorine is needed.

Raymond Davis, Jr., and his colleagues at Brookhaven National Laboratory placed a tank containing nearly 400,000 liters of cleaning fluid 1.5 km beneath the Earth's surface in a gold mine at Lead, South Dakota. A mine was chosen so that the surrounding material of the Earth would keep cosmic rays (high-energy particles from space) from reaching the cleaning fluid and creating false signals. Cosmic-ray particles are stopped by thick layers of Earth, but neutrinos find them of no significance. Calculations show that solar neutrinos should produce about one atom of radioactive argon in the tank each day.

Davis's experiment, begun in 1970, detected only about one third as many neutrinos as predicted by solar models. This was a shocking result because astronomers thought they had a pretty good understanding of both neutrinos and the Sun's interior. This demonstrates how important it is not to rest on your laurels in science, but to continue doing new experiments whenever possible. Other experiments have now confirmed the fact that we do indeed detect fewer neutrinos than predicted by the best solar models. Davis received the 2002 Nobel Prize in physics for his work.

Does this mean that the best models we have of the Sun aren't good enough? To explain the missing neutrinos, astronomers tried tinkering with the solar models. If, say, the temperature in the core were slightly lower than the models predict or if the composition of the interior were somewhat different from the surface, maybe fewer neutrinos would be produced. Before the measurements of solar oscillations, some very ingenious models were devised that seemed almost to solve the problem. However, the measurements of oscillations really do tell us quite accurately what the interior of the Sun must be like, and we now know that the various models inspired by the shortage of neutrinos cannot be right.

If our original solar model was right all along, then what about the "missing" neutrinos? Do they behave in some unexpected way so that they "get lost" en route to South Dakota from the center of the Sun? Fusion in the Sun produces only one type of neutrino, the so-called electron neutrino, and the initial experiments to detect solar neutrinos were, quite sensibly, designed to detect this one type. Physicists have shown, however, that there are actually three types of neutrinos. Suppose the electron neutrinos produced in the Sun change to a different type during their journey from the center of the Sun to the Earth. Then the initial experiments would have been unable to detect them. Such a transformation, by the way, is called a *neutrino oscillation*.

A recent experiment at the Sudbury Neutrino Observatory in Canada was the first one designed to capture all three types of neutrinos. The experiment is located in a mine 2 km underground. The neutrino detector consists of a 12-m-diameter transparent acrylic plastic sphere, which contains 1000 metric tons of heavy water (Figure 15.15 and the opening figure for this chapter). Remember that an ordinary water nucleus contains two hydrogen atoms and one oxygen atom. Heavy water instead contains two deuterium atoms and one oxygen atom, and incoming neutrinos can occasionally break up the loosely bound proton and neutron that make up the deuterium nucleus. The sphere of heavy water is surrounded by a shield of 1700 metric tons of very pure water, which in turn is surrounded by 9600 photomultipliers, devices that detect flashes of light produced after neutrinos interact with the heavy water.

To the enormous relief of astronomers who make models of the Sun, the Sudbury experiment detects about 1 neutrino per hour and has shown that the *total* number of neutrinos reaching the heavy water is just what solar models predict. Only one third of these, however, are electron neutrinos. It appears that two thirds of the electron neutrinos produced by the Sun transformed themselves into one of the other types of neutrinos as they made their way from the core of the Sun to the Earth. This is why the earlier experiments saw only one-third the number of neutrinos we expected.

Although it is not intuitively obvious, such neutrino oscillations can happen only if the mass of the electron neu-

Courtesy of SNO

■ FIGURE 15.15

Sudbury Neutrino Detector   An artist's drawing of the Sudbury neutrino detector. The cavity, more than 2 km (6800 ft) underground, is the size of a ten-story building. The blue sphere holds 1000 metric tons of heavy water.

trino is not zero. Other experiments indicate that its mass is tiny (even compared to the electron). But the fact that the neutrino has mass at all has deep implications for both physics and astronomy. For example, we will look at the role that neutrinos play in the inventory of the mass of the universe in Chapter 28.

Isn't it amazing how a series of experiments that began with enough cleaning fluid to fill a swimming pool being brought down the shafts of an old gold mine is now teaching us about the energy source of the Sun and the properties of matter! This is a good example of how experiments in astronomy, coupled with the best models we can devise, can still lead to fundamental changes in our understanding of nature.

SURFING THE WEB

Albert Einstein Online: www.westegg.com/einstein/
S. Morgan Friedman of the University of Pennsylvania has assembled this very useful Web site that has a link to just about every other Web site about Einstein. Includes information about his life, books and papers, collections of quotes, pictures, related science, and more.

Princeton Plasma Physics Laboratory Site: ippex.pppl.gov/
About the quest for controlled fusion on Earth.

GONG Project Site: www.gong.noao.edu/index.html
The Global Oscillations Network Group (GONG) is an international collaboration for helioseismology.

SUMMARY

15.1 The Sun produces an enormous amount of energy every second. The Earth is 4.5 billion years old, so the Sun must have been shining for at least that long. Neither chemical burning nor gravitational contraction can account for the energy radiated by the Sun during all this time.

15.2 Solar energy is produced by interactions of **elementary particles**—that is, protons, neutrons, electrons, **positrons,** and **neutrinos.** Specifically, the source of the Sun's energy is the **fusion** of hydrogen to form helium. The series of reactions required to convert hydrogen to helium is called the **proton-proton cycle.** A helium atom is about 0.71 percent less massive than the four hydrogen atoms that combine to form it, and that lost mass is converted to energy (with the amount of energy given by the formula $E = mc^2$).

15.3 Even though we cannot see inside the Sun, it is possible to calculate what its interior must be like. As input for these calculations, we use what we know about the Sun. It is made entirely of hot gas. Apart from some very tiny changes, the Sun is neither expanding nor contracting (it is in **hydrostatic equilibrium**) but instead puts out energy at a constant rate. Fusion of hydrogen occurs in the center of the Sun, and the energy generated is carried to the surface by **radiation** and **convection.** A solar model describes the structure of the Sun's interior. Specifically, it describes how pressure, temperature, mass, and luminosity depend on distance from the center of the Sun.

15.4 Studies of solar oscillations (**solar seismology**) and neutrinos can provide observational data about the Sun's interior. The technique of solar seismology has so far shown that the composition of the interior is much like that of the surface (except in the core, where some of the original hydrogen has been converted to helium), and that the convection zone extends about 30 percent of the way from the Sun's surface to its center. Solar seismology can also detect active regions on the far side of the Sun and provide better predictions of solar storms that may affect the Earth. A recent experiment has shown that solar models predict accurately the number of electron neutrinos produced by nuclear reactions in the core of the Sun. However, two thirds of these neutrinos are converted to different types of neutrinos during their long journey from the Sun to the Earth, a result that also indicates that neutrinos are not massless particles.

INTER-ACTIVITY

A The text states that meteorites falling into the Sun could not be the source of the Sun's energy because the necessary increase in the mass of the Sun would lengthen the Earth's orbital period by 2 seconds per year. Have your group discuss what effects this would cause as the centuries went on.

B Solar astronomers can learn more about the Sun's interior if they can observe the oscillations 24 hours each day. That means that they cannot have their observations interrupted by the day/night cycle. Such an experiment, called the GONG (Global Oscillation Network Group) project, has been set up. To save money, this experiment was designed to make use of the minimum possible number of telescopes. It turns out that if the sites are selected carefully, the Sun can be observed all but about 10 percent of the time with only six observing stations. What factors have to be taken into consideration in selecting the observing sites? Can you suggest six general geographic locations that would optimize the amount of time that the Sun can be observed? Check your answer by looking at the GONG Web site.

C What would it be like if we actually manage to get controlled fusion on Earth to be economically feasible? If the hydrogen in water becomes the fuel for releasing enormous amounts of energy, have your group discuss how this would affect the world economy and international politics. (Think of the role that oil and natural gas deposits now play on the world scene.)

D Your group is a delegation sent to the city council of a small mining town to explain why the government is putting a swimming-pool-sized vat of commercial cleaning fluid down one of the shafts of an old gold mine. How would you approach this meeting? Assuming that the members of the city council do not have much science background, how would you explain the importance of the project to them? Suggest some visual aids you could use.

1. How do we know the age of the Sun?

2. Explain how we know that the Sun's energy is not supplied either by chemical burning, as in fires here on Earth, or by gravitational contraction (shrinking).

3. What is the ultimate source of energy that makes the Sun shine?

4. What is the *p-p* chain or proton-proton cycle? Give the three steps and explain what happens in each step?

5. How is a neutrino different from a neutron? List all the ways you can think of.

6. Describe in your own words what is meant by the statement that the Sun is in hydrostatic equilibrium.

7. Two astronomy students travel to South Dakota. One stands on the Earth's surface and enjoys some sunshine. At the same time, the other descends into a gold mine where neutrinos are detected, arriving in time to measure the creation of a new radioactive argon nucleus. Although the photon at the surface and the neutrinos in the mine arrive at the same time, they have had very different histories. Describe the differences.

8. Why do measurements of the number of neutrinos emitted by the Sun tell us about conditions deep in the solar interior?

9. Do neutrinos have mass? Describe how the answer to this question has changed over time and why.

THOUGHT QUESTIONS

10. The Sun is much larger and more massive than the Earth. Do you think the density of the Sun is larger or smaller than that of the Earth? Write down your answer before you look up the densities. Now find the values of the densities elsewhere in this text. Were you right? Explain clearly the meanings of density and mass.

11. A friend who has not had the benefit of an astronomy course suggests that the Sun must be full of burning coal to shine as brightly as it does. List as many arguments as you can against this hypothesis.

12. Which of the following transformations is (are) fusion and which is (are) fission? (See Appendix 13 for a list of the elements.)

a. helium to carbon
b. carbon to iron
c. uranium to lead
d. boron to carbon
e. oxygen to neon

13. Why is a higher temperature required to fuse hydrogen to helium by means of the CNO cycle than is required by the process that occurs in the Sun, which involves only isotopes of hydrogen and helium?

14. The Earth's atmosphere is in hydrostatic equilibrium. What this means is that the pressure at any point in the atmosphere must be high enough to support the weight of air above it. How would you expect the pressure on Mt. Everest to differ from the pressure in your classroom? Explain why.

15. Explain what it means when we say that the Earth's oceans are in hydrostatic equilibrium. Now suppose you are a scuba diver. Would you expect the pressure to increase or decrease as you dive below the surface to a depth of 200 ft? Why?

16. What mechanism transfers heat away from the surface of the Moon? If the Moon is losing energy in this way, why does it not simply become colder and colder?

17. Suppose you are standing a few feet away from a bonfire on a cold fall evening. Your face begins to feel hot. What is the mechanism that transfers heat from the fire to your face? (*Hint:* Is the air between you and the fire hotter or cooler than your face?)

18. Give some everyday examples of the transport of heat by convection and by radiation.

19. Suppose the proton-proton cycle in the Sun were to slow down suddenly and generate energy at only 95 percent of its current rate. Would an observer on the Earth see an immediate decrease in the Sun's brightness? Would she immediately see a decrease in the number of neutrinos emitted by the Sun?

20. Do you think that nuclear fusion takes place in the atmospheres of stars? Why or why not?

21. Why is fission not an important energy source in the Sun?

22. Why do you suppose so great a fraction of the Sun's energy comes from its central regions? Within what fraction of the Sun's radius does practically all of the Sun's luminosity originate? (See Figure 15.12.) Within what radius of the Sun has its original hydrogen been partially used up? Discuss what relationship the answers to these questions bear to one another.

FIGURING FOR YOURSELF

You can use what you have learned in this chapter to estimate how long the Sun will keep shining, and then you can tell your friends just how soon they need to start worrying. To do this, first you need to estimate just how much energy the Sun can generate over its lifetime. Then you need to know how fast the Sun is radiating its energy away. The total energy available divided by the rate at which it is being spent (radiated away) will tell you how long the energy will last. (A human analogy would be the case of someone who decided to stop working and live off the money she had in the bank. The amount of money available divided by the rate at which it is being spent per day would determine how long this time off from work would last.) The problems below will guide you through this process step by step:

23. Assume that the mass of the Sun is 75 percent hydrogen and that all of this mass could be converted to energy according to Einstein's equation $E = mc^2$. How much total energy could the Sun generate? If m is in kilograms and c is in km/s, then E will be expressed in joules. (The mass of the Sun is given in Appendix 6.)

24. In fact, the conversion of mass to energy in the Sun is not 100 percent efficient. As we have seen in the text, the conversion of four hydrogen atoms to one helium atom results in the conversion of about 0.02862 times the mass of a proton to energy. How much energy in joules does one such reaction produce? (See Appendix 6 for the mass of the hydrogen atom, which for all practical purposes is the mass of a proton.)

25. Now suppose that all of the hydrogen atoms in the Sun were converted to helium. How much total energy would be produced? (To calculate the answer, you will have to estimate how many hydrogen atoms are in the Sun. This will give you good practice with scientific notation, since the numbers involved are very large! See Appendix 4 for a review of scientific notation.)

26. Models of the Sun indicate that only about 10 percent of the total hydrogen in the Sun will participate in nuclear reactions, since it is only the hydrogen in the central regions that is at a high enough temperature. Find the total energy radiated per second by the Sun in this chapter, and use that along with the result of problem 23 and the information given here to estimate the lifetime of the Sun. (*Hint:* Make sure you keep track of the units; if the luminosity is the energy radiated per second, your answer will also be in seconds. You should convert the answer to something more meaningful, such as years.)

27. Show that the statement in the text is correct—namely, that roughly 600 million tons of hydrogen must be converted to helium in the Sun each second to explain its energy output. (*Hint:* Recall Einstein's most famous formula, and remember that for each kilogram of hydrogen, 0.0071 kg of mass is converted to energy.) How long will it be before 10 percent of the hydrogen is converted to helium? Does this answer agree with the lifetime you calculated in problem 26?

28. Every second, the Sun converts 4 million tons of matter to energy. How long will it take the Sun to reduce its mass by 1 percent? Compare your answer with the lifetime of the Sun so far.

SUGGESTIONS FOR FURTHER READING

Badash, L. "The Age of the Earth Debate" in *Scientific American*, Aug. 1989, p. 90.

Bahcall, J. "How the Sun Shines" in *Mercury* (the magazine of the Astronomical Society of the Pacific), Sept./Oct. 2001, p. 30.

Davies, P. "Particle Physics for Everybody" in *Sky & Telescope*, Dec. 1987, p. 582.

Fischer, D. "Closing In on the Solar Neutrino Problem" in *Sky & Telescope*, Oct. 1992, p. 378.

Goldsmith, D. *The Astronomers*. 1991, St. Martin's Press. Chapter 6 is "Why Stars Shine."

Hathaway, D. "Journey to the Heart of the Sun" in *Astronomy*, Jan. 1995, p. 38.

Kennedy, J. "GONG: Probing the Sun's Hidden Heart" in *Sky & Telescope*, Oct. 1996, p. 20. Update on helioseismology.

LoPresto, J. "Looking Inside the Sun" in *Astronomy*, Mar. 1989, p. 20. On helioseismology.

Park, R. *Voodoo Science*. 2000, Oxford U. Press. Chapters 1 and 5 tell the full story of the "cold fusion" uproar.

Rousseau, D. "Case Studies in Pathological Science" in *American Scientist*, Jan./Feb. 1992, p. 54. Explores the story of "cold fusion."

Sutton, C. *Spaceship Neutrino*. 1992, Cambridge U. Press. Definitive book on neutrinos.

Trefil, J. "How Stars Shine" in *Astronomy*, Jan. 1998, p. 56.

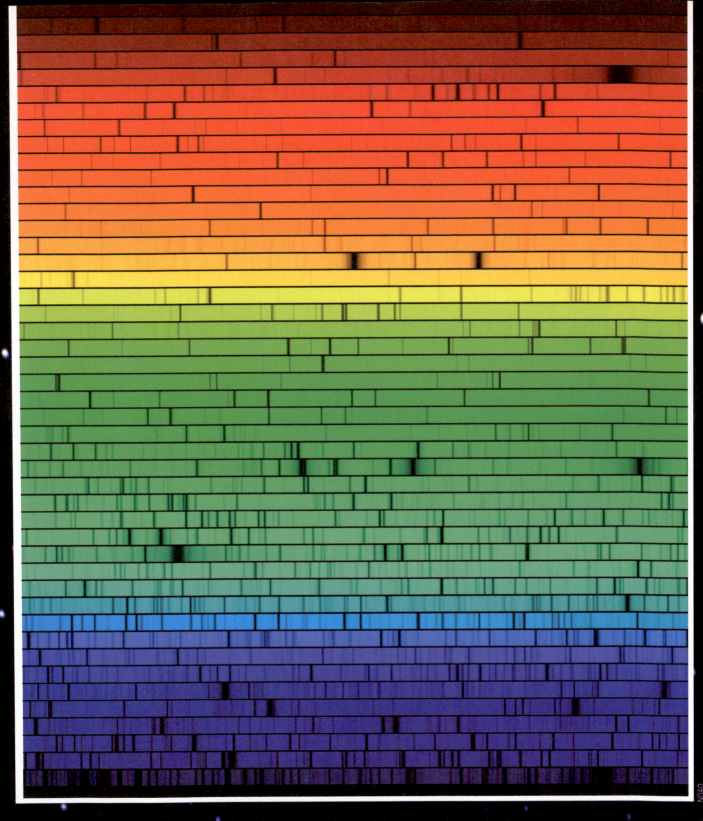

The Spectrum of the Sun This image shows the spectrum of the Sun in the visible-light region of the spectrum (400–700 nm). As the colors show, wavelengths increase from left to right and then from bottom to top. Imagine that the individual strips from bottom to top were once pasted together into one long strip. The dark lines crossing the spectrum are produced by absorption of light by atoms in the solar photosphere. Analysis of the spectra of the Sun and stars can tell us their temperature, what they are made of, how fast they are rotating, and how fast they are moving toward or away from us. This information in turn helps us to piece together the life story of stars from birth to death.

16 Analyzing Starlight

Twinkle, twinkle little
* star,*
I don't wonder what
* you are,*
For by spectroscopic ken
I know that you are
* hydrogen.*

Anonymous, quoted by D. Bush in *Science and English Poetry* (1950)

THINKING AHEAD

When you look at the sky at night, all of the stars appear to be mere points of light. You may be surprised to learn that they also look like points through the largest telescopes. They are simply too far away for us to make out any structure.[1] Everything we know about stars—how they are born, what they are made of, how long they live, how they will ultimately die—must be learned by decoding the messages contained in the feeble light that reaches the Earth. Suppose you were given the challenge of learning all about the stars. What questions would you ask first? How would you go about finding the answers?

One of the major scientific achievements of the 20th century was mapping out the life history of stars. In this chapter we describe the techniques that scientists have used to study these objects, which lie far beyond our grasp.

The first thing scientists do when faced with the task of studying a new class of objects is to try to find the right pigeonholes in which to put them. Whether the objects are butterflies, human diseases, or subatomic particles, scientists first try to measure some basic properties that characterize the objects. Stars, it turns out, can be classified according to what their temperatures are, how much material they contain (their mass), and how much energy they produce. As we learn more about the stars, we will use these characteristics to begin assembling clues to the main problems we are interested in solving: How do stars form? How long do they survive? What is their ultimate fate?

Virtual Laboratories

The Spectral Sequence and H-R Diagram

[1] The disks of a few especially large stars can now be studied with special techniques (see Chapter 21). But most stars look like points no matter what we do.

National Optical Astronomy Observatories

■ **FIGURE 16.1**

The Colors of Stars A time exposure showing star colors in the constellation Orion as it sets over Kitt Peak National Observatory. The camera shutter was left open for a long exposure, during which the Earth's turning motion drew out the light of each star into a long line. To make the photo more interesting, the photographer ended with a short exposure, giving the dots at the right. The colors of the various stars are caused by their different temperatures.

We begin our voyage to the stars by describing how astronomers use light to measure the characteristics of stars. Very simple observations show that not all stars are alike. You can make some of these observations yourself, especially if you get away from city lights or use a pair of binoculars. Take a good look at the night sky. It is obvious that all stars do not appear to be equally bright. They are not all the same color either. Although most are too dim to excite our color vision, a few brighter stars are obviously red while others appear white or bluish. In the winter sky, a good constellation for seeing star colors is Orion, the hunter (Figure 16.1).

Color is a good indication of the temperature of a hot glowing gas, with red being the coolest, and blue and violet being the hottest (see Chapter 4). Since stars are giant balls of glowing gas, we can immediately guess that those that look blue-white are hotter than those that appear to be red (Figure 16.2). Quantitative measurements show that this guess is correct. The Sun looks yellow, a color that is in the middle of the visible spectrum. Therefore, we can predict that the Sun's temperature is also somewhere in the middle, between the temperatures of the hottest and coolest stars, and again we will find out that this prediction is correct.

Other questions about stars can be answered only by means of careful observations with large telescopes. Since the stars are suns, it should not surprise you that the same techniques, including spectroscopy, used to study the Sun can also be used to find out what stars are like.

16.1 THE BRIGHTNESS OF STARS

Perhaps the most important characteristic of a star is its **luminosity**—the total amount of energy at all wavelengths that it emits per second. In Chapter 15 we saw that the Sun (pictured as a big light bulb) puts out a tremendous amount of radiation every second. (And there are stars far more luminous than the Sun out there.) Later, we will see that if we can measure how much energy a star emits and we also know its mass, then we can calculate how long it can continue to shine before it exhausts its nuclear energy and begins to die.

Ace⊗Astronomy™ Log into AceAstronomy and select this chapter to see the Active Figure called "Apparent Brightness," and Astronomy Exercise "Apparent Brightness."

16.1.1 Standardizing Brightness

Note that luminosity is how much energy a star *gives off* each second, not how much energy ultimately reaches our eyes or telescope on Earth. Stars are very democratic in the way they emit energy: The same amount goes in every direction in space. Only a minuscule fraction of the energy given off by a star actually reaches an observer on Earth. We call the amount of a star's energy that reaches a given area (say, 1 square meter) each second here on Earth its **apparent brightness.** If you look at the night sky, you see a wide range of apparent brightnesses among the stars. Most stars, in fact, are so dim that you need a telescope to detect them.

If all stars were the same luminosity—if they were *standard bulbs*—we could use the difference in their apparent brightnesses to tell us something we very much want to know: how far away they are. Imagine you are in a big concert hall or ballroom that is dark except for a few dozen 25-watt (W) bulbs placed in fixtures around the walls. Since they are all 25-W bulbs, their luminosity (energy output) is the same. But from where you are standing in one corner, they do *not* have the same apparent brightness. Those close to you look brighter (more of their light reaches your eye), while those far away look dimmer (their light has spread out more before reaching you), so you can tell which bulbs are closest to you. In the same way, if all the stars had the same luminosity, we could immediately say that the bright-looking stars were close by and the dim-looking ones were far away.

■ FIGURE 16.2

Stars in Sagittarius This image, which was taken by the Hubble Space Telescope, shows stars in the direction toward the center of the Milky Way Galaxy. The bright stars glitter like colored jewels on a black velvet background. The color of a star indicates its temperature. Blue-white stars are much hotter than the Sun, while red stars are cooler. On average, the stars in this field are at a distance of about 25,000 light years (which means it takes light 25,000 years to traverse the distance from them to us) and the width of the field is about 13.3 light years.

To pin this idea down more precisely, recall from Chapter 4 that we already know exactly how light fades with increasing distance. The energy we receive is inversely proportional to the square of the distance. If, for example, we have two stars of the same brightness and one is twice as far away as the other, it will look four times dimmer than the closer one. If it is three times farther away, it will look nine (three squared) times dimmer, and so forth.

Alas, the stars do not have the courtesy to come in one standard luminosity. (Actually, we are pretty glad about that, because having many different types of stars makes the universe a much more interesting place.) But this means that if a star looks dim in the sky, we cannot tell whether it appears dim because it has a low luminosity but is relatively nearby, or because it has a high luminosity but is very far away. (By the way, although stars in general did not turn out to be "standard bulbs," astronomers have not given up on the idea of finding some types of objects in the cosmos that all come with the same luminosity "built in." We will return to our quest for standard bulbs in future chapters as our voyages take us into more distant regions of the universe.)

Because the stars are not standard bulbs, we cannot just read off their luminosities from their apparent brightnesses. We must first compensate for the dimming effects of distance on light, and to do that we must know how far away they are. Distance is among the most difficult of all astronomical measurements. We will return to how it is determined after we have learned more about the stars. For now, we turn to the easier problem of measuring apparent brightness.

16.1.2 The Magnitude Scale

Because the brightness of a star is one of its most obvious characteristics, the first measurements were made long ago. Modern astronomers have attempted to retain the historic system even as measurement techniques have changed, and as a result we are cursed with an overly complex system for describing stellar brightness. Because this system (called **magnitudes**) is used for star maps, by amateur astronomers, and often appears in news stories, we will explain it here—even though your authors are sorely tempted to stop teaching this old-fashioned system!

The process of measuring the apparent brightness of stars is called *photometry* (from the Greek *photo*, "light," and *-metry*, "to measure"). As we saw in Chapter 1, astronomical photometry began with Hipparchus. Around 150 B.C.E. he erected an observatory on the island of Rhodes in the Mediterranean. There he prepared a catalog of nearly 1000 stars that included not only their positions but also estimates of their apparent brightness.

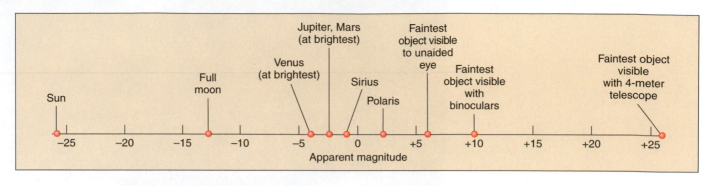

FIGURE 16.3

Apparent Magnitudes of Well-Known Objects The faintest magnitudes that can be detected by the eye, binoculars, and a large telescope are also shown.

Hipparchus did not have photographic plates or any instruments that could measure brightness accurately, so he simply made estimates with his eye. He sorted the stars into six brightness categories, which he called magnitudes. He referred to the brightest stars in his catalog as first-magnitude stars, while those so faint he could barely see them were sixth-magnitude stars.

During the 19th century, astronomers attempted to make the scale more precise by establishing exactly how much the apparent brightness of a sixth-magnitude star differs from that of a first-magnitude star. Measurements showed that we receive about 100 times more light from a star of the first magnitude than from one of the sixth. Based on this measurement, astronomers then defined an accurate magnitude system in which a difference of five magnitudes corresponds exactly to a brightness ratio of 100:1.

So what number is it that, when multiplied together five times, gives you this factor of 100? Play on your calculator, and see if you can get it. The answer turns out to be 2.512, which is the fifth root of 100. Let's round it off to 2.5 among friends. This means stars that differ by about 1 magnitude differ in brightness by a factor of about 2.5. Thus a fifth-magnitude star gives us about 2.5 times as much light as one of the sixth magnitude, a fourth-magnitude star gives 2.5 times as much light as a fifth-magnitude star, and so forth.

Here are a few rules of thumb that might help those new to this strange system. If two stars differ in brightness by 4 magnitudes, they differ by a factor of about 40 in brightness. If they are 2.5 magnitudes apart, they differ in brightness by a factor of 10; and 0.7 magnitude corresponds to a difference in brightness of a factor of 2.

The brightest stars, those that were traditionally referred to as first-magnitude stars, actually turned out (when measured accurately) not to be identical in brightness. For example, the brightest star in the sky, Sirius, sends us about ten times as much light as the average first-magnitude star. On the modern magnitude scale, Sirius has been assigned a magnitude of −1.5. (An average first-magnitude star would have a magnitude of 1, and, as we just saw, a factor of 10 is equivalent to a difference of 2.5 magnitudes). Sev-

eral of the planets appear even brighter. Venus at its brightest is of magnitude −4.4, while the Sun has a magnitude of −26.2.

Figure 16.3 shows the range of observed magnitudes from the brightest to the faintest, along with the actual magnitudes of several well-known objects. The important fact to remember when using magnitudes is that the system goes backward: The *larger* the magnitude, the *fainter* the object you are observing!

Today, many astronomers (and astronomy students) wish we had a less cumbersome system, and many researchers have stopped using magnitudes altogether. But tradition plays a strong role in astronomy, as it does in numerous other areas of human endeavor, and so the magnitude system survives.

16.1.3 Other Units of Brightness

Although the magnitude scale is still used for visual astronomy, it is not used at all in newer branches of the field. In radio astronomy, for example, no equivalent of the magnitude system has been defined (thank goodness!). Rather, the radio astronomer measures the amount of energy being collected each second by each square meter of a radio telescope and expresses the brightness of each source in terms of, for example, watts per square meter.

Similarly, most researchers in the fields of infrared, x-ray, and gamma-ray astronomy use energy per area per second rather than magnitudes to express the results of their measurements. Nevertheless, astronomers in all fields are careful to distinguish between the total *luminosity* of the source (even when that luminosity is all in x rays) and the amount of energy that happens to reach us on Earth. After all, the luminosity is a really important characteristic that tells us a lot about the object in question, whereas the energy that reaches the Earth is an accident of cosmic geography.

To make the comparison among stars easy, in this text we avoid the use of magnitudes as much as possible and we express the luminosity of other stars in terms of the Sun's luminosity. For example, the luminosity of Sirius is 23 times that of the Sun. We use the symbol L_{Sun} to denote

FIGURING FOR YOURSELF: MAGNITUDES

Even scientists can't take fifth roots in their heads, so astronomers have summarized the above discussion in an equation to help calculate the difference in brightness for stars with different magnitudes. If m_1 and m_2 are the magnitudes of two stars, then we can calculate the ratio of their brightness (b_2/b_1) using this equation:

$$m_1 - m_2 = 2.5 \log\left(\frac{b_2}{b_1}\right) \quad \text{or} \quad \frac{b_2}{b_1} = 2.5^{m_1 - m_2}$$

Few students today are familiar with logs—mathematical tools for keeping track of exponents that are not whole numbers. Mostly such calculations are done with calculators or on computers that have these functions.

Here is another way to write this equation:

$$\frac{b_2}{b_1} = (100^{1/5})^{m_1 - m_2}$$

Let's do a real example, just to show how this works. Imagine that an astronomer has just discovered something special about a dim star (magnitude 8.5) and she wants to tell her students how much dimmer the star is than Sirius. Star 1 in the equation will be our dim star and star 2 will be Sirius. Remember, Sirius has a magnitude of -1.5. In that case:

$$\frac{b_2}{b_1} = (100^{1/5})^{8.5 - (-1.5)} = (100^{1/5})^{10}$$

$$= (100)^2 = 100 \times 100 = 10,000$$

the Sun's luminosity; hence, that of Sirius can be written as $23L_{\text{Sun}}$.

16.2 COLORS OF STARS

Stars are not all the same color because they do not all have identical temperatures. Wien's law is the equation that relates the color of a star to its temperature (see Chapter 4). Blue colors dominate the light output of very hot stars, while cool stars emit most of their light energy at red wavelengths. To define *color* precisely, astronomers have devised quantitative methods for characterizing the color of a star and then using those colors to determine stellar temperatures. Unfortunately, these quantitative methods involve those pesky magnitudes again. We describe how it is done in this section, but in the chapters that follow we will always give the temperature rather than the color of the stars we are describing.

16.2.1 Color and Temperature

The color of a star provides a measure of its intrinsic or true temperature (apart from the effects of reddening by interstellar dust, which will be discussed in Chapter 19) and does not depend on where the observer happens to be. This idea should make sense to you. After all, you know from everyday experience that the color of an object appears the same no matter how far away it is. If we could somehow take a star, observe it, and then move it much farther away, its apparent brightness (magnitude) would change. But this change in brightness is the same for all wavelengths, and so its color would remain the same.

The hottest stars (nuclei of planetary nebulae described in Chapter 21) have temperatures of 100,000 degrees Kelvin (K), and the coolest true stars have temperatures of about 2000 K. (A class of "failed stars," called brown dwarfs, which are described later in this chapter, are cooler still.) Our Sun's surface temperature is about 6000 K; its dominant color is a slightly greenish yellow. It looks more yellow seen from the Earth's surface because our planet's air molecules scatter some of the green light out of the beams of sunlight that reach us, leaving more yellow behind.

16.2.2 Color Indices

In order to find the exact color of a star, astronomers normally measure the stellar brightness through filters, each of which transmits only the light from a particular narrow band of wavelengths (colors). A crude example of a filter in everyday life is a piece of red cellophane, which, when held in front of your eyes, lets only the red colors of light through.

One commonly used set of filters in astronomy measures stellar brightness at three wavelengths corresponding to ultraviolet, blue, and yellow light. The filters are given names: U (ultraviolet), B (blue), and V (visual—for yellow). These filters transmit light near the wavelengths of 360 nanometers (nm), 420 nm, and 540 nm, respectively. The brightness measured through each filter is usually expressed in magnitudes. The difference between any two of these magnitudes—say, between the blue and the visual magnitudes (B − V)—is called a *color index*.

By agreement among astronomers, the ultraviolet, blue, and visual magnitudes of the UBV system are adjusted to give a color index of 0 to a star with a surface temperature of about 10,000 K. The B − V color indices of stars range from −0.4 for stars with temperatures of about 50,000 K to +2.0 for the reddest, with temperatures of about 2000 K. The (B − V) for the Sun is about +0.62.

Why use color index if it ultimately implies temperature? Because the brightness of a star through a filter is what astronomers actually measure, and we are always more comfortable when our statements have to do with measurable quantities.

16.3 THE SPECTRA OF STARS

Measuring colors is only one way of analyzing starlight. Instead of filters we can use a spectrograph to spread out the light into a spectrum (see Chapters 4 and 5). As early as 1823, the German physicist Joseph Fraunhofer observed that the spectra of stars have dark lines crossing a continuous band of colors (see the spectrum at the beginning of this chapter). In 1864, an English astronomer, Sir William Huggins (Figure 16.4), succeeded in identifying some of the lines in stellar spectra as those of known elements on Earth, showing that the same chemical elements found in the Sun and planets exist in the stars. Since then, astronomers have worked hard to perfect experimental techniques for obtaining and measuring spectra, and they have developed a theoretical understanding of what can be learned from spectra. Today spectroscopic analysis is one of the cornerstones of astronomical research.

16.3.1 Formation of Stellar Spectra

When the spectra of different stars were first observed, astronomers found that they were not all identical. Since the dark lines are produced by the chemical elements present in the stars, astronomers first thought that the spectra differ from one another because stars are not all made of the same chemical elements. This hypothesis turned out to be wrong. *The primary reason that stellar spectra look differ-*

■ **FIGURE 16.4**

William Huggins (1824–1910) Huggins was the first to identify the lines in the spectrum of a star other than the Sun; he also took the first spectrogram, or photograph of a stellar spectrum.

ent is that stars have different temperatures. Most stars have nearly the same composition as the Sun, with only a few exceptions.

Hydrogen, for example, is by far the most abundant element in all stars (except, as we will see in later chapters, those in advanced stages of their lives). However, lines of hydrogen are not seen in the spectra of some types of stars. In the atmospheres of the hottest stars, hydrogen atoms are completely ionized. Because the electron and the proton are separated, ionized hydrogen cannot produce absorption lines. (Recall from Chapter 4 that the lines are the result of electrons in orbit around a nucleus changing energy levels.)

In the atmospheres of the coolest stars, hydrogen atoms have their electrons attached and can switch energy levels to produce lines. However, practically all of the hydrogen atoms are in the lowest energy state (unexcited) in these stars and thus can absorb only those photons able to lift an electron from that first energy level to a higher level. The photons absorbed in this way produce a series of absorption lines that lie in the ultraviolet part of the spectrum and hence cannot be studied from the Earth's surface. What this means is that if you observe the spectrum of a very hot or very cool star with a typical telescope on the surface of the Earth, the most common element in that star will not show any lines.

The hydrogen lines in the visible part of the spectrum (called *Balmer lines;* see Chapter 4) are strongest in stars with intermediate temperatures—not too hot and not too cool. Calculations show that the optimum temperature for producing visible hydrogen lines is about 10,000 K. At this temperature, an appreciable number of hydrogen atoms are excited to the second energy level. They can then absorb additional photons, rise to still higher levels of excitation, and produce a dark absorption line. Hydrogen lines are less conspicuous in the spectra of both hotter and cooler stars, even though hydrogen is just about equally abundant in all the stars. Similarly, every other chemical element, in each of its possible stages of ionization, has a characteristic temperature at which it is most effective in producing absorption lines in any particular part of the spectrum.

16.3.2 Classification of Stellar Spectra

Because a star's temperature determines which absorption lines are present in its spectrum, we can use the spectrum to measure its surface temperature. Until recently, astronomers sorted stars according to the patterns of lines seen in them into only seven principal **spectral classes.** From hottest to coldest, these seven spectral classes are designated O, B, A, F, G, K, and M. Recently, astronomers have added two additional classes for even cooler objects—L and T.

At this point you are probably looking at these letters with amazement and asking yourself why astronomers didn't call the spectral types A, B, C, and so on. It's a long story (partly discussed in a moment) in which tradition won

Ace✺Astronomy™ Log into AceAstronomy and select this chapter to see Astronomy Exercise "Stellar Atomic Absorption Lines."

■ **FIGURE 16.5**

Absorption Lines in Stars of Different Temperatures A graph showing the strength of absorption lines of different elements as we move from hot (left) to cool (right) stars along the sequence of spectral types.

out over common sense. To help them remember this crazy order of letters, generations of astronomy students have used the mnemonic "Oh be a fine girl, kiss me." (Today, when many astronomy students are women, you can easily substitute "guy" for "girl.") Other mnemonics, which we hope will not be relevant for you, include "Oh brother, astronomers frequently give killer midterms" and "Oh boy, an F grade kills me!" With the new letters, the mnemonic might be "Oh be a fine girl (guy); kiss my lips tenderly."

Each of these spectral classes, except possibly for the T class, which is still being defined, is further subdivided into ten subclasses designated by the numbers 0 through 9. A B0 star is the hottest type of B star; a B9 star is the coolest type of B star and is only slightly hotter than an A0 star. And just one more item of vocabulary: For historical reasons, astronomers call all elements heavier than helium "metals," even though most of these elements do not show metallic properties. (If you are getting annoyed at the strange jargon astronomers use, just bear in mind that every field of human knowledge develops its own peculiar vocabulary. Try reading a credit card agreement or loan document these days without training in the law!)

As Figure 16.5 shows, in the hottest O stars (those with temperatures over 28,000 K), only lines of ionized helium and highly ionized atoms of other elements are conspicuous. Hydrogen lines are strongest in A stars with atmospheric temperatures of about 10,000 K. Ionized metals provide the most conspicuous lines in stars with temperatures from 6000 to 7500 K (spectral type F). In the coolest M stars (below 3500 K), absorption bands of titanium oxide and other molecules are very strong. The sequence of spectral classes is summarized in Table 16.1. The spectral class assigned to the Sun is G2.

To see how spectral classification works, let's use Figure 16.5. Suppose you have a spectrum in which the hydrogen lines are about half as strong as those seen in an A star. Looking at the red lines in our figure, you see that the star could be either a B star or a G star. But if the spectrum also contains helium lines, then it is a B star, whereas if it

contains lines of ionized iron and other metals, it must be a G star.

The differences in spectra are quite obvious once you get familiar with them. If you look at Figure 16.7, you can see that you too could assign a spectral class to a star whose type was not already known. All you have to do is match the pattern of spectral lines to a standard star, like the ones shown in the figure, whose type has already been determined.

Much of the pioneering work on the classification of stellar spectra was carried out at Harvard University in the first decades of the 20th century. The basis for these studies was a monumental collection of nearly a million photographic spectra of stars, obtained from many years of observations made at Harvard College Observatory in Massachusetts as well as at its remote observing stations in South America and South Africa.

Working with this data base, Annie Cannon (see the Voyagers in Astronomy box) personally measured the spectra of some 400,000 stars and assigned them to spectral classes. At the time, nobody knew that temperature is the primary factor in determining the appearance of a stellar spectrum, and so the Harvard workers at first simply named the spectral classes with letters of the alphabet according to the complexity of the spectra, with type A being the simplest. It was Annie Cannon who initially glimpsed a better classification system, and she focused on just a few letters from the original system. Now that we understand that spectral classes depend on the temperature of the stars, we list the types not in order of how many lines are seen, but rather from hottest to coolest: O, B, A, F, G, K, M, L, and T.

Both colors and spectral types can be used to estimate the temperature of a star. Spectra are harder to measure because the light must be spread out into all the colors of the rainbow, and detectors must be sensitive enough to respond to individual wavelengths. In order to measure colors, the detectors need only respond to the many wavelengths that pass simultaneously through the colored filters that have been chosen—that is, to *all* the blue light or *all* the yellow-green light.

TABLE 16.1 *Spectral Classes for Stars*

Spectral Class	Color	Approximate Temperature (K)	Principal Features	Examples
O	Violet	>28,000	Relatively few absorption lines. Lines of doubly ionized nitrogen, triply ionized silicon, and other highly ionized atoms.	10 Lacertae
B	Blue	10,000–28,000	Lines of neutral helium, singly and doubly ionized silicon, singly ionized oxygen, and magnesium. Hydrogen lines more pronounced than in O-type stars.	Rigel Spica
A	Blue	7500–10,000	Strong lines of hydrogen. Lines of singly ionized magnesium, silicon, iron, titanium, calcium, and others. Lines of some neutral metals show weakly.	Sirius Vega
F	Blue to white	6000–7500	Hydrogen lines weaker than in A-type stars but still conspicuous. Lines of singly ionized calcium, iron, and chromium, plus lines of neutral iron and chromium, are present, as are lines of other neutral metals.	Canopus Procyon
G	White to yellow	5000–6000	Lines of ionized calcium are the most conspicuous spectral features. Many lines of ionized and neutral metals are present. Hydrogen lines are weaker than in F-type stars. Bands of the molecule CH are strong.	Sun Capella
K	Orange to red	3500–5000	Lines of neutral metals predominate. The CH bands are still present.	Arcturus Aldebaran
M	Red	2000–3500	Strong lines of neutral metals and molecular bands of titanium oxide dominate.	Betelgeuse Antares
L	Infrared	1300–2000	Lines of steam, metallic hydrides, carbon monoxide, neutral sodium, potassium, cesium, and rubidium.	Teide 1
T	Infrared°	700–1300	Methane lines.	Gliese 229B

° Absorption by methane molecules makes T dwarfs a bit less red than L dwarfs.

University of Massachusetts and the Infrared Processing and Analysis Center/NASA

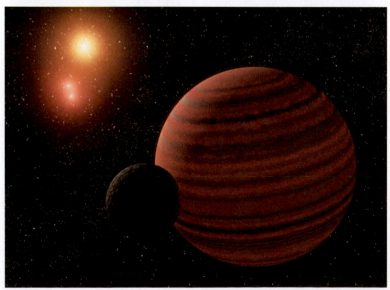

Dr. Robert Hurt

■ **FIGURE 16.6**

A Cool T Dwarf This brown dwarf, called Gliese 570D, has an estimated temperature of 750 K, which definitely makes it spectral type T. It is part of a quadruple system with three cool stars—one yellow and two red—about 19 light years away. We see an image of the brown dwarf without its companion stars on the left (arrow) and the planet Neptune with its moon Triton on the right taken by the 2MASS infrared survey. Note that they have similar colors in the infrared due to the presence of methane in their atmospheres. Below is an artist's conception of the whole system, with the three stars in the upper left and the brown dwarf in the foreground. The artist has imagined an even smaller companion object orbiting 570D.

Annie Cannon: Classifier of the Stars

Annie Jump Cannon was born in Delaware in 1863. In 1880 she went to Wellesley College, one of the new breed of U.S. colleges opening up to educate young women. Only five years old at the time, Wellesley had the second student physics lab in the country and provided excellent training in basic science. After college, Cannon spent a decade with her parents but was very dissatisfied, longing to do scientific work. With her mother's death in 1893, she was able to return to Wellesley as a teaching assistant and also to take courses at Radcliffe, the women's college associated with Harvard.

This was a time when the director of the Harvard Observatory, Edward C. Pickering, needed lots of help with his ambitious program of classifying stellar spectra. Pickering quickly discovered that educated young women could be hired as assistants for one-third or one-fourth the salary paid to men, and they would often put up with working conditions and repetitive tasks that men with the same education would not tolerate. (We should emphasize that astronomers were not alone in reaching such conclusions about the relatively

Annie Jump Cannon
(1863–1941)

new idea of women working outside the home: Women were exploited and underestimated in many fields. This is a legacy from which our society is just beginning to emerge.)

Cannon wanted to do astronomy, and she was quickly hired by Pickering to help with the classification of spectra. After a while she became so good at it that she could visually examine and determine the spectral types of several hundred stars per hour (dictating her conclusions to an assistant). She made many discoveries while in-

vestigating the Harvard photographic plates, including 300 variable stars (stars whose luminosity changes). But her main legacy is a marvelous catalog of spectral types for hundreds of thousands of stars, which served as a foundation for much of 20th-century astronomy.

In 1911 a visiting committee of astronomers reported that "she is the one person in the world who can do this work quickly and accurately" and urged Harvard to give Cannon an official appointment in keeping with her skill and renown. Not until 1938, however, did Harvard appoint her an astronomer at the university; she was then 75 years old.

Cannon received the first honorary degree from Oxford awarded to a woman, and she became the first woman to be elected an officer of the American Astronomical Society, the main professional organization of astronomers in this country. She generously donated the money from one of the major prizes she had won to found a special award for women in astronomy, now known as the Annie Jump Cannon Prize. True to form, she continued classifying stellar spectra almost to the very end of her life in 1941.

16.3.3 Spectral Classes L and T

The scheme devised by Annie Cannon worked well until 1995, when astronomers began to discover objects even cooler than M9. We use the word *object* because many of the new discoveries are not true stars. A star is defined as an object that during some part of its lifetime derives 100 percent of its energy from the same process that makes the Sun shine—the conversion of hydrogen nuclei (protons) to helium. Objects with masses less than about 7.2 percent of the mass of our Sun ($0.072M_{Sun}$) do not become hot enough for the fusion of protons to take place. Even before the first such object was found, this class of objects, with masses intermediate between stars and planets, was given the name **brown dwarfs.**

Brown dwarfs are very difficult to observe because they are extremely faint and cool, and they put out most of

their light in the infrared part of the spectrum. It was only after the construction of very large telescopes, like the Keck telescopes in Hawaii, and the development of very sensitive infrared detectors that the search for brown dwarfs succeeded. The first brown dwarf was discovered in 1995, and there are now more than 100 candidate objects that may turn out to be brown dwarfs when they are studied more thoroughly.

Initially, brown dwarfs were given spectral classes like M10+ or "much cooler than M9," but so many are now known that it is possible to begin assigning spectral types. The hotter brown dwarfs are given types L0–L9 (temperatures in the range 2000–1300 K), while still cooler objects are called T dwarfs (see Figure 16.6). In L stars, the lines of titanium oxide, which are strong in M stars, have disappeared. This is because the L stars are so cool that atoms and molecules can gather together into dust particles in their atmospheres; the titanium is locked up in the dust grains rather than being available to form molecules of titanium oxide. Lines of steam (hot water vapor) are present, along with lines of carbon monoxide and neutral sodium, potassium, cesium, and rubidium. Methane (CH_4) is strong in T dwarfs, as it is in the atmosphere of the giant planets in our own solar system.

Most brown dwarfs start out with atmospheric temperatures and spectra like those of M dwarfs, even though the brown dwarfs are not hot enough in their interiors to fuse protons. (Nuclear reactions involving deuterium do take place in some brown dwarfs.) Brown dwarfs, unlike M dwarfs, cool steadily throughout their lifetimes, passing through the L dwarf stage and ultimately becoming T dwarfs after 1–2 billion years. Since galaxies like our own are much older than this, most of the brown dwarfs ever formed are currently T dwarfs.

16.4 SPECTROSCOPY: THE KEY TO THE UNIVERSE

Analyzing the spectrum of a star can teach us all kinds of things in addition to its temperature. We can measure its detailed chemical composition as well as the pressure in its atmosphere. From the pressure, we get clues about its size. We can also measure its motion toward or away from us and estimate its rotation.

16.4.1 Clues to the Size of a Star

As we shall see in the next chapter, stars come in a wide variety of sizes. At some periods in their lives, stars can expand to enormous dimensions. Stars of such exaggerated size are called **giants**. Luckily for the astronomer, stellar spectra can be used to distinguish giants from run-of-the-mill stars (such as our Sun).

Suppose you want to determine whether a star is a giant. By definition, a giant star has a large extended photosphere. Because it is so large, its atoms are spread over a large volume, the density of atoms is low, and the pressure is therefore also low. The ionization of atoms in the star's outer layers is caused mainly by photons, and the amount of energy carried by photons is determined by temperature. But how long atoms *stay* ionized depends in part on pressure. Compared with what happens in the Sun (with its relatively dense photosphere), ionized atoms in a giant star's photosphere are less likely to pass close enough to electrons to interact, combine with each other, and thereby become neutral again. To see why this is so, think of automobile traffic. Collisions are much more likely during rush hour, when the density of cars is high.

Low-density gases, therefore, maintain a higher average degree of ionization than do high-density gases of the same temperature. The difference is large enough to affect the strengths of spectral lines, and for this reason careful study of spectra can tell which of two stars at the same tem-

perature has higher pressure (and is thus more compressed) and which has lower pressure (and thus must be extended.)

16.4.2 Abundances of the Elements

Dark lines of a majority of the known chemical elements have now been identified in the spectra of the Sun and stars. If we see lines of iron in a star's spectrum, for example, then we know immediately that the star must contain iron.

Note that the *absence* of an element's spectral lines does not necessarily mean that the element itself is absent. As we saw, the temperature and pressure in a star's atmosphere will determine what types of atoms are able to produce absorption lines. Only if the physical conditions in a star's photosphere are such that lines of an element *should* (according to calculations) be there can we conclude that the absence of observable spectral lines implies low abundance of the element.

Suppose two stars have identical temperatures and pressures, but the lines of, say, sodium are stronger in one than in the other. Stronger lines mean that there are more atoms in the stellar photosphere absorbing light. Therefore, we know immediately that the star with stronger sodium lines contains more sodium (Figure 16.7). Difficult calculations are required to determine exactly how much more, but those calculations can be done for any element observed in any star with any temperature and pressure. It is this kind of detailed analysis that has shown that the relative abundances of the different chemical elements in the Sun (see Table 14.2) and in most stars are approximately the same.

Of course, astronomy textbooks such as ours always make these things sound a bit easier than they really are. If you look at the solar spectrum at the beginning of this chapter, you may get some feeling for how hard it is to decode all of the information contained in the thousands of absorption lines. First of all, it has taken many years of careful laboratory work on Earth to determine the precise wavelengths at which hot gases of each element have their spectral lines. Long books and computer databases have been compiled to show the lines that can be seen at each temperature.

Second, stellar spectra usually have many lines from a number of elements, and we have to be careful to sort them out correctly. Sometimes nature is unkind, and lines of different elements have identical wavelengths, thereby adding to the confusion. And third, as we saw in Chapter 4, the motion of the star changes the location of each of the lines, and the observed wavelengths may not match laboratory measurements exactly. So, in practice, analyzing stellar spectra is a demanding, sometimes frustrating task that requires both training and skill.

Studies of stellar spectra have shown that hydrogen makes up about three quarters of the mass of most stars. Hydrogen and helium together make up from 96 to 99 percent of the mass; in some stars they amount to more than 99.9 percent. Among the 4 percent or less of "heavy elements," neon, oxygen, nitrogen, carbon, magnesium, argon, silicon, sulfur, iron, and chlorine are among the most abun-

■ **FIGURE 16.7**

Spectra of Stars with Different Spectral Classes The spectral type assigned to each of these stellar spectra is listed at the left of the picture. The strongest four lines seen at spectral type A1 (one in the red, one in the blue green, and two in the blue) are Balmer lines of hydrogen. Note how these lines weaken at both higher and lower temperatures, as the graph in Figure 16.5 also indicates. The strong pair of closely spaced lines in the yellow in the cool stars is due to neutral sodium (one of the neutral "metals" in Figure 16.5).

NOAO/AURA/NSF

dant. Generally, but not invariably, the elements of lower atomic weight are more abundant than those of higher atomic weight.

Take a careful look at this list of elements. Two of the most abundant are hydrogen and oxygen (which make up water); add carbon and nitrogen and you are starting to write the prescription for the chemistry of an astronomy student. We are made of elements that are common in the universe—just mixed together in a far more sophisticated form (and a much cooler environment) than is a star.

Appendix 13 lists how common each element is in the universe (compared to hydrogen); these estimates are based primarily on investigation of the Sun, which is a typical star. Some very rare elements, however, have not been detected in the Sun. Estimates of the amounts of these elements in the universe are based on laboratory measurements of their abundance in primitive meteorites, which are considered representative of unaltered material condensed from the solar nebula (see Chapter 13).

16.4.3   Radial Velocity

When we measure the spectrum of a star, we determine the *wavelengths* of each of its lines. If the star is not moving with respect to the Sun, then the wavelengths corresponding to each element will be the same as those we measure in a laboratory here on Earth. But if stars are moving toward or away from us, we must consider the *Doppler*

Ace ☉ Astronomy™ Log into AceAstronomy and select this chapter to see Astronomy Exercise "Doppler Shift."

effect (see Section 4.6). We should see all the spectral lines of moving stars shifted either toward the red end of the spectrum, if the star is moving away from us, or toward the blue (violet) end, if it is moving toward us. The greater the shift, the faster the star is moving. Such motion, along an imaginary line connecting the star and the observer, is called **radial velocity** and is usually measured in kilometers per second.

William Huggins, pioneering yet again, made the first radial velocity determination of a star in 1868. He observed the Doppler shift in one of the hydrogen lines in the spectrum of Sirius and found that this star is moving toward the solar system. Today the radial velocity can be measured for any star bright enough for its spectrum to be observed. As we will see in the next chapter, radial velocity measurements of double stars are crucial in deriving stellar masses.

16.4.4   Proper Motion

For the sake of completeness, we should note that there is another type of motion that cannot be detected with stellar spectra. Stars can move across the sky (perpendicular to our line of sight) as well as toward or away from us. Motion across the sky is referred to as *proper motion*. If stars have some motion in this perpendicular direction, we see it as a change in the relative positions of the stars on the "dome" of the sky (Figure 16.8). These changes are very slow. Even the star with the largest proper motion takes 200 years to

Yerkes Observatory

■ **FIGURE 16.8**

Large Proper Motion Two photographs of Barnard's star, the star with the largest known proper motion, showing how this faint star has moved over a period of 22 years.

Astronomy and Philanthropy

Throughout the history of astronomy, contributions from wealthy patrons of the science have made an enormous difference for building new instruments and carrying out long-term research projects. Edward Pickering's stellar classification project, which was to stretch over several decades, was made possible by major donations from Anna Draper. She was the widow of Henry Draper, a physician who was one of the most accomplished amateur astronomers of the 19th century and the first man to successfully photograph the spectrum of a star. Anna Draper gave several hundred thousand dollars (a lot more money then than today!) to Harvard Observatory. As a result, the great spectroscopic survey is still known as the Henry Draper Memorial, and many stars are still referred to by their "HD" numbers in that catalog (such as HD 209458).

In the 1870s, the eccentric piano builder and real estate magnate James Lick decided to leave some of his fortune to build the world's largest telescope. (This was actually his second plan for a memorial; fortunately he had been talked out of his first plan, which was to build the world's largest pyramid in downtown San Francisco as a monument to himself.) When, in 1887, the pier to house the telescope was finished, Lick's body was entombed in it. Atop the foundation rose a

James Lick

Mary Lea Shane Archives of the Lick Observatory

36-in. refractor, for many years the main instrument at the Lick Observatory near San Jose.

The Lick telescope remained the largest in the world until 1897, when George Ellery Hale persuaded railroad millionaire Charles Yerkes to build a 40-in. telescope near Chicago (see Chapter 5).

More recently Howard Keck, whose family made its fortune in the oil business, gave $70 million from his family foundation to the California Institute of Technology to help build the world's largest telescope atop the 14,000-ft peak of Mauna Kea in Hawaii (see Chapter 5). The Keck Foundation was so pleased with what is now called the Keck telescope that they gave $74 million more to build "Keck II," another 10-m reflector on the same volcanic peak.

Now if any of you reading this book become millionaires or billionaires and astronomy has sparked your interest, do keep an astronomical instrument or project in mind as you make out your will. But frankly, private philanthropy could not possibly support the full enterprise of scientific research in astronomy. Much of our exploration of the universe is financed by such federal agencies as the National Science Foundation and NASA in the United States and by government agencies in other countries. In this way, all of us, through a very small share of our tax dollars, get to be philanthropists for astronomy.

---

change its position in the sky by an amount equal to the width of the full Moon, and the motions of the other stars are smaller yet.

For this reason we do not notice any change in the positions of the bright stars during the course of a human lifetime. If we could live long enough, however, the changes would become obvious. For example, some 50,000 years from now, terrestrial observers (if humans haven't wiped themselves out by then) will find the handle of the Big Dipper unmistakably more bent than it is now (Figure 16.9).

We measure proper motion in terms of how far the star moves in the sky in arcseconds per year. That is, the measurement of proper motion tells us only by how much of an angle in the sky a star has changed its position. In order to convert this angular motion to a real speed, we need to know how far away the star is. If two stars at different distances are moving at the same speed perpendicular to our line of sight, the closer one will appear to move farther

across the sky in a year's time. As an analogy, imagine you are standing at the edge of a freeway. Cars will appear to whiz past you. If you then watch the traffic from a vantage point half a mile away, the cars will move much more slowly across your field of vision.

To know the true velocity of a star—that is, its total speed and the direction in which it is moving through space—we must know its radial velocity, proper motion, and distance. The motions of stars also cause their distances from us to change. Again, the changes occur slowly but over several hundred thousand years can be large enough to cause significant changes in brightness for nearby stars. For example, 300,000 years ago, Aldebaran, the bright star in Taurus, the bull, was the brightest star in the sky. Sirius is currently the brightest one, but 300,000 years from now it will have moved away and faded somewhat, and Vega, the bright blue star in Lyra, will take over its place of honor as the brightest star in Earth's skies.

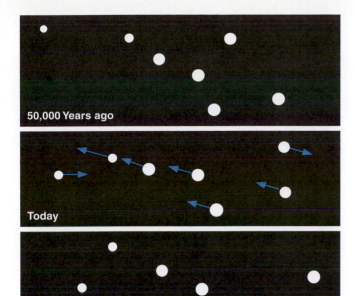

50,000 Years ago

Today

50,000 Years hence

■ **FIGURE 16.9**

Changes in the Big Dipper The change in the appearance of the Big Dipper due to proper motion of the stars over 100,000 years.

Ace★Astronomy™ Log into AceAstronomy and select this chapter to see Astronomy Exercise "Stellar Rotation."

16.4.5 Rotation

We can use the Doppler effect to measure how fast a star rotates. If an object is rotating, then (unless its axis of rotation happens to be pointed exactly toward us) one of its sides is approaching us while the other is receding. This is clearly the case for the Sun or a planet; we can observe the light from either the approaching or receding edge of these nearby objects and measure directly the Doppler shifts that arise from the rotation.

Stars, however, are so far away that they all appear as unresolved points. The best we can do is analyze the light from the entire star at once. Even though we cannot distinguish them with our eyes or photographs, the lines in the light that comes from the side of the star rotating toward us are shifted to shorter wavelengths and the lines in the light from the opposite edge of the star are shifted to longer wavelengths. You can think of each spectral line that we observe as a sum or composite of spectral lines originating from different points on the star's disk, which are moving at different speeds with respect to us. Each point has its own Doppler shift, so the absorption line we see from the whole star is actually much wider than it would be if the star were not rotating. If a star is rotating rapidly, all its spectral lines should be quite broad. In fact, astronomers call this effect *line broadening,* and the amount of broadening can tell us the speed at which the star rotates (Figure 16.10).

■ **FIGURE 16.10**

Using a Spectrum to Determine Stellar Rotation The rotation of a star broadens its spectral lines.

Measurements of the widths of spectral lines show that many stars hotter than the Sun rotate in periods of only a day or two. The Sun, with its rotation period of about a month, rotates rather slowly. The rotation of most stars cooler than the Sun is slower still and often cannot be measured with our present techniques.

As you can see, spectroscopy is an extremely powerful technique that helps us learn all kinds of information about stars that we simply could not gather any other way. We will see in later chapters that these same techniques can also teach us about galaxies, which are the most distant objects that we can observe. Without spectroscopy, we would know next to nothing about the universe beyond the solar system.

SURFING THE WEB

Spectral Classification:
zebu.uoregon.edu/~imamura/208/jan18/mk.html
A brief tutorial on the spectral classification system, with graphic examples of what the spectra of different types of stars look like.

Tutorial on Magnitudes:
www.physics.uq.edu.au/people/ross/ph227/survey.mags.htm
An introduction to magnitudes and color indices.

2Mass Page: www.ipac.caltech.edu/2MASS
These pages describe the work of the 2 Micron All Sky Survey, the cataloging of infrared sources that led to the discovery of the L- and T-type stars. The site combines technical and nontechnical information.

Women in Astronomy Web Site:
www.astrosociety.org/education/resources/womenast_bib.html
This page brings together written and Web resources on the contributions of women to astronomy, with general readings, references about specific women, and links.

SUMMARY

16.1 The total energy emitted per second by a star is called its **luminosity.** How bright a star looks to us is called its **apparent brightness.** For historical reasons, the apparent brightnesses of stars are often expressed in terms of **magnitudes.** If one star is 5 magnitudes brighter than another, it emits 100 times more energy. Since the apparent brightness of a star depends on its luminosity and distance, determination of apparent brightness and measurement of the distance to a star provide enough information to calculate its luminosity.

16.2 Stars have different colors, which are indicators of temperature. The hottest stars tend to appear blue or blue-white, while the coolest stars are red. The color index of a star is the difference in the magnitudes measured at any two wavelengths.

16.3 The differences in the spectra of stars are principally due to differences in temperature, not composition. The spectra of stars are described in terms of **spectral classes.** In order of decreasing temperature, these spectral classes are O, B, A, F, G, K, M, L, and T. The classes L and T have been added recently to describe newly discovered objects—mainly **brown dwarfs**—that are cooler than M9.

16.4 Spectra of stars of the same temperature but different atmospheric pressure have subtle differences, so spectra can be used to determine whether a star has a large radius and low atmospheric pressure (a **giant** star) or a small radius and high atmospheric pressure. Stellar spectra can also be used to determine the chemical composition of stars; hydrogen and helium make up most of the mass of all stars (just as they do in the Sun). Measurements of line shifts produced by the Doppler effect indicate the **radial velocity** of a star. Broadening of spectral lines by the Doppler effect is a measure of rotational velocity.

INTER-ACTIVITY

 A The Voyagers in Astronomy box on Annie Cannon discusses some of the difficulties women who wanted to do astronomy faced in the first half of the 20th century. What does your group think about the situation for women today? Do men and women have an equal chance to become scientists? Discuss whether, in your experience, boys and girls were equally encouraged to do science and math where you went to school.

B After you have done activity A, here is a little minidrama your group can try. One member of the group plays a well-known woman astronomer, who has some observing time on a large telescope (scheduled 8 months ago) and has to fly there with her graduate students tomorrow. Her husband works in a large corporation and has an important meeting with a client tomorrow. They have two children, ages 4 and 10. The grandparents do not live near them. The person who helps them with child care has just called in sick for tomorrow. The two parents are having a discussion about what they should do. If you prefer, all the people in the group can take the two roles in turn.

C Suppose you could observe a star that has only one spectral line. Can you tell what element that spectral line comes from? Make a list of reasons with your group about why you answered yes or no.

D A very wealthy alumnus of your college decides to give $50 million to the astronomy department to build a world-class observatory for learning more about the characteristics of stars. Have your group discuss what kind of equipment they would put in the observatory. Where should this observatory be located? Justify your answers. (You may want to refer back to Chapter 5 on telescopes and to revisit this question as you learn more about the stars in future chapters.)

E For some astronomers, introducing a new spectral type for the stars (like the types L and T discussed in the text) is similar to having to introduce a new area code for telephone calls. No one likes to disrupt the old system, but sometimes it is simply necessary. Have your group make a list of steps an astronomer would have to go through to persuade colleagues that a new spectral class is needed.

REVIEW QUESTIONS

Ace⊙Astronomy™ Assess your understanding of this chapter's topics with additional quizzing and animations at **http://ace.brookscole.com/voyages**

1. What two factors determine how bright a star appears to be in the sky?
2. Explain why color is a measure of a star's temperature.
3. What is the main reason that the spectra of all stars are not identical? Explain.

4. What elements are stars mostly made of? How do we know this?
5. What did Annie Cannon contribute to the understanding of stellar spectra?
6. Name three characteristics of a star that can be determined by measuring its spectrum. Explain how you would use a spectrum to determine these characteristics.
7. How do objects of spectral types L and T differ from those of the other spectral types?

THOUGHT QUESTIONS

8. If the star Sirius emits 23 times more energy than the Sun, why does the Sun appear brighter in the sky?

9. Draw a picture showing how two stars of equal luminosity—one blue and the other red—would appear on two images, one taken through a filter that passes mainly blue light and the other through a filter that transmits mainly red light.

10. Table 16.1 lists the temperature ranges that correspond to the different spectral types. What part of the star do these temperatures refer to? Why?

11. Suppose you are given the task of measuring the colors of the stars in Appendix 11 through three filters. The first transmits blue light, the second transmits yellow light, and the third transmits red light. The way colors are defined, if you observe the star Vega, it will appear equally bright through each of the three filters. Which stars will appear brighter through the blue filter than through the red filter? Which stars will appear brighter through the red filter? Which star is likely to have colors most nearly like those of Vega?

12. Star X has lines of ionized helium in its spectrum, and star Y has bands of titanium oxide. Which is hotter? Why? The spectrum of star Z shows lines of ionized helium and also molecular bands of titanium oxide. What is strange about this spectrum? Can you suggest an explanation?

13. The spectrum of the Sun has hundreds of strong lines of nonionized iron but only a few, very weak lines of helium. A star of spectral type B has very strong lines of helium but very weak iron lines. Do these differences mean that the Sun contains more iron and less helium than the B star? Explain.

14. What are the approximate spectral classes of stars with the following characteristics?

a. Balmer lines of hydrogen are very strong; some lines of ionized metals are present.
b. The strongest lines are those of ionized helium.
c. Lines of ionized calcium are the strongest in the spectrum; hydrogen lines show with only moderate strength; lines of neutral and ionized metals are present.
d. The strongest lines are those of neutral metals and bands of titanium oxide.

15. Look at Appendix 13. Can you identify any relationship between the abundance of an element and its atomic weight? Are there any obvious exceptions to this relationship? (In Chapters 21 and 22, we will learn why this pattern exists.)

16. Appendix 10 lists the nearest stars. Are most of these stars hotter or cooler than the Sun? Do any of them emit more energy than the Sun? If so, which ones?

17. Appendix 11 lists the stars that appear brightest in our sky. Are most of these hotter or cooler than the Sun? Can you suggest a reason for the difference between this answer and the answer to question 16? (*Hint:* Look at the luminosities.) Is there any tendency for a correlation between temperature and luminosity? Are there exceptions to the correlation? (The tendency toward the correlation and the reason for exceptions will be discussed in the next chapter.)

18. In Appendix 11, how much more luminous is the brightest of the stars than the faintest?

19. What is the brightest star in the sky (other than the Sun)? The second brightest? What color is Betelgeuse? Use Appendix 11 to find the answers.

20. Suppose people of 1 million years ago had left behind maps of the night sky. Would these maps represent accurately the sky that we see today? Why or why not?

21. Why can only a lower limit to the rate of stellar rotation be determined from line broadening rather than the actual rotation rate? (Refer to Figure 16.10.)

22. Why do you think astronomers have suggested two different spectral types (L and T) for the brown dwarfs? Why was one not enough?

FIGURING FOR YOURSELF

Use the equations relating magnitude and apparent brightness given in Section 16.1.2 and the box on page 361 for problems 23–28.

23. Verify that if two stars have a difference of 5 magnitudes, this corresponds to a factor of 100 in the ratio (b_2/b_1); that 2.5 magnitudes corresponds to a factor of 10; and that 0.7 magnitude corresponds to a factor of 2.

24. As seen from the Earth, the Sun has an apparent magnitude of about −26. What is the apparent magnitude of the Sun as seen from Saturn, about 10 astronomical units (AU) away? (Remember that 1 AU is the distance from the Earth to the Sun and that the brightness decreases as the inverse square of the distance.) Would the Sun still be the brightest star in the sky?

25. An astronomer is investigating a faint star that has recently been discovered on very sensitive surveys of the sky. The star has a magnitude of 16. How much less bright is it than Antares, a star with magnitude roughly equal to 1?

26. The center of a faint but active galaxy has magnitude 26. How much less bright does it look than the very faintest star that our eyes can see, roughly magnitude 6?

27. You have enough information from this chapter to estimate the distance to Alpha Centauri, the second nearest star, which has an apparent magnitude of −1. Since it is a G2 star, like the Sun, assume it has the same luminosity as the Sun and the difference in magnitudes is a result only of the difference in distance. Estimate how far away Alpha Centauri is. Describe the necessary steps in words and then do the calculation. (As we will see in Chapter 18, this method—namely, assuming that stars with identical spectral types emit the same amount of energy—actually is used to estimate distances to stars.) If you assume the distance to the Sun is in AU, your answer will come out in AU.

28. Do problem 27 again, this time using the information that the Sun is 93,000,000 million miles away. You will get a very large number of miles as your answer. To get a better feeling for how the distances compare, try calculating the time it takes light at a speed of 186,000 mi/s to travel from the Sun to the Earth and from Alpha Centauri to the Earth. For Alpha Centauri, figure out how long the trip will take in years as well as in seconds. (As we will see in Chapter 18, astronomers express large distances in terms of light travel time.)

SUGGESTIONS FOR FURTHER READING

Berman, B. "Magnitude Cum Laude" in *Astronomy,* Dec. 1998, p. 92. On how we measure the apparent brightnesses of stars.

Fraknoi, A. and Freitag, R. "Women in Astronomy: A Resource Guide" in *Mercury* (the magazine of the Astronomical Society of the Pacific), Jan./Feb. 1992, p. 27. Part of an issue devoted to a discussion of women in astronomy.

Hearnshaw, J. "Origins of the Stellar Magnitude Scale" in *Sky & Telescope,* Nov. 1992, p. 494. A good history of how we have come to have this cumbersome system.

Hearnshaw, J. *The Analysis of Starlight.* 1986, Cambridge U. Press. A history of spectroscopy in astronomy.

Hirshfeld, A. "The Absolute Magnitude of Stars" in *Sky & Telescope,* Sept. 1994, p. 35.

Kaler, J. "Stars in the Cellar: Classes Lost and Found" in *Sky & Telescope,* Sept. 2000, p. 39. An introduction to spectral types and the new classes L and T.

Kaler, J. "Origins of the Spectral Sequence" in *Sky & Telescope,* Feb. 1986, p. 129.

Kaler, J. *Stars.* 1992, Scientific American Library/W. H. Freeman. Good modern review of our understanding of stars.

Kaler, J. *Stars and Their Spectra.* 1989, Cambridge U. Press. A detailed introduction to the field of spectroscopy and what it can tell us about the stars.

Kidwell, P. "Three Women of American Astronomy" in *American Scientist,* May/June 1990, p. 244. Focuses on Annie Cannon and Cecilia Payne.

Skrutskie, M. "2MASS: Unveiling the Infrared Universe" in *Sky & Telescope,* July 2001, p. 34. On the all-sky survey at 2 microns.

Sneden, C. "Reading the Colors of the Stars" in *Astronomy,* Apr. 1989, p. 36. Discusses what we learn from spectroscopy.

Steffey, P. "The Truth about Star Colors" in *Sky & Telescope,* Sept. 1992, p. 266. About the color index, and how the eye and film "see" colors.

Tomkins, J. "Once and Future Celestial Kings" in *Sky & Telescope,* Apr. 1998, p. 59. Calculating the motion of stars and determining which stars were, are, and will be brightest in the sky.

SUN

Low-mass star

Brown dwarf

Jupiter

Earth

Gemini Observatory/Artwork by Jon Lomberg

This artist's conception shows the relative sizes of some astronomical objects. In this chapter we discuss how to determine the masses and diameters of stars. The sizes of stars hold clues to their evolution.

*Raving politics, never
    at rest—
as this poor earth's pale
    history runs—
What is it all but a trouble
    of ants
in the gleam of a million
    million of suns?*

**Alfred Lord Tennyson
in the poem
"Vastness" (1889)**

THINKING AHEAD

How do stars form? How long do they live? And how do they die? Stop and think how hard it is to answer these questions. After all, in your lifetime, you have not seen any of the visible stars change significantly, let alone turn on or off. In fact, the human race has not been recording observations of the heavens long enough to see any star go through all the stages of its life cycle.

To help you see how you might go about tackling the problem of determining the life story of the stars, imagine for a moment that you are the captain of a well-equipped starship, sent from a civilization that orbits another star to study alien life-forms. (You happen to resemble an intelligent cauliflower with arms and legs, but you have some excellent scientists among your crew.) You have arrived on Earth, and your schedule gives your crew exactly one Earth-day to learn all they can about the life cycle of the dominant life-form on this new planet.[1] What would be your advice to the landing party?

The crew could pursue many strategies, after all. One might be for each cauliflower to watch a single member of the human species for a whole day, to see if any of them went through important life-cycle changes. For a random sample of humans, this would typically turn out to be a disappointingly unproductive strategy. A better approach might be for the crew to make a widespread

Virtual Laboratories

 The Spectral Sequence and H-R Diagram

Binary Stars, Accretion Disks and Kepler's Laws

[1] An interesting question is whether a new visitor to Earth would immediately deduce that humans are the dominant life-form here. Some commentators have suggested that a more reasonable conclusion would be that cars rule the Earth. After all, they seem to swallow and disgorge humans (and other animals) at will, and humans are clearly their servants—washing, feeding, and polishing them with domestic regularity.

Ace✹Astronomy™ The AceAstronomy icon throughout the text indicates an opportunity for you to test yourself on key concepts and to explore animations and interactions on the AceAstronomy website at **http://ace.brookscole.com/voyages**

(a)

(b)

■ **FIGURE 17.1**

Diversity in People and Stars   (a) Stars, like people, come in a variety of colors. (b) This image shows a cluster of stars whose catalog number is NGC 2420. The cluster contains about 1000 stars packed into a roughly spherical volume of space that is 30 LY in diameter. The stars are about one-third the age of the Sun, or about 1.7 billion years old. The different colors and luminosities of these stars provide clues about their life stories.

survey, cataloging as many different humans as possible and then trying to determine which of the differences among humans provide clues about how they develop.

If you don't know anything about human beings to begin with (being an intelligent cauliflower, after all), those human characteristics that are important (and those that are simply random variations from one specimen to the next) would not be at all obvious. For example, you might note that humans come in different colors and hypothesize that they start life a rich, dark brown, getting lighter and lighter with age. The humans with the darkest skin would be the youngest, according to this theory. It's an interesting idea, but it would turn out to be completely wrong. Or you might note that humans come with a bit of fuzzy stuff on top of their heads, and make the very reasonable suggestion that the length of this fuzz is a good measure of age. This also would not get you very far down the road toward the cauliflower Nobel Prize.

On the other hand, there are characteristics that (properly understood) might help you pin down the human life cycle. For the first part of a human life, body mass and height are reasonably good indicators of age. And the smoothness of our skin tends to decrease as we get older. To understand such subtle indicators would require a broad sampling of human characteristics and many years of careful study (even though your data would be collected in a single day).

Astronomers faced a similar problem with stars (Figure 17.1). Stars live such a long time that nothing much can be gained from staring at one for a human lifetime. It was necessary to measure the characteristics of many stars (to take a celestial census, in effect) and then determine which characteristics help us understand the stars' life stories. Like the cauliflowers of our example, astronomers tried a variety of hypotheses about stars until they came up with the right approach to comprehending their development. But the key was making a thorough census of the stars around us.

17.1   A STELLAR CENSUS

Before we can make our own survey, we need to agree on a unit of distance appropriate to the objects we are studying. The stars are all so far away that kilometers (and even astronomical units) would be very cumbersome to use; so—as discussed in the Prologue—astronomers use a much larger "measuring stick" called the *light year* (LY). A light year is the distance that light (the fastest signal we know) travels in 1 year. Since light covers an astounding 300,000 km per second, and since there are a lot of seconds in 1 year, a light year is a very large quantity: 9.5 trillion (9.5×10^{12}) km to be exact. (Bear in mind that the light year is a unit of *distance* even though the term *year* appears in it.) If you drove at the legal U.S. speed limit without stopping for food or rest, you would not arrive at the end of a light year in space until roughly 12 million years had passed. And the closest star is more than 4 LY away.

Notice that we have not yet said much about how such enormous distances can be measured. That is a complicated question, to which we will return in the next chapter. For now, let us assume that distances have been measured for stars in our cosmic vicinity so that we can proceed with our census.

17.1.1   Small Is Beautiful— Or at Least More Common

When we do a census of people in the United States, we count the inhabitants by neighborhood. We can try the same approach for our stellar census and begin with our own immediate neighborhood. As we shall see, we run into two problems—just as we do with a census of human beings. First, it is hard to be sure we have counted *all* the inhabitants; second, our local neighborhood may not contain all possible types of people.

Figure 17.2 shows an estimate of the stars in our own local neighborhood—within 26 LY of the Sun. (The Milky Way Galaxy, in which we live, is 100,000 LY in diameter, so this figure really applies to a *very* local neighborhood, one that contains a *tiny* fraction of all the billions of stars in the Milky Way.) The sizes of the balls in the figure are an indication of their masses. You can see that there are many more small stars than big ones. Only five of the stars in our local neighborhood are significantly more massive than the Sun. This is truly a case where small triumphs over large— at least in terms of numbers. The Sun is more massive than the vast majority of stars in our vicinity.

Along with the L and T dwarfs already discovered in our neighborhood, Figure 17.2 shows (as magenta balls on the right) some of the hundreds of T dwarfs that astronomers still expect to find. Many of these are even cooler than the coolest currently known T dwarf. The reason the lowest mass dwarfs are so hard to find is that they put out very little light—ten thousand to a million times less light than the Sun. Only recently has our technology progressed to the point that we can detect these dim, cool objects.

To put all this in perspective, we note that stars fainter than the Sun cannot be seen with the unaided eye unless they are *very* nearby. For example, stars with luminosities ranging from 1/100 to 1/10,000 the luminosity of the Sun (L_{Sun}) are very common, but a star with a luminosity of $1/100\ L_{Sun}$ would have to be within 5 LY to be visible to the naked eye—and only three stars (all in one system) are this close to us. The nearest of these three stars, Proxima Centauri, still cannot be seen without a telescope because it has such a low luminosity.

Astronomers are working hard these days to complete the census of our local neighborhood. For example, a dozen new M stars within 33 LY were reported at the beginning of 2002. Many of the coolest T dwarfs that we think must be nearby are, however, too faint to be seen even with the best ongoing surveys and will not be discovered for years.

17.1.2 Bright Does Not Necessarily Mean Close

If we confine our census to the local neighborhood, we will miss many of the most interesting kinds of stars. After all, the neighborhood in which you live does not contain all the types of people—distinguished according to age, education, income, race, and so on—that live in the entire country. For example, a few people do live to be over 100 years old, but there may be no such individual within several miles of where you live. In order to sample the full range of the human population, you would have to extend your census to a much larger area. Similarly, some types of stars simply are not found nearby.

A clue that we are missing something in our stellar census comes from the fact that only six of the 20 stars that appear brightest in our sky—Sirius, Vega, Altair, Alpha Centauri, Fomalhaut, and Procyon—are found within 26 LY of

Dr. Robert Hurt of the Infrared Processing and Analysis Center/NASA

■ **FIGURE 17.2**

Stars and Brown Dwarfs near the Sun This whimsical painting shows an estimate of the objects within 26 LY of the Sun and visible in the northern hemisphere, along with an estimate of the objects that remain to be discovered. Stars are shown on the left pan, brown dwarfs on the right. There are 4 A stars (blue); one F star (green); 5 G stars (yellow), one of which is the Sun; 22 K stars (orange); 87 M stars (red); and 9 white dwarfs on the left. There are at least as many brown dwarfs on the right pan as there are stars on the left, but—as the scale shows—they contribute far less mass to the neighborhood. Astronomers also estimate that there are 220 extremely cool T dwarfs that have not yet been found. Astronomers estimate that, when we have discovered all our neighbors, about two thirds of the objects will probably turn out to be L and T dwarfs.

the Sun (see Figure 17.2 and Appendix 11). Why are we missing most of the brightest stars when we take our census of the local neighborhood?

The answer, interestingly enough, is that the stars that appear brightest are *not* the ones closest to us. The brightest stars look the way they do because they emit a very large amount of energy—so much, in fact, that they do not have to be nearby to look brilliant. You can confirm this by looking at Appendix 11, which gives distances for the 20 stars that appear brightest from Earth. The most distant of these stars is more than *3000 light years* from us. In fact, it turns out that *most* of the stars visible without a telescope are hundreds of light years away and many times more luminous than the Sun. Among the 6000 stars visible to the unaided eye, only about 50 are intrinsically fainter than the Sun. Note also that several of the stars in Appendix 11 are spectral type B, a type that is completely missing from Figure 17.2.

The most luminous of the bright stars listed in Appendix 11 emit 100,000 times more energy than does the Sun. These highly luminous stars are missing from the solar neighborhood because they are very rare. None of them happens to be in the tiny volume of space immediately surrounding the Sun, and only this small volume was surveyed to get the data shown in Figure 17.2.

For example, let's consider the most luminous stars—those 100 or more times as luminous as the Sun. Although such stars are rare, they are visible to the unaided eye even when hundreds to thousands of light years away. A star with a luminosity 10,000 times greater than the Sun's can be seen without a telescope out to a distance of 5000 LY. The volume of space included within a distance of 5000 LY, however, is enormous; so even though highly luminous stars are intrinsically rare, many of them are readily visible to our unaided eye.

The contrast between these two samples of stars—those that are close to us and those that can be seen with the unaided eye—is an example of a *selection effect*. When a population of objects (stars in this example) includes a great variety of different types, we must be careful what conclusions we draw from an examination of any particular subgroup. Certainly we would be fooling ourselves if we assumed that the stars visible to the unaided eye are characteristic of the general stellar population; this subgroup is heavily weighted to the most luminous stars. It requires much more effort to assemble a complete data set for the nearest stars, since most are so faint that they can be observed only with a telescope. However, it is only by doing so that astronomers are able to work out the properties of the vast majority of the stars, which are actually much smaller and fainter than our own Sun. Let's look at how we measure some of these properties.

17.2 MEASURING STELLAR MASSES

The mass of a star—how much material it contains—is one of its most important characteristics. If we know a star's mass, as we shall see, we can estimate how long it will shine and what its ultimate fate will be. Yet the mass of a star is very difficult to measure directly. Somehow we need to put a star on the cosmic equivalent of a scale.

Luckily, not all stars live like the Sun, in isolation from other stars. About half the stars are **binary stars**—two stars that orbit each other, bound together by gravity. Masses of binary stars can be calculated from measurements of their orbits, just as the mass of the Sun can be derived by measuring the orbits of the planets around it (see Chapter 2).

17.2.1 Binary Stars

Before we discuss in more detail how mass can be measured, we will take a closer look at stars that come in pairs. The first binary star was discovered in 1650, less than half a century after Galileo began to observe the sky with a telescope. John Baptiste Riccioli, an Italian astronomer, noted that the star Mizar, in the middle of the Big Dipper's handle, appeared through his telescope as two stars. Since that discovery, thousands of binary stars have been cataloged. (Astronomers call any pair of stars close to each other in the sky *double stars,* but not all of these form a true binary; that is, not all of them are physically associated. Some are just chance alignments of stars that are actually at different distances from us.) Although stars most commonly come in pairs, there are also triple and quadruple systems.

One well-known binary star is Castor, located in the constellation of Gemini. By 1804 astronomer William Herschel, who also discovered the planet Uranus, had noted that the fainter component of Castor had slightly changed its position relative to the brighter component. (We use the term *component* to mean a member of a star system.) Here was evidence that one star was moving around another. It was the first evidence that gravitational influences exist outside the solar system. The orbital motion of the binary star Kruger 60 is shown in Figure 17.3. Binary star systems in which both of the stars can be seen with a telescope are called **visual binaries.**

Edward C. Pickering (at Harvard) discovered a second class of binary stars in 1889—a class in which only one of the stars is actually seen directly. He was examining the spectrum of Mizar and found that the dark absorption lines

■ **FIGURE 17.3**

Revolution of a Binary Star Three photographs covering a period of about 12 years show the mutual revolution of the two stars that make up the nearby star system Kruger 60.

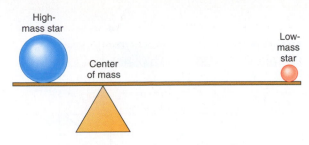

FIGURE 17.4

Binary Star System   In a binary star system, both stars orbit their center of mass. A star with higher mass will be balanced closer to the center of mass, while a star with lower mass will be balanced farther from it.

Ace⟳Astronomy™   Log into AceAstronomy and select this chapter to see the Active Figure called "Center of Mass."

in the brighter star's spectrum were usually double. Not only were there two lines where astronomers normally saw only one, but the spacing of the lines was constantly changing. At times the lines even became single. Pickering correctly deduced that the brighter component of Mizar, called Mizar A, is itself really two stars that revolve about each other in a period of 104 days. Stars like Mizar A, which appear as single stars when photographed or observed visually through the telescope, but which spectroscopy shows really to be double stars, are called **spectroscopic binaries.**

Mizar, by the way, is a good example of just how complex such star systems can be. Mizar has been known for centuries to have a faint companion called Alcor, which can be seen without a telescope. Mizar and Alcor form an *optical double*—a pair of stars that appear close together in the sky but do not orbit each other. Through a telescope, as Riccioli discovered in 1650, Mizar can be seen to have another, closer companion that does orbit it; Mizar is thus a *visual binary.* The two components that make up this visual binary, known as Mizar A and Mizar B, are both spectroscopic binaries. So, Mizar is really a quadruple system of stars.

Strictly speaking, it is not correct to describe the motion of a binary star system by saying that one star orbits the other. Gravity is a *mutual* attraction. Each star exerts a gravitational force on the other, with the result that both stars orbit a point between them called the *center of mass.* Imagine that the two stars are seated at either end of a seesaw. The point at which the fulcrum would have to be located in order for the seesaw to balance is the center of mass, and it is always closer to the more massive star (Figure 17.4).

Figure 17.5 shows two stars (A and B) moving around their center of mass, along with one line in the spectrum of each star that we observe from the system at different times. When one star is approaching us relative to the center of mass, the other star is receding from us. In the top illustration, star A is moving toward us, so the line in its spectrum is Doppler-shifted toward the blue end of the spectrum. Star B is moving away from us, so its line shows a red shift. When we observe the composite spectrum of the two stars, the line appears double. When the two stars are both moving across our line of sight (neither away from nor toward us), however, they both have the same radial

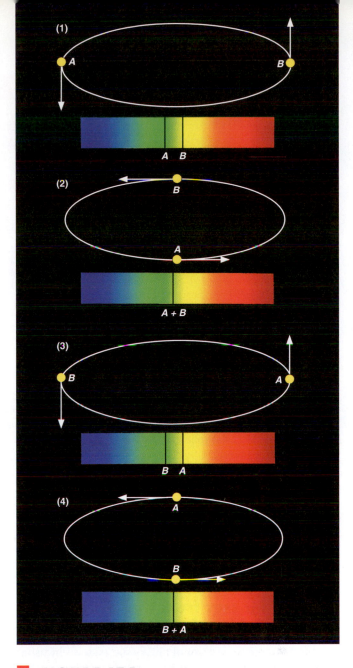

FIGURE 17.5

Motions of Two Stars Orbiting Each Other   The true orbit is tipped slightly with respect to the page on which it is printed (or if it were a real system seen in the sky, the orbit would be tilted with respect to our line of sight). If the orbit were exactly in the plane of the paper (or the sky), then it would look nearly circular, but we would see no change in radial velocity. If the orbit were perpendicular to the plane of the paper, then the stars would appear to move back and forth in a straight line, and we would see the largest possible radial-velocity variations. We see changes in velocity because when one star is moving toward the Earth, the other is moving away; half a cycle later the situation is reversed. Doppler shifts cause the spectral lines to move back and forth. At times, lines from both stars can be seen well separated from each other. When the two stars are moving perpendicular to our line of sight (that is, they are not moving either toward or away from us), the two lines are exactly superimposed, and so we see only a single spectral line.

Ace⟳Astronomy™   Log into AceAstronomy and select this chapter to see Astronomy Exercise "Spectroscopic Binaries."

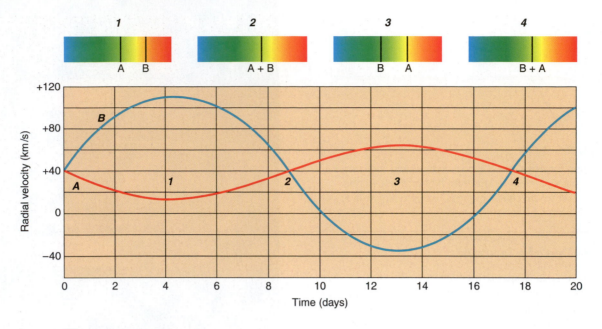

FIGURE 17.6

Radial Velocities in a Spectroscopic Binary System These curves plot the radial velocities of two stars in a spectroscopic binary system, showing how the stars alternately approach and recede from the Earth. The positions on the curve corresponding to the illustrations in Figure 17.5 are marked.

velocity (that of the pair's center of mass); hence the spectral lines of the two stars come together. This is shown in the second and fourth illustrations in Figure 17.5. A plot showing how the velocities of the stars change with time is called a *radial-velocity curve;* the curve for the binary system in Figure 17.5 is shown in Figure 17.6.

17.2.2 Masses from the Orbits of Binary Stars

We can estimate the masses of binary star systems by using Newton's reformulation of Kepler's third law (discussed in Section 2.3). Kepler found that the time a planet takes to go around the Sun is related by a specific mathematical formula to its distance from the Sun. In our binary star situation, if two objects are in mutual revolution, then the period (P) with which they go around each other is related to the semimajor axis (D) of the orbit of one with respect to the other, according to the equation

$$D^3 = (M_1 + M_2)P^2$$

where D is in astronomical units, P is measured in years, and $M_1 + M_2$ is the sum of the masses of the two stars in units of the Sun's mass. Thus, if we can observe the size of the orbit and the period of mutual revolution of the stars in a binary system, we can calculate the sum of their masses.

Most spectroscopic binaries have periods ranging from a few days to a few months, with separations of usually less than 1 AU between their member stars. Recall that an AU is the distance from the Earth to the Sun, so this is a small

separation and very hard to see at the distances of stars. This is why many of these systems are known to be double only through careful study of their spectra.

We can analyze a radial-velocity curve (such as the one in Figure 17.6) to determine the masses of the stars in a spectroscopic binary. This is complex in practice but not hard in principle. We measure the speeds of the stars from the Doppler effect. We then determine the period—how long the stars take to go through an orbital cycle—from the velocity curve. Knowing how fast the stars are moving and how long they take to go around tells us the circumference of the orbit and hence the separation of the stars in kilometers or astronomical units. From Kepler's law, the period and the separation allow us to calculate the sum of the stars' masses.

Of course, knowing the sum of the masses is not as useful as knowing the mass of each star separately. But the relative orbital speeds of the two stars can tell us how much of the total mass each star has. The more massive star is closer to the center of gravity and therefore has a smaller orbit. Hence it moves more slowly to get around in the same time. In practice we also need to know how the binary system is oriented in the sky to our line of sight; but if we do, and the just-described steps are carried out carefully, the result is a calculation of the masses of each of the two stars in the system.

In short, a good measurement of the motion of two stars around a common center of mass, combined with the laws of gravity, allows us to determine the masses of stars in such systems. These mass measurements are absolutely

■ **FIGURE 17.7**

Brown Dwarfs in Orion These images show the region surrounding the Trapezium star cluster inside a star-forming region called the Orion Nebula. No brown dwarfs are seen in the visible light image on the left, both because they put out very little light in the visible and because they are hidden within the clouds of dust in this region. The right-hand image was taken in infrared light, which can make its way to us through the dust. The faintest objects in this image are brown dwarfs with masses between 10 and 80 times the mass of Jupiter.

crucial to developing a theory of how stars evolve. One of the best things about this method is that it is independent of the location of the binary system. It works as well for stars 1000 LY away as for those in our immediate neighborhood.

17.2.3 The Range of Stellar Masses

How large can the mass of a star be? Stars more massive than the Sun are rare. None of the stars within 30 LY of the Sun has a mass greater than four times that of the Sun. Searches at large distances from the Sun have led to the discovery of a few stars with masses up to about 100 times that of the Sun, and a handful of stars (a few out of several billion) may have masses as large as 200 solar masses. However, most stars are smaller than the Sun.

According to theoretical calculations, the smallest mass that a true star can have is about 1/12 that of the Sun. By a "true" star, astronomers mean one that becomes hot enough to fuse protons to form helium (as discussed in Chapter 15). Objects with masses between roughly 1/100 and 1/12 that of the Sun may produce energy for a brief time by means of nuclear reactions involving deuterium, but they do not become hot enough to fuse protons. Such objects are intermediate in mass between stars and planets and, as we described in Chapter 16, have been given the name **brown**

dwarfs (Figure 17.7). Brown dwarfs are similar to Jupiter in radius but have masses from approximately 10–15 to 80 times larger than the mass of Jupiter.[2]

Objects with still smaller masses (less than about 1/100 the mass of the Sun or 10-15 Jupiter masses) are called planets. They may radiate energy produced by the radioactive elements that they contain, and they may also radiate heat generated by slowly compressing under their own weight (a process called gravitational contraction). However, their interiors will never reach temperatures high enough for any nuclear reactions, not even those involving deuterium, to take place. Jupiter, whose mass is about 1/1000 the mass of the Sun, is unquestionably a planet, for example. Until the 1990s we could only detect planets in our own solar system, but now we have begun to detect them elsewhere as well. (We will discuss these exciting observations in Chapter 20.)

---

[2] Exactly where to put the dividing line between planets and brown dwarfs is a subject of some debate among astronomers as we write this book (as is, in fact, the exact definition of each of these objects). Even those who accept deuterium fusion as the crucial defining issue for brown dwarfs concede that, depending on the composition of the star and other factors, the lowest mass for such a dwarf could be anywhere from 10 to 15 Jupiter masses.

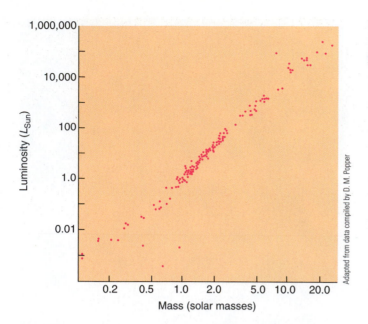

Adapted from data compiled by D. M. Popper

■ FIGURE 17.8

The Mass-Luminosity Relation The plotted points show the masses and luminosities of stars. The three points lying below the sequence of points are all white dwarf stars.

Ace⊘Astronomy™ Log into AceAstronomy and select this chapter to see Astronomy Exercise "Mass-Star Luminosity Relation."

17.2.4 The Mass-Luminosity Relation

Now that we have measurements of the characteristics of many different types of stars, we can search for relationships among the characteristics. For example, we can ask whether the mass and luminosity of a star are related. It turns out that for most stars, they are: The more massive stars are generally also the more luminous. This relationship, known as the **mass-luminosity relation,** is shown graphically in Figure 17.8. Each point represents a star whose mass and luminosity are both known. Horizontal position on the graph shows the star's mass, given in units of the Sun's mass, and vertical position shows its luminosity in units of the Sun's luminosity.

In mathematical terms, we can say that

$$L/L_{Sun} \sim (M/M_{Sun})^{3.9}$$

It's a reasonably good estimate to say that luminosity varies as the fourth power of the mass. If two stars differ in mass by a factor of 2, then the more massive one will be 16 times brighter; if one star is 1/3 the mass of another, it will be 81 times less luminous.

Notice how good this relationship is. Most stars fall along a line running from the lower left (low mass, low luminosity) corner of the diagram to the upper right (high mass, high luminosity) corner. About 90 percent of all stars obey the mass-luminosity relation illustrated in Figure 17.8. Later we will explore why such a relationship exists and what we can learn from the roughly 10 percent of stars that "disobey" it.

17.3 DIAMETERS OF STARS

It is easy to measure the diameter of the Sun. Its angular diameter—that is, its apparent size on the sky—is about 1/2°. If we know the angle the Sun takes up in the sky and how far away it is, we can calculate its true (linear) diameter, which is 1.39 million km, or about 109 times the diameter of the Earth.

Unfortunately, the Sun is the only star whose angular diameter is easily measured. All the other stars are so far away that they look like pinpoints of light through even the largest telescopes. (They often seem to be bigger, but that is merely distortion introduced by turbulence in the Earth's atmosphere.) Luckily, there are several techniques that astronomers can use to estimate the sizes of stars. One method makes use of the Stefan-Boltzmann law for the relationship between energy radiated and temperature (see Chapter 4); it is described in the Figuring for Yourself section at the end of the chapter. Let us look at two other methods that astronomers regularly use.

17.3.1 Stars Blocked by the Moon

One technique, which gives very precise diameters but can be used for only a few stars, is to observe the dimming of light that occurs when the Moon passes in front of a star. What astronomers measure (with great precision) is the time required for the star's brightness to drop to zero as the edge of the Moon moves across it. Since we know how rapidly the Moon moves in its orbit around the Earth, it is possible to calculate the angular diameter of the star. If the distance to the star is also known, we can calculate its diameter in kilometers. This method works only for fairly bright stars that happen to lie along the zodiac, where the Moon (or, much more rarely, a planet) can pass in front of them as seen from the Earth.

17.3.2 Eclipsing Binary Stars

Accurate sizes for a large number of stars come from measurements of **eclipsing binaries,** and so we must make a brief detour from our main story to examine this type of star system. Some binary stars are lined up in such a way that, when viewed from the Earth, each star passes in front of the other during every revolution (Figure 17.9). When one star blocks the light of the other, preventing it from reaching the Earth, the luminosity of the system decreases, and astronomers say that an eclipse has occurred.

The discovery of the first eclipsing binary helped solve a long-standing puzzle in astronomy. The star Algol, in the constellation of Perseus, changes its brightness in an odd but regular way. Normally Algol is a second-magnitude star, but at intervals of 2 days, 20 hours, 49 minutes it fades to

FIGURE 17.9

Light Curve of an Eclipsing Binary We see the light output of an eclipsing binary star with total eclipses (one star passes directly in front of and behind the other). The numbers indicate parts of the light curve corresponding to various positions of the smaller star in its orbit. When the small star, which is hotter, goes behind the larger one, its light is completely blocked, and there is a strong dip in the light curve. When the smaller star goes in front of the larger one, it blocks the light from a portion of the cooler star's surface. The dip is smaller because the cooler star emits less energy over this area than does the hotter star (remember Wien's law).

Ace **Astronomy**™ Log into AceAstronomy and select this chapter to see Astronomy Exercise "Eclipsing Binaries."

one third of its regular brightness. After a few hours it brightens to normal again. This effect is easily seen even without a telescope if you know what to look for; try observing Algol if you have access to clear skies without too much light pollution. (See Making Connections: *Astronomy and Mythology.*)

In 1783 a young English astronomer named John Goodricke made a careful study of Algol (see the Voyagers in Astronomy box in Chapter 18). Even though Goodricke could neither hear nor speak, he made a number of major discoveries in the 21 years of his brief life. He suggested that Algol's unusual brightness variations might be due to an invisible companion that regularly passes in front of the brighter star and blocks its light. Unfortunately, Goodricke had no way to test this idea, since it was not until about a century later that equipment became good enough to measure Algol's spectrum.

In 1889 the German astronomer Hermann Vogel demonstrated that, like Mizar, Algol is a spectroscopic binary. The spectral lines of Algol were not observed to be double because the fainter star of the pair gives off too little light compared with the brighter star for its lines to be conspicuous in the composite spectrum. Nevertheless, the periodic shifting back and forth of the brighter star's lines gave evidence that it was revolving about an unseen companion. (The lines of both components need not be visible for a star to be recognized as a spectroscopic binary.)

The discovery that Algol is a spectroscopic binary verified Goodricke's hypothesis. The plane in which the stars revolve is turned nearly edgewise to our line of sight, and each star passes in front of the other during every revolution. (The eclipse of the fainter star in the Algol system is not very noticeable because the part of it that is covered contributes little to the total light of the system. This second eclipse can, however, be detected by careful measurements.) Any binary star produces eclipses if viewed from the proper direction, near the plane of its orbit, so that one star passes in front of the other (see Figure 17.9). But from our vantage point on Earth, only a few binary star systems are oriented in this way.

17.3.3 Diameters of Eclipsing Binary Stars

We now turn back to the main thread of our story to discuss how all this can be used to measure the sizes of stars. The technique involves making a *light curve* of an eclipsing binary, a graph that plots how the brightness changes with time. Let us consider a hypothetical binary system in which the stars are very different in size, like those illustrated in Figure 17.10. To make life easy, we will assume that the orbit is viewed exactly edge-on. Even though we cannot see the two stars separately in such a system, the light curve can tell us what is happening. When the small star just starts to pass behind the large star (a point we call *first contact*), the brightness begins to drop. The eclipse becomes total (the small star is completely hidden) at the point called *second contact*. At the end of the total eclipse (*third contact*), the small star begins to emerge. When the small star has reached *last contact*, the eclipse is completely over.

To see how this allows us to measure diameters, look carefully at Figure 17.10. During the time interval between first and second contacts, the small star has moved a distance equal to its own diameter. During the time interval from first to third contacts, the small star has moved a distance equal to the diameter of the large star. If the spectral lines of both stars are visible in the spectrum of the binary, then the speed of the small star with respect to the large one can be measured from the Doppler shift. But knowing the speed with which the small star is moving and how long it took to cover some distance can tell us the span of that distance—in this case, the diameters of the stars. The speed multiplied by the time interval from first to second contact gives the diameter of the small star. We multiply it by the time between the first and third contacts to get the diameter of the large star.

In actuality, the orbits are generally not seen exactly edge-on, and the light from each star may be only partially blocked by the other. Furthermore, binary star orbits, just like the orbits of the planets, are ellipses, not circles.

Astronomy and Mythology: Algol the Demon Star and Perseus the Hero

The name Algol comes from the Arabic *Ras al Ghul,* meaning "the demon's head." The word *ghoul* in English has the same derivation. As discussed in Chapter 1, many of the bright stars have Arabic names because during the long dark ages in medieval Europe, it was Arabic astronomers who preserved and expanded the Greek and Roman knowledge of the skies. The reference to the demon is part of the ancient Greek legend of the hero Perseus, who is commemorated by the constellation in which we find Algol and whose adventures involve many of the characters associated with the northern constellations.

Perseus was one of the many half-god heroes fathered by Zeus (Jupiter in the Roman version), the king of the gods in Greek mythology. Zeus had, to put it delicately, a roving eye and was always fathering somebody or other with a human maiden who caught his fancy. (Perseus derives from *Per Zeus,* meaning "fathered by Zeus.") Set adrift with his mother by an (understandably) upset stepfather, Perseus grew up on an island in the Aegean Sea. The king there, taking an interest in Perseus' mother, tried to get rid of the young man by assigning him an extremely difficult task.

In a moment of overarching pride, a beautiful young woman named Medusa had compared her golden hair to that of the goddess Athena (Minerva for the Romans). The Greek gods did not take kindly to being compared to mere mortals, and Athena turned Medusa into a "Gorgon," a hideous, evil creature with writhing snakes for hair and a face that turned anyone who looked at it into stone. Perseus was given the task of slaying the demon, which seemed like a pretty sure way to get him out of the way forever.

But because Perseus had a god for a father, some of the other gods gave him tools for the job, including Athena's reflective shield and the winged sandals of Hermes (Mercury in the Roman story). By flying over her and looking only at her reflection, Perseus was able to cut off Medusa's head without ever looking at her directly. Taking her head (which could still turn onlookers to stone even without being attached to her body) with him, Perseus continued on to other adventures.

He next came to a rocky seashore, where boasting had gotten another family into serious trouble with the gods. Queen Cassiopeia had dared to compare her own beauty to that of the Nereids, sea nymphs who were daughters of Poseidon (Neptune in Roman mythology), the god of the sea. Poseidon was so offended that he created a sea-monster named Cetus to devastate the kingdom. King Cepheus, Cassiopeia's beleaguered husband, consulted the oracle, who told him that he must sacrifice his beautiful daughter Andromeda to the monster.

When Perseus came along and found Andromeda chained to a rock near the sea, awaiting her fate, he rescued her by turning the monster to stone. (Scholars of mythology actually trace the essence of this story back to far older legends from ancient Mesopotamia, in which the god-hero Marduk vanquishes a monster named Tiamat. Symbolically, a hero like Perseus or Marduk is usually associated with the Sun, the monster with the power of night, and the beautiful maiden with the fragile beauty of dawn, which the Sun releases after its nightly struggle with darkness.)

Many of the characters in these Greek legends can be found as constellations in the sky, not necessarily resembling their namesakes but serving as reminders of the story. (See the star map for October after the Appendices.) For example, vain Cassiopeia is sentenced to be very close to the celestial pole, rotating perpetually around the sky and hanging upside down every winter. The ancients imagined Andromeda still chained to her rock (it is much easier to see the chain of stars than to recognize the beautiful maiden in this star grouping). Perseus is next to her with the head of Medusa swinging from his belt. Algol represents this gorgon head and has long been associated with evil and bad fortune in such tales. Some commentators have speculated that the star's change in brightness (which can be observed with the unaided eye) may have contributed to its unpleasant reputation, with the ancients regarding such a change as an evil sort of "wink."

However, all these effects can be sorted out from very careful measurements of the light curve.

17.3.4 Stellar Diameters

The results of stellar size measurements confirm that most nearby stars are roughly the size of the Sun—with typical diameters of a million kilometers or so. Faint stars, as we might have expected, are generally smaller than more luminous stars. However, there are some dramatic exceptions to this simple generalization.

A few of the very luminous stars, those that are also red (indicating relatively low surface temperatures), turn out to be truly enormous. These stars are called, appropriately enough, *giants* or *supergiants*. An example is Betelgeuse, the second brightest star in the constellation of Orion and

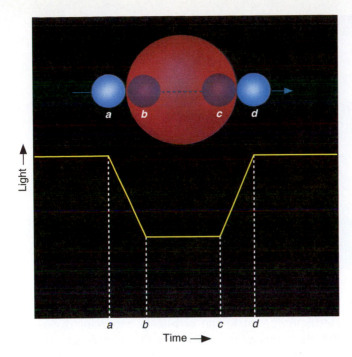

■ **FIGURE 17.10**

Light Curve of an Edge-On Eclipsing Binary Here we see the light curve of a hypothetical eclipsing binary star whose orbit we view exactly edge-on, in which the two stars fully eclipse each other. From the time intervals between contacts it is possible to estimate the diameters of the two stars.

one of the dozen brightest stars in our sky. Its diameter is greater than 10 AU, large enough to fill the entire inner solar system almost as far out as Jupiter. In Chapter 21 we will look in detail at the evolutionary process that leads to the formation of giant and supergiant stars.

17.4 THE H-R DIAGRAM

In this chapter and the previous one we have described some of the characteristics by which we might classify stars, and how those characteristics are measured. These ideas are summarized in Table 17.1. We have also given an example of a relationship between two of these characteristics in the mass-luminosity relation. When the characteristics of large numbers of stars were measured at the beginning of the 20th century, astronomers were able to begin a deeper search for patterns and relationships in these data.

To help understand what sorts of relationships might be found, let us return briefly to our intelligent cauliflowers who are trying to make sense of their data about human beings. Being good scientists, they might try plotting their data in different ways. Suppose they make a plot of the heights of a large sample of humans against their weights (which is a measure of their mass). Such a plot is shown in Figure 17.11 and it has some interesting features. In the way we have chosen to present our data, height increases

TABLE 17.1 *Measuring the Characteristics of Stars*

Characteristic	Technique
Surface temperature	1. Determine the color (very rough). 2. Measure the spectrum and get the spectral type.
Chemical composition	Determine which lines are present in the spectrum.
Luminosity	Measure the apparent brightness and compensate for distance.
Radial velocity	Measure the Doppler shift in the spectrum.
Rotation	Measure the width of spectral lines.
Mass	Measure the period and radial-velocity curves of spectroscopic binary stars.
Diameter	1. Measure the way a star's light is blocked by the Moon. 2. Measure the light curves and Doppler shifts for eclipsing binary stars.

upward, while weight increases to the left. Notice that humans are not randomly distributed in that graph. Most points fall along a sequence that goes from the upper left to the lower right.

We can conclude from this graph that generally speaking, taller human beings weigh more, while shorter ones weigh less. This makes sense if you are familiar with the structure of human beings. Typically, if we have bigger bones, we have more flesh to fill out our larger frame. It's not mathematically exact—there is a wide range of

■ **FIGURE 17.11**

Height Versus Weight The plot of the heights and weights of a representative group of human beings. Most points lie along a "main sequence" representing normal people, but there are a few exceptions.

Sterrewacht Leiden and Princeton University Archives

■ **FIGURE 17.12**

Hertzsprung and Russell
Ejnar Hertzsprung (1873–1967) and Henry Norris Russell (1877–1957) independently discovered the relationship between the luminosity and surface temperature of stars that is summarized in what is now called the H-R diagram.

variation—but it's not a bad overall rule. And, of course, there are some dramatic exceptions. You occasionally see a short human who is very overweight and would thus be more to the bottom left of our diagram than the average sequence of people. Or you might have a very tall, skinny fashion model with great height but relatively small weight, who would be found near the upper right of the figure.

A similar diagram has been found extremely useful for understanding the lives of stars. In 1913 American astronomer Henry Norris Russell plotted the luminosities of stars against their spectral classes (a way of denoting their surface temperatures). This investigation, and a similar in-

dependent study in 1911 by Danish astronomer Ejnar Hertzsprung, led to the extremely important discovery that the temperature and luminosity of stars are related (Figure 17.12).

17.4.1 Features of the H-R Diagram

Following Hertzsprung and Russell, let us plot the temperature (or spectral class) of a selected group of nearby stars against their luminosity and see what we find (Figure 17.13). Such a plot is frequently called the **Hertzsprung-Russell diagram,** abbreviated **H-R diagram.** It is one of

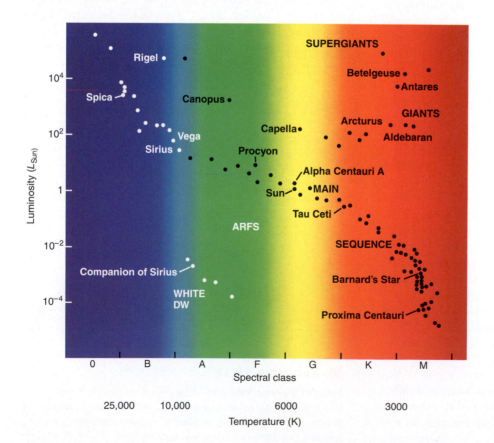

Ace Astronomy™ Log into AceAstronomy and select this chapter to see the Active Figure called "Animated HR Diagram" and Astronomy Exercise "3D HR Diagram."

■ **FIGURE 17.13**

H-R Diagram for a Selected Sample of Stars In such diagrams, luminosity is plotted along the vertical axis. Along the horizontal axis we can plot either temperature or spectral type. Several of the brightest stars are identified by name. Most stars fall on the main sequence.

Henry Norris Russell

When Henry Norris Russell graduated from Princeton University, his work had been so brilliant that the faculty decided to create a new level of honors degree beyond "summa cum laude" for him. His students later remembered him as a man whose thinking was three times faster than just about anybody else's. His memory was so phenomenal, he could correctly quote an enormous number of poems and limericks, the entire Bible, tables of mathematical functions, and almost anything he had learned about astronomy. He was nervous, active, competitive, critical, and very articulate; he tended to dominate every meeting he attended. In outward appearance he was an old-fashioned product of the 19th century, who wore high-top black shoes and high starched collars and carried an umbrella every day of his life. His 264 papers were enormously influential in many areas of astronomy.

Born in 1877, the son of a Presbyterian minister, Russell showed early promise. When he was 12, his family sent him to live with an aunt in Princeton so he could attend a top preparatory school. He lived in the same house in that town until his death in 1957 (interrupted only by a brief stay in Europe for graduate work). He was fond of recounting that both his mother and his maternal grandmother had won prizes in mathematics, and that he probably inherited his talents in that field from their side of the family.

Before Russell, American astronomers devoted themselves mainly to surveying the stars and making impressive catalogs of their properties— especially their spectra (as described in Chapter 16). Russell began to see that interpreting the spectra of stars required a much more sophisticated understanding of the physics of the atom, a subject that was being developed by European physicists in the 1910s and 1920s. Russell embarked on a lifelong quest to ascertain the physical conditions inside stars from the clues in their spectra; his work inspired, and was continued by, a generation of astronomers, many trained by Russell and his collaborators.

Russell also made important contributions in the study of binary stars and the measurement of star masses, the origin of the solar system, the atmospheres of planets, and the measurement of distances in astronomy, among other fields. He was an influential teacher and popularizer of astronomy, writing a column on astronomical topics for *Scientific American* magazine for more than 40 years. He and two colleagues wrote a textbook for college astronomy classes that helped train astronomers and astronomy enthusiasts over several decades. That book set the scene for the kind of textbook you are now reading, which not only lays out the facts of astronomy but also explains how they fit together. Russell gave lectures around the country, often emphasizing the importance of understanding modern physics in order to grasp what was happening in astronomy.

Harlow Shapley, director of the Harvard College Observatory, called Russell "the dean of American astronomers." Russell was certainly regarded as the leader of the field for many years and was consulted on many astronomical problems by colleagues from around the world. Today, one of the highest recognitions that an astronomer can receive is an award from the American Astronomical Society called the Russell Prize, set up in his memory.

the most important and widely used diagrams in astronomy, with applications that extend far beyond the purposes for which it was originally developed nearly a century ago.

It is customary to plot H-R diagrams in such a way that temperature increases toward the left and luminosity toward the top. Notice the similarity to our plot of height and weight for people. Stars, like people, are not distributed over the diagram at random, as they would be if they exhibited all combinations of luminosity and temperature. Instead, we see that the stars cluster into certain parts of the H-R diagram. The great majority are aligned along a narrow sequence running from the upper left (hot, highly luminous) to the lower right (cool, less luminous). This band of points is called the **main sequence.** It represents a relationship between *temperature* and *luminosity* that is followed by most stars. We can summarize this relationship by saying that hotter stars are more luminous than cooler ones.

A number of stars, however, lie above the main sequence on the H-R diagram, in the upper right (cool, high luminosity) region. How can a star be at once cool, meaning each square meter on the star does not put out all that much energy, and yet very luminous? The only way is for the star to be enormous—to have so many square meters on its surface that the *total* energy output is still large. These stars must be *giants* or *supergiants*, the stars of huge diameter we discussed above.

The stars in the lower left (hot, low luminosity) corner of the diagram, on the other hand, have high surface temperatures, so that each square meter on a given star puts

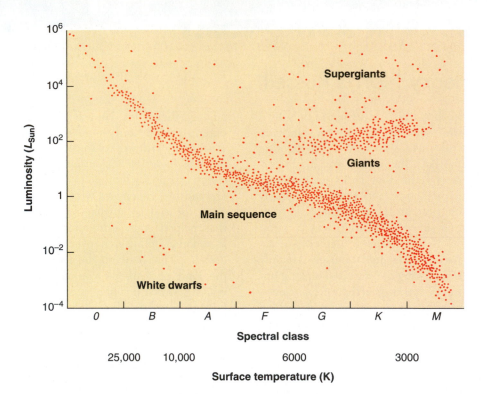

Ace✺Astronomy™ Log into AceAstronomy and select this chapter to see the Active Figure called "Animated HR Diagram."

■ **FIGURE 17.14**

Schematic H-R Diagram for Many Stars Ninety percent of all stars on such a diagram fall along a narrow band called the main sequence. A minority of stars are found in the upper right; they are both cool and bright and must be giants. Some stars fall in the lower left of the diagram; they are both hot and dim and must be dwarfs.

out a lot of energy. How then can the overall star be dim? It must be that it has a very small total surface area; such stars are known as **white dwarfs** (white because the colors blend together to make them look bluish-white). We will say more about these puzzling objects in a moment. Figure 17.14 is a schematic H-R diagram for a large sample of stars, drawn to make the different types more apparent.

Now think back to our discussion of star surveys. It is difficult to plot an H-R diagram that is truly representative of all stars because most stars are so faint that we cannot see those outside our immediate neighborhood. The stars plotted in Figure 17.13 were selected because their distances are known. This sample omits many intrinsically faint stars that are nearby but have not had their distances measured, so it shows fewer faint main-sequence stars than a "fair" diagram would. To be truly representative of the stellar population, an H-R diagram should be plotted for all stars within a certain distance. Unfortunately, our knowledge is reasonably complete only for stars within 10 to 20 LY of the Sun, among which there are no giants or supergiants. Still, from many surveys (and more can now be done with new, more powerful telescopes) we estimate that overall about 90 percent of the true stars (excluding brown dwarfs) in our part of space are main-sequence stars, about 10 percent are white dwarfs, and fewer than 1 percent are giants or supergiants.

These estimates can be used directly to understand the lives of stars. Permit us another quick analogy with people. Suppose the intelligent cauliflowers return to your town and this time focus their attention on the location of young people, ages 6 to 18. Their survey teams fan out and take data about where such youngsters are found at all times during a 24-hour day. Some are found in the pizza parlor, others at home, others at the movies, many in school. After surveying a very large number of young people, the teams determine that, averaged over the course of the 24 hours, one third of all youngsters are found in school.

How can they interpret this result? Does it mean that two thirds of students are truants and the remaining one third spend all their time in school? No, we must bear in mind that the survey teams counted youngsters throughout the full 24-hour day. Some survey teams worked at night, when most youngsters were at home asleep, and others worked in the late afternoon, when most youngsters were on their way home from school (and likely to be enjoying a pizza). If the survey was truly representative, we *can* conclude, however, that if an average of one third of all youngsters are found in school, then humans ages 6 to 18 must spend about one third *of their time* in school.

We can do something similar for stars. We find that on average 90 percent of all stars are located on the main sequence of the H-R diagram. If we can identify some activity or life stage with the main sequence, then it follows that stars must spend 90 percent of their lives in that activity or life stage.

17.4.2 Understanding the Main Sequence

In Chapter 15 we discussed the Sun as a representative star. We saw that what stars such as the Sun "do for a living" is to convert protons to helium deep in their interiors via the process of nuclear fusion, thus producing energy. The fusion of protons to helium is an excellent, long-lasting source of energy for a star because the bulk of every star consists of hydrogen atoms, whose nuclei are protons.

Our theoretical models (constructed on computers as described in Chapter 15) of how stars evolve over time show us that a typical star will spend about 90 percent of its life fusing the abundant hydrogen in its core into helium. This then is a good explanation of why 90 percent of all stars are found on the main sequence in the H-R diagram. But if all the stars on the main sequence are doing the same thing (fusing hydrogen), why are they distributed along a sequence of points? That is, why do they differ in luminosity and surface temperature (which is what we are plotting on the H-R diagram)?

To help us understand how main-sequence stars differ, we can use one of the most important results from our studies of model stars. Astrophysicists have been able to show that the structure of stars that are in equilibrium and derive all their energy from nuclear fusion is completely and uniquely determined by just two quantities: *total mass* and *composition*. This fact provides an interpretation of many features of the H-R diagram.

Imagine a cluster of stars forming from a cloud of interstellar "raw material" whose chemical composition is similar to the Sun's. (We'll describe this process in more detail in Chapter 20, but for now the details will not concern us.) In such a cloud, all condensations that become stars will begin with the same chemical composition and will differ from one another only in mass. Now suppose that we compute a model of each of these stars for the time at which it becomes stable and derives its energy from nuclear reactions, but before it has time to alter its composition appreciably as a result of these reactions.

The models calculated for these stars allow us to determine their luminosities, temperatures, and sizes. If we plot the results from the models—one point for each model star—on the H-R diagram, we get something that looks just like the main sequence we saw for real stars.

And here is what we find when we do this. The model stars with the largest masses are the hottest and most luminous, and they are located at the upper left of the diagram.

The least-massive model stars are the coolest and least luminous, and they are placed at the lower right of the plot. The other model stars all lie along a line running diagonally across the diagram. *The main sequence turns out to be a sequence of stellar masses.*

This makes sense if you think about it. The most massive stars have the most weight and can thus compress their centers to the greatest degree. This means they are the hottest inside and the best at generating energy from nuclear reactions deep within. As a result, they shine with the greatest luminosity and have the hottest surface temperatures. The stars with lowest mass, in turn, are the coolest inside and the least effective in generating energy. Thus they are the least luminous and the coolest on the surface. Our Sun lies somewhere in the middle of these extremes (as you can see in Figure 17.13). The characteristics of representative main-sequence stars (excluding brown dwarfs, which are not true stars) are listed in Table 17.2.

Note that this is exactly what we found earlier, in Section 17.2, when we examined the mass-luminosity relation (see Figure 17.8). We observed that 90 percent of all stars seem to follow the relationship; these are the 90 percent of all stars that lie on the main sequence in our H-R diagram. Our models and our observations agree.

What about the other stars on the real H-R diagram—the giants and supergiants and the white dwarfs? As we will see in the next few chapters, these are what main-sequence stars turn into as they age; they are the later stages in a star's life. As a star consumes its nuclear fuel, its source of energy changes, as do its chemical composition and interior structure. These changes cause the star to alter its luminosity and surface temperature so that it no longer lies on the main sequence on our diagram.

17.4.3 Extremes of Stellar Luminosities, Diameters, and Densities

We can use the H-R diagram to explore the extremes in size, luminosity, and density found among the stars. Such extreme stars are not only interesting to fans of the *Guinness*

TABLE 17.2 *Characteristics of Main-Sequence Stars*

Spectral Type	Mass (Sun = 1)	Luminosity (Sun = 1)	Temperature	Radius (Sun = 1)
O5	40	7×10^5	40,000 K	18
B0	16	2.7×10^5	28,000 K	7
A0	3.3	55	10,000 K	2.5
F0	1.7	5	7,500 K	1.4
G0	1.1	1.4	6,000 K	1.1
K0	0.8	0.35	5,000 K	0.8
M0	0.4	0.05	3,500 K	0.6

NASA/SAO/CXC

■ **FIGURE 17.15**

Sirius's White Dwarf Companion in X Rays This image of the Sirius star system was taken with the Chandra x-ray telescope. The bright object in the center of the image is the white dwarf companion. Sirius is the faint object next to the white dwarf; what we are seeing from Sirius is probably not x-ray radiation but rather ultraviolet light that has leaked into the detector. Note that x-ray intensities of these two objects are completely reversed from the situation in visible light. Sirius almost completely swamps the visible light from the white dwarf, which is about 8200 times fainter and is very difficult to detect.

Book of World Records; they can teach us a lot about how stars work. For example, we saw that the most massive main-sequence stars are the most luminous ones. We know of a few extreme stars that are a million times more luminous than the Sun, with masses that exceed 100 times the Sun's mass. These superluminous stars, which are at the upper left of the H-R diagram, are exceedingly hot, very blue stars of spectral type O. These are the stars that would be the most conspicuous at vast distances in space.

The cool supergiants in the upper right corner of the H-R diagram are as much as ten thousand times as luminous as the Sun. In addition, these stars have diameters very much larger than that of the Sun. As discussed above, some supergiants are so large that if the solar system could be centered in one, the star's surface would lie beyond the orbit of Mars. We will have to ask, in coming chapters, what process can make a star swell up to such an enormous size, and how long these "swollen" stars can last in their distended state.

In contrast, the very common red, cool, low-luminosity stars at the lower end of the main sequence are much smaller and more compact than the Sun. An example of such a red dwarf is Ross 614B, with a surface temperature of 2700 K and only 1/2000 of the Sun's luminosity. We call such a star a dwarf because its diameter is only 1/10 that of the Sun. A star with such a low luminosity also has a low mass (about 1/12 that of the Sun). This combination of mass and diameter means that the star has an average density about 80 times that of the Sun. Its density must be higher, in fact, than that of any known solid found on the surface of the Earth. (Despite this, the star is made of gas throughout because its center is so hot.)

The faint red main-sequence stars are not the stars of the most extreme densities, however. The white dwarfs, at the lower left corner of the H-R diagram, have densities many times greater still.

17.4.4 The White Dwarfs

The first white dwarf star was detected in 1862. Called Sirius B, it forms a binary system with Sirius A, the brightest-appearing star in the sky. It eluded discovery and analysis for a long time because its faint light tends to be lost in the glare of nearby Sirius A (Figure 17.15). (Since Sirius is often called the Dog Star—being located in the constellation of Canis Major, the big dog—Sirius B is sometimes nicknamed the Pup.)

We have now found hundreds of white dwarfs. Figure 17.2 shows that about 7 percent of the true stars (spectral types O–M) in our local neighborhood are white dwarfs. A good example of a typical white dwarf is the nearby star 40 Eridani B. Its surface temperature is a relatively hot 12,000 K, but its luminosity is only $1/275 \, L_{Sun}$. Calculations show that its radius is only 1.4 percent of the Sun's, or about the same as that of the Earth, and its volume is 2.5×10^{-6} that of the Sun. Its mass, however, is 0.43 times the Sun's mass, just a little less than half. To fit such a substantial mass into so tiny a volume, the star's density must be about 170,000 times the density of the Sun, or more than 200,000 g/cm^3. A teaspoonful of this material would have a mass of some 50 tons! At such densities, matter cannot exist in its usual state; we will examine the peculiar behavior of this type of matter in Chapter 22. For now, we just note that white dwarfs are dying stars, reaching the end of their productive lives and ready for their stories to be over.

The British astrophysicist (and science popularizer) Arthur Eddington described the first known white dwarf this way: "The message of the companion of Sirius, when decoded, ran: 'I am composed of material three thousand times denser than anything you've ever come across. A ton of my material would be a little nugget you could put in a matchbox.' What reply could one make to something like that? Well, the reply most of us made in 1914 was, 'Shut up; don't talk nonsense.'" Today, however, astronomers not only accept that stars as dense as white dwarfs exist but—as we will see—have found even denser objects in their quest to understand the evolution of different types of stars.

SURFING THE WEB

Stars of the Week Site:
www.astro.uiuc.edu/~kaler/sow/sow.html
Astronomer and author James Kaler hosts this site, featuring "biographical summaries" of famous stars—not the Hollywood type, but ones in the real sky. You can learn more about each star's place in ancient legends, its characteristics, companions, and so on.

How Big Is That Star Activity:
imagine.gsfc.nasa.gov/docs/teachers/lessons/star_size/star _size_cover.html
An elementary activity on using actual eclipsing binary star data (in x rays) to measure star diameters.

Eclipsing Binary Stars Site:
www.physics.sfasu.edu/astro/binstar.html
Dan Bruton at Austin State University has put a series of animations, articles, and links here showing how astronomers use eclipsing binary light curves.

Henry Norris Russell's Work:
mondrian.princeton.edu/cgi-bin/mfs/05/Companion/russell _henry.html
A brief summary of Russell's career is part of this site devoted to Princeton University history.

SUMMARY

17.1 To understand the properties of stars, we must make wide-ranging surveys. We find the stars that appear brightest to our eyes are bright primarily because they are intrinsically very luminous, not because they are the closest to us. Most of the nearest stars are intrinsically so faint that they can be seen only with the aid of a telescope. The luminosity of stars ranges from more than $10^6 L_{Sun}$ for the most luminous stars to $10^{-6} L_{Sun}$ for brown dwarfs. Stars with low mass and low luminosity are much more common than stars with high mass and high luminosity. Most of the brown dwarfs in the local neighborhood have not yet been discovered.

17.2 The masses of stars can be determined by analysis of the orbits of **binary stars**—two stars that orbit a common center of mass. In **visual binaries** the two stars can be seen separately in a telescope, while in a **spectroscopic binary** only the spectrum reveals the presence of two stars. Stellar masses range from about 1/12 to (rarely) more than 100 times the mass of the Sun. Objects with masses smaller than required to fuse protons but large enough to fuse deuterium are called **brown dwarfs.** Objects in which no nuclear reactions can take place are planets. The most massive stars are, in most cases, also the most luminous, and this correlation is known as the **mass-luminosity relation.**

17.3 The diameters of stars can be determined by measuring the time it takes an object (the Moon, a planet, or a companion star) to pass in front of it and block its light. Diameters of members of **eclipsing binary** systems (where the stars pass in front of each other) can be determined through analysis of their orbital motions.

17.4 The **Hertzsprung-Russell diagram,** or **H-R diagram,** is a plot of stellar luminosity as a function of surface temperature. Most stars lie on the **main sequence,** which extends diagonally across the H-R diagram from high temperature and high luminosity to low temperature and low luminosity. The position of a star along the main sequence is determined by its mass. High-mass stars emit more energy and are hotter than low-mass stars on the main sequence. Main-sequence stars derive their energy from the fusion of protons to helium. About 90 percent of the stars lie on the main sequence. Only about 10 percent of the stars are **white dwarfs,** and fewer than 1 percent are *giants* or *supergiants*.

INTER-ACTIVITY

A Two stars are seen close together in the sky, and your group is given the task of determining whether they are a visual binary or whether they just happen to be seen in nearly the same direction. Make a list of the types of measurements you would make to determine whether they orbit each other.

B Your group is given information about five main-sequence stars that are among the brightest-appearing stars in the sky and yet are pretty far away. Where would these stars be on the H-R diagram and why?

Next your group is given information about five main-sequence stars that are typical of the stars closest to us. Where would these stars be on the H-R diagram and why?

C A very wealthy (but eccentric) alumnus of your college donates a lot of money for a fund that will help in the search for more brown dwarfs. Your group is the committee in charge of this fund. How would you spend the money? (Be as specific as you can, listing instruments and observing programs.)

REVIEW QUESTIONS

Ace✸Astronomy™ Assess your understanding of this chapter's topics with additional quizzing and animations at **http://ace.brookscole.com/voyages**

1. How does the mass of the Sun compare with that of other stars in our local neighborhood? Brown dwarfs are the most numerous stars in our neighborhood. Explain why they do not contribute the most mass.
2. Name and describe the three types of binary systems.
3. Describe two ways of determining the diameter of a star.

4. What are the largest and smallest known values of the mass, luminosity, surface temperature, and diameter of stars (roughly)?
5. You are able to take spectra of both stars in an eclipsing binary system. List all properties of the stars that can be measured from their spectra and light curves.
6. Sketch an H-R diagram. Label the axes. Show where cool supergiants, white dwarfs, the Sun, and main-sequence stars are found.

THOUGHT QUESTIONS

7. Is the Sun an average star? Why or why not?

8. Suppose you want to determine the average educational level of people throughout the nation. Since it would be a great deal of work to survey every citizen, you decide to make your task easier by asking only the people on your campus. Will you get an accurate answer? Will your survey be distorted by a selection effect? Explain.

9. Why do most known visual binaries have relatively long periods and most spectroscopic binaries have relatively short periods?

10. Figure 17.10 shows the light curve of a hypothetical eclipsing binary star in which the light of one star is completely blocked by another. What would the light curve look like for a system in which the light of the smaller star

is only partially blocked by the larger one? Assume the smaller star is the hotter one. Sketch the relative positions of the two stars that correspond to various portions of the light curve.

11. There are fewer eclipsing binaries than spectroscopic binaries. Explain why. Within 50 LY of the Sun, visual binaries outnumber eclipsing binaries. Why? Which is easier to observe at large distances—a spectroscopic binary or a visual binary?

12. The eclipsing binary Algol drops from maximum to minimum brightness in about 4 hours, remains at minimum brightness for 20 minutes, and then takes another 4 hours to return to maximum brightness. Assume that we view this system exactly edge-on, so that one star crosses directly in front of the other. Is one star much larger than the other, or are they fairly similar in size? (*Hint:* Refer to the diagrams of eclipsing binary light curves.)

13. Consider the following data on five stars:

Star	Apparent Magnitude	Spectrum
1	12	G, main sequence
2	8	K, giant
3	12	K, main sequence
4	15	O, main sequence
5	5	M, main sequence

a. Which is the hottest?
b. Coolest?
c. Most luminous?
d. Least luminous?
e. Nearest?
f. Most distant?

In each case, give your reasoning. (Recall that apparent magnitude is a measure of apparent brightness, where the larger the number, the dimmer the star appears to us.)

14. Which changes by the largest factor along the main sequence from spectral types O to M—mass or luminosity? Suppose two main-sequence stars differ in mass by a factor of 3. Which star will be more luminous? By how much?

15. Suppose you want to search for brown dwarfs using a space telescope. Will you design your telescope to detect light in the ultraviolet or the infrared part of the spectrum? Why?

16. An astronomer discovers a type-M star with a large luminosity. How is this possible? What kind of star is it?

17. Approximately 6000 stars are bright enough to be seen without a telescope. Are any of these white dwarfs? Use the information given in this chapter to explain your reasoning.

18. Use the data in Appendix 11 to plot an H-R diagram for the brightest stars. Use the data from Table 17.2 to show where the main sequence lies. Do 90 percent of the brightest stars lie on or near the main sequence? Explain why or why not.

19. Use the diagram you have drawn for question 18 to answer the following questions: Which star is more massive—Sirius or Alpha Centauri? Rigel and Regulus have nearly the same spectral type. Which is larger? Rigel and Betelgeuse have nearly the same luminosity. Which is larger? Which is redder?

20. Use the data in Appendix 10 to plot an H-R diagram for this sample of nearby stars. How does this plot differ from the one for the brightest stars (question 18)? Why?

FIGURING FOR YOURSELF

Based on what you have learned in this chapter, you can now estimate the diameters, the masses, and the densities of stars. When you do so, you are using basically the same methods that professional astronomers use to obtain rough estimates of these quantities. More accurate answers can be obtained from detailed modeling, but astronomers often use these simple methods just to make sure they have not made any large mistakes in their more complicated calculations.

Let's start by looking at Appendix 11, which provides information about the 20 stars that appear the brightest in our skies. This table gives their luminosities relative to the Sun and their spectral types. Let's see how much we can learn from just these two pieces of information.

21. In order to calculate, we need numbers. Spectral types are an indicator of temperature, but we need to turn that indicator into at least a rough number. For the first ten stars in Appendix 11, estimate their temperatures from their spectral types. Use information in the figures and/or tables in this chapter and describe how you made the estimates.

22. We can estimate the masses of most of the stars in Appendix 11 from the mass-luminosity relation in Figure 17.8. However, remember this relationship works only for main-sequence stars. Determine which of the first ten stars in Appendix 11 are main-sequence stars. Use one of the figures in this chapter. Make a table of the stars' masses.

23. A much more accurate way to calculate masses is to use Kepler's third law:

$$D^3 = (M_1 + M_2)P^2$$

where D is the separation of two stars in AU, M is their mass, and P is the period in years. Sirius is one of the few binary stars in Appendix 11 for which we have enough information to apply this equation. In this case, the two stars are separated by about 20 AU and have an orbital period of about 50 years. What is the sum of the masses of the two stars? Since Sirius is a normal main-sequence A star, estimate its mass from the mass-luminosity relation (see problem 22). What then is the mass of its faint companion?

24. Sirius' companion star is about 8200 times fainter than Sirius itself and yet has about the same temperature. We can use this information to calculate the ratio of their sizes (i.e., the ratio of their radii). Remember the Stefan-Boltzmann law from Chapter 4: The energy flux (energy emitted per second per square meter by a blackbody, like the Sun) is given by

$$F = \sigma T^4$$

where σ is a constant and T is the temperature. The surface area of a sphere is given by

$$A = 4\pi R^2$$

where R is the radius of the star. The luminosity of a star is given by its surface area in square meters times the energy flux ($A \times F$). We know that Sirius has a luminosity 8200 times greater than its companion. How much bigger is Sirius than its companion? (*Hint:* Set up a ratio of the luminosities of the two stars.)

25. We now know the difference in the radii of the two stars in the Sirius system, but let's express the absolute ra-

dius of the companion in terms of something we know, like the radius of the Sun. Assume the temperature of the Sun is 5800 K and the temperature of Sirius is 10,000 K. Find the luminosity of Sirius from the data in this book.

26. Now calculate the radius of Sirius relative to that of the Sun. Then calculate the radius of the companion to Sirius. How does this radius compare with that of the Earth?

27. From the above, you should have deduced that the companion to Sirius contains about the same mass as the Sun crammed into a volume roughly the size of the Earth. Now calculate the density of the companion relative to the density of the Sun. (Density equals mass/volume, and the volume of a sphere is given by $(4/3)\pi R^3$.) How does this density compare with that of water and other materials listed in this text? Can you see why astronomers were so surprised and puzzled when they first determined the orbit of the companion to Sirius?

28. How much would you weigh if you were suddenly transported to the companion of Sirius? You may use your own weight (or if you don't want to own up to what it is, assume you weigh 70 kg or 150 pounds). In this case, assume that the companion to Sirius has a mass equal to that of the Sun and a radius equal to that of the Earth. Remember Newton's law of gravity:

$$F = GM_1M_2/R^2$$

and that your weight is proportional to the force that you feel. What kind of star should you travel to if you want to *lose* weight (and not gain it)?

29. The star Betelgeuse has a temperature of 3000 K and a luminosity of 100,000 L_{Sun}. Calculate the radius of Betelgeuse relative to the Sun.

SUGGESTIONS FOR FURTHER READING

Croswell, K. "The Grand Illusion: What We See Is Not Necessarily Representative of the Universe" in *Astronomy*, Nov. 1992, p. 44.

Davis, J. "Measuring the Stars" in *Sky & Telescope*, Oct. 1991, p. 361. Explains direct measurements of stellar diameters.

DeVorkin, D. "Henry Norris Russell" in *Scientific American*, May 1989.

Henry, T. "Brown Dwarfs: Revealed at Last" in *Sky & Telescope*, Apr. 1996, p. 24.

Kaler, J. *Stars*. 1992, Scientific American Library/W. H. Freeman. Good introduction.

Kaler, J. *The One Hundred Greatest Stars*. 2002, Springer-Verlag. Guided tour of the characteristics of the 100 most interesting stars.

Kaler, J. "Journeys on the H-R Diagram" in *Sky & Telescope*, May 1988, p. 483.

Kopal, Z. "Eclipsing Binary Stars: Algol and Its Celestial Relations" in *Mercury* (the magazine of the Astronomical Society of the Pacific), May/June 1990, p. 88.

McAllister, H. "Twenty Years of Seeing Double" in *Sky & Telescope*, Nov. 1996, p. 28. An update on modern studies of binary stars.

Nielsen, A. "E. Hertzsprung—Measurer of Stars" in *Sky & Telescope*, Jan. 1968, p. 4.

Parker, B. "Those Amazing White Dwarfs" in *Astronomy*, July 1984, p. 15. Focuses on the history of their discovery.

Phillip, A. and Green, L. "Henry N. Russell and the H-R Diagram" in *Sky & Telescope*, April 1978, p. 306.

Roth, J. and Sinnott, R. "Our Studies of Celestial Neighbors" in *Sky & Telescope*, Oct. 1996, p. 32. On finding the nearest stars.

Tanguay, R. "Observing Double Stars for Fun and Science" in *Sky & Telescope*, Feb. 1999, p. 116. An amateur astronomer discusses how you can contribute to this branch of astronomy.

The Globular Cluster M 80 (NGC 6093) This beautiful image shows a giant cluster of stars called Messier 80, located about 28,000 light years from Earth. Such crowded groups, which astronomers call globular clusters, contain hundreds of thousands of stars, including some of the RR Lyrae variables discussed in this chapter. Especially obvious in this picture are the bright red giants, which are stars similar to the Sun in mass that are nearing the ends of their lives.

18

Celestial Distances

Ace Astronomy™ The AceAstronomy icon throughout the text indicates an opportunity for you to test yourself on key concepts and to explore animations and interactions on the AceAstronomy website at **http:// ace.brookscole.com /voyages**

THINKING AHEAD

How large is the universe? What is the most distant thing we can see? These are among the biggest questions astronomers can ask, and we certainly want to answer them as we continue our voyages in this book. But just as babies must crawl before they can take their first halting steps, so too we have to start with a more modest question: How far away are the stars? And even this question proves to be very hard to answer. After all, stars are mere points of light. Suppose you see a point of light in the darkness when you are driving on a country road late at night. How can you tell whether it is a nearby firefly, an oncoming motorcycle some distance away, or the porch light of a house much farther down the road? It's not so easy, is it? Astronomers faced an even more difficult problem when they tried to estimate how far away the stars are.

The determination of astronomical distances is central to understanding the nature of stars, but measuring such distances accurately is very difficult. After all, we cannot send a graduate student out to lay a tape measure between the Sun and even the nearest star. Over the years, astronomers have developed a variety of clever techniques for estimating the vast distances that separate us from the stars. For nearby stars, we can use methods similar to the ones surveyors use here on Earth. For more distant objects, we have to apply some of the information about stars described in the preceding two chapters. We shall see that nature also provides a kind of cosmic guidepost in the form of a special type of star that varies in brightness.

Virtual Laboratories

 The Astronomical Distance Scale

Astronomers now have a chain of methods for measuring cosmic distances, one that stretches from the Earth to the stars to the farthest reaches of the universe. One of the characteristics of that chain is that its links depend on one another: The measurement of distances to remote galaxies depends on the measurement of distances to the stars within our own Galaxy, which in turn depends on the accuracy of measurements within the solar system. The entire chain of cosmic distances is only as strong as its weakest link, and so it is important that every link be as accurate as possible.

In this chapter we begin with the fundamental definitions of distances on Earth and then extend our reach outward to the stars.

18.1 FUNDAMENTAL UNITS OF DISTANCE

The first measures of distances were based on human dimensions—the inch as the distance between knuckles on the finger, or the yard as the span from the extended index finger to the nose of the British king. Later, the requirements of commerce led to some standardization of such units, but each nation tended to set up its own definitions. It was not until the middle of the 18th century that any real efforts were made to establish a uniform, international set of standards.

18.1.1 The Metric System

One of the enduring legacies of the Napoleonic era was the establishment of the *metric system* of units, officially adopted in France in 1799 and now used in most countries around the world. The fundamental metric unit of length is the *meter,* originally defined as one ten-millionth of the distance along the Earth's surface from the equator to the pole. French astronomers of the 17th and 18th centuries were pioneers in determining the dimensions of the Earth, so it was logical to use their information as the foundation of the new system.

Practical problems exist with a definition expressed in terms of the size of the Earth, since anyone wishing to determine the distance from one place to another can hardly be expected to go out and remeasure the planet. Therefore, an intermediate standard meter consisting of a bar of platinum–iridium metal was set up in Paris. In 1889, by international agreement, this bar was defined to be exactly 1 m in length, and precise copies of the original meter bar were made to serve as standards for other nations.

Other units of length are derived from the meter. Thus 1 kilometer (km) equals 1000 m, 1 centimeter (cm) equals 1/100 m, and so on. Even the old British and American units, such as the inch and the mile, are now defined in terms of the metric system.

18.1.2 Modern Redefinitions of the Meter

In 1960 the official definition of the meter was changed again. As a result of improved technology for generating spectral lines of precisely known wavelength, the meter was redefined to equal 1,650,763.73 wavelengths of a particular atomic transition in the element krypton-86. The advantage of this redefinition is that anyone with a suitably equipped laboratory can reproduce a standard meter, without reference to any particular metal bar.

In 1983 the meter was redefined once more, this time in terms of the velocity of light. Light in a vacuum can travel a distance of 1 meter in 1/299,792,458.6 second. This then defines the speed of light in a vacuum to be 299,792,458.6 m/s. Today, therefore, light travel time provides our basic unit of length. Put it another way, a distance of *one light second* (LS) (the amount of space light covers in 1 second) is defined to be 299,792,458.6 m. We could just as well use the light second as the fundamental unit of length, but for practical reasons (and to respect tradition), we have defined the meter as a small fraction of the light second.

18.1.3 Distances Within the Solar System

The work of Copernicus and Kepler established the *relative* distances of the planets—that is, how far from the Sun one planet is compared to another (see Chapters 1 and 2). But their work could not establish the *absolute* distances (in light seconds or meters or other standard units of length). This is like knowing the height of all the students in your class only as compared to the height of your astronomy instructor, but not in inches or centimeters. Somebody's height has to be measured directly.

Similarly, to establish absolute distances, astronomers had to measure one distance in the solar system directly. Estimates of the distance to Venus were made as Venus crossed the face of the Sun in 1761 and 1769, and an international campaign was organized to estimate the distance to the asteroid Eros in the early 1930s, when its orbit brought it close to Earth. By the way, a crossing (or *transit*) of Venus across the Sun is happening again in 2004 and 2012.

The key to our modern determination of solar system dimensions is *radar,* a type of radio wave that can bounce off solid objects (Figure 18.1). As discussed in several earlier chapters, by timing how long a radar beam (traveling at the speed of light) takes to reach another world and return, we can measure the distance involved very accurately. In 1958, radar signals were bounced off Venus for the first time, providing a direct measurement of the distance from Earth to Venus in terms of light seconds (from the round-trip travel time of the radar signal). Subsequently, radar has also been used to determine the distances to Mercury, Mars, the satellites of Jupiter, the rings of Saturn, and several asteroids. Note, by the way, that it is not possible to use

FIGURE 18.1

A Radar Telescope This dish-shaped antenna, part of the NASA Deep Space Network in California's Mojave Desert, can send and receive radar waves and thus measure the distances to planets, satellites, and asteroids.

radar to measure the distance to the Sun directly because the Sun does not reflect radar very efficiently. But we can measure the distance to many other solar system objects and use Kepler's laws to give us the distance to the Sun.

From the various (related) solar system distances, astronomers selected the average distance from the Earth to the Sun as our standard "measuring stick" within the solar system; this average distance is called the *astronomical unit* (AU). We then express all the other distances in the solar system in terms of the astronomical unit. Years of painstaking analyses of radar measurements have led to a determination of the length of the astronomical unit to a precision of about one part in a billion. The length of 1 AU can be expressed in light travel time as 499.004854 LS, or about 8.3 light minutes (LM). If we use the definition of the meter given previously, this is equivalent to 1 AU = 149,597,892,000 m.

These distances are, of course, given here to a much higher level of precision than is normally needed. In this text, we are usually content to express numbers to a couple of significant places and leave it at that. For our purposes it will be sufficient to round off these numbers:

Speed of light: $c = 3 \times 10^8$ m/s $= 3 \times 10^5$ km/s
Length of light second: LS $= 3 \times 10^8$ m
$\qquad\qquad\qquad\qquad = 3 \times 10^5$ km
Astronomical unit: AU $= 1.50 \times 10^{11}$ m
$\qquad\qquad\qquad = 1.50 \times 10^8$ km $= 500$ LS

We now know the absolute distance scale within our own solar system with fantastic accuracy. This is the first link in the chain of cosmic distances.

SURVEYING THE STARS

It is an enormous step to go from the planets to the stars. The nearest star is hundreds of thousands of astronomical units from the Earth. Yet in principle we can survey distances to the stars using the same technique that a civil engineer employs to survey the distance to an inaccessible mountain or tree—the method of *triangulation*.

18.2.1 Triangulation in Space

A practical example of triangulation is your own depth perception. As you are pleased to discover every morning when you look in the mirror, your two eyes are located some distance apart. You therefore view the world from two different vantage points, and it is this dual perspective that allows you to get a general sense of how far away objects are.

To see what we mean, take a pen and hold it a few inches in front of your face. Look at it first with one eye (closing the other) and then switch eyes. Note how the pen seems to shift relative to objects across the room. Now hold the pen at arm's length: The shift is less. If you play with the pen for a while, you will notice that the farther away you hold it, the less it seems to shift. Your brain automatically performs such comparisons and gives you a pretty good sense of how far away things in your immediate neighborhood are.

If your arms were made of rubber, you could stretch the pen far enough away from your eyes that the shift would become imperceptible. This is because our depth perception fails for objects more than a few tens of meters away. It would take a larger distance between viewing perspectives than the spacing between the eyes to see the shift of an object a city block or more from you.

Let's see how surveyors take advantage of the same idea. Suppose you are trying to measure the distance to a tree across a deep river (Figure 18.2). You set up two observing stations some distance apart. That distance (line AB in Figure 18.2) is called the *baseline*. Now the direction to the tree (C in the figure) in relation to the baseline is observed from each station. Note that C appears in different directions from the two stations. This apparent change in direction of the remote object due to a change in vantage point of the observer is called **parallax.**

The parallax is also the angle that lines AC and BC make—in mathematical terms, the angle subtended by the baseline. A knowledge of the angles at A and B and the length of the baseline, AB, allows the triangle ABC to be solved for any of its dimensions—say, the distance AC or BC. The solution could be reached by constructing a scale drawing or by using trigonometry to make a numerical calculation. If the tree were farther away, the whole triangle would be longer and skinnier, and the parallax angle would be smaller. Thus, we have the general rule that the smaller the parallax, the more distant the object we are measuring must be.

In practice, the kinds of baselines surveyors use for measuring distances on Earth are completely useless when we try to gauge distances in space. The farther away an astronomical object lies, the longer the baseline has to be to give us a reasonable chance of making a measurement. Un-

fortunately, nearly all astronomical objects are very far away. To measure their distances requires either a very large baseline or highly precise angular measurements, or both. The Moon is the only object near enough that its distance can be found fairly accurately with measurements made without a telescope. Ptolemy determined the distance to the Moon correctly to within a few percent. He used the Earth itself as a baseline, measuring the position of the Moon relative to the stars at two different times of night.

With the aid of telescopes, later astronomers were able to measure the distances to the nearer planets or asteroids by using the Earth's diameter as a baseline. This is how the astronomical unit was first established. To reach for the stars, however, requires a much longer baseline for triangulation, and extremely sensitive measurements. Such a baseline is provided by the Earth's annual trip around the Sun.

18.2.2 Distances to Stars

As the Earth travels from one side of its orbit to the other, it graciously provides us with a baseline of 2 AU, or about 300 million km. Although this is a much bigger baseline than the diameter of the Earth, the stars are so far away that the resulting parallax shift is *still* not visible to the naked eye—not even for the closest stars.

In Chapter 1 we discussed how this dilemma perplexed the ancient Greeks, some of whom had actually suggested that the Sun might be the center of the solar system, with the Earth in motion around it. Aristotle and others argued, however, that the Earth could not be revolving about the Sun. If it were, they said, we would observe the parallax of the nearer stars against the background of more distant objects as we viewed the sky from different parts of the Earth's orbit (Figure 18.3). Tycho Brahe advanced the

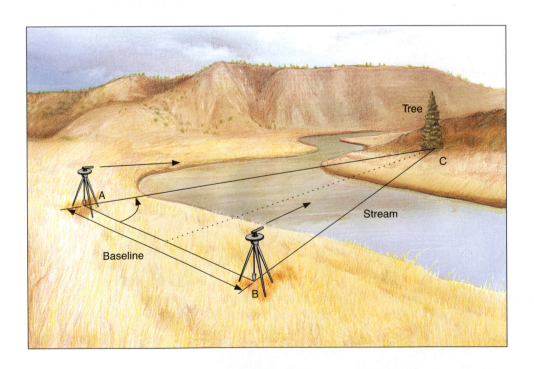

FIGURE 18.2

Triangulation Triangulation allows us to measure distances to inaccessible objects. By getting the angle to a tree from two different vantage points, we can calculate the properties of the triangle they make and thus the distance to the tree.

Ace●Astronomy™ Log into AceAstronomy and select this chapter to see Astronomy Exercise "Parallax II."

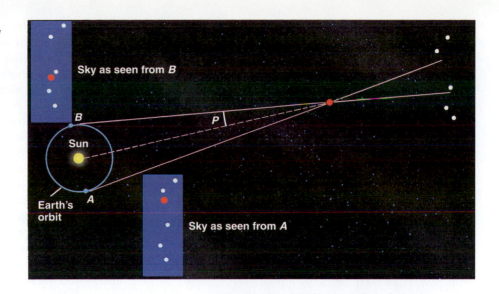

Sky as seen from *B*

B

Sun

P

Earth's orbit

A

Sky as seen from *A*

■ FIGURE 18.3

Parallax As the Earth revolves around the Sun, the direction in which we see a nearby star varies with respect to distant stars. We define the parallax of the nearby star to be one half of the total change in direction, and we usually measure it in arcseconds.

same argument nearly 2000 years later, when his careful measurements of stellar positions with the unaided eye revealed no such shift.

These early observers did not realize how truly distant the stars were and how small the change in their positions therefore was, even with the entire orbit of the Earth as a baseline. The problem was that they did not have tools to measure parallax shifts too small to be seen with the human eye.

By the 18th century, when there was no longer serious doubt about the Earth's revolution, it became clear that the stars must be extremely distant. Astronomers equipped with telescopes began to devise instruments capable of measuring the tiny shifts of nearby stars relative to the background of more distant (and thus unshifting) celestial objects. This was a significant technical challenge, since, even for the nearest stars, parallax angles are usually only a fraction of a second of arc. Recall that one second of arc is an angle of only 1/3600 of a degree (see Chapter 1). A coin the size of a U.S. quarter would appear to have a diameter of 1 arcsec if you were viewing it from a distance of about 3 miles, or 5 km! Think about how small an angle that is. No wonder it took astronomers a while before they could measure such tiny shifts.

The first successful detections of stellar parallax were in the year 1838, when Friedrich Bessel in Germany (Figure 18.4), Thomas Henderson, a Scottish astronomer working at the Cape of Good Hope, and Friedrich Struve in Russia independently measured the parallaxes of the stars 61 Cygni, Alpha Centauri, and Vega, respectively. Even the closest star, Alpha Centauri, showed a total displacement of only about 1.5 arcsec during the course of a year.

Figure 18.3 shows how such measurements work. Seen from opposite sides of the Earth's orbit, a nearby star shifts position when compared to a pattern of more distant stars. Astronomers actually define parallax to be *one-half* the angle that a star shifts when seen from opposite sides of the Earth's orbit (the angle labeled *P* in Figure 18.3). The

■ FIGURE 18.4

Bessel, First Measurer of Parallax Friedrich Wilhelm Bessel (1784–1846) made the first authenticated measurement of the distance to a star (61 Cygni) in 1838, a feat that had eluded many dedicated astronomers for almost a century.

reason for this is just that they prefer to deal with a baseline of 1 AU instead of 2 AU.

18.2.3 Units of Stellar Distance

With a baseline of 1 AU, how far away would a star have to be to have a parallax of 1 arcsec? The answer turns out to be 206,265 AU, or 3.26 light years. This is 3.1×10^{13} km (in words, 31 million million kilometers; see Figuring for Yourself at the end of the chapter). We give this unit a special name, the **parsec** (abbreviated **pc**)—derived from "the distance at which we have a **par**allax of one **sec**ond." The distance of a star in parsecs (D) is just the reciprocal of its parallax (p) in arcseconds; that is,

$$D = \frac{1}{p}$$

Thus a star with a parallax of 0.1 arcsec would be found at a distance of 10 pc, and one with a parallax of 0.05 arcsec would be 20 pc away.

Back in the days when most of our distances came from parallax measurements, a parsec was a useful unit of distance, but it is not as intuitive as the **light year,** which we defined in Section 17.1. One advantage of the light year as a unit is that it emphasizes the fact that as we look out into space, we are also looking back into time. The light that we see from a star 100 LY away left that star 100 years ago. What we study is not the star as it is now, but rather as it was in the past. The light that reaches our telescopes today from distant galaxies left them before the Earth even existed.

In this text, we will use light years as our unit of distance, but many astronomers still use parsecs when they write technical papers or talk with each other at meetings. To convert between the two distance units, just bear in mind: 1 pc = 3.26 LY, and 1 LY = 0.31 pc.

ASTRONOMY BASICS
Naming Stars

You may be wondering why stars have such a confusing assortment of names. Just look at the first three stars to have their parallaxes measured: 61 Cygni, Alpha Centauri, and Vega. Each of these names comes from a different tradition of designating stars.

The brightest stars have names that derive from the ancients. Some are from the Greek, such as Sirius, which means "the scorched one"—a reference to its brilliance. A few are from Latin, but many of the best-known names are from Arabic because, as discussed in Chapter 1, much of Greek and Roman astronomy was "rediscovered" in Europe after the dark ages by means of Arabic translations. Vega, for example, means "swooping Eagle," and Betelgeuse (pronounced "Beetle-juice") means "right hand of the central one."

In 1603 the German astronomer Johann Bayer introduced a more systematic approach to naming stars. For each constellation, he assigned a Greek letter to the brightest stars, roughly in order of brightness. In the constellation of Orion, for example, Betelgeuse is the brightest star, so it got the first letter in the Greek alphabet—alpha—and is known as Alpha Orionis. (*Orionis* is the possessive form of *Orion,* so Alpha Orionis means "the first of Orion.") A star called Rigel, being the second brightest in that constellation, is called Beta Orionis (Figure 18.5). Since there are 24 letters in the Greek alphabet, this system allows the labeling of 24 stars in each constellation, but constellations usually have many more stars than that.

In 1725 the English Astronomer Royal John Flamsteed introduced yet another system, in which the brighter stars eventually got a number in each constellation in order of their location in the sky or, more precisely, their right ascension. The system of sky coordinates that includes right ascension was discussed in Chapter 3. In this system Betelgeuse is called 58 Orionis and 61 Cygni is the 61st star in the constellation of Cygnus, the swan.

It gets worse! As astronomers began to understand more and more about stars, they drew up a series of specialized star catalogs, and fans of those catalogs began calling stars by their catalog numbers. If you look at Appendix 10—our list of the nearest stars (many of which are much too faint to get an ancient name, Bayer letter, or Flamsteed number)—you will see references to some of these catalogs. An example is a set of stars labeled with a BD number, for "Bonner Durchmusterung." This was a mammoth catalog of over 324,000 stars in a series of zones in the sky, organized at the Bonn Observatory in the 1850s and 1860s. Keep in mind that this catalog was made before photography or computers came into use, so the position of each star had to be measured (at least twice) by eye, a daunting undertaking.

There is also a completely different system for keeping track of stars whose luminosity varies, and another for stars that brighten explosively at unpredictable times. Astronomers have gotten used to the many different star-naming systems, but students often find them bewildering and wish astronomers would settle down to one. Don't hold your breath: In astronomy, as in many fields of human thought, tradition holds a powerful attraction. Still, with high-speed computer databases to aid human memory, names may become less and less necessary. Today astronomers often refer to stars by their precise locations in the sky rather than by their names or various catalog numbers.

■ ■ ■ ■ ■ ■ ■ ■ ■ ■ ■ ■ ■

18.2.4 The Nearest Stars

No known star (other than the Sun) is within 1 LY or even 1 pc of the Earth. The stellar neighbors nearest the Sun are three stars that make up a multiple system in the constellation of Centaurus. To the unaided eye the system appears as a single bright star, called Alpha Centauri, only 30° from the south celestial pole and hence not visible from the mainland United States. Alpha Centauri itself is a binary star—two stars in mutual revolution, too close together to be distinguished without a telescope. These two stars are 4.4 LY from us. Nearby is the third member of the system, a faint star known as Proxima Centauri. Proxima, with a distance of 4.3 LY, is slightly closer to us than the other two stars. (By the way, a few astronomers have started to question whether Proxima is actually bound to the Alpha Cen-

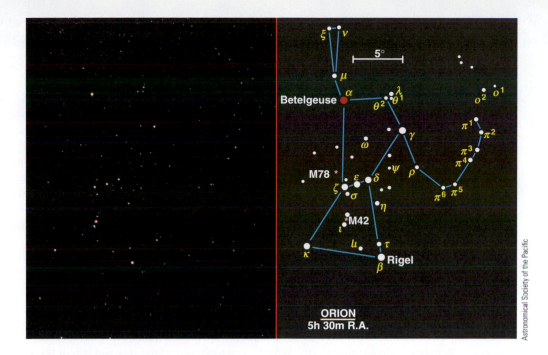

■ FIGURE 18.5

Objects in Orion Side by side are a photograph and a diagram of the brightest objects in or near the star pattern of Orion, the hunter (of Greek mythology), in the constellation of Orion. The Greek letters in Bayer's system are shown in yellow. The objects denoted M42 and M78 are not stars but nebulae—clouds of gas and dust; these numbers come from a list of "fuzzy objects" made by Charles Messier in 1781.

tauri pair; some lines of evidence show that it may simply be passing close to Alpha Centauri temporarily.)

The nearest star visible without a telescope from most parts of the United States is the brightest-appearing of all the stars, Sirius, which has a distance of a little more than 8 LY. As we saw in the previous chapter, it too is a binary system, composed of a faint white dwarf orbiting a bluish-white main-sequence star. It is interesting to note that light reaches us from the Sun in about 8 minutes and from Sirius in about 8 years.

The measurements of stellar parallax have been revolutionized by the launch of the spacecraft Hipparcos, which measured distances out to about 300 LY with an accuracy of 10 to 20 percent (see Making Connections: *Parallax and Space Astronomy*). However, even 300 LY is less than 1 percent the size of our Galaxy's main disk. If we are going to reach very far away from our own neighborhood with our chain of methods for measuring cosmic distances, we need some completely new techniques.

18.3 **VARIABLE STARS: ONE KEY TO COSMIC DISTANCES**

18.3.1 Standard Bulbs Revisited

Let's briefly review the key reasons that measuring distances to the stars is such a struggle. As discussed in Section 16.1, our problem is that stars come in a bewildering variety of intrinsic luminosities. (If stars were light bulbs, we'd say they come in a wide range of wattages.) Suppose, instead, that all stars had the same wattage or luminosity. In

that case the more distant ones would always look dimmer, and we could tell how far away a star was simply by how dim it appeared. In the real universe, however, when we look at a star in our sky (with eye or telescope) and measure its apparent brightness, we cannot know whether it looks dim because it's a low-wattage bulb or because it is far away, or perhaps some of each.

Astronomers need to discover something else about the star that allows us to "read off" its intrinsic luminosity—in effect, to know what the star's true wattage is. With this information, we can then attribute how dim it looks from Earth to its distance. Recall that the apparent brightness of an object decreases with the square of the distance to that object. If two objects have the same luminosity but one is three times farther than the other, the more distant one will look nine times fainter. Therefore, if we know the luminosity of a star and its apparent brightness, we can calculate how far away it is.

Astronomers have long searched for techniques that would somehow allow us to determine the luminosity of a star, and it is to these techniques that we turn next.

18.3.2 Variable Stars

The breakthrough in measuring distances to remote parts of our own Galaxy, and to other galaxies as well, came from the study of *variable stars*. Most stars are constant in their luminosity, at least to within a percent or two. Like the Sun, they generate a steady flow of energy from their interiors. However, some stars are seen to vary in brightness and for this reason are called variable stars. Many such stars vary on a regular cycle, like the flashing bulbs that decorate stores and homes during the winter holidays.

Parallax and Space Astronomy

One of the most difficult things about precisely measuring the tiny angles of parallax shifts from Earth is that you have to observe the stars through our planet's atmosphere. As we saw in Chapter 5, the effect of the atmosphere is to spread out the points of starlight into fuzzy disks, making exact measurements of their positions more difficult. Astronomers have long dreamed of being able to measure parallaxes from space, and an orbiting observatory turned this dream into reality.

The name of the Hipparcos satellite, launched in 1989 by the European Space Agency, is both an abbreviation for **Hi**gh **P**recision **Par**allax **Co**llecting **S**atellite and a tribute to Hipparchus, the pioneering Greek astronomer whose work we discussed in Chapter 1. The satellite was designed to make the most accurate parallax measurements in history from 36,000 km above the Earth. However, its on-board rocket motor failed to fire, which meant it did not get the needed boost to reach the desired altitude. Hipparcos ended up spending its four-year life in an elliptical orbit that varied from 500 to 36,000 km high. In this orbit, the satellite plunged into the Earth's radiation belts every five hours or so, which finally took its toll on the solar panels that provide energy to power the instruments.

Nevertheless, the mission was successful. Two catalogs of Hipparcos data have now been published. One gives po-sitions of 120,000 stars to an accuracy of one-thousandth of an arcsecond—about the diameter of a golf ball in New York as viewed from Europe. The second catalog contains information for more than a million stars, whose positions have been measured to thirty-thousandths of an arcsecond. We now have accurate parallax measurements of stars out to distances of about 300 LY. With ground-based tele-scopes, accurate measurements were feasible out to only about 60 LY.

About 200 new nearby stars have also been discovered, the nearest only about 18 LY away. Several hundred stars originally thought to be within 75 LY are now known to be much farther away. Because estimates of the age as well as the size of the universe depend on the chain of cosmic dis-tances, the Hipparcos results also affect our estimates of the size and age of the universe. They suggest that the old-est stars in the universe may be only 12–14 billion years old—younger than the 18 billion years estimated from ear-lier ground-based observations.

For a long time, the measurement of parallaxes and ac-curate stellar positions was a backwater of astronomical re-search—mainly because the accuracy of measurements did not improve much for about 100 years. However, the ability to make measurements from space has revolution-ized this field of astronomy. The European Space Agency

---

Let's define some tools to help us keep track of how a star varies. A graph that shows how the brightness of a variable star changes with time is called a **light curve** (Figure 18.6). The *maximum* is the point of the light curve where the star has its greatest brightness; the *minimum* is the point where it is faintest. If the light variations re-peat themselves periodically, the interval between the two maxima is called the *period* of the star. (If this kind of graph looks familiar, it is because we introduced it in Sec-tion 17.3.)

18.3.3 Pulsating Variables

There are two special types of variable stars for which—as we will see—measurements of the light curve give us ac-curate distances. These are called **cepheids** and the **RR Lyrae** variables, both of which are **pulsating variable stars.** Such a star actually changes its diameter with time—periodically expanding and contracting, as your chest does when you breathe. We now understand that these stars are going through a brief unstable stage late in their lives.

Ace*Astronomy*™ Log into AceAstronomy and select this chapter to see Astronomy Exercise "Cepheid Variable."

■ **FIGURE 18.6**

Cepheid Light Curve
This light curve shows how the brightness changes with time for a typical cepheid variable.

is now studying the feasibility of a spacecraft that would measure the positions of hundreds of millions of stars with an accuracy of a few ten-millionths of an arcsecond. Not only would we then have a three-dimensional map of a large fraction of our own Milky Way Galaxy, but we would also have an unbreakably strong link in the chain of cosmic distances.

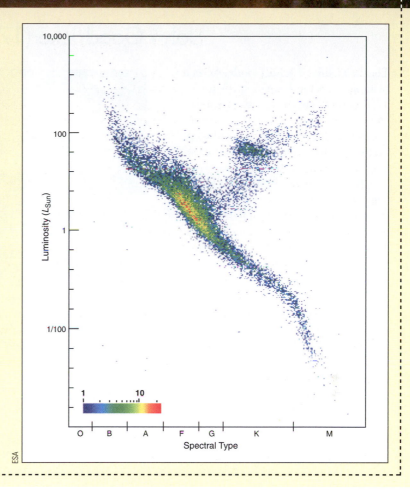

An H-R diagram of stars measured by Hipparcos. This plot includes 16,631 stars for which the parallaxes have an accuracy of 10 percent or better. The colors indicate the numbers of stars at each point of the diagram, with red corresponding to the largest number and blue to the lowest. Note that even this set of measurements, which brings together the best parallax data available, does not reach far enough out in space to include any supergiants (compare this diagram with Figures 17.13 and 17.14).

The expansion and contraction of pulsating variables can be measured by using the Doppler effect. The lines in the spectrum shift toward the blue as the surface of the star moves toward us and then shift to the red as it shrinks back. As the star pulsates, it also changes its overall color, indicating that its temperature is also varying. And, most important for our purposes, the luminosity of the pulsating variable also changes in a regular way as it expands and contracts.

18.3.4 Cepheid Variables

Cepheids are large, yellow, pulsating stars named for the first-known star of the group, Delta Cephei. This, by the way, is another example of how confusing naming conventions get in astronomy; here a whole class of stars is named after the constellation in which the first one happened to be found. We textbook authors can only apologize to our students for the whole mess!

The variability of Delta Cephei was discovered in 1784 by the young English astronomer John Goodricke (see Voyagers in Astronomy: *John Goodricke*). The star rises rather rapidly to maximum light and then falls more slowly to minimum light, taking a total of 5.4 days for one cycle. The curve in Figure 18.6 represents the light variation of Delta Cephei.

Several hundred cepheid variables are known in our Galaxy. Most cepheids have periods in the range of 3 to 50 days and luminosities that are about 1000 to 10,000 times greater than that of the Sun. Their variations in luminosity range from a few percent to a factor of ten.

Polaris, the North Star, is a cepheid variable that for a long time varied by one tenth of a magnitude, or by about 10 percent in visual luminosity, in a period of just under four days. Recent measurements indicate that the amount by which the brightness of Polaris changes is decreasing and that sometime in the future this star will no longer be a pulsating variable. This is just one more piece of evidence that stars really do evolve and change in fundamental ways as they age.

Ace🌀**Astronomy**™ Log into AceAstronomy and select this chapter to see Astronomy Exercise "Cepheid Variable I & II."

18.3.5 The Period-Luminosity Relation

The importance of cepheid variables lies in the fact that their periods and average luminosities turn out to be directly related. The longer the period (the longer the star

VOYAGERS IN ASTRONOMY

John Goodricke (1764–1786)

The brief life of John Goodricke is a testament to the human spirit under adversity. Born deaf and unable to speak, Goodricke nevertheless made a number of pioneering discoveries in astronomy through patient and careful observations of the heavens.

Born in Holland where his father was on a diplomatic mission, Goodricke was sent back to England at age 8 to study at a special school for the deaf. He did sufficiently well to enter Warrington Academy, a secondary school that offered no special assistance for students with handicaps. His mathematics teacher there inspired an interest in astronomy, and in 1781, at age 17, Goodricke began observing the sky at his family home in York, England. Within a year he had discovered the brightness variations of the star Algol (discussed in Chapter 17) and suggested that an unseen companion star was causing the changes, a theory that waited over 100 years for proof. His paper on the subject was read before the Royal Society (the main British group of scientists) in

A portrait of John Goodricke by artist J. Scouler, which now hangs in the Royal Astronomical Society in London. There is some controversy about whether this is actually what Goodricke looked like or whether the painting was much retouched to please his family.

Courtesy of the San Diego State University special collections library

1783 and won him a medal from that distinguished group.

In the meantime, Goodricke had discovered two other stars that varied regularly, Beta Lyrae and Delta Cephei, both of which continued to interest astronomers for years to come. Goodricke shared his interest in observing with his older cousin, Edward Pigott, who went on to discover other variable stars during his much longer life. But Goodricke's time was quickly drawing to a close; at age 21, only two weeks after he was elected to the Royal Society, he caught a cold while making astronomical observations and never recovered.

Today the University of York has a building named Goodricke Hall and a plaque that honors his contributions to science. Yet if you go to the churchyard cemetery where he is buried, an overgrown tombstone has only the initials "J. G." to show where he lies. Astronomer Zdenek Kopal, who has looked carefully into Goodricke's life, has speculated on why the marker is so modest: Perhaps the rather staid Goodricke relatives were ashamed of having a "deaf-mute" in the family and could not sufficiently appreciate how much a man who could not hear could nevertheless see.

takes to vary), the greater the luminosity. This **period-luminosity relation** was a remarkable discovery, one for which astronomers still (pardon the expression) thank their lucky stars. The period of such a star is easy to measure—a good telescope and a good clock are all you need. Once you have the period, the relationship (which can be put into precise mathematical terms) will give you the luminosity of the star. Astronomers can then compare this intrinsic brightness with the apparent brightness of the star. As we saw, the difference between the two allows them to calculate the distance.

The relation between period and luminosity was discovered in 1908 by Henrietta Leavitt (Figure 18.7), a staff member at the Harvard College Observatory (one of a number of women working for low wages assisting Edward Pickering, the observatory's director; see Voyagers in Astronomy: *Annie Cannon* in Chapter 16). Some hundreds of cepheid variables had been discovered in the Large and Small Magellanic Clouds (Figure 18.8), two great star sys-

tems that are actually neighboring galaxies (although they were not known to be galaxies then).

These systems presented a wonderful opportunity to study the behavior of variable stars independent of their distance. For all practical purposes, the Magellanic Clouds are so far away that astronomers can assume that all the stars in them are at roughly the same distance from us. (In the same way, all the suburbs of Los Angeles are roughly the same distance from New York City. Of course, if you are *in* Los Angeles, you will notice distances between the suburbs, but compared to how far away New York City is, the differences seem small.) If all the variable stars in the Magellanic Clouds are at roughly the same distance, then any difference in their apparent brightnesses must be a reflection of differences in their intrinsic luminosities.

Leavitt found that the brighter-appearing cepheids always have the longer periods of light variation. Thus, she reasoned, the period must be related to the luminosity of the stars. When Leavitt did this work, the distance to the

FIGURE 18.7

Henrietta Leavitt Henrietta Swan Leavitt (1868–1921) worked at the Harvard College Observatory as one of a number of underpaid women "computers." Studying photographs of the Magellanic Clouds, she found over 1700 variable stars, including 20 cepheids. Since all the cepheids in these systems were at roughly the same distance, she was able to compare their luminosities and periods of variation, and she thus discovered a relationship between these characteristics that led to a new and much better way of estimating cosmic distances.

Harvard College Observatory Archives

Magellanic Clouds was not known, so she was only able to show that luminosity was related to period. She could not determine exactly what the relationship is. To define the period–luminosity relation with actual numbers (to *calibrate* it), astronomers first had to measure the actual distances to a few nearby cepheids in another way. (This was accomplished by finding cepheids associated in clusters with other stars whose distances could be estimated from their spectra, as discussed in the next section of this chapter.) But once the relation was thus defined, it could give us the distance to any cepheid, wherever it might be located (Figure 18.9).

Here at last was the technique astronomers had been searching for to break the confines of distance that parallax imposed on them. Cepheids can be observed and monitored, it turns out, in many parts of our own Galaxy and in other galaxies as well. Astronomers, including Ejnar Hertzsprung and Harvard's Harlow Shapley, immediately saw the potential of the new technique; they and many others set to work exploring more distant reaches of space using cepheids as signposts. As we will see, this work still continues, as the Hubble Space Telescope and other modern instruments try to identify and measure individual cepheids in galaxies farther and farther away (Figure 18.10).

FIGURE 18.8

Large Magellanic Cloud
The Large Magellanic Cloud (so named because Magellan's crew were the first Europeans to record it) is a small irregular-shaped galaxy near our own Milky Way. It was in this galaxy that Henrietta Leavitt discovered the cepheid period–luminosity relation.

National Optical Astronomy Observatories

Find a Cepheid variable star and measure its period

Use the Period-Luminosity Law to calculate the star's luminosity

Measure the star's apparent brightness

Compare the luminosity with the apparent brightness to calculate the distance

■ **FIGURE 18.9**

Using a Cepheid to Measure Distance

Wendy Freedman, Carnegie Institution of Washington, and NASA

■ **FIGURE 18.10**

A Really Distant Cepheid   This image of part of the galaxy called M100 was taken with the Hubble Space Telescope in 1994. The insets show a single cepheid in the galaxy going through its cycle of brightness variations. What makes these images remarkable is that M100 is 51 million LY away, one of the most distant galaxies in which individual cepheids have been identified and measured. This series of faint smudges also shows just how hard it is to measure such distant stars.

■ FIGURE 18.11

Period-Luminosity Relation for Cepheid Variables
In this class of variable stars, the time the star takes to go through a cycle of luminosity changes is related to the average luminosity of the star. The longer the star takes to vary, the brighter it is. Also shown are the period and luminosity for RR Lyrae stars.

18.3.6 RR Lyrae Stars

A related group of stars, whose nature was understood somewhat later than that of the cepheids, are called RR Lyrae variables, named for the star RR Lyrae, the best-known member of the group. More common than the cepheids but less luminous, thousands of these pulsating variables are known in our Galaxy. The periods of RR Lyrae stars are always less than one day, and their changes in brightness are typically less than about a factor of two.

Astronomers have observed that the RR Lyrae stars occurring in any particular star cluster all have about the same apparent brightness. Since stars in a cluster are all at approximately the same distance, it follows that RR Lyrae variables must all have nearly the same intrinsic luminosity, which turns out to be about $50\,L_{Sun}$ (in this sense, RR Lyrae stars are a little bit like standard light bulbs). Figure 18.11 displays the ranges of periods and luminosities for both the cepheids and the RR Lyrae stars.

RR Lyrae stars can be detected out to a distance of about 2 million LY, and cepheids to about 60 million LY. Compare these limits with parallaxes that even from space will probably not be measured for stars more distant than a few hundred LY. You can see from this comparison just how important the discovery of the period–luminosity relation for variable stars was in enabling astronomers to extend their measurements of cosmic distances.

18.4 THE H-R DIAGRAM AND COSMIC DISTANCES

18.4.1 Distances from Spectral Types

As satisfying and productive as variable stars have been for distance measurement, these stars are rare and are not found near all the objects to which we wish to measure distances. Suppose, for example, we need the distance to a star that is not varying, or to a group of stars, none of which is a variable. In this case, the H–R diagram can come to our rescue.

If we can observe the spectrum of a star, we can estimate its distance from our understanding of the H–R diagram. As discussed in Chapter 16, a detailed examination of a stellar spectrum allows astronomers to classify the star into one of the *spectral types* indicating surface temperature. (The types are O, B, A, F, G, K, M, L, and T; each of these can be divided into numbered subgroups.) In general, however, the spectral type alone is not enough to allow us to estimate luminosity. Look again at Figure 17.14. A G2 star could be a main-sequence star with a luminosity of $1\,L_{Sun}$, or it could be a giant with a luminosity of $100\,L_{Sun}$, or even a supergiant with a still higher luminosity.

But we can learn more from a star's spectrum than just its temperature. Remember, for example, that we can detect pressure differences in stars from the details of the spectrum (see Section 16.4). This knowledge is very useful because giant stars are larger (and have lower pressures) than main-sequence stars, and supergiants are still larger than giants. If we look in detail at the spectrum of a star, we can determine whether it is a main-sequence star, a giant, or a supergiant.

Suppose, for example, that the spectrum, color, and other properties of a distant G2 star match those of the Sun exactly. It is then reasonable to conclude that this distant star is likely to be a main-sequence star just like the Sun, and to have the same luminosity as the Sun.

The most widely used system of star classification divides stars of a given spectral class into six categories called **luminosity classes.** These luminosity classes are denoted by Roman numerals as follows:

Ia: Brightest supergiants
Ib: Less luminous supergiants
II: Bright giants
III: Giants
IV: Subgiants (intermediate between giants and main-sequence stars)
V: Main-sequence stars

The full spectral specification of a star includes its luminosity class. For example, a main-sequence star with spectral class F3 is written as F3 V. The specification for an M2 giant is M2 III. Figure 18.12 illustrates the approximate

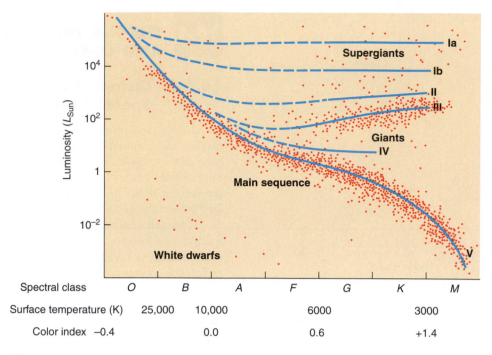

FIGURE 18.12

Luminosity Classes Stars of the same temperature (or spectral class) can fall into different luminosity classes on the Hertzsprung–Russell diagram. By studying details of the spectrum for each star, astronomers can determine which luminosity class they fall in (whether they are main sequence stars, giant stars, or supergiant stars).

positions of stars of various luminosity classes on the H–R diagram. The dashed portions of the lines represent regions with very few or no stars.

With both its spectral and luminosity classes known, a star's position on the H–R diagram is uniquely determined. Since the diagram plots luminosity versus temperature, this means we can now read off the star's luminosity (once its spectrum has helped us place it on the diagram). As before, if we know how luminous the star really is and see how dim it looks, the difference allows us to calculate its distance. (For historical reasons, astronomers sometimes call this method of distance determination *spectroscopic parallax,* even though the method has nothing to do with parallax.)

18.4.2 A Few Words About the Real World

Introductory textbooks such as ours work hard to present the material in a straightforward and simplified way. In doing so, we sometimes do our students a disservice by making scientific techniques seem too clean and painless. In the real world, the techniques we have just described turn out to be messy and difficult, and often give astronomers headaches that last long into the day!

For example, the relationships we have described—such as the period–luminosity relation for certain variable stars—aren't exactly straight lines on a graph. The points representing many stars scatter widely when plotted, and thus the distances derived from them also have a certain built-in scatter or uncertainty.

The distances we measure with the methods we have discussed are therefore only accurate to within a certain percentage of error—sometimes 10 percent, sometimes 25 percent, sometimes as much as 50 percent or more. A 25 percent error for a star estimated to be 10,000 LY away means it could be anywhere from 7500 to 12,500 LY away. This would be an unacceptable uncertainty if you were loading fuel into a spaceship for a trip to the star, but it is not a bad first figure to work with if you are an astronomer stuck on planet Earth.

Nor is the construction of H–R diagrams as easy as you might think at first. To make a good diagram, one needs to measure the characteristics of many stars, which can be a time-consuming task. Since our own solar neighborhood is already well mapped, the stars astronomers want to study to advance our knowledge are likely to be far away and faint. It may take hours of observing to obtain a single spectrum. Observers and their graduate students may have to spend many nights at the telescope (and many days back home working with their data) before they get their distance measurement.

Nevertheless, with these tools—parallaxes for the nearest stars, RR Lyrae variable stars and the H–R diagram for clusters of stars in our own and nearby galaxies, and cepheids out to distances of 60 million LY—we can measure distances throughout our own Galaxy and beyond to a good number of neighboring stellar systems. We can combine such distances with measurements of composition, luminosity, and temperature made with the techniques described in Chapters 16 and 17. These make up the arsenal of information we need to trace the evolution of stars from birth to death, the subject to which we turn in the chapters that follow.

SURFING THE WEB

The ABC's of Distances:
www.astro.ucla.edu/~wright/distance.htm
Astronomer Ned Wright (UCLA) gives a concise primer on many different methods of obtaining distances. This site is at a higher level than our textbook, but it is an excellent review for those with some background in astronomy.

The Transit of Venus and the Solar Parallax:
www.dsellers.demon.co.uk/venus/ven_ch1.htm
David Sellers recounts the story of how observations of Venus moving across the face of the Sun helped astronomers to measure the astronomical unit. Uses algebra and trigonometry to show how the calculations are done.

F. W. Bessel Site:
www-groups.dcs.st-and.ac.uk/~history/Mathematicians/Bessel.html
A brief site about the first person to detect stellar parallax, with references and links.

Hipparcos Mission Site: astro.estec.esa.nl/Hipparcos/
Background, results, catalogs of data, and educational resources from the mission to observe parallaxes from space. Some sections are technical, but others are accessible to students.

American Association of Variable Star Observers:
www.aavso.org
This organization of amateur astronomers helps to keep track of the behavior of variable stars. Its site has some background material, observing instructions, data tables, and links to organizations and programs elsewhere.

Hubble Project Measuring Distant Cepheids:
hubblesite.org/newscenter/archive/1999/19/
As explained further in Chapter 25, using cepheid variables to get distances to groups of galaxies beyond our own also allows us to measure some of the fundamental properties of the universe. Here is a progress report on this work by an international team of astronomers using the Hubble Space Telescope.

SUMMARY

18.1 Early measurements of length were based on human dimensions, but today we use worldwide standards that specify lengths in units such as the meter. Distances within the solar system are now determined by timing how long it takes radar signals to travel from the Earth to the surface of a planet or other body and then return.

18.2 For stars that are relatively nearby, we can "triangulate" the distance from a baseline created by the Earth's annual motion around the Sun. Half the shift in a nearby star's position relative to very distant background stars, as viewed from opposite sides of the Earth's orbit, is called the **parallax** of that star and is a measure of its distance. The units used to measure stellar distance are the **light year**

(LY), the distance light travels in one year, and the **parsec (pc),** the distance of a star with a parallax of 1 arcsec (1 parsec = 3.26 LY). The first successful measurements of stellar parallaxes were reported in 1838. Parallax measurements are a fundamental link in the chain of cosmic distances. The Hipparcos satellite has allowed us to measure accurate parallaxes for stars out to about 300 LY.

18.3 Cepheids and **RR Lyrae** stars are two types of **pulsating variable stars. Light curves** of these stars show that their luminosities vary with a regularly repeating period. Both types of variables obey a **period-luminosity relation,** so measuring their periods can tell us their luminosities. Then we can calculate their distances by comparing their luminosities with their apparent brightnesses, and this can allow us to measure distances to these stars out to over 60 million LY.

18.4 Stars with identical temperatures but different pressures (and diameters) have slightly different spectra. Spectral classification can therefore be used to estimate the **luminosity class** of a star as well as its temperature. As a result, a spectrum can allow us to pinpoint where the star is located on an H–R diagram and establish its luminosity. This, with the star's apparent brightness, again yields its distance.

INTER-ACTIVITY

A In this chapter, we explain the various measurements that have been used to establish the size of a standard meter. Your group should discuss why we change the definitions of our standard units of measurement in science from time to time. What factors in our modern society contribute to the growth of technology? Does technology "drive" science, or does science "drive" technology? Or do you think the two are so intertwined that it's impossible to say which is the driver?

B Cepheids are scattered throughout our own Milky Way Galaxy, but the period–luminosity relation was discovered from observations of the Magellanic Clouds, a satellite galaxy now known to be about 160,000 LY away. What reasons can you give to explain why the relation was not discovered from observations of cepheids in our own Galaxy? Would your answer change if there were a small cluster in our own Galaxy that contained 20 cepheids?

C You want to write a proposal to use the Hubble Space Telescope to look for the brightest cepheids in the galaxy M100 and estimate their luminosities. What observations would you need to make? Make a list of all the reasons such observations are harder than it first might appear.

D Why does your group think so many different ways of naming stars developed through history? (Think back to the days before e-mail and the Web connected everyone so easily.) Are there other fields where things are named confusingly and arbitrarily? How do stars differ from other phenomena that science and other professions tend to catalog?

E Although cepheid and RR Lyrae variable stars tend to change their brightness pretty regularly (while they are in that stage of their lives), some variable stars are unpredictable or change their behavior even during the course of a single human lifetime. Amateur astronomers all over the world follow such variable stars patiently and persistently, sending their nightly observations to huge databases that are being kept on the behavior of many thousands of stars. None of the hobbyists who do this get paid for making such painstaking observations. Have your group discuss why they do it. Would you ever consider a hobby that involves so much work, long into the night, often on work nights? If observing variable stars doesn't "turn you on," is there something you could do as a volunteer that does get you excited? Why?

F In the figure in the Making Connections box, the highest concentration of stars occurs in the middle of the main sequence. Can your group give reasons why this might be so? Why are there fewer very hot stars and fewer very cool stars on this diagram?

REVIEW QUESTIONS

Ace⟲Astronomy™ Assess your understanding of this chapter's topics with additional quizzing and animations at **http://ace.brookscole.com/voyages**

1. Explain how parallax measurements can be used to determine distances to stars. Why can we not make accurate measurements of parallax beyond a certain distance?

2. Make up a table relating the following units of astronomical distance: kilometer, Earth radius, solar radius, astronomical unit, light year, and parsec.

3. Suppose you have discovered a new RR Lyrae variable star. What steps would you take to determine its distance?

4. Explain how you would use the spectrum of a star to estimate its distance.

5. Which method would you use to obtain the distance to each of the following?

a. An asteroid crossing the Earth's orbit

b. A star astronomers believe to be no more than 50 LY from the Sun

c. A tight group of stars in the Milky Way Galaxy that includes a significant number of variable stars

d. A star that is not variable but for which you can obtain a clearly defined spectrum

THOUGHT QUESTIONS

6. What would be the advantage of making parallax measurements from Pluto rather than from Earth? Would there be a disadvantage?

7. Parallaxes are measured in fractions of an arcsecond. One arcsecond equals 1/60 arcmin; an arcminute is in turn 1/60°. To get some idea of how big 1° is, go outside at night and find the Big Dipper. The two pointer stars at the end of the bowl are 5.5° apart. The two stars across the top of the bowl are 10° apart. (Ten degrees is also about the width of your fist when held at arm's length and projected against the sky.) Mizar, the second star from the end of the Big Dipper's handle, appears double. The fainter star, Alcor, is about 12 arcmin from Mizar. For comparison, the diameter of the full moon is about 30 arcmin. The belt of Orion is about 3° long. Why did it take until 1838 to make parallax measurements for even the nearest stars?

8. For centuries, astronomers wondered whether comets were true celestial objects, like the planets and stars, or a phenomenon that occurred in the atmosphere of the Earth. Describe an experiment to determine which of these two possibilities is correct.

9. The Sun is much closer to the Earth than are the nearest stars, yet it is not possible to measure accurately the parallax of the Sun relative to the stars by measuring its position directly. Explain why.

10. Parallaxes of stars are sometimes measured relative to the positions of galaxies or distant objects called quasars. Why is this a good technique?

11. Figure 18.6 is the light curve for the prototype cepheid variable Delta Cephei. How does the luminosity of this star compare with that of the Sun?

12. Look at Appendices 10 and 11. What percentage of the stars in each list are main-sequence stars (remember the luminosity classes)? Why is this percentage so different for the two lists of stars?

13. Suppose you measure the temperature of a star to be identical to that of the Sun. Is this enough information to determine its distance? Explain.

14. Which of the following can you determine about a star without knowing its distance: radial velocity, temperature, apparent magnitude, luminosity? Explain.

15. A G2 star has a luminosity 100 times that of the Sun. What kind of star is it? How does its radius compare with that of the Sun?

16. A star has a temperature of 10,000 K and a luminosity of $10^{-2} L_{Sun}$. What kind of star is it?

FIGURING FOR YOURSELF

As we saw, estimating accurate distances to remote objects that we can never visit is one of the most challenging problems that astronomers face. As we move to ever larger distances, we must change the techniques that we use. Let's do some calculations with several of these techniques:

Direct measurements of distance are possible within the solar system. We can bounce radar signals off the nearest planets and asteroids that pass near the Earth and measure the round-trip transit time. We can send commands to a spacecraft and wait to see how long it takes for the spacecraft to acknowledge that it has received the commands. In both cases, the equation is one we have used before: distance = velocity times time, or $D = v \times t$.

17. A radar astronomer who is new at the job claims she beamed radio waves to Jupiter and received an echo exactly 48 min later. Do you believe her? Why?

18. A light year is the distance light travels in 1 year. Given that light travels at a speed of 300,000 km/s, how many kilometers are there in a light year? (Check to make sure you are using consistent units and not mixing, for example, years and seconds.)

19. Demonstrate that 1 pc equals 3.086×10^{13} km and that it also equals 3.26 LY. Show your calculations.

For nearby stars, we can measure the apparent shift in their positions as the Earth orbits the Sun. We wrote earlier that an object must be 206,265 astronomical units distant to have a parallax of 1 second of arc. This must seem like a very strange number, but you can figure out why this is the right value. We will start by estimating the diameter of the Sun and then apply the same idea to a star with a parallax of 1 arcsecond. Make a sketch that has a round circle to represent the Sun, place the Earth some distance away, and put an observer on it. Draw two lines from the point where the observer is standing, one to each side of the Sun. Sketch a circle centered at the Earth with its circumference passing through the center of the Sun. Now think about proportions.

20. The Sun spans about half a degree on the sky. A full circle has 360°. The circumference of the circle centered on the Earth and passing through the Sun is given by:

circumference = $2\pi \times 93,000,000$ miles

Then the following two ratios are equal:

$$\frac{0.5°}{360°} = \frac{\text{diameter of Sun}}{2\pi \times 93,000,000}$$

Calculate the diameter of the Sun. How does your answer compare to the actual diameter?

21. Now apply this idea to calculating the distance to a star that has a parallax of 1 arcsec. Draw a picture similar to the one we suggested above and calculate the distance in AU. Did you get 206,265 AU? How many light years is this?

22. The best parallaxes are those obtained with Hipparcos and have an accuracy of 0.001 arcsec. If you want to measure the distance to a star with an accuracy of 10 percent,

its parallax must be ten times larger than the typical error. How far away can you obtain a distance that is accurate to 10 percent with Hipparcos data? Our Galaxy is 100,000 LY in diameter. What fraction of the diameter of the Galaxy is the distance for which we can measure accurate parallaxes?

23. Astronomers are always making comparisons between measurements in astronomy and something that might be more familiar. For example, the Hipparcos Web site tells us that the measurement accuracy of 0.001 arcsec is equivalent to the angle made by a golf ball viewed from across the Atlantic Ocean, or to the angle made by the height of a person on the Moon as viewed from Earth, or to the length of growth of a human hair in 10 seconds as seen from 10 meters away. Use the ideas in question 20 to verify one of the first two comparisons. For extra credit, estimate how much a human hair grows in 10 seconds from the data you have.

24. As you found from question 22, even with the amazing accuracy of Hipparcos, we can measure parallaxes for only a tiny fraction of the stars in the Galaxy. To reach still larger distances, we need a new technique. That is why cepheids are so valuable: We can see them to very large distances. How far? To answer this question, you need to know three things: (1) a difference of five magnitudes corresponds to a difference of a factor of 100 in energy output or in brightness; (2) the brightness of a cepheid varies as the inverse square of its distance; and (3) about the faintest cepheid for which you can measure a light curve with a ground-based telescope has an apparent magnitude of 25. Let's use these ideas to estimate the distance to which you can see a cepheid that is 10,000 times brighter than the Sun.

a. How bright would this cepheid appear to be if it were at the same distance as the Sun? Assume the Sun has an apparent magnitude of −26 (see Figure 16.3).

b. What is the magnitude difference between our cepheid, which was at a distance of 1 AU, and the faintest cepheid that can be seen with the largest telescope?

c. Round your answer in the previous question to 60 magnitudes. What are the relative brightnesses of the two stars? (*Hint:* Remember that each five-magnitude interval corresponds to a factor of 100 in brightness.) What are the relative distances?

d. If 1 parsec corresponds to 206,265 AU, how many parsecs away can you see this cepheid? How many light years is this? How does this compare with the size of the Galaxy? (You should get an answer that says that with cepheids you can measure distances well beyond the boundary of our Galaxy.)

SUGGESTIONS FOR FURTHER READING

Adams, A. "The Triumph of Hipparcos" in *Astronomy,* Dec. 1997, p. 60. Brief introduction.

Ferris, T. *Coming of Age in the Milky Way.* 1988, Morrow. A history of how we established the scale of the cosmos.

Hirshfeld, A. "The Absolute Magnitude of Stars" in *Sky & Telescope,* Sep. 1994, p. 35. Good review of how we measure luminosity, with charts.

Hirshfeld, A. "The Race to Measure the Cosmos" in *Sky & Telescope,* Nov. 2001, p. 38. On parallax.

Hodge, P. "How Far Away Are the Hyades?" in *Sky & Telescope,* Feb. 1988, p. 138. A history of how we measure distance to this important cluster.

Marschall, L. et al. "Parallax You Can See" in *Sky & Telescope,* Dec. 1992, p. 626. On measuring parallax for yourself.

Maunder, M. and Moore, P. *Transit: When Planets Cross the Sun.* 1999, Springer-Verlag. On the history and observation of transits.

Reddy, F. "How Far the Stars" in *Astronomy,* June 1983, p. 6. Nice summary of the entire chain of distances.

Rowan-Robinson, M. *The Cosmological Distance Ladder.* 1985, W. H. Freeman. Somewhat technical introduction to measuring distances in the universe.

Trefil, J. "Puzzling Out Parallax" in *Astronomy,* Sep. 1998, p. 46. On the concept and history of parallax.

Turon, C. "Measuring the Universe" in *Sky & Telescope,* July 1997, p. 28. On the Hipparcos mission and its results.

Webb, S. *Measuring the Universe: The Cosmological Distance Ladder.* 1999, Praxis/Springer-Verlag.

Zimmerman, R. "Polaris: The Code-Blue Star" in *Astronomy,* Mar. 1995, p. 45. On the famous Cepheid variable and how it is changing.

The Pleiades star cluster contains hundreds of stars (only a few are visible here) and is located about 400 LY from the Sun. The blue nebulosity is starlight reflected by interstellar dust in a cloud that the cluster happens to be passing through at the present time. One important way we become aware of material between the stars is to observe the effects of a star's radiation on nearby gas and dust.

19

Between the Stars: Gas and Dust in Space

[P]hotographs] showed the entire group of stars [the Pleiades cluster] with an entangling system of nebulous matter, which seemed to bind together the different stars with misty wreaths and streams of filmy light . . . all of which is entirely beyond the keenest vision and the most powerful telescope.

E. E. Barnard, writing in *Popular Astronomy*, vol. 6, p. 439 (1898)

THINKING AHEAD

Where do stars come from? We already know that stars die because ultimately they exhaust their nuclear fuel. We might therefore make the hypothesis that new stars come into existence to replace the ones that die. In order to form new stars, however, we need the raw material to make them. What kind of material would you look for? How would you detect it, especially if it is not yet in the form of stars and cannot generate its own energy?

One of the most exciting discoveries of 20th-century astronomy was that our Galaxy contains not only stars but also vast quantities of this "raw material"—atoms or molecules of gas and tiny solid dust particles that are found between the stars. Studies of this raw material help us understand how new stars form and give us important clues about our own origins billions of years ago.

Ace Astronomy™ The AceAstronomy icon throughout the text indicates an opportunity for you to test yourself on key concepts and to explore animations and interactions on the AceAstronomy website at **http://ace.brookscole.com/voyages**

Virtual Laboratories

 Cosmic Rays

19.1 THE INTERSTELLAR MEDIUM

Astronomers refer to all the material between stars as **interstellar matter;** the entire collection of interstellar matter is called the *interstellar medium.* Some interstellar material is concentrated into giant clouds, which are called **nebulae** (Latin for "clouds"). The best-known nebulae are the ones that we can see glowing or reflecting visible light; there are many pictures of these in this chapter. Interstellar clouds do not last for the lifetime of the universe, but instead collide with each other, coalesce, and grow. Some form stars within them that then inject heat into the cloud material and disperse it. When stars die, they, in turn, eject some of their material into interstellar space. This material can then form new clouds and begin the cycle over again.

About 99 percent of the material between the stars is in the form of a *gas*—that is, it consists of individual atoms or molecules. The most abundant elements in this gas are hydrogen and helium (which we saw are also the most abundant elements in the stars), but this gas also includes other elements. Some of the gas is in the form of molecules—combinations of atoms. The remaining 1 percent is solid—frozen particles consisting of many atoms and molecules that are called **interstellar grains** or **interstellar dust** (Figure 19.1).

If all the interstellar gas were spread out smoothly, there would be about one atom of gas per cubic centimeter in interstellar space. (In contrast, the air in the room where you

Photo by David Malin, © Royal Observatory, Edinburgh

■ FIGURE 19.1

Various Types of Interstellar Matter The reddish nebulae on this spectacular photograph glow with light emitted by hydrogen atoms. The darkest areas are clouds of dust that block the light from stars behind them. The upper part of the picture is filled with the bluish glow of light reflected from hot stars embedded in the outskirts of a huge, cool cloud of dust and gas. The cool supergiant star Antares is in the lower left part of the picture. This star is shedding some of its outer atmosphere and is surrounded by a cloud of its own making that reflects the red light of the star. The red nebula at the middle right partially surrounds the star Sigma Scorpii. (To the right of Antares is M4, a much more distant cluster of extremely old stars.)

are reading this book has roughly 10^{19} atoms per cubic centimeter.) The dust grains are even scarcer. A cubic kilometer of space would contain only a few hundred to a few thousand tiny grains, each typically less than one ten-thousandth of a millimeter in diameter. These numbers are just averages, however, because the gas and dust are distributed in a patchy and irregular way, much as water vapor in the Earth's atmosphere is often concentrated into clouds.

In some interstellar clouds the density of gas and dust may exceed the average by as much as a thousand times or more, but even this density is more nearly a vacuum than any we can make on Earth. To show what we mean, let's imagine a vertical tube of air reaching from the ground to the top of the Earth's atmosphere with a cross section of 1 square meter. Now let us extend the same size tube from the top of the atmosphere all the way to the edge of the observable universe—over 10 billion light years (LY) away. Long though it is, the second tube would still contain fewer atoms than the one in our planet's atmosphere!

While the density of interstellar matter is very low, the volume of space in which such matter is found is huge, and so its *total* mass is substantial. To see why, we must bear in mind that stars occupy only a tiny fraction of the volume of the Milky Way Galaxy. For example, it takes light only about 2 seconds to travel a distance equal to the radius of the Sun, but more than four *years* to travel from the Sun to the nearest star. Even if the spaces among the stars are sparsely populated, there's just a lot of space out there.

Astronomers estimate that the total mass of gas and dust in the Milky Way Galaxy is equal to about 20 percent of the mass contained in stars. This means that the mass of the interstellar matter in our Galaxy amounts to about 7 billion times the mass of the Sun. There is plenty of raw material in the Galaxy to make many generations of new stars and planets (and perhaps even astronomy students).

ASTRONOMY BASICS
Naming the Nebulae

As you look at the captions for some of the spectacular photographs in this and the next chapter, you will notice the variety of names given to the nebulae. A few, which in small telescopes look like something recognizable, are sometimes named after the creatures or objects they resemble. Examples include the Crab, Tarantula, and Keyhole Nebulae. But most have only numbers that are entries in a catalog of astronomical objects.

Perhaps the best-known catalog of nebulae (as well as star clusters and galaxies) was compiled by the French astronomer Charles Messier (1730–1817). Messier's passion was discovering comets, and his devotion to this cause earned him the nickname "The Comet Ferret" from King Louis XV. When comets are first seen coming toward the Sun, they look like little fuzzy patches of light; in small telescopes, they are easy to confuse with nebulae or with group-

ings of many stars so far away that their light is all blended together. Time and again, Messier's heart leapt as he thought he had discovered one of his treasured comets, only to find that he had "merely" observed a nebula or cluster.

In frustration, Messier set out to catalog the position and appearance of over 100 objects that could be mistaken for comets. For him, this list was merely a tool in the far more important work of comet hunting. He would be very surprised if he returned today to discover that no one recalls his comets anymore, but his catalog of "fuzzy things that are not comets" is still widely used. When Figure 19.1 refers to M4, it denotes the fourth entry in Messier's list.

A far more extensive listing was compiled under the title of the *New General Catalog (NGC) of Nebulae and Star Clusters* in 1888 by John Dreyer, working at the observatory in Armagh, Ireland. He based his compilation on the work of William Herschel and his son John, plus many other observers who followed them. With the addition of two further listings (called the *Index Catalogs*), Dreyer's compilation eventually included 13,000 objects. Astronomers today still use his NGC numbers when referring to most nebulae and star groups.

■ ■ ■ ■ ■ ■ ■ ■ ■ ■ ■ ■

19.2 INTERSTELLAR GAS

Interstellar gas, depending on where it is located, can be as cold as a few degrees above absolute zero or as hot as a million degrees or more. We will begin our voyage through the interstellar medium by exploring the gas.

19.2.1 Ionized Hydrogen (H II) Regions—Gas Near Hot Stars

Some of the most spectacular astronomical photographs show interstellar gas located near hot stars (Figure 19.2). The strongest line in the visible region of the hydrogen spectrum is the red Balmer line (see Section 4.5), and this emission line accounts for the characteristic red glow in images like Figure 19.2.

Hot stars heat nearby gas to temperatures close to 10,000 degrees Kelvin (K). The ultraviolet radiation from the stars also ionizes the hydrogen (remember that during ionization, the electron is stripped completely away from the proton). Such a detached proton won't remain alone forever when attractive electrons are around; it will capture a free electron, becoming neutral hydrogen once more. However, such a neutral atom can then absorb ultraviolet radiation again and start the cycle over. At a typical moment, most of the atoms near a star are in the ionized state.

Since hydrogen is the main constituent of interstellar gas, we often characterize a region of space according to whether its hydrogen is neutral or ionized. A cloud of ionized hydrogen is called an **H II region.** (Scientists who

FIGURE 19.2

Dusty Nebulae The red glow that dominates this region in the constellation of Sagittarius is produced by the first line in the Balmer series of hydrogen. Hydrogen emission indicates that there are hot young stars nearby that ionize these clouds of gas. When electrons then recombine with protons and move back down to the lowest energy orbit, emission lines are produced. The blue color seen at the edges of some of the clouds is produced by small particles of dust, which scatter the light from the hot stars. Dust can also be seen silhouetted against the glowing gas and where it blocks the light from old, yellowish stars, which are seen in great numbers in the direction of the center of our Galaxy.

work with spectra use the Roman numeral I to indicate that an atom is neutral; successively higher Roman numerals are used for each higher stage of ionization. H II thus refers to hydrogen that has lost its one electron; Fe III is iron with two electrons missing.)

The electrons that are captured by the hydrogen nuclei cascade down through the various energy levels of the hydrogen atoms on their way to the lowest level, or ground state. During each transition downward, they give up energy in the form of light (as we explained in Chapter 4). This process of converting ultraviolet radiation into visible light is called *fluorescence*. Interstellar gas contains other elements besides hydrogen. Many of them are also ionized in the vicinity of hot stars; they then capture electrons and emit light, just as hydrogen does, which allows them to be observed by astronomers. But generally, the red hydrogen line is the strongest, and that is why H II regions look red.

A fluorescent light on Earth works using the same principles as a fluorescent H II region. When you turn on the current, electrons collide with atoms of mercury vapor in the tube. The mercury is excited to a high-energy state

because of these collisions. When the electrons in the mercury atoms return to lower energy levels, some of the energy they emit is in the form of ultraviolet photons. These, in turn, strike a phosphor-coated screen on the inner wall of the light tube. The atoms in the screen absorb the ultraviolet photons and emit visible light as they cascade downward among the energy levels. (The difference is that these atoms give off a wider range of light colors, which mix to give the characteristic white glow of fluorescent lights, while the hydrogen atoms in an H II region give off a more limited set of colors.)

19.2.2 Neutral Hydrogen Clouds

The very hot stars required to produce H II regions are rare, and only a small fraction of interstellar matter is close enough to such hot stars to be ionized by them. How do we go about looking for cold interstellar matter?

Unfortunately, cold hydrogen—far from the radiation of hot stars—does not produce any spectral lines in the visible part of the spectrum, and so for a long time astronomers had no way to detect it. Fortunately, however, other elements do produce strong spectral lines when they are cold. Some of the strongest lines in cold interstellar gas are produced by calcium and sodium. Since the gas is cold, what we see are dark absorption lines produced when atoms of calcium, sodium, and certain other elements absorb some of the light from stars that lie behind them (as discussed in Chapter 4).

The first evidence for absorption by interstellar clouds came from the analysis of a spectroscopic binary star (see Chapter 17). While most of the lines in the spectrum of this binary shifted alternately from longer to shorter wavelengths and back again, as we would expect from the Doppler effect for stars in orbit around each other, a few lines in the spectrum remained fixed in wavelength. Since both stars are moving in a binary system, lines that showed no motion puzzled astronomers. Subsequent work demonstrated that these lines were not formed in the star's atmosphere at all, but rather in a cold cloud of gas located between the Earth and the binary star.

Even though hydrogen is still the most abundant element in the colder regions, hydrogen atoms do not produce observable lines in the visible part of the spectrum. The direct detection of cold hydrogen gas had to await the development of sensitive radio telescopes. (See Section 5.4 for a discussion of how to detect long-wavelength radio radiation.)

Dutch astronomers predicted that hydrogen would produce a strong line at a wavelength of 21 centimeters (cm). That's quite a long wavelength, implying that the wave has a low frequency and low energy and does not come from electrons jumping between the energy levels (as we discussed in Chapter 4). Instead, energy is emitted when the electron does a flip, something like an acrobat in a circus flipping upright after standing on his head.

The flip works like this. A hydrogen atom acts as if its components were spinning. There is a tiny amount of rotational energy associated with the spin of the electron around

(a)   (b)

■ **FIGURE 19.3**

Formation of the 21-cm Line   When the electron in a hydrogen atom is in the orbit
closest to the nucleus, the proton and the electron may be spinning either (a) in the same
direction or (b) in opposite directions. When the electron flips over, the atom gains or loses
a tiny bit of energy by either absorbing or emitting electromagnetic energy with a wavelength of 21 cm.

its axis and with its orbital motion around the nucleus (proton). In addition, the proton has a spin of its own. If the proton and electron are spinning in opposite directions, the
atom as a whole has a very slightly lower energy than if the
two spins are aligned (Figure 19.3). If an atom in the lower
energy state (spins opposed) acquires a small amount of energy, then the spins of the proton and electron can be
aligned, leaving the atom in a slightly *excited state.* If the
atom then loses that same amount of energy again, it returns to its ground state. The amount of energy involved
corresponds to a wave with a wavelength of 21 cm.

Neutral hydrogen atoms can acquire small amounts of
energy through collisions with electrons and other atoms.
Such collisions are extremely rare in the sparse gases of interstellar space. An individual atom may wait many years
before such an encounter aligns the spins of its proton and
electron. Nevertheless, over many millions of years a significant fraction of the hydrogen atoms are excited by a collision. (Out there in cold space, that's about as much excitement as an atom typically experiences.)

An excited atom can later lose its excess energy either
by colliding with another particle or by giving off a radio
wave with a wavelength of 21 cm. If there are no collisions,
an excited hydrogen atom will wait an average of about 10
million years before emitting a photon and returning to its
state of lowest energy.

Equipment sensitive enough to detect the *21-cm line*
of neutral hydrogen became available in 1951. Dutch astronomers had built an instrument to detect the 21-cm
waves that they had predicted, but a fire destroyed it. As a
result, two Harvard physicists, Harold Ewen and Edward
Purcell, made the first detection (Figure 19.4), soon followed by confirmations from the Dutch and a group in
Australia. Observations showed that the 21-cm line is produced in clouds that have temperatures of about 100 K.
(Remember that a temperature of 0 K is −273°C. So 100 K
is −173°C—very cold in human terms.)

Most of the cold hydrogen is confined to an extremely
flat layer, less than 300 LY thick, that extends throughout

the disk of the Milky Way Galaxy. Since the detection of the
21-cm line, many other radio lines produced by both atoms
and molecules have been discovered (as we will discuss in
a moment), and these have allowed astronomers to map out
the cold gas throughout our home galaxy.

19.2.3   Ultra-Hot Interstellar Gas

Not all of the gas that lies far from hot stars is cold. Some of
the interstellar gas is at a temperature of a *million* degrees,
even though there is no visible source of heat nearby. The
discovery of this ultra-hot interstellar gas was a big surprise.
Before the launch of astronomical observatories into space,
astronomers assumed that most of the region between stars
was filled with cool hydrogen. But telescopes launched
above the Earth's atmosphere obtained ultraviolet spectra
that contained interstellar lines produced by oxygen atoms
that have been ionized five times. To strip five electrons
from their orbits around an oxygen nucleus requires a lot of
energy. In fact, these observations can be explained only if
the temperature of the interstellar matter where these
atoms occur is approximately a million degrees.

Theorists have now shown that the source of energy
producing these remarkable temperatures is the explosion
of massive stars at the ends of their lives (Figure 19.5). Such
explosions, called *supernovae,* will be discussed in detail in
Chapter 22. For now we'll just say that some stars, nearing
the ends of their lives, become unstable and literally explode. These explosions send high-temperature gas, moving at velocities of thousands (and even tens of thousands)
of kilometers per second, out into interstellar space, where
it has a tremendous heating effect on the neighborhood gas
atoms it encounters.

Astronomers estimate that one supernova explodes
roughly every 25 years somewhere in the Galaxy. On
average, hot gas from a supernova will sweep through any
given point in the Galaxy about once every 2 million years.
At this rate, the sweeping action is continuous enough to
keep much of the space between colder clouds filled with

Photos courtesy of E. M. Purcell and Harvard University

■ FIGURE 19.4

Ewen and Purcell   We see Harold Ewen in 1952 working with the horn antenna, atop the Lyman Physics Laboratory at Harvard, that made the first detection of interstellar 21-cm hydrogen radiation. The inset shows Edward Purcell, winner of the 1952 Nobel Prize in physics, a few years later.

gas at a temperature of a million degrees. As you might expect, there is also interstellar gas at temperatures of a few thousand degrees, especially in regions that mark the transition between the very cold and the very hot gas.

19.2.4   Interstellar Molecules

A few simple molecules out in space, such as CN and CH, were discovered decades ago because they produce absorption lines in the visible-light spectra of stars behind them. When more sophisticated equipment for obtaining spectra in radio and infrared wavelengths became available, astronomers—to their surprise—found much more complex molecules in interstellar clouds as well.

Just as atoms leave their "fingerprints" in the spectrum of visible light, so the vibration and rotation of atoms within molecules can leave spectral fingerprints in radio and infrared waves. If we spread out the radiation at such longer wavelengths, we can detect emission or absorption lines in the spectrum that are characteristic of specific molecules.

Over the years, experiments in Earth laboratories have shown us the exact wavelengths associated with changes in the rotation and vibration of many common molecules—giving us a template of possible lines against which we can now compare our observations of interstellar matter.

Over 120 kinds of molecules have been identified in interstellar space, most in giant clouds that also contain substantial amounts of gas and dust. Among the simpler molecules are H_2 (molecular hydrogen), water, and ammonia (whose smell you recognize in strong home cleaning products). The more complex molecules astronomers have found are mostly combinations of hydrogen, oxygen, carbon, nitrogen, and sulfur atoms.

Many of these molecules are *organic* (those associated with organic carbon chemistry on Earth) and include formaldehyde (used to preserve living tissues), alcohol (see Making Connections: *Cocktails in Space*), and antifreeze. In 1996, astronomers discovered acetic acid (the prime ingredient of vinegar) in a cloud lying in the direction of the constellation of Sagittarius. To balance the sour

■ **FIGURE 19.5**

The Vela Supernova Remnant
About 12,000 years ago, a dying star in the constellation Vela exploded, becoming about as bright as the full Moon in the Earth's skies. This photograph shows a portion of the shell of gas that is expanding outward from the site of the explosion.

with the sweet, a simple sugar (glycolaldehyde) has also been found. Relatively heavy molecules such as HC_9N and $HC_{11}N$, which have long chains of carbon atoms, have been detected in some cold clouds. And astronomers have recently found benzene, a molecule that is a ring of six carbon and six hydrogen atoms and is thought to be a likely stepping stone in the formation of more complex organic molecules. Other features in infrared spectra hint at complex rings of as many as 50 carbon atoms. See Table 19.1 for a list of a few of the more interesting molecules that have been found so far.

The cold interstellar clouds also contain cyanoacetylene (HC_3N) and acetaldehyde (CH_3CHO), generally regarded as starting points for *amino acid* formation. These are the building blocks of proteins, which are among the fundamental chemicals from which living organisms on Earth are constructed. The presence of these organic molecules does not imply that life exists in space, but it does show that the chemical building blocks of life can form under a wide range of conditions in the universe. As we learn

TABLE 19.1 *Some Interesting Interstellar Molecules*

Name	Chemical Symbol	Use on Earth
Ammonia	NH_3	Household cleansers
Formaldehyde	H_2CO	Embalming fluid
Acetylene	HC_2H	Fuel for a welding torch
Acetic acid	$C_2H_2O_4$	The essence of vinegar
Ethyl alcohol	CH_3CH_2OH	End-of-semester parties
Ethylene glycol	$HOCH_2CH_2OH$	Antifreeze ingredient
Benzene	C_6H_6	Carbon ring, ingredient in varnishes and dyes

Cocktails in Space

Among the molecules astronomers have identified in interstellar clouds is alcohol, which comes in two varieties: methyl (or wood) alcohol and ethyl alcohol (the kind you find in cocktails). Ethyl alcohol is a pretty complex molecule, written by chemists as C_2H_5OH. It is quite plentiful in space (relatively speaking). In clouds where it has been identified, we detect up to one molecule for every cubic meter. The largest of the cold clouds (which can be several hundred light years across) have enough ethyl alcohol to make 10^{28} fifths of liquor.

Spouses of future interstellar astronauts, however, need not fear that their wives or husbands will become interstellar alcoholics. Even if a spaceship were equipped with a giant funnel 1 km across and could scoop it through such a cloud at the speed of light, it would take about a thousand years to gather up enough alcohol for one standard martini!

Furthermore, the very same clouds also contain water (H_2O) molecules. Your scoop would gather them up as well, and there are a lot more of them because they are simpler and thus easier to form. For the fun of it, one astronomical paper actually calculated the proof of a typical cloud. *Proof* is the ratio of alcohol to water in a drink, where 0-proof means all water, 100-proof means half alcohol and half water, and 200-proof means all alcohol. The proof of the interstellar cloud was only 0.2, hardly enough to blur your memory after a big astronomy exam.

---

more about how complex molecules are produced in interstellar clouds, we gain an increased understanding of the kinds of processes that preceded the beginnings of life on Earth billions of years ago.

Once such molecules have formed in space, surviving is a challenge. Most of the more complex molecules would be *dissociated* (torn apart into individual atoms) by the higher-energy radiation that comes from stars. Therefore, these molecules can survive only where they are shielded from ultraviolet starlight. Fortunately for the molecules, there are dense, dark, giant clouds that contain dust, which acts like a thick blanket of interstellar smog and keeps ultraviolet starlight from penetrating the interior of the cloud (although the radio waves that signal the presence of the molecules readily get out). It is in these clouds that we find

molecules; in fact, these giant clumps of interstellar matter are called *molecular clouds*.

Within these giant clouds of interstellar material, we also see new stars forming and beginning the stellar life cycle that is our subject in the next several chapters. But let us first learn more about the dust these clouds contain.

19.3 COSMIC DUST

Figure 19.6 shows a striking example of what is actually a common sight in large telescopes: a dark region on the sky that appears nearly empty of stars. For a long time astronomers debated whether these dark regions were empty

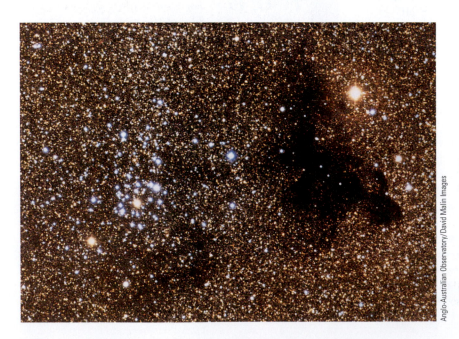

■ **FIGURE 19.6**

A Dark Dusty Cloud We see a star cluster (NGC 6520) next to a dark cloud of interstellar matter (Barnard 86). Old stars in our Galaxy are yellowish and form the brightest part of the Milky Way. The dark cloud is seen in front of these background stars, visible only because it blocks out the light from the stars beyond. The small cluster of young blue stars may have begun, millions of years ago, as just such a dark cloud.

Anglo-Australian Observatory/David Malin Images

VOYAGERS IN ASTRONOMY

Edward Emerson Barnard

Born in 1857 in Nashville, Tennessee, two months after his father died, Edward Barnard grew up in such poor circumstances that he had to drop out of school at age 9 to help support his ailing mother. He soon became an assistant to a local photographer, where he learned to love both photography and astronomy, destined to become the dual passions of his life. He worked as a photographer's aide for 17 years, studying astronomy on his own. In 1883 he obtained a job as an assistant at the Vanderbilt University Observatory, which enabled him at last to take some astronomy courses.

Married in 1881, Barnard built a house for his family that he could ill afford. But as it happened, a patent-medicine manufacturer offered a $200 prize (a lot of money in those days) for the discovery of any new comet. With the determination that became characteristic of him, Barnard spent every clear night searching for comets. He discovered seven of them between 1881 and 1887, earning enough money to make the payments on his home; this "Comet House" later became a local attraction. (By the end of his life, Barnard had found 17 comets through diligent observation.)

In 1887 Barnard got a position at the newly founded Lick Observatory, where he soon locked horns with the director, Edward Holden, a blustering

E. E. Barnard at the Lick Observatory 36-in. refractor.

Mary Lea Shane Archives of the Lick Observatory

administrator who made Barnard's life miserable. (To be fair, Barnard soon tried to do the same for him.) Despite being denied the telescope time that he needed for his photographic work, Barnard in 1892 managed to discover the first new satellite found around Jupiter since Galileo's day, a stunning observational feat that earned him world renown. Now in a position to demand more telescope time, he perfected his photographic techniques and soon began to publish the best images of the Milky Way taken up to that time. It was during the course of this work that he began to examine the dark regions among the crowded star lanes of the Galaxy, and to realize

that they must be vast clouds of obscuring material (rather than "holes" in the distribution of stars).

Astronomer-historian Donald Osterbrock has called Barnard an "observaholic": His daily mood seemed to depend entirely on how clear the sky promised to be for his night of observing. He was a driven, neurotic man, concerned about his lack of formal training, fearful of being scorned, and afraid that he might somehow slip back into the poverty of his younger days. He had difficulty taking vacations and lived for his work: Only serious illness could deter him from making astronomical observations.

In 1895 Barnard, having had enough of the political battles at Lick, accepted a job at the Yerkes Observatory near Chicago, where he remained until his death in 1923. He continued his photographic work, publishing compilations of his images that became classic photographic atlases, and investigating the varieties of nebulae revealed in his photographs. He also made measurements of the sizes and features of planets, participated in observations of solar eclipses, and carefully cataloged dark nebulae (see Figure 19.6). In 1916 he discovered the star with the largest proper motion, the second-closest star system to our own (see Chapter 16). It is now called Barnard's Star in his honor.

"tunnels" through which we looked beyond the stars of the Milky Way Galaxy into intergalactic space, or clouds of some dark material that blocked the light of the stars beyond. American astronomer E. E. Barnard is generally credited with showing from his extensive series of nebula photographs that the latter interpretation is the correct one (see the Voyagers in Astronomy box).

Dust betrays its presence in several ways: by blocking the light from distant stars, by emitting energy in the infrared part of the spectrum, by reflecting the light from nearby stars, and by making distant stars look redder than they really are.

19.3.1 Detecting Dust

The dark cloud (called Barnard 86) seen in Figure 19.6 blocks the light of the many stars that lie behind it; note how the regions in other parts of the photograph are crowded with stars. The only stars visible in the direction of Barnard 86 are those that happen to lie in front of the dark cloud from our perspective. Barnard 86 is an example of a relatively dense cloud or *dark nebula* containing tiny solid dust grains. Such opaque clouds are conspicuous on any photograph of the Milky Way (see the figures in Chapter 24). The "dark rift," which runs lengthwise down a long part of

T. A. Rector (NOAO/AURA/NSF) and Hubble Heritage Team (STScI/AURA/NASA)

ESA/ISO/CAM, L. North (Stockholm Observatory)

■ **FIGURE 19.7**

Visible and Infrared Images of the Horsehead Nebula in Orion This dark cloud is one of the best-known images in astronomy, probably because it really does resemble a horse's head. The horse-head shape is an extension of a large cloud of dust that fills the lower part of the picture. The top image shows this region in visible light; dust clouds are especially easy to see when they lie in front of a bright background such as this one. The bottom image shows infrared radiation from the region of the Horsehead (as recorded by the European Infrared Space Observatory). Note how the regions that appear dark in visible light appear bright in the infrared. The dust is heated by nearby stars and re-radiates this heat in the infrared. Only the top of the horse's head shows in the infrared image. Bright dots seen in the nebula below and to the left of the Horsehead and at the top of the Horsehead are young, newly formed stars. The insets in the bottom image show the horsehead and the bright nebula in more detail.

the summer Milky Way and appears to split it in two, is produced by a collection of such obscuring clouds.

While dust clouds are too cold to radiate a measurable amount of energy in the visible part of the spectrum, they glow brightly in the infrared (Figure 19.7). Small dust grains absorb visible light and ultraviolet radiation very efficiently.

The grains are heated by the absorbed radiation, typically to temperatures from 20 to about 500 K, and re-radiate this heat at infrared wavelengths.

We can use Wien's law (see Section 4.2) to estimate where in the electromagnetic spectrum this radiation falls. For a temperature of 100 K, the maximum is at about 30 mi-

IPAC/JPL/NASA

■ **FIGURE 19.8**

Infrared Cirrus This infrared image shows emission from infrared cirrus in a 12.5 × 12.5-degree field near the south celestial pole. The cirrus consists of microscopic dust grains, which are heated by radiation from stars and re-radiate their energy in the infrared. This image was taken by the IRAS telescope and is a composite of 12-micron (color-coded blue), 25-micron (color-coded green), and 60-micron (color-coded red) wavelengths. Stars are bright at 12 microns and thus appear as blue dots in the picture.

crons (1 micron = 10^{-6} m = 10^{-4} cm), while grains as cold as 20 K will radiate most strongly near 150 microns. The Earth's atmosphere is opaque to radiation at these wavelengths, so emission by interstellar dust is best measured from space.

Observations from above the Earth's atmosphere by IRAS (the Infrared Astronomical Satellite) showed that dust clouds are present throughout the plane of the Milky Way (Figure 19.8). The bright patches of emission have been given the name **infrared cirrus** because their appearance resembles the cirrus clouds in our own atmosphere. The closest infrared cirrus clouds are about 300 LY away.

Some dense clouds of dust are close to luminous stars and scatter enough starlight to become visible. Such a cloud of dust, illuminated by starlight, is called a *reflection nebula,* since the light we see is starlight reflected off the grains of dust. One of the best-known examples is the nebulosity around each of the brightest stars in the Pleiades cluster (see the figure at the beginning of this chapter). The dust grains are small, and such small particles turn out to scatter light with blue wavelengths more efficiently than light at red wavelengths. A reflection nebula, therefore, usually appears bluer than its illuminating star (Figure 19.9).

Gas and dust are generally intermixed in space, although the proportions are not exactly the same every-

NASA and the Hubble Heritage Team (STScI)

■ **FIGURE 19.9**

Scattering by Interstellar Dust The bluish light in this image is an example of a reflection nebula (NGC 1999). Like fog around a street lamp, a reflection nebula shines only because the dust within it scatters light from a nearby bright source. This nebula is illuminated by a recently formed star, which is visible just to the left of center. The jet black T-shaped structure in the center of the nebula is a cold cloud of gas, molecules, and dust, which is so dense it blocks all of the light behind it. This region is close to the Orion Nebula and is about 1500 LY from Earth.

where. The presence of dust is apparent on many photographs of emission nebulae; Figure 19.10 shows a beautiful image of the Trifid Nebula in the constellation of Sagittarius, where we see an H II region surrounded by a blue reflection nebula. Which type of nebula appears brighter depends on the kind of stars that cause the gas and dust to glow. Stars cooler than about 25,000 K have so little ultraviolet radiation of wavelengths shorter than 91.2 nanometers (nm)—which is the wavelength required to ionize hydrogen—that the reflection nebulae around such stars outshine the emission nebulae. Stars hotter than 25,000 K emit enough ultraviolet energy that the emission nebulae produced around them generally outshine the reflection nebulae.

19.3.2 Interstellar Reddening

The tiny interstellar dust grains absorb some of the starlight they intercept. But at least half of the starlight that interacts with a grain is merely scattered; that is, it is redirected helter skelter in all directions. Since neither the absorbed nor the scattered starlight reaches us directly, both absorption and scattering make stars look dimmer. The effects of both processes are called **interstellar extinction.**

Astronomers first came to understand interstellar extinction about 70 years ago as the explanation of a puzzling observation. In the early part of the 20th century, astronomers discovered that some stars are red even though their spectral lines indicate that they must be extremely hot (and thus should look blue). The solution to this seeming contradiction turned out to be that the light from these hot stars is not only dimmed but also **reddened** by interstellar dust.

Dust does not interact with all the colors of visible light the same way. Much of the violet, blue, and green light from these stars has been scattered or absorbed by dust, so it does not reach the Earth. Some of their orange and red light, with longer wavelengths, on the other hand, more easily penetrates the intervening dust and completes its long journey through space to enter Earth-based telescopes (Figure 19.11). Thus the star looks redder from Earth than it would if you could see it from nearby. (Strictly speaking, *reddening* is not the most accurate term for this process, since no red color is added; instead, blues and related colors are subtracted, so it should more properly be called "deblueing.")

We have all seen an example of reddening. The Sun appears much redder at sunset than it does at noon. The lower the Sun is in the sky, the longer the path its light must travel through the atmosphere. Over this greater distance, there is a greater chance that sunlight will be scattered. Since red light is less likely to be scattered than blue light, the Sun appears more and more red as it approaches the horizon.

By the way, scattering of sunlight is also what causes our sky to look blue, even though the gases that make up the Earth's atmosphere are transparent. As sunlight comes in, it scatters from the molecules of air. The small size of the molecules means that the blue colors scatter much more efficiently than the greens, yellows, and reds. Thus the blue in sunlight is scattered out of the beam and all over the sky. The light from the Sun that comes to your eye, on the other

Ace⊙Astronomy™ Log into AceAstronomy and select this chapter to see Astronomy Exercise "Why Is the Sky Blue?"

■ **FIGURE 19.10**

The Trifid Nebula In the reddish H II region, the hydrogen is ionized by nearby hot stars and glows through the process called fluorescence. The surrounding blue region is a reflection nebula. The Trifid Nebula (M20) is about 30 LY in diameter and about 3000 LY from the Sun. This beautiful image was processed by astronomical photographer David Malin using special photographic techniques to bring out faint details.

Ace⊙Astronomy™ Log into AceAstronomy and select this chapter to see the Active Figure called "Scattering."

■ **FIGURE 19.11**

Scattering of Light by Dust Interstellar dust scatters blue light more efficiently than red light, thereby making distant stars appear redder and giving clouds of dust near stars a bluish hue. Here a red ray of light from a star comes straight through to the observer, while a blue ray is shown scattering. A similar scattering process makes the Earth's sky look blue.

hand, is missing some of its blue, so the Sun looks a bit yellower, even when it is high in the sky, than it would from space.

The fact that starlight is reddened by interstellar dust means that long-wavelength radiation is transmitted through the Galaxy more efficiently than short-wavelength radiation. Consequently, if we wish to see farther in a direction with considerable interstellar material, we should look at long wavelengths. This simple fact provides one of the motivations for the development of infrared astronomy. In the infrared region at 2 microns (2000 nm), for example, the obscuration is only one-sixth as great as in the visible region (500 nm), and we can therefore study stars that are more than twice as distant before their light is blocked by interstellar dust. This ability to see farther by observing in the infrared portion of the spectrum represents a major gain for astronomers trying to understand the structure of our Galaxy or probing its puzzling center (see Chapter 24).

19.3.3 Interstellar Grains

Why doesn't *gas* redden distant stars like dust does? Except for specific spectral lines, atomic or molecular gas is almost transparent. Consider the Earth's atmosphere. Despite its very high density compared with that of interstellar gas, it is so transparent as to be practically invisible. The quantity of *gas* required to produce the observed absorption of light in interstellar space would have to be enormous. The gravitational attraction of so great a mass of gas would affect the motions of stars in ways that could easily be detected. Such motions are not observed, and thus the interstellar absorption cannot be the result of gases.

Although gas does not absorb much light, we know from everyday experience that tiny solid or liquid particles can be very efficient absorbers. Water vapor in the air is quite invisible. When some of that vapor condenses into tiny water droplets, however, the resulting cloud is opaque. Dust storms, smoke, and smog offer familiar examples of the efficiency with which solid particles absorb light. On the basis of arguments like these, astronomers have concluded that widely scattered *solid* particles in interstellar space are responsible for the observed dimming of starlight. What are these particles made of? And how did they form?

Observations like the pictures in this chapter show that a great deal of this dust exists; hence it must be primarily composed of elements that are abundant in the universe (and in interstellar matter). After hydrogen and helium, the most abundant elements are oxygen, carbon, and nitrogen. These three elements, along with magnesium, silicon, iron, and perhaps hydrogen itself, turn out to be the most important components of interstellar dust. Many of the dust particles can be characterized as sootlike (i.e., rich in carbon) or sandlike (containing silicon and oxygen). Grains of interstellar dust are found in meteorites and can be identified because the abundances of certain isotopes are different from what we see in other solar system material. Several different interstellar dust substances have been identified in this way in the laboratory, including graphite and dia-

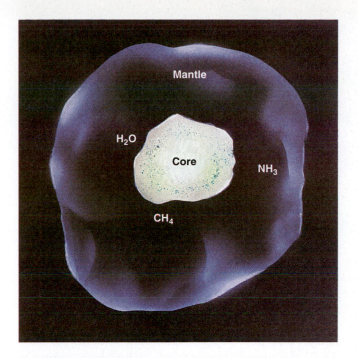

■ **FIGURE 19.12**

Model of an Interstellar Dust Grain A typical interstellar grain is thought to consist of a core of rocky material (silicates) or graphite, surrounded by a mantle of ices. Typical grain sizes are 10^{-8} to 10^{-7} m.

monds. (Don't get excited; these diamonds are only a billionth of a meter in size and would hardly make an impressive engagement ring!)

The most widely accepted model pictures the grains with rocky cores that are either like soot (rich in carbon) or like sand (rich in silicates). These cores are covered by icy mantles (Figure 19.12). The most common ices in the grains are water (H_2O), methane (CH_4), and ammonia (NH_3)—all built out of atoms that are especially abundant in the realm of the stars.

Typical individual grains must be just slightly smaller than the wavelength of visible light. If the grains were a lot smaller, they would not block the light efficiently, as Figure 19.9 and other images in this chapter show that it does. For an analogy, imagine a wall of bowling balls, which are considerably smaller than the wavelength of radio waves. Such a wall could not keep radio signals from reaching us inside a bowling alley.

On the other hand, if the dust grains were much larger than the wavelength of light, then starlight would not be reddened. Again, our wall of bowling balls, which are much larger than the wavelength of light, would block both blue and red light with equal efficiency. In this way we can deduce that a characteristic interstellar dust grain contains 10^6 to 10^9 atoms and has a diameter of 10^{-8} to 10^{-7} m (10 to 100 nm). This is actually more like the specks of solid matter in cigarette smoke than the larger grains of dust you find under your desk when you are too busy studying astronomy to clean properly.

19.4 COSMIC RAYS

In addition to gas and dust, a third class of particles, noteworthy for the high speeds with which they travel, is found in interstellar space. **Cosmic rays** were discovered in 1911 by an Austrian physicist, Victor Hess, who flew simple instruments aboard balloons and showed that high-speed particles arrive at the Earth from space (Figure 19.13). These particles have nearly the same composition as ordinary interstellar gas. Their behavior, however, is radically different from the gas we have discussed so far.

19.4.1 The Nature of Cosmic Rays

Cosmic rays are mostly high-speed atomic nuclei and electrons. Velocities equal to 90 percent of the speed of light are typical. Most cosmic rays are hydrogen nuclei (protons) stripped of their accompanying electron, but helium and heavier nuclei constitute about 9 percent of the particles. Ten to 20 percent of the cosmic rays with masses equal to the mass of the electron carry positive charge rather than the negative charge that characterizes electrons. Such particles are called **positrons** and are a form of antimatter (see Section 15.2.2).

The abundances of various atomic nuclei in cosmic rays mirror the abundances in stars and interstellar gas,

■ FIGURE 19.13

Cosmic Ray Pioneer Victor Hess returns from a 1912 balloon flight that reached an altitude of 5⅓ km. It was on such balloon flights that Hess discovered cosmic rays.

Photo courtesy of Martin Pomerantz

with one important exception. The light elements lithium, beryllium, and boron are far more abundant in cosmic rays than in the Sun and stars. These light elements are formed when high-speed cosmic ray nuclei of carbon, nitrogen, and oxygen collide with protons in interstellar space and break apart. (By the way, if you, like most readers, have not memorized all the elements and want to see how any of those we mention fit into the sequence of elements, you will find them all listed in Appendix 13 in order of the number of protons they contain.)

Cosmic rays reach the Earth in substantial numbers, and we can determine their properties either by capturing them directly or by observing the reactions that occur when they collide with atoms in our atmosphere. The total energy deposited by cosmic rays in the Earth's atmosphere is only about one-billionth the energy received from the Sun, but it is comparable to the total energy received in the form of starlight. Some of the cosmic rays come to the Earth from the surface of the Sun, but most come from outside the solar system.

19.4.2 Where Do They Come From?

There is a serious problem in identifying the source of cosmic rays. Since light travels in straight lines, we can tell where it comes from simply by looking. Cosmic rays are charged particles, and their direction of motion can be changed by magnetic fields. The paths of cosmic rays are curved both by magnetic fields in interstellar space and by the Earth's own field. Calculations show that low-energy cosmic rays may spiral many times around the Earth before entering the atmosphere where we can detect them. If an airplane circles an airport many times before landing, it is impossible to determine the direction and city from which it came. So, too, after a cosmic ray circles the Earth several times, it is impossible to know where its journey began.

There are a few clues, however, about where cosmic rays might be generated. We know, for example, that magnetic fields in interstellar space are strong enough to keep all but the most energetic cosmic rays from escaping the Galaxy. It therefore seems likely that they are produced somewhere inside the Galaxy. The only likely exceptions are those with the very highest energy. Such cosmic rays move so rapidly that they are not significantly influenced by interstellar magnetic fields, and thus they could escape our Galaxy. By analogy, they could escape other galaxies as well, so some of the highest-energy cosmic rays that we detect may have been created in some distant galaxy. Still, most cosmic rays must have their source inside the Milky Way.

We can also estimate how far typical cosmic rays travel before striking the Earth. The light elements lithium, beryllium, and boron hold the key. Since these elements are formed when carbon, nitrogen, and oxygen strike interstellar protons, we can calculate how long, on the average, cosmic rays must travel through space in order to experience enough collisions to account for the amount of lithium and the other light elements that they contain. It turns out that the required distance is about 30 times around the Galaxy.

At speeds near the speed of light, it takes perhaps 3 million to 10 million years for the average cosmic ray to travel this distance. This is only a small fraction of the age of the Galaxy or the universe, so cosmic rays must have been created fairly recently on a cosmic timescale.

The best candidates for a source of cosmic rays are the *supernova* explosions, which mark the deaths of massive stars (and which we will discuss in Chapter 22). There are enough explosions, and the explosions generate enough energy, to account for the observed number of cosmic rays. What we do not yet know is what precise mechanism in these explosions accelerates protons and other atomic nuclei to the fast speeds we see. Some collapsed stars (including star remnants left over from supernova explosions) may, under the right circumstances, also serve as accelerators of particles. In any case, we again find that the raw material of the Galaxy is enriched by the life cycle of stars. Let's now look at this enrichment process in more detail.

19.5 RECYCLING COSMIC MATERIAL

Just as materials from living (and nonliving) things on Earth are recycled over long periods of time, so the cosmos recycles its atoms over the generations of stars.

19.5.1 The Life Cycle of Interstellar Matter

Much of the interstellar material in our Galaxy has been ejected into space by old and dying stars. We will talk more about this process in Chapters 21 and 22, but for now just bear in mind what we learned in Chapter 15. What stars "do for a living" is to fuse heavier elements from lighter ones, producing energy in the process. As stars mature, they begin to lose some of the newly made elements to the reservoir of interstellar matter. Astronomers estimate that about 17 percent of the mass formed into stars is ultimately returned to the interstellar medium.

So, for example, dust forms when grains can condense in regions where gas is dense and cool. One place where the right conditions are found is in the winds from dying stars (the red giants and supergiants we discussed in Chapter 17).

The two kinds of dust grains are formed in different kinds of stars. Sootlike carbon-rich grains are expelled from stars that have synthesized carbon in their interiors, dredged it up into their atmospheres, and then expelled it in a stellar wind. Sandlike silicate grains are expelled in winds from stars that have atmospheric compositions more like that of the Sun when they develop their winds.

The most massive stars do not fade quietly away, but end their lives with the giant explosions called supernovae. A supernova drives an expanding spherical shell of stellar material, shock waves, and energy out into space. In many cases, stars form close together in space, and when several explode at nearly the same time (i.e., within a few million years), they form a *superbubble* that expands and fills more and more of space. Because of the huge amount of energy involved, the material inside this superbubble is very hot—with temperatures on the order of 1 million K. About 50 percent of the space between stars is filled with this very hot gas.

The important thing to bear in mind about the raw material of the Galaxy is that, over long timescales, the gas and dust clouds are not static; their number and arrangement change with time. Sometimes the expanding bubbles may overrun cooler interstellar clouds, which then find themselves inside the hot regions (Figure 19.14). The expanding shells of the bubbles also sweep up interstellar matter and

Diagram adapted from work published in *The Astrophysical Journal* by C. McKee and J. Ostriker

■ **FIGURE 19.14**

Distribution of Interstellar Matter   Interstellar clouds are embedded in hot, low-density gas heated to temperatures as high as 1 million K by supernova explosions. In the upper right, a supernova remnant is shown sweeping through interstellar space.

concentrate it at their outer boundaries. Superbubbles can run into other bubbles, combining into tunnels of hot gas surrounded by walls of cooler, denser gas that is not ionized. Some of this gas breaks up into cooler pockets and forms new clouds. These, in turn, can collide with each other and merge to form the giant molecular clouds where most new stars are formed.

These molecular clouds contain the dust grains originally expelled in the stellar winds, and additional dust particles can form in their cool, dense interiors. The dust grains slowly gather atoms—perhaps only one each day. But over the millions and billions of years, the numbers can add up.

The surfaces of the dust grains (see Section 19.3)—which would seem very large if you were an atom—provide "nooks and crannies" where these atoms can stick long enough to find partners and form molecules. (Think of the dust grains as "interstellar social clubs" where lonely atoms can meet and form meaningful relationships.) Eventually the dust grains become coated with ices. The presence of the dust shields the molecules inside the clouds from ultraviolet radiation and cosmic rays that would break them up.

When stars finally begin to form within the cloud, they heat the grains and evaporate the ices. The gravitational attraction of the newly forming stars also increases the density of the surrounding cloud material. Many more chemical reactions take place in the gas surrounding the newly forming stars, and this is where organic molecules are formed. These molecules can be incorporated into newly formed planetary systems, and, in fact, the early Earth may have been seeded in just such a way.

Indeed, scientists speculate that the water on the Earth may have come in part from interstellar grains. Recent observations from space have shown that water is abundant in dense interstellar clouds. Since stars are formed from this material, water must be present when solar systems, including our own, come into existence. The water in our oceans and lakes may have come initially from water locked into the rocky material that accreted to form the Earth. Alternatively, the water may have been brought to Earth when asteroids and comets impacted it. Scientists estimate that one comet impact every thousand years during Earth's first billion years would have been enough to account for the water we see today.

Any interstellar grains that are incorporated into newly forming stars will be destroyed by their high temperatures. But eventually, each new generation of stars will evolve to become red giants, with stellar winds of their own. Some of these stars will become supernovae and explode. Thus the process of recycling cosmic material can start all over again.

19.5.2 Interstellar Matter Around the Sun

We want to conclude our discussion of interstellar matter by asking how such material is organized in our immediate neighborhood. The Sun is located in a region where the density of interstellar matter is very low—so low that the region is referred to as the Local Bubble. There are probably two reasons for the low density. First, the Sun is located in a region between the spiral arms of our Galaxy, and interstellar matter tends to be concentrated in the spiral arms (see Chapter 24). Second, we are apparently located inside a hot superbubble like the ones we have described above. Indeed, the temperature of the low-density gas in the solar neighborhood is about a million degrees, just like the other superbubbles that spread throughout our Galaxy. Because there is so little hot material (only 0.005 atom per cubic centimeter), this high temperature does not affect the stars or planets in the area in any way.

There is a region in the direction of the constellations Scorpius and Centaurus where a lot of star formation took place about 15 million years ago. Winds from these stars and supernovae explosions (that occurred when the most massive stars exhausted their fuel supply) combined to blow the interstellar matter that initially surrounded these stars outward (similar to Figure 19.15). This drove the gas and dust into the low-density interarm region where the Sun is located. The rim of this expanding superbubble reached the Sun about 7.6 million years ago and now lies more than 200 LY past the Sun in the general direction of the constellations of Orion, Perseus, and Auriga.

Some scientists even speculate that a supernova explosion in this region may have caused the extinction of some species on Earth, although the number of species affected would have been much smaller than in the worldwide extinction that led to the elimination of the dinosaurs. As we saw, supernova explosions will produce high-speed cosmic rays. If enough cosmic rays impinge on the Earth's atmosphere, they could destroy our planet's protective ozone layer. Increased ultraviolet radiation could then reach the upper layers of the oceans, destroying the phytoplankton, microscopic plants found near the surface. This would create a famine for those species that feed on them.

It is well established that about 2 million years ago, there was an abrupt extinction of many bivalve species (creatures with two valves, like clams and oysters) in the tropical and temperate oceans. Astronomers are now refining their calculations to establish whether or not cosmic rays from a supernova at a distance of 120 LY (the location of the Scorpius-Centaurus cluster 2 million years ago) could reach the Earth in sufficient numbers to cause significant damage to the ozone layer.

A few clouds do exist within the Local Bubble. The Sun itself seems to have entered a cloud about 10,000 years ago. This cloud is warm (with a temperature of about 7500 K) and has a density of 0.1 hydrogen atom per cubic centimeter—higher than most of the Local Bubble but so tenuous that it is referred to as *Local Fluff*. (Aren't these astronomical names fun sometimes?) While this is a pretty thin cloud, it contributes 50 to 100 times more particles than the solar wind to the diffuse material between the planets in our solar system. These interstellar particles have been detected and their numbers counted by the spacecraft traveling between the planets. Perhaps someday

■ **FIGURE 19.15**

An Expanding Bubble A cluster of extremely hot stars is located at the center of this bubble-like expanding shell of gas. Winds of energetic particles, mainly hydrogen nuclei, are blowing away from the surfaces of these stars at speeds that are close to 4000 km/s. These outward-flowing streams of particles plow into stationary interstellar gas, building up the density around the outer rim and releasing energy in the form of hydrogen emission. The bubble is about 400 LY in diameter and is located in a neighbor galaxy called the Large Magellanic Cloud.

© Anglo-Australian Observatory/David Malin Images

scientists will devise a way to collect them without destroying them and to return them to Earth, and so that we can touch—or at least study in our laboratories—these messengers from distant stars.

SURFING THE WEB

 The Messier Catalog: www.seds.org/messier/
Brings together information, images, background, current research issues, and web links for each object in Messier's catalog of nebulae and galaxies.

Sources of Great Nebula Images on the Web:

- *Anglo-Australian Observatory Image Collection:* www.aao.gov.au/images.html

- *European Southern Observatory Images:* www.eso.org/outreach/gallery/

- *National Optical Astronomy Observatories Image Gallery:* www.noao.edu/image_gallery/

- *Hubble Space Telescope Images:* hubblesite.org/newscenter/archive

The Web Nebulae: www.seds.org/billa/twn
Another good site where you can look at images of and information about some of the well-known nebulae.

Interstellar Medium Online Tutorial: www-ssg.sr.unh.edu/tof/Outreach/Interstellar/index.html
?Interdepth.html
Questions and answers about the interstellar medium (ISM) and how we study it.

3-D Mapping of the Local Interstellar Medium: spacsun.rice.edu/~twg/lism.html
Interesting artistic "impressions" of the gas in our local neighborhood, including the Local Bubble, with some occasionally technical information.

Cosmic Rays Learning Site: helios.gsfc.nasa.gov/cosmic.html
Part of the "Cosmic and Heliospheric Learning Center" at NASA, this site explains about cosmic rays, keeps track of recent news items, and provides links.

19.1 About 20 percent of the visible matter in the Galaxy is in the form of gas and dust, which serve as the raw material for new stars. About 99 percent of this **interstellar matter** is in the form of gas—individual atoms or molecules. The most abundant elements in the interstellar gas are hydrogen and helium. About 1 percent of the interstellar matter is in the form of solid **interstellar dust grains.**

19.2 Interstellar gas may be hot or cold. Gas found near hot stars emits light by *fluorescence;* that is, light is emitted when an electron is captured by an ion and cascades down to lower energy levels. Glowing clouds (**nebulae**) of ionized hydrogen are called **H II regions** and have temperatures of about 10,000 K. Most hydrogen in interstellar space is not ionized and can best be studied by radio measurements of the 21-cm line. Some of the gas in interstellar space is at a temperature of a million degrees, even though it is far away from hot stars; this ultra-hot gas is probably heated when rapidly moving gas ejected in supernova explosions sweeps through space. Some of the gas inside large clouds is in the form of molecules; more than 120 different molecules have been found in space, including the basic building blocks of proteins, which in turn are fundamental to life as we know it here on Earth.

19.3 Interstellar dust can be detected: (1) when it blocks the light of stars behind it, (2) when it scatters the light from nearby stars, and (3) because it makes distant stars look both redder and fainter. These effects are called **reddening** and **interstellar extinction.** Dust can also be detected in the infrared because it emits heat radiation. Much of the dust is found in clouds called **infrared cirrus.** The dust particles are about the same size as the wavelength of light and consist of rocky cores that are either sootlike (carbon-rich) or sandlike (silicates) with mantles made of ices such as water, ammonia, and methane.

19.4 **Cosmic rays** are particles that travel through interstellar space at a typical speed of 90 percent the speed of light. The most abundant elements in cosmic rays are the nuclei of hydrogen and helium, but **positrons** are also found. It is likely that many cosmic rays are produced in supernova explosions.

19.5 Much of the interstellar matter in our Galaxy was ejected by dying stars either in the form of stellar winds or when the stars became supernovae. This material then expanded into space, in some cases forming giant *superbubbles* that can sweep up additional interstellar matter, eventually forming clouds where new stars can be born. Thus gas and dust are recycled through successive generations of stars. The Sun is located at the edge of a low-density cloud called the *Local Fluff.* The Sun and this cloud are located within the *Local Bubble,* a region extending to at least 300 LY from the Sun, within which the density of interstellar material is extremely low.

INTER-ACTIVITY

A The Sun is located in a region where the density of interstellar matter is low. Suppose that instead it were located in a dense cloud about 20 LY in diameter that dimmed the visible light from stars lying outside it by a factor of 100. Have your group discuss how this would have affected the development of civilization on Earth. For example, would it have presented a problem for early navigators?

B Your group members should look through the pictures in this chapter. How big are the nebulae you see in the images? Are there any clues either in the images or in the captions? Are the clouds they are part of significantly bigger than the nebulae we can see? How can we determine the sizes?

C How do the members of your group think astronomers are able to estimate the distances of such nebulae in our own Galaxy? (*Hint:* Look at the images. Can you see anything between us and the nebula in

some cases. Review Chapter 18, if you need to remind yourself about methods of measuring distances.)

D The text suggests that a tube of air extending from the surface of the Earth to the top of the atmosphere contains more atoms than a tube of the same diameter extending from the top of the atmosphere to the edge of the observable universe. Scientists often do what they call "back of the envelope calculations," in which they make very rough approximations just to see whether statements or ideas are true. Try doing such a "quick and dirty" estimate for this statement with your group. What are the steps in comparing the numbers of atoms contained in the two different tubes? What information do you need to make the approximations? Can you find it in this text? And is the statement true?

E If your astronomy course has involved learning about the solar system before you got to this chapter, have your group discuss where else besides interstellar clouds astronomers have been discovering organic molecules (the chemical building blocks of life). How might the discoveries of such molecules in our own solar system be related to the molecules in the clouds discussed in this chapter?

F Two stars both have a reddish appearance in telescopes. One star is actually red; the other's light has been reddened by interstellar dust on its way to us.

Have your group make a list of the observations you could perform to determine which star is which.

G You have been asked to give a talk to your little brother's middle school class on astronomy, and you decide to talk about how nature recycles gas and dust. Have your group discuss what images from this book you would use in your talk. In what order? What is the one big idea you would like the students to remember when the class is over?

REVIEW QUESTIONS

Ace⟲Astronomy™ Assess your understanding of this chapter's topics with additional quizzing and animations at **http://ace.brookscole.com/voyages**

1. Identify several dark nebulae in photographs in this book (don't restrict yourself to this chapter alone). Give the figure numbers of the photographs, and specify where the dark nebulae are to be found on them.
2. Why do nebulae near hot stars look red? Why do dust clouds near stars usually look blue?
3. Describe the characteristics of the various types of interstellar gas clouds.

4. Prepare a table listing the different ways in which (a) dust and (b) gas can be detected in interstellar space.
5. Describe how the 21-cm line of hydrogen is formed. Why is this line such an important tool for understanding the interstellar medium?
6. Describe the properties of the dust grains found in the space between stars.
7. Why is it difficult to determine where cosmic rays come from?
8. What causes reddening of starlight? Explain how the reddish color of the Sun's disk at sunset is caused by the same process.

THOUGHT QUESTIONS

9. Figure 19.1 shows a reddish glow around the star Antares, and yet the caption states that this is a dust cloud. What observations would you make to determine whether the red glow is actually produced by dust or whether it is produced by an H II region?

10. If the red glow around Antares is indeed produced by reflection of the light from Antares by dust, what does its red appearance tell you about the likely temperature of Antares? Look up the spectral type of Antares in Appendix 11. Was your estimate of the temperature about right? In most of the images in this chapter, a red glow is associated with ionized hydrogen. Would you expect to find an H II region around Antares? Explain your answer.

11. Even though neutral hydrogen is the most abundant element in interstellar matter, it was detected first with a radio telescope, not a visible-light telescope. Explain why. (The explanation given in Chapter 16 for the fact that hydrogen lines are not strong in stars of all temperatures may be helpful.)

12. The terms H II and H_2 are both pronounced "H two." What is the difference in meaning of these two terms? Suppose someone told you that she had discovered H II around

the star Aldebaran. Would you believe her? Can there be such a thing as H III? Why or why not?

13. Describe the spectrum of each of the following:
a. starlight reflected by dust
b. a star behind invisible interstellar gas
c. an emission nebula

14. According to the text, a star must be hotter than about 25,000 K to produce an H II region. Both the hottest white dwarfs and main-sequence O stars have temperatures hotter than 25,000 K. Which type of star can ionize more hydrogen? Why?

15. From the comments in the text about which kinds of stars produce emission nebulae and which kinds are associated with reflection nebulae, what can you say about the temperatures of the stars in the Pleiades? (See the image that begins this chapter.)

16. One way to calculate the size and shape of the Galaxy is to estimate the distances to faint stars from their observed apparent magnitudes, and to note the distance at which stars are no longer observable. The first astronomers to try this experiment did not know that starlight is

dimmed by interstellar dust. Their estimates of the size of the Galaxy were much too small. Explain why.

17. Short-wavelength light is absorbed more efficiently by the Earth's atmosphere than is long-wavelength light. Sunburns are caused primarily by sunlight with wavelengths between 280 and 320 nm. The heat that we feel, however, is produced mainly by infrared radiation. Use this information to explain in general terms why it is easier to get sunburned at noon than in the late afternoon, even though the Sun feels nearly as hot at, say, 4:00 P.M. as it does at noon.

(*Hint*: Draw diagrams showing the Earth with its atmosphere, and where the Sun is relative to you.)

18. New stars form in regions where the density of gas and dust is high. Suppose you wanted to search for some recently formed stars. Would you more likely be successful if you observed at visible wavelengths or at infrared wavelengths? Why?

19. In big cities, you can see much farther on days without smog. Why?

FIGURING FOR YOURSELF

You can estimate how much interstellar mass our Galaxy contains. All you need to know is how big the Galaxy is and what the average density is using the formula:

total mass = volume × density

You have to remember to use consistent units—say, meters and kilograms. We will assume that our Galaxy is shaped like a cylinder; the volume of a cylinder equals the area of its base times its height:

$$V = \pi r^2 h$$

where r is the radius of the cylinder and h is its height. Use the appendices to find the constants you need, like how many meters there are in a light year or what the mass of a hydrogen atom is, to answer the following question.

20. Suppose that the average density of hydrogen gas in our Galaxy is one atom per cubic centimeter. If the Galaxy is a cylinder with a diameter of 100,000 LY and a height of 200 LY, how many hydrogen atoms are in the interstellar gas? What is the mass of this quantity of hydrogen? Presumably you have calculated this number in units of kilograms or grams, and the number is so large it is meaningless. What is the ratio of this mass to the mass of the Sun? How does your answer compare with the value given in the text, which is based on much more detailed observations and calculations?

21. H II regions can exist only if there is a nearby star hot enough to ionize hydrogen. Hydrogen is ionized only by radiation with wavelengths shorter than 91.2 nm. What is the temperature of a star that emits its maximum energy at 91.2 nm? (Use Wien's law from Chapter 4.) Based on this result, what are the spectral types of those stars likely to provide enough energy to produce H II regions?

22. Dust was originally discovered because the stars in certain clusters seemed to be fainter than expected. Suppose a star is behind a cloud of dust that dims its brightness

by a factor of 100. Suppose you do not realize the dust is there. How much in error will your distance estimate be? Can you think of any measurement you might make to detect the dust?

Clouds of interstellar matter and the million-degree gas between them must be at approximately the same pressure. If the cloud pressure were higher, the cloud would expand until its pressure matched that of its environment. If, on the other hand, the pressure of the hot gas were greater than that of a cloud embedded in it, the hot gas would compress the cloud and force it to shrink until its pressure became high enough to resist further compression. Imagine a balloon as an analogy. If you add air to it, thereby increasing its pressure, the balloon will expand. If you squeeze it with your hands, you can make the balloon smaller because of the greater external pressure.

The pressure in a gas depends on both the temperature and the number of particles per cubic centimeter (its density):

$$P = nkT$$

where P is the pressure, n is the number of particles per unit volume (e.g., per cubic centimeter), and T is the temperature. You can see why this formula is plausible. If the gas particles are hotter, then their energy of motion is greater, and they collide harder with one another. If the density is greater, there are more collisions and more pressure. As an analogy, think about a crowded room. You can fit more people into a room when they are not moving around very much—say, at a formal cocktail party. But if they want to do the waltz, there will be many more collisions. Now let's apply these ideas to interstellar matter.

23. How would the density inside a cold cloud ($T \sim 10$ K) compare with the density of the ultra-hot interstellar gas ($T \sim 10^6$ K) if they were in perfect pressure equilibrium? (It takes a large cloud to be able to shield its interior from heating so that it can be at such a low temperature.) Which region do you think is more suitable for the creation of new stars? Why?

24. The text states that the Local Fluff, which surrounds the Sun, has a temperature of 7500 K and a density of 0.1 atom per cm^3. The Local Fluff is embedded in hot gas with a temperature of 10^6 K and a density of 0.005 atom per cm^3. Are they in equilibrium? What is likely to happen to the Local Fluff?

SUGGESTIONS FOR FURTHER READING

Bowyer, S. et al. "Observing a Partly Cloudy Universe" in *Sky & Telescope*, Dec. 1994, p. 36. Review of the Extreme Ultraviolet Explorer mission and what it has shown us about the ISM.

Friedlander, M. *A Thin Cosmic Rain: Particles from Outer Space.* 2000, Harvard U. Press. Good introduction to cosmic rays.

Goodman, A. "Recycling in the Universe" in *Sky & Telescope*, Nov. 2000, p. 44. Review of how stellar evolution, ISM, and supernovae all work together to recycle cosmic material.

Helfand, D. "Fleet Messengers from the Cosmos" in *Sky & Telescope*, Mar. 1988, p. 265. Excellent summary of history and current understanding of cosmic rays.

Knapp, G. "The Stuff Between the Stars" in *Sky & Telescope*, May 1995, p. 20.

Nadis, S. "Searching for the Molecules of Life in Space" in *Sky & Telescope*, Jan. 2002, p. 32. Recent observations of water in the ISM by satellite telescopes.

Reynolds, R. "The Gas Between the Stars" in *Scientific American*, Jan. 2002, p. 34. On the interstellar medium.

Sheehan, W. *The Immortal Fire Within: The Life and Work of E. E. Barnard.* 1995, Cambridge U. Press. An excellent biography.

Shore, L. and Shore, S. "The Chaotic Material Between the Stars" in *Astronomy,* June 1988, p. 6.

Teske, R. "The Star That Blew a Hole in Space" in *Astronomy,* Dec. 1993, p. 31. On the Local Bubble.

Trefil, J. "Discovering Cosmic Rays" in *Astronomy,* Jan. 2001, p. 36. Concise introduction to history and current understanding.

Verschuur, G. "Interstellar Molecules" in *Sky & Telescope,* Apr. 1992, p. 379.

Verschuur, G. "Barnard's Dark Dilemma" in *Astronomy,* Feb. 1989, p. 30.

Verschuur, G. *Interstellar Matters.* 1989, Springer-Verlag. Essays on dust and the interstellar medium. Combines good historical material and reviews of modern topics.

Whitman, A. "Seeking Summer's Dark Nebulae" in *Sky & Telescope,* Aug. 1998, p. 114. On finding some interstellar dust clouds with small telescopes.

Wynn-Williams, G. "Bubbles, Tunnels, Onions, Sheets: The Diffuse Interstellar Medium" in *Mercury* (the magazine of the Astronomical Society of the Pacific), Jan./Feb. 1993, p. 2.

Wolfgang Brandner (JPL/IPAC), Eva K. Grebel (U. of Washington), You-Hua Cu (U. of Illinois, Urbana-Champaign), & NASA

The Eagle Nebula (M16) Stars form in great clouds of gas and dust, and this image shows a large region of such cosmic raw material. The gas is visible because, about 2 million years ago, the cloud produced a cluster of bright stars, whose light ionizes the hydrogen gas nearby and causes it to glow. The cluster can be seen just above and to the right of the darker columns of dust at the center of the image. The dark columns or "elephant trunks" of material are seen in much more detail in Figures 20.1 and 20.2. This false-color image was created by combining images taken through filters that select lines of hydrogen alpha (green), oxygen (blue), and sulfur (red).

20

The Birth of Stars and the Discovery of Planets Outside the Solar System

All these illustrious worlds, and many more, Which by the tube astronomers explore . . . Are suns, are centers, whose superior sway, Planets of various magnitudes obey.

Sir Richard Blackmore in the poem *The Creation*, Book II, quoted by A. Meadows in *The High Firmament* (1969, Leicester University Press)

THINKING AHEAD

Do planets circle other stars, or is ours the only planetary system? Scientists and philosophers have speculated about the answer to this question for centuries. In the past decade, observational techniques have finally become good enough to allow us to answer the question directly. Astronomers have now detected many more planets outside our solar system than in it, and the number is growing apace. However, even before planets were detected, astronomers had predicted that planetary systems were likely to be frequent by-products of the star-formation process. In this chapter we look at how interstellar matter is transformed into stars and planets.

The formation of stars is not a topic of mere historical interest; star formation is a continuous process that is going on *right now*. Astronomers estimate that every year in our Galaxy on average 3 solar masses of interstellar gas and dust are converted into stars. This may sound like a small amount of mass, but 3 solar masses of material every year for, say, a billion years add up to a lot of stars.

As we begin our study of the birth of stars, it may be useful to remember some of the key things we've learned so far. Table 20.1 summarizes several important ideas about stars that were established in earlier chapters.

Ace✸Astronomy ™ The AceAstronomy icon throughout the text indicates an opportunity for you to test yourself on key concepts and to explore animations and interactions on the AceAstronomy website at **http://ace.brookscole.com/voyages**

Virtual Laboratories

 Extrasolar Planets

TABLE 20.1	*Basics About Stars from Earlier Chapters*

- Stable (main-sequence) stars such as the Sun maintain equilibrium by producing energy through nuclear fusion in their cores. The ability to generate energy by fusion defines a star. (Sections 15.2, 15.3)

- Each second in the Sun approximately 600 million tons of hydrogen undergo fusion into helium, with about 4 million tons turning to energy in the process. This rate of hydrogen use means that eventually the Sun (and all other stars) will run out of central fuel. (Section 15.2)

- Stars come with many different masses, ranging from 1/12 M_{Sun} to roughly 100 M_{Sun}. There are far more low-mass than high-mass stars. (Section 17.2)

- The most massive main-sequence stars (spectral type O) are also the most luminous and have the highest surface temperature. The lowest-mass stars on the main sequence (spectral type M or L) are the least luminous and the coolest. (Section 17.4)

- A galaxy of stars such as the Milky Way contains enormous amounts of gas and dust—enough to make billions of stars like the Sun. (Section 19.1)

20.1 STAR FORMATION

If we want to find stars still in the process of formation, we must look in places that have plenty of the raw material from which stars are assembled. Since stars are made of gas, we focus our attention (and our telescopes) on the dense clouds of gas that dot the Milky Way (see the opening image in this chapter and Figures 20.1 and 20.2).

20.1.1 Molecular Clouds: Stellar Nurseries

As we saw in Chapter 19, the most massive reservoirs of interstellar matter—and some of the most massive objects in the Milky Way Galaxy—are the **giant molecular clouds.** Their name reflects the coldness of their interiors; with characteristic temperatures of only 10–20 degrees Kelvin (K), most of their gas atoms are bound into molecules. These clouds are the birthplaces of most stars in our Galaxy.

The masses of molecular clouds range from a thousand times the mass of the Sun to about 3 million solar masses. Molecular clouds have a complex filamentary structure, much like cirrus clouds on Earth, but the molecular cloud

Jeff Hester and Paul Scowen, Arizona State University, and NASA

■ FIGURE 20.1

Pillars of Dust in M16 This Hubble Space Telescope image of the central regions of M16 shows huge columns of cool gas, including molecular hydrogen (H_2), and dust. These columns are of higher density than the surrounding regions and have resisted evaporation by the ultraviolet radiation from a cluster of hot stars just beyond the upper right corner of this image. The tallest pillar is about 1 LY long, and the M16 region is about 7000 LY away from us.

Jeff Hester and Paul Scowen, Arizona State University, and NASA

■ **FIGURE 20.2**

Dense Globules in M16 This close-up of one of the pillars in Figure 20.1 shows some very dense globules, many of which harbor embryonic stars. Astronomers coined the term *evaporating gas globules* for these structures, in part so they could say we found *e.g.g.s* inside the Eagle Nebula! It is possible that because these embryos or e.g.g.s are exposed to the relentless action of the radiation from nearby hot stars, some may not yet have collected enough material to make a star and may thus be "still-born."

filaments can be up to 1000 light years (LY) long. Within the clouds are high-density regions called *clumps*, where most of the star formation takes place.

The conditions in these clumps—low temperature and high density—are just what is required to make new stars. Remember what we learned in Chapter 15—the essence of the life story of any star is the ongoing competition between two forces: *gravity* and *pressure*. The force of gravity, pulling inward, tries to make a star collapse. Internal pressure produced by the motions of the gas atoms, pushing outward, tries to force the star to expand. When a star is first forming, low temperature (and hence low pressure) and high density (and hence greater gravitational attraction) both work to give gravity the advantage. In order to form a star—that is, a dense, hot ball of matter capable of starting nuclear reactions deep within—we need a clump of interstellar atoms and molecules to shrink in radius and increase in density by a factor of nearly 10^{20}. It is the force of gravity that produces this drastic collapse.

20.1.2 The Orion Molecular Cloud

The closest and best-studied stellar nursery is in the constellation of Orion, the hunter, about 1500 LY away (Figure 20.3a). This grouping of stars is easy to recognize by the conspicuous "belt" of three stars that mark the hunter's waist. The Orion molecular cloud is a truly impressive structure. In its long dimension it stretches over a distance of about 100 LY. The total quantity of molecular gas is about 200,000 times the mass of the Sun. Most of the cloud does not glow with visible light but betrays its presence by the radiation it gives off at infrared and radio wavelengths (Figure 20.3b).

A wave of star formation that began about 12 million years ago at one edge of this molecular cloud, near the western shoulder of the Orion star figure, has slowly moved through the cloud, leaving behind groups of newly formed stars. The stars in Orion's belt are about 8 million years old, and the stars near the middle of the "sword" hanging from Orion's belt are only 300,000 to 1 million years old. These stars can be seen easily with binoculars, and a small telescope shows the glowing gas that still surrounds them—called the Orion Nebula.

The region in the sword where star formation is still taking place is called the Orion Nebula cluster or sometimes the Trapezium cluster (Figure 20.4). About 2200 stars are found in a region only slightly larger than a dozen light years in diameter. Compare this with our own solar neighborhood, where the typical spacing between stars is about 3 LY. Only a small number of stars in the Orion cluster can

VISIBLE LIGHT

(a)

INFRARED

(b)

Infrared Processing and Analysis Center/JPL

■ **FIGURE 20.3**

Orion in Visible and Infrared (a) The Orion star group was named after the legendary hunter in Greek mythology. Three stars close together in a line mark Orion's belt. The ancients imagined a sword hanging from the belt; the object at the end of the blue line in this sword is the Orion Nebula. (b) This wide-angle infrared view of the same area was taken with the Infrared Astronomical Satellite. Heated dust clouds dominate in this false-color image, and many of the stars that stood out on part (a) are now invisible. An exception is the cool red-giant star Betelgeuse, which can be seen as a yellowish point at the left vertex of the blue triangle (at Orion's left armpit). The large yellow ring to the right of Betelgeuse is the remnant of an exploded star. The infrared image lets us see how large the Orion molecular cloud really is. The two colorful regions in Figure 20.4 are the two bright yellow splotches at the left end of and below Orion's belt. The lower one is the Orion Nebula and the higher one is the region of the horsehead dust nebula.

Anglo-Australian Observatory Board

■ **FIGURE 20.4**

Star-Forming Regions in Orion This wide-angle view includes the belt and sword of Orion. The Orion Nebula is the fuzzy white object near the bottom. A cloud of dust that resembles a horsehead can be seen against the reddish glow of hydrogen in the upper left. The bright, overexposed star above the horsehead and the one at the very top center of the image are two of the stars in Orion's belt. All the visible features of gas and dust are evidence of the much larger molecular cloud that lies behind them and that (at radio wavelengths) would fill most of the field shown here.

(a)

(b)

Anglo-Australian Observatory/David Malin Images

2MASS, IPAC & University of Massachusetts

■ **FIGURE 20.5**

The Central Region of the Orion Nebula The Orion Nebula harbors some of the youngest stars in the solar neighborhood. At the heart of the nebula is the Trapezium cluster, which includes four very bright stars that provide much of the energy that causes the nebula to glow so brightly. In these images, we see a section of the nebula in visible light (a) and infrared (b). The four bright stars in the center of the visible-light image are the Trapezium stars. Notice that most of the stars seen in the infrared are completely hidden by dust in the left image.

be seen with visible light, but infrared images—which penetrate the dust better—detect more than 2000 stars (Figure 20.5).

The typical clump within which star formation takes place contains only about one-thousandth the mass of the parent molecular cloud and only enough material to make a few hundred to a thousand stars. Individual stars form within their own smaller concentrations of material, which are called *cores*. Star formation is not a very efficient process. In Orion, less than 1 percent of the material in the cloud has been turned into stars. That is why we still see a substantial amount of gas and dust in the region of the Trapezium stars. The leftover material is eventually heated, either by the radiation and winds from the hot stars that form or by explosions of the most massive stars. (We will see in later chapters that the most massive stars go through their lives very quickly and end by exploding.) Whether gently or explosively, the material in the neighborhood of the new stars is blown away into interstellar space (as previewed in Chapter 19). Older groups or clusters of stars can therefore be easily observed in visible light because they are no longer shrouded in dust and gas (Figure 20.6). We will discuss these clusters in the next chapter.

Because of the correlation between stellar ages and position in the Orion region, we know that star formation has been moving progressively through this molecular cloud. Although we do not know what initially caused stars to begin forming in Orion, there is good evidence that the first generation of stars triggered the formation of addi-

tional stars, which in turn led to the formation of still more stars (Figure 20.7).

The basic idea is this: When a massive star is formed, it emits a large amount of ultraviolet radiation and also ejects high-speed particles in the form of a stellar wind. This injection of energy heats the surrounding gas in the molecular cloud and causes it to expand. When massive stars exhaust their supply of fuel, they explode, and the energy of the explosion also heats the gas. The hot gases pile into the surrounding cold cloud, compressing the material in it and increasing its density. If this increase in density is large enough, gravity will overcome pressure and stars will begin to form in the compressed gas. Such a chain reaction—where the brightest and hottest stars of one area become the cause of star formation "next door"—seems to have occurred not only in Orion but also in many other molecular clouds.

20.1.3 The Birth of a Star

Although regions such as Orion give us clues about how star formation begins, the subsequent stages are still shrouded in mystery (and a lot of dust). There is an enormous difference between the density of a molecular cloud core and the density of the youngest stars that can be detected. Direct observations of this collapse to higher density are nearly impossible for three reasons. First, the dust-shrouded interiors of molecular clouds where stellar births take place cannot be observed with visible light. It is only with the

T. A. Rector, B. A.Wolpa, M. Hanna, and NOAO/AURA/NSF

■ **FIGURE 20.6**

The Rosette Nebula A cluster of stars formed recently in the center of this nebula. Stellar winds and pressure produced by the radiation from these hot stars have blown the gas and dust away from the cluster so that the newly formed stars are easily seen in visible light. The nebula still contains many globules of dust. This star-formation region covers an area on the sky that is six times larger than the area covered by the full moon. The colors in this image are not what your eyes would see; our picture was produced by combining images taken in the emission lines of hydrogen alpha (red), oxygen (green), and sulfur (blue).

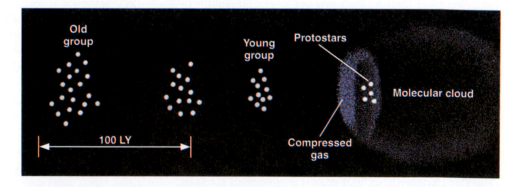

■ **FIGURE 20.7**

Propagating Star Formation This schematic diagram shows how star formation can move progressively through a molecular cloud. The oldest group of stars lies to the left of the diagram and has expanded because of the motions of individual stars. Eventually the stars in the group will disperse and no longer be recognizable as a cluster. The youngest group of stars lies to the right, next to the molecular cloud. This group of stars is only 1 to 2 million years old. The pressure of the hot, ionized gas surrounding these stars compresses the material in the nearby edge of the molecular cloud and initiates the gravitational collapse that will lead to the formation of more stars.

Based on drawings by F. Shu, F. Adams, and S. Lizano

■ FIGURE 20.8

The Formation of a Star (a) Dense cores form within a molecular cloud. (b) A proto-star with a surrounding disk of material forms at the center of a dense core, accumulating additional material from the molecular cloud through gravitational attraction. (c) A stellar wind breaks out but is confined by the disk to flow out along the two poles of the star. (d) Eventually this wind sweeps away the cloud material and halts the accumulation of additional material, and a newly formed star, surrounded by a disk, becomes observable. These sketches are not drawn to the same scale. The diameter of a typical accreting enve-lope is about 5000 AU. The typical diameter of the disk is about 100 AU or slightly larger than the diameter of the orbit of Pluto.

new techniques of infrared and millimeter radio astronomy that we are able to make any measurements at all. Second, the timescale for the initial collapse—thousands of years—is very short, astronomically speaking. Since each star spends such a tiny fraction of its life in this stage, relatively few stars are going through the collapse process at any given time. Third, the collapse of a new star occurs in a region so small (0.3 LY) that in most cases we cannot resolve it with existing techniques.

Nevertheless, through a combination of theoretical calculations and the limited observations available, as-tronomers have pieced together a picture of what the ear-liest stages of stellar evolution are likely to be.

The first step in the process of creating stars is the for-mation within a clump of gas and dust—through a process we do not fully understand—of the dense cores we dis-cussed earlier (Figure 20.8a). Each dense core then begins to build a star by attracting additional matter from the sur-rounding cloud material. Eventually, the gravitational force of the infalling gas becomes strong enough to overwhelm the pressure exerted by the cold material that forms the dense cores. The material then undergoes a rapid collapse, and the density of the core increases greatly as a result. During the time a dense core is contracting to become a true star, but before the fusion of protons to produce he-lium begins, we call the object a **protostar.**

The natural turbulence inside a clump tends to give any portion of it some initial spinning motion (even if it is

very slow). As a result, each collapsing core is expected to spin. According to the law of conservation of angular mo-mentum (discussed in Chapter 2), a rotating body spins more rapidly as it decreases in size. In other words, if the object can turn its material around a smaller circle, it can move that material more quickly—like a figure skater spin-ning more rapidly as she brings her arms in tight to her body. This is exactly what happens when a core contracts to form a protostar: As it shrinks, its rate of spin increases.

But all directions on a spinning sphere are not created equal. As the protostar rotates, it is much easier for mate-rial to fall right onto the poles (which spin most slowly) than onto the equator (where material moves around most rapidly). Therefore, gas and dust falling in toward the pro-tostar's equator are "held back" by the rotation and form a whirling extended disk around the equator (Figure 20.8b). You may have observed this same "equator effect" on the amusement park ride in which you stand with your back to a cylinder that is spun faster and faster. As you spin really fast, you are pushed against the wall so strongly that you can not possibly fall toward the center of the cylinder. Gas can, however, fall onto the protostar easily from directions away from the star's equator.

The protostar and disk at this stage are embedded in an envelope of dust and gas, from which material is still falling onto the protostar. This dusty envelope blocks visible light, but infrared radiation can get through. As a result, in this phase of its evolution, the protostar itself is observable only in the infrared region of the spectrum. Once almost all of the available material has been accreted and the central protostar has reached nearly its final mass, it is given a special name. It is now called a *T Tauri star*, named after one of the best studied and brightest members of this class of stars, which was discovered in the constellation of Taurus. (Astronomers have a tendency to name types of stars after the first example they discover or come to understand. It's not an elegant system, but it works for us.)

20.1.4 Winds and Jets

Observations show that T Tauri stars next go through a stage involving a powerful outflow of particles from the surface. This **stellar wind** consists mainly of protons (hydrogen nuclei) and electrons streaming away from the star at speeds of a few hundred kilometers per second (several hundred thousand miles per hour). When the wind first starts up, the disk of material around the star's equator blocks the wind in this direction. Where the wind particles *can* escape most effectively is in the direction of the star's poles (Figure 20.8c). Astronomers have actually seen evidence of these beams of particles shooting out in opposite directions from the polar regions of newly formed stars. In many cases, these beams point back to the location of a protostar that is still so completely shrouded in dust that we cannot yet see it.

Such double beams (or *jets*) of outflowing material are sometimes quite broad, but on other images we see a remarkably narrow set of jets (Figure 20.9). On occasion, the jets collide with a somewhat denser lump of material nearby, excite its atoms, and cause them to emit light. These glowing regions, called *Herbig–Haro* (or HH) *objects* after the two astronomers who first identified them, allow us to trace the progress of the jet to a distance of a light year or more from the star that produced it. Figure 20.10 shows two spectacular images of Herbig–Haro objects taken with the Hubble Space Telescope. As always happens in the real world, the images also show a remarkable amount of complex structure in the jets and the clouds of material they energize.

The wind from a forming star will ultimately sweep away the material that remains in the obscuring envelope of dust and gas, leaving behind the naked disk and protostar, which can now be seen with visible light (Figure 20.8d). We should note that at this point, the protostar itself is still contracting slowly and has not yet reached the *main-sequence* stage on the H–R diagram (a concept introduced in Chapter 17). The disk can be detected directly when observed at infrared wavelengths or when it is seen silhouetted against a bright background (Figure 20.11).

This description of a protostar surrounded by a rotating disk of gas and dust sounds very much like what happened when the Sun and planets formed. Indeed, one of the most important discoveries from the study of star formation in the last decade of the 20th century was that disks are an inevitable by-product of the process of creating stars. The next question that astronomers set out to answer was: Do the disks around protostars also form planets? We will return to this question at the end of the chapter.

To keep life simple, we have described the formation of single stars. Many stars are members of binary or triple systems, where several stars are born together. In this case, the stars form in nearly the same way. Widely separated binaries may each have their own disk; close binaries may share a single disk (see Figure 20.11).

20.2 THE H-R DIAGRAM AND THE STUDY OF STELLAR EVOLUTION

One of the best ways to summarize all of these details about how a star or protostar changes with time is to use a Hertzsprung–Russell (H–R) diagram. As a star goes through the stages of its life, its luminosity and temperature change. Thus its position on the H–R diagram, in which luminosity is plotted against temperature, also changes. As a star ages, we must replot it in different places on the diagram. Therefore astronomers often speak of a star *moving* on the H–R diagram, or of its evolution tracing out a path on the diagram. Of course, in this context, "tracing out a path" has nothing to do with the star's motion through space; this is just a shorthand way of saying that its temperature and luminosity change as it evolves. (We can also think of people "moving" in the plot of height versus weight shown in Figure 17.11. How would you expect the position of the point that represents the height and weight of a person to change from the time the person is born to the time she is very old?)

To estimate just how much the luminosity and temperature of a star change as it ages, we must resort to calculations. Theorists compute a series of *models* for a star, with each successive model representing a later point in time. Stars may change for a variety of reasons. Protostars, for example, change in size because they are contracting, and their temperature and luminosity change as they do so. After nuclear fusion begins in the star's core (see Chapter 15), main-sequence stars change because they are using up their nuclear fuel.

Given a model that represents a star at one stage of its evolution, we can calculate what it will be like at a slightly later time. At each step, the model predicts the luminosity and size of the star, and from these values we can figure out its surface temperature. A series of points on an H–R diagram, calculated in this way, allows us to follow the life changes of a star and hence is called its *evolutionary track*.

20.2.1 Evolutionary Tracks

Let's now use these ideas to follow the evolution of protostars that are on their way to becoming main-sequence stars. The evolutionary tracks of newly forming stars with a range

The vertical text on the right side of the image reads: A. Watson, K. Stapelfeldt, J. Krist, and C. Burrows & NASA

■ **FIGURE 20.9**

Hubble Images of Gas Jets Flowing Away from a Protostar   Here we see
a protostar, known to us as HH 30 because it is a Herbig–Haro object. The star is about
450 LY away and only about 1 million years old. Light from the star itself is blocked by
a disk, which is larger than 60 billion km in diameter and is seen almost edge-on. Thinner
material above and below the central part of the disk reflects light toward us. Jets are seen
emerging in opposite directions and perpendicular to the disk. The material in these jets
is flowing outward at speeds up to 960,000 km/h. The series of three images shows
changes during a period of six years. Every few months a compact clump of gas is
ejected, and its motion outward can be followed. The changes in the brightness of the
disk may be due to motions of clouds within the disk that alternately block some of the
light and then let it through. This image corresponds to the stage in the life of a protostar
shown in Figure 20.8c.

J. Morse & NASA

HH47

C. Burrows, J. Morse, and J. Hester & NASA

■ **FIGURE 20.10**

Outflows from Protostars These images were taken with the Hubble Space Telescope and show jets flowing outward from newly formed stars. In the top image of HH 47, a protostar 1500 LY away (invisible inside a dusty disk at the left edge of the image) produces a very complicated jet. The star may actually be wobbling, perhaps because it has a companion. Light from the star illuminates the white region at the left because light can emerge perpendicular to the disk (just as the jet does). At right the jet is plowing into existing clumps of interstellar gas, producing a shock wave that resembles an arrowhead. The white bar is 1000 AU (1000 times the distance between Earth and Sun). The bottom image (HH 1 and 2) shows a classic double-beam jet emanating from a protostar (hidden in a dust disk in the center) in the constellation of Orion. Tip to tip, these jets are more than 1 LY long. The bright regions (first identified by Herbig and Haro) are places where the jet is slamming into a clump of interstellar gas.

of stellar masses are shown in Figure 20.12. These young stellar objects are not yet producing energy by nuclear reactions, but they derive their energy from gravitational contraction—by the sort of process proposed for the Sun by Helmholtz and Kelvin in the last century (see Chapter 15).

Initially, a protostar remains fairly cool, with a very large radius and a very low density. It is transparent to infrared radiation, and the heat generated by gravitational contraction can be radiated away freely into space. Because heat builds up slowly inside the protostar, the gas pressure remains low and the outer layers fall almost unhindered toward the center. Thus the protostar undergoes very rapid collapse, a stage that corresponds to the roughly vertical lines at the right of Figure 20.12. As the star shrinks, its surface area gets smaller, and so its total luminosity decreases.

The rapid contraction stops only when the protostar becomes dense and opaque enough to trap the heat released by gravitational contraction.

When the star begins to retain its heat, the contraction becomes much slower, and changes inside the contracting star keep the luminosity of stars like our Sun roughly constant. The surface temperature starts to build up and the star "moves" to the left in the H–R diagram. Stars first become visible only after the stellar wind described earlier clears away the surrounding dust and gas. This can happen during the rapid-contraction phase for low-mass stars, but high-mass stars remain shrouded in dust until they reach the **main sequence** (see the dashed line in Figure 20.12).

To help you keep track of the various stages that stars go through in their lives, it can be useful to compare the de-

CoKu Tau1

500 AU

DG Tau B

Haro 6-5B

IRAS 04016+2610

IRAS 04248+2612

IRAS 04302+2247

D. Padgett, W. Brandner, and K. Stapelfeldt; IPAC/Caltech/JPL & NASA

■ **FIGURE 20.11**

Disks Around Protostars   These Hubble Space Telescope infrared images show disks around young stars in the constellation of Taurus, in a region about 450 LY away. In some cases, we can see the central star (or stars—some are binaries). In other cases, the dark horizontal bands indicate regions where the dust disk is so thick that even infrared radiation from the star embedded within it cannot make its way through. The bright glowing regions are starlight reflected from the upper and lower surfaces of the disk, which are less dense than the central regions.

velopment of a star to that of a human being. (Clearly, you will not find an exact correspondence, but thinking through the stages in human terms may help you remember some of the ideas we are trying to emphasize.) Protostars might be compared to human embryos—as yet unable to sustain themselves but drawing resources from their environment as they grow. Just as the birth of a child is the moment it is called upon to produce its own energy (through eating and breathing), so astronomers say that a star is born when it is able to sustain itself through nuclear reactions.

When the star's central temperature becomes high enough (about 10 million K) to fuse hydrogen into helium, we say that the star has *reached the main sequence* on the H–R diagram. (Astronomers call such a star a main-sequence star even when the discussion does not concern the H–R diagram.) It is now a full-fledged star, more or less in equilibrium, and its rate of change slows dramatically. Only the

gradual depletion of hydrogen as it is transformed into helium in the core slowly changes the star's properties.

The mass of a star determines exactly where it falls on the main sequence. As Figure 20.12 shows, massive stars on the main sequence have high temperatures and high luminosities. Low-mass stars have low temperatures and low luminosities.

Objects of extremely low mass never achieve high enough central temperatures to ignite nuclear reactions. The lower end of the main sequence stops where stars have a mass just barely great enough to sustain nuclear reactions at a sufficient rate to stop gravitational contraction. This critical mass is calculated to be about 0.072 times the mass of the Sun. As we discussed in Chapter 17, objects below this critical mass are called either brown dwarfs or planets. At the other extreme, the upper end of the main sequence terminates at the point where the energy radiated by the

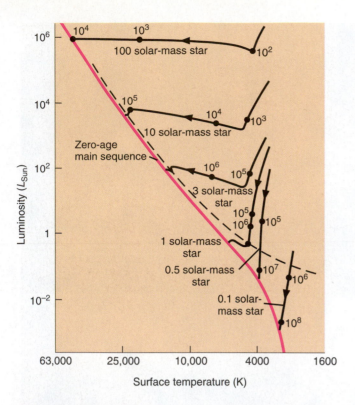

■ FIGURE 20.12

Evolutionary Tracks for Contracting Protostars
Tracks are plotted on the H–R diagram to show how stars of different masses change during the early parts of their lives. The number next to each dark point on a track is the rough number of years it takes an embryo star to reach that stage. You can see that the more mass a star has, the shorter time it takes to go through each stage. Stars that lie above the dashed line are typically still surrounded by infalling material and are hidden by it.

Ace◉Astronomy™ Log into AceAstronomy and select this chapter to see the Active Figure called "Animated HR Diagram."

newly forming massive star becomes so great that it halts the accretion of additional matter. The upper limit of stellar mass is between 100 and 200 solar masses.

20.2.2 Evolutionary Timescales

How long it takes a star to form depends on its mass. The numbers that label the points on each track in Figure 20.12 are the times, in years, required for the embryo stars to reach the stages we have been discussing. Stars of mass much higher than the Sun's reach the main sequence in a few thousand to a million years. The Sun required millions of years before it was born. Tens of millions of years are required for stars of lower mass to evolve to the lower main sequence. (We will see that, in fact, massive stars go through *all* stages of evolution faster than low-mass stars do.)

We will take up the subsequent stages in the life of a star in the next chapter, examining what happens after stars arrive on the main sequence and begin a "prolonged ado-

lescence" of fusing hydrogen to form helium. But now we want to examine the connection between the formation of stars and planets.

20.3 EVIDENCE THAT PLANETS FORM AROUND OTHER STARS

Having developed on one and finding it essential to our existence, we have a special interest in *planets*. Yet planets outside the solar system are extremely difficult to detect. Recall that we see planets in our own system only because they reflect sunlight and are close by. When we look to the other stars, we find that the amount of light a planet reflects is a depressingly tiny fraction of the light its star gives off. Furthermore, from a distance, planets are lost in the glare of their much brighter parent stars.

You might compare a planet orbiting a star to a mosquito flying around one of those giant spotlights at a shopping center opening. From close up, you might spot the mosquito, especially if it has an oily reflective belly. But now imagine viewing the scene from some distance away—say, from an airplane. You could see the spotlight just fine, but what are your chances of catching the light reflected from the mosquito?

Even though astronomers have not yet succeeded in *seeing* a planet around another star, they have detected planets elsewhere *indirectly*—by studying the disks of material from which planets might be condensing and by observing a planet's effects on its parent star.

20.3.1 Disks Around Protostars: Planetary Systems in Formation?

It is a lot easier to detect the spread-out raw material from which planets might be assembled than to detect planets after they are fully formed. From our study of the solar system, we understand that planets form by the gathering together of gas and dust particles in orbit around a newly created star. Each dust particle is heated by the young protostar and radiates in the infrared region of the spectrum. Before any planets form, we can detect such radiation from all of the individual dust particles that are destined to become parts of planets. We can also detect the silhouette of the disk if it blocks bright light coming from a source behind it (Figure 20.13).

However, once the dust particles gather together and form a few planets (and maybe some moons), the overwhelming majority of the dust is hidden in the interiors of the planets where we cannot see it. All we can now detect is the radiation from the outside surfaces, which cover a drastically smaller area than the huge dusty disk from which they formed. The amount of infrared radiation is therefore great-

FIGURE 20.13

Disks Around Protostars   These Hubble Space Telescope images show four disks around young stars in the Orion Nebula. The dark, dusty disks are seen silhouetted against the bright backdrop of the glowing gas in the nebula. The size of each image is about 30 times the diameter of our planetary system; this means the disks we see here range in size from two to eight times the orbit of Pluto. The red glow at the center of each disk is a young star, no more than a million years old. These images correspond to the stage in the life of a protostar shown in Figure 20.8d.

M. McCaughrean and C. R. O'Dell & NASA

est before the dust particles combine into planets. For this reason, our search for planets begins with a search for radiation from the material required to make them.

A disk of gas and dust appears to be an essential part of star formation. Observations show that nearly all very young protostars have disks and that the disks range in size from 10 to 1000 astronomical units, or AU. (For comparison, the average diameter of the orbit of Pluto—that is, the rough size of our own planetary system—is 80 AU.) The mass contained in these disks is typically 1 to 10 percent of the mass of our own Sun, which is more than the mass of all the planets in our solar system put together. Observations therefore demonstrate that a large fraction of stars begin their lives with enough material in the right place to form a planetary system.

20.3.2   The Time Required to Form Planets

We can use observations of how the disks change with time to estimate how long it might take for planets to form. If we measure the temperature and luminosity of a protostar, then we can place it in an H–R diagram like the one shown in Figure 20.12. By comparing the real star with our models of how protostars should evolve with time, we can estimate its age. We can then look at how the disks we observe change with the ages of the stars that they surround.

What the observations show is that if a protostar is less than about 1 to 3 million years old, its disk extends all the way

from very close to the surface of the star out to tens or hundreds of astronomical units away. In older stars, we find disks with outer parts that still contain large amounts of dust, but the inner regions have lost most of their dust. In these objects, the disk looks like a donut with the protostar centered in its hole. The inner dense parts of most disks have disappeared by the time the stars are 10 million years old.

Calculations show that the formation of one or more planets could produce such a donut-like distribution of dust. Suppose a planet forms a few astronomical units away from the protostar, presumably due to the gathering together of matter from the disk. As the planet grows in mass, the process clears out a dust-free region in its immediate neighborhood. Calculations also show that any small dust particles and gas that were initially located in the region between the protostar and the planet, and that are not swept up by the planet, will then fall onto the star very quickly—in about 50,000 years.

Matter lying outside the planet's orbit, in contrast, is prevented from moving into the hole by the gravitational forces exerted by the planet. (We saw something similar in Saturn's rings, where the action of the shepherd moons keeps the material near the edge of the rings from spreading out.) If the formation of a planet is indeed what produces and sustains the holes in the disks that surround very young stars, then planets must form in 3 to 30 million years. This is a short period compared with the lifetimes of most stars and shows that the formation of planets may be a quick by-product of the birth of stars.

20.3.3 Debris Disks and Shepherd Planets

The dust around newly formed stars is gradually either incorporated into the growing planets in the newly forming solar system or ejected through gravitational interactions with the planets into space. The dust will disappear after about 30 million years unless the disk is continually supplied with new material. Comets and asteroids are the source of new dust. As the planet-sized bodies grow, they stir up the orbits of smaller objects. These small bodies collide at high speeds, shatter, and produce tiny particles of silicate dust and ices that keep the disk supplied with the debris from these collisions.

Over several hundred million years, the comets and asteroids will gradually be reduced in number, the frequency of collisions will go down, and the supply of dust will diminish. Remember that the heavy bombardment in the early solar system ended when the Sun was only about 500 million years old. Observations show that the dusty debris disks around other stars also become undetectable by the time the stars reach an age of 400 to 500 million years. It is likely, however, that some small amount of cometary material will remain in orbit, much like our Kuiper belt, which is a flattened disk of comets outside the orbit of Neptune in the solar system.

These debris disks provide evidence for the existence of planets outside the solar system. Although we cannot see the planets themselves, we can see their effects on the dusty debris. Just as the moons of Saturn shepherd the particles in the rings, so too newly formed planets can concentrate the dust particles into clumps and arcs.

Debris disks have now been found around nearly a dozen stars, including one called HR 4796A, which is estimated to be about 10 million years old (Figure 20.14). The narrow ring of dust is about 70 AU from its star but only about 17 AU wide. In some stars, the brightness of the rings varies with position; around other stars, there are bright arcs and gaps in the rings. The brightness indicates the relative concentration of dust, since what we are seeing is infrared (heat radiation) from the dust particles in the rings. More dust means more radiation.

An example of a clumpy disk is the one around the main-sequence star Epsilon Eridani, which, at a distance of only 10 LY, is one of the stars closest to us. Epsilon Eridani is slightly cooler than the Sun (its spectral type is K2 V), and it is only about one-tenth as old as our star. Observations at millimeter wavelengths show that Epsilon Eridani is surrounded by a donut-shaped ring of dust that contains some bright patches (Figure 20.15). The diameter of the disk around Epsilon Eridani is about the same as the diameter of the Kuiper belt.

The bright spots in the ring around Epsilon Eridani might be warmer dust trapped around a planet that formed inside the donut. Alternatively, they could be a concentration of dust brought together by the gravitational influence of a planet orbiting just inside the ring. Inside the disk, there

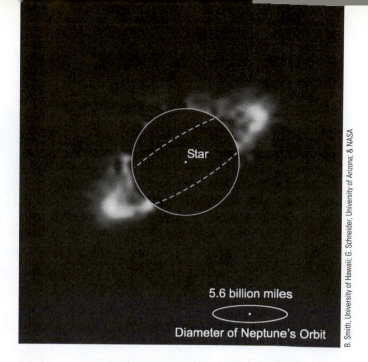

B. Smith, University of Hawaii; G. Schneider, University of Arizona; & NASA

■ **FIGURE 20.14**

Dust Ring Around a Young Star This near-infrared image from the Hubble Space Telescope shows a narrow ring of dust around the very young star HR 4796A, which lies about 220 LY away in the constellation of Centaurus. The ring is very narrow, spanning the same distance that separates Mars from Uranus in our solar system. (The ring, however, is much farther from its star, lying at what would be about twice the distance of Pluto from our Sun.) The image was taken with a *coronagraph*, a device that covers the bright star itself and allows faint structures around it to be recorded.

is very little material, probably because this region has been swept clean by planets that have formed from the dust.

Several groups have made observations that show a planet that orbits Epsilon Eridani at a distance of about 3.3 AU (a distance that would be between Mars and Jupiter in our solar system and much closer to the star than the dust ring). This inner planet has a mass that is at least 0.9 Jupiter mass. It is one of about a hundred such planets that have been found around other stars, as we will discuss in the next section.[1] This planet, however, is too far away from the ring to explain its structure.

To explain the structure of its ring of dust, theorists have calculated that Epsilon Eridani must have a planet with a mass about 0.2 times the mass of Jupiter orbiting the star at a distance of about 55–65 AU. If this prediction is right, then astronomers should see the bright spot revolve around the star at a rate of slightly more than 1° per year. This planet could not be detected with the planet-finding techniques

[1] The reason this planet is harder to confirm than some of the others we will be discussing in a moment is that Epsilon Eridani is a young, rapidly rotating star with a strong magnetic field. It has active starspots (like the sunspots on our own star) and its spectrum is more difficult to interpret as a result.

Declination

-9:27:00
26:50
10
20
30
40
50
28:00
10

3:32:50 58   57   56   55   54   53
Right ascension

Size of
Pluto's orbit

J. Greaves, et al., Joint Astronomy Centre

■ **FIGURE 20.15**

The Disk Around Epsilon Eridani   This image was taken at millimeter wavelengths (using the 15-m Maxwell telescope at the Mauna Kea Observatory in Hawaii) and shows a disk of dust surrounding the nearby star Epsilon Eridani. The bright knot at the 8 o'clock position in the disk might indicate the presence of a planet. The white star indicates where the star is located; such a star is not very bright at these wavelengths. The ring lies about 60 AU from the star, which is estimated to be about half a billion to a billion years old.

now in use (see the next section), but the predicted change in position of the bright spot is large enough to be measurable within a few years, and so this prediction is testable.

The second brightest star in the northern hemisphere, Vega, also has a disk, which has one clump of dust about 60 AU to the southwest of the star and a second about 75 AU to the northeast. Calculations show that both clumps could be explained if Vega is circled by a planet with a mass several times larger than that of Jupiter and a semimajor axis of about 30 AU. Someday astronomers may develop the techniques needed to image this planet if it is really there.

20.4 PLANETS BEYOND THE SOLAR SYSTEM: SEARCH AND DISCOVERY

Astronomers have tried a variety of techniques to search for evidence of planets around mature stars that have lost their disks, with dramatic recent success. Since planets are so difficult to image, the kinds of experiments possible with current technology look for the effects of planets *on their stars*. In particular, astronomers are measuring subtle changes in the stars' motion through space caused by the pull of surrounding planets.

20.4.1 Search for Orbital Motion

To understand how this approach works, consider a single Jupiter-like planet in orbit about a star. Both the planet and the star in such systems actually revolve about their *common center of mass*. Remember from our discussion in Chapter 2 that gravity is a mutual attraction. The star and the planet each exert a force on the other, and we can find a stable point between them about which both objects move. The smaller the mass of a body in such a system, the larger the orbit. So a massive star barely swings around the center of mass, while a low-mass planet makes a much larger "tour."

Suppose the planet is like Jupiter and has a mass about one-thousandth that of the star; in this case, the size of the star's orbit is one-thousandth the size of the planet's. To get a sense of how difficult observing such motion might be, let's see how hard Jupiter would be to detect from the distance of a nearby star. Consider alien astronomers trying to observe our own system from Alpha Centauri, the closest star system to our own (about 4.3 LY away). Even if they had the best telescopes we presently use, these alien astronomers could not detect faint Jupiter directly. But there are two ways they could try to detect the orbital motion of the Sun. One way is to look for changes in its position on the sky. The second is to use the Doppler effect to look for changes in its velocity.

The diameter of Jupiter's apparent orbit viewed from Alpha Centauri is 10 seconds of arc (arcsec) and the diameter of the Sun's orbit is 0.010 arcsec. (A second of arc is 1/3600 degree.) If the aliens could measure the apparent position of the Sun (which is bright and easy to detect) to sufficient precision, they would see it describe an orbit of diameter 0.010 arcsec with a period equal to that of Jupiter, which is 12 years. In other words, if they watched the Sun for 12 years, they would see it wiggle back and forth in the sky by this minuscule fraction of a degree. From the observed motion and the period of the "wiggle," they could deduce the mass of Jupiter and its distance using Kepler's laws.

Measuring positions in the sky this accurately is extremely difficult, and so far Earth-based astronomers have not made any confirmed detections of planets using this technique. New NASA missions are being planned, however, which will someday make such measurements from space.

As the star and planet orbit each other, part of their motion will be in our line of sight (i.e., toward us or away from us). Such motion (as discussed in Chapter 4) can be measured using the *Doppler effect* and the star's spectrum. As the star moves back and forth in orbit around the system's center of mass in response to the gravitational tug of an orbiting planet, the lines in its spectrum will shift back and forth (Figure 20.16).

Let's again consider the example of the Sun. Its *radial velocity* (motion toward or away from us) changes by about 13 meters/second (m/s) with a period of 12 years because of the gravitational pull of Jupiter. (The change becomes slightly more, 15 m/s, if the effects of Saturn are also included.) This corresponds to about 30 miles per hour —

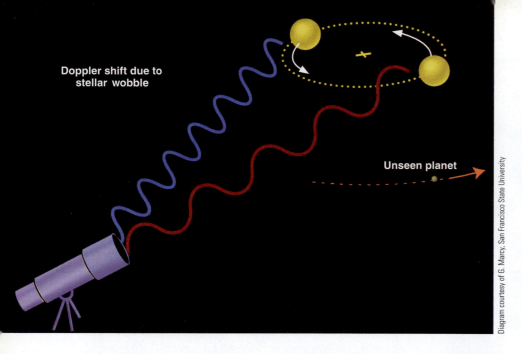

Doppler shift due to stellar wobble

Unseen planet

Diagram courtesy of G. Marcy, San Francisco State University

■ **FIGURE 20.16**

Doppler Method of Detecting Planets
The motion of a star and a planetary companion around a common center of mass can be detected as a cyclical change in the Doppler shift of the star. When the star is moving away from us, the lines in its spectrum show a tiny redshift; when it is moving toward us, they show a tiny blueshift. The change in color (wavelength) has been exaggerated here for educational purposes. In reality, the Doppler shifts we measure are extremely small and require sophisticated equipment to be detected.

roughly the speed at which many of us drive around town. Detecting motion at this level in a star's spectrum presents an enormous technical challenge, but several groups of astronomers around the world, using specialized spectrographs designed for this purpose, have recently succeeded.

20.4.2 The Discovery of Planets

Michel Mayor and Didier Queloz of the Geneva Observatory used precise Doppler measurements of velocity to discover the first planet in 1995 (Figure 20.17). This planet orbits a star resembling our Sun called 51 Pegasi, about 40 LY away. (The star can be found in the sky near the great square of Pegasus, the flying horse of Greek mythology, one of the easiest-to-find star patterns.) To everyone's surprise, the planet takes a mere 4.2 days to orbit the star. Remember that Mercury, the innermost planet in our solar system, takes 88 days to go around once.

The planet must be very close to 51 Pegasi, circling it about 7 million km away (Figure 20.18). At that distance, the energy of the star should heat the planet's surface to a temperature of a few thousand degrees C (a bit hot for future tourism). From the planet's motion, astronomers calculate that it has at least half the mass of Jupiter, making it clearly a jovian and not a terrestrial planet.

Since that initial discovery, the rate of progress has been breathtaking. As we write this paragraph, over 140 giant planets have been discovered orbiting other stars. Most of the discoveries are due to Geoffrey Marcy, Paul Butler, Steven Vogt, and their collaborators, primarily based on observations with the 3-meter telescope at the Lick Observatory in California and the 10-meter Keck telescope in Hawaii (Figure 20.19). There is about one new discovery each month, and so we recommend that you refer to the Web sites listed at the end of this chapter for up-to-date information. But stop for a minute and think about what we are discussing. Less than a decade after the first discovery,

M. Mayor and D. Queloz

■ **FIGURE 20.17**

European Planet-Hunting Team Michel Mayor and Didier Queloz of the Geneva Observatory were the first to discover a giant planet around a Sun-like star. They are continuing their work using telescopes in Europe and in Chile and have found several other planets since 1995.

we now know fifteen times more planets outside our solar system than within it!

Many of the initial planets discovered are massive (Jupiter-mass or larger) planets orbiting close to their stars—astronomers have called these "hot Jupiters." The existence of giant planets so close to their stars was a surprise, and these observations have forced us to reassess our ideas about how planetary systems form (something we'll discuss in a moment). But for now, bear in mind that the Doppler-shift method—which relies on the pull of a planet making its star "wiggle" back and forth around the center of mass—is most effective at finding planets that are close to their stars and massive. These, after all, cause the biggest "wiggles" in the motion of their stars and take the shortest

FIGURE 20.18

Giant Planet Close to a Sun-Like Star This artist's concept shows what the giant planet discovered orbiting the star 51 Pegasi might look like close up. As explained in the chapter, this planet was the first of more than a dozen jovian planets found around other stars whose orbits turned out to be smaller than the orbit of Mercury in our own system. The planet around 51 Pegasi is only about 7 million km from its star and takes only 4.2 days to complete its orbit. At that distance, the surface temperature of the planet would be so hot that (as astronomer Geoff Marcy has quipped) "a roast chicken might be done in a microsecond." The artist has shown prominences and sunspots on 51 Pegasi, evidence of an active atmosphere that might extend a significant way to the giant planet. She also shows the planet with bands like Jupiter, although our measurements can allow us to estimate only the mass of the planet, not its density, and thus we have no idea what sorts of materials the planet is made of.

Painting by Lynette Cook

time to complete a full orbit. So if such planets exist, we would expect to be finding them first. (Scientists call this a *selection effect*—where our technique of discovery selects certain kinds of objects as "easy finds"). As we spend more time watching target stars and as our ability to measure smaller Doppler shifts improves, this technique can reveal more distant and less massive planets, too. It is only very recently that astronomers have found a handful of solar systems like our own—that is, systems with Jupiter-sized planets taking a year or longer to orbit their stars.

Table 20.2 summarizes the characteristics of the 110 planets discovered through March 2004; some of these systems are shown on Figure 20.20. (If the terms describing orbits in the table are not familiar, see Chapter 2.) The masses of the planets in Figure 20.20 range from about a tenth the mass of Jupiter to more than 13 Jupiter masses. By the end of 2004, astronomers had discovered even more planets, including three with masses like Neptune's (only about 0.05 Jupiter masses or 18 Earth masses). We note that all of the masses we list are minimum masses. To determine the exact mass using the Doppler shift and Kepler's laws, we must also know the angle at which the planet's orbit is oriented to our view—something we don't have an independent way of knowing in most cases. Still, if the minimum mass is as large as the ones listed in the table, we

San Francisco State University

FIGURE 20.19

American Planet-Hunting Team Paul Butler and Geoff Marcy were both at San Francisco State University when they confirmed the discovery of the planet around 51 Pegasi and went on to discover most of the planets that have been found around other stars so far. They later separated, with Butler going to the Anglo-Australian Observatory to look for planets in the Southern Hemisphere and Marcy continuing the search in the Northern Hemisphere.

Masses and Orbital Radii of Extrasolar Planets Known So Far

Star

HD83443	0.34 M_J
HD46375	0.25 M_J
HD187123	0.53 M_J
HD179949	0.92 M_J
HD209458	0.83 M_J
BD-103166	0.48 M_J
HD75289	0.45 M_J
TauBoo	4.7 M_J
HD75289	0.79 M_J
51Peg	0.45 M_J
HD19874	0.77 M_J
Ups And	0.88 M_J, 1.9 M_J, 4.2 M_J
HD188748	0.24 M_J
HD88988	7.8 M_J
HD182020	7.3 M_J
HD217107	7.2 M_J
HD130322	1.1 M_J
GJ86	4.2 M_J
55 Cancrf	0.8 & 0.3 M_J, 4 M_J
HD38529	0.79 M_J, 12.7 M_J
HD195019	3.5 M_J
HD8434	0.48 M_J
GJ876	0.6 & 1.9 M_J
BD-CCRB	0.99 M_J
HD741580	1.5 M_J
HD188443	7.8 M_J, 17 M_J
HD121504	0.89 M_J
HD178911	8.4 M_J
HD18141	0.22 M_J
HD114782	10 M_J
HD80808	3.4 M_J
70Vir	7.4 M_J
HD52265	1.1 M_J
HD1237	3.4 M_J
HD37124	0.88 M_J, 1.00 M_J
HD73526	2.8 M_J
HD82943	0.88 M_J, 1.6 M_J
HD8574	2.2 M_J
HD202206	14 M_J
HD150706	0.95 M_J
HD10979	3.7 M_J
HD134967	7.8 M_J
HD12661	2.2 M_J, 1.54 M_J
HD189830	2.9 M_J
HD89744	7.1 M_J
HD17051	2.1 M_J
HD92788	3.8 M_J
HD142	1.3 M_J
HD128311	2.8 M_J
HD28185	5.8 M_J
HD108874	1.8 M_J
HD1203	1.8 M_J
HD177630	1.2 M_J
HD210277	1.2 M_J
HD27442	1.9 M_J
HD829430	1.8 M_J
HD114783	0.99 M_J
HD147513	7.0 M_J
HD20367	7.7 M_J
HD75458	8.8 M_J
HD222582	5.7 M_J
HD23079	2.7 M_J
HD141937	9.6 M_J
HD180691	1.7 M_J
HD114388	1.0 M_J
HD1208	0.80 M_J
16 Ccrb	1.8 M_J
HD213240	4.4 M_J
HD114729	0.87 M_J
47 UMS	2.5 M_J, 0.76 M_J
HD10697	8.0 M_J
HD2039	4.7 M_J
HD190226	5.0 M_J
HD50554	3.7 M_J
HD218437	1.9 M_J
HD136118	11 M_J
HD198050	28 M_J
HD108252	8.7 M_J
HD218435	1.2 M_J
HD33838	27 M_J
HD30177	7.8 M_J
HD23598	8.0 M_J
HD145875	3.8 M_J
HD72859	2.5 M_J
HD74158c	7.4 M_J
HD39091	10 M_J
GJ777a	1.3 M_J

Orbital Semimajor Axis (AU)

Distance of planet from star

■ FIGURE 20.20

Masses and Orbital Radii of Known Extrasolar Planets

A plot showing the planetary systems discovered by summer 2002, along with their masses in terms of the mass of Jupiter and the sizes of their orbits measured in astronomical units. Note that nine of the stars have more than one planetary companion. Source: Adapted from the California & Carnegie Planet Search website: exoplanet.org (Marcy, Butler, et al.)

TABLE 20.2 *Some Properties of the First 110 Extrasolar Planets Discovered*

Number Found	Characteristic
2	System of triple planets
7	Systems containing two planets
17	Planets with periods shorter than ten days
35	Planets closer to their stars than Mercury
57	Planets closer to their stars than 1 AU, which is the semimajor axis of Earth's orbit
54	Planets with periods longer than one year
60	Planets with eccentricities larger than Pluto, which is the largest in the solar system
7	Planets with minimum mass less than that of Saturn; all are closer to their Sun than Mercury

know for sure that we are dealing with planets significantly more massive than Earth.

The most massive companions probably burned deuterium early in their evolution and according to the definition used in this textbook might more properly be called brown dwarfs. Some astronomers argue for another definition of a planet—namely, an object that forms in a circumstellar disk through the aggregation of planetesimals into a rocky central core. However, the vast majority of the objects listed in Figure 20.20 never were hot enough to achieve nuclear fusion and are true planets by any definition.

The first triple system of planets was discovered orbiting Upsilon Andromedae, a star that is about 3 billion years old and about 44 LY away. The first quadruple system discovered orbits 55 Cancri, a sunlike star about 41 LY from Earth. By 2002, three planets had been found around this star, taking from 15 days to almost 15 years to orbit. Later measurements, however, showed that the velocity pattern of the star (its "wiggle") was even more complex. A fourth planet, with a mass only 18 times Earth's mass (one of the smallest yet found) and an orbital period of only about 3 days must also be circling 55 Cancri. A few other systems with multiple planets are also known and more are being discovered each year.

Another interesting planet is the one orbiting the star HD 209458. This planet passes directly in front of its parent star for about 3 hours every 3.5 days as we view it from Earth. A careful measurement of the tiny bit of starlight that is blocked as the planet crosses the face of the star then tells us how large the planet is. (This is similar to how astronomers use eclipsing binaries to measure the sizes of stars, as explained in Section 17.3.) In fact, using the Hubble Space Telescope, astronomers are now able to rule out any moons around this planet larger than 1.2 Earths, since such a moon would also have caused some measurable dimming. They also detect the planet's hot atmosphere evaporating.

The planet around HD 209458 has about 70 percent the mass of Jupiter, but its radius is about 35 percent larger than Jupiter's. Perhaps the planet is bloated because it is so

hot. Orbiting in less than 4 days, it is so close to the star that the temperature of the planet is predicted to be higher than 1500°C. In any case, these measurements confirm that the planet is a gas giant. When the planet passes in front of HD 209458, the atoms in the planet's atmosphere absorb starlight. Observations of this absorption were first made at the wavelengths of the yellow sodium lines and show that the atmosphere of the planet contains sodium. Searches for other elements, such as methane, water vapor, and potassium, will be made in the future.

Our search shows that planets with masses close to that of Jupiter are much more common than planets with larger masses (see problem 17). Since it is much easier to detect planets with large masses, this result is not a selection effect but a real trend. Nature seems to prefer its planets not to grow too big, for reasons we are still trying to understand.

As more Doppler observations are made and more time passes, astronomers expect to find planets with still lower masses, at greater distances from their parent stars, and hence with periods more like those of the giant planets in our own solar system. To detect a planet with an orbit and a mass like Jupiter, astronomers would have to make measurements with an accuracy of about 3 m/s for more than a decade, and most of the 2000 stars currently being surveyed for planets have been monitored for only half a dozen years or less.

20.4.3 Explaining the Planets We Found

The orbits of these newly discovered planets shocked astronomers. Before these discoveries, most astronomers expected that other planetary systems would be much like our own—planets following circular orbits, with massive planets found several astronomical units from their parent star. Yet most of the newly discovered planets are very different from those in our own solar system. As we saw, many are "hot Jupiters," orbiting closer to their suns than Mercury does. The name sounds funny, but it makes an important point: In the solar system, we think of Jupiters as cold worlds, far from their stars! And there was another surprise: Most of the planets that lie more than a few tenths of an astronomical unit from their star have highly elliptical (i.e., noncircular) orbits.

Traditionally, we have assumed that the planets in our solar system formed at about their current distances from the Sun and have remained there ever since. The first step in the formation of a giant planet is to build up a solid core, which happens when *planetesimals* (small chunks of solid matter—ice and dust particles) collide and stick. Eventually this core becomes massive enough to begin sweeping up gaseous material in the disk, thereby building the gas giants Jupiter and Saturn.

But this model works only if the giant planets are formed far from the central star (about 5–10 AU), where the disk is cold enough to have a fairly high density of solid matter. It cannot explain the hot Jupiters, which are located inside the orbit of Mercury, where any rocky matter would be completely vaporized. It also cannot explain elliptical orbits because the orbit of a protoplanet, whatever its initial shape, will quickly become circular through interactions with the surrounding disk of material and will remain that way as the planet grows by sweeping up additional matter.

So we have two choices: Either we find a new model for forming planets close to the searing heat of the parent star, or we find a way to change the orbits of planets so that cold Jupiters can travel inward after they form. Most researchers favor the latter. Calculations show that if a planet forms while a substantial amount of gas remains in the disk, then some of the planet's orbital angular momentum can be transferred to the disk. As it loses momentum (in a process that reminds us of the effects of friction), the planet will spiral inward. This process can transport giant planets, initially formed in cold regions of the disk, close to the central star—thereby explaining the hot Jupiters.

Some of the migrating planets very likely fall into the star and could enrich the star in heavy elements. Most of the stars that have been found to have planets so far do indeed contain more heavy elements than the Sun. Astronomers are now trying to determine whether this enrichment is due to the stars having a number of planets "for lunch" or whether planets are simply more likely to form in disks that are initially enriched in heavy elements. The two possibilities predict slightly different abundances, and observational tests are in progress as we write.

This process of forming planets is a "bottoms-up" model. Planets start out with a small core and build themselves up gradually through accretion. The problem is that this process takes 10 to 20 million years, after which the disk must last long enough to enable the inward migration. Some theorists argue that this is just too long, since (as we discussed earlier) observations show that inner parts of most disks disappear in 10 million years or less. As an alternative, researchers are exploring "top-down" planet formation, in which the matter in the disk becomes unstable and forms large clumps of material, which collapse and form planets in only a hundred thousand years.

What about the elliptical orbits of the giant planets found farther out? Calculations show that if several giant planets form around the same star, all relatively close together, then they will interact gravitationally and their orbits will change as a result. Some planets will be sent shooting out into space and will leave the neighborhood of the star altogether. Free-floating planets with masses between 8 and 15 times the mass of Jupiter have been found in star clusters in the area of the Orion giant molecular cloud, but we can't tell whether these planets were ejected from a planetary system or formed in isolation, as stars do.

The planets left behind around the star will have elliptical orbits rather than the circular ones they started out with. By the time all of this happens, most of the disk material will already have been swept up to form the multiple gas giants, and there won't be enough gas remaining to make orbits circular again.

The new picture then is that the formation of planetary systems is much more chaotic than we used to think. Think of planets when they form as being like skaters in a rink. The original model, when we had only our own solar system as a guide, assumed that the planets behaved like polite skaters, all obeying the rules and all moving in nearly the same direction, following roughly circular paths. The new picture corresponds more to a roller derby, where the skaters crash into one another, change directions, and sometimes are thrown entirely out of the rink.

Some of this chaotic behavior may even have affected our own solar system. Uranus and Neptune probably did not form at their present distances from the Sun but rather much closer to where Jupiter and Saturn are now. The reason for this idea is that the density in the disk of matter surrounding the Sun at the time the planets formed was so low outside the orbit of Saturn that it would take several billion years to build up Uranus and Neptune. Yet we know from observing other stars that the disks around protostars survive for only a few million years. Therefore scientists are developing models to calculate whether Uranus and Neptune could have formed near the current locations of Jupiter and Saturn, and then been kicked out to larger distances through gravitational interactions with their neighbors.

All these wonderful new observations illustrate how dangerous it can be to draw conclusions about a phenomenon in science (in this case how planetary systems form and arrange themselves) when you have only one example to work with. Until 1995, we knew of only one planetary system—our own. Now that over 100 star systems with planets are known, and more are being discovered all the time, our understanding and our hypotheses are growing much more sophisticated.

20.4.4 Habitable Planets Circling Other Stars

If giant planets can spiral in from several astronomical units away and be swallowed by their stars in the early days of a planetary system, then terrestrial planets that are formed even closer to the parent star are even more likely to be destroyed. This certainly makes astronomers wonder whether systems like our own and, in particular, planets like Earth might be rare. Were Jupiter and Saturn formed just as the material in the disk surrounding the protosun was cleared away, so that they could remain in their original orbits and not migrate inward? Can habitable terrestrial planets therefore survive only in those systems where the timing of early events is just right?

We can't answer such questions until we can detect planets around other stars with masses comparable to that of Earth. Astronomers are already planning new instruments and techniques for making the kinds of measurements it would take to show us Earths around other stars.

One approach is to use an indirect technique called *transit photometry*. In any large sample of stars with planets circling them, a few of the systems will be aligned so that the planets pass in front of their star on each orbit, as seen from Earth. One example, HD 209458, was described earlier in this chapter. If we measure the brightness of the star with great precision, we can detect the tiny drop in light as the planet transits (moves across the stellar disk) and blocks a bit of its radiation. Observing from space (where very high precision can be obtained), we might someday use this approach to detect planets as small as Earth and thus determine the distributions of planetary sizes and orbits.

The best possible evidence for an Earth-like planet elsewhere would be an image. After all, "seeing is believing" is a very human prejudice. But imaging a distant planet is a formidable challenge indeed. Suppose, for example, you were a great distance away and wished to detect reflected light from the Earth. The Earth intercepts and reflects less than one billionth of the Sun's radiation, so its apparent brightness in visible light is less than a billionth that of the Sun! But the faintness of potential planets is not the biggest problem.

The real difficulty is that the faint light from a planet is swamped by the blaze of radiation from its parent star. If you are nearsighted, try looking at streetlights at night with your glasses off. (If your vision is good, you can achieve the same effect by squinting.) You will see a halo of light surrounding every light. Bright stars seen through a telescope also appear to be surrounded by a halo of light. In this case, the problem is not that the telescope is nearsighted, but rather that slight imperfections in its optics and atmospheric blurring prevent the star's light from coming into focus in a completely sharp point. Planets, if any, would lie within this halo, and their faint light could not be seen in the glare.

Overcoming this problem is of one of NASA's major goals for the 21st century. One technique is to build infrared interferometers in space (see Chapter 5 for a discussion of radio interferometers). Again, we need to go into space to escape the blurring effects of Earth's atmosphere. The infrared is the optimum wavelength range in which to observe because planets get brighter in the infrared while stars get fainter, thereby making it easier to detect a planet against the glare of its star. Interferometry is an efficient way to obtain *high resolution* (to make out finer detail), which is what we need to observe the star and the nearby planet as two separate objects. Special techniques can be used to artificially suppress the light from the central star and make it easier to see the planet itself.

Even if we go into space, it is not clear whether we can successfully obtain such images from the Earth's immediate neighborhood. We are still deep within the Sun's *zodiacal cloud*—a thin layer of dust mainly between the Sun and Jupiter. This cloud tends to scatter and spread out starlight, making it hard to take really sharp images. It may be necessary to make our observations far from the Sun—say, out at the orbit of Jupiter—to minimize the scattering of light.

Once astronomers actually image an Earth-like planet, the next step would be to measure its spectrum and thus determine the composition of its atmosphere. The spectrum might even indicate whether life is present. In our own atmosphere, oxygen is produced by photosynthesis and methane by the decay of organic matter. If life were absent, neither element would be present in our atmosphere. So the discovery of methane and oxygen in the atmosphere of an Earth-like planet would be strong evidence that life as we know it is present. (Of course, forms of life that we cannot yet imagine might produce other gases, so the absence of oxygen and methane may not mean that life is absent.)

It is amazing to think that astronomers are beginning to develop realistic plans for instruments in space that can look for life on distant worlds. The next century should bring an answer to the question of whether habitable planets are also common, and perhaps even to the question of whether there are other planets teeming with life. The discovery of planets outside the solar system lends a new spirit of optimism to the search for life elsewhere, a subject to which we will return in Chapter 29.

SURFING THE WEB

Sources of more information on Herbig–Haro objects:

- *Catalog of HH Objects:*
  www-astro.phast.umass.edu/catalogs/HHcat/HHintro.html

- *Hubble Images:*
  hubblesite.org/newscenter/archive/1995/24/

Hubble Images: hubblesite.org/newscenter/archive
For images of dust disks around other stars, check the gallery of Hubble Space Telescope images.

Extra-Solar Planets Encyclopedia:
www.obspm.fr/encycl/encycl.html
Jean Schneider of the Paris Observatory keeps a detailed catalog of confirmed and unconfirmed new planets, with basic data, useful background information, and references.

Searching for Extra-Solar Planets:
exoplanets.org/exoplanets_pub.html
A site from Geoff Marcy's team about the search for planets with charts, summaries, papers, popular articles, links, and an excellent graphic summary.

Lists of Planet Search Programs Worldwide:

- *Darwin Site:* ast.star.rl.ac.uk/darwin/searches.html

- *Extra-Solar Encyclopedia Site:*
  www.obspm.fr/encycl/searches.html
  Excellent annotated lists of links to all the research programs that are working on or are proposed for working on the search for planets outside the solar system, including many we have not had room to discuss in the chapter.

20.1 Most stars form in **giant molecular clouds** that have masses as large as 3×10^6 times the mass of the Sun. Their longest dimension can range from 50 to 1000 LY. The best-studied molecular cloud is Orion, where star formation began about 12 million years ago and is moving progressively through the cloud. Recently formed hot stars are exposing many stages of the process of star formation to our view in Orion. Stars typically form in groups of a few hundred inside a *clump* of gas and dust that has a higher density than the surrounding molecular cloud. The formation of a star inside one of these clumps begins with a dense *core* of material that accretes matter and collapses due to gravity. The accumulation of material halts when the **protostar** develops a strong **stellar wind.** A turbulent cloud forms a rotating star with an equatorial disk of material. The wind tends to emerge more easily in the direction of a protostar's poles, leading to jets of material being observed from the star. These can collide with the material around the star and produce regions that emit light in the form of emission lines; these regions are called Herbig–Haro objects after the astronomers who first described them.

20.2 The evolution of a star can be described in terms of its temperature and luminosity changes, which can best be followed by plotting them on an H–R diagram. Protostars generate energy through gravitational contraction. The initial gravitational collapse takes several thousand years, and then a slow contraction typically continues for millions of years, until the star reaches the **main sequence** and nuclear reactions begin. The higher the mass of a star, the shorter the time it spends in each stage of evolution. The masses of stars range from about 0.072 to 100–200 times the mass of the Sun.

20.3 There is observational evidence that most protostars are surrounded by disks with large enough diameters and enough mass (as much as 10 percent that of the Sun) to form planets. Although we don't yet know what fraction of these disks actually form planets, the properties of these disks do change systematically with time. Initially, an opaque disk extends all the way to the surface of the protostar. After a few million years the inner part of the disk is cleared of dust, and the disk is then shaped like a donut with the protostar centered in the hole. The development of such a hole can be explained if a large planet has formed at its outer boundary. After 30 million years or so, all of the original cloud of material from which the star formed has disappeared from the vicinity of the star, because the material has been accreted by the star, incorporated into its planets, or expelled from the system altogether. Around a few still older stars, we see disks formed from the debris produced when small bodies (comets and asteroids) collide with each other. These debris disks are in narrow rings, and the density of material varies with its position in the rings. The distribution of matter in the rings is probably determined by shepherd planets, just as Saturn's shepherd moons affect the orbits of the material in its rings.

20.4 At present we can search for planets around nearby stars only by looking for the star's motion around the star–planet system's common center of mass. This can be done by looking for changes either in the star's position on the sky over time or in its radial velocity (as seen in the Doppler shift of the lines in its spectrum). Radial velocity searches in the past five years have been stunningly successful. We now know of about ten times more planets outside our own solar system than within it, and more are being found at the rate of about one per month. To our surprise, most of these planets are not at all like the ones in our own solar system. "Hot Jupiters" are found very close to their parent stars, and many planets at distances of more than a few astronomical units from the central star have elliptical orbits. Astronomers now think that planets may migrate after they are formed, with some moving closer to the star or even falling into it, and others being ejected into space through gravitational interactions with other nearby planets. Ambitious space experiments are now being planned that might make it possible to image Earth-like planets outside the solar system and even to obtain their spectra. Oxygen would be an unmistakable indicator of biological activity.

INTER-ACTIVITY

 A The search for planets around other stars is an example of a research problem in which selection effects must be taken into account to determine whether the conclusions are valid. "Selection effect" is a fancy way of saying that the observations are in some way biased or not truly representative of the actual situation. Have your group discuss how selection effects are relevant in answering the following questions:

    1. No terrestrial (Earth-like) planets have yet been discovered through measurements of radial velocities. Therefore, a friend of yours asserts that there must be no habitable planets in the universe other than Earth. Is your friend right?

    2. So far no brown dwarfs (objects with masses in the range of, say, 20–100 times the mass of Jupiter) have been found with orbits less than about 5 AU

around other stars. A friend of yours says that therefore brown dwarfs are not likely to be found in close orbits around other stars. Is your friend right?

B Your group is a subcommittee of scientists examining whether any of the "hot Jupiters" (giant planets closer to their stars than Mercury is to the Sun) could have life on or near them. Can you come up with places on or near such planets where life could develop or where forms of life might survive?

C A wealthy couple who are alumni of your college or university leaves the astronomy program several million dollars in their will, to spend in the best way possible to search for "infant stars in our section of the Galaxy." Your group has been assigned the happy task of advising the dean on how best to spend the money. What kind of instruments and search program would you recommend, and why?

D Even before Earth-like planets have been found, some people consider the discovery of any planets (even hot Jupiters) around other stars one of the most

important events in the history of astronomical research. Some astronomers have been surprised that the public is not more excited about the planet discoveries. One reason that has been suggested for this lack of public surprise and excitement is that science fiction stories have long prepared us for there being planets around other stars. (The Starship Enterprise on the *Star Trek* TV series found some in just about every weekly episode!) What does your group think? Did you know about the discovery of planets around other stars before taking this course? Do you consider it exciting? Were you surprised to hear about it? Are science fiction movies and books good or bad tools for astronomy education in general, do you think?

E What if future space instruments reveal an Earth-like planet with oxygen and methane in its atmosphere orbiting another star. Suppose the star is 50 LY away. What does your group suggest astronomers do next? How much effort and money would you recommend be put into finding out more about this planet and why?

REVIEW QUESTIONS

1. Give several reasons the Orion molecular cloud is such a useful "laboratory" for studying the stages of star formation.
2. Why is star formation more likely to occur in cold molecular clouds than in regions where the temperature of the interstellar medium is several hundred thousand degrees?
3. Why have we learned a lot more about star formation since the invention of detectors sensitive to infrared radiation?
4. Describe what happens when a star forms. Begin with a

dense core of material in a molecular cloud, and trace the evolution up to the time the newly formed star reaches the main sequence.
5. Why is it so hard to see planets around other stars and so easy to see them around our own?
6. What techniques have been used to search for planets around other stars?
7. How were the first planets around stars like the Sun discovered?
8. Explain the current ideas about how "hot Jupiters" formed around other stars.
9. Explain why taking an image of a planet around another star is so difficult, and how astronomers hope to overcome these difficulties in the future.

THOUGHT QUESTIONS

10. A friend of yours, who did not do well in her astronomy class, tells you she believes that all stars are old and that none could possibly be born today. What arguments would you use to persuade her that stars are being born somewhere in the Galaxy during our lifetime?

11. Look at the four stages in the birth of a star shown in Figure 20.8. In which stage(s) can we see the star in vis-

ible light? In infrared radiation? In which stage(s) is the star generating energy by converting hydrogen to helium?

12. Observations suggest that it takes more than 3 million years for the dust to begin clearing out of the inner regions of the disks surrounding protostars. Suppose this is the minimum time required to form a planet. Would you expect

to find a planet around a 10-solar-mass star? (Refer to Figure 20.12.)

13. The evolutionary track for a star of 1 solar mass remains nearly vertical in the H–R diagram for a while (see Figure 20.12). How is its luminosity changing during this time? Its temperature? Its radius? What is its source of energy?

14. Suppose you wanted to image a planet around another star. Would you try to observe in visible light or in the infrared? Why? Would the planet be easier to see if it were at 1 AU or 5 AU?

15. Why were giant planets close to their stars the first ones to be discovered? Why has the same technique not been used yet to discover giant planets at the distance of Saturn?

FIGURING FOR YOURSELF

16. Use the data in this chapter about the rate at which interstellar matter is being converted into stars and the data from Chapter 19 about how much interstellar matter is contained in the Milky Way Galaxy to estimate how much longer star formation can continue. Compare this time with the age of the universe (about 10 to 15 billion years). Can you think of any factors that would extend the time over which star formation might continue? These results argue that we live in a special time as far as star formation is concerned. In a few billion years, astronomers could not study this process directly, and observations of galaxies, which we will discuss later in this text, suggest that star formation was much more active a few billion years ago.

17. Construct a histogram (bar graph) that uses the data from Figure 20.20 to show how many planets have masses that range from the largest observed to less than 1 Jupiter mass. Put all the planets between 2 and 3 Jupiter masses

into one bin, 3 and 4 into another, and so on. Describe the histogram you got. Is the trend real or the result of a selection effect? Are we likely to have missed a large number of massive planets? Why? Are we likely to have missed planets with small masses? Why? In general, how would you expect this histogram to change as we obtain better observations over the next decade?

18. When astronomers found the first giant planets with orbits of only a few days, they did not know whether those planets were gaseous and liquid like Jupiter or rocky like Mercury. The observations of HD 209458 settled this question because observations of the eclipse of the star by this planet made it possible to determine the radius of the planet. Use the data given in the text to estimate the density of this planet, and then use that information to explain why it must be a gas giant.

SUGGESTIONS FOR FURTHER READING

Caillault, J. et al. "The New Stars of M42" in *Astronomy,* Nov. 1994, p. 40. On studies of circumstellar disks in Orion.

Frank, A. "Starmaker: The New Story of Stellar Birth" in *Astronomy,* July 1996, p. 52.

Jayawardhana, R. "Spying on Stellar Nurseries" in *Astronomy,* Nov. 1998, p. 62. On protoplanetary disks.

Lada, C. "Deciphering the Mysteries of Stellar Origins" in *Sky & Telescope,* May 1993, p. 18. Fine beginning article.

MacRobert, A. "A Star Hop in the Heart of Orion" in *Sky & Telescope,* Jan. 1998, p. 90. On observing this region of star formation with a small telescope.

O'Dell, C. R. "Exploring the Orion Nebula" in *Sky & Telescope,* Dec. 1994, p. 20. Good review with Hubble results.

Reipurth, B. and Heathcote, S. "Herbig–Haro Objects and the Birth of Stars" in *Sky & Telescope,* Oct. 1995, p. 38.

Stahler, S. "The Early Life of Stars" in *Scientific American,* July 1991, p. 48.

On the Search for Planets Elsewhere

Boss, A. *Looking for Earths: The Race to Find New Solar Systems.* 1998, Wiley.

Doyle, L. "Searching for Shadows of Other Earths" in *Scientific American,* Sept. 2000, p. 58.

Fischer, D. "Prowling for Planets" in *Mercury* (the magazine of the Astronomical Society of the Pacific), July/Aug. 2000, p. 13. Review by the first woman in history to discover a planet.

Kaisler, D. "The Puzzles of Planethood" in *Sky & Telescope,* Aug. 2002, p. 33. On what differentiates planets from brown dwarfs.

Marcy, G. and Butler, R. "The Diversity of Planetary Systems" in *Sky & Telescope,* Mar. 1998, p. 30. A progress report on planet discoveries by the leading team. (See also their "Hunting Planets Beyond" in *Astronomy,* Mar. 2000, p. 43.)

McInnis, D. "Wanted: Life-Bearing Planets" in *Astronomy,* Apr. 1998, p. 38. On future instruments that could find planets more like Earth.

Schilling, G. "The Race to Epsilon Eridani" in *Sky & Telescope,* June 2001, p. 34. On the star and its possible planet.

Stephens, S. "Planet Hunters" in *Astronomy,* July 1998, p. 58. Profile of Geoff Marcy and Paul Butler, who have found the largest number of extrasolar planets.

During the later phases of stellar evolution, stars expel some of their mass, which returns to the interstellar medium to form new stars. Here is a beautiful image of a star losing mass taken with the Hubble Space Telescope. This planetary nebula is known as Mz 3 or the ant nebula and is about 3000 LY away from the Sun. We see a central star that has ejected mass preferentially in two opposite directions. The object is about 1.6 LY long. The image is color coded so that red corresponds to an emission line of sulfur, green to nitrogen, blue to hydrogen, and blue/violet to oxygen.

21

Stars: From Adolescence to Old Age

The universe at large would suffer as little, in its splendor and variety, by the destruction of our planet, as the verdure and sublime magnitude of a forest would suffer by the fall of a single leaf.

Thomas Chalmers in *Discourses on the Christian Revelation Viewed in Connection with the Modern Astronomy* (1817)

THINKING AHEAD

We know that the Sun cannot last forever. Sooner or later (much, much later, we hope!) it will exhaust its nuclear fuel and cease to shine. But how will it change during its long lifetime? And what do these changes mean for the future of the Earth?

We now turn from the birth of stars and planets to the rest of their life stages. This is not an easy task, since stars live much longer than astronomers. Thus we cannot hope to see the life story of any single star unfold before our eyes or telescopes. Like the rushed crew of our imaginary starship in Chapter 17 (with only one day to study the lives of the Earth's human inhabitants), we must survey as many of the stellar inhabitants in the Galaxy as possible. If we are lucky (and thorough), we can catch at least a few of them in each possible stage of their lives.

Surveys in our own galactic neighborhood and beyond show that stars have many different characteristics (Figure 21.1). Some of the differences come about because stars have different masses and thus different temperatures and luminosities. But others are the result of changes that occur as stars age. Through a combination of observation, theory, and clever detective work, we can use these differences to piece together the life story of a star.

© Anglo-Australian Telescope Board

■ **FIGURE 21.1**

Stars in the Constellation of Sagittarius   This snapshot shows a yellowish population of older stars with a small cluster of younger bluish stars (NGC 6520) as well as a dark cloud (Barnard 86) that blocks the light from stars behind it.

21.1   EVOLUTION FROM THE MAIN SEQUENCE TO GIANTS

One of the best ways to get a "snapshot" of a group of stars is by plotting their properties on an H–R diagram. We have already used the H–R diagram to follow the evolution of protostars up to the time they reach the main sequence. Now let's see what happens next.

Once a star has reached the main-sequence stage of its life, it derives its energy almost entirely from the conversion of hydrogen to helium via the process of nuclear fusion (see Chapter 15). Since hydrogen is the most abundant element in stars, this process can maintain the star's equilibrium for a long time. Thus all stars remain on the main sequence for most of their lives. Some astronomers like to call the main-sequence phase the star's "prolonged adolescence" or "adulthood" (continuing our analogy to the stages in a human life).

The left-hand edge of the main-sequence band in the H–R diagram is called the **zero-age main sequence** (see Figure 17.14). We use the term *zero-age* to mark the time when each star reaches the main sequence and its hydro-gen fusion reactions begin. The zero-age main sequence is a continuous line in the H–R diagram that shows where stars of different masses but similar chemical composition begin to fuse hydrogen.

Since only 0.7 percent of the hydrogen used in fusion reactions is converted into energy, the star does not change its *total* mass appreciably during this long period. It does, however, change the chemical composition in its central regions, where the nuclear reactions occur: Hydrogen is gradually depleted and helium accumulates. This change of composition changes the luminosity, temperature, size, and interior structure of the star. When a star's luminosity and temperature begin to change, the point that represents the star on the H–R diagram moves away from the zero-age main sequence.

Calculations show that the temperature and density in the inner region slowly increase as helium accumulates in the center of a star. As the temperature gets hotter, each proton acquires more energy of motion on average; this means it is more likely to interact with other protons and as a result the rate of fusion also increases. (For the proton–proton cycle described in Chapter 15, the rate of fusion goes up roughly as the temperature to the fourth power. If the temperature were to double, then, the rate of fusion would increase by a factor of 2^4, or 16 times.)

If the rate of fusion goes up, the rate at which energy is being generated also increases and the luminosity of the star gradually rises. Initially, however, these changes are small, and stars remain within the main-sequence band on the H-R diagram for most of their lifetimes.

21.1.1 Lifetimes on the Main Sequence

How many years a star remains in the main-sequence band depends on its mass. You might think that a more massive star, having more fuel, would last longer, but it's not that simple. The lifetime of a star in a particular stage of evolution depends on how much nuclear fuel it has and on *how fast* it uses up that fuel. (In the same way, how long people can keep spending money depends not only on how much money they have but also on how quickly they spend it. This is why many lottery winners who go on spending sprees quickly wind up poor again.) In the case of stars, more massive stars use up their fuel much more quickly than stars of low mass.

The reason massive stars are such spendthrifts is that the rate of fusion depends *very* strongly on the star's core temperature. And what determines how hot a star's central regions get? It is the *mass* of the star—the weight of the overlying layers—that determines how high the pressure in the center must be: Higher mass requires higher pressure to balance it. Higher pressure in turn is produced by higher temperature. The higher the temperature in the central regions, the faster the star races through its storehouse of central hydrogen. Although massive stars have more fuel, they burn it so prodigiously that their lifetimes are much shorter than those of their low-mass counterparts. You can also understand now why the most massive main-sequence stars are also the most luminous. Like new rock stars with their first platinum album, they spend their resources at an astounding rate.

The main-sequence lifetimes of stars of different masses are listed in Table 21.1. This table shows that the most massive stars spend only a few million years on the main sequence. A star of 1 solar mass remains there for roughly 10 billion years, while a star of about 0.4 solar mass has a main-sequence lifetime of some 200 billion years—longer than the current age of the universe. (Bear in mind, however, that every star spends *most* of its total lifetime on the main sequence. Stars devote an average of 90 percent of their lives to peacefully fusing hydrogen into helium.)

These results are not merely of academic interest. Human beings developed on a planet around a G-type star whose stable main-sequence lifetime was so long that it afforded life on Earth plenty of time to evolve. If we were to search for intelligent life like our own on planets around other stars, it would be a pretty big waste of time to search around O- or B-type stars. These stars remain stable for such a short time that the development of creatures complicated enough to take astronomy courses is very unlikely.

Spectral Type	Mass (Mass of Sun = 1)	Lifetime on Main Sequence (years)
O5	40	1 million
B0	16	10 million
A0	3.3	500 million
F0	1.7	2.7 billion
G0	1.1	9 billion
K0	0.8	14 billion
M0	0.4	200 billion

TABLE 21.1 *Lifetimes of Main-Sequence Stars*

21.1.2 From Main-Sequence Star to Red Giant

Eventually all the hydrogen in the stellar core, where it is hot enough for fusion reactions, is used up. The core now contains only helium, "contaminated" by whatever small percentage of heavier elements the star had to begin with. The helium in the core can be thought of as the accumulated "ash" from the nuclear "burning" of hydrogen during the main-sequence stage.

Energy can no longer be generated by hydrogen fusion because the hydrogen is all gone and, as we will see, the fusion of helium requires much higher temperatures. Since the central temperature is not yet high enough to fuse helium, there is no nuclear energy source to supply heat to the central region of the star. Gravity again takes over and, after a long period of stability, the core begins to contract. Once more the star's energy is partially supplied by gravitational energy, in the way described by Kelvin and Helmholtz (see Section 15.1). As the star's core shrinks, the energy of the inward-falling material is converted to heat.

The heat generated in this way, like all heat, flows outward to where it is a bit cooler. In the process, the heat raises the temperature of the hydrogen that spent the whole long main-sequence time just outside the core. Like an understudy waiting in the wings of a hit play for a chance at fame and glory, this hydrogen was almost (but not quite) hot enough to undergo fusion and take part in the main action that sustains the star. Now, the additional heat produced by the shrinking core puts this hydrogen "over the limit," and a shell of hydrogen nuclei just outside the core becomes hot enough for hydrogen fusion to begin.

New energy produced by fusion now pours outward from this shell and begins to heat up layers of the star farther out, causing them to expand. Meanwhile, the helium core continues to contract, producing more heat right around it. This leads to more fusion in the shell of fresh hydrogen outside the core. The additional fusion produces more energy, which also flows out into the upper layers of the star.

Most stars actually generate more energy when they are fusing hydrogen in the shell surrounding the core than

■ **FIGURE 21.2**

Relative Sizes of Stars This computer-generated graphic shows the size of the Sun (yellow) compared with the sizes of Delta Boötis, a giant star (orange), and Xi Cygni, a supergiant (red). The sizes of the other stars were measured using the Palomar Testbed Interferometer, an instrument that allows astronomers to make out much finer detail (greater resolution) by connecting together two telescopes 110 meters apart.

they did when hydrogen fusion was confined to the central part of the star; thus they increase in luminosity. With all the new energy pouring outward, the outer layers of the star begin to expand, and the star eventually grows and grows until it reaches enormous proportions (Figure 21.2).

When you take the lid off a pot of boiling water, the steam can expand and it cools down. In the same way, the expansion of the outer layers of a star causes the temperature at the surface to decrease. As it cools, the star's overall color becomes redder. (We saw in Chapter 4 that red corresponds to cooler temperature.) So the star becomes simultaneously more luminous and cooler; on the H–R diagram the star therefore leaves the main-sequence band and moves upward and to the right. The star becomes one of the *red giants* or supergiants first discussed in Chapter 17. You might say that these stars have "split personalities": Their cores are contracting while their outer layers are expanding. (Note that the giant stars would not all look deep red; they simply get *redder* as they get cooler. A giant star's overall color will always depend on its surface temperature at a given time.)

Just how different are these red giants and supergiants? Table 21.2 compares the Sun with the red supergiant Betelgeuse, visible above Orion's belt as the bright red star that marks the hunter's armpit. Relative to the Sun, the supergiant has a much larger radius, a much lower average density, a cooler surface, and a much hotter core.

These giant stars can become so large that if we were to replace the Sun with one of them, its outer atmosphere would extend to the orbit of Mars or even beyond (Figure 21.3). This is the next stage in the life of a star as it moves (to continue our analogy to human lives) from its long period of youth into middle age. (After all, many human beings today also see their outer layers expand a bit during middle age!)

21.1.3 Models for Evolution to the Giant Stage

As we discussed earlier, astronomers can construct computer models of stars with different masses and compositions to see how stars change throughout their lives. Figure

TABLE 21.2 *Comparing a Supergiant with the Sun*

Property	Sun	Betelgeuse
Mass (2×10^{33} g)	1	16
Radius (km)	700,000	500,000,000
Surface temperature (K)	5,800	3,600
Core temperature (K)	15,000,000	160,000,000
Luminosity (4×10^{26} W)	1	46,000
Average density (g/cm^3)	1.4	1.3×10^{-7}
Age (million years)	4,500	10

Source: Data from research by G. J. Mathews, G. Herczeg, and D. Dearborn

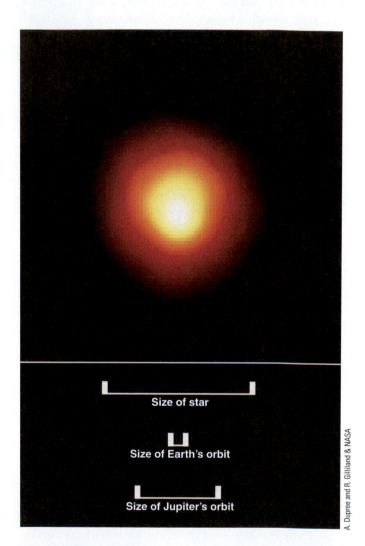

Size of star

Size of Earth's orbit

Size of Jupiter's orbit

■ **FIGURE 21.3**

The Supergiant Betelgeuse This star is in the constellation of Orion (see Figure 20.3). Here we see an image taken in ultraviolet light with the Hubble Space Telescope—the first direct image ever made of the surface of another star. As shown by the bars at the bottom, Betelgeuse has an extended atmosphere so large that, if it were at the center of our solar system, it would stretch past the orbit of Jupiter.

21.4, based on theoretical calculations by University of Illinois astronomer Icko Iben, shows an H–R diagram with several tracks of evolution from the main sequence to the giant stage. Tracks are shown for stars with different masses and with chemical compositions similar to that of the Sun. The red line is the initial or zero-age main sequence. The numbers along the tracks in Figure 21.4 indicate the times, in years, required for the stars to reach those points in their evolution after leaving the main sequence. Once again, you can see that the more massive a star is, the more quickly it goes through each stage in its life.

Note that the most massive model in this diagram has a mass similar to that of Betelgeuse, and so its evolutionary track shows approximately what the history of Betelgeuse has been. The track for a 1-solar-mass star shows that the Sun is still in the main-sequence phase of evolution, since it is only about 4.5 billion years old. It will be billions of years before the Sun begins its own "climb" away from the main sequence—the expansion of its outer layers that will make it a red giant.

21.2 STAR CLUSTERS

The preceding description of stellar evolution is based on calculations. No star completes its main-sequence lifetime or its evolution to a red giant quickly enough for us to observe these structural changes as they happen. Fortunately, nature has provided us with a way to test our calculations.

Instead of observing the evolution of a single star, we can look at a group or *cluster* of stars. We look for a group of stars that is very close together in space, held together by gravity, often moving around a common center. Then it is reasonable to assume that the individual stars in the group all formed at nearly the same time, from the same cloud, and with the same composition. We expect that these stars will differ only in mass. And mass is what determines how quickly they go through each stage of their lives.

Since stars with higher masses evolve more quickly, we can find clusters in which massive stars have already completed their main-sequence phase of evolution and become red giants while stars of lower mass in the same cluster are still on the main sequence, or even undergoing pre-main-sequence gravitational contraction. We can see many stages

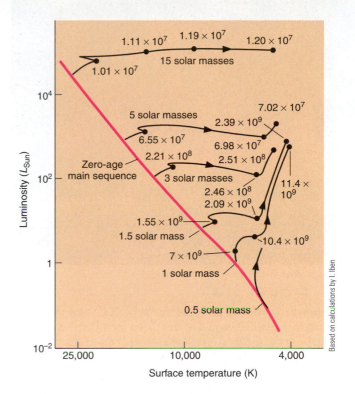

FIGURE 21.4

Evolutionary Tracks for Stars of Different Masses
The solid black lines show the predicted evolution from the main sequence through the red-giant or supergiant stage on the H–R diagram. Each track is labeled with the mass of the star that it is describing. The numbers show how many years each star takes to become a giant after arriving on the main sequence. The red line is the zero-age main sequence.

Ace⊛Astronomy™ Log into AceAstronomy and select this chapter to see the Active Figure called "Animated HR Diagram" and Astronomy Exercise "Mass-Star Lifetime Relation."

of stellar evolution among the members of a single cluster, and we can see whether our models can explain why the H–R diagrams of clusters of different ages look the way they do.

The three basic types of clusters are globular clusters, open clusters, and stellar associations. Their properties are summarized in Table 21.3. As we will see, globular clusters contain only very old stars, whereas open clusters and associations contain young stars.

TABLE 21.3 *Characteristics of Star Clusters*

	Globular Clusters	Open Clusters	Associations
Number in Galaxy	150	Thousands	Thousands
Location in Galaxy	Halo and nuclear bulge	Disk (and spiral arms)	Spiral arms
Diameter (LY)	50–450	<30	100–500
Mass (solar masses)	10^4–10^6	10^2–10^3	10^2–10^3
Number of stars	10^4–10^6	50–1000	10^2–10^4
Color of brightest stars	Red	Red or blue	Blue
Luminosity of cluster (L_{Sun})	10^4–10^6	10^2–10^6	10^4–10^7

Photo by David Malin; © Anglo-Australian Observatory

■ FIGURE 21.5

Omega Centauri   Located about 17,000 LY away, the globular cluster Omega Centauri is the most massive globular cluster in our Galaxy. It contains several million stars.

21.2.1 Globular Clusters

The **globular clusters** were given their name because they are nearly symmetrical round systems of, typically, hundreds of thousands of stars. The most massive globular in our Galaxy is Omega Centauri, which is about 17,000 LY distant and contains several million stars (Figure 21.5). Note that the brightest stars in this cluster, which are giants that have already completed the main-sequence phase of their evolution, are pale yellow. These stars have typical temperatures of about 2500 K and about the same color as a tungsten light bulb.

What would it be like to live inside a globular cluster? In the dense central regions, the stars would be roughly a million times closer together than in our neighborhood (Figure 21.6). If the Earth orbited one of the inner stars in a globular cluster, the nearest stars would be *light months*, not light years, away. They would still appear as points of light but would be brighter than any of the stars we see in our own sky. The Milky Way would probably be impossible to see through the bright haze of starlight produced by the cluster.

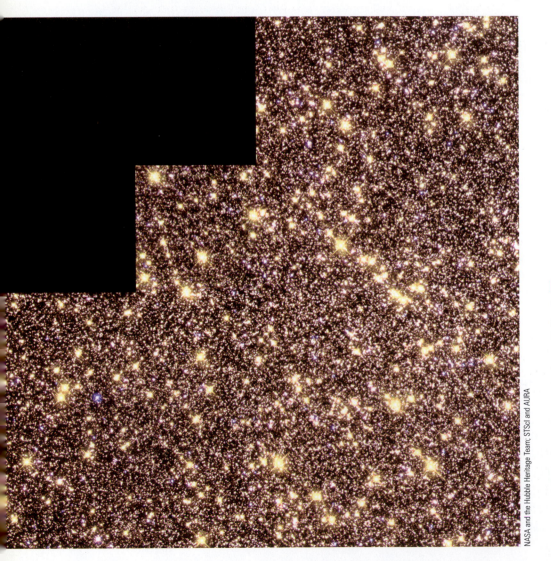

NASA and the Hubble Heritage Team; STScI and AURA

■ FIGURE 21.6

The Central Region of the Globular Cluster Omega Centauri   The Hubble Space Telescope is needed to see the individual stars in the densely packed central regions of Omega Centauri. This image contains about 50,000 stars in a region about 7 LY across. (The full cluster is about 450 LY across.) Most of the stars in this image are faint yellow main-sequence stars like our Sun. The bright yellow-orange stars are red giants that have begun to exhaust their nuclear fuel and have expanded to diameters about 100 times larger than the Sun. The faint blue stars are past the red-giant phase and are in the process of evolving to become white dwarfs.

About 150 globular clusters are known in our Galaxy. Most of them are in a spherical halo (or cloud) surrounding the flat disk formed by the majority of the Galaxy's stars. All the globular clusters are very far from the Sun, and some are found at distances of 60,000 LY or more from the galactic plane. The diameters of globular star clusters range from 50 LY to more than 450 LY.

21.2.2 Open Clusters

Open clusters are found in the disk of the Galaxy, often associated with interstellar matter. Open clusters are smaller than globular clusters, usually having diameters of less than 30 LY, and they typically contain only several dozen to several hundred stars (Figure 21.7). The stars in open clusters usually appear well separated from one another, even in the central regions, which explains why they are called "open." Our Galaxy contains thousands of open clusters, but we can see only a small fraction of them. Interstellar dust, which is also concentrated in the disk, dims the light of more distant clusters so much that they are undetectable (see Chapter 19).

Several open clusters are visible to the unaided eye. Most famous among them is the Pleiades (see the image that opens Chapter 19), which appears as a tiny group of six stars (some people can see even more) arranged like a dipper in the constellation of Taurus, the bull. A good pair of binoculars shows dozens of stars in the cluster, and a telescope reveals hundreds. (A car company, Subaru, takes its name from the Japanese term for this cluster; you can see the star group on the rear lock cover of each Subaru.)

The Hyades is another famous open cluster in Taurus. To the naked eye, it appears as a V-shaped group of faint stars marking the face of the bull. Telescopes show that the Hyades actually contains more than 200 stars.

21.2.3 Stellar Associations

An **association** is a group of extremely young stars and typically contains 5 to 50 hot, bright O and B stars scattered over a region of space some 100–500 LY in diameter. Associations also contain hundreds to thousands of low-mass stars, but these are much fainter and less conspicuous. The presence of hot, luminous stars indicates that star formation in the association has occurred in the last million years or so. Since O stars go through their lives in only about a million years, they would not still be around unless star formation has occurred recently. It is therefore not surprising that associations are found in regions rich in the gas and dust required to form new stars. Because associations, like ordinary open clusters, lie in regions occupied by dusty interstellar matter, most are hidden from our view.

21.3 CHECKING OUT THE THEORY

Open clusters are younger than globular clusters, and associations are typically somewhat younger still. We know this because the stars in these different types of clusters are found in different places in the H–R diagram, and we can use their locations in combination with theoretical calculations to estimate their ages.

21.3.1 H–R Diagrams of Young Clusters

What does theory predict for the H–R diagram of a cluster whose stars have recently condensed from an interstellar cloud? After a few million years ("recently" for astronomers), the most massive stars should have completed their contraction phase and be on the main sequence, while the less massive ones should be off to the right, still on their way to the main sequence. These ideas are illustrated in Figure 21.8, which shows the H–R diagram calculated by R. Kippenhahn and his associates at Munich for a hypothetical cluster with an age of 3 million years.

■ **FIGURE 21.7**

The Jewel Box (NGC 4755) This open cluster of young, bright stars is about 8000 LY away from the Sun. Note the contrast in color between the bright yellow supergiant and the hot blue main-sequence stars. The name comes from its 19th-century description by John Herschel as "a casket of variously colored precious stones."

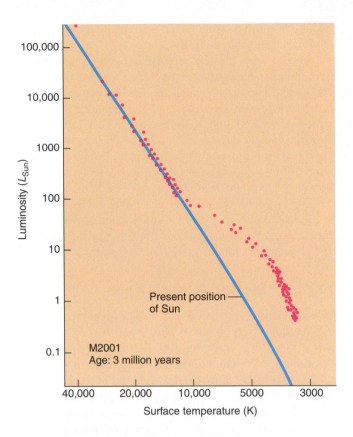

Young Cluster H–R Diagram   We see an H–R diagram for a hypothetical cluster 3 million years old. Note that the high-mass (high-luminosity) stars have already arrived at the main-sequence stage of their lives, while the lower-mass (lower-luminosity) stars are still to the right of the zero-age main sequence, not yet hot enough to begin fusion of hydrogen.

There are real star clusters that fit this description. The first to be studied (in about 1950) was NGC 2264, which is still associated with the region of gas and dust from which it was born (Figure 21.9). Its H–R diagram is shown in Figure 21.10. The cluster in the middle of the Orion Nebula (shown in Figures 20.4 and 20.5) is in a similar stage of evolution.

As clusters get older, their H–R diagrams begin to change. After a short time—less than a million years after they reach the main sequence—the most massive stars use up the hydrogen in their cores and evolve off the main sequence to become red giants. As more time goes on, stars of lower and lower mass begin to leave the main sequence and make their way to the upper right of the H–R diagram. Figure 21.11 is a photograph of NGC 3293, a cluster that is about 10 million years old. The dense clouds of gas and dust are gone. One massive star has evolved to become a red giant and stands out as an especially bright orange member of the cluster.

Figure 21.12 shows the H–R diagram of the open cluster M41, which is roughly 100 million years old; by this time, a significant number of stars have moved off to the

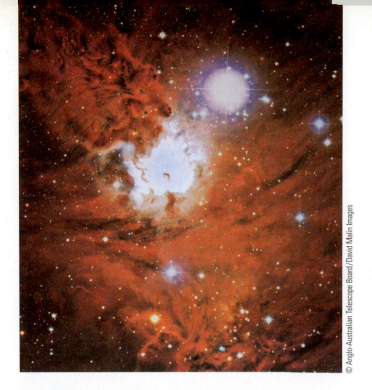

The Young Cluster NGC 2264   Located about 2500 LY from us, this region of newly formed stars is a complex mixture of red hydrogen gas ionized by hot embedded stars, dark obscuring dust lanes, and brilliant young stars.

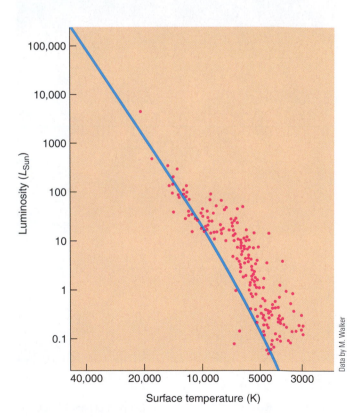

H–R Diagram for Cluster NGC 2264   Compare with Figure 21.8; although the points scatter a bit more here, the theoretical and observational diagrams are remarkably—and satisfyingly—similar.

■ **FIGURE 21.11**

The Open Star Cluster NGC 3293
All the stars in such clusters form at about the same time. The most massive stars, however, exhaust their nuclear fuel more rapidly and hence evolve more quickly than stars of low mass. As stars evolve, they become redder. The bright orange star in NGC 3293 is the member of the cluster that has evolved most rapidly.

right and become red giants. Note the gap that appears in this H–R diagram between the stars near the main sequence and the red giants. A gap does not necessarily imply that stars avoid a region of temperatures and luminosities. In this case, it simply represents a domain of temperature and luminosity through which stars evolve very quickly. We see a gap for M41 because at this particular moment we have not caught a star in the process of scurrying across this part of the diagram.

21.3.2 H-R Diagrams of Older Clusters

After 4 billion years have passed, many more stars—including stars only a few times more massive than the Sun—have left the main sequence (Figure 21.13). This means that no stars are left near the top of the main sequence; only the low-mass stars near the bottom remain. The

■ **FIGURE 21.12**

H–R Diagram for Cluster M41   This cluster is older than NGC 2264 (see Figure 21.10) and contains several red giants. Some of its more massive stars are no longer close to the zero-age main sequence (blue line).

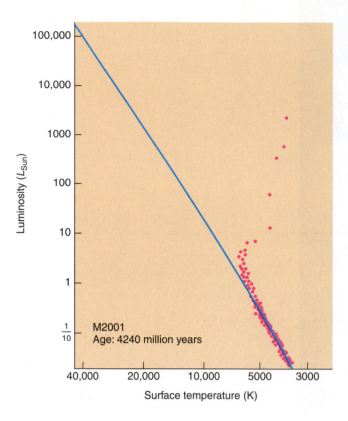

FIGURE 21.13

H–R Diagram for an Older Cluster We see the H–R diagram for a hypothetical cluster at an age of 4.24 billion years. Note that most of the stars on the upper part of the main sequence have now turned off toward the red-giant region.

Ace Astronomy™ Log into AceAstronomy and select this chapter to see the Active Figure called "Cluster Turnoff."

Data by J. Hesser and collaborators

FIGURE 21.14

H–R Diagram for Cluster 47 Tucanae This H–R diagram is for the globular cluster 47 Tucanae (which you can see in Figure 28.3). Note that the scale of luminosity is different from those of the other H–R diagrams in this chapter. We are focusing on the lower portion of the main sequence, the only part where stars still remain in this old cluster.

older the cluster, the lower the point on the main sequence where stars begin to move toward the red giant region.

The oldest clusters of all are the globulars. Figure 21.14 is the H–R diagram of the globular cluster 47 Tucanae. Notice that the luminosity and temperature scales are different from the other H–R diagrams in this chapter. In Figure 21.13, for example, the luminosity scale on the left side of the diagram goes from 1/10 to 100,000 times the Sun's luminosity. But in Figure 21.14, the luminosity scale has been significantly reduced in extent. So many stars in this old cluster have had time to turn off the main sequence that only the very bottom of the main sequence remains.

Just how old are the different clusters we have been discussing? To get actual ages (in years), we must compare the appearance of our calculated H–R diagrams for different ages with actual H–R diagrams. Astronomers use the position of the top of the main sequence (i.e., the luminosity at which stars begin to move off the main sequence to become red giants) as a measure of the age of a cluster. Compare the luminosities of the brightest stars still on the main sequence in Figures 21.10 and 21.13, for example.

Some associations and open clusters are as young as 1 million years old, and others are several billion years old. Once all of the interstellar matter surrounding a cluster has been used to form stars or has dispersed and moved away from the cluster, star formation ceases, and stars of progressively lower mass move off the main sequence, as shown in Figures 21.10, 21.12, and 21.13.

Even the youngest of the globular clusters in our Galaxy is older than the oldest open cluster. All of the globular clusters have main sequences that turn off at a luminosity less than that of the Sun. Star formation in these systems ceased billions of years ago, and no new stars are coming on to the main sequence to replace the ones that have turned off.

Indeed, the globular clusters are the oldest structures in our Galaxy (and in other galaxies as well). The oldest have ages of about 13 billion years. Since these are the oldest objects we know of, this estimate is one of the best limits we have on the age of the universe itself—it must be at least 13 billion years old. We will return to the fascinating question of determining the age of the entire universe in Chapter 28.

21.4 FURTHER EVOLUTION OF STARS

The "life story" we have related so far applies to all stars: Every one of them starts as a contracting protostar, then lives most of its life as a stable main-sequence star, and eventually moves off the main sequence toward the red-giant region. The pace at which each star goes through the stages depends, as you now know, on its mass, with more massive stars evolving more quickly. But after this point, the life stories of stars of different masses diverge, with a wider range of behavior possible according to mass, composition, and the presence of any nearby companion stars.

Because we have written this book for nonscience students who are taking their first astronomy course, we will recount a somewhat simplified version of what happens to stars as they move toward the final stages in their lives. We will (perhaps to your heartfelt relief) not delve into all the possible ways stars can behave. Instead, we will focus on only the key stages in the evolution of single stars and show how the evolution of high-mass stars differs from that of low-mass stars (such as our Sun).

21.4.1 Helium Fusion

Let's begin by considering stars whose *initial* masses are comparatively low—no more than about two to three times the mass of our Sun. (That may not sound all that low, but stars with masses less than this all behave in a fairly similar fashion. We will see what happens to more massive stars in the next section.) Because there are many more low-mass stars than high-mass stars (see Chapter 17), the vast majority of stars—including our Sun—follow the scenario we are about to relate. By the way, we carefully used the term *initial masses* of stars because, as we will see, stars can lose quite a bit of mass in the process of aging and dying.

Remember that red giants start out with a helium core where no energy generation is taking place, surrounded by a shell where hydrogen is undergoing fusion. The core, however, is shrinking and growing hotter. Once it reaches a temperature of 100 million K (but not before), three helium atoms can fuse to form a single carbon nucleus. This process is called the **triple-alpha process,** so named because physicists call the nucleus of the helium atom an alpha particle.

When the triple-alpha process begins in the low-mass stars, our calculations show that the entire core is ignited in a quick burst of fusion called the **helium flash.** As soon as the temperature becomes high enough to start the triple-alpha process, the extra energy released is transmitted quickly through the entire core, producing a rapid heating of all the helium there. The heating speeds up the nuclear reactions, which provides more heating, which ac-

celerates the nuclear reactions. We have runaway generation of energy, which reignites the entire core in a flash.

You might wonder why the next major step in nuclear fusion in stars involves *three* helium nuclei and not just two. Although it is a lot easier to get two heliums to collide, the product of this collision is not stable and falls apart very quickly. It takes three helium nuclei coming together to make a stable nuclear structure. Given that each helium nucleus has two positive protons and such protons all repel one another, you can begin to see the problem. It takes a temperature of 100 million K to slam helium nuclei together so that their six protons can overcome their natural repulsion and stick. But when they do, the star has produced a carbon nucleus!

Ace Astronomy™ Log into AceAstronomy and select this chapter to see the Active Figure called "Stellar Evolution of High and Low Mass Stars."

ASTRONOMY BASICS
Stars in Your Little Finger

Stop reading for a moment and look at your little finger. It's full of carbon atoms because carbon is a fundamental chemical building block for life on Earth. Every one of those carbon atoms was once inside a red giant star and was fused from helium nuclei in the triple-alpha process. All the carbon on Earth—in you, in the charcoal you use for barbecuing, and in the diamonds you might exchange with your sweetheart—was "cooked" by previous generations of stars. How the carbon atoms (and other elements) made their way from inside some of those stars to become part of the Earth is something we will discuss in the next chapter. For now, we want to emphasize that our description of stellar evolution is, in a very real sense, the story of our own cosmic "roots"—the history of how our own atoms originated among the stars. We are made of "star-stuff."

■ ■ ■ ■ ■ ■ ■ ■ ■ ■ ■ ■ ■

21.4.2 Becoming a Giant Again

After the helium flash, the star, having survived the "energy crisis" that followed the end of the main-sequence stage, finds its balance again. As the star readjusts to the release of energy from the triple alpha process in its core, its internal structure changes once more. Its surface temperature increases and its overall luminosity decreases. The point that represents the star on the H–R diagram thus moves to a new position to the left of and somewhat below its place as a red giant (Figure 21.15). The star then continues to fuse the helium in its core for a while, returning to the kind of equilibrium between pressure and gravity that characterized the main-sequence stage.

During this time, a newly formed carbon nucleus can sometimes be joined by another helium nucleus to produce a nucleus of oxygen.

However, at a temperature of 100 million degrees, the inner core is converting its helium fuel to carbon (and oxygen) at a rapid rate. Thus, the new period of stability cannot last very long: It is far shorter than the main-sequence stage. Soon all the helium hot enough for fusion will be used up, and again the inner core will not be able to generate energy via fusion. Once more, gravity will take over.

The star's situation is analogous to the end of the main-sequence stage (when the central hydrogen got used up), but the star now has a somewhat more complicated structure. Again the star's core begins to collapse under its own weight. Heat released by the shrinking of the carbon and oxygen core flows into a shell of helium just above the core. This helium, which had not been hot enough for fusion into carbon earlier, is heated just enough for fusion to begin and to generate a new flow of energy.

Farther out in the star another shell forms where fresh hydrogen has been heated enough to form helium. As energy flows outward from the two shells, once again the outer regions of the star begin to expand. Its brief period of stability over, the star moves back to the red giant domain on the H–R diagram for a short time (see Figure 21.15). But this is a brief and final burst of glory.

Recall that the last time the star was in this predicament, helium fusion came to its rescue. The temperature at the star's center eventually became hot enough for the *product* of the previous step of fusion (helium) to become the *fuel* for the next step (helium fusing into carbon). But the step after the fusion of helium nuclei requires a temperature so hot that the kinds of lower-mass stars we are discussing simply cannot compress their cores to reach it. No further types of fusion are possible for such a star.

In stars with masses similar to that of the Sun, the formation of a carbon–oxygen core thus marks the end of the generation of nuclear energy at the center of the star. The star must now confront the fact that its death is near. We will discuss the death of stars in the next chapter, but in the meantime Table 21.4 summarizes the stages discussed so far in the life of a star with the mass of the Sun. One thing that gives us confidence in our calculations of stellar evolu-

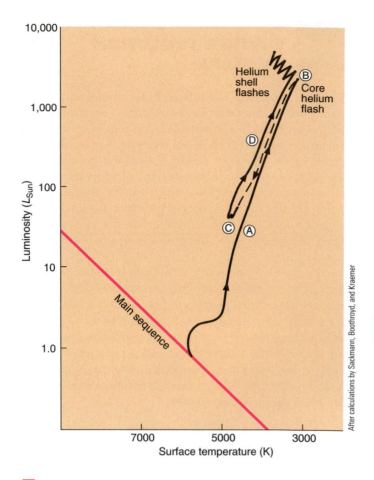

After calculations by Sackmann, Boothroyd, and Kraemer

■ **FIGURE 21.15**

Evolution of a Star Like the Sun on an H–R Diagram
Each stage in the star's life is labeled with a letter. (A) The star evolves from the main sequence to be a red giant, decreasing in surface temperature and increasing in luminosity. (B) A helium flash occurs at this point, leading to a readjustment of the star's internal structure and to (C) a brief period of stability during which helium is fused to carbon and oxygen in the core (in the process the star becomes hotter and less luminous than it was as a red giant). (D) After the central helium is exhausted, the star becomes a giant again and moves to higher luminosity and lower temperature. By this time, however, the star has exhausted its inner resources and will soon begin to die. Where the evolutionary track becomes a dashed line, the changes are so rapid that they are difficult to model.

TABLE 21.4 *The Evolution of a Star with the Sun's Mass*

Stage	Time in This Stage (years)	Surface Temperature (K)	Luminosity (L_{Sun})	Diameter (diameter of Sun = 1)
Main sequence	11 billion	6000	1	1
Becomes red giant	1.3 billion	Down to 3100	Up to 2300	165
Helium fusion	100 million	4800	50	10
Giant again	20 million	3100	5200	180

tion is that when we make H–R diagrams of older clusters, we see stars in each of the stages we have been discussing.

21.4.3 Mass Loss from Giant Stars and the Formation of Planetary Nebulae

When stars become giants, they begin to lose a substantial fraction of their mass into space. Astronomers estimate that by the time a star like the Sun reaches the point of the helium flash, for example, it will lose as much as 25 percent of its mass. And it can lose still more mass when it ascends the red-giant branch for the second time. As a result, aging stars are surrounded by one or more expanding shells of gas, each containing as much as 0.1 or 0.2 M_{Sun} of material (10–20 percent of the Sun's mass).

When nuclear energy generation in the carbon–oxygen core ceases, the star's core begins to shrink again and to heat up as it gets more and more compressed. The whole star follows along, shrinking and also becoming very hot—reaching surface temperatures of 100,000 K. Such hot stars are very strong sources of stellar winds and ultraviolet radiation, which sweep outward into the shells of material ejected when the star was a red giant. The winds and the ultraviolet radiation heat the shells, ionize them, and set them aglow (just as ultraviolet radiation from hot, young stars produces H II regions—see Chapter 19).

The result is the creation of some of the most beautiful objects in the cosmos (see the gallery in Figure 21.16 plus the image that opens this chapter). These objects were given an extremely misleading name when first found: **planetary nebulae.** The name is derived from the fact that a few planetary nebulae, when viewed through a small telescope, bear a superficial resemblance to planets. Actually, they have nothing to do with planets, but once names are put into regular use in astronomy, it is extremely difficult to change them. There are tens of thousands of planetary nebulae in our own Galaxy, although many are hidden from view because their light is absorbed by interstellar dust.

As Figure 21.16 shows, sometimes a planetary nebula appears to be a simple ring. Others have faint shells surrounding the bright ring, which are evidence of multiple episodes of mass loss when the star was a red giant (see Figure 21.16d). In a few cases, we see two lobes of matter flowing in opposite directions. Many astronomers think that all planetary nebulae are basically the same but that the shape we see depends on the viewing angle (Figure 21.17). According to this idea, the dying star is surrounded by a very dense donut-shaped disk of gas. (Theorists do not yet have a good explanation for why the dying star should produce this ring, but since observers see it, the star has managed somehow.)

As the star continues to lose mass, any less dense gas that leaves the star cannot penetrate the thick donut. Where it *can* flow outward is in directions perpendicular to the disk. If we look perpendicular to the direction of outflow, we see the disk and both of the outward flows (Figure 21.16b). If we look "down the barrel" and into the flows, we see a ring (Figure 21.16a). At intermediate angles, we may see wonderfully complex structures.

Planetary nebula shells usually expand at speeds of 20–30 km/s, and a typical planetary nebula has a diameter of about 1 LY. If we assume that the gas shell has expanded at a constant speed, we can calculate that the shells of all the planetary nebulae visible to us were ejected within the past 50,000 years or so. After this amount of time, the shells have expanded so much that they are too thin and tenuous to be seen. That's a pretty short time that each planetary nebula can be observed (when compared to the whole lifetime of the star). Given the number of such nebulae we nevertheless see, we must conclude that a large fraction of all stars evolve through the planetary nebula phase. This confirms our view of planetary nebulae as a sort of "last gasp" of low-mass star evolution.

21.4.4 Cosmic Recycling

The loss of mass by dying stars is a key step in the gigantic cosmic recycling scheme we discussed in Chapter 19. Remember that stars form from vast clouds of gas and dust. As they end their lives, stars return part of themselves to the galactic reservoirs of raw material. Eventually, some of the expelled material from aging stars will participate in the formation of new star systems.

However, the atoms returned to the Galaxy by an aging star are not necessarily the same ones it received initially. The star, after all, has fused new elements over the course of its life. And during the red-giant stage, material from the star's central regions is dredged up and mixed with its outer layers. As a result, the winds that blow outward from such stars include atoms that were "newly minted" inside the stars' cores. (As we will see, this mechanism is even more effective for high-mass stars, but it does work for stars with masses like that of the Sun.) In this way, the raw material of the Galaxy not only is resupplied but also receives infusions of new elements. You might say this allows the universe to get more "interesting" all the time.

21.5 THE EVOLUTION OF MORE MASSIVE STARS

If what we have described so far were the whole story of the evolution of stars and elements, we would have a big problem on our hands. We will see in later chapters that in our best models of the first few minutes of the universe, everything starts with the two simplest elements—hydrogen and helium (plus a tiny bit of lithium). All the predictions of the models imply that no heavier elements were produced at the beginning. Yet when we look around us on Earth, we see lots of other elements besides hydrogen and helium.

(a)

(b)

(c)

(d)

■ **FIGURE 21.16 A Gallery of Planetary Nebulae**

A Gallery of Planetary Nebulae We show off the capabilities of the Hubble Space Telescope (HST) with a series of beautiful images depicting some intriguing planetary nebulae.

(a) Perhaps the best known planetary nebula is the Ring Nebula (M57), located about 2000 LY away in the constellation of Draco. The ring is about 1 LY in diameter, and the central star has a temperature of about 120,000°C. Careful study of this image has shown scientists that, instead of looking at a spherical shell around this dying star, we may be looking down the barrel of a tube or cone. Blue isolates emission from very hot helium, which is located very close to the star; red shows emission from ionized nitrogen, which is radiated by the coolest gas farthest from the star; and green represents oxygen emission, which is produced at intermediate temperatures and is at an intermediate distance from the star.

(b) In this planetary nebula, the central star (which is part of a binary system) has ejected mass preferentially in two opposite directions. In other images, a disk, perpendicular to the two long streams of gas, can be seen around the two stars in the middle. The stellar outburst that resulted in the expulsion of matter occurred about 1200 years ago. Neutral oxygen is shown in red, once-ionized nitrogen in green, and twice-ionized oxygen in blue. The planetary nebula is about 2100 LY away in the constellation Ophiuchus.

(c) In this image of the planetary nebula NGC 6751, blue regions mark the hottest gas, which forms a ring around the central star. Orange and red show the locations of cooler gas. The origin of these cool streamers is not known, but their shape indicates that they are affected by radiation and stellar winds from the hot star at the center. The temperature of the star is about 140,000°C. The diameter of the nebula is about 600 times larger than the diameter of our own solar system. The nebula is about 6500 LY from us and in the constellation of Aquila.

(d) This image of the planetary nebula NGC 7027 shows several stages of mass loss. Faint blue concentric shells surrounding the central region identify mass that was shed slowly from the surface of the star when it became a red giant. Somewhat later, the remaining outer layers were ejected but not in a spherically symmetric way. The dense clouds formed by this late ejection produce the bright inner regions. The hot central star can be seen faintly near the center of the nebulosity. NGC 7027 is about 3000 LY away in the direction of the constellation Cygnus.

■ **FIGURE 21.17**

A Model to Explain the Different Shapes of Planetary Nebulae It may be that the many different shapes we see in a gallery of planetary nebulae can be explained by a single model viewed from different directions. In this model, the hot central star is surrounded by a thick torus (or donut) of gas. The star's wind cannot flow out into space in the direction of the torus but can escape freely in directions perpendicular to it. If we look along the direction of the flow, we see a spherical shell of gas (like looking directly down into an empty ice cream cone). If we look along the equator of the torus, we see both outflows. At in between angles, we may see a very complex shape.

The Red Giant Sun and the Fate of the Earth

How will the evolution of the Sun affect conditions on Earth? Although the Sun has appeared reasonably steady in size and luminosity over recorded human history, that brief span means nothing compared with the timescales we have been discussing. Let's examine the long-term prospects for our planet.

The Sun took its place on the zero-age main sequence approximately 4.5 billion years ago. At that time it emitted only about 70 percent of the energy that it radiates today. One might expect that Earth would have been a lot colder than it is now, with the oceans frozen solid. But if this were the case, it would be hard to explain why simple life forms existed when the Earth was less than a billion years old. Scientists now think that much more carbon dioxide was present in the Earth's atmosphere when it was young, and that a much stronger greenhouse effect kept the Earth warm. (In the greenhouse effect, a gas like carbon dioxide allows the Sun's light to come in but does not allow the infrared radiation from the ground to escape back into space. So the temperature near the Earth's surface increases.)

Carbon dioxide has since steadily declined as the Sun has increased in luminosity. As the brighter Sun increases the temperature of the Earth, rocks weather faster and react with carbon dioxide, removing it from the atmosphere. The warmer Sun and the weaker greenhouse effect have kept the Earth at a nearly constant temperature for most of its life. This remarkable coincidence, which has resulted in fairly stable climactic conditions, has been key to the development of complex life-forms on our planet.

As a result of changes caused by the buildup of helium in its core, the Sun will continue to increase in luminosity

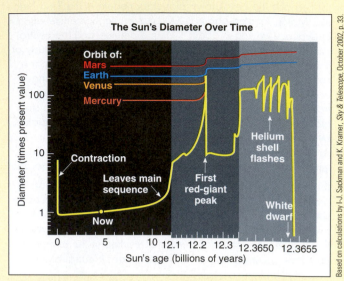

Relationship Between the Diameter and Age of the Sun This diagram shows what our best models predict about how the diameter of the Sun will change over long periods of time. Shown on the same scale are the sizes of the orbits of the terrestrial planets. Note that the timescale along the bottom changes as we move from section to section of the diagram. The planet orbits grow larger as the Sun loses mass. Earth remains outside the Sun, while Mercury and Venus are swallowed up.

as it grows older. More and more radiation will reach the Earth. For a while, the amount of carbon dioxide will continue to decrease. (Note that this effect counteracts increases in carbon dioxide from human activities, but on much too slow a timescale to undo the changes in climate that are likely to occur in the next 100 years.)

These elements must have been made somewhere in the universe, *and the only place hot enough to make them is inside stars.* One of the fundamental discoveries of 20th-century astronomy is that the stars are the source of all the chemical richness that characterizes our world and our lives.

We have already seen that carbon and oxygen are the results of fusion inside stars that become red giants. But where do the heavier elements we know and love (such as the silicon and iron inside the Earth, and the gold and silver in our jewelry) come from? The kind of stars we have been discussing so far never get hot enough to make them.

It turns out that such heavier elements can be formed only late in the lives of more massive stars.

21.5.1 Making New Elements in Massive Stars

Massive stars evolve in much the same way that the Sun does (but always more quickly)—up to the formation of a carbon–oxygen core. One difference is that for stars with more than about twice the mass of the Sun, helium begins fusion more gradually, rather than with a sudden flash. Also, when more massive stars become red giants, they become

Eventually, the heating of the Earth will melt the polar caps and increase the evaporation of the oceans. Water vapor is also an efficient greenhouse gas and will more than compensate for the decrease in carbon dioxide. Sooner or later—atmospheric models are not yet good enough to say exactly when, but estimates range from 500 million to 2 billion years—the increased water vapor will cause a runaway greenhouse effect.

About 3.5 billion years from now, the Earth will lose its water vapor. In the upper atmosphere, sunlight breaks water vapor down into hydrogen and oxygen, and the fast-moving hydrogen atoms escape into outer space. Like Humpty Dumpty, the water molecules cannot be put back together again. The Earth will start to resemble the Venus of today, and temperatures will become much too high for life as we know it.

All of this will happen before the Sun even becomes a red giant. Then the bad news really starts. Most astronomers used to think that the expanding Sun would engulf and incinerate the Earth, thereby sealing its fate. In fact, the Sun will swallow Mercury and probably Venus, and friction with our star's outer atmosphere will make these planets spiral inward until they are completely vaporized. Earth will probably escape this fate because of the mass loss that will occur when the Sun becomes a red giant. The gravitational attraction of the Sun decreases when it loses mass. The result is that the diameter of the Earth's orbit will increase (remember Kepler's third law). Models indicate that the mass loss is just enough to keep the Earth above the surface of the Sun (see the figure). However, that close to our star, all life on Earth will surely be incinerated.

What then are the prospects for preserving Earth life as we know it? The first strategy you might think of would be to move humanity to a more distant and cooler planet. However, calculations indicate that there are long periods of time (several hundred million years) when no planet is habitable. For example, Earth becomes far too warm for life long before Mars warms up enough.

A better alternative may be to move the Earth progressively farther from the Sun. The idea is to use gravity in the same way NASA has used it to send spacecraft to distant planets. When a spacecraft flies near a planet, the planet's motion can be used to speed up the spacecraft, slow it down, or redirect it. Calculations show that if we were to redirect an asteroid so that it follows just the right orbit between Earth and Jupiter, it could transfer orbital energy from Jupiter to Earth and move Earth slowly outward, pulling us away from the expanding Sun on each flyby. Since we have hundreds of millions of years to change Earth's orbit, the effect of each flyby need not be large. Of course, the people directing the asteroid had better get the orbit exactly right and not hit Earth!

It may seem crazy to think about projects to move an entire planet to a different orbit. But remember that we are talking about the distant future. If, by some miracle, human beings are able to get along for all that time and don't blow ourselves to bits, our technology is likely to be far more sophisticated than it is today. It may also be that if humans survive for hundreds of millions of years, we may spread to planets or habitats around other stars. Indeed Earth, by then, might be a museum world to which youngsters from other planets return to learn about the origin of our species. It is also possible that evolution will by then have changed us in ways that allow us to survive in very different environments. Wouldn't it be exciting to see how the story of the human race turns out after all those years?

---

so bright and large that we call them *supergiants*. Such stars can expand until their outer regions become as large as the orbit of Jupiter, which is precisely what the Hubble Space Telescope has shown for the star Betelgeuse (see Figure 21.3). They also lose mass very effectively, producing dramatic winds and outbursts as they age. Figure 21.18 shows a wonderful image of the very massive star Eta Carinae, with a great deal of ejected material clearly visible.

But the crucial way that massive stars diverge from the story we have outlined is that they can start further kinds of fusion in their centers. The outer layers of a star with a mass greater than about 8 solar masses weigh enough to compress the carbon–oxygen core until it becomes hot enough to ignite carbon. Carbon can fuse into still more oxygen as well as neon, sodium, magnesium, and finally silicon. After each of the possible sources of nuclear fuel is exhausted, the core contracts until it reaches a temperature high enough to lead to the fusion of still heavier nuclei.

Theorists have now found mechanisms whereby virtually all chemical elements of atomic weights up to that of iron can be built up by this **nucleosynthesis** (the making of new atomic nuclei) in the centers of the more massive red giant stars. This still leaves the question of where

■ FIGURE 21.18

Eta Carinae With a mass at least 100 times that of the Sun, the hot supergiant Eta Carinae is one of the most massive stars known. This highly computer-processed image from the Hubble Space Telescope records the two giant lobes and equatorial disk of material it has ejected in the course of its evolution. The pink outer region is material ejected in an outburst seen in 1843, the largest such mass-loss event that any star is known to have survived. Moving away from the star at a speed of about 1000 km/s, the material is rich in nitrogen and other elements formed in the interior of the star. The inner blue-white region is material ejected at lower speeds and thus still closer to the star. It appears blue-white because it contains dust and reflects the light of Eta Carinae, whose luminosity is 4 million times that of our Sun.

elements *heavier* than iron come from. We will see in the next chapter that when massive stars finally exhaust their nuclear fuel, they most often die in a spectacular explosion. Heavier elements can be synthesized in the unspeakable violence of such explosions.

It is remarkable that our theories of nucleosynthesis inside stars are able to predict the relative abundances with which the elements occur in nature. That is, the way stars build up elements during various nuclear reactions can really explain why some elements are common and others quite rare.

21.5.2 The Elements in Globular Clusters and Open Clusters Are Not the Same!

The fact that elements are made in stars over time explains an important difference between globular and open clusters. Hydrogen and helium, the most abundant elements in stars in the solar neighborhood, are also the most abundant constituents of stars in both kinds of clusters. However, the abundances of the elements heavier than helium are very different.

In the Sun and most of its neighboring stars, the combined abundance (by mass) of the elements heavier than hydrogen and helium is between 1 and 4 percent of the star's mass. Spectra show that most open-cluster stars also have 1 to 4 percent of their matter in the form of heavy elements. Globular clusters, however, are a different story. The heavy-element abundance of stars in typical globular clusters is found to be only 1/10 to 1/100 that of the Sun.

These differences in chemical composition are a direct consequence of when the cluster stars were formed. The very first generation of stars initially contained only hydrogen and helium. We have seen that these stars, to generate energy, created heavier elements in their interiors. In the last stages of their lives, they ejected matter, now enriched in heavy elements, into the reservoirs of raw material between the stars. Such matter was then incorporated into a new generation of stars.

This means that the relative abundance of the heavy elements must be less and less as we look further into the past. We saw that the globular clusters are much older than the open clusters. Since globular-cluster stars formed much earlier than those in open clusters, they have only a relatively small abundance of elements heavier than hydrogen and helium.

As time passes, the proportion of heavier elements increases. This means that the first generation of stars that formed in our Galaxy was unlikely to have been accompanied by a planet like Earth, full of silicon, iron, and many other heavy elements. The Earth (and the astronomy students who live on it) was possible only after stars had a chance to make and recycle their heavier elements.

Ace Astronomy™ Log into AceAstronomy and select this chapter to see the Active Figure called "Future of the Sun."

21.5.3 Approaching Death

Compared with the main-sequence lifetimes of stars, the events that characterize the last stages of stellar evolution pass very quickly. As the star's luminosity increases, its rate of nuclear fuel consumption goes up rapidly—just at that point in its life when its fuel supply is beginning to run down. It is as if a person suddenly did everything possible to hasten death—overeating, overdrinking, smoking like a chimney, and so on.

After the prime fuel, hydrogen, is exhausted in a star's core, other sources of nuclear energy are available to the star—in the fusion first of helium and then of other, more complex elements. But the energy yield of these reactions is much less than that of the fusion of hydrogen to helium. And to trigger these reactions, the central temperature must be higher than that required for the fusion of hydro-gen to helium, leading to even more rapid consumption of fuel. Clearly this is a losing game, and very quickly the star reaches its end. As it does so, however, some remarkable things can happen—as we will see in the next chapter.

SURFING THE WEB

 A Tour of the Orion Region:
www.gb.nrao.edu/~rmaddale/Education/OrionTourCenter/index.htm
Astronomer Ronald Maddalena conducts this Web tour of what the Orion molecular cloud region looks like at many different wavelengths.

 Recent Work on the Supergiant Star Betelgeuse:

- hubblesite.org/newscenter/archive/1996/04/
  (Hubble observations and measurements)

- www.nrao.edu/pr/betel/
  (observations with an array of radio telescopes)

- cfa-www.harvard.edu/cfa/ep/pressrel/betel97.htm
  (on how the star pulsates)

 Recent Work on the Supergiant Eta Carinae:

- www.etacarinae.iag.usp.br
  (information and images, some technical, some quite accessible)

- hubblesite.org/newscenter/archive/1996/23/
  (dramatic Hubble observations)

- chandra.harvard.edu/press/99_releases/press_100899.html
  (observations with the Chandra X-ray Observatory)

 Messier Catalog Clusters:
www.seds.org/messier/objects.html#cluster
The wonderful Messier catalog site we recommended in Chapter 19 includes lists of and information about the best-known open and globular clusters, and many useful links

 Galleries and Discussions of Planetary Nebula Images:

- hubblesite.org/newscenter/archive/category/nebula/planetary/
  (Hubble images)

- hubblesite.org/newscenter/archive/1997/38/
  (some of the best Hubble images, plus background information)

- www.noao.edu/image_gallery/planetary_nebulae.html
  (images from telescopes on the ground and in space)

- www.blackskies.com
  (a page by amateur astronomer Doug Snyder on observing planetary nebulae with smaller telescopes)

 Planetary Nebulae and the Future of the Solar System:
www.astro.washington.edu/balick/WFPC2
Astronomer Bruce Balick (U. of Washington), part of a team studying planetary nebulae with the Hubble Space Telescope and other instruments, has put together an informative set of pages describing the team's work and the implications for understanding the future of our Sun.

SUMMARY

21.1 When stars first begin to fuse hydrogen to helium, they lie on the **zero-age main sequence.** The amount of time a star spends in the main-sequence stage depends on its mass. More massive stars complete each stage of evolution more quickly than low-mass stars. The fusion of hydrogen to form helium changes the interior composition of a star, which in turn results in changes in its temperature, luminosity, and radius. Eventually, as stars age, they evolve away from the main sequence to become red giants or supergiants. The core of a red giant is contracting, but the outer layers are expanding as a result of fusion in a shell outside the core. The star gets larger and redder as it expands and cools.

21.2 Calculations that show what happens as stars age can be checked by measuring the properties of stars in clusters. The members of a given cluster were formed at about the same time and have the same composition, so they differ mainly in mass and thus in the stage they have reached in their lives. There are three types of star clusters. **Globular clusters** have diameters of 50–450 LY, contain hundreds of thousands of old stars, and are distributed in a halo around the Galaxy. **Open clusters** typically contain hundreds of young to middle-aged stars, are located in the plane of the Galaxy, and have diameters less than 30 LY. **Associations** are found in regions of gas and dust and contain extremely young stars.

21.3 The H–R diagram of stars in a cluster changes systematically as the cluster grows older. The most massive stars evolve the most rapidly. In the youngest clusters and associations, highly luminous blue stars are on the main sequence; the stars with the lowest masses lie to the right of the main sequence and are still contracting toward it. With passing time, stars of progressively lower mass evolve away from (turn off) the main sequence. In globular clusters, which have a typical age of about 13 billion years, there are no luminous blue stars at all. Astronomers can use the turn-off point from the main sequence to determine the age of a cluster.

21.4 After stars become red giants, their cores eventually become hot enough to produce energy by fusing helium to form carbon and oxygen. The fusion of three helium nuclei produces carbon through the **triple-alpha process.** The rapid onset of helium fusion in the core of a low-mass star is called the **helium flash.** After this, the star becomes stable and reduces its luminosity and size briefly. In stars with masses about twice the mass of the Sun or less, fusion stops after the helium in the core has been exhausted. Fusion of hydrogen and helium in shells around the contracting core makes the star a bright giant again, but only temporarily. When the star is a red giant, it can shed its outer layers and thereby expose hot inner layers. **Planetary nebulae** (which have nothing to do with planets) are shells of gas ejected by such stars, set glowing by the ultraviolet radiation of the dying central star.

21.5 In stars with masses greater than about 8 solar masses, nuclear reactions involving carbon, oxygen, and still heavier elements can build up nuclei as heavy as iron. Even heavier nuclei can be created during supernova explosions. This creation of elements is called **nucleosynthesis.** The late stages of evolution occur very quickly. Ultimately all stars must use up all of their available energy supplies. In the process of dying, most stars eject some matter, enriched in heavy elements, into interstellar space where it can be used to form new stars. Each succeeding generation of stars therefore contains a larger proportion of elements heavier than hydrogen and helium. This progressive enrichment explains why the stars in open clusters contain more heavy elements than do those in globular clusters, and it tells us where most of the atoms on Earth and in our bodies come from.

INTER-ACTIVITY

A Have your group take a look at the list of the brightest stars in the sky in Appendix 11. What fraction of them are past the main-sequence phase of evolution? The text says that stars spend 90 percent of their lifetimes in the main-sequence phase of evolution. That suggests that if we have a fair sample of stars, 90 percent of them should be main-sequence stars. Your group should brainstorm why 90 percent of the brightest stars aren't in the main-sequence phase of evolution.

B Reading an H–R diagram can be tricky. Suppose your group is given the H–R diagram of a star cluster. Stars above and to the right of the main sequence could be either red giants that had evolved away from the main sequence or very young stars that are still evolving toward the main sequence. Discuss how you would decide which they are.

C In Chapter 29, we discuss some of the efforts now underway to search for radio signals from possible intelligent civilizations around other stars. Our present resources for carrying out such searches are very limited and there are many stars in the Galaxy. Your group is a committee set up by the International Astronomical Union to come up with a list of the best possible stars with which such a search should begin. Make a list of criteria for choosing the stars on the list, and explain the reasons behind each entry.

D Make a list of the reasons a star that formed at the very beginning of the universe (soon after the big bang) could not have a planet with astronomy students reading astronomy textbooks (even if it is the same mass as our Sun).

E Since we are pretty sure that when the Sun becomes a giant star, all life on Earth will be wiped out, should we start making preparations of any kind? Let's suppose that a political leader who fell asleep during large parts of his astronomy class suddenly hears about this problem from a large donor and appoints your group as a task force to make suggestions for how to prepare for the end of the Earth. Make a list of arguments for why such a task force is not really necessary.

REVIEW QUESTIONS

Ace⊜Astronomy™ Assess your understanding of this chapter's topics with additional quizzing and animations at **http://ace.brookscole.com/voyages**

1. Compare the following stages in the lives of a human being and a star: prenatal, birth, prolonged adolescence, middle age, old age. What does a star with the mass of our Sun do in each of these stages?

2. What is the main factor that determines where a star falls along the main sequence?

3. What happens when a star exhausts the hydrogen in its core and stops the generation of energy by the nuclear fusion of hydrogen to helium?

4. Describe the evolution of a star with a mass similar to that of the Sun, from the protostar stage to the time it first becomes a red giant. Give the description in words and then sketch the evolution on an H–R diagram.

5. Describe the evolution of a star with a mass similar to that of the Sun, from just after it first becomes a red giant to the time it exhausts the last type of fuel its core is capable of fusing. After describing the stages in words, sketch them on an H–R diagram.

6. Suppose you have discovered a new star cluster. How would you determine whether it is an open or a globular cluster? List several characteristics that might help you decide.

7. Explain how an H–R diagram can be used to determine the age of a cluster of stars.

8. Where did the carbon atoms in the trunk of a tree on your college campus come from originally? Where did the neon in the fabled "neon lights of Broadway" come from originally?

9. What is a planetary nebula? Will we have one around the Sun? Roughly when?

THOUGHT QUESTIONS

10. Use star charts to identify at least one open cluster visible at this time of year. (Such charts can be found in *Sky & Telescope* and *Astronomy* magazines each month.) The Pleiades and Hyades are good autumn subjects, and Praesepe is good for springtime viewing. Go out and look at these clusters with binoculars and describe what you see.

11. Is the Sun on the zero-age main sequence? Explain your answer.

12. Which of the planets in our solar system have orbits that are smaller than the photospheric radius of Betelgeuse listed in Table 21.2?

13. Would you expect to find an Earth-like planet around a very low-mass star that formed right at the beginning of a globular cluster's life? Explain.

14. In the H–R diagrams for some young clusters, stars of both very low and very high luminosity are off to the right of the main sequence, whereas those of intermediate luminosity are on the main sequence. Can you offer an explanation? Sketch an H–R diagram for such a cluster.

15. If the Sun were a member of the cluster NGC 2264, would it be on the main sequence yet? Why?

16. If all the stars in a cluster have nearly the same age, why are clusters useful in studying evolutionary effects (different stages in the lives of stars)?

17. Suppose a star cluster were at such a large distance that it appeared as an unresolved spot of light through the telescope. What would you expect the overall color of the spot to be if it were the image of the cluster immediately after it was formed? How would the color differ after 10^{10} years? Why?

18. Suppose an astronomer known for joking around told you she had found a type-O main-sequence star that contained no elements heavier than helium. Would you believe her? Why?

One of the beautiful things about science is that you don't have to take someone else's word for anything. You can check and verify the results for yourself (although sometimes it may take very expensive equipment and a lot of training). If a result can't be verified independently, then it isn't science. You can check some of the assertions in this chapter with only a pencil and paper and what you have already learned. See if what we said was right!

19. The text says that a star does not change its mass by very much during the course of its main-sequence lifetime. While it is on the main sequence, a star converts about 10 percent of the hydrogen initially present into helium. Look in earlier chapters to find out what percentage of the hydrogen mass involved in fusion is lost because it is converted to energy. So how much does the mass of the whole star change as a result of fusion? Were we correct to say that the mass of a star does not change significantly while it is on the main sequence?

20. The text states that massive stars have shorter lifetimes than low-mass stars. Even though massive stars have more fuel to burn, they use it up faster than low-mass stars. You can check and see whether this statement is true. The lifetime of a star is directly proportional to the amount of mass (fuel) it contains and inversely proportional to the rate at which it uses up that fuel (i.e., to its luminosity). Since the lifetime of the Sun is about 10^{10} years, we have the following relationship:

$$T = 10^{10}\frac{M}{L}\text{ years}$$

where T is the lifetime of a main-sequence star, M is its mass measured in terms of the mass of the Sun, and L is its luminosity measured in terms of the Sun's luminosity.

(a) Explain in words why this equation works.
(b) Use the data in Table 17.2 to calculate the ages of the main-sequence stars listed.

(c) Do low-mass stars have longer main-sequence lifetimes?
(d) Do you get the same answers as those in Table 21.1?

21. You can use the equation in problem 20 to estimate the approximate ages of the clusters in Figures 21.10, 21.12, 21.13, and 21.14. Use the information in the figures to determine the luminosity of the most massive star still on the main sequence. Now use the data in Table 17.2 to estimate the mass of this star. Then calculate the age of the cluster. This method is similar to the procedure used by astronomers to obtain the ages of clusters, except they use actual data and model calculations rather than simply making estimates from a drawing. How do your ages compare with the ages in the text?

22. Detailed model calculations indicate that when the Sun reaches the top of the giant branch a second time, it will have a mass of only $0.68M_{Sun}$. Its diameter will be 172 million km.

(a) Use Kepler's third law

$$(M_1 + M_2)P^2 = D^3$$

to calculate the size of Earth's orbit. (You can find our explanation of this formula in Chapter 2.) Is this larger or smaller than the diameter of the Sun at this time?
(b) What would your conclusion be about the fate of Earth if the Sun had no mass loss and thus Earth remained in its present orbit?

23. You can estimate the age of the planetary nebula in Figure 21.16c. The diameter of the nebula is 600 times the diameter of our own solar system, or about 0.8 LY. The gas is expanding away from the star at a rate of about 25 mi/s. Remember that distance = velocity × time, and calculate how long ago the gas left the star if its speed has been constant the whole time. Make sure you use consistent units for time, speed, and distance.

SUGGESTIONS FOR FURTHER READING

Chaboyer, B. "Rip Van Twinkle: The Oldest Stars Have Been Growing Younger" in *Scientific American,* May 2001, p. 44. On determining the age of the oldest stars in globular clusters.

Darling, D. "Breezes, Bangs, and Blowouts: Stellar Evolution Through Mass Loss" in *Astronomy,* Sept. 1985, p. 78; Nov. 1985, p. 94.

Davidson, K. "Crisis at Eta Carinae?" in *Sky & Telescope,* Jan. 1998, p. 36.

Djorgovsky, G. "The Dynamic Lives of Globular Clusters" in *Sky & Telescope,* Oct. 1998, p. 38. Cluster evolution and blue straggler stars.

Frank, A. "Angry Giants of the Universe" in *Astronomy,* Oct. 1997, p. 32. On luminous blue variables like Eta Carinae.

Garlick, M. "The Fate of the Earth" in *Sky & Telescope,* Oct. 2002, p. 30. What will happen when our Sun becomes a red giant.

Iben, I. and Tutukov, A. "The Lives of the Stars: From Birth to Death and Beyond" in *Sky & Telescope,* Dec. 1997, p. 36.

Kaler, J. *Stars.* 1992, Scientific American Library/W. H. Freeman. Good modern introduction to stellar evolution.

Kaler, J. "The Largest Stars in the Galaxy" in *Astronomy,* Oct. 1990, p. 30. On red supergiants.

Kwok, S. *Cosmic Butterflies: The Colorful Mysteries of Planetary Nebulae.* 2001, Cambridge U. Press. Introduction with over 100 images.

Kwok, S. "What Is the Real Shape of the Ring Nebula?" in *Sky & Telescope,* July 2000, p. 33. On seeing planetary nebulae from different angles.

Kwok, S. "Stellar Metamorphosis" in *Sky & Telescope,* Oct. 1998, p. 30. How planetary nebulae form.

Mullan, D. "Caution! High Winds Beyond This Point" in *Astronomy,* Jan. 1982, p. 74. On stellar winds and mass loss.

Wolfgang Brandner, JPL/IPAC; Eva K. Grebel, University of Washington; You-Hua Chu, University of Illinois at Urbana–Champaign; and NASA

The Stellar Life Cycle   This remarkable picture of NGC 3603, taken with the Hubble Space Telescope, captures the life cycle of massive stars from birth to near-death. In the upper right corner, we see small, dense, dark clouds that contain enough dust to block the light from behind them. It is likely that star formation will occur at some time in the future within these clouds. Near the center of the image is a cluster that contains massive hot stars that are only a few million years old. A torrent of ionizing radiation and fast stellar winds from these massive stars has blown a large cavity around the cluster. Near the bottom of the image, left of center, there are two small, tadpole-shaped emission nebulae. What we are seeing here is probably gas and dust evaporating from protoplanetary disks. The conspicuous bluish features in the upper left of center are associated with a blue supergiant, which is located at the center of a ring of glowing gas. This ring and the two blobs of glowing blue gas perpendicular to it were ejected by the star and are rich in heavy elements built inside the star by nuclear fusion reactions. The ring closely resembles that around SN 1987A, a star that exploded in 1987, and perhaps this star, too, will someday become a supernova.

The Death of Stars

He told me to look at my hand, for a part of it came from a star that exploded too long ago to imagine.

Paul Zindel in his play
The Effect of Gamma Rays on Man-in-the-Moon Marigolds

THINKING AHEAD

Do stars die with a bang or a whimper? In the preceding two chapters we have followed the life story of stars, from the process of birth to the brink of death. Now we are ready to explore the ways that stars end their lives. Sooner or later each star exhausts its store of nuclear energy. Without an internal energy source, every star eventually gives way to the inexorable pull of gravity and collapses under its own weight.

Following the rough distinction made in the last chapter, we will discuss the end-of-life evolution of stars of lower and higher mass separately. What determines the outcome—bang or whimper—is the mass of the star *when it is ready to die,* not the mass it was born with. As we noted in the last chapter, stars can lose a significant amount of mass in their middle and old age.

Ace◉**Astronomy**™   The AceAstronomy icon throughout the text indicates an opportunity for you to test yourself on key concepts and to explore animations and interactions on the AceAstronomy website at **http://ace.brookscole.com/voyages**

Virtual Laboratories

 White Dwarfs, Novae, and Supernovae

 Neutron Stars and Pulsars

22.1 THE DEATH OF LOW-MASS STARS

Let's begin with those stars whose final mass just before death is less than about 1.4 times the mass of the Sun. (We will explain why this is the crucial dividing line in a moment.) Note that most stars in the universe fall into this category. The number of stars decreases as mass increases (see Chapter 17); as in the music business, only a few stars get to be superstars. Furthermore, many stars with an initial mass much greater than 1.4 solar masses (M_{Sun}) will be reduced to that level by the time they die. Stars that start out with masses of at least 7.5 M_{Sun} and possibly as much as 10 M_{Sun} manage to lose enough mass to fit into this category (an accomplishment Jenny Craig customers would surely envy!)

22.1.1 A Star in Crisis

In the last chapter, we left the life story of a star with a mass like the Sun's just after it had climbed up to the red-giant region of the H–R diagram for a second time and had shed some of its outer layers to form a planetary nebula. Recall that during this time the *core* of the star was undergoing an "energy crisis." Earlier in its life, during a brief stable period, helium in the core had gotten hot enough to fuse into carbon (and oxygen). But after this helium was exhausted, the star's core had once more found itself without a source of pressure to balance gravity, and so had begun to contract.

This collapse is the final event in the life of the core. Because the star's mass is relatively low, it cannot push the temperature inside high enough to begin another round of fusion (the way larger-mass stars can). The core continues to shrink until it reaches a density equal to nearly a million times the density of water! At this extreme density, a new and different way for matter to behave kicks in and helps the star achieve a final state of equilibrium. In the process, what remains of the star becomes one of the strange *white dwarfs* that we met in Chapter 17.

22.1.2 Degenerate Stars

Because white dwarfs are far more dense than any substance on Earth, the matter inside them behaves in a very strange way—unlike anything we know from everyday experience. At this high density, *electrons* resist being squeezed closer together and set up a powerful pressure inside the core. This pressure is the result of the fundamental rules that govern the behavior of electrons. According to these rules, which have been verified by studying how electrons behave under laboratory conditions, no two electrons can be in the same place at the same time doing the same thing. We specify the *place* of an electron by its precise position in space, and we specify what it is doing by its motion and the way it is spinning.

The temperature in the interior of a star is always so high that the atoms are stripped of virtually all their electrons. For most of a star's life, the density of matter is also relatively low and the electrons in the star are moving rapidly. It is very unlikely that any two of them will be in the same place moving in exactly the same way at the same time. But this all changes when a star exhausts its store of nuclear energy and begins its final collapse.

As the star's core contracts, electrons are squeezed closer and closer together. Eventually a star like the Sun becomes so dense that further contraction would require two or more electrons to violate the rule against occupying the same place and moving in the same way. Such a hot, dense gas is said to be **degenerate** (a term coined by physicists and not related to the electron's moral character). The electrons in a degenerate gas resist further crowding with overwhelming pressure. (It's as if the electrons said, "You can press inward all you want, but there is simply no room for any other electrons to squeeze in here without violating the rules of our existence.")

The degenerate electrons do not require an input of heat to maintain the pressure they exert, and so a star with this kind of structure, if nothing disturbs it, can last essentially forever. (Note that the repulsive force between degenerate electrons is different from, and much stronger than, the normal electrical repulsion between charges that have the same sign.)

The electrons in a degenerate gas do move about, as do particles in any gas, but not with a lot of freedom. A particular electron cannot change position or momentum until another electron in an adjacent state gets out of the way. The situation is much like that in the parking lot after a big football game. Vehicles are closely packed, and a given car cannot move until the one in front of it moves, leaving an empty space to be filled. Of course the dying star also has atomic nuclei in it, not just electrons, but in this stage it is the pressure of the electrons that halts the collapse of the core.

22.1.3 White Dwarfs

White dwarfs, then, are stars with degenerate electron cores that cannot contract any further. Calculations showing that this is the likely end state of low-mass stars were first carried out by the Indian-American astrophysicist S. Chandrasekhar (see the Voyagers in Astronomy box). He was able to show how much a star will shrink before the degenerate electrons stop its contraction and hence what its final diameter will be (Figure 22.1). Note that the larger the mass of the star, the *smaller* its radius.

A white dwarf with a mass like that of the Sun has a diameter about the same as that of the Earth. This is a remarkably small size for a star and means that its material is very dense. A *teaspoonful* of white-dwarf material (if it could be brought to Earth in its compressed state) would weigh more than a full garbage truck.

VOYAGERS IN ASTRONOMY

Subrahmanyan Chandrasekhar

Born in 1910 in Lahore, India, Sub-rahmanyan Chandrasekhar (known as Chandra to his friends and colleagues) grew up in a home that encouraged scholarship and an interest in science. His uncle, C. V. Raman, was a physicist who won the 1930 Nobel Prize. A precocious student, Chandra tried to read as much as he could about the latest ideas in physics and astronomy, although obtaining technical books was not easy in India at the time. He finished college at age 19 and won a scholarship to study in England. It was during the long boat voyage to get to graduate school that he first began doing calculations about the structure of white dwarf stars.

Chandra developed his ideas during and after his studies as a graduate student, showing—as we have discussed—that white dwarfs with masses greater than 1.4 times the mass of the Sun cannot exist, and that the theory predicts the existence of other kinds of stellar corpses. He wrote later that he felt very shy and lonely during this period, isolated from other students, afraid to assert himself, and sometimes waiting for hours to speak with some of the famous professors he had read about in India. His calculations soon brought him into conflict with certain distinguished astronomers, including Sir

S. Chandrasekhar
(1910–1995)

Arthur Eddington, who publicly ridiculed Chandra's ideas. At a number of meetings of astronomers, such leaders in the field as Henry Norris Russell refused to give Chandra the opportunity to defend his ideas, while allowing his more senior critics lots of time to criticize them.

Yet Chandra persevered, writing books and articles elucidating his theories, which turned out not only to be correct, but to lay the foundation for much of our modern understanding of the death of stars. In 1983, he received the Nobel Prize in physics for this early work.

In 1937 Chandra came to America and joined the faculty at the University of Chicago, where he remained for the rest of his life. There he devoted himself to research and teaching, making major contributions to many fields of astronomy—from our understanding of the motions of stars through the Galaxy to the behavior of the bizarre objects called black holes (see Chapter 23). In 1999, NASA named its sophisticated orbiting x-ray telescope (designed in part to explore such stellar corpses) the Chandra Observatory.

Chandra spent a great deal of time with his graduate students, supervising the research of more than 50 PhDs during his life. He took his teaching responsibilities very seriously: During the 1940s, while based at the Yerkes Observatory, he willingly drove the more than 100-mile trip to the university each week to teach a class of only a few students.

Chandra also had a deep devotion to music, art, and philosophy, writing articles and books about the relationship between the humanities and science. He once wrote that "one can learn science the way one enjoys music or art . . . Heisenberg had a marvelous phrase 'shuddering before the beautiful' . . . that is the kind of feeling I have."

When Chandrasekhar made his calculations about white dwarfs, he found something very surprising, which is clear in Figure 22.1 as well. The size of a white dwarf shrinks as the mass in the star increases. According to the best theoretical models, a white dwarf with a mass of about 1.4 M_{Sun} or larger would have a radius of zero! What the calculations are telling us is that even the force of degenerate electrons cannot stop the collapse of a star with more mass than this. The maximum mass that a star can have and still become a white dwarf—1.4 M_{Sun}—is called the **Chandrasekhar limit.** Stars with masses that exceed this limit

have a different kind of end in store—one that we will explore in the next section.

22.1.4 The Ultimate Fate of White Dwarfs

If the birth of a main-sequence star is defined by the onset of fusion reactions, then we must consider the end of all fusion reactions to be the time of a star's death. As the core is stabilized by degeneracy pressure, a last shudder of fusion passes through the outside of the star, consuming the little

Courtesy of Emilio Segrè Visual Archives, Physics Today Collection

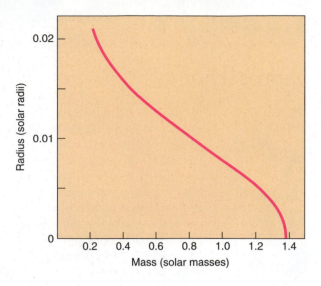

FIGURE 22.1

Relating Masses and Radii of White Dwarfs Models of white-dwarf structure predict that as the mass of the star increases (toward the right), it gets smaller and smaller.

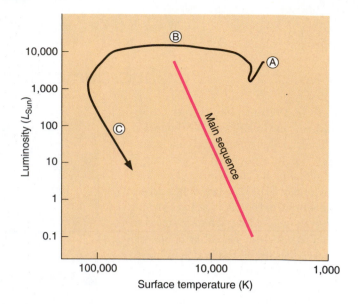

FIGURE 22.2

Evolutionary Track for a Star Like the Sun This diagram shows the changes in luminosity and surface temperature for a star with a mass like the Sun's as it nears the end of its life. After the star becomes a giant again (point A on the diagram), it will lose more and more mass as its core begins to collapse. The mass loss will expose the hot inner core, which will appear at the center of a planetary nebula. In this stage the star moves across the diagram to the left as it becomes hotter and hotter during its collapse (point B). At first the luminosity remains nearly constant, but as the star begins to cool off, it becomes less and less bright (point C). It is now a white dwarf and will continue to cool slowly for billions of years until all of its remaining store of energy is radiated away. (This model assumes the Sun will lose about 46 percent of its mass during the giant stages—based on calculations by Sackmann, Boothroyd, and Kraemer.)

hydrogen still remaining. Now the star is a true white dwarf: It has no further source of energy. (Figure 22.2 shows the path of a star like the Sun on the H–R diagram during its final stages.)

Since the white dwarf can no longer contract (or produce energy through fusion), its only energy source is the heat represented by the motions of the atomic nuclei in its interior. The light it emits comes from this internal stored heat, which is substantial. Gradually, however, the white dwarf radiates away all its heat into space. After many billions of years, the nuclei will be moving much more slowly, and the white dwarf will no longer shine. It will then be a *black dwarf*—a cold stellar corpse with the mass of a star and the size of a planet. It will be composed mostly of carbon and oxygen, the products of the most advanced fusion reactions of which the star was capable.

We have one final surprise as we leave our low-mass star in the stellar graveyard. Calculations show that as a degenerate star cools, the atoms inside it in essence "solidify" into a giant, highly compact lattice (organized rows of atoms just like in a crystal). When carbon is compressed and crystallized in this way, it becomes a *diamond*. A white dwarf star is the most impressive engagement present you could ever see, although any attempt to mine the diamond-like material inside would crush an ardent lover instantly!

22.1.5 Evidence That Stars Can Shed a Lot of Mass as They Evolve

Whether or not a star will become a white dwarf depends on how much mass is lost in the red-giant and earlier phases of evolution. All stars that have masses below the

Chandrasekhar limit when they run out of fuel will become white dwarfs, no matter what mass they were born with. But which stars shed enough mass to reach this limit?

One strategy for answering this question is to look in young, open clusters. For example, detailed observations of NGC 1818, a young cluster in our neighbor galaxy, the Large Magellanic Cloud, show that member stars with masses of 7.5 M_{Sun} are still on the main sequence. That means that only stars with masses greater than this amount have had time to exhaust their supplies of nuclear energy and complete their evolution to the white-dwarf stage. Yet this cluster contains at least one white dwarf (Figure 22.3). This means that the star that turned into the white dwarf must have had a main-sequence mass of more than 7.5 M_{Sun}, since stars with lower masses have not yet had time to use up their stores of nuclear energy. It must, therefore, have gotten rid of more than 6 M_{Sun} so that its mass at the time nuclear energy generation ceased was less than 1.4 M_{Sun}. This shows that even stars that originally make our

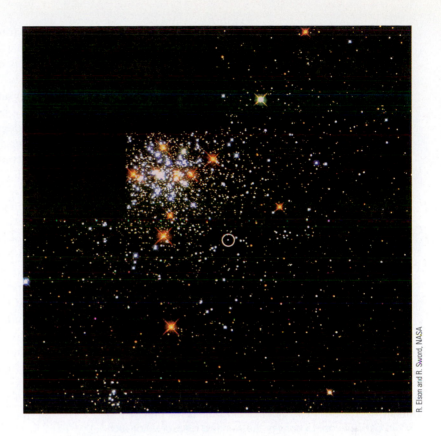

R. Elson and R. Sword, NASA

■ FIGURE 22.3

White Dwarf in a Young Cluster   The cluster of stars shown in this Hubble image is in a neighboring galaxy called the Large Magellanic Cloud, at a distance of about 160,000 LY. The cluster is about 40 million years old. When astronomers construct its H–R diagram, they find that stars with masses of about 7.5 M_{Sun} are just beginning to evolve away from the main sequence. Yet the circled star is a white dwarf that is a member of the cluster. Since a star can become a white dwarf only if it has less than 1.4 M_{Sun}, this star must have lost more than 6 M_{Sun} between the time it left the main sequence and the time it completed the giant phases of its evolution.

Sun look like a lightweight can wind up with so little mass by the end that they will die in the same way our Sun will.

<h2>22.2 EVOLUTION OF MASSIVE STARS: AN EXPLOSIVE FINISH</h2>

Thanks to mass loss, then, stars with masses up to at least 7.5 M_{Sun} (and perhaps even more) probably end their lives as white dwarfs. But we know stars can have masses as large as 150 M_{Sun}. These are the stars that die with a bang.

22.2.1 Nuclear Fusion of Heavy Elements

After the helium in its core is exhausted (see Section 21.5), the evolution of a massive star takes a significantly different course from that of lower-mass stars. In a massive star, the weight of the outer layers is sufficient to force the carbon and oxygen core to contract until it becomes hot enough to fuse carbon into oxygen, neon, and magnesium. This cycle of contraction, heating, and the ignition of another nuclear fuel repeats several more times. After each of the possible nuclear fuels is exhausted, the core contracts again, until it reaches a temperature high enough to fuse still heavier nuclei. The products of carbon burning can be further con-

verted to silicon, sulfur, calcium, and argon. And these elements, when heated to a still higher temperature can combine to produce iron. Massive stars go through these stages very, very quickly. In really massive stars, some fusion stages toward the very end can take only months or even days!

At this late stage of its evolution, a massive star resembles an onion with an iron core. As we get farther from the center, we find shells of decreasing temperature in which nuclear reactions involve nuclei of progressively lower mass—silicon and sulfur, oxygen, neon, carbon, helium, and finally hydrogen (Figure 22.4).

But there is a limit to how long this process of peacefully building up elements by fusion can go on. The fusion of silicon into iron is the last step in the sequence of peaceful element production. Up to this point, each fusion reaction has *produced* energy because the nucleus of each fusion product has been a bit more stable than the nuclei that formed it. As discussed in Chapter 15, light nuclei give up some of their binding energy in the process of fusing into more tightly bound, heavier nuclei. It is this released energy that maintains the pressure in the core so that the star does not collapse. But of all the nuclei known, iron is the most tightly bound and thus the most stable.

You might think of the situation like this: All smaller nuclei want to "grow up" to be like iron, and they are willing to pay (*produce* energy) to move toward that goal. But iron is a mature nucleus with good self-esteem, perfectly

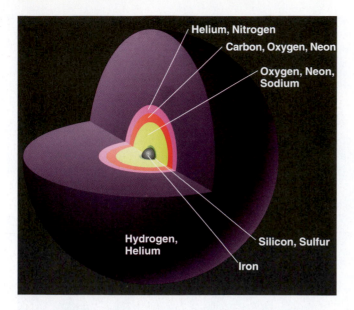

■ FIGURE 22.4

Structure of an Old Massive Star Just before its final gravitational collapse, a massive star resembles an onion. The iron core is surrounded by layers of silicon and sulfur, oxygen, neon, carbon mixed with some oxygen, helium, and finally hydrogen. (Note that this diagram is not precisely to scale but is meant to convey the general idea of what such a star would be like.)

content being iron; it requires payment (must *absorb* energy) to change its stable nuclear structure. This is the exact opposite of what has happened in each nuclear reaction so far: Instead of providing energy to balance the inward pull of gravity, any nuclear reactions involving iron would remove some energy from the core of the star.

Unable to generate energy, the star now faces catastrophe.

22.2.2 Collapse into a Ball of Neutrons

When nuclear reactions stop, the core of a massive star is supported by degenerate electrons, just as a white dwarf is. For stars that begin their evolution with masses of at least $10\ M_{Sun}$, this core is made of iron. For stars with initial masses in the range 8 to $10\ M_{Sun}$, the core is made of oxygen, neon, and magnesium because the star never gets hot enough to form elements as heavy as iron.

While no energy is being generated within the white-dwarf core of the star, fusion does still occur in the shells that surround the core. As the shells finish their fusion reactions and stop producing energy, the ashes of the last reaction fall onto the white-dwarf core, increasing its mass. As Figure 22.1 shows, higher mass means a smaller core. The core can contract because even a degenerate gas is still mostly empty space. Electrons and atomic nuclei are, after

all, extremely small. The electrons and nuclei in a stellar core may be crowded compared to the air in your room, but there is still lots of space between them.

The electrons at first resist being crowded closer together and so the core shrinks only a small amount. Ultimately, however, the iron core reaches a mass so large that even degenerate electrons can no longer support it. When the density reaches 4×10^{11} g/cm^3 (400 billion times the density of water), some electrons are actually squeezed into the atomic nuclei, where they combine with protons to form neutrons. (This can happen only at the outrageous densities found in collapsing stars.) Some of the electrons are now gone, so the core can no longer resist the crushing mass of the star's overlying layers. The core begins to shrink rapidly. More and more electrons are pushed into the atomic nuclei, which ultimately become so saturated with neutrons that they cannot hold onto them.

At this point the neutrons are squeezed out of the nuclei and can exert a new force. As is true for electrons, it turns out that the neutrons strongly resist being in the same place and moving in the same way. The force that can be exerted by such *degenerate neutrons* is much greater than that produced by degenerate electrons, so they can ultimately resist the collapse. Calculations show that the upper limit on the mass of stars made only of neutrons is about $3\ M_{Sun}$.

In other words, if the collapsing core has less mass than $3\ M_{Sun}$, it can reach a stable state as a crushed ball made mainly of neutrons, which astronomers call a **neutron star.** However, if the mass of the core is greater than this limit, then even neutron degeneracy cannot stop the core from collapsing and the star becomes something unbelievably compressed called a *black hole.* Black holes are the subject of the next chapter; for now, we restrict ourselves to those stars in which the collapse of the core is in fact halted by degenerate neutrons.

22.2.3 Collapse and Explosion

When the collapse is stopped by the degenerate neutrons, the core is saved from further destruction, but the rest of the star is literally blown apart. Here's how it happens.

The collapse that takes place when electrons are absorbed into the nuclei is very rapid. In less than a second, the core, which originally was approximately the same diameter as the Earth, collapses to a diameter of less than 20 km. The speed with which material falls inward reaches one fourth the speed of light. The collapse halts only when the density of the core exceeds the density of an atomic nucleus (which is the densest form of matter we know). A typical neutron star is so compressed that to duplicate its density we would have to squeeze all the people in the world into a single *raindrop*. This would give us one raindrop's worth of a neutron star.

The neutron degenerate core strongly resists further compression, abruptly halting the collapse. The shock of the abrupt jolt initiates a shock wave that starts to propagate outward. However, this shock alone is not enough to create a star explosion. The energy produced by the bounce is quickly absorbed by atomic nuclei in the dense overlying layers of gas, where it breaks up the nuclei into individual neutrons and protons.

Our understanding of nuclear processes indicates that each time an electron and a proton in the star's core merge to make a neutron, the merger releases a *neutrino*. These ghostly subatomic particles, introduced in Chapter 15, carry away some of the nuclear energy. It is their presence that launches the final disastrous explosion of the star. The total energy contained in the neutrinos is huge. In the first second of the explosion, the power carried by the neutrinos (10^{46} watts) is greater than the power put out by all the stars in all the galaxies that we can see!

While neutrinos ordinarily do not interact very much with ordinary matter (we accused them of being downright antisocial in Chapter 15), matter near the center of a collapsing star is so dense that the neutrinos do interact with it. They deposit some of this energy in the layers of the star just outside the core. This huge, sudden input of energy reverses the infall of these layers and drives them explosively outward. Huge amounts of star-stuff are spewed into space.

The resulting explosion is called a **supernova** (Figure 22.5). When these explosions happen close by, they can be among the most spectacular celestial events, as we will discuss in the next section. (Actually, there are at least two different types of supernova explosions; the kind we have been describing, which is the collapse of a massive star, are called for historical reasons Type II supernovae. We will describe how the types differ in Section 22.5.)

Ace **Astronomy**™   Log into AceAstronomy and select this chapter to see the Active Figure called "End State of Stars."

| **TABLE 22.1** | *The Ultimate Fate of Stars with Different Masses* |

Initial Mass (Mass of Sun = 1)	Final State at End of Its Life
<0.01	Planet
0.01 to 0.08	Brown dwarf
0.08 to 0.25	White dwarf made mostly of helium
0.25 to 8–10	White dwarf made mostly of carbon and oxygen
8–10 to 12°	White dwarf made of oxygen–neon–magnesium
12 to 40	Supernova explosion that leaves a neutron star
>40	Supernova explosion that leaves a black hole

° Stars in this mass range may produce a type of supernova different from the one we have discussed.

Anglo-Australian Telescope Board

■ **FIGURE 22.5**

A Star Explodes   We see before and after pictures of the field around Supernova 1987A in the Large Magellanic Cloud. An arrow points to the star that exploded. It's easy to put such an arrow there after the explosion; astronomers can only wish such arrows appeared ahead of time so we could know which star was next. The difference in image quality between these pictures is an effect of the Earth's atmosphere, which was steadier when the plates used to make the pre-supernova picture were taken.

Table 22.1 summarizes the discussion so far about what happens to stars of different initial masses at the ends of their lives. Like so much of our scientific understanding, this list represents a progress report: It is the best we can do with our present models and observations. The mass limits corresponding to various outcomes may change somewhat as models improve. There is much we do not yet understand about the details of what happens when stars die.

Supernovae in History

Although many supernova explosions in our own Galaxy have gone unnoticed, a few were so spectacular that they were clearly seen and recorded by sky-watchers and historians at the time. We can use these records, going back two millennia, to help us pinpoint where the exploding stars were and thus where to look for their remnants today.

The most dramatic supernova was observed in the year 1006. It appeared in May as a brilliant point of light visible during the daytime, perhaps 100 times brighter than the planet Venus. It was bright enough to cast shadows on the ground during the night and was recorded with awe and fear by observers all over Europe and Asia. No one had seen anything like it before; Chinese astronomers, noting that it was a temporary spectacle, called it a "guest star."

Astronomers David Clark and Richard Stephenson have scoured records from around the world to find more than 20 reports of the 1006 supernova. This has allowed them to determine with some accuracy where in the sky the explosion occurred. They place it in the modern constellation of Lupus; at roughly the position they have determined we do find a supernova remnant, now quite faint. From the way its filaments are expanding, it indeed appears to be about 1000 years old.

Another guest star was clearly recorded in Chinese records in July 1054. The remnant of that star, called the Crab Nebula, is shown in Figure 22.11 on page 503. It is a marvelously complex object, whose study has been a key to understanding the death of massive stars. We estimate that it was about as bright as the planet Jupiter, nowhere near as dazzling as the 1006 event but still quite dramatic to anyone who kept track of objects in the sky. There is some evidence that Native Americans in New Mexico, for example, recorded the new star with the crescent moon nearby in a cave painting at a ceremonially important location. Another fainter supernova was seen in 1181.

The next supernova became visible in November 1572 and, being brighter than the planet Venus, was quickly spotted by a number of observers, including the young Tycho Brahe (see Chapter 2). His careful measurements of the star over a year and a half showed that it was not a comet or something in the Earth's atmosphere, since it did not move relative to the stars. He correctly deduced that it must be a phenomenon belonging to the realm of the stars, not of the solar system. The remnant of Tycho's Supernova (as it is now called) can still be detected in many different bands of the electromagnetic spectrum.

Not to be outdone, Johannes Kepler, Tycho Brahe's scientific heir, found his own supernova in 1604. Fainter than Tycho's, it nevertheless remained visible for about a year. Kepler wrote a book about his observations that was read by many with an interest in the heavens, including Galileo.

22.2.4 The Supernova Giveth and the Supernova Taketh Away

After the supernova explosion, the life of a massive star comes to an end. But the death of each massive star is an important event in the history of its galaxy. The elements built up by fusion during the star's life are now "recycled" into space by the explosion, making them available to form new stars and planets. Without mass loss from supernovae and planetary nebulae, neither the authors nor the readers of this book would exist.

But the supernova explosion has one more creative contribution to make, one we alluded to in the preceding chapter when we asked where the atoms in your jewelry came from. The supernova explosion produces a flood of energetic neutrons that barrel through the expanding material. These neutrons can be absorbed by iron and other nuclei, where they can turn into protons. Thus they build up elements that are more massive than iron, including such terrestrial favorites as gold and silver. This is the only place we know where such atoms as gold or uranium can be made. Next time you wear some gold jewelry (or give some to your sweetheart), bear in mind that every one of those gold atoms was once part of an exploding star!

When supernovae explode, these elements (as well as the ones the star made during more stable times) are ejected into the existing gas between the stars and mixed with it. Thus supernovae play a major role in building up the supply of chemical elements in the universe (Figure 22.6).

Supernovae are also thought to be the source of the high-energy *cosmic-ray* particles discussed in Section 19.4. Trapped by the magnetic field of the Galaxy, the particles from exploded stars continue to circulate around the vast spiral of the Milky Way. Scientists speculate that high-speed cosmic rays hitting the genetic material of Earth organisms over billions of years may have contributed to the

Several Views of Tycho Brahe's 1572 Supernova (a) A broadsheet about the "new star" from 1573. (b) A modern map of the radio emission from the expanding shell of gas ejected by the supernova explosion. The different colors correspond to different intensities of radio emission produced by extremely energetic electrons gyrating in a magnetic field. No central source has been found; apparently nothing remains of the star that exploded. (c) As the debris from the explosion expands, it barrels into the surrounding interstellar gas, producing very high temperatures. The shocked material glows in x rays, as shown here in an image recorded by the Chandra x-ray satellite. Red, green, and blue represent low, medium, and high energies, respectively. The image is cut off at the bottom because the southernmost region of the remnant fell outside the field of view of the detector.

No supernova has been spotted in our Galaxy for the past 300 years. Since the explosion of a visible supernova is a chance event, there is no way to say when the next one might occur. Around the world, dozens of professional and amateur astronomers keep a sharp lookout for "new" stars that appear overnight, hoping to be the first to spot the next guest star in our sky and make a little history themselves.

steady *mutations*—subtle changes in the genetic code—that drive the evolution of life on our planet. In all these ways, supernovae have played a part in the development of new generations of stars, planets, and life.

But supernovae also have a dark side. Suppose a lifeform has the misfortune to develop around a star that happens to lie near a massive star destined to become a supernova. Such life-forms may find themselves snuffed out when the harsh radiation and high-energy particles from the neighbor star's explosion reach their world. If, as some astronomers speculate, life can develop on many planets around long-lived (lower-mass) stars, then the suitability of that life's *own star* and planet may not be all that matters for its long-term evolution and survival. Life may well have formed around a number of pleasantly stable stars only to be wiped out because a massive nearby star went supernova. Just as children born in a war zone may find themselves the unjust victims of their violent neighborhood, life

too close to a star that goes supernova may fall prey to having been born in the wrong place at the wrong time.

What is a safe distance to be from a supernova explosion? A lot depends on the violence of the particular explosion, what type of supernova it is (see Section 22.5), and what level of destruction we are willing to accept. Calculations suggest that a supernova less than 50 light years (LY) away from us would certainly end all life on Earth, and that even one 100 LY away would have drastic consequences for the radiation levels here. One minor extinction of sea creatures may actually have been caused by a supernova at a distance of about 120 LY (see Section 19.5).

The good news is that there are at present no massive stars that promise to become supernovae within 50 LY of the Sun. (This is in part because the kinds of massive stars that become supernovae are overall quite rare.) The massive star closest to us, Spica (in the constellation of Virgo), is about 260 LY away, probably a safe distance.

■ **FIGURE 22.6**

Cygnus Loop Supernova Remnant
These loops of gas are a small section of the remnant of a supernova that exploded more than 15,000 years ago in the direction of the constellation of Cygnus, the swan. The material from the supernova explosion, which is moving from left to right across our field of view, is blasting into interstellar gas in the neighborhood. The shock of the collision causes the material to glow. The supernova material is enriched in heavy elements created by nuclear processes inside the star that exploded. Eventually this gas will combine with other material in interstellar space and be incorporated into new generations of stars. The colors in this image are not what your eyes would see; each shows the glow of one specific type of atom: Blue is doubly ionized oxygen, red is singly ionized sulfur, and green is hydrogen.

22.3 SUPERNOVA OBSERVATIONS

Supernovae were discovered long before astronomers realized that these spectacular cataclysms mark the death of stars (see Making Connections: *Supernovae in History*). The word *nova* means "new" in Latin; before telescopes, when a star too dim to be seen with the unaided eye suddenly flared up in a brilliant explosion, observers concluded it must be a brand-new star. Twentieth-century astronomers reclassified the explosions with the greatest luminosity as *super*novae.

From historical records, from studies of the remnants of supernova explosions in our own Galaxy, and from analyses of supernovae in other galaxies, we estimate that, on average, one supernova explosion occurs somewhere in the Milky Way Galaxy every 25 to 100 years. Unfortunately, however, no supernova explosion has been detected in our Galaxy since the invention of the telescope. Either we have been exceptionally unlucky or, more likely, recent explosions have taken place in parts of the Galaxy where interstellar dust blocks light from reaching us.

At their maximum brightness, the most luminous supernovae have about 10 billion times the luminosity of the Sun. For a brief time, a supernova may outshine the entire galaxy in which it appears. After maximum brightness, the star fades in light and disappears from telescopic visibility within a few months or years. At the time of their outbursts, supernovae eject material at typical velocities of 10,000 km/s (and speeds twice that high have been observed). A speed of 20,000 km/s corresponds to about 44 million mi/h, truly an indication of unimaginably great cosmic violence.

22.3.1 Supernova 1987A

Our most detailed information about what happens when a supernova occurs comes from an event that was observed in 1987. Before dawn on February 24, Ian Shelton, a Canadian astronomer working at an observatory in Chile, pulled a photographic plate from the developer. Two nights earlier he had begun a survey of the Large Magellanic Cloud, a small galaxy that is one of the Milky Way's nearest neighbors in space. Where he expected to see only faint stars, he saw a large bright spot. Concerned that his photograph was flawed, Shelton went outside to look at the Large Magellanic Cloud—and saw that a new object had indeed appeared in the sky (see Figure 22.5). He soon realized he had discovered a supernova, one that could be seen with the unaided eye even though it was about 160,000 LY away.

Now known as SN 1987A, since it was the first supernova discovered in 1987, this brilliant newcomer to the southern sky gave astronomers their first opportunity to

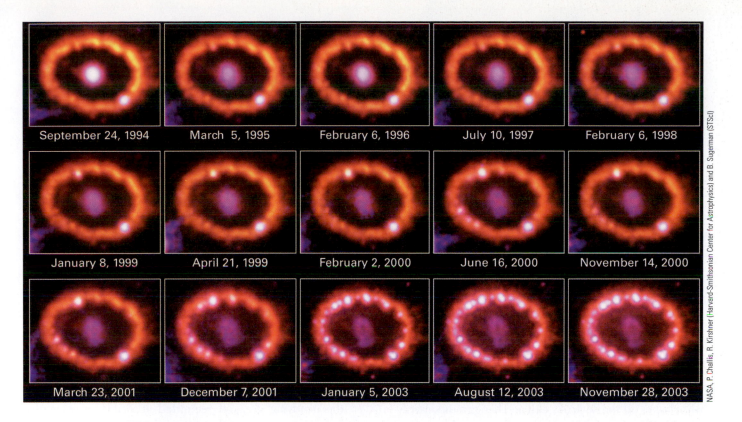

Credit line (vertical, right margin): NASA, P. Challis, R. Kirshner (Harvard-Smithsonian Center for Astrophysics) and B. Sugerman (STScI)

Image labels (by row):

Row 1: September 24, 1994 | March 5, 1995 | February 6, 1996 | July 10, 1997 | February 6, 1998

Row 2: January 8, 1999 | April 21, 1999 | February 2, 2000 | June 16, 2000 | November 14, 2000

Row 3: March 23, 2001 | December 7, 2001 | January 5, 2003 | August 12, 2003 | November 28, 2003

■ **FIGURE 22.7**

The Ring Around Supernova 1987A   About 30,000 years ago, when the star that exploded in 1987 was a red supergiant, it expelled a ring of gas. This ring is now about 1 LY across. New gas ejected at high speed by the 1987 supernova explosion has been racing toward this ring and is now blasting into it. The shock wave produced by the collision compresses and heats the gas in the ring and causes it to glow brightly. The first bright spot was seen by the Hubble Space Telescope in 1996, and in subsequent years many more hot spots appeared as the blast wave reached more distant parts of the irregularly shaped ring. The elongated and expanding object in the middle of the ring is slower debris from the supernova explosion.

study the death of a relatively nearby star with modern instruments. It was also the first time astronomers had observed a star *before* it became a supernova. The star that blew up had been included in earlier surveys of the Large Magellanic Cloud, and as a result we know the star was a blue supergiant just before the explosion.

By combining theory and observations at many different wavelengths, astronomers have reconstructed the life story of the star that became SN 1987A. Formed about 10 million years ago, it originally had a mass of about 20 M_{Sun}. For 90 percent of its life it lived quietly on the main sequence, converting hydrogen to helium. At this time its luminosity was about 60,000 times that of the Sun (L_{Sun}), and its spectral type was O. When hydrogen was exhausted in the center of the star, the core contracted and ultimately became hot enough to fuse helium. By this time, the star was a red supergiant emitting about 100,000 times more energy than the Sun. While in this stage, the star lost some of its mass. This material has actually been detected by observations with the Hubble Space Telescope (Figure 22.7).

The gas driven out into space by the subsequent supernova explosion is just now beginning to collide with the material lost by the star when it was a red supergiant.

Helium fusion lasted only about 1 million years. When the helium was exhausted at the center of the star, the core contracted again, the surface also decreased in radius, and the star became a blue supergiant with a luminosity still about equal to 100,000 L_{Sun}. This is what it still looked like on the outside when it exploded.

The stages of evolution of the star that became SN 1987A, including the ones following helium exhaustion, are listed in Table 22.2. While we don't expect you to remember these numbers, note the patterns in the table. Each stage of evolution happens more quickly than the preceding one, the temperature and pressure in the core increase, and progressively heavier elements are the source of fusion energy. Once iron was created, the collapse began. It was a catastrophic collapse, lasting only a few tenths of a second; the speed of infall in the outer portion of the iron core reached 70,000 km/s, about one-fourth the speed of light.

TABLE 22.2 *Evolution of the Star That Exploded as SN 1987A*

Phase	Central Temperature (K)	Central Density (g/cm^3)	Time Spent in This Phase
Hydrogen fusion	40×10^6	5	9×10^6 years
Helium fusion	170×10^6	900	10^6 years
Carbon fusion	700×10^6	150,000	10^3 years
Neon fusion	1.5×10^9	10^7	Several years
Oxygen fusion	2.1×10^9	10^7	Several years
Silicon fusion	3.5×10^9	10^8	Days
Core collapse	200×10^9	2×10^{14}	Tenths of a second

In the meantime, the outer shells of neon, helium, and hydrogen in the star did not yet know about the collapse. Information about the physical movement of different layers travels through a star at the speed of sound and cannot reach the surface in the few tenths of a second required for the core collapse to occur. Thus the surface layers of our star hung briefly suspended, much like a cartoon character who dashes off the edge of a cliff and hangs momentarily in space before realizing that he is no longer held up by anything.

The collapse of the core continued until the densities rose to several times that of an atomic nucleus. The resistance to further collapse then became so great that the core rebounded. Infalling material ran into the "brick wall" of the rebounding core and was thrown outward with a great shock wave. Neutrinos poured out of the core, helping the shock wave blow the star apart. The shock reached the surface of the star a few hours later, and the star began to brighten into the supernova Ian Shelton observed in 1987.

22.3.2 The Synthesis of Heavy Elements

The variations in the brightness of SN 1987A, which are shown in Figure 22.8, helped confirm our ideas about heavy element production. In a single day the star soared in brightness by a factor of about 1000 and became just visible without a telescope. The star then continued to increase slowly in brightness until it was about the same apparent magnitude as the stars in the Little Dipper. Up until about day 40 after the outburst, the energy being radiated away was produced by the explosion itself. But then SN 1987A did not continue to fade away, as we might have expected the light from the explosion to do. Instead, SN 1987A remained bright, as energy from newly created radioactive elements came into play.

One of the elements formed in a supernova explosion is radioactive nickel, with an atomic mass of 56 (that is, the total number of protons plus neutrons in its nucleus is 56). Nickel-56 is unstable and changes spontaneously, with a half-life of about 6 days, to cobalt-56. (Recall that a half-life is the time it takes for half the nuclei in a sample to undergo radioactive decay.) Cobalt-56 in turn decays with a half-life of about 77 days to iron-56, which is stable. Energetic gamma rays are emitted when these radioactive nuclei decay. Those gamma rays then serve as a new source of energy for the expanding layers. The gamma rays are absorbed in the overlying gas and reemitted at visible wavelengths, keeping the remains of the star bright.

As you can see in Figure 22.8, astronomers did observe brightening due to radioactive nuclei in the first few months following the supernova's outburst and then saw the extra

■ **FIGURE 22.8**

The Change in the Brightness of SN 1987A with Time Note how the rate of decline of the supernova's light slowed between days 40 and 500. During this time the brightness was mainly due to the energy emitted by newly formed (and fast-decaying) radioactive elements.

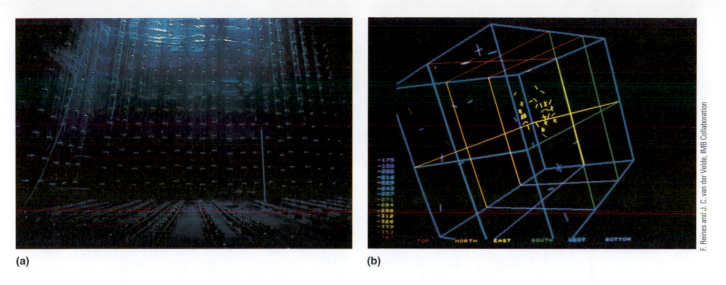

(a)

(b)

F. Reines and J. C. van der Velde, IMB Collaboration

■ FIGURE 22.9

Neutrino Detector (a) The neutrino detector under Lake Erie has 2048 light-sensitive tubes distributed around a tank that holds 8000 tons of pure water. (b) A computer-generated display shows one of the neutrino detections on February 23, 1987. The yellow crosses and slashes near the center show which tubes were triggered by the passage of the neutrino.

light die away as more and more of the radioactive nuclei decayed to stable iron. The gamma-ray heating was responsible for virtually all of the radiation detected from SN 1987A after day 40. Some gamma rays also escaped directly without being absorbed. These were detected by Earth-orbiting telescopes at the wavelengths expected for the decay of radioactive nickel and cobalt, clearly confirming our theory that new elements were formed in the crucible of the supernova.

22.3.3 Neutrinos from SN 1987A

If there had been any human observers in the Large Magellanic Cloud about 160,000 years ago, the explosion we call SN 1987A would have been a brilliant spectacle in their skies. Yet we now know that less than 1/10 of 1 percent of the energy of the explosion appeared as visible light. About 1 percent of the energy was required to destroy the star, and the rest was carried away by neutrinos. The overall energy in these neutrinos was truly astounding. In the first second, as we noted earlier in our general discussion of supernovae, their total luminosity exceeded the luminosity of all the stars in all the galaxies in the part of the universe that we can observe. And the supernova generated this energy in a volume less than 50 km in diameter! Supernovae are by far the most violent events in the universe, and their light turns out to be only the tip of the iceberg in revealing that violence.

The neutrinos from SN 1987A were detected by two instruments, which might be called "neutrino telescopes," almost a full day before Shelton's observations. (This is because the neutrinos get out of the exploding star more easily than light does and also because you don't need to wait until nightfall to catch a "glimpse" of them.) Both neutrino telescopes, one in a deep mine in Japan and the other under Lake Erie, consist of several thousand tons of purified water surrounded by several hundred light-sensitive detectors. Incoming neutrinos interact with the water to produce positrons and electrons, which move rapidly through the water and emit deep blue light (Figure 22.9).

Altogether, 19 neutrinos were detected. Since the neutrino telescopes were in the Northern Hemisphere and the supernova occurred in the Southern Hemisphere, the detected neutrinos had already passed through the Earth and were on their way back out into space when they were captured!

Only a few neutrinos were detected because the probability that they will interact with ordinary matter is very, very low. It is estimated that the supernova actually released 10^{58} neutrinos. A tiny fraction of these, about 50 billion, eventually passed through each square centimeter of the Earth's surface. About a million people actually experienced a neutrino interaction within their bodies as a result of the supernova. This interaction happened to only a single nucleus in each person and thus had absolutely no biological effect; it went completely unnoticed by everyone concerned.

Since the neutrinos come directly from the heart of the supernova, their energies provide a measure of the temperature of the core as the star was exploding. The central temperature was about 200 billion K, a stunning figure to which no earthly analogy can bring much meaning. With neutrino telescopes we are peering into the final moment in the life stories of massive stars and observing conditions beyond all human experience. Yet we are also seeing the unmistakable hints of our own origins.

Courtesy, AIP Emilio Segrè Visual Archives, Weber Collection

■ **FIGURE 22.10**
Antony Hewish and Jocelyn Bell

22.4 PULSARS AND THE DISCOVERY OF NEUTRON STARS

After the supernova explosion fades away, all that is left behind is the neutron star. Neutron stars are the densest objects in the universe; the force of gravity at their surface is 10^{11} times greater than what we experience at the Earth's surface. The interior of a neutron star is composed of about 95 percent neutrons, with a small number of protons and electrons mixed in. In effect, a neutron star is a giant atomic nucleus, with a mass about 10^{57} times the mass of a proton. Its diameter is more like the size of a small town or an asteroid than a star. (Table 22.3 compares the properties of neutron stars and white dwarfs.) Because it is so small, a neutron star probably strikes you as the object least likely to be observed from thousands of light years away. Yet neutron stars do manage to signal their presence across vast gulfs of space.

22.4.1 The Discovery of Neutron Stars

In 1967 Jocelyn Bell, a research student at Cambridge University, was studying distant radio sources with a special detector that had been designed and built by her advisor Antony Hewish to find rapid variations in radio signals (Figure 22.10). The project computers spewed out reams of paper showing where the telescope had surveyed the sky, and it was the job of Hewish's graduate students to go through it all, searching for interesting phenomena. In September 1967 Bell discovered what she called "a bit of scruff"— a strange radio signal unlike anything seen before.

What Bell had found, in the constellation of Vulpecula, was a source of rapid, sharp, intense, and extremely regular pulses of radio radiation. Like the regular ticking of a clock, the pulses arrived precisely every 1.33728 seconds. Such

TABLE 22.3	Properties of a Typical White Dwarf and a Neutron Star	
Property	**White Dwarf**	**Neutron Star**
Mass (Sun = 1)	1.0 (always <1.4)	Always >1.4 and <3
Radius	5000 km	10 km
Density	5×10^5 g/cm^3	10^{14} g/cm^3

exactness led the scientists to speculate that perhaps they had found signals from an intelligent civilization. Radio astronomers even half-jokingly dubbed the source "LGM" for "little green men." Soon, however, three similar sources were discovered in widely separated directions in the sky.

When it became apparent that this type of source was fairly common, astronomers concluded that they were highly unlikely to be signals from other civilizations. By today more than a thousand such sources have been discovered; they are now called **pulsars,** short for pulsating radio sources.

The pulse periods of different pulsars range from a little longer than 1/1000 second to nearly 10 seconds. At first, the pulsars seemed particularly mysterious because nothing could be seen at their location on visible-light photographs. But then a pulsar was discovered right in the center of the Crab Nebula, a cloud of gas produced by a supernova that was recorded by the Chinese in 1054 A.D. (Figure 22.11). The energy from the Crab Nebula pulsar arrives in sharp bursts that occur 30 times each second—with a regularity that would be the envy of a Swiss watchmaker. In addition to pulses of radio energy, we can observe pulses of visible light and x rays from the Crab Nebula.

The Crab Nebula is a fascinating object. The whole nebula glows with radiation at many wavelengths, and its overall energy output is more than 100,000 times that of the Sun—not a bad trick for the remnant of a supernova that exploded almost a thousand years ago. Astronomers

■ **FIGURE 22.11**

The Crab Nebula A composite picture showing x-ray (blue) and visible-light (red) emission from the Crab Nebula, which is about 6000 LY away. The pulsar is the bright spot at the center of the concentric rings. The x rays extend a shorter distance from the pulsar than the light because the high-energy electrons that produce the x rays lose their energy quickly as they move away from the pulsar. Data taken over about a year show that particles stream away from the inner ring at about half the speed of light. Apparently the inner ring is where the high-speed wind from the pulsar really accelerates the particles in the nebula. The jet that is perpendicular to the rings is a stream of matter and antimatter electrons also moving at half the speed of light!

NASA/HST/ASU/J. Hester et al.

began to look for a connection between the pulsar and the large energy output of the surrounding nebula.

22.4.2 A Spinning Lighthouse Model

By applying a combination of theory and observation, astronomers concluded that pulsars must be *spinning neutron stars.* According to this model, a neutron star is something like a lighthouse on a rocky coast. To warn ships in all directions and yet not cost too much to operate, the light in a modern lighthouse turns, sweeping its beam across the dark sea. From the vantage point of a ship, you see a pulse of light each time the beam points in your direction. In the same way, radiation from a small region on a neutron star sweeps across the oceans of space, giving us a pulse of radiation each time the beam points toward the Earth.

Neutron stars are ideal candidates for such a job because the collapse has made them so small that they can turn very rapidly. Recall the principle of the *conservation of angular momentum* from Section 2.2: If an object gets smaller, it can spin more rapidly. Even if the parent star was rotating very slowly when it was on the main sequence, its rotation had to speed up as it collapsed to form a neutron star. With a diameter of only 10 to 20 km, a neutron star can complete one full spin in only a fraction of a second. This is just the sort of time period we observe between pulsar pulses.

Any magnetic field that existed in the original star will be highly compressed when the core collapses to a neutron star. At the surface of the neutron star, protons and electrons are caught up in this spinning field and accelerated nearly to the speed of light. In only two places—at the north and south magnetic poles—can the trapped particles escape the strong hold of the magnetic field (Figure 22.12). The same effect can be seen (in reverse) on Earth, where charged particles from space are *kept out* by our planet's magnetic field everywhere except near the poles. As a result, Earth's auroras (caused when charged particles hit the atmosphere at high speed) are seen only near the poles.

Note that in a neutron star, the magnetic north and south poles do not have to be anywhere close to the north and south poles defined by the star's rotation. In the same way, the magnetic poles on the planets Uranus and Neptune are not lined up with the poles of the planets' spin. Figure 22.12 shows the poles of the magnetic field perpendicular to the poles of rotation, but the two kinds of poles could make any angle.

At the two magnetic poles, the particles from the neutron star are focused into a narrow beam and come streaming out of the whirling magnetic region at enormous speeds. They emit energy over a broad range of the electromagnetic spectrum. The radiation itself is also confined to a narrow beam, which explains why the pulsar acts like a lighthouse. As the rotation carries first one and then the other magnetic pole of the star into our view, we see a pulse of radiation each time.

22.4.3 Tests of the Model

This explanation of pulsars in terms of beams of radiation from highly magnetic and rapidly spinning neutron stars is a very clever idea. But what evidence do we have that it is

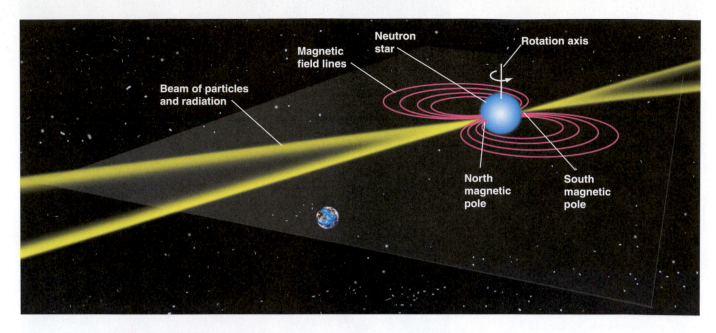

■ FIGURE 22.12

Model of a Pulsar A diagram showing how beams of radiation at the magnetic poles of a neutron star can give rise to pulses of emission as the star rotates. As each beam sweeps over the Earth, like a lighthouse beam sweeping over a distant ship, we see a short pulse of radiation. This model requires that the magnetic poles be located in different places from the rotation poles.

Ace⊙Astronomy™ Log into AceAstronomy and select this chapter to see the Active Figure called "Neutron Star."

the correct model? First, we can measure the masses of some pulsars, and they do turn out be in the range of 1.4 to 1.8 times that of the Sun—just what theorists predict for neutron stars. The masses are found using Kepler's laws for those few pulsars that are members of binary star systems.

But there is an even better confirming argument, which brings us back to the Crab Nebula and its vast energy output. When the high-energy charged particles from the neutron star pulsar hit the slower-moving material from the supernova, they energize this material and cause it to "glow" at many different wavelengths—just what we observe from the Crab Nebula (see Figure 22.11). The pulsar beams are a power source that "lights up" the nebula long after the initial explosion of the star that made it.

Who "pays the bills" for all the energy we see coming out of a remnant like the Crab Nebula? After all, when energy emerges from one place, it must be depleted in another. The ultimate energy source in our model is the rotation of the neutron star, which propels charged particles outward and spins its magnetic field at enormous speeds. As its rotational energy is used to excite the Crab Nebula year after year, the pulsar inside the nebula slows down. As it slows, the pulses come a little less often; more time elapses before the slower neutron star brings its beam back around.

Several decades of careful observations have now shown that the Crab Nebula pulsar is not a perfectly regular clock as we originally thought. Instead it is gradually slowing down. Having measured how much the pulsar is slowing down, we can calculate how much rotation energy

the neutron star is losing. Remember that it is very densely packed and spins amazingly fast. Even a tiny slowing down can mean an immense loss of energy.

To the satisfaction of astronomers, the rotational energy lost by the pulsar turned out to be the same as the amount of energy emerging from the nebula surrounding it. In other words, the slowing down of a rotating neutron star can explain precisely why the Crab Nebula is glowing with the amount of energy we observe.

22.4.4 The Evolution of Pulsars

From observations of the pulsars discovered so far, astronomers have concluded that one new pulsar is born somewhere in the Galaxy every 25 to 100 years, the same rate at which supernovae are estimated to occur. Calculations suggest that the typical lifetime of a pulsar is about 10 million years; after that the neutron star no longer rotates fast enough to produce significant beams of particles and energy, and it is then no longer observable. We estimate that there are about 100 million neutron stars in our Galaxy.

The Crab pulsar is rather young (only about 900 years old) and has a short period, while other, older pulsars have already slowed to longer periods. Pulsars thousands of years old have lost too much energy to emit appreciably in the visible and x-ray wavelengths, and they are observed only as radio pulsars; their periods are a second or longer.

There is one other reason we can see only a fraction of the pulsars in the Galaxy. Consider our lighthouse model

Touched by a Neutron Star

On August 27, 1998, at about 3:22 A.M. Pacific Daylight Time, the Earth was bathed with a stream of x-ray and gamma-ray radiation from a neutron star about 20,000 LY from the Earth. What made this event so remarkable was that, despite the distance of the source, this tidal wave of radiation had measurable effects on the Earth's atmosphere. The high-energy surge lasted about 5 minutes and at the upper edge of the atmosphere was about a tenth as strong as the radiation associated with a dental x ray.

The primary effect of the radiation was on a layer high in the Earth's atmosphere called the *ionosphere*. At night the ionosphere is normally at a height of about 85 km, but during the day energy from the Sun ionizes more molecules and lowers the boundary of the ionosphere to a height of about 60 km. The pulse of x-ray and gamma-ray radiation produced about the same level of ionization as the daytime Sun! It also caused some sensitive satellites above the atmosphere to shut down their electronics.

Measurements by telescopes in space indicate that the source of radiation was a special type of fast-spinning neutron star called a *magnetar*. Astronomers Robert Duncan

and Christopher Thomson gave them this name because their magnetic fields are stronger than that of any other type of astronomical source—in this case, about 800 trillion times stronger than the magnetic field of the Earth!

A magnetar is thought to consist of a superdense core of neutrons surrounded by a rigid crust of atoms about a mile deep with a surface made of iron. The magnetar's field is so strong that it creates huge stresses inside that can sometimes crack open the hard crust, causing a starquake. The vibrating crust produces an enormous blast of radiation. An astronaut 0.1 LY from this particular magnetar would have received a fatal dose from the blast in less than a second.

Fortunately, we were far enough away to be safe. Could a magnetar ever present a real danger to Earth? To produce enough energy to disrupt the ozone layer, a magnetar would have to be located within the cloud of comets that surround the solar system, and we know no magnetars are that close. Nevertheless, it is a fascinating discovery that events on distant star corpses can have measurable effects on the Earth.

again. On Earth, all ships approach on the same plane—the surface of the ocean—so the lighthouse can be built to sweep its beam over that surface. But in space, objects can be anywhere in three dimensions. As a given pulsar's beam sweeps over a circle in space, there is absolutely no guarantee that this circle will include the direction of Earth. In fact, if you think about it, many more circles in space will *not* include the Earth. Thus we estimate that we are unable to observe a large number of neutron stars because their pulsar beams miss us entirely.

At the same time, it turns out that only a few of the pulsars discovered so far are embedded in the visible clouds of gas that mark the remnant of a supernova. This might at first seem mysterious, since we know that supernovae give rise to neutron stars and that we should expect each pulsar to have begun its life in a supernova explosion. But the lifetime of a pulsar turns out to be about 100 times longer than the length of time required for the expanding gas of a supernova remnant to disperse into interstellar space. Thus most pulsars are found with no other trace left of the explosion that produced them.

In addition, some pulsars are ejected by a supernova explosion that is not the same in all directions (some astronomers call this "getting a birth kick"). We know such kicks happen because we see a number of young supernova remnants in nearby galaxies where the pulsar is to one side

of the remnant and racing away at several hundred miles per second (Figure 22.13).

22.5 THE EVOLUTION OF BINARY STAR SYSTEMS

The discussion of the life stories of stars presented so far has suffered from a bias—what we might call "single-star chauvinism." Because the human race developed around a star that goes through life alone, we tend to think of most stars in isolation. But as we saw in Chapter 17, it now appears that as many as half of all stars may develop in *binary* systems—those in which two stars are born in each other's gravitational embrace and go through life orbiting a common center of mass.

For these stars, the presence of a close-by companion can have a profound influence on their evolution. Under the right circumstances, stars can exchange material, especially during the stages when one of them swells up into a giant or supergiant or has a strong wind. When this happens and the companion stars are sufficiently close, material can flow from one star to another, decreasing the mass of the donor and increasing the mass of the recipient. Such

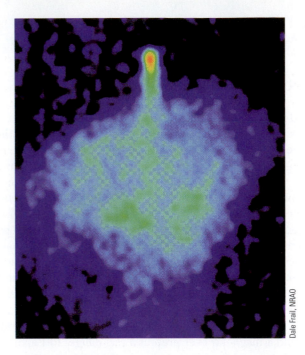

Dale Frail, NRAO

■ **FIGURE 22.13**

Speeding Pulsar   This radio map shows a distant supernova remnant in our Galaxy, nicknamed the Duck Nebula. We see a speeding pulsar (in the yellow region) whose rapid motion has overtaken the expanding material from the explosion. The map was made with the Very Large Array.

mass transfer can be especially dramatic when the recipient is a stellar remnant such as a white dwarf or a neutron star. While the detailed story of how such binary stars evolve is beyond the scope of our book, we do want to mention a few examples of how the stages of evolution described in this chapter may change when there are two stars in a system.

22.5.1   White-Dwarf Explosions: The Mild Kind

Let's consider a system of two stars: one has become a white dwarf and the other is gradually transferring material onto it. As fresh hydrogen from the outer layers of its companion accumulates on the surface of the hot white dwarf, it begins to build up a layer of hydrogen. As more and more hydrogen accumulates and heats up on the surface of the degenerate star, the new layer eventually reaches a temperature that causes fusion to begin in a sudden, explosive way, blasting much of the new material away. The white dwarf quickly (but only briefly) becomes quite bright. To observers before the invention of the telescope, it seemed that a new star suddenly appeared, and they called it a **nova.** Novae fade away in a few months to a few years.

Hundreds of novae have been observed, each of them occurring in a binary star system, and each later showing a shell of expelled material. A number of stars have more

than one nova episode, as more material accumulates on the white dwarf and the whole process repeats. As long as the episodes do not increase the mass of the white dwarf beyond the Chandrasekhar limit (by transferring too much mass), the dense white dwarf itself remains pretty much unaffected by the explosions on its surface.

22.5.2   White-Dwarf Explosions: The Violent Kind

If a white dwarf accumulates matter from a companion star at a much faster rate, it can be pushed over the Chandrasekhar limit. The evolution of such a binary system is shown in Figure 22.14. When its mass exceeds 1.4 M_{Sun}, such an

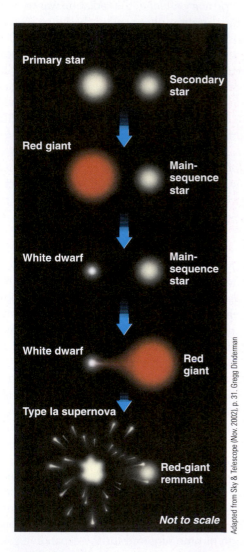

Adapted from Sky & Telescope (Nov. 2002), p. 31. Gregg Dinderman

■ **FIGURE 22.14**

The Evolution of a Binary System   The more massive star evolves first to become a red giant and then a white dwarf. The white dwarf then begins to attract material from its companion, which in turn evolves to become a red giant. Eventually the white dwarf acquires so much mass that it is pushed over the Chandrasekhar limit and becomes a Type Ia supernova.

object can no longer support itself as a white dwarf and it begins to collapse. As it does so it heats up, and new nuclear reactions begin in the degenerate core. In less than a second, an enormous amount of fusion takes place; the energy released is so great that it completely destroys the white dwarf. Gases are blown out into space at velocities of several thousand kilometers per second, and no central star remains behind.

Such an explosion is also called a supernova, since, like the destruction of a high-mass star, it produces a huge amount of energy in a very short time, destroying the star in the process. We call these overfed white dwarfs Type Ia supernovae (distinguished from the Type IIs discussed earlier, which signal the deaths of massive stars). Tycho's Supernova (see Making Connections: *Supernovae in History*) appears to have been caused by such a white dwarf with an overly generous companion. (Very recent observations suggest that some mechanism allows white dwarfs with masses less than 1.4 M_{Sun} in binary systems to explode violently as well, although astronomers do not yet have a detailed model of how this happens.)

22.5.3 Neutron Stars with Companions

It is possible that a binary system can survive the explosion of one of its members. In that case, an ordinary star can share a system with a neutron star. If material is transferred from the "living" star to its "dead" (and highly compressed) companion, this material will be pulled in by the strong gravity of the neutron star. Such infalling gas will be compressed and heated to incredible temperatures. It will quickly become so hot that it will experience an explosive burst of fusion. The energies involved are so great that we would expect much of the radiation from the burst to emerge as x rays. And indeed, high-energy observatories above the Earth's atmosphere (see Chapter 5) have recorded many objects that undergo just these types of x-ray *bursts*.

If the neutron star and its companion are positioned the right way, a significant amount of material can be transferred to the neutron star and can set it spinning faster (as spin energy is also transferred). Astronomers have found several pulsars in binary systems that are spinning at almost *1000 times per second!* Such a rapid spin could not have come from the birth of the neutron star—it must have been externally caused. (Recall that the Crab Nebula pulsar, one of the youngest pulsars known, was spinning 30 times per second.) Indeed, some of the fast pulsars are observed to be part of binary systems, while others may be alone only because they may have "fully consumed" their former partner stars through the mass transfer process. (These have sometimes been called "black widow pulsars.")

We have now reached the end of our description of the final states of stars, yet one piece of the story remains to be filled in. We saw that stars whose masses are less than 1.4 M_{Sun} at the time they run out of fuel end their lives as white dwarfs. Dying stars with masses between 1.4 and about 3 M_{Sun} become neutron stars. But there are stars whose masses are greater than 3 M_{Sun} when they exhaust their fuel supplies. What becomes of them? The truly bizarre result of the death of such massive stellar cores is the subject of our next chapter.

SURFING THE WEB

💻 *The Crab Nebula Story:*
xrtpub.harvard.edu/xray_sources/crab/crab.html
A short, colorfully written introduction to the history and science involving the best known supernova remnant.

💻 *Crab Nebula: The Movie:*
hubblesite.org/newscenter/archive/2002/24
A sequence of Hubble and Chandra Space Telescope images of the central regions of the Crab Nebula have been assembled into a very brief movie accompanied by animation showing how the pulsar affects its environment. It comes with some useful background material.

💻 *Hubble Observations of Supernova 1987A:*
hubblesite.org/newscenter/archive/2000/11/ and
hubblesite.org/newscenter/archive/1998/08/
Hubble images of the 1987 supernova in a neighbor galaxy, with a history of major discoveries, background information, and new results.

💻 *Michael's Supernova Page:*
stupendous.rit.edu/richmond/sne/sne.html
Michael Richmond of the Rochester Institute of Technology keeps one of the most informative Web sites about supernovae. It includes lists of all recent supernovae observed in other galaxies as

well as historical supernovae in our Galaxy, links to other sites, and a clear explanation of the different types of supernovae. Of special interest is his somewhat technical essay on "Will a Nearby Supernova Endanger Life on Earth?" at: stupendous.rit.edu/richmond/answers/snrisks.txt

 Introduction to Neutron Stars:
www.astro.umd.edu/~miller/nstar.html
Coleman Miller of the University of Maryland keeps this site, which goes from easy to hard as you get into it, but it has lots of good information about the corpses of massive stars.

Magnetars Page:
solomon.as.utexas.edu/~duncan/magnetar.html
Robert Duncan, one of the originators of the idea of magnetars, assembled this useful site in 2000 with background information, a short history of the idea, and an explanation of the observations that support it.

SUMMARY

22.1 During the course of their evolution, stars shed their outer layers and lose a significant fraction of their initial mass. Stars with masses of at least $7.5\,M_{Sun}$ can lose enough mass to become white dwarfs, which have masses less than the **Chandrasekhar limit** (about $1.4\,M_{Sun}$). A typical white dwarf has a mass about the same as that of the Sun and a diameter comparable to that of the Earth. The pressure exerted by **degenerate electrons** keeps white dwarfs from contracting to still smaller diameters. Eventually, white dwarfs cool off to become black dwarfs, stellar remnants made mainly of carbon and oxygen.

22.2 In a massive star, hydrogen fusion in the core is followed by several other fusion reactions involving heavier elements. Just before it exhausts all sources of energy, a massive star has an iron core surrounded by shells of silicon and sulfur, oxygen, neon, carbon, helium, and hydrogen. The fusion of iron requires energy (rather than releasing it). If the mass of a star's iron core exceeds the Chandrasekhar limit (but is less than $3\,M_{Sun}$), the core collapses until its density exceeds that of an atomic nucleus, forming a **neutron star** with a typical diameter of 20 km. The core rebounds and transfers energy outward, blowing off the outer layers of the star in a Type II supernova explosion.

22.3 A supernova occurs on average once every 25 to 100 years in the Milky Way Galaxy. Despite the odds, however, only one bright supernova has been observed since the invention of the telescope, and this supernova (SN 1987A) oc-

curred in a neighboring galaxy, the Large Magellanic Cloud. The star that became SN 1987A began its life as a blue supergiant, evolved to become a red supergiant, and returned to being a blue supergiant at the time it exploded. Studies of SN 1987A have detected neutrinos and confirmed theoretical calculations of what happens during such explosions, including the formation of elements beyond iron. Supernovae are a main source of high-energy cosmic rays and can be dangerous for any living organisms in nearby star systems.

22.4 At least some supernovae leave behind a highly magnetic, rapidly rotating neutron star, which can be observed as a **pulsar** if its beam of escaping particles and focused radiation is pointing toward us. Pulsars emit rapid pulses of radiation at regular intervals; their periods are in the range of 0.001 to 10 seconds. The rotating neutron star acts like a lighthouse, sweeping its beam in a circle and giving us a pulse of radiation when the beam sweeps over the Earth. Pulsars lose energy as they age, the rotation slows, and their periods increase.

22.5 When a white dwarf or neutron star is a member of a close binary star system, its companion star can transfer mass to it. Material falling *gradually* onto a white dwarf can explode in a sudden burst of fusion and make a **nova**. If material falls *rapidly* onto a white dwarf, it can push it over the Chandrasekhar limit and cause it to explode completely as a Type Ia supernova. Material falling onto a neutron star can cause powerful bursts of x-ray radiation.

INTER-ACTIVITY

A Someone in your group uses a large telescope to observe an expanding shell of gas. Discuss what measurements you could make to determine whether you have discovered a planetary nebula or the remnant of a supernova explosion.

B The star Sirius (the brightest star in our northern skies) has a white-dwarf companion. Sirius has a mass of about $2\,M_{Sun}$ and is still on the main sequence, while its companion is already a star corpse. Remember that a white dwarf can't have a mass greater than

1.4 M_{Sun}. Assuming that the two stars formed at the same time, your group should discuss how Sirius could have a white-dwarf companion. *Hint:* Was the initial mass of the white-dwarf star larger or smaller than that of Sirius?

C What would people today do if a brilliant star suddenly became visible during the daytime? What kind of fear and superstition might result from a supernova that was really bright in our skies? Have your group invent some headlines that the supermarket tabloids would feature.

D Suppose a supernova exploded only 40 LY from Earth. Have your group discuss what effects there may be on the Earth when the radiation reaches us and later when the particles reach us. Would there be any way to protect people from the supernova effects?

E When pulsars were discovered, the astronomers involved with the discovery talked about finding "little green men." If you had been in their shoes, what tests would you have performed to see whether such a pulsating source of radio waves was natural or the result of an alien intelligence? Today, several groups around the world are actively searching for possible radio signals from intelligent civilizations. How might you expect such signals to differ from pulsar signals?

F Your little brother, who has not had the benefit of an astronomy course, reads about white dwarfs and neutron stars in a magazine and decides it would be fun to go near them or even try to land on them. Is this a good idea for future tourism? Have your group make a list of reasons it would not be safe for children (or adults) to go near a white dwarf and a neutron star.

REVIEW QUESTIONS

1. How does a white dwarf differ from a neutron star? How does each form? What keeps each from collapsing under its own weight?

2. Describe the evolution of a star with a mass like that of the Sun, from the main-sequence phase of its evolution until it becomes a white dwarf.

3. Describe the evolution of a massive star (say, 20 times the mass of the Sun) up to the point at which it becomes a supernova. How does the evolution of a massive star differ from that of the Sun? Why?

4. How do the two types of supernovae discussed in this chapter differ? What kind of star gives rise to each type?

5. A star begins its life with a mass of 5 M_{Sun} but ends its life as a white dwarf with a mass of 0.8 M_{Sun}. List the stages in the star's life during which it most likely lost some of the mass it started with. How did mass loss occur in each stage?

6. If the formation of a neutron star leads to a supernova explosion, explain why only three out of the hundreds of known pulsars are found in supernova remnants.

7. How can the Crab Nebula shine with the energy of something like 100,000 Suns when the star that formed the nebula exploded almost 1000 years ago? Who "pays the bills" for much of the radiation we see coming from the nebula?

8. How is a nova different from a Type I supernova? How does it differ from a Type II supernova?

THOUGHT QUESTIONS

9. Arrange the following stars in order of their evolution:
a. A star with no nuclear reactions going on in the core, which is made primarily of carbon and oxygen.
b. A star of uniform composition from center to surface; it contains hydrogen but has no nuclear reactions going on in the core.
c. A star that is fusing hydrogen to form helium in its core.
d. A star that is fusing helium to carbon in the core and hydrogen to helium in a shell around the core.
e. A star that has no nuclear reactions going on in the core but is fusing hydrogen to form helium in a shell around the core.

10. Would you expect to find any white dwarfs in the Orion Nebula? (See Chapter 20 to remind yourself of its characteristics.) Why or why not?

11. Suppose no stars more massive than about 2 M_{Sun} had ever formed. Would life as we know it have been able to develop? Why or why not?

12. Would you be more likely to observe a Type II supernova (the explosion of a massive star) in a globular cluster or in an open cluster? Why?

13. Astronomers believe there are something like 100 million neutron stars in the Galaxy, yet we have only found about

a thousand pulsars in the Milky Way. Give several reasons these numbers are so different. Explain each reason.

14. Would you expect to observe every supernova in our own Galaxy? Why or why not?

15. The Large Magellanic Cloud has about one tenth the number of stars found in our own Galaxy. Suppose the mix of high- and low-mass stars is exactly the same in both galaxies. Approximately how often does a supernova occur in the Large Magellanic Cloud?

16. Look at the list of the nearest stars in Appendix 10. Would you expect any of these to become supernovae? Why or why not?

17. If most stars become white dwarfs at the ends of their lives and the formation of white dwarfs is accompanied by the production of a planetary nebula, why are there more white dwarfs than planetary nebulae in the Galaxy?

18. Which are more numerous in our Galaxy: white dwarfs or neutron stars? Explain your answer.

FIGURING FOR YOURSELF

How Much Would You Weigh on a Stellar Corpse?

Let's calculate what it's really like near a white dwarf or neutron star.

First, let's ask how strongly gravity accelerates you on a body (which is a measure of how strongly you are pulled to the surface). Recall from Chapter 2 that the force of gravity, F, between two bodies is calculated as

$$F = \frac{GM_1M_2}{R^2}$$

where G is the gravitational constant—6.67×10^{-11} Nm^2/kg^2, M_1 and M_2 are the masses of the two bodies, and R is their separation. Also, from Newton's second law,

$$F = M \times a$$

where a is the acceleration of a body with mass M. So let's consider the situation of a mass—say, you—standing on a body, such as the Earth or a white dwarf. You are M_1 and the body you are standing on is M_2. The distance between you and the center of gravity of the body on which you stand is just its radius, R. The force exerted on you is

$$F = M_1 \times a = \frac{GM_1M_2}{R^2}$$

The acceleration of gravity on that world, which is given the letter g, is $(G \times M)/R^2$. We can now calculate the acceleration of gravity for planet Earth (see Appendix 6 for the numbers):

$$g = \frac{6.67 \times 10^{-11}\,Nm^2/kg^2 \times 6 \times 10^{24}\,kg}{(6.4 \times 10^6\,m)^2}$$

$$= \frac{4.0 \times 10^{14}}{4.1 \times 10^{13}} = 9.8\,m/s^2$$

[You might wonder how the units worked out; where did the Newtons (N) go? A Newton = $kg \times m/s^2$.]

The speed you need to get away from a body is called its *escape velocity*. It is given by the formula

$$v_{esc}^2 = \frac{2GM}{R}$$

For the Earth, $v_{esc}^2 = 2 \times (6.67 \times 10^{-11}\,Nm^2/kg^2 \times 6 \times 10^{24}\,kg) / 6.4 \times 10^6\,m = 1.25 \times 10^8$. Now remember, this is the square of the escape velocity. Taking the square root gives $1.12 \times 10^4\,m/s = 11.2\,km/s$. This is the speed (roughly 25,000 mph) that you have to go to escape Earth's gravity.

The density of a body is its mass divided by its volume. The volume of a sphere is $V = \frac{4}{3}\pi R^3$, so

$$density = \frac{M}{(4/3)\pi R^3}$$

For the Earth, we find the average density is

$$\frac{6 \times 10^{24}\,kg}{1.33 \times 3.14 \times (6.4 \times 10^6\,m)^3} = 5500\,kg/m^3$$

Recall that the density of water is $1000\,kg/m^3$; so the Earth has 5.5 times the density of water.

19. What is the acceleration of gravity at the surface of the Sun? (See Appendix 6 for the Sun's key characteristics.) How much greater is this than g at the surface of the Earth? Calculate what you would weigh on the surface of the Sun. Your weight would be your Earth weight multiplied by the ratio of the acceleration of gravity on the Sun to the acceleration of gravity on the Earth. (OK, OK, we know that the Sun does not have a solid surface to stand on and that you would be vaporized if you were at the Sun's photosphere. Humor us for the sake of doing these calculations.)

20. What is the escape velocity from the Sun? How much greater is it than the escape velocity from the Earth?

21. What is the average density of the Sun? How does it compare to the average density of the Earth?

22. Say that a particular white dwarf has the mass of the Sun but the radius of the Earth. What is the acceleration of gravity at the surface of the white dwarf? How much greater is this than g at the surface of the Earth? What would you weigh at the surface of the white dwarf (again granting us the dubious notion that you could survive there)?

23. What is the escape velocity from the white dwarf in problem 22? How much greater is it than the escape velocity from Earth?

24. What is the average density of the white dwarf in problem 22? How does it compare to the average density of the Earth?

25. Now take a neutron star that has twice the mass of the Sun but a radius of 10 km. What is the acceleration of gravity at the surface of the neutron star? How much greater is this than g at the surface of the Earth? What would you weigh at the surface of the neutron star (provided you could somehow not become a puddle of protoplasm)?

26. What is the escape velocity from the neutron star in problem 25? How much greater is it than the escape velocity from Earth?

27. What is the average density of the neutron star in problem 25? How does it compare to the average density of the Earth?

A scientific hypothesis that explains one set of observations can often be tested by determining whether it fits other independent observations. The text says low-mass stars evolve from the red-giant phase to become central stars of planetary nebulae, and then these stars simply cool off over time to become white dwarfs. Such stars are supported by degenerate electrons, and we would not expect much change in size as the central star of a planetary nebula makes the transition to the white-dwarf state. So let's see how the size of a typical star inside a planetary nebula compares with the size of a white dwarf.

28. One way to calculate the radius of a star (see *Figuring for Yourself* in Chapter 17, problem 24) is to use its luminosity and its temperature and assume that the star radiates approximately like a blackbody. Astronomers have measured the characteristics of central stars of planetary nebulae and have found that a typical central star is 16 times as luminous and 20 times as hot (about 110,000 K) as the Sun. Find the radius in terms of the Sun's. How does this radius compare with that of a typical white dwarf?

29. According to a model described in the text, a neutron star has a radius of about 10 km. The pulses occur once per rotation. According to Einstein's theory of relativity, nothing can move faster than the speed of light. Check to make sure that this pulsar model does not violate relativity. Calculate the rotation speed of the Crab Nebula pulsar at its equator, given its period of 0.033 second. (Remember that distance equals velocity × time and that the circumference of a circle is given by $2\pi R$.)

30. Do the same calculations as in problem 29, but for a pulsar that rotates 1000 times per second.

SUGGESTIONS FOR FURTHER READING

See some of the references for the last chapter, plus:

Filippenko, A. "A Supernova with an Identity Crisis" in *Sky & Telescope*, Dec. 1993, p. 30. Good review of supernovae in general and of supernova 1993J in M81.

Graham-Smith, F. "Pulsars Today" in *Sky & Telescope*, Sept. 1990, p. 240.

Greenstein, G. "Neutron Stars and the Discovery of Pulsars" in *Mercury*, Mar./Apr. 1985, p. 34; May/June 1985, p. 66.

Gribbin, J. *Stardust: Supernovae and Life—the Cosmic Connection.* 2000, Yale U. Press. On the origin of the elements and how they are recycled by supernovae.

Iben, I. and Tutukov, A. "The Lives of Binary Stars: From Birth to Death and Beyond" in *Sky & Telescope*, Jan. 1998, p. 42.

Irion, R. "Pursuing the Most Extreme Stars" in *Astronomy*, Jan. 1999, p. 48. On pulsars.

Kaler, J. "The Smallest Stars in the Universe" in *Astronomy*, Nov. 1991, p. 50. On white dwarfs, neutron stars, and pulsars.

Kawaler, S. and Winget, D. "White Dwarfs: Fossil Stars" in *Sky & Telescope*, Aug. 1987, p. 132.

Kirshner, R. "Supernova: The Death of a Star" in *National Geographic*, May 1988, p. 618. Excellent introduction for beginners to SN 1987A.

Kirshner, R. "Supernova 1987A: The First Ten Years" in *Sky & Telescope*, Feb. 1997, p. 35.

Marschall, L. *The Supernova Story*, 2nd ed. 1994, Princeton U. Press. The introduction of choice to supernovae and SN 1987A.

Maurer, S. "Taking the Pulse of Neutron Stars" in *Sky & Telescope*, Aug. 2001, p. 32. Review of recent ideas and observations of pulsars.

Nadis, S. "Neutron Stars with Attitude" in *Astronomy*, Mar. 1999, p. 52. On magnetars.

Naeye, R. "Ka-Boom: How Stars Explode" in *Astronomy*, July 1997, p. 44. On modeling supernovae.

Robinson, L. "Supernovae, Neutrinos, and Amateur Astronomers" in *Sky & Telescope*, Aug. 1999, p. 31. On how you can help find supernovae.

Wheeler, J. C. *Cosmic Catastrophes: Supernovae, Gamma-ray Bursts, and Adventures in Hyperspace.* 2000, Cambridge U. Press. Binary star evolution, supernovae, white dwarfs, black holes, and what happens when they flare up.

Zimmerman, R. "Into the Maelstrom" in *Astronomy*, Nov. 1998, p. 44. About the Crab Nebula.

A Cannibal Black Hole This is an artist's conception of what happens in the vicinity of a black hole, which is all that is left when a very massive star implodes at the end of its life. In this drawing, gas from a companion star (not shown) is attracted by the black hole and swirls around it before falling inside the event horizon—never to be seen again. As the gas swirls closer to the black hole, it becomes hotter, reaching temperatures of millions of degrees at the inner edge. At this temperature, the gas emits x rays. Note that the inner edge of the disk is white on the right side and red on the left side. This is because the gas on the right side is moving toward the observer, so the radiation is shifted to higher energies; conversely, the gas on the left side is moving away from the observer, so its radiation is shifted to lower energies. The inset depicts gas very close to the black hole; this material cannot stay in orbit and eventually falls into the black hole. Very close to the event horizon, the gravitational redshift (the slowing of time in the vicinity of a large gravitational field) shifts the radiation to longer (redder) wavelengths. Inside the event horizon (black area), the extreme curvature of spacetime bends the light rays back into the black hole, so no light escapes.

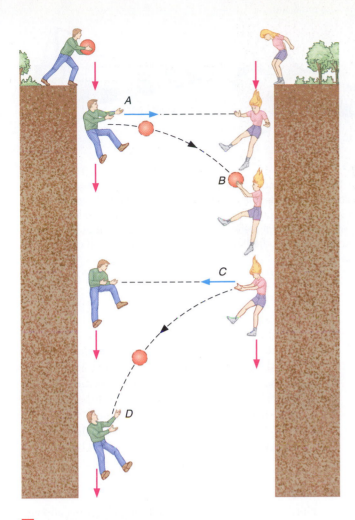

FIGURE 23.2

Free Fall A brave couple is playing catch as they descend into a bottomless abyss. Since the boy, girl, and ball all fall at the same speed, it appears to them that they can play catch by throwing the ball in a straight line between them. Within their world, there appears to be no gravity.

FIGURE 23.3

Astronauts in Space When the Space Shuttle is in free-fall orbit around the Earth, everything stays put or moves uniformly because no apparent gravity acts inside the spacecraft.

uations are completely *equivalent*. The idea that life in a freely falling laboratory is indistinguishable from, and hence equivalent to, life with no gravity is called the **equivalence principle**.

23.1.1 Gravity or Acceleration?

This simple idea has *big* consequences. Let's begin by considering what happens if a foolhardy boy and girl simultaneously jump from opposite banks into a bottomless chasm (Figure 23.2). If we ignore air friction, then we can say that while they fall, they both accelerate downward at the same rate and feel no external force acting on them. They can throw a ball back and forth, always aiming it straight at each other, as if there were no gravity. The ball falls at the same rate they do, so it can always remain in a line between them.

Such a game of catch is very different on the surface of the Earth. Everyone who grows up feeling gravity knows that a ball, once thrown, falls to the ground. Thus, in order to play catch with someone, you must aim the ball upward so that it follows an arc—rising and then falling as it moves forward—until it is caught at the other end.

Now suppose we isolate our freely falling boy, girl, and ball inside in a large box that is falling with them. No one inside the box is aware of any gravitational force. If the children let go of the ball, it doesn't fall to the bottom of the box or anywhere else, but merely stays there or moves in a straight line, depending on whether it is given any motion.

Astronauts in the Space Shuttle orbiting the Earth live in an environment just like that of the boy and girl sealed in a freely falling box (Figure 23.3). The shuttle in orbit is "falling" freely around the Earth (as explained in Section 2.3). While in free fall, the astronauts live in a magical world where there seems to be no gravitational force. One can give a wrench a shove, and it moves at constant speed across the orbiting laboratory. A pencil set in midair remains there, as if no force were acting on it.

Appearances are misleading, however. There is a force in this situation. Both the Shuttle and the astronauts continually fall around the Earth, pulled by its gravity. But since all fall together—shuttle, astronauts, wrench, and pencil—inside the shuttle all gravitational forces appear to be absent.

Thus the shuttle provides an excellent example of the principle of equivalence—how local effects of gravity can be completely compensated by the right acceleration. To the astronauts, falling around the Earth creates the same effects as being far off in space, remote from all gravitational influences.

■ **FIGURE 23.4**

Curved Light Path In a spaceship moving to the left (in this figure) in its orbit about a planet, light is beamed from the rear, A, toward the front, B. Meanwhile the ship is falling out of its straight path (the amount it falls is exaggerated here). We might therefore expect the light to strike at B′, above the target in the ship. Instead, the light, bent by gravity, follows a curved path and strikes at C.

23.1.2 The Paths of Light and Matter

Einstein postulated that the equivalence principle is a fundamental fact of nature, and that there is *no* experiment inside the spacecraft by which an astronaut can ever distinguish between being weightless in remote space and being in free fall near a planet like the Earth. This would apply to experiments done with beams of light as well. But the minute we use light in our experiments, we are led to some very disturbing conclusions—and it is these conclusions that lead us to general relativity.

It is a fundamental observation from everyday life that beams of light travel in straight lines. Imagine that the Space Shuttle is moving through empty space far from any gravity. Send a laser beam from the back of the ship to the front, and it will travel in a nice straight line and land on the front wall exactly opposite the point from which it left the rear wall. If the equivalence principle really applies universally, then this same experiment performed in free fall around the Earth should give us the exact same result.

Imagine that the astronauts again shine a beam of light along the length of their ship. But, as shown in Figure 23.4, the orbiting shuttle falls a bit between the time the light leaves the back wall and the time it hits the front wall. (The amount of the fall is grossly exaggerated in Figure 23.4 to illustrate the effect.) Therefore, if the beam of light follows a straight line but the ship's path curves downward, then the light should strike the front wall at a point higher than the point from which it left.

However, this would violate the principle of equivalence—the two experiments would give different results. We are thus faced with giving up one of our two assumptions. Either the principle of equivalence is not correct, or light does not always travel in straight lines. Instead of dropping what probably seemed at the time like a ridiculous idea, Einstein worked out what would happen if light sometimes does *not* follow a straight path.

Let's suppose the principle of equivalence is right. Then the light beam must arrive directly opposite the point

from which it started in the ship. The light, like the ball the children were throwing back and forth, *must fall with the ship* if that ship is in orbit around the Earth (see Figure 23.4). This would make its path curve downward, like the path of the ball, and thus the light would hit the front wall exactly opposite the spot from which it came.

Thinking this over, you might well conclude that it doesn't seem like such a big problem; why *can't* light fall the way balls do? But, as discussed in Chapter 4, light is profoundly different from balls. Balls have mass, while light does not.

Here is where Einstein's intuition and genius allowed him to make a profound leap. He gave physical meaning to the strange result of our thought experiment. Einstein suggested that the light curves down to meet the front of the shuttle because the Earth's gravity actually bends the *fabric of space and time.* This radical idea—which we will explain next—keeps the behavior of light the same in both empty space and free fall, but it changes some of our most basic and cherished ideas about space and time. The reason we take Einstein's suggestion seriously is that, as we will see, experiments now clearly show his intuitive leap was correct.

23.2 SPACETIME AND GRAVITY

Is light actually bent from its straight-line path by Earth's gravity? Einstein preferred to think that it is space and time that are affected by gravity; light beams, and everything else that travels through space and time, then find their paths affected. Light always follows the shortest path—but that path may not always be straight. This is true for human travel as well. On the curved surface of planet Earth, if you want to fly from Chicago to Rome, the shortest distance is not a straight line but the arc of a *great circle* (which we defined at the beginning of Chapter 3).

23.2.1 Linkages: Mass, Space, and Time

To show what Einstein's insight really means, let's first consider how we locate an event in space and time. For example, imagine you have to describe to worried school officials the fire that broke out in your room when your roommate tried making shish kebob in the fireplace. You explain that your dorm is at 6400 College Avenue, which locates the room in a left–right direction; you are on the 5th floor, which tells where you are in the up–down direction; and you are the 6th room back from the elevator, which tells where you are in the forward–backward direction. Then you explain that the fire broke out at 6:23 P.M. (but was soon brought under control). *Any* event in the universe, whether nearby or far away, can be pinpointed using the three dimensions of space and the one dimension of time.

Newton considered space and time completely independent, and that continued to be the accepted view until the beginning of the 20th century. But Einstein showed that there was an intimate connection between space and time, and that only by considering the two together—in what we call **spacetime**—can we build up a correct picture of the physical world. We examine spacetime a bit more closely in the next subsection.

The gist of Einstein's general theory is that the presence of mass (gravity) curves or warps the fabric of spacetime. When something else—a beam of light, an electron, or the Starship Enterprise—enters such a region of distorted spacetime, the path of that something else will be different from what it would have been in the absence of the mass. This idea is often summarized in the following way: Matter tells spacetime how to curve, and the curvature of spacetime tells other matter how to move.

The amount of distortion in spacetime depends on the mass involved and on how concentrated and compact it is. Terrestrial objects, such as the book you are reading, have far too little mass to introduce any noticeable distortion. (If, like most people, you have never heard of distorted spacetime, that's the reason; it plays no role in life on Earth. Newton's view of gravity is just fine for building bridges, skyscrapers, or amusement-park rides.) It takes a mass like a star to produce measurable distortions, and a white dwarf produces more distortion just above its surface than does a red giant with the same mass. So you see, we *are* eventually going to talk about collapsing stars again, but not before discussing Einstein's ideas (and the evidence for them) in more detail.

23.2.2 Spacetime Examples

How can we understand the distortion of spacetime by the presence of some (significant) amount of mass? Let's try the following analogy. You may have seen maps of New York City that squeeze the full three dimensions of this towering metropolis onto a flat sheet of paper and still have enough information so tourists will not get lost. Let's do something similar with diagrams of spacetime.

■ FIGURE 23.5

A Spacetime Diagram This diagram shows the progress of a motorist traveling east across the Kansas landscape. Distance traveled is plotted along the horizontal axis. The time elapsed since the motorist left the starting point is plotted along the vertical axis.

Figure 23.5, for example, shows the progress of a motorist driving to the east on a stretch of road in Kansas where the countryside is absolutely flat. Since our motorist is traveling only in the east–west direction and the terrain is flat, we can ignore the other two dimensions of space. The amount of time elapsed since he left home is shown on the vertical axis, and the distance traveled eastward is shown on the horizontal axis. From A to B he drove at a uniform speed; unfortunately, it was too fast a uniform speed and a police car spotted him. From B to C he stopped to receive his ticket and made no progress through space, only through time. From C to D he drove more slowly because the police car was behind him.

In the same way, we might (in our imaginations) squeeze the four dimensions of spacetime in the universe onto a flat sheet. To show the distortion, however, we will make it a rubber sheet that can stretch or warp if we put objects on it. Let's imagine we borrow one from a local hospital and stretch it taut on four posts.

To complete the analogy, we need something that normally travels in a straight line (as light does). Suppose we have an extremely intelligent ant that has been trained by some skilled psychologists to walk in a straight line. We reward the ant with a piece of sugar at the end of each run to reinforce the desired behavior.

We begin with just the rubber sheet and the ant, simulating empty space with no mass in it. We put the ant on one side of the sheet and it walks in a beautiful straight line over to the other side (Figure 23.6a). We next put a small grain of sand on the rubber sheet. The sand does distort the sheet a tiny bit, but this is not a distortion that we or the ant can measure. If we send the ant so it goes close to, but not

■ **FIGURE 23.6**

Three-Dimensional Analogy for Spacetime (a) On a flat rubber sheet, a trained ant has no trouble walking in a straight line. (b) When a massive object creates a big depression in the sheet, the ant, who must walk where the sheet takes it, finds its path changed (warped) dramatically.

on top of, the sand grain, it has little trouble continuing to walk in a straight line and earning its sugary reward.

Now we grab something with a little more mass—say, a small pebble. It bends or distorts the sheet just a bit right around its position. If we send the ant into this region, it finds its path slightly altered by the distortion of the sheet. The distortion is not large, but if we follow the ant's path carefully, we notice it deviating slightly from a straight line.

The effect gets more noticeable as we increase the mass of the object we put on the sheet. Let's say we now use a massive paperweight. Such a heavy object distorts or warps the rubber sheet very effectively, putting good sag in it. From our point of view, we can see that the sheet near the paperweight is no longer straight!

Now let's again send the ant on a journey that takes it close to, but not on top of, the paperweight (Figure 23.6b). Far away from the paperweight, the ant has no trouble doing its walk, which looks straight to us. As it nears the paperweight, however, the ant is forced down into the sag. It must then climb up the other side before it can return to walking on an undistorted part of the sheet. All this while, the ant is following the shortest path it can, but through no

fault of its own (after all, ants can't fly, so it has to stay on the sheet) this path is curved by the distortion of the sheet itself.

In the same way, according to Einstein's theory, light always follows the shortest path through spacetime. But large concentrations of mass distort spacetime, and the shortest, most direct paths are no longer straight lines, but curves.

How large does a mass have to be before we can measure the distortion of spacetime? In 1916, when Einstein first proposed his theory, no distortion had been detected at the surface of the Earth (so the Earth might have played the role of the grain of sand in our analogy). Something with a mass like our Sun's was necessary to detect the effect Einstein was describing. We will discuss how this effect was measured using the Sun in the next section.

The paperweight in our analogy might be a white dwarf or a neutron star. The distortion of spacetime is much more noticeable around these compact, massive objects than around the Sun. And when, to return to the situation described at the beginning of the chapter, a star core with more than three times the mass of the Sun collapses forever, the distortions of spacetime can become truly mind-boggling!

23.3 TESTS OF GENERAL RELATIVITY

What Einstein proposed was nothing less than a major revolution in our understanding of space and time. It was a new model of gravity, in which gravity was seen not as a force, but as something that changes the geometry of spacetime. Like all new ideas in science, no matter who advances them, Einstein's theory had to be tested by comparing its predictions against the experimental evidence. This was no easy matter because the effects of the new view of gravity were apparent only when the mass was quite large. (For smaller masses, it required measuring techniques that would not become available until decades later.)

When the distorting mass is small, the predictions of general relativity must agree with those of Newton's theory, which, after all, has served us admirably in our technology and in guiding space probes to the other planets. In familiar territory, therefore, the differences between the predictions of the two theories are subtle and difficult to detect. Nevertheless, Einstein was able to demonstrate one proof of his theory that could be found in existing data and to suggest another one that would be tested just a few years later.

23.3.1 The Motion of Mercury

Of the planets in our solar system, Mercury orbits closest to the Sun and is thus most affected by the distortion of spacetime produced by the Sun's gravity. Einstein wondered if the distortion might produce a noticeable difference in the motion of Mercury that was not predicted by Newton's theory. It turned out that the difference was subtle, but it was definitely there. Most important, it had already been measured.

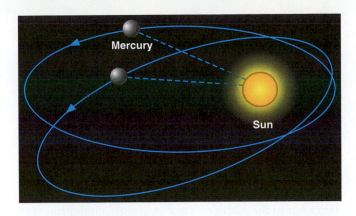

■ FIGURE 23.7

Mercury's Wobble The major axis of the orbit of a planet, such as Mercury, rotates in space slightly because of various perturbations. In Mercury's case the amount of rotation is a bit larger than can be accounted for by the gravitational forces exerted by other planets; this difference is precisely explained by the theory of general relativity. Mercury, being the closest planet to the Sun, has its orbit most affected by the warping of spacetime near the Sun. The change from orbit to orbit has been significantly exaggerated on this diagram.

Mercury has a highly elliptical orbit, so that it is only about two thirds as far from the Sun at perihelion as it is at aphelion. (These terms were defined in Chapter 2 and, like all our astronomical vocabulary, are also explained in the glossary at the end of the book. This has been a friendly reminder from your authors.) It can be calculated that the gravitational effects (perturbations) of the other planets on Mercury should produce an advance of its perihelion. What this means is that each successive perihelion occurs in a slightly different direction as seen from the Sun (Figure 23.7).

According to Newtonian theory, the gravitational forces exerted by the planets will cause Mercury's perihelion to advance by about 531 seconds of arc (arcsec) per century. In the last century, however, it was observed that the actual advance is 574 arcsec per century. The discrepancy was first pointed out by Urbain Leverrier, the codiscoverer of Neptune. Just as discrepancies in the motion of Uranus allowed astronomers to discover the presence of Neptune, so it was thought that the discrepancy in the motion of Mercury could mean the presence of an undiscovered inner planet. Astronomers searched for this planet near the Sun, even giving it a name—Vulcan, after the Roman god of fire. (The name would later be used for the home planet of a famous character on a popular television show about future space travel.)

But no planet has ever been found near the Sun, and the discrepancy was still bothering astronomers when Einstein was doing his calculations. General relativity, however, predicts that due to the curvature of spacetime there should be a tiny additional push on Mercury, over and above that predicted by Newtonian theory, at each perihelion. The result is to make the major axis of Mercury's orbit rotate slowly in space. The prediction of relativity is that the direction of perihelion should change by 43 arcsec per century. This is remarkably close to the observed discrepancy, and it gave Einstein a lot of confidence as he advanced his theory. The relativistic advance of perihelion was later also observed in the orbits of several asteroids that come close to the Sun.

23.3.2 Deflection of Starlight

Einstein's second test was something that had not been observed before and would thus provide an excellent proof of his theory. Since spacetime is more curved in regions where the gravitational field is strong, we would expect light passing very near the Sun to appear to follow a curved path (Figure 23.8), just like the ant in our analogy. Einstein calculated from general relativity theory that starlight just

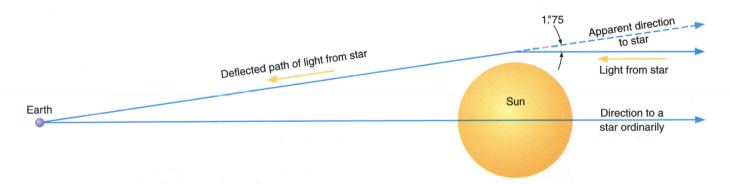

■ FIGURE 23.8

Curvature of Light Paths Near the Sun Starlight passing near the Sun is deflected slightly by the "warping" of spacetime. Before passing by the Sun, the light from the star was traveling parallel to the bottom edge of the page. When it passed near the Sun, the path was altered slightly. When we see the light, we assume the light beam has been traveling in a straight path throughout its journey, and so we measure the position of the star to be slightly different from its true position. If we were to observe the star at another time, when the Sun is not in the way, we would measure its true position.

grazing the Sun's surface should be deflected by an angle of 1.75 arcsec. Could such a deflection be observed?

We encounter a small "technical problem" when photographing starlight coming very close to the Sun: The Sun is an outrageously bright source of starlight itself! But during a total solar eclipse, much of the Sun's light is blocked out, allowing the stars near the Sun to be photographed. In a paper published during World War I, Einstein (writing in a German journal) suggested that observation during an eclipse could detect the deflection of light passing near the Sun.

The technique involves taking a photograph six months earlier of the stars in front of which the eclipsed Sun will appear and measuring the position of all the stars accurately. Then the same stars are photographed during the eclipse. This is when the starlight has to travel to us by skirting the Sun and moving through measurably warped spacetime. Thus its path is no longer a straight line. As seen from Earth, the stars closest to the Sun will be "out of place"—slightly away from their regular positions.

A single copy of that paper, passed through neutral Holland, reached the British astronomer Arthur S. Eddington, who noted that the next suitable eclipse was on May 29, 1919. The British organized two expeditions to observe it; one on the island of Principe, off the coast of West Africa, and the other in Sobral, in northern Brazil. Despite some problems with the weather, both expeditions obtained successful photographs. The stars seen near the Sun were indeed displaced, and to the accuracy of the measurements, which was about 20 percent, the shifts were consistent with the predictions of relativity. More modern experiments with radio waves have confirmed that the actual displacements are within 1 percent of what general relativity predicts.

The confirmation of the theory by the eclipse expeditions in 1919 was a triumph that made Einstein a world celebrity.

23.4 TIME IN GENERAL RELATIVITY

General relativity theory makes various predictions about the behavior of space and time. One of these predictions, put in everyday terms, is that *the stronger the gravity, the slower the pace of time.* Such a statement goes very much counter to our intuitive sense of time as a flow we all share. Time has always seemed the most democratic of concepts: All of us, regardless of wealth or status, appear to move together from the cradle to the grave in the great current of time.

But Einstein argued that it only seems this way to us because all humans so far have lived and died in the gravitational environment of the Earth. We have had no chance to test the idea that the pace of time might depend on the strength of gravity because we have not experienced radically different gravities. And, the differences in the flow of time are extremely small until truly large masses are involved. Nevertheless, Einstein's prediction can now be tested, both on Earth and in space.

23.4.1 The Tests of Time

An ingenious experiment in 1959 used the most accurate atomic clock known to compare time measurements on the ground floor and the top floor of the physics building at Harvard University. For a clock, the experimenters used the frequency (the number of cycles per second) of gamma rays emitted by radioactive cobalt. Einstein's theory predicts that such a cobalt clock on the ground floor, being a bit closer to the Earth's center of gravity, should run very slightly slower than the same clock on the top floor. This is precisely what the experimenters observed. Later, atomic clocks were taken up in high-flying aircraft and even on one of the Gemini space flights. In each case, the clocks farther from the Earth ran a bit faster.

The effect is more pronounced if the gravity involved is the Sun's and not the Earth's. If stronger gravity slows the pace of time, then it will take longer for a light or radio wave that passes very near the edge of the Sun to reach the Earth than we would expect on the basis of Newton's law of gravity. (It also takes longer because spacetime is curved in the vicinity of the Sun.) The smaller the distance between the ray of light and the edge of the Sun at closest approach, the longer the delay in the arrival time.

In November 1976, when the two Viking spacecraft were operating on its surface, Mars, as seen from Earth, went behind the Sun (Figure 23.9). Scientists had preprogrammed Viking to send a radio wave toward the Earth that would go extremely close to the outer regions of the Sun. According to general relativity, there would be a delay because the radio wave would be passing through a region where time ran more slowly. The experiment was able to confirm Einstein's theory to within 0.1 percent.

23.4.2 Gravitational Redshift

What does it mean to say that time runs more slowly? When light enters a region of strong gravity and time slows down, the light experiences a change in its frequency and wavelength. To understand what happens, let's recall that a wave of light is a repeating phenomenon—crest follows trough follows crest with great regularity. In this sense, each light wave is a little clock, keeping time with its wave cycle. If stronger gravity slows down the pace of time, then the rate at which crest follows crest must slow down—the waves become less *frequent*. This means that there must be more space between the crests; in other words, the wavelength of the wave must increase. This kind of increase (when caused by the motion of the source) is just what we called a *redshift* in Chapter 4. Here, because it is gravity and not motion that produces the longer wavelengths, we call the effect a *gravitational redshift*.

The advent of space-age technology made it possible to measure gravitational redshift with very high accuracy.

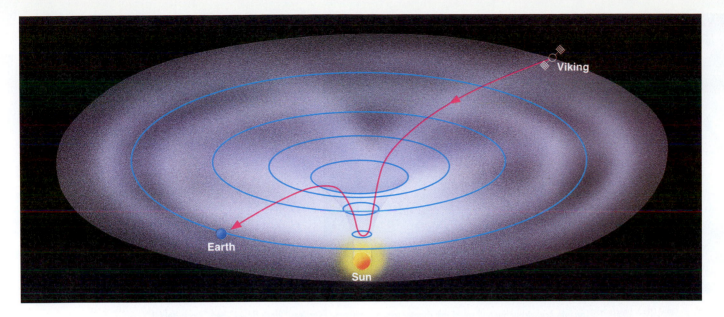

FIGURE 23.9

Time Delays for Radio Waves Near the Sun   Radio signals from the Viking lander on Mars are delayed when they pass near the Sun, where spacetime is curved relatively strongly. In this picture, spacetime is pictured as a two-dimensional rubber sheet.

In the mid-1970s a hydrogen *maser,* a device (akin to a laser) that produces a microwave radio signal at a particular wavelength, was carried by a rocket to an altitude of 10,000 km. The rocket-borne maser was used to detect the radiation from a similar maser on the ground. The radiation showed a redshift due to the Earth's gravity that confirmed the relativity predictions to within a few parts in 10,000.

So far, then, each prediction of general relativity that can be tested has been confirmed within the accuracy of the experiments. Today, general relativity is accepted as our best description of gravity and is used by astronomers and physicists to understand the behavior of active galaxies, the beginning of the universe, and the subject with which we began the chapter—the death of truly massive stars.

23.5   BLACK HOLES

Let's now apply what we have learned about gravity and spacetime curvature to the collapsing core in a very massive star. We saw that if the core's mass is greater than about $3\,M_{\text{Sun}}$, theory says that nothing can stop the core from collapsing forever. We will examine this situation from two perspectives: first from a pre-Einstein point of view, and then with the aid of general relativity.

23.5.1   Classical Collapse

Let's begin with a thought experiment. We want to know what speeds are required to escape from the gravitational pull of different objects. A rocket must be launched from the

surface of the Earth at a very high speed if it is to escape the pull of the Earth's gravity. In fact, any object—rocket, ball, astronomy book—that is thrown into the air with a velocity less than 11 km/s will soon fall back to the Earth's surface. Only those objects launched with a velocity greater than this *escape velocity* can get away from the Earth.

The escape velocity from the surface of the Sun is higher yet—618 km/s. Now imagine that we begin to compress the Sun, forcing it to shrink in diameter. Recall that the pull of gravity depends on both the mass that is pulling you and your distance from the center of gravity of that mass. If the Sun is compressed, its *mass* will remain the same, but the *distance* between a point on the Sun's surface and the center will get smaller and smaller. Thus, as we compress the star, the pull of gravity for an object on the shrinking surface will get stronger and stronger (Figure 23.10).

When the shrinking Sun reaches the diameter of a neutron star (less than 100 km), the velocity required to escape its gravitational pull will be about half the speed of light. Suppose we continue to compress the Sun to a smaller and smaller diameter. (We saw this can't happen to our Sun in the real world because of electron degeneracy, but it will happen to the core of the supermassive star with which our chapter started.) Ultimately, as the Sun shrinks, the escape velocity would exceed the speed of light. If the speed you need to get away is faster than the fastest possible speed in the universe, then nothing, not even light, is able to escape. An object with such large escape velocity emits no light, and anything that falls into it can never return.

In modern terminology, we call an object from which light cannot escape a *black hole,* a name suggested by the American scientist John Wheeler in 1969 (Figure 23.11).

■ **FIGURE 23.10**

Formation of a Black Hole   At left, an (imaginary) astronaut floats near the surface of a massive star-core about to collapse. As the same mass falls into a smaller sphere, the gravity at the surface goes up, making it harder for anything to escape from the stellar surface. Eventually the mass collapses into so small a sphere that the escape velocity exceeds the speed of light and nothing can get away. Note that the size of the astronaut has been exaggerated so you can see him. In the last picture, the astronaut is just outside the event horizon and is stretched and squeezed by the strong gravity.

NO ESCAPE

Ace ◯ Astronomy™   Log into AceAstronomy and select this chapter to see Astronomy Exercises "Black Hole I" and "Black Hole II."

The idea that such objects might exist is, however, not a new one. Cambridge professor and amateur astronomer John Michell wrote a paper in 1783 about the possibility that stars with escape velocities exceeding that of light might exist. And in 1796 the French mathematician Pierre Simon, the Marquis de Laplace, made similar calculations using Newton's theory of gravity; he called the resulting objects "dark bodies."

While these early calculations provided a hint that something strange should be expected if very massive objects collapse under their own gravity, we really need general relativity theory to give an adequate description of what happens in such a situation.

Photo courtesy Roy Bishop

■ **FIGURE 23.11**

John Wheeler   This brilliant physicist did much pioneering work in general relativity theory and coined the term *black hole* in 1969.

23.5.2   Collapse with Relativity

General relativity tells us that gravity is really a curvature in spacetime. As gravity increases (as in the collapsing Sun of our thought experiment), the curvature gets larger and larger. Eventually, if the Sun could shrink down to a diameter of about 6 km, only light beams sent out perpendicular to the surface would escape. All others would fall back onto the star (Figure 23.12). If the Sun could then shrink just a little more, even that one remaining light beam would no longer be able to escape.

Keep in mind that gravity is not pulling on the light. Gravity has curved spacetime, and light (like the trained ant of our earlier example), "doing its best" to go in a straight line, is now confronted with a world in which straight lines have become curved paths. The collapsing star is a *black hole*, in this view, because the very concept of "out" has no geometrical meaning. The star has become trapped in its own little pocket of spacetime, from which there is no escape.

The star's geometry cuts off communication with the rest of the universe at precisely the moment when, in our earlier picture, the escape velocity becomes equal to the speed of light. The size of the star at this moment defines a surface that we call the **event horizon.** It's a wonderfully descriptive name: Just as objects that sink below our horizon cannot be seen on Earth, so any event inside the event horizon can no longer affect the rest of the universe.

Imagine a future spacecraft foolish enough to land on the surface of a massive star just as it begins to collapse in the way we have been describing. Perhaps the captain is asleep at the gravity meter, and before the crew can say "Albert Einstein," they have collapsed with the star inside the event horizon. Frantically, they send an escape pod straight outward. But paths outward twist around to become paths

Light Paths Near a Massive Object Suppose a person could stand on the surface of a normal star with a flashlight. The light leaving the flashlight travels in a straight line no matter where the flashlight is pointed. Now consider what happens if the star collapses so that it is just a little larger than a black hole. All of the light paths except the one straight up curve back to the surface. When the star shrinks inside the event horizon and becomes a black hole, even a beam directed straight up returns.

Ace ✺ Astronomy™ Log into AceAstronomy and select this chapter to see the Active Figure called "Schwarzschild Radius."

inward, and the pod turns around and falls toward the center of the black hole. In desperation now, they send a radio message to their loved ones, bidding good-bye. But radio waves, like light, must travel through spacetime, and curved spacetime allows nothing to get out. Their woeful message remains unheard. Events inside the event horizon can never again affect events outside it.

The characteristics of an event horizon were first worked out by astronomer and mathematician Karl Schwarzschild (Figure 23.13). A member of the German army in World War I, he died in 1916 of an illness he contracted while doing artillery shell calculations on the Russian front. His paper on the theory of event horizons was among the last things he finished as he was dying; it was the first exact solution to Einstein's equations of general relativity. The radius of the event horizon is called the *Schwarzschild radius* in his memory.

The event horizon is the boundary of the black hole; calculations show that it does not get smaller once the whole star has collapsed inside it. It is the region that separates the things trapped inside it from the rest of the universe. Anything coming from the outside is also trapped

Karl Schwarzschild This German scientist was the first to demonstrate mathematically that a black hole is possible.

once it comes inside the event horizon. The horizon's size turns out to depend only on the mass inside it. For a black hole of 1 M_{Sun}, the Schwarzschild radius is about 3 km; thus the entire black hole is about one third the size of a neutron star of that same mass. Feed the black hole some mass, and the horizon will grow—but not very much. Doubling the mass will make a black hole 6 km in radius, still very tiny on the cosmic scale.

The event horizons of more massive black holes have larger radii. For example, if a globular cluster of 100,000 stars (solar masses) could collapse to a black hole, it would be 300,000 km in radius, a little less than half the radius of the Sun. If the entire Galaxy could collapse to a black hole, it would be only about 10^{12} km in radius—about a tenth of a light year. Smaller masses have correspondingly smaller horizons: For the Earth to become a black hole, it would have to be compressed to a radius of only 1 cm—about the size of a grape. A typical asteroid, if crushed to a small enough size to be a black hole, would have the dimensions of an atomic nucleus!

23.5.3 A Black Hole Myth

Much of the modern folklore about black holes is misleading. One idea you may have heard is that black holes are monsters that go about sucking things up with their gravity. Actually, it is only close to a black hole that the strange effects we have been discussing come into play. The gravitational attraction far away from a black hole is the same as that of the star that collapsed to form it.

Remember that the gravity of any star some distance away acts as if all of its mass were concentrated at a point in the center, which we call the center of gravity. For real stars, we merely *imagine* that all the mass is concentrated

Gravity and Time Machines

Time machines are one of the favorite devices of science fiction. Such a device would allow you to move through time at a different pace or in a different direction from everyone else. General relativity suggests that it is possible, in theory, to construct a time machine using gravity that could take you into the future.

Let's imagine a place where gravity is terribly strong, such as near a black hole. General relativity predicts that the stronger the gravity, the slower the pace of time (as seen by a distant observer). So imagine a future astronaut, with a fast spaceship, who volunteers to go on a mission to such a high-gravity environment. The astronaut leaves in the year 2200, just after graduating from college at age 22. She takes, let's say, exactly ten years to get to the black hole. Once there, she orbits some distance from it, taking care not to get pulled in.

She is now in a high-gravity realm where time passes much more slowly than it does on Earth. This isn't just an effect on the mechanism of her clocks—*time itself* is running slowly. That means that every way she has of measuring time will give the same slowed-down reading when compared to time passing on Earth. Her heart will beat more slowly, her hair will grow more slowly, her antique wristwatch will tick more slowly, and so on. She is not aware of this slowing down because all her readings of time, whether made by her own bodily functions or with mechanical equipment, are measuring the same—slower—time. Meanwhile, back on Earth, time passes as it always does.

Our astronaut now emerges from the region of the black hole, her mission of exploration finished, and returns to Earth. Before leaving, she carefully notes that (according to her timepieces) she spent about two weeks around the black hole. She then takes exactly ten years to return to Earth. Her calculations tell her that since she was 22 when she left the Earth, she will be 42 plus two weeks when she returns. So the year on Earth, she figures, should be 2242, and her classmates should now be approaching their midlife crises.

But our astronaut should have paid more attention in her astronomy class! Because time slowed down near the black hole, much less time passed for her than for the people on Earth. While her clocks measured two weeks spent near the black hole, more than 2000 weeks (depending on how close she got) could well have passed on Earth. That's equal to 40 years, meaning her classmates will be senior citizens in their 80s when she (a mere 42-year-old) returns. On Earth it will be not 2242, but 2282—and she will say that she has arrived *in the future*.

Is this scenario real? Well, it has a few practical challenges: We don't think any black holes are close enough for

---

there; for black holes, all the mass *really is* concentrated at a point in the center. So if you are a star or distant planet orbiting around a star that becomes a black hole, your orbit may not be significantly affected by the collapse of the star (although it may be affected by any mass loss that precedes the collapse). If, on the other hand, you venture close to the event horizon, it would be very hard for you to resist the "pull" of the warped spacetime near the black hole.

If another star or a spaceship were to pass one or two solar radii from a black hole, Newton's laws would be adequate to describe what would happen to it. Only very near the surface of a black hole is its gravitation so strong that Newton's laws break down. A massive star coming into our neighborhood would be far, far safer to us as a black hole than it would have been earlier in its life as a brilliant, hot star.

23.5.4 A Trip into a Black Hole

The fact that they cannot see inside black holes has not kept scientists from trying to calculate what they are like. One of the first things these calculations showed was that the formation of a black hole obliterates nearly all information about the star that collapsed to form it. Physicists like to say "black holes have no hair," meaning that nothing sticks out of a black hole to give us clues about what kind of star produced it. The only information a black hole can reveal about itself is its mass, its spin (rotation), and whether it has any electrical charge.

What happens to the collapsing star-core that made the black hole? Our best calculations predict that the material will continue to collapse under its own weight, forming an infinitely *squozen* point—a place of zero volume and infinite density—to which we give the name **singularity.** At the singularity, spacetime ceases to exist. The laws of physics as we know them break down. We do not yet have the physical understanding or the mathematical tools to describe the singularity itself. From the outside, however, the entire structure of a basic black hole (one that is not rotating) can be described as a singularity surrounded by an event horizon. In comparison with humans, black holes are really very simple objects!

Scientists have also calculated what would happen if a daring (perhaps we should say suicidal) astronaut were to fall into a black hole. Let's take up an observing position a

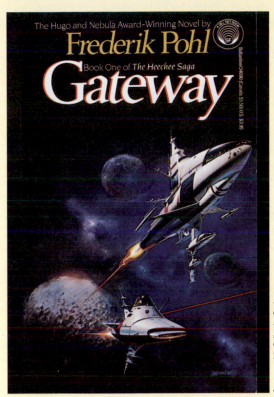

The Hugo and Nebula Award-Winning Novel by
Frederik Pohl
Book One of *The Heechee Saga*
Gateway

Cover copyright Ballantine Books

Gateway by F. Pohl

us to reach in ten years, and we don't think any spaceship or human can survive near a black hole. But the key point, about the slowing down of time, is a natural consequence of Einstein's general theory of relativity, and we saw that its predictions have been confirmed by experiment after experiment.

Science-fiction writers who pay attention to developments in astronomy and physics have seized upon this idea for wonderful plot devices. In Larry Niven's novel *World Out of Time* (1976, Ballantine Books), the protagonist travels around an extremely massive black hole and arrives back on Earth some 3 million years in the future. And in Fred Pohl's *Gateway* (1977, Ballantine Books), the "hero" has to push his girlfriend into a black hole so he can escape its strong gravity. Not only can they never speak or touch again, but he must also deal with the additional guilt of knowing that, now that he is far from the black hole, his time is passing much more quickly than hers. To him she seems almost suspended in time, and (if he had a good enough telescope) he could see her there for the rest of his life, looking back at him in disbelief that he could have done so horrible a thing. (Indeed, the "hero" spends much of the book talking to a robot psychiatrist named Albert about how guilty he feels.)

long, safe distance away from the event horizon and watch this astronaut fall toward it. At first he falls away from us, moving ever faster, just as though he were approaching any massive star. However, as he nears the event horizon of the black hole, things change. The strong gravitational field around the black hole will make his clocks run more slowly, when seen from our outside perspective.

If, as he approaches the event horizon, he sends out a signal once per second according to his clock, we will see the spacing between signals grow longer and longer until it becomes infinitely long when he reaches the event horizon. (If the infalling astronaut uses a blue light to send his signals every second, we will see the light get redder and redder until its wavelength is nearly infinite.) As the spacing between clock ticks approaches infinity, it will appear to us that the astronaut is slowly coming to a stop, frozen in time at the event horizon.

In the same way, all matter falling into a black hole will also appear to an outside observer to stop at the event horizon, frozen in place and taking an infinite time to fall through it. But don't think that matter falling into a black hole will therefore be easily visible at the event horizon. The

tremendous redshift will make it very difficult to observe any radiation from the "frozen" victims of the black hole.

This, however, is only how we, located far away from the black hole, see things. To the astronaut, his time goes at its normal rate and he falls right on through the event horizon into the black hole. (Remember, this horizon is not a physical barrier, but only a region in space where the curvature of spacetime makes escape impossible.)

You may have trouble with the idea that you (watching from far away) and the astronaut (falling in) have such very different ideas about what has happened. This is the reason Einstein's ideas about space and time are called theories of *relativity*. What each observer measures about the world depends on (is relative to) his or her frame of reference. The observer in strong gravity measures time and space differently from the one sitting in weaker gravity. When Einstein proposed these ideas, many scientists also had difficulty with the idea that two such very different views of the same event could be correct, each in its own "world," and they tried to find a mistake in the calculations. There were no mistakes: We and the astronaut really would see the fall into a black hole very differently.

For the astronaut, there is no turning back. Once inside the event horizon, the astronaut, any signals from his radio transmitter, and any cries of regret are doomed to remain hidden forever from the universe outside. He will, however, not have a long time (from his perspective) to feel sorry for himself as he approaches the black hole. Suppose he is falling feet first. The force of gravity exerted by the singularity on his feet will be slightly greater than on his head, and he will be stretched slightly. Since the singularity is a point, the left side of his body will be pulled slightly toward the right, and the right side slightly toward the left, bringing each side closer to the singularity. The astronaut will therefore be slightly squeezed in one direction and stretched in the other.

The Earth exerts similar *tidal forces* on an astronaut performing a spacewalk. In the case of the Earth, the tidal forces are so small that they pose no threat to the health and safety of the astronaut. Not so in the case of a black hole. Sooner or later, as the astronaut approaches the black hole, the tidal forces will become so great that the astronaut will be ripped apart. His legs will be ripped from his torso, his ankles from his legs, his toes from his feet, and so forth. Soon it will be only the individual atoms from his shredded body that will continue their inexorable fall into the singularity. (Jumping into a black hole is definitely a once-in-a-lifetime experience!)

23.6 EVIDENCE FOR BLACK HOLES

Theory tells us what black holes are like. But do they actually exist? And how do we go about looking for something that is many light years away, only about 10 km across, and completely black? It turns out that the trick is not to look for the black hole itself but instead to look for what it does to a nearby companion star.

As we saw, when very massive stars collapse, they leave behind their gravitational influence. What if a member of a double-star system becomes a black hole, and its companion manages to survive the death of the massive star? While the black hole disappears from our view, we may be able to deduce its presence from the things it does to its companion.

23.6.1 Requirements for a Black Hole

So here is a prescription for finding a black hole: First look for a star whose motion (determined from the Doppler shift of its spectral lines) shows it to be a member of a binary star system. If both stars are visible, neither can be a black hole, so focus your attention on just those systems where only one star of the pair is visible even with our most sensitive telescopes.

Being invisible is not enough, however, because a relatively faint star might be hard to see next to the glare of a brilliant companion or in a shroud of dust. And even if the star really is invisible, it could be a neutron star. Therefore, we must also have evidence that the unseen star has a mass too high to be a neutron star and that it is a collapsed object—an extremely small one.

We can use Kepler's law (see Chapter 2) and our knowledge of the visible star to measure the mass of the invisible member of the pair. If the mass is greater than about $3\,M_{Sun}$, then we are likely seeing (or, more precisely, not seeing) a black hole—as long as we can make sure the object really is a collapsed star.

If matter falls toward a compact object of high gravity, the material is accelerated to high speed. Near the event horizon of a black hole, matter is moving at velocities that approach the speed of light. As the atoms whirl chaotically toward the event horizon, they rub against each other; internal friction can heat them to temperatures of 100 million K or more. Such hot matter emits radiation in the form of flickering x rays. The last part of our prescription, then, is to look for a source of x rays associated with the binary system. Since x rays do not penetrate the Earth's atmosphere, such sources must be found using x-ray telescopes in space.

The infalling gas that produces the x-ray emission comes from the black hole's companion star. As we saw in Chapter 22, stars in close binary systems can exchange mass, especially as one of the members expands into a red giant. Suppose that one star in a double-star system has evolved to a black hole and that the second star begins to expand. If the two stars are not too far apart, the outer layers of the expanding star may reach the point where the black hole exerts more gravitational force on them than do the inner layers of the red giant to which the atmosphere belongs. The outer atmosphere has then passed through the point of no return between the stars and falls toward the black hole.

The mutual revolution of the giant star and the black hole causes the material falling toward the black hole to spiral around it rather than flowing directly into it. The infalling gas whirls around the black hole in a pancake of matter called an *accretion disk*. It is in the inner part of this disk that matter is revolving about the black hole so fast that internal friction heats it up to x-ray–emitting temperatures (see the image at the beginning of this chapter).

Another way to form an accretion disk in a binary star system is to have a powerful stellar wind come from the black hole's companion. Such winds are a characteristic of several stages in a star's life. Some of the ejected gas in the wind will then flow close enough to the black hole to be captured by it into the disk (Figure 23.14).

We should point out that, as often happens, the measurements we have been discussing are not quite as simple as they are described in introductory textbooks. In real life, Kepler's law allows us to calculate only the combined mass of the two stars in the binary system. We must learn more about the visible star of the pair and its history to disentangle its mass from the unseen companion's. Also, the calculations are affected by how the orbit of the two stars is tilted

■ **FIGURE 23.14**

Black Hole Accretion Disk in a Binary Star System   Mass lost from a giant star through a stellar wind streams toward a black hole and swirls around it before finally falling in. In the inner portions of the accretion disk, the matter is revolving so fast that internal friction heats it to very high temperatures and x rays are emitted.

toward the Earth, something we can rarely measure. And neutron stars can also have accretion disks that produce x rays, so astronomers must study the properties of these x rays carefully when trying to determine what kind of object is at the center of the disk. Nevertheless, several systems that clearly contain black holes have now been found.

23.6.2   The Discovery of Black Holes

Because the x rays are such an important clue that we are dealing with a black hole that is having some of its companion for lunch, the search for black holes had to await the launch of sophisticated x-ray telescopes into space. These instruments must have the resolution to distinguish the location of x-ray sources and allow us to match them to the positions of binary star systems.

The first black hole binary system to be discovered is called Cygnus X-1. The visible star in this binary system is spectral type O. Measurements of the Doppler shifts of the O star's spectral lines show that it has an unseen companion. The x rays flickering from it strongly indicate that the companion is a small collapsed object. The mass of the invisible collapsed companion is at least six times that of the Sun. The companion is therefore too massive to be either a white dwarf or a neutron star.

A number of other binary systems also meet all the conditions for containing a black hole. Table 23.1 lists the characteristics of some of the best examples.

23.6.3   Feeding a Black Hole

After an isolated star, or even one in a binary star system, becomes a black hole, it probably won't be able to grow much larger. Out in the suburban regions of the Milky Way Galaxy where we live (see Chapter 24), stars and star systems are much too far apart for other stars to provide "food" to a hungry black hole. After all, material must approach close to the event horizon before the gravity is any different from that of the star that became the black hole.

But, as we will see, the central regions of galaxies are quite different from the outer parts. Here stars and raw material can be quite crowded together, and they can interact much more frequently with each other. Therefore, black holes in the centers of galaxies may have a much better opportunity to find mass close enough to their event horizons to pull it in. Black holes are not particular about what they "eat": They are happy to consume other stars, asteroids, gas, dust, and even other black holes. (If two black holes merge, you just get a black hole with more mass and a larger event horizon.)

TABLE 23.1 *Some Black Hole Candidates in Binary Star Systems*

Name/Catalog Designation°	Companion Star Spectral Type	Orbital Period (days)	Black Hole Mass Estimates (M_{Sun})
LMC X-1	O giant	3.0	4–10
Cygnus X-1	O supergiant	5.6	6–15
SAX J1819.3–2525 (V4641 Sgr)	B giant	2.8	6–7
LMC X-3	B main sequence	1.7	4–11
4U1543-47 (IL Lup)	A main sequence	1.1	4–7
GRO J1655 − 40 (Nova Sco 1994)	F subgiant	2.4	4–15
GRS1915 + 105	K giant	33.5	9–14
GS202 + 1338 (V404 Cyg)	K giant	6.5	> 6
XTE J1550–564	K giant	1.5	8–11
A0620-00 (V616 Mon)	K main sequence	7.8	4–15
H1705 − 250 (Nova Oph 1977)	K main sequence	0.52	4–15
GS1124 − 683 (Nova Mus 1991)	K main sequence	0.43	4–15
GS2000 + 25 (QZ Vul)	K main sequence	0.35	4–15
GRS1009 − 45 (Nova Velorum 1993)	K dwarf	0.29	4–8
XTE J1118 + 480	K dwarf	0.17	6–7
XTE J1859 + 226	K dwarf	0.38	8–12
GRO J0422 + 32	M dwarf	0.21	4–15

°As you can tell, there is no standard way of naming these candidates. The chain of numbers is the location of the source in right ascension and declination (the longitude and latitude system on the sky); the letter(s) preceding them designate the satellite that discovered the candidate—A for Ariel, G for Ginga, and so on. The notations in parentheses are those used by astronomers who study binary star systems or novae.

Source: Data based on information provided by Jeffrey McClintock (CfA) and Jerome Orosz (San Diego State).

As a result, black holes in crowded regions can grow, eventually swallowing thousands or even millions of times the mass of the Sun. Observations with the Hubble Space Telescope have recently shown dramatic evidence for the existence of black holes in the centers of many galaxies. These black holes can contain more than a billion solar masses. The feeding frenzy of such supermassive black holes may be responsible for some of the most energetic phenomena in the universe (see Chapters 24 and 26). And evidence from x-ray observations is also starting to indicate the existence of "middle-weight black holes," whose masses are dozens to thousands of times the mass of the Sun. The crowded inner regions of the globular clusters we described in Chapter 21 may be just the right breeding grounds for such intermediate black holes.

Over the past decade, many observations, especially with the Hubble Space Telescope and with x-ray satellites, have been made that can be explained only if black holes really do exist. Furthermore, the observational tests of Einstein's general theory of relativity have convinced even the most skeptical scientists that his picture of warped or curved spacetime is indeed our best description of the effects of gravity.

23.7 GRAVITATIONAL WAVE ASTRONOMY

There is another part of Einstein's ideas about gravity that can be tested as a way of checking the theory that underlies black holes. According to general relativity, the geometry of spacetime depends on where matter is located. Any rearrangement of matter—say, from a sphere to a sausage shape—creates a disturbance in spacetime. This disturbance is called a **gravitational wave,** and relativity predicts that it should spread outward at the speed of light. The big problem with trying to study such waves is that they are tremendously weaker than electromagnetic waves and correspondingly difficult to detect.

23.7.1 Proof from a Pulsar

We now know that gravitational waves exist. In 1974, astronomers Joseph Taylor and Russell Hulse discovered a pulsar (with the designation PSR1913 + 16) orbiting another neutron star. Pulled by the powerful gravity of its companion, the pulsar is moving at about one tenth the speed of light in its orbit.

According to general relativity, this system of star corpses should be radiating gravitational energy at a high enough rate to cause the pulsar and its companion to spiral closer together. If this is correct, then the orbital period should decrease (according to Kepler's third law) by one ten-millionth of a second per orbit. Continuing observations showed that the period is decreasing by precisely this amount. Such a loss of energy in the system can be due only to the radiation of gravitational waves—thus confirming their existence. Taylor and Hulse shared the 1993 Nobel Prize in physics for this work.

23.7.2   Direct Observations

Although such an indirect proof convinced physicists that gravitational waves exist, they would still like to detect the waves directly. What we need are phenomena that are powerful enough to produce a lot of gravitational waves, so that we can actually measure them. Here are some examples:

- the coalescence of two neutron stars in a binary system that spiral together until they merge
- the swallowing of a neutron star by a black hole
- the coalescence (merger) of two black holes
- the implosion of a massive star to form a neutron star or a black hole
- the first "shudder" when space and time came into existence and the universe began

What might a gravitational-wave "telescope" look like? Scientists have designed two experiments to detect gravitational waves. Both rely on the fact that when a gravitational wave passes by the Earth, the ripple in spacetime will cause objects to move ever so slightly. Since the waves are so weak, the trick is to devise a system that will detect these slight motions. And when we say *slight,* we do mean slight! Suppose two test masses are suspended from wires 4 km apart and are free to move in response to a gravitational wave from a typical source. The expected change in distance between the two masses is only one hundred-millionth the diameter of a hydrogen atom! Obviously, such a tiny change can be detected only if the test masses are isolated from all other disturbances—including motions of gas molecules in the air.

LIGO (Laser Interferometer Gravitational-Wave Observatory) is a ground-based facility that started operation in 2002. LIGO has two observing stations, one in Louisiana and the other in the state of Washington. The effects of gravitational waves are so small that a detection will be believed only if it is reported by two widely separated detectors. Local events that might cause small motions of the test masses and mimic gravitational waves—such as small earthquakes, ocean tides, and even traffic—should affect the two sites differently.

Each of the LIGO stations consists of two 4-km-long, 4-ft-diameter vacuum pipes arranged in an L-shape. A test mass with a mirror on it is suspended by wire at each of the four ends of the pipes. Ultra-stable laser beams traverse the pipes and are used to measure any change in the distance between the mirrors (Figure 23.15).

Laser Interferometer Gravitational Wave Observatory/Caltech

■ **FIGURE 23.15**

Gravitational-Wave Telescope   An aerial view of the LIGO facility at Hanford, Washington.

LISA (Laser Interferometer Space Antenna) is designed to carry out the same experiment in space. In this case, three spacecraft fly in formation, maintaining the shape of an equilateral triangle with sides 5 million kilometers long. Each spacecraft will contain two free-floating test masses, and lasers will be used to measure the distances between the test masses on different spacecraft. The goal is to measure the 5-million-km distance with an accuracy of about one tenth the size of an atom. This mission could fly as early as 2008, provided it receives funding.

The ideas discussed in this chapter may seem strange and overwhelming, especially the first time you read them. The consequences of the general theory of relativity take some getting used to. But you have to admit that they make the universe more interesting and bizarre than you probably thought before you took this class.

SURFING THE WEB

The History of General Relativity:
www-groups.dcs.st-and.ac.uk/~history/HistTopics/General_relativity.html
A concise history of the development of the theory of general relativity, with links to biographies of the main participants.

Spacetime Wrinkles Website:
www.ncsa.uiuc.edu/Cyberia/NumRel/NumRelHome.html
An intriguing and well-produced "exhibit on line" about Einstein's theory of relativity and its astronomical implications, including some movies that simulate situations such as the collision of two black holes.

Frequently Asked Questions about Black Holes:

- *Center for Particle Astrophysics at Berkeley:* cosmology.berkeley.edu/Education/Bhfaq.html

- *NASA's Ask the Space Scientist site:* image.gsfc.nasa.gov/poetry/ask/abholes.html

Cambridge Black Holes Page:
www.amtp.cam.ac.uk/user/gr/public/bh_home.html
The Cambridge University Relativity Group has a short set of pages on black holes and calculations relating to them. You can go exploring from there to learn about cosmology, cosmic strings, quantum gravity, and other frontier topics.

Black Hole Animations:
Using high-speed computers, several groups of physicists have simulated the behavior of black holes and what it might be like to fall into one. The following sites have Web movies from such simulations:

- *Movies from the Edge of Spacetime:* www.ncsa.uiuc.edu/Cyberia/Expo/Theater_Img/NumRel.html

- *Numerical Relativity Exhibitions:* jean-luc.ncsa.uiuc.edu/Exhibits/

LIGO site: www.ligo.caltech.edu
The full story on the first major gravitational-wave observatory.

SUMMARY

23.1 Einstein proposed the **equivalence principle** as the foundation of the theory of **general relativity.** According to this principle, there is no way that anyone or any experiment in a sealed environment can distinguish between free fall and the absence of gravity.

23.2 By considering the consequences of this principle, Einstein concluded that we live in a curved **spacetime.** The distribution of matter determines the curvature of spacetime; other objects (and even light) entering a region of spacetime must follow its curvature.

23.3 In weak gravitational fields, the predictions of general relativity agree with the predictions of Newton's theory of gravity. However, general relativity predicts that starlight or radio waves will be deflected when they pass near the Sun, and that the position where Mercury is at perihelion would change by 43 arcsec per century even if there were no other planets in the solar system to perturb its orbit. These predictions have been verified by experiment.

23.4 General relativity predicts that the stronger the gravity, the more slowly time must run. Experiments on Earth and with spacecraft have confirmed this prediction with remarkable accuracy. When light or other radiation emerges from a compact stellar corpse, it shows a gravitational redshift due to the slowing of time.

23.5 Theory suggests that stars with stellar cores more massive than three times the mass of the Sun at the time they exhaust their nuclear fuel will collapse to become **black holes.** The surface surrounding a black hole, where the escape velocity equals the speed of light, is called the **event horizon,** and the radius of the surface is called the *Schwarzschild radius.* Nothing, not even light, can escape through the event horizon from the black hole. At its center, each black hole has a **singularity,** a point of infinite density and zero volume. Matter falling into a black hole appears, as viewed by an outside observer, to freeze in position at the event horizon. However, if we were riding on the infalling matter, we would pass through the event horizon. As we approach the singularity, the tidal forces would tear our bodies apart.

23.6 The best evidence that stellar-mass black holes actually do exist comes from binary star systems that have (a) one star of the pair that is invisible, (b) x-ray emission characteristic of an *accretion disk* around a compact object, and (c) an indication from the orbit and characteristics of the visible star that the mass of its invisible companion is greater than $3 M_{Sun}$. Several systems with these characteristics have been found. Black holes with masses greater than a billion solar masses are found in the centers of large galaxies.

23.7 General relativity also predicts that the rearrangement of matter in space should produce **gravitational waves.** The existence of such waves was confirmed in observations of a pulsar in orbit around another neutron star whose orbits were spiraling closer and losing energy in the form of gravitational waves. New experiments are under way using freely suspended test masses to detect the passing of gravitational waves directly.

INTER-ACTIVITY

A A computer-science major takes an astronomy course like the one you are taking and becomes fascinated with black holes. Later in life, he founds his own Web company and becomes very wealthy when it goes public. He sets up a foundation to support the search for black holes in our Galaxy. Your group is the allocation committee of this foundation; how would you distribute money each year to increase the chances that more black holes will be found?

B Suppose for a minute that stars evolve *without* losing any mass at any stage of their lives. Your group is given a list of binary star systems. Each binary contains one main-sequence star and one invisible companion. The spectral types of the main-sequence stars range from spectral type O to M. Your job is to determine whether any of the invisible companions might be black holes. Which ones are worth observing? Why? (*Hint:* Remember that in a binary star system, the two stars form at the same time, but the pace of their evolution depends on the mass of each star.)

C You live in the far future, and you have been convicted (falsely) of high treason. The method of execution is to send everyone in your group into a black hole, but you get to pick which one. Since you are doomed to die, you would at least like to see what the inside of a black hole is like—even if you can't tell anyone outside about it. Would you choose a black hole with a mass equal to that of Jupiter or one with a mass equal to that of an entire galaxy? Why? What would happen to you as you approached the event horizon in each case? (*Hint:* Consider the difference in force on your feet and your head as you cross over the event horizon.)

D General relativity is one of the areas of modern astrophysics where we can clearly see the frontiers of human knowledge. We have started to learn about black holes and warped spacetime only recently, and we are humbled by how much we still don't know. Research in this field is supported mostly by grants from government agencies. Have your group discuss what reasons there are for our tax dollars to support such "far-out" work. Can you make a list of "far-out" (seemingly impractical) areas of research in past centuries that later led to practical applications? What if general relativity does not have many practical applications? Do you think a small part of society's funds should still go to exploring theories about the nature of space and time?

E Once you all have read this chapter, work with your group to come up with a plot for a science fiction story that uses the properties of black holes.

REVIEW QUESTIONS

1. How does the equivalence principle lead us to suspect that spacetime might be curved?

2. If general relativity offers the best description of what happens in the presence of gravity, why do physicists still make use of Newton's equations in describing gravitational forces on Earth (when building a bridge, for example)?

3. Einstein's general theory of relativity made predictions about the outcome of several experiments that had not yet been carried out at the time the theory was first published. Describe three experiments that verified the predictions of the theory.

4. If a black hole emits no radiation, what evidence do astronomers and physicists today have that the theory of black holes is correct?

5. What characteristics must a binary star have to be a good candidate for a black hole? Why is each of these characteristics important?

6. Suppose the Earth were to fall into a black hole. What would happen to it?

7. A student becomes so excited by the whole idea of black holes that he decides to jump into one.
   a. What is the trip like for him?
   b. What is it like for the rest of the class, watching from afar?

8. What is a gravitational wave and why is it so hard to detect?

9. What are some strong sources of gravitational waves that astronomers hope to detect in the future?

THOUGHT QUESTIONS

10. Imagine that you have built a large room around the boy and girl in Figure 23.2, and that this room is falling at exactly the same rate as they are. Galileo showed that, if there is no air friction, light and heavy objects that are dropping due to gravity will fall at the same rate. Suppose this were not true and that instead heavy objects fall faster. Also suppose that the boy in Figure 23.2 is twice as massive as the girl. What would happen? Would this violate the equivalence principle?

11. A monkey hanging from a tree branch sees a hunter aiming a rifle directly at him. The monkey then sees a flash and knows that the rifle has been fired. Reacting quickly, the monkey lets go of the branch and drops so that the bullet can pass harmlessly over his head. Does this act save the monkey's life? Why?

12. During the 1970s, the United States had in orbit a small space station called Skylab. Some of the astronauts exercised by running around the inside wall of their cylindrical vehicle. How could they stay against the wall while running rather than float aimlessly inside Skylab? What physical principles were involved?

13. Why would we not expect to detect x rays from a disk of matter about an ordinary star?

14. Look elsewhere in this book for the necessary data, and indicate what the final stage of evolution—white dwarf, neutron star, or black hole—will be for each of these kinds of stars.
   a. spectral type-O main-sequence star
   b. B main-sequence star
   c. A main-sequence star
   d. G main-sequence star
   e. M main-sequence star

15. Which is likely to be more common in our Galaxy: white dwarfs or black holes? Why?

16. If the Sun could suddenly collapse to a black hole, how would the period of the Earth's revolution about it differ from what it is now?

FIGURING FOR YOURSELF

The size of the event horizon of a black hole depends on the mass of the black hole. The greater the mass, the larger the radius of the event horizon. General relativity calculations show that the formula for the radius of the event horizon is

$$R = \frac{2GM}{c^2}$$

where c is the speed of light, G is the gravitational constant, and M is the mass of the black hole. Note that in this formula, 2, *G*, and *c* are all constants; only the mass changes from black hole to black hole.

17. Look up *G*, *c*, and the mass of the Sun in Appendix 6 of this text, and calculate the radius of a black hole that has the same radius as the Sun. (Note that this is only a theoretical calculation. The Sun does not have enough mass to become a black hole.)

18. Suppose you wanted to know the size of black holes with masses that are larger or smaller than the Sun. You could go through all the steps in problem 17, wrestling with a lot of large numbers with large exponents. You could be clever, however, and evaluate all the constants in the equation once and then simply vary the mass. You could even express the mass in terms of the Sun's mass and make future calculations really easy. Show that the event horizon equation is equivalent to saying that the radius of the event horizon is equal to 3 km times the mass of the black hole in units of the Sun's mass.

19. Use the result from problem 18 and calculate the radius of a black hole with a mass equal to (a) the Earth, (b) a B0-type main sequence star, (c) a globular cluster, and (d) the Milky Way Galaxy. Look elsewhere in this text and the appendices for tables that provide data on the mass of these four objects.

20. Since the force of gravity outside the event horizon of a black hole is the same as that of an ordinary object of the same mass, Kepler's third law is valid. Suppose that the Earth collapsed to the size of a golf ball. What would be the period of revolution of the Moon, orbiting at its current distance of 400,000 km? Use Kepler's third law to calculate: (a) the period of revolution of a spacecraft orbiting at a distance of 6000 km and (b) the period of revolution of a miniature spacecraft orbiting at a distance of 0.1 m.

21. Calculate the orbital speed of the mini-spaceship in problem 20 and compare it with the speed of light.

22. The first binary system discovered that provided strong evidence that black holes really do exist was Cygnus X-1. You could now make this discovery yourself if you had access to the same data that astronomers had when the discovery was first made. The binary system contains a supergiant of spectral type O with a mass of roughly 30 M_{Sun}. The companion is invisible. The orbital period of the binary system is 5.6 days, and the orbital velocity of the O star is about 350 km/s. Use Kepler's third law to calculate the mass of the companion. Is your result consistent with the possibility that the companion is a black hole? What other evidence would you want before announcing your discovery to the world?

SUGGESTIONS FOR FURTHER READING

Begelman, M. and Rees, M. *Gravity's Fatal Attraction: Black Holes in the Universe.* 1996, Scientific American Library. Nicely done introduction to the astronomy of black holes.

Charles, P. and Wagner, R. "Black Holes in Binary Stars: Weighing the Evidence" in *Sky & Telescope*, May 1996, p. 38. An excellent review.

Jayawardhana, R. "Beyond Black" in *Astronomy*, June 2002, p. 28. On finding evidence of the existence of event horizons and thus black holes.

McClintock, J. "Do Black Holes Exist?" in *Sky & Telescope*, Jan. 1988, p. 28.

Overbye, D. "God's Turnstile: The Work of John Wheeler and Stephen Hawking" in *Mercury* (the magazine of the Astronomical Society of the Pacific), July/Aug. 1991, p. 98.

Parker, B. "Where Have All the Black Holes Gone?" in *Astronomy*, Oct. 1994, p. 36. On candidates within our Galaxy.

Rees, M. "To the Edge of Space and Time" in *Astronomy*, July 1998, p. 48. About black holes.

Thorne, K. *Black Holes and Time Warps.* 1994, W. W. Norton. The definitive introduction to black holes.

Wheeler, J. *A Journey into Gravity and Spacetime.* 1990, Scientific American Library/W. H. Freeman. A brilliant introduction by one of the foremost scientists of our time.

Wheeler, J. "Of Wormholes, Time Machines, and Paradoxes" in *Astronomy*, Feb. 1996, p. 52. On some of the "far-out" consequences of general relativity.

Will, C. *Was Einstein Right?—Putting General Relativity to the Test.* 1986, Basic Books. Superb guide to the many tests.

Zirker, J. "Testing Einstein's General Relativity During Eclipses" in *Mercury*, July/Aug. 1985, p. 98.

On Gravitational Waves

Bartusiak, M. "Catch a Gravity Wave" in *Astronomy*, Oct. 2000, p. 54.

Bartusiak, M. *Einstein's Unfinished Symphony.* 2000, Joseph Henry Press.

Frank, A. "Teaching Einstein to Dance: The Dynamic World of General Relativity" in *Sky & Telescope*, Oct. 2000, p. 50.

Gibbs, W. "Ripples in Spacetime" in *Scientific American*, Apr. 2002, p. 62.

Sanders, G. and Beckett, D. "LIGO: An Antenna Tuned to the Songs of Gravity" in *Sky & Telescope*, Oct. 2000, p. 41.

Anglo-Australian Observatory/David Malin Images

The Southern Milky Way   The brightest part of the Milky Way is located in the southern hemisphere. The constellations in this region are Centaurus, the Southern Cross, and Carina. The bright stars Alpha and Beta Centauri in the lower left of the image point up and to the right to the Southern Cross, which is located next to a dark dust cloud called the Coal Sack. The reddish glows are produced by clouds of ionized hydrogen.

Across the height of heaven there runs a road, Clear when the night is bare, the Milky Way, Famed for its sheen of white . . .

Ovid, *Metamorphoses*, quoted by E. C. Krupp in *Beyond the Blue Horizon* (1991, HarperCollins)

THINKING AHEAD

Today we know that our Sun is just one of many billions of stars that make up the huge cosmic island we call the Milky Way Galaxy. How can we "weigh" such an enormous system of stars and measure its total mass?

O ne of the most striking features you can see in a truly dark sky is a band of faint white light, called the Milky Way, which stretches from one horizon to the other. The name comes from ancient legends that compare its faint white splash of light to a stream of spilled milk. But folktales differ from culture to culture: One East African tribe thought of it as the smoke of ancient campfires, several Native American stories tell of a path across the sky traveled by sacred animals, and in Siberia it was known as the seam of the tent of the sky. In the Southern Hemisphere, the Milky Way looks even brighter— so bright, in fact, that native people in South America gave names to various portions of it.

In 1610 Galileo made the first telescopic survey of the Milky Way and discovered that it is composed of a multitude of individual stars. We know that our Sun is just one of many billions of stars that make up a huge cosmic pinwheel called the **Milky Way Galaxy.** In this chapter, we begin our voyage beyond the neighborhood of the Sun by exploring our own Galaxy. What we learn will then help us interpret the more feeble light emitted by other galaxies far beyond our own.

Ace◐Astronomy™   The AceAstronomy icon throughout the text indicates an opportunity for you to test yourself on key concepts and to explore animations and interactions on the AceAstronomy website at **http://ace.brookscole.com/voyages**

Virtual Laboratories

 Dark Matter

24.1 THE ARCHITECTURE OF THE GALAXY

The Milky Way Galaxy surrounds us, and you might think it is easy to study because it is so close. However, the very fact that we are embedded within it presents a difficult challenge. Suppose you were given the task of mapping New York City. You could do a much better job from a helicopter flying over the city than you could if you were standing in Times Square and forbidden to move. Similarly, it would be easier to map our Galaxy if we could only get a little way outside it, but instead we are trapped inside and, worse yet, we are out in its suburbs—far from the galactic equivalent of Times Square.

24.1.1 Herschel Measures the Galaxy

In 1785 William Herschel (Figure 24.1) made the first important discovery about the architecture of the Milky Way Galaxy. By counting stars in different directions, he found that most of the stars he could see lay in a narrow band, the Milky Way, circling the sky, and that the numbers of stars were about the same in any direction around this band. He therefore concluded that the stellar system to which the Sun belongs has the shape of a wheel (he might have called it a Frisbee except Frisbees hadn't been invented yet), and that the Sun must be near the hub of the wheel (Figure 24.2).

To understand why Herschel reached this conclusion, imagine that you are a member of the band standing in formation during halftime at a football game. If you count the band members you see in different directions and get about the same number each time, you can conclude that the band has arranged itself in a circular pattern with you at the

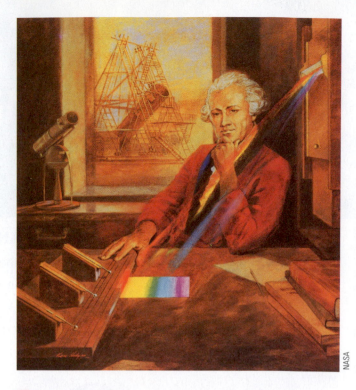

■ FIGURE 24.1

William Herschel (1738–1822) Herschel was a German musician who emigrated to England and took up astronomy in his spare time. He discovered the planet Uranus, built several large telescopes, and made measurements of the Sun's place in the Galaxy, the Sun's motion through space, and the comparative brightnesses of stars. This painting shows how Herschel discovered infrared radiation. After dispersing sunlight into a spectrum, he placed thermometers at the position of each color and measured the increase in temperature. He found that the temperature rose even when the thermometers were placed beyond the red end of the spectrum, thereby demonstrating the existence of invisible heat radiation.

■ FIGURE 24.2

Herschel's Diagram of the Milky Way Herschel constructed this cross section of the Galaxy by counting stars in various directions. The large circle shows the location of the Sun.

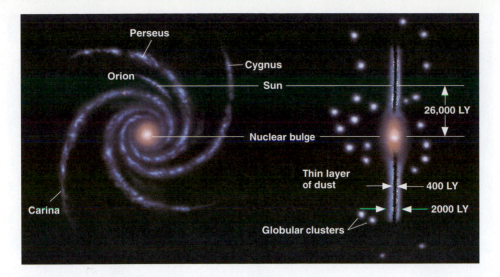

A Schematic Representation of the Galaxy The left image shows the view looking down on the spiral disk; the right image shows the view looking "sideways" along the disk. The major spiral arms are labeled. The Sun is located on the inside edge of the short Orion spur.

Ace◯**Astronomy**™ Log into AceAstronomy and select this chapter to see the Active Figure called "Explorable Milky Way," and Astronomy Exercise "Milky Way Galaxy."

center. Since you see no band members above you or underground, you know that the circle made by the band is much flatter than it is wide.

We now know that Herschel was right about the shape of our system, but wrong about where the Sun lies within the disk. As we saw in Chapter 19, we live in a dusty Galaxy. Because interstellar dust absorbs the light from stars, Herschel could see only those stars within about 6000 light years (LY) of the Sun. Today we know that this is a very small section of the entire disk of stars that makes up the Galaxy.

In the same way, if we stand on the streets of a large urban area on a smoggy day, we can see only a small section of the city. The dust and pollution in the air block the light from more distant neighborhoods. Herschel, confined to using visible light for his observations, had no way of penetrating the dust between the stars and so did not discover what we know today—namely, that the Sun is actually located fairly far from the center of the Galaxy.

24.1.2 Disks and Haloes

With modern instruments, astronomers can now penetrate the "smog" of the Milky Way. Just as you can listen to your favorite radio station whether or not it is smoggy in the city, astronomers can "tune in" on radio and infrared radiation from distant parts of the Galaxy. Measurements at these wavelengths (as well as observations of other galaxies like ours) have given us a good idea of what the Milky Way would look like if we *could* observe it from a distance.

Figure 24.3 sketches what we would see if we could view the Galaxy face-on and edge-on. The brightest part of the Galaxy consists of a thin, circular, rotating disk of stars distributed across a region about 100,000 LY in diameter and about 1000 LY thick. (Given how thin the disk is, perhaps a French crepe is a more appropriate analogy than a wheel.) In addition to stars, the dust and gas from which stars form are also found mostly in the thin disk of the Galaxy. The mass of the interstellar matter is about 20 percent of the mass of the stars in this disk.

As the diagram in Figure 24.3 shows, the stars, gas, and dust are not spread evenly throughout the disk but are concentrated in a series of spiral arms. We will describe the spiral structure in more detail shortly. The Sun is located about halfway between the center of the Galaxy and the edge of the disk and only about 70 LY above its central plane.

Our thin disk of young stars, gas, and dust is embedded in a thicker disk of older stars; this thicker disk extends about 3000 LY above and below the mid-plane of the thin disk and contains only about 5 percent as much mass as the thin disk.

Close in to the galactic center (within about 12,000 LY), the stars are no longer confined to the disk but form a **nuclear bulge.** When we use ordinary visible radiation, we can only glimpse the stars in the bulge in those rare directions where there happens to be relatively little interstellar dust (Figure 24.4). The first picture that actually succeeded in showing the bulge as a whole was taken at infrared wavelengths (Figure 24.5).

The fact that much of the bulge is obscured by dust makes its shape difficult to determine. For a long time, astronomers assumed it was spherical. However, infrared images and other data indicate that the bulge is about two

■ FIGURE 24.4

The Sky Toward the Center of the Galaxy Baade's window is a region with little interstellar dust, so we can see stars in the bulge of the Galaxy. In this direction we also see a rich concentration of stars along with a number of open clusters (M6, M7, M18, M21, M23, M24, M25) and globular clusters (M9, M22, M28, M54, M69, M70).

■ FIGURE 24.5

The Inner Part of the Milky Way Galaxy This picture was obtained in 1990 by an experiment aboard the Cosmic Background Explorer Satellite (COBE). Taken in the near-infrared part of the spectrum, the image permits us to see, for the first time, the bulge of old stars that surround the center of our Galaxy. Redder colors correspond to regions with more dust, and this dust forms a thin disk of material that shows the plane of the Milky Way. The same dust, building up into a kind of interstellar smog, prevents us from seeing the inner parts of our Galaxy with visible light.

■ **FIGURE 24.6**

Spiral Galaxies That Resemble the Milky Way   The image on the left shows M83, which is a galaxy that is very much like our own. This galaxy is at a distance of about 20 million LY and has a concentration of older, yellow stars in its round nuclear bulge, with younger, blue stars and patchy red clouds of glowing gas and dark dust lanes in the spiral arms. For a long time, it was thought that the nuclear bulge of the Milky Way was also spherical, like the one in M83. Now, however, it appears that the bulge is elongated, more like NGC 1365, which is shown in the right-hand image. This galaxy is about 40 million LY away. Note the bright bar of light, with the nuclear bulge as its hub. The bulge and the bar both appear yellowish because the bright stars in them are old red giants. Slender curved arms with blue stars and red patches of glowing gas project from the ends of the bar.

times longer than it is wide, rather like a peanut. We know many other spiral galaxies that also have bar-shaped concentrations of stars in their central regions; for that reason they are called *barred spirals* (Figure 24.6). At the center of the nuclear bulge is a tremendous concentration of matter, which we will discuss in Section 24.4.

The thin and thick disks and the nuclear bulge are embedded in a spherical **halo** of very old, faint stars that extends to a distance of at least 150,000 LY from the galactic center. The globular clusters are also found in this halo. The whole of the Milky Way, including the halo of old stars,

is surrounded by a bubble of hot gas. The temperature of this gas is about 1 million degrees, but it is very tenuous (less than 1 particle per 10,000 cm^3). It extends out at least to the distance of the small neighbor galaxies we call the Magellanic Clouds (160,000 LY). The hot gas is probably a relic of the formation of the Milky Way. Theories of galaxy formation predict that hot gas should be present around galaxies when they first form and that it would take longer than the age of the universe for this gas to cool.

The mass in the Milky Way extends even farther out, well beyond the boundary of the luminous stars to a dis-

Harlow Shapley: Mapmaker to the Stars

Until the early 1900s, astronomers generally accepted Herschel's conclusion that the Sun is near the center of the Galaxy. The discovery of the Galaxy's true size and our actual location came about largely through the efforts of Harlow Shapley. In 1917, he was studying RR Lyrae variable stars (see Section 18.3) in globular clusters. By comparing the known intrinsic luminosity of these stars to how bright they appeared, Shapley could calculate how far away they are. (Recall that it is distance that makes the stars look dimmer than they would be "up close," and that the brightness fades as the distance squared.) Knowing the distance to any star in a cluster then tells us the distance to the cluster itself.

Globular clusters are found in regions that are free of interstellar dust and can be seen at very large distances. When Shapley used the distances and directions of 93 globular clusters to map out their positions in space, he found that the clusters are distributed in a spherical volume, which has its center not at the Sun but at a distant point along the Milky Way in the direction of Sagittarius (see Figure 24.4). Shapley then made the bold assumption, since verified by many other observations, that the point on

Harlow Shapley (1885–1972)

Harvard College Observatory Archives

which the system of globular clusters is centered is also the center of the entire Galaxy.

Shapley's work showed once and for all that our star has no special place in the Galaxy. We are in a nondescript region of the Milky Way, one of at least 100 billion stars that circle the distant center of our Galaxy.

Born in 1885 on a farm in Missouri, Harlow Shapley at first dropped out of school with only the equivalent of a fifth-grade education. He studied at home and at age 16 got a job as a newspaper reporter covering crime stories. Frustrated by the lack of op-

portunities for someone who had not finished high school, Shapley went back and completed a six-year high school program in only two years, graduating as class valedictorian.

In 1907, at age 22, he went to the University of Missouri, intent on studying journalism, but found that the School of Journalism would not open for a year. Leafing through the college catalog (or so he told the story later), he chanced to see "Astronomy" among the subjects beginning with "A." Recalling his boyhood interest in the stars, he decided to study astronomy for the next year (and the rest, as the saying goes, is history).

Upon graduation Shapley received a fellowship for graduate study at Princeton and began to work with the brilliant Henry Norris Russell (see Voyagers in Astronomy: *Henry Norris Russell* in Chapter 17). For his PhD thesis, Shapley made major contributions to the methods of analyzing the behavior of eclipsing binary stars. He was also able to show that cepheid variable stars are not binary systems, as some people thought at the time, but individual stars that pulsate with striking regularity.

Impressed with Shapley's work, George Ellery Hale offered him a position at the Mount Wilson Observa-

tance of at least 600,000 LY from the center of the Galaxy. This **dark matter halo** has been detected only because it affects the orbits of other stars and galaxies, but it hasn't actually been seen. This mysterious outer region will be the subject of Section 24.3.

Some vital statistics and a sketch of the thin and thick disks and the stellar halo are given in Table 24.1. Note particularly how the ages of stars correlate with where they are found. As we shall see, this information holds important clues to how the Milky Way formed.

Establishing this overall picture of the Galaxy from our suburban, dust-shrouded vantage point has been one of the great achievements of modern astronomy (and one

that took decades of effort by astronomers working with a wide range of telescopes). One thing that helped enormously was the discovery that our Galaxy is not unique in its characteristics. There are many other such spiral-shaped islands of stars in the universe. For example, the Milky Way resembles the Andromeda Galaxy, which at a distance of about 2.3 million LY is one of our nearest neighbors (see Figure 24.13 later in the chapter). Just as you can get a much better picture of yourself if someone else takes the photo from some distance away, pictures (and other data) from nearby galaxies that resemble ours have been vital to our understanding the properties of the Milky Way.

tory, where the young man took advantage of the clear mountain air and the 60-inch reflector to do his pioneering study of variable stars in globular clusters.

Shapley subsequently accepted the directorship of the Harvard College Observatory, and over the next 30 years he and his collaborators made contributions to many fields of astronomy, including the study of neighboring galaxies, the discovery of dwarf galaxies, a survey of the distribution of galaxies in the universe, and much more. He wrote a series of nontechnical books and articles, and became known as one of the most effective popularizers of astronomy. Shapley enjoyed giving lectures around the country, including at many smaller colleges where students and faculty rarely got to interact with scientists of his caliber.

During World War II, Shapley helped rescue many scientists and their families from Eastern Europe; later he helped found UNESCO, the United Nations Educational, Scientific, and Cultural Organization. He wrote a little pamphlet called *Science from Shipboard* for men and women in the armed services who had to spend many weeks on board transport ships to Europe. And during the diffi-

Globular Clusters Outline the Milky Way Harlow Shapley's diagram shows the location of globular clusters, with the position of the Sun also marked. The black area shows Herschel's old diagram, centered on the Sun, approximately to scale (see Figure 24.2).

cult period of the 1950s when congressional committees began their "witch hunts" for communist sympathizers (including such liberal leaders as Shapley), he spoke out forcefully and fearlessly in defense of the freedom of thought. A man of many interests, he was fascinated by the behavior of ants, and wrote scientific papers about them as well as about galaxies.

By the time he died in 1972, Shapley was acknowledged as one of the pivotal figures of modern astronomy, a "20th-century Copernicus" who mapped the Milky Way and showed us our place in the Galaxy.

Property	Thin Disk	Thick Disk	Stellar Halo (Excludes Dark Matter)
Stellar mass	$4 \times 10^{10}\,M_{Sun}$	A few percent of the thin disk mass	$10^{10}\,M_{Sun}$
Luminosity	$3 \times 10^{10}\,L_{Sun}$	?	$8 \times 10^{8}\,L_{Sun}$
Typical Age of Stars	1 million to 10 billion years	11 billion years	13 billion years
Heavier-element abundance	High	Intermediate	Very low
Rotation	High	Moderately high	None

TABLE 24.1 *Characteristics of the Milky Way Galaxy*

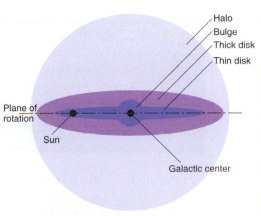

The Major Parts of the Galaxy

Reprinted with permiss on from *Science, Vol 287.* © 2000 AAS

24.2 SPIRAL STRUCTURE

Astronomers were able to make tremendous progress in mapping the spiral structure of the Milky Way after the discovery of the 21-cm line from cold hydrogen atoms (see Chapter 19). Remember that interstellar dust prevents us from seeing stars at large distances in the disk. However, radio waves at 21 cm pass right through the dust, and since the stars and the interstellar matter like to live together, mapping interstellar gas can tell us where the stars are as well. After decades of surveying this gas, we are beginning to pin down the precise layout of our Galaxy.

24.2.1 The Arms of the Milky Way

The Galaxy has four major **spiral arms** with some smaller spurs (shown in Figure 24.3). The Sun is near the inner edge of a short arm or spur called the *Orion arm*, which is about 15,000 LY long and contains such conspicuous features as the Cygnus Rift (the great dark nebula in the summer Milky Way) and the Orion Nebula.

More distant, and therefore less conspicuous, are the *Carina* and *Perseus arms,* located, respectively, about 6000 LY inside and outside the Sun's position relative to the galactic center. Both of these arms and the more distant *Cygnus arm* are about 80,000 LY long. The fourth major arm is unnamed and difficult to detect because it runs along the other side of the Galaxy. Radiation from it is mingled with radiation from the central regions of the Galaxy, as seen from our vantage point.

24.2.2 Formation of Spiral Structure

At the Sun's distance from its center, the Galaxy does not rotate like a solid wheel. Instead, each star follows its own orbit around the Galaxy's center. Individual stars obey Kepler's third law, just as planets in our solar system do. Re-

member that Pluto takes longer than the Earth to complete one full circuit around the Sun. In just the same way, stars in larger orbits in the Galaxy trail behind those in smaller ones. This effect is called *differential galactic rotation.*

Differential rotation explains why so much of the material in the disk of the Milky Way is concentrated into elongated features that resemble spiral arms. No matter what the original distribution of the material might be, the differential rotation of the Galaxy can stretch it out into spiral features. Figure 24.7 shows the development of spiral arms from two irregular blobs of interstellar matter, as the portions of the blobs closest to the galactic center move fastest, while those farther out trail behind.

But this picture of spiral arms presents astronomers with an immediate problem. If that's all there were to the story, differential rotation—over the roughly 13-billion-year history of the Galaxy—would have wound the Galaxy's arms tighter and tighter until all semblance of structure had disappeared. Somehow, the graceful spiral arms must maintain themselves over time, even as the material in them turns and turns around the center of the Milky Way.

We can think of the spiral arms as regions where matter is more densely concentrated. The arms are defined by a concentration of stars and interstellar matter, but individual stars and gas clouds may move into and out of the arms. To see what we mean, consider this analogy: Imagine that you are driving on a highway and that several cars with very nervous student drivers are moving unusually slowly in two of the lanes ahead of you. Traffic behind these cars will be forced to slow down, and the density of cars around them will increase. (So will the noise level, as people honk their horns and make rude remarks.)

Over time, individual cars like yours will manage to get past the slowly moving cars, but others will take their place in the traffic jam. Viewed from high above the highway, there will be a clump of cars moving along the freeway at the same speed as the two slowly moving cars. The clump will be moving more slowly than the traffic both in front of and behind the jam. The cars stuck in the clump will be dif-

■ **FIGURE 24.7**

Formation of Spiral Arms This sketch shows how spiral arms might form from irregular clouds of interstellar material stretched out by the different rotation rates throughout the Galaxy. The regions farthest from the galactic center rotate most slowly.

ferent as time passes; each driver will be caught in the traffic jam only for a while.

Calculations show that, in the same way, the "traffic" of stars and interstellar matter in the Galaxy slows down and clumps when it meets the denser regions of the spiral arms. Gravity causes objects moving around the Galaxy to slow down slightly in the regions of the spiral arms and linger there longer than elsewhere in their orbits. The regions of the concentrations of higher density rotate more slowly than does the actual material in the Galaxy, so that individual stars, clusters, gas, and dust eventually pass through the spiral arms—just as the cars ultimately make it through the traffic jam.

This theory for how spiral arms form is called the **spiral density wave** model. Spiral density waves have been observed directly in the rings of Saturn, another region that contains a disk with many objects circling a common center.

As gas and dust clouds approach the inner boundaries of an arm and encounter the higher density of slowly moving matter, they crash into it. It is just where the shock of the collision occurs that the compression required for star formation is most likely to take place. This explains why the spiral arms are sites of vigorous star formation and the home of the youngest stars. In some other galaxies, whose spiral arms we can view face-on, we see young stars, along with the densest dust clouds, occupying the inner boundaries of spiral arms—just as this theory predicts.

The spiral density model does not explain everything about the structure of galaxies. Even our Galaxy has shorter spurs or segments between spiral arms that are not so easy to explain. Other spiral galaxies have even more chaotic or complicated patterns. Theorists have found other ways, unrelated to spiral density waves, to produce elongated structures that look like spiral arms. For example, great chain reactions of star formation may move progressively through molecular clouds and produce extended regions of young stars that mimic spiral arms (see Section 20.1). It is likely that more than one mechanism contributes to building up the specific design of galaxies such as the Milky Way.

24.3 THE MASS OF THE GALAXY

When we described the sections of the Milky Way, we said that the stars are now known to be surrounded by a much larger halo of invisible matter. Let's see how this surprising discovery was made.

24.3.1 Kepler Helps Weigh the Galaxy

The Sun, like all the other stars in the Galaxy, orbits the center of the Milky Way. Our star's orbit is nearly circular and lies in the Galaxy's disk. The speed of the Sun in its orbit is about 220 km/s, which means it takes us approxi-

mately 225 million years to go once around. We call the period of the Sun's revolution the *galactic year*. It is a long time compared to human timescales; during the entire lifetime of the Earth, only about 20 galactic years have passed. And we have gone a mere fraction of the way around the Galaxy in all the time that humans have gazed into the sky.

We can use the information about the Sun's orbit to estimate the mass of the Galaxy (just as we could "weigh" the Sun by monitoring the orbit of a planet around it—see Chapter 2). Let's assume that the Sun's orbit is circular and that the Galaxy is roughly spherical, both of which are fairly accurate assumptions. Long ago, Newton showed that if you have matter distributed in the shape of a sphere, then it is simple to calculate the pull of gravity on some object just outside that sphere: You can assume gravity acts as if all the matter were concentrated at a point in the center of the sphere. So, for our purposes we have a point in the center of the Galaxy, containing all the mass that lies inward of the Sun's position, and we have the Sun orbiting that point about 26,000 LY away.

This is the sort of situation to which Kepler's third law (as modified by Newton) can be directly applied. Plugging numbers into Kepler's formula (which we do in the Figuring for Yourself section at the end of this chapter), we can calculate the sum of the masses of the Galaxy and the Sun. However, the mass of the Sun is completely trivial compared to the mass of the Galaxy. Thus, for all practical purposes, the result (about 100 billion solar masses) is the mass of the Milky Way. More sophisticated calculations based on complicated models give a similar result.

This estimate tells us how much mass is contained in the volume inside the Sun's orbit. It is a good estimate for the *total* mass of the Galaxy only if hardly any mass is outside the Sun's orbit. For many years astronomers thought this assumption was reasonable. The number of bright stars and the amount of *luminous matter* (meaning any material from which we can detect electromagnetic radiation) both drop off dramatically at distances of more than about 30,000 LY from the galactic center.

24.3.2 A Galaxy of Mostly Invisible Matter

In science, what seems to be a reasonable assumption can later turn out to be wrong (which is why we continue to do observations and experiments every chance we get). There is a lot more to the Milky Way than meets the eye (or our instruments)! While there is relatively little luminous matter beyond 30,000 LY, a lot of *invisible matter* exists at great distances from the galactic center.

We can understand how astronomers detected this invisible matter by remembering that, according to Kepler's third law, objects orbiting at large distances from a massive object move more slowly than do objects closer to that central mass. In the case of the solar system, for example, the outer planets move more slowly in their orbits than the planets close to the Sun.

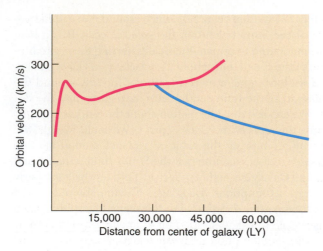

■ FIGURE 24.8

Rotation Curve of the Galaxy   The orbital speed of carbon monoxide (CO) gas at different distances from the center of the Milky Way Galaxy is shown in red. The blue curve shows what the rotation curve would look like if all of the matter in the Galaxy were located inside a radius of 30,000 LY. Instead of going down, the speed of gas clouds farther out goes up, indicating a great deal of mass beyond the Sun's orbit.

There are a few objects, including globular clusters and some nearby small satellite galaxies, that lie well outside the luminous boundary of the Milky Way. If most of the mass of our Galaxy were concentrated within the luminous region, then these very distant objects should travel around their galactic orbits at lower speeds than, for example, that of the Sun.

It turns out, however, that the few objects seen at large distances from the luminous boundary of the Milky Way Galaxy are *not* moving more slowly than the Sun. There are some globular clusters and RR Lyrae stars between 30,000 and 150,000 LY from the center of the Galaxy, and their orbital velocities are even greater than the Sun's (Figure 24.8).

What do these higher speeds mean? Kepler's third law tells us how fast objects must orbit a source of gravity if they are neither to fall in (because they move too slowly) nor to escape (because they move too fast). If the Galaxy had only the mass calculated in the previous section, then the high-speed outer objects should long ago have escaped the grip of the Milky Way. The fact that they have not done so means that our Galaxy must have more gravity than can be supplied by the luminous matter—in fact, a *lot* more gravity. The high speed of these outer objects tells us that the source of this gravity must extend outward from the center far beyond the Sun's orbit.

If the gravity were supplied by stars or by something else that gives off radiation, we should have spotted this outer material long ago. We are therefore forced to the reluctant conclusion that this matter is invisible and has, except for its gravitational pull, gone entirely undetected!

Studies of the motions of the most remote globular clusters and the small galaxies that orbit our own show that the total mass of the Galaxy is at least $2 \times 10^{12} \, M_{Sun}$, about twenty times greater than the amount of luminous matter. The **dark matter** (as astronomers have come to call the invisible material) extends to a distance of at least 600,000 LY from the center of the Galaxy. Theoretical arguments about how galaxies form suggest that this halo of dark matter is spherical, but observations are not yet good enough to confirm or refute this prediction.

The obvious question is: What is the dark matter made of? Let's look at a list of "suspects" taken from our study of astronomy so far. Since this matter is invisible, it clearly cannot be in the form of ordinary stars. And it cannot be gas in any form (remember that there has to be a lot of it). If it were neutral hydrogen gas, its 21-cm radiation would have been detected. If it were ionized hydrogen, it should be hot enough to emit visible radiation. If a lot of hydrogen atoms out there had combined into hydrogen molecules, these should produce dark features in the ultraviolet spectra of objects lying beyond the Galaxy, but such features have not been seen. Nor can the dark matter consist of interstellar dust, since dust in the required quantities would clearly block the light from distant galaxies.

What are our other possibilities? The dark matter cannot be a huge number of black holes (of stellar mass) or old neutron stars, since interstellar matter falling onto such objects would produce more x rays than are observed. Also, recall that the formation of black holes, neutron stars, and white dwarfs is preceded by mass loss, which scatters heavy elements into space to be incorporated into subsequent generations of stars. If the dark matter consisted of an enormous number of any of these objects, they would have blown off and recycled a lot of heavier elements over the history of the Galaxy. In that case, the young stars we observe in our Galaxy today would contain much larger abundances of heavy elements than they actually do.

Brown dwarfs and loose Jupiter-like planets have also been ruled out. As we learned in Chapter 23, the general theory of relativity predicts that the path traveled by light is changed when it passes near a concentration of mass. It turns out that when the two objects appear close enough together on the sky, the image of the more distant object also becomes significantly brighter. By looking for the temporary brightening that occurs when a dark matter object in our own Galaxy moves across the path traveled by light from stars in the Magellanic Clouds, astronomers have shown that the dark matter cannot be made up of a lot of small objects with masses between one-millionth and one-tenth the mass of the Sun.

What's left? One possibility is that the dark matter is composed of exotic subatomic particles of a type not yet detected on Earth. Very sophisticated (and difficult) experiments are now under way to look for such particles.

The problem of dark matter is by no means confined to the Milky Way. Observations show that dark matter must also be present in other galaxies (whose outer regions also orbit too fast "for their own good"). As we will see, it even exists in great clusters of galaxies, whose members move around under the influence of far more gravity than can be accounted for by luminous matter alone. Chapter 27 contains a summary of the recent results of searches for dark matter.

Stop a moment and consider how astounding the conclusion we have reached really is. Perhaps as much as 95 percent of the mass in our Galaxy (and many other galaxies) is invisible, and we do not even know what it is made of! The stars and raw material we can observe may be merely the tip of the cosmic iceberg (to use an old cliché); underlying it all may be other matter, perhaps familiar, perhaps startlingly new. Understanding the nature of this dark matter is one of the great challenges of astronomy today.

24.4 THE CENTER OF THE GALAXY

At the beginning of this chapter, we hinted that the core of our Galaxy contains a large concentration of mass. In fact, we now have evidence that the very center contains a black hole with enough mass inside to make several million Suns. This amazing conclusion needs some explanation. We cannot see a black hole directly because by definition it radiates no energy. And we cannot even see into the center of the Galaxy in visible light because of absorption by the interstellar dust that lies between us and the galactic center.

Light from the central region of the Galaxy is dimmed by a factor of a trillion (10^{12}) by this dust.

Fortunately, we are not blind at other wavelengths. Infrared and radio radiation, which have long wavelengths compared with the sizes of the interstellar dust grains, flow around the dust particles and reach our telescopes. In fact, the very bright radio source in the nucleus, known as Sagittarius A*, was the first cosmic radio source astronomers discovered.

24.4.1 A Journey Toward the Center

We will journey to the dark heart of our Galaxy in stages, stopping along the way to see just what a rich assortment of objects is present. Figure 24.9 is a radio image of a region about 1500 LY across, centered on Sagittarius A*, the bright radio source that marks the nucleus of the Galaxy. Much of the radio emission comes from hot gas heated either by clusters of hot stars (the stars themselves do not produce radio emission and can't be seen in the picture) or by supernova blast waves. Most of the hollow circles you see on the image are supernova remnants. The other main source of radio emission is electrons moving at high speed in regions with strong magnetic fields. The bright thin arcs and "threads" show us where this type of emission is produced.

Figure 24.10 shows the x-ray emission from a region 400 LY across and 900 LY wide centered on Sagittarius A*.

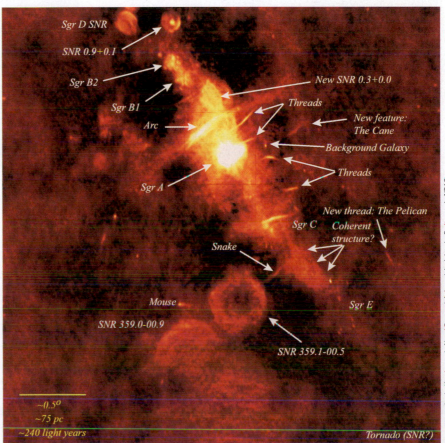

■ **FIGURE 24.9**

Radio Image of Galactic Center Region This radio map of the center of the Galaxy (at a wavelength of 90 cm) was constructed from data obtained with the Very Large Array radio telescope. Brighter regions are more intense in radio waves. The galactic center is inside the region labeled Sgr A. Sgr B1 and B2 are regions of active star formation. Many filaments or threadlike features are seen, as well as a number of shells (labeled SNR), which are supernova remnants. The bar at bottom left is about 240 LY long, to give you a sense of scale.

Sgr D SNR
SNR 0.9+0.1
Sgr B2
Sgr B1
Arc
Sgr A
New SNR 0.3+0.0
Threads
New feature: The Cane
Background Galaxy
Threads
New thread: The Pelican
Sgr C
Coherent structure?
Snake
Mouse
Sgr E
SNR 359.0-00.9
SNR 359.1-00.5
Tornado (SNR?)

~0.5°
~75 pc
~240 light years

Image processing by Kassim et al., Naval Research Labs; original data by Pedlar et al., NRAO

NASA/U. of Massachusetts/D. Wang et al.

■ **FIGURE 24.10**

The Galactic Center in X Rays This mosaic of 30 images taken with the Chandra x-ray satellite shows a region 400 by 900 LY in extent and centered on Sagittarius A*, the bright white source in the center of the picture. The point x-ray sources are white dwarfs, neutron stars, and black holes. The blue-green object at the far left is a supernova remnant. The diffuse "haze" is emission from gas at a temperature of 10 million degrees. This hot gas is flowing away from the center out into the rest of the Galaxy.

Seen in this picture are hundreds of white dwarfs, neutron stars, and black holes. The diffuse haze in the picture is emission from gas that lies among the stars and is at a temperature of 10 million degrees.

Within the central few light years of the Galaxy, we find the supermassive black hole Sagittarius A°, which is surrounded by a cluster of stars, some of which are hot and massive and formed in the last few million years. There is also a ring of dust and molecular gas that is revolving around the center of the Galaxy, along with some ionized gas streamers that are heated by the hot stars (Figure 24.11).

24.4.2 The Dark Heart of the Galaxy

If we zero in on the inner few light years of the Galaxy with an infrared telescope equipped with adaptive optics, we see a region crowded with individual stars (Figure 24.12). These stars have now been observed for almost a decade, and astronomers have detected their orbital motions around the very center of the Galaxy. These orbits provide convincing evidence that the stars are moving around a supermassive black hole at the center of our Galaxy.

As we saw in Chapter 23, proving that a black hole exists is a challenge because the hole itself emits no radiation. What astronomers must do is prove that a small region in space contains far more mass than could be accounted for by a very dense cluster of stars or something else made of ordinary matter. For example, the radius of a black hole with a mass of about 3 million M_{Sun} would only be about the size of the Sun. The corresponding density within this region of space would be much higher than that of any star cluster or any other ordinary astronomical object.

The first clue that a black hole *might* lie at the center of the Galaxy came from radio astronomy. As matter spirals inward toward the event horizon of a black hole, it is heated in a whirling *accretion disk* and produces radio radiation. (Such accretion disks were explained in Chapter 23.) Mea-

NRAO/AUI

■ **FIGURE 24.11**

The Central 10 Light Years of the Galaxy This image shows the radio emission from hot, ionized gas in the center of the Milky Way. The intensity of radio waves increases from blue to red. Sagittarius A*, which is not marked in this picture, is located near where the vertical bar of the anchor-shaped area of strong radio emission (red) intersects the right-hand cross piece of the anchor.

surements with the Very Large Array (VLA) radio telescope show that the diameter of the radio source Sagittarius A° is no larger than the diameter of Jupiter's orbit (10 AU), and it may even be smaller—close to the size of the event horizon.

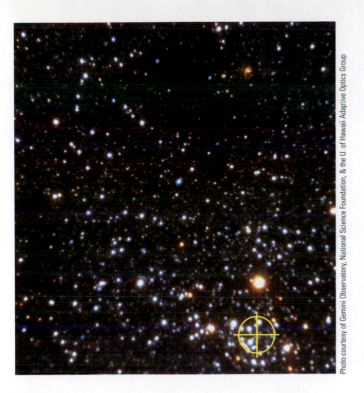

Photo courtesy of Gemini Observatory, National Science Foundation, & the U. of Hawaii Adaptive Optics Group

■ **FIGURE 24.12**

Infrared Image of the Stars in the Center of the Galaxy The density of stars near the center of the Galaxy is about 300,000 times higher than the density in the region near the Sun. This image, taken with the Gemini North telescope, shows a region about 3 LY across. Sagittarius A* is located at the center of the Galaxy, which is indicated by cross hairs near the bottom right of the image.

However, we have to establish that Sagittarius A* is not only small but also has a very large mass. That is where the orbiting stars come in. If we combine observations of their periods and the size of their orbits with Kepler's third law, we can estimate the mass of the object that keeps them in their orbits. One of the stars has been observed for about ten years, and in that time it has completed about two thirds of its journey around Sagittarius A*. Its closest approach takes it to a distance of only 124 astronomical units (AU) or about 17 light hours from the black hole. This orbit, when combined with observations of other stars close to the galactic center, indicates that a mass of 2.6 million M_{Sun} must be concentrated inside the orbit—that is, within 17 light hours of the center of the Galaxy.

Since we can't see this mass at any wavelength, it must be dark. If it were anything other than a supermassive black hole—low-mass stars that emit very little light or neutron stars or a very large number of small black holes—then calculations show that these objects would be so densely packed that they would collapse to a single black hole within a hundred thousand years. This is a very short time compared with the age of the Galaxy, which is 13 to 14 billion years old. Since it seems very unlikely that we would

have caught a cluster of objects just before it collapsed, the evidence for a supermassive black hole at the center of the Galaxy is convincing indeed.

24.4.3 Finding the Source

Where did this black hole come from? The answer is a matter of ongoing debate among astronomers. One possibility is that a large cloud of gas near the center of the Milky Way collapsed directly to form a black hole. Since we find large black holes at the centers of most other galaxies (see Chapter 26), even ones that are very young, this collapse probably took place when the Milky Way was just beginning to take shape. The mass of this black hole might have been only a few hundred thousand solar masses. Over billions of years, as it devoured nearby stars and gas clouds, the black hole grew to its current size.

At the present time, gas and dust are falling into the galactic center at the rate of about 1 M_{Sun} per thousand years. If matter had been falling in at the same rate for about 5 billion years, then it would have easily been possible to accumulate the matter needed to form a black hole with a mass of several million solar masses. Stars are also on the black hole's menu. The density of stars near the galactic center is high enough that we would expect a star to pass near the black hole and be swallowed by it every ten thousand years or so. As this happens, some of the energy of infall is released as radiation. As a result, the center of the Galaxy might flare up and briefly outshine all the stars in the Milky Way.

24.5 STELLAR POPULATIONS IN THE GALAXY

In the first section of this chapter, we described the thin disk, thick disk, and stellar halo. Look back at Table 24.1 and note some of the patterns: Young stars lie in the thin disk, are rich in metals, and orbit the Galaxy's center at high speed. The stars in the halo are old, have low abundances of elements heavier than hydrogen and helium, and barely revolve at all. The stars in the thick disk are intermediate between these two extremes. Let's first see why age and heavier-element abundance are correlated and then see what these correlations tell us about the origin of our Galaxy.

24.5.1 Two Kinds of Stars

The discovery that there are two different kinds of stars was first made by Walter Baade during World War II. As a German national, Baade was not allowed to do war research as many other scientists were, so he was able to make regular use of the Mount Wilson telescopes in southern California. His observations were aided by the darker skies that resulted from the wartime blackout of Los Angeles.

Photo by Tony Hallas/Inset: Caltech Archives

■ **FIGURE 24.13**

The Galaxy in Andromeda (M31) This neighboring spiral looks very much like our
own Galaxy. Note the bulge of older, yellowish stars in the center, the bluer and younger
stars in the outer regions, and the dust in the disk that blocks some of the light from the
bulge. *Inset:* Walter Baade (1893–1960).

Ace⊘Astronomy™ Log into AceAstronomy and select this chapter to see the Active Figure called "M31 at Many Wavelengths."

Baade was impressed by the similarity of the mainly
reddish stars in the Andromeda galaxy's nuclear bulge (Figure 24.13) to those in our own Galaxy's globular clusters
and the halo, and by the difference in color between all of
these stars and the bluer stars found in the spiral arms near
the Sun. On this basis, he called all the stars in the halo and
globular clusters **population II,** and the bright blue stars
in the spiral arms **population I.**

We now know that the populations differ not only in
their locations in the Galaxy but also in their chemical composition, age, and orbital motions around the center of the
Galaxy. Population I stars, found only in the disk, are especially concentrated in the spiral arms and follow nearly circular orbits around the galactic center. Examples are bright
supergiant stars, main-sequence stars of high luminosity
(spectral classes O and B), and members of young open star
clusters. Interstellar matter and molecular clouds are
found in the same places as population I stars.

Population II stars show no correlation with the location of the spiral arms. These objects are found throughout
the Galaxy. Some are in the disk, but many others follow el-

liptical orbits that carry them high above the galactic disk
into the halo. Examples include planetary nebulae and RR
Lyrae variables. The stars in globular clusters, found almost entirely in the halo, are also classified as population II.

Today we know much more about stellar evolution
than did astronomers in the 1940s, and we can determine
the ages of stars. Population I includes stars with a range of
ages. While some are as old as 10 billion years, others are
still forming today. Most important, population I includes
all of the massive young stars that formed in the last few
million years. The Sun, which is about 5 billion years old, is
a population I star. Population II, on the other hand, consists entirely of old stars that formed very early in the history of the Galaxy; typical ages are 11 to 13 billion years.

We also now have good measurements of the compositions of stars. Nearly all stars appear to be composed mostly
of hydrogen and helium, but their abundances of the heavier elements differ. In the Sun and other population I stars,
the heavy elements (those heavier than hydrogen and helium) account for about 1 to 4 percent of the total stellar
mass. Population II stars in the outer galactic halo and in

globular clusters have much lower abundances of the heavy elements—often less than one-tenth or even one-hundredth the concentrations found in the Sun.

As we discussed in earlier chapters, heavy elements are created in stars. They return to the Galaxy's reserves of raw material when stars die, only to be recycled into new generations of stars. Thus, as time goes on, stars are born with larger and larger supplies of heavy elements. Population II stars formed when the abundance of elements heavier than hydrogen and helium was low. We might call them the senior citizens of the Milky Way; like many human seniors, they lived during simpler times (at least as far as the abundances of the elements are concerned). Population I stars formed later, after mass lost by dying members of the first generations of stars had seeded the interstellar medium with elements heavier than hydrogen and helium.

Just recently, European astronomers using the Very Large Telescope discovered a star on the outskirts of the Galaxy with an abundance of iron that is 20 times less than the most primitive star that had been known before. It has just one iron atom for every 7 billion atoms of hydrogen. For comparison, the Sun has one iron atom for every 275,000 atoms of hydrogen. The star, whose catalog number is HE0107-5240, is estimated to be more than 12 billion years old and must have formed very early in the history of the Milky Way. Like finding a person who is older than 110, the discovery of this star gives us hope that we can delve even further back in galactic history than we had earlier thought.

24.5.2   The Real World

With rare exceptions, we should never trust any theory that divides the world into just two categories. Since Baade's pioneering work, we have learned that the notion that all stars can be characterized as either older and poor in heavy elements, or younger and rich in heavy elements, is an oversimplification. The idea of two populations helps organize our initial thoughts about the Galaxy, but we now know it cannot explain everything we observe.

For example, most of the stars in the nuclear bulge of the Galaxy are more than 10 billion years old. Yet the abundance of heavy elements in these stars is about the same as in the Sun. Astronomers think that star formation in the crowded nuclear bulge occurred very rapidly just after the Milky Way Galaxy formed. After a few million years, the first generation of massive short-lived stars then expelled heavy elements in supernova explosions and enriched subsequent generations of stars. Thus even stars that formed in the bulge more than 10 billion years ago started with a good supply of heavy elements.

Exactly the opposite occurred in the Small Magellanic Cloud, a small galaxy that orbits our own. Even the youngest stars in this galaxy are deficient in heavy elements, presumably because star formation has occurred so slowly that there have been, so far, relatively few supernova explosions. (Smaller galaxies also have more trouble holding onto the gas produced in supernova explosions long enough to recycle it. Low-mass galaxies exert only a modest gravitational force, and the high-speed gas ejected by supernovae can easily escape from them.)

The elements with which a star is endowed depend not only on when it forms in the history of its galaxy, but also on how many stars in its part of the galaxy have already completed their lives by the time the star is ready to form.

24.6   THE FORMATION OF THE GALAXY

Information about stellar populations holds vital clues to how our Galaxy was built over time. The flattened disk shape of the Galaxy suggests that it formed through a process similar to the one that leads to the formation of a protostar (see Chapter 20). Building on this idea, early models assumed that the Galaxy formed from a single rotating cloud. But, as we shall see, this turns out to be only part of the story.

24.6.1   The Protogalactic Cloud

Since the oldest stars—those in the halo and in globular clusters—are distributed in a sphere centered on the nucleus of the Galaxy, it makes sense to assume that the protogalactic cloud was roughly spherical. The oldest stars in the halo have ages of about 13 billion years, and so we estimate that the formation of the Galaxy began about that long ago. Then, just as in the case of star formation, the protogalactic cloud collapsed and formed a thin rotating disk. Stars born before the cloud collapsed did not participate in the collapse, but continue to orbit in the halo to the present day (Figure 24.14).

Gravitational forces caused the gas in the thin disk to fragment into clouds or clumps with masses like those of star clusters. These individual clouds then fragmented further to form stars. Since the oldest stars in the disk are nearly as old as the stars in the halo, the collapse must have been rapid (astronomically speaking), requiring perhaps no more than a few hundred million years.

24.6.2   Collision Victims

Recently, astronomers have learned that the evolution of the Galaxy has not been quite as peaceful as this traditional model suggests. In 1994 astronomers discovered a small new galaxy in the direction of the constellation of Sagittarius. The Sagittarius dwarf galaxy is only 50,000 LY away from the nucleus of our Galaxy and is the closest galaxy known (Figure 24.15). It is very elongated, and its shape indicates that it is being torn apart by the Galaxy's gravitational force—just as Comet Shoemaker–Levy was torn apart when it passed too close to Jupiter in 1992. The Sagittarius galaxy is a small one with about 150,000 stars, all of which seem doomed to end up in the bulge and halo of our

■ FIGURE 24.14

Formation of the Galaxy The Milky Way Galaxy initially formed from a rotating cloud of gas that collapsed due to gravity. Halo stars and globular clusters either formed prior to the collapse or were formed elsewhere and attracted by gravity to the Galaxy early in its history. Stars in the disk formed late, and the gas from which they were made was contaminated with heavy elements produced in early generations of stars. Stars were added to the Milky Way by collisions and mergers with other small galaxies.

own Galaxy. But don't sound the funeral bells quite yet; the destruction of the Sagittarius dwarf will take another 100 million years or so.

Calculations indicate that the Galaxy's thick disk may be a product of one or more such collisions with other galaxies. Accretion of a satellite galaxy would stir up the orbits of the stars and gas clouds originally in the thin disk and cause them to move higher above and below the mid-plane of the Galaxy. If such a collision happened about 10 billion years ago, then any gas in the two galaxies that had not yet formed into stars would have had plenty of time to settle back down into the thin disk. The gas could then have begun forming subsequent generations of population I stars. This timing is also consistent with the typical ages of stars in the thick disk.

There is evidence of other past collisions. When a small galaxy ventures too close, the force of gravity exerted by our Galaxy tugs harder on the near side than on the far side. The net effect is that the stars that originally belonged to the small galaxy are spread out into a long stream that or-

bits through the halo of the Milky Way (Figure 24.16). This stream can maintain its identity for billions of years. Astronomers have now identified several such streams originating from different small galaxies that were swallowed by the much larger Milky Way. It has recently been suggested that large globular clusters, like Omega Centauri, were actually dense nuclei of cannibalized dwarf galaxies. The globular cluster M54 (see Figure 24.4) is the nucleus of the Sagittarius dwarf that is currently merging with the Milky Way. The stars in the outer regions of such galaxies are stripped off by the Milky Way, but the central dense regions manage to survive.

The Milky Way has more collisions in store. The Large and Small Magellanic Clouds (see Figure P.14 in the Prologue), two nearby satellite galaxies, are spiraling ever closer to our Galaxy and, according to calculations, will merge with it in the distant future. Already there is evidence that streams of matter have been torn out of the Clouds during past close encounters with the Milky Way. The Milky Way Galaxy has eight other small satellite galaxies, and at least

Diagram from Gerry Gilmore, Institute of Astronomy, Cambridge

■ **FIGURE 24.15**

The Sagittarius Dwarf   In 1994 British astronomers discovered a galaxy in the constellation of Sagittarius, located only about 50,000 LY from the center of the Milky Way and falling into our Galaxy. This image covers a region approximately 70° × 50° and combines a black-and-white view of the disk of our Galaxy with a red contour map showing the brightness of the dwarf galaxy. This galaxy lies on the other side of the galactic center from us. The white stars in the red region mark the locations of several globular clusters. The cross marks the galactic center. The horizontal line corresponds to the galactic plane. The blue outline on either side of the galactic plane corresponds to the infrared image in Figure 24.5. The boxes mark regions where detailed studies of individual stars led to the discovery of this galaxy.

Paul Harding/Case Western Reserve University

■ **FIGURE 24.16**

Streams in the Galactic Halo   When a small galaxy is swallowed by the Milky Way, its member stars are stripped away and form streams of stars in the galactic halo. This image is based on calculations of what these streams might look like if the Milky Way swallowed 50 dwarf galaxies over the past 10 billion years. Although color coding helps us to sort out streams from different galaxies in this computer plot, in real life the stars in the streams are mixed in with other stars in the halo and are hard to identify.

John Dubinski, University of Toronto

■ **FIGURE 24.17**

The Collision of the Milky Way Galaxy with Andromeda   In about 3 billion years the Milky Way and the Andromeda Galaxies will collide, separate, and then come back together to form an elliptical galaxy. The interaction will take about a billion years. On these computer plots, the Milky Way is shown face-on, while Andromeda is tilted. The Milky Way moves from the bottom upward to the left of Andromeda and then to the upper right before returning to complete the merger.

some of these dwarfs appear to have been split off from the Magellanic Clouds when they passed close to the Milky Way.

In about 3 billion years, the Milky Way itself will be swallowed up, since it and the Andromeda Galaxy are on a collision course. Our computer models show that, after a complex interaction, the two will merge to form a larger, more rounded galaxy (Figure 24.17).

We are thus coming to realize that "environmental influences" (and not just its original characteristics) play an important role in determining the properties and development of our Galaxy. In future chapters we will see that collisions and mergers are a major factor in the evolution of many other galaxies as well.

SUMMARY

24.1 The **Milky Way Galaxy** consists of a thin disk containing dust, gas, and young stars; a spherical **halo** containing very old stars, globular clusters, and RR Lyrae variables; a thick disk, with stars that have properties intermediate between those in the thin disk and the halo; a bar-shaped **nuclear bulge** of stars around the center; and a supermassive black hole at the very center. The Sun is located in the outskirts of the Milky Way about 26,000 LY from the center.

24.2 The Galaxy has four main **spiral arms** and several short spurs; the Sun is located on one of these spurs. Measurements show that the Galaxy does not rotate as a solid body, but instead its stars follow Kepler's laws; those closer to the galactic center complete their orbits more quickly. The **spiral density wave** theory is one way to account for the spiral arms. Calculations show that the gravitational forces within the Galaxy cause stars and gas clouds to slow down in the vicinity of the spiral arms, thereby leading to higher densities of material. When molecular clouds attempt to pass through these regions of higher density, the gas within them is compressed and star formation is triggered.

24.3 The Sun revolves completely around the galactic center in about 225 million years (sometimes called a *galactic year*). The mass of the Galaxy can be determined by measuring the orbital velocities of stars or interstellar matter. The total mass of the Galaxy is about 10^{12} M_{Sun}, and about 95 percent of this mass consists of **dark matter** that emits no electromagnetic radiation and can be detected only because of the gravitational force it exerts on visible stars and interstellar matter. This dark matter is located mostly in the Galaxy's halo; its nature is not understood at present.

24.4 A supermassive black hole is located at the center of the Galaxy. Measurements of the velocities of stars located within a few light days of the center show that the mass inside their orbits around the center is about 2.6 million M_{Sun}. The density of this matter concentration exceeds that of the densest known star clusters by a factor of nearly a million; the only known object with such a high density is a black hole.

24.5 We can roughly divide the stars in the Galaxy into two categories. Old stars with few heavy elements are referred to as **population II** stars and are found in the halo and in globular clusters. **Population I** stars contain more heavy elements than globular cluster and halo stars, are younger and found in the disk, and are especially concentrated in the spiral arms. The Sun is a member of population I. Population I stars formed after previous generations of stars produced heavy elements and ejected them into the interstellar medium. The bulge stars, which are also typically more than 10 billion years old, do have substantial amounts of heavy elements, presumably because there were many massive first-generation stars in this dense region and these quickly contaminated the next generations of stars with the products of nucleosynthesis.

24.6 The Galaxy formed about 13 billion years ago. Models suggest that the stars in the halo and globular clusters formed first, while the Galaxy was spherical. The gas, somewhat enriched in heavy elements by the first generation of stars, then collapsed from a spherical distribution to a disk-shaped distribution. Stars are still forming today from the gas and dust that remain in the disk. Star formation occurs most rapidly in the spiral arms, where the density of interstellar matter is highest. The Galaxy captured (and is still capturing) additional stars and globular clusters from small galaxies that venture too close to the Milky Way. In about 3 billion years, the Galaxy will collide with the Andromeda Galaxy, and after about a billion years, the two galaxies will merge to form a larger, more rounded galaxy.

INTER-ACTIVITY

A You are captured by space aliens who take you inside a complex cloud of interstellar gas, dust, and a few newly formed stars. To escape, you need to make a map of the cloud. Luckily, the aliens have a complete astronomical observatory with equipment for measuring all the bands of the electromagnetic spectrum. Use what you have learned in this chapter, and have your group discuss what kinds of maps you would make of the cloud to plot your most effective escape route.

B The diagram that Herschel made of the Milky Way has a very irregular outer boundary (see Figure 24.2). Can your group think of a reason for this?

C Suppose that for your final exam in this course, your group is assigned telescope time to observe a star selected for you by your professor. The professor tells you the position of the star on the sky (its right ascension and declination) but nothing else. You can make any observations you wish. How would you go about determining whether the star is a member of population I or population II?

D The existence of dark matter comes as a great surprise and its nature is a great mystery today. Some day astronomers will know a lot more about it, and it will be a routine part of textbooks like ours. Can your group make a list of earlier astronomical observations that began as a surprise and mystery, but wound up (with more observations) as well-understood parts of introductory textbooks?

E Physicist Gregory Benford has written a series of science-fiction novels that take place near the center of the Milky Way Galaxy in the far future. Suppose your group were writing such a story. How would the environment near the galactic center differ from the environment in the "galactic suburbs," where the Sun is located? Would life as we know it have an easier or harder time surviving on planets that orbit stars near the center (and why)?

F These days, in most urban areas, city lights completely swamp the faint light of the Milky Way in our skies. Have each member of your group go home and survey ten friends or relatives, explaining what the Milky Way is and then asking if they have ever seen it. Report back to your group and discuss your reactions to the survey. How important is it that many kids growing up on Earth today never (or rarely) get to see our home galaxy in the sky?

REVIEW QUESTIONS

Ace Astronomy™ Assess your understanding of this chapter's topics with additional quizzing and animations at **http://ace.brookscole.com/voyages**

1. Explain why we see the Milky Way as a faint band of light stretching across the sky.
2. Explain where (and why) in a spiral galaxy you would expect to find globular clusters, molecular clouds, and atomic hydrogen.
3. Describe several characteristics that distinguish population I from population II stars.
4. Briefly describe the main parts of our Galaxy.
5. Describe the evidence indicating that a black hole may be at the center of our Galaxy.
6. Explain why the abundances of heavy elements in stars correlate with their positions in the Galaxy.
7. What will be the long-term future of our Galaxy?

8. Suppose the Milky Way were a band of light extending only halfway around the sky (that is, in a semicircle). What, then, would you conclude about the Sun's location in the Galaxy? Give your reasoning.

9. The globular clusters revolve around the Galaxy in highly elliptical orbits. Where would you expect the clusters to spend most of their time? (Think of Kepler's laws.) At any given time, would you expect most globular clusters to be moving at high or low speeds with respect to the center of the Galaxy? Why?

10. Shapley used the positions of globular clusters to determine the location of the Galactic center. Could he have used open clusters? Why or why not?

11. Consider the following five kinds of objects: (1) open cluster, (2) giant molecular cloud, (3) globular cluster, (4) group of O and B stars, and (5) planetary nebulae.
a. Which occur only in spiral arms?
b. Which occur only in the parts of the Galaxy other than the spiral arms?
c. Which are thought to be very young?
d. Which are thought to be very old?
e. Which have the hottest stars?

12. The dwarf galaxy in Sagittarius is the one closest to the Milky Way, yet it was discovered only in 1994. Can you think of a reason it was not discovered earlier? (*Hint:* Think about what else is in its constellation.)

13. Suppose three stars lie in the disk of the Galaxy at distances of 20,000 LY, 25,000 LY, and 30,000 LY from the galactic center, and suppose that all three are lined up in such a way that it is possible to draw a straight line through them and on to the center of the Galaxy. How will the relative positions of these three stars change with time? Assume that their orbits are all circular and lie in the plane of the disk.

14. Why does star formation occur primarily in the disk of the Galaxy?

15. Where in the Galaxy would you expect to find Type II supernovae, which are the explosions of massive stars that go through their lives very quickly? Where would you expect to find Type I supernovae, which involve the explosions of white dwarfs?

16. Suppose that stars evolved without losing mass—that once matter was incorporated into a star, it remained there forever. How would the appearance of the Galaxy be different from what it is now? Would there be population I and population II stars? What other differences would there be?

17. Use the data in Appendix 12 to identify the galaxies that are satellites of the Milky Way. How far away are they on average? How do their sizes compare with that of our own Galaxy?

FIGURING FOR YOURSELF

Kepler's third law (see Section 2.3), which was so useful in determining masses in the solar system and in calculating masses of binary stars, also comes in handy when we study the Milky Way. It might seem as if we were making a big jump to take a law that was derived by watching the planets and apply it to the entire Galaxy. But every test we have been able to make indicates that the physical laws that work in our own local neighborhood are valid throughout the universe.

To use the formula in Section 2.3, we must put the distance in astronomical units and the period in years. The Sun orbits the center of the Galaxy in 225 million years at a distance of 26,000 LY. There are 6.3×10^4 AU in 1 LY, so 26,000 LY becomes 1.6×10^9 AU. If it takes about 2.25×10^8 years for the Sun to orbit the center of the Milky Way, Kepler's law gives:

$$M_{\text{Galaxy}} + M_{\text{Sun}} = \frac{a^3}{P^2} = \frac{(1.6 \times 10^9)^3}{(2.25 \times 10^8)^2}$$

$$= \text{approximately } 10^{11} M_{\text{Sun}}$$

As we said in the chapter, since the Sun's mass is negligible compared to that of the entire Galaxy, this is essentially the mass of the Milky Way inside the Sun's orbit. (Only the part of the Galaxy's mass within our orbit affects the motion, but since we are out in the suburbs, this is not a bad first estimate of the mass of the ordinary matter in the Galaxy. As we saw in the chapter, it's a whole different story if we want to count the dark matter.)

18. Assume that the Sun orbits the center of the Galaxy at a speed of 220 km/s and a distance of 26,000 LY from the center.

a. Calculate the circumference of the Sun's orbit, assuming it to be approximately circular. (Remember that the circumference of a circle is given by $2\pi R$, where R is the radius of the circle. Be sure to use consistent units. The conversion from light years to kilometers/second can be found in an appendix, or you can calculate it for yourself: The speed of light is 300,000 km/s, and you can determine the number of seconds in a year.)

b. Calculate the Sun's period, the "galactic year." Again, be careful with the units. Does it agree with the number we gave above?

19. Suppose the Sun orbited a little farther out, but the mass of the Galaxy inside its orbit remained the same as we calculated above. What would be its period at a distance of 30,000 LY?

20. We have said that the Galaxy rotates differentially; that is, stars in the inner parts complete a full 360° orbit around the center of the Galaxy more rapidly than stars farther out. Use Kepler's third law and the mass we derived above to calculate the period of a star that is only 5000 LY from the center. Now do the same calculation for a globular cluster at a distance of 50,000 LY. Suppose the Sun, this star, and the globular cluster all fall on a straight line through the center of the Galaxy. Where will they be relative to each other after the Sun completes one full journey around the center of the Galaxy? (Assume that all of the mass in the Galaxy is concentrated at its center.)

21. If our solar system is 4.6 billion years old, how many galactic years has planet Earth been around?

22. Suppose the average mass of a star in the Galaxy is one third of a solar mass. Use the value for the mass of the Galaxy that we calculated above, and estimate how many stars are in the Milky Way. Give some reasons it is reasonable to assume that the mass of an average star is less than the mass of the Sun.

23. The first clue that the Galaxy contains a lot of dark matter was the observation that the orbital velocities of stars did not decrease with increasing distance from the center of the Galaxy. Construct a rotation curve for the solar system by using the orbital velocities of the planets, which can be found in Appendix 7. How does this curve differ from the rotation curve for the Galaxy? What does it tell you about where most of the mass in the solar system is concentrated?

24. The best evidence for a black hole at the center of the Galaxy also comes from the application of Kepler's third law. Suppose a star at a distance of 5 light days from the center of the Galaxy has an orbital speed of 2500 km/s. How much mass must be located inside its orbit?

25. The next step in deciding whether the object in problem 24 is a black hole is to estimate the density of this mass. Assume that all of the mass is spread uniformly throughout a sphere with a radius of 5 light days. What is the density? [Remember that the volume of a sphere is given by $V = (4/3)\pi R^3$.] Explain why the density might be even higher than the value you have calculated. How does this density compare with that of the Sun or other objects we have talked about in this book?

26. Suppose the Sagittarius dwarf galaxy merges completely with the Milky Way and adds 150,000 stars to it. Estimate the percentage change in the mass of the Milky Way. Will this be enough mass to affect the orbit of the Sun around the galactic center? Assume that all of the Sagittarius Galaxy's stars end up in the nuclear bulge of the Galaxy and explain your answer.

SUGGESTIONS FOR FURTHER READING

Binney, J. "The Evolution of Our Galaxy" in *Sky & Telescope*, Mar. 1995, p. 20.

Croswell, K. *Alchemy of the Heavens*. 1995, Doubleday/Anchor. A popular-level review of current Milky Way research.

Duncan, A. and Haynes, R. "A New Look at the Milky Way" in *Sky & Telescope*, Sept. 1997, p. 46. Brief look at new radio maps of our Galaxy.

Ferris, T. *Coming of Age in the Milky Way*. 1988, Morrow. Historical survey.

Henbest, N. and Couper, H. *The Guide to the Galaxy*. 1994, Cambridge U. Press. Illustrated history of the growth of our understanding, plus a tour of the Milky Way.

Irion, R. "A Crushing End for Our Galaxy" in *Science*, Jan. 7, 2000, p. 62. On the role of mergers in the evolution of the Milky Way.

Laughlin, G. and Adams, F. "Celebrating the Galactic Millenium" in *Astronomy*, Nov. 2001, p. 39. The long-term future of the Milky Way in the next 90 billion years.

Mateo, M. "Searching for Dark Matter" in *Sky & Telescope*, Jan. 1994, p. 20.

Schulkin, B. "Does a Monster Lurk Nearby?" in *Astronomy*, Sept. 1997, p. 42. On what lies at the galactic center.

Szpir, M. "Passing the Bar Exam" in *Astronomy*, Mar. 1999, p. 46. On evidence that our Galaxy is a barred spiral.

Trimble, V. and Parker, S. "Meet the Milky Way" in *Sky & Telescope,* Jan. 1995, p. 26. Overview of our Galaxy.

van den Bergh, S. and Hesser, J. "How the Milky Way Formed" in *Scientific American,* Jan. 1993, p. 72.

Verschuur, G. "In the Beginning" in *Astronomy,* Oct. 1993, p. 40. On globular clusters and the Galaxy.

Verschuur, G. "Journey into the Galaxy" in *Astronomy,* Jan. 1993, p. 32. Excellent tour.

Whitt, K. "The Milky Way from the Inside" in *Astronomy,* Nov. 2001, p. 58. Fantastic panorama image of the Galaxy, with finder charts and explanations.

Zimmerman, R. "Heart of Darkness" in *Astronomy,* Oct. 2001, p. 42. On the center of the Galaxy and the black hole there.

The Hubble Deep Field North   This historic photograph is one of the deepest images ever taken of the universe of galaxies. Two hundred seventy-six exposures, taken with the Hubble Space Telescope in December 1995 over 150 consecutive orbits, were combined to make an image that showed fainter galaxies than had ever been glimpsed before, including some that were a record-breaking 4 billion times fainter than the human eye can see. About 2500 galaxies are visible when the image is examined closely. Some are so far away that we are seeing them as they were only a billion years or so after the universe began its expansion. The region photographed, a tiny patch of sky just above the Big Dipper (about the size that a grain of sand would look if held at arm's length), was selected because it is far from the disk of the Milky Way and contains very few nearby galaxies.

25 Galaxies

There is nothing like astronomy to pull the stuff out of man.
His stupid dreams and red-rooster importance: let him count the star-swirls.

From the poem *Star-Swirls* by Robinson Jeffers (1924)

THINKING AHEAD

Exploration of a new land often raises more questions than it answers. In the last chapter we explored our own Galaxy. But is it the only one? If there are others, are they like the Milky Way? How far away are they? As we shall see, some galaxies turn out to be so far away that their light has been journeying toward us for billions of years. These remote galaxies can tell us what the universe was like when it was young.

In this chapter we start our exploration of the vast realm of the galaxies. In fact, there are so many galaxies that a modern telescope allows us to see millions of them just in the bowl of the Big Dipper (see the image opposite). Like tourists from a small town making their first visit to the great cities of the world, we will be awed by the beauty and variety of the galaxies. And yet we will recognize much of what we see as being familiar from our experiences back home, and we will be impressed by how much we can learn by looking at structures built long ago.

We begin our voyage with a guide to the properties of galaxies, much as a tourist begins with a guidebook to the cities on the itinerary. In later chapters we will look more carefully at the history of galaxies, how they have changed over time, and how they acquired their many different forms.

The very idea that other galaxies exist was controversial for a long time. As late as 1920, many astronomers, including Harlow Shapley, thought the Milky Way encompassed all there was in the universe. The proof in 1924 that our Galaxy is not unique was one of the great scientific advances of the 20th century.

Virtual Laboratories

 The Astronomical Distance Scale

25.1 THE GREAT NEBULA DEBATE

Growing up at a time when the Hubble Space Telescope orbits above our heads and giant telescopes are springing up on the great mountaintops of the world, you may be surprised to learn that we were not sure about the existence of other galaxies for such a long time. But today's giant telescopes and electronic detectors are rather recent additions to the astronomer's toolbox.

With the telescopes available in earlier centuries, galaxies looked like small fuzzy patches of light that were difficult to distinguish from the star clusters and gas-and-dust clouds that are part of our own Galaxy. All objects that were not sharp points of light were given the same name, *nebulae,* the Latin word for "clouds." Because even their precise shapes were often hard to make out and no techniques had yet been devised for measuring their distances, the nature of the nebulae was the subject of much debate.

As early as the 18th century, the philosopher Immanuel Kant (1724–1804) suggested that some of the nebulae might be distant systems of stars (other Milky Ways), but proof of this suggestion was beyond the capabilities of the telescopes of that time. By the early 20th century, some nebulae had been correctly identified as star clusters and others (such as the Orion Nebula) as gaseous nebulae. Most of them, however, looked faint and indistinct even with the best telescopes, and their distances remained unknown.

(For more on how such nebulae are named, by the way, see the Astronomy Basics box in Chapter 19.) If these nebulae were nearby, with distances comparable to those of observable stars, they were most likely clouds of gas or groups of stars within our Galaxy. If, on the other hand, they were remote, far beyond the edge of the Galaxy, they could be other systems containing billions of stars.

To determine what the nebulae are, astronomers had to find a way of measuring the distances to at least some of them. When the 100-in. (2.5-m) telescope on Mount Wilson in southern California went into operation, astronomers finally had the large telescope they needed to settle the controversy.

Working with the 100-in. telescope, Edwin Hubble was able to resolve individual stars in several of the brighter spiral-shaped nebulae, including M31, the great spiral in Andromeda mentioned in the previous chapter (see the Voyagers in Astronomy box). Among these stars he discovered some faint variables that—when he analyzed their light curves—turned out to be cepheids. Here were reliable indicators that Hubble could use to measure the distances to the nebulae using the technique pioneered by Henrietta Leavitt (see Chapter 18). After painstaking work, he estimated that the Andromeda Galaxy was about 900,000 LY away from us. This meant that it was a separate galaxy of stars located well outside the boundaries of the Milky Way (Figure 25.1). Today we know it is actually slightly more than twice as distant as Hubble's first estimate, but his conclusion about its true nature remains unchanged.

■ **FIGURE 25.1**

The Andromeda Galaxy Also known by its catalog number M31, this large spiral galaxy is very similar in appearance to, and slightly larger than, our own Galaxy. Two companion galaxies are seen in the lower center (M32) and the upper right (NGC 205). At a distance of about 2.2 million LY, Andromeda is the spiral galaxy that is nearest our own in space. The area shown in this image is quite large on the sky, covering about five times the area of the full Moon. This true-color image was created by combining images taken in seven filters: U (violet), B (blue), V (green), R (orange), I (red), Hydrogen-Alpha (red), and Oxygen [OIII] (green). The small reddish regions show where ionized hydrogen is located; the ionization is produced by radiation from recently formed hot, massive stars.

T. A. Rector and B. A. Wolpa/NOAO/AURA/NSF

Edwin Hubble: Expanding the Universe

The son of a Missouri insurance agent, Edwin Hubble graduated from high school at age 16. He excelled in sports, winning letters in track and basketball at the University of Chicago, where he studied both science and languages. Both his father and grandfather wanted him to study law, however, and he gave in to family pressure. He received a prestigious Rhodes scholarship to Oxford University in England, where he studied law with only middling enthusiasm. Returning to the United States, he spent a year teaching high school physics and Spanish as well as coaching basketball, while trying to determine his life's direction.

The pull of astronomy eventually proved too strong to resist, and so Hubble went back to the University of Chicago for graduate work. Just as he was about to get his degree and accept an offer to work at the soon-to-be-completed 100-in. telescope, the United States entered World War I and Hubble enlisted as an officer. Although the war had ended by the time he arrived in Europe, he received more officer's training abroad and enjoyed a brief time of further astro-

Edwin Hubble
(1889–1953)

J. Stokley/A.S.P. Archives

nomical study at Cambridge before being sent home.

In 1919, at age 30, he joined the staff at Mount Wilson and began working with the world's largest telescope. Ripened by experience, energetic, disciplined, and a skillful observer, Hubble soon established some of the most important ideas in modern astronomy. He showed that other galaxies existed, classified them on the

basis of their shapes, found a pattern to their motion (and thus put the notion of an expanding universe on a firm observational footing), and began a lifelong program to study the distribution of galaxies in the universe. Although a few others had glimpsed pieces of the puzzle, it was Hubble who put it all together and showed that an understanding of the large-scale structure of the universe was feasible.

His work brought Hubble much renown and many medals, awards, and honorary degrees. As he became better known (he was the first astronomer to appear on the cover of *Time* magazine), he and his wife enjoyed and cultivated friendships with movie stars and writers in southern California. Hubble was instrumental (if you'll pardon the pun) in the planning and building of the 200-in. telescope on Palomar Mountain and he had begun to use it for studying galaxies when he was felled by a stroke in 1953. When astronomers built a space telescope that would allow them to extend Hubble's work to distances he could only dream about, it seemed natural to name it in his honor.

No one in human history had ever measured a distance so great. When Hubble's paper on the distances to nebulae was read before a meeting of the American Astronomical Society on the first day of 1925, the entire room erupted in a standing ovation. A new era had begun in the study of the universe, and a new scientific field—extragalactic astronomy—had just been born.

1920s, when a single photograph or spectrum of a galaxy could take a full night of tireless observation. In recent decades, larger telescopes and electronic detectors have made this task less difficult, although observing the most distant galaxies (those that show us the universe in its earliest phases) still requires enormous effort. (See this chapter's opening image, which required ten days of observation with the Hubble Space Telescope.)

The first step in trying to understand a new type of object is often simply to describe it. Remember, the first step in understanding stellar spectra was simply to sort them according to appearance (see Chapter 16). As it turns out, the biggest and most luminous galaxies come in one of two basic shapes: Either they have spiral arms, like our own Galaxy, or they appear to be elliptical (something like the Goodyear blimp). Many small galaxies, in contrast, have an irregular shape.

25.2 TYPES OF GALAXIES

Having established the existence of other galaxies, Hubble and others then began to observe them more closely—noting their shapes, their contents, and as many other properties as they could measure. This was a daunting task in the

25.2.1 Spiral Galaxies

Our own Galaxy and M31, the Andromeda Galaxy, which is believed to be much like it, are typical large **spiral galaxies** (see Figures 24.13 and 25.1). They consist of a nuclear bulge, a halo, a disk, and spiral arms. Interstellar material is usually spread throughout the disks of spiral galaxies. Bright emission nebulae and hot young stars are present, especially in the spiral arms (Figure 25.2), showing that new star formation is still going on. The disks are often dusty, which is especially noticeable in those systems that we view almost edge-on (Figure 25.3).

In galaxies that we see face-on, the bright stars and emission nebulae make the arms of spirals stand out like those of a Fourth of July pinwheel (see Figure 25.2). Open star clusters can be seen in the arms of nearer spirals, and globular clusters are often visible in their halos. Spiral galaxies contain a mixture of young and old stars, just as the Milky Way does. All spirals rotate, and the direction of their spin is such that the arms appear to trail, like the coattails of a swift runner.

About two thirds of the nearby spiral galaxies have boxy or peanut-shaped bars of stars running through their nuclei (Figure 25.4). Showing great originality, astronomers call these galaxies *barred spirals*. As we noted in the previous chapter, the Milky Way has a modest bar, too (Figure 25.5). The spiral arms usually begin from the ends of the bar. The fact that bars are so common suggests that they are long-lived; it may be that most spiral galaxies form a bar at some point during their evolution.

In both normal and barred spirals, we observe a range of different shapes. At one extreme the nuclear bulge is large and luminous, the arms are faint and tightly coiled, and bright emission nebulae and supergiant stars are inconspicuous. At the other extreme, the nuclear bulge is small—indeed almost absent—and the arms are loosely wound. In these latter galaxies, luminous stars and emission nebulae are very prominent. Our Galaxy and M31 are both intermediate between the two extremes. Photographs of spiral galaxies, illustrating the different types, are shown in Figures 25.6 and 25.7.

Ace Astronomy™ Log into AceAstronomy and select this chapter to see the Active Figure called "Galaxy Types."

■ **FIGURE 25.2**

Hot Stars in a Spiral Galaxy Star formation is very active in the Whirlpool Galaxy (M 51). This image shows its inner region with many bright clusters of recently formed stars. These clusters are associated with reddish clouds of ionized hydrogen, and both outline the galaxy's spiral arms. Note also the complex structures in the dust clouds in this galaxy. This galaxy is about 31 million LY distant, and the region shown is about 30,000 LY across.

C. Howk, JHU; B. Savage, University of Wisconsin; N. A. Sharp
NOAO/WIYN/NOAO/NSF

■ FIGURE 25.3

A Spiral Galaxy Viewed Edge-on
NGC 4013 is a spiral galaxy similar to
the Milky Way. It is at a distance of 55
million LY. We view this galaxy almost
exactly edge-on, and from this angle we
can clearly see the dust in the plane of
the galaxy; it appears dark because it
absorbs the light from the stars in the
galaxy. Most of the dust lies in a plane
that is about 500 LY thick.

The luminous parts of spiral galaxies range in diameter
from about 20,000 to more than 100,000 LY. There is also
considerable dark matter in them, just as there is in the
Milky Way; we deduce its presence from how fast stars in
the outer parts of the galaxy are moving in their orbits. From
the observational data available, spiral galaxy masses are estimated to range from a billion to a thousand billion Suns
(10^9 to 10^{12} M_{Sun}). The total luminosities of most spirals fall
in the range of a hundred million to a hundred billion times
the luminosity of our Sun (10^8 to 10^{11} L_{Sun}). Our Galaxy and
M31 are relatively large and massive, as spirals go.

25.2.2 Elliptical Galaxies

Elliptical galaxies consist almost entirely of old stars
and have shapes that are spheres or ellipsoids (somewhat squashed spheres). They contain no trace of spiral
arms. Their light is dominated by older reddish stars (the

Photo by David Malin; © Anglo-Australian Observatory

■ FIGURE 25.4

A Barred Spiral NGC 3351 is a barred spiral galaxy
about 25 million LY away. Note that the spiral arms begin at
the ends of the bar.

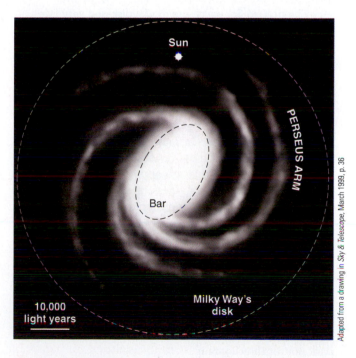

Adapted from a drawing in Sky & Telescope, March 1999, p. 36

■ FIGURE 25.5

The Milky Way Bar Space-based infrared observations
show that the Milky Way Galaxy has a modest bar in its
central regions. Here we see the main spiral arms and the
slight elongation of the central bulge of our Galaxy's disk.

NGC 1201   Type S0   NGC 2841   Type Sb

NGC 2811   Type Sa   NGC 3031 M81   Type Sb

NGC 488   Type Sab   NGC 628 M74   Type Sc

Palomar Observatory, Caltech

NGC 2859   Type SB0   NGC 2523   Type SBb(r)

NGC 175   Type SBab(s)   NGC 1073   Type SBc(sr)

NGC 1300   Type SBb(s)   NGC 2525   Type SBc(s)

Palomar Observatory, Caltech

■ **FIGURE 25.6**

An Album of Spirals   The different types of spiral galaxies range from S0, which are mostly bulge and very little disk, to Sc, which are mostly disk and very little bulge.

■ **FIGURE 25.7**

An Album of Barred Spirals   Here we see the types of barred spirals, paralleling the types of regular spirals. The majority of nearby spiral galaxies turn out to have a barlike structure in them.

population II stars discussed in the previous chapter). In the larger nearby ellipticals, many globular clusters can be identified (Figure 25.8). Dust and emission nebulae are not conspicuous in elliptical galaxies, but many do contain a small amount of interstellar matter.

Elliptical galaxies show various degrees of flattening, ranging from systems that are approximately spherical to those that approach the flatness of spirals (Figure 25.9). The rare *giant ellipticals* (for example, M87 in Figure 25.8) reach luminosities of $10^{11} L_{\text{Sun}}$. The mass in a giant elliptical can be as great as $10^{13} M_{\text{Sun}}$. The diameters of these large galaxies extend over at least several hundred thousand light years and are considerably larger than the largest spirals. Although individual stars orbit around the center of an elliptical galaxy, the orbits are not all in the same direction as they are in spirals. Therefore, ellipticals don't rotate in a systematic way, and it is hard to estimate how much dark matter they contain.

We find that elliptical galaxies range all the way from the giants, just described, to dwarfs, which may be the most common kind of galaxy. *Dwarf ellipticals* escaped our notice for a long time because they are very faint and difficult to see. An example of a dwarf elliptical is the And VI system shown in Figure 25.10. There are so few bright stars in

Photo by David Malin; © Anglo-Australian Telescope Board

■ **FIGURE 25.8**

Globular Clusters Around an Elliptical   This image, prepared with special photographic masking techniques, shows a swarm of globular clusters surrounding the giant elliptical galaxy M87. A jet of material is shooting out of the galaxy's core, which is not visible here. M87 is an active galaxy, with strong radio and x-ray emissions; such galaxies are discussed in the next chapter.

Palomar Observatory, Caltech

■ **FIGURE 25.9**

Types of Elliptical Galaxies Elliptical galaxies are classified by whether they are spherical (E0) or more elongated.

this galaxy that even its central regions are resolved (i.e., we can see individual stars). However, the total number of stars (most of which are too faint to show in our photo) is probably at least several million. The luminosity of this typical dwarf is about equal to that of the brightest globular clusters.

Intermediate between the giant and dwarf elliptical galaxies are systems such as M32 and NGC 205, two companions of M31. They can be seen in the photographs of M31 in Figures 24.13 and 25.1.

25.2.3 Irregular Galaxies

Hubble classified galaxies that do not have the regular shapes associated with the categories we just described in the catchall bin of **irregular galaxies,** and we continue to

use his term. Typically, irregular galaxies have lower masses and luminosities than spiral galaxies. Irregular galaxies often appear chaotic, and many are undergoing relatively intense star formation activity. They contain both young population I stars and old population II stars.

The two best-known irregular galaxies are the Large and Small Magellanic Clouds (Figures 25.11 and 25.12), which are a little more than 160,000 LY away and are among our nearest extragalactic neighbors. Their name reflects the fact that Ferdinand Magellan and his crew, making their round-the-world journey, were the first European travelers to notice them. Although not visible from the United States and Europe, these two systems are prominent from the Southern Hemisphere, where they look like wispy clouds in the night sky. They are only about one tenth as distant as the Andromeda spiral. The Large Cloud contains the 30 Doradus complex (also known as the Tarantula Nebula), one of the largest and most luminous groups of supergiant stars and associated gas known in any galaxy.

The Small Magellanic Cloud is considerably less massive than the Large Cloud and is six times longer than it is wide. This narrow wisp of material points directly toward our Galaxy like an arrow. The Small Cloud was probably the victim of a near collision with the Large Cloud some 200 million years ago. It is now being pulled apart by the gravity of the Milky Way.

25.2.4 Do Galaxy Types Evolve?

Encouraged by the success of the H–R diagram for stars (see Chapter 17), astronomers studying galaxies hoped to find some sort of comparable scheme, where differences in appearance could be tied to different evolutionary stages. Wouldn't it be nice if every elliptical galaxy evolved into a spiral, for example, just as every main-sequence star evolves into a red giant? Several simple ideas of this kind were tried, some by Hubble himself, but none stood the test of time (and observation).

■ **FIGURE 25.10**

The Dwarf Galaxy And VI This galaxy is about 4 million LY away, which is very near for a galaxy, but it was discovered only in 1999. Dwarf galaxies are probably the most common type, but they are very difficult to detect because they are so faint. Most of the stars are slightly red because they have already evolved to become red giants. The few blue stars in this image are probably foreground stars in our own Milky Way. The right panel is an enlarged high-resolution view of the center of And VI showing the tight packing of stars.

George Jacoby/WIYN/NOAO/NSF

National Optical Astronomy Observatories

■ **FIGURE 25.11**

The Large Magellanic Cloud
This satellite of our own Galaxy is visible to the naked eye from the Southern Hemisphere. The large red nebula (called the Tarantula) is the site of active star formation and contains many young supergiant stars.

Because no simple scheme for evolving one type of galaxy into another could be found, astronomers then tended to the opposite point of view. For a while, most astronomers thought that all galaxies formed very early in the history of the universe and that the differences between them had to do with the rate of star formation. Ellipticals were those galaxies in which all the interstellar matter was converted rapidly to stars. Spirals were galaxies in which star formation occurred slowly over the entire lifetime of the galaxy. This idea turned out to be too simple as well.

Today we understand that at least some galaxies have changed types over the billions of years since the universe began. As we shall see in later chapters, collisions and mergers of smaller galaxies, including spirals, in the centers of dense galaxy clusters can build up the massive ellipticals such as M87. Even isolated spirals can change their appearance over time. As they consume their gas, the rate of star formation will slow down and the spiral arms will gradually become less conspicuous. Over long periods, spirals therefore begin to look more like the galaxies at the top left of Figures 25.6 and 25.7 (which astronomers refer to as S0 types). Over the past two decades, the study of how galaxies evolve over the lifetime of the universe has become one of the most active fields of astrophysical research. We will discuss the evolution of galaxies in more detail in Chapter 27, but let's first see in a little more detail just what different galaxies are like.

25.3 PROPERTIES OF GALAXIES

National Optical Astronomy Observatories

■ **FIGURE 25.12**

The Small Magellanic Cloud This dwarf irregular galaxy is another satellite of the Milky Way.

The technique for deriving the masses of galaxies is basically the same as that used to estimate the mass of the Sun, the stars, and our own Galaxy: We measure how fast objects in the outer regions of the galaxy are orbiting the center, and then we use this information along with Kepler's third law to calculate how much mass is inside that orbit.

25.3.1 Masses of Galaxies

Astronomers can measure the rotation speed in spiral galaxies by obtaining spectra of either stars or gas and looking for wavelength shifts produced by the Doppler effect.

Such observations of M31 (our spiral neighbor in Andromeda), for example, show it to have a mass (within the main visible part of the galaxy, out to a distance of 100,000 LY from the center) of about $4 \times 10^{11} M_{Sun}$, which is about the same as the mass of our own Galaxy. The *total* mass of M31 is higher than $4 \times 10^{11} M_{Sun}$, however, because we have not included the material that lies more than 100,000 LY from the center. Like our own Galaxy, Andromeda appears to have a large amount of dark matter beyond its luminous boundary.

Elliptical galaxies do not rotate; for them we must use a slightly different technique to measure mass. Their stars are still moving in orbit around the galactic center, but not in the organized way that characterizes spirals. Since elliptical galaxies contain stars that are billions of years old, we can assume that the galaxies themselves are not flying apart. Therefore, if we can measure the various speeds with which the stars are moving in their orbits around the center of the galaxy, we can calculate how much mass the galaxy must contain in order to hold the stars within it.

In practice, the spectrum of a galaxy is a composite of the spectra of its many stars, whose different motions produce different Doppler shifts (some red, some blue). The result is that the lines we observe from the entire galaxy contain the combination of many Doppler shifts; they therefore look much wider in a spectrum than would the same lines in a hypothetical galaxy in which the stars had no orbital motion. Astronomers call this phenomenon *line broadening*. The amount by which each line broadens indicates the range of speeds at which the stars are moving with respect to the center of the galaxy. The range of speeds depends, in turn, on the force of gravity that holds the stars within the galaxies. With information about the speeds, it is possible to calculate the mass of the elliptical galaxy.

Table 25.1 summarizes the range of masses (and other properties) of the various types of galaxies. The most massive galaxies are the giant ellipticals, but the lowest mass galaxies are ellipticals as well. On average, irregular galaxies have less mass than spirals.

25.3.2 Mass-to-Light Ratio

A useful way of characterizing a galaxy is by noting the ratio of its mass (in units of the Sun's mass) to its light output (in units of the Sun's luminosity). This single number tells us roughly what kind of stars make up most of the luminous population of the galaxy, and it also tells us whether a lot of dark matter is present. For stars like the Sun, the **mass-to-light ratio** is 1 by our definition.

Galaxies are not, of course, composed entirely of stars that are identical to the Sun. The overwhelming majority of stars are less massive and less luminous than the Sun, and usually these stars contribute most of the mass of a system without accounting for very much light. The mass-to-light ratio for the low-mass stars is greater than 1 (you can verify this using the data in Table 17.2). Therefore, a galaxy's mass-to-light ratio is also generally greater than 1, with the exact value depending on the ratio of high-mass stars to low-mass stars.

Galaxies in which star formation is still occurring have many massive stars, and their mass-to-light ratios are usually in the range of 1 to 10. Galaxies that consist mostly of an older stellar population, such as ellipticals, in which the massive stars have already completed their evolution and have ceased to shine, have mass-to-light ratios of 10 to 20.

But these figures refer only to the inner, conspicuous parts of galaxies (Figure 25.13). In Chapter 24 we discussed the evidence for dark matter in the outer regions of our own Galaxy, extending much farther from the galactic center than do the bright stars and gas. Recent measurements of the rotation speeds of the outer parts of nearby galaxies, such as the Andromeda spiral, suggest that they, too, have extended distributions of dark matter around the

TABLE 25.1 *Characteristics of the Different Types of Galaxies*

Characteristic	Spirals	Ellipticals	Irregulars
Mass (M_{Sun})	10^9 to 10^{12}	10^5 to 10^{13}	10^8 to 10^{11}
Diameter (thousands of LY)	15 to 150	3 to 600	3 to 30
Luminosity (L_{Sun})	10^8 to 10^{11}	10^6 to 10^{11}	10^7 to 2×10^9
Populations of stars	Old and young	Old	Old and young
Interstellar matter	Gas and dust	Almost no dust; little gas	Much gas; some have little dust, some much dust
Mass-to-light ratio in the visible part	2 to 10	10 to 20	1 to 10
Mass-to-light ratio for total galaxy	100	100	?

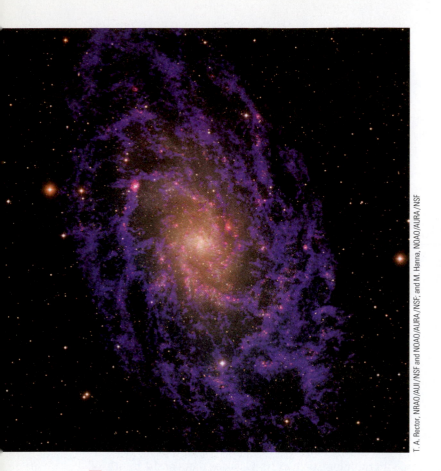

■ FIGURE 25.13

M33, a Nearby Spiral Galaxy This unusual image was created by combining visible-light and radio information. M33 is more than 30,000 LY across, and more than 2 million LY distant. The visible-light data in this image show the many stars within the galaxy, as well as reddish star-forming regions that are filled with hot hydrogen gas. The radio data (shown in blue-violet) reveal cool hydrogen gas that cannot be seen with a visible-light telescope. What neither the optical nor the radio measurements can show is any dark matter that this galaxy contains.

visible disk of stars and dust. This largely invisible matter adds to the mass of the galaxy while contributing nothing to its luminosity, thus increasing the mass-to-light ratio. If invisible dark matter is present in a galaxy, its mass-to-light ratio can be as high as 100. The two different mass-to-light ratios measured for various types of galaxies are given in Table 25.1.

These measurements of other galaxies support the conclusion already reached from studies of the rotation of our own Galaxy—namely, that most of the material in the universe cannot at present be observed directly in any part of the electromagnetic spectrum. An understanding of the properties and distribution of this invisible matter is crucial to our understanding of galaxies. Through the gravitational force it exerts, dark matter plays a dominant role in galaxy formation and early evolution. As we will see in Chapter 28, it may also help determine the ultimate fate of the universe.

There is an interesting parallel here between our time and the time during which Edwin Hubble was receiving his training in astronomy. By 1920 many scientists were aware that astronomy stood on the brink of important breakthroughs if only the nature and behavior of the nebulae could be settled with better observations. In the same way, many astronomers today feel we may be closing in on a far more sophisticated understanding of the large-scale structure of the universe if only we can learn more about the nature and properties of dark matter. If you follow astronomy in the newspapers or news magazines (as we hope you will), you should be hearing more about dark matter in the years to come.

25.4 THE EXTRAGALACTIC DISTANCE SCALE

To determine many of the properties of a galaxy, such as its luminosity or its size, we must first know how far away it is. If we know the distance to a galaxy, we can convert how bright the galaxy appears to us in the sky into its true luminosity because we know the precise way light is dimmed by distance. (The same galaxy ten times farther away, for example, would look a hundred times dimmer.) But the measurement of galaxy distances is one of the most difficult problems in modern astronomy: All the galaxies are far away, and most are so distant we cannot even make out individual stars in them.

For decades after Hubble's initial work, the techniques used to measure galaxy distances were relatively inaccurate, and different astronomers derived distances that differed by as much as a factor of two. Imagine if the distance between your home or dorm and your astronomy class were this uncertain; it would be difficult to make sure you got to class on time. In the past few years, however, astronomers have devised new techniques for measuring distances to galaxies; all of them give the same answer to within an accuracy of about 10 percent. As we will see, this means we may finally be able to make reliable estimates of the scale of the universe.

25.4.1 Variable Stars

Before astronomers could measure distances to other galaxies, they first had to establish the scale of cosmic distances using objects in our own Galaxy. We described the chain of these distance methods in Chapter 18 (and we recommend that you review that chapter if it has been a while since you've read it). Astronomers were especially delighted when they discovered that they could measure distances using certain kinds of intrinsically luminous *variable stars*, such as cepheids, which can be seen at very large distances.

After the variables in nearby galaxies had been used to make distance measurements for a few decades, Walter Baade (whose work on stellar populations was discussed in

T. A. Rector, NRAO/AUI/NSF and NOAO/AURA/NSF; and M. Hanna, NOAO/AURA/NSF the previous chapter) showed that there were actually two kinds of cepheids and that astronomers had been unwittingly mixing them up. As a result, in the early 1950s all the distances to the galaxies had to be increased by about a factor of two. We mention this because we want you to bear in mind, as you read on, that science is always a progress report. Our first tentative steps in such difficult investigations are always subject to future revision as our techniques become more sophisticated.

The amount of work involved in finding cepheids and measuring their periods can be enormous. Hubble, for example, obtained 350 long-exposure photographs of the nearby spiral M31 over a period of 18 years and identified only 40 cepheids. Even though cepheids are fairly luminous stars, they can be detected in only about 30 of the nearest galaxies with the world's largest ground-based telescopes. As mentioned in Chapter 18, one of the main projects carried out during the first years of operation of the Hubble Space Telescope was the measurement of cepheids in more-distant galaxies to improve the accuracy of the extragalactic distance scale (see Figure 18.9). Recently, astronomers working with the Hubble have extended such measurements out to 108 million LY—a triumph of technology and determination.

Nevertheless, we can use cepheids only to measure distances within a small pocket of the universe of galaxies. After all, to use this method we must be able to resolve single stars and follow their subtle variations. Beyond a certain distance, even our finest space telescopes cannot help us do this. Fortunately, there are other ways to measure the distances to galaxies.

25.4.2 Standard Bulbs

We discussed in Chapter 18 the great frustration that astronomers felt when they realized that the stars in general were not *standard bulbs*. If every light bulb in a huge auditorium is a standard 100-watt bulb, then bulbs that look brighter to us must be closer, while those that look dimmer must be farther away. If every star were a standard luminosity (or wattage), then we could similarly "read off" their distances based on how bright they appear to us. Alas, as we have seen, neither stars nor galaxies come in one standard-issue luminosity. Nonetheless, astronomers have been searching for objects out there that do act in some way like a standard bulb—that have the same intrinsic (built-in) brightness wherever they are.

A number of suggestions have been made for what sorts of objects might be standard bulbs, including the brightest supergiant stars, planetary nebulae (which give off a lot of ultraviolet radiation), and the average globular cluster in a galaxy. One object turns out to be particularly useful: Type I supernovae—those that involve the explosion of a white dwarf in a binary system (see Section 22.5). Observations show that supernovae of this type all reach nearly the same luminosity (about $10^{10} L_{Sun}$) at maximum light. With such tremendous luminosities, these super-

A. Riess et al./Harvard–Smithsonian Center for Astrophysics

■ **FIGURE 25.14**

Supernovae as Distance Indicators Four images of Type I supernovae (shown with the arrows) in different galaxies, taken with the 1.2-m telescope at the Whipple Observatory. If Type I supernovae are all the same maximum luminosity, then those in more-distant galaxies will appear dimmer and can be used to estimate distances.

novae have been detected out to a distance of more than 8 billion LY (Figure 25.14).

At very large distances we can use the total light emitted by an entire galaxy as a standard bulb. This technique, however, will not work for a single isolated galaxy because galaxies span an enormous range in intrinsic luminosity (see Table 25.1) and we cannot tell from appearance alone what the luminosity of any one galaxy really is. (For example, ellipticals with different luminosities often look similar when seen in isolation.) We can only tell which galaxies are highly luminous, and which are not, if we see a collection of them with various brightnesses side by side. Luckily, galaxies (as we will see in Chapter 27) are social creatures and come in groups and clusters. Thus we can use the apparent brightness of the brightest elliptical members (say, the average of the brightest five) in a large cluster to estimate the distance to the cluster.

There is much debate among astronomers about how good these various standard bulbs are. It's as if, to continue our analogy, standard 100-watt bulbs came from a slipshod manufacturer who made some of them 90 watts and others 110 watts. In this case you could only use the brightness of any given bulb to estimate its brightness plus or minus 10 percent. The universe is not in the business of making its objects into exact standard bulbs either, so there is still some variation in the luminosities of the best of these distance indicators. We must use such methods with a knowledge of their fallibility and compare them to other independent methods carefully every chance we get.

25.4.3 New Techniques

Some powerful new techniques for measuring distances have recently been developed. All start with observations of nearby galaxies whose distances have been measured through the use of cepheids and standard bulbs. The characteristics of these nearby galaxies are carefully calibrated and used to measure distances to galaxies so far away that individual stars and clusters can no longer be detected. Let's examine two of these techniques briefly.

The first method makes use of an interesting relationship noticed in the late 1970s by Brent Tully of the University of Hawaii and Richard Fisher of the National Radio Astronomy Observatory. They discovered that the *luminosity* of a spiral galaxy is related to its *rotational velocity* (how fast it spins). Why is this true? The more mass a galaxy has, the faster the objects in its outer regions must orbit. A more-massive galaxy has more stars in it and is thus more luminous (ignoring dark matter for a moment). Using our terminology from the previous section, we can say that if the mass-to-light ratios for various spiral galaxies are pretty similar, then we can estimate the luminosity of a spiral galaxy by measuring its mass and we can estimate its mass by measuring its rotational velocity.

Tully and Fisher used the 21-cm radiation from cold hydrogen to determine how rapidly material in spiral galaxies is orbiting around their centers. Since 21-cm radiation from stationary atoms comes in a nice narrow line, the width of the 21-cm line produced by a whole galaxy tells us the range of orbital velocities of the galaxy's hydrogen gas. The broader the line, the faster the gas is orbiting in the galaxy, and the more massive and luminous the galaxy turns out to be.

It is somewhat surprising that this technique works, since much of the mass associated with galaxies is dark matter, which does not contribute at all to the luminosity but does affect the rotation speed. There is also no obvious reason that the mass-to-light ratios should be similar for all spiral galaxies. Nevertheless, observations of nearer galaxies show that measuring the rotational velocity of a galaxy provides an accurate estimate of its intrinsic luminosity. Once we know how luminous the galaxy really is, we can compare the luminosity to the apparent brightness and use the difference to calculate its distance.

Another new technique involves measuring fluctuations in the apparent surface brightness of elliptical galaxies. This technique has been explored by John Tonry at the University of Hawaii. Elliptical galaxies contain mostly very old stars and very little gas or dust. A perfectly sharp picture of an elliptical galaxy would look much like that of a globular cluster, with many individual stars appearing as discrete points of light. Even with the blurring caused by the Earth's atmosphere, the image of an elliptical galaxy does not appear perfectly smooth. Rather, it is mottled or bumpy due to the naturally lumpy distribution of light emitted by the individual stars belonging to that galaxy (Figure 25.15).

The amount of bumpiness depends on the distance to the galaxy. In a nearby galaxy we can see individual stars or

Nearby galaxy

Same galaxy at three times the distance

Galaxy star field

What the CCD sees

Adapted from information by J. Tonry in *Sky & Telescope* magazine.

■ **FIGURE 25.15**

A Method for Measuring the Distances to Galaxies These drawings show the appearance of a galaxy on a modern electronic charge-coupled device (CCD). Such detectors are made up of many small, sensitive areas called *pixels* (shown as little boxes on our figure), each of which measures all the light from a small area of the sky. If the galaxy being observed is nearby (*left images*), some pixels will measure the light from one or a few stars, others will measure blank sky, and the brightness measured by the individual pixels will differ by large amounts. If the galaxy is farther away (*right images*), every pixel will detect light from some stars, which now appear crowded close together on the sky. This means there will be less difference in the light that neighboring pixels measure. In this way, the smoothness of the distribution of light in the image can be used to estimate the distance to the galaxy.

TABLE 25.2 *Some Methods for Estimating Distances to Galaxies*

Method (SB = standard bulb)	How Reliable	Galaxy Type	Approximate Distance Range (millions of LY)
Cepheid variables	Very	Spirals, irregulars	0–110
Brightest stars (SB)	Moderate	Spirals, irregulars	0–150
Planetary nebulae (SB)	Very	All	0–70
Globular clusters (SB)	Moderate	All	0–100
Surface brightness fluctuations	Very	Ellipticals	0–100
Tully–Fisher method (21-cm line widths)	Very	Spirals, irregulars	0–300
Type I supernovae (SB)	Very	All	0–11,000
Brightest galaxy in cluster (SB)	Very	Ellipticals in clusters	70–13,000
Redshifts (Hubble's law)	Very	All	300–13,000

clusters of stars, and the image has many bumps of varying brightness. In a very distant galaxy the stars cannot be resolved at all, and the image is smooth.

The distance to an elliptical galaxy can therefore be estimated by measuring the degree of bumpiness in the light distribution. This technique will not work for spiral galaxies, whose disks contain large amounts of dust that can also cause fluctuations in brightness distribution. But the method seems to give good distances for elliptical galaxies, which are the very ones for which the Tully–Fisher technique does not work.

Table 25.2 lists the type of galaxy for which each of the distance techniques is useful, the range of distances over which the technique can be applied, and the reliability of the distance estimates derived with each technique.

25.5 THE EXPANDING UNIVERSE

We now come to one of the most important discoveries ever made in astronomy—the fact that the universe is *expanding*. Before we describe how the discovery was made, we should point out that the first steps in the study of galaxies came at a time when the techniques of spectroscopy were also making great strides. Astronomers using large telescopes could record the spectrum of a faint star or galaxy on photographic plates, guiding their telescopes so they remained pointed to the same object for many hours. The resulting spectra of galaxies contained a wealth of information about the composition, mass, and motion of these great star systems.

25.5.1 Slipher's Pioneering Observations

Curiously, the discovery of the expansion of the universe began with the search for Martians and other solar systems. In 1894 the controversial (and wealthy) astronomer Percival

Lowell established an observatory in Flagstaff, Arizona, to study the planets and to search for life in the universe. Lowell thought that the spiral nebulae might be solar systems in the process of formation. He therefore asked one of the observatory's young astronomers, Vesto M. Slipher (Figure 25.16), to photograph the spectra of some of the spiral nebulae to see if their spectral lines might show chemical compositions like those expected for newly forming planets.

The Lowell Observatory's major instrument was a 24-in. refracting telescope, which was not at all well suited to observations of faint spiral nebulae. With the technology available in those days, photographic plates had to be exposed for 20 to 40 hours to produce a good spectrum (in

Lowell Observatory

■ **FIGURE 25.16**

Vesto M. Slipher, 1875–1969 V. M. Slipher spent his entire career at the Lowell Observatory, where he discovered the large radial velocities of galaxies.

which the positions of the lines could reveal a galaxy's motion). This often meant continuing to expose the same photograph over several nights. Beginning in 1912 and making heroic efforts over a period of about 20 years, Slipher managed to photograph the spectra of more than 40 nebulae.

To his surprise, the spectral lines of most galaxies showed astounding **redshifts.** By "redshift" we mean that the lines in the spectra are displaced toward longer wavelengths (toward the red end of the visible spectrum). Recall from Chapter 4 that a redshift is seen when the source of the waves is moving away from us. Slipher's observations showed that most spirals are racing away at huge speeds; the highest velocity he measured was 1800 km/s. Only a few spirals, such as M31, now known to be our close neighbors, turned out to be approaching us. All the other galaxies were moving away. Slipher first announced this discovery in 1914, years before Hubble showed that these objects were other galaxies and before anyone knew how far away they were.

Ace ◉ Astronomy™ Log into AceAstronomy and select this chapter to see the Active Figure called "Hubble."

25.5.2 The Hubble Law

The profound implications of Slipher's work became apparent only during the 1920s, when Hubble found ways of estimating the distances of the spiral nebulae. Hubble carried out the key observations in collaboration with a remarkable man, Milton Humason (Figure 25.17), who dropped out of school in the 8th grade and began his astronomical career by driving a mule train up the trail on Mount Wilson to the observatory. In those early days, supplies had to be brought up that way; even astronomers hiked up to the mountaintop for their turns at the telescope. Humason became interested in the work of the astronomers and, after marrying the daughter of the observatory's electrician, took a job as janitor there. After a time he became a night assistant, helping the astronomers run

■ **FIGURE 25.17**

Milton Humason, 1891–1972 Milton Humason worked with Hubble to establish the expansion of the universe.

the telescope and take data. Eventually he made such a mark that he became a full astronomer at the observatory.

By the late 1920s Humason was collaborating with Hubble by photographing the spectra of faint galaxies with the 100-in. telescope. (By then there was no question that the spiral nebulae were in fact galaxies.) When Hubble laid his own distance estimates next to measurements of the recession velocities (the speed with which the galaxies were moving away), he found something stunning: There was a relationship between distance and velocity for galaxies. *The more distant the galaxy, the faster it was receding from us.*

In 1931 Hubble and Humason jointly published a classic paper in the *Astrophysical Journal,* the foremost technical publication read by U.S. astronomers. They compared the distances and velocities of remote galaxies moving away from us at speeds as high as 20,000 km/s and were able to show that the recession velocities of galaxies are directly proportional to their distances from us (Figure 25.18).

Ace ◉ Astronomy™ Log into AceAstronomy and select this chapter to see Astronomy Exercise "Hubble Relation."

(a)

(b)

■ **FIGURE 25.18**

The Hubble Law (a) Hubble's original velocity–distance relationship, adapted from his 1929 paper in the *Proceedings of the National Academy of Sciences.* (b) Hubble and Humason's velocity–distance relationship, adapted from their 1931 paper in the *Astrophysical Journal.* The red dots at the lower left are the points in the diagram in the 1929 paper. Comparison of the two graphs shows how rapidly the determination of galactic distances and redshifts progressed in the two years between these publications.

We now know that this relationship holds for every galaxy except a few of the nearest ones! The few galaxies approaching us turn out to be part of the Milky Way's own group of galaxies, which have their own individual motions, just as birds flying in a swarm may fly in slightly different directions at slightly different speeds even though the entire flock travels through space together.

Written as a formula, the relationship between velocity and distance is

$$v = H \times d$$

where v is the recession speed, d is the distance, and H is a number called the **Hubble constant.** This equation is now known as the **Hubble law.**

ASTRONOMY BASICS
Constants of Proportionality

Mathematical relationships such as the Hubble law are pretty common in life. To take a simple example, suppose your college or university hires you to call rich alumni and ask for donations. You are paid $1.50 for each call; the more calls you squeeze in between studying astronomy and other courses, the more money you take home. We can set up a formula that connects p, your pay, and n, the number of calls:

$$p = A \times n$$

where A is the Alumni constant, with a value of 1.50 dollars. If you make 20 calls, you will earn 1.50 dollars times 20, or $30.

Suppose your boss forgets to tell you what you will get paid for each call. You can calculate the Alumni constant that governs your pay by keeping track of how many calls you make and noting your gross pay each week. If you make 100 calls the first week and are paid $150, you can deduce that the constant is 1.5 (in units of dollars per call). Hubble, of course, had no higher agent to tell him what his constant would be; he had to calculate its value from the measurements of distance and velocity.

■ ■ ■ ■ ■ ■ ■ ■ ■ ■ ■ ■ ■ ■

Astronomers express the value of Hubble's constant in units that relate to how they measure speed and velocity for galaxies. In this book, we will use km/s per million LY as that unit. For many years, estimates of the value of the Hubble constant have been in the range of 15 to 30 km/s per million LY. The most recent work appears to be converging on a value near 20 km/s per million LY. If H is 20 km/s per million LY, then a galaxy moves away from us at a speed of 20 km/s for every million light years of its distance. As an example, a galaxy 100 million LY away is moving away from us at a speed of 2000 km/s.

The Hubble law tells us something fundamental about the universe. Since all but the nearest galaxies appear to be in motion *away* from us, with the most distant ones moving the fastest, we must be living in an *expanding universe.* We will explore the implications of this idea shortly, as well as in the final chapters of this text. For now, we will just say that Hubble's observation underlies all our theories about the origin and evolution of the universe.

25.5.3 Hubble's Law and Distances

The regularity expressed in the Hubble law has a built-in bonus: It gives us a new way to determine the distances to remote galaxies. First we must reliably establish Hubble's constant by measuring *both* the distance and the velocity of many galaxies in many directions, to be sure the Hubble law is truly a universal property of galaxies. But once we have this constant and are satisfied that it applies everywhere, much more of the universe opens up for distance determination. Basically, if we can obtain a spectrum of a galaxy, we can immediately tell how far away it is.

The procedure works like this. We use the spectrum to measure the speed with which the galaxy is moving away from us. If we then put this speed and the Hubble constant into the Hubble law equation, we can solve for the distance. See the Figuring for Yourself box at the end of the chapter for examples.

This is a very important technique because, as we have seen, most of our methods for determining galaxy distances don't take us much beyond 600 million LY (and they have many uncertainties). But to use the Hubble law as a distance indicator, all we need to do is get a spectrum of a galaxy and measure the Doppler shift, something astronomers are now very good at.

With large telescopes and modern spectrographs, spectra can be taken of extremely faint galaxies. As we will see in the chapters to come, astronomers can now measure galaxies with redshifts that imply a velocity away from us of more than 90 percent the speed of light. At such high velocities, lines in the ultraviolet region of the spectrum—which normally cannot be observed from the ground—are shifted to yellow and red wavelengths and can be photographed as ordinary light.

25.5.4 Models for an Expanding Universe

At first, thinking about Hubble's law and being a fan of Copernicus and Shapley, you might be shocked. Are all the galaxies really moving *away from us?* Is there, after all, something special about our position in the universe? Worry not; the fact that galaxies obey the Hubble law only shows that the universe is expanding uniformly. A uniformly expanding universe is one that is expanding at the same rate everywhere. In such a universe, *we and all other observers,* no matter where they are located, must observe a proportionality between the velocities and distances of remote galaxies.

■ FIGURE 25.19

Stretching a Ruler As explained in the text, ants on a stretching ruler see other ants moving away from them. The speed with which another ant moves away is proportional to its distance.

To see why, first imagine a ruler made of flexible rubber, with the usual lines marked off at each centimeter. Now suppose someone with strong arms grabs each end of the ruler and slowly stretches it so that, say, it doubles in length in 1 min (Figure 25.19). Consider an intelligent ant sitting on the mark at 2 cm—a point that is not at either end or in the middle of the ruler. He measures how fast other ants, sitting at the 4-, 7-, and 12-cm marks, move away from him as the ruler stretches.

The ant at 4 cm, originally 2 cm away from our ant, has doubled its distance in 1 min; it therefore moved away at a speed of 2 cm/min. The one at the 7-cm mark, which was originally 5 cm away from our ant, is now 10 cm away; it thus had to move at 5 cm/min. The one that started at the 12-cm mark, which was 10 cm away from the ant doing the counting, is now 20 cm away, meaning it must have raced away at a speed of 10 cm/min. Ants at different distances move away at different speeds, and their speeds are proportional to their distances (just as the Hubble law states for galaxies). Yet all the ruler was doing was stretching uniformly.

Now let's repeat the analysis, but put the intelligent ant on some other mark—say, on 7 or 12. We discover that, as long as the ruler stretches uniformly, this ant also finds every other ant moving away at a speed proportional to its distance. In other words, the kind of relationship expressed by the Hubble law can be explained by a uniform stretching of the "world" of the ants. And *all* the ants in our simple diagram will see *the other ants* moving away from them as the ruler stretches.

For a three-dimensional analogy, let's look at the loaf of raisin bread in Figure 25.20. The chef has put too much yeast in the dough, and when she sets the bread out to rise, it doubles in size during the next hour, causing all the raisins to move farther apart. On the figure, we again pick a representative raisin (one that is not at the edge or the center of the loaf) and show the distances from it to several others in the figure (before and after the loaf expands).

Measure the increases in distance and calculate the speeds for yourself. Since each distance doubles during the hour, each raisin moves away from our selected raisin at a speed proportional to its distance. The same is true no matter which raisin you start with. Each raisin (if it could talk) would say, "Look! All the other raisins are expanding away from me. And their speeds are proportional to their distances."

Ace Astronomy™ Log into AceAstronomy and select this chapter to see the Active Figure called "Raisin Bread."

■ FIGURE 25.20

Expanding Raisin Bread As the raisin bread bakes, the raisins see other raisins moving away. More distant raisins move away faster in a uniformly expanding bread.

Our two analogies are useful for clarifying our thinking, but you must not take them literally. On both the ruler and the raisin bread, there are points that are at the end or edge. You can use these to pinpoint the middle of the ruler and the loaf. While our models of the universe have some resemblance to the properties of the ruler and the loaf, we will see that they do not have a center or an edge.

What is useful to notice about both the ants and the raisins is that they themselves did not "cause" their motion. It isn't as if the raisins decided to take a trip away from each other and then put on roller skates to get away. No, in both our analogies it was the stretching of the medium (the ruler or the bread) that moved the ants or the raisins farther apart. In the same way, we will see that the galaxies don't have rocket motors propelling them away from each other.

Instead, they are passive participants in the *expansion of space*. As space stretches, the galaxies are carried farther and farther apart.

The expansion of the universe, by the way, does not imply that the individual galaxies and clusters of galaxies themselves are expanding. The raisins in our analogy do not grow in size as the loaf expands. Similarly, gravity holds galaxies and clusters together, and they merely get farther away from each other—without themselves changing in size—as the universe expands.

In the final chapter of our book, we will explore some of the implications of this expansion for the past history and ultimate fate of the universe. But first we turn to some of the most intriguing and energetic objects in the expanding universe.

SURFING THE WEB

Introduction to Galaxies:
crux.astr.ua.edu/goodies/data_resources/galaxies.text
Clear introduction to galaxies, with good background information and a list of easy galaxies for amateur observing.

Edwin Hubble:
antwrp.gsfc.nasa.gov/diamond_jubilee/d_1996/sandage_hubble.html
An article on the life and work of Hubble by his student and successor, Allan Sandage.

Measuring the Expansion Rate of the Universe:
hubblesite.org/newscenter/archive/1999/19/
If you look at the background material here, you'll find a nice chronology of how we discovered and measured the expansion of the universe.

The Hubble Deep Field Images:

- hubblesite.org/newscenter/archive/1996/01

- hubblesite.org/newscenter/archive/1998/41

- hubblesite.org/newscenter/archive/1998/32
  In 1995 and 1998, the Hubble Space Telescope took the two really deep images of the universe of galaxies (one in the northern skies and one looking south). In 1998, astronomers also took an even deeper infrared image of part of the northern field. View or download the famous images, read captions and background information, and learn about some of the research they have led to.

Collections of Galaxy Images:
Each of the following sites contains a wonderful album of galaxy images taken with the world's largest telescopes:

- *Anglo-Australian Observatory:*
  www.aao.gov.au/images/general/galaxy_frames.html

- *Hubble Space Telescope:*
  hubblesite.org/newscenter/archive/

- *National Optical Astronomy Observatories:*
  www.noao.edu/image_gallery/galaxies.html

- *William Keel's Collection at the University of Alabama:*
  crux.astr.ua.edu/choosepic.html

- *Messier Catalog Site:*
  www.seds.org/messier/objects.html#galaxy

25.1 Faint star clusters, clouds of glowing gas, dust clouds reflecting starlight, and galaxies all appeared as faint patches of light (or nebulae) in the telescopes available at the beginning of the 20th century. It was only when Hubble measured the distance to the Andromeda Galaxy using cepheid variables in 1924 that the existence of other galaxies similar to the Milky Way in size and content was established.

25.2 The majority of bright galaxies are either **spirals** or **ellipticals.** Spiral galaxies contain both old and young stars, as well as interstellar matter, and have typical masses in the range of 10^9 to 10^{12} M_{Sun}. Our own Galaxy is a large spiral. Ellipticals are spheroidal or slightly elongated systems that consist almost entirely of old stars, with very little interstellar matter. Elliptical galaxies range in size from giants, more massive than any spiral, down to dwarfs, with masses of only about 10^6 M_{Sun}. A small percentage of galaxies with more chaotic shapes are classified as **irregulars.** Collisions and mergers of spiral galaxies can result in the formation of a giant elliptical galaxy.

25.3 The masses of spiral galaxies are determined from measurements of their rates of rotation. The masses of elliptical galaxies are estimated from analyses of the motions of the stars within them. Galaxies can be characterized by their **mass-to-light ratios.** The luminous parts of galaxies with active star formation typically have mass-to-light ratios in the range of 1 to 10; the luminous parts of elliptical galaxies, which contain only old stars, typically have mass-to-light ratios of 10 to 20. The mass-to-light ratios of whole galaxies, including their outer regions, are as high as 100, indicating the presence of a great deal of dark matter.

25.4 Astronomers determine the distances to galaxies using a variety of methods, including the period–luminosity relationship for cepheid variables; objects such as Type I supernovae, which appear to be standard bulbs; and the Tully–Fisher method, which connects the line broadening of 21-cm radiation to the luminosity of spiral galaxies. Each method has limitations in terms of its precision, the kinds of galaxies it can be used with, and the range of distances over which it can be applied.

25.5 The universe is expanding. Observations show that the spectral lines of distant galaxies are **redshifted,** and that their recession velocities are proportional to their distances from us, a relationship known as the **Hubble law.** The rate of recession, called the **Hubble constant,** is approximately 20 km/s per million LY. We are not at the center of this expansion: An observer in any other galaxy would see the same pattern of expansion that we do.

INTER-ACTIVITY

A Throughout much of this century, the 100-in. telescope on Mount Wilson (completed in 1917) and the 200-in. telescope on Palomar Mountain (completed in 1948) were the only ones large enough to obtain spectra of faint galaxies. Only a handful of astronomers (all men, since until the 1960s women were not given time on these two telescopes) were allowed to use these facilities, and in general the observers did not compete with each other by working on the same problems. Now there are many other telescopes, and several different groups do often work on the same problem. For example, two different groups have independently developed the techniques for using supernovae to determine the distances to galaxies at high redshifts. Which approach do you think is better for the science? Which is more cost effective? Why?

B A distant relative, whom you invite to dinner so you can share all the exciting things you have learned in your astronomy class, says he does not believe that other galaxies are made up of stars. You come back to your group and ask them to help you respond. What kinds of measurements would you make to show that other galaxies *are* composed of stars?

C Look at Figure 25.1 with your group. What does the difference in color between the spiral arms and the nuclear bulge of Andromeda tell you about the difference in the types of stars that populate these two regions of the galaxy? Which side of the galaxy is closer to us? Why?

D The members of your group should examine the Hubble Deep Field, the image with which this chapter opens. (Or see the Deep Field Web sites suggested in Surfing the Web.) Imagine you are the astronomers who obtained this image and you are just getting your first look at it, seeing deeper into the universe of galaxies than anyone else ever had. Which objects are galaxies and which are stars? How do the galaxies in the image differ? In general, how can you tell which galaxies are farther away?

E What is your reaction to reading about the discovery of the expanding universe? Discuss how the members of the group feel about a universe "in motion." Do you share Einstein's original sense that this is not the kind of universe you feel comfortable with? What do you think could have caused space to be expanding?

F In science fiction, characters sometimes talk about visiting other galaxies. Discuss with your group how realistic this idea is. Even if we had fast spaceships (traveling close to the speed of light, the speed limit of the universe) how likely are we to be able to reach another galaxy? Why?

REVIEW QUESTIONS

Ace Astronomy™ Assess your understanding of this chapter's topics with additional quizzing and animations at **http://ace.brookscole.com/ voyages**

1. Describe the main distinguishing features of spiral, elliptical, and irregular galaxies.
2. Why did it take so long for the existence of other galaxies to be established? What finally convinced astronomers that they do exist?
3. Explain what the mass-to-light ratio is and why it is smaller in regions of star formation in spiral galaxies than in the central regions of elliptical galaxies.
4. If we now realize that dwarf ellipticals are the most common type of galaxy, why did they escape our notice for so long?

5. Describe the best ways to measure the distance to:
   a. a nearby spiral galaxy
   b. a nearby elliptical galaxy
   c. a distant isolated elliptical galaxy
   d. a distant isolated spiral galaxy
   e. a distant elliptical galaxy that is a member of a galaxy cluster
6. Why is the Hubble law considered one of the most important discoveries in the history of astronomy?
7. What does it mean to say that the universe is expanding? For example, is your astronomy classroom expanding? Is the solar system? Why or why not?

THOUGHT QUESTIONS

8. In 1920 Harlow Shapley and another astronomer, Heber Curtis, held a debate. Although it began as a discussion of one issue (the size of our Galaxy), it eventually came to involve the larger question of the existence of other galaxies. Use the resources in your library or the World Wide Web to look up information about this debate. Summarize the points that each participant made, and compare to what we know today. (*Hint:* You can start by looking in our Surfing the Web section for Chapter 24 or the Suggestions for Further Reading on the next page.)

9. Where might the gas and dust (if any) in an elliptical galaxy come from?

10. Why can we not determine distances to galaxies by the same method used to measure the parallaxes of stars?

11. Which is redder: a spiral galaxy or an elliptical galaxy?

12. Suppose the stars in an elliptical galaxy all formed within a few million years shortly after the universe began. Suppose these stars have a range of masses, just as the stars in our own Galaxy do. How would the color of the elliptical

galaxy change over the next several billion years? How would its luminosity change? Why?

13. Starting with the determination of the size of the Earth, outline the sequence of steps necessary to obtain the distance to a remote cluster of galaxies. (*Hint:* Review Chapter 18.)

14. Suppose that the Milky Way Galaxy were truly isolated and that no other galaxies existed within 100 million LY. Suppose that galaxies were observed in larger numbers at distances greater than 100 million LY. Why would it be more difficult to determine accurate distances to those galaxies than if there were also galaxies relatively close by?

15. Suppose you were Hubble and Humason, working on the distances and Doppler shifts of the galaxies. What sorts of things would you have to do to convince yourself (and others) that the relationship you were seeing between the two quantities was a real feature of the behavior of the universe? (For example, would data from two galaxies be enough to demonstrate the Hubble law?)

FIGURING FOR YOURSELF

The Hubble law ($v = H \times d$) is a beautiful relationship because this simple equation allows us to calculate the distance to any galaxy. Here is an example to show how we use it in practice.

Suppose we have measured Hubble's constant to be 20 km/s per million LY. This means that if a galaxy is 1 million LY farther away, it will move away 20 km/s faster. So, if we find a galaxy that is moving away at a speed of 20,000 km/s, the Hubble law tells us the galaxy must be at a distance of

$$d = \frac{v}{H} = \frac{20{,}000\,\text{km/s}}{(20\ \text{km/s})/(1{,}000{,}000\ \text{LY})}$$

$$= \frac{(20{,}000)(1{,}000{,}000\ \text{LY})}{20} = 1{,}000{,}000{,}000\ \text{LY}$$

Note that the km/s in the numerator and denominator cancel, and that the factor of 1,000,000 LY in the denominator of the constant must be divided correctly before we get our distance of 1 billion LY.

16. Suppose a supernova explosion occurred in a galaxy at a distance of 10^8 LY. If we are only now detecting it, how long ago did the supernova actually occur? According to the Hubble law, what is the velocity with which this galaxy is moving away from us? (Assume a Hubble constant of 20 km/s per million LY.)

17. A cluster of galaxies is observed to have a recession velocity of 60,000 km/s. Find the distance to the cluster. (Assume the same Hubble constant as in problem 16).

18. Suppose we could measure the distance to a galaxy using one of the distance techniques listed in Table 25.2 and it turns out to be 200 million LY. The galaxy's redshift tells us its recession velocity is 5000 km/s. What is the Hubble constant?

19. As in previous chapters, Kepler's third law comes to our aid and makes it possible to calculate the mass of any galaxy for which we can measure the rotational velocity. Measurements of the Andromeda Galaxy show that it is rotating at a speed of 230 km/s, at a distance of about 100,000 LY or 6×10^9 AU from the center. What is the mass you derive from this rotation? Is the true mass of the galaxy likely to be larger or smaller than this estimate? Why?

20. Calculate the mass-to-light ratio for each of the stars listed in Table 17.2.

21. Calculate the mass-to-light ratio for a globular cluster with a luminosity of $10^6\,L_{\text{Sun}}$ and 10^5 stars. (Assume that the average mass of a star in such a cluster is $1\,M_{\text{Sun}}$.)

22. Calculate the mass-to-light ratio for a superluminous star of $100\,M_{\text{Sun}}$ that has the same luminosity of $10^6\,L_{\text{Sun}}$ as the globular cluster in problem 21.

23. Now let's try calculating the mass-to-light ratio for galaxies. Use the data in Table 25.1 to calculate the mass-to-light ratio for the most luminous elliptical galaxies. Do the same for the most luminous spirals. What do these numbers tell you about the differences in the types of stars that inhabit these two different types of galaxies?

24. Suppose you discovered a galaxy with a mass-to-light ratio of 200. Could you devise any combination of stars that would produce this ratio? Explain. What would you conclude about the galaxy you have discovered?

SUGGESTIONS FOR FURTHER READING

Bartusiak, M. "What Makes Galaxies Change" in *Astronomy,* Jan. 1997, p. 37. Nice overview of galaxy evolution.

Bothun, G. "Beyond the Hubble Sequence" in *Sky & Telescope,* May 2000, p. 36. History and updating of Hubble's classification scheme.

Christianson, G. *Edwin Hubble: Mariner of the Nebulae.* 1995, Farrar, Straus, & Giroux. The definitive biography of Hubble.

Christianson, G. "Mastering the Universe" in *Astronomy,* Feb. 1999, p. 60. Brief introduction to Hubble's life and work.

Dalcanton, J. "The Overlooked Galaxies" in *Sky & Telescope,* Apr. 1998, p. 28. On low-brightness galaxies, which have been easy to miss.

Dressler, A. *Voyage to the Great Attractor.* 1994, Knopf. Outstanding book by a noted astronomer on how modern extragalactic astronomy is done.

Eicher, D. "Candles to Light the Night" in *Astronomy,* Sept. 1994, p. 33. Introduction to standard bulbs and the cosmic distance scale.

Freedman, W. "The Expansion Rate and Size of the Universe" in *Scientific American,* Nov. 1992, p. 76.

Hartley, K. "Elliptical Galaxies Forged by Collision" in *Astronomy,* May 1989, p. 42.

Hodge, P. "The Extragalactic Distance Scale: Agreement at Last?" in *Sky & Telescope,* Oct. 1993, p. 16.

Impey, C. "Ghost Galaxies of the Cosmos" in *Astronomy*, June 1996, p. 40. Another article on low-brightness galaxies.

Kaufmann, G. and van den Bosch, F. "The Life Cycle of Galaxies" in *Scientific American*, June 2002, p. 46. On galaxy evolution and how it leads to the different types of galaxies.

Martin, P. and Friedli, D. "At the Hearts of Barred Galaxies" in *Sky & Telescope*, Mar. 1999, p. 32. On barred spirals.

Osterbrock, D. "Edwin Hubble and the Expanding Universe" in *Scientific American*, July 1993, p. 84.

Parker, B. "The Discovery of the Expanding Universe" in *Sky & Telescope*, Sept. 1986, p. 227.

Russell, D. "Island Universes from Wright to Hubble" in *Sky & Telescope*, Jan. 1999, p. 56. A history of our discovery of galaxies.

Skiff, B. "Exploring the Hubble Sequence by Eye" in *Sky & Telescope*, May 2000, p. 120. A guide for small-telescope observers to finding different types of galaxies.

Smith, R. "The Great Debate Revisited" in *Sky & Telescope*, Jan. 1983, p. 28. On the Shapley–Curtis debate concerning the extent of the Milky Way and the existence of other galaxies.

Trefil, J. "Galaxies" in *Smithsonian*, Jan. 1989, p. 36. Nice long review article.

A Gravitational Lens The brightest spiral and elliptical galaxies in this Hubble Space Telescope image are members of a cluster of galaxies called Abell 2218, which is 2 billion LY from Earth. Abell 2218 is so massive that its gravitational field deflects light rays passing through it, much as a lens bends light to form an image. The "arcs" in this image are the distorted images of very distant galaxies, which lie 5 to 10 times farther away than the cluster.

*Quasar, quasar, burning
    bright
In the forests of the night,
What immortal hand
    or eye
Could frame thy fearful
    luminosity?*

**With profound
apologies to
William Blake**

THINKING AHEAD

We have already seen that our own Galaxy harbors a giant black hole at its center. Do other galaxies also have such monsters at their cores, and, if so, how would we go about searching for them? How big can such a central black hole get? And what can black holes tell us about the formation and evolution of their host galaxies?

During the first half of the 20th century, astronomers viewed the universe of galaxies as a mostly peaceful place. They assumed that galaxies formed billions of years ago and then evolved slowly as the populations of stars within them formed, aged, and died. That placid picture completely changed in the last few decades of the 20th century.

Today astronomers can see that the universe is often shaped by violent events, including cataclysmic explosions of supernovae, collisions of whole galaxies, and the tremendous outpouring of energy as matter interacts in the environment surrounding massive black holes. The key event that began to change our view of the universe was the discovery of a new class of objects—the quasars.

Ace⟳**Astronomy**™ The AceAstronomy icon throughout the text indicates an opportunity for you to test yourself on key concepts and to explore animations and interactions on the AceAstronomy website at **http://ace.brookscole.com/voyages**

Virtual Laboratories

 General Relativity, Black Holes, and Gravitational Lensing

 Dark Matter

26.1 THE QUASARS

The story of the quasars begins in 1960, when astronomers discovered that what looked like two faint blue stars were strong sources of radio waves (Figure 26.1). The Sun does not radiate much energy at radio wavelengths, and astronomers expected that other stars would also be quiet in the radio region of the spectrum. The spectra of these radio "stars" only deepened the mystery: They had emission lines, but astronomers at first could not identify them with any known substance. By the 1960s, astronomers had a century of experience in identifying elements and compounds in the spectra of stars. Elaborate tables had been published showing the lines that each element would produce under a wide range of conditions. A "star" with unidentifiable lines in the ordinary visible light spectrum had to be something completely new.

26.1.1 Redshifts: The Key to the Quasars

The breakthrough came in 1963. At Caltech's Palomar Observatory, Maarten Schmidt (Figure 26.2) was puzzling over the spectrum of one of the radio stars, which was named 3C 273 because it was the 273rd entry in the third Cambridge catalog of radio sources (Figure 26.3). There were

■ **FIGURE 26.2**

Quasar Pioneers   Maarten Schmidt (left), who solved the puzzle of the quasar spectra in 1963, shares a joke in this 1987 photo with Allan Sandage, who took the first spectrum of a quasar. Sandage has also been instrumental in measuring the value of the Hubble constant.

strong emission lines in the spectrum, and Schmidt recognized that they had the same spacing between them as the Balmer lines of hydrogen (see Chapter 4). But the lines in 3C 273 were shifted far to the red of the wavelengths at which the Balmer lines are normally located. Indeed, these lines were at such long wavelengths that if the redshifts were attributed to the Doppler effect, 3C 273 was receding from us at a speed of 45,000 kilometers/second (km/s), or about 15 percent of the speed of light! Since stars don't show Doppler shifts this large, no one had thought of considering high redshifts to be the cause of the strange spectra.

The puzzling emission lines in other starlike radio sources were then reexamined to see whether they, too, might be well-known lines with large redshifts. Such proved to be the case, but the other objects were found to be receding from us at even greater speeds. Their astounding speeds showed that the radio "stars" could not possibly be stars in our own Galaxy. Any true star moving faster than a few hundred kilometers per second would be able to overcome the gravitational force of the Galaxy and completely escape from it. (As we shall see later in this chapter, astronomers eventually discovered that there was also more to these "stars" than just a point of light.)

It turns out that these high-velocity objects look like stars only because they are compact and very far away. Since they superficially resemble stars but do not share their properties, they were given the name *quasi-stellar radio sources*. Later, astronomers discovered objects with large redshifts that appear starlike but have no radio emission. Today, all these objects are referred to as *quasi-stellar objects (QSOs)* or, as they are more popularly known, **quasars.** (The name was soon appropriated by a manufacturer of home electronics.)

Thousands of QSOs have now been discovered, and spectra are available for a representative sample. All the

■ **FIGURE 26.1**

A Typical Quasar   The bars in this image mark the quasar known by its catalog number, PKS 1117–248. Note that nothing in this image distinguishes the quasar from an ordinary star. Its spectrum, however, shows that it is moving away from us at a speed of 36 percent of the speed of light, or 67,000 miles per second. The maximum speed observed for any star is only a few hundred miles per second.

■ **FIGURE 26.3**

The Quasar 3C 273 This Chandra x-ray image shows 3C 273 and the powerful jet shooting away from it. High-powered jets are often seen being driven outward from quasars, often at velocities very close to the speed of light. As we shall discuss later in this chapter, the energy emitted from the jet in 3C 273 probably comes from gas that falls toward a supermassive black hole at the center of the quasar but is redirected into a collimated jet. The quasar's jet is about 150,000 LY long. 3C 273 is the brightest quasar in our sky and was the first to be identified.

spectra show large to very large redshifts (and none show blueshifts). The largest correspond to shifts in wavelength that are greater than $\Delta\lambda/\lambda = 6$, where $\Delta\lambda$ is the difference between the wavelength that a spectral line would have if the source were stationary and the wavelength that it actually has. The stationary or laboratory wavelength is indicated by the Greek letter λ (lambda).

In the record-holding quasars, the first Lyman series line of hydrogen, with a laboratory wavelength of 121.5 nanometers (nm) in the ultraviolet portion of the spectrum, is shifted all the way through the visible region to beyond 800 nm. At such high redshifts, the simple formula for converting a Doppler shift to speed must be modified to take into account the effects of the theory of relativity (see Section 4.6). If we apply the relativistic form of the Doppler shift formula, we find that these redshifts correspond to velocities of about 96 percent of the speed of light! (See the Figuring for Yourself box at the end of the chapter for more on how to do calculations for such large redshifts.)

26.1.2 Quasars Obey the Hubble Law

The first question astronomers asked was whether the large redshifts could be caused by something other than the Doppler effect. Strong gravity causes a redshift, as we saw

in Chapter 23. However, a gravitational force strong enough to produce the very large redshifts of the quasars would leave other evidence in the spectrum, and that evidence is simply not there. Could the quasars be high-speed projectiles shot out from galaxies by some unknown process? That might explain their high speeds, but then some galaxy somewhere would surely be shooting out a projectile that happened to come *toward* us and show a blueshift. Yet only redshifts have been observed.

So let's test the hypothesis that the most obvious explanation is correct—namely, that quasars are distant objects, like galaxies, and thus participating in the expansion of the universe. If this idea is right, then quasars must obey the Hubble law (see Chapter 25). In order to determine whether this is the case, we need to know their velocities, which can be measured from their spectra. We also need to measure their distances, but we must do so without assuming that the Hubble law applies to quasars (which is what we are trying to test)! How could we do this? Two techniques have turned out to work. Both involve finding galaxies that are associated with quasars and measuring the redshifts of the galaxies instead. These methods work because we know that redshifts *do* give the correct distances to *galaxies*.

As we will discuss in more detail in the next chapter, galaxies throughout the universe are found in groups and clusters. We know that these groups are physically associated not only because they are close together in the sky but also because their member galaxies show the same redshift. Now suppose we see a quasar in the same area of the sky. Such an association does not prove that the quasar is at the same distance as the cluster. After all, it could be a coincidence—like seeing a bird and an airplane together, only to realize that the plane is really much farther away.

Astronomers, however, have now found many examples of a quasar that is seen in the same direction as a cluster of galaxies and has the same redshift as the cluster. One or two examples might be written off as an unlucky coincidence. Coincidences do occasionally happen, but as gamblers trying to outsmart the house in Las Vegas can attest, they do not happen very often. As more and more clusters were found with quasars inside the cluster boundaries and possessing the same redshift as all the cluster members, it became clear that the redshifts of quasars must be caused by the same effect that causes the redshifts of ordinary galaxies—the expanding universe.

Observations with the Hubble Space Telescope provided even stronger evidence. Quasars turn out to be located at the centers of galaxies. Hints that this is true had been obtained with ground-based telescopes, but space observations were required to make a convincing case. The reason is that quasars outshine whole galaxies by factors of 10 to 100 or even more. When this light passes through the Earth's atmosphere, it is blurred by turbulence and drowns out the faint light from the surrounding galaxy—much as the bright headlights from an oncoming car at night make it difficult to see anything close by.

The Hubble Space Telescope, however, is not affected by atmospheric turbulence and can detect the faint glow

J. Bahcall, M. Disney & NASA

■ **FIGURE 26.4**

Quasar Host Galaxies   The Hubble Space Telescope reveals the much fainter "host" galaxies around quasars. The top left-hand image shows a quasar at the heart of a spiral galaxy 1.4 billion LY from Earth. The bottom left-hand image shows a quasar at the heart of an elliptical galaxy some 1.5 billion LY from us. The middle images show remote pairs of interacting galaxies, one of which harbors a quasar. Each of the right-hand images shows long tails of gas and dust streaming away from a galaxy that contains a quasar. Such tails are produced when one galaxy collides with another.

from the galaxies that host quasars (Figure 26.4). Quasars have been found in the cores of both spiral and elliptical galaxies. Many quasar host galaxies are involved in a collision with a second galaxy, providing, as we shall see, an important clue to the source of their prodigious energy output.

26.1.3   The Size of the Energy Source

Given their great distances away, the quasars have to be extremely luminous to be visible to us at all—far brighter than any normal galaxy. In visible light alone, most are far more energetic than the brightest elliptical galaxies. But quasars also emit energy at x-ray and ultraviolet wavelengths, and some are radio sources as well. When all their radiation is added together, some QSOs have total luminosities as large as $10^{14} L_{\text{Sun}}$, which is 10 to 100 times the brightness of the brighter elliptical galaxies.

Finding a mechanism to produce the large amount of energy emitted by a quasar would be difficult under any circumstances. But there is an additional problem. When astronomers began monitoring quasars carefully, they found that some vary in luminosity on timescales of months, weeks, or even, in some cases, days. This variation is irregular and can change the brightness of a quasar by a few tens of percent in both its light and radio output.

Think about what such a change in luminosity means. A quasar at its dimmest is still more brilliant than any normal galaxy. Now imagine that the brightness increases by 30 percent in a few weeks. Whatever mechanism is responsible must be able to release new energy at rates that stagger our imaginations. The most dramatic changes in quasar brightness are equivalent to the energy released by 100 trillion Suns. To produce this much energy we would have to convert the total mass of about ten Earths into energy every minute!

FIGURE 26.5

Time to Vary Limits Size This diagram shows why light variations from a large region in space appear to last for an extended period of time as viewed from Earth. Suppose all the stars in this cluster, which is 10 LY across, brighten simultaneously and instantaneously. From Earth, star *A* will appear to brighten five years before star *B*, which in turn will appear to brighten five years earlier than star *C*. It will take ten years for an Earth observer to get the full effect of the brightening.

Moreover, because the fluctuations occur in such short times, the part of a quasar that is varying must be smaller than the distance light travels in the time it takes the variation to occur—typically a few months. To see why this must be so, let's consider a cluster of stars 10 LY in diameter at a very large distance from the Earth (Figure 26.5). Suppose every star in this cluster somehow brightens simultaneously and remains bright. When the light from this event arrives at the Earth, we would first see the brighter light from stars on the near side; five years later we would see increased light from stars at the center. Ten years would pass before we detected more light from stars on the far side.

Even though all stars in the cluster brightened at the same time, the fact that the cluster is 10 LY wide means that ten years must elapse before the light from every part of the cluster reaches us. From Earth we would see the cluster get brighter and brighter, as light from more and more stars began to reach us. Not until ten years after the brightening began would we see the cluster reach maximum brightness. In other words, if an extended object suddenly flares up, it will seem to brighten over a period of time equal to the time it takes light to travel across the object from its far side.

We can apply this idea to brightness changes in quasars to estimate their diameters. Because QSOs typically vary (get brighter and dimmer) over periods of a few months, the region where the energy is generated can be no larger than a few light months across. If it were larger, it would take longer than a few months for the light from the far side to reach us.

How large is a region of a few light months? Pluto, usually the outermost planet in our solar system, is about 5.5 light hours from us, while the nearest star is 4 LY away. Clearly a region a few light months across is tiny relative to the size of an entire galaxy. And some quasars vary even more rapidly, which means their energy is generated in an even smaller region. Whatever mechanism powers the quasars must be able to generate more energy than that produced by an entire galaxy in a volume of space that, in some cases, is not much larger than our solar system!

26.2 ACTIVE GALAXIES

How do quasars produce so much energy in such a tiny region? The first clues to the answer came from the study of some rather peculiar galaxies that turned out to have properties intermediate between those of ordinary galaxies and quasars. Let's look first at these observations, and then we will describe a single model that can explain the bewildering variety of peculiar galaxies and quasars.

The peculiar galaxies that we are concerned with are almost as luminous as quasars and share many of their properties, though to a less spectacular degree. Since abnormal amounts of energy are produced in their centers, they are said to have **active galactic nuclei** and are often referred to as *AGNs*. The galaxies that house these active nuclei are called **active galaxies.** (We now understand that such active galaxies actually have smaller versions of quasars embedded in their centers.)

26.2.1 Seyfert Galaxies

A good example of an active galaxy type is a group called *Seyfert galaxies,* which are spirals with pointlike energetic nuclei (Figure 26.6). They are named after the astronomer who discovered the first examples. Like quasars, Seyferts have strong, broad emission lines in their spectra. Since stars typically produce *absorption* lines in the spectrum of a galaxy, *emission* lines indicate that there are hot gas clouds near the Seyfert nuclei. The width of the lines indicates that the glowing gas is moving at speeds up to thousands of kilometers per second.

Some Seyferts, like quasars, are radio or x-ray sources or both, and all emit strongly in the infrared. Some show brightness variations over a period of a few months. As we concluded for quasars, the region from which the radiation comes can therefore be no more than a few light months across.

Studies of many such objects show that Seyfert and other peculiar galaxies tend to be more luminous than

■ **FIGURE 26.6**

Seyfert Galaxy NGC 1566 This galaxy is about 50 million LY distant and appears to be a normal spiral. However, observations show that it has a very luminous nucleus with many of the characteristics of a quasar, though it is much less energetic than a quasar. The active region at the center of NGC 1566 varies on a timescale of less than a month, indicating that it is extremely compact. Spectra show that hot gas near the tiny nucleus is moving at an abnormally high velocity.

■ **FIGURE 26.7**

The Jet of M87 Streaming out like a cosmic searchlight from the center of the galaxy M87 is one of nature's most amazing phenomena, a huge jet of electrons and other particles traveling at nearly the speed of light. In this Hubble Space Telescope image, the blue of the jet contrasts with the yellow glow from the combined light of billions of unseen stars and yellow, pointlike globular clusters that make up this galaxy. As we shall see later in this chapter, the jet, which is several thousand light years long, originates in a disk of superheated gas swirling around a giant black hole at the center of M87. The light that we see is produced by electrons twisting along magnetic field lines in the jet, a process known as *synchrotron radiation*, which gives the jet its bluish tint.

normal galaxies but less luminous than quasars. The bright but pointlike nuclei indicate that Seyferts, like quasars, produce enormous amounts of energy in a small region at their centers. Seyferts are close enough that we can study them in detail, and we can see that a significant fraction of their power comes from a source other than the energy output of individual stars.

26.2.2 Active Elliptical Galaxies

Spirals are not alone in showing peculiar activity in their cores. Some elliptical galaxies also have powerful central energy sources. Many giant elliptical galaxies that appear comparatively normal when seen with visible light turn out to be powerful emitters of radio energy; hence we call them *radio galaxies*. The galaxy M87, discussed in the previous chapter, is a fine example; Figure 25.8 shows it looking deceptively peaceful. But Hubble Space Telescope (Figure 26.7) and radio and x-ray images (Figure 26.8) allow us to "look" deeper inside the galaxy. There we find an

intense source of radio emission in the center, and a complicated jet of material extending from it for more than 6000 LY. Recent measurements show that knots of material in this jet are moving outward at speeds approaching two-thirds the speed of light. We will have more to say about what's happening at the center of M87 in the next section.

In some radio galaxies, most of the radio emission comes from small regions near their centers (reminding us of quasars). In others, we find not only bright sources in the nucleus but also larger extended regions of radio emission surrounding the galaxy. In about three quarters of the radio galaxies, the radio source is double, with most of the radiation coming from extended regions on opposite sides of each galaxy (Figure 26.9). Typically, the two emitting regions (called *radio lobes*) are far larger than the galaxy itself, and their centers are a few hundred thousand light years away from it. Often, observations also reveal two narrow jets of radio radiation pointing away from the galaxy toward the large extended sources. These jets are similar to the one in Figure 26.9 but can be more than 1 million LY long.

X ray: NASA /CXC /MIT/H. Marshall et al.; Radio: F. Zhou; F. Owen, NRAO; J. Biretta, STScl.
Visible light: NASA /STScl /UMBC/E. Perlman et al.

■ **FIGURE 26.8**

Multiwavelength Images of the Jet in M87 The jet in M87 can be observed in visible light, radio waves, and x rays. At the extreme left of the image, the bright galactic nucleus harboring a super-massive black hole shines. The radiation at each wavelength is produced by synchrotron radiation.

It appears that in all these objects hot ionized gases are "shot out" along the jets into the radio lobes by an extremely intense energy source in the center of the galaxy. Some mechanism in the center is able to direct the flow of energetic particles into narrowly focused beams that spread out and interact with colder neutral gas as they get beyond the confines of the galaxy. With new telescopes and detectors, similar jets can now be seen in more than half of all quasars.

Ace⟲Astronomy™ Log into AceAstronomy and select this chapter to see the Active Figure called "Jet Deflection."

D. Clark et al.; National Radio Astronomy Observatory

■ **FIGURE 26.9**

Jets Ejected by the Seyfert Galaxy 3C 219 To construct this unusual picture, a visible-light image (in blue) has been combined with a radio image (in red). You can see that the lobes of radio emission extend far beyond the visible parts of the galaxy (which is the blimp-shaped blue region in the center); a short jet connects the lobes and the galaxy.

26.2.3 A Variety of Active Galactic Nuclei

Our examples barely scratch the surface of the many different types of active galaxies astronomers have uncovered in the last few decades. Although at first the variety in their appearance was confusing, today we understand that the differences are primarily due to the angles at which we see their jets or lobes. (The distances of the galaxies and the environments in which their activity takes place also play a role.)

Remember that galaxies can be oriented in every possible way to our line of sight from Earth. Sometimes we look right down into a jet, as into the barrel of a gun; in these cases we see a brilliant spot—a pointlike source of energy. The jets of other galaxies shoot out to the side from our perspective, giving us the best view. Many times we see the central source and the jets or the lobes at an angle, which makes the appearance more complex. And, of course, the farther away the object is, the more challenging is the task of observing it. For example, some of the quasars that looked like points of light with the technology of the 1960s are now seen to have jets coming from them or to be surrounded by a galaxy.

Based on many such observations and decades of constructing theoretical models, astronomers now view all of these objects as part of the same basic phenomenon. A powerful source of energy at the centers of active galaxies is responsible for all the different forms of activity, including the outpouring of energy by quasars. And that source of energy turns out to be a massive black hole.

26.3 BLACK HOLES AT THE CENTERS OF GALAXIES

As we saw in Chapter 24, our own Galaxy has a black hole at its center and energy is emitted from a compact region around the black hole. Astronomers now think that all elliptical galaxies and all spirals with nuclear bulges have black holes at their centers. The amount of energy emitted by material near the black hole depends on two things: the mass of the black hole and the amount of matter that is falling into it.

If a black hole with a billion Suns' worth of mass inside (10^9 M_{Sun}) accretes (gathers) even a relatively modest amount of additional material—say, about 10 M_{Sun} per year—then (as we shall see) it can produce as much energy as a thousand normal galaxies. This is enough to account for the total energy of a quasar. On the other hand, if the mass of the black hole is smaller than a billion solar masses or the accretion rate is low, then the amount of energy emitted can be much smaller, as it is in the case of the Milky Way.

26.3.1 Observational Evidence for Black Holes

In order to prove that a black hole is present at the center of a galaxy, we must demonstrate that so much mass is crammed into so small a volume that no normal objects—massive stars or clusters of stars—could possibly account for it (see Chapter 24). We use Kepler's third law to estimate the mass at the center from the speed of material whirling around it. To apply this law we must find either gas or stars orbiting very close to the object whose mass we are trying to calculate. The Hubble Space Telescope (HST) is especially well suited to this task because it is above the blurring of the Earth's atmosphere and can obtain spectra very close to the bright central regions of active galaxies. The Doppler effect is then used to measure radial velocities of the orbiting material and so derive the speed with which it moves around.

One of the first galaxies to be studied with the HST is our old favorite, the giant elliptical M87. HST images showed that there is a disk of hot (10,000 K) gas swirling around the center of M87 (Figure 26.10). While it was surprising to find hot gas in an elliptical galaxy, the discovery was extremely useful for pinning down the existence of the black hole. Astronomers measured the Doppler shift of spectral lines emitted by this gas, found its speed of rotation, and then used the speed to derive the amount of mass inside the disk, applying Kepler's third law.

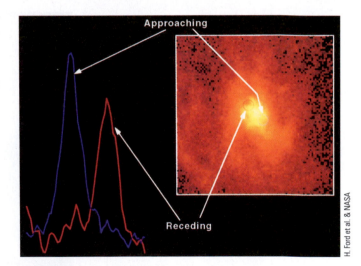

■ FIGURE 26.10

Evidence for a Black Hole at the Center of M87
This disk of whirling gas was discovered at the center of the giant elliptical galaxy M87 with the Hubble Space Telescope. Observations made on opposite sides of the disk show that one side is approaching us (the spectral lines are blueshifted by the Doppler effect) while the other is receding (lines redshifted), a clear indication that the disk is rotating. The rotation speed is about 550 km/s, or 1.2 million mi/h. Such a high rotation speed is evidence that there is a massive black hole at the center of M87.

■ FIGURE 26.11

Another Galaxy with a Black-Hole Disk The smaller image at the upper left shows an elliptical galaxy called NGC 7052 located in the constellation of Vulpecula, almost 200 million LY from Earth. At the galaxy's center (larger image) is a dust disk roughly 3700 LY in diameter. The disk rotates like a giant merry-go-round: Gas in the inner part (186 LY from the center) whirls around at a speed of 155 km/s (341,000 mi/h). From these measurements and Kepler's third law, it is possible to estimate that the disk whirls around a central black hole with a mass of 300 million Suns.

R. P. van der Marel, F. C. van den Bosch & NASA

These measurements showed that there is a mass of about 2.5 billion M_{Sun} concentrated in a tiny region at the very center of M87. So much mass in such a small volume of space must be a black hole. Let's stop for a moment and take in this figure—a single black hole that has swallowed enough material to make 2.5 billion stars like the Sun! Few astronomical measurements have ever led to so mind-boggling a result. What a strange environment the neighborhood of such a supermassive black hole must be.

Another example is shown in Figure 26.11. Here we see a disk of dust that surrounds a 300-million-M_{Sun} black hole in the center of an elliptical galaxy. (The bright spot in the center is produced by the combined light of stars that have been pulled close together by the gravitational force of the black hole.) The mass of the black hole was again derived from measurements of the rotational speed of the disk. The dust is moving around at 155 km/s (341,000 mi/h) at a distance of only 186 LY from its center. Given the pull of the mass at the center, we expect that the whole dust disk should be swallowed by the black hole in several billion years.

There is now evidence for black holes in many more galaxies. Indeed observations of stellar velocities near the centers of galaxies suggest that all galaxies that have a spherical concentration of stars—either elliptical galaxies or spiral galaxies with nuclear bulges—harbor black holes at their centers. Among the latter is our neighbor the Andromeda Galaxy. The mass of the central black holes in the galaxies studied so far is typically about 0.5 percent (1/200) of the total mass of the surrounding elliptical galaxy or nuclear bulge. The most massive galaxies, like the giant elliptical M87, thus have the most massive black holes (Figure 26.12).

Very recently, black holes have been found in two globular clusters, and again the mass of the black hole is about 0.5 percent of the total mass of the cluster. Since the masses of globular clusters are much smaller than the masses of galaxies, the masses of their central black holes are smaller as well—only 4000 and 20,000 M_{Sun} in the two cases studied so far. More observations are needed to determine whether all globular clusters contain black holes.

Many astronomers think that central black holes are an inevitable consequence of the formation of a dense spherical concentration of stars (Figure 26.13). That dense concentration of stars might be as small as a globular cluster or as large as a giant elliptical galaxy. As we shall see in the next chapter, theories suggest that galaxies formed from collapsing clouds of gas. Some of the gas formed stars. But it is likely that some of the gas settled to the center, where it became so concentrated that it formed a black hole.

Once formed, black holes will continue to grow by feeding on gas and stars that pass too close to it. Also, as we will see in the next chapter, galaxies sometimes collide and merge with close neighbors. The black holes at the center of each galaxy will then merge to form a larger black hole (Figure 26.14). Such mergers may explain why ellipticals like M87 can grow so large. Interstellar material and stars from one or more of the victims of these collisions may fall toward the black hole and add to its mass.

26.3.2 Energy Production Around a Black Hole

By now you may be willing to entertain the idea that huge black holes lurk at the centers of active galaxies. But we still need to answer the question of how such a black hole can account for one of the most powerful sources of energy in the universe. As we saw in Chapter 23, a black hole itself can radiate no energy. Any energy we detect from it must come from material very close to the black hole but not inside its event horizon.

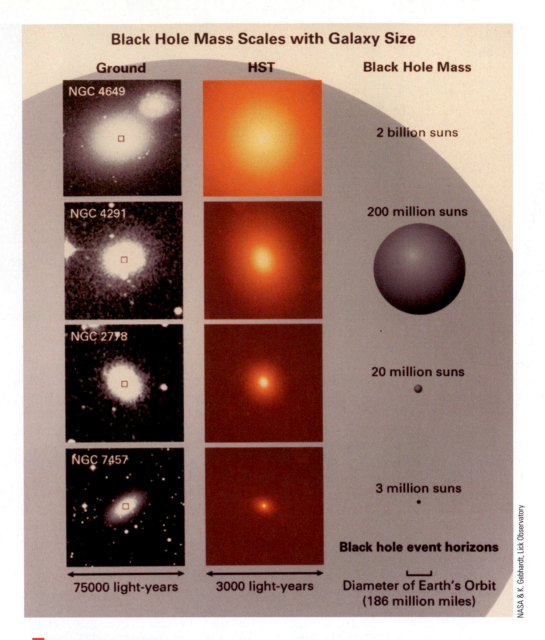

Black Hole Mass Scales with Galaxy Size

Ground	HST	Black Hole Mass
NGC 4649		2 billion suns
NGC 4291		200 million suns
NGC 2778		20 million suns
NGC 7457		3 million suns
		Black hole event horizons
75000 light-years	3000 light-years	Diameter of Earth's Orbit (186 million miles)

NASA & K. Gebhardt, Lick Observatory

■ **FIGURE 26.12**

Relationship Between Black-Hole Mass and the Mass of Host Galaxy
Observations show that there is a close correlation between the mass of the black hole at the center of a galaxy and the mass of the spherical distribution of stars that surrounds the black hole. That spherical distribution may be in the form of either an elliptical galaxy or the nuclear bulge of a spiral galaxy.

In a galaxy, a central black hole attracts matter—stars, dust, and gas—orbiting in the dense nuclear regions. This matter spirals in toward the spinning black hole and forms an *accretion disk* of material around it (see Chapter 23). As the material spirals ever closer to the black hole, it accelerates and becomes compressed, heating up to temperatures of millions of degrees. Such hot matter can radiate prodigious amounts of energy as it falls in toward the black hole.

To convince yourself that falling into a region with strong gravity can release a great deal of energy, imagine dropping your astronomy textbook out the window of the ground floor of the library. It will land with a thud, and maybe give a surprised pigeon a nasty bump, but the energy released by its fall will not be very great. Now take the same book up to the 15th floor of a tall building and drop it from there. For anyone below, astronomy could suddenly become a deadly subject; the book hits the ground with a great deal of energy. Dropping things from far away into the much stronger gravity of a black hole is much more effective. Just as the falling book can heat up the air, shake the ground, or produce sound energy that can be heard

1. Primordial hydrogen cloud collapses around small 'seed' black hole.

2. Infalling gas feeds the hole with more mass and forms stars.

3. Collapse yields a giant elliptical galaxy. Black hole growth stops.

1. Two disk galaxies with central black holes fall toward each other.

2. The galaxies collide, and their cores begin to merge along with their black holes.

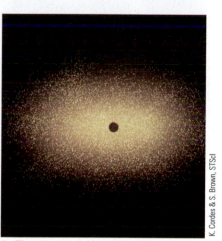

3. The merger yields a giant elliptical galaxy with a central black hole that has grown proportionally more massive.

K. Cordes & S. Brown, STScI

■ **FIGURE 26.13**

Two Ways to Make a Black Hole A black hole may be built up during the initial collapse of a cloud of gas to form a galaxy or through the merger of central black holes when two galaxies collide.

some distance away, so the energy of material falling toward a black hole can be converted to significant amounts of electromagnetic radiation.

The exact way in which the energy of infalling material is converted to radiation near a black hole is far more complicated than our simple example. To cite just one factor, the material does not go straight in but whirls around in a chaotic accretion disk. In fact, to understand what happens in the "rough and tumble" region around a massive black hole, astronomers and physicists must resort to computer simulations (and they require supercomputers, fast machines capable of awesome numbers of calculations per second). The details of these models are beyond the scope of our book, but they look very promising.

26.3.3 Connecting the Model and the Observations

A number of the phenomena that we observe in active galaxies and quasars can be explained naturally in terms of the black hole model. First, reasonable amounts of material falling into a very massive black hole can produce the vast quantities of energy that are emitted by quasars and active galactic nuclei. Detailed calculations show that about 10 percent of the mass of matter falling into a black hole can be converted to energy. This is a very efficient process: Remember that during the entire course of the evolution of a star like the Sun, less than 1 percent of its mass is converted to energy by nuclear fusion.

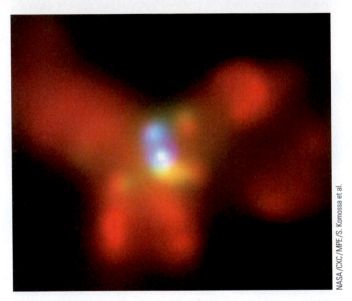

NASA/CXC/MPE/S. Komossa et al.

■ **FIGURE 26.14**

Colliding Galaxies with Two Black Holes We compare Hubble visible-light (*top*) and Chandra x-ray (*bottom*) images of the central regions of NGC 6240, a galaxy about 400 million LY away. It is a prime example of a galaxy in which stars are forming, evolving, and exploding at an exceptionally rapid rate due to a relatively recent merger (30 million years ago). The Chandra image shows two bright x-ray sources, each produced by hot gas surrounding a black hole. Over the course of the next few hundred million years, the two supermassive black holes, which are about 3000 LY apart, will drift toward each other and merge to form an even larger black hole. This detection of a binary black hole supports the idea that black holes grow to enormous masses in the centers of galaxies by merging with other black holes.

The event horizon of a black hole is very small; recall from Chapter 23 that a black hole with a mass of 1 billion M_{Sun} would have a radius of approximately 3 billion km, about the distance between the Sun and the planet Uranus. The emission produced by infalling matter comes from a small volume of space closely surrounding the black hole. The small size of the radiating region is exactly what we need to explain the fact that quasars vary on timescales of weeks to months.

We saw that quasar spectra are dominated by strong emission lines. Such lines form when a hot glowing gas is present, which is just what we would expect to see around such an energetic black hole. Our models suggest that the broad emission lines are formed in relatively dense clouds within about half a light year of the black hole. These clouds are located farther out than the accretion disk and may provide additional fuel for the black hole in the future.

26.3.4 Radio Jets

So far, our model seems to explain the central energy source in quasars and active galaxies. But, as we have seen, quasars and other active galaxies also have long jets that glow with radio waves, light, and sometimes even x rays, and that extend far beyond the limits of the parent galaxy. Can we find a way for our black hole and its accretion disk to produce these jets of energetic particles as well?

Many different observations have now traced these jets to within 3 to 30 LY of the parent quasar or galactic nucleus. While the black hole and accretion disk are typically smaller than 1 LY, we nevertheless presume that if the jets come this close, they probably originate in the vicinity of the black hole.

Why are energetic electrons and other particles ejected into jets, and often into two oppositely directed jets, rather than in all directions? Again, we must use theoretical models and supercomputer simulations of what happens when a lot of material whirls inward in a crowded black hole accretion disk. There is no agreement on exactly how jets form, but it has become clear that any material escaping from the neighborhood of the black hole has an easier time doing so perpendicular to the disk.

In some ways the inner regions of black-hole accretion disks resemble a baby that is just learning to eat by herself. As much food can sometimes wind up being spit out in various directions as going into the baby's mouth. In the same way, some of the material whirling inward toward a black hole finds itself under tremendous pressure and orbiting with tremendous speed. Under such conditions, our simulations show that a significant amount of material can be flung outward—not back along the disk, where more material is crowding in, but above and below the disk. If the disk is thick (as it tends to be when a lot of material falls in quickly), it can channel the outrushing material into narrow beams perpendicular to the disk (Figure 26.15).

Figure 26.16 shows observations of an elliptical galaxy that behaves in exactly this way. At the center of this active galaxy, there is a ring of dust and gas about 400 LY in diameter, surrounding a 1.2-billion-M_{Sun} black hole. Radio observations show that two jets emerge in a direction perpendicular to the ring, just as the model predicts.

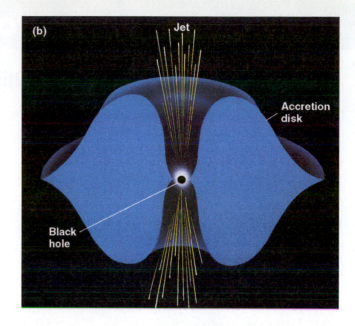

FIGURE 26.15

Models of Accretion Disks These schematic drawings show what accretion disks might look like around large black holes. (a) A thin accretion disk. (b) A "fat" disk—the type needed to account for channeling the outflow of hot material into narrow jets oriented perpendicular to the disk.

| 380 Arc Seconds | 17 Arc Seconds |
| 88,000 Light-Years | 400 Light-Years |

W. Jaffe, H. Ford & NASA

FIGURE 26.16

Jets and Disk in an Active Galaxy The picture on the left shows the active elliptical galaxy NGC 4261, which is located in the Virgo cluster at a distance of about 100 million LY. The galaxy itself—the white circular region in the center—is shown the way it looks in visible light, while the jets are seen at radio wavelengths. A Hubble Space Telescope image of the central portion of the galaxy is shown on the right. It contains a ring of dust and gas about 2500 LY in diameter, surrounding a supermassive black hole. Note that the jets emerge from the galaxy in a direction perpendicular to the plane of the ring, consistent with the models shown in Figure 26.15.

Quasars and the Attitudes of Astronomers

The discovery of quasars in the early 1960s was the first in a series of surprises astronomers had in store. Within another decade they would find neutron stars (in the form of pulsars), the first hints of black holes (in binary x-ray sources), and even the radio echo of the Big Bang itself. Many more new discoveries lay ahead.

As Maarten Schmidt reminisced in 1988, "This had, I believe, a profound impact on the conduct of those practicing astronomy. Before the 1960s, there was much authoritarianism in the field. New ideas expressed at meetings would be instantly judged by senior astronomers and rejected if too far out." We saw a good example of this in the trouble Chandrasekhar had in finding acceptance for his ideas about the death of stars with cores greater than $1.4\,M_{\text{Sun}}$.

"The discoveries of the 1960s," Schmidt continued, "were an embarrassment, in the sense that they were totally unexpected and could not be evaluated immediately. In reaction to these developments, an attitude has evolved where even outlandish ideas in astronomy are taken seriously. Given our lack of solid knowledge in extragalactic astronomy, this is probably to be preferred over authoritarianism."[1]

This is not to say that astronomers (being human) don't continue to have prejudices and preferences. For example, a small group of astronomers who thought that the redshifts of quasars were not connected with their distances (which was definitely a minority opinion) often felt excluded from meetings or from access to telescopes in the 1960s and 1970s. It's not so clear that they actually *were* excluded, as much as that they felt the very difficult pressure of knowing that most of their colleagues strongly disagreed with them. As it turned out, the evidence—which must ultimately decide all scientific questions—was not on their side either.

But today, as better instruments bring solutions to some problems and starkly illuminate our ignorance about others, the entire field of astronomy seems more open to discussing unusual ideas. Of course, before any hypotheses become accepted, they must be tested—again and again—against the evidence that nature itself reveals. Still, the many strange proposals published about what dark matter might be (we'll mention just a few in the next chapter) attest to the new openness that Schmidt described.

[1] From M. Schmidt: "The Discovery of Quasars" in *Modern Cosmology in Retrospect,* ed. B. Bertotti et al. (1990, Cambridge University Press).

With this black hole model, we have come a long way toward understanding the quasars and active galaxies that seemed very mysterious only a few decades ago. As often happens in astronomy, a combination of better instruments (making better observations) and improved theoretical models enabled us to make significant progress on a puzzling aspect of the cosmos.

 ## 26.4 QUASARS AS PROBES OF EVOLUTION OF THE UNIVERSE

The quasars' brilliance and large distance make them ideal probes of the far reaches of the universe and its remote past. Recall that when introducing quasars, we mentioned that they generally tend to be far away. When we see extremely distant objects, we are seeing them as they were long ago. Radiation from a quasar 8 billion LY away is telling us what that quasar was like 8 billion years ago, much closer to the time that the galaxy that surrounds it first formed. Astronomers have now detected light emitted

from quasars when the universe was less than 10 percent as old as it is now.

26.4.1 The Evolution of Quasars

Quasars provide compelling evidence that we live in an evolving universe—one that changes with time. They tell us that astronomers living billions of years ago would have seen a universe that is very different from the universe of today. Counts of the number of quasars at different redshifts (and thus distances) show us how dramatic these changes are (Figure 26.17). The number of quasars was the greatest at the time the universe was only 10–20 percent of its present age.

In order to explain this result, we make use of our model of the energy source of the quasars—namely, that quasars are black holes with enough fuel to make a brilliant accretion disk right around them. The fact that there were more quasars long ago (far away) than there are today (nearby) could be explained if there was more material to be accreted early in the history of the universe. If that fuel were mostly consumed in the first few billion years after the universe began its expansion, then later in its life, a

FIGURE 26.17

The Number of Quasars as We Look Back in Time
This graph shows the number of quasars at earlier and earlier times (that is, farther and farther away). An age of 0 corresponds to the beginning of the universe; an age of 1 corresponds to the present time. Note that quasars were most abundant when the universe was about 10–20 percent of its current age.

hungry black hole would have very little left with which to light up the galaxy's central regions.

In other words, if matter in the accretion disk is continually being depleted by falling into the black hole or being blown out from the galaxy in the form of jets, then a quasar can continue to radiate only as long as new gas is available to replenish the accretion disk. Where does this "food" for the black hole come from originally and how might it be replenished?

Observations indicate that the collision of galaxies is a primary source of fuel for the black holes at their centers. If two galaxies collide and merge, then gas and dust from one may come close enough to the black hole in the other to be devoured by it and so provide the necessary fuel. Astronomers have found that collisions were much more common early in the history of the universe than they are today. Therefore, we should generally see more quasars long ago (far away) than we do today (nearby)—as we in fact do.

The swallowing of one galaxy by another may even restart a quasar or active galaxy after it has been quiescent. If we see a *nearby* quasar or active galaxy, we can check to see whether it has been involved in any cosmic traffic accidents recently. And, indeed, many of the relatively nearby, still active quasars, including 3C 273, appear to be embedded in galaxies that have experienced collisions with other galaxies. Gas and dust from these "victim" galaxies have ap-

parently been swept up by a dormant black hole and have provided the new source of fuel required to rekindle it.

In addition to a merger with another galaxy, another possible source of fuel for the black hole is matter from the host galaxy itself. Interstellar matter near the center of the galaxy and stars that venture too close to the black hole may be accreted by it. We expect to find more gas and dust in the central regions early in a galaxy's life than later on. This again means that we would expect that the number and luminosity of quasars powered in this way would decline with time. Both ellipticals and the nuclear bulges of spirals today have very little raw material left to serve as a source of fuel for the black hole.

Most of the black holes in nearby galaxies, including in our own Milky Way, are now dark and quiet—mere shadows of their former selves. Without large quantities of fresh "food," they glow only weakly as bits of local material spiral inward toward the black hole.

26.4.2 Quasars and the Evolution of Elliptical Galaxies

Many of the closest quasars are found in the centers of elliptical galaxies. This suggests that distant quasars are located in the ancestors of today's elliptical galaxies, and that by studying the host galaxies of distant quasars we can learn what elliptical galaxies were like when the universe was young. As we saw, quasars are useful signposts to distant galaxies because they are so bright and their redshifts are easily measured. However, the superb, sharp images of the Hubble Space Telescope are required to study the host galaxies. The Earth's atmosphere tends to blur the light of the quasar and the galaxy together, making it impossible for ground-based telescopes to study the light from the galaxy alone.

The Hubble observations show that massive elliptical galaxies were already present around quasars when the universe was only about 4 to 5 billion years old—that is, when it was only one-third its current age. However, these galaxies seem to be bluer than nearby ellipticals, which indicates that they have more hot, young stars. As we look back in time, it appears we are seeing massive ellipticals in the process of formation.

In the next chapter, we will explore in more detail what astronomers have learned about the evolution of galaxies in general.

26.4.3 Quasars at the Dawn of Time

Quasars are so luminous that they can be seen at greater distances than even the brightest elliptical galaxies. In the past few years, there has been a kind of contest among astronomers to find the most distant quasar. Several are now known that are roughly 13 billion LY away (Figure 26.18). This means that they emitted their light when the universe was only about a billion years old, or only about 7 percent as old as it is now. These results suggest that supermassive

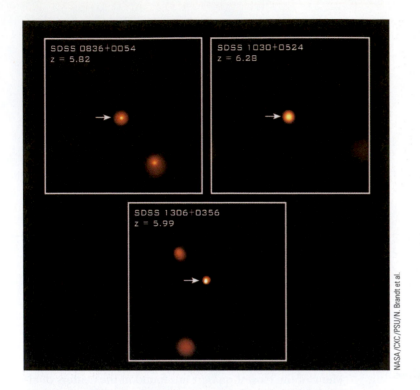

NASA/CXC/PSU/N. Brandt et al.

■ **FIGURE 26.18**

Three Quasars at a Distance of 13 Billion Light Years   These three quasars, recently discovered at visible wavelengths by the Sloan Digital Sky Survey, are shown here in x rays. They are examples of the most distant known quasars; the x rays Chandra detected were emitted when the universe was only a billion years old—about 7 percent of the present age of the universe. Estimates put the mass of the black holes that cause the x-ray emission at somewhere between 1 and 10 billion M_{Sun}. The implication is that at least some supermassive black holes formed very quickly after the universe began its expansion.

black holes—that is, black holes with masses of a billion solar masses or more—formed very quickly in the early universe. Unfortunately, these quasars are so far away that we cannot see what kind of galaxy, if any, surrounds them.

How so much material might have been collected together in so short a time has puzzled theorists. It is as if someone set out to build the Empire State Building in a single day. To avoid this problem, some theorists have suggested that the brightness of these distant quasars is an optical illusion. If the light from a distant quasar passes near a concentration of matter, like a cluster of galaxies, then according to general relativity the light path may be bent in a variety of ways. In the process, relativity theory says the light may be amplified, thereby making the quasar appear brighter. Since the light from a quasar at a distance of 13 billion LY must travel a very long distance to reach us, it is likely that somewhere in that journey it will pass near a massive galaxy or cluster of galaxies and be affected by such **gravitational lensing.** Quasars at a distance of only 1 billion LY are much less likely to be gravitationally lensed.

Lensing can enhance the brightness of a quasar by a factor of 10 or more. Since the mass of the black hole is estimated from the amount of energy emitted, lensing may fool us into an overestimate. If we know a quasar has been lensed, we can decrease our estimate of the intrinsic luminosity by a factor of 10. This will lower the mass the black hole is required to have, thereby making assembly in a billion years easier to explain.

NASA, ESA, Richard Ellis, Caltech; and Jean-Paul Kneib, Observatoire Midi-Pyrenees, France

■ **FIGURE 26.19**

Gravitational Lensing of a Distant Object   Gravitational lensing can cause objects to appear brighter, thereby making visible distant objects that would otherwise be too faint to observe. Here we see a small faint galaxy whose light has been amplified by the intervening massive cluster and also split into two. (The cluster is shown in the opening image of this chapter.) The galaxy (indicated by the arrows) is 13.4 billion LY distant (if we assume the age of the universe is 14 billion years), and so we may be seeing one of the early building blocks of the larger galaxies that we see in today's universe.

FIGURE 26.20

Gravitational Lensing This drawing shows how a gravitational lens can make two images, as in Figure 26.21. Two light rays from a distant quasar are bent while passing a foreground galaxy; they then arrive together at Earth. Although the two beams of light contain the same information, they now appear to come from two different points on the sky. This sketch is oversimplified and not to scale, but it gives a rough idea of the lensing phenomenon.

It turns out that gravitational lensing can also produce multiple images (see the next section). Astronomers are now imaging bright, distant quasars with the Hubble Space Telescope to determine whether they see multiple images, as would be expected if gravitational lensing has increased the apparent brightness of the most distant quasars.

Although the brightening caused by gravitational lensing may confuse the interpretation of observations of distant quasars, it can make otherwise invisible objects bright enough to detect. Figure 26.19 shows an example of a very distant faint galaxy that has become observable only because its light path passes through a large concentration of massive galaxies.

26.4.4 Gravitational Lensing and Multiple Images

The general theory of relativity predicts that light will be deflected in the vicinity of a strong gravitational field (see Chapter 23). Calculations show that, like a reflection in a funhouse mirror, such bending can produce multiple or twisted images (Figure 26.20). The first example of a gravitationally lensed quasar was discovered in 1979. Astronomers noticed that a pair of quasars, very close to each other in the sky and known by the single name QSO 0957 + 561 (the numbers

give their coordinates in the sky), had identical spectra and redshifts (Figure 26.21). The astronomers suggested that the two quasars might actually be only one, and that an intervening galaxy was serving as a gravitational lens.

Subsequent observations showed that there is a dim galaxy lying just above the lower quasar in Figure 26.21. In fact, the galaxy turns out to be a member of a galaxy cluster that is much closer to us than is the quasar. The geometry and estimated mass of the galaxy are just right to produce gravitational lensing.

Gravitational lenses can produce not only double images but also multiple images, arcs, or rings (as shown in the gallery of odd shapes in Figure 26.22). A remarkable example is shown in this chapter's opening image, in which

FIGURE 26.21

The Double Quasar QSO 0957+561 A massive galaxy acting as a gravitational lens forms multiple images of a quasar lying far beyond it. Here we see two images of a single distant quasar. The two images are the bright pointlike objects that appear in a nearly vertical line near the center of the picture. The foreground galaxy that is producing the images appears between the two quasars.

W. C. Keel

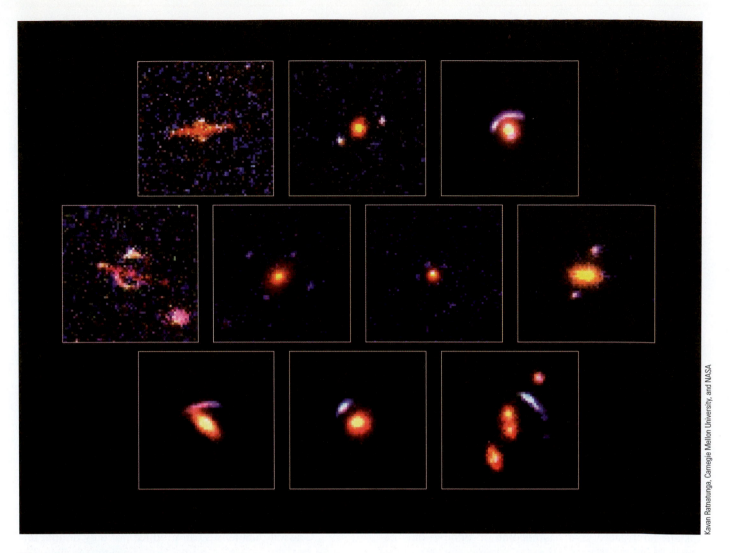

Kavan Ratnatunga, Carnegie Mellon University, and NASA

■ **FIGURE 26.22**

A Gallery of Gravitational Lenses   The number and shape of gravitationally lensed images depend on how the distant source and the closer lensing galaxy are lined up. These ten images taken with the Hubble Space Telescope show some of the possibilities. Strictly speaking, we would need spectra to prove that the various images in any one of these frames are of the same distant object, but it is likely that most of these are examples of gravitational lensing.

(*top left*) A lensed pair on either side of the disk of an edge-on lensing galaxy.

(*top center*) A pair of bluer lensed images around a red spherical elliptical lensing galaxy. Two much fainter images can barely be seen above and below the galaxy, which might make this a quadruple system.

(*top right*) A very good lens candidate, in which a bluer lensed source is seen as an extended arc about the redder elliptical lensing galaxy.

(*middle far left*) Here the lensed image is stretched nearly into a ring. It has been nicknamed "the London Underground" because it resembles the logo of the subway system in England's capital.

(*middle near left*) This is the first and brightest lens system discovered with the Hubble Space Telescope, and it has now been confirmed spectroscopically using large ground-based telescopes. The elliptical lensing galaxy is 7 billion LY away, while the lensed quasar (seen as four blue dots in a shape called an Einstein cross) is about 4 billion LY farther away.

(*middle near right*) The second quadruple lens candidate discovered with Hubble is similar to the first, but appears smaller and fainter.

(*middle far right*) A pair of bluish lensed images above and below a brighter, redder galaxy.

(*bottom left*) The image of an edge-on disklike galaxy (blue arc) has been significantly distorted by the redder lensing elliptical galaxy.

(*bottom center*) A blue arc in the Hubble Deep Field image (see the opening image for Chapter 25).

(*bottom right*) A blue arc caused by light traveling through a small group of four galaxies.

we see the images of background galaxies distorted into gracefully curving arcs by the foreground (yellow) cluster of galaxies.

We should note that visible galaxies are not the only possible gravitational lenses. Dark matter can also reveal its presence by producing this effect. The many ovals, arcs, and small portions of arcs seen in the image at the beginning of this chapter can be used to map the total mass in the foreground cluster—including dark matter. The resulting maps show that the foreground galaxy cluster contains much more matter than is actually seen in visible light. If the dark matter were not there, we would see significantly fewer distorted images of the background galaxies. Astronomers are currently observing lensed images all over the sky to determine where the dark matter is located and how much of it exists. Already these observations have confirmed the idea we discussed in Chapter 24—that most of the mass in the universe is in the form of dark matter.

SURFING THE WEB

Quasars and Active Galactic Nuclei:
www.astr.ua.edu/keel/agn/
An annotated gallery of images showing the wide range of activity in galaxies. There is also an introduction, a glossary, and background information.

Quasars: The Light Fantastic:
hubblesite.org/newscenter/archive/1996/35/astrofile
This brief "backgrounder" from the public information office at the Hubble gives a bit of the history of the discovery and understanding of quasars.

Quasar & Active Galaxy Tutorial:
casswww.ucsd.edu/public/tutorial/Quasars.html
Astronomer Gene Smith reviews the essentials of this topic, with illustrations.

M87 Information & Images:
www.seds.org/messier/m/m087.html
Part of a Messier catalog site, these pages explore the giant elliptical galaxy M87 and its environment (the Virgo Cluster).

Gravitational Lenses:

- *Primer on Gravitational Lenses:*
  www.scibridge.sdsu.edu/coursemats/introsci/
  sysinteractions/grav_lenses/index.html
  NASA's Joe Dolan gives a brief introduction and pictorial tour of gravitational lensing, with links to other sources of information.

- *Lens an Astrophysicist:*
  theory2.phys.cwru.edu/~pete/GravitationalLens/
  Purely for fun, you can pick an image of astronomers working in this field and put their photos through a gravitational lens. You can import images of others and subject them to the same indignity.

Some Hubble Space Telescope Results with Gravitational Lenses:

- *Lens in Galaxy Cluster Abell 2218:*
  hubblesite.org/newscenter/archive/1995/14/

- *Using a Lens to Discover a Distant Galaxy:*
  hubblesite.org/newscenter/archive/1997/25/

- *More Lenses:*
  hubblesite.org/newscenter/archive/1999/18/

- *Using Lenses to Find Galaxy Building Blocks:*
  hubblesite.org/newscenter/archive/2001/32/

SUMMARY

26.1 The first **quasars** discovered looked like stars but had strong radio emission. The quasar spectra obtained so far show redshifts ranging from 15 to more than 96 percent of the speed of light. The evidence shows that quasars obey the Hubble law and are at the great distances implied by these redshifts. To be noticeable at such great distances, they must have 10 to 100 times the luminosity of the brighter normal galaxies. Their variations show that this tremendous energy output is generated in a small volume—in some cases, in a region not much larger than our own solar system. Some quasars are members of small groups or clusters with the same redshift as the quasar. Observations with the HST show that quasars lie at the centers of galaxies. Both spirals and ellipticals can harbor quasars.

26.2 Astronomers now view quasars as the most extreme example of a class of peculiar galaxies that generate large amounts of energy in a small **active galactic nucleus.** Seyferts are an example of such **active galaxies.** Some giant elliptical galaxies are also strong radio sources. In many cases, jets of radio emission extend from the active center to large radio-emitting regions located on either side of the galaxy or quasar. Active galaxies have luminosities intermediate between those of normal galaxies and quasars.

26.3 Both active galactic nuclei and quasars derive their energy from material falling toward, and forming a hot accretion disk around, a massive black hole. The mass of the black hole is typically about 1/200 the mass of the stars in the elliptical galaxy or the nuclear bulge that surrounds it. This model can account for the large amount of energy emitted and for the fact that the energy is produced in a relatively small volume of space. Less massive black holes have been found in some globular clusters as well.

26.4 Quasars were much more common billions of years ago than they are now, and astronomers speculate that they mark an early stage in the formation of galaxies. Quasars were more likely to be active when the universe was young and fuel for the accretion disk and jets was more available. Quasar activity can apparently be retriggered by a collision between two galaxies, which provides a new source of fuel to feed the black hole. Quasars can be used as probes of the distant reaches of the universe. Galaxies that surrounded quasars when the universe was only about one-third of its current age are bluer and therefore have larger populations of hot, young stars than today's ellipticals. This may be an early stage in the evolution of galaxies. Observations of distant quasars demonstrate that the **gravitational lensing** effects predicted by general relativity actually do occur. Gravitational lenses can produce various types of images, including double or multiple images and even arcs and rings. Analysis of the effects of gravitational lensing can be used to determine where dark matter is located and how much dark matter exists. Such observations support the idea that most of the mass in the universe is in the form of dark matter.

INTER-ACTIVITY

A When quasars were first discovered and the source of their great energy was unknown, some astronomers searched for evidence that quasars are much nearer to us than their redshifts imply. (That way, they would not have to produce so much energy to look as bright as they do.) One way was to find a "mismatched pair"—a quasar and a galaxy with different redshifts that lie in very nearly the same direction in the sky. Suppose you do find one and only one galaxy with a quasar very close by, and the redshift of the quasar is six times larger than that of the galaxy. Have your group discuss whether you could then conclude that the two objects are at the same distance and that redshift is *not* a reliable indicator of distance. Why? Suppose you found three such pairs, each with different mismatched redshifts? Suppose *every* galaxy has a nearby quasar with a different redshift. How would your answer change and why?

B Large ground-based telescopes typically can grant time to only one out of every four astronomers who apply for observing time. One prominent astronomer tried for several years to establish that the redshifts of quasars do not indicate their distances. At first he was given time on the world's largest telescope, but eventually it became clearer that quasars were just the centers of active galaxies and that their redshifts really did indicate distance. At this point, he was denied observing time by the committee of astronomers who reviewed such proposals. Suppose your group had been the committee. What decision would you have made? Why? (In general, what criteria should astronomers use for allowing astronomers whose views completely disagree with the prevailing opinion to be able to pursue their research?)

C You get so excited about gravitational lensing that you talk about it with an intelligent friend who has not yet taken an astronomy course. After hearing you out, this friend starts to worry. She says, "If gravitational lenses can distort quasar images, sometimes creating multiple, or ghost, images of the same object, then how can we trust any point of light in the sky to be real? Maybe many of the stars we see are just ghost images or lensed images, too!" Have your group discuss how to respond. (*Hint:* Think about the path that the light of a quasar took on its way to us and the path that the light of a typical star takes.)

D Based on the information in this chapter and in Chapter 23, have your group discuss what it would be like near the event horizon of a supermassive black hole in a quasar or active galaxy. Make a list of all the reasons a trip to that region would not be good for your health. Be specific.

REVIEW QUESTIONS

Ace◓Astronomy™ Assess your understanding of this chapter's topics with additional quizzing and animations at **http://ace.brookscole.com/ voyages**

1. Describe some differences between quasars and normal galaxies.
2. Describe the arguments supporting the idea that quasars are at the distances indicated by their redshifts.
3. In what ways are active galaxies like quasars but different from normal galaxies?

4. Describe the process by which the action of a black hole can explain the energy radiated by quasars.
5. What is a gravitational lens? Show with a diagram how it is possible to see two quasars in the sky when in reality there is only one.
6. Describe the observations that convinced astronomers that M87 is an active galaxy.
7. How do gravitational lenses help astronomers learn about dark matter in the universe?

THOUGHT QUESTIONS

8. Suppose you observe a starlike object in the sky. How can you determine whether it is a star or a quasar?

9. Why don't any of the methods for establishing distances to galaxies, described in Chapter 25 (other than Hubble's law itself), work for quasars?

10. One of the early hypotheses to explain the high redshifts of quasars was that these objects had been ejected at very high velocities from galaxies. This idea was rejected because no quasars with large blueshifts have been found. Explain why we would expect to see quasars with both blueshifted and redshifted lines if they were ejected from nearby galaxies.

11. If we see a double image of a quasar produced by a gravitational lens and can obtain a spectrum of the galaxy that is acting as the gravitational lens, we can then put limits on the distance to the quasar. Explain how.

12. A friend of yours who has watched many *Star Trek* episodes says, "I thought that black holes pulled everything into them. Why then do astronomers think that black holes can explain the great *outpouring* of energy from quasars?" How would you respond?

13. The opening image of this chapter shows a cluster of yellow galaxies that produces several images of blue galaxies through gravitational lensing. Which are more distant: the blue galaxies or the yellow galaxies? The light in the galaxies comes from stars. How do the temperatures of the stars that dominate the light of the cluster galaxies differ from the temperatures of the stars that dominate the light of the blue-lensed galaxy? Which galaxy's light is dominated by young stars?

FIGURING FOR YOURSELF

In Section 4.6, we gave the formula for the Doppler shift, which astronomers denote by the letter z, as:

$$z = \frac{\Delta\lambda}{\lambda} = \frac{v}{c}$$

where λ is the wavelength emitted by a source of radiation that is not moving, $\Delta\lambda$ is the difference between that wavelength and the wavelength we measure, v is the speed with which the source moves away, and c (as usual) is the speed of light.

A line in the spectrum of a galaxy is at a wavelength of 393 nanometers (nm, or 10^{-9} m) when the source is at rest. Let's say the line is measured to be longer than this value (redshifted) by 7.86 nm. Then its redshift $z =$ 7.86 nm/393 nm = 0.02, so its speed away from us is 2 percent of the speed of light ($v/c = 0.02$).

This formula is fine for galaxies that are relatively nearby and are moving away from us slowly in the expansion of the universe. But the quasars and distant galaxies we discussed in this chapter are moving away at speeds close to the speed of light. In that case, converting a Doppler shift (redshift) to a distance must include the effects of the special theory of relativity, which explains how measurements of space and time change when we see things moving at high speeds. The details of how this is done are way beyond the level of this text, but we can share with you the relativistic formula for the Doppler shift:

$$\frac{v}{c} = \frac{(z+1)^2 - 1}{(z+1)^2 + 1}$$

Let's do an example. Suppose a distant quasar has a redshift of 5. At what fraction of the speed of light is the quasar moving away? We calculate as follows:

$$\frac{v}{c} = \frac{(5+1)^2 - 1}{(5+1)^2 + 1} = \frac{36-1}{36+1} = \frac{35}{37} = 0.946$$

The quasar is thus receding from us at about 95 percent the speed of light.

14. Show that no matter how big a redshift (z) we measure, v/c will never be greater than 1. (In other words, no galaxy we observe can be moving away faster than the speed of light.)

15. If a quasar has a redshift of 3.3, at what fraction of the speed of light is it moving away from us?

16. If a quasar is moving away from us at $v/c = 0.8$, what is the measured redshift?

17. In Section 26.1.1, we discuss that the largest redshifts found so far are greater than 6! Suppose we find a quasar with a redshift of 6.1. With what fraction of the speed of light is it moving away from us?

18. Rapid variability in quasars indicates that the region in which the energy is generated must be small. You can show why this is true. Suppose, for example, that the region in which the energy is generated is a transparent sphere 1 LY in diameter. Suppose that in 1 s this region brightens by a factor of 10 and remains bright for two years, after which it returns to its original luminosity. Draw its light curve (a graph of its brightness over time) as viewed from Earth.

19. Large redshifts move the positions of spectral lines to longer wavelengths and change what can be observed from the ground. For example, suppose a quasar has a redshift of $\Delta\lambda/\lambda = 4.1$. At what wavelength would you make observations in order to detect its Lyman line of hydrogen, which has a laboratory or rest wavelength of 121.6 nm? Would this line be observable with a ground-based telescope in a quasar with zero redshift? Would it be observable from the ground in a quasar with a redshift of $\Delta\lambda/\lambda = 4.1$?

20. Once again in this chapter, we see the use of Kepler's third law to estimate the mass of supermassive black holes. This chapter supplied the result of the calculation of the mass of the black hole in NGC 4261. In order to get this answer, astronomers had to measure the velocity of particles in the ring of dust and gas that surrounds the black hole. How high were these velocities? Turn Kepler's third law around and use the information given in this chapter about the galaxy NGC 4261—the mass of the black hole at its center and the diameter of the surrounding ring of dust and gas—to calculate how long it would take a dust particle in the ring to complete a single orbit around the black hole. Assume that the only force acting on the dust particle is the gravitational force exerted by the black hole. Calculate the velocity of the dust particle in km/s.

SUGGESTIONS FOR FURTHER READING

Bartusiak, M. "A Beast in the Core" in *Astronomy,* July 1998, p. 42. On supermassive black holes at the centers of galaxies.

Bartusiak, M. "Gravity's Rainbow" in *Astronomy,* Aug. 1997, p. 44. On gravitational lenses.

Bechtold, J. "Shadows of Creation: Quasar Absorption Lines and the Genesis of Galaxies" in *Sky & Telescope,* Sept. 1997, p. 29.

Croswell, K. "Have Astronomers Solved the Quasar Enigma?" in *Astronomy,* Feb. 1993, p. 29.

Disney, M. "A New Look at Quasars" in *Scientific American,* June 1998, p. 52.

Djorgovski, S. "Fires at Cosmic Dawn" in *Astronomy,* Sept. 1995, p. 36. On quasars and what we can learn from them.

Finkbeiner, A. "Active Galactic Nuclei: Sorting Out the Mess" in *Sky & Telescope,* Aug. 1992, p. 138. Good introduction by a science writer.

Ford, H. and Tsvetanov, Z. "Massive Black Holes at the Hearts of Galaxies" in *Sky & Telescope,* June 1996, p. 28. Nice overview.

Nadis, S. "Here, There, and Everywhere" in *Astronomy,* Feb. 2001, p. 34. On Hubble observations showing how common supermassive black holes are in galaxies.

Olson, S. "Black Hole Hunters" in *Astronomy,* May 1999, p. 48. Profiles four astronomers who search for "hungry" black holes at the centers of active galaxies.

Petersen, C. "The Universe through Gravity's Lens" in *Sky & Telescope,* Sept. 2001, p. 32. An eight-page review of lensing.

Preston, R. "Beacons in Time: Maarten Schmidt and the Discovery of Quasars" in *Mercury,* Jan./Feb. 1988, p. 2.

Voit, G. "The Rise and Fall of Quasars" in *Sky & Telescope,* May 1999, p. 40. Good overview of how quasars fit into cosmic history.

Wambsganss, J. "Gravity's Kaleidoscope" in *Scientific American,* Nov. 2001, p. 65. On gravitational lenses and microlensing phenomena.

Wright, A. and Wright, H. *At the Edge of the Universe.* 1989, Ellis Horwood/John Wiley. Introductory book on extragalactic astronomy.

Merging Neutron Stars In this imaginative painting, artist Don Dixon
shows a pair of neutron stars about to merge, with their "solid" crusts breaking
up. With temperatures around a million degrees, the neutron stars are shown
as they might appear to x-ray–sensitive eyes. Such mergers may be responsible
for some of the mysterious gamma-ray bursts described in this Interlude.

Interlude:
The Mystery of the Gamma-Ray Bursts

Everybody loves a good mystery story, and astronomers are no exception. A mystery that astronomers have been trying to unravel for the last three decades now seems—at last—to be close to a solution. It's a good story, one that involves cold-war politics, telescopes in space, neutron stars and black holes, and some new-fashioned detective work. It is also a good case study of how astronomers—just like detectives—often piece together clues from many sources to find a perpetrator.

The mystery was uncovered in the mid-1960s, not by an astronomical project, but as a result of a search for the tell-tale signs of nuclear-weapon explosions. The U.S. Defense Department launched a series of Vela satellites to make sure that no country was violating a treaty that banned the detonation of nuclear weapons in space. (The satellite name comes from the Spanish word *velar*, which can mean "to watch over.") Since nuclear explosions produce the most energetic form of electromagnetic waves called *gamma rays* (see Chapter 4), the Vela satellites contained detectors to search for this type of radiation. The satellites did not detect any events from human activities, but they did—to everyone's surprise—detect short bursts of gamma rays coming from random directions in the universe.

News of the discovery was first published (after extensive analysis) in 1973; however, the origin of the bursts remained a mystery. No one knew what produced the brief flashes of gamma rays or how far away the sources were. They could have been as close as the outskirts of our own solar system or as distant as the edge of the observable universe. Of course, lots of intriguing theories were proposed about what and where the bursts might be, but good evidence to decide among these theories was lacking.

With the launch of the Compton Gamma-Ray Observatory by NASA in 1991 (Figure I.1), astronomers began to identify many more bursts and to learn more about them. Approximately once per day, the NASA satellite detected a flash of gamma rays somewhere in the sky that lasted from a fraction of a second to several hundred seconds. The sources of the bursts were distributed *isotropically;* that is, they could appear anywhere in the sky (Figure I.2). Almost never did a second burst come from the same location. Before the Compton measurements, astronomers had expected that the most likely place for the bursts to come from was the main disk of our own (frisbee-shaped) Galaxy. If this had been the case, however, more bursts would have been seen in the crowded plane of the Milky Way than above or below it.

For several years, astronomers actively debated whether the burst sources were relatively nearby or very far away—the two possibilities for isotropic bursts. Nearby locations might include the cloud of comets that surrounds

■ **FIGURE I.1**

The Compton Observatory The deployment of the gamma-ray observatory in 1991 from the space shuttle Atlantis. Weighing more than 16 tons, it was one of the largest scientific payloads ever launched into space.

the solar system or the halo of our Galaxy, which is large and spherical and also surrounds us in all directions (see Section 24.1). If, on the other hand, the bursts occurred at very large distances, they could come from other galaxies, which are also distributed uniformly in all directions.

Both the very local and the very distant hypotheses required something strange to be going on. If the bursts were coming from the cold outer reaches of our own solar system, then astronomers had to hypothesize some new kind of physical process that could produce unpredictable flashes of high-energy gamma rays in an otherwise quiet region of space. And if the bursts came from galaxies millions or billions of light years away, then they must be extremely powerful to be observable at such a distance; indeed they had to be the biggest explosions in the universe.

I.2 GETTING RESOLUTION

The problem with trying to figure out the source of the gamma-ray bursters was that our instruments for detecting gamma rays could not pinpoint the exact place in the sky where the burst was happening. Gamma-ray telescopes can show us an area only a few degrees of arc on a side where the burst is located; within this area, visible-light telescopes can detect vast numbers of possible "perpetrators." There was simply no way to know which of the many candidate objects (if any) was the source of the gamma-ray burst. It was like the problem of a flash camera going off at a crowded news conference. Unless you happen to be looking exactly at the camera at the right moment, you cannot tell which photographer was responsible.

Yet astronomers suspected (perhaps we should say hoped) that if they *could* pinpoint the exact position of one of these rapid bursts, then a lingering "afterglow" in other wavelengths might be able to tell more about the source. As discussed in Chapter 4 and throughout the text, light and other waves often contain information that we can decode to learn more about the source of the waves. For example, lines in the spectrum of light can tell us about the Doppler shift of the source (see Section 4.6). If the source is a galaxy, we can use Hubble's law to tell us how far away it is.

The breakthrough came with the launch of the Italian–Dutch BeppoSAX x-ray satellite in 1996 (Figure I.3). If you are wondering about the name, Beppo was the nickname of a noted Italian physicist and SAX is *Satellite per Astronomia X.* It turns out that some gamma-ray bursts can in fact be detected by their x-ray afterglow, and BeppoSAX carried just the instruments needed to measure the position of such an x-ray source accurately. At any given time, BeppoSAX could observe about 5 percent of the sky. If there is about one gamma-ray event per day over the whole sky, then just by accident astronomers expected that about once every 20 days BeppoSAX would be pointed to the "right" part of the sky and would detect the x rays that follow a burst of gamma rays. BeppoSAX could then determine a position accurate to about 1 arcminute, or about 1/30 the

2704 BATSE Gamma-Ray Bursts

+90

+180

−180

−90

■ **FIGURE 1.2**

Distribution of Gamma-Ray Bursts in the Sky A map of where gamma-ray bursts have been found in the sky by the Compton Gamma-Ray Observatory. There are 2704 bursts shown. The map is oriented so that the disk of the Milky Way would stretch across the center line (or equator) of the oval. Note that the bursts show no preference at all for the plane of the Milky Way, as many other types of objects in the sky do. Colors indicate the total energy in the burst: Red dots indicate long-duration, bright bursts; blue and purple dots show short, weaker bursts.

diameter of the full Moon. With that kind of accuracy, a visible-light telescope could be pointed to the same part of the sky and look for something that was fading in brightness.

We should emphasize that this technique would not work if the x-ray and visible-light afterglows were as short as the gamma-ray bursts. A fraction of a second—or even several hundred seconds—is just not long enough for a satellite like BeppoSAX to pinpoint the source and to alert telescopes around the world to move to the right position and make observations. Fortunately, at least some of the gamma-ray bursts have a longer afterglow at x-ray, visible, and radio wavelengths. Think of how a bonfire dies away slowly—with the orange-red flames fading and being replaced with a dull red glow that lingers and then with heat radiation (infrared). Similarly, we can still observe some gamma-ray-burst sources at longer wavelengths for weeks to months after the gamma rays themselves have disappeared.

1.3 THE FIRST OBSERVATIONS

Two crucial BeppoSAX burst observations in 1997 helped to resolve the mystery of the gamma-ray bursts. The first burst came on February 28 from the direction of the constellation of Orion. Within eight hours, astronomers working with the satellite had found the x-ray afterglow, reoriented the spacecraft to focus the narrow-field x-ray camera on the source, and determined a position they could hand over to astronomers around the world.

That very night the 4.2-m William Herschel Telescope on the Canary Islands was able to find the fading visible-light glow, and within a week the world's largest telescopes could see something fuzzy right where the fading point of light had been. When the Hubble Space Telescope was pointed in the right direction in March (and again in September), it found that the burst seemed to have occurred

■ **FIGURE 1.3**

BeppoSAX The BeppoSAX satellite was launched by the Italian Space Agency in 1996 and can pinpoint the direction from which x rays from gamma-ray-burst sources are coming.

Andrew Fruchter, Space Telescope Science Institute; Elena Pian, ITSRE-CNR; and NASA

■ **FIGURE I.4**

Hubble Space Telescope View of a Gamma-Ray Burst This false-color image, taken on September 5, 1997, shows the fading afterglow of the February 28, 1997 burst fireball, as well as the fuzzy object in which the source appears to be embedded. Using the HST, astronomers were able to detect the afterglow six months after the initial burst. The top left view shows the region of the burst (and is repeated smaller at the bottom left, but with the addition of a box showing the area that is enlarged in the right-hand views). The top right-hand enlargement then shows the burst source and what appears to be its host galaxy. The bottom right-hand box has them labeled for clarity. Note that the gamma-ray source is not in the center of the galaxy.

near the edge of a faint galaxy (Figure I.4). Now, it is possible that the burst source was much closer to us and just happened to be aligned with a more distant galaxy, so this observation was suggestive, but not conclusive.

But then came a burst on May 8 from the direction of the constellation Camelopardalis. In a coordinated international effort, BeppoSAX again took only hours to fix a reasonably precise position, and almost immediately a telescope on Kitt Peak in Arizona was able to catch the visible-light afterglow. Within two days, the largest telescope in the world (the Keck in Hawaii) collected enough light to record a good spectrum. What that spectrum showed was clear evidence that the gamma-ray burst had taken place in a distant galaxy.

In Chapter 25, we described how we can determine the distance to galaxies by measuring their redshifts. The space between the galaxies contains clouds of gas, and these clouds leave tell-tale absorption lines in the spectrum of distant galaxies whose light we see going through them. Since these clouds participate in the expansion of the universe (just as galaxies do), their absorption lines also show a redshift. We can use Hubble's law to tell us how far away the clouds are. The May gamma-ray-burst afterglow spectrum showed absorption features from just such a cloud that was 4 billion light years (LY) from the Sun. Since the light from the burst source passed through the cloud on its way to us, the gamma-ray burst must have occurred at an even greater distance!

■ **FIGURE I.5**

The ROTSE Experiment ROTSE team members Jim Wren (left) and Robert Kehoe pose with the ROTSE-1 camera array that was used to capture the light from the January 23, 1999 gamma-ray burst. The instrument was built inexpensively from telephoto lenses, electronic light detectors, and other parts designed for amateur astronomers.

Astronomers have planned several more satellites to detect gamma ray burst sources. The newest of these is Swift, designed to be agile and quick, like the bird of the same name. It was launched in November 2004, and can relay the positions of bursts within seconds of discovery. Even during its first month of checkout, Swift detected several gamma ray bursts, and many more discoveries are expected.

I.4 NETWORKING TO CATCH MORE BURSTS

Swift and other satellites make it possible to locate gamma-ray bursts precisely, and we know that the afterglows at visible and radio wavelengths last long enough to be observed. But the "window of opportunity" for follow-up observations while the burst sources are brightest is only a few minutes. Therefore, astronomers have established an automated system to notify observers worldwide that a burst has occurred.

Now when one of our orbiting high-energy telescopes discovers a burst, its rough location is immediately transmitted to a *Gamma Ray Burst Coordinates Network,* based at NASA's Goddard Space Flight Center. From there, computers use the Internet to alert observers on the ground within a few seconds to look for the visible-light afterglow. Specially designed automated telescopes lie in wait for an alert from space and then move rapidly, point themselves to the right position, and start snapping images of the sky.

The first success with such a telescope was achieved by a team of astronomers from the University of Michigan and the Livermore and Los Alamos National Laboratories, who designed an automated device they called the *Robotic Optical Transient Search Experiment (ROTSE).* The system demonstrated its capabilities with a dramatic observation on January 23, 1999 (Figure I.5). The Compton Gamma-

Ray Observatory recorded a gamma-ray burst in the constellation of Boötes during the early morning hours (U.S. time). After just 22 seconds, the information reached the ROTSE telescope in Los Alamos, New Mexico, which began to take pictures of the region of the burst. Amazingly, the automated ROTSE images caught the visible-light afterglow still brightening and then starting to fade. The process was so fast that the visible light was actually caught while the 110-second gamma-ray burst was still going on — the first time a burst had been "sighted" in real time.

Armed with an accurate position from the optical information, a team of astronomers from several institutions used the Keck telescope in Hawaii the next night to find that the fading light of the burst was inside a faint galaxy. Taking a spectrum, they found the galaxy to be about 9 billion LY away. On February 8 and 9, the Hubble Space Telescope took an image of the "bursting" object and managed to capture its light even though it had meanwhile faded to some 4 million times fainter than it had been at maximum.

For the light to be detectable from a distance of 9 billion LY, the event that produced the burst must have been astoundingly energetic. If the source was radiating energy in all directions, then the energy released during the burst would have been equivalent to the luminosity of ten million billion (10^{16}) Suns. Other bursts are estimated to have luminosities as high as 10^{20} Suns.

I.5 TO BEAM OR NOT TO BEAM

For a source to flare up to those kinds of luminosities is a real challenge. However, there is one way to reduce the power required of the "mechanism" that makes gamma-ray bursts. So far, our discussion has assumed that the source of the gamma rays behaves "democratically"—like a household light bulb. That is to say, when it "lights up," it can be seen equally well from all directions.

But as we have learned in discussing pulsars in Chapter 22 and quasars in Chapter 26, not all sources of radiation in the universe are like this. Some produce thin beams of radiation or jets of particles that are concentrated into only one or two directions. A laser pointer and a lighthouse on the ocean are examples of such beamed sources on Earth. If, when a burst occurs, the gamma rays come out in only one or two beams, then our estimates of the luminosity of the source can be reduced and the bursts may be easier to explain. In that case, however, the beam has to sweep over the Earth for us to be able to see the burst.

If many sources are "cheap" and put their energy only into beams, then we must also accept that not every source will be pointing toward us. Only a fraction of the bursters will thus be detectable from Earth. This means that there may be even more bursters out there, many of which we will never detect, but which our explanation of what causes the bursts must still account for.

I.6 THE SOURCE OF THE ENERGY

After identifying and following up dozens of gamma-ray bursts, astronomers have begun to piece together clues about what the source of energy is likely to be. Our observations have given us the most important clue—the "smoking gun" for our mystery: The bursts that have visible afterglows are found in star-forming galaxies. If we are really talking about sources billions of light years away, we must look for a mechanism that can produce large amounts of energy very quickly. Most of the suggested mechanisms use the deaths or the "corpses" of massive stars. You will remember from Chapters 22 and 23 that these corpses are either spinning neutron stars or black holes.

Before we tell you about one mechanism for which good evidence is accumulating, we should clarify that not all bursts are necessarily powered by the same mechanism. One crucial distinction may be how long a burst lasts. Astronomers have divided gamma-ray bursts into two categories: short-duration ones (lasting less than 4 seconds) and long-duration ones (lasting more than 4 seconds).

Particular progress has been made in understanding those long-duration bursts that have afterglows at lower energies. In this model, the burst itself is generally thought to be connected with an event at the end of the life of a massive star, but it is an event that only a tiny fraction of massive stars experience. The event astronomers have in mind is the collapse of the core of the star to form a spinning, magnetic black hole. The rest of the star explodes in the way we discussed in Chapter 22. Sometimes such an event is called a *hypernova* to distinguish it from the more routine *supernova*. Because the star corpse is both magnetic and spinning rapidly, its sudden collapse is complex and can produce swirling jets of particles and beams of radiation.

Astronomers have several models about what happens during the collapse process, but all the models assume that a small amount of mass (less than 0.00001 M_{Sun}) is ejected

at speeds close to that of light. The ejected particles are so energetic that collisions among them produce electron–positron pairs—particles of matter and antimatter. Therefore, in a region 100 to 1000 km across (smaller than the diameter of the Earth), there is suddenly a dense sea of radiation, electrons, positrons, neutrons, and protons—a kind of cosmic fireball that simply dwarfs any explosion we can create on Earth.

The expanding blast wave from the fireball soon plows into the interstellar matter in the neighborhood. As the high-speed particles are slowed, they gradually lose energy and emit radiation at longer and longer wavelengths. This accounts for the afterglow of x rays, visible light, and radio waves. Where does the material that the fireball hits come from? Some of it was ejected by the high-mass star during its explosion and in earlier stages of its life. But some of it is cosmic raw material that lives in the neighborhood. Remember that massive stars have short lifetimes and would thus still be embedded in the interstellar matter from which they formed.

What evidence do we have to support this model for the longer gamma-ray bursts? First of all, in several cases we find that the source of the gamma-ray burst is not at the center of its galaxy but farther out. A massive star about to die would not have to be near the galaxy's core; in fact, it would be more likely in regions with raw materials and active star formation. Second, some afterglow spectra appear to have lines of iron and other elements produced in the late stages of evolution of a massive star. Third, and perhaps most intriguing, the light level in some of the burst afterglows does not simply decline with time but levels off or even brightens slightly. If the fireball from the formation of the black hole (acting alone) simply continues to expand after the burst, we would expect its light to decrease with time. But if the rest of the star explodes in a supernova, then its light would soon be added to the afterglow and could keep the light output high for a much longer time.

In some cases, the supernova explosion and the collapse to a black hole occur at the same time. In at least one case, however, the explosion may have occurred before the final collapse. There is one gamma-ray source in which x-ray lines of iron have been observed in the afterglow, and it appears that the iron is located some distance away from the site where the burst gamma rays were produced. One possible explanation is that a massive star first collapsed to become a fast-spinning neutron star, producing a supernova and creating an expanding shell of iron-rich material in the process. The neutron star contains enough mass to become a black hole but does not collapse immediately because it is held back by its very fast rotation. When its rotation gradually slows, about 10 years after the initial supernova explosion, the neutron star finally does collapse into a black hole, producing the gamma-ray burst that then illuminates the surrounding iron-rich shell of gas (Figure I.6).

What about the shorter gamma-ray bursts, for which we do not have afterglows, and which make up about a third of the bursts we have observed? One possibility is that these are produced by the *merger* of two neutron stars, two

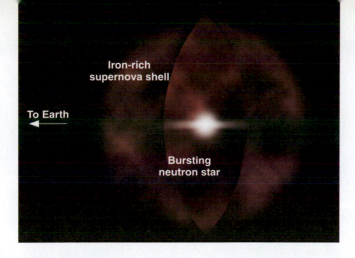

FIGURE I.6

Model of a Gamma-Ray Burst An artist's conception of a two-stage explosion involving a supernova and a gamma-ray burst. First a massive star collapses to form a rapidly rotating neutron star, shedding its shell of gas rich in iron and other heavy elements. Ten years later the neutron star has slowed enough to collapse to a black hole, producing a burst of gamma rays in the process.

black holes, or a neutron star and a black hole (see the image that opens this Interlude). Although systems with two stellar corpses are not common, we presume such double systems will form in every galaxy. Theories suggest that such mergers can indeed produce enormous energy bursts in very short times. But until afterglows from such shorter bursts are detected at other wavelengths, we cannot do a test to pick among our models.

Still, astronomers are pleased that at least the long-duration bursts have been connected with the deaths of massive stars in distant galaxies. Gamma-ray bursts are far brighter than quasars, although only for a short time. The bursts are so bright that they could easily be seen at redshifts that correspond to a few hundred million years after the expansion of the universe began, which is when theorists think that the first generation of stars formed. Some theories predict that the first stars are likely to be massive and complete their evolution in only a million years or so. If this turns out to be the case, then gamma-ray bursts may provide us with the best way of probing the universe when stars and galaxies first began to form.

SURFING THE WEB

 Gamma-Ray Bursts Background Information: imagine.gsfc.nasa.gov/docs/introduction/bursts.html Nontechnical overview of gamma-ray burst observations and models at two different levels of difficulty.

Gamma-Ray Instrument and Experiment Sites:

- *ROTSE Experiment:* www.umich.edu/~rotse Information and pictures about the automated camera system that caught the January 23, 1999, burst. Some parts are a bit technical.

- *HETE Spacecraft:* space.mit.edu/HETE/

- *INTEGRAL Spacecraft:* astro.estec.esa.nl/ SA-general/Projects/Integral/integral.html

- *Swift Spacecraft:* swift.gsfc.nasa.gov/docs/swift/ swiftsc.html (also you can see gamma-ray bursts in real time at: grb.sonoma.edu/)

Hubble Space Telescope Observations of Gamma-Ray Bursts:

- *January 23, 1999, burst:* hubblesite.org/newscenter/archive/1999/09/

- *December 14, 1997, burst:* hubblesite.org/newscenter/archive/1998/17/

- *February 28, 1997, burst:* hubblesite.org/newscenter/archive/1997/30/

SUGGESTIONS FOR FURTHER READING

Katz, J. *The Biggest Bangs.* 2002, Oxford U. Press. A book devoted to gamma-ray bursts by a scientist in the field.

Leonard, P. and Bonnell, J. "Gamma-ray Bursts of Doom" in *Sky & Telescope*, Feb. 1998, p. 28.

Schilling, G. *Flash: The Hunt for the Biggest Explosions in the Universe.* 2002, Cambridge U. Press. By a leading science journalist.

Schilling, G. "Stalking Cosmic Explosions" in *Astronomy*, Feb.

2003, p. 48. Profile of Jan Van Paradijs and the hunt for what the bursts are.

Schilling, G. "Gamma-ray Bursts Caught Holding Supernova Debris" in *Sky & Telescope*, Feb. 2001, p. 22.

Wheeler, J. C. *Cosmic Catastrophes: Supernovae, Gamma-ray Bursts, and Adventures in Hyperspace.* 2000, Cambridge U. Press. See especially Chapter 11. Despite the title, this is a fine introduction to cosmic violence by an astronomer.

Composite

Infrared

Visible

NASA/JPL-Caltech/Z. Wang (Harvard-Smithsonian CfA); Visible: M.Rushing/NOAO

Colliding Galaxies . Collisions and mergers of galaxies strongly influence their evolution. These images show the dramatic merger of two galaxies, NGC 4038 and 4039 (sometimes nicknamed the Antennae). They are located around 68 million light-years away and have been merging for about the last 800 million years. A tremendous burst of star formation has been triggered by this collision, particularly at the site where the two galaxies overlap (the region covered by these pictures.) The two nuclei, or centers, of the merging galaxies can be seen as yellow-white areas, one above the other on all three images.

The lower right panel shows the galaxies in visible light, while an infrared image taken by the Spitzer space telescope is at the upper right. The infrared picture is dominated by clouds of dust heated by newly born stars hidden within them (the red regions); many of them crowd the region between the galaxies, to the left of the two nuclei. These intense star forming regions are buried in the dust and thus not seen in the visible light picture. The main image (at left) is a false-color composite, showing visible-light features as blue and green and infrared features as red. As usual, looking in more than one band of the electromagnetic spectrum can help us learn much more about what is going on in these galaxies.

27

The Evolution and Distribution of Galaxies

*B*ut I want to make sure of our whereabouts and whenabouts," said Van, "it is a philosophical need."

Vladimir Nabokov in *Ada* (1969, Vintage Books)

THINKING AHEAD

Were galaxies the same billions of years ago as they are now? Or can we find evidence that whole galaxies, just like the stars within them, have evolved over time? If so, what determines whether a galaxy will "grow up" to be a spiral or an elliptical? And what is the role of nature versus nurture? That is to say, how much of a galaxy's development is determined by what it looks like when it is born and how much is influenced by its environment?

*S*top and think about the opening quote. As prudent voyagers, we want to know our "whereabouts and whenabouts" in the larger scheme of things. Fortunately, we now have the tools we need to explore the universe almost back to the time it began. The huge new telescopes and sensitive detectors built in the last decade make it possible to obtain both images and spectra of really distant and faint galaxies. Their spectra tell us their redshifts and thus we know how far away they are. The spectra also allow us to measure what their stars are made of and to estimate how old they are from how many heavier elements are present in them.

Nevertheless, the challenges of studying the evolution of galaxies are formidable. First, galaxies are made up of stars, and it was only after astronomers understood the evolution of individual stars that they could begin to explore how whole systems of stars change with time. Fifty years ago, any attempt to

Virtual Laboratories

 Dark Matter

Large Scale Structure

describe the evolution of galaxies would have been pointless: We simply did not know enough about the life histories of stars.

A second difficulty in the study of distant galaxies is that they look very, very dim. Until recently it was extremely hard, even with large telescopes, to determine the shapes of the most remote galaxies. So measuring how they change with time became possible only when our instruments could show us the details of faint galaxies with greater resolution. Today large telescopes on the ground and instruments like the Hubble Space Telescope (HST) in space are finally making such a task possible.

We do have one remarkable asset in studying galactic evolution. As we have seen, the universe itself is a kind of time machine that permits us to observe remote galaxies as they were long ago. For the closest galaxies, the time the light takes to reach us is a few hundred thousand to a few million years. Typically not much changes over times that short. But when we observe a galaxy that is 13 billion light years (LY) away, we are seeing it as it was when the light left it 13 billion years ago. By observing ever more distant objects, we look ever further back toward a time when both galaxies and the universe were young (Figure 27.1).

The distance to a galaxy is derived from its redshift—that is, by how much the lines in its spectrum are shifted to the red because of the expansion of the universe. The conversion of redshift to distance depends on the Hubble constant, which in turn depends on certain properties of the universe, including how much mass it contains. We will describe the currently accepted model of the universe in the next chapter. For the purposes of this chapter, it is enough to know that the current best estimate for the age of the universe is 14 billion years. In that case, if we see an object at a distance of 6 billion LY, we are seeing it as it was when the universe was 8 billion years old. If we see something 13 billion LY away, we are seeing it as it was when the universe was only a billion years old.

Let's begin by exploring what we know about the evolution of galaxies. Then we will look at how galaxies are distributed in space, so that we know what the universe looks like on the large scale. Finally, we will complete our inventory of the contents of the universe by returning to the question of the mysterious dark matter—how much there is, where it is, and what it might be.

27.1 OBSERVATIONS OF DISTANT GALAXIES

Astronomy is one of the few sciences in which all measurements must be made at a distance. Geologists can take samples of the objects they are studying; chemists can conduct experiments in their laboratories to determine what a substance is made of; archeologists can use carbon dating to determine how old something is. But astronomers can't pick up and play with a star or galaxy. If they want to know what galaxies are made of and how they have changed over the lifetime of the universe, they must decode the messages carried by the small number of photons that reach the Earth.

27.1.1 Spectra, Colors, and Shapes

Fortunately (as we hope you have learned in this course) electromagnetic radiation is a rich source of information. In addition to distance, studies of the Doppler shifts of a galaxy's spectral lines can tell us how fast the galaxy is rotating and hence how massive it is. Detailed analysis of such lines can also indicate what types of stars inhabit a galaxy and whether it contains large amounts of interstellar matter.

Unfortunately, many galaxies are so faint that collecting enough light to produce a measurable spectrum is impossible. Astronomers thus have to use a much rougher guide to estimate what kinds of stars inhabit the faintest galaxies—their overall colors. Look again at Figure 27.1 and notice that some of the galaxies are blue and others are red-orange. Now remember that hot luminous blue stars are massive and have lifetimes of only a few million years. If we see a blue galaxy, we know that it must contain many hot luminous blue stars and that star formation must have

R. Windhorst et al. & NASA

■ FIGURE 27.1

Astronomical Time Travel This true-color, long-exposure image, made during 48 orbits of the Earth with the HST, shows a small area in the direction of the constellation of Hercules. We can see galaxies an estimated 3 to 8 billion LY away, including a series of faint blue galaxies that were much more common in that earlier time than they are today. These galaxies appear blue because they are undergoing active star formation and making hot, bright blue stars.

occurred in the past few million years. A yellow or red galaxy, on the other hand, contains mostly old stars, which typically formed billions of years before the light that we now see was emitted.

Another important clue to the nature of a galaxy is its shape. Spiral galaxies can be distinguished from elliptical galaxies by their shape. Spiral galaxies contain young stars and large amounts of interstellar matter, while elliptical galaxies have mostly old stars and very little interstellar matter. Elliptical galaxies turned most of their interstellar matter into stars many billions of years ago, but star formation has continued until the present day in spiral galaxies.

If we can count the number of galaxies of each type during each epoch of the universe, it will help us understand how the pace of star formation changes with time. As we will see later in this chapter, galaxies in the distant universe—that is, young galaxies—look very different from the older galaxies that we see nearby in the present-day universe.

27.1.2 The First Generation of Stars

Since stars are the source of nearly all the light emitted by galaxies, we must learn about the evolution of galaxies by studying the stars within them. What we find is that nearly all galaxies contain at least some very old stars. For example, our own Galaxy contains globular clusters with stars that are at least 13 billion years old, and some may be even older than that. Since a galaxy must be at least as old as the oldest stars in it, the Milky Way must have been born at least 13 billion years ago.

As we will discuss in the next chapter, astronomers have traced the expansion of the universe backward in time and discovered that the universe itself is only about 14 billion years old. Thus it appears that the globular-cluster stars in the Milky Way must have formed not more than a billion years after the expansion began.

Several other observations also establish that star formation in the cosmos began very early. Astronomers have used spectra to determine the composition of some elliptical galaxies that are so far away that the light we see from them left when the universe was only half as old as it is now. Yet these ellipticals contain old red stars, which must have formed billions of years earlier still. Quantitative modeling indicates that star formation in elliptical galaxies began less than a billion years or so after the universe started its expansion, and new stars continued to form for a few billion years. But then star formation apparently stopped. When we compare distant elliptical galaxies with ones nearby, we find that ellipticals have not changed very much since the universe reached about half its current age.

Observations of the most luminous galaxies take us even further back in time. Recently, as we have already noted, astronomers have discovered a few galaxies that are so far away that the light we see now left them only a billion years or so after the beginning (Figure 27.2). Yet the spectra of some of these galaxies contain lines of heavy ele-

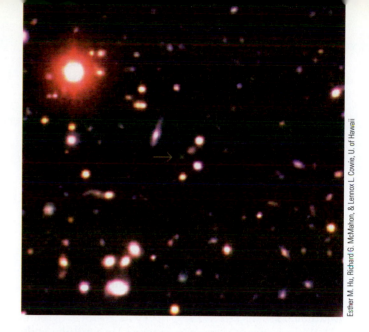

■ FIGURE 27.2

A Very Distant Galaxy This image was taken with the Keck 10-m telescope and shows the field around a luminous galaxy at a distance of about 13 billion LY (indicated by the arrow). Long exposures in the far red and at infrared wavelengths were combined to make the image. The very distant galaxy was detected because it has a strong emission line of hydrogen, which is formed in regions where star formation is taking place.

ments, including carbon, silicon, aluminum, and sulfur. These elements were not present when the universe began but had to be manufactured in the interiors of stars. This means that when the light from these galaxies was emitted, an entire generation of stars had already been born, lived out their lives, and died—spewing out the new elements made in their interiors—even before the universe was a billion years old. And it wasn't just a few stars in each galaxy that got started this way. Enough had to live and die to affect the overall composition of the galaxy, in a way we can still measure in the spectrum from far away.

Observations of quasars support this conclusion. We can measure the abundances of heavy elements in the gas near quasar black holes. The composition of this gas in quasars at distances of about 12.5 billion LY is like that of the Sun. This means that a large portion of the gas surrounding the black holes must already have been cycled through stars during the first 1.5 billion years after the expansion of the universe began. If we allow time for this cycling, then their first stars must have formed when the universe was only a few hundred million years old.

27.1.3 Clues to the Evolution of Galaxies

The observation that all galaxies contain some old stars led to the hypothesis, popular when your authors were in graduate school, that galaxies were born fully formed near the time when the universe began its expansion. It is as if human beings were born as adults and did not have to pass through the various stages of development from infancy through the teens. If this hypothesis were correct, the most distant galaxies should have shapes and sizes very much like the galaxies we see nearby. After they formed, galaxies should change only slowly, as successive generations of stars within them formed, evolved, and died. As the interstellar matter was slowly used up and fewer new stars formed, the galaxies would gradually become dominated by fainter, older stars and look dimmer and dimmer.

Thanks to the HST and the new generation of large, ground-based telescopes, we now know that this picture of galaxies evolving peacefully and in isolation from one another is *completely wrong*. Galaxies in the distant universe do *not* look like the Milky Way and nearby galaxies such as Andromeda.

Why were astronomers so wrong? Up until the early 1990s, the most distant normal galaxy that had been observed was at a distance of 8 billion LY. During the last 8 billion years, many galaxies—and particularly the giant ellipticals, which are the most luminous and therefore the easiest to see at large distances—did evolve peacefully and slowly and in isolation. The HST and other powerful new telescopes that came on line in the 1990s made it possible to pierce the 8-billion-LY barrier, however. We now know

of more than a thousand galaxies at distances greater than 8 billion LY (and some more than 13 billion LY).

Much of the work on the evolution of galaxies has taken place in the Hubble Deep Fields. These are two small regions of the sky, one in the northern hemisphere and the other in the southern hemisphere, where the HST took extremely long exposures in order to detect very faint, very distant, and therefore very young galaxies (Figure 27.3 and the opening image in Chapter 25). To be precise, the HST took many exposures of the same spot in the sky orbit after orbit, and these were carefully added together to produce the deep views we have. Just how hard this kind of research is can be illustrated by the fact that the HST can make images of galaxies that are 100 times fainter than the spectroscopic limits of today's giant ground-based telescopes. That turns out to mean that we can obtain spectra needed to determine redshifts for fewer than 5 percent of the galaxies in these images.

Although we do not have spectra for most of the faint galaxies, the HST is especially well suited to studying their *shapes* because the images taken in space are not blurred by the Earth's atmosphere. To the surprise of astronomers, the distant galaxies did not fit Hubble's classification scheme at all. Remember that Edwin Hubble found that nearly all nearby galaxies could be classified into a few categories, depending on whether they were ellipticals or spi-

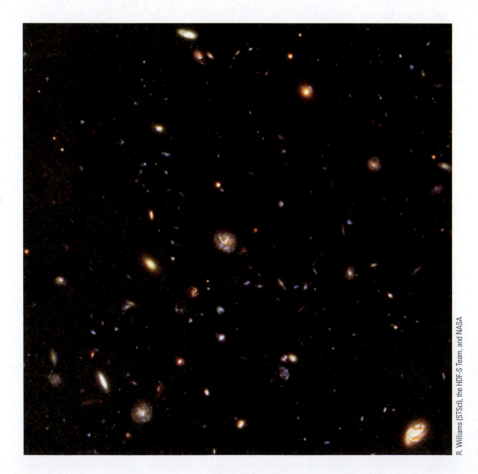

R. Williams (STScI), the HDF-S Team, and NASA

■ **FIGURE 27.3**

Hubble Deep Field South This image is part of the result of a 10-day observation with the HST of a region that is rich in galaxies (located toward the constellation Tucana near the south celestial pole). We see some pinwheel-shaped spiral galaxies, which are like the Milky Way. We also find a variety of peculiar-shaped galaxies that are in collision with companion galaxies. Elliptical galaxies, which contain mostly old stars, appear as reddish blobs.

z = 0.75   z = 0.95

z = 0.96   z = 1.01

z = 1.36   z = 2.01

z = 2.27   z = 2.80

Ferguson, Dickinson, and Williams at STScI

■ **FIGURE 27.4**

Chaotic Shapes of Distant Galaxies   Selected galaxies from the Hubble Deep Field North viewed at visible-light (left) and near-infrared wavelengths. The redshift (z) of each galaxy is shown below its images. Note the peculiar shapes. These galaxies do not fit the Hubble classification scheme of elliptical and spiral galaxies with uncomplicated shapes.

rals. The distant galaxies observed by the HST look very different from present-day galaxies, without identifiable spiral arms, disks, and bulges (Figure 27.4). In other words, the shapes of galaxies have changed with time. In fact, we now know that the Hubble scheme works well only for the last half of the age of the universe. Before then, galaxies were much more chaotic.

It's not just the shapes that are different. Nearly all the galaxies farther than 11 billion LY away—that is, galaxies that are less than 3 billion years old—are extremely blue, indicating that they contain a lot of young stars and that star formation is occurring at a faster rate than in nearby galaxies. Stars that formed more than 11 billion years ago will be very old stars today, and indeed we find mostly old stars in the nuclear bulges of nearby spirals and in elliptical galaxies. When we look at galaxies at distances of 11 to 12 billion LY, we are seeing the ancestors of elliptical galaxies and of the nuclear bulges of spirals.

Another surprise: Very distant galaxies turn out to be systematically smaller on average than nearby galaxies. Relatively few galaxies at distances greater than 8 billion LY have masses greater than $10^{11}\,M_{\text{Sun}}$, one-twentieth the mass of the Milky Way, if we include its dark matter halo. At a distance of 11 billion LY, only a few galaxies have masses greater than $10^{10}\,M_{\text{Sun}}$. What we see instead are fragments of today's galaxies. For example, the gravitationally lensed red object in Figure 26.19, which is 13.4 billion LY away, is only about 500 LY across, compared with the Milky Way's diameter of 100,000 LY. The Milky Way is about 100,000 times more massive than this fragment.

What the observations are showing us is that galaxies have grown in size as the universe has aged. Not only were galaxies smaller several billion years ago, but there were more of them; gas-rich galaxies, particularly the less luminous ones, were much more numerous then than they are today.

Now let's stop and think: What could be the reason that today's galaxies are much larger and fewer in number than the galaxies seen 10 billion or more years ago?

27.2 THE EVOLUTION OF GALAXIES

If you guessed that the answer to the question we just posed is that little galaxies come together to build big galaxies, you are exactly right. (But we made it easy by offering you all the clues in the last section. In real life, it took astronomers several decades to arrive at this conclusion.) One of the most important discoveries in the past few years is that collisions and mergers of whole galaxies play a crucial role in determining how galaxies acquired the shapes and sizes we see today. Only a few of the nearby galaxies are currently involved in collisions, but detailed studies of those tell us what to look for when we seek evidence of mergers in very distant and very faint galaxies. Let's examine in more detail what happens when two galaxies collide.

27.2.1 Mergers and Cannibalism

The figure that opens this chapter shows a very beautiful picture of two galaxies that are colliding. The stars themselves in this pair of galaxies will not be affected much by this cataclysmic event. Since there is a lot of space between the stars, a direct collision between two stars is very unlikely. However, the *orbits* of many of the stars will be changed as the two galaxies move through each other, and the change in orbits can totally alter the appearance of the interacting galaxies. A gallery of interesting colliding galaxies is shown in Figure 27.5. Great rings, huge tendrils of stars and gas, and other complex structures can form. Indeed, these strange

shapes are the signposts that astronomers use to identify colliding galaxies.

■ ■ ■ ■ ■ ■ ■ ■ ■ ■ ■ ■ ■

The details of galaxy collisions are complex and the process can take hundreds of millions of years. Thus collisions are best simulated on a large computer (Figure 27.6). The calculations show that if the collision is slow, the colliding galaxies may coalesce to form a single galaxy. When two galaxies of equal size are involved, we call such an interaction a **merger** (the term applied in the business world to two equal companies that join forces). But small galaxies can also be swallowed by larger ones—a process astronomers have called, with some relish, **galactic cannibalism** (Figure 27.7).

The very large elliptical galaxies we discussed in Chapter 25 probably form by cannibalizing smaller galaxies in their clusters. These "monster" galaxies frequently possess more than one nucleus and have probably acquired their

(a)

K. Borne & NASA

(b)

NOAO

(c)

W. Keel and ESO

(d)

NASA and Hubble Heritage Team (STScI)

■ **FIGURE 27.5**

A Gallery of Interacting Galaxies   (a) The Cartwheel Galaxy, seen in this HST image, is the result of a head-on collision. The spiral galaxy at left collided with another galaxy (probably one of the two at the right), to produce a ring of vigorous star formation (seen in blue). The ring is about 150,000 LY across and contains several billion new stars. (b) NGC 4676 A and B are nicknamed "The Mice." This is a visible-light image that has been computer processed and color coded to show subtle details in the levels of light. You can see the long, narrow tails of stars pulled away from the galaxies by the interactions of the two spirals. (c) NGC 6240 shows two galaxy nuclei quite close together in the center, with tails of material indicating that the two spiral galaxies must have been involved in a collision. Observations with the IRAS have shown that this interacting pair puts out a tremendous amount of energy in the infrared. This is consistent with vigorous star formation in the center warming vast quantities of dust (which hides much of this activity from our view in the visible region of the spectrum). (d) In this HST image, two spiral galaxies are in the process of colliding. The gravitational force exerted by the larger galaxy on the left (NGC 2207) has already distorted the shape of its smaller victim (IC2163). Stars and gas have been flung out into long streamers that extend 100,000 LY toward the right-hand edge of the picture. Billions of years from now the two galaxies will merge to become one.

(a)

(b)

(c)

(d)

Computer images courtesy of Josh Barnes, University of Hawaii

■ **FIGURE 27.6**

Computer Simulation of a Galaxy Collision This computer simulation produces a structure that strongly resembles NGC 4038/39, the pair of interacting galaxies shown in the opening image of this chapter. The sequence shows the galaxies at (a) 60 million years, (b) 185 million years, (c) 310 million years, and (d) 435 million years after the interaction began.

Carl Grillmair (California Institute of Technology) and NASA

■ **FIGURE 27.7**

Galactic Cannibalism This HST image shows the eerie silhouette of dark dust clouds against the glowing nucleus of the elliptical galaxy NGC 1316. Elliptical galaxies normally contain very little dust. These clouds are probably the remnant of a small companion galaxy that was cannibalized by NGC 1316 about 100 million years ago.

■ **FIGURE 27.8**

A Galaxy with Multiple Nuclei   This false-color image shows the distribution of light intensity in the galaxy NGC 6166 in the constellation of Hercules. The three peaks in the central region indicate that this giant elliptical galaxy has probably consumed several companion galaxies. NGC 6166 lies at the center of a galaxy cluster, where there are many potential victims.

unusually high luminosities by swallowing nearby galaxies. The multiple nuclei are the remnants of their victims (Figures 27.8 and 26.14). Many of the peculiar galaxies that we observe also owe their chaotic shapes to past interactions. As we will discuss in the next section, slow collisions and mergers can even transform spiral galaxies into elliptical galaxies.

A change in shape is not all that happens when galaxies collide. If either galaxy contains interstellar matter, that is where the real action takes place. Interstellar gas clouds are large and likely to experience direct impacts with other clouds. These violent collisions compress the gas in the clouds, and the increased density of the gas can increase the rate at which stars are being formed by as much as a factor of 100.

Astronomers call this abrupt increase in the number of stars being formed a *starburst,* and the galaxies in which the increase occurs are termed *starburst galaxies* (Figures 27.9 and 27.10). In some interacting galaxies, star formation is so intense that all the available gas is exhausted in only a few million years; the burst of star formation is clearly only a temporary phenomenon. While a starburst is going on, however, the galaxy where it is taking place becomes much brighter and much easier to detect at large distances.

When astronomers finally had the tools to examine a significant number of galaxies at distances of 11 to 12 billion LY, they found that these very young galaxies resemble nearby starburst galaxies that are involved in mergers. They often have multiple nuclei and peculiar shapes, just as nearby galaxies involved in mergers do. These distant young

■ **FIGURE 27.9**

Starburst Galaxy   Most galaxies form new stars at a fairly slow rate, but members of a rare class known as starburst galaxies blaze with extremely active star formation. The galaxy NGC 3310 is one such starburst galaxy, and this HST image shows that it is forming bright clusters of new stars at a prodigious rate. This galaxy is 59 million LY away and has a diameter of about 52,000 LY. You can see bright blue star clusters in the galaxy's spiral arms, each with as many as a million young stars in it.

The image credit reads vertically: NASA, J. English (U. Manitoba), S. Hunsberger (PSU), Z. Levay (STScI), S. Gallagher (PSU), J. Charlton (PSU)

■ **FIGURE 27.10**

Starburst Associated with Colliding Galaxies Two of the galaxies in the small group known as Stephan's Quintet have recently collided. Like a diamond necklace around the galaxy at the top of this image, and scattered elsewhere in the field, you can see more than 100 bright star clusters containing millions of bright young stars. The ages of the clusters range from 2 million to 1 billion years, suggesting that there have been several different collisions within this group of galaxies, each leading to bursts of star formation. Stephan's Quintet is located at a distance of 270 million LY.

galaxies also have faster rates of star formation than nearby single galaxies, and they contain lots of blue, young stars, as nearby merging galaxies do. The young galaxies may also contain older stars, but the light from any old stars that may be present is completely swamped by the light of the hot, blue O and B stars.

Galaxy mergers in today's universe are rare. Only about 5 percent of nearby galaxies are currently involved in interactions. Interactions were much more common billions of years ago (Figure 27.11). It is not an exaggeration to say that today the typical nearby galaxy is single, but 10 billion years ago the typical galaxy was double. Clearly, interactions of galaxies play a crucial role in their evolution.

Let's now put together all of these clues to trace the life history of galaxies. What follows is the current consensus picture, but research in this field is moving rapidly and some of these ideas may be modified as new observations are made.

27.2.2 The Evolution of Galaxies

Since we see quasars at distances of more than 13 billion LY—that is, when the universe was less than a billion years old—we know that large condensations of matter had begun to form that early. Remember that quasars are powered by black holes and that the mass of the black hole is typically about 0.5 percent of the mass of the surrounding spherical concentration of stars. It seems likely that the black holes and the spherical distribution of ordinary matter around them formed at the same time. Regions of higher density formed larger black holes and more massive concentrations of surrounding matter.

We saw in Chapter 26 that many quasars are found in the centers of elliptical galaxies. This means that some of the first concentrations of matter must have evolved into the elliptical galaxies that we see in today's universe. There have been two models for how this evolution took place. The first, suggested in the 1960s, proposes that elliptical galaxies formed in a single, rapid collapse of gas, that virtually all of the gas was turned quickly into stars, and that the galaxies have subsequently changed slowly as the stars evolve. The alternative model proposes that today's giant ellipticals were formed mostly through mergers of smaller galaxies that had already converted at least some of their gas into stars. In other words, astronomers are debating whether giant ellipticals formed most of their stars in the large galaxy that we see today or in separate small galaxies that subsequently merged.

Since we see some luminous quasars at distances of 13 billion LY or more, it is likely that a few giant ellipticals formed very early through the collapse of a single cloud. However, as we write this, the best evidence seems to show that mature giant elliptical galaxies were rare before the universe was about 6 billion years old and that they are much more common today than when the universe was young. This observation strongly favors the hypothesis that most giant ellipticals were formed through mergers of smaller galaxies.

Observations indicate that most of the gas in elliptical galaxies was converted to stars by the time the universe was about 3 billion years old. Certainly, starbursts caused by mergers would have helped turn gas into stars more quickly. While slow star formation continued for a while after that, it appears that elliptical galaxies have not formed many new stars in the past 8 billion years.

The situation with spiral galaxies is very different. The nuclear bulges of these galaxies formed early, like the elliptical galaxies (Figure 27.12). However, the disks formed later (remember that the stars in the disk of the Milky Way

Pieter van Dokkum, Marijn Franx (U. of Groningen / Leiden), ESA, NASA

■ FIGURE 27.11

Collisions of Galaxies in a Distant Cluster   The large picture on the left shows the HST image of a cluster of galaxies at a distance of about 8 billion LY. Among the 81 galaxies that have been examined in some detail, 13 are the result of recent collisions of pairs of galaxies. The eight smaller images on the right are close-ups of some of the colliding galaxies. The merger process typically takes a billion years or so.

are younger than the stars in the bulge and the halo) and still contain gas and dust. Since the rate of star formation depends on the density of interstellar gas and dust, star formation in the less densely packed disks of spiral galaxies has continued to the present day. However, the rate of star formation today is about ten times lower than it was 8 billion years ago. The number of stars being formed drops as the gas is used up.

The shapes of spiral galaxies have also continued to change until recent time (cosmologically speaking). We now estimate that spirals with well-defined, long spiral arms developed only in the last 4 billion years or so. The ancestors of today's spiral galaxies were much more likely to be involved in mergers earlier than about 4 billion years ago, and these interactions probably prevented the formation of long-lived spiral arms.

Why did ellipticals form stars at a much faster rate than spiral galaxies did? Astronomers are still working on theories of galaxy formation, and so we have only partial answers. Key factors seem to be a galaxy's rotation and density. Ellipticals rotate more slowly than spirals and their mass densities are high—both factors that favor more rapid star formation.

27.3 THE DISTRIBUTION OF GALAXIES IN SPACE

In the preceding section, we emphasized the role of mergers in shaping the evolution of galaxies. In order to collide, galaxies must be fairly close together. To estimate how often collisions occur, astronomers need to know how galaxies are distributed in space. Are most of them isolated from one another or do they congregate in groups? If so, how large are the groups? And how, in general, are galaxies and their groups arranged in the cosmos? Are there as many in one direction of the sky as in any other, for example?

RAPID COLLAPSE

1. Primordial hydrogen cloud.

2. Cloud collapses under gravity.

3. Large bulge of ancient stars dominates galaxy.

ENVIRONMENTAL EFFECTS

1. Disk galaxy and companion.

2. Smaller galaxy falls into disk galaxy.

3. Bulge inflates with addition of young stars and gas.

NASA

■ **FIGURE 27.12**

Growth of Spiral Bulges   The nuclear bulges of some spiral galaxies formed through the collapse of a single protogalactic cloud (top row). Others grew over time through mergers with other smaller galaxies (bottom row).

Edwin Hubble found answers to some of these questions only a few years after he first showed that other galaxies exist. As he examined galaxies all over the sky, Hubble made two discoveries that are crucial for studies of the evolution of the universe.

27.3.1   The Cosmological Principle

Hubble made his observations with what were then the world's largest telescopes—the 100-in. and 60-in. reflectors on Mount Wilson. These telescopes have small fields of view; they can only see a small part of the heavens at a time. To photograph the entire sky with the 100-in. telescope, for example, would take thousands of years. So instead Hubble sampled the sky in many regions, much as Herschel did with

his star gauging (see Section 24.1). In the 1930s, Hubble photographed 1283 sample areas, and on each print he carefully counted the numbers of galaxy images.

The first discovery Hubble made was that the number of galaxies visible in each area of the sky is about the same. (Strictly speaking this is true only if the light from distant galaxies is not absorbed by dust in our own Galaxy, but Hubble made corrections for this absorption.) He also found that the numbers of galaxies increase with faintness, as we would expect if the density of galaxies is about the same at all distances from us.

To understand what we mean, imagine you are taking snapshots in a crowded stadium during a sold-out concert. The people sitting near you look big, so only a few of them will fit into a photo. But if you focus on the people sitting in seats on the far side of the stadium, they look so small that many more will fit into your picture. If all parts of the stadium have the same seat arrangements, then as you look farther and farther away, your photo will get more and more crowded with people. In the same way, as Hubble looked at fainter and fainter galaxies, he saw more and more of them.

Hubble's findings are enormously important because they indicate that the universe is both **isotropic** and **homogeneous:** It looks the same in all directions, and a large volume of space at any given redshift or distance is much like any other volume at that redshift. If that's so, it doesn't matter what section of the universe we observe (as long as it's a sizable portion)—any section will look the same as any other.

In other words, Hubble's results suggest not only that the universe is about the same everywhere, apart from changes with time, but also that the part we can see around us, aside from small-scale local differences, is representative of the whole. The idea that the universe is the same everywhere is called the **cosmological principle** and is the starting assumption for nearly all theories that describe the entire universe (see the next chapter).

Without the cosmological principle, we could make no progress at all in studying the universe. Suppose our own local neighborhood were unusual in some way. Then we could no more understand what the universe is like than if we were marooned on a warm south-sea island without outside communication and were trying to understand the geography of the Earth. From our limited vantage point, we could not know that some parts of the planet are covered with snow and ice, or that large continents exist with a much greater variety of terrain than that found on our island.

Hubble merely counted the numbers of galaxies in various directions without knowing how far away most of them were. With modern instruments, astronomers have measured the velocities and distances of thousands of galaxies, and so built up a meaningful picture of the large-scale structure of the universe. In the rest of this section we describe what we know about the distribution of galaxies, beginning with those that are nearby.

27.3.2 The Local Group

The region of the universe for which we have the most detailed information is, as you would expect, our own local neighborhood. It turns out that the Milky Way Galaxy is a member of a small group of galaxies called, not too imaginatively, the **Local Group.** It is spread over about 3 million LY and contains more than 40 members. There are three large spiral galaxies (our own, the Andromeda Galaxy, and M33), two intermediate ellipticals, and many dwarf ellipticals and irregular galaxies. A partial list of members of the Local Group is given in Appendix 12, and from that list you can get an idea of what these galaxies are like.

New members of the Local Group are still being discovered. We mentioned in Chapter 24 that a dwarf galaxy only about 80,000 LY from the Earth and about 50,000 LY from the center of the Galaxy was recently found in the constellation of Sagittarius. (This dwarf is actually venturing too close to the much larger Milky Way and will eventually be consumed by it.) Several new dwarf galaxies have also been found near the Andromeda Galaxy. Such dwarf galaxies are difficult to find because they typically contain relatively few stars and it is hard to distinguish them from the foreground stars in our own Milky Way.

Figure 27.13 is a rough sketch showing where the brighter members of the Local Group are located. The average of the motions of all the galaxies in the group indicates that its total mass is about $5 \times 10^{12} \, M_{Sun}$—about two to three times the mass of the Milky Way. A substantial amount of this mass is in the form of dark matter.

27.3.3 Neighboring Groups and Clusters

Small groups like ours are hard to notice at larger distances. Fortunately, there are much more substantial groups called **galaxy clusters.** Such clusters are described as *poor* or *rich* depending on how many galaxies they contain. Rich clusters have thousands or even tens of thousands of galaxies, although many of them are faint and hard to detect.

The nearest moderately rich galaxy cluster is called the Virgo cluster, after the constellation in which it is seen. It is about 50 million LY away and contains thousands of members, of which a few are shown in Figure 27.14. The giant elliptical (and very active) galaxy M87, which you came to know and love in the previous chapter, belongs to the Virgo cluster. Although M87 is not shown in Figure 27.14, two other giant ellipticals in the cluster are.

A good example of a cluster that is much larger than the Virgo complex is the Coma cluster, with a diameter of at least 10 million LY (you can see it in the opening figure of our Prologue). About 250 to 300 million LY distant, this cluster is centered on two giant ellipticals whose luminosities equal about 400 billion Suns. Thousands of galaxies have been observed in Coma, but the galaxies we see are almost certainly only part of what is really there. Dwarf galaxies are too faint to be seen at the distance of Coma, but we expect they are part of this cluster just as they are part of nearer ones. If so, then Coma likely contains tens of thousands of galaxies. The total mass of this cluster is about $4 \times 10^{15} \, M_{Sun}$ (enough mass to make 4 million billion stars like the Sun!).

Let's pause here for a moment of perspective. We are now discussing numbers that overwhelm even astronomers. The Coma cluster may have 10, 20, or 30 thousand galaxies, and each galaxy has billions and billions of stars. If you were traveling at the speed of light, it would still take you more than 10 million years (longer than the history of the human species) to cross this giant swarm of galaxies. And if you lived on a planet on the outskirts of one of these galaxies, many other members of the cluster would be close enough to be noteworthy sights in your nighttime sky.

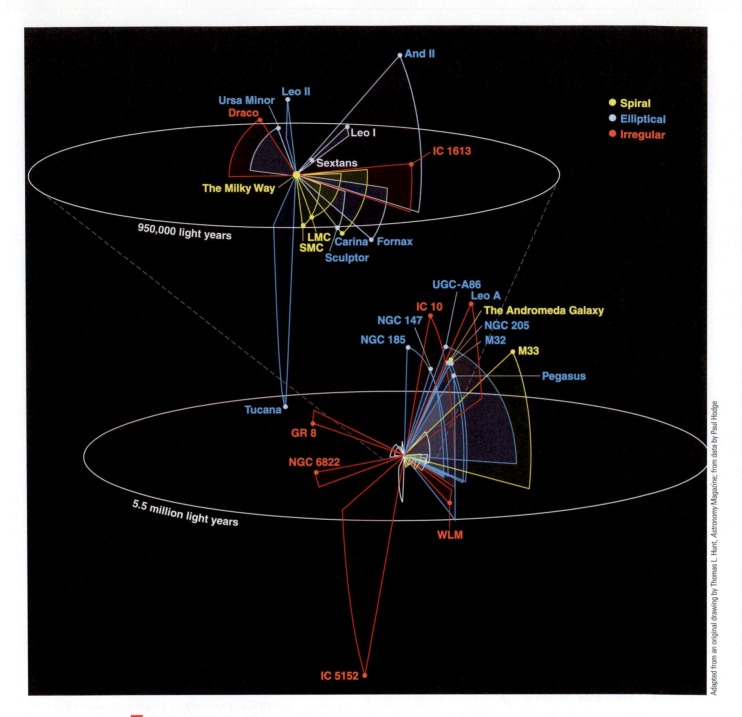

■ FIGURE 27.13

The Local Group This is a three-dimensional view of some members of the Local Group of galaxies, with our Milky Way at the center. The exploded view at the top shows the region closest to the Milky Way. The three largest galaxies among the three dozen or so members of the Local Group are all spirals; the others are small irregular galaxies and dwarf ellipticals. A number of new members of the group have been found since this map was made.

Really rich clusters such as Coma usually have a high concentration of galaxies near the center. We can see giant elliptical galaxies in these central regions but few if any spiral galaxies. The spirals that do exist are generally on the outskirts of clusters. We might say that ellipticals are highly "so-cial": They are often found in groups and very much enjoy "hanging out" with other ellipticals in crowded situations. It is precisely in such crowds that collisions are most likely, and as we have discussed earlier, we think that most large ellipticals are built through mergers of smaller galaxies. Spirals,

■ FIGURE 27.14

Central Region of the Virgo Cluster
Virgo is the nearest rich cluster and is about
50 million LY away. It contains hundreds of
bright galaxies. In this picture you can see
only part of the cluster, including two giant
elliptical galaxies, M84 and M86, plus a
number of spirals.

on the other hand, are more "shy": They are more likely to
be found in poor clusters or on the edges of rich clusters
where collisions are less likely to disrupt the spiral arms or
strip out the gas needed for continued star formation.

27.3.4 Superclusters and Voids

After astronomers discovered clusters of galaxies, they naturally wondered whether there were still larger structures
in the universe. Do the clusters of galaxies gather together?
To answer this question, we must be able to map large parts
of the universe in three dimensions. We must know not
only the position of a galaxy *on* the sky (that's two dimensions) but also its distance from us (the third dimension).

This means we must be able to measure the redshift of
each galaxy in our map. Taking a spectrum of each individual galaxy to do this is a much more time-consuming task
than simply counting galaxies seen in different directions
on the sky, as Hubble did. Today, astronomers have found

ways to get the spectra of many galaxies in the same field of
view (sometimes hundreds at a time) to cut down the time
it takes to finish their maps.

Another challenge astronomers faced in deciding how
to go about constructing a map of the universe is similar to
the task confronted by the first team of explorers in a huge,
uncharted territory on Earth. Since there is only one band
of explorers and an enormous amount of land, they have to
make choices about where to go first. One strategy might
be to strike out in a straight line in order to get a sense of
the terrain. They might, for example, cross some mostly
empty prairies and then hit a dense forest of trees. As they
make their way through the forest, they learn how thick it
is in the direction they are traveling, but not its width to
their left or right. Then a river crosses their path; as they
wade across, they can measure its width but learn nothing
about its length. Still, as they go on in their straight line, they
begin to get some sense of what the landscape is like and
can make at least part of a map. Other explorers, striking

Margaret Geller: Cosmic Surveyor

Born in 1947, Margaret Geller is the daughter of a chemist who encouraged her interest in science and helped her visualize the three-dimensional structure of molecules as a child. (It was a skill that would come in very handy for visualizing the three-dimensional structure of the universe.) She remembers being bored in elementary school but was encouraged to read on her own by her parents. Her recollections also include subtle messages from teachers that mathematics (her strong early interest) was not a field for girls, but she did not allow herself to be deterred.

Geller obtained a B.A. in physics from the University of California at Berkeley and became the second woman to receive a PhD in physics from Princeton. There, while working with James Peebles, one of the world's leading cosmologists (see Chapter 28), she became interested in problems relating to the large-scale structure of the universe. In 1980 she accepted a research position at the Harvard–Smithsonian Center for Astrophysics, one of the most dynamic institutions for astronomy research in the country. She saw that to make progress in understanding how galaxies and clusters are organized, a far more intensive series of surveys was required. Although it would not bear fruit for many years, Geller and her collaborators began the long, arduous task of mapping the galaxies.

Her team was fortunate to be given access to a telescope that could

Dr. Margaret Geller

be dedicated to their project, the 60-in. reflector on Mount Hopkins, near Tucson, Arizona, where they and their

■ ■ ■ ■ ■ ■ ■ ■ ■ ■ ■ ■ ■

Geller and her team point their telescope at a determined position in the sky and then let the rotation of the Earth bring new galaxies into view.

■ ■ ■ ■ ■ ■ ■ ■ ■ ■ ■ ■ ■

assistants continue to take spectra to determine galaxy distances. To get a slice of the universe, they point their telescope at a determined position in

the sky and then let the rotation of the Earth bring new galaxies into their field of view. In this way they have measured the positions and redshifts of more than 18,000 galaxies and made a wide range of interesting maps to display their data. Their surveys now include "slices" in both the Northern and Southern Hemispheres.

As news of her important work spread beyond the community of astronomers, Geller received a MacArthur Foundation Fellowship in 1990. These fellowships, popularly called "the MacArthur genius awards," are designed to recognize truly creative work in a wide range of fields. Geller continues to have a strong interest in visualization and has (with filmmaker Boyd Estus) made several award-winning videos explaining her work to nonscientists (one is entitled *So Many Galaxies, So Little Time!*). She has appeared on a variety of national news and documentary programs, including the *MacNeil/Lehrer Newshour, The Astronomers,* and *The Infinite Voyage.* Energetic and outspoken, she has given talks on her work to many audiences around the country and works hard to find ways to explain the significance of her pioneering surveys to the public.

"It's exciting to discover something that nobody's seen before. [To be] one of the first three people to ever see that slice of the universe . . . [was] sort of being like Columbus. . . . Nobody expected such a striking pattern!"—Margaret Geller

out in other directions, will someday help fill in the remaining parts of that map.

Astronomers must make the same sort of choices. We cannot explore the universe in every direction: There are far too many galaxies and far too few telescopes (and graduate students) to do the job. But we can pick a single direction or a small slice of the sky and start mapping the galaxies. Margaret Geller (see the Voyagers in Astronomy box), John Huchra, and their students at the Harvard–Smithsonian Center for Astrophysics pioneered this technique, and several other groups have extended their work to cover larger volumes of space. One survey, called the Sloan Digital Sky Survey, is profiled in the Making Connections box for this chapter.

Astronomy and Technology: The Sloan Digital Sky Survey

In Edwin Hubble's day (and even when the authors of your text were first learning astronomy), spectra of galaxies had to be taken one at a time. The faint light of a distant galaxy gathered by a large telescope was put through a slit, and then a spectrometer was used to separate the colors and record the spectrum. This is a laborious process, ill-suited to the demands of making large-scale maps that require the redshifts of many thousands of galaxies.

But new technology has come to the rescue of astronomers who would like to see good three-dimensional maps of the universe of galaxies before they retire. The most ambitious survey of the sky is now under way in the Lincoln National Forest atop the Sacramento Mountains of New Mexico. Called the Sloan Digital Sky Survey, after the foundation that has provided a large part of the funding, the program uses a 2.5-m telescope (about the same aperture as the HST) as a wide-angle astronomical camera. During a projected multiyear mapping program, astronomers are using electronic light detectors to take images of more than 100 million objects, covering one quarter of the celestial sphere. Like many large projects in modern science, the Sloan Survey involves scientists and engineers from many different institutions, ranging from universities to national laboratories.

When the night is crisp and clear, astronomers use the instrument to make images recording the position and brightness of celestial objects in long strips of the sky. The information in each strip is digitally recorded as a record for future generations. But when the seeing (recall this term from Chapter 5) is only adequate, the telescope is used for taking spectra of galaxies and quasars—it does so for as many as 640 objects *at a time!*

The key to the success of the project is a series of *optical fibers*, thin tubes of flexible glass that can channel light from a source to an electronic chip that records the spectrum. After taking images of a part of the sky and identifying which objects are galaxies, project scientists drill an aluminum plate with holes for attaching fibers at the location of each galaxy. The telescope is then pointed at the right section of the sky, and the fibers lead the light of each galaxy to the spectrometer for individual recording.

About an hour is sufficient for each set of spectra, and the predrilled aluminum plates can be switched quickly. Thus it is possible to take as many as 5000 spectra in one night (provided the weather cooperates).

After the survey is completed, we will have spectra (and thus redshifts and distances) for nearly a million galaxies and 100,000 quasars. Already, studies of the data obtained to date have led to important discoveries, including quasars at distances so large we are seeing them as they were when the universe was only 5% as old as it is now; several gravitationally-lensed quasars; a faint dwarf galaxy in the Local Group; and several remnants of small galaxies cannibalized by our own Milky Way.

The information recorded by the Sloan Survey staggers the imagination. The data come in at 8 megabytes per second (this means 8 million individual numbers or characters every second). Over the course of the project, scientists expect to record more than 15 terabytes or 15 thousand billion bytes, which they estimate is comparable to the information contained in the Library of Congress! (And you think you have a lot to absorb for your astronomy final.) Organizing and sorting this volume of data are a formidable challenge, even in our information age.

Fermilab Visual Media Services

The Sloan Digital Sky Survey Telescope at Apache Point Observatory is seen here with the protective building rolled off (note the tracks on either side of the white fence) and the boxy wind baffle protecting the telescope structure.

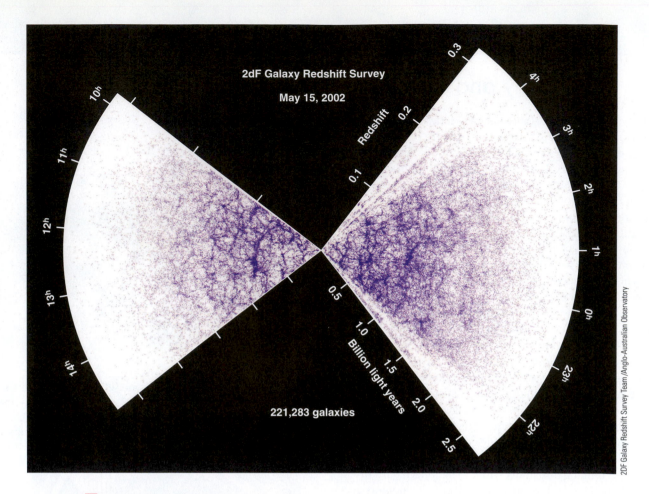

2dF Galaxy Redshift Survey

May 15, 2002

221,283 galaxies

2dF Galaxy Redshift Survey Team./Anglo-Australian Observatory

■ **FIGURE 27.15**

Two Slices of the Universe   This map shows the distribution of galaxies in space as seen from our location, which is in the middle where the slices come together. The figure includes points for 221,283 galaxies for which redshifts have been measured. Note the concentration of galaxies in narrow bands or lanes, with large voids (some as wide as 150 million LY) between them. Some of the longer filamentary concentrations of galaxies extend over a distance of several hundred million light years. We would obtain a similar distribution if we took a slice through a sponge.

A plot of the galaxies measured by the largest of these surveys, which was conducted at the Anglo-Australian Observatory, is shown in Figure 27.15. To the surprise of astronomers, maps like the one in the figure showed that clusters of galaxies are not arranged uniformly throughout the universe, but are found in huge filamentary **superclusters** that look like great arcs of inkblots splattered across a page. The superclusters resemble an irregularly torn sheet of paper or a pancake in shape—they can extend for hundreds of millions of light years in two dimensions but are only 10 to 20 million LY thick in the third dimension. Detailed study of some of these structures shows that their masses are greater than $10^{16} M_{Sun}$, which is 10,000 times more massive than the Milky Way Galaxy.

Separating the filaments and sheets are the **voids,** which look like huge empty bubbles walled in by the great arcs of galaxies. They have typical diameters of 150 million

LY, with the clusters of galaxies concentrated along their walls. The whole arrangement of filaments and voids reminds us of a sponge, the inside of a honeycomb, or a hunk of Swiss cheese with very large holes. If you take a good slice or cross section through any of these, you will see something that looks roughly like Figure 27.15.

Before these voids were discovered, most astronomers would probably have predicted that the regions between giant clusters of galaxies were filled with many small groups of galaxies, or even with isolated individual galaxies. Careful searches within these voids have found few galaxies of any kind. Apparently, 90 percent of the galaxies occupy less than 10 percent of the volume of space.

At this point, if you have been thinking about our discussions of the expanding universe in Chapter 25, you may be wondering what exactly in Figure 27.15 is expanding. We know that the galaxies and clusters of galaxies are held together by gravity and do not expand as the universe does. However, the voids do grow larger and the filaments move farther apart as space stretches.

After learning *how* galaxies are distributed, astronomers began to try to explain *why* they are distributed that way. The existence of such large filaments of galaxies and voids is a puzzle because we have evidence (to be discussed in the next chapter) that the universe was extremely smooth a few hundred thousand years after it formed. The challenge for theoreticians is to understand how a featureless universe changed into the complex and lumpy one that we see today.

27.4.1 Forming Voids and Filaments

For example, confronted by data such as that plotted in Figure 27.15, which shows that galaxies are distributed on the walls of bubble-like structures, some astronomers argued that matter used to be inside the bubbles and then was somehow cleaned out and pushed toward the walls. But how could this happen? One possibility is that there were giant explosions during the first few billion years of the universe that swept the gas into bubble-like shells, and that galaxies subsequently formed in those shells. Unfortunately, despite considerable ingenuity, astronomers have yet to devise a way to produce sufficiently energetic explosions to account for the largest voids.

Observations with the HST have found evidence for a massive supercluster at a distance of about 8 billion LY (Figure 27.16), which means that large concentrations of galaxies had already come together when the universe was less than half as old as it is now. Some theorists argue that these concentrations formed much earlier—when the universe was only a few hundred thousand years old. At that time, no galaxies were yet present, and so if this idea is right, then the initial concentrations must have consisted of great quantities of gas and probably of dark matter.

As we shall see in the next chapter, when the universe was a few hundred thousand years old, *everything* was at a temperature of a few thousand degrees. Theorists suggest that at this early time, all of the hot gas was pulsing, much like sound waves set up pulses in the air of a night club with an especially loud band. This pulsing could have concentrated matter into high-density peaks and created emptier spaces between them. When the universe cooled, the concentrations of matter were "frozen in," and galaxies ultimately formed from the matter in these high-density regions.

If all this structure exists, you may want to ask whether the universe can really be described as homogeneous and isotropic, as we claimed earlier. The answer is probably yes,

Megan Donahue (STScI)/NASA

■ FIGURE 27.16

Distant Cluster of Galaxies   This HST image shows a cluster of galaxies at a distance of 8 billion LY. The cluster contains thousands of galaxies, and its mass is several thousand times larger than the mass of the Milky Way Galaxy. It's amazing to have images like this, showing us light that left its galaxies before the Sun and the Earth ever formed.

provided we consider regions of the universe large enough to include a number of superclusters and voids. This is similar to making a meaningful survey of small towns and wide-open spaces in the American Midwest. If you live in a town and look only at the houses nearest you, your perspective will be biased. Similarly, if you live on an isolated farm and look only at the nearest few acres, you will miss all the things that happen exclusively in urban areas. A representative survey must catch enough people or territory to include cities, villages, farms, and open country.

27.4.2 The Big Picture

Let's put all these ideas together to sketch how the universe came to look the way it does. Initially, as we said, the distribution of matter was nearly—but not quite—smooth and uniform. That "not quite" is the key to everything. Here and there were regions where the density was ever so slightly higher than average (some of this increased density was provided by dark matter). As in the case of star formation, the fate of these higher-density regions depended on the balance between pressure (the outward push of expansion) and the inward pull of gravity.

Initially, each individual region expanded because the whole universe was expanding. However, as the universe continued to expand, the regions of higher density acquired still more mass because they exerted a slightly larger than average gravitational force on surrounding material. If the inward pull of gravity was high enough, the individual region ultimately stopped expanding. It then began to collapse into an irregularly shaped blob. In each region the collapse was more rapid in one direction, and so the concentrations of matter were not spherical but came to resemble giant pancakes.

These pancakes occurred throughout the early universe, oriented in different directions and collapsing at different rates. The pancakes provided the framework for the filamentary and bubble-like structures that we see preserved in the universe today.

The universe then proceeded to "build itself" from the bottom up. Within the pancakes, smaller structures formed first and then merged to build larger ones. The first dense concentrations of matter that collapsed in the pancakes were the size of small galaxies or globular clusters. These fragments then gradually assembled to build clusters and ultimately superclusters of galaxies.

According to this picture, small galaxies and large star clusters first formed in the highest density regions of all—where the pancakes intersect—when the universe was about 2 percent of its current age. Some stars may have formed even before the first star clusters and galaxies came into existence.

Clusters of galaxies then formed as individual galaxies congregated, drawn together by their mutual gravitational attraction (Figure 27.17). First a few galaxies came together to form groups, much like our own Local Group. Then the groups began combining to form clusters and eventually superclusters. This model predicts that clusters and super-clusters should still be in the process of gathering together, and observations do in fact suggest that clusters are still gathering up their flocks of galaxies.

Most giant elliptical galaxies formed through the collision and merger of many smaller fragments. A collision of two systems of gas and stars tends to stir up the orbits of the individual stars within each system and also to strip away from the system's outer regions any matter that is not held as strongly by gravity as the inner material. The result of many collisions is thus to build round systems that lack extended disks. Some spiral galaxies formed in relatively isolated regions from a single cloud of gas, which collapsed to make a flattened disk. Others acquired additional stars through collisions, and these stars now populate their halos and nuclear bulges. As we have seen, our own Milky Way is still capturing small galaxies and adding them to its halo.

27.5 A UNIVERSE OF (MOSTLY) DARK MATTER?

So far this chapter has focused almost entirely on matter that radiates electromagnetic energy. But, as we have pointed out in several earlier chapters, it is now clear that galaxies contain large amounts of dark matter as well. There is much more dark matter, in fact, than matter we can see—which means it would be foolish to ignore the effect of this unseen material in our theories about the structure of the universe. (As many a ship captain in the polar seas found out too late, the part of the iceberg visible above

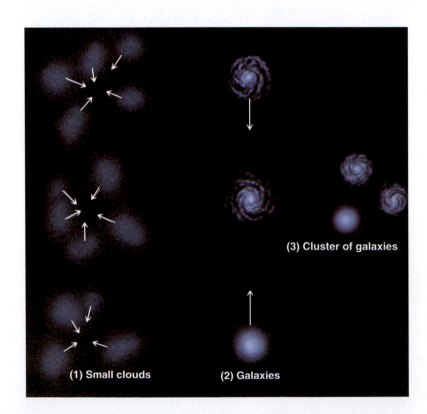

(3) Cluster of galaxies

(1) Small clouds

(2) Galaxies

■ FIGURE 27.17

Bottom-up Formation of Galaxies This schematic diagram shows how galaxies might have formed if small clouds formed first and then congregated to form galaxies and then clusters of galaxies.

the ocean's surface was not necessarily the only part he needed to pay attention to.) The dark matter is extremely important in determining the evolution of galaxies and of the universe as a whole.

The idea that much of the universe is filled with dark matter may seem bizarre, but we can cite a historical example of "dark matter" much closer to home. In the mid-19th century, measurements showed that the planet Uranus did not follow exactly the orbit predicted if one added up the gravitational forces of all the known objects in the solar system. Its orbital deviations were attributed to the gravitational effects of an (at the time) invisible planet. Calculations showed where that planet had to be, and Neptune was discovered just about in the predicted location.

In the same way, astronomers are now trying to determine the location and amount of dark matter in galaxies by measuring its gravitational effects on objects we can see. And, by measuring the way that galaxies move in clusters, scientists are discovering that dark matter may play an important role in galaxy evolution as well. It appears that dark matter makes up most of the matter in the universe. Let's look at the search for dark matter and the quest to determine what it might be made of.

27.5.1 Dark Matter in the Local Neighborhood

The first place we might look for dark matter is in our own solar system. Astronomers have examined the orbits of the known planets and of spacecraft as they journey to the outer planets and beyond. No deviations have been found from the orbits predicted on the basis of the objects already discovered in our solar system, and so no evidence exists for large amounts of nearby dark matter.

Astronomers have also looked for evidence of dark matter in the region of the Milky Way Galaxy that lies within a few hundred light years of the Sun. In this vicinity most of the stars are restricted to a thin disk. It is possible to calculate how much mass the disk must contain in order to keep the stars from wandering far above or below it. The total matter that must be in the disk is less than twice the amount of luminous matter. This means that no more than half the mass near the Sun is dark.

27.5.2 Dark Matter Around Galaxies

In contrast with our local neighborhood, there is (as we saw in Chapter 24) evidence suggesting that 90 percent of the mass in the entire Galaxy may be in the form of a halo of dark matter. In other words, there is apparently about nine times more dark matter than visible matter. The stars in the outer region of the Milky Way are revolving very rapidly around its center. The mass contained in all the stars and all the interstellar matter in the Galaxy does not exert enough gravitational force to explain how those stars remain in their orbits and do not fly away. Only by having large

Wm. Keel & the Astronomical Society of the Pacific

■ FIGURE 27.18

Rotation Indicates Dark Matter   We see the edge-on spiral galaxy NGC 5746, with a graph that shows the velocity of rotation at a series of points spanning the width of the galaxy. As is true of the Milky Way, the speed of rotation does not decrease with distance from the center. This indicates the presence of a halo of dark matter outside the boundary of the luminous matter. This dark matter causes the outer regions to rotate faster than the observed matter alone could explain.

amounts of unseen matter could the Galaxy be holding on to all those fast-moving outer stars. The same result is found for other spiral galaxies as well.

Analyzing the rotation of spiral galaxies suggests that the dark matter is found in a large halo surrounding the luminous parts of each galaxy (Figure 27.18). The radius of this halo may be as large as 300,000 LY, much larger than the visible size of these galaxies.

27.5.3 Dark Matter in Clusters of Galaxies

Galaxies in clusters also move around; they orbit the cluster's center of mass. It is not possible for us to follow a galaxy around its entire orbit. For example, it takes 10 billion years or longer for the Andromeda and Milky Way Galaxies to complete a single orbit around each other. It is possible, however, to measure the velocities with which galaxies in a cluster are moving and then estimate what the total mass in the cluster must be to keep the individual galaxies from flying off into space. The observations indicate that the total amount of dark matter in clusters probably exceeds the mass within the galaxies themselves, indicating that dark matter exists *between* galaxies as well as inside them.

As we saw, the universe is expanding, but this expansion is not perfectly uniform—thanks to the interfering hand of gravity. Suppose, for example, that a galaxy lies outside but relatively close to a rich cluster of galaxies. The gravitational force of the cluster will tug on that neighboring galaxy and slow down the rate at which it moves away from the cluster due to the expansion of the universe.

Consider the Local Group of galaxies, lying on the outskirts of the Virgo supercluster. The mass concentrated at the center of the Virgo cluster exerts a gravitational force on the Local Group. As a result, the Local Group is moving away from the center of the Virgo cluster at a velocity a few hundred kilometers per second slower than the Hubble law predicts. By measuring such deviations from a smooth expansion, astronomers can estimate the total amount of mass contained in large clusters.

Astronomers have now measured accurate distances and velocities for thousands of galaxies within about 150 million LY of the Milky Way. Superimposed on the more local motions we find a new and surprising trend. These galaxies tend to be flowing toward an enormous concentration of mass, which has been named the *Great Attractor*. The mass of the Great Attractor is estimated to be $3 \times 10^{16} M_{Sun}$, equivalent to tens of thousands of galaxies. This mass is much larger than the amount of luminous matter seen in this direction, and so most of the matter in the Great Attractor must be dark. Don't let the fun name astronomers gave this concentration of mass impress you too much, however. There are likely to be many other such regions (as there are many emptier voids) in the vastness of the universe.

27.5.4 Mass-to-Light Ratio

Section 25.3 described the use of the mass-to-light ratio to characterize the matter in galaxies or clusters of galaxies. For systems that contain mostly old stars, the mass-to-light ratio is typically 10 to 20, where mass and light are measured in units of the Sun's mass and luminosity. Mass-to-light ratios of 100 or more are a signal that a substantial amount of dark matter is present. Table 27.1 summarizes the results of measurements of mass-to-light ratios for various classes of objects. Very large mass-to-light ratios are found for all systems of galaxy size and larger, which indicates that dark matter is present in all these types of objects. This is why we say that dark matter apparently makes up most of the total mass of the universe. Astronomers currently estimate that the typical ratio of dark matter to luminous matter in a large volume of space is about 7 to 1.

TABLE 27.1 *Mass-to-Light Ratios*

Type of Object	Mass-to-Light Ratio
Sun	1
Matter in vicinity of Sun	2
Total mass in Milky Way	10
Small groups of galaxies	50–150
Rich clusters of galaxies	250–300

The clustering of galaxies can be used to derive the total amount of mass in a given region of space, while visible radiation is a good indicator of where the luminous mass is. Studies show that the dark matter is distributed in the same way as the luminous matter. The dark matter halos do extend beyond the luminous boundaries of the galaxies that they surround. However, where there are large clusters of galaxies, there are also large amounts of dark matter. Voids in the galaxy distribution are also voids in the distribution of dark matter.

27.5.5 What Is the Dark Matter?

How do we go about figuring out what the dark matter consists of? The technique we might use depends on its composition. Some of the dark matter could be made up of normal particles—protons and neutrons and electrons. If these protons, neutrons, and electrons were assembled into black holes, brown dwarfs, or even white dwarfs, they would be invisible to us. The latter two types of objects do emit some radiation but have such low luminosities that they cannot be seen at distances greater than a few thousand light years.

We can, however, look for such compact objects because they can act as gravitational lenses. (See Section 26.4 for more on gravitational lenses.) Suppose the dark matter in the halo of the Milky Way were made up of black holes, brown dwarfs, and white dwarfs. These objects have been whimsically dubbed MACHOs (MAssive Compact Halo Objects). If an invisible MACHO passes directly between a distant star and the Earth, it acts as a gravitational lens, focusing the light from the distant star. This causes it to appear to brighten over a time interval of several days before returning to its normal brightness. Since we can't predict when any given star might brighten this way, we have to monitor huge numbers of stars to catch one in the act. There are not enough astronomers to keep monitoring so many stars, but today's automated telescopes and computer systems can do it for us.

Research teams making observations of millions of stars in the nearby galaxy called the Large Magellanic Cloud have recently reported several examples of the type of brightening expected if MACHOs are present in the halo of the Milky Way. However, there are simply not enough such events to account for most of the dark matter in the halo.

This result is a bit disappointing, since it means that we still have to figure out the nature of the majority of dark matter. And, as we shall discover in the next chapter, a variety of experiments lead us to conclude that the types of matter we are familiar with can make up only a tiny portion of the dark matter. The rest must therefore be composed of some type of particle we have yet to detect in our laboratories here on Earth.

Solving the dark matter problem is one of the biggest challenges facing astronomers. After all, we can hardly understand the evolution of the universe without understanding what the dark matter is. What role did dark matter play in forming the higher-density "seeds" that led to the formation of galaxies? And if many galaxies have large halos made of dark matter, how does this affect their interactions with one another and the shapes and types of galaxies that their collisions create?

Astronomers armed with various theories are working hard to produce models of galaxy structure and evolution that take dark matter into account in just the right way. Unfortunately, since we don't know what the dark matter is, we don't know how it behaves and therefore we cannot know what that "right way" is. Here is a dramatic demonstration of what we have tried to emphasize throughout this book: Science is always a "progress report" and we often encounter areas where we have more questions than answers.

SURFING THE WEB

 Specific Groups or Clusters:

- *The Local Group Page:*
  bozo.lpl.arizona.edu/messier/more/local.html

- *Virgo Cluster of Galaxies:*
  bozo.lpl.arizona.edu/messier/more/virgo.html

 Some Redshift Surveys (Maps of Regions of the Universe):

- *Sloan Digital Sky Survey Page:* www.sdss.org

- *2DF Galaxy Redshift Survey:*
  www.mso.anu.edu.au/2dFGRS/

- *Center for Astrophysics Redhsift Survey:*
  cfa-www.harvard.edu/~huchra/zcat

- *Deep Extragalactic Evolutionary Probe:*
  deep.ucolick.org

 Galaxy Transformations:
www.ifa.hawaii.edu/faculty/barnes/transform.html
Josh Barnes of the University of Hawaii, one of the astronomers simulating galaxy mergers on high-speed computers, gives a taste of his work with MPEG movies on this site.

Dark Matter:

- *Queen's University Dark Matter Tutorial:*
  www.astro.queensu.ca/~dursi/dm-tutorial/dm0.html

- *Essay by Joseph Silk:*
  astron.berkeley.edu/~mwhite/darkmatter/essay.html

- *The MACHO Project (to search for lensing effects):*
  wwwmacho.mcmaster.ca

SUMMARY

27.1 When we look at distant galaxies, we are looking back in time. We have now seen galaxies as they were when the universe was about 1 billion years old—less than one-tenth as old as it is now. The universe now is about 14 billion years old. The color of a galaxy is an indicator of the age of the stars that populate it. Blue galaxies must contain a lot of hot, massive, young stars. Galaxies that contain only old stars tend to be yellow-red. The first generation of stars formed when the universe was only a few hundred million years old. Galaxies observed when the universe was only a few billion years old tend to be smaller, to have more irregular shapes, and to have more rapid star formation than the galaxies we see nearby in today's universe.

27.2 When galaxies of comparable size collide and coalesce, we call it a **merger,** but when a small galaxy is swallowed by a much larger one, we use the term **galactic cannibalism.** Collisions play an important role in the evolution of galaxies. If the collision involves at least one galaxy rich in interstellar matter, the resulting compression of the gas will result in a burst of star formation. Mergers were much more common when the universe was young, and many of the most distant galaxies that we see are starburst galaxies that are involved in collisions. Since we have seen some luminous quasars at a distance of 13 billion LY, it is likely that some large elliptical galaxies were forming then. However, most giant elliptical galaxies were formed later through mergers of smaller galaxies that had already formed stars. There has been very little star formation in ellipticals in the last 8 billion years or so. The disks of spiral galaxies formed later than either ellipticals or the nuclear bulges of spirals and were still acquiring their distinctive shapes 4 billion years ago. The rate of star formation has decreased by about a factor of ten during the past 8 billion years.

27.3 Counts of galaxies in various directions establish that the universe on the large scale is **homogeneous** and **isotropic** (the same everywhere and in all directions, apart from evolutionary changes with time). The sameness of the universe everywhere is referred to as the **cosmological principle.** Galaxies are grouped together in clusters. The Milky Way Galaxy is a member of the **Local Group,** which contains at least 40 member galaxies. Rich clusters (such as Virgo and Coma) contain thousands or tens of thousands of galaxies. **Galaxy clusters** often group together with other clusters to form large-scale structures called **superclusters,** which can extend over distances of several hundred million light years. Clusters and superclusters are found in filamentary structures that fill only a small fraction of space. Most of space consists of **voids** between superclusters, with nearly all galaxies confined to less than 10 percent of the total volume.

27.4 Initially matter in the universe was distributed almost—but not quite—uniformly. The challenge for galaxy formation theories is to show how this "not quite" smooth distribution of matter developed the structures—galaxies and galaxy clusters—that we see today. It is likely that the filamentary distribution of galaxies and voids was built in near the beginning, before stars and galaxies began to form. The first condensations of matter were about the mass of a large star cluster or a small galaxy. These smaller structures then merged over cosmic time to form large galaxies, clusters of galaxies, and superclusters of galaxies. Superclusters today are still gathering up more galaxies.

27.5 The visible matter in the universe does not exert a large enough gravitational force to hold stars in their orbits within galaxies or to hold galaxies in their orbits around other galaxies, which requires about seven times more dark matter than luminous matter. Although some of the dark matter may be made up of ordinary matter—protons and neutrons, perhaps in the form of very faint stars or black holes—most of it probably consists of some totally new type of particle not yet detected on Earth. Gravitational lensing effects on distant objects have been used to search the outer region of our Galaxy for any dark matter in the form of compact, dim stars or star remnants, but not enough such objects have been found to account for all the dark matter.

INTER-ACTIVITY

A Suppose you developed a theory to account for the evolution of New York City. Have your group discuss whether it would resemble the development of structure in the universe (as we have described it in this chapter): In other words, would it be assembled from the bottom up out of smaller components?

B Your group is appointed the data management team for the Sloan Digital Sky Survey (see the Making Connections box). Data are coming into the project at a rate of 8 megabytes per second (where a megabyte is a million individual pieces of information). Over five years you expect to get 10 thousand billion bytes of information, including positions and brightnesses for 100 million celestial objects, and distances to more than a million galaxies. Discuss how you would store, publish, and display such a wealth of information today. How do you think things will change by the time the survey is finished in about 2005?

C Most astronomers believe that dark matter exists and is a large fraction of the total matter in the universe. At the same time, most astronomers do not believe that UFOs are evidence that we are being visited by aliens from another world. Yet astronomers have never actually seen either dark matter or a UFO. Why do you think one idea is widely accepted by scientists and the other is not? Which idea do you think is more believable? Give your reasoning.

D Someone in your group describes the redshift surveys of galaxies to a friend, who says she's never heard of a bigger waste of effort. Who cares, she asks, about the large-scale structure of the universe? What is your group's reaction, and what reasons could you come up with for putting money into figuring out how the universe is organized?

E The leader of a small but very wealthy country is obsessed by maps. He has put together a fabulous collection of Earth maps, purchased all the maps of other planets that astronomers have assembled, and now wants to commission the best possible map of the entire universe. Your group is selected to advise him. What sort of instruments and surveys should he invest in to produce a good map of the cosmos? Be as specific as you can.

REVIEW QUESTIONS

1. How are distant (young) galaxies different from the galaxies that we see in the universe today?
2. What is the evidence that star formation began when the universe was only a few hundred million years old?
3. Describe the evolution of an elliptical galaxy. How does the evolution of a spiral galaxy differ from that of an elliptical?
4. Explain what we mean when we call the universe homogeneous and isotropic. Would you say that the distribution of elephants on the Earth is homogeneous and isotropic? Why?
5. Describe the organization of galaxies into groupings, from the Local Group to superclusters.
6. What is the evidence that a large fraction of the matter in the universe is invisible?

THOUGHT QUESTIONS

7. Describe how you might use the color of a galaxy to determine something about what kinds of stars it contains.

8. Suppose a galaxy formed stars for a few million years and then stopped. What would be the most massive stars on the main sequence after 500 million years? After 10 billion years? How would the color of the galaxy change over this time span? (Refer to Table 21.1.)

9. Given the ideas presented here about how galaxies form, would you expect to find a giant elliptical galaxy in the Local Group? Why or why not? Is there in fact a giant elliptical in the Local Group?

10. Can an elliptical galaxy evolve into a spiral? Explain your answer. Can a spiral turn into an elliptical? How?

11. Suppose you are standing in the center of a large, densely populated city that is exactly circular, surrounded by a ring of suburbs with lower-density population, surrounded in turn by a ring of farmland. Would you say the population distribution is isotropic? Homogeneous?

12. Use the data in Appendix 12 to determine which is more common in the Local Group: large luminous galaxies or small faint galaxies. Which is more common: spirals or ellipticals?

13. Based on data in Appendix 12, would you describe the Milky Way Galaxy as a typical member of the Local Group? Why or why not?

14. Astronomers have been making maps by observing a slice of the universe and seeing where the galaxies lie within that slice. If the universe is isotropic and homogeneous, why do they need more than one slice? Suppose they now want to make each slice extend farther into the universe. What do they need to do?

FIGURING FOR YOURSELF

To determine the distribution of galaxies in three-dimensional space, astronomers have to measure their positions and their redshifts. The larger the volume of space you survey, the more likely you are to be measuring a fair sample of the universe as a whole. However, the work involved increases *very* rapidly as you increase the volume covered by the survey.

Let's do some quick calculations to see why this is so. Remember that the volume of a sphere, V, is given by the formula:

$$V = \left(\frac{4}{3}\right)\pi R^3$$

where R is the radius of the sphere. If you go out twice as far, you will be increasing the volume of space you have to survey, not by a factor of 2 but by 2^3, which is 8. If you triple the distance you want to cover, the volume goes up by a factor of 81!

15. Suppose that you have completed a survey of all the galaxies within 30 million LY and you now want to survey

to 60 million LY. What volume of space is covered by your second survey? How much larger is this volume than the volume of your first survey?

16. How much fainter an object will you have to be able to measure in order to include the same kinds of galaxies in your second survey? Remember that the brightness of an object varies as the inverse square of the distance.

17. If galaxies are distributed homogeneously, how many times more of them would you expect to count on your second survey?

18. How much longer will it take you to do your second survey?

19. Suppose you want to increase your complete survey to 120 million LY (four times the original distance from us). How much larger a volume would your survey cover? How much longer would it take than the original survey?

Perhaps now you can see why astronomers sample regions of the sky instead of the entire volume of space when they set out to measure the distribution of galaxies at large distances from the Earth.

20. Galaxies are found in the "walls" of huge voids; very few galaxies are found in the voids themselves. The text says that the structure of filaments and voids has been present in the universe since shortly after the expansion began 14 billion years ago. In science, we always have to check whether some conclusion is contradicted by any other information we have. In this case, we can ask whether the voids would have filled up with galaxies in 14 billion years. Observations show that, in addition to the motion associated with the expansion of the universe, the galaxies in the walls of the voids are moving in random directions at typical speeds of 300 km/s. At least some of them will be moving into the voids. How far into the void will a galaxy move in 14 billion years? Is it a reasonable hypothesis that the voids have existed for 14 billion years?

In this chapter, and throughout the book, we have expressed the distance to galaxies in light years. To obtain distances in light years, it is necessary to measure the redshift, use the Doppler formula to convert the redshift to a velocity, and then use the Hubble law (see Chapter 25) to convert the velocity to a distance.

Astronomers often indicate the relative distances to objects in terms of the redshift (z) itself, which is defined to be

$$z = \frac{\Delta\lambda}{\lambda}$$

where λ is the wavelength a spectral line would have in, for example, a laboratory source that is not moving relative to the observer, and $\Delta\lambda$ is the difference between the laboratory wavelength and the wavelength observed in the galaxy. The advantage of using z is that it can be measured directly on a spectrum; it will not change even if estimates of the Hubble constant should change.

21. Calculate the velocity of the most distant galaxies in Figure 27.15 using the Hubble constant given in this text and the distance in light years given in the diagram. Then estimate whether the astronomers who constructed the diagram in Figure 27.15 used the same Hubble constant that we have adopted for the text. To do this, note that the distances to the galaxies in Figure 27.15 are labeled in both light years and z. Use the data in the figure to estimate what Hubble constant was adopted. Remember the Doppler formula for velocity ($v = c \times \Delta\lambda/\lambda$) and the Hubble law ($v = H \times d$, where d is the distance to a galaxy). For these low velocities, you can neglect relativistic effects (which were discussed in the Figuring for Yourself section in Chapter 26).

22. If the Coma cluster of galaxies is about 275 million LY away, what will be the redshift of a typical galaxy in the cluster (if we ignore the motion of the galaxies around the cluster's center and just worry about the expansion of the universe)?

SUGGESTIONS FOR FURTHER READING

Abrams, B. and Stecker, M. *Structures in Space.* 2000, Springer-Verlag. Book and CD-ROM about structure at all scales, but mostly the galactic and extragalactic.

Barnes, J., et al. "Colliding Galaxies" in *Scientific American*, Aug. 1991, p. 40.

Bartusiak, M. "Outsmarting the Early Universe" in *Astronomy*, Oct. 1998, p. 54. Profile of Chuck Steidel and his work on finding distant galaxies to probe galaxy evolution.

Bartusiak, M., et al. "The New Dark Age of Astronomy" in *Astronomy*, Oct. 1996, p. 36. A special issue focusing on the theory and observations of dark matter.

Benningfield, D. "Galaxies Colliding in the Night" in *Astronomy*, Nov. 1996, p. 37. About galaxy collisions and mergers.

Croswell, K. "To Kill a Galaxy" in *Astronomy*, Dec. 1996, p. 36. On how collisions with smaller neighbors have shaped the Milky Way.

Dressler, A. "The Journey Back to the Source" in *Sky & Telescope*, Oct. 1998, p. 46. On looking out (and back) to the birth of galaxies.

Dressler, A. *Voyage to the Great Attractor.* 1994, A. Knopf. A noted astronomer describes how we find large-scale structure.

Finkbeiner, A. "Invisible Astronomers Give Their All to the Sloan" in *Science,* 25 May 2001, p. 1472. The story of the Sloan Digitial Sky Survey and the astronomers behind it.

Geller, M. and Huchra, J. "Mapping the Universe" in *Sky & Telescope,* Aug. 1991, p. 134.

Henry, J. "The Evolution of Galaxy Clusters" in *Scientific American,* Dec. 1998, p. 52.

Henry, J., et al. "The Evolution of Galaxy Clusters" in *Scientific American,* Dec. 1998, p. 52.

Hodge, P. "Our New Improved Cluster of Galaxies" in *Astronomy,* Feb. 1994, p. 26. On the Local Group.

Jayawardhana, R. "Our Galaxy's Nearest Neighbor" in *Sky & Telescope,* May 1998, p. 42. On the Sagittarius dwarf galaxy.

Keel, W. "Before Galaxies Were Galaxies" in *Astronomy,* July 1997, p. 58. On how galaxies formed and evolved in the early universe.

Knapp, G. "Mining the Heavens: The Sloan Digital Sky Survey" in *Sky & Telescope,* Aug. 1997, p. 40.

Kron, R. and Butler, S. "Stars and Strips Forever" in *Astronomy,* Feb. 1999, p. 48. On the Sloan Digital Survey.

Larson, R. and Bromm, V. "The First Stars in the Universe" in *Scientific American,* Dec. 2001, p. 64. On the dark ages and the birth of the first stars.

Macchetto, F. and Dickinson, M. "Galaxies in the Young Universe" in *Scientific American,* May 1997, p. 92. On learning about galaxy formation and evolution from deep images.

MacRobert, A. "Mastering the Virgo Cluster" in *Sky & Telescope,* May 1994, p. 42. How to observe the galaxies through small telescopes.

Parker, S. and Roth, J. "The Hubble Deep-Field" in *Sky & Telescope,* May 1996, p. 48. A first analysis of the deepest image of the universe yet taken.

Roth, J. "When Galaxies Collide" in *Sky & Telescope,* Mar. 1998, p. 48. Nice introduction to the theory and observations of merging galaxies.

Schramm, D. "Dark Matter and the Origin of Cosmic Structure" in *Sky & Telescope,* Oct. 1994, p. 28.

Tytell, D. "A Wide Deep Field: Getting the Big Picture" in *Sky & Telescope,* Sept. 2001, p. 42. On the NOAO survey of deep sky objects.

West, M. "Galaxy Clusters: Urbanization of the Cosmos" in *Sky & Telescope,* Jan. 1997, p. 30. Useful overview.

The Space Telescope of the Future   A drawing of the James Webb Space Telescope, which is currently planned for launch in 2010. The blue sunshade shadows the primary mirror and science instruments. The primary mirror is expected to be at least 20 feet in diameter. Before and during launch, the mirror will be folded up. After the telescope is placed in its orbit, ground controllers will command it to unfold the mirror petals. To see distant galaxies whose light has been shifted to long wavelengths, the telescope will carry several instruments for taking infrared images and spectra.

28

The Big Bang

Making a model of the universe is like trying to pitch a tent on a moonless night in a howling Arctic wind. The tent is theory. The wind is experiment. Progress is made whenever a tent peg proves sturdy enough to hold.

Timothy Ferris in "Minds and Matter," a brief article on cosmology in *The New Yorker*, May 15, 1995

THINKING AHEAD

As we look farther and farther out in the universe, we look farther and farther back in time. What lies at the beginning of time? How much of the early evolution of the universe can we actually "see" with modern instruments? And where do our instruments become so blind that we must rely on theory?

We are now ready to complete our voyage through the universe by asking the largest questions that astronomers can ask. How did the entire universe come into being? How has it changed since the beginning? What will its ultimate fate be? For most of human history, ideas about the origin and evolution of the universe were considered the realm of philosophy and religion. Ancient people often told very beautiful and poetic stories about creation—made up to account for the unknowable and unexplainable. All of that has changed. Today new instruments allow us to do observations to probe both the past and the future of the cosmos.

Indeed, our telescopes and detectors have become so sensitive that we can actually observe directly what the universe was like from the time it was a few hundred thousand years old. We have detected galaxies at a time when the universe was only about a billion years old, and even larger telescopes are now being planned with the goal of detecting newly forming galaxies and quasars when the universe was younger still (opening figure and Figure 28.1).

Ace⊛**Astronomy**™ The AceAstronomy icon throughout the text indicates an opportunity for you to test yourself on key concepts and to explore animations and interactions on the AceAstronomy website at **http://ace.brookscole.com/voyages**

Virtual Laboratories

 Dark Matter

 Cosmology and Cosmic Background Radiation

■ **FIGURE 28.1**

A 30-meter Telescope This drawing shows what a 30-m ground-based infrared telescope might look like. Note that the mirror is made of many small segments, and therefore the telescope is referred to as the giant segmented mirror telescope (GSMT). It is likely that a telescope like this will be built in the next 15 years or so to study the spectra of really distant galaxies and determine when and how they formed.

The study of the universe as a whole is called **cosmology.** Astronomers now talk about our entering an era of *precision cosmology*—a time when we can make measurements that tell us quantitatively how the universe came to have the properties that we observe today. Cosmology is an enormous field, making rapid progress on many fronts, and it has been the topic of more popular books than any other astronomical subject. In this chapter we can give you only an overview of the highlights of current cosmological thinking. But at the end you will find a list of our favorite nontechnical books and Web sites on the subject; we urge you to consult them if our discussion whets your appetite for more.

Let us begin by reviewing, in Table 28.1, some of the observational discoveries about the universe as a whole that have already been covered in this book. For example, we know from observations of redshifts and the Hubble

<table>
<tr><td colspan="2">**TABLE 28.1** *Some Observed Characteristics of the Universe*</td></tr>
<tr><td>1. All galaxies show a redshift proportional to distance, implying that the universe is expanding (Chapter 25).</td></tr>
<tr><td>2. The distribution of galaxies on the largest scale is isotropic and homogeneous (the cosmological principle; Chapter 27).</td></tr>
<tr><td>3. The contents of the universe evolve with time: Hydrogen and helium are changed into heavier elements inside stars (Chapters 15, 21, 22).</td></tr>
</table>

law that the universe is expanding. Thus we do not need to consider any theory of the universe that does not account in a natural way for this expansion. In the discussion that follows, we build on this fundamental observation (and others) to construct the best model we can currently make of the cosmos.

28.1 THE AGE OF THE UNIVERSE

With hindsight, it is surprising that scientists in the 1920s and 1930s were so shocked to discover that the universe is expanding. In fact, our theories of gravity demand that the universe *must* be either expanding or contracting. To show what we mean, let's begin with a universe of finite size—say, a giant ball of a thousand galaxies. All these galaxies attract each other because of their gravity. If they were initially stationary, they would inevitably begin to move closer together and eventually collide. They could avoid this collapse only if for some reason they happened to be moving away from each other at high speeds. In just the same way, only if a rocket is launched at high enough speed can it avoid falling back to Earth.

The problem of what happens in an infinite universe is harder to solve, but Einstein used his theory of general relativity to show that even infinite universes cannot be static. Since astronomers at that time did not yet know the universe was expanding (and Einstein himself was philosophically unwilling to accept a universe in motion), he changed his equations by introducing an arbitrary new term (we might call it a fudge factor) called the **cosmological constant.** It represented a hypothetical force of repulsion that could balance gravitational attraction on the largest scales and permit galaxies to remain at fixed distances from one another.

About a decade later Hubble and his co-workers reported that the universe was expanding, so that no mysterious balancing force was needed. Einstein is reported to have said that the introduction of the cosmological constant was "the biggest blunder of my life." As we shall see later in this chapter, however, new observations indicate that the expansion is *accelerating.* In a way, Einstein may have been right after all.

Ace Astronomy™ Log into AceAstronomy and select this chapter to see Astronomy Exercise "Age of the Universe."

28.1.1 The Hubble Time

If we had a movie of the expanding universe and ran the film *backward,* what would we see? The galaxies, instead of moving apart, would move *together* in our movie—getting closer and closer all the time. Eventually, we would find that all matter was once concentrated in an infinitesimally small volume. Astronomers identify this time as the *beginning of the universe.* The explosion of that concentrated universe at the beginning of time is called the **big bang** (not a bad term, since you can't have a bigger bang than one that creates the entire universe!). But when did this bang occur?

We can make a reasonable estimate of the time since the expansion began. To see how astronomers do this, let's begin with an analogy. Suppose your astronomy class decides to have a party (a kind of "big bang") at someone's home to celebrate the end of the semester. Unfortunately, everyone is celebrating with so much enthusiasm that the neighbors call the police, who arrive and send everyone away at the same moment. You get home at 2 A.M., still somewhat upset about the way the party ended, and realize you forgot to look at your watch to see what time the police got there. You use a map to find out that the distance between the party and your house is 40 km. And you also remember that you drove the whole trip at a steady speed of 80 km/h (since you were worried about the police cars following you). Therefore, you can calculate how long the trip must have taken:

$$\text{time} = \frac{\text{distance}}{\text{velocity}} = \frac{40 \text{ km}}{80 \text{ km/h}} = 0.5 \text{ h}$$

So the party must have broken up at 1:30 A.M.

No humans were around to look at their watches when the universe began, but we can use the same technique to estimate when the galaxies began moving away from each other. (Remember that, in reality, it is space that is expanding, not the galaxies that are somehow traveling on their own!) If we can measure how far apart the galaxies have gotten and how fast they are moving, then we can figure out how long the trip has been.

Let's call the age of the universe measured in this way T_0. The time it has taken a galaxy to move a distance, d, away from the Milky Way (remember that at the beginning the galaxies were all together in a very tiny volume) is (as in our example)

$$T_0 = \frac{d}{v}$$

where v is the velocity of the galaxy. Since individual galaxies have their own local motions, we want to make measurements not for just one galaxy but for a good sample of them. If we can measure the speed with which many different galaxies are moving away and also the distances between them, we can establish how long ago the expansion began.

Making such measurements should sound very familiar. This is just what Edwin Hubble and many astronomers after him needed to do in order to establish the Hubble law and the Hubble constant. We learned in Chapter 25 that a galaxy's distance and its velocity in the expanding universe are related by

$$v = H \times d$$

where H is the Hubble constant. Combining these two expressions gives us

$$T_0 = \frac{d}{v} = \frac{d}{H \times d} = \frac{1}{H}$$

We see, then, that the work of calculating this time was already done for us when astronomers measured the Hub-

ble constant. The age of the universe estimated in this way turns out to be just the *reciprocal of the Hubble constant* (that is, $1/H$). This age estimate is sometimes called the *Hubble time*. For a Hubble constant of 20 km/s per million LY, the Hubble time is about 15 billion years.

To make numbers easier to remember, we have done some rounding here. Estimates for the Hubble constant are actually closer to 21 or 22 km/s per million LY, which makes the age closer to 14 billion years. But there is still about a 10 percent uncertainty in the Hubble constant, which means the age of the universe estimated this way is also uncertain by about 10 percent. To put these uncertainties in perspective, however, you should know that 20 years ago, the uncertainty was a factor of 2. Remarkable progress toward pinning these numbers down has been made in the last few years! (You may see the Hubble constant quoted in the press and elsewhere in terms of km/s per million *parsecs*. These are the units used by professional astronomers, and in these units the Hubble constant is equal to 70 km/s per million parsecs, again with an uncertainty of about 10 percent.)

28.1.2 The Role of Deceleration

The Hubble time is the right age for the universe only if the expansion rate has been constant throughout the time since the universe began to expand (Figure 28.2). Continuing with our end-of-the-semester-party analogy, this is equivalent to assuming that you traveled home from the party at a constant rate, when in fact this may not have been the case. At first, angry about having to leave, you may have driven fast, but then as you calmed down—and thought about police cars on the highway—you may have begun to slow down until you were driving at a more socially acceptable speed (such as 80 km/h). In this case, given that you were driving faster at the beginning, the trip home would have taken less than a half-hour.

In the same way, in calculating the Hubble time, we have assumed that H has been constant throughout all of time. This may not be a good assumption. Matter creates gravity, whereby all objects pull on all other objects. This mutual attraction will slow the expansion as time goes on, which means that if gravity were the only force acting (a big *if*, as we shall see in the next section), then the rate of expansion must have been faster in the past than it is today. In this case, we would say the universe has been *decelerating* since the beginning. How much it has decelerated depends on the importance of gravity in slowing the expansion.

If the universe were nearly empty, the role of gravity would be minor. Then the deceleration would be close to zero and the universe would have been expanding at a constant rate. But in a universe with any significant density of matter, the deceleration means that the expansion should be slower now than it used to be. In this case, the age of the universe is less than our earlier estimate, assuming a constant expansion rate. If we use the current rate of expansion to estimate how long it took the galaxies to reach their current separations, we will overestimate the age of

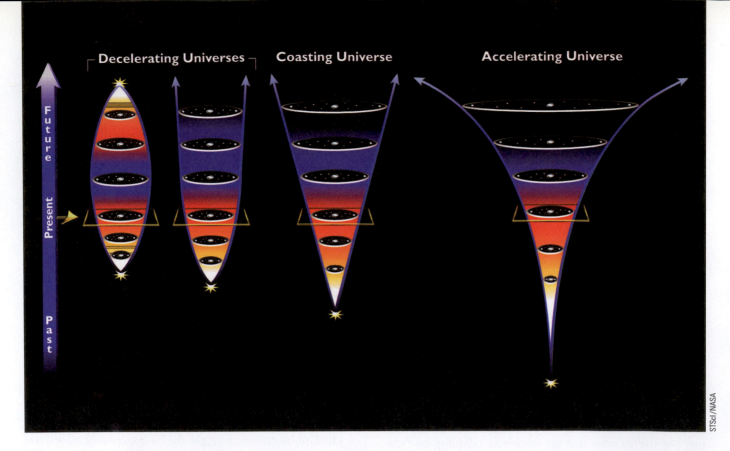

STScI /NASA

■ **FIGURE 28.2**

Expansion Rates and the Age of the Universe These drawings show four possible models of the universe. The yellow rectangle marks the present in all four cases, and for all four the Hubble constant is equal to the same value at the present time. Time is measured in the vertical direction. The two universes on the left are ones in which the rate of expansion slows over time. The one on the far left will eventually slow, come to a stop, and reverse, ending up in a "big crunch," while the one next to it will continue to expand forever, but more slowly as time passes. Since these universes were expanding more rapidly in the past than they are today, it has taken them less time to reach the current size of the universe than it has for the "coasting" universe, which has expanded at a constant rate given by the Hubble constant throughout all of cosmic time. The accelerating universe on the right expanded more slowly in the past than today and has taken the longest time to reach the current size of the universe.

the universe—just as we may have overestimated the time it took you to get home from the party.

28.1.3 A Universal Acceleration

But what if Einstein's initial thought was right, and there is a cosmological constant of some kind? Recently, astronomers have begun using Type Ia supernovae to measure the rate of expansion of the universe. Remember that this type of supernova occurs when a white dwarf gathers enough matter to be pushed over the Chandrasekhar limit and to explode. There is good evidence that all of these supernovae have very nearly the same brightness at maximum light (and astronomers know how to correct for the slight differences among them). Thus, such supernovae can serve as "standard bulbs" (as discussed in our chapter on galaxies).

If we detect a Type Ia supernova in a distant galaxy, we therefore immediately know its distance very accurately. From the redshift in its spectrum, we can measure how fast that galaxy is moving away from us. If we substitute the value into Hubble's law, we can get the Hubble constant and in this way determine whether the expansion rate of the universe (which is what the Hubble constant tells us) at the time the supernova explosion occurred is the same as the rate today.

Supernovae of Type Ia are extremely bright, and we can detect them at distances of several billion LY. Two groups of astronomers have been searching for these supernovae and have observed more than a hundred of them so far. Both groups conclude that the universe is expanding at a faster rate today than it was at the time the supernova explosions occurred—that is, when the universe was about a third of its current age. The universe, according to these results, is *accelerating*. This means that something is pushing galaxies apart at an increasing rate—just the effect that Einstein's cosmological constant would add to our recipe for the universe. (Einstein had "tuned" his constant to balance gravity exactly. In reality, it appears that nature tuned it to be a bit stronger, so it can speed up the expansion.)

The energy associated with this mysterious "something" is not possessed by matter or radiation but by "empty" space. One interpretation is that tiny elementary particles flicker in and out of existence and give space a kind of springiness that pushes it apart. Attempts have been made to calculate how large this effect might be, but so far the predictions do not agree with the acceleration of the expansion derived from the supernova observations. Since we don't know what the source of the energy is, it has been given the name **dark energy.** Note that this new component of the universe is not the dark matter we talked about in earlier chapters, but something else that we have also not detected in our laboratories here on Earth.

At this point, you can perhaps see why the claim that the universe is accelerating makes many scientists uncomfortable. Pushing things apart at an increasing rate requires energy; speeding up the entire universe requires, as you might imagine, a lot of energy! To understand how much, let's recall that energy and mass are, according to Einstein, equivalent. Let's say that the total amount of matter in the universe plus the amount of dark energy converted to its equivalent in mass add up to 100 percent. Then calculations show that true matter contributes only about 30 percent of the mass and the dark energy contributes the remaining 70 percent!

Because this observation is so unsettling, many astronomers initially were very skeptical about the result and tried to find another explanation. What we actually observe is that the most distant supernovae are about 20 percent fainter than they would be if the expansion rate were constant. An accelerating expansion provides a natural explanation for this observation. If the universe is expanding faster now than in the past, our motion away from the distant supernovae has speeded up since the light left them, sweeping us farther away. The light then has to travel a longer distance to reach us than if the expansion rate were constant. Because the amount of radiation that reaches us decreases by the inverse square of the distance, the supernovae look fainter than predicted for a constant expansion rate.

But just like the skeptical astronomers, you should be asking at this point whether there might be other factors that would make distant supernovae look dimmer without resorting to something quite so drastic as an accelerating universe. For example, if dust were between the supernovae and us, light would be absorbed by this dust, and the supernovae would look fainter. However, the types of dust we know about make objects look redder as well as fainter, and there is no evidence of reddening of distant supernovae.

Another possibility is that supernova explosions were intrinsically fainter when the universe was young. After all, the composition of the dying stars might have been some-

what different long ago because there had been less time to build up elements heavier than hydrogen and helium. That's an interesting idea; however, the spectra of distant supernovae look identical to the spectra of nearby supernovae; astronomers see no evidence for a significant difference in composition.

Scientists are looking for other possibilities for dimming the light of distant supernovae. Because this result is so important—namely, that we don't know what makes up 70 percent of the universe—a great deal of effort is being expended on trying to find out whether anything is wrong with either the observations or their interpretation. However, so far no one has found a convincing challenge to the evidence that the expansion is accelerating. And, as we shall see later in this chapter, there is now evidence from observations that do not involve supernovae that this dark energy does indeed exist.

The possibility that the universe is accelerating makes our estimates of the age of the universe rather uncertain. As we have seen, if there is no acceleration or deceleration, then, for a Hubble constant of 20 km/s per million LY, the age of the universe is 15 billion years. If the universe is decelerating—moving slower now than it did in the beginning—the universe is even younger. If the expansion is accelerating, it could be older than 15 billion years.

If, as we currently think (see the next section), the universe decelerated for a time near the beginning and then began to accelerate, then the Hubble constant alone is not enough to determine the age. We need to know just how much matter and dark energy there are. It turns out, however, that if we use the best current estimates of the amount of mass and dark energy (30 percent matter and 70 percent dark energy) to calculate the age of the universe, we get very nearly the same result as if we use the simple formula given above: $T_0 = 1/H$.

28.1.4 Comparing Ages

How else can we estimate the age of the universe besides measuring the rate at which it expands? One way is to find the oldest objects whose ages we can measure. After all, the universe has to be at least as old as the oldest objects in it. In our Galaxy and others, the oldest stars are found in the globular clusters (Figure 28.3), which can be dated using the methods described in Section 21.3.

The accuracy of the age estimates of the globular clusters have improved markedly in recent years for two reasons. First, models of the interiors of globular cluster stars have been improved, mainly through better information about how atoms absorb radiation as it makes its way from the center of a star out into space. Second, the Hipparcos satellite (see Chapter 18) has improved the accuracy of the distance calculations to globular clusters. The conclusion is that the oldest stars formed about 12 to 13 billion years ago.

This age estimate has recently been confirmed by the study of the spectrum of uranium in the stars. The isotope uranium-238 is radioactive and decays (changes into

European Southern Observatory

■ FIGURE 28.3

A Rich Globular Cluster Globular clusters, such as 47 Tucanae seen here, are among the oldest objects in our Galaxy and can be used to estimate its age. This image was taken with the 1-m Schmidt telescope at La Silla in Chile.

another element) over time. (Uranium-238 gets its designation because it has 92 protons and 146 neutrons.) We know (from how stars and supernovae make elements) how much uranium-238 is generally made compared to other elements. Suppose we measure the amount of uranium relative to nonradioactive elements in a very old star and in our own Sun and compare the abundances. With those pieces of information, we can estimate how much longer the uranium has been decaying in the very old star because we know from our own Sun how much uranium decays in 4.5 billion years.

The line of uranium is very weak even in the Sun, but it has now been measured in one extremely old star using the European Very Large Telescope. Comparing the abundance with that in the solar system, whose age we know, astronomers estimate that the star is 12.5 billion years old, with an uncertainty of about 3 billion years. While the uncertainty is large, it is important confirmation of the ages estimated by studies of the globular cluster stars because the uranium age estimate is completely independent. It does not depend either on the measurement of distances or on models of the interiors of stars.

As we shall see later in this chapter, the globular cluster stars probably did not form until the expansion of the universe had been under way for almost a billion years. Accordingly, based on their ages, we estimate that the age of the universe is about 14 billion years. This age is certainly compatible with the range of possible ages allowed by the Hubble constant. In this text, we have adopted 14 billion years as the age of the universe.

Having examined a few of the key observations, we are now ready to look at how these ideas might be applied to develop a model for evolution of the universe as a whole. All our models begin with the facts listed in Table 28.1. Then they go on to make predictions about how the universe has evolved so far and what will happen to it in the future.

28.2.1 The Expanding Universe

Every model of the universe must include the expansion we observe. Furthermore, the cosmological principle tells us that the universe is homogeneous. As a result, the expansion rate must be uniform (the same everywhere during any epoch of cosmic time). If so, we don't need to think about the entire universe when we think about the expansion, we can just look at any portion of it.

In Chapter 25, we hinted that when we think of the expansion of the universe, it is more correct to think of space itself stretching, rather than of galaxies moving through static space. Nevertheless, we have since been discussing the redshifts of galaxies as if they resulted from the motion of the galaxies themselves.

Now, however, it is time to finally put such childish notions behind us and take a more grown-up look at the cosmic expansion. Recall from our discussion of Einstein's theory of general relativity in Chapter 23 that space (or, more precisely, spacetime) is not a mere backdrop to the action of the universe, as Newton thought. Rather, it is an active participant—affected by and in turn affecting the matter and energy in the universe.

Since the expansion of the universe is the stretching of all spacetime, all points in the universe are stretching together. Thus, the expansion began *everywhere at once!* Unfortunately for tourist agencies of the future, there is no location you can visit where the stretching of space began or the big bang happened.

To describe just how space stretches, we say the cosmic expansion causes the universe to undergo a uniform change in *scale* over time. By scale we mean, for example, the distance between two clusters of galaxies. It is customary to represent the scale by the factor R; if R doubles, then the distance between the clusters has doubled. Since the universe is expanding at the same rate everywhere, the change in R tells us how much it has expanded (or contracted) at any given time. For a static universe, R would be constant as time passes. In an expanding universe, R increases with time.

If it is space that is stretching and not the galaxies that are moving through space, then why do the galaxies show redshifts in their spectra? When you were young and naive—three chapters ago—it was fine to discuss the redshifts of distant galaxies as if they resulted from their motion away from us. But now that you are an older and wiser student of cosmology, this view will simply not do!

■ **FIGURE 28.4**

Expansion and Redshift As an elastic surface expands, a wave on its surface stretches. For light waves, the increase in wavelength is seen as a redshift.

A more accurate view of the redshifts of galaxies is that the waves are stretched by the stretching of the space they travel through. Think about the light from a remote galaxy. As it moves away from its source, the light has to travel through space. If space is stretching during all the time the light is traveling, the light waves will be stretched as well. As we saw in Chapter 4, a redshift is a stretching of waves; the wavelength of each wave increases (Figure 28.4). Light from more distant galaxies has more space to cover and so has stretched more than light from closer ones and thus shows a greater redshift.

The simplest scenario of an expanding universe would be one in which R increases with time at a constant rate. But you already know that life is not so simple! The universe contains a great deal of mass and its gravity decelerates the expansion—by a large amount if the universe contains a lot of matter, or by a negligible amount if the universe is nearly empty. Then there is the problem of the force associated with the dark energy, which accelerates the expansion.

We can make different models for different amounts of mass and for different values of the cosmological constant. In some models—as we will see—the universe expands forever. In others, it stops expanding and starts to contract. If we could measure the precise amount by which the rate of expansion is changing, we could select the correct model and collect our Nobel Prize.

Unfortunately, it is very difficult to measure directly whether the rate of expansion is changing with time and in what way. One method might be to measure the distances and speeds of very distant galaxies. We are seeing these as they were long ago and so we could see how much faster (or slower) they were moving when the universe was young. However, think about how we usually measure the distance to very remote groups of galaxies. We use the Hubble law, which allows for only one rate of expansion, not for a rate that changes with time. When we use this method to estimate how far the galaxies are from us, we are already *assuming* that we live in a uniformly expanding universe.

To check on how much the expansion rate is changing, we need an *independent* way of measuring the distance. For example, if a certain type of galaxy were a standard bulb (see Chapter 25), we could use its apparent brightness to tell us how far away each galaxy is. Alas, not only are galaxies not standard bulbs at any *given* time in history, but we also know that their brightness can change over time. As we have seen, galaxies evolve, experience collisions and mergers, and sometimes undergo bursts of star formation that make them unusually bright for several million years.

The only standard bulb identified so far for large distances is Type I supernovae. They can be seen out to distances of several billion light years, and they offer the best method of estimating whether the rate of expansion changes with time (Figure 28.5). As we have already learned, measurements of supernovae seem to show that the universe is

NASA, Adam Riess (STScI)

■ **FIGURE 28.5**

Supernova in a Very Distant Galaxy These three images show a supernova that exploded in a galaxy 10 billion LY away. Astronomers discovered this incredibly faint supernova by comparing images taken about two years apart. The large image shows a portion of the Hubble Deep Field North. The white box shows the area blown up in the lower left image. The arrow points to the galaxy that hosted the supernova. It is elliptical and its red color indicates that its light is dominated by old stars. The picture on the lower right is the difference between the images taken two years apart. All of the sources that had the same brightness in both images disappear when we subtract one from the other, leaving only a white dot, which is the supernova.

accelerating, thus ruling out any simple gravity-only models. Gravity does still play a role, however, and we need to know just how large its effect is.

28.2.2 The Effects of Gravity

Let's start by ignoring the dark energy and consider the effects of gravity alone. We can estimate how much gravity slows the expansion rate by estimating how much material the universe contains and what gravitational force is exerted by that material. Here is where the cosmological principle really comes in handy. Since the universe is the same all over, we only need to measure how much material exists in a (large) representative sample of it. (Such sampling is how pollsters can describe how we feel about political issues without asking each and every person in the country.)

What astronomers look at in practice, to see the effects of gravity, is the *average density* of the universe. This is the mass of matter (including the equivalent mass of energy)[1] that would be contained in each unit of volume (say, 1 cm^3) if all the stars, galaxies, and other objects were taken apart, atom by atom, and if all those particles, along with the light and other energy, were distributed throughout all of space with absolute uniformity. (For the time being, we will not include any effects due to dark energy in these calculations, but we will return to the dark energy soon.) If the average density is low, the universe will not have much gravity; it will not decelerate very much and can expand forever. High average density, on the other hand, means that a lot of gravity is pulling the galaxies together, and so the expansion will eventually stop.

Note that the redshift tells us how fast the galaxies are moving away from each other (or, more precisely, how rapidly the scale factor of space is increasing). This means that we can calculate the **critical density** for the universe—the mass per unit volume that will be just enough to slow the expansion to zero at some time infinitely far in the future. If the actual density is higher than this critical density, then the expansion will ultimately reverse and the universe will begin to contract. If the actual density is lower, then the universe will expand forever.

These various possibilities are illustrated in Figure 28.6. Time increases to the right, and the scale, R, increases upward in the figure. Today, at the point marked "present" along the time axis, R is increasing in each model. We are still "early" in the history of the universe, so the galaxies are expanding no matter which model is right. (The same situation holds for a baseball thrown high into the air. Although it may eventually fall back down, at the beginning of the throw it moves upward.) The straight dashed line corresponds to the empty universe with no deceleration; it intercepts the time axis at a time T_0 (the Hubble time) in the past. The curves below the dashed line represent models with no cosmological constant and with varying amounts of

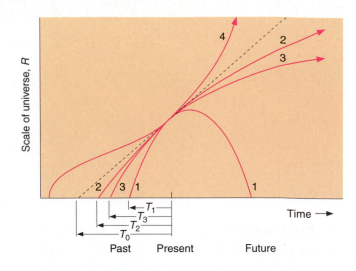

■ FIGURE 28.6

Models of the Universe   A plot of R, the scale of the universe, against time for various cosmological models. Curve 1 represents a universe where the density is greater than the critical value, curve 2 represents a universe with a density lower than critical, and curve 3 is a critical-density universe. Curve 4 represents a universe with a positive cosmological constant and a mass density that is less than the critical density. The dashed line is for an empty universe, one in which the expansion is not slowed by gravity or accelerated by the cosmological constant. Note that the age of the universe equals the Hubble time only for this (unlikely) possibility.

deceleration, starting from the big bang at shorter times in the past. The curve above the dashed line shows what happens if there is a cosmological constant and the expansion is accelerating. Let's take a look at the future according to the different models.

Let's start with curve 1 in Figure 28.6. In this case, the actual density of the universe is higher than the critical density. This universe will stop expanding some time in the future and begin contracting. This model is called a "closed universe." Eventually the scale drops to zero, which means that space will have shrunk down to infinitely small size. The noted physicist John A. Wheeler (who gave black holes their name) called this the "big crunch" because matter, energy, space, and time would all be crushed out of existence. Note that the big crunch is the opposite of the big bang; it is an *implosion* and not an explosion.

It is tempting to speculate that another big bang might follow the crunch, giving rise to a new expansion phase and then another contraction—perhaps oscillating between successive big bangs and big crunches indefinitely in the past and future. Such speculation is sometimes referred to as the *oscillating theory* of the universe, but it is not really a theory because we know of no mechanism that can produce another big bang. General relativity predicts, instead, that at the crunch the universe would collapse into a singularity; all the mass of the universe would be contained within a point of zero volume and infinite density.

---

[1] By equivalent mass we mean the mass that would result if the energy were turned into mass using Einstein's formula, $E = mc^2$.

What Might It Be Like in the Distant Future?

Some say the world will end in fire,
Some say in ice.
From what I've tasted of desire
I hold with those who favor fire.

—From the poem "Fire and Ice"
by Robert Frost (1923)

Given the destructive power of impacting asteroids, expanding red giants, and nearby supernovae, our species may not be around in the remote future. Nevertheless, you might enjoy speculating about what it would be like, in the various models, to live in a much, much older universe.

The far future in a universe that ultimately contracts to the big crunch would be exciting but not very healthy for living things. As space contracted (R gets smaller and smaller), the galaxies would see each other's light blueshifted and not redshifted; that is, the compression of space would compress the waves. But shorter waves have more energy, and thus the temperature of the cosmic background radiation would increase with time (see Section 28.4).

As the universe got smaller and denser, it would become easier to circumnavigate. Rays of light following curved paths in spacetime could eventually make it around the universe and come back to where they started. In theory, you could see a star die in one direction and then see the light of its birth come around from the other direction. "Ghost galaxies"—images of galaxies whose light had gone completely around the universe—would become visible, adding to the number of galaxies we observed.

Ultimately, as space shrank and the temperature of the background radiation increased, the temperature of space would become higher than that of any planets. Heat would then flow from space to a planet, instead of the other way around. (Just think how much we now take for granted that space is a cold place into which our excess heat flows.)

As a planet heated up, any life-forms on its surface would be "cooked" by the heat of space. As the radiation of space got even hotter, heat would eventually flow from space into the stars, breaking them apart. Their end could be mourned only briefly, however, because soon all matter and energy would be crushed out of existence by the closing down of spacetime in the big crunch.

Continued expansion is unsettling in a completely different way. If the universe expands forever (R increases without limit), the clusters of galaxies would spread ever farther apart with time. As the eons passed, the universe would get thinner, colder, and darker.

Within each galaxy, stars would continue to go through their lives, eventually becoming white dwarfs, neutron stars, and black holes. Low-mass stars might take a long time to finish their evolution, but in this model we would literally have all the time in the world! Ultimately, even the white dwarfs would cool down to be black dwarfs; any neutron stars that revealed themselves as pulsars would slowly stop spinning; and black holes with accretion disks would one day complete their "meals." The final states of stars would all be dark and difficult to observe.

The light that now reveals galaxies to us would eventually go out. Even if a small pocket of raw material were left in one unsung corner of a galaxy, ready to be turned into a fresh cluster of stars, we would only have to wait until the time that their evolution was also done. And time is one thing this model of the universe has plenty of. There would surely come a time when all the stars were out, galaxies were as dark as space, and no source of heat remained to help living things survive. Then the lifeless galaxies would just continue to move apart in their lightless realm.

If these views of the future seem discouraging (from a human perspective), you might take heart from the knowledge that science is always a progress report. The most advanced ideas about the universe from a hundred years ago now strike us as rather primitive. It may well be that our best models of today will in a hundred or a thousand years also seem rather childish and that there are other factors determining the ultimate fate of the universe of which we are still completely unaware.

The oscillating model is more a philosophical idea than a scientific one; scientific hypotheses, after all, must be tested with experiments. There is no way to test whether another cycle of the universe could arise from the singularity because nothing—not matter, not energy, not space, not time—could survive from the universe that existed before the big crunch.

If the density of the universe is less than the critical density (curve 2 in Figure 28.6), gravity is never important enough to stop the expansion, and so the universe expands forever. Such a universe is infinite and this model is called an "open universe." Time and space begin with the big bang, but they have no end; the universe simply continues expanding, always a bit more slowly as time goes on. Groups of galaxies eventually get so far apart that it would be difficult for observers in any of them to see the others. (See the Making Connections box for more about the distant future in the closed and open universe models.)

At the critical density (curve 3) the universe can just barely expand forever. The critical-density universe has an age of exactly two-thirds T_0, where T_0 is the age of the empty universe. Universes that will someday begin to contract have ages less than two-thirds T_0.

28.2.3 The Cosmic Tug of War

We might summarize our discussion so far by saying that a "tug of war" is going on between the forces that push everything apart and gravity, which pulls everything together. Who will win this tug of war is one of the greatest questions in astronomy, for the answer will tell us what the ultimate fate of the universe will be.

What happens now if we put the effects of dark energy into our model of the universe? Curve 4 corresponds to a universe with a cosmological constant (an accelerating expansion) and a matter density that is less than the critical density. The curve has a complicated shape because in the beginning, when the matter is all very close together, the rate of expansion is most influenced by gravity. The pressure force represented by the cosmological constant (the dark energy we discussed earlier) acts only over large scales and becomes more important relative to the force of gravity as the universe grows larger and the matter begins to thin out. Thus, at first the universe slows down, but as space stretches, the acceleration plays a greater role and the universe speeds up.

If there is a cosmological constant and the matter density is less than the critical density, then the expansion will continue forever. This is the oldest of the model universes illustrated in Figure 28.6. It is even older than the flat universe and started longer ago than T_0. The various models we have described are summarized in Table 28.2.

So now let's look at what our observations can tell us about all these choices. Since we know the universe is accelerating, we already know that gravity is not winning the tug of war. However, our measurements can tell us something more. Even in the absence of dark energy, the expansion would continue forever because the amount of matter turns out to be less than the critical density. The critical density depends on H_0. If the Hubble constant is around 20 km/s per million LY, the critical density is about 10^{-26} kg/m^3 (see the Figuring for Yourself section at the end of the chapter). Let's see how this value compares with the actual density of the universe.

There are several methods by which we can try to determine the average density of matter in space. One way is to count all the galaxies out to a given distance and use estimates of their masses, including dark matter, to calculate the average density. Such estimates indicate a density of 1 to 2×10^{-27} kg/m^3 (10 to 20 percent of critical), which by itself is too small to stop the expansion. A lot of the dark matter lies outside the boundaries of galaxies, so our inventory is not yet complete. But even if we add an estimate of the dark matter outside galaxies, our total won't rise beyond about 30 percent of the critical density. We'll pin these numbers down more precisely in Section 28.5, where we will also include the effects of dark energy.

In any case, things do not look good for fans of the closed universe (big crunch) model. Between the lack of sufficient matter and the effect of the dark energy, it looks like we live in a forever expanding universe.

28.2.4 Ages of Distant Galaxies

In Chapter 25 we discussed how we can use the Hubble law to measure the distance to a galaxy. But that simple method works only with galaxies that are not moving too fast (that is, are not too far away). Once we get to large distances, we are looking so far into the past that we must take changes in the rate of the expansion of the universe into account. Since we cannot measure these changes directly, we must assume one of the models of the universe so that we can convert large redshifts into distances.

This is why astronomers squirm when reporters and students ask them exactly how far away some newly discovered distant quasar or galaxy is. We really can't give an answer without first explaining the model of the universe we are assuming in calculating it (by which time a reporter or student is long gone or asleep!). In Chapter 27, where we did give distances in light years, we assumed a mass density of 0.3 times the critical density and a dark energy equivalent to 0.7 times the critical density. We will refer to these values as the "standard model of the universe," and the evidence for them will be explained later in this chapter.

TABLE 28.2 *Some Models of the Universe*

Kind of Universe (model)	Age (billions of years)*	Ultimate Fate of Universe
Mass alone exceeds critical density	Less than 10	Stop expanding and contract
Mass alone equals critical density	10	Barely expand forever
Mass is less than critical and there is no dark energy	Greater than 10	Expand forever
Mass = 30% and dark energy = 70% of critical density	14 billion years	Expand forever at an accelerating rate

*The numbers here assume a Hubble constant of 20 km/s per million LY. If the Hubble constant is smaller than this, the ages increase; if it is larger, they decrease.

TABLE 28.3 Ages of the Universe at Different Redshifts

Redshift	Percent of Current Age of Universe When the Light Was Emitted (mass = critical density)	Percent of Current Age of Universe When the Light Was Emitted (mass = 0.3 critical density; dark energy = 0.7 critical density)
0	100 (now)	100 (now)
0.5	55	63
1.0	35	42
2.0	19	25
3.0	12	16
4.0	9	12
4.5	8	11
4.9	7	9
Infinite	0	0

Once we assume a model, we can use it to calculate the age of the universe at the time an object emitted the light we see. As an example, Table 28.3 lists the times that light was emitted by objects at different redshifts as fractions of the current age of the universe. The times are given for two models so you can see that the difference is not large. The first model assumes that the universe has a critical density of matter and no cosmological constant. The second model is the standard model described in the preceding paragraph. The first column in the table is the redshift, which is given by the equation $z = \Delta\lambda/\lambda_0$ and is a measure of how much the wavelength of light has been stretched by the expansion of the universe on its long journey to us. The numbers are not so important, but notice that as we find objects with higher and higher redshifts, we are looking back to smaller and smaller fractions of the age of the universe.

We have already learned some important things by observing objects at large redshifts. For example, we saw that quasars were most abundant when the universe was roughly 20 percent of its current age (see Figure 26.17). At a still earlier time, quasars were exceedingly rare. Perhaps in most regions of space it takes a few billion years to build massive black holes. Whatever the explanation, these observations provide clear evidence that our universe is evolving with time.

28.3 THE BEGINNING OF THE UNIVERSE

As we look farther and farther back, we see the galaxies and quasars thin out, and we approach the era when matter had still not settled into the structures we observe today. What were things like when the universe was young and space had not yet stretched very significantly? In other words, what was it like just after the big bang?

28.3.1 The History of the Idea

It is one thing to say the universe had a beginning (as the equations of general relativity imply) and quite another to describe that beginning. The Belgian priest and cosmologist Georges Lemaître (1894–1966) was probably the first to propose a specific model for the big bang itself (Figure 28.7). He envisioned all the matter of the universe starting in one great bulk he called the *primeval atom,* which then broke into a tremendous number of pieces. Each of these pieces continued to fragment further until they became the present atoms of the universe, created in a vast nuclear

Yerkes Observatory

■ FIGURE 28.7

Abbé Georges Lemaître (1894–1966) This Belgian cosmologist studied theology at Mechelen and mathematics and physics at the University of Leuven. It was there that he began to explore the expansion of the universe and postulated its explosive beginning. He actually predicted the Hubble law two years before its verification, and he was the first to consider seriously the physical processes by which the universe began.

■ FIGURE 28.8

George Gamow and Collaborators This composite image shows George Gamow emerging like a genie from a bottle of *ylem*, a Greek term for the original substance from which the world formed. Gamow revived the term to describe the material of the hot big bang. Flanking him are Robert Herman (left) and Ralph Alpher, with whom he collaborated in working out the physics of the big bang. The modern composer Karlheinz Stockhausen was inspired by Gamow's ideas to write a piece of music called *Ylem,* in which the players actually move away from the stage as they perform, simulating the expansion of the universe.

fission. In a popular account of his theory, Lemaître wrote, "The evolution of the world could be compared to a display of fireworks just ended—some few red wisps, ashes and smoke. Standing on a well-cooled cinder we see the slow fading of the suns and we try to recall the vanished brilliance of the origin of the worlds."

Physicists today know much more about nuclear physics than was known in the 1920s and have shown that the primeval fission model cannot be correct. Yet Lemaître's

vision was in some respects prophetic. We still believe that everything was together at the beginning; it was just not in the form of matter as we now know it.

In the 1940s the American physicist George Gamow (Figure 28.8) suggested a universe with a different kind of beginning—involving nuclear fusion instead of fission. He worked out the details with Ralph Alpher, and they published the results in 1948. (Gamow, who had a quirky sense of humor, decided at the last minute to add the name of physicist Hans Bethe to their paper, so that the co-authors would be Alpher, Bethe, and Gamow, a pun on the first three letters of the Greek alphabet: alpha, beta, and gamma.) Gamow's universe started with fundamental particles that built up the heavy elements by fusion in the big bang.

Gamow's ideas were close to our modern view, except we now know that the early universe only remained hot enough for fusion for only a short while. Thus only the three lightest elements—hydrogen, helium, and a small amount of lithium—were formed in appreciable abundances at the beginning. The heavier elements formed later in stars. Since the 1940s, many astronomers and physicists have worked on a detailed theory of what happened in the early stages of the universe.

28.3.2 The First Few Minutes

Three basic ideas hold the key to tracing the changes that occurred during the first few minutes after the universe began. The first is that the universe cools as it expands, much as gas cools when sprayed from an aerosol can. Figure 28.9 shows how the temperature changes with the passage of time. In the first fraction of a second, the universe was unimaginably hot. By the time 0.01 s had elapsed, the temperature had dropped to 100 billion (10^{11}) K. After about 3 min, it had fallen to about 1 billion (10^9) K, still some 70 times hotter than the interior of the Sun. After a few hundred thousand years, the temperature was down to a mere 3000 K, and the universe has continued to cool since that time.

All of these temperatures but the last are derived from theoretical calculations, since (obviously) no one was there to measure them directly. As we will see, however, we have

■ FIGURE 28.9

The Temperature of the Universe This graph shows how the temperature of the universe varies with time as predicted by the standard model of the big bang. Note that both the temperature (vertical axis) and the time in seconds (horizontal axis) change over vast scales on this compressed diagram.

actually detected the feeble glow of radiation emitted at a time when the universe was a few hundred thousand years old. Indeed, the fact that we have done so is one of the strongest arguments in favor of the big bang model.

The second idea in understanding the evolution of the universe is that at very early times the universe was so hot that it contained mostly radiation (and not the matter that we find dominating today). The photons—the packets of pure electromagnetic energy described in Chapter 4— that filled the universe could collide and produce material particles; that is, under the conditions just after the big bang, energy could turn into matter (and matter could turn into energy). We can calculate how much mass is produced from a given amount of energy by using Einstein's formula $E = mc^2$ (see Chapter 15).

The idea that energy can turn into matter is a new one for many students because it is not part of our everyday experience. That's because when we compare the universe today to what it was like right after the big bang, we live in cold, hard times! The photons in the universe today typically have far less energy than the amount required to make new matter. In Chapter 15 we briefly mentioned that when subatomic particles of matter and *antimatter* collide, they turn into pure energy. But the reverse, energy turning into matter and antimatter, is equally possible. This process has been observed in particle accelerators around the world. If we have enough energy, under the right circumstances, new particles of matter (and antimatter) are indeed created—and the conditions were right during the first few minutes after the expansion of the universe began.

Our third key point is that the hotter the universe the more energetic the photons available to make matter and antimatter (see Figure 28.9). To take a specific example, at a temperature of 6 billion (6×10^9) K, the collision of two typical photons can create an electron and its antimatter counterpart, a positron. If the temperature exceeds 10^{14} K, much more massive protons and antiprotons can be created.

28.3.3 The Evolution of the Early Universe

Keeping these three ideas in mind, we can trace the evolution of the universe from the time it was about 0.01 s old and had a temperature of about 100 billion K. Why not begin at the very beginning? There are as yet no theories that allow us to penetrate to a time before about 10^{-43} s. (If you are not very familiar with scientific notation, this number is a decimal point followed by 42 zeros and then a one. It is so small that we cannot relate it to anything in our everyday experience, but when it comes to understanding the universe, our minds must be willing to roam over territories both very large and very small.) When the universe was that young, its density was so high that the theory of general relativity is not adequate to describe it, and even the concept of time breaks down.

Scientists, by the way, have been somewhat more successful in describing the universe when it was older than 10^{-43} s but still less than about 0.01 s old. During that time

the universe was filled with energy and strongly interacting subatomic particles. Although the theory of these particles is difficult to deal with, recently theoretical physicists have begun to speculate about what things may have been like during this very, very early time. We will look at some of their speculations later in this chapter.

By the time the universe was 0.01 s old, it was filled with radiation and with types of matter that are still present in our world today. At that time, the universe consisted of a soup of matter and radiation; the matter included protons and neutrons, leftovers from an even younger and hotter universe. Each particle collided rapidly with other particles. The temperature was no longer high enough to allow colliding photons to produce neutrons or protons, but it was sufficient for the production of electrons and positrons (Figure 28.10a). There was probably also a sea of exotic

(a) 10^{-2} Seconds

(b) 3 Minutes

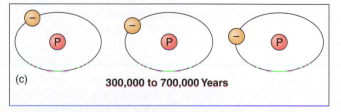

(c) 300,000 to 700,000 Years

■ **FIGURE 28.10**

Particle Interactions in the Early Universe (a) In the first fractions of a second, when the universe was very hot, energy was converted to particles and antiparticles. The reverse reaction also happened: A particle and an antiparticle could collide and produce energy. (b) As the temperature of the universe decreased, the energy of typical photons became too low to create matter. Instead, existing particles fused to create such nuclei as deuterium and helium. (c) Later it became cool enough for electrons to settle down with nuclei and make neutral atoms. Most of the universe was still hydrogen.

subatomic particles that would later play a role as dark matter. All the particles jiggled about on their own; it was still much too hot for protons and neutrons to combine to form the nuclei of atoms.

Our picture is of a seething cauldron of a universe, with photons colliding and interchanging energy, sometimes being destroyed to create a pair of particles. The particles in the universe also collide with one another. Frequently, a matter particle and an antimatter particle meet and turn each other into a burst of gamma-ray radiation.

Among the particles created in the early phases of the universe was the ghostly neutrino (see Chapter 15), which today interacts only rarely with ordinary matter. In the crowded conditions of the very early universe, however, neutrinos ran into so many electrons and positrons that they experienced frequent interactions despite their "antisocial" natures.

By the time the universe was a little more than 1 s old, the density had dropped to the point where neutrinos no longer interacted with matter but simply traveled freely through space. In fact, these neutrinos should now be all around us. Since they have been traveling through space unimpeded (and hence unchanged) since the universe was 1 s old, measurement of their properties would offer one of the best tests of the big bang model. Unfortunately, the very characteristic that makes them so useful—they interact so weakly with matter that they have survived unaltered for all but the first second of time—also renders them undetectable, at least with present techniques. Perhaps someday someone will devise a way to capture these elusive messengers from the past.

28.3.4 Atomic Nuclei Form

When the universe was about 3 min old and its temperature was down to about 900 million K, protons and neutrons could combine. At higher temperatures, these atomic nuclei had immediately been blasted apart by interactions with high-energy photons and so could not survive. But at the temperatures and densities reached between 3 and 4 min after the beginning, deuterium (a proton and neutron) lasted long enough that collisions could convert some of it to helium, which consists of two protons and two neutrons (Figure 28.10b). In essence, the entire universe was acting the way centers of stars do today—fusing new elements from simpler components. In addition, a little bit of element 3, lithium, could also form.

This burst of cosmic fusion was only a brief interlude, however. The universe was expanding and cooling down. This meant that no elements beyond lithium could form (it got too cool too fast), and even the light elements stopped forming after a few minutes. In the cool universe we know and love today, the fusion of new elements is limited to the centers of stars and the explosions of supernovae.

Still, the fact that the big bang model allows the creation of a good deal of helium is the answer to a long-standing mystery in astronomy. Put simply, there is just too much helium in the universe to be explained by what happens inside stars. All the generations of stars that have produced helium since the big bang cannot account for the quantity of helium we observe. Furthermore, even the oldest stars and the most distant galaxies show significant amounts of helium. These observations find a natural explanation in the synthesis of helium by the big bang itself during the first few minutes of time. We estimate that *ten times more helium* was manufactured in the first 3 min of the universe than in all the generations of stars during the succeeding 10 to 15 billion years.

28.3.5 Learning from Deuterium

We can learn even more from the way the early universe made atomic nuclei. It turns out that all of the deuterium in the universe was formed during the first 3 min. In stars, any region hot enough to fuse two protons to form a deuterium nucleus is also hot enough to change it further—either by destroying it through a collision with an energetic photon or by converting it to helium through nuclear reactions.

The amount of deuterium produced in the first 3 min of creation depends on the density of the universe at the time deuterium was formed. If the density was relatively high, nearly all the deuterium would have been converted to helium through interactions with protons, just as it is in stars. If the density was relatively low, then the universe expanded and thinned out rapidly enough that some deuterium survived. The amount of deuterium we see today thus gives us a clue to the density of the universe when it was about 3 min old. Theoretical models can relate the density then to the density now; thus, measurements of the abundance of deuterium today can give us an estimate of the current density of the universe.

The deuterium measurements indicate that the present-day density of matter is about 4×10^{-28} kg/m, consistent with earlier estimates that the total mass in the universe is less than the critical density. This is, however, a lower limit on the amount of mass that might be present. The deuterium abundance is determined by the density of protons and neutrons alone, since these are the particles that interact to form it. From the deuterium abundance we know that not enough protons and neutrons are present, by a factor of about 25, to produce a critical-density universe.

If, however, there are dark matter particles of some other kind that are not involved in nuclear reactions, then the density could be higher than what is predicted for ordinary matter alone. As you have read, we have evidence that more mass is in the form of dark matter than in the form of the types of matter familiar to us here on Earth. The dark matter is apparently made of some exotic, unknown kind of particle, and not combinations of protons and neutrons like the readers of this book.

28.4 THE COSMIC BACKGROUND RADIATION

Although fusion stopped after a few minutes, the universe continued to resemble the interior of a star for a few hundred thousand years: It remained hot and opaque, with radiation being scattered from one particle to another. It was still too hot for electrons to "settle down" and become associated with a particular nucleus. And electrons are especially effective at scattering photons, thus ensuring that no radiation ever got very far in the early universe without having its path changed. In a way, the universe was like an enormous crowd right after a popular concert; if you get separated from a friend, even if he is wearing a flashing button, it is impossible to see through the dense crowd to spot him. Only after the crowd clears is there a path for the light from his button to reach you.

Ace✪Astronomy™ Log into AceAstronomy and select this chapter to see the Active Figure called "Cosmic Microwave Background Radiation."

28.4.1 The Universe Becomes Transparent

Not until a few hundred thousand years after the big bang, when the temperature had dropped to about 3000 K and the density of atomic nuclei to about 1000 per cm^3, did the electrons and nuclei combine to form stable atoms of hydrogen and helium (see Figure 28.10c). With no free electrons to scatter photons, the universe became transparent for the first time in cosmic history. From this point on, matter and radiation interacted much less frequently; we say that they *decoupled* from each other and evolved separately. Suddenly, electromagnetic radiation could really travel, and it has been traveling through the universe ever since. If we are to detect the radiation of the early universe, it will come from this *decoupling time* when radiation was first allowed to move over significant distances (Figure 28.11).

One billion years after the big bang, stars and galaxies had begun to form. Deep in the interiors of stars matter

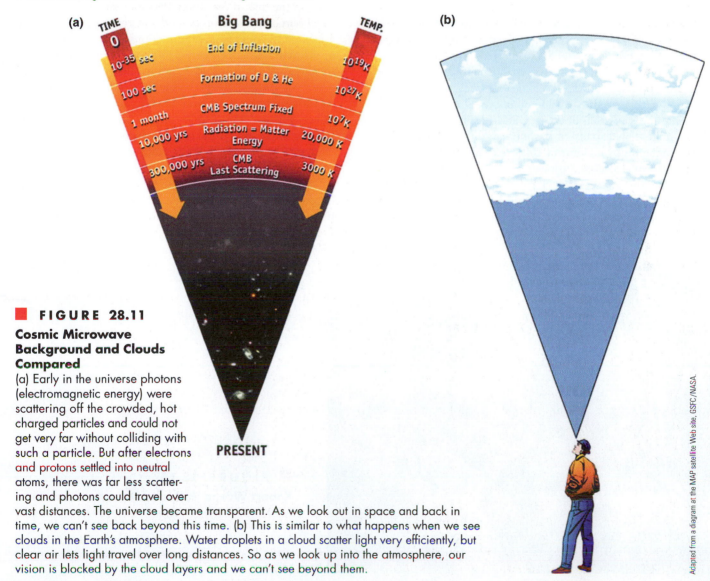

(a)

TIME | Big Bang | TEMP.

0
10^{-35} sec — End of Inflation — 10^{19} K
100 sec — Formation of D & He — 10^{27} K
1 month — CMB Spectrum Fixed — 10^7 K
10,000 yrs — Radiation = Matter Energy — 20,000 K
300,000 yrs — CMB Last Scattering — 3000 K

PRESENT

(b)

■ **FIGURE 28.11**

Cosmic Microwave Background and Clouds Compared
(a) Early in the universe photons (electromagnetic energy) were scattering off the crowded, hot charged particles and could not get very far without colliding with such a particle. But after electrons and protons settled into neutral atoms, there was far less scattering and photons could travel over vast distances. The universe became transparent. As we look out in space and back in time, we can't see back beyond this time. (b) This is similar to what happens when we see clouds in the Earth's atmosphere. Water droplets in a cloud scatter light very efficiently, but clear air lets light travel over long distances. So as we look up into the atmosphere, our vision is blocked by the cloud layers and we can't see beyond them.

Adapted from a diagram at the MAP satellite Web site, GSFC/NASA.

was reheated, nuclear reactions were ignited, and the more gradual synthesis of the heavier elements began. In the meantime, the radiation from the decoupling time continued to cool as space stretched and all radiation became redshifted (thus carrying less and less energy). As billions of years passed, the afterglow of the big bang faded away.

We conclude this quick tour of our model of the early universe with a reminder. You must not think of the big bang as a *localized* explosion *in space,* like an exploding superstar. There were no boundaries and no site of the explosion. It was an explosion *of space* (and time and matter and energy) that happened everywhere in the universe. All matter and energy that exist today, including the particles of which you are made, came from the big bang. We were and still are in the midst of the big bang; it is all around us.

28.4.2 Discovery of the Cosmic Background Radiation

In the late 1940s, Ralph Alpher and Robert Herman, working with George Gamow (see Figure 28.8), realized that just before the universe became transparent, it must have been radiating like a blackbody at a temperature of 3000 K (see Chapter 4). If we could have seen that radiation just after neutral atoms formed, it would have resembled radiation from a reddish star. It was as if a giant fireball filled all of the universe.

But that was nearly 14 billion years ago, and in the meantime, the scale of the universe has increased a thousandfold. This expansion has increased the wavelength of the radiation by a factor of 1000, and, according to Wien's law, has correspondingly lowered the temperature by a factor of 1000 (see Section 4.2). Alpher and Herman predicted that the glow from the fireball should now be at radio wavelengths and should resemble the radiation from a

blackbody at a temperature only a few degrees above absolute zero. Since the fireball was everywhere throughout the universe, the radiation left over from it should also be everywhere. If our eyes were sensitive to radio wavelengths, the whole sky would appear to glow very faintly. However, our eyes are not sensitive to these wavelengths, and at the time Alpher and Herman made their prediction there were no instruments that could detect the glow. Over the years, their prediction was forgotten.

In the mid-1960s in Holmdel, New Jersey, Arno Penzias and Robert Wilson of AT&T's Bell Laboratories were using a delicate microwave antenna (Figure 28.12) to measure the intensity of radio radiation all around the sky (to check on possible sources of radiation that might interfere with communications satellites). They were plagued with some unexpected background noise, just like faint static on a radio, that they could not get rid of. The puzzling thing about this radiation was that it seemed to be coming from all directions at once. This is very unusual in astronomy; after all, most radiation has a specific direction where it is strongest—the direction of the Sun, or a supernova remnant, or the disk of the Milky Way, for example.

Penzias and Wilson at first thought any radiation that appeared to come from all directions must originate from inside their telescope, so they took everything apart to look for the source of the noise. They even found that some pigeons had roosted inside the big horn and had left (as Penzias delicately put it) "a layer of white, sticky, dielectric substance coating the inside of the antenna." However, nothing the scientists did could reduce the background radiation to zero, and they reluctantly came to accept that it must be real and coming from space.

Penzias and Wilson were not cosmologists, but as they began to discuss their puzzling discovery with other scientists, they were quickly put in touch with a group of as-

AT&T Bell Laboratories

■ **FIGURE 28.12**
Robert Wilson (left) and Arno Penzias These two scientists are posing in front of the horn-shaped antenna with which they discovered the cosmic background radiation. The photo was taken in 1978, just after they received the Nobel Prize in physics.

tronomers and physicists at Princeton University (a short drive away). These astronomers had—as it happened—been redoing the calculations of Gamow's group from the 1940s and realized that the radiation from the decoupling time should be detectable as a faint afterglow of radio waves. Their work predicted that the temperature corresponding to this **cosmic background radiation (CBR)** should be about 3. K. Penzias and Wilson found the intensity of the radiation they had discovered matched a blackbody with just that temperature.

Many other experiments on Earth and in space soon confirmed the discovery: The radiation was indeed coming from all directions (it was isotropic) and matched the predictions of the big bang theory with remarkable precision. Penzias and Wilson had inadvertently observed the glow from the primeval fireball. They received the Nobel Prize for their work in 1978. And just before his death in 1966, Lemaître learned that his "vanished brilliance" had been discovered and confirmed.

28.4.3 Properties of the Background Radiation

The first accurate measurements of the CBR were made with a satellite orbiting the Earth. Named the Cosmic Background Explorer (COBE), it was launched by NASA on November 18, 1989 (Figure 28.13). The data it received quickly showed that the CBR closely matches the radiation expected from a blackbody with a temperature of 2.73 K (Figure 28.14). This is exactly the result expected if the CBR is indeed redshifted radiation emitted by a hot gas that filled all of space shortly after the universe began.

The first important conclusion from measurements of the CBR, therefore, is that the universe we have today has evolved from a hot, uniform state. This observation also provides direct support for the idea that we live in an evolving universe, since the universe is cooler today than it was in the beginning.

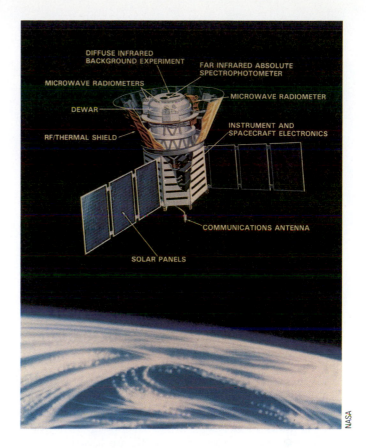

NASA

■ **FIGURE 28.13**

The Cosmic Background Explorer An artist's conception of COBE, the satellite designed to explore the cosmic background radiation at infrared and microwave wavelengths. Various instruments aboard the probe are marked.

A second result (seen with other instruments and now confirmed by COBE) is that the CBR appears to be slightly hotter in one direction than in the exact opposite direction in the sky. This difference comes about because of our own motion through space. If you approach a blackbody, its

■ **FIGURE 28.14**

The Cosmic Background Radiation
The solid line shows how the intensity of radiation should change with wavelength for a blackbody with a temperature of 2.73 K. The boxes show the intensity of the cosmic background radiation as measured at various wavelengths by COBE's instruments. The fit is perfect. When this graph was first shown at a meeting of astronomers, they gave it a standing ovation.

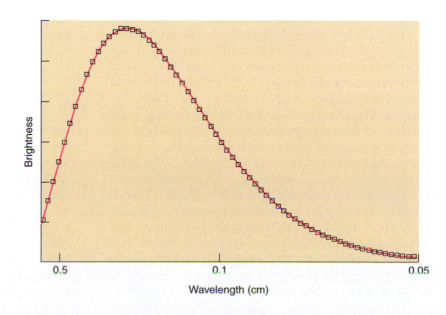

Brightness

Wavelength (cm)

0.5   0.1   0.05

radiation is slightly Doppler-shifted to shorter wavelengths and therefore resembles radiation from a slightly hotter blackbody. If you move away from a blackbody, its radiation is redshifted and resembles the radiation from a slightly cooler object. Since the CBR fills the universe, we can determine how we are moving by noting the directions in which it is redshifted and blueshifted (cooler and hotter) as well as the amount of the shift.

The small temperature differences of the CBR indicate that the Sun, the Milky Way, and the whole Local Group of galaxies are moving at a speed of about 500 km/s in the general direction of the constellation Hydra. Note that this motion is in addition to the motion of these galaxies resulting from the overall expansion of the universe. As we saw in Chapter 27, the extra motion is probably caused by the gravitational attraction of an unusually dense concentration of mostly dark matter (the Great Attractor) that pulls the Local Group toward Hydra.

28.4.4 Small Differences in the CBR

It was known even before the launch of COBE that the CBR is extremely *isotropic*. In fact, its uniformity is one of the best confirmations of the cosmological principle. Ground-based measurements showed that if we look at places in the sky that differ in direction by less than a degree, any fluctuations in the intensity of the CBR are less than a few parts in ten thousand.

According to our theories, however, the temperature could not have been *perfectly* uniform when the CBR was emitted. After all, the CBR is radiation that was scattering from the particles in the universe at the time of decoupling. If the radiation were completely smooth, then all those particles must have been distributed through space absolutely evenly. Yet it is those particles that have become all the galaxies and stars (and astronomy students) that now inhabit the cosmos. Had they been completely smoothly distributed, they could not have formed all the large-scale structure now present in the universe—the clusters and superclusters of galaxies discussed in the last few chapters.

The early universe must have had tiny density fluctuations from which such structure could evolve. Regions of higher than average density would have attracted additional matter and eventually grown into the galaxies and clusters that we see today. For complicated reasons, these denser regions would appear to us to be hot spots; that is, they would have higher than average temperatures. Therefore, if the seeds of present-day galaxies existed at the time the CBR was emitted, we should see some slight changes in the CBR temperature as we look in different directions in the sky.

Scientists working with the data from the COBE satellite did indeed detect very subtle temperature differences in the CBR. The temperature variations are typically only 10 millionths of a degree K. The regions of higher than average temperature come in a variety of sizes, but even the smallest is far too large to be the precursor of an individual galaxy, or even a supercluster of galaxies. This is because

the COBE instrument had "blurry vision" (poor resolution) and could measure only large patches of the sky.

That situation has now been remedied by instruments flown using balloons. One such experiment, launched at the South Pole, was given the clever name of BOOMERANG (balloon observations of millimetric extragalactic radiation and geomagnetics). The first BOOMERANG balloon and its instruments circumnavigated Antarctica, riding on circumpolar winds and remaining aloft for ten days before, true to their name, they were retrieved (Figure 28.15). BOOMERANG had a resolution about 35 times sharper than COBE and could see regions comparable to the full moon in size; that is, it could see areas about 30 arcminutes across. These areas are the right size to be the precursors of clusters and superclusters of galaxies.

Theoretical calculations show that the size of the hot and cold spots in the CBR depends on the total density of the universe. (It's not at all obvious that it should do so, and it takes some pretty fancy calculations—way beyond the level of our text—to make the connection. But having such a dependence is very useful.) The total density we are discussing here includes both the amount of mass in the universe and the value of the cosmological constant. That is, we must add together mass and energy: ordinary matter, dark matter, and the dark energy that is speeding up the expansion.

If the mass density of all three combined is equal to the critical density, then the hot and cold spots should typically be about a degree in size. If we live in a universe with a density greater than critical, then the typical sizes will be larger than one degree. If the universe has a density less than critical, then the structures will appear smaller. In Figure 28.16 you can do the comparison yourself. The BOOMERANG observations indicate that we live in a critical-density universe.

More detailed measurements of the properties of the CBR can tell us even more about the properties of the early universe. In February 2003, the first results from the Wilkinson microwave anisotropy probe (WMAP) spacecraft were reported. Named after David Wilkinson, one of the pioneers in mapping the CBR, the probe was designed to make full-sky maps of the CBR with much greater accuracy and precision than earlier experiments. This experiment confirmed the picture of the universe that had been built up from many separate experiments over the past decade, as summarized in the earlier sections of this chapter, but it improved the precision of the various numbers that describe what the universe is like.

The age of the universe according to the WMAP data is 13.7 billion years, with an uncertainty of only 200,000 years. WMAP confirmed that we live in a critical-density universe. Normal matter (protons and neutrons) make up 4.4 percent of the total density. Dark matter plus normal matter add up to 27 percent of the total density. Dark energy contributes the remaining 73 percent. The Hubble constant is 21.8 km/s per million LY (71 km/s per million parsecs). The age of the universe at decoupling—that is,

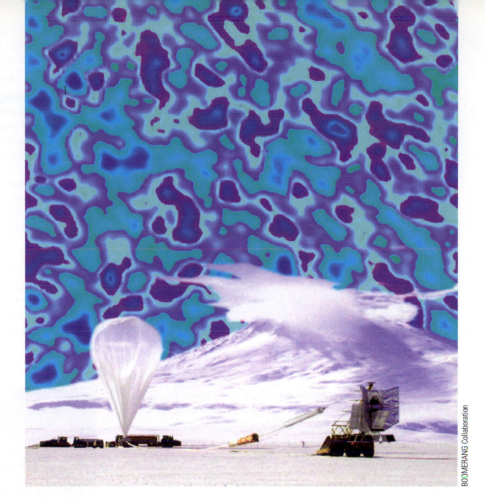

BOOMERANG Collaboration

■ **FIGURE 28.15**

The CMB Sky over Mt. Erebus
In this composite picture, the cosmic microwave background (CMB) sky is seen behind the BOOMERANG balloon, which is being prepared for launch. The BOOMERANG images of the CBR have been overlaid onto the sky to indicate what size the fluctuations would appear to be if our eyes were sensitive to millimeter radiation. The colors in the BOOMERANG image correspond to differences in temperature, with the largest differences being about 10 millionths of a degree K.

25°

BOOMERANG

BOOMERANG Collaboration

■ **FIGURE 28.16**

BOOMERANG Measurements Compared with Predictions of Model Universes   If we live in a critical-density universe (center image in bottom row), then the hot and cold spots will have typical sizes of about one degree. If the density exceeds the critical density, then the structures will appear to be larger than one degree (lower left). If the density is lower than the critical density, then the structures will appear to be smaller than one degree (lower right). As you can see, the BOOMERANG observations are matched best by a critical-density universe.

when the CBR was emitted—was 379,000 years. A new result from WMAP is that the first stars appeared only 200 million years after the big bang.

Perhaps the most surprising result of the WMAP experiment is that there were no surprises. The WMAP results were announced after the rest of this chapter had been written, and if we round the WMAP values for the content and age of the universe to be consistent with the lower accuracy of the earlier estimates, we do not need to change a single number in what was already written. In other words, the very strange universe that we have been describing, with only 4 percent of its contents being made up of the kinds of matter we are familiar with here on Earth, is really the universe we live in.

28.5 **WHAT IS THE UNIVERSE REALLY MADE OF?**

Stop and think about the last section. It should shock you. What we are saying is that 96 percent of the stuff of the universe is either dark matter or dark energy—neither of which has ever been detected in a laboratory on Earth. This whole text, which has focused mostly on objects that emit electromagnetic radiation, has been ignoring 96 percent of what is out there. Who says there aren't big mysteries yet to solve in science!

MAP/NASA

Changing Estimates of the Contents of the Universe This diagram shows the changes in our understanding of the contents of the universe over the past three decades. In the 1970s, we suspected that most of the matter in the universe was invisible, but we thought it might be ordinary matter (protons, neutrons, and so on) that was simply not producing electromagnetic radiation. By the 1980s, it was becoming likely that most of the dark matter was made of something we had not yet detected on Earth. By the late 1990s, a variety of experiments had shown that we live in a critical-density universe and that dark energy contributes about 70 percent of what is required to reach critical density. Note how the estimate of the relative importance of ordinary luminous matter (shown in yellow) has diminished over time.

Figure 28.17 shows how our ideas of the composition of the universe have changed over the past three decades. The fraction of the universe that we think is made of the same particles as astronomy students has been decreasing steadily.

28.5.1   Where Is the Matter?

Let's assume that WMAP and other recent measurements are right and we really do live in a critical-density universe. Have we found all the matter and energy that contribute to it?

As we discussed earlier in this chapter, we can estimate the amount of ordinary matter in the universe from the deuterium abundance. A comparison of observed deuterium abundances with theory indicates that ordinary matter should add up to about 4 percent of the critical density. Note that this value is just what is derived independently from measurements of the CBR. The fact that they agree is a *very* strong indication that our models are on the right track.

Observations so far have detected enough mass in the galaxies and stars to account for about 1 percent of the critical density. Is there more ordinary matter hiding somewhere? Recent observations suggest this is probably the case. Measurements with the Hubble Space Telescope have discovered very hot clouds that lie in the space between galaxies. Light from a very distant quasar was used as a searchlight. The beam from the quasar penetrated through several intergalactic gas clouds on its long journey to Earth, and atoms of oxygen that had lost five of their electrons absorbed some of the light from the quasar (Figure 28.18). This showed as absorption lines in the quasar's spectrum.

Oxygen can lose this many electrons only if its temperature is at least 100,000 degrees. At such a high temperature, every hydrogen atom will have lost its single electron, and so there is no way to detect the hydrogen that is also present in

John Godfrey (STScI)/NASA

Discovering Hot Gas Clouds Between Galaxies The light from a distant quasar passes through intervening clouds on its long journey to Earth. These clouds contain ionized oxygen atoms that absorb some of the light from the quasar. The light from the particular quasar we discuss in the text passed through four separate clouds as it traveled a distance of several billion light years to reach Earth.

these clouds. If, however, we assume that there are about 10,000 hydrogen atoms for every 6 oxygen atoms, which is what we observe in stars, we can then estimate the amount of hydrogen in the hot clouds. It turns out that the amount of matter in these intergalactic clouds is about equal to the amount of luminous matter in stars and galaxies.

Where did all these hot clouds come from? Theories of galaxy formation suggest that the early universe was filled with a web of gas filaments. Clusters of galaxies form in the higher density regions where the filaments intersect. When

TABLE 28.4 *The Contributions of Different Objects to the Density of the Universe*

Object	Density as a Percent of Critical Density
Luminous matter (stars, etc.)	<1
Ordinary matter	4 (some may be in the form of hot intergalactic gas)
Total matter, bright and dark	27
Equivalent mass density of the dark energy	73

"I CAN'T TELL YOU WHAT'S IN THE DARK MATTER SANDWICH. NO ONE KNOWS WHAT'S IN THE DARK MATTER SANDWICH."

large numbers of massive stars in these galaxies exploded as supernovae, they spewed out oxygen atoms synthesized in the centers of stars. The fast-moving oxygen atoms (and other atoms) then left their home galaxies and mixed with hydrogen that had been left behind in the filaments (and was not incorporated into stars or galaxies). The gas was heated by shock waves from colliding filaments, much as shells from supernovae collide with and heat the interstellar medium in our own Galaxy (see Chapter 19).

Table 28.4 summarizes the current best estimates of the amount of matter estimated to be present in various astronomical objects. Luminous matter in galaxies contributes less than 1 percent of the mass required to reach critical density. Another 1 percent may be found in hot gas in intergalactic space. Perhaps the remaining 2 percent of ordinary matter lies hidden in intergalactic space.

So now we have 4 percent. Where is the rest? In Chapter 27, we said that maps of galaxy clusters indicate that there is about seven times as much dark matter as light matter. That puts us up to 32 percent (4 + 28) of the critical density. Given the uncertainties in estimating the ratio of dark to light matter, this is very close to the CBR estimates that matter of all kinds makes up 27 percent of the critical density. The remainder of the critical density is in the form of the cosmological constant or "dark energy" that's been lurking in the background of this whole chapter.

Many astronomers find the situation we have described very satisfying. Several independent experiments now agree on the type of universe we live in and the inventory of what it contains. We seem to be very close to having a cosmological model that explains nearly everything. Others are not yet ready to jump on the bandwagon. They say, "Show me the 96 percent of the universe we can't detect directly—for example, find me some dark matter!"

28.5.2 Dark-Matter Possibilities

How should astronomers go about looking for dark matter? The techniques depend on what we think it might be made of. We described the attempt to look for MACHOs, compact objects that might look dark and yet contain ordinary matter, in Section 27.5.5. Recall that such searches could account for only a small fraction of the dark matter.

Another possible form that dark matter can take is some type of elementary particle that we have not yet detected on Earth—a particle that has mass and exists in sufficient abundance to contribute 23 percent of the critical density. Some physics theories predict the existence of such particles. One class of these has been given the name WIMPs, which stands for *weakly interacting massive particles*. (The name MACHOs was selected to fit well with WIMPs; it's science humor.) Since these particles do not participate in nuclear reactions that lead to the production of deuterium, the deuterium abundance puts no limits on how many WIMPs might be in the universe. (A number of other exotic particles have also been suggested as prime constituents of dark matter, by the way, but we will confine our discussion to WIMPs as a useful example.)

If large numbers of WIMPs do exist, then some of them should be passing through our physics laboratories right now. The trick is to catch them. Since by definition they interact only weakly (infrequently) with other matter, the chances that they will have a measurable effect are small. We also don't know anything about these particles, and so we can't predict what kind of effect they might produce. Nevertheless, more than 20 experiments are now under way to try to determine whether these elusive WIMPs actually exist.

One type of detector is based on the idea that a WIMP might occasionally collide with an atomic nucleus and cause it to move. This motion would then be transferred to other particles in the detector, causing a very(!) small change in temperature. Estimates are that the change might be about one millionth of a degree, which could be measured if the detector is at a temperature very close to absolute zero. Another approach is to make a detector out of material that emits a (very faint) flash of light when one of its nuclei or electrons recoils after being struck by another particle. Yet another idea is to look for microscopic trails of destruction when a WIMP travels through a piece of mica. We have pieces of mica that are more than a billion years old. Over that period of time there should have been at least a few occasions when a WIMP collided with an atom's nucleus, knocking it out of place enough to push other neighboring atoms out of place and leaving a detectable trail of damage.

If it is really true that most of the matter in the universe is made of some type of particle that we have not yet discovered, then we must accept the challenge of trying to detect it. The search is on—in huge accelerators, in university laboratories around the world, and in deep underground mines, where scientists are trying to trap elusive dark matter particles just as they once succeeded in capturing neutrinos. Stay tuned!

28.5.3 Dark Matter and the Formation of Galaxies

Elusive as dark matter may be in today's universe, galaxies would probably never have formed without it. As we have seen in Chapters 25 and 27, galaxies grew from density fluctuations in the early universe. The observations with COBE and other experiments give us information on the size of those fluctuations. It turns out that the density variations are too small, at least according to our current theories, to have formed galaxies in the first billion years or so after the big bang. Yet observations indicate that galaxies were indeed formed that early.

Our instruments that measure the CBR, however, give us information about density fluctuations only for the type of matter that interacts with radiation. Suppose there is a type of matter that does not interact with light at all— namely, dark matter. This matter could have much greater variations in density and might form a kind of gravitational trap that could have begun to attract ordinary matter immediately after the universe became transparent. As ordinary matter became increasingly concentrated, it could have turned into galaxies more quickly thanks to these traps.

For an analogy, imagine a boulevard with traffic lights every half-mile or so. Suppose you are part of a motorcade of cars accompanied by police who lead you past each light, even if it is red. So, too, when the early universe was opaque, radiation carried ordinary matter with it, sweeping past the concentrations of dark matter. Now suppose the police leave the motorcade and it meets some red lights. The lights act as traffic traps; approaching cars now have to stop, and so they bunch up. Likewise, after the early universe became transparent, ordinary matter interacted with radiation only occasionally and so could fall into the dark-matter traps.

The size of the gravitational traps depends on the nature of the dark matter. Suppose it is moving near the speed of light; astronomers call this **hot dark matter.** Neutrinos are a good example of hot dark matter. In this case, any small-scale fluctuations in density would have been smoothed out by the rapid streaming of particles, as they move from high- to low-density regions. With smaller

dense regions not being able to survive, large-scale structure would form first.

If, on the other hand, the dark matter moves slowly— we call this **cold dark matter**—then the particles do not have time to move far enough to smooth out small-scale density fluctuations. In this case, relatively small structures, the size of globular clusters or individual galaxies, are likely to form first.

Neither hot nor cold dark matter is entirely successful in explaining the distribution of galaxies discussed in Chapter 27. Hot dark matter models predict that all galaxies should be found in large sheetlike structures, which are not seen. Cold dark matter cannot produce voids, walls, and long structures such as the Great Attractor. Now theories are being developed that contain both hot and cold dark matter. Even though current models are not adequate to explain how galaxies form, the important point is that galaxies are difficult to form at all unless a substantial amount of dark matter of some kind is present.

28.5.4 The Universe in a Nutshell

Figure 28.19 summarizes the entire history of the observable universe in a single diagram. The universe was very hot when it began to expand. We have fossil remnants of the very early universe in the form of neutrons, protons, electrons, and neutrinos and the atomic nuclei that formed when the universe was about 3 min old: deuterium, helium, and a small amount of lithium. The dark matter also remains, but we do not yet know what form it is in.

The universe gradually cooled, and when it was about four hundred thousand years old and at a temperature of about 3000 K, electrons combined with protons to form hydrogen atoms. At this point the universe became transparent to light, and we have detected the CBR emitted at this time. The universe still contained no stars or galaxies, and so it entered what astronomers have begun to call "the dark ages" (since stars were not lighting up the darkness). During the next several hundred million years, small fluctuations in density of the dark matter grew and formed gravitational traps that concentrated the ordinary matter, which began to form stars and galaxies.

By the time the universe was about two hundred million years old, it had begun its own renaissance; it was again blazing with radiation, but this time from newly formed stars, star clusters, and small galaxies. Over the next several billion years, small galaxies merged to form the giants we see today. Clusters and superclusters of galaxies began to grow, and the universe eventually began to resemble the structure we described in Chapter 27.

During the next 20 years, astronomers plan to build giant new telescopes both in space and on the ground to explore even farther back in time. In 2012, we hope to launch a successor to the Hubble Space Telescope, a 6.5-m telescope that will be assembled in space. The predictions are that with this powerful instrument (see the image that

FIGURE 28.19

The History of the Universe This image summarizes the changes that have occurred in the universe during the last 14 billion years. Protons, deuterium, helium, and some lithium were produced in the initial fireball. A few hundred thousand years after the Big Bang, the universe became transparent to electromagnetic radiation for the first time. COBE and other instruments have been used to study the radiation emitted at that time and still visible today (the cosmic background radiation). The universe was then dark (except for this background radiation) until the first stars began to form about 200 million years after the big bang. The James Webb Space Telescope is being designed to look back to the time when the first galaxies formed. Existing space and ground-based telescopes have made substantial progress in studying the subsequent evolution of galaxies.

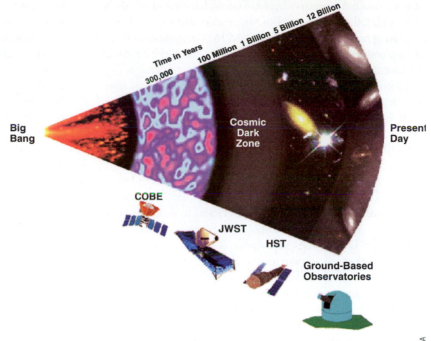

opens this chapter) we should be able to look back far enough to observe the formation of the first galaxies.

28.6 THE INFLATIONARY UNIVERSE

The hot big bang model that we have been describing is remarkably successful. It accounts for the expansion of the universe, explains the observations of the CBR, and correctly predicts the abundances of the light elements. As it turns out, this model also predicts that there should be exactly three types of neutrinos in nature, and this prediction has been confirmed by experiments with high-energy accelerators. The theory is not complete, however.

28.6.1 Problems with the Standard Big Bang Model

There are a number of observed facts about the universe that the standard big bang model does not explain. For example, we are unsure about the origin of the density fluctuations that ultimately grew into galaxies. And we are puzzled by the remarkable *uniformity* of the universe. The CBR is the same, no matter in which direction we look, to an accuracy of about one part in 100,000. This sameness might be expected if all the parts of the visible universe were in contact at some point and had time to come to the same temperature. In the same way, if we put some ice into a glass of lukewarm water and wait a while, the ice will melt and the water will cool down until they are the same temperature.

If we accept the standard big bang model, however, all parts of the visible universe were *not* in contact at any time. The fastest information can travel from one point to another is the speed of light. There is a maximum distance that light can have traveled from any point since the time the universe began. This distance is called that point's *horizon distance* because anything farther away is "below its horizon"—unable to make contact with it. One region of space separated from another by more than the horizon distance has been completely isolated from it through the entire history of the universe.

If we measure the CBR in two opposite directions in the sky, we are observing regions that were significantly beyond each other's horizon distance at the time the CBR was emitted. We can see both regions, but *they* can never have seen each other! Why, then, are their temperatures so precisely the same? According to the standard big bang model, they have never been able to exchange information, and there is no reason they should have identical temperatures. (It's a little like seeing the clothes that all the students wear at two schools in different parts of the world become identical, without the students ever having been in contact.) The only explanation is simply that the universe somehow *started out* being absolutely uniform (which is like saying all students were born liking the same clothes). Scientists are always uncomfortable when they must appeal to a special set of initial conditions to account for what they see.

Another problem with the standard big bang model is that it does not explain why the density of the universe is equal to the critical density. The mass density could have been, after all, so low and the effects of dark energy so high that the expansion would have been too rapid to form any galaxies at all. Alternatively, there could have been so much matter that the universe would have already begun to contract long before now. The standard big bang model offers no explanation of why the actual density should be poised so close to the critical value.

Before we can discuss some intriguing new ideas that seek to explain these characteristics of the universe, we must first digress and talk about the forces acting on subatomic particles. Then we will return to discussing the grand picture of how the universe might have evolved.

28.6.2 Grand Unified Theories

In physical science, the term *force* is used to describe anything that can change the motion of a particle or body (see Chapter 2). One of the remarkable discoveries of modern science is that all known physical processes can be described through the action of four forces: gravity, electromagnetism, the strong nuclear force, and the weak nuclear force (Table 28.5).

Although gravity is perhaps the most familiar to you, and certainly appears strong if you jump off a tall building, the force of gravity between two elementary particles—say, two protons—is by far the weakest of the four forces. Electromagnetism, which includes both magnetic and electrical forces, holds atoms together and produces the electromagnetic radiation that we use to study the universe. The weak nuclear force is weak only in comparison to its strong "cousin" but is in fact much stronger than gravity.

Both the weak and strong nuclear forces differ from the first two forces in that they act only over very small distances—those comparable to the size of an atomic nucleus or less. The weak force is involved in radioactive decay and in reactions that result in the production of neutrinos. The strong force holds protons and neutrons together in an atomic nucleus (as we saw in Chapter 15).

Physicists have wondered why there are four forces in the universe—why not 300 or, preferably, just one? An important hint comes from the name *electromagnetic* force. For a long time scientists thought that the forces of electricity and magnetism were separate, but James Clerk Maxwell (see Chapter 4) was able to *unify* these forces, to show that they are aspects of the same phenomenon. In the same way, many scientists (including Einstein) have wondered

whether the four forces we now know could also be unified. Physicists have developed models, called **grand unified theories (GUTs),** that unify three of the four forces.

In these theories, the strong, weak, and electromagnetic forces are not three independent forces but instead are different manifestations or aspects of what is, in fact, a single force. The theories predict that at high enough temperatures there would be only one force. At lower temperatures (like the ones in the universe today), however, this single force has changed into three different forces (Figure 28.20). Just as different gases freeze at different temperatures, we can say that the different forces "froze out" of the

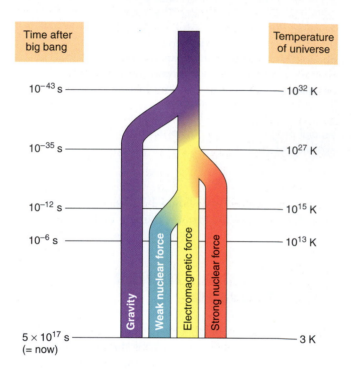

■ **FIGURE 28.20**

The Four Forces That Govern the Universe The strength of the four forces depends on the temperature of the universe. This diagram (inspired by some grand unified theories) shows that at very early times, when the temperature of the universe was very high, all four forces were indistinguishable. As the universe cooled, the forces took on separate and distinctive characteristics.

TABLE 28.5	*The Forces of Nature*		
Force	**Relative Strength Today**	**Range**	**Important Applications**
Gravity	1	Whole universe	Motions of planets, stars, galaxies
Electromagnetism	10^{36}	Whole universe	Atoms, molecules, electricity, magnetic fields
Weak nuclear force	10^{33}	10^{-17} m	Radioactive decay
Strong nuclear force	10^{38}	10^{-15} m	The existence of atomic nuclei

unified force at different temperatures. Unfortunately, the temperatures at which the three forces were one are so high that they cannot be reached in any terrestrial laboratory. Only the early universe, at times prior to 10^{-35} s, was hot enough to unify these forces. (Many physicists think that gravity is also unified with the other forces at still higher temperatures.)

28.6.3 The Inflationary Hypothesis

Some forms of the GUTs predict that a remarkable event occurred when the universe was about 10^{-35} s old and the forces were starting to separate. The equations of general relativity, combined with the special state of matter at that time, predict that gravity could briefly have been a repulsive force. In our own time, gravity is an attractive force that slows the expansion of the universe, but for a brief instant near 10^{-35} s after the expansion began, gravity could actually have accelerated the expansion. It is as if the cosmological constant had, for a brief instant, been very large and produced a tremendous repulsive force in the universe.

A model universe in which this rapid, early expansion occurs is called an **inflationary universe.** The inflationary universe is identical to the big bang universe for all time after the first 10^{-30} s. Prior to that, there was a brief period of extraordinarily rapid expansion or inflation during which the scale of the universe increased by a factor of about 10^{50} times more than predicted by standard big bang models (Figure 28.21). As the universe expanded, its temperature dropped below the critical value at which all three forces behave in a symmetrical fashion. In the cooler, asymmetrical universe, the nuclear forces dominated the electromagnetic force; they continue to do so in our world today.

Prior to the inflation, all the parts of the universe that we can now see were so small and close together that they could exchange information; that is, the horizon distance included all of the universe that we can now observe. Before inflation occurred, there was adequate time for the observable universe to homogenize itself and come to the same temperature. Then inflation expanded those regions tremendously, so that many parts of the universe are now beyond each other's horizon.

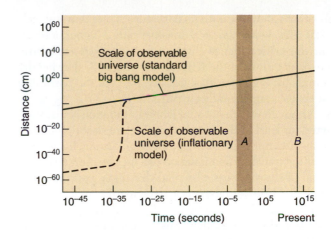

■ **FIGURE 28.21**

Expansion of the Universe   This graph shows how the scale factor of the observable universe changes with time for the standard big bang model (solid line) and for the inflationary model (dashed line). (Note that the time scale at the bottom is extremely compressed.) During inflation, regions that were very small and in contact with each other are suddenly blown up to be much larger and outside each other's horizon distance. The two models are the same for all times after 10^{-30} s. Electrons, positrons, and the lightest atomic nuclei are formed during the time interval labeled A. The universe becomes transparent to radiation at the time designated B. "Now" is at the right edge of the figure.

Another appeal of the inflationary model is its prediction that the density of the universe should be exactly equal to the critical density. To see why this is so, remember that the curvature of spacetime is intimately linked to the density of matter. (We discussed in Chapter 23 that the stronger the gravity, the greater the curvature or warping of spacetime.) If the universe began with some curvature in its spacetime, one analogy for it might be the skin of a balloon. The period of inflation was equivalent to blowing up the balloon to a tremendous size. The universe became so big that from our vantage point no curvature should be visible (Figure 28.22). In the same way, the Earth's surface is so big that it looks flat to us in any location. Calculations

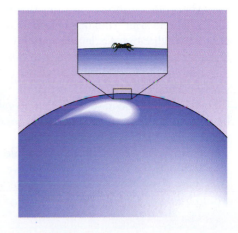

■ **FIGURE 28.22**

Analogy for Inflation   During a period of rapid inflation, a curved balloon grows so large that to any local observer it looks flat.

show that a universe with no curvature is one that is at critical density. Universes with densities either higher or lower than the critical density would show marked curvature. The observations of the CBR in Figure 28.15, showing the universe having the critical density, rule out the possibility that space is significantly curved.

Although we are not sure whether inflation actually occurred (at least in the simple form we are describing), scientists are impressed by the number of known characteristics of the universe that inflation can explain. Yet, if you are typical of the students who read this book, you may have found this brief discussion of dark matter, inflation, and cosmology a bit frustrating. We have offered glimpses of theories and observations but have been unable to provide satisfying answers to some of the problems we have raised. These ideas are at the forefront of modern science, where questions are more numerous than answers, and much more work is needed before we can see clearly. Bear in mind that less than a century has passed since Hubble demonstrated the existence of other galaxies. The quest to understand just how the universe of galaxies came to be will keep astronomers busy for a long time to come.

28.7 THE ANTHROPIC PRINCIPLE

Despite our uncertainties, we must admit that the picture we have developed about the evolution of our universe is a remarkable one. With new telescopes we have begun to collect enough observational evidence that we can describe how the universe evolved from a mere fraction of a second after the expansion began. Although this is an impressive achievement, we still cannot explain some characteristics of the universe—and yet, it turns out that if these characteristics were any different, we would not be here to ask about them. Let's look at some of these "lucky accidents," beginning with the observations of the cosmic background radiation (CBR).

28.7.1 Lucky Accidents

As we described in this chapter, the CBR is radiation that was emitted when the universe was a few hundred thousand years old. Observations show that the temperature of the radiation varies from one region to another typically by about 10 millionths of a degree, and these temperature differences signal small differences in density. But suppose the tiny, early fluctuations in density had been much smaller. Then calculations show that the pull of gravity near them would have been so small that no galaxies would ever have formed. What if the fluctuations in density had been much larger? Then it is possible that very dense regions would have condensed, and these would simply have collapsed directly to black holes without ever forming galaxies and stars. Even if galaxies had been able to form, space would have been filled with intense x rays and gamma rays,

and it would have been difficult for life-forms to develop and survive. The density of stars within galaxies would be so high that interactions and collisions among them would be frequent. In such a universe, planetary systems could rarely survive long enough for life to develop.

Another lucky accident is that the universe is so finely balanced between expansion and contraction. It is expanding, but very slowly. If the expansion had been at a much faster rate, all of the matter would have thinned out before galaxies could form. If everything were expanding at a much slower rate, then—even with so little matter in the universe—the expansion would have reversed and all of the matter would have recollapsed, probably into a black hole—again, no stars, no planets, no life.

The development of life-forms on Earth depends on still more lucky coincidences. Had matter and antimatter been present initially in exactly equal proportions, then all matter would have been annihilated and turned into pure energy. We owe our existence to the fact that there was slightly more matter than antimatter. (After most of the matter made contact with an equal amount of antimatter, turning into energy, a small amount of additional matter must have been present. We are all descendants of that bit of "unbalanced" matter.)

If nuclear fusion reactions occurred at a somewhat faster rate than they actually do, then at the time of the initial fireball, all of the matter would have been converted from hydrogen to helium to carbon and all the way to iron (the most stable nucleus). That would mean that no stars would have formed, since the existence of stars depends on there being light elements, which can undergo fusion and make the stars shine. In addition, the structure of atomic nuclei had to be just right to make it possible for three helium atoms to come together easily to fuse carbon, which is the basis of life. If the triple alpha process we discussed in Chapter 21 were too unlikely, not enough carbon would have formed to lead to biology as we know it. At the same time, it had to be hard enough to fuse carbon to oxygen that a large amount of carbon survived for billions of years.

Neutrinos have to interact with matter at just the right infrequent rate. Supernova explosions occur when neutrinos escape from the cores of collapsing stars, deposit some of their energy in the surrounding stellar envelope, and cause it to blow out and away into space. The heavy elements that are ejected in such explosions are essential ingredients of life on Earth. If neutrinos did not interact with matter at all, they would escape from the cores of collapsing stars without causing the explosion. If neutrinos interacted strongly with matter, they would remain trapped in the stellar core. In either case, the heavy elements would remain locked up inside the collapsing star.

If gravity were a much stronger force than it is, stars could form that contain much smaller masses, and their lifetimes would be measured in years rather than billions of years. Chemical processes, on the other hand, would not be speeded up if gravity were a stronger force, and so there would be no time for life to develop while stars were so short-lived. Even if life did develop in a stronger-gravity universe, life-forms would have to be tiny or they could not stand up or move around.

28.7.2 What Had to Be, Had to Be

In summary, a specific set of rules and conditions in the universe has allowed complexity and life on Earth to develop. As yet we have no theory to explain why this "right" set of conditions occurred. For this reason, many scientists are beginning to accept an idea we call the **anthropic principle**—namely, that the physical laws must be what they are precisely because these are the only laws that allow for the existence of humans.

Some scientists speculate that our universe is but one of countless universes, each with a different set of physical laws. Some of those universes might be stillborn, collapsing before any structure forms. Others may expand so quickly that they remain essentially featureless with no stars and galaxies. In other words, there may be a much larger universe that contains our own. This larger universe (existing perhaps in more dimensions that we can become aware of)
is infinite and eternal; it generates many, many inflating regions, each of which evolves into a separate universe, which may be completely unlike any of the other separate universes. Our universe is then the way it is because it is the only way it could be and have humans like ourselves in it to discover its properties.

It is difficult to know how to test these ideas because we can never make contact with any other universe. For some scientists, our discussion in this section borders on the philosophical and metaphysical. Perhaps someday our understanding of physics will develop to the point that we can know why the gravitational constant is as strong as it is, why the universe is expanding at exactly the rate it is, and why all of the other "lucky accidents" happened—why they were inevitable and could be no other way. Then this anthropic idea would no longer be necessary. No one knows, however, whether we will ever have an explanation for why this universe works the way it does.

We have come a long way in our voyage through the universe. We have learned a remarkable amount about *how* and *when* the cosmos came to be, but the question of *why* the universe is the way it is remains as elusive as ever.

SURFING THE WEB

Cosmology: A Research Briefing:
www.nap.edu/readingroom/books/cosmology
A 1995 report by the Panel on Cosmology of the National Research Council, this long document is a mostly nontechnical summary of the key areas of cosmological research and plans for the future.

Cosmology for Beginners:
epunix.biols.susx.ac.uk/Home/John_Gribbin/
Science writer John Gribbin (who has a PhD in astrophysics) explains selected topics in modern cosmological research, including inflation, dark matter, and the work of Gamow.

Ned Wright's Cosmology Tutorial:
www.astro.ucla.edu/~wright/cosmolog.htm
Wright, a professor of astronomy at UCLA, keeps a reliable (if somewhat technical) summary of cosmological ideas and developments on this site. It is especially useful for students who have had a bit of algebra and physics.

Introduction to the Cosmic Microwave Background:
www.astro.ubc.ca/people/cmb_intro.html
Astronomers Douglas Scott and Martin White give background information, frequently asked questions, and recent results about the CBR.

Penzias and Wilson Nobel Prize site:
www.nobel.se/physics/laureates/1978/index.html
Includes information on the scientists and their work (see also: www.bell-labs.com/user/apenzias).

Specific Missions and Projects to Investigate the CBR:

- *COBE Satellite:*
  lambda.gsfc.nasa.gov/product/cobe/c_overview.html

- *BOOMERANG Project:*
  cmb.phys.cwru.edu/boomerang

- *Microwave Anisotropy Project Home Page:*
  map.gsfc.nasa.gov/
  MAP is a NASA satellite that takes off where COBE left off.

For sites about dark matter, see the listings for Chapter 27.

SUMMARY

28.1 **Cosmology** is the study of the organization and evolution of the universe. The universe is expanding, and this is one of the key observational starting points for modern cosmological theories. Before Hubble showed that the universe was expanding, Einstein introduced a **cosmological constant** into his equations to produce an outward pressure to counterbalance gravity. In an expanding universe, this constant represents a kind of **dark energy** that accelerates the expansion. Measurements of the distant supernovae indicate that the expansion of the universe is indeed accelerating. To estimate the age of the universe we must allow for changes in the rate of expansion. Gravity acts to slow the expansion, while the cosmological constant or some form of dark energy acts as an accelerator. After allowing for these effects, astronomers estimate that all of the matter within the universe was concentrated in an infinitesimally small volume roughly 14 billion years ago, a time we call the **big bang.** This estimate agrees with age estimates of the oldest stars and of the abundance of uranium recently measured in a very old star.

28.2 The big bang happened throughout space everywhere at once, and the universe has been expanding ever since. That means it undergoes a change in scale with time; space stretches and distances grow larger by the same factor everywhere at a given time. In theory, the universe in the future might contract to a singularity if the density of matter is high enough to reverse the expansion. Alternatively, a low-density universe will expand forever. Observations show that we live in a universe with a low density of matter. When combined with the effects of dark energy, this low density means that the expansion will continue to accelerate and galaxies will move ever farther apart.

28.3 The universe cools as it expands. The energy of photons is determined by their temperature, and calculations show that in the hot, early universe photons had so much energy that when they collided with one another, they could produce material particles. As the universe expanded and cooled, protons and neutrons formed first; then came electrons and positrons. Next fusion reactions produced deuterium, helium, and lithium nuclei.

28.4 When the universe became cool enough to form neutral hydrogen atoms, the universe became transparent to radiation. Scientists have detected the **cosmic background radiation (CBR)** from the hot, early universe. Measurements with the COBE satellite show that the CBR is a blackbody with a temperature of 2.73 K. Tiny fluctuations in the CBR may show us the seeds of large-scale structure in the universe. Detailed measurements of the these fluctuations show that we live in a critical density universe and that the critical density is composed of 27 percent matter, including dark matter, and 73 percent dark energy. Ordinary matter—the kinds of elementary particles we find on Earth—make up only 4 percent of the critical density. CBR measurements also indicate that the universe is 13.7 billion years old and that the first stars formed 200 million years after the expansion began.

28.5 Twenty-three percent of the critical density is composed of dark matter. To explain so much dark matter, some physics theories predict that additional types of particles should exist. One type has been given the name of WIMPs (weakly interacting massive particles). Dark matter plays an essential role in forming galaxies. Since by definition these particles interact only very weakly (if at all) with radiation, they could have congregated while the universe was still very hot and filled with radiation. They would thus have formed gravitational traps that quickly attracted and concentrated ordinary matter after the universe became transparent and matter and radiation decoupled. This rapid concentration of matter enabled galaxies to form by the time the universe was a billion years old.

28.6 The big bang model does not explain why the CBR has the same temperature in all directions. Neither does it explain why the density of the universe is so close to the critical density. New **grand unified theories (GUTs),** which predict a period of very rapid expansion, or **inflation,** when the universe was 10^{-35} s old, are being developed to try to explain these observations.

28.7 Recently, many cosmologists have noted that the existence of humans depends on the fact that certain properties of the universe—the size of density fluctuations in the early universe, the strength of gravity, the structure of atoms—were just right. The idea that physical laws must be the way they are because otherwise we could not be here to measure them is called the **anthropic principle.**

INTER-ACTIVITY

A One of the most exciting things about taking a modern astronomy class is being able to make a list like Table 28.1, giving the characteristics of the universe as a whole. Have your group look at the table after you have read Chapter 28 and talked about it in class. Would you add anything to the list? In other words, can you come up with other known facts or principles that apply to the entire universe?

B This chapter deals with some pretty big questions and ideas. Some belief systems teach us that there are questions to which "we were not meant to know" the answers. Other people feel that if our minds and instruments are capable of exploring a question, then it becomes part of our birthright as thinking human beings. Have your group discuss your personal reactions to discussing questions like the beginning of time and space and the ultimate fate of the universe. Does it make you nervous to hear about scientists discussing these issues? Or is it exciting to know that we can now gather scientific evidence about the origin and fate of the cosmos? (In discussing this, you may find that members of your group strongly disagree; try to be respectful of each others' points of view.)

C A popular model of the universe in the 1950s and 1960s was the so-called steady-state cosmology. In this model, the universe was not only the same everywhere and in all directions (homogeneous and isotropic) but also the same *at all times*. We know the universe is expanding and the galaxies are thinning out, and so this model hypothesized that new matter was continually coming into existence to fill in the space between galaxies as they moved farther apart. If so, the infinite universe did not have to have a sudden beginning but could simply exist forever in a steady state. Have your group discuss your reaction to this model. Do you find it more appealing philosophically than the big bang model? Can you cite some evidence that indicates that the universe was not the same billions of years ago as it is now—that it is *not* in a steady state?

D One of the lucky accidents that characterizes our universe is that the timescale for the development of intelligent life on Earth and the lifetime of the Sun are comparable. Have your group discuss what would happen if the two timescales were very different. Suppose, for example, that the time for intelligent life to evolve was ten times longer than the main-sequence lifetime of the Sun. Would our civilization have ever developed? Now suppose the time for intelligent life to evolve was ten times shorter than the main-sequence lifetime of the Sun. Would we be around? (This latter discussion requires considerable thought, including such ideas as what the early stages in the Sun's life were like and how much the early Earth was bombarded by asteroids and comets.)

REVIEW QUESTIONS

Ace Astronomy™ Assess your understanding of this chapter's topics with additional quizzing and animations at **http://ace.brookscole.com/voyages**

1. What are the basic observations about the universe that any theory of cosmology must explain?
2. Describe some possible futures for the universe. What property of the universe determines which of these possibilities is the correct one?
3. Which formed first in the early universe: protons and neutrons, or electrons and positrons? Why?
4. Which formed first: hydrogen nuclei or hydrogen atoms? Explain the sequence of events that led to each.

5. Describe at least two characteristics of the universe that are explained by the standard big bang model.
6. Describe two properties of the universe that are not explained by the standard big bang model (without inflation). How does inflation explain these two properties?
7. Why do astronomers believe there must be dark matter that is not in the form of atoms with protons and neutrons?
8. What is dark energy and what evidence do astronomers have that it is an important component of the universe?
9. Describe the anthropic principle. What are three properties of the universe that make it "ready" to have life-forms like you in it?

THOUGHT QUESTIONS

10. What is the most useful probe of the early evolution of the universe: a giant elliptical galaxy or an irregular galaxy such as the Large Magellanic Clouds? Why?

11. What are the advantages and disadvantages of using quasars to probe the early history of the universe?

12. Describe the evidence that the expansion of the universe is accelerating. Would acceleration occur if the universe were composed entirely of matter (that is, if there were no dark energy)?

13. Suppose the universe expands forever. Describe what will become of the radiation from the primeval fireball. What will the future evolution of galaxies be like? Could life as we know it survive forever in such a universe? Why?

14. Some theorists expected that observations would show that the density of matter in the universe is just equal to the critical density. Do the current observations support this hypothesis?

15. Summarize the evidence for the existence of dark matter in the universe.

16. In this text, we have discussed numerous motions of the Earth as it travels through space with the Sun. Describe as many of these as you can.

17. There are a variety of ways of estimating the ages of various objects in the universe. Describe some of these ways, and indicate how well they agree with one another and with the age of the universe itself as estimated by its expansion.

18. Since the time of Copernicus, each revolution in astronomy has moved humans farther from the center of the universe. Now it appears that we may not even be made of the most common form of matter. Trace the changes in scientific thought about the central nature of the Earth, the Sun, and our Galaxy on a cosmic scale. Explain how the notion that most of the universe is made of dark matter continues this "Copernican tradition."

19. The anthropic principle says that in some sense we are observing a special universe; if the universe were different, we could never have come to exist. Comment on how this fits with the Copernican tradition described in question 18.

20. Construct a time line for the universe, and indicate when various significant events occurred, from the beginning of the expansion to the formation of the Sun to the appearance of humans on Earth.

FIGURING FOR YOURSELF

As we saw in the chapter, the *critical density* is that combination of matter and energy that brings the universe coasting to a stop at time infinity. Einstein's equations lead to the following expression for the critical density (ρ_{crit}):

$$\rho_{crit} = \frac{3H^2}{8\pi G}$$

where H is the Hubble constant and G is the universal constant of gravity (6.67×10^{-11} Nm^2/kg^2). Let's substitute our best values and see what we get. Take an $H = 22$ km/s per million LY. We need to convert both km and LY to meters for consistency. A million LY $= 10^6 \times 9.5 \times 10^{15}$ m $= 9.5 \times 10^{21}$ m. And 22 km/s $= 2.2 \times 10^4$ m/s. That makes $H = 2.3 \times 10^{-18}$/s and $H^2 = 5.36 \times 10^{-36}$/s^2. So,

$$\rho_{crit} = \frac{3 \times 5.36 \times 10^{-36}}{8 \times 3.14 \times 6.67 \times 10^{-11}}$$

$$= 9.6 \times 10^{-27} \text{ kg/m}^3$$

which we can round off to the 10^{-26} kg/m^3 given in Section 28.2.3. (To make the units work out, you have to know that N, the unit of force, is the same as kg $\times$ m/s^2.)

Now we can compare densities we measure in the universe to this critical value. Note that density is mass per unit volume, but energy has an equivalent mass of $m = E/c^2$ (from Einstein's equation $E = mc^2$).

21. Suppose the Hubble constant is not 22 but 33 km/s per million LY. Then what would the critical density be?

22. Assume that the average galaxy contains $10^{11} M_{Sun}$ and that the average distance between galaxies is 10 million LY. Calculate the average density of matter (mass per unit volume) in galaxies. What fraction is this of the critical density we calculated above?

Does the cosmic microwave background radiation (CBR) contribute significantly to the density of the universe? Since mass and energy are equivalent, we can convert the energy of the photons in the CBR to the equivalent amount of mass and compare it with the critical density. Let's walk through the steps in the following problems, and see what the answer is.

23. The CBR contains roughly 400 million photons per m^3. The energy of each photon depends on its wavelength. So the first step is to calculate the typical wavelength of a CBR photon. The CBR is blackbody radiation at a temperature of 2.73 K. According to Wien's law, the peak wavelength in nanometers is given by $\lambda_{max} = 3 \times 10^6/T$. Calculate the wavelength at which the CBR is a maximum and, to make the units consistent, convert this wavelength from nanometers to meters.

24. The next step is to calculate the energy of a typical photon. Assume for this approximate calculation that each photon has the wavelength calculated in problem 23. The energy of a photon is given by $E = hc/\lambda$, where h is Planck's constant and is equal to 6.626×10^{-34} joule $\times$ s, c is the speed of light in m/s, and λ is the wavelength in meters.

25. To get the energy in a cubic meter of space, multiply the energy per photon calculated in problem 24 by the number of photons per cubic meter given above.

26. To convert this energy to an equivalent in mass, use Einstein's equation $E = mc^2$. Divide the energy per m^3 calculated in problem 25 by the speed of light squared. Check your units; you should have an answer in kg/m^3. Now compare this answer with the critical density. Your answer should be several powers of ten smaller than the critical density. In order words, the contribution of the CBR photons to the overall density of the universe is much, much smaller than the contribution made by stars and galaxies.

27. There is still some uncertainty in the Hubble constant. Current estimates range from about 19.9 km/s per million LY to 23 km/s per million LY. Assume that the Hubble constant has been constant since the big bang. What is the possible range in the ages of the universe? Use the equation in the text, $T_0 = 1/H$, and make sure you use consistent units. Twenty years ago, estimates for the Hubble constant ranged from 50 to 100 km/s per megaparsec. What are the possible ages for the universe from those values? Can you rule out some of these possibilities on the basis of other evidence?

28. It is possible to derive the age of the universe given the value of the Hubble constant and the distance to a galaxy, again with the assumption that the value of the Hubble constant has not changed since the big bang. Consider a galaxy at a distance of 400 million LY receding from us at a velocity, v. If the Hubble constant is 20 km/s per million LY, what is its velocity? How long ago was that galaxy right next door to our own Galaxy if it has always been receding at its present rate? Express your answer in years. Since the universe began when all galaxies were very close together, this number is a rough estimate for the age of the universe.

SUGGESTIONS FOR FURTHER READING

Good Popular-Level Cosmology Books

Croswell, K. *The Universe at Midnight: Observations Illuminating the Cosmos.* 2001, Free Press. Updates on cosmology. Lots on dark matter, the accelerating universe, and so on.

Davies, P. *The Last Three Minutes.* 1994, Basic Books. Introduction to the ultimate fate of the universe.

Ferris, T. *The Whole Shebang: A State-of-the-Universe Report.* 1997, Simon & Schuster. A distinguished science journalist reports on cosmology today.

Goldsmith, D. *The Runaway Universe: The Race to Find the Future of the Cosmos.* 2000, Perseus. A first report on the supernova results and cosmological implications.

Guth, A. *The Inflationary Universe.* 1997, Addison-Wesley. One of the key scientists responsible for the inflationary hypothesis describes how it came about.

Hogan, C. *The Little Book of the Big Bang.* 1998, Copernicus. A concise introduction by an astronomer active in the field.

Krauss, L. *Quintessence: The Mystery of Missing Mass in the Universe.* 2000, Basic Books. On dark matter and its cosmological implications.

Livio, M. *The Accelerating Universe.* 2000, John Wiley. Poetic introduction to the new cosmology by a scientist who works with the Hubble Space Telescope.

Mather, J. and Boslaugh, J. *The Very First Light.* 1996, Basic Books/HarperCollins. A scientist and science writer recount the science and politics of the COBE satellite.

Overbye, D. *Lonely Hearts of the Cosmos.* 1991, Harper Collins. Wonderful introduction to cosmology in the 1980s, with a focus on the people involved.

Rees, M. *Before the Beginning: Our Universe and Others.* 1997, Helix. The Astronomer Royal of England, and one of the leading astronomers of our time, explains many well-established and speculative aspects of cosmology.

Silk, J. *The Big Bang.* 3rd ed. 2001, Freeman. Somewhat more technical.

Articles on Cosmology

Adams, F. and Laughlin, G. "The Future of the Universe" in *Sky & Telescope,* Aug. 1998, p. 32. A look into the distant future in an open universe. (Also see their "Embracing the End" in *Astronomy,* Oct. 2000, p. 48.)

Bucher, M. and Spergel, D. "Inflation in a Low-Density Universe" in *Scientific American,* Jan. 1999, p. 62. On new and improved inflation theories.

Caldwell, R. and Kamionkowski, M. "Echoes from the Big Bang" in *Scientific American,* Jan. 2001, p. 38. On studying the details of the CMB to learn about the universe.

Chaboyer, B. "Rip Van Twinkle: The Oldest Stars Have Been Growing Younger" in *Scientific American,* May 2001, p. 44. On determining the age of the oldest stars, and how it fits with the Hubble time.

Davies, P. "Everyone's Guide to Cosmology" in *Sky & Telescope,* March 1991, p. 250. Good introductory article.

Falk, D. "An Interconnected Universe: Exploring the Topology of the Cosmos" in *Sky & Telescope,* July 1999, p. 45.

Ferris, T. "Inflating the Cosmos" in *Astronomy,* July 1997, p. 38. On the inflationary hypothesis.

Fienberg, R. "COBE Confronts the Big Bang" in *Sky & Telescope,* July 1992, p. 34. Good summary of the temperature fluctuations discovery.

Finkbeiner, A. "Cosmic Yardsticks: Supernovae and the Fate of the Universe" in *Sky & Telescope,* Sept. 1998, p. 38. Clear introduction to using supernovae as standard bulbs and some intriguing results.

Grimes, K. and Boyle, A. "The Universe Takes Shape" in *Astronomy,* Oct. 2002, p. 34. On experiments to determine the geometry of the universe.

Hogan, C., et al. "Surveying Spacetime with Supernovae" in *Scientific American,* Jan. 1999, p. 46. On hints of an accelerating universe. (See also: Krauss, L. "Cosmological Antigravity" in the same issue on the theory behind it.)

Larson, R. and Bromm, V. "The First Stars in the Universe" in *Scientific American,* Dec. 2001, p. 64. On the dark ages and the birth of the first stars.

Livio, M. "Moving Right Along" in *Astronomy,* July 2002, p. 34. On the accelerating universe and some suggested explanations for it.

Luminet, J., et al. "Is Space Finite?" in *Scientific American,* Apr. 1999, p. 90. On understanding and measuring the geometry of spacetime.

Nadis, S. "Cosmic Inflation Comes of Age" in *Astronomy,* Apr. 2002, p. 28. Science writer describes current status and versions of the inflationary cosmology.

Odenwald, S. "Space-time: The Final Frontier" in *Sky & Telescope,* Feb. 1996, p. 24. On where the big bang may have come from.

Ostriker, J. and Steinhardt, P. "The Quintessential Universe" in *Scientific American,* Jan. 2001, p. 47. On new ideas on the cause of the acceleration of the universe.

Roth, J. "Dating the Cosmos: A Progress Report" in *Sky & Telescope,* Oct. 1997, p. 42. Update on comparing the expansion age to the ages of the oldest stars.

Astrobiology: The Road to Life in the Universe   In this fanciful montage produced by a NASA artist, we see one roadmap for discovering life in the universe. Learning more about the origins and domains of life on Earth, we can explore Mars, which may have had life billions of years ago when conditions were warmer. Next is Jupiter's satellite Europa, under whose icy surface a liquid ocean may harbor some kind of life. The road winds onward to other stars, which may well have planets around them that are hospitable to some kind of life.

29 Life in the Universe

*We shall not cease from
exploration,
And the end of all our
exploring,
Will be to arrive where
we started,
And know the place for
the first time.*

**T. S. Eliot, *Little
Gidding* (from *The
Four Quartets* in
*The Collected Poems
of T. S. Eliot*, 1934,
1936, Harcourt Brace
& World)**

THINKING AHEAD

Suppose that one day we receive a message from a civilization around another star. In a series of pictures transmitted across space, they tell us a bit about their world and themselves. Do we reply? And if we do, will all humanity speak with a single voice, or will we send a bewildering variety of answers—from different countries, religions, and ethnic groups? Who speaks for Earth?

Our voyages have taken us through billions of light years of space and billions of years of time. As we have learned more about the universe, we have naturally wondered whether there might be other forms of life out there—perhaps even other creatures who also take astronomy courses and think about worlds beyond their own.

In this final chapter, we want to consider how life began on Earth, to ask whether the same processes could have led to life on other worlds, and then to suggest some strategies for finding out about life elsewhere. These are all topics that are central to the science of life in the universe, variously called exobiology, bioastronomy, or astrobiology. (We'll use the last term in our book.)

Sometimes the media confuse the various topics that we discuss in this chapter. Remember that the search for life on other planets is not the same as the search for *intelligent* life, which (if it exists) is surely much more rare. Even on Earth today, most life is microbial, and humans have existed on our planet

Ace☉Astronomy™   The AceAstronomy icon throughout the text indicates an opportunity for you to test yourself on key concepts and to explore animations and interactions on the AceAstronomy website at **http://ace.brookscole.com/voyages**

for only 0.1 percent of the age of the planet. Also, it is important to distinguish between the scientific search for life, including intelligent life, and "belief in UFOs." As we will discuss below, most scientists think it is likely that the universe is teeming with life, but none of them find the evidence for UFOs and extraterrestrial visitors convincing.

29.1 THE COSMIC CONTEXT FOR LIFE

We saw that the universe was born in the big bang about 14 billion years ago. After the initial hot, dense fireball of creation cooled sufficiently for atoms to exist, all matter consisted of hydrogen and helium (with a very small amount of lithium). As the universe aged, processes within stars created the other elements, including those that make up the Earth (such as iron, silicon, magnesium, and oxygen) and those of prime biological interest (such as carbon, oxygen, and nitrogen). These and other elements combined in space to produce a wide variety of compounds, including complex organic chemicals that form the basis of life on Earth. While we do not understand the details of the origin of life, it is clear that these events took place within the context of the chemical evolution of the universe.

Ace ◐ Astronomy™   Log into AceAstronomy and select this chapter to see the Active Figure called "Earth Calendar."

29.1.1 Where Were the Atoms in Your Body Billions of Years Ago?

Let's review what we have learned in our voyages about the history of the cosmos by taking a look at the history of the atoms in your body, starting long before they became a part of you.

The universe began as hydrogen and helium, and the "first generation" stars contained little else. The hydrogen atoms in the water in your body formed at this early time; they are the oldest atoms that are part of you. But it is not possible to make an organism as complex and interesting as you with only these elements. More complex atoms had to be "cooked" in the only places in the universe hot enough to do the job—the centers of stars. (Fusing together simpler nuclei to make more complex ones is what stars "do for a living" and also the way they produce the energy we see coming from them.) The most massive stars not only produced the greatest variety of new nuclei, but then had the courtesy to explode, scattering the newly minted atoms into space (Figure 29.1).

Astronomers know from observations of distant parts of the universe that stars must have formed, lived out their lives, and exploded within the first billion years or so, because spectra of distant (and therefore ancient) galaxies already show the presence of some of the heavier elements. Over the years, thanks to supernova explosions (and mass loss in all kinds of stars), the gas between the stars became increasingly enriched with heavier elements. In the cooler outer layers of old stars, atoms frequently combined into solid particles that we call interstellar dust. The next generations of stars and planets then formed from reservoirs of enriched gas and dust. They thus contained atoms of carbon, nitrogen, silicon, iron, and the rest of the familiar elements. One of the most remarkable discoveries of modern astronomy is that life on Earth is mostly composed of just those elements that stars find easiest to make.

About 5 billion years ago, a cloud of gas and dust in this cosmic neighborhood began to collapse under its own weight. Out of this cloud formed the Sun and its planets, together with all the smaller bodies that also orbit the Sun (Figure 29.2). The third planet from the Sun, as it cooled,

■ **FIGURE 29.1**

A Star That Exploded   The Crab Nebula is the remnant of an exploding star, first seen in 1054. Now almost 11 LY across, this object still glows with tremendous energy in many bands of the electromagnetic spectrum. It is powered inside by the compressed whirling remnant (corpse) of the original star, whose beams of energy still stir and excite the atoms thrown out by the explosion. The left-hand image, taken with a ground-based telescope, shows the full nebula. The right-hand image, taken with the Hubble Space Telescope, is a close-up of the Crab's central region. The stellar corpse can be seen as the left member of a pair of stars near the center of the frame. HST was able to observe wisps of material streaming away from it at half the speed of light! Such exploding stars recycle new elements made during the star's life into the general supply of raw material from which new generations of stars and planets then form.

J. Hester, P. Scowen, and NASA

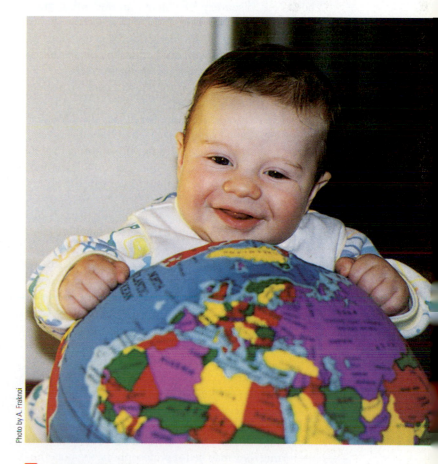

■ **FIGURE 29.2**

A Comet Comet Hyakutake captured in 1996 by amateur astrophotographer Robert Provin of California State University, Northridge. This 12-min exposure was taken on a clear night in the Mojave Desert with a camera and telephoto lens mounted on a telescope tripod. As the moving comet was held steady in the camera, the stars appeared to streak.

developed an atmosphere that served to moderate temperature extremes and allow the formation of large quantities of liquid water on its surface. The chemicals available on the cooling Earth were probably further enriched by the addition of complex molecules frozen in the nuclei of comets that collided with our planet.

The chemical variety and moderate conditions on Earth eventually led to the formation of self-reproducing molecules and the beginnings of life. Over the billions of years of Earth history, life slowly evolved and became more complex. The course of evolution was punctuated by occasional planetwide changes caused by collisions with those smaller bodies that did not make it into the Sun or one of its accompanying worlds. Mammals may owe their domination of the Earth's surface to just such a collision 65 million years ago.

Through many twisting turns, the course of evolution on Earth produced a creature with self-consciousness, able to ask questions about its own origins and place in the cosmos (Figure 29.3). Like most of the Earth, this creature is composed of atoms that were forged in earlier generations of stars—in this case, assembled rather cleverly into brains, kidneys, fingers, and faces. We might say that through the thoughts of human beings, the matter in the universe can become aware of itself.

Think about those atoms in your body for a minute. They are merely on loan to you from the lending library of atoms that make up our local corner of the universe. Atoms of many kinds circulate through your body and then leave it with the breath you exhale and the food you excrete. Even the atoms that take up more permanent residence in your tissues will not be part of you much longer than you

■ **FIGURE 29.3**

A Young Human Human beings have the intellect to wonder about their planet and what lies beyond it. Through them, the universe becomes aware of itself.

are alive. Ultimately, you will return your atoms to the vast reservoir of the Earth, where they will be incorporated into other structures and even other living things in the millennia to come.

This picture of *cosmic evolution,* of our descent from the stars, has been obtained through the efforts of scientists in many fields over many decades. Some of its details are still tentative and incomplete, but we feel reasonably confident in its broad outlines. It is remarkable how much we have been able to learn in the short time we have had the instruments to probe the physical nature of the universe.

29.1.2 The Copernican Principle

Our study of astronomy has taught us that we have always been wrong in the past whenever we have claimed that the Earth is somehow unique. Copernicus and Galileo showed us that the Earth is not the center of the solar system but merely one of a number of objects orbiting the Sun. Our study of the stars has demonstrated that the Sun itself is a rather undistinguished star, living through its long main-sequence stage like so many billions of others. There seems nothing special about our position in the Milky Way Galaxy either, and nothing surprising about our Galaxy's position in either its own group or its supercluster.

The recent discovery of planets around other stars confirms our idea that the formation of planets is probably a natural consequence of the formation of many kinds of stars. While our current techniques of finding planets allow us to identify only giant planets, there is no reason to believe that other systems could not contain smaller planets like the Earth as well.

Philosophers of science sometimes call the idea that there is nothing special about our place in the universe the *Copernican principle.* Although it may be tempting to consider ourselves the central and unique focus of all creation, no evidence for such a belief is found in any of the observations discussed in this book.

Most scientists, therefore, would be surprised if life were limited to our planet and had started nowhere else. There are billions of stars in our Galaxy old enough for life to have developed on a planet around them, and there are billions of other galaxies as well. Astronomers and biologists have long conjectured that a series of events similar to those on the early Earth probably led to living organisms on many planets around other stars. And, where conditions are right, such life might even have evolved to become what we would call intelligent—that is, aware of and interested in its own cosmic history. (In this sense, we conclude—with tongue firmly in cheek—that taking an astronomy class is the supreme example of intelligent behavior in the universe!)

Such arguments from the Copernican principle, however interesting they may be for philosophers, are nonetheless insufficient for scientists. Science demands data. We would like to find actual evidence for the existence of life (and perhaps even of intelligent life) elsewhere. Despite the sensationalistic claims of unidentified flying objects (UFOs) and alien abductions in the tabloid media, no such evidence has yet been found. But because many scientists feel that such a discovery would be a defining moment in the history of the human species, we are beginning to search seriously for evidence of life beyond the Earth.

29.1.3 So Where Are They?

If the Copernican principle is applied to life, then biology may be rather common among planets. Taken to its logical limit, the Copernican principle also suggests that intelligent life might be common. Intelligence like ours has some very special properties, including an ability to progress rapidly through the application of technology. Indeed, some philosophers today think that we will soon construct artificial intelligence that is superior to that of humans, in which case future evolution may be dominated by "creatures" composed of silicon, manipulating increasingly sophisticated systems of data and software. It seems entirely possible that such super-intelligent machines might decide to move beyond their own planetary systems and explore the Galaxy, or even beyond. Or organic life around other (older) stars may have started a billion years earlier than we did on Earth and may thus have had a lot more time to send out probes.

Faced with such a prospect, physicist Enrico Fermi asked the question half a century ago that is now called the *Fermi paradox:* Where are they? If life and intelligence are common and have such tremendous capacity for growth, why is there not a network of galactic civilizations whose presence extends even into a backwater solar system like ours?

Several solutions have been suggested to the Fermi paradox. Perhaps life is common but intelligence (or at least technological civilization) is rare. Perhaps such a network will come about in the future but has not yet had the time to develop. Maybe there are invisible streams of data flowing past us all the time that we are not advanced enough or sensitive enough to detect. Or maybe advanced species make it a practice not to interfere with immature, developing consciousness such as our own. We do not yet know whether any advanced life is out there and, if it is, why we are not aware of it. Still, you might want to think about these issues as you read the rest of this chapter.

29.2 ASTROBIOLOGY

In the last decade of the 20th century, many discoveries in both astronomy and biology stimulated the development of **astrobiology.** Astrobiologists study the origin, distribution, and ultimate fate of life in the universe. The field brings together astronomers, planetary scientists, biochemists, environmentalists, geologists, geneticists, and microbiologists to work on the same problems from their own perspectives.

Among the issues that astrobiologists are grappling with are the conditions under which life arose on the Earth

■ FIGURE 29.4

A Cloud of Gas and Dust This cloud of gas and dust in the constellation of Scorpius is the sort of region where complex molecules are found. It is also the sort of cloud where new stars form from the reservoir of gas and dust in the cloud. Radiation from a group of hot stars (off the picture to the bottom left) called the Scorpius OB Association is "eating into" the cloud, sweeping it into an elongated shape and causing the reddish glow seen at its tip.

and the reasons for the extraordinary adaptability of life on our planet. They are also involved in planning the continuing search for evidence of life on Mars and Jupiter's moon Europa, and in trying to understand where habitable zones might be found around other stars. Let's look at some of these issues in more detail.

29.2.1 The Building Blocks of Life

While no unambiguous evidence for life has yet been found anywhere beyond the Earth, life's chemical building blocks have been detected in a wide range of extraterrestrial environments. Meteorites, rocks that fall to Earth from space, have yielded many amino acids (the molecular building blocks of proteins) whose chemical structures mark them as having an extraterrestrial origin. When we examine the gas and dust around comets, we also find a number of **organic molecules**—those carbon-based compounds that on Earth are associated with the chemistry of life.

One of the most interesting results of modern radio astronomy has been the discovery of organic molecules in giant clouds of gas and dust between stars. More than 100 different molecules have been identified in these reservoirs of cosmic raw material, including formaldehyde ("embalming fluid"), alcohol, and others we know as important stepping stones in the development of life on Earth. Using radio telescopes and radio spectrometers, astronomers can measure the abundances of various chemicals in these clouds. We find organic molecules most readily in regions where the interstellar dust is most abundant, and it turns out these are precisely the regions where star formation (and probably planet formation) happens most easily (Figure 29.4).

Starting in the early 1950s, scientists have tried to duplicate in their laboratories the chemical pathways that led to life on our planet. In a series of experiments pioneered by Stanley Miller and Harold Urey at the University of Chicago, biochemists have simulated conditions on the early Earth and have been able to produce many of the funda-

Photo by David Malin, © Anglo-Australian Observatory

APPARATUS USED BY MILLER FOR THE SYNTHESIS
OF AMINO ACIDS BY ELECTRIC DISCHARGE

Electrodes

CH₄
NH₃
H₂O
H₂

Gases

Spark discharge

Water out

Condenser

Water in

Water containing
organic compounds

Boiling water

Trap

F. Drake and the Astronomical Society of the Pacific

■ **FIGURE 29.5**

Simulating the Early Earth The Miller–Urey experiment, performed in 1953, studies chemical reactions that might have taken place on the early Earth. An "atmosphere" consisting of methane, ammonia, water vapor, and hydrogen was subjected to electrical sparks to simulate lightning. Water at the bottom of the apparatus provided an "ocean" into which materials synthesized in the atmosphere could fall. When its contents were analyzed, the "ocean" was found to contain a variety of amino acids, the building blocks of proteins. Today we understand that the atmosphere of the early Earth probably had more carbon dioxide and much less hydrogen than Miller and Urey used, and this reduces the number of organic molecules formed in such experiments.

mental building blocks of life, including those that go into forming proteins and nucleic acids (Figure 29.5).

Although that's encouraging, there are problems with the experiments. The most interesting chemistry from a biological perspective takes place with hydrogen-rich or *reducing* gases, such as ammonia and methane. However, the early atmosphere of the Earth was probably dominated by carbon dioxide (as Venus's atmosphere still is today). and never contained abundant reducing gases. Thus, while we know it is relatively easy to make complex carbon compounds of biological interest, we are not sure that the conditions on Earth were quite right for this to happen in sufficient quantities. It is more likely that some of the building blocks of life did not form on our planet at all, but rather in some chemically more favorable locale (such as the outer parts of the solar nebula from which the planets formed). It is even conceivable that life itself originated elsewhere and was seeded onto our planet.

29.2.2 The Origin of Life

The carbon compounds that form the chemical basis of life may be common in the universe, but it is still a giant step from these building blocks to a living cell. Even the simplest molecules of the genetic material in a cell contain millions of molecular units, each arranged in a precise sequence. Furthermore, even the most primitive life required two special capabilities: a means of extracting energy from its environment and a means of reproducing itself. Biologists can see ways that either of these capabilities might have formed in a natural environment, but we are still a long way from knowing how the two came together in the first life-forms.

Unfortunately, we have little evidence about the actual origin of life on our planet. No fossils remain from the critical first few million years after the end of the heavy bombardment by asteroids and comets about 3.9 billion years ago. Large impacts would have sterilized the surface layers of the Earth, wiping out any life that had started and perhaps requiring life to start again and again.

When the large impacts ceased, the scene was set for a more peaceful environment on our planet. If the oceans of Earth contained accumulated organic material from impacts with icy planetesimals (like the comets of today), the ingredients were available to make living organisms, even though we do not understand the sequence of events in any detail. For tens of millions of years after its formation, this earliest life probably extracted energy from chemicals in the ocean and used the accumulated organic material there. Eventually, however, this "free lunch" was depleted and life was forced to find other ways to survive. The evolution of more complex organisms then began.

One of the key early steps along that evolutionary pathway was the development of DNA (deoxyribonucleic acid)—the twisted molecular strands of nucleic acid that encode the genomic information in terrestrial life-forms. It is DNA that tells each cell how to reproduce and grow. It also controls the chemical "factory" within each cell that utilizes chemical energy and manufactures the proteins and other chemicals that life requires. All life on Earth today uses the same fundamental DNA-based chemistry. Everything can be traced back to a common ancestor that emerged in the oceans of Earth perhaps 3.8 billion years ago. But events before that time are murky at best. Probably the earliest life—none of which survives today—had a much simpler chemical structure, based on RNA or even simpler precursors.

Another innovation of great importance was the development of **photosynthesis:** life's ability to extract chemical energy from sunlight. Previously life had to make do with dissolved chemicals in the environment, a much weaker source of usable energy. Photosynthesis is a complex, multistep chemical process, but it allows life to draw directly on the vast energy of sunlight that strikes the Earth every day. There is some evidence—still rather controversial—that

■ **FIGURE 29.6**

Microfossil of Ancient Terrestrial Life A fossil microbe from 3.5-billion-year-old sediments of Western Australia. A portion of the filament is covered by an overlaid image, which shows the results of experiments that probed for the presence of carbon. The white areas contain carbon and thus indicate the survival of organic matter. Some scientists think that such microbes might already have developed photosynthesis.

the earliest fossils of microbial life that we have found on Earth are the remains of photosynthetic bacteria (called blue-green algae or *cyanobacteria*). If this is so, then photosynthesis must already have been active 3.5 billion years ago (Figure 29.6). Other scientists think that photosynthesis arose later, but there is general agreement that the extraction of energy from sunlight was going strong by 3.0 billion years ago.

Photosynthesis releases oxygen as a waste product. The free oxygen produced by photosynthesis began accumulating in our atmosphere about 2.2 billion years ago. As the amount of this gas increased, three oxygen atoms could combine to form ozone. As a result, a layer of ozone developed high in the Earth's atmosphere, providing protection from the Sun's ultraviolet radiation. This allowed life to col-

onize the landmasses of our planet instead of remaining only in the ocean.

The rise in oxygen levels was deadly to some microbes (since oxygen breaks apart a number of organic molecules). But it allowed other kinds of life to thrive, because combining with oxygen is an especially effective way for cells to get energy from organic materials that plants produce. (This happens every time you eat a good salad for lunch.) Energy produced through *oxydation* made possible a great proliferation of organisms, which continued to evolve in an oxygen-rich environment.

The details of that evolution are properly the subject of biology courses, but we note that the process of evolution by natural selection (survival of the fittest) provides a clear explanation for the development of Earth's remarkable variety of life-forms. It does not, however, directly solve the mystery of life's earliest beginnings. We hypothesize that life will arise whenever conditions are appropriate, and there is no evidence to contradict this assumption. But this hypothesis is another form of the Copernican principle: Something that happened on Earth is likely to have happened elsewhere in the universe. Until we actually find extraterrestrial life, we cannot really know how well our hypothesis fits the actual workings of nature.

29.2.3 Habitable Environments

To understand the role of life in the universe, we must explore the range of environments that might support living things. Scientists approach this problem in two ways. In this section we discuss habitability from the perspective of the basic properties of carbon-based life. In the next section we will look at the diversity of environmental conditions on Earth where life is found.

Why do we limit ourselves to carbon chemistry? The simplest answer is that the only life we know, and the only organic chemistry we can experiment with, is based on carbon. It's not that alternative chemistries haven't been considered. However, a careful examination of the properties of different elements shows that carbon is unique in its ability to form complex chemical bonds with a wide variety of other elements, including both oxygen and hydrogen. The carbon-based molecules can also combine to form very long chains (called *biopolymers*) and even to build up the double helix of the DNA molecule, with its millions of components. No other element has these properties.

Other important biological elements that combine with carbon, hydrogen, and oxygen are nitrogen, sulfur, and phosphorus. As we saw, all of these are among the most abundant elements in the universe, ensuring that the components for carbon-based life are widely available.

Much of organic chemistry is enabled by the presence of liquid water. Water is the best solvent, which is why we use it for washing. It also turns out that the range of temperatures in which water is a liquid (from 0°C to 100°C) is precisely the range where much of carbon-based chemistry

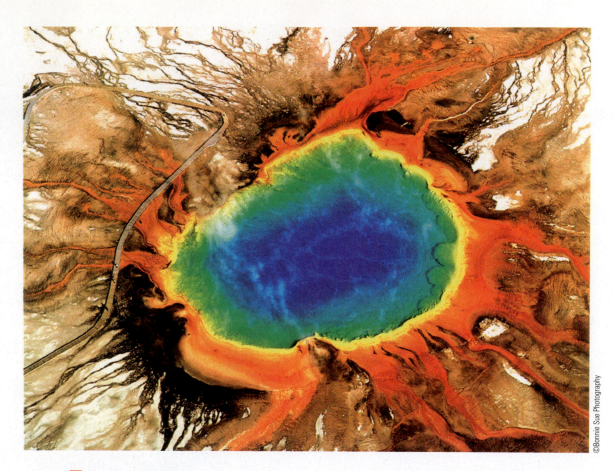

■ **FIGURE 29.7**

Yellowstone Hot Spring with Colorful Microbial Life   Grand Prismatic Spring in Yellowstone National Park supports a wide range of microbial life under conditions that we would consider extreme. The rings of color mark the habitats of microbes with different temperature tolerances. The water, which wells up near the center at temperatures close to boiling (100°C), cools as it spreads from the center and overflows the lip of the pool. You can see people on the boardwalk winding around at the left.

is active. At temperatures higher than 100°C, the larger carbon molecules start to come apart. This is why boiling water kills most microbes.

Although H_2O is common in the universe, liquid water is much less so. Most water is in the form of either vapor (in the atmospheres of cool stars) or solid (in ice grains within clouds of interstellar gas and dust). Liquid water requires a pressure greater than 0.006 bar; at lower pressures H_2O can have only two states: solid and gas. (Recall that 1 bar is the pressure at sea level on Earth.) Liquid water is thus associated with *planets,* on their surfaces or in their interiors. This limits us in seeking habitable environments.

In addition to liquid water and organic chemicals, life requires a source of extractable energy. Energy is needed to manufacture the complex molecules of life as well as to undertake such functions as locomotion and thinking in more complicated creatures. Living things use the energy from their surroundings to manufacture the complex

chemistry of life; if the energy source is removed, we "die" and these complex molecules break down and return to the environment.

Simple or primitive life on Earth extracts energy from dissolved chemicals through fermentation and other reactions. This chemical energy derives ultimately from geothermal sources, such as vents of hot, mineral-rich water deep under the sea or hot springs like those in Yellowstone National Park (Figure 29.7). In contrast, photosynthesis enables life to tap the much greater energy of sunlight itself. Photosynthesis also yields carbohydrates as a by-product, and these can be used as an energy source by other organisms, such as animals. Most terrestrial life is part of a food chain that is ultimately dependent on green-plant photosynthesis.

A habitable environment, then, seems to require the presence of three things: abundant raw material in the form of carbon compounds, liquid water (which points us toward planets), and an energy source based on either chemistry or sunlight.

29.2.4 Life in Extreme Conditions

On Earth, life has evolved to fill many ecological niches, some of them quite different from our everyday experience. Two obviously incompatible environments on Earth are the air and the sea. Creatures that thrive in the sea will die if placed in the air, while those of us who prefer air will quickly drown if submerged in water. But there are even stranger environments on our planet where specialized life has managed to survive and even flourish. We call these **extremophiles,** meaning that their environments seem extreme to us; like most life on Earth, they are mostly microbes. Note, however, that these extremophiles are not necessarily simple or primitive; in fact, a great deal of evolutionary adaptation was required for them to learn to function in these environments.

Most life works best at temperatures between about 15°C and 60°C. Microbes that prefer lower temperature are called *psychrophiles,* and those that prefer heat are called *thermophiles* (where "phile" is the Greek suffix meaning to like or love, as in "anglophile," someone who likes the British). At the low end, life can often survive at temperatures even below 0°C in the presence of some kind of natural "antifreeze," although metabolism slows down and often stops. Microbes that were frozen and dormant for tens of thousands of years in the Antarctic ice have been revived in our laboratories.

At high temperatures, thermophiles have developed mechanisms to make the chemical repairs that are needed as carbon-based compounds begin to come apart. Unlike the dormant state at low temperatures, adaptation to high temperatures requires active chemical intervention. Many microbes flourish in the Yellowstone hot springs at temperatures up to 100°C. The record for a thermophile is 113°C at deep-sea vents, where the pressure of the water above the vent is so great that 113°C is still below the boiling point of water (Figure 29.8).

Three related environmental extremes involve moisture, salt, and acid. Many microbes can survive extremely dry conditions in much the way they do low temperatures—by going into a dormant state and waiting for better conditions to return. Some microbes are so tolerant of high salinity that they can survive even in the waters of the Dead Sea. The range of acidity in which life has been found goes from pH less than 0 to greater than 9. The numbers in pH, which stands for potential of hydrogen, measure whether a substance is acidic (0–7), neutral (exactly 7), or alkaline (7–14). One example of an acid environment is the Rio Tinto (Red River) of southern Spain, which originates in a region of extensive mineral deposits and maintains a steady pH of 2.5 all the way to its mouth (Figure 29.9). A rich microbe community inhabits this river, and indeed

Peter Ryan /Scripps/Science Photo Library/Photo Researchers, Inc.

■ FIGURE 29.8

"Black Smoker" on the Ocean Floor Starting in 1977, marine geologists and biologists have been discovering that hot, mineral-rich water can emerge from vents on the ocean floor. As you can see here, when the hot vent water mixes with the much colder ocean water, the dissolved minerals can fall out of solution and make a kind of "black smoke." The regions around these hydrothermal vents are full of life; more than 350 new species have been discovered close to them. In the pitch darkness of the ocean floor, these forms of life derive their energy not from sunlight but from the chemistry of the vents.

some of the microbes help to maintain the constant pH of 2.5 because that is the environment they like.

One of the most surprising cases of tolerance to extremes is exhibited by the bacterium called *Deinococcus radiodurans,* which is found (among other places) in the cooling water of nuclear reactors. (It is called an atomophile!) With its highly developed chemical repair mechanisms, *radiodurans* can survive ultraviolet or particle radiation up to 6000 rads per hour, a thousand times more than you would be able to tolerate. It is also resistant to many

David Morrison

■ **FIGURE 29.9**
Rio Tinto, the Acid River of Spain
The waters of Rio Tinto are acidic, with a pH of 2.5. Its reddish color comes from dissolved minerals. Yet the river supports microbes that are adapted to the acidic conditions and actually help maintain the river at a constant pH of 2.5 from its source to its mouth.

unpleasant industrial chemicals, and it is commonly found in toxic waste dumps. No one knows what evolutionary pathways enabled *radiodurans* to develop this amazing tolerance for otherwise lethal conditions.

So we see that life has learned to survive on Earth in a remarkable range of environments on the surface, in the water and air, and even deep underground. Nearly every conceivable ecological niche seems to be filled, although in many cases the rate of metabolism and other biological processes is very slow. But there are some exceptions. No organism has learned to extract the water it needs from ice: The ice sheets of Greenland are not green! There are also no organisms that carry out their life cycle entirely in the air. Life (as we know it) requires liquid water and something substantial such as the land or ocean to make a home.

29.3 SEARCHING FOR LIFE BEYOND EARTH

Astronomers and planetary scientists are on the forefront of the search for life in the universe. In this section we discuss two kinds of searches. First is the direct exploration of planets within our own solar system, especially Mars and Europa. Second is the even more difficult task of searching for evidence of life—*biomarkers*—on planets circling other stars. Finally, in Section 29.4, we will examine SETI, the search for extraterrestrial *intelligence*. As you will see,

the approaches taken in these three cases are very different, even though the goal of each is the same—to determine whether life on Earth is unique in the universe.

29.3.1 Life on Mars

In 1996, an enormous amount of public and media attention was focused on a suggestion by an interdisciplinary team of scientists that they had found evidence of life on Mars. They had done detailed laboratory analysis of an ancient martian rock sample recovered in Antarctica, a 4-billion-year-old meteorite called ALH 84001. Their work showed that this bit of martian crust (knocked into space by a powerful impact; Figure 29.10) had experienced wet conditions about 3.5 billion years ago. At that time, Mars's atmosphere was thicker and water could flow—leaving traces of carbonate minerals and some organic compounds embedded in the potato-sized piece of rock that eventually fell to Earth. The team further hypothesized that tiny structures seen in the rock at extremely high magnification were the fossilized remains of ancient microbial life.

If there are indeed fossils in ancient martian rocks, this would be one of the most spectacular discoveries in the history of science. As we write in 2003, it appears to most scientists that this meteorite was contaminated with terrestrial chemicals during the millennia it lay trapped in the Antarctic ice and that the "fossils" are too small to represent anything once living. But even if the rock's specific features are not organic in origin, the work on ALH 84001

■ **FIGURE 29.10**

Martian Impact This painting shows how a large rock impacting Mars might have ejected some smaller chunks from the red planet. After millions of years of orbiting the Sun, some fragments intersected the Earth and wound up in the Antarctic, where scientists later found them.

Painting by Don Davis, © SETI Institute

■ **FIGURE 29.11**

Potential Mars Rover Landing Site The most promising sites in the search for fossil microbes on Mars are places where water once pooled, such as ancient ponds or lakes. One of the places the Mars Exploration Rover mission plans to land is at Gusev Crater, an impact basin about 160 km in diameter, seen at the top of this image. A 600-km-long channel called Ma'adim Vallis drains into it, and there is speculation, based on the appearance of the crater floor, that significant water may have accumulated in Gusev in the past.

USGS/NASA/JPL

verified that wet, apparently Earth-like conditions once existed on Mars. And the energetic debate about martian meteorites has stimulated thinking about how we might detect fossil life on other planets.

Much of the exploration of Mars by spacecraft has focused on either the environmental conditions there or the search for evidence of life. There is abundant geological evidence that Mars once had a warmer climate and liquid water. The 1976 Viking landers carried instruments designed specifically to detect microbial life in the soil, but the results were negative. Although Mars has the most Earth-like environment of any planet in the solar system, the surface is too dry, too cold, and too affected by solar ultraviolet radiation to meet the requirements for habitability.

However, there is no reason to think that life could not have begun on Mars about 4 billion years ago, since Earth and Mars apparently had similar surface conditions then. This brings us back to ALH 84001 and the arguments about fossil microbes. Astrobiologists today are learning how to identify fossil microbes in ancient Earth rocks and preparing to apply these lessons to Mars. There are plans for future missions that will include the return of samples selected from sedimentary rocks at sites (such as ancient lake beds or hot springs) that once held water (Figure 29.11). The most powerful searches for martian life (past or present) will thus be carried out in our laboratories here on Earth.

If there is evidence from these rocks or elsewhere that Mars once had life, then scientists will accelerate their search for possible survivors. They will look for life-forms that may have evolved to deal with the deteriorating climate of Mars, perhaps by finding some oasis or refuge that is warmer and wetter than most of the martian surface. NASA's theme in searching for life is "follow the water." The most likely source of liquid water on Mars today is deep below the surface, where extensive *aquifers* (layers containing groundwater) may exist. Perhaps someday astronauts on Mars will drill deep wells down to this layer of liquid water and finally encounter living alien life.

One interesting twist on the search for life is emphasized by the presence of ALH 84001 and other Mars rocks on Earth. Mars and Earth are close enough together that they have exchanged material in this way throughout their history (although most of the traffic has been from Mars to Earth, as a consequence of the lower surface gravity on Mars). It is entirely possible that some of these rocks may have contained viable microorganisms, not just fossils. Thus Mars might have seeded Earth, or the two planets could have exchanged biological material. It is thus conceivable that if we eventually find living things on Mars they will be genetically similar to terrestrial life, for the good reason that they are our distant cousins.

29.3.2 Life in the Outer Solar System

Another location of great interest to astrobiologists is Jupiter's satellite Europa. Galileo spacecraft data on this Moon-sized world indicate the presence of a global ocean of liquid water beneath a thick ice crust. As the only location in the solar system beyond Earth where there is known to be a large body of liquid water, Europa beckons us to further exploration (Figure 29.12)

Life requires an energy source, and sunlight does not penetrate below the thick icy crust of Europa. Thus any life-forms there are unlikely to have evolved photosynthesis. What other energy sources might be present on Europa? To remain liquid, the global ocean must be warmed by heat generated by tides and now escaping from the interior of Europa. Hot (or at least warm) springs might be active there, analogous to those we have discovered in the deep oceans of the Earth. Europa might therefore support life that derives all its energy from the mineral-laden water in such springs.

Some people have suggested that Europa is the most likely place beyond the Earth to find life in the solar system. On the other hand, we might question whether life could originate in a dark ocean heated only by hot springs. Since we do not know exactly how life formed on Earth, it is impossible to evaluate this possibility for Europa. One thing is clear, however: If there is life in the europan oceans, it is likely to be unrelated to terrestrial life. There is no exchange of rocks between Earth and Europa to provide the possibility of cross-contamination, as is the case for Mars. Thus Europa holds out the tantalizing prospect of a clear *second genesis*—an independent origin of life. If so, we can hardly guess what that life might be like. Will it be carbon based? Will it utilize protein chemistry? Will it possess a genetic material something like DNA or RNA? Or will it be more exotic than we can now imagine?

29.3.3 Habitable Planets Orbiting Other Stars

Although our robot probes are still confined to the worlds of our own solar system, we certainly do not want to limit our search for life to our own neighborhood. How can we

JPL/NASA

■ FIGURE 29.12

Icy Surface of Europa The Galileo spacecraft demonstrated that Jupiter's satellite Europa, which is about the same size as our Moon, has a global ocean of liquid water roughly 100 km deep. Unfortunately this ocean lies beneath a thick crust of ice, and no sunlight penetrates into the cold dark waters beneath. This picture illustrates the cracked surface of Europa (top) on the same scale as the San Francisco peninsula (bottom). The round, flat, smooth area at left on the Europa image is probably an area where some fluid coming up from below buried the cracks and ridges that crisscross the satellite's surface.

gather evidence about the possibility of life among the stars?

As we have discussed previously, one of the most important recent developments in astronomy has been the ability to detect planets circling distant solar-type stars—with more than 100 such planets discovered by the end of 2002. So far, we are able to detect only giant planets (like Jupiter and Saturn) and these are unlikely as the home of life. Still, their existence holds out the prospect of smaller worlds (either Earth-like planets or satellites orbiting the giants) that might support liquid water and the other conditions necessary for biology.

In evaluating the prospect for life in distant planetary systems, astrobiologists have developed the idea of a **habitable zone**—a region around a star where suitable conditions might exist for life. Since the focus is on the presence of liquid water, the usual definition of a habitable zone is the distance where water will be liquid on the surface of a terrestrial-type planet (that is, a planet with roughly the mass of the Earth).

Obviously the Earth is in the habitable zone for the solar system, although there are periods in the past when most of the oceans froze over (we call these times "snowball Earth"). Note, however, that the Earth is warmer than the freezing temperature of water only because the greenhouse effect in our atmosphere raises the average temperature by about 25°C over the value it would have if there were no greenhouse gases in the atmosphere. Thus, we must consider the nature of any atmosphere as well as the distance from the star in evaluating the range of habitability.

The luminosity of stars like the Sun changes over their main-sequence lifetime, and this further complicates the picture. Calculations indicate that the power output of the Sun has increased by at least 30 percent over the past 4 billion years.[1] Yet geological evidence suggests that liquid water has been present on the Earth's surface for that entire time, and indeed the coldest periods (the snowball Earth times) took place within the past billion years, not earlier when the Sun was fainter. Apparently the Earth had a larger greenhouse effect early on, and the declining greenhouse effect has roughly compensated for the increasing luminosity of the Sun. Such changes must be factored into our estimates of where a star's habitable zone might lie at any given time.

Within the solar system, Venus has evolved through a runaway greenhouse effect into an oven where life is impossible, but it was once inside the habitable zone. Mars today is too cold and dry for surface life, but in the past it had a thicker atmosphere and apparently supported surface water (although perhaps its lakes and seas were ice-covered). Today Mars seems to be outside the habitable zone, but if the Earth (with its greater ability to retain an atmosphere) were in the orbit of Mars, it might still be relatively warm. In other words, the current inhospitable nature of Mars is as much a consequence of its small mass as its distance from the Sun.

This is all very complicated, and scientists still differ in what they consider to be a habitable zone. Most would agree that for the solar system, habitability means having a mass at least as great as that of Mars (11 percent that of the Earth) and orbiting between about 0.9 and 1.2 AU. Stars that have lower luminosity than the Sun will have a smaller habitable zone that is closer in to the star. Stars with somewhat higher luminosities will have larger zones, but their ultraviolet output will be greater and may pose a hazard for life. Stars that have really large masses and luminosities

won't last long enough for life to develop around them, so we won't worry about them.

29.3.4 Biomarkers

Our prime candidate worlds in the search for life beyond the solar system are terrestrial-type planets within the habitable zones of their stars. In addition, we require that the stars themselves be reasonably constant in brightness (no variable stars allowed) and that the orbits of the planets have low eccentricity (look more like circles than ellipses), so that their surface temperatures are stable. We are unable to detect such planets with our current technology, but within a decade or so, space missions should allow us to determine how common such habitable planets are and to identify nearby candidate systems for further study.

The fact that a planet is within the habitable zone and is therefore considered able to support life does not ensure, of course, that it is actually inhabited. Indeed, one of the most important questions in astrobiology is just that: Will life arise naturally when the environmental conditions are correct? It is thus very important to consider how we might recognize the signature of life on a distant planet.

Even with the largest space-based telescope we can contemplate, we will never be able to obtain an image of a distant planet, as we do the worlds in our own solar system, let alone visit with robot spacecraft. Astronomers therefore need a global **biomarker**—something distinctive to separate a live world from a dead one. To be detectable, these biomarkers should involve changes in atmospheric or surface chemistry that can *only* be the result of life.

If we observed Earth from a great distance and took sensitive visible-light and infrared spectra, we might just see such biomarkers. The most easily detectable evidence is the presence of abundant free oxygen in the atmosphere, which produces distinctive features in near infrared spectra (Figure 29.13). Because oxygen is a highly reactive element, any planet with lots of it must have a way of producing fresh supplies regularly. On Earth, oxygen is the by-product of photosynthesis; if life on Earth should cease, the oxygen in the atmosphere would disappear within a few thousand years. The oxygen, therefore, is a product of life, which in effect uses sunlight to manufacture atmospheric oxygen. A similar atmospheric gas is methane, produced by microbes in swamps (it used to be called swamp gas) and in the digestive tracts of some mammals. Without the presence of life, methane would be quickly oxidized and disappear from the atmosphere. In fact, the most distinctive biomarker for Earth is the simultaneous presence of these two mutually inconsistent gases: oxygen and methane.

The only problem with using oxygen and methane as biomarkers is that for more than half the history of the Earth, there was no free oxygen. Even after the biological invention of photosynthesis, it took a long time before oxygen achieved measurable concentrations. Thus, for 2 billion years Earth was a living planet but without the oxygen/methane biomarker in its atmosphere. Today scientists are

---

[1] The Sun is very gradually brightening because it is converting more and more of its core hydrogen into helium through fusion. When four hydrogen atoms are turned into one helium atom, the core pressure is reduced (fewer particles to do the pressing). This means the outer layers of the Sun can squeeze harder on the core. When squeezed, the core gets hotter and more hydrogen atoms fuse. This very gradually raises the energy output of our star.

■ **FIGURE 29.13**
Spectrum of Planet Earth
This shows what the spectrum of our planet in infrared radiation would look like from another world. Note the absorption feature from ozone (a form of oxygen).

David DesMarais, NASA, Ames

■ **FIGURE 29.14**

Microbial Mats   Microbes that can do photosynthesis have played the dominant role in producing the Earth's oxygen for 3 billion years. This photo shows a cross-section of a mat of such organisms living today in the salt flats of Baja California. Hundreds of different species of microbes "work together" in a miniature ecosystem that is less than a centimeter thick. A coin is shown to give you a sense of scale.

beginning to look in more detail at the way early life functioned on the Earth and to try to identify other biomarkers that would be useful in the study of a faint and distant world (Figure 29.14). Alternatively, we can look for one quite dramatic biomarker of an intelligent technological life-form: the broadcast of radio signals.

29.4 THE SEARCH FOR EXTRATERRESTRIAL INTELLIGENCE

For the first time in human history, we know that our planetary system is not unique. Given all the developments discussed in this chapter, it may also be that life has developed on many planets around other stars. If that life is microbial, or even if it includes metazoans (large animals), such as

those that have populated the Earth for the most recent several hundred million years, we do not yet have the technology to discover it. But suppose in a few cases intelligent, technical civilizations have arisen, as has happened on Earth in the most recent blink of cosmic time. If there is such an intelligent species out there, how could we make contact with it?

This problem is similar to making contact with people who live in a remote part of the Earth. If students in the United States want to converse with students in Australia, for example, they have two choices. Either one group gets on an airplane and travels to meet the other, or they communicate via some message medium (today probably by telephone, fax, or e-mail). Given how expensive airline tickets are, most students would probably select the message route.

In the same way, if we want to get in touch with intelligent life around other stars, we can travel or we can try to exchange messages. Because of the great distances involved, interstellar space travel is very slow and very expensive. The fastest spacecraft the human species has built so far would take almost 80,000 years to get to the nearest star. While we could certainly design a faster craft, the more quickly we require it to travel, the greater the energy cost involved. To reach neighboring stars in less than a human life span, we would have to travel close to the speed of light. In that case, however, the expense would become truly astronomical.

29.4.1   Interstellar Travel

Bernard Oliver, then Vice President of the Hewlett-Packard Corporation and an engineer with an abiding interest in life elsewhere, made a revealing calculation about the costs of rapid space travel. Since we do not know what sort of technology we (or other civilizations) might someday develop, Oliver considered a trip to the nearest star in a spaceship with a "perfect engine"—one that would convert its fuel into energy with 100 percent efficiency. (No future technology can possibly do better than this. In reality, nature is unlikely to yield efficiency even close to the perfect value.) Even with a perfect engine, the energy cost of a single round-trip journey at 70 percent the speed of light turns out to be equivalent to about 500,000 years' worth of total U.S. electrical energy consumption!

In case you are wondering why this figure is so high, you must remember that the voyagers could not depend on

■ FIGURE 29.15

Interstellar Message This is the image engraved on the plaques aboard the Pioneer 10 and 11 spacecraft. The human figures are drawn in proportion to the spacecraft, which is shown behind them. The Sun and planets in the solar system can be seen at the bottom, with the trajectory that the spacecraft followed. The lines and markings in the left center show the positions and pulse periods for a number of pulsars, which might help locate the spacecraft's origins in space and time.

finding "gas stations" open at their destination; therefore, they would have to carry the fuel for the return legs of the journey with them, and getting that huge mass of fuel up to 70 percent the speed of light would be very expensive. The important thing about Oliver's calculation is that it does not depend on present-day technology (since it assumes a perfect engine) but only on the known laws of science. What it shows is that no matter who does the traveling, it is very expensive to go fast enough to get to the stars within the course of a single human life.

This is one reason astronomers are so skeptical about claims that UFOs are spaceships from extraterrestrial civilizations. Given the distance and expense involved, it seems unlikely that the dozens of UFOs (and even UFO abductions) claimed each year could all be visitors from other stars so fascinated by Earth civilization that they are willing to expend fantastically large amounts of energy or time to reach us. Nor does it seem credible that these visitors have made this long and expensive journey and then systematically avoided contacting our governments or political and intellectual leaders.

In fact, a sober evaluation of UFO reports often converts them to IFOs (identified flying objects) or NFOs (not-at-all flying objects). While some are hoaxes, others are natural phenomena, such as ball lightning, fireballs, bright planets, or even flocks of birds with reflective bellies. Still others are human craft, such as private planes with some lights missing or secret military aircraft. It is also interesting that the group of people who most avidly look at the night sky, the amateur astronomers, have never reported UFO sightings. Further, not a single UFO has ever

left behind any physical evidence that can be tested in a laboratory and shown to be of nonterrestrial origin.[2]

29.4.2 Messages on Spacecraft

While space travel by living creatures seems very difficult, robot probes can travel over long distances and over long periods of time without needing a good cup of coffee. Four spacecraft built by the human species—two Pioneers and two Voyagers—having finished their program of planetary exploration, are now leaving the solar system. At their coasting speeds, they will take hundreds of thousands or millions of years to get anywhere close to another star. They are the first products of human technology to go beyond our home system, however, so we wanted to put messages on board to show where they came from.

Each Pioneer carries a plaque with a pictorial message engraved on a gold-anodized aluminum plate (Figure 29.15). The Voyagers, launched in 1977, have audio and video records attached (Figure 29.16), which allowed the inclusion of more than 100 photographs and a selection of music from around the world. (Included among the excerpts from Bach, Beethoven, folk music, tribal chants, and others, is one piece of rock—"Johnny B. Goode" by Chuck

[2] If you are interested in pursuing the topic of what UFOs are and aren't, we recommend the following books: P. Klass, *UFO Abductions: A Dangerous Game* (1988, Prometheus Books); C. Peebles, *Watch the Skies: A Chronicle of the Flying Saucer Myth* (1994, Smithsonian Institution Press); and R. Shaeffer, *UFO Sightings: The Evidence* (1998, Prometheus Books). Also see: www.astrosociety.org/education/resources/pseudobib.html.

(a)

(b)

NASA

■ **FIGURE 29.16**

The Voyager Record (a) Encoded onto a gold-coated copper disk, the Voyager record contains 118 photographs, 90 minutes of music from around the world, greetings in almost 60 languages, and other audio material. It is a summary of the sights and sounds of Earth. (b) One of the images encoded onto the record. Originally, the team devising the record wanted to send a photograph from a medical book of a nude man and a pregnant woman. However, NASA was concerned about offending some people on Earth, and so artist Jon Lomberg drew a silhouette version, which allowed a fetus inside the woman to be shown as well.

Berry.) Given the enormity of the space between stars in our section of the Galaxy, it is very unlikely that these messages will ever be received by anyone. They are more like a note in a bottle thrown into the sea by a shipwrecked sailor, with no realistic expectation of its being found soon but a slim hope that perhaps someday, somehow, someone will know of the sender's fate.

Ace☯Astronomy™ Log into AceAstronomy and select this chapter to see Active Figure called "Interstellar Communications."

29.4.3 Communicating with the Stars

If direct visits to stars are unlikely, we must turn to the alternative for making contact—exchanging messages. Here the news is a lot better. We already know (and have learned to use) a messenger—electromagnetic radiation—that

The Voyager Message

An excerpt from the Voyager record:

We cast this message into the cosmos. It is likely to survive a billion years into our future, when our civilization is profoundly altered. . . . If [another] civilization intercepts Voyager and can understand these recorded contents, here is our message:

This is a present from a small, distant world, a token of our sounds, our science, our images, our music, our thoughts, and our feelings. We are attempting to survive our time so we may live into yours. We hope, someday, having solved the problems we face, to join a community of galactic civilizations. This record represents our hope and our determination, and our good will in a vast and awesome universe.

—Jimmy Carter, President of the United States, June 16, 1977 moves through space at the fastest speed in the universe. Traveling at the speed of light, radiation reaches the nearest star in only four years and does so at a tiny fraction of the cost of sending material objects. These advantages are so clear and obvious that we assume they will occur to any other species of intelligent beings who develop technology.

However, we have access to a wide spectrum of electromagnetic radiation, ranging from the longest-wavelength radio waves to the shortest-wavelength gamma rays. Which would be the best for interstellar communication? It would not be smart to select a wavelength that is easily absorbed by interstellar gas and dust, or one that is unlikely to penetrate the atmosphere of a planet like ours. Nor would we want to pick a wave that has lots of competition for attention in our neighborhood.

One final criterion makes the selection easier: We want the radiation to be inexpensive to produce in large quantities. When we consider all these requirements, radio waves turn out to be the best answer. Being the lowest-frequency (and lowest-energy) band of the spectrum, they are not very expensive to produce (permitting us to use them extensively for communications on the Earth). They are not significantly absorbed by interstellar dust and gas. With some exceptions, they easily pass through the Earth's atmosphere and through the atmospheres of the other planets we are acquainted with.

29.4.4 The Cosmic Haystack

For these reasons, many astronomers have decided that the radio band is probably the best place in the spectrum for communication among intelligent civilizations. In particular, we are focusing on the part of the radio band called microwaves, where both the universe and the atmosphere of planets turn out to produce the least amount of static. Having made such a decision, however, we still have many questions and a daunting task ahead of us. Shall we *send* a message, or try to *receive* one? Obviously, if every civilization decides to receive only, then no one will be sending, and everyone will be disappointed. On the other hand, it may be appropriate for us to *begin* by listening, since we are likely to be among the most primitive civilizations in the Galaxy who are interested in exchanging messages.

We do not make this statement to insult the human species (which, with certain exceptions, we are rather fond of). Instead, we base it on the fact that humans have had the ability to receive (or send) a radio message across interstellar distances for only a few decades. Compared to the ages of the stars and the Galaxy, this is a mere instant. If there are civilizations out there that are ahead of us in development by even a short time (in the cosmic sense), they are likely to have a head start of many, many years. If there are civilizations behind us, chances are they are sufficiently far behind that they have not yet developed radio communications.

In other words, we, who have just started, may well be the "youngest" species in the Galaxy with this capability

(see the Figuring for Yourself section at the end of the chapter). Just as the youngest members of a community are often told to be quiet and listen to their elders for a while before they say something foolish, so we may want to begin our exercise in extraterrestrial communication by listening.

Even restricting our activities to listening, however, leaves us with an array of challenging questions. For example, you know from your own experience with radio transmissions that a typical signal comes in on only one channel (which means it is carried by one small frequency band of radio waves). The owners of your favorite station are confident—because there aren't that many channels on the radio dial—that you will find them despite this (although a few actually transmit over two different frequencies, one on the AM band and one on the FM band). Many of us, when we first arrive in a new city, scan up and down the radio band until we find the stations we like, but if we have only an AM radio in our car, and the stations that play our favorite music are all FM, then we are out of luck.

In the same way, it would be very expensive (and perhaps even ill mannered) for an extraterrestrial civilization to broadcast on a huge number of channels. Most likely, they select one or a few channels for their particular message. (Communicating on a narrow band of channels also helps distinguish an artificial message from the radio static that comes from natural cosmic processes.) But the radio band of the electromagnetic spectrum contains an astronomically large number of possible channels. How can we know in advance which one they have selected, and how they have coded their message into the radio signal?

There is another problem that may make it hard for you to find your favorite type of radio station. If your radio has a poor antenna, then it may not pick up the signal from a weak station some distance away. You may not learn about the existence of that station (and others like it) until you buy better equipment. The same will be true for interstellar transmissions. If an extraterrestrial civilization's signal is just too weak for our present-day radio telescopes, they may be broadcasting their little alien hearts out, but we will miss it completely.

Table 29.1 summarizes these and other factors that scientists must grapple with when they try to tune into radio messages from distant civilizations. Because their success depends on either guessing right about so many factors or searching through all the possibilities for each factor, some scientists have compared their quest to looking for a needle in a haystack. Thus, they like to say that the list of factors in Table 29.1 defines the *cosmic haystack problem*.

29.4.5 Radio Searches

Although the cosmic haystack problem seems daunting, many other research problems in astronomy also require a large investment of time, equipment, and patient effort. And, as several astronomers have pointed out, if we don't search, we're sure not to find anything. Therefore, several groups of radio astronomers have undertaken searches for extraterrestrial messages during the last four decades.

The very first search was conducted by astronomer Frank Drake in 1960, using the 85-ft antenna at the National Radio Astronomy Observatory (Figure 29.17). Called Project Ozma, after the queen of the exotic Land of Oz in the children's stories of L. Frank Baum, Drake's experiment involved looking at about 7200 channels and two nearby stars over a period of 200 hours. Although he found nothing, Drake demonstrated the feasibility of such a search and set the stage for the more sophisticated projects that followed. (It is interesting to note that what took 200 hours in 1960 can be done with today's automated systems in about a thousandth of a second.)

More than 60 additional radio searches have been carried out by scientists around the world, each exploring a minuscule region of the cosmic haystack. Although a number of initially interesting signals emerged, all either have been traced to local interference or have lasted such a short time that the signals could not be verified. Scientists are continuing several of the searches, always trying to improve their equipment and beat the odds against finding that elusive needle of a message.

In 1992, NASA began the most comprehensive SETI observations ever undertaken, only to have Congress cut the funding for their project after less than a year. Using private donations, the nonprofit SETI Institute has undertaken to continue the search. They now call it Project Phoenix, since the program has risen from the ashes of its funding crisis.

Using modern electronics and computers, the Phoenix system can tune to tens of millions of channels simultaneously (Figure 29.18). Its software searches for promising signals (such as continuous waves or pulses) and alerts the experimenters if an interesting signal persists on any one channel or moves between channels because of Doppler shift. The intent of the project (which resumed in 1995) is

National Radio Astronomy Observatory

■ **FIGURE 29.17**

Project Ozma   A 25th-anniversary photo of some members of the Project Ozma team standing in front of the 85-ft radio telescope with which the 1960 search for extraterrestrial messages was performed. Frank Drake is in the back row, second from the right.

(a)

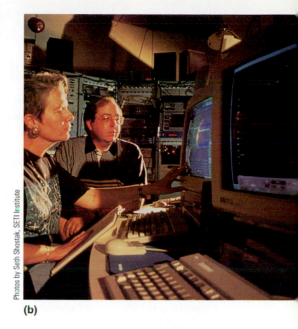

Photos by Seth Shostak, SETI Institute

(b)

■ **FIGURE 29.18**

Project Phoenix (a) The 64-m radio telescope at Parkes, Australia, was used in 1995 to search for radio signals from possible extraterrestrial civilizations around 200 stars. (b) Phoenix Project scientists Jill Tarter and Peter Backus are shown at the telescope controls during their observing run.

to search 2 billion channels for each of about 1000 nearby stars. To make it onto their list, a star must be roughly similar to the Sun and at least 3 billion years old. That means it could have had enough time for intelligent life to develop on any Earth-like planets around it. The majority of the initial candidate stars have now been observed. Other SETI programs search only a selected range of channels but sweep the sky in the hope of encountering a message. So far, no signal has been found; the search will continue using other large radio telescopes and even better equipment.

Receivers are constantly improving, and the sensitivity of SETI programs advances rapidly. One of the next steps will be to build much larger radio telescopes that are optimized for SETI. The SETI Institute is constructing the Allen Telescope Array, which uses many relatively small and inexpensive radio dishes linked together to provide the equivalent of 1 hectare of collecting area (Figure 29.19). A hectare is an area of 10,000 square meters, or about 2.5 acres. There is hope of expanding this concept to a 100-hectare or 1-square-kilometer area, which would be a powerful telescope for radio astronomy as well as SETI.

What kind of signals do we hope to pick up? We on Earth are inadvertently sending out a flood of radio signals, dominated by high-power television and military radar systems. This is a kind of leakage signal, similar to the wasted light energy that is beamed upward by poorly designed street lights and advertising signs. Could we detect a similar leakage of radio signals from another civilization? The answer is just barely, but only for the nearest stars. For the most part,

SETI Institute

■ **FIGURE 29.19**

Allen Telescope Array The most ambitious expansion of SETI today, the Allen Telescope Array will eventually consist of 350 small antennas linked together. This photo shows the first three antennas, which began test operations in California in December 2002.

therefore, current radio SETI searches are looking for beacons, not trying to eavesdrop on someone else's communication system. Our prospects for success depend on how often civilizations arise, how long they last, and how patient they are about broadcasting their locations to the cosmos (see the Figuring for Yourself box at the end of the chapter).

Jill Tarter: Trying to Make Contact

Jill Tarter

1997 was quite a year for Jill Cornell Tarter, one of the world's leading scientists in the SETI field. The SETI Institute announced that she was the recipient of its first endowed chair (the equivalent of an endowed research professorship) named in honor of Bernard Oliver. The National Science Foundation approved a proposal by a group of scientists and educators she headed to develop an innovative hands-on high school curriculum based on the ideas of cosmic evolution (the topics of this chapter). And, at roughly the same time, she was besieged with requests for media interviews as news reports identified her as the model for Ellie Arroway, the protagonist of *Contact*, Carl Sagan's bestselling novel about SETI. The book had been made into a high-budget science-fiction film starring Jodie Foster, who had actually talked with Tarter before taking the role.

Tarter is quick to point out that "Carl Sagan wrote a book about a woman who does what I do, not about me." Still, as the only woman in such a senior position in the small field of SETI, she was the center of a great deal of public attention. (However, colleagues and reporters pointed out that this was nothing compared to what would happen if her search for radio signals from other civilizations recorded a success.)

Being the only woman in a group is not a new situation for Tarter, who was often the only woman in her advanced science or math classes. She had been encouraged by her father in both her interest in science and her "tinkering," and as an undergraduate at Cornell University, she majored in engineering physics. That training now comes in very handy as she puts together and maintains the complex devices that automatically scan for signals from other civilizations.

Switching to astrophysics for her graduate studies, Tarter wrote a PhD thesis that, among other topics, considered the formation of failed stars—those whose mass was not sufficient for the nuclear reactions that power the stars to begin inside them. Tarter coined the term *brown dwarf* for these small, dim objects, and it has remained the name astronomers use. Later, her postgraduate work involved thinking about "dark matter," so-far unseen material that seems (from its gravitational pull) to be a very substantial component of galaxies and groups of galaxies. Now that she is working full time searching for extraterrestrial civilizations, Tarter likes to joke that her entire career seems to have involved searching for the most elusive things in the universe.

It was while she was still in graduate school that Stuart Bowyer, one of her professors at Berkeley, gave her a study on the possibilities of SETI (chaired, as it turns out, by Bernard Oliver). Bowyer asked her if she wanted to be involved in a small experiment to siphon off a bit of radiation from a radio telescope as astronomers used it year in and year out and see whether there was any hint of an intelligently coded radio message buried in the radio noise. Her engineering and computer programming skills became essential to the project, and soon she was hooked on the search for life elsewhere. Years later, as Tarter plans for the Allen Telescope Array, she may one day (like Ellie Arroway in the film) be the first person on Earth to know that humanity is not alone in the cosmos.

29.4.6 SETI Outside the Radio Realm

For the reasons discussed above, most SETI programs search for signals at radio wavelengths. It seems reasonable to us that if another civilization were broadcasting a beacon for purposes of interstellar communication, it would use microwaves. But maybe the civilization isn't broadcasting on purpose. In that case, is there any other way we could pick up other evidence for its existence?

Recently technology has allowed astronomers to expand the search into the domain of visible light. You might think it would be hopeless to try to detect a flash of visible light from a planet, given the brilliance of the star it orbits. This is why we cannot measure the reflected light of planets around other stars. The feeble light of the planet is simply swamped by the "big light" in the neighborhood. So another civilization would need a mighty strong beacon to compete with their star.

However, in recent years, human engineers have learned how to make flashes of light brighter than the Sun. The trick is to "turn on" the light for a very brief time, so that the costs are manageable. But ultra-bright, ultra-short laser pulses (operating for periods of a billionth of a second) can pack a lot of energy and can be coded to carry a message. We also have the technology to detect such short pulses—not with human senses, but with special detectors that can be "tuned" to hunt automatically for such short bursts of light from nearby stars.

Why would any civilization try to outshine its own star in this way? It turns out that the cost of sending an ultra-short laser pulse in the direction of a few promising stars can be less than the cost of sweeping a continuous radio message across the whole sky. Or perhaps they too have a special fondness for light messages because one of their senses evolved using light. Several programs are now experimenting with "optical SETI," which can be done with only a modest telescope (Figure 29.20). (The term *optical* here means using visible light.)

If we let our imaginations expand, we might think of other possibilities. What if a truly advanced civilization should decide to (or need to) renovate its planetary system to maximize the area for life? It could do so by breaking apart some planets or moons and building a sphere that

■ **FIGURE 29.20**

Optical SETI at the University of California
Undergraduate physics major Shelley Wright designed and built one of the first optical SETI systems, designed to detect extremely short-duration flashes of visible light. She is shown here with the Lick Observatory 1.1-m telescope where the observations are being made.

completely encloses the star and intercepts all its light. The star would then become invisible, but the huge artificial sphere might glow very brightly at infrared wavelengths as all the starlight was eventually converted to heat and re-radiated into space. In fact, we might imagine engineering on an even larger scale, capturing the energy from many stars. When astronomers find new and mysterious objects in the universe, there is always the temptation to ask ourselves whether these could be the constructions of distant, powerful minds with knowledge and skills almost beyond our imagining. It is a wonderful idea, but so far no evidence of such a civilization has been discovered.

29.4.7 What If We Succeed?

No one can predict when or whether SETI searches will be successful. It may well be that civilizations technologically far in advance of our own use other forms of communication that we are not yet aware of. After all, 150 years ago we did not have an inkling of the possibilities of radio communication, while today it is difficult to imagine our civilization without it. On the other hand, we would never dream of giving a preschooler a book like the one you are reading. Young children learning to read are given simple books until they have mastered the basics. We hope that advanced civilizations remember their own youth and send out messages that even youngsters like us can find and interpret.

What will happen if we do find a radio signal that is unambiguously the product of an extraterrestrial intelligence? The existence of the signal itself will be of tremendous philosophical importance, demonstrating that we are not alone in the cosmos. But unless we are able to interpret the message, there may not be much practical value in the discovery. If we can eventually work out a method of mutual communication, however, an interesting question will arise: Who will speak for planet Earth?

Suppose we know that a star 35 LY away has a technological civilization around it and that they have given us a kind of "code" by which we can make ourselves understood to each other. (An easy way to begin interacting might be to send pictures.) Who on Earth decides what to send? Does the whole planet try to agree on one set of messages, or can any individual or group send a separate communication? There may be many countries, religious groups, cultural organizations, corporations, and individuals who can afford a radio antenna and are interested in getting their message out. Among ourselves, we rarely speak with one voice; should we try to do so in addressing the universe? Confronting such questions may be a good test of whether there is intelligent life on Earth.

Whether or not we ultimately turn out to be the only intelligent species in our part of the Galaxy, our exploration of the cosmos will surely continue. An important part of that exploration will still be the search for biomarkers from inhabited planets that have not produced technological creatures that send out radio signals. After all, creatures like butterflies and dolphins may never build radio antennas, but we are happy to share our planet with them and would be delighted to find their counterparts on other worlds.

A humble acknowledgment of how much we have left to learn about the universe is one of the fundamental hallmarks of science. This should not, however, prevent us from feeling exhilarated about how much we have already managed to discover, and curious about what else we might find out.

Our progress report on the ideas of astronomy ends here, but we hope that your interest in the universe does not. We hope you will keep up with developments in astronomy through the media, or by going to an occasional public lecture by a local scientist. Who, after all, can even guess all the amazing things that future research projects will reveal about both the universe and our connection with it?

SURFING THE WEB

The Astrobiology Web: www.astrobiology.com
This private site, maintained by a space enthusiast, collects a vast amount of information about many aspects of astrobiology.

NASA Astrobiology Site: astrobiology.arc.nasa.gov
NASA's Ames Research Center is taking a leading role in astrobiology and its site can point you to NASA's plans for missions, as well as useful background information.

SETI Institute Home Page: www.seti.org
The main scientific organization searching for signals from extraterrestrial civilizations (which also works to understand the formation and prevalence of life beyond the Earth).

The Planetary Society's Search:
planetary.org/html/UPDATES/seti/index.html
This membership and advocacy organization helps pay for several SETI projects.

Drake Equation Calculators:

- www.seti.org/seti/seti_science/drake_calculator.html

- www.activemind.com/Mysterious/Topics/SETI/drake_equation.html

SETI@Home: setiathome.ssl.berkeley.edu
This screen saver allows anyone with a home computer to analyze a tiny bit of SETI data to see if there are any messages buried in it.

SUMMARY

29.1 The Earth and its life-forms are a product of cosmic evolution. The universe began with only the very simplest elements. More complex nuclei are produced via nuclear fusion at the hot centers of stars; some stars explode at the end of their lives and thus recycle the newly formed elements into space. Our solar system formed about 5 billion years ago from a cloud of gas and dust enriched by several generations of heavier element production in stars. The *Copernican principle,* which suggests that there is nothing special about our place in the universe, implies that if life could develop on Earth, it should be able to develop in other places as well. The Fermi paradox asks why more advanced life-forms have not visited us, if life is common.

29.2 The study of life in the universe, including its origin on Earth, is called **astrobiology.** Carbon-based or **organic molecules** are abundant in space, but the actual origin of life—of cells that can extract energy from their environment and reproduce themselves—remains mysterious. Proliferation of life on Earth is possible because of **photosynthesis,** the ability to extract chemical energy from sunlight. Life has evolved to occupy many diverse ecological niches, with **extremophiles** living in conditions of heat, cold, dryness, salinity, acidity, or high radiation that seem extreme to us. The basic requirements of liquid water, organic compounds, and an energy source point us toward searching for life on Earth-like or terrestrial planets.

29.3 Astronomers and astrobiologists are searching for life within our solar system but are also beginning to turn their attention toward planets around other stars. Mars is the most promising site, and the hypothesis of fossil life in the Mars rock ALH 84001 has stimulated additional interest. The strategy of "follow the water" also leads us to Europa, Jupiter's moon that has an ocean of liquid water. The same requirement of liquid water is part of the concept of a **habitable zone** around other stars—regions where life may be able to develop. When future telescopes are able to detect the light or infrared from Earth-like planets within the habitable zones of other stars, we will search for **biomarkers**—global changes (such as oxygen in the atmosphere) that provide evidence of life.

29.4 Astronomers are also engaged in the search for extraterrestrial *intelligent* life (**SETI**). Because the stars are so far away, traveling to them is either very slow or extremely expensive (in terms of energy required). Despite the many UFO reports and tremendous media publicity, there is no evidence that any of these are related to extraterrestrial spacecraft. The spacecraft we have sent beyond the planets (two Pioneers and two Voyagers) each contain a plaque or recording with information about humanity. Scientists have determined that the best way to communicate with any intelligent civilizations out there is by using electromagnetic waves, and radio waves seem best suited to the task. Since 1960, more than 60 projects have searched for radio messages from civilizations around other stars. So far, they have only begun to comb the many different possible stars, frequencies, signal types, and other factors that make up what we call the *cosmic haystack problem.* Another approach, called optical SETI, searches for extremely strong, but very brief pulses of light. If we do find a signal someday, deciding whether to answer and what to answer may be two of the greatest challenges humanity will face.

INTER-ACTIVITY

A If one of the rocks from Mars does turn out to have unambiguous signs of ancient life that formed on Mars, what does your group think would be the implications of such a discovery for science and for our view of life elsewhere? Would such a discovery have any long-term effects on your own thinking?

B Suppose we receive a radio message from a civilization around another star (which shows clear evidence of intelligence, even if we can't quite decipher all the details of what they send). What does your group think the implications of this discovery would be? How would your own thinking or philosophy be affected by such a discovery?

C A radio message has been received from a civilization around a star 40 LY away, which contains (in pictures) quite a bit of information about the beings that sent the message. The president of the United States has appointed your group a high-level commission to advise whether humanity should answer the message (which was not particularly directed at us but comes from a beacon that, like a lighthouse, sweeps out a circle in space). How would you advise the president? Does your group agree, or do you have a majority or minority view to present?

D If there is no evidence that the so-called UFOs are extraterrestrial visitors, why does your group think that television shows, newspapers, and movies spend so much time and effort publicizing the point of view that UFOs are craft from other worlds? Who stands to gain by exaggerating stories of unknown lights in the sky or simply fabricating stories that alien visitors are already here?

E Do you think scientists should just ignore all the media publicity about UFOs or should they try to respond? Why?

F Suppose your group had been the team planning the sights and sounds of Earth for the Voyager spacecraft audio/video record. What would you have put on the record to represent our planet to another civilization?

G Let's suppose Earth civilization has decided to broadcast a radio (or television) message announcing our existence to other possible civilizations among the stars. Your group is part of a large task force of scientists, communications specialists, and people from the humanities charged with deciding the form and content of our message. What would you recommend?

H If we do send messages into space to announce our presence, would it be useful to start with a reading of one of Shakespeare's plays coded as a very powerful FM broadcast? Why or why not?

I Think of examples of contact with aliens you have seen in movies and on TV. Discuss with your group how realistic these have been, given what you have learned in this class. Why do you think Hollywood does so many shows and films that are not scientifically valid?

REVIEW QUESTIONS

Ace⊙Astronomy™ Assess your understanding of this chapter's topics with additional quizzing and animations at **http://ace.brookscole.com/ voyages**

1. What is the Copernican principle? Make a list of scientific discoveries that confirm it.
2. Where in the solar system (and beyond) have scientists found evidence of organic molecules?
3. Give a short history of the atoms that are now in your little finger, going back to the beginning of the universe.
4. What is a biomarker? Give some possible examples of biomarkers we might look for beyond the solar system?
5. Why are Mars and Europa the top targets for the study of astrobiology? Explain.
6. Why is traveling between the stars difficult?
7. What are the advantages to using radio waves for communication between civilizations that live around different stars? List as many as you can.
8. What is the "cosmic haystack problem"? List as many of its components as you can think of.
9. In what ways is Project Phoenix an improvement over the first search ever conducted for radio signals from extraterrestrial civilizations?
10. What is optical SETI? Why is starlight not an overwhelming problem for such searches?

11. Would a human have been possible during the first generation of stars that formed right after the big bang? Why or why not?

12. How would you demonstrate that a rock found in Antarctica is from Mars and not from Earth?

13. If we do find life on Mars, what might be some ways to check whether it formed separately from Earth life, or whether exchanges of material between the two planets meant that the two forms of life have a common origin?

14. What kind of evidence do you think would convince astronomers that an extraterrestrial spacecraft has landed on the Earth? (*Hint:* See question 12.)

15. What are some reasons that more advanced civilizations might want to send out some messages to other star systems?

16. What are some answers to the Fermi paradox? Can you think of some that are not discussed in this chapter?

FIGURING FOR YOURSELF: THE DRAKE EQUATION

Ace Astronomy™ Log into AceAstronomy and select this chapter to see the Active Figure called "Drake Equation."

The first scientific meeting devoted to SETI (the search for extraterrestrial intelligence) was held in 1961 at the U.S. National Radio Astronomy Observatory, where Frank Drake had just completed his pioneering Project Ozma observations. To provide some structure to the wide-ranging discussions about life in the universe, Drake wrote an equation on the blackboard that took the difficult question of estimating the number of civilizations in the Galaxy and broke it down into a series of smaller, more manageable questions. Ever since then, both astronomers and students have used this Drake equation as a means of approaching the most challenging question: Are we alone? Since this is at present an unanswerable question, astronomer Jill Tarter has called the Drake equation a "way of organizing our ignorance."

The form of the Drake equation is very simple. To estimate the number of communicating civilizations that currently exist in the Galaxy (we will define these terms more carefully in a moment), we multiply the rate of formation of such civilizations (number per year) by their average lifetime (in years). In symbols,

$$N = R(\text{total}) \times L$$

This problem is similar to estimating the number of frogs in a fresh pond. If 50 new frogs are born in the pond each year (the rate of formation) and the frogs each live an average of 8 years, then, when things settle down after a decade or so, the number of frogs in the pond at any given time should be 50×8, or 400.

To make his formula easier to use (and more interesting!), however, Drake separated the rate of formation $R(\text{total})$ into a series of probabilities:

$$R(\text{total}) = R(\text{star}) \times f_p \times f_e \times f_l \times f_i \times f_c$$

$R(\text{star})$ is the rate of formation of stars like the Sun in our Galaxy, which is about 10 stars per year. Each of the other terms is a fraction or probability (less than or equal to 1.0), and the product of all these probabilities is itself the total probability that each star will have an intelligent, technological, communicating civilization that we might want to talk to. We have:

f_p = the fraction of these stars with planets (thanks to recent discoveries, this is now known to be between 0.1 and 1.0)

f_e = the fraction of the planetary systems that include Earth-like or habitable planets (maybe all of them if they are like the solar system, but maybe a smaller fraction if hot Jupiters spiral inward in some systems)

f_l = the fraction of habitable planets that actually support life (from the Copernican principle, you might say it's near 1.0, but we just don't know)

f_i = the fraction of inhabited planets that develop advanced intelligence (it happened on Earth, but only after more than 4 billion years of evolution; it's hard to say how likely it is without having other examples)

f_c = the fraction of these intelligent civilizations that develop science and the technology to build radio telescopes and transmitters for communication (something that has happened on Earth only in the last few decades and could probably happen only on worlds where the inhabitants can mine metals)

Each of these factors can be discussed and perhaps evaluated, but we must guess at many of the values. In particular, we don't know how to calculate the probability of something that happened once on Earth but has not been observed elsewhere—and these include the development of life, of intelligent life, and of technological life (the last three factors in the equation).

Even if we don't know the answers, you will find that it is instructive to make some guesses and calculate the resulting number N. Let's start with the optimism implicit in the Copernican principle and set the last three terms equal to 1.0. If R is 10, f_p is 0.1, and f_e is 1.0, the equation becomes

$$N = R(\text{total}) \times L = L$$

Now we see the importance of the term *L*, the lifetime of a communicating civilization. We have had this capability for only a few decades. Suppose we assume that this stage in our history lasts only one century; then, with our optimistic assumptions about the other factors, $L = 100$ years and $N = 100$ such civilizations in the entire Galaxy. In that case there are so few other civilizations like ours that we are unlikely to detect any signals in a SETI search. But suppose the average lifetime is a million years; in that case there are a million such civilizations in the Galaxy and several are likely to be close enough for communication. (You may be amused to read that Frank Drake drives a car with license plate "NEQLSL.")

The most important conclusion from this calculation is that, even if we are extremely optimistic about the probabilities, the only way we can expect success from SETI is if other civilizations are much older (and hence probably much more advanced) than ours. We must be the youngest kid on the block (in our corner of the Galaxy). Most other civilizations are likely to be much older and perhaps much smarter (or at least more experienced) than we. It's a sobering thought. If we had made less optimistic assumptions, this conclusion would be even stronger.

17. Using the information from this and previous chapters, pick values that make sense to you for the probabilities in the Drake equation. What is the resulting value of *N*, the number of communicating civilizations?

18. Select the most optimistic and pessimistic values that you think are reasonable (and this choice requires some judgments, since we really don't know the values of these numbers) and calculate the corresponding range in values of *N*.

19. Suppose astronomers discover a radio message from a civilization whose planet orbits a star 35 LY away. Their message encourages us to send a radio answer, which we decide to do. Suppose our governing bodies take two years to decide whether and how to answer. When our answer arrives there, their governing bodies take one year to frame an answer to us. How long after we get their first message can we hope to get their reply to ours? (Once communication gets going, should we continue to wait for a reply before we send the next message?)

20. Think of our Milky Way Galaxy as a flat disk of diameter 100,000 LY. Suppose we are one of 1000 civilizations, randomly distributed through the disk, interested in communicating via radio waves. How far away would the nearest such civilization be from us (on average)?

SUGGESTIONS FOR FURTHER READING

Croswell, K. "Interstellar Trekking" in *Astronomy,* June 1998, p. 46. Science perspective on how to travel to the stars.

Darling, D. *Life Everywhere: The Maverick Science of Astrobiology.* 2001, Basic Books. Short overview of the science of astrobiology.

Davies, P. *The Fifth Miracle: The Search for the Origin and Meaning of Life.* 1999, Simon & Shuster. Scientific and philosophical issues associated with the origin of life and the search for life elsewhere.

deDuve, C. *Vital Dust: The Origin and Evolution of Life on Earth.* 1995, Basic Books. A Nobel laureate recounts our modern understanding of how life emerged on our planet.

Gross, M. *Life on the Edge.* 1998, Plenum. A British biologist discusses how life can thrive under remarkably extreme conditions, and how we can search for such life elsewhere.

Jakosky, B. *The Search for Life on Other Planets.* 1998, Cambridge University Press. Popular account of the prospects for life in our solar system and beyond by one of the pioneers of astrobiology.

Krauss, L. *Atom: An Odyssey from the Big Bang to Life on Earth . . . and Beyond.* 2001, Little, Brown. Traces the cosmic history of an oxygen atom.

Lemonick, M. *Other Worlds: The Search for Life in the Universe.* 1998, Simon & Schuster. *Time* magazine senior science correspondent summarizes the new ideas and discoveries.

LePage, A. and MacRobert, A. "SETI Searches Today" in *Sky & Telescope,* Dec. 1998, p. 44. On the scientific searches for signals from other civilizations.

Nadis, S. "Using Lasers to Detect ET" in *Astronomy,* Sept. 2002, p. 44. On optical SETI, with profiles of the main players.

Pendleton, Y. and Farmer, J. "Life: A Cosmic Imperative?" in *Sky & Telescope,* July 1997, p. 42. On the rise of life and environments that make it possible or impossible.

Shilling, G. "The Chance of Finding Aliens: Re-evaluating the Drake Equation" in *Sky & Telescope,* Dec. 1998, p. 36.

Shostak, S. "The Future of SETI" in *Sky & Telescope,* Apr. 2001, p. 42. New searches, new strategies; nice long review.

Shostak, S. "When ET Calls Us" in *Astronomy,* Sept. 1997, p. 36. What do we do when we receive a message?

Shostak, S. *Sharing the Universe: Perspectives on Extraterrestrial Life.* 1998, Berkeley Hills. Wide-ranging and often humorous discussions by one of the most popular speakers and writers on SETI.

Tarter, J. and Chyba, C. "Is There Life Elsewhere in the Universe?" in *Scientific American,* Dec. 1999, p. 118. On future search for life in the solar system and beyond.

Treiman, A. "Microbes in a Martian Meteorite: The Evidence Has Grown More Cloudy" in *Sky & Telescope,* Apr. 1999, p. 52.

Webb, S. *Where Is Everybody?* 2002, Copernicus/Praxis. On 50 solutions to the Fermi paradox.

1 Astronomy on the World Wide Web

In just a few short years, the World Wide Web has become a planet-wide library, full of information, images, opinion, and argument. Astronomers, always eager to use technological tools like computers, jumped on the Web bandwagon early, and thousands of sites are dedicated to astronomical ideas.

This great wealth of astronomy pages also presents a problem: Unlike a textbook that is written by (and checked by) experts in their fields, Web sites can be posted by anyone. Thus many Web sites contain misinformation, exaggeration, misunderstanding, and downright nonsense about the universe. Some of these sites are offered by well-meaning but inexperienced enthusiasts, while others are by charlatans or hoaxers out to make a quick buck or promote a hidden cause.

Throughout *Voyages* we have recommended reliable Web sites for each chapter and topic as we went along. In this appendix, we list some more general sites that are especially good at guiding astronomy beginners to helpful information, the latest images, or reliable organizations. Most of the sites we list include links to other sites that their Webmasters have found useful.

Note that Web addresses and sites can change faster than any mere mortal can follow. For the latest information on Web sites we recommend, check the *Voyages* Web site at: info.brookscole .com/voyages

1. Magazines That Cover Astronomy

Astronomy **Magazine** (www.astronomy.com) has the largest circulation of any magazine devoted to the universe and is designed especially for astronomy hobbyists and armchair astronomers. The site features many sections on the hobby of astronomy and on things you can do to become involved.

Sky & Telescope **Magazine** (skyandtelescope.com) is an older and somewhat higher-level magazine for astronomy hobbyists. The site is especially rich in observational guides, software and equipment reviews, and lists of astronomy clubs around the country.

Scientific American **Magazine** (www.sciam.com) offers one astronomy article about every second issue. These articles, a number of which are reproduced on their Web site, are at a slightly higher level than *Voyages,* but—often written by the astronomers who have done the work being described—are authoritative and up-to-date.

Mercury **Magazine** (www.astrosociety.org/pubs/mercury/ mercury.html) is a popular-level magazine published by the Astronomical Society of the Pacific that features articles and columns for astronomers, educators, and astronomy enthusiasts. The site often deals with more philosophical issues and the influence of astronomy on other areas of human thought.

Other Magazines That Cover Astronomy

- *Science News:* www.sciencenews.org
- *Griffith Observer:* www.griffithobs.org/Observer.html
- *Astronomy Now:* www.astronomynow.com

2. Organizations That Deal with Astronomy for the Public

The Astronomical Society of the Pacific (www.astrosociety .org) is the largest and oldest national astronomy organization in the United States. The membership of the society, founded in 1889, includes astronomers, educators, and astronomy enthusiasts. The site features a nonprofit catalog of interesting astronomy items (posters, videos, software, etc.) and also a large quantity of information related to astronomy education.

The Planetary Society (www.planetary.org), founded by the late Carl Sagan, works to encourage planetary exploration and the search for life elsewhere. Its Web site has many pages of information in these two areas.

The Astronomical League (www.astroleague.org) is the umbrella organization of astronomy clubs in the United States. The site has good lists of astronomy clubs around the country, activities and hints for astronomy hobbyists, and information on regional and national programs.

The Royal Astronomical Society of Canada (www.rasc.ca) is a membership organization that unites professional and amateur astronomers in Canada and has 26 centers around that country with local activities. Specific calendars for each location can be found on the site.

NASA (www.nasa.gov) has an astronomical amount of information on its many Web sites; the trick is to find what you need. Their site has improved in recent years, with better navigation and search functions. Most projects and NASA centers have their own Web sites and we have recommended many of these throughout the chapters.

3. Sites for Astronomy and Space Science News

- www.skyandtelescope.com
- www.astronomy.com
- space.com
- www.universe.com
- www.spacedaily.com
- science.nasa.gov

Each of these sites is run by people with a science background and tries to feature (and explain) new developments as they occur.

4. Selected Sources for the Latest Astronomy Images

Anglo-Australian Observatory Image Collection: www.aao.gov.au/images.html
A marvelous library of images (focusing on nebulae and galaxies) taken using large Australian telescopes. Many are by David Malin, who is acknowledged to be one of the finest astronomical photographers of our time. The images include captions and ordering information.

Hubble Space Telescope Images: hubblesite.org/newscenter/archive
All the magnificent Hubble Space Telescope images are here, with captions; many have detailed background information and new research results included. You can check the latest images, browse the ones the staff considers the Hubble's "greatest hits," or search for objects of interest to you. Best of all, the images can all be downloaded in various formats completely free.

National Optical Astronomy Observatories Image Gallery: www.noao.edu/image_gallery/
NOAO includes a number of major telescopes in the United States and the Southern Hemisphere. Some of the best images from NOAO instruments are collected and organized at this site.

Planetary Photojournal: photojournal.jpl.nasa.gov
This site, run by the Jet Propulsion Laboratories and the U.S. Geological Survey Branch of Astrogeology in Flagstaff, is one of the most useful resources on the Web. It features thousands of the best NASA images from planetary exploration, with detailed captions and excellent indexing, and more are being added all the time. You can dial up images by world, feature name, date, or catalog number, and download images in a number of popular formats.

Astronomy Picture of the Day: antwrp.gsfc.nasa.gov/apod/astropix.html
Astronomers Robert Nemiroff and Jerry Bonell feature one relatively new celestial image each day with a brief nontechnical caption. Over the years, some of the best astronomical images have been featured here and an index is available on site.

European Southern Observatory: www.eso.org/outreach/gallery/astro
This growing album contains images from the large telescopes in the Southern Hemisphere run by a consortium of European countries. With the advent of the Very Large Telescope, there is an increasing number of important new images and results on this site.

Infrared Image Gallery: www.ipac.caltech.edu/Outreach/Gallery
This site collects infrared images from space missions and telescopes on the ground and also offers an intriguing set of comparisons showing images at visible and invisible wavelengths.

Other Ways to Find Astronomy and Space Images

- *Google image search:* www.google.com/advanced_image_search?hl=en
- *NASA Image Exchange:* nix.nasa.gov
- *Great Images at NASA (mostly historical):* grin.hq.nasa.gov/
- *Index to NASA Image Sites:* www.nasa.gov/gallery/photo/index.html

5. Debunking Pseudo-science

Astronomical Pseudo-science: A Skeptic's Resource List: www.astrosociety.org/education/resources/pseudobib.html
An annotated list of written and Web resources for dealing with astrology, UFOs, the Face on Mars, ancient astronauts, and many similar fringe topics.

Debunking Pseudo-science (in general): www.csicop.org
For students or instructors who want assistance in dealing with topics that mix up science with new-age philosophy, paranormal beliefs, conspiracy theories, and utter silliness, the premier organization is the awkwardly named, but very effective, Committee for the Scientific Investigation of Claims of the Paranormal (CSICOP). Composed of scientists, educators, magicians, legal experts, doctors, and others who are tired of the uncritical acceptance of pseudo-science by the media and the public, CSICOP seeks to provide the skeptical, rational side of many controversial topics, such as UFOs, astrology, psychic power, the Bermuda Triangle, crop circles, and so on. The site is a treasure trove of information, anecdotes, experiments, and links.

6. Sources and Reviews of Astronomical Software

Sky & Telescope's Software Pages: skyandtelescope.com/resources/software
Lists and explains a wide range of astronomical freeware and shareware, programs you can download and copy.

John Mosley's Software Directory: www.griffithobs.org/software.html
The editor of the magazine for planetarium educators lists a wide range of astronomical software by publisher, plus references to published reviews.

Dan Bruton's Software List: www.physics.sfasu.edu/astro/software.html
Includes listings and links to over 200 pieces of astronomical software.

Bill Arnett's Software List: www.seds.org/billa/astrosoftware.html
A long listing of both commercial and noncommercial packages with links.

7. Miscellaneous Sites

Science Fiction Stories with Good Astronomy and Physics: www.astrosociety.org/education/resources/scifi.html
A list of novels and stories organized by topic in astronomy, with a brief description of each.

Astronomical Observing Tips: skyandtelescope.com/howto/
This series of excellent articles from *Sky & Telescope* magazine introduces beginners to observational astronomy. It includes good advice on getting started with just your eyes, binoculars, small telescopes, star wheels, photography, and much more.

History of Astronomy Site: www.astro.uni-bonn.de/~pbrosche/astoria.html
Professor Wolfgang Dick maintains this comprehensive series of links to sites about all aspects of astronomical history.

Among its best features is the ability to select the name of a historical astronomer and find links that provide instant biographical information.

AstroWeb: Astronomy/Astrophysics on the Internet:
www.cv.nrao.edu/fits/www/astronomy.html

This site, maintained by a consortium of astronomy institutions, is a collection of thousands of astronomy-related links, organized by subject. Some links are to highly technical sites, some to pages that are just right for beginners. You can check whether an astronomer has a personal Web page, look up what astronomy departments or observatories have put on the Web, and delve further into many specialized areas.

Space Calendar: wwwjpl.nasa.gov/calendar

Ron Baalke at JPL keeps this amazing listing of space- and astronomy-related events, including upcoming launches, significant anniversaries, and things to see in the sky. Most entries in the calendar have a Web link where you can find more information.

Sources of Astronomical Information

With the growth of the World Wide Web, even print magazines and all organizations dedicated to working with the public have set up Web sites of basic information. Many of their Web sites are given in Appendix 1. Still, you may want to contact some of the magazines and organizations by mail to subscribe or ask for specific materials. Here we list a few of the best-known public information providers in astronomy.

1. Popular-level Astronomy Magazines

Astronomy, Kalmbach Publishing, P.O. Box 1612, Waukesha, WI 53187. The astronomy publication with the largest circulation in the world; it is very colorful and basic.

Griffith Observer, Griffith Observatory, 2800 E. Observatory Rd., Los Angeles, CA 90027. A small magazine specializing in historical topics, edited with humor and verve.

Mercury, Astronomical Society of the Pacific, 390 Ashton Ave., San Francisco, CA 94112. The magazine of the largest general membership astronomy organization in the United States, with many interesting features, particularly about the influence of astronomy on other fields, astronomy teaching, and the stories behind the new discoveries.

Planetary Report, Planetary Society, 65 N. Catalina Ave., Pasadena, CA 91106. The magazine of an organization that promotes the exploration of the solar system and the search for life elsewhere.

Stardate, McDonald Observatory, RLM 15.308, University of Texas, Austin, TX 78712. Accompanies the popular radio program on astronomy.

Sky & Telescope, P.O. Box 9111, Belmont, MA 02178. Found in most libraries, this is the popular astronomy magazine "of record," with many excellent articles on astronomy and amateur astronomy.

Sky News, Box 10, Yarker, Ontario K0K 3N0 Canada; www.skynewsmagazine.com. This is Canada's popular-level astronomy magazine, with astronomy news and features for hobbyists.

The following magazines (found in most libraries) also contain reports on astronomical developments: *Discover, National Geographic, Science News,* and *Scientific American.*

2. Organizations for Astronomy Enthusiasts

American Association of Variable Star Observers, 25 Birch St., Cambridge, MA 02138 (617-354-0484); www.aavso.org. An organization of amateur astronomers devoted to making serious observations of stars whose brightness changes.

Association of Lunar and Planetary Observers, c/o Donald Parker, 12911 Lerida St., Coral Gables, FL 33156; www.lpl.arizona.edu/alpo. An organization of amateur astronomers who monitor and report on objects in our solar system.

Astronomical League, 9201 Ward Pkwy., Kansas City, MO 64114; www. astroleague.org. Umbrella organization of amateur astronomy clubs in the United States, with many activities. If you write to this all-volunteer organization, be sure to enclose a stamped, self-addressed envelope.

Astronomical Society of the Pacific, 390 Ashton Ave., San Francisco, CA 94112 (415-337-1100); www.astrosociety.org. This international organization brings together scientists, teachers, and people with a general interest in astronomy; its name is merely a reminder of its origins on the West Coast of the United States in 1889. Write for the catalog of interesting astronomy materials for all levels.

Committee for the Scientific Investigation of Claims of the Paranormal (CSICOP), P.O. Box 703, Buffalo, NY 14226 (716-636-1425); www.csicop.org. An organization of scientists, educators, magicians, and other skeptics that seeks to inform the public about the rational perspective on such pseudo-sciences as astrology, UFOs, and psychic power. The committee publishes *The Skeptical Inquirer* magazine, full of great debunking articles, and holds meetings and workshops around the world.

International Dark Sky Association, 3225 N. First Ave., Tucson, AZ 85719; www.darksky.org. Small non-profit organization devoted to fighting light pollution and educating politicians, lighting engineers, and the public about the importance of not spilling light where it will interfere with astronomical observations.

The Planetary Society, 65 N. Catalina Ave., Pasadena, CA 91106 (818-793-1675); www.planetary.org. Large national membership organization was founded by Carl Sagan and others. It lobbies for more planetary exploration and SETI, and publishes a colorful magazine.

The Royal Astronomical Society of Canada, 136 Dupont St., Toronto, Ontario M5R 1V2 (416-924-7973); www.rasc.ca. The main organization of amateur astronomers in Canada, with local centers in each province. Write for information on contacting the center near you.

Note: The American Astronomical Society is the professional organization for astronomers in the United States. The journals and activities are mostly for research scientists, but the Web site does have an excellent introduction for high school students to becoming an astronomer. See: http://www.aas.org/education/career.html

3

Glossary

Note: The number in parentheses is the chapter where the term is first defined or explained (P is for the Prologue; I for the Interlude).

absolute brightness (absolute magnitude) A measure of the luminosity of an object like a star. *See* luminosity.

absolute zero A temperature of $-273°C$ (or 0 K), where all molecular motion stops. (4)

absorption spectrum Dark lines superimposed on a continuous spectrum. (4)

accelerate To change velocity; to speed up, slow down, or change direction. (1)

accretion Gradual accumulation of mass, as by a planet forming from colliding particles in the solar nebula or gas falling into a black hole. (13)

accretion disk A disk of matter spiraling in toward a massive object; the disk shape is the result of conservation of angular momentum. (23)

active galactic nucleus A galaxy is said to have an active nucleus if unusually violent events are taking place in its center, emitting large quantities of electromagnetic radiation. Seyfert galaxies and quasars are examples of galaxies with active nuclei. (26)

active region An area on the Sun where magnetic fields are concentrated; sunspots, prominences, and flares all tend to occur in active regions. (14)

adaptive optics Systems used with telescopes that change to compensate for distortions in an image introduced by the atmosphere, thus producing sharper images. (5)

albedo The fraction of sunlight reaching it that a planet or asteroid reflects; reflectivity.

alpha particle The nucleus of a helium atom, consisting of two protons and two neutrons. (4)

amplitude The range in variability, as in the light from a variable star.

angular diameter Angle subtended by the diameter of an object.

angular momentum A measure of the momentum associated with motion about an axis or fixed point. (2)

antapex (solar) Direction away from which the Sun is moving with respect to the local standard of rest.

anthropic principle The idea that physical laws must be the way they are because otherwise we could not be here to measure them. (28)

Antarctic Circle Parallel of latitude 66°30′S; at this latitude the noon altitude of the Sun is 0° on the date of the summer solstice. (3)

antimatter Matter consisting of antiparticles: antiprotons (protons with negative rather than positive charge), positrons (positively charged electrons), and antineutrons. (15)

aperture The diameter of an opening, or of the primary lens or mirror of a telescope. (5)

apex (solar) The direction toward which the Sun is moving with respect to the local standard of rest.

aphelion Point in its orbit where a planet (or other orbiting object) is farthest from the Sun. (2)

apogee Point in its orbit where an Earth satellite is farthest from the Earth. (2)

apparent brightness A measure of the observed light received from a star or other object at the Earth—that is, how bright an object appears in the sky, as contrasted with its luminosity. (4)

apparent solar time Time as measured by the position of the Sun in the sky (the time that would be indicated by a sundial). (3)

Arctic Circle Parallel of latitude 66°30′N; at this latitude the noon altitude of the Sun is 0° on the date of the winter solstice. (3)

array (interferometer) A group of several telescopes that are connected together to make observations at high angular resolution. (5)

association A loose group of young stars whose spectral types, motions, and positions in the sky indicate that they probably had a common origin. (21)

asterism An especially noticeable star pattern in the sky, such as the Big Dipper. (1)

asteroid A stony or metallic object orbiting the Sun that is smaller than a major planet but that shows no evidence of an atmosphere or of other types of activity associated with comets. (6)

asteroid belt The region of the solar system between the orbits of Mars and Jupiter in which most asteroids are located. The main belt, where the orbits are generally the most stable, extends from 2.2 to 3.3 AU from the Sun. (12)

astrobiology The multidisciplinary study of life in the universe: its origin, evolution, distribution, and fate; similar terms are *exobiology* and *bioastronomy*. (29)

astrology The pseudoscience that deals with the supposed influences on human destiny of the configurations and locations in the sky of the Sun, Moon, and planets; a primitive belief system that had its origin in ancient Babylonia. (1)

astronomical unit (AU) Originally meant to be the semimajor axis of the orbit of the Earth; now defined as the semimajor axis of the orbit of a hypothetical body with the mass and period that Gauss assumed for the Earth. The semimajor axis of the orbit of the Earth is actually 1.000000230 AU. (2)

atom   The smallest particle of an element that retains the properties that characterize that element. (P)

atomic mass unit   *Chemical:* one-sixteenth of the mean mass of an oxygen atom. *Physical:* one-twelfth of the mass of an atom of the most common isotope of carbon. The atomic mass unit is approximately the mass of a hydrogen atom, 1.67×10^{-27} kg.

atomic number   The number of protons in each atom of a particular element.

atomic weight   The mean mass of an atom of a particular element in atomic mass units.

aurora   Light radiated by atoms and ions in the ionosphere, mostly in the magnetic polar regions. (7)

autumnal equinox   The intersection of the ecliptic and celestial equator where the Sun crosses the equator from north to south. A time when every place on Earth has 12 hours of daylight and 12 hours of night. (3)

axis   An imaginary line about which a body rotates. (1)

azimuth   The angle along the celestial horizon, measured eastward from the north point to the intersection of the horizon with the vertical circle passing through an object. (1)

Balmer lines   Emission or absorption lines in the spectrum of hydrogen that arise from transitions between the second (or first excited) and higher energy states of the hydrogen atom. (4)

bands (in spectra)   Emission or absorption lines, usually in the spectra of chemical compounds, so numerous and closely spaced that they coalesce into broad emission or absorption bands. (4)

bar   A force of 100,000 newtons acting on a surface area of 1 square meter. The average pressure of the Earth's atmosphere at sea level is 1.013 bars. (7)

barred spiral galaxy   Spiral galaxy in which the spiral arms begin from the ends of a "bar" running through the nucleus rather than from the nucleus itself. (25)

barycenter   The center of mass of two mutually revolving bodies. (20)

basalt   Igneous rock produced by the cooling of lava. Basalts make up most of Earth's oceanic crust and are also found on other planets that have experienced extensive volcanic activity. (7)

big bang theory   A theory of cosmology in which the expansion of the universe began with a primeval explosion. (28)

binary stars   Two stars that revolve about each other. (17)

binding energy   The energy required to separate completely the constituent parts of an atomic nucleus. (15)

bioastronomy   *See* astrobiology.

biomarker   Evidence of the presence of life, especially a global indication of life on a planet that could be detected remotely (such as an unusual atmospheric composition). (29)

blackbody   A hypothetical perfect radiator, which absorbs and reemits all radiation incident upon it. (4)

black dwarf   The final state of evolution for a low-mass star, in which all of its energy sources are exhausted and it no longer emits significant radiation. (22)

black hole   A collapsed massive star (or other collapsed body) whose velocity of escape is equal to or greater than the speed of light; thus no radiation can escape from it. (23)

Bohr atom   A particular model of an atom, invented by Niels Bohr, in which the electrons are described as revolving about the nucleus in circular orbits. (4)

brown dwarf   An object intermediate in size between a planet and a star. The approximate mass range is from about 1/100 of the mass of the Sun up to the lower mass limit for self-sustaining nuclear reactions, which is 0.072 the mass of the Sun. (17)

caldera   Volcanic depression or crater, usually produced by the collapse of a volcano's summit. (9)

carbon-nitrogen-oxygen (CNO) cycle   A series of nuclear reactions in the interiors of stars involving carbon as a catalyst, by which hydrogen is transformed to helium. (15)

carbonaceous meteorite   A primitive meteorite made primarily of silicates but including chemically bound water, free carbon, and complex organic compounds. Also called carbonaceous chondrites. (13)

carbonate   Chemical compound that contains CO_2, such as calcium carbonate ($CaCO_3$); carbonates can "store" carbon dioxide in solid form. (13)

Cassegrain focus   An optical arrangement in a reflecting telescope in which light is reflected by a second mirror to a point behind the primary mirror. (5)

CBR   *See* cosmic background radiation.

CCD   *See* charge-coupled device.

celestial equator   A great circle on the celestial sphere 90° from the celestial poles; where the celestial sphere intersects the plane of the Earth's equator. (1)

celestial meridian   An imaginary line on the celestial sphere passing through the north and south points on the horizon and through the zenith.

celestial poles   Points about which the celestial sphere appears to rotate; intersections of the celestial sphere with the Earth's polar axis. (1)

celestial sphere   Apparent sphere of the sky; a sphere of large radius centered on the observer. Directions of objects in the sky can be denoted by their position on the celestial sphere. (1)

center of gravity   Center of mass.

center of mass   The average position of the various mass elements of a body or system, weighted according to their distances from that center of mass; that point in an isolated system that moves with constant velocity according to Newton's first law of motion. (17)

cepheid variable   A star that belongs to a class of yellow supergiant pulsating stars. These stars vary periodically in brightness, and the relationship between their periods and luminosities is useful in deriving distances to them. (18)

Chandrasekhar limit   The upper limit to the mass of a white dwarf (equals 1.4 times the mass of the Sun). (22)

charge-coupled device (CCD)   An array of high-sensitivity electronic detectors of electromagnetic radiation, used at the focus of a telescope (or camera lens) to record an image or spectrum. (5)

chemical condensation sequence   The calculated order of chemical compounds and minerals that would form at different temperatures in a cooling gas of cosmic composition; used to infer the composition of grains that formed in the solar nebula at different distances from the protosun. (13)

chondrite   A primitive stony meteorite

chondrule   Small spheres (roughly the size of a pea or smaller) of once-melted rock that are found in most chondrite meteorites.

chromosphere   That part of the solar atmosphere that lies immediately above the photospheric layers. (14)

circular satellite velocity The critical speed that a revolving body must have in order to follow a circular orbit. (2)

circumpolar zone Those portions of the celestial sphere near the celestial poles that are either always above or always below the horizon. (1)

circumstellar disk Thin disk of gas and dust surrounding a very young star or protostar out of which a planetary system may form; similar to the solar nebula that gave birth to our solar system (6, 13)

closed universe A model of the universe in which the curvature of space is such that straight lines eventually curve back upon themselves; in this model, the universe expands from a big bang, stops, and then contracts to a big crunch. (28)

cluster of galaxies A system of galaxies containing several to thousands of member galaxies. (27)

color index Difference between the magnitudes of a star or other object measured in light of two different spectral regions—for example, blue minus visual (B − V) magnitudes. (16)

coma (of comet) The cloud of evaporated gas and dust around a comet nucleus. (12)

comet A small body of icy and dusty matter that revolves about the Sun. When a comet comes near the Sun, some of its material vaporizes, forming a large head of tenuous gas and often a tail. (6)

compound A substance composed of two or more chemical elements.

conduction The transfer of energy by the direct passing of energy or electrons from atom to atom. (15)

conservation of angular momentum The law that the total amount of angular momentum in a system remains the same (in the absence of any force not directed toward or away from the point or axis about which the angular momentum is referred). (2)

constellation One of 88 sectors into which astronomers divide the celestial sphere; many constellations are named after a prominent group of stars within them that represents a person, animal, or legendary creature from ancient mythology. (P)

continental drift A gradual movement of the continents over the surface of the Earth due to plate tectonics. (7)

continuous spectrum A spectrum of light composed of radiation of a continuous range of wavelengths or colors rather than only certain discrete wavelengths. (4)

convection The transfer of energy by moving currents in a fluid. (7)

core (of a planet) The central part of a planet, consisting of higher-density material. (7)

corona (of Sun) Outer atmosphere of the Sun. (3)

corona (on Venus) Distinctive large tectonic feature on Venus, apparently caused by a rising plume of hot magma in the mantle of the planet (9)

coronal hole A region in the Sun's outer atmosphere that appears darker because there is less hot gas there. (14)

cosmic background radiation (CBR) The microwave radiation coming from all directions that is the redshifted glow of the big bang. (28)

cosmic rays Atomic nuclei (mostly protons) that are observed to strike the Earth's atmosphere with exceedingly high energies. (19)

cosmological constant A term in the equations of general relativity that represents a repulsive force in the universe. *See also* dark energy. (28)

cosmological principle The assumption that, on the large scale, the universe at any given time is the same everywhere—isotropic and homogeneous. (27)

cosmology The study of the organization and evolution of the universe. (28)

crater A circular depression (from the Greek word for "bowl"), generally of impact origin. (8)

crescent moon One of the phases of the Moon when it appears less than half full. (3)

critical density In cosmology, the density that provides enough gravity to bring the expansion of the universe just to a stop after infinite time. (28)

crust The outer layer of a terrestrial planet. (6)

cryovolcanism Geological processes on cold worlds (such as the outer-planet satellites) that mimic volcanic activity but involve fluids such as water rather than silicate magma. (11)

dark energy The energy that is causing the expansion of the universe to accelerate. Its existence is inferred from observations of distant supernovae. (28)

dark matter Nonluminous mass, whose presence can be inferred only because of its gravitational influence on luminous matter. The composition of the dark matter is not known. (24)

dark nebula A cloud of interstellar dust that obscures the light of more distant stars and appears as a small opaque region in the sky. (19)

declination Angular distance north or south of the celestial equator. (3)

degenerate gas A gas in which the allowable states for the electrons have been filled; it behaves according to different laws from those that apply to "perfect" gases and resists further compression. (22)

density The ratio of the mass of an object to its volume. (2)

deuterium A "heavy" form of hydrogen, in which the nucleus of each atom consists of one proton and one neutron. (15)

differential galactic rotation The rotation of the Galaxy, not as a solid wheel but so that parts adjacent to one another do not always stay close together. (24)

differentiated meteorite Meteorite derived from a differentiated parent body and hence not primitive. (13)

differentiation (geological) Gravitational separation of materials of different density into layers in the interior of a planet or satellite. (6)

disk (of Galaxy) The central plane or "wheel" of our Galaxy, where most of the luminous mass is concentrated. (24)

dispersion Separation of the different wavelengths of white light through being refracted by different amounts. (4)

Doppler effect Apparent change in wavelength or frequency of the radiation from a source due to its relative motion away from or toward the observer. (4)

dust tail (of comet) Dust loosened by the sublimation of ice in a comet that is then pushed by photons from the Sun into a curved stream. (12)

Earth-approaching asteroid An asteroid with an orbit that either crosses the Earth's orbit or will at some time cross the Earth's orbit as it evolves under the influence of the planets' gravity. *See also* near-Earth object. (12)

eccentricity (of ellipse) Ratio of the distance between the foci to the major axis. (2)

eclipse The cutting off of all or part of the light of one body by another; in planetary science, the passing of one body into the shadow of another. (3)

eclipsing binary star A binary star in which the plane of revolution of the two stars is nearly edge-on to our line of sight, so that the light of one star is periodically diminished by the other passing in front of it. (17)

ecliptic The apparent annual path of the Sun on the celestial sphere. (1)

effective temperature *See* temperature (effective).

ejecta Material excavated from an impact crater, such as the blanket of material surrounding lunar craters and crater rays. (8)

electromagnetic force One of the four fundamental forces or interactions of nature; the force that acts between charges and binds atoms and molecules together. (28)

electromagnetic radiation Radiation consisting of waves propagated through regularly varying electric and magnetic fields; includes radio, infrared, visible light, ultraviolet, x rays, and gamma rays. (4)

electromagnetic spectrum The whole array or family of electromagnetic waves, from radio to gamma rays. (4)

electron A negatively charged subatomic particle that normally moves around the nucleus of an atom. (4)

element A substance that cannot be decomposed by chemical means into simpler substances.

elementary particle One of the basic particles of matter. The most familiar elementary particles are the proton, neutron, and electron. (15)

ellipse A curve for which the sum of the distances from any point on the ellipse to two points inside (called the foci) is always the same. (2)

elliptical galaxy A galaxy whose shape is an ellipse and that contains no conspicuous interstellar material. (25)

ellipticity The ratio (in an ellipse) of the major axis minus the minor axis to the major axis.

emission line A discrete bright line in the spectrum. (4)

emission nebula A gaseous nebula that derives its visible light from the fluorescence of atoms using ultraviolet light from a star in or near the nebula. (19)

emission spectrum A spectrum consisting of emission lines. (4)

energy flux The amount of energy passing through a unit area (for example, 1 square meter) per second. The units of flux are $J\,m^{-2}\,s^{-1}$. (4)

energy level A particular level, or amount, of energy possessed by an atom or ion above the energy it possesses in its least energetic state; also used to refer to the states of energy an electron can have in an atom. (4)

epicycle A circular orbit of a body in the Ptolemaic system, the center of which revolves about another circle (the deferent). (1)

equator A great circle on the Earth, 90° from (or equidistant from) each pole. (1)

equinox One of the intersections of the ecliptic and celestial equator; one of the two times during the year when the lengths of the day and the night are the same. (3)

equivalence principle Principle that a gravitational force and a suitable acceleration are indistinguishable within a sufficiently local environment. (23)

escape velocity The velocity a body must achieve to break away from the gravity of another body and never return to it. (2)

eucrite meteorite One of a class of basaltic meteorites believed to have originated on the asteroid Vesta.

event horizon The surface through which a collapsing star passes when its velocity of escape is equal to the speed of light—that is, when the star becomes a black hole. (23)

excitation The process of giving an atom or an ion an amount of energy greater than it has in its least-energy state. (4)

exclusion principle *See* Pauli exclusion principle.

exobiology *See* astrobiology

extinction Reduction of the light from a celestial body produced by the Earth's atmosphere, or by interstellar absorption. *See also* mass extinction. (19)

extragalactic Beyond our own Milky Way Galaxy.

extrasolar planet A planet orbiting a star other than the Sun.

extremophile An organism (usually a microbe) that tolerates or even thrives under conditions that we and most of the life around us would consider hostile, such as very high or low temperature or acidity. (29)

eyepiece A magnifying lens used to view the image produced by the objective of a telescope. (5)

fall (of meteorites) Meteorites seen in the sky and recovered on the ground. (13)

fault In geology, a crack or break in the crust of a planet along which slippage or movement can take place, accompanied by seismic activity. (7)

Fermi paradox The question why, if there are many advanced civilizations in the Galaxy, none of them are visiting us or have left indications of their existence. (29)

field A mathematical description of the effect of forces, such as gravity, that act on distant objects. For example, a given mass produces a gravitational field in the space surrounding it, which produces a gravitational force on objects within that space. (4)

find (of meteorites) A meteorite that has been recovered but was not seen to fall. (13)

fireball A spectacular meteor seen for longer than an instant in the sky. (13)

fission The breakup of a heavy atomic nucleus into two or more lighter ones. (15)

flare A sudden and temporary outburst of light from an extended region of the Sun's surface. (14)

fluorescence The absorption of light of one wavelength and the reemission of it at another wavelength; especially the conversion of ultraviolet into visible light. (19)

flux The rate at which energy or matter crosses a unit area of a surface; so, energy per unit time per unit area. (4)

focal length The distance from a lens or mirror to the point where light converged by it comes to a focus. (5)

focus (of ellipse) One of two fixed points inside an ellipse from which the sum of the distances to any point on the ellipse is a constant. (2)

focus (of telescope) Point where the rays of light converged by a mirror or lens meet. (5)

forbidden lines Spectral lines that are not usually observed under laboratory conditions because they result from atomic transitions that are highly improbable on Earth.

force That which can change the momentum of a body; numerically, the rate at which the body's momentum changes. That which causes masses to accelerate. (2)

Fraunhofer line An absorption line in the spectrum of the Sun or of a star. (4)

Fraunhofer spectrum The array of absorption lines in the spectrum of the Sun or a star. (4)

frequency Number of vibrations per unit time; number of waves that cross a given point per unit time (in radiation). (4)

fusion The building up of heavier atomic nuclei from lighter ones. (15)

galactic cannibalism   The process by which a larger galaxy strips material from or completely swallows a smaller one. (27)

galactic cluster   An "open" cluster of stars located in the spiral arms or disk of the Galaxy. (21)

galaxy   A large assemblage of stars; a typical galaxy contains millions to hundreds of billions of stars. (P)

Galaxy   The galaxy to which the Sun and our neighboring stars belong; the band of light in the sky we call the Milky Way is light from many remote stars in the disk of the Galaxy. (24)

gamma rays   Photons (of electromagnetic radiation) of energy higher than those of x rays; the most energetic form of electromagnetic radiation. (4)

gamma-ray burst   A brief pulse of gamma rays; primarily originating in distant galaxies. (I)

general relativity theory   Einstein's theory relating acceleration, gravity, and the structure (geometry) of space and time. (23)

geocentric   Centered on the Earth. (1)

giant (star)   A star of large luminosity and radius. (16)

giant molecular cloud   Large, cold interstellar clouds with diameters of dozens of light years and typical masses of 10^5 solar masses; found in the spiral arms of galaxies, these clouds are where massive stars form. (20)

giant planet   *See* jovian planet.

globular cluster   One of about 150 large spherical star clusters that form a system of clusters centered on the center of the Galaxy. (21)

grand unified theories (GUTS)   Physical theories that attempt to describe the four interactions (forces) of nature as different manifestations of a single force. (28)

granite   The type of igneous silicate rock that makes up most of the continental crust of the Earth. (7)

granulation   The rice-grain-like structure of the solar photosphere; granulation is produced by upwelling currents of gas that are slightly hotter, and therefore brighter, than the surrounding regions, which are flowing downward into the Sun. (14)

gravity   The mutual attraction of material bodies or particles. (2)

gravitational energy   Energy that can be released by the gravitational collapse, or partial collapse, of a system—that is, by particles that fall in toward the center of gravity. (15)

gravitational lens   A massive object, such as a galaxy, whose gravity deflects the light (as predicted by Einstein's general theory of relativity) of an object behind it, thus showing us distorted or multiple images of the distant source. (26)

gravitational redshift   The redshift of electromagnetic radiation caused by a gravitational field. The slowing of clocks in a gravitational field. (23)

gravitational waves   Disturbances in the curvature of spacetime caused by changes in how matter is distributed; gravitational waves propagate at (or near) the speed of light. (23)

great circle   Circle on the surface of a sphere that is the curve of intersection of the sphere with a plane passing through its center. (3)

greenhouse effect   The blanketing (absorption) of infrared radiation near the surface of a planet—for example, by carbon dioxide in its atmosphere. (7)

ground state   The lowest energy state of an atom. (4)

H or H₀   *See* Hubble constant.

HI region   Region of neutral hydrogen in interstellar space. (19)

HII region   Region of ionized hydrogen in interstellar space. (19)

habitable zone   Region around a star in which liquid water could exist on the surface of terrestrial-size planets, hence the most probable place to look for life. (29)

half-life   The time required for half of the radioactive atoms in a sample to disintegrate. (6)

halo (of galaxy)   The outermost extent of our Galaxy or another, containing a sparse distribution of stars and globular clusters in a more or less spherical distribution. More recently, astronomers also speak of a halo of dark matter around galaxies. (24)

heavy elements   In astronomy, usually those elements of higher atomic number than helium.

helio-   Prefix referring to the Sun.

heliocentric   Centered on the Sun. (1)

helium flash   The nearly explosive ignition of helium in the triple-alpha process in the dense core of a red giant star. (21)

Herbig-Haro (HH) object   Luminous knots of gas in an area of star formation, which are set to glow by jets of material from a protostar. (20)

hertz   A unit of frequency: one cycle per second. Named for Heinrich Hertz, who first produced radio radiation. (4)

Hertzsprung-Russell (H-R) diagram   A plot of luminosity against surface temperature (or spectral type) for a group of stars. (17)

highlands   The lighter, heavily cratered regions of the Moon, which are generally several kilometers higher than the maria; also similar heavily cratered regions on Mars. (8)

homogeneous   Having a consistent and even distribution of matter that is the same everywhere. (27)

horizon (astronomical)   A great circle on the celestial sphere 90° from the zenith; more popularly, the circle around us where the dome of the sky meets the Earth. (1)

horoscope   A chart used by astrologers, showing the positions along the zodiac and in the sky of the Sun, Moon, and planets at some given instant and as seen from a particular place on Earth—usually corresponding to the time and place of a person's birth. (1)

Hubble constant   Constant of proportionality between the velocities of remote galaxies and their distances. (25)

Hubble law (or the law of the redshifts)   The rule that the radial velocities of remote galaxies are proportional to their distances from us. (25)

hydrostatic equilibrium   A balance between the weights of various layers, as in a star or the Earth's atmosphere, and the pressures that support them. (15)

hypothesis   A tentative supposition or scientific model advanced to explain certain facts or phenomena and subject to further tests and verification. (P)

igneous rock   Any rock produced by cooling from a molten state. (7)

inclination (of an orbit)   The angle between the orbital plane of a revolving body and some fundamental plane—usually the plane of the celestial equator or of the ecliptic.

inertia   The property of matter that requires a force to act on it to change its state of motion; the tendency of objects to continue doing what they are doing in the absence of outside forces. (2)

inertial system   A system of coordinates that is not itself accelerated but that either is at rest or is moving with constant velocity.

inflationary universe   A theory of cosmology in which the universe is assumed to have undergone a phase of very rapid expansion during the first 10^{-30} s. After this period of rapid expansion, the standard big bang and inflationary models are identical. (28)

infrared cirrus Patches of interstellar dust that emit infrared radiation and look like cirrus clouds on infrared images of the sky. (19)

infrared radiation Electromagnetic radiation of wavelength longer than the longest (red) wavelengths that can be perceived by the eye, but shorter than radio wavelengths. (4)

interference A phenomenon of waves that mix together such that their crests and troughs can alternately reinforce and cancel one another. (5)

interferometer An instrument that combines electromagnetic radiation (or gravitational waves) from one or more telescopes to obtain a resolution equivalent to what would be obtained with a single telescope with a diameter equal to the baseline separating the individual separate telescopes. (5)

international date line An arbitrary line on the surface of the Earth near longitude 180° across which the date changes by one day. (3)

interstellar dust Tiny solid grains in interstellar space, thought to consist of a core of rocklike material (silicates) or graphite surrounded by a mantle of ices. Water, methane, and ammonia are probably the most abundant ices. (19)

interstellar extinction The attenuation or absorption of light by dust in the interstellar medium. (19)

interstellar medium or interstellar matter Gas and dust between the stars in a galaxy. (19)

inverse-square law (for light) The amount of energy (light) flowing through a given area in a given time (flux) decreases in proportion to the square of the distance from the source of energy or light. (4)

ion An atom that has become electrically charged by the addition or loss of one or more electrons. (4)

ion tail (of comet) A stream of ionized particles evaporated from a comet and then swept away from the Sun by the solar wind. (12)

ionization The process by which an atom gains or loses electrons. (4)

ionosphere The upper region of the Earth's atmosphere in which many of the atoms are ionized. (7)

iron meteorite A meteorite composed primarily of iron and nickel. (13)

irregular galaxy A galaxy without rotational symmetry; neither a spiral nor an elliptical galaxy. (25)

irregular satellite A planetary satellite with an orbit that is retrograde, or of high inclination or eccentricity.

isotope Any of two or more forms of the same element whose atoms all have the same number of protons but different numbers of neutrons. (4)

isotropic The same in all directions. (27)

joule The metric unit of energy; the work done by a force of 1 newton (N) acting through a distance of 1 m.

jovian planet or giant planet Any of the planets Jupiter, Saturn, Uranus, and Neptune in our solar system, or planets of roughly that mass and composition in other planetary systems. (6)

KBO *See* Kuiper belt object.

kinetic energy Energy associated with motion.

Kuiper belt A region of space beyond Neptune that is dynamically stable (like the asteroid belt); the source region for most short-period comets. (12)

Kuiper belt object (KBO) An object in the Kuiper belt beyond Neptune; sometimes a comet that is derived from the Kuiper belt. (12)

laser An acronym for light amplification by stimulated emission of radiation; a device for amplifying a light signal at a particular wavelength into a coherent beam.

latitude A north-south coordinate on the surface of the Earth; the angular distance north or south of the equator measured along a meridian passing through a place. (3)

law of areas Kepler's second law: The radius vector from the Sun to any planet sweeps out equal areas in the planet's orbital plane in equal intervals of time. (2)

law of the redshifts *See* Hubble law.

leap year A calendar year with 366 days, inserted approximately every four years to make the average length of the calendar year as nearly equal as possible to the tropical year. (3)

light or visible light Electromagnetic radiation that is visible to the eye. (4)

light curve A graph that displays the time variation of the light from a variable or eclipsing binary star or, more generally, from any other object whose radiation output changes with time. (17)

light year The distance light travels in a vacuum in one year; $1 \text{ LY} = 9.46 \times 10^{12}$ km, or about 6×10^{12} mi. (P)

line broadening The phenomenon by which spectral lines are not precisely sharp but have finite widths. (16)

Local Group The small cluster of galaxies to which our Galaxy belongs. (27)

local standard of rest A coordinate system that shares the average motion of the Sun and its neighboring stars about the galactic center.

long-period comet A comet with orbital period longer than about a century; mostly derived from the Oort comet cloud. (12)

longitude An east-west coordinate on the Earth's surface; the angular distance, measured east or west along the equator, from the Greenwich meridian to the meridian passing through a place. (3)

luminosity The rate at which a star or other object emits electromagnetic energy into space; the total power output of an object. (16)

luminosity class A classification of a star according to its luminosity within a given spectral class. Our Sun, a G2 V star, has luminosity class V. (18)

luminous matter Material that gives off light or other electromagnetic radiation (as opposed to dark matter).

lunar eclipse An eclipse of the Moon, in which the Moon moves into the shadow of the Earth. (3)

Lyman lines A series of absorption or emission lines in the spectrum of hydrogen that arise from transitions to and from the lowest-energy states of the hydrogen atom. (4)

Magellanic Clouds Two neighboring galaxies visible to the naked eye from southern latitudes. (24)

magma High-temperature molten state of rock, usually of silicate mineral composition and with dissolved gases and other volatiles. (7)

magnetic field The region of space near a magnetized body within which magnetic forces can be detected.

magnetic pole One of two points on a magnet (or a planet) at which the greatest density of lines of force emerge. A compass needle aligns itself along the local lines of force on the

Earth and points more or less toward the magnetic poles of the Earth.

magnetosphere The region around a planet in which its intrinsic magnetic field dominates the interplanetary field carried by the solar wind; hence, the region within which charged particles can be trapped by the planetary magnetic field. (7)

magnitude A system of measuring the amount of light flux received from a star or other luminous object. The higher the magnitude, the less radiation we receive from the object. (1)

main sequence A sequence of stars on the Hertzsprung–Russell diagram, containing the majority of stars, that runs diagonally from the upper left to the lower right. (17)

major axis (of ellipse) The maximum diameter of an ellipse. (2)

mantle (of Earth) The greatest part of the Earth's interior, lying between the crust and the core. (7)

mare (pl. maria) Latin for "sea"; name applied to the dark, relatively smooth features that cover 17 percent of the Moon's surface. (8)

mass A measure of the total amount of material in a body; defined either by the inertial properties of the body or by its gravitational influence on other bodies (2).

mass extinction The sudden disappearance in the fossil record of a large number of species of life, to be replaced by new species in subsequent layers. Mass extinctions are indications of catastrophic changes in the environment, such as might be produced by a large impact on the Earth. (7)

mass-to-light ratio The ratio of the total mass of a galaxy to its total luminosity, usually expressed in units of solar mass and solar luminosity. The mass-to-light ratio gives a rough indication of the types of stars contained within a galaxy and whether or not substantial quantities of dark matter are present. (25)

mass-luminosity relation An empirical relation between the masses and luminosities of many (principally main-sequence) stars. (17)

Maunder Minimum The interval from 1645 to 1715 when solar activity was very low. (14)

mean solar day Average length of the apparent solar day. (3)

mean solar time Time based on the rotation of the Earth; mean solar time passes at a constant rate, unlike apparent solar time. (3)

merger (of galaxies) When galaxies (of roughly comparable size) collide and form one combined structure. (27)

meridian (celestial) The great circle on the celestial sphere that passes through an observer's zenith and the north (or south) celestial pole. (3)

meridian (terrestrial) The great circle on the surface of the Earth that passes through a particular place and the North and South Poles of the Earth. (3)

Messier catalog A catalog of nonstellar objects compiled by Charles Messier in 1787 (includes nebulae, star clusters, and galaxies). (19)

metals In general, any element or compound whose electron structure makes it a good conductor of electricity. Among astronomers, the term is often used to mean all elements beyond hydrogen and helium. (16)

metamorphic rock Any rock produced by physical and chemical alteration (without melting) under high temperature and pressure. (7)

meteor A flash of light observed when a small piece of solid matter enters the Earth's atmosphere and burns up; popularly called a "shooting star." (6)

meteor shower Many meteors appearing to radiate from one point in the sky; produced when the Earth passes through a cometary dust stream. (13)

meteorite A portion of a meteoroid that survives passage through the atmosphere and strikes the ground. (6)

meteorite shower The breakup of an object entering the Earth's atmosphere at high speed to produce multiple meteorites falling over an area of many square kilometers. (13)

meteoroid A particle or chunk of typically rocky or metallic material in space before any encounter with the Earth.

micron Old term for micrometer (10^{-6} m).

microwave Electromagnetic radiation at radio wavelengths, somewhat longer than the longest infrared waves. (4)

Milky Way The band of light encircling the sky, which is due to the many stars and diffuse nebulae lying near the plane of the Galaxy. (24)

minerals The solid compounds (often primarily silicon and oxygen) that form rocks.

model atmosphere or photosphere The result of a theoretical calculation of the run of temperature, pressure, density, and so on through the outer layers of the Sun or a star.

molecule A combination of two or more atoms bound together; the smallest particle of a chemical compound or substance that exhibits the chemical properties of that substance.

momentum A measure of the inertia or state of motion of a body; the momentum of a body is the product of its mass and velocity. In the absence of a force, momentum is conserved. (2)

near-Earth object (NEO) A comet or asteroid whose path intersects the orbit of the Earth. (12)

nebula Cloud of interstellar gas or dust; the term is typically used for clouds that are seen to glow with visible light or infrared. (19)

neutrino A fundamental particle that has little or no rest mass and no charge but that does have spin and energy. Neutrinos rarely interact with ordinary matter. (15)

neutron A subatomic particle with no charge and with mass approximately equal to that of the proton.

neutron star A star of extremely high density composed almost entirely of neutrons. (12)

nonthermal radiation *See* synchrotron radiation.

north celestial pole *See* celestial poles.

nova A star that experiences a sudden outburst of radiant energy, temporarily increasing its luminosity by hundreds to thousands of times. (22)

nuclear Referring to the nucleus of the atom.

nuclear bulge Central part of our or another galaxy. (24)

nuclear transformation The change of one atomic nucleus into another, as in nuclear fusion. (15)

nucleosynthesis The building up of heavy elements from lighter ones by nuclear fusion. (21)

nucleus (of atom) The heavy part of an atom, composed mostly of protons and neutrons, and about which the electrons revolve. (4)

nucleus (of comet) The solid chunk of ice and dust in the head of a comet. (12)

nucleus (of galaxy) Concentration of matter at the center of a galaxy. (24)

occultation The passage of an object of large angular size in front of a smaller object, such as the Moon in front of a distant star or the rings of Saturn in front of the Voyager spacecraft. *Compare with* transit. (11)

Oort comet cloud The large spherical region around the Sun from which most "new" comets come; a reservoir of objects with aphelia at about 50,000 AU. (12)

opacity Absorbing power; capacity to impede the passage of light. (15)

open cluster A comparatively loose cluster of stars, containing from a few dozen to a few thousand members, located in the spiral arms or disk of the Galaxy; sometimes referred to as a galactic cluster. (21)

open universe A model of the universe in which gravity is not strong enough to bring the universe to a halt; it expands forever. In this model the geometry of spacetime is such that if you go in a straight line, not only can you never return to where you started, but even more space opens up than you would expect from Euclidean geometry. (28)

optical In astronomy, relating to the visible-light band of the electromagnetic spectrum. Optical observations are those made with visible light. (4)

optical double star Two stars at different distances that are seen nearly lined up in projection so that they appear close together, but that are not really gravitationally associated. (17)

orbit The path of an object that is in revolution about another object or point. (2)

organic compound A compound containing carbon, especially a complex carbon compound; not necessarily produced by life. (29)

organic molecule A molecule that contains carbon, especially complex hydrocarbons. *See also* organic compound. (29)

oscillation A periodic motion; in the case of the Sun, a periodic or quasi-periodic expansion and contraction of the whole Sun or some portion of it. (15)

ozone A heavy molecule of oxygen that contains three atoms rather than the more normal two; designated O_3. (7)

parabola A conic section of eccentricity 1.0; the curve of the intersection between a circular cone and a plane parallel to a straight line in the surface of the cone.

parallax An apparent displacement of a nearby star that results from the motion of the Earth around the Sun; numerically, the angle subtended by 1 AU at the distance of a particular star. (18)

parsec A unit of distance in astronomy, equal to 3.26 light years. At a distance of 1 parsec, a star has a parallax of one arc-second. (18)

Pauli exclusion principle Quantum mechanical principle by which no two electrons (or similar particles) can have the same position and momentum.

peculiar velocity The velocity of a star with respect to the local standard of rest—that is, its space motion, corrected for the motion of the Sun with respect to our neighboring stars.

penumbra The outer, not completely dark part of a shadow; the region from which the source of light is not completely hidden. (3)

perfect radiator or blackbody A body that absorbs and subsequently reemits all radiation incident upon it. (4)

periastron The place in the orbit of a star in a binary star system where it is closest to its companion star.

perigee The place in the orbit of an Earth satellite where it is closest to the center of the Earth. (2)

perihelion The place in the orbit of an object revolving about the Sun where it is closest to the Sun's center. (2)

period-luminosity relation An empirical relation between the periods and luminosities of certain variable stars. (18)

perturbation The small disturbing effect on the motion of a body produced by a third body or other external agent. (2)

phase A particular point in the time of a cycle (for example, the time after maximum for a variable star) or any distinct time period in a sequence of events (as in the phases of the Moon). (3)

photochemistry Chemical changes caused by electromagnetic radiation. (10)

photometry The measurement of light (or other electromagnetic wave) intensities. (5)

photon A discrete unit of electromagnetic energy. (4)

photosphere The region of the solar (or a stellar) atmosphere from which continuous radiation escapes into space. (14)

photosynthesis Complex sequence of chemical reactions through which some living things can use sunlight to manufacture products that store energy (such as carbohydrates), releasing oxygen as one by-product. (7)

pixel An individual picture element in a detector—for example, a particular silicon diode in a CCD.

plage A bright region of the solar surface observed in the light of some spectral line. (14)

Planck's constant The constant of proportionality relating the energy of a photon to its frequency. (4)

planet Any of the nine largest objects revolving about the Sun, or any similar objects that orbit other stars. Unlike stars, planets do not (for the most part) give off their own light but only reflect the light of their parent star. (1)

planetarium An optical device for projecting on a screen or domed ceiling the stars and planets and their apparent motions in the sky. (1)

planetary nebula A shell of gas ejected by and expanding away from an extremely hot low-mass star that is nearing the end of its life. (22)

planetesimals Objects, from tens to hundreds of kilometers in diameter, that formed in the solar nebula as an intermediate step between tiny grains and the larger planetary objects we see today. The comets and some asteroids may be leftover planetesimals. (6)

plasma A hot ionized gas.

plate tectonics The motion of segments or plates of the outer layer of the Earth over the underlying mantle. (7)

populations I and II Two classes of stars (and systems of stars) classified according to their spectral characteristics, chemical compositions, radial velocities, ages, and locations in the Galaxy. (24)

positron An electron with a positive rather than negative charge; an antielectron. (15)

potential energy Stored energy that can be converted into other forms; especially gravitational energy.

power The rate at which work is done (energy used per second)—that is, the rate at which energy is transferred or absorbed. The metric unit of power is the watt, and 1 watt is equivalent to 1 joule/s.

precession (of Earth) A slow, conical motion of the Earth's axis of rotation caused principally by the gravitational pull of the Moon and Sun on the Earth's equatorial bulge. (1)

precession of the equinoxes Slow westward motion of the equinoxes along the ecliptic that results from precession. (1)

pressure Force per unit area; expressed in units of atmospheres or pascals.

prime focus   The point in a telescope where the objective focuses the light. (5)

prime meridian   The terrestrial meridian passing through the site of the old Royal Greenwich Observatory; longitude 0°. (3)

primitive   In planetary science and meteoritics, an object or rock that is little changed, chemically, since its formation. Also used to refer to the chemical composition of an atmosphere that has not undergone extensive chemical evolution. (7)

primitive meteorite   A meteorite that has not been greatly altered chemically since its condensation from the solar nebula; called in meteoritics a chondrite (either ordinary chondrite or carbonaceous chondrite). (13)

primitive rock   Any rock that has not experienced great heat or pressure and therefore remains representative of the original condensates from the solar nebula. (7)

principle of equivalence   Principle that a gravitational force and a suitable acceleration are indistinguishable within a sufficiently local environment. (23)

prism   A wedge-shaped piece of glass that is used to disperse white light into a spectrum. (5)

prominence   A phenomenon in the solar corona that commonly appears like a flame above the edge of the Sun. (14)

proper motion   The angular change per year in the direction of a star as seen from the Sun. (16)

proton   A heavy subatomic particle that carries a positive charge; one of the two principal constituents of the atomic nucleus. (4)

proton–proton cycle   A series of thermonuclear reactions by which nuclei of hydrogen are built up into nuclei of helium. (15)

protoplanet or -star or -galaxy   A very young planet (or a star or galaxy) still in the process of formation. (20)

pulsar   A variable radio source of small angular size that emits very rapid radio pulses in very regular periods that range from fractions of a second to several seconds. (22)

pulsating variable   A variable star that pulsates in size and luminosity. (18)

quantum mechanics   The branch of physics that deals with the structure of atoms and their interactions with one another and with radiation. (4)

quasar   An object of very high redshift that looks like a star but is extragalactic and highly luminous; an active galactic nucleus where the galaxy appears faint from Earth. (26)

radar   The technique of transmitting radio waves to an object and then detecting the radiation that the object reflects back to the transmitter; used to measure the distance to, and motion of, a target object or to form images of it. (5)

radial velocity   The component of relative velocity that lies in the line of sight; motion toward or away from the observer. (4)

radial-velocity curve   A plot of the variation of radial velocity with time for a binary or variable star. (17)

radiant (of meteor shower)   The point in the sky from which the meteors belonging to a shower seem to radiate. (13)

radiation   A mode of energy transport whereby energy is transmitted through a vacuum; also the transmitted energy itself. (4, 15)

radiation pressure   The transfer of momentum carried by electromagnetic radiation to a body that the radiation impinges upon.

radio galaxy   A galaxy that emits greater amounts of radio radiation than average. (26)

radio telescope   A telescope designed to make observations in radio wavelengths. (5)

radioactive dating   The technique of determining the ages of rocks or other specimens by the decay of certain radioactive elements contained therein. (6)

radioactivity (radioactive decay)   The process by which certain kinds of atomic nuclei naturally decompose, with the spontaneous emission of subatomic particles and gamma rays. (6)

red giant   A large, cool star of high luminosity; a star occupying the upper right portion of the Hertzsprung–Russell diagram. (21)

reddening (interstellar)   The reddening of starlight passing through interstellar dust because dust scatters blue light more effectively than red. (19)

redshift   A shift to longer wavelengths of light, typically a Doppler shift caused by the motion of the source or a shift due to the expansion of space. (25)

reducing   In chemistry, referring to conditions in which hydrogen dominates over oxygen, so that most other elements form compounds with hydrogen. In very reducing conditions, free hydrogen (H_2) is present and free oxygen (O_2) cannot exist. (6)

reflecting telescope   A telescope in which the principal light collector is a concave mirror. (5)

reflection nebula   A relatively dense dust cloud in interstellar space that is illuminated by reflected starlight. (19)

refracting telescope   A telescope in which the principal light collector is a lens or system of lenses. (5)

refraction   The bending of light rays passing from one transparent medium (or a vacuum) to another. (3)

relativistic particle (or electron)   A particle (electron) moving at nearly the speed of light.

relativity   A theory formulated by Einstein that describes the relationships between measurements of physical phenomena by two different observers who are in relative motion at constant velocity (the special theory of relativity) or that describes how a gravitational field can be replaced by a curvature of spacetime (the general theory of relativity). (23)

resolution   The detail in an image; specifically the smallest angular (or linear) features that can be distinguished. (5)

resonance   An orbital condition in which one object is subject to periodic gravitational perturbations by another; most commonly arising when two objects orbiting a third have periods of revolution that are simple multiples or fractions of each other. (11)

retrograde (rotation or revolution)   Backward with respect to the common direction of motion in the solar system; clockwise as viewed from the north, and going from east to west rather than from west to east. (1)

retrograde motion   An apparent westward motion of a planet on the celestial sphere or with respect to the stars. (1)

revolution   The motion of one body around another.

rift zone   In geology, a place where the crust is being torn apart by internal forces, generally associated with the injection of new material from the mantle and with the slow separation of tectonic plates. (7)

right ascension   A coordinate for measuring the east–west positions of celestial bodies; the angle measured eastward along the celestial equator from the vernal equinox to the hour circle passing through a body. (3)

Roche limit   *See* tidal stability limit.

rotation   Turning of a body about an axis running through it.

RR Lyrae variable   One of a class of giant pulsating stars with periods shorter than one day. (18)

runaway greenhouse effect   An irreversible process whereby the heating of a planet leads to an increase in its atmospheric greenhouse effect and thus to further heating, thereby significantly altering the composition of its atmosphere and the temperature of its surface. (9)

satellite   An object that revolves about a planet.

Schwarzschild radius   *See* event horizon.

scientific method   The procedure scientists follow to understand the natural world: (1) the observation of phenomena or the results of experiments; (2) the formulation of hypotheses that describe these phenomena and that are consistent with the body of knowledge available; (3) the testing of these hypotheses by noting whether or not they adequately predict and describe new phenomena or the results of new experiments; and (4) the modification or rejection of hypotheses that are not confirmed by observations or experiment. (P)

sedimentary rock   Any rock formed by the deposition and cementing of fine grains of material. (7)

seeing   The unsteadiness of the Earth's atmosphere, which blurs telescopic images. Good seeing means the atmosphere is steady. (5)

seismic waves   Vibrations that travel through the interior of the Earth or any other object; on Earth these are generally caused by earthquakes. (7)

seismology (solar)   The study of small changes in the radial velocity of the Sun as a whole or of small regions on the surface of the Sun. Analyses of these velocity changes can be used to infer the internal structure of the Sun. (15)

seismology (terrestrial)   The study of earthquakes, the conditions that produce them, and the internal structure of the Earth as deduced from analyses of seismic waves. (7)

semimajor axis   Half the major axis of a conic section, such as an ellipse. (2)

SETI   The search for extraterrestrial intelligence, usually applied to searches for radio signals from other civilizations. (29)

Seyfert galaxy   A galaxy with an active galactic nucleus; one whose nucleus shows bright emission lines; first described by C. Seyfert. (26)

shepherd satellite   Informal term for a satellite that is thought to maintain the structure of a planetary ring through its close gravitational influence. (11)

short-period comet   A comet with orbital period shorter than about a century; most are derived from the Kuiper belt and are gravitationally linked to Jupiter (called Jupiter-family comets). (12)

sidereal day   The Earth's rotation period as defined by the positions of the stars in the sky; the time between successive passages of the same star through the meridian. (3)

sidereal period   The period of revolution of one body about another measured with respect to the stars.

sidereal time   Time on Earth measured with respect to the stars rather than the Sun; the local hour angle of the vernal equinox. (3)

sidereal year   Period of the Earth's revolution about the Sun with respect to the stars. (3)

sign (of zodiac)   Astrological term for any of 12 equal sections along the ecliptic, each of length 30°. Because of precession, these signs today are no longer lined up with the constellations from which they received their names. (1)

singularity   A theoretical point of zero volume and infinite density to which any object that becomes a black hole must collapse, according to the general theory of relativity. (23)

solar activity   Phenomena that cause changes in the appearance or energy output of the solar atmosphere, such as sunspots, plages, and flares. (14)

solar antapex   Direction away from which the Sun is moving with respect to the local standard of rest.

solar apex   The direction toward which the Sun is moving with respect to the local standard of rest.

solar day   The Earth's rotation period as defined by the position of the Sun in the sky; the time between successive passages of the Sun through the meridian. (3)

solar eclipse   An eclipse of the Sun by the Moon, caused by the passage of the Moon in front of the Sun. Solar eclipses can occur only at the time of the new moon. (3)

solar motion   Motion of the Sun, or the velocity of the Sun, with respect to the local standard of rest.

solar nebula   The cloud of gas and dust from which the solar system formed. *See also* circumstellar disk. (6)

solar seismology   The study of pulsations or oscillations of the Sun in order to determine the characteristics of the solar interior. (15)

solar system   The system of the Sun and the planets, their satellites, asteroids, comets, KBOs, and other objects revolving around the Sun. (6)

solar time   A time based on the Sun; usually the hour angle of the Sun plus 12 h. (3)

solar wind   A flow of hot charged particles leaving the Sun. (14)

solstice   Either of two points on the celestial sphere where the Sun reaches its maximum distances north and south of the celestial equator; time of the year when the daylight is the longest or the shortest. (3)

south celestial pole   *See* celestial poles.

space velocity or space motion   The velocity of a star with respect to the Sun.

spacetime   A system of one time and three spatial coordinates, with respect to which the time and place of an event can be specified. (23)

spectral class (or type)   The classification of stars according to their temperatures using the characteristics of their spectra; the types are O, B, A, F, G, K, M, with L and T added recently for cooler stars that recent surveys have revealed. (16)

spectral line   Radiation at a particular wavelength of light produced by the emission or absorption of energy by an atom. (4)

spectral sequence   The sequence of spectral classes of stars arranged in order of decreasing temperatures of stars of those classes; O B A F G K M L T. (16)

spectrometer   An instrument for obtaining a spectrum; in astronomy, usually attached to a telescope to record the spectrum of a star, galaxy, or other astronomical object. (5)

spectroscopic binary star   A binary star in which the components are not resolved but whose binary nature is indicated by periodic variations in radial velocity, indicating orbital motion. (17)

spectroscopic parallax   A parallax (or distance) of a star that is derived by comparing the apparent magnitude of the star with its absolute magnitude as deduced from its spectral characteristics. (18)

spectroscopy The study of spectra.

spectrum The array of colors or wavelengths obtained when light (or other radiation) from a source is dispersed, as in passing it through a prism or grating. (4)

speed The rate at which an object moves without regard to its direction of motion; the numerical or absolute value of velocity. (2)

spicule A jet of rising material in the solar chromosphere. (14)

spiral arms Arms (or long denser regions) of interstellar material and young stars that wind out in a plane from the central nucleus of a spiral galaxy. (24)

spiral density wave A mechanism for the generation of spiral structure in galaxies; a density wave interacts with interstellar matter and triggers the formation of stars. Spiral density waves are also seen in the rings of Saturn. (11, 24)

spiral galaxy A flattened, rotating galaxy with pinwheel-like arms of interstellar material and young stars winding out from its nucleus. (25)

spring tide The highest tidal range of the month, produced when the Moon is near either the full or the new phase. (3)

standard bulb An astronomical object of known luminosity; such an object can be used to determine distances; also sometimes called a standard candle. (16, 25)

star A sphere of gas shining under its own power; massive enough to support the fusion of hydrogen to helium in the interior. (P)

star cluster An assemblage of stars held together by their mutual gravity. (21)

Stefan-Boltzmann law A formula from which the rate at which a blackbody radiates energy can be computed; the total rate of energy emission from a unit area of a blackbody is proportional to the fourth power of its absolute temperature. (4)

stellar evolution The changes that take place in the characteristics of stars as they age. (21)

stellar model The result of a theoretical calculation of the physical conditions in the different layers of a star's interior.

stellar parallax See parallax.

stellar wind The outflow of gas, sometimes at speeds as high as hundreds of kilometers per second, from a star. (20)

stony-iron meteorite A type of differentiated meteorite that is a blend of nickel-iron and silicate materials. (13)

stony meteorite A meteorite composed mostly of stony material, either primitive or differentiated. (13)

stratosphere The layer of the Earth's atmosphere above the troposphere (where most weather takes place) and below the ionosphere. (7)

strong nuclear force or strong interaction The force that binds together the parts of the atomic nucleus. (15, 28)

subduction zone In terrestrial geology, a region where one crustal plate is forced under another, generally associated with earthquakes, volcanic activity, and the formation of deep ocean trenches. (7)

summer solstice The point on the celestial sphere where the Sun reaches its farthest distance north of the celestial equator; the day with the longest amount of daylight. (3)

Sun The star about which the Earth and other planets revolve.

sunspot A temporary cool region in the solar photosphere that appears dark by contrast against the surrounding hotter photosphere. (14)

sunspot cycle The semiregular 11-year period with which the frequency of sunspots fluctuates. (14)

supercluster A large region of space (more than 100 million LY across) where groups and clusters of galaxies are more concentrated; a cluster of clusters of galaxies. (27)

supergiant A star of very high luminosity and relatively low temperature. (17)

supernova An explosion that marks the final stage of evolution of a star. A Type I supernova occurs when a white dwarf accretes enough matter to exceed the Chandrasekhar limit, collapses, and explodes. A Type II supernova marks the final collapse of a massive star. (22)

surface gravity The weight of a unit mass at the surface of a body.

synchrotron radiation The radiation emitted by charged particles being accelerated in magnetic fields and moving at speeds near that of light. (10)

tail (of a comet) See dust tail or ion tail.

tangential (transverse) velocity The component of a star's space velocity that lies in the plane of the sky.

tectonic Geological features that result from stresses and pressures in the crust of a planet. Tectonic forces can lead to earthquakes and motion of the crust. (7)

temperature A measure of how fast the particles in a body are moving or vibrating in place; a measure of the average heat energy in a body.

temperature (Celsius) Temperature measured on scale where water freezes at 0° and boils at 100°.

temperature (color) The temperature of a star as estimated from the intensity of the stellar radiation at two or more colors or wavelengths.

temperature (effective) The temperature of a blackbody that would radiate the same total amount of energy that a particular object, such as a star, does.

temperature (excitation) The temperature of a star as estimated from the relative strengths of lines in its spectrum that originate from atoms in different stages of excitation.

temperature (Fahrenheit) Temperature measured on a scale where water freezes at 32° and boils at 212°.

temperature (ionization) The temperature of a star as estimated from the relative strengths of lines in its spectrum that originate from atoms in different stages of ionization.

temperature (Kelvin) Absolute temperature measured in Celsius degrees, with the zero point at absolute zero ($-273°C$).

temperature (radiation) The temperature of a blackbody that radiates the same amount of energy in a given spectral region as does a particular body.

terrestrial planet Any of the planets Mercury, Venus, Earth, or Mars; sometimes the Moon is included in the list. Also any planet with mass roughly in the interval from 10 Earths to 1/10 Earth, which orbits another star. (6)

theory A set of hypotheses and laws that have been well demonstrated to apply to a wide range of phenomena associated with a particular subject.

thermal energy Energy associated with the motions of the molecules or atoms in a substance. (15)

thermal equilibrium A balance between the input and outflow of heat in a system. (15)

thermal radiation The electromagnetic radiation emitted by any object or gas that is not at absolute zero.

thermonuclear energy Energy associated with thermonuclear reactions or that can be released through thermonuclear reactions. (15)

thermonuclear reaction A nuclear reaction or transformation that results from encounters between particles that are given high velocities (by heating them). (15)

tidal force A differential gravitational force acting on two sides of an object that tends to deform that object. (3)

tidal stability limit The distance—approximately 2.5 planetary radii from the center—within which differential gravitational forces (or tides) are stronger than the mutual gravitational attraction between two adjacent orbiting objects. Within this limit, fragments are not likely to accrete or assemble themselves into a larger object. Also called the Roche limit. (11)

tide Deformation of a body by the differential gravitational force exerted on it by another body; in the Earth, the deformation of the ocean surface by the differential gravitational forces exerted by the Moon and Sun. (3)

transit Passage of an object of small angular size in front of another of larger size, as in the transit of a planet in front of a star. *Compare with* occultation. (11)

transition region The region in the Sun's atmosphere where the temperature rises very rapidly from the relatively low temperatures that characterize the chromosphere to the high temperatures of the corona. (14)

triple-alpha process A sequence of two nuclear reactions by which three helium nuclei are built up into one carbon nucleus. (21)

Trojan asteroid An asteroid with the same orbital period as Jupiter (12 years) that is part of a cluster either 60 degrees in front of or 60 degrees behind Jupiter. (12)

Tropic of Cancer The parallel (circle) of latitude 23.5° N. (3)

Tropic of Capricorn The parallel (circle) of latitude 23.5° S. (3)

tropical year Period of revolution of the Earth about the Sun with respect to the vernal equinox. (3)

troposphere Lowest level of the Earth's atmosphere, where most weather takes place. (7)

turbulence Random motions of gas masses, as in the atmosphere of a star.

21-cm line A line in the spectrum of neutral hydrogen at the radio wavelength of 21 cm. (19)

ultraviolet radiation Electromagnetic radiation of wavelengths shorter than the shortest visible wavelengths; radiation of wavelengths in the approximate range 10 to 400 nm. (4)

umbra The central, completely dark part of a shadow. (3)

universe The totality of all matter, radiation, and space; everything accessible to our observations.

variable star A star that varies in luminosity. (18)

velocity The speed and the direction a body is moving—for example, 44 km/s toward the north galactic pole. (2)

velocity of escape The speed with which an object must move in order to enter a parabolic orbit about another body (such as the Earth) and hence move permanently away from the vicinity of that body. (2)

vernal equinox The point on the celestial sphere where the Sun crosses the celestial equator passing from south to north; a time in the course of the year when the day and night are roughly equal. (3)

very long baseline interferometry A technique of radio astronomy whereby signals from telescopes thousands of kilometers apart combine to obtain very high resolution by letting waves from different sites interfere with each other. (5)

visual binary star A binary star in which the two components are telescopically resolved. (17)

void A region between clusters and superclusters of galaxies that appears relatively empty of galaxies. (27)

volatile materials Materials that are gaseous at fairly low temperatures. This is a relative term, usually applied to the gases in planetary atmospheres and to common ices (H_2O, CO_2, and so on), but it is also sometimes used for elements such as cadmium, zinc, lead, and rubidium that form gases at temperatures up to 1000 K. (These are called volatile elements, as opposed to refractory elements.) (7)

volume A measure of the total space occupied by a body. (2)

watt A unit of power (energy per unit time; a joule per second).

wavelength The spacing of the crests or troughs in a wave (4).

weak nuclear force or weak interaction The nuclear force involved in radioactive decay. The weak force is characterized by the slow rate of certain nuclear reactions, such as the decay of the neutron, which occurs with a half-life of 11 min.

weight A measure of the force due to gravitational attraction.

white dwarf A low-mass star that has exhausted most or all of its nuclear fuel and has collapsed to a very small size; such a star is near its final stage of life. (17)

Wien's law Formula that relates the temperature of a blackbody to the wavelength at which it emits the greatest intensity of radiation. (4)

winter solstice Point on the celestial sphere where the Sun reaches its farthest distance south of the celestial equator; the time of the year with the shortest amount of daylight. (3)

x rays Photons of wavelengths intermediate between those of ultraviolet radiation and gamma rays. (4)

year The period of revolution of the Earth around the Sun. (1)

Zeeman effect A splitting or broadening of spectral lines due to magnetic fields. (14)

zenith The point on the celestial sphere opposite to the direction of gravity; the point directly above the observer. (1)

zero-age main sequence Main sequence on the H-R diagram for a system of stars that have completed their contraction from interstellar matter and are now deriving all their energy from nuclear reactions, but whose chemical composition has not yet been altered by nuclear reactions. (21)

zodiac A belt around the sky about 18° wide centered on the ecliptic. (1)

zone of avoidance A region near the Milky Way where obscuration by interstellar dust is so heavy that few or no exterior galaxies can be seen.

4

Powers-of-Ten Notation

In astronomy (and other sciences), it is often necessary to deal with very large or very small numbers. In fact, when numbers become truly large in everyday life, such as the U.S. national debt, we call them astronomical. Among the ideas astronomers must routinely deal with is that the Earth is 150,000,000,000 m from the Sun, and the mass of the hydrogen atom is 0.00000000000000000000000000167 kg. No one in his or her right mind would want to continue writing so many zeros!

Instead, scientists have agreed on a kind of shorthand notation, which not only is easier to write but (as we shall see) also makes multiplication and division of large and small numbers much less difficult. If you have never used this powers-of-ten notation or scientific notation, it may take a bit of time to get used to it, but you will soon find it much easier than keeping all those zeros.

Writing Large Numbers

The convention in this notation is that we generally have only one number to the left of the decimal point. If a number is not in this format, it must be changed. The number 6 is already in the right format because for integers, we understand there to be a decimal point to the right of them. So 6 is really 6., and there is indeed only one number to the left of the decimal point. But the number 465 (which is 465.) has three numbers to the left of the decimal point and is thus ripe for conversion.

To change 465 to proper form, we must make it 4.65 and then keep track of the change we have made. (Think of the number as a weekly salary and suddenly it makes a lot of difference whether we have $465 or $4.65.) We keep track of the number of places we moved the decimal point by expressing it as a power of ten. So 465 becomes 4.65×10^2, or 4.65 multiplied by ten to the second power. The small raised 2 is called an exponent, and it tells us how many times we moved the decimal point to the left.

Note that 10^2 also designates 10 squared or 10×10, which equals 100. And 4.65×100 is just 465, the number we started with. Another way to look at scientific notation is that we separate out the messy numbers out front, and leave the smooth units of ten for the exponent to denote. So a number like 1,372,568 becomes 1.372568 times a million (10^6), or 1.372568 times 10 multiplied by itself 6 times. We had to move the decimal point six places to the left (from its place after the 8) to get the number into the form where there is only one digit to the left of the decimal point.

The reason we call this powers-of-ten notation is that our counting system is based on increases of ten; each place in our numbering system is ten times greater than the place to the right of it. As you have probably learned, this got started because human beings have ten fingers and we started counting with them.

It is interesting to speculate that if we ever meet intelligent life-forms with only eight fingers, their counting system would probably be a powers-of-eight notation!

So, in the example we started with, the number of meters from the Earth to the Sun is 1.5×10^{11}. Elsewhere in the book, we mention that a string a light year long would fit around the Earth's equator 236 million or 236,000,000 times. In scientific notation, this becomes 2.36×10^8. Now if you like expressing things in millions, as the annual reports of successful companies do, you might like to write this number as 236×10^6. However, the usual convention is to have only one number to the left of the decimal point.

Writing Small Numbers

Now take a number like 0.00347, which is also not in the standard (agreed-to) form for scientific notation. To put it into that format, we must make the first part of it 3.47 by moving the decimal point three places to the right. Note that this motion to the right is the opposite of the motion to the left that we discussed above. To keep track, we call this change negative and put a minus sign in the exponent. Thus 0.00347 becomes 3.47×10^{-3}.

In the example we gave at the beginning, the mass of the hydrogen atom would then be written as 1.67×10^{-27} kg. In this system, 1 is written as 10^0, a tenth as 10^{-1}, a hundredth as 10^{-2}, and so forth. Note that any number, no matter how large or how small, can be expressed in scientific notation.

Multiplication and Division

The powers-of-ten notation is not only compact and convenient, it also simplifies arithmetic. To multiply two numbers expressed as powers of ten, you need only multiply the numbers out front and then add the exponents. If there are no numbers out front, as in $100 \times 100,000$, then you just add the exponents (in our notation, $10^2 \times 10^5 = 10^7$). When there are numbers out front, you do have to multiply them, but they are much easier to deal with than numbers with many zeros in them. Here are two examples:

$$3 \times 10^5 \times 2 \times 10^9 = 6 \times 10^{14}$$

$$0.04 \times 6,000,000 = 4 \times 10^{-2} \times 6 \times 10^6$$

$$= 24 \times 10^4 = 2.4 \times 10^5$$

Note in the second example that when we added the exponents, we treated negative exponents as we do in regular arithmetic (-2 plus 6 equals 4). Also, notice that the first result we got had a 24 in it, which was not in the acceptable form, having two places to the left of the decimal point, and we therefore changed it to 2.4.

To divide, you divide the numbers out front and subtract the exponents. Here are three examples:

$$1,000,000 \div 1000 = 10^6 \div 10^3 = 10^{6-3} = 10^3$$

$$9 \times 10^{12} \div 2 \times 10^3 = 4.5 \times 10^9$$

$$2.8 \times 10^2 \div 6.2 \times 10^5 = 0.452 \times 10^{-3} = 4.52 \times 10^{-4}$$

If this is the first time you have met scientific notation, we urge you to practice using it (you might start by solving the exercises below). Like any new language, the notation looks complicated at first, but gets easier as you practice it.

Exercises

1. On April 8, 1996, the Galileo spacecraft was 775 million kilometers from Earth. Convert this number to scientific notation. How many astronomical units is this? (An astronomical unit is the distance from the Earth to the Sun; see above, but remember to keep your units consistent!)

2. During the first six years of its operation, the Hubble Space Telescope circled the Earth 37,000 times, for a total of 1,280,000,000 km. Use scientific notation to find the number of kilometers in one orbit.

3. In a college cafeteria, a soybean-vegetable burger is offered as an alternative to regular hamburgers. If 489,875 burgers were eaten during the course of a school year, and 997 of them were veggie-burgers, what fraction of the burgers does this represent?

4. In a June 1990 Gallup poll, 27 percent of adult Americans thought that alien beings have actually landed on Earth. The number of adults in the United States in 1990 was (according to the census) about 186,000,000. Use scientific notation to determine how many adults believe aliens have visited the Earth.

5. In 1995, 1.7 million degrees were awarded by American colleges and universities. Among these were 41,000 PhD degrees. What fraction of the degrees were PhDs? Express this number as a percent. (Now find jobs for all those PhDs!)

6. A star 60 light years away has been found to have a large planet orbiting it. Your uncle wants to know the distance to this planet in old-fashioned miles. If light travels 186,000 miles per second, and there are 60 seconds in a minute, 60 minutes in an hour, 24 hours in a day, and 365 days in a year, how many miles away is that star?

5

Units Used in Science

In the American system of measurement (originally developed in England), the fundamental units of length, weight, and time are the foot, pound, and second, respectively. There are also larger and smaller units, which include the ton (2240 lb), the mile (5280 ft), the rod (16½ ft), the yard (3 ft), the inch (1/12 ft), the ounce (1/16 lb), and so on. Such units, whose origins in decisions by British royalty have been forgotten by most people, are inconvenient for conversion or doing calculations.

In science, therefore, it is more usual to use the *metric system,* which has been adopted in virtually all countries except the United States. Its great advantage is that every unit increases by a factor of ten, instead of the strange factors in the American system. The fundamental units of the metric system are

length: 1 meter (m)

mass: 1 kilogram (kg)

time: 1 second (s)

A meter was originally intended to be 1 ten-millionth of the distance from the equator to the North Pole along the surface of the Earth. It is about 1.1 yd. A kilogram is the mass that on Earth results in a weight of about 2.2 lb. The second is the same in both metric and American units.

The most commonly used quantities of length and mass of the metric system are listed next.

Length

1 kilometer (km) = 1000 meters = 0.6214 mile

1 meter (m) = 0.001 km = 1.094 yards = 39.37 inches

1 centimeter (cm) = 0.01 meter = 0.3937 inch

1 millimeter (mm) = 0.001 meter = 0.1 cm

1 micrometer (μm) = 0.000001 meter = 0.0001 cm

1 nanometer (nm) = 10^{-9} meter = 10^{-7} cm

To convert from the American system, use these helpful factors:

1 mile = 1.61 km

1 inch = 2.54 cm

Mass

Although we don't make the distinction very carefully in everyday life on Earth, strictly speaking the kilogram is a unit of mass (measuring how many atoms a body has) and the pound is a unit of weight (measuring how strongly the Earth's gravity pulls on a body).

1 metric ton = 10^6 grams = 1000 kg (and it produces a weight of 2.205×10^3 lb on Earth)

1 kg = 1000 grams (and it produces a weight of 2.2046 lb on Earth)

1 gram (g) = 0.0353 oz (and the equivalent weight is 0.002205 lb)

1 milligram (mg) = 0.001 g

A weight of 1 lb is equivalent on Earth to a mass of 0.4536 kg, while a weight of 1 oz is produced by a mass of 28.35 g.

Temperature

Three temperature scales are in general use:

1. Fahrenheit (F); water freezes at 32°F and boils at 212°F.

2. Celsius or centigrade° (C); water freezes at 0°C and boils at 100°C.

3. Kelvin or absolute (K); water freezes at 273 K and boils at 373 K.

All molecular motion ceases at about −459°F = −273°C = 0 K, a temperature called *absolute zero.* Kelvin temperature is measured from this lowest possible temperature. It is the temperature scale most often used in astronomy. Kelvins have the same value as centigrade or Celsius degrees, since the difference between the freezing and boiling points of water is 100 degrees in each. (Note that we just say "Kelvins," not Kelvin degrees.)

On the Fahrenheit scale, the difference between the freezing and boiling points of water is 180 degrees. Thus, to convert Celsius degrees or Kelvins to Fahrenheit degrees, it is necessary to multiply by 180/100 = 9/5. To convert from Fahrenheit degrees to Celsius degrees or Kelvins, it is necessary to multiply by 100/180 = 5/9. The full conversion formulas are:

K = °C + 273

°C = 0.555 × (°F − 32)

°F = (1.8 × °C) + 32

° *Celsius* is the correct term; it has a more modern standardization but differs from the old centigrade scale by less than 0.1°.

Some Useful Constants for Astronomy

Physical Constants

speed of light $(c) = 2.9979 \times 10^8$ m/s

gravitational constant $(G) = 6.672 \times 10^{-11}$ N m^2/kg^2

Planck's constant $(h) = 6.626 \times 10^{-34}$ joule $\cdot$ s

mass of a hydrogen atom $(m_H) = 1.673 \times 10^{-27}$ kg

mass of an electron $(m_e) = 9.109 \times 10^{-31}$ kg

Rydberg constant $(R) = 1.0974 \times 10^7$ per m

Stefan-Boltzmann constant $(\sigma) = 5.670 \times 10^{-8}$ joule/(s $\cdot$ m^2 deg^4)
    [deg stands for degrees Celsius or Kelvins]

constant in Wien's law $(\lambda_{max} T) = 2.898 \times 10^{-3}$ m deg

electron volt (energy) (eV) $= 1.602 \times 10^{-19}$ joules

energy equivalent of 1 ton TNT $= 4.3 \times 10^9$ joules

Astronomical Constants

astronomical unit (AU) $= 1.496 \times 10^{11}$ m

light year (LY) $= 9.461 \times 10^{15}$ m

parsec (pc) $= 3.086 \times 10^{16}$ m $= 3.262$ LY

sidereal year (yr) $= 3.158 \times 10^7$ s

mass of Earth $(M_E) = 5.977 \times 10^{24}$ kg

equatorial radius of Earth $(R_E) = 6.378 \times 10^6$ m

obliquity of ecliptic $= 23° 27'$

surface gravity of Earth $(g) = 9.807$ m/s^2

escape velocity of Earth $(v_E) = 1.119 \times 10^4$ m/s

mass of Sun $(M_{Sun}) = 1.989 \times 10^{30}$ kg

equatorial radius of Sun $(R_{Sun}) = 6.960 \times 10^8$ m

luminosity of Sun $(L_{Sun}) = 3.83 \times 10^{26}$ watts

solar constant (flux of energy received at Earth)
    $S = 1.37 \times 10^3$ watts/m^2

Hubble constant (H_0) = approximately 20 km/s per million LY
    or approximately 70 km/s per megaparsec

7

Physical and Orbital Data for the Planets

Physical Data for the Planets

Planet	Diameter (km)	Diameter (Earth = 1)	Mass (Earth = 1)	Mean Density (g/cm³)	Rotation Period (days)	Inclination of Equator to Orbit (°)	Surface Gravity (Earth = 1)	Velocity of Escape (km/s)
Mercury	4,878	0.38	0.055	5.43	58.6	0.0	0.38	4.3
Venus	12,104	0.95	0.82	5.24	−243.0	177.4	0.91	10.4
Earth	12,756	1.00	1.00	5.52	0.997	23.4	1.00	11.2
Mars	6,794	0.53	0.107	3.9	1.026	25.2	0.38	5.0
Jupiter	142,800	11.2	317.8	1.3	0.41	3.1	2.53	60
Saturn	120,540	9.41	94.3	0.7	0.43	26.7	1.07	36
Uranus	51,200	4.01	14.6	1.3	−0.72	97.9	0.92	21
Neptune	49,500	3.88	17.2	1.7	0.67	29	1.18	24
Pluto	2,300	0.18	0.0025	2.0	−6.387	118	0.09	1

Orbital Data for the Planets

Planet	Semimajor Axis AU	Semimajor Axis 10⁶ km	Sidereal Period Tropical Years	Sidereal Period Days	Mean Orbital Speed (km/s)	Orbital Eccentricity	Inclination of Orbit to Ecliptic (°)
Mercury	0.3871	57.9	0.24085	88	47.9	0.206	7.004
Venus	0.7233	108.2	0.61521	225	35.0	0.007	3.394
Earth	1.0000	149.6	1.000039	365	29.8	0.017	0.0
Mars	1.5237	227.9	1.88089	687	24.1	0.093	1.850
(Ceres)	2.7671	414	4.603		17.9	0.077	10.6
Jupiter	5.2028	778	11.86		13.1	0.048	1.308
Saturn	9.538	1427	29.46		9.6	0.056	2.488
Uranus	19.191	2871	84.07		6.8	0.046	0.774
Neptune	30.061	4497	164.82		5.4	0.010	1.774
Pluto	39.529	5913	248.6		4.7	0.248	17.15

8

Selected Satellites of the Planets

Note: As this book goes to press, more than 100 satellites are now known in the solar system and more are being discovered on a regular basis. Many of these are small bodies, most likely captured asteroids. It's no longer practical to list every known satellite in such an appendix, and we therefore restrict ourselves to the more significant bodies that orbit each planet. See the Web sites listed for Chapters 6 and 11 for information on all the known satellites.

Planet	Satellite Name	Discovery	Semimajor Axis (km × 1000)	Period (days)	Diameter (km)	Mass (10^{20} kg)	Density (g/cm^3)
Earth	Moon	—	384	27.32	3476	735	3.3
Mars	Phobos	Hall (1877)	9.4	0.32	23	1×10^{-4}	2.0
	Deimos	Hall (1877)	23.5	1.26	13	2×10^{-5}	1.7
Jupiter	Metis	Voyager (1979)	128	0.29	20	—	—
	Adrastea	Voyager (1979)	129	0.30	40	—	—
	Amalthea	Barnard (1892)	181	0.50	200	—	—
	Thebe	Voyager (1979)	222	0.67	90	—	—
	Io	Galileo (1610)	422	1.77	3630	894	3.6
	Europa	Galileo (1610)	671	3.55	3138	480	3.0
	Ganymede	Galileo (1610)	1,070	7.16	5262	1482	1.9
	Callisto	Galileo (1610)	1,883	16.69	4800	1077	1.9
	Leda	Kowal (1974)	11,090	239	15	—	—
	Himalia	Perrine (1904)	11,480	251	180	—	—
	Lysithea	Nicholson (1938)	11,720	259	40	—	—
	Elara	Perrine (1905)	11,740	260	80	—	—
	Ananke	Nicholson (1951)	21,200	631 (R)	30	—	—
	Carme	Nicholson (1938)	22,600	692 (R)	40	—	—
	Pasiphae	Melotte (1908)	23,500	735 (R)	40	—	—
	Sinope	Nicholson (1914)	23,700	758 (R)	40	—	—
Saturn	Unnamed	Voyager (1985)	118.2	0.48	15?	3×10^{-5}	—
	Pan	Voyager (1985)	133.6	0.58	20	3×10^{-5}	—
	Atlas	Voyager (1980)	137.7	0.60	40	—	—
	Prometheus	Voyager (1980)	139.4	0.61	80	—	—
	Pandora	Voyager (1980)	141.7	0.63	100	—	—
	Janus	Dollfus (1966)	151.4	0.69	190	—	—
	Epimetheus	Fountain, Larson (1980)	151.4	0.69	120	—	—

(Table continues)

Planet	Satellite Name	Discovery	Semimajor Axis (km × 1000)	Period (days)	Diameter (km)	Mass (10²⁰ kg)	Density (g/cm³)
Saturn	Mimas	Herschel (1789)	186	0.94	394	0.4	1.2
	Enceladus	Herschel (1789)	238	1.37	502	0.8	1.2
	Tethys	Cassini (1684)	295	1.89	1048	7.5	1.3
	Telesto	Reitsema et al. (1980)	295	1.89	25	—	—
	Calypso	Pascu et al. (1980)	295	1.89	25	—	—
	Dione	Cassini (1684)	377	2.74	1120	11	1.4
	Helene	Lecacheux, Laques (1980)	377	2.74	30	—	—
	Rhea	Cassini (1672)	527	4.52	1530	25	1.3
	Titan	Huygens (1655)	1,222	15.95	5150	1346	1.9
	Hyperion	Bond, Lassell (1848)	1,481	21.3	270	—	—
	Iapetus	Cassini (1671)	3,561	79.3	1435	19	1.2
	Phoebe	Pickering (1898)	12,950	550 (R)	220	—	—
Uranus	Cordelia	Voyager (1986)	49.8	0.34	40?	—	—
	Ophelia	Voyager (1986)	53.8	0.38	50?	—	—
	Bianca	Voyager (1986)	59.2	0.44	50?	—	—
	Cressida	Voyager (1986)	61.8	0.46	60?	—	—
	Desdemona	Voyager (1986)	62.7	0.48	60?	—	—
	Juliet	Voyager (1986)	64.4	0.50	80?	—	—
	Portia	Voyager (1986)	66.1	0.51	80?	—	—
	Rosalind	Voyager (1986)	69.9	0.56	60?	—	—
	Belinda	Voyager (1986)	75.3	0.63	60?	—	—
	Unnamed	Voyager (1986)	73.5	0.63	40?	—	—
	Puck	Voyager (1985)	86.0	0.76	170	—	—
	Miranda	Kuiper (1948)	130	1.41	485	0.8	1.3
	Ariel	Lassell (1851)	191	2.52	1160	13	1.6
	Umbriel	Lassell (1851)	266	4.14	1190	13	1.4
	Titania	Herschel (1787)	436	8.71	1610	35	1.6
	Oberon	Herschel (1787)	583	13.5	1550	29	1.5
	Caliban	Gladman et. al. (1997)	7,000	562	60?	—	—
	Sycorax	Nicholson et. al. (1997)	12,000	1,261	120?	—	—
Neptune	Naiad	Voyager (1989)	48	0.30	50	—	—
	Thalassa	Voyager (1989)	50	0.31	90	—	—
	Despina	Voyager (1989)	53	0.33	150	—	—
	Galatea	Voyager (1989)	62	0.40	150	—	—
	Larissa	Voyager (1989)	74	0.55	200	—	—
	Proteus	Voyager (1989)	118	1.12	400	—	—
	Triton	Lassell (1846)	355	5.88 (R)	2720	220	2.1
	Nereid	Kuiper (1949)	5,511	360	340	—	—
Pluto	Charon	Christy (1978)	19.7	6.39	1200	—	—

(R) = Retrograde orbit

Upcoming (Total) Eclipses

1 Total Eclipses of the Sun

Date	Duration of Totality (min)	Where Visible
2003 Nov. 23	2.0	Antarctica
2005 April 8	0.7	South Pacific
2006 March 29	4.1	Africa, western Asia, Russia
2008 Aug. 1	2.4	Arctic Ocean, Siberia, China
2009 July 22	6.6	India, China, South Pacific
2010 July 11	5.3	South Pacific
2012 Nov. 13	4.0	Northern Australia, South Pacific
2013 Nov. 3	1.7	Atlantic, Central Africa
2015 March 20	4.1	North Atlantic, Arctic
2016 March 9	4.5	Indonesia, Pacific
2017 Aug. 21	2.7	Pacific, United States, Atlantic
2019 July 2	4.5	South Pacific, South America
2020 Dec. 14	2.2	South Pacific, South America, South Atlantic
2021 Dec. 4	1.9	Antarctica
2023 April 20	4.5	South Pacific, Mexico, eastern United States
2026 Aug. 12	2.3	Arctic, Greenland, North Atlantic, Spain
2027 Aug. 2	6.4	North Africa, Arabia, Indian Ocean
2028 July 22	5.1	Indian Ocean, Australia, New Zealand
2030 Nov. 25	3.7	South Africa, Indian Ocean, Australia

2 Total Eclipses of the Moon

2003 Nov. 9
2004 May 4
2004 Oct. 28
2007 March 3
2007 Aug. 28
2008 Feb. 21
2010 Dec. 21
2011 June 15
2011 Dec. 10
2014 Apr. 15

Note: Lunar eclipses are much more "democratic" than solar ones. They are visible over the entire hemisphere of the Earth where the Moon can be seen (see Chapter 3).

Name (catalog number)	Distance (LY)	Spectral Type	Location[1] RA	Location[1] Dec	Luminosity (Sun = 1)
Sun	—	G2V	—	—	1.0
Proxima Centauri	4.2	M5V	14 30	−62 41	6×10^{-6}
Alpha Centauri A	4.4	G2V	14 40	−60 50	1.5
Alpha Centauri B	4.4	KOV	14 40	−60 50	0.5
Barnard's Star (Gliese 699)	6.0	M4V	17 58	+04 42	4×10^{-4}
Wolf 359 (Gliese 406)	7.8	M6V	10 56	+07 03	2×10^{-5}
Lalande 21185 (HD 95735)	8.3	M2V	11 03	+35 58	5×10^{-3}
Sirius A	8.6	A1V	06 45	−16 43	24
Sirius B	8.6	w.d.[2]	06 45	−16 43	3×10^{-3}
Luyten 726-8 A (Gliese 65A)	8.7	M5V	01 39	−17 57	6×10^{-5}
Luyten 726-8 B (UV Ceti)	8.7	M6V	01 39	−17 58	4×10^{-5}
Ross 154 (Gliese 729)	9.7	M4V	18 50	−23 50	5×10^{-4}
Ross 248 (Gliese 905)	10.3	M6V	23 42	+44 11	1×10^{-4}
Epsilon Eridani (Gliese 144)	10.5	K2V	03 33	−09 27	0.3
Lacaille 9352 (Gliese 887)	10.7	M1V	23 06	−35 51	1×10^{-2}
Ross 128 (Gliese 447)	10.9	M4V	11 48	+00 48	3×10^{-4}
Luyten 789-6 A (Gliese 866A)	11.3	M5V[3]	22 39	−15 18	1×10^{-4}
Luyten 789-6 B	11.3	—	22 39	−15 18	—
Luyten 789-6 C	11.3	—	22 39	−15 18	—
Procyon A	11.4	F51V	07 39	+05 13	7.7
Procyon B	11.4	w.d.[2]	07 39	+05 13	6×10^{-4}
61 Cygni A (Gliese 820A)	11.4	K5V	21 07	+38 45	8×10^{-2}
61 Cygni B	11.4	K7V	21 07	+38 45	4×10^{-2}
Gliese 725 A	11.5	M3V	18 43	+59 38	3×10^{-3}
Gliese 725 B	11.5	M4V	18 43	+59 38	2×10^{-3}
Gliese 15 A	11.6	M1V	00 18	+44 01	$6 \times 10-3$
Gliese 15 B	11.6	M3V	00 18	+44 01	4×10^{-4}
Epsilon Indi (Gliese 845)	11.8	K5V	22 03	−56 47	0.14
GJ 1111 (DX Cancri)	11.8	M7V	08 30	+26 47	1×10^{-5}
Tau Ceti (Gliese 71)	11.9	G8V	01 44	−15 56	0.45

[1] Location (right ascension and declination) given for Epoch 2000.0
[2] White dwarf star
[3] The stars in this system are so close to each other that it is not possible to measure their spectral types and luminosities separately.

(Table continues)

| Name (catalog number) | Distance (LY) | Spectral Type | Location[1] | | Luminosity (Sun = 1) |
			RA	Dec	
GJ 1061	11.9	M5V	03 36	−44 31	8×10^{-5}
Luyten 725-32 (YZ Ceti)	12.1	M5V	01 12	−16 60	3×10^{-4}
Gliese 273 (Luyten's Star)	12.4	M4V	07 27	+05 14	1×10^{-3}
Gliese 191 (Kapteyn's Star)	12.8	M1V	05 12	−45 01	4×10^{-3}
Gliese 825 (AX Microscopium)	12.9	M0V	21 17	−38 52	3×10^{-2}

[1] Location (right ascension and declination) given for Epoch 2000.0

With many thanks to Dr. Todd Henry and the RECONS team for updated information. For the latest data on nearby star systems, see their Web page at http://joy.chara.gsu.edu/RECONS/TOP100.htm

The Brightest Stars

Note: These are the stars that appear the brightest visually, as seen from our vantage point on Earth. They are not necessarily the stars that are intrinsically the brightest.

Name[1]	Luminosity (Sun = 1)	Distance[2] (LY)	Spectral Type	Proper Motion (arcsec/yr)	Right Ascension (Epoch 2000.0) (h)	(m)	Declination (Epoch 2000.0) (deg)	(min)
Sirius (α CMa)	24	8.6	A1V	1.34	06	45.1	−16	43
Canopus (α Car)	7.3×10^3	228	F0II	0.02	06	24.0	−52	42
Alpha Centauri	2	4	G2V	3.68	14	39.6	−60	50
Arcturus (α Boo)	187	37	K1.5II	2.28	14	15.7	+19	11
Vega (α Lyr)	50	25	A0V	0.35	18	36.9	+38	47
Capella (α Aur)	145	42	G8III	0.43	05	16.7	+46	00
Rigel (β Ori)	6×10^4	772	B8Ia	0.00	05	14.5	−08	12
Procyon (α Cmi)	7	11	F5IV-V	1.26	07	39.3	+05	13
Betelgeuse (α Ori)	7×10^5	427	M1Iab	0.03	05	55.2	+07	24
Achernar (α Eri)	2800	144	B3V	0.10	01	37.7	−57	14
Beta Centauri	6.5×10^4	525	B1III	0.04	14	03.8	−60	22
Altair (α Acll)	10	17	A7V	0.66	19	50.8	+08	52
Aldebaran (α Tau)	450	65	K5III	0.20	04	35.9	+16	31
Spica (α Vir)	1.2×10^4	262	B1III	0.05	13	25.2	−11	10
Antares (α Sco)	8.5×10^5	604	M1.5Ib	0.03	16	29.4	−26	26
Pollux (β Gem)	40	34	K0III	0.62	07	45.3	+28	02
Fomalhaut (α PsA)	16	25	A3V	0.37	22	57.6	−29	37
Deneb (α Cyg)	2.4×10^5	3228	A2Ia	0.00	20	41.4	+45	17
Beta Crucis	1.6×10^4	352	B0.5IV	0.05	12	47.7	−59	41
Regulus (α Leo)	230	77	B7V	0.25	10	08.3	+11	58

[1] The brightest stars typically have names from antiquity (although most fainter stars are not given names but merely catalog designations). Next to each star's ancient name, we have put its name in the system originated by Bayer (see the Astronomy Basics box in Chapter 18). The abbreviations of the constellations are given in Appendix 14.
[2] The distances of the more remote stars are estimated from their spectral types and apparent brightnesses and are only approximate. The luminosities for those stars are approximate to the same degree.

12

The Brightest Members of the Local Group of Galaxies

Galaxy	Type[1]	Right Ascension[2] (h)	(m)	Declination (degrees)	Distance[3] (1000 LY)	Luminosity (L_{Sun})	Apparent Magnitude	Diameter (1000 LY)	Mass[4] ($10^6 M_{Sun}$)
Milky Way	SBb	17	46	−29	—	1.3×10^{10}	—	100	10^6
Andromeda (M31, NGC224)	Sb	00	43	+41	2900	2.6×10^{10}	3.4	160	4×10^5
M33 (NGC598)	Sc	01	34	31	3000	3.4×10^9	5.7	60	25,000
Large Magellanic Cloud	Irr	05	20	−69	179	2.1×10^9	0.1	35	20,000
Small Magellanic Cloud	Irr	00	52	−73	210	3.7×10^8	2.3	18	6,000
IC10	Irr	00	20	+59	4200	9.3×10^7	10.3	9	
NGC205	E5pec	00	41	+42	2900	2.3×10^8	8.5	17	10,000
M32 (NGC221)	E2	00	43	+41	2900	3.4×10^8	8.1	10	3,000
NGC6822	Irr	19	45	−15	1700	5.3×10^7	9		
WLM	Irr	00	02	−15	4200	5.6×10^7	10.9	15	
NGC185	E3pec	00	39	+48	2500	9.3×10^7	9.2	11	
IC1613	Irr	01	05	+02	2900	1.2×10^8	9.2	17	
NGC147	E5	00	33	+48	2400	6.4×10^7	9.5	11	
Leo A	Irr	09	59	+31	7000				
Pegasus	Irr	23	29	+15	6000				
Fornax	E2	02	40	−34	530	1.2×10^7	8.1	2	
DD0210	Irr	20	47	−13	2000	7.7×10^5	13.9	1	
Sagittarius Dwarf[5]	DwE7	18	55	−31	80				
Sagittarius	Irr	19	30	−18	2000				
Sculptor	E3	01	00	−34	300	4.1×10^5	10.5		
Andromeda I	E3	00	46	+38	2900	3.4×10^6	13.2		
Andromeda III	E2	00	35	+36	2900	2.3×10^6	13.5		
Andromeda II	E0	01	16	+33	2900	3.7×10^6	13		
Pisces (LGS3)	Irr	01	04	+22	3000	4.4×10^5	15.4	2	
Leo I	E3	10	09	+12	880	7.0×10^6	9.8		
Leo II	E0	11	14	+22	800				
Ursa Minor	E4	15	09	+67	240	1.8×10^5	10.9	3	
Draco	E0	17	20	+58	280	2.6×10^5	10.9	4	

(Table continues)

Galaxy	Type[1]	Right Ascension[2] (h)	Right Ascension[2] (m)	Declination (degrees)	Distance[3] (1000 LY)	Luminosity (L_{Sun})	Apparent Magnitude	Diameter (1000 LY)	Mass[4] ($10^6 M_{Sun}$)
Carina	E3	06	42	−51	300				
Andromeda V	DwE	01	10	+48	2900	1.2×10^6			
Phoenix	Irr	01	51	−44	1600	6.4×10^5	—		
Sextans	DwE3	10	13	−02	300	4.9×10^5			
Tucana	DwE5	22	42	−64	3000	5.3×10^5			
Andromeda VI	DwE	23	52	+25	2900	—			

[1] S means spiral, SB means barred spiral, E means elliptical, Irr means irregular, Dw means dwarf, pec means peculiar; the numbers represent subgroups into which Hubble and others divided these broad categories.

[2] Coordinates are given for Epoch 2000.0.

[3] Many of the distances (and therefore the diameters calculated from them) are only approximate.

[4] Masses have been measured for only a few of the brightest and/or nearest galaxies. Determining masses for irregular galaxies requires the measurement of spectra of a large number of faint stars, and this work has not yet been done for most of the stars in the Local Group.

[5] This close neighbor galaxy is so extended on the sky that giving a magnitude would not make sense.

13 The Chemical Elements

Element	Symbol	Atomic Number	Atomic Weight* (chemical scale)	Number of Atoms per 10^{12} Hydrogen Atoms
Hydrogen	H	1	1.0080	1×10^{12}
Helium	He	2	4.003	8×10^{10}
Lithium	Li	3	6.940	2×10^{3}
Beryllium	Be	4	9.013	3×10^{1}
Boron	B	5	10.82	9×10^{2}
Carbon	C	6	12.011	4.5×10^{8}
Nitrogen	N	7	14.008	9.2×10^{7}
Oxygen	O	8	16.00	7.4×10^{8}
Fluorine	F	9	19.00	3.1×10^{4}
Neon	Ne	10	20.183	1.3×10^{8}
Sodium	Na	11	22.991	2.1×10^{6}
Magnesium	Mg	12	24.32	4.0×10^{7}
Aluminum	Al	13	26.98	3.1×10^{6}
Silicon	Si	14	28.09	3.7×10^{7}
Phosphorus	P	15	30.975	3.8×10^{5}
Sulfur	S	16	32.066	1.9×10^{7}
Chlorine	Cl	17	35.457	1.9×10^{5}
Argon	Ar(A)	18	39.944	3.8×10^{6}
Potassium	K	19	39.100	1.4×10^{5}
Calcium	Ca	20	40.08	2.2×10^{6}
Scandium	Sc	21	44.96	1.3×10^{3}
Titanium	Ti	22	47.90	8.9×10^{4}
Vanadium	V	23	50.95	1.0×10^{4}
Chromium	Cr	24	52.01	5.1×10^{5}
Manganese	Mn	25	54.94	3.5×10^{5}
Iron	Fe	26	55.85	3.2×10^{7}
Cobalt	Co	27	58.94	8.3×10^{4}
Nickel	Ni	28	58.71	1.9×10^{6}
Copper	Cu	29	63.54	1.9×10^{4}
Zinc	Zn	30	65.38	4.7×10^{4}
Gallium	Ga	31	69.72	1.4×10^{3}
Germanium	Ge	32	72.60	4.4×10^{3}
Arsenic	As	33	74.91	2.5×10^{2}
Selenium	Se	34	78.96	2.3×10^{3}

* Where average atomic weights have not been well determined, the atomic mass numbers of the most stable isotopes are given in parentheses.

(*Table continues*)

Element	Symbol	Atomic Number	Atomic Weight* (chemical scale)	Number of Atoms per 10^{12} Hydrogen Atoms
Bromine	Br	35	79.916	4.4×10^2
Krypton	Kr	36	83.80	1.7×10^3
Rubidium	Rb	37	85.48	2.6×10^2
Strontium	Sr	38	87.63	8.8×10^2
Yttrium	Y	39	88.92	2.5×10^2
Zirconium	Zr	40	91.22	4.0×10^2
Niobium (Columbium)	Nb(Cb)	41	92.91	2.6×10^1
Molybdenum	Mo	42	95.95	9.3×10^1
Technetium	Tc(Ma)	43	(99)	—
Ruthenium	Ru	44	101.1	68
Rhodium	Rh	45	102.91	13
Palladium	Pd	46	106.4	51
Silver	Ag	47	107.880	20
Cadmium	Cd	48	112.41	63
Indium	In	49	114.82	7
Tin	Sn	50	118.70	1.4×10^2
Antimony	Sb	51	121.76	13
Tellurium	Te	52	127.61	1.8×10^2
Iodine	I (J)	53	126.91	33
Xenon	Xe(X)	54	131.30	1.6×10^2
Cesium	Cs	55	132.91	14
Barium	Ba	56	137.36	1.6×10^2
Lanthanum	La	57	138.92	17
Cerium	Ce	58	140.13	43
Praseodymium	Pr	59	140.92	6
Neodymium	Nd	60	144.27	31
Promethium	Pm	61	(147)	—
Samarium	Sm(Sa)	62	150.35	10
Europium	Eu	63	152.00	4
Gadolinium	Gd	64	157.26	13
Terbium	Tb	65	158.93	2
Dysprosium	Dy(Ds)	66	162.51	15
Holmium	Ho	67	164.94	3
Erbium	Er	68	167.27	9
Thulium	Tm(Tu)	69	168.94	2
Ytterbium	Yb	70	173.04	8
Lutecium	Lu(Cp)	71	174.99	2
Hafnium	Hf	72	178.50	6
Tantalum	Ta	73	180.95	1
Tungsten	W	74	183.86	5
Rhenium	Re	75	186.22	2
Osmium	Os	76	190.2	27
Iridium	Ir	77	192.2	24
Platinum	Pt	78	195.09	56
Gold	Au	79	197.00	6
Mercury	Hg	80	200.61	19
Thallium	Tl	81	204.39	8

* Where average atomic weights have not been well determined, the atomic mass numbers of the most stable isotopes are given in parentheses.

Element	Symbol	Atomic Number	Atomic Weight°(chemical scale)	Number of Atoms per 10^{12} Hydrogen Atoms
Lead	Pb	82	207.21	1.2×10^2
Bismuth	Bi	83	209.00	5
Polonium	Po	84	(209)	—
Astatine	At	85	(210)	—
Radon	Rn	86	(222)	—
Francium	Fr(Fa)	87	(223)	—
Radium	Ra	88	226.05	—
Actinium	Ac	89	(227)	—
Thorium	Th	90	232.12	1
Protactinium	Pa	91	(231)	—
Uranium	U(Ur)	92	238.07	1
Neptunium	Np	93	(237)	—
Plutonium	Pu	94	(244)	—
Americium	Am	95	(243)	—
Curium	Cm	96	(248)	—
Berkelium	Bk	97	(247)	—
Californium	Cf	98	(251)	—
Einsteinium	E	99	(254)	—
Fermium	Fm	100	(253)	—
Mendeleevium	Mv	101	(256)	—
Nobelium	No	102	(253)	—
Lawrencium	Lr	103	(262)	—
Rutherfordium	Rf	104	(261)	—
Dubnium	Db	105	(262)	—
Seaborgium	Sg	106	(263)	—
Bohrium	Bh	107	(262)	—
Hassium	Hs	108	(264)	—
Meitnerium	Mt	109	(266)	—
Ununnilium	Uun	110	(269)	—
Unununium	Uuu	111	(272)	—
Ununbium	Uub	112	(277)	—
Ununquadium	Uuq	114	(285)	—
Ununhexium	Uuh	116	(289)	—
Ununoctium	Uuo	118	(293)	—

° Where average atomic weights have not been well determined, the atomic mass numbers of the most stable isotopes are given in parentheses.

14 The Constellations

Constellation (Latin name)	Genitive Case Ending[1]	English Name or Description	Abbre- viation	Approximate Position	
				α	δ
				h	°
Andromeda	Andromedae	Princess of Ethiopia	And	1	+40
Antila	Antilae	Air pump	Ant	10	−35
Apus	Apodis	Bird of paradise	Aps	16	−75
Aquarius	Aquarii	Water bearer	Aqr	23	−15
Aquila	Aquilae	Eagle	Aql	20	+5
Ara	Arae	Altar	Ara	17	−55
Aries	Arietis	Ram	Ari	3	+20
Auriga	Aurigae	Charioteer	Aur	6	+40
Boötes	Boötis	Herdsman	Boo	15	+30
Caelum	Caeli	Engraving tool	Cae	5	−40
Camelopardalis	Camelopardalis	Giraffe	Cam	6	+70
Cancer	Cancri	Crab	Cnc	9	+20
Canes Venatici	Canum Venaticorum	Hunting dogs	CVn	13	+40
Canis Major	Canis Majoris	Big dog	CMa	7	−20
Canis Minor	Canis Minoris	Little dog	CMi	8	+5
Capricornus	Capricorni	Sea goat	Cap	21	−20
Carina°	Carinae	Keel of the Argonauts' ship	Car	9	−60
Cassiopeia	Cassiopeiae	Queen of Ethiopia	Cas	1	+60
Centaurus	Centauri	Centaur	Cen	13	−50
Cepheus	Cephei	King of Ethiopia	Cep	22	+70
Cetus	Ceti	Sea monster (whale)	Cet	2	−10
Chamaeleon	Chamaeleontis	Chameleon	Cha	11	−80
Circinus	Circini	Compasses	Cir	15	−60
Columba	Columbae	Dove	Col	6	−35
Coma Berenices	Comae Berenices	Berenice's hair	Com	13	+20
Corona Australis	Coronae Australis	Southern crown	CrA	19	−40
Corona Borealis	Coronae Borealis	Northern crown	CrB	16	+30
Corvus	Corvi	Crow	Crv	12	−20

[1] When a constellation name is by itself in a sentence, we use the form in the first column (for example, the comet last night was seen in Crux). But when the constellation is part of the name of an object, we use the form in the second column (for example, the brightest star in that constellation is Alpha Crucis).

° The four constellations Carina, Puppis, Pyxis, and Vela originally formed the single constellation Argo Navis.

Constellation (Latin name)	Genitive Case Ending	English Name or Description	Abbre-viation	Approximate Position α h	δ °
Crater	Crateris	Cup	Crt	11	−15
Crux	Crucis	Cross (southern)	Cru	12	−60
Cygnus	Cygni	Swan	Cyg	21	+40
Delphinus	Delphini	Porpoise	Del	21	+10
Dorado	Doradus	Swordfish	Dor	5	−65
Draco	Draconis	Dragon	Dra	17	+65
Equuleus	Equulei	Little horse	Equ	21	+10
Eridanus	Eridani	River	Eri	3	−20
Fornax	Fornacis	Furnace	For	3	−30
Gemini	Geminorum	Twins	Gem	7	+20
Grus	Gruis	Crane	Gru	22	−45
Hercules	Herculis	Hercules, son of Zeus	Her	17	+30
Horologium	Horologii	Clock	Hor	3	−60
Hydra	Hydrae	Sea serpent	Hya	10	−20
Hydrus	Hydri	Water snake	Hyi	2	−75
Indus	Indi	Indian	Ind	21	−55
Lacerta	Lacertae	Lizard	Lac	22	+45
Leo	Leonis	Lion	Leo	11	+15
Leo Minor	Leonis Minoris	Little lion	LMi	10	+35
Lepus	Leporis	Hare	Lep	6	−20
Libra	Librae	Balance	Lib	15	−15
Lupus	Lupi	Wolf	Lup	15	−45
Lynx	Lyncis	Lynx	Lyn	8	+45
Lyra	Lyrae	Lyre or harp	Lyr	19	+40
Mensa	Mensae	Table Mountain	Men	5	−80
Microscopium	Microscopii	Microscope	Mic	21	−35
Monoceros	Monocerotis	Unicorn	Mon	7	−5
Musca	Muscae	Fly	Mus	12	−70
Norma	Normae	Carpenter's level	Nor	16	−50
Octans	Octantis	Octant	Oct	22	−85
Ophiuchus	Ophiuchi	Holder of serpent	Oph	17	0
Orion	Orionis	Orion, the hunter	Ori	5	+5
Pavo	Pavonis	Peacock	Pav	20	−65
Pegasus	Pegasi	Pegasus, the winged horse	Peg	22	+20
Perseus	Persei	Perseus, hero who saved Andromeda	Per	3	+45
Phoenix	Phoenicis	Phoenix	Phe	1	−50
Pictor	Pictoris	Easel	Pic	6	−55
Pisces	Piscium	Fishes	Psc	1	+15
Piscis Austrinus	Piscis Austrini	Southern fish	PsA	22	−30
Puppis°	Puppis	Stern of the Argonauts' ship	Pup	8	−40

[1] When a constellation name is by itself in a sentence, we use the form in the first column (for example, the comet last night was seen in Crux). But when the constellation is part of the name of an object, we use the form in the second column (for example, the brightest star in that constellation is Alpha Crucis).

° The four constellations Carina, Puppis, Pyxis, and Vela originally formed the single constellation Argo Navis.

(Table continues)

Constellation (Latin name)	Genitive Case Ending	English Name or Description	Abbreviation	Approximate Position α (h)	δ (°)
Pyxis° (= Malus)	Pyxidus	Compass of the Argonauts' ship	Pyx	9	−30
Reticulum	Reticuli	Net	Ret	4	−60
Sagitta	Sagittae	Arrow	Sge	20	+10
Sagittarius	Sagittarii	Archer	Sgr	19	−25
Scorpius	Scorpii	Scorpion	Sco	17	−40
Sculptor	Sculptoris	Sculptor's tools	Scl	0	−30
Scutum	Scuti	Shield	Sct	19	−10
Serpens	Serpentis	Serpent	Ser	17	0
Sextans	Sextantis	Sextant	Sex	10	0
Taurus	Tauri	Bull	Tau	4	+15
Telescopium	Telescopii	Telescope	Tel	19	−50
Triangulum	Trianguli	Triangle	Tri	2	+30
Triangulum Australe	Trianguli Australis	Southern triangle	TrA	16	−65
Tucana	Tucanae	Toucan	Tuc	0	−65
Ursa Major	Ursae Majoris	Big bear	UMa	11	+50
Ursa Minor	Ursae Minoris	Little bear	UMi	15	+70
Vela°	Velorum	Sail of the Argonauts' ship	Vel	9	−50
Virgo	Virginis	Virgin	Vir	13	0
Volans	Volantis	Flying fish	Vol	8	−70
Vulpecula	Vulpeculae	Fox	Vul	20	+25

[1] When a constellation name is by itself in a sentence, we use the form in the first column (for example, the comet last night was seen in Crux). But when the constellation is part of the name of an object, we use the form in the second column (for example, the brightest star in that constellation is Alpha Crucis).

° The four constellations Carina, Puppis, Pyxis, and Vela originally formed the single constellation Argo Navis.

15

The Messier Catalog of Nebulae and Star Clusters

M	NGC or (IC)	Right Ascension (2000)		Declination (2000)		Apparent Visual Magnitude	Description
		h	m	(deg)	(min)		
1	1952	5	34.5	+22	01	8.4	"Crab" nebula in Taurus; remains of SN1054
2	7089	21	33.5	−0	50	6.4	Globular cluster in Aquarius
3	5272	13	42.2	+28	23	6.3	Globular cluster in Canes Venatici
4	6121	16	23.6	−26	32	5.9	Globular cluster in Scorpius
5	5904	15	18.6	+2	05	5.8	Globular cluster in Serpens
6	6405	17	40.1	−32	13	4.2	Open cluster in Scorpius
7	6475	17	53.9	−34	49	3.3	Open cluster in Scorpius
8	6523	18	03.8	−24	23	5.1	"Lagoon" nebula in Sagittarius
9	6333	17	19.2	−18	31	7.9	Globular cluster in Ophiuchus
10	6254	16	57.1	−4	06	6.7	Globular cluster in Ophiuchus
11	6705	18	51.1	−6	16	5.8	Open cluster in Scutum
12	6218	16	47.2	−1	57	6.6	Globular cluster in Ophiuchus
13	6205	16	41.7	+36	28	5.9	Globular cluster in Hercules
14	6402	17	37.6	−3	15	7.6	Globular cluster in Ophiuchus
15	7078	21	30.0	+12	10	6.4	Globular cluster in Pegasus
16	6611	18	18.8	−13	58	6.6	Open cluster with nebulosity in Serpens
17	6618	18	20.8	−16	11	7.5	"Swan" or "Omega" nebula in Sagittarius
18	6613	18	19.9	−17	08	6.9	Open cluster in Sagittarius
19	6273	17	02.6	−26	16	6.9	Globular cluster in Ophiuchus
20	6514	18	02.3	−23	02	8.5	"Trifid" nebula in Sagittarius
21	6531	18	04.6	−22	30	5.9	Open cluster in Sagittarius
22	6656	18	36.4	−23	54	5.1	Globular cluster in Sagittarius
23	6494	17	56.8	−19	01	5.5	Open cluster in Sagittarius
24	6603	18	16.9	−18	29	4.6	Star field with open cluster in Sagittarius
25	(4725)	18	31.6	−19	15	4.6	Open cluster in Sagittarius
26	6694	18	45.2	−9	24	8.0	Open cluster in Scutum
27	6853	19	59.6	+22	43	8.1	"Dumbbell" planetary nebula in Vulpecula
28	6626	18	24.5	−24	52	6.9	Globular cluster in Sagittarius
29	6913	20	23.9	+38	32	7.0	Open cluster in Cygnus
30	7099	21	40.4	−23	11	7.5	Globular cluster in Capricornus
31	224	0	42.7	+41	16	3.5	Andromeda galaxy

(Table continues)

M	NGC or (IC)	Right Ascension (2000) h	Right Ascension (2000) m	Declination (2000) (deg)	Declination (2000) (min)	Apparent Visual Magnitude	Description
32	221	0	42.7	+40	52	8.2	Elliptical galaxy; companion to M31
33	598	1	33.9	+30	39	5.7	Spiral galaxy in Triangulum
34	1039	2	42.0	+42	47	5.2	Open cluster in Perseus
35	2168	6	08.9	+24	20	5.1	Open cluster in Gemini
36	1960	5	36.1	+34	08	6.5	Open cluster in Auriga
37	2099	5	52.4	+32	33	5.6	Open cluster in Auriga
38	1912	5	28.7	+35	50	6.4	Open cluster in Auriga
39	7092	21	32.2	+48	26	4.6	Open cluster in Cygnus
40	—	12	22.4	+58	05	8.0	Close double star in Ursa Major
41	2287	6	47.0	−20	44	4.5	Loose open cluster in Canis Major
42	1976	5	35.4	−5	27	4.0	Orion nebula
43	1982	5	35.6	−5	16	9.0	Northeast portion of Orion nebula
44	2632	8	40.1	+19	59	3.1	Praesepe; open cluster in Cancer
45	—	3	47.0	+24	07	1.2	The Pleiades; open cluster in Taurus
46	2437	7	41.8	−14	49	6.1	Open cluster in Puppis
47	2422	7	36.6	−14	30	4.4	Loose group of stars in Puppis
48	2548	8	13.8	−5	48	5.8	"Cluster of very small stars"
49	4472	12	29.8	+8	00	8.5	Elliptical galaxy in Virgo
50	2323	7	03.2	−8	20	6.0	Loose open cluster in Monoceros
51	5194	13	29.9	+47	12	8.4	"Whirlpool" spiral galaxy in Canes Venatici
52	7654	23	24.2	+61	35	6.9	Loose open cluster in Cassiopeia
53	5024	13	12.9	+18	10	7.7	Globular cluster in Coma Berenices
54	6715	18	55.1	−30	29	7.7	Globular cluster in Sagittarius
55	6809	19	40.0	−30	58	7.0	Globular cluster in Sagittarius
56	6779	19	16.6	+30	11	8.3	Globular cluster in Lyra
57	6720	18	53.6	+33	02	9.0	"Ring" nebula; planetary nebula in Lyra
58	4579	12	37.7	+11	49	9.8	Spiral galaxy in Virgo
59	4621	12	42.0	+11	39	9.8	Spiral galaxy in Virgo
60	4649	12	43.7	+11	33	8.8	Elliptical galaxy in Virgo
61	4303	12	21.9	+4	28	9.7	Spiral galaxy in Virgo
62	6266	17	01.2	−30	07	6.6	Globular cluster in Scorpius
63	5055	13	15.8	+42	02	8.6	Spiral galaxy in Canes Venatici
64	4826	12	56.7	+21	41	8.5	Spiral galaxy in Coma Berenices
65	3623	11	18.9	+13	05	9.3	Spiral galaxy in Leo
66	3627	11	20.2	+12	59	9.0	Spiral galaxy in Leo; companion to M65
67	2682	8	50.4	+11	49	6.9	Open cluster in Cancer
68	4590	12	39.5	−26	45	8.2	Globular cluster in Hydra
69	6637	18	31.4	−32	21	7.7	Globular cluster in Sagittarius
70	6681	18	43.2	−32	18	8.1	Globular cluster in Sagittarius
71	6838	19	53.8	+18	47	8.0	Globular cluster in Sagittarius
72	6981	20	53.5	−12	32	9.3	Globular cluster in Aquarius
73	6994	20	59.0	−12	38	9.1	Faint open cluster in Aquarius
74	628	1	36.7	+15	47	9.3	Spiral galaxy in Pisces
75	6864	20	06.1	−21	55	8.6	Globular cluster in Sagittarius

M	NGC or (IC)	Right Ascension (2000) h	Right Ascension (2000) m	Declination (2000) (deg)	Declination (2000) (min)	Apparent Visual Magnitude	Description
76	650	1	42.4	+51	34	11.4	Planetary nebula in Perseus
77	1068	2	42.7	0	01	8.9	Spiral galaxy in Cetus
78	2068	5	46.7	0	03	8.3	Small emission nebula in Orion
79	1904	5	24.5	−24	33	7.8	Globular cluster in Lepus
80	6093	16	17.0	−22	59	7.3	Globular cluster in Scorpius
81	3031	9	54.2	+69	04	7.0	Spiral galaxy in Ursa Major
82	3034	9	55.8	+69	41	8.4	Irregular galaxy in Ursa Major
83	5236	13	37.0	-29	52	7.6	Spiral galaxy in Hydra
84	4374	12	25.1	+12	53	9.3	Elliptical galaxy in Virgo
85	4382	12	25.4	+18	11	9.3	Elliptical galaxy in Coma Berenices
86	4406	12	26.2	+12	57	9.2	Elliptical galaxy in Virgo
87	4486	12	30.8	+12	24	8.7	Elliptical galaxy inVirgo
88	4501	12	32.0	+14	25	9.5	Spiral galaxy in Coma Berenices
89	4552	12	35.7	+12	33	9.8	Elliptical galaxy in Virgo
90	4569	12	36.8	+13	10	9.5	Spiral galaxy in Virgo
91	4548	12	35.4	+14	30	10.2	Spiral galaxy in Coma Berenices [1]
92	6341	17	17.1	+43	08	6.4	Globular cluster in Hercules
93	2447	7	44.6	-23	52	6.5	Open cluster in Puppis
94	4736	12	50.9	+41	07	8.2	Spiral galaxy in CanesVenatici
95	3351	10	44.0	+11	42	9.7	Barred spiral galaxy in Leo
96	3368	10	46.8	+11	49	9.2	Spiral galaxy in Leo
97	3587	11	14.8	+55	01	11.1	"Owl" nebula; planetary nebula in Ursa Major
98	4192	12	13.8	+14	54	10.1	Spiral galaxy in Coma Berenices
99	4254	12	18.8	+14	25	9.9	Spiral galaxy in Coma Berenices
100	4321	12	22.9	+15	49	9.4	Spiral galaxy in Coma Berenices
101	5457	14	03.2	+54	21	7.9	Spiral galaxy in Ursa Major
102	5866(?)	15	06.5	+55	46	10.5	Spiral galaxy (identification in doubt) [1]
103	581	1	33.2	+60	42	7.4	Open cluster in Cassiopeia
104 °	4594	12	40.0	−11	37	8.3	Spiral galaxy in Virgo
105 °	3379	10	47.8	+12	35	9.3	Elliptical galaxy in Leo
106 °	4258	12	19.0	+47	18	8.4	Spiral galaxy in Canes Venatici
107 °	6171	16	32.5	−13	03	8.2	Globular cluster in Ophiuchus
108 °	3556	11	11.5	+55	40	10.0	Spiral galaxy in Ursa Major
109 °	3992	11	57.6	+53	23	9.8	Spiral galaxy in Ursa Major
110 °	205	0	40.4	+41	41	8.1	Elliptical galaxy; companion to M31

[1] According to some versions of Messier's list, M91 is actually a repeat of M58 and M102 is the same as M101, but here we give the galaxies that modern lists usually cite.

° Not in Messier's original (1781) list; added later by others.

Credits

This page constitutes an extension of the copyright page. We have made every effort to trace the ownership of all copyrighted material and to secure permission from copyright holders. In the event of any question arising as to the use of any material, we will be pleased to make the necessary corrections in future printings. Thanks are due to the following authors, publishers, and agents for permission to use the material indicated.

Prologue. Opener: Courtesy of William Baum and NASA. **P.1:** USGS. **P.2:** European Southern Observatory. **P.3:** C.R. O'Dell and NASA. **P.4:** NASA. **P.5:** ESA. **P.8:** JPL/NASA. **P.11:** Eastern Southern Observatory. **P.12:** Roger Angel, Steward Observatory/University of Arizona. **P.13:** David Malin/Anglo-Australian Telescope Board/David Malin Images. **P.14:** Photo taken with the U.K. Schmidt Telescope; © Anglo-Australian Telescope Board/David Malin Images. **P.15:** Tony Hallas. **P.16:** Photo taken with the U.K. Schmidt Telescope; © Anglo-Australian Telescope Board/David Malin Images.

Chapter 1. CO1: U. of Toronto. **Fig.1.3:** "National Optical Astronomy Observatories". **Fig.1.7:** "J.M. Pasachoff and the Chapin Library". **Fig.1.8:** Kent Wood. **Fig.1.14:** Yerkes Observatory. **Fig.1.15:** "Crawford Collection, Royal Observatory, Edinburgh". **Fig.1.17:** Yerkes Observatory. **Fig.1.18:** "Instituto e Museo di Storia della Scienza de Florenza".

Chapter 2. CO2: NASA. **Fig.2.1:** Granger Collection, New York. **Fig.2.2:** AIP/Niels Bohr Library. **Fig.2.6:** AIP/Niels Bohr Library. **Fig.2.7:** NASA. **Fig.2.9:** NASA. **Fig.2.11b:** "Crawford Collection, Royal Observatory, Edinburgh". **Fig.2.12:** US Space Command, NORAD. **Fig.2.13:** NASA/ARC. **Fig.2.14a:** Yerkes Observatory. **Fig.2.14b:** Corbis/Bettmann.

Chapter 3. CO3: NASA. **Fig.3.2:** © 1993 Hal Berol/Visuals Unlimited. **Fig.3.3:** © Bob Emott, Photographer. **Fig.3.11:** David Morrison. **Fig.3.12:** David Morrison. **Fig.3.17a:** Courtesy Nova Scotia Tourism. **Fig.17b:** Courtesy Nova Scotia Tourism. **Fig.3.21:** © 1991 Stephen J. Edberg. **Voyagers:** © Royal Society.

Chapter 4. CO4: Martha Haynes and Riccardo Giovanelli, Cornell University. **Fig.4.1:** American Institute of Physics, Niels Bohr Library. **Fig.4.2:** © 1993 Comstock, Inc. **Fig.4.6:** Max Planck Institute for Extraterrestrial Physics. **Fig.4.10:** National Solar Observatory/NOAO.

Chapter 5. CO5: Johnson Space Ctr/NASA. **Fig.5.1a:** Infrared Processing and Analysis Center/JPL. **Fig.5.1b:** Max Planck Institut fur Extraterrestrische Physik. **Fig.5.1c:** JPL/NASA. **Fig.5.2a:** David Morrison. **Fig.5.2b:** David Morrison. **Fig.5.7:** Gemini Observatory. **Fig.5.8a:** California Institute of Technology. **Fig.5.8b:** Gemini Observatory. **Fig.5.9:** California Association for Research in Astronomy. **Fig.5.10:** ESO. **Fig.5.11:** © William Keck Observatory. **Fig.5.13:** Bell Laboratories. **Fig.5.14:** NRAO/AUI (Nat'l Radio Astronomy Observatory). **Fig.5.15:** NRAO. **Fig.5.16:** NRAO. **Fig.5.17:** National Astronomy and Ionosphere Center. **Fig.5.18:** NASA/CXC/SAO. **Fig.5.19:** ESO. **Voyagers:** G.E. Hale: Caltech. The Yerkes telescope: Yerkes Observatory.

Chapter 6. CO6: JPL/NASA. **Fig.6.1:** NASA/JSC. **Fig.6.3:** NASA. **Fig.6.4:** NASA. **Fig.6.5:** NASA/JPL. **Fig.6.6:** NASA/JPL. **Fig.6.7:**

E. Kakorchka, U. of Arizona; and NASA. **Fig.6.8:** JPL/NASA. **Fig.6.9:** STScI/NASA. **Fig.6.10:** NASA. **Fig.6.12:** Art by planetary astronomer William Hartmann. **Fig.6.13a:** C.R. O'Dell/Rice University/NASA. **Fig.6.13b:** C.R. O'Dell/Rice University/NASA. **Fig.6.13c:** C.R. O'Dell/Rice University/NASA. **Voyagers:** unknown.

Chapter 7. CO7: G. Briggs/USGS. **Fig.7.1:** NASA. **Fig.7.3:** J. Frawley and NASA. **Fig.7.5:** David Morrison. **Fig.7.8:** David Porter/Science Photo Library/Photo Researchers, Inc. **Fig.7.9:** David Morrison. **Fig.7.11:** NASA. **Fig.7.12:** NASA/ARC, courtesy of David DesMarais. **Fig.7.16:** Novosty. **Fig.7.17:** Meteor Crater, Northern Arizona. **Fig.7.18:** Painting by Don Davis. **Voyagers:** Historical Pictures Service, Chicago.

Chapter 8. CO8: NASA. **Fig.8.1:** NASA. **Fig.8.2:** NASA. **Fig.8.3:** NASA. **Fig.8.4:** NASA. **Fig.8.5:** USGS. **Fig.8.6:** NASA. **Fig.8.7:** NASA. **Fig.8.8:** NASA. **Fig.8.9:** NASA. **Fig.8.10:** NASA. **Fig.8.11:** NASA. **Fig.8.14:** NASA. **Fig.8.16a-d:** Courtesy of A.G.W. Cameron, Center for Astrophysics. **Fig.8.17:** University of Arizona, Courtesy of Robert Strom. **Fig.8.19:** Mosaic courtesy of the Astrogeology Team, USGS, Flagstaff. **Fig.8.20:** NASA/JPL. **Fig.8.21:** NASA/JPL.

Chapter 9. CO9: NASA/USGS, courtesy of Alfred McEwen. **Fig.9.1:** NASA/ARC. **Fig.9.2:** Steve Larson, University of Arizona. **Fig.9.3:** Lowell Observatory. **Fig.9.4:** NASA/JPL. **Fig.9.5a:** NASA/JPL. **Fig.9.5b:** NASA/JPL. **Fig.9.6:** NASA/JPL. **Fig.9.7:** NASA/JPL. **Fig.9.8:** NASA/JPL. **Fig.9.9:** USSR Academy of Science and Carle Pieters, Brown University. **Fig.9.11:** Space Telescope Science Institute/NASA. **Fig.9.12:** NASA/JPL/Caltech. **Fig.9.13:** NASA/MOLA. Team led by David Smith at Goddard Space Flight Center. **Fig.9.14:** "Kees Veenenbos/veenen@tiscali.nl." **Fig.9.15:** NASA/USGS. **Fig.9.16a:** NASA/JPL. **Fig.9.16b:** NASA/JPL. **Fig.9.16c:** NASA/JPL. **Fig.9.17:** NASA/JPL. **Fig.9.18:** Malin Space Science Systems/NASA. **Fig.9.19:** NASA/JPL/Malin Space Science Systems. **Fig.9.2:** NASA/JSC. **Fig.9.21:** NASA/JPL/Malin Space Science Systems. **Fig.9.22:** NASA/from Mars Digital Image Map, processing by Brian Fessler/LPI. **Fig.9.23:** NASA/Malin Space Science Systems. **Fig.9.24:** NASA/JPL. **Fig.9.25a:** Malin Space Science Systems/NASA. **Fig.9.25b:** NASA/JPL/Malin Space Science Systems. **Fig.9.26:** Malin Space Science Systems/NASA. **Voyagers:** Lowell Observatory. **Making Connections (1):** JPL/NASA. **Making Connections (2):** NASA/Malin Space Science Systems.

Chapter 10. CO10: NASA/JPL. **Fig.10.1:** NASA/JPL/University of Arizona. **Fig.10.2:** NASA/JPL. **Fig.10.3:** NASA/ARC. **Fig.10.4:** NASA/JPL. **Fig.10.5:** Erich Karkoschka, Univ. of Arizona and NASA. **Fig.10.8:** Imke dePater/NRAO. **Fig.10.9:** NASA/JPL. **Fig.10.10:** Space Telescope Science Institute/NASA. **Fig.10.11:** NASA/JPL. **Fig.10.13:** NASA/JPL. **Fig.10.14a:** Andrew Ingersoll and Don Davis, courtesy of Sky Publishing Corp. **Fig.10.14b:** NASA/JPL. **Fig.10.15:** NASA/JPL. **Fig.10.16a:** Caltech/NASA. **Fig.10.16b:** Caltech/NASA. **Fig.10.16c:** Cornell University/Caltech/NASA. **Voyagers:** Photo by D. Hakanson.

Chapter 11. CO11: DLR/NASA. **Fig.11.1:** NASA/JPL. **Fig.11.2:** NASA/JPL. **Fig.11.3:** Arizona State Univ./JPL/NASA. **Fig.11.4:** Galileo Imaging Team/JPL/NASA. **Fig.11.5:** NASA/JPL. **Fig.11.6:** NASA/JPL. **Fig.11.7:** NASA/JPL. **Fig.11.8:** NASA/JPL. **Fig.11.9:** NASA/JPL. **Fig.11.10:** NASA/JPL. **Fig.11.11:** DLR (German Aerospace Center)/

NASA. **Fig.11.12:** NASA/USGS. **Fig.11.13:** NASA/JPL. **Fig.11.14:** University of Arizona/NASA. **Fig.11.15:** University of Arizona/NASA. **Fig.11.16:** University of Arizona/NASA. **Fig.11.17:** NASA/JPL. **Fig.11.19:** NASA/Image processing at USGS. **Fig.11.20:** Lowell Observatory. **Fig.11.21:** A. Stern, M. Buie, NASA and ESA. **Fig.11.23:** NASA/JPL. **Fig.11.24:** NASA/JPL. **Fig.11.25:** NASA/JPL. **Fig.11.26:** NASA/JPL. **Fig.11.27:** NASA/JPL. **Voyagers:** Lowell Observatory.

Chapter 12. CO12: Tony Hallas, AstroPhoto. **Fig.12.1:** Yerkes Observatory. **Fig.12.2:** Edward Bowell, Lowell Observatory. **Fig.12.4:** NASA. **Fig.12.5:** "NASA/JPL." **Fig.12.6:** NASA/JPL. **Fig.12.7:** Malin Space Systems/JPL/NASA. **Fig.12.9:** NASA/JPL, courtesy of Steven Ostro. **Fig.12.10:** NASA/JHUAPL. **Fig.12.11:** NASA/JHUAPL. **Fig.12.12:** NASA/JHUAPL. **Fig.12.13:** Anglo-Australian Observatory/David Malin Images. **Fig.12.15:** Donald Brownlee, University of Washington. **Fig.12.16a:** Max Planck Institut fur Aeronomie, and Ball Aerospace Corporation, courtesy of Harold Reitsema. **Fig.12.16b:** NASA/USGS. **Fig.12.19:** Caltech/Palomar Observatory. **Fig.12.20:** Eastern Southern Observatory. **Fig.12.21a:** University of Hawaii. **Fig.12.21b:** NASA, Harold Weaver (Johns Hopkins University), and the HST Comet Linear Investigation. **Fig.12.22:** NASA/STScI. **Fig.12.23:** NASA/STScI. **Voyagers:** Yerkes Observatory, University of Chicago.

Chapter 13. CO13: Painting by Don Dixon. **Fig.13.1:** P.Parviainen, Polar Image. **Fig.13.2:** Photo © 1986 John Sanford. **Fig.13.4:** unknown. **Fig.13.5:** J.O. Wheeler; courtesy Department of Library Services, American Museum of Natural History. **Fig.13.6:** NASA. **Fig.13.7:** Chip Clark. **Fig.13.10:** NASA, M. Clampin (STScI), H. Ford (JHU), G. Illingworth (UCO/Lick), J. Krist (STScI), D. Ardila (JHU), D. Golimowski (JHU) the ACS Science Team and ESA. **Fig.13.11:** Geoffrey Bryden/JPL/NASA. **Fig.13.15:** NASA/USGS, courtesy of Alfred McEwen. **Fig.13.16:** JPL/NASA. **Making Connections:** © 1982, 1992 Dan Haar/ The Hartford Courant.

Chapter 14. CO14: SOHO/NASA. **Fig.14.2:** Harvard University Archives. **Fig.14.3:** SOHO/NASA. **Fig.14.4:** David Alexander; NASA/Yohkoh. **Fig.14.6:** High Altitude Observatory/NCAR. **Fig.14.7:** "TRACE/ Stanford-Lockheed Institute for Space Research." **Fig.14.8:** "W. Livingston, National Solar Observatories/NOAO." **Fig.14.9:** National Solar Observatory/NOAO. **Fig.14.10a:** LMATC/NSO/NASA. **Fig.14.10b:** LMATC/NSO/NASA. **Fig.14.11:** NOAO. **Fig.14.12:** Data courtesy of J. Harvey, National Solar Observatory (Tucson/Kitt Peak, AZ). **Fig.14.13:** SOHO/NASA. **Fig.14.14:** Solar & Heliospheric Observatory/NASA. **Fig.14.15:** LMATC/NSO/NASA. **Fig.14.17:** "Kunsthistorisches Museum, Vienna; Photo by Erich Lessing. Art Resource." **Voyagers:** Courtesy of Stanford University. **Making Connections:** SOHO/NASA.

Chapter 15. CO15: Courtesy of the KamLAND Collaboration. **Fig.15.1:** Doug Sokell/Visuals Unlimited. **Fig.15.2a:** "Smithsonian Institution, courtesy AIP Emilio Segre Visual Archives." **Fig.15.2b:** "Smithsonian Institution, courtesy AIP Emilio Segre Visual Archives." **Fig.15.13:** NOAO. **Fig.15.14:** SOHO/NASA. **Fig.15.15:** "Courtesy of Sudbury Neutrino Observatory/SNO." **Voyagers:** "Permission granted by the Albert Einstein Archives. The Hebrew University of Jerusalem, Israel." **Making Connections:** Published with kind permission of ITER.

Chapter 16. CO16: NOAO. **Fig.16.1:** NOAO. **Fig.16.2:** "Hubble Heritage Team using data collected by John Trauger (Jet Propulsion Laboratory), Jon Holtzman (New Mexico State University), & collaborators." **Fig.16.4:** "Mary Lea Shane Archives of the Lick Observatory." **Fig.16.6a:** Courtesy of the University of Massachusetts and the Infrared Processing and Analysis Center. **Fig.16.6b:** Courtesy of the University of Massachusetts and the Infrared Processing and Analysis Center. **Fig.16.6c:** Dr. Robert Hurt, the Infrared Processing and Analysis Center. **Fig.16.7:** NOAO/AURA/NSF. **Fig.16.8a:** Yerkes Observatory. **Fig.16.8b:** Yerkes Observatory. **Voyagers:** Harvard College Observatory Archives. **Making Connections:** "Mary Lea Shane Archives of the Lick Observatory."

Chapter 17. CO17: Artwork by Jon Lomberg, © Gemini Observatory. **Fig.17.1a:** Jeff Greenberg/Photo Researchers, Inc. **Fig.17.1b:** WIYN/ NOA/NSF. **Fig.17.2:** "Art rendered by Dr. Robert Hurt of the Infrared Processing and Analysis Center/NASA." **Fig.17.3:** Yerkes Observatory. **Fig.17.7:** "NASA and K. Luhman (Harvard-Smithsonian Center for Astrophysics)." **Fig.17.12a:** "Sterrewacht Leiden and Princeton University Archives." **Fig.17.12b:** "Sterrewacht Leiden and Princeton University Archives." **Fig.17.15a:** NASA/SAO/CXC.

Chapter 18.CO18: Hubble Heritage Team/AURA/STScI/NASA. **Fig.18.1:** NASA/JPL. **Fig.18.4:** Yerkes Observatory. **Fig.18.5a:** Astronomical Society of the Pacific. **Fig.18.5:** Richard Norton, Science Graphics. **Fig.18.7:** Harvard College Observatory Archives. **Fig.18.8:** NOAO. **Fig.18.10:** "Wendy Freedman, Carnegie Institution of Washington, and NASA." **Voyagers:** "Courtesy of the San Diego State University special collections library."

Chapter 19. CO19: "Royal Observatory Edinburgh/Anglo-Australian Telescope Board/David Malin Images." **Fig.19.1:** "© Royal Observatory, Edinburgh/David Malin Images." **Fig.19.2:** "Anglo-Australian Observatory/ Royal Observatory, Edinburgh/David Malin Images." **Fig.19.4a:** E.M. Purcell and Harvard University. **Fig.19.4b:** E.M. Purcell and Harvard University. **Fig.19.5:** "© 1979 Royal Observatory, Anglo-Australian Telescope Board/David Malin Images." **Fig.19.6:** "Anglo-Australian Observatory/ David Malin Images." **Fig.19.7(lft):** "T. A. Rector (NOAO/AURA/NSF) and Hubble Heritage Team (STScI/AURA/NASA)." **Fig.19.7(rgt):** ESA/ ISO, CAM, L. Nordh (Stockholm Observatory). **Fig.19.8:** IPAC/JPL/ NASA. **Fig.19.9:** NASA and Hubble Heritage Team (STScI). **Fig.19.10:** Anglo-Australian Observatory. **Fig.19.13:** Photo courtesy of Martin Pomerantz. **Fig.19.15:** Anglo-Australian Observatory. **Voyagers:** "Mary Lea Shane Archives of the Lick Observatory."

Chapter 20. C020: "Wolfgang Brandner (JPL/IPAC), Eva K. Grebel (Univ. of Washington), You-Hua Chu (Univ. of Illinois Urbana-Champaign), and NASA." **Fig.20.1:** Jeff Hester and Paul Scowen, Arizona State U. and NASA. **Fig.20.2:** "Jeff Hester and Paul Scowen, Arizona State U. and NASA." **Fig.20.3a:** Infrared Processing and Analysis Center/JPL. **Fig.20.3b:** Infrared Processing and Analysis Center/JPL. **Fig.20.4:** Anglo-Australian Observatory/David Malin Images. **Fig.20.5a:** "Anglo-Australian Observatory/David Malin Images." **Fig.20.5b:** "Infrared Processing and Analysis Center & University of Massachusetts." **Fig.20.6:** "T.A. Rector, B.A. Wolpa, M. Hanna, KPNO 0.9-m Mosaic, NOAO/AURA/NSF." **Fig.20.9:** "NASA, Alan Watson (Instituto de Astronomia, UNAM, Mexico), Karl Stapelfeldt (JPL), John Krist (STSI), and Chris Burrows (ESA/ STSI)." **Fig.20.10a:** J. Morse and NASA. **Fig.20.10b:** C. Burrows, J. Morse, J. Hester, and NASA. **Fig.20.11:** "D. Padgett (IPAC/Caltech), W. Brandner (IPAC), K. Stapelfeldt (JPL), and NASA." **Fig.20.13a:** M. McCaughrean, C.R. O'Dell, and NASA. **Fig.20.13b:** M. McCaughrean, C.R. O'Dell, and NASA. **Fig.20.13c:** M. McCaughrean, C.R. O'Dell, and NASA. **Fig.20.13d:** M. McCaughrean, C.R. O'Dell, and NASA. **Fig.20.14:** "B. Smith, U. of Hawaii, G. Schneider, U. of Arizona; and NASA." **Fig.20.15:** J. Greaves, et al., Joint Astronomy Centre. **Fig.20.17:** M. Mayor and D. Queloz. **Fig.20.18:** Painting by Lynette Cook. **Fig.20.19:** San Francisco State University.

Chapter 21. CO21: NASA, ESA, and the Hubble Heritage Team (STScI/ AURA). **Fig.21.1:** "© Anglo-Australian Telescope Board/David Malin Images." **Fig.21.2:** G. van Belle/JPL/NASA. **Fig.21.3:** A. Dupree, R. Gilliland, and NASA. **Fig.21.5:** "© Anglo-Australian Telescope Board/ David Malin Images." **Fig.21.6:** NASA and the Hubble Heritage Team (STScI/AURA). **Fig.21.7:** "© Anglo-Australian Telescope Board/David Malin Images." **Fig.21.9:** "© Anglo-Australian Telescope Board/David Malin Images." **Fig.21.11:** "© Anglo-Australian Telescope Board/David Malin Images." **Fig.21.16a:** "Hubble Heritage Team/AURA/STScI/ NASA." **Fig.21.16b:** "Bruce Balick (Univ of Washington), Vincent Icke (Leiden Univ, The Netherlands), Garrelt Mellema (Stockholm Univ), and NASA." **Fig.21.16c:** NASA, The Hubble Heritage Team (STScI/AURA). **Fig.21.16d:** H. Bond (STScI) and NASA. **Fig.21.17a(inset):** "Eastern Southern Observatory (original photo from Fraknoi)." **Fig.21.17b(inset):** "Massimo Stiavelli (STScI), Inge Heyer (STScI) et al. & The Hubble Heritage Team AURA/STScI/NASA)." **Fig.21.17c(inset):** "Hubble Heritage Team/AURA/STScI/NASA." **Fig.21.18:** J. Morse, U. of Colorado, and NASA.

Chapter 22. CO22: Wolfgang Brandner (JPL/IPAC), Eva Grebel (Univ. Washington), You-Hua Chu (Univ. Illinois, Urbana-Champaign) and NASA. **Fig.22.3:** R. Elson and R. Sword, NASA. **Fig.22.5a-b:** "Anglo-Australian Telescope Board/David Malin Images." **Fig.22.6:** Space Telescope Science Institute/NASA. **Fig.22.7:** Peter Challis (Harvard University) and the SInS collaboration. **Fig.22.9a:** "F. Reines and J. C. van der Velde, IMB Collaboration." **Fig.22.9b:** "F. Reines and J. C. van der Velde, IMB Collaboration." **Fig.22.10a:** "Courtesy of AIP Emilio Segre Visual Archives, Weber Collection." **Fig.22.10b:** "Courtesy of AIP Emilio Segre Visual Archives, Weber Collection." **Fig.22.11:** NASA/HST/ASU/J. Hester et al. **Fig.22.13:** "© 1979 Royal Observatory, Anglo-Australian Telescope Board/David Malin Images." **Voyagers:** "Courtesy of Emilio Segre Visual Archives, Physics Today Collection." **Making Connections(a-b):** National Radio Astronomy Observatory/AUI. **Making Connections(c):** NASA/CXC/SAO.

Chapter 23. CO23: Harvard-Smithsonian Center for Astrophysics/SAO/NASA. **Fig.23.1:** Photo courtesy of the archives, Caltech. **Fig.23.3:** NASA. **Fig.23.11:** Photo courtesy Roy Bishop. **Fig.23.13:** Yerkes Observatory. **Fig.23.15:** LIGO/Caltech. **Making Connections:** cover copyright Ballantine Books.

Chapter 24. CO24: Anglo-Australian Observatory/David Malin Images. NASA. **Fig.24.4:** W. Keel, U. Alabama and CTIO. **Fig.24.5:** NASA. **Fig.24.6:** "Anglo-Australian Telescope Board/David Malin Images." **Fig.24.9:** "Image processing by Kassim et al., Naval Research Labs; original data by Pedlar et al., NRAO." **Fig.24.10:** NASA/UMass/D. Wang et al. **Fig.24.11:** NRAO/AUI. **Fig.24.12:** Photo courtesy of Gemini Observatory, National Science Foundation and the University of Hawaii Adaptive Optics Group. **Fig.24.13:** Photo by Tony Hallas. **Fig.24.13(inset):** Caltech Archives. **Fig.24.15:** "Diagram from Gerry Gilmore, Institute of Astronomy, Cambridge." **Fig.24.16:** Paul Harding/Case Western Reserve University. **Fig.24.17a:** John Dubinski/Univ. of Toronto/Canadian Institute of Theoretical Astrophysics. **Fig.24.17b:** John Dubinski/Univ. of Toronto/Canadian Institute of Theoretical Astrophysics. **Fig.24.17c:** John Dubinski/Univ. of Toronto/Canadian Institute of Theoretical Astrophysics. **Voyagers:** Harvard College Observatory Archives.

Chapter 25. CO25: R. Williams, the Hubble Deep Field Team, and NASA. **Fig.25.1:** T.A. Rector and B.A. Wolpa/NOAO/AURA/NSF. **Fig.25.2:** NASA and the Hubble Heritage Team/STScI/AURA. **Fig.25.3:** C. Howk (JHU), B. Savage (U. Wisconsin), N.A. Sharp (NOAO/WIYN/NOAO/NSF. **Fig.25.4:** Anglo-Australian Observatory/David Malin Images. **Fig.25.6:** Palomar Observatory, Caltech. **Fig.25.7:** Palomar Observatory, Caltech. **Fig.25.8:** Anglo-Australian Observatory/David Malin Images. **Fig.25.9:** Palomar Observatory, Caltech. **Fig.25.10:** George Jacoby/WIYN/NOAO/NSF. **Fig.25.11:** NOAO. **Fig.25.12:** NOAO. **Fig.25.13:** "T.A. Rector (NRAO/AUI/NSF & NOAO/AURA/NSF) & M. Hanna (NOAO/AURA/NSF)." **Fig.25.14:** "A. Reiss et al./Harvard-Smithsonian Center for Astrophysics." **Fig.25.16:** Lowell Observatory. **Fig.25.17:** Caltech Archives. **Voyagers:** Photo by J. Stokley/A.S.P. Archives.

Chapter 26. CO26: NASA, Andrew Fruchter & the ERO Team (Sylvia Baggett (STScI), Richard Hook (ST-ECF), Zoltan Levay (STScI). **Fig.26.1:** W. C. Keel, Univ. of Alabama. **Fig.26.2:** Andrew Fraknoi. **Fig.26.3:** NASA/CXC/SAO/H. Marshall et al. **Fig.26.4:** John Bahcall (Institute for Advanced Study, Princeton), Mike Disney (University of Wales) and NASA. **Fig.26.6:** "Anglo-Australian Telescope Board/David Malin Images." **Fig.26.7:** NASA & the Hubble Heritage Team, STScI/AURA. **Fig.26.8:** "X-ray: NASA/CXC/MIT/H. Marshall et al. Radio: F. Zhou, F. Owen (NRAO), J. Biretta (STScI) Optical: NASA/STScI/UMBC/E. Perlman et al." **Fig.26.9:** D. Clark et al, National Radio Astronomy Observatory. **Fig.26.10:** H. Ford et al & NASA. **Fig.26.11(lft):** R.P. van der Marel, F.C. van den Bosch & NASA. **Fig.26.11(rgt):** R.P. van der Marel, F.C. van den Bosch & NASA. **Fig.26.12:** NASA & K. Gebhardt, Lick Observatory. **Fig.26.13:** K. Cordes & S. Brown, STScI. **Fig.26.14:** X-ray: NASA/CXC/MPE/S. Komossa et al.; Optical: NASA/STScI/R.P. van der Marel, J. Gerssen. **Fig.26.16(lft):** W. Jaffe, H. Ford, & NASA. **Fig.26.16(rgt):** W. Jaffe, H. Ford, & NASA. **Fig.26.18:** NASA/CXC/PSU/N. Brandt et al. **Fig.26.19:** NASA, ESA, Richard Ellis (Caltech) and Jean-Paul Kneib (Observatoire Midi-Pyrenees, France). **Fig.26.21:** W.C. Keel, Univ. of Alabama. **Fig.26.22:** Kavan Ratnatunga (Carnegie Mellon Univ.) and NASA.

Interlude. COI: Painting © by Don Dixon. **I.1:** NASA. **I.2:** BATSE/NASA. **I.3:** Italian Space Agency. **I.4:** "Andrew Fruchter, Space Telescope Science Institute; Elena Pian, ITSRE-CNR; and NASA." **I.5:** Carl Akerlof, U. of Michigan.

Chapter 27. CO27: B. Whitmore (STScI) & NASA. **Fig.27.1:** R. Windhorst et al. & NASA. **Fig.27.2:** Esther M. Hu, Richard G. McMahon, & Lennox L. Cowie, Univ. Hawaii. **Fig.27.3:** R. Williams (STScI), the HDF-S Team and NASA. **Fig.27.4:** Ferguson, Dickinson, Williams at STScI. **Fig.27.5a:** K. Borne & NASA. **Fig.27.5b:** NOAO. **Fig.27.5c:** W. Keel and ESO. **Fig.27.5d:** NASA and Hubble Heritage Team (STScI). **Fig.27.6(a-d):** "Computer image courtesy of Josh Barnes, University of Hawaii." **Fig.27.7:** Carl Grillmair (California Institute of Technology) and NASA.R. Schild. **Fig.27.9:** NASA and The Hubble Heritage Team (STScI/AURA). **Fig.27.10:** "NASA, J. English (U. Manitoba), S. Hunsberger (PSU), Z. Levay (STSI), S. Gallagher (PSU), J. Charlton (PSU)." **Fig.27.11:** Pieter van Dokkum, Marijn Franx (U. Groningen/Leiden), ESA, NASA. **Fig.27.12:** NASA. **Fig.27.14:** "© Anglo-Australian Telescope Board/David Malin Images." **Fig.27.15:** "Anglo-Australian Observatory/David Malin Images." **Fig.27.16:** Megan Donahue (STScI)/NASA. **Fig.27.18:** William Keel & Astronomical Society of the Pacific. **Voyagers:** CfA. **Making Connections:** Fermilab Visual Media Services.

Chapter 28. CO28: NASA. **Fig.28.1:** NOAO. **Fig.28.2:** NASA. **Fig.28.3:** European Southern Observatory. **Fig.28.5:** NASA, Adam Riess (STScI). **Fig.28.7:** Yerkes Observatory. **Fig.28.8:** Courtesy of Ralph Alpher. **Fig.28.11:** NASA. **Fig.28.12:** AT&T Bell Laboratories. **Fig.28.13:** NASA. **Fig.28.15:** The BOMERANG Collaboration. **Fig.28.16:** The BOMERANG Collaboration. **Fig.28.17:** NASA/GSFC. **Fig.28.18:** John Godfrey (STScI)/NASA. **Fig.28.19:** NASA. **Cartoon:** © Ted Goff. Reprinted by permission.

Chapter 29. CO29: NASA; artist: Roger Arno. **Fig.29.1(lft):** J. Hester, P. Scowen, and NASA. **Fig.29.1(rgt):** J. Hester, P. Scowen, and NASA. **Fig.29.2:** "Robert Provin, California State University, Northridge." **Fig.29.3:** A. Fraknoi. **Fig.29.4:** "© Anglo-Australian Telescope Board/David Malin Images." **Fig.29.6:** William Schopf, UCLA. **Fig.29.7:** © Bonnie Sue Photography. **Fig.29.8:** "Peter Ryan/Scripps/Science Photo Library/Photo Researchers, Inc." **Fig.29.9:** David Morrison. **Fig.29.10:** Painting by Don Davis, © SETI Institute. **Fig.29.11:** USGS/NASA/JPL. **Fig.29.12:** NASA/JPL. **Fig.29.14:** David DesMarais, NASA, Ames. **Fig.29.15:** NASA. **Fig.29.16(a-b):** NASA. **Fig.29.17:** NRAO. **Fig.29.18(a-b):** Photos by Seth Shostak, SETI Institute. **Fig.29.19:** SETI Institute. **Fig.29.20:** Seth Shostak, SETI Institute. **Voyagers:** Seth Shostak, SETI Institute.

Index

dark rift in, 425–426
definition of, 706
differential rotation of, 542
disk of stars in, 537, *537*
dusk clouds in, 427, *427*
formation of, 549–552, *550*
full-sky view of, *12*
globular clusters in, 540, *541*
halo of stars in, 539–540
Herschel's measurements of, 536–537, *536*
illustration of major parts of, *541*
interstellar dust in, 10, 12
invisible matter in, 543–545
Local Group and, 625, *626*
mass of, 543–545
nuclear bulge in, 537, *538*, 539
rotation curve of, *544*
scale examples of, 10, *11*
spiral structure of, 542–543, *542, 563*
stellar populations in, 547–549
supernova explosions in, 498
types of stars in, 547–549
Web sites about, 553
See also galaxies
Miller, Coleman, 508
Miller, Stanley, 677
Mimas, 261
Minas Basin, 75
minerals, 708
mirrors, telescope, 112–113, *112, 114, 114*
Mizar, 378–379
models, scientific, 4
modulation, 122
molecular clouds, 424, 440–443
    Orion constellation and, 441–443, *442*
    star formation and, 440–441
molecules, 14, 708
    interstellar, 422–424
    organic, 296–297, 677
momentum
    angular, 49, *49*, 76
    change of, 48
    conservation of, 47, 503
    definition of, 708
Moon (of Earth), 71–79, 176–186
    age of, 145
    apparent motion of, 25
    basic properties of, 176, *176*
    composition and structure of, 178–179
    craters on, *144*, 181–184, *184*, 272
    distance from Earth to, 8
    eclipses of, 77, 79, *79*
    exploration of, *136, 174*, 176–178, *177*, 181
    geological activity on, 180, 303
    highlands of, 179, *179*
    history of, 179–181
    impact basins on, 180, *180*
    maria of, 179–180, *180*
    mountains on, 181, *181*, 304
    observation of, 183
    ocean tides and, 74–77
    origin of, 185–186, *186*
    phases of, 71–73, *72*, 81, *183*
    revolution and rotation of, 73–74, *74*
    rocks collected from, 145, 176, *177, 180*
    size of, *245*
    solar eclipses and, 77–79
    south pole of, *178*
    stellar diameters and, 382
    surface of, 179–181
    viewed from Earth, *176*
    Web sites about, 81, 190
moons. *See* satellites (natural)
morning star, 38
Morrison, David, 170
motion
    Kepler's laws of, 46–47

Newton's laws of, 47
    proper, 367–368, *367, 369*
    retrograde, 30, *30*
    See also planetary motion
mountains
    on Earth, 160, *161*, 304–305, *304*
    on Mars, 205, *206*, 304–305, *304*
    on the Moon, 181, *181*, 304
    on terrestrial planets, 304–305, *304*
    on Venus, 200, 304, *304*, 305
Mount Wilson Observatory, 121, 540–541, 560, 624
Multi-Mirror (MMT) telescope, *115*
Mundi Systematic, De (Newton), *54*
Murchison meteorite, 296
Musgrave, Story, *108*
mutations, 497
mutual gravitation, 55
mythology, 384

Nabokov, Vladimir, 613
names
    of asteroids, 141, 268
    of comets, 141
    of nebulae, 419
    of planets, 141
    of stars, 402
Nanedi Valles, 212
NASA
    Ames Research Center, 55, 125, 262, 694
    Deep Space Network, *399*
    Goddard Space Flight Center, 127, 609
    Jet Propulsion Laboratory, 273
    Johnson Space Flight Center, 190
    Planetary Photojournal, 149
    Web sites, 149, 190, 698
natal astrology, 31
National Optical Astronomy Observatories (NOAO), 16, 699
National Radio Astronomy Observatory, 690, *690*
National Science Foundation, 692
National Space Science Data Center, 149
nature
    forces of, 15, 664–665, *664*
    laws of, 5
neap tides, 76, 77
near-Earth objects (NEOs), 272–273, *272*, 708
nearest stars, 722–723
NEAR-Shoemaker spacecraft, *140*, 271, 274, *274*
nebulae, 418
    dark, 425–426, *426*
    definition of, 708
    galaxies and, 560–561
    ionized hydrogen and, 419, *420*
    Messier catalog of, 733–735
    naming of, 419
    planetary, 477, *478, 479*
    reflection, 427, *427*
    Web sites about, 433, 507
Nemiroff, Robert, 17
Neptune
    appearance of, 228
    atmosphere of, *234*, 235, *235*, 237
    basic properties of, 227–228, *228*
    composition and structure of, 229, *230*
    discovery of, 55–56
    Great Dark Spot on, 237
    internal energy source of, 231
    Kuiper belt objects and, 281
    magnetic field of, 231
    orbit of, 255–256
    rings of, 245, 260, *261*
    rotation of, *228*
    satellites of, 245, 254–255, *255, 364*, 720
    seasons on, 228

spacecraft explorations of, *225, 226, 227*
    storms on, 237
    Web sites about, 239
    winds and weather on, 236, *236*
neutral hydrogen clouds, 420–421
neutrinos, 341
    definition of, 708
    detection of, *336*, 351–352, *352*
    early universe and, 654
    oscillation of, 352
    solar, 343, 348, *348*, 351–352
    stellar explosions and, 495
    Supernova 1987A and, 501, *501*
    Web sites about, 353
neutrons, 15, 98, *341*, 494, 708
neutron stars, 494, 708
    companions to, 507
    discovery of, 502
    gamma-ray bursts and, *604*, 610–611, *611*
    merging of, *604*
    properties of, *502*
    radiation from, 505
    rotation of, 503–504, *504*
    Web sites about, 508
    See also pulsars
New Astronomy, The (Kepler), 45
New General Catalog (NGC) of Nebulae and Star Clusters (Dreyer), 419
new moon, 72
New Sinai, The (Clough), 52
Newton, Isaac, 5, *47*, 276, 514, 517
    comets studied by, 275
    law of gravity, 50
    laws of motion, 47–48, 51
    prism experiment, 91
    reflecting telescope, 113
    Web sites about, 57
Newton Crater, *213*
Newtonian focus, *113*
NGC 1232 galaxy, *11*
NGC 1365 galaxy, 539
NGC 1566 galaxy, 586
NGC 1818 star cluster, 492, *493*
NGC 1999 nebula, 427
NGC 2264 star cluster, 472, *472*
NGC 2420 star cluster, *376*
NGC 3293 star cluster, 472, *473*
NGC 3351 galaxy, 563
NGC 3603 nebula, 488
NGC 4013 galaxy, 563
NGC 4038 galaxy, 612, 620
NGC 4039 galaxy, 612, 620
NGC 4261 galaxy, 593
NGC 5746 galaxy, 633
NGC 6240 galaxy, 592, 619
NGC 6520 star cluster, 424
NGC 6751 nebula, 478, 479
NGC 7027 nebula, 478, 479
NGC 7052 galaxy, 589
Niven, Larry, 525
Nobeyama Cosmic Radio Observatory, *124*
North America Nebula, 266
north celestial pole, 21–22, *22, 23*
north circumpolar zone, 22
novae, 506, 708
nuclear bulge, 537, *538*, 539, 708
nuclear fission, 342, *342*
nuclear forces, 664, *664*
nuclear fusion, 35, 342–346, *342*
    controlled production of, 344
    heat required for, 342–343
    heavy elements and, 493–494, 500–501
    interior of the Sun and, 343
    proton-proton cycle and, *343*, 345, *345*
    stellar evolution and, 475
    Web sites about, 352
nuclear transformation, 708

THE NIGHT SKY IN JANUARY

Latitude of chart is 34°N, but it is
practical throughout the continental
United States.

To use: Hold chart vertically and turn
it so the direction you are facing
shows at the bottom.

Chart time (Local Standard):

10 p.m. First of month

9 p.m. Middle of month

8 p.m. Last of month

NORTHERN HORIZON

EASTERN HORIZON

WESTERN HORIZON

SOUTHERN HORIZON

THE NIGHT SKY IN FEBRUARY

Latitude of chart is 34° N, but it is practical throughout the continental United States.

To use: Hold chart vertically and turn it so the direction you are facing shows at the bottom.

Chart time (Local Standard):
10 p.m. First of month
9 p.m. Middle of month
8 p.m. Last of month

Star Chart from *GRIFFITH OBSERVER*, Griffith Observatory, Los Angeles

THE NIGHT SKY IN MARCH

Latitude of chart is 34°N, but it is
practical throughout the continental
United States.

To use: Hold chart vertically and turn
it so the direction you are facing
shows at the bottom.

Chart time (Local Standard):

10 p.m. First of month

9 p.m. Middle of month

8 p.m. Last of month

Star Chart from *GRIFFITH OBSERVER*, Griffith Observatory, Los Angeles.

NORTHERN HORIZON

CEPHEUS
CASSIOPEIA
DRACO
PERSEUS
VEGA
POLARIS 'NORTH STAR'
HERCULES
CAPELLA
URSA MINOR "LITTLE DIPPER"
AURIGA
TAURUS
CORONA BOREALIS
BOOTES
URSA MAJOR "BIG DIPPER"
SERPENS
ALDEBARAN
CASTOR
GEMINI
POLLUX
ARCTURUS
CANCER
BETELGEUSE
RIGEL
LEO
ORION
REGULUS
PROCYON
CANIS MINOR
VIRGO
SIRIUS
LIBRA
SPICA
CORVUS
CANIS MAJOR
HYDRA

EASTERN HORIZON

WESTERN HORIZON

SOUTHERN HORIZON

THE NIGHT SKY IN APRIL

Latitude of chart is 34°N, but it is
practical throughout the continental
United States.

To use: Hold chart vertically and turn
it so the direction you are facing
shows at the bottom.

Chart time (Local Standard):
10 p.m. First of month
9 p.m. Middle of month
8 p.m. Last of month

Star Chart from *GRIFFITH OBSERVER*, Griffith Observatory, Los Angeles

THE NIGHT SKY IN MAY

Latitude of chart is 34°N, but it is practical throughout the continental United States.

To use: Hold chart vertically and turn it so the direction you are facing shows at the bottom.

Chart time (Local Standard):
10 p.m. First of month
9 p.m. Middle of month
8 p.m. Last of month

Star Chart from *GRIFFITH OBSERVER*, Griffith Observatory, Los Angeles

NORTHERN HORIZON

CASSIOPEIA

CEPHEUS

POLARIS "NORTH STAR"

CYGNUS "NORTHERN CROSS"

DENEB

LITTLE DIPPER

URSA MINOR

DRACO

DELPHINUS

VEGA

LYRA

SAGITTA

HERCULES

URSA MAJOR "BIG DIPPER"

CASTOR

POLLUX

GEMINI

CANCER

EASTERN HORIZON

AQUILA

ALTAIR

CORONA BOREALIS

BOOTES

ARCTURUS

LEO

REGULUS

WESTERN HORIZON

SERPENS

OPHIUCHUS

SERPENS

VIRGO

SAGITTARIUS

ANTARES

LIBRA

SPICA

CORVUS

HYDRA

SCORPIUS

SOUTHERN HORIZON

THE NIGHT SKY IN JUNE

Latitude of chart is 34°N, but it is
practical throughout the continental
United States.

To use: Hold chart vertically and turn
it so the direction you are facing
shows at the bottom.

Chart time (Local Standard):

10 p.m. First of month

9 p.m. Middle of month

8 p.m. Last of month

Star Chart from *GRIFFITH OBSERVER*, Griffith Observatory, Los Angeles

NORTHERN HORIZON

CASSIOPEIA

CEPHEUS

POLARIS NORTH STAR

URSA MINOR
"LITTLE DIPPER"

URSA MAJOR "BIG DIPPER"

DRACO

CYGNUS NORTHERN CROSS

DENEB

LEO

REGULUS

PEGASUS

VEGA

LYRA

CORONA BOREALIS

ARCTURUS

BOOTES

AQUARIUS

DELPHINUS

ALTAIR

SAGITTA

HERCULES

OPHIUCHUS

VIRGO

CORVUS

AQUILA

SERPENS

SPICA

CAPRICORNUS

SERPENS

LIBRA

SAGITTARIUS

ANTARES

SCORPIUS

EASTERN HORIZON

WESTERN HORIZON

SOUTHERN HORIZON

THE NIGHT SKY IN JULY

Latitude of chart is 34°N, but it is
practical throughout the continental
United States.

To use: Hold chart vertically and turn
it so the direction you are facing
shows at the bottom.

Chart time (Local Standard):

10 p.m. First of month

9 p.m. Middle of month

8 p.m. Last of month

Star Chart from *GRIFFITH OBSERVER*, Griffith Observatory, Los Angeles

NORTHERN HORIZON

EASTERN HORIZON

WESTERN HORIZON

SOUTHERN HORIZON

THE NIGHT SKY IN AUGUST

Latitude of chart is 34°N, but it is
practical throughout the continental
United States.

To use: Hold chart vertically and turn
it so the direction you are facing
shows at the bottom.

Chart time (Local Standard):

10 p.m. First of month

9 p.m. Middle of month

8 p.m. Last of month

Star Chart from *GRIFFITH OBSERVER*, Griffith Observatory, Los Angeles

THE NIGHT SKY IN SEPTEMBER

Latitude of chart is 34°N, but it is practical throughout the continental United States.

To use: Hold chart vertically and turn it so the direction you are facing shows at the bottom.

Chart time (Local Standard):

10 p.m. First of month

9 p.m. Middle of month

8 p.m. Last of month

Star Chart from *GRIFFITH OBSERVER*, Griffith Observatory, Los Angeles

NORTHERN HORIZON

EASTERN HORIZON

WESTERN HORIZON

SOUTHERN HORIZON

THE NIGHT SKY IN OCTOBER

Latitude of chart is 34°N, but it is
practical throughout the continental
United States.

To use: Hold chart vertically and turn
it so the direction you are facing
shows at the bottom.

Chart time (Local Standard):
10 p.m. First of month
9 p.m. Middle of month
8 p.m. Last of month

Star Chart from *GRIFFITH OBSERVER*, Griffith Observatory, Los Angeles

THE NIGHT SKY IN NOVEMBER

Latitude of chart is 34°N, but it is
practical throughout the continental
United States.

To use: Hold chart vertically and turn
it so the direction you are facing
shows at the bottom.

Chart time (Local Standard):

10 p.m. First of month

9 p.m. Middle of month

8 p.m. Last of month

Star Chart from *GRIFFITH OBSERVER*, Griffith Observatory, Los Angeles

THE NIGHT SKY IN DECEMBER

Latitude of chart is 34°N, but it is practical throughout the continental United States.

To use: Hold chart vertically and turn it so the direction you are facing shows at the bottom.

Chart time (Local Standard):

10 p.m. First of month

9 p.m. Middle of month

8 p.m. Last of month

Star Chart from *GRIFFITH OBSERVER*, Griffith Observatory, Los Angeles